This Book Has Been
Supported by the

SEELEY W. MUDD
Memorial Fund

Colonel Seeley W. Mudd won lasting recognition in the mining profession for the vast mineral riches he discovered and developed for the benefit of mankind. Yet when he died in 1926, his friends and colleagues reserved their highest tributes for the unstinting help he had given to young members of the profession. In administering the Fund in memory of Colonel Mudd, AIME has sought to perpetuate his lifelong interest in newcomers to the mining fraternity by using the Fund to produce publications for the profession and to present each new Junior Member of AIME with a book of his choice. The Seeley W. Mudd Memorial Fund was entrusted to AIME on Sept. 9, 1929, by Colonel Mudd's widow, Della M. Mudd, and his two sons, Harvey S. Mudd and Seeley G. Mudd.

Economics of the Mineral Industries

4th Edition

Economics
of the Mineral Industries

(A Series of Articles by Specialists)

4th Edition

Edited by

William A. Vogely
Professor, The Pennsylvania State University

Carla Sydney Stone and Richard T. Newcomb
Associate Editors

W. Carey Hardy and Joe E. Wirsching
(Editorial Board)

Sponsored by the Seeley W. Mudd Memorial Fund
Fourth Edition Completely Revised and Rewritten

A Volume in the Seeley W. Mudd Series published by
American Institute of Mining, Metallurgical, and Petroleum Engineers, Inc.
New York • 1985

Printed in the United States of America
by Guinn Printing Incorporated, Hoboken, N.J.

ISSN 0-89520-438-X
Library of Congress Catalog Card Number 85-70525

HARVEY SEELEY MUDD
1888–1955

Harvey Mudd, to whom this volume is fittingly dedicated, was one of his generation's outstanding mining engineers. His quality in this respect was somewhat obscured by his unusual modesty and the fact that much of his business activity was within closely held organizations. Nevertheless, he was accorded such professional honors as the presidency of AIME (1945), the Egleston Medal for distinguished engineering achievement (Columbia University, 1949), and honorary doctorates from three universities (California, Columbia, and Loyola).

He was born in Leadville, Colo. Attending Stanford and Columbia Universities, he was granted the degree of Engineer of Mines from the latter in 1912. His principal business activity was with Cyprus Mines Corp., organized to operate copper pyrites properties on the Island of Cyprus. The deposits had been worked extensively in ancient times and were rediscovered after 17 centuries of idleness by C. Godfrey Gunther, with the backing of Seeley W. Mudd and Philip Wiseman. Early results were disappointing, and it took considerable skill and perseverance to bring the operation through initial stages, postwar copper price deflation, and the depression of the thirties. Harvey Mudd was active in this development from 1918 onward, being president of the company for most of the intervening years until shortly before his death. To him goes most of the credit for its growth into an unusually successful enterprise, with enlightened employee relations, sound diplomatic policies, technical capability, and a balance sheet such as is rarely seen. Harvey Mudd was also active in several other mining developments, including the Johnson Camp, Old Dick, and Pima properties in Arizona; the Cactus Queen and Afterthought in California; and the Marcona iron mine in Peru. He was also a director of Texas Gulf Sulphur Co., Mesabi Iron Co., Southern Pacific, and Pacific Mutual Life Insurance Co.

His activities and honors as a leading citizen and farsighted philanthropist were almost too numerous to mention. During the First World War he was assistant secretary of the War Minerals Board. He was a member of the Board of Consulting Engineers of the Los Angeles Water District. In 1944 he was chairman of the Citizens Advisory Committee to the State Reconstruction and Employment Commission. He was a leader in civic fund raising for various charitable purposes and gave generously of both time and money to the support of artistic, cultural, and educational institutions. In 1935 he was cited as the Los Angeles citizen who had given the community the most valuable and unselfish service.

To all of these activities Harvey Mudd brought the rare combination of qualities which assured his success as an engineer and administrator—objectivity, an encyclopedic memory, uncompromising ethics, studious attention, frankness, tolerance, enterprise, and sound judgment. As an engineer he was impregnable because he combined an unusual appetite for details with amazing capacity to perceive the essentials of a problem. The young mineral technologist seeking guidance can go a long way toward guaranteeing his own professional and economic success by remembering and emulating the above mentioned qualities of Harvey S. Mudd.

—*Evan Just*

CONTENTS

FOREWORD

The fourth edition of the *Economics of the Mineral Industries* is completely rewritten to address the state of the mineral sectors and of mineral economic principles as of 1984. Since the third edition was written in the early 1970s there have been major changes both in the structure of these industries and in the tools of analysis of them, as well as in public policy. Also, resources have become a much more popular subject for study at universities and colleges in the United States and in many other areas of the world. There are now four formal Ph.D. programs in mineral economics in the United States: The Pennsylvania State University, Colorado School of Mines, West Virginia University, and University of Arizona, in order of establishment. There are courses at many other institutions, and several masters level degree programs, often as part of economics, mining or business administration departments. Faculty from these programs are represented as authors in this volume.

This edition has three objectives:

1. To provide a book suitable for use in a wide variety of courses at universities and colleges.

2. To provide to the members of AIME a reference to which they can turn for understanding of the principles and applications of economics, geostatistics, modeling, and other analytical techniques to the mineral industries.

3. To provide an objective review of the public policy issues concerning minerals that the United States faces.

This edition differs in scope from the previous editions. First, principles, trends and interrelationships rather than specific commodity problems are addressed. Statistical data are largely omitted as their inclusion would tend to date the volume quickly, and such data are easily obtained from government or industry sources on a current basis. The focus is also, except for Part 1, on the United States, necessary to keep the volume at a reasonable length.

Reader's Guide:

Part 1. Minerals and the Economy contains three chapters. Chapter 1.1 is an up-to-date presentation of the behavior of mineral prices in real terms over the past 110 years. The hypothesis of no increase in mineral scarcity over this period is sustained. The effects of energy price increases, flowing from market control, and of environmental regulation are apparent in the 1970–1980 price behavior. Chapter 1.2 develops the role of minerals in the developing economics of the world. Chapter 1.3 stresses the international nature of the mineral industries. All of Part 1 is written at a level suitable for use by nonspecialists, and an introductory material for students.

Part 2. Here the formal economic theory of mineral production demand and resource appraisal are presented. The level of these chapters is much more rigorous and they require knowledge of economic theory and some mathematical background for understanding. This part is designed for reference by trained people and for instruction in formal mineral economic principles.

Part 3 contains chapters on analysis applied to the mineral sectors. Chapter 3.7 covers investment and financial analysis. This chapter is designed for reference by those concerned with the evaluation of mineral deposits for their economic viability, and for use in evaluation courses. Chapter 3.8 and .9 describes the current state of the art in mineral and energy modeling.

Part 4 describes the major mineral sectors—the metals, oil and gas and the solid fuels—with emphasis on aspects that are of special importance to each sector.

Finally, Part 5 addresses public policy. Chapter 5.13 stresses the contribution of economic principles to the development of efficient public policy, Chapter 5.14 reviews the policy discussions during the 1970s on nonfuel minerals, and Chapter 5.15 covers environmental policy with respect to minerals in the United States. This part is designed for reference by those interested in mineral policy issues and analysis.

As editor of this edition, I was greatly assisted by the associate editors, who aided me in refereeing the draft chapters. I am also grateful to the authors who contributed their effort, and met strict deadline schedules. I thank my many colleagues who also served as chapter reviewers. Associate editor Carla Sydney Stone, has prepared an Introduction to discuss an issue of importance.

WILLIAM A. VOGELY, EDITOR
July 1984

INTRODUCTION

It is appropriate that the fourth edition of *Economics of the Mineral Industries* was planned to be published in 1984. While George Orwell's predictions of a world ruled by a technological monster called "Big Brother" have not come true; computer, communications, and information processing technologies have advanced to a degree that not even Mr. Orwell could have predicted.

Just as the characters in *1984* sat mesmerized before a large television screen, so do much of the American public watch the numbers and words which appear on the millions of video display terminals in corporations and universities. However, the information which is displayed across the screens is not a product which can be consumed or used in the manufacture of other products, but a tool to be used by managers and policy makers in reaching decisions about the effects of various methods of production.

In reaction to increased competition from abroad in basic industries such as mining, steelmaking, automobile and equipment manufacturing and high domestic energy and environmental costs, many American policy makers have declared that the future of this country rests in the hands of information processors, computer manufacturers, and other "clean, high-tech" industries. In doing so, these governmental and corporate officials ignore the larger economic effects of a purely service society or "high-tech" economy.

The United States presently supplies much of its own needs for agricultural products, copper, steel, industrial equipment and even oil. It is virtually self sufficient in coal, building materials, fertilizers and food. It exports not only its technological know-how but also most of the forementioned raw materials and goods in addition to millions of consumer products each year. As Professor Henry McCarl points out in his chapter, the United States is an integral supplier to the world's flow of minerals and manufactured goods. Should the United States cease to be a producer and become solely a consumer of raw materials, not only would this country's economy be affected, but so would the world's.

It is legitimate for American officials to want to reduce the dependence of the economy on the cyclicallity of raw materials prices. It also is reasonable for these officials to want to put unemployed miners and metal workers to work in industries in which the United States is a significant exporter rather than consumer. In order to attract new high technology companies to their part of the country, local government bodies have offered generous tax benefits, promised low labor costs, new industrial "parks" and other incentives. Unfortunately, many of these new companies are just as subject to competition and price instability in the goods and services they sell as the basic industries that they were to replace.

Worse still, mineral production, basic industry, and agriculture no longer seem to be considered a vital part of America's future. The great strides made by the mining and manufacturing industries in productivity; cost, safety, and environmental control; and energy use during the 1970s were all but ignored by government and corporate economists as they grappled with the problems posed by the explosive increase in the price of oil. Although thousands of pages were written about the "dire" consequences facing those countries which were dependent on oil imports, little if any attention, then or now, was given to the strategic and economic problems facing the United States should it be transformed into a service or "high-tech" economy.

The best thinkers in the field of mineral economics have contributed to this fourth edition of *Economics of Mineral Industries*. It is much more than a handbook for practicing engineers or a college textbook for future mineral economists. Rather, this volume provides the foundation and models for theoreticians and practitioners to use in solving the basic supply and demand questions facing the world's minerals industries. It is an example of how the new computer and information processing technologies can be used to aid the production of raw materials and basic industrial products. No author pretends that the charts, graphs and formulae contained in this volume are directly consumable products. These chapters present the scope and breadth of theories, models, and mathematical methods available as aids to industry and governmental managers in reaching crucial decisions about the role of mineral production today and in the future.

CARLA SYDNEY STONE
Wilmington, Delaware
July, 1984

Part 1

MINERALS AND THE ECONOMY

1.1

Minerals and Economic Growth

John G. Myers* and Harold J. Barnett†

INTRODUCTION, CONCEPTS, AND MEASUREMENT

Introduction

Periods in the history of man, from his earliest appearance to the present, are often characterized by the principal materials used. In nearly all cases, the material or fuel is of mineral origin—stone, copper, bronze, iron, coal, and so forth—indicating the principal ingredient of the tools or source of power of the period. Each of these periods represented great improvement over the preceding one in terms of human well-being, as measured by population growth and other indicators (Cipolla, 1978). A major reason for the improvements in production and well-being was the nature of the material or fuel, each successively superior to the preceding, which made it possible to produce more food, fuel, shelter, and clothing with a given amount of labor.

If we accept this well-documented description of economic progress as accurate, we may then turn to the consideration of two related questions concerning this development. The first, easily answered, is why man spent long periods of time (millenia, in many cases) making tools from inferior materials, such as copper, instead of passing directly to superior materials, such as iron and steel. The second, closely related question, concerns the relation between the rate of economic growth and the quantity of mineral reserves available at any time.

For the first question, the answer is that the technology required to discover, extract, and process each successive material had to be discovered and developed before the material could be used. Once the technology was available, the new material could be and was exploited widely, raising productivity—output per unit of labor—and consequent consumption per person.

This familiar idea underlies a fundamental principle of the economics of resources, namely, at any time, available resources depend upon available technology. No mineral or other raw material is a resource unless the technology has been developed to utilize it. But once the technology is at hand, the new resources associated with it provide the means for economic growth and rising levels of living.

This principle is subject to certain conditions. For example, governments or other institutions may prevent a mineral from being used, even though the technology is available. And, if a mineral is located in a remote place, transportation charges may make it uneconomic to exploit. Such minor qualifications do not, however, change the thrust of the argument that technology is the overriding determinant.

With respect to the second question, it might seem apparent that the greater the quantity of reserves available, up to some point, the more economic growth will be facilitated. This line of reasoning leads to the conclusion that as reserves are consumed the quantity available will be reduced and economic growth will eventually be restrained. Reasoning of this sort underlies much popular (and scholarly) thought, in which depletion of mineral and other resources is assumed to lead to constraints on economic growth, through higher materials costs and absolute limitations on materials use.

Two important qualifications must be made to this explanation, however, which together

*Southern Illinois University at Carbondale
†Washington University, St. Louis

often result in a development over time which is the exact opposite of that predicted by the depletion and increasing scarcity hypothesis. The first and most important qualification is changing technology; this is basically the same consideration as that given in answering the first question, above. As technology changes—new methods of discovery, extraction, and processing are developed—the prices of materials derived from minerals not only may not rise, as expected from depletion alone, but may fall, either absolutely or relative to non-extractive materials and finished products.

The second qualification is that the quantity of reserves known and available may be a function of demand. Exploration and initial development required to create reserves are costly. They are undertaken only as needed to maintain an inventory of reserves considered adequate relative to the current and anticipated rates of extraction. Exploration is also affected by improvements in technology, so the cost of finding new mineral deposits normally rises less than would be true if depletion of the most easily found deposits were the only determinant.

In this brief introduction, some of the highlights of the relationship between minerals and economic growth have been touched upon. In order to test the hypothesis of increasing scarcity, a number of concepts must first be discussed and measurement problems explored. Then developments in the United States during the last century will be studied.

Depletion, Increasing Cost, and Technical Change

Much economic analysis is carried out in a framework which specifically excludes technical change. This is particularly true of studies of exhaustible resources, of which studies of minerals exploitation form a large part.

Studies of minerals exploitation go back at least to the early 19th Century. A notable example is a study by Jevons (1906), which pointed out that the rapid rise in the economic growth of Great Britain during the preceding 100 years had been based in large part on the use of coal. Given that the initial quantity of coal in the nation was fixed, Jevons deduced that growth in industrial output and advantageous foreign trade could not be maintained at past rates as these coal resources were depleted. The possibilities of substitute fuels, such as petroleum, were not considered.

An analytic model, which describes the consumption over time of a fixed amount of a resource, was given by Hotelling (1931). Under conditions of constant costs of extraction, the price of a mineral will rise over time owing to the necessity for reimbursement to the owner of reserves for postponing production. Subsequent researchers have elaborated the Hotelling theory in many ways (Devarajan and Fisher, 1981). Most theoretical analyses have supported the Hotelling finding of rising prices in the long run. (The reader is referred to Chapter 2.4 of this volume, ''The Production of Mineral Commodities,'' for a more complete discussion.)

In the following sections, four influences on mineral prices and costs are discussed which bear on the topic of this chapter, minerals and economic growth. These influences are: (a) technical progress in discovery and extraction; (b) discoveries of new reserves; (c) deviations from a competitive price path; and (d) externalities.

Technical change was discussed above. Its impact is to lower the cost of discovery, extraction, or processing of minerals, and thus offset the rise in price brought about by depletion. Discovery of new reserves will result in a drop in the price of a resource, but eventually this will be followed by a rise in price from the new level. A series of such discoveries will result in a ratchet-like path for the resource. If the discoveries are unanticipated, a similar, ratchet-like path will result, except that the price declines and increases will be steeper. These ratchet-like paths can follow a rising or falling trend, of course, depending on the size of the successive discoveries. Increasing size will lead to a falling trend, and vice versa (Dasgupta and Heal, 1979).

Noncompetitive Price Behavior

A wide variety of market organizations can be examined under this heading—monopoly, monopsony, and many others. Under fairly general static conditions, however, a similar result is obtained: the price rises as reserves are depleted, but the initial price is higher and the rate of price increase is slower than under competitive conditions.

The noncompetitive market form that has had the greatest impact in the minerals area is the cartel. Here a group of producers combine to seek monopolistic control over the market for a mineral. In recent times, the most outstanding example is OPEC (the Organization of Petroleum Exporting Countries). This organization, formed in 1960, accounts for less than one-half of world production of oil, but much larger fractions of world trade in oil and world reserves.

A cartel can operate in a number of ways in attempting to achieve market control: (a) a minimum price can be set for cartel members that is higher than the competitive price; (b) production quotas for members can be set such that the sum of the quotas is less than the total output of the group that would result under competitive conditions; (c) exclusive market areas can be allocated to members within which monopoly control can be exercised; or (d) some combination of (a), (b), and (c) can be used. In all cases, the essence of cartel operation is to restrict output and thereby increase price. The choice of operating method is often dictated by ease of enforcement; how best can the cartel prevent cheating (which reduces the market power of the cartel) among its members?

OPEC began by setting a minimum oil price (with premiums and discounts for different qualities and locations), but later added production quotas in an attempt to strengthen control over its members. The success of OPEC in raising the price of oil, from $3 per barrel in 1973 to $34 in early 1983, is by far the most spectacular in modern times. Although OPEC reduced its base price to about $29 during March 1983 in response to a "glut" of oil brought about by a deep recession in western, industrialized nations and increasing production by non-OPEC nations, it still exercises significant market power.

The essential fact, for the relation of minerals to economic growth, is that the exercise of monopoly power over a widely used resource, such as oil, can hinder economic growth to an alarming degree. Events during 1974–5 and 1980–3 illustrate this problem, as does the entire 1973–83 period.

Externalities and the Environment

A simple definition of an externality is an effect that does not enter into a market transaction, but results from it. If two parties, A and B, conduct a private transaction, but a third party, C, is harmed, an externality is associated with the transaction.

Two examples will help to make this concept clear. A blast furnace in an integrated steel mill emits particulates (soot), sulfur oxides, and other pollutants into the air. The steel mill carries on market transactions with sellers of minerals (coal, limestone, iron ore, etc.) and with purchasers of steel products, but other persons, not parties to the transactions, are adversely affected by the emissions. A second example is of a coal strip mine which makes a permanent change in the appearance of a natural landscape. Private transactions are carried out between the mine operator and the owner of the mineral rights, and between the mine operator and the purchaser of the coal. Other persons who enjoyed the appearance of the landscape before it was strip mined are adversely affected by the market transactions to which they are not parties.

Externalities affect the relation between minerals and economic growth. The price of a mineral or mineral product may not reflect the true cost to society of its production. The true cost may be higher than the market price, and the two may rise or fall at different rates over time. The result is that the mineral may be over-consumed, and the general welfare reduced in comparison with a situation in which the mineral price reflected its true cost to society (Fisher, 1981).

Federal, state, and local governments have long regulated the effects of some externalities on third parties. Until the late 1960s, however, such regulation applied mainly to a limited number of clear-cut cases, such as obvious contamination of drinking water, or the creation of hazards to navigation. Beginning with the National Environmental Policy Act of 1969, the United States has passed a number of laws, and created a large body of related regulations implementing these laws, that have the effect of reducing the negative externalities of minerals production and use. Pollution of the air, water, and land has been increasingly restricted by these laws and regulations. The creation of "disamenities" such as noise, unsightly landscapes, and the destruction of the wilderness has also been increasingly restricted.

Regulations of this type have the effect of raising the prices of minerals and products made from them, such as metals, relative to all other commodities combined. While many nonminerals have also been strongly affected by the regulations, the prices of minerals and their products have received a greater relative impact than those of "all others." To the extent that the regulations are "cost effective," that is, pollution and disamenities are reduced in a reasonably efficient manner, the general welfare is enhanced. In the terminology of environmental economics, the regulations internalize the costs of minerals production and use; costs that were external, imposed on third parties, are now properly incorporated in the cost of production.

The process of internalizing these costs began, for the main part, in the late 1960s and is still in progress. Thus, domestic prices of minerals and minerals products have been experiencing upward pressure over this period. If the costs of all externalities are eventually included in product prices by the regulation process, these new prices will then reflect the true cost to society of the product use. Since the new prices will generally be higher than before external costs were included, use of the associated products will be less than they would be at lower prices.

In summary, it can be said that the absence of pollution and other costs from minerals prices kept these prices too low and led to overconsumption of minerals. The gradual incorporation of these costs in minerals prices, since the late 1960s, has raised minerals prices and reduced the use of minerals and their production.

Common Property Resources and Irreversibility

One form of externality frequently associated with minerals extraction is an irreversible change of a common property resource. An example of such an irreversibility is the change in a natural landscape brought about by strip mining of coal mentioned in the preceding section. The common property aspect is that before appropriation, the landscape, wilderness area, area of the ocean, clean air, and so forth, belong to us all, but property rights are not specifically assigned. In cases where the landscape is changed, the wilderness destroyed, or a biological species driven to extinction, the change is irreversible (Fisher, Krutilla, and Ciccheti, 1972).

As the term implies, an irreversibility differs from other externalities in that the services formerly yielded by the environment, termed amenities, are no longer available to society. Many cases of pollution are reversible, in that the absorptive abilities of the environment will eventually make the air fit to breathe and the water fit to drink or swim in, if the pollution is reduced. These amenities are then once again available to society. This is not the case when an irreversible change occurs.

As minerals extraction is expanded, irreversible changes often take place. Legislative protection of natural environments to prevent such changes has a long history; the creation of the national park system is a prime example. The Wilderness Act of 1964 and the Wild and Scenic Rivers Act of 1968 are recent government actions designed to prevent irreversible changes in some locations (Smith, 1974). The critical aspects for the topic of this chapter are that irreversible changes have taken place and are currently taking place in the extraction of minerals, and that there is a difference in kind between reversible and irreversible environmental damage.

When a mineral, or other natural resource, is depleted, it may rise in price reflecting increased scarcity. Of course, technical change may offset this effect in whole or part. When extraction of the mineral leads to an irreversible change in a natural environment, however, the amenities produced by the natural environment are permanently reduced, and technical change cannot restore them.

Natural environments, as noted above, yield services without which life on earth would be impossible. The quantity of these services cannot be increased, however, which distinguishes them from privately produced goods. The services of the atmosphere, bodies of water, wilderness areas, and other common properties cannot be enhanced by technical change. As the production of private goods increases over time, the services of common property resources become relatively more scarce. Because common property resource services are not traded in markets and consequently are not priced, their costs are not included in the production costs of private goods.

This discussion may be summed up as follows. The common property concept leads to a similar conclusion as that reached in the previous section, namely, that the prices of many privately-produced goods do not reflect the cost to society of their production. As governments intervene to protect common property resources, the costs of some privately-produced goods will rise relative to others. The largest relative price rises will be experienced by those goods that use the greatest volume of the services of common property resources; many goods produced from, or with the aid of, minerals are among these.

The irreversibility concept adds a new dimension to the discussion. The destruction of a natural environment reduced the services of common property resources. Insofar as minerals production leads to an irreversible change, a permanent reduction in these services results.

Measurement of Minerals Scarcity and Economic Growth

In order to establish the extent to which the increasing relative scarcity of minerals will impede economic growth, and vice versa, concepts of increasing relative scarcity and economic growth must now be stated more precisely and their implications explored. Only then can empirical analysis be meaningfully undertaken.

Minerals Scarcity: With the one major exception of external costs, a rise in the ratio of the price of a mineral to the general price level means that the mineral has become relatively more scarce. The relative price of a mineral may rise from (a) depletion, (b) slower than average technical change, or (c) noncompetitive forces. Whatever the reason for the rise in relative price, the fact remains that purchasers must pay more for the minerals. This hinders production of the commodities for which the mineral is an input. Production costs increase, quantity demanded falls, and output is reduced. Conversely, a decline in the relative price of a mineral enhances the production of commodities for which the mineral is an output. An unchanged relative price is neutral.

The exclusion of external costs means that the price does not reflect the full cost to society of producing the mineral. Historically, when government actions to internalize costs were negligible, the *levels* of the relative prices of minerals were clearly too low. Whether the change in the relative price of a mineral over time was distorted by the omission of external costs depends on the behavior of the ratio of external costs to the mineral price. If the ratio was stable, then the movement of the relative price of the mineral is a good indicator of relative scarcity.

There is presumptive evidence, however, that the ratios of external costs to prices of minerals rose up to about 1970. This is derived from the discussion in the preceding section. Services of common property resources cannot be expanded, and are therefore becoming relatively more scarce. Since minerals production consumes an above average amount of the common property services, these external costs presumably rose relative to minerals prices. Beginning about 1970, however, an increasing fraction of these costs have been internalized, so the ratio of external costs to minerals prices has probably fallen.

A number of other measures of changing relative scarcity have been examined (Barnett and Morse, 1963). These include various productivity indices and their reciprocals, often termed "real costs." Examples of productivity indices are output per unit of labor input, per unit of other inputs, or per unit of a composite input (such as labor and capital in some fixed proportion). Another measure of scarcity that has been suggested is the cost of discovering new reserves. In this chapter, for the reasons outlined above, relative price changes will be used as measures of changing scarcity, subject to the qualification of external costs.

Economic Growth: This is usually measured by changes in net national product (NNP), in the aggregate or per capita, after correction for price changes. The result, "real" NNP, represents the physical volume of production of goods and services, during some period, that is available for consumption or investment in productive capacity. In practice, the gross national product (GNP) is more commonly used, because there are difficulties in the measurement of capital consumption (mainly depreciation), which must be subtracted from GNP to obtain NNP.

The aggregate price level, mentioned above, is the ratio of GNP in current prices to real GNP. Changes in that ratio indicate changes in the general price level. Changes in the ratio of mineral price to the general price level indicate changes in the *relative* price of that mineral.

A qualification to the usefulness of changes in real NNP as a measure of economic growth comes from the concept of external costs. To the extent that external costs are not deducted from the value of production, changes in real NNP overstate economic growth (Nordhaus and Tobin, 1972). A striking case of this occurs when an irreversible change in a common property resource takes place. As a result, the services of common property resources are reduced for all future periods.

A satisfactory measure of national welfare, real production net of external costs, is not available. However, external costs, since the 1960s, have been increasingly internalized, so real NNP has been approaching a more satisfactory measure of national welfare.

CASE STUDY: U.S. DEVELOPMENTS, 1870–1980

Relevance of Case Study

In 1870, the United States was largely an agricultural nation; 48% of all employed persons worked in agriculture, and many more were employed in closely related areas (Fig. 1.1.1). By 1980, an extraordinary change had taken place. Most of the population lived in urban places and only three percent of the labor force was employed in agriculture. This change was the result, of course, of a very rapid and thorough industrialization of the nation. This industrialization has a vast number of far-reaching consequences. The most important of these, for the topic of this chapter, was the enormous rise in quantity of minerals consumed.

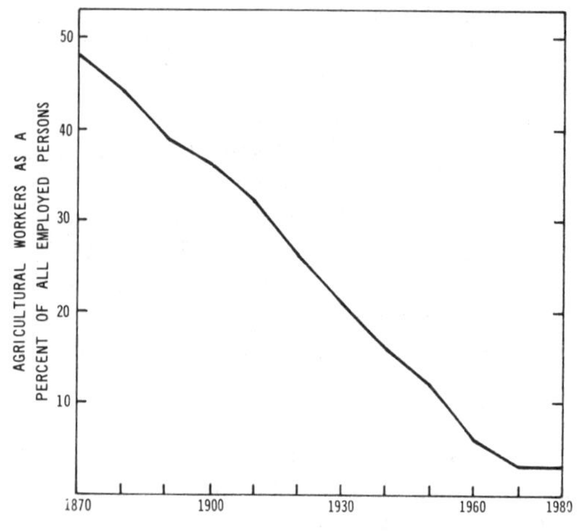

Fig. 1.1.1—Employment in Agriculture, 1870–1980 Source: U.S. Bureau of the Census, 1975.

In 1870, only modest amounts of ores were used in the nation, and the principal fuel used was wood. All that changed rapidly, however, as manufacturing consumed more and more minerals, for use as raw materials and fuel, and other sectors of the economy—transportation, households, services, and government—consumed increasing amounts of mineral fuels. By 1980, the United States was a highly-industrialized nation, making enormous demands on minerals, both domestically produced and imported. (Table 1.1.1)

In addition, the United States has become by far the leading agricultural nation in the world. This is also based on minerals use. Mineral fertilizers have been substituted for agricultural land, petroleum-driven engines in place of animal power, and metal farm machinery and implements for those of wood.

The extraordinary demands made on minerals by this development would have resulted in sharp increases in the relative prices of minerals, had the simple static model of exhaustible resources held sway. In the remainder of this section the statistical evidence of the 110-year period will be examined with respect both to the simple static model and to the additional influences on price discussed in the previous section. First, however, a brief outline of data sources and

Table 1.1.1—Mineral Consumption, 1870–1970, Selected Years
Physical Quantities Valued at 1967 Prices
(Millions of Dollars)

Year	Total	Fuels	Metals	Non-metal, Non-fuel Minerals
1870	456	258	64	134
1880	964	510	172	283
1890	1,878	995	349	535
1900	2,635	1,551	584	500
1910	4,936	3,104	1,143	689
1920	6,716	4,533	1,511	673
1930	7,860	5,426	1,450	985
1940	9,669	6,627	2,136	906
1950	14,419	9,748	3,021	1,650
1960	18,669	13,181	2,788	2,700
1970	27,326	19,651	3,691	3,984
	Annual Average Rate of Change (%)			
1870–1970	+4.2	+4.4	+4.1	+3.5

Source: Manthy, 1978.

adjustment is presented.

Data Sources and Adjustment

Two overlapping periods are examined, 1870 to 1970, and 1950 to 1980. The purpose of this treatment is the availability of data and the nature of the 1970 to 1980 period. For the earlier period, a carefully prepared set of minerals data is available in a volume which extends only to 1973 (Manthy, 1978). The 1970 to 1980 period is unique in so many respects that coverage of this decade is important. The unique aspects of this period can be seen only in comparison with earlier years, so two data sets which extend from 1950 to 1980 are analyzed: one for fuels (Energy Information Administration, 1983) and the other for metals and ores (Myers et al.).

The 1870 to 1970 period is analyzed in an attempt to discover durable trends. Earlier long-term studies include Potter and Christy (1962), Barnett and Morse (1963), Smith (1978), and Slade (1982). A study of the Post-WW II period using a methodology similar to that of this chapter is Hall and Hall (forthcoming).

Prices of individual minerals and groups of minerals are expressed relative to the GNP implicit price deflator to seek an answer to the question of whether minerals are becoming more scarce relative to other goods and services. The procedure is to divide the minerals price by the GNP deflator. If the resulting relative (deflated) price rises over time, increasing scarcity is indicated, and vice versa.

Relative Price Changes 1870 to 1970

For the entire 100-year period, the relative price of minerals definitely fell. For each of the three main groups shown in Table 1.1.2, fuels, metals, and other minerals, prices rose less than the GNP deflator. When the 100-year period is divided into five 20-year subperiods, some variation is found. During the first subperiod, 1870 to 1890, the relative price of all minerals, as well as of each main group, fell. The next three subperiods reveal no significant changes in the average relative price of all minerals, although some groups show definite trends. During the last 20 years, 1950 to 1970, the total and each subgroup experienced significant declines in relative price.

Fuels: Within the fuels group, divergent price patterns are found. The price of petroleum fell substantially from 1870 to 1910, in both absolute and relative terms, as is typical of new commodities. It rose from 1910 to 1950, again both absolutely and relatively, and then fell in relative terms from 1950 to 1970. The post-war decline resulted principally from rising imports

Table 1.1.2—Changes in Relative Prices of Minerals, 1870 to 1970, Minerals Prices Divided by GNP Deflator
(Average Annual Rates of Change, %)

Mineral Category	Period					
	1870-1970	1870-1890	1890-1910	1910-1930	1930-1950	1950-1970
All minerals	−0.7†	−5.0†	−0.6	−0.3	0.1	−1.0†
Fuels	−0.4†	−5.4†	−0.5	1.2	0.6	−1.0†
Metals	−0.7†	−1.9†	0.5	−3.3†	1.1†	−0.5†
Non-fuel, non-metal	−1.5†	−4.9†	−1.3†	−3.1†	−2.1†	−1.3†
Fuels*						
Petroleum	−0.4†	−6.0†	−0.6	1.4	1.2†	−1.4†
Natural Gas	n.a.	n.a.	n.a.	n.a.	−4.4†	2.4†
Bituminous Coal	0.1	−1.5†	0.3	0.0	2.8†	−2.2†
Metals*						
Pig Iron	−0.4†	−2.0†	0.5	−2.1†	1.4†	−1.0†
Copper	−1.0†	−2.1†	0.5	−4.0†	1.2	0.7
Lead	−0.4†	−0.6	0.2	−0.8	3.2†	−2.7†
Zinc	−0.6†	−0.9	0.5	−2.8	3.0†	−2.1†
Molybdenum	n.a.	n.a.	n.a.	n.a.	−2.1†	1.2†

†Significantly different from zero at the 5% probability level.
*Only selected components are shown separately.
n.a. = Not available
Source: Minerals prices from Manthy, 1978; GNP deflator from U.S. Bureau of the Census, 1975, and U.S. Department of Commerce, 1981.

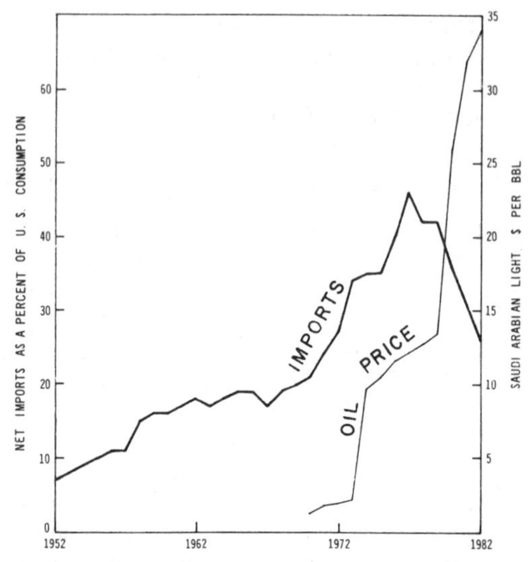

Fig. 1.1.2—U.S. Petroleum Imports and OPEC Price, 1952–1982 Source: Energy Information Administration, 1983.

of low-priced petroleum and products (Fig. 1.1.2). It is noteworthy that the absolute price per barrel of petroleum in 1970 was about the same as it had been in 1871; in relative terms, the petroleum price had declined substantially, despite an enormous increase in production and use.

Representative national natural gas prices are not available prior to the end of World War I. The relative price of natural gas fell substantially until the end of World War II; after that it rose rapidly. Natural gas was transformed from a waste by-product of petroleum production to a commodity prized for its desirable characteristics; this occurred mainly after WW II.

Coal prices generally fell in relative terms from 1870 to 1910, and again from 1950 to 1970. Only during the 20 years from 1930 to 1950 did the price of coal rise relative to the general price level.

Metals: The relative prices of metals fell for the entire 100-year period, and for three subperiods (1870 to 1890, 1910 to 1930, and 1950 to 1970), but rose significantly from 1930 to 1950. From 1930 to 1950, pig iron, lead, and zinc rose sharply, but from 1950 to 1970, each of these metals fell in relative price. In the case of molybdenum, which has become more important in recent years, relative price fell from 1930 to 1950, and then rose from 1950 to 1970.

Nonfuel, Nonmetal Minerals: This group is composed of a variety of minerals, including crushed and broken stone, dimension stone, clays, etc. The relative price of the group fell for the entire 100-year period, as well as for each subperiod.

Summary of Period: Despite an extraordinary rise in consumption of minerals, their relative prices generally fell during the 100-year period. As the United States grew from a largely rural, predominantly agricultural nation into a great industrial power, heavy demands were made on minerals resources. Yet minerals prices show no evidence of increasing scarcity; indeed the contrary is true, because each major minerals category reveals a decline in relative price for the entire period.

By 1970, however, a number of developments had appeared that raised questions regarding the validity of price trends as indicators of changes in relative scarcity, and in the continuation of the price patterns themselves. These developments include an increasing awareness of the impact of minerals extraction and use, and an increase in the strength of cartels among suppliers of imported minerals. The effects of these developments came to be felt in the 1970 to 1980 period, which is examined in the next section of this chapter.

The Post-World-War II Period

During the two decades following World War II, the United States experienced rapid economic growth accompanied by moderate inflation (Table 1.1.3). By 1950, the economy had completed the post-war adjustment and was growing at an average rate of 3.5% per year (real GNP). Several recessions were experienced during the 1950s and 1960s, but all were mild by historical standards and growth was only briefly interrupted. The general price level rose in every year after 1949, but the rate of increase was modest by recent standards. From 1950 to 1970, the average annual increase was 2.2%; the year-to-year change exceeded six percent only once, during the Korean War (1950 to 1951).

The pattern during the 1970 to 1980 decade was quite different. Growth in real GNP was slower (3.1% per year, on average) and was interrupted by two severe recessions (1973 to 1975, and 1980). Inflation reached rates that had not been seen since World Wars I and II. For the decade of the 1970s, the average rise in the GNP deflator was 6.7%, and year-to-year changes exceeded eight percent four times.

Many influences combined to produce the economic difficulties of the 1970s. During the late 1960s, inflation began to rise as a result of the heavy demands made on the economy by the Vietnam War. The higher inflation rate attained by the end of the 1960s was not reduced significantly in the 1970s, despite a variety of policy undertakings (fiscal, monetary, and wage and price control) by the federal government. Also during the late 1960s, the overall rate of productivity growth (real GNP per hour worked) slowed, and remained below the 1950–65 rate throughout the 1970s. Slower productivity growth, combined with a high rate of increase in wage rates, added to inflationary pressures.

Other influences which contributed to the unfortunate record of the 1970s are directly related to minerals extraction and consumption. During the late 1960s and throughout the 1970s, laws designed to protect the environment from degradation were passed and regulations put into force. In addition, the Mine Safety and Health Act of 1969 and the Occupational Safety and Health Act of 1970 were made into law, and a series of regulations followed; these were designed to prevent injury and other damage to the health of miners and other workers.

Table 1.1.3—Economic Growth and Inflation, 1950 to 1980
Real GNP (1972 Prices) and GNP Deflator (1972 = 100)

Year	Real GNP (Billions of 1972 dollars)	Implicit Price Deflator (1972 = 100)
1950	534.8	53.56
1951	579.4	57.09
1952	600.8	57.92
1953	623.6	58.82
1954	616.1	59.55
1955	657.5	60.84
1956	671.6	62.79
1957	683.8	64.93
1958	680.9	66.04
1959	721.7	67.60
1960	737.2	68.70
1961	756.6	69.33
1962	800.3	70.61
1963	832.5	71.67
1964	876.4	72.77
1965	929.3	74.36
1966	984.8	76.76
1967	1.011.4	79.06
1968	1,058.1	82.54
1969	1,087.6	86.79
1970	1,085.6	91.45
1971	1,122.4	96.01
1972	1,185.9	100.00
1973	1.254.3	105.75
1974	1,246.3	115.08
1975	1,231.6	125.79
1976	1,298.2	132.34
1977	1,369.7	140.05
1978	1,438.6	150.42
1979	1,479.4	163.42
1980	1,475.0	178.42
	Average Annual Rate of Change (%)	
1950–1970	**3.5†**	**2.2†**
1970–1980	**3.1†**	**6.7†**

†Significantly different from zero at the 5% probability level.
Source: U.S. Department of Commerce, 1981, 1982, and 1983.

Both types of regulations, environmental and occupational, were intended to redress past failures of the market system and have had measurably beneficial effects, but they have brought higher costs. Manufacturing plants, mines, and other establishments required to reduce emissions to the air and water achieved the necessary goals at substantial cost in terms of capital and operating expenditures. Similarly, the protection of workers from accidents and health hazards raised capital and operating costs of the establishments affected.

The major point for this discussion is that minerals producing and processing establishments were among those most strongly affected by the new regulations. It is fair to say that the production of basic materials, of which minerals and their products form a large part, bore the brunt of the impact of these regulations. As a result, the prices of many minerals during the 1970s rose relative to the general price level. The price increases in basic materials caused increases in the general price level as well, thus increasing the inflation rate.

A second strong influence on minerals prices was the sharp rise in oil prices, brought about by the actions of OPEC (Fig. 1.1.2). Net petroleum imports, although higher than during the prior 20 years, accounted for only 35%, on average, of U.S. consumption during the 1970–80 period. Nevertheless, the sharp rise in OPEC prices spread through the domestic economy, affecting the prices of domestically produced petroleum, other fuels and electricity, the production costs of heavy energy users, and the overall rate of inflation.

In addition to the impact of inflation, the increase in regulations and fuels prices caused major difficulties for producers. Instead of concentrating their efforts on the customary tasks of producing goods and services, large amounts of time and expenditures were taken up by the necessity of dealing with these new problems. National productivity and production suffered as a result. And, as mentioned above, these effects were most pronounced in the production of basic materials, which include minerals and their products.

Hypothesis of Increasing Minerals and Scarcity in the 1970s: The preceding discussion provides a rationale for formulating the following hypothesis: the relative decline in minerals prices during the 1950 to 1970 period was reversed during the 1970s, which would be a distinct break from the evidence of the preceding 100 years, as outlined earlier.

A test of this hypothesis can be carried out with a fairly simple trend analysis.

Let Y_t = relative price of a mineral in year t,
t = year—1949, for years from 1950 to 1980
D = dummy variable with value of zero from 1950 to 1969 and value of one thereafter,
a_0, a_1, b, c = constant coefficients.

A regression estimate of the model

$$log_e Y = a_0 + a_1 D + bt + cDt$$

will permit a test of the hypothesis for a mineral or group. An estimate of the trend during the 1950 to 1970 period is given by the value of "d." If the estimate value of "c" is significantly different from zero, it indicates that the trend changed during the 1970s from the 1950 to 1970 period. If the estimates of both "b" and "c" are statistically significant, then their sum measures the trend during the 1970s. This test is applied first to fuels and then to other minerals in the following sections.

Fuels: The results of the estimation procedure, shown in Table 1.1.4, indicate a sharp change in the trends of relative prices during the 1970s. Petroleum and coal prices, which, in relative terms, had been falling during the 1950 to 1970 period, rose strongly thereafter. A similar pattern is found for the composite measure. (The trend for the 1950 to 1970 period in Table 1.1.4 differs from that in Table 1.1.2, for total fuels, because the basic data sources used different methods to aggregate fuel prices.) Natural gas prices, which had been rising from 1950 to 1970, show a greatly accelerated rise during the 1970s.

All the trend rates for fuels are significantly different from zero, so the hypothesis of increasing relative prices during the 1970s cannot be rejected for this group.

Ores and Metals: For the 1950 to 1970 period, no general trend is discernable. Four categories show statistically significant downward trends, two show significant rising trends, and four show no definite direction. A definite change appears for the following period: seven categories show significant positive changes. As a result, seven of the 10 ores and metals listed in

Table 1.1.4—Comparison of Relative Price Changes of Minerals, 1950 to 1970 and 1970 to 1980
Minerals Prices Divided by GNP Deflator
(Average Annual Rates of Change, %)

Mineral Category	Trend 1950 to 1970 "b"	Change* in Trend "c"	Trend 1970 to 1980 "b" + "c"
Fuels#	−1.7†	13.8†	12.1†
Petroleum	−1.4†	12.5†	11.0†
Natural Gas	2.6†	15.0†	17.6†
Bituminous Coal	−2.2†	11.7†	9.4†
Ores and Metals#	−0.2	2.5	2.3
Iron Ore	−1.1†	1.5	0.4
Steel	−0.3	4.7†	4.5†
Manganese	−6.2†	10.6†	4.4†
Aluminum	−0.1	4.8†	4.7†
Copper	0.3	−1.7	−1.3
Lead	−2.8†	+8.8†	6.0†
Mercury	3.9†	−14.1†	−10.3†
Nickel	1.1†	2.5†	3.6†
Tin	0.6	9.1†	9.8†
Zinc	−2.1†	4.9†	2.7†

*See text for explanation.
†Significantly different from zero at the 5% probability level.
#Only selected components are shown separately.
Source: Fuels prices from Energy Information Administration, 1983; Ores and metals prices from Myers et al.; GNP deflator from U.S. Department of Commerce, 1981, 1982, and 1983.

Table 1.1.4 show definite rising trends during the 1970 to 1980 period, and only one shows a definite decline. The weight of the evidence, therefore, supports the hypothesis of increasing relative prices during the 1970s.

Meaning of Test Results: If the test results are accepted as valid, the apparent meaning is that minerals became relatively more scarce during the decade of the 1970s, which is a reversal of the trend found for the preceding 100 years. To the extent that this change in trend is the result of the price rises brought about by OPEC, the rising scarcity conclusion is valid. It is, however, subject to a different interpretation than is often presented. That is, depletion of exhaustible resources does not explain the pattern of the 1970s; instead, noncompetitive market behavior is the basic explanation. In this case, the 1980s will furnish an interesting test case: as OPEC's market power weakens and inflation subsides, will the recent upward trend in the relative price of minerals—fuels and metals—be reversed? Although some current evidence points in this direction, more data are needed in order to reach a firm conclusion.

The other important set of influences on minerals prices, discussed earlier in this chapter, arises from environmental, and worker safety and health regulations. To the extent that minerals price increases caused by these regulations represent the gradual inclusion in production costs of negative impacts formerly imposed on others (externalities), the price increases are not valid indicators of relative scarcity. In other words, the increases in the relative prices of minerals in the Seventies may reflect the gradual inclusion of costs which should have been included in minerals prices in the past. Once the inclusion process has been completed, the relative prices may rise or fall, but the trend should be different than during the transition. Inasmuch as the inclusion process is continuing during the 1980s (proposed acid rain legislation is an example), it will be some time before a test of this explanation can be made.

Conclusions of Case Study

During the 100 years from 1870 to 1970, the United States was transformed from a small, primarily agricultural nation to a large highly industrialized nation. Mineral consumption at the end of the period was 60 times the amount at the beginning (Table 1.1.2).

The following 10 years, 1970 to 1980, reveal a distinct change in this pattern, however. The relative price of every member of the fuels group examined rose sharply. And, of 10 ores and metals examined, seven show statistically significant price increases while only one shows a significant fall (Table 1.1.4); the remaining two members of this group show no definite trend.

Two explanations offered for the change in the 1970s are the actions of OPEC and the rise in regulations (for environmental control and for worker safety and health). Only the first of these, namely, OPEC's price hikes, represents true increasing scarcity and is not the result of the depletion of exhaustible resources but rather arises from the exercise of non-competitive market power.

The second explanation, increasing regulation, makes the interpretation of relative price changes difficult. Insofar as past minerals prices did not reflect the true cost to society of their production and use, while the new regulations are causing them to do so, the resulting price changes give no clear evidence. That is, the prices changes resulting from increasing regulation can overstate or understate changes in the relative scarcity of minerals. For the United States, price changes prior to these new regulations are imperfect indicators of relative scarcity, because only part of total costs were reflected in minerals prices.

Developments during the 1980s may make it possible to measure the separate importance of each of these explanations. If OPEC's power wanes or rises, or if the regulation of minerals stabilizes or increase, it may be possible to make a definite judgment regarding the question: Has there been a substantial change in the long-term trend? And if so, what are the causes?

Additional insight into the nature of the developments of the 1970s may be drawn from a comparison of the post-war behavior of prices in Europe with those in the United States.

EUROPEAN POSTWAR EXPERIENCE

Coverage, Data Sources, and Data Adjustment

To facilitate comparison with the U.S. and avoid tedious repetition, combined data for the 19 European members of the OECD (Organization for Economic Cooperation and Development) are analyzed. The 19 nations, market economies mostly located in western Europe, are the following: Austria, Belgium, Denmark, Finland, France, the Federal Republic of Germany, Greece, Iceland, Ireland, Italy, Luxembourg, The Netherlands, Norway, Portugal, Spain, Sweden, Switzerland, Turkey, and the United Kingdom.

Minerals price data were taken from unpublished materials (Myers et al.). The deflator used to convert the price series into relative terms was computed by dividing the combined national products of these nations in current prices by the combined values in constant prices.

Relative Price Changes, 1950 to 1970

The minerals shown in Table 1.1.5 are almost identical with those in Table 1.1.4. A comparison of the figures from these two tables for the 1950 to 1970 period reveals that the direction of the trend was the same for the two major groups (fuels, and ores and metals) and for nine of the 12 individual categories (Table 1.1.6). The difference is that the magnitude of negative trend found for both the United States and OECD-Europe is much larger in the latter. Only one metal, mercury, rose in relative price in OECD-Europe and at a lesser rate than it did in the United States. All other categories fell more in relative price in OECD-Europe than in the United States.

An explanation for this difference may be drawn from the comparison of deflators shown in Table 1.1.7. The rate of inflation was more

Table 1.1.5—Western Europe, Relative Price Changes of Minerals, 1950 to 1970 and 1970 to 1980 Minerals Prices Divided by GNP Deflator (Average Annual Rates of Change, %)

Mineral Category	1950 to 1970	1970 to 1980
Fuels*	−3.7	16.6
Petroleum	−4.2	18.3
Natural Gas	n.a.	6.6
Coal	−3.0	2.9
Ores and Metals*	−1.9	−2.6
Iron Ore	−3.0	−4.4
Steel	−2.0	0.8
Manganese	−7.8	−0.6
Aluminum	−1.9	−0.2
Copper	−0.4	−7.5
Lead	−5.2	1.5
Mercury	1.8	−14.1
Nickel	−0.4	−1.3
Tin	−1.2	5.4
Zinc	−4.5	−3.8

*Only selected components are shown separately.
n.a. = Not available.
Source: Myers et al.

Table 1.1.6—Relative Price Changes of Minerals, OECD-Europe and U.S., 1950 to 1980 (Average Annual Rates of Change, %)

Mineral Category	1950-70		1970-80	
	Europe	U.S.	Europe	U.S.
Fuels	−3.7	−1.7	16.6	12.1
Petroleum	−4.2	−1.4	18.3	11.0
Natural Gas	n.a.	2.6	6.6	17.6
Coal	−3.0	−2.2	2.9	9.4
Ores and Metals	−1.9	−0.2	−2.6	2.3
Iron Ore	−3.0	−1.1	−4.4	0.4
Steel	−2.0	−0.3	0.8	4.5
Manganese	−7.8	−6.2	−0.6	4.4
Aluminum	−1.0	−0.1	−0.2	4.7
Copper	−0.4	0.3	−7.5	−1.3
Lead	−5.2	−2.8	1.5	6.0
Mercury	1.8	3.9	−14.1	−10.3
Nickel	−0.4	1.1	−1.3	3.6
Tin	−1.2	0.6	5.4	9.8
Zinc	−4.5	−2.1	−3.8	2.7

Source: Tables 4 and 5.

Table 1.1.7—Comparison of Gross Product Deflators, U.S. and OECD-Europe, 1950 to 1980 (Average Annual Rates of Change, %)

	1950 to 1970	1970 to 1980
U.S.	2.2	6.7
OECD-Europe	4.8	11.7

Source: U.S. Department of Commerce, 1981, 1982, and 1983; Myers et al.

than twice as great in Europe than in the United States from 1950 to 1970. This accounts for nearly all the difference in relative price trends between the United States and Europe from 1950 to 1970; the price changes before deflation, for the categories shown in these tables, are lower (less increase or greater decrease) in the United States in nine out of 14 cases, higher in five (Table 1.1.8).

The meaning of this is clear: inflation was greater in Europe than in the U.S., but it was predominantly in non-mineral goods, and in services. In the United States, price trends in minerals were more similar to those in other commodities.

Table 1.1.8—OECD-Europe and U.S., Minerals Price Changes, 1950 to 1980 (Not Deflated) Average Annual Rates of Change, %)

Mineral Category	1950–70		1970–80	
	Europe	U.S.	Europe	U.S.
Fuels	0.6	1.1	30.3	20.6
Petroleum	0.1	1.2	32.2	21.1
Natural Gas	n.a.	4.6	19.1	25.2
Coal	1.3	1.6	15.0	15.7
Ores and Metals	2.5	2.1	8.8	9.6
Iron Ore	1.3	1.0	6.8	7.6
Steel	2.4	2.0	12.6	12.0
Manganese	−3.7	−4.1	11.1	11.9
Aluminum	2.4	2.1	11.5	12.2
Copper	4.1	3.0	3.3	5.7
Lead	−1.0	−0.4	13.4	13.7
Mercury	6.3	5.9	−4.0	−3.4
Nickel	4.0	3.7	10.3	11.0
Tin	3.2	3.0	17.8	18.1
Zinc	−0.2	0.2	7.5	10.1

Source: See Tables 4 and 5.

The Decade of the 1970s

Inflation was much greater in Europe than in the United States during the 1970s, as well (Table 1.1.7). This alone, however, will not explain the differences in relative price trends in minerals in the 1970s between Europe and the U.S. (Table 1.1.6). Two additional influences must be taken into account in order to understand the differences between the two sets of figures. These are price control of fuels in the United States and movements in the exchange rate of the U.S. dollar for European currencies. The price control effect is discussed immediately below. Exchange rate movements apply mainly to ores and metals, so they will be discussed in a later section.

Fuels: Two of the three principal fuels, petroleum and natural gas, were subject to price control in the United States during the 1970s. The price of domestically-produced petroleum was frozen on August 15, 1971, and then permitted to rise gradually during the rest of the decade. Price controls on petroleum and its products were not completely removed until 1981. The result was that the price of petroleum in the United States lagged behind the world price, which was rising rapidly as a result of the actions of OPEC. The relative price of petroleum in Europe reflected the world price during the 1970s, while that in the United States did not (Table 1.1.6). As a result, the petroleum price rise was greater in Europe, and, because of the great importance of petroleum among fuels, the price rise of the fuels group was also greater in Europe.

The price of natural gas in interstate commerce in the United States has been controlled since the late 1950s. During the 1970s, price controls were reduced substantially, and prices rose rapidly as they moved from a very low base toward a level similar to that of petroleum prices. In Europe, natural gas, only recently developed in the Netherlands, did not surge in price as a result of decontrol.

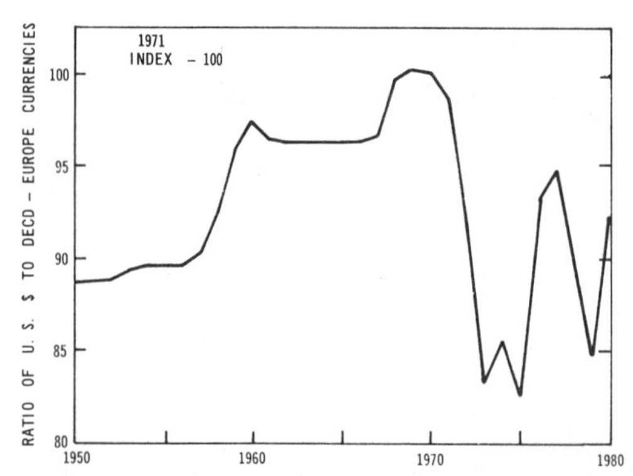

Fig. 1.1.3—Value of U.S. $ in Terms of OECD-Europe Currencies, 1950–1980 Source: Myers et al.

Ores and Metals: These minerals are extensively traded among nations, so the price in an importing nation varies with the nation's exchange rate. In 1971 and again in 1972, major realignments in the exchange rates among OECD members took place. The principal changes were to lower the value of the U.S. dollar relative to the currencies of most of its trading partners, in particular relative to OECD-Europe (Fig. 1.1.3). During the rest of the decade, exchange rates fluctuated widely, but the U.S. dollar remained well below the 1970 level, in comparison with OECD-Europe.

The effect of this exchange rate development was to cause the prices of ores and metals to rise more in the United States than in Europe. As shown in Table 1.1.7, the average price for the ores and metals group, and the prices of nine of the 10 individual categories, increased more in the U.S. This was a reversal of the pattern of the 1950 to 1970 period, when the exchange rate of the dollar was rising in comparison with OECD-Europe (Fig. 1.1.3).

In summary, the relative price of ores and metals in the United States rose, in part, because of a weakened exchange rate while the reverse occurred in Europe. The most important point for the topic of this chapter is that no increase in the relative scarcity of ores and metals became apparent for Europe in the 1970s.

CONCLUSIONS OF CHAPTER

A principal idea put forth in this chapter is that economic growth is aided if the prices of minerals are falling relative to the prices of other goods, and hindered if they are rising. A great deal of popular and theoretical discussion predicts that the relative prices of minerals will rise over time as the result of depletion of the limited quantities of minerals in the earth's crust. The result of these price increases will be to slow economic growth.

An examination of 100 years of minerals prices in the United States up to 1970 reveals a relative decline, both for the period as a whole and for most subperiods. This is contrary to the hypothesis of increasing relative scarcity. It is well-known, however, that some of the costs of producing and using these minerals were not reflected in their prices. This came about through the failure to account for the costs of environmental damage in minerals prices.

During the 1970s, most minerals prices in the U.S. rose relative to the general price level. The causes for this reversal of past trends are not as yet clear. OPEC greatly increased the prices at which it sells petroleum. The exchange rate moved against the U.S. dollar, making imported minerals more costly. In addition, government regulations had the effect of raising minerals prices as the result of efforts to protect the environment. In Europe, the relative prices of nonfuel minerals continued to fall during the 1970s, aided by a strengthened exchange rate, and a high rate of inflation in other products.

It appears to be too early to conclude that declining historical price trends in minerals have been reversed. The durability and strength of OPEC or similar cartels, the full but as yet unknown cost of environmental protection, and exchange rate developments will all affect the outcome. As of 1983, however, there is not convincing evidence that minerals are truly becoming relatively more scarce.

References

1. Barnett, Harold J., and Morse, Chandler, 1963, *Scarcity and Growth*, Johns Hopkins University Press, Baltimore.
2. Cipolla, Carl, 1978, *The Economic History of World Population*, 7th ed., Penguin, Harmondsworth, Middlesex, England.
3. Dasgupta, P.S., and Heal, G.M., 1979, *Economic Theory and Exhaustible Resources*, James Nisbet, Welwyn, Herts., and Cambridge University Press.
4. Devarajan, Shantayanan, and Fisher, Anthony C., 1981, "Hotelling's 'Economics of Exhaustible Resources': Fifty Years Later," *Journal of Economic Literature*, Vol. 19, No. 1, March, pp. 65–73.
5. Energy Information Administration, 1983, U.S. Department of Energy, *1982 Annual Energy Review*, Washington, D.C.
6. Fisher, Anthony C., 1981, *Resource and Environmental Economics*, Cambridge University Press.
7. Fisher, Anthony C., Krutilla, J.V., and Ciccheti, C.J., 1972, "The Economics of Environmental Preservation: A Theoretical and Empirical Analysis," *American Economic Review*, Vol. 62, No. 4, September, pp. 605–19.
8. Hall, Darwin C. and Hall, Jane V., (Forthcoming), "Concepts and Measures of Natural Resource Scarcity with a Summary of Recent Trends," *Journal of Environmental Economics and Management*.
9. Hotelling, Harold, 1931, "The Economics of Exhaustible Resources," *Journal of Political Economy*, Vol. 39, No. 2, April, pp. 137–75.
10. Jevons, W. Stanley, 1906, *The Coal Question*, A.W. Flux, ed., 3rd ed., Augustus M. Kelley, New York, 1965 reprint.
11. Manthy, Robert S., 1978, *Natural Resource Commodities—A Century of Statistics*, Johns Hopkins University Press, Baltimore.
12. Myers, John G., Barnett, Harold J., and van Muiswinkel, Gerard M., "Unpublished Materials."
13. Nordhaus, William, and Tobin, James, 1972, "Is Growth Obsolete," in *Economic Growth*, Fiftieth Anniversary Colloquium V, National Bureau of Economic Research, New York.
14. Potter, Neal, and Christy, Francis T., Jr., 1962, *Trends in Natural Resource Commodities*, Johns Hopkins University Press, Baltimore.
15. Slade, Margaret E., 1982, "Trends in Natural-Resource Commodity Prices: An Analysis of the Time Domain," and "Cycles in Natural-Resource Commodity Prices: An Analysis of the Frequency Domain," *Journal of Environmental Economics and Management*, Vol. 9, pp. 122–37 and 138–48.
16. Smith, V. Kerry, 1974, *Technical Change, Relative Prices, and Environmental Resource Evaluation*, Resources for the Future, Washington, D.C.
17. Smith, V. Kerry, 1978, "Measuring Natural Resource Scarcity: Theory and Practice," *Journal of Environmental Economics and Management*, Vol. 5, pp. 150–71.
18. U.S. Bureau of the Census, 1975, *Historical Statistics of the United States, Colonial Times to 1970, Bicentennial Edition, Part 1*, Washington, D.C.
19. U.S. Department of Commerce, 1981, *The National Income and Product Accounts of the United States, 1929–76 Statistical Tables*, Washington, D.C.
20. U.S. Department of Commerce, 1982, *Survey of Current Business*, Vol. 62, No. 7, July, pp. 23, 99, and 131.
21. U.S. Department of Commerce, 1983, *Survey of Current Business*, Vol. 63, No. 7, July, pp. 23 and 80.

1.2

Minerals and the Developing Economies

Charles J. Johnson and William S. Pintz*

INTRODUCTION

This chapter on the role of minerals in developing economies is oriented toward economic policy issues. The discussion and analysis cut across the technical, economic, legal, and policy fields. The goal of this chapter is to provide relevant background information important to individuals involved with the minerals sector in developing countries. Where possible, trends are shown to provide a basis for understanding the evolving role of minerals in developing economies.

This chapter is divided into the following five sections: (1) characteristics of developing countries (which presents selected statistical comparisons); (2) government objectives in developing countries; (3) government policy formulation in developing countries; (4) determinants of future trends, and (5) a summary of the chapter's major points combined with the authors' view.

CHARACTERISTICS OF DEVELOPING COUNTRIES

There are 155 nations in the world categorized as "developing countries." In total these nations represent about one third of the value of the world's mineral production, three quarters of the world's population, and over half of the world's land surface area. The group of so-called developing countries is quite diverse and there are exceptions to almost any general observation about the group as a whole. Most definitions of the term "developing country" are based on levels of per capita income. However, if per capita income is the sole criterion used in categorization, then several oil-exporting countries would jump from the category of developing country to the category of developed country.[a] Obviously, this results in a poor fit with respect to other measures. In general, the majority of developing countries can be characterized by relatively low per capita income, high population growth rates, high levels of unemployment, low levels of education, short average life expectancies, income distribution skewed toward the highest 20% of the population, and a large share of exports represented by primary commodities.[b]

For the purposes of this chapter a slightly modified version of the widely used World Bank system of classification (shown in Table 1.2.1) is used. As summarized in Table 1.2.1, developing countries are divided into three main groups based on estimated gross national product (GNP) per capita. It should be noted that the high-income, oil-exporting countries have characteristics of both developing and industrialized countries; however they are included for most statistical comparisons within the group of developing countries.

Of the 155 developing countries in the world, approximately half produce minerals other than common building materials, and only 32 account for more than five percent of world pro-

*Resources Systems Institute, East-West Center, Honolulu, Hawaii

[a]In this chapter "developed country" is synonymous with "industrialized country." These phrases are used interchangeably.

[b]It should be emphasized that there are numerous exceptions to this generalization. For example, among the low-income developing countries, Sri Lanka and Vietnam stand out because of their high literacy rates; and among lower-middle income developing countries Cuba, Jamaica, and Costa Rica stand out because of their life expectancies.

duction of any single mineral commodity. The 32 developing countries shown in Table 1.2.2 account for 85% of the value of nonfuel minerals and about 60% of the value of fuel minerals produced by all developing countries. In the context of total world production the share of total value accounted for by these 32 countries is roughly one third (37% of nonfuel and 29% of fuel minerals production.)

In constant 1978 dollars the total value of all minerals produced in 1983 is estimated at roughly 550 billion dollars—compared to 234 billion dollars in 1973. The share of this total accounted for by fuel increased from 77% in 1973 to about 93% in 1983, largely as a result of the major petroleum price increases.

Table 1.2.3 shows the distribution of world minerals production by value and country group for 1978. If high-income, oil-exporting countries are added to the group of developing countries, then approximately half (48%) of the value of world minerals production was accounted for by developing countries. If nonfuel and fuel minerals are separated then developing coun-

Table 1.2.1—Classification of World Economies

Country Group	Population (millions)	Area ('000 km²)	Approximate GNP per capita range (1981 $)	Representative Countries
Developing Countries low-income	2,248.1	35,947	80–400	Bangladesh, Zaire, Tanzania
lower-middle income	430.7	15,225	400–1,700	Zambia, Papua New Guinea, Indonesia
upper-middle income	525.9	23,885	1,700–5,700	Malaysia, Brazil, South Africa, Venezuela
High-Income, Oil-Exporters*	14.4	4,241	8,000–25,000	Libya, Saudi Arabia, Kuwait
Developed (Market Economies)	669.3	31,154	5,200–17,500	United Kingdom, Japan, United States
East European Nonmarket Economies	371.8	23,422	2,000–5,700	Romania, Poland, USSR

*These countries have the characteristics of both developing and industrialized countries.
Source: World Bank, 1980.

Table 1.2.3—Percentage Share of World Minerals Production (by value) in 1978

Country Group	Nonfuel	Fuel	Total Minerals
Developing Countries			
low-income	9.4	9.8	9.8
lower-middle income	11.8	4.6	5.5
upper-middle income	22.2	16.7	17.4
High-Income, Oil-Exports	0.0	17.8	15.5
Subtotal	43.4	48.9	48.2
Developed (Market Economies)	32.6	27.8	28.4
East European Nonmarket Economies	24.1	23.4	23.5
Total*	100.1	100.1	100.1

*Total percentages do not add to 100 because of independent rounding.
Adapted from: Callot, 1981.

tries accounted for 43% and 49% respectively of the value of world nonfuel and fuel minerals production in 1978. In contrast, in 1973 developing countries accounting for 46% and 45% respectively of the value of world nonfuel and fuel minerals production. The increase in the share of fuel minerals production between 1973 and 1978 in developing countries is as expected; however, the decrease in the share of nonfuel minerals production is not considered indicative of the long-term trend.

Between countries minerals production varies widely even for countries within the same per capita income group. However, when the averages for each country group are compared there is a substantial correlation between the value of minerals production per square kilometer and the level of gross domestic product (GDP) per capita. This correlation is shown in Fig. 1.2.1, where the value of minerals production per square kilometer increases as GDP increases to the 3000 to 4000 dollars per capita range in 1978 (thereafter increasing only slowly). Not shown is the high-income, oil-exporting group which is a ma-

Table 1.2.2—Developing Countries with Substantial Minerals Production (5% or more of any mineral commodity)

Country	Commodity*	Percent Production (by value)	Country	Commodity	Percent Production (by value)
Angola	Diamonds	6		Lead	5
Bolivia	Antimony	16		Silver	15
	Tin	13	Mongolia	Fluorspar	7
	Tungsten	6	Morocco	Phosphate	19
Botswana†	Diamonds	15	Namibia	Diamonds	20
Brazil	Iron Ore	11		Uranium	5
	Manganese	9	New Caledonia	Nickel	5
Chile	Copper	13	Peru	Barite	7
	Molybdenum	13		Lead	5
China	Antimony	18		Silver	11
	Barite	5		Zinc	7
	Coal	23	Philippines	Cobalt	5
	Fluorspar	6	Saudi Arabia	Oil	14
	Iron Ore	7	Sierre Leone	Diamonds	5
	Magnesite	11	South Africa	Antimony	15
	Pyrite	10		Asbestos	7
	Talc	5		Chromite	32
	Tin	9		Diamonds	26
	Tungsten	20		Fluorspar	7
Cuba	Cobalt	7		Gold	58
Gabon	Manganese	8		Manganese	20
Guinea	Bauxite	19		Platinum	46
India	Barite	5		Uranium	9
	Iron Ore	5		Vanadium	39
	Manganese	6	Surinam	Bauxite	6
	Mica	33	Thailand	Antimony	6
				Barite	6
Indonesia	Tin	12		Tin	13
Iran	Oil	9		Tungsten	7
Korea, Dem.	Magnesite	17	Turkey	Boron	28
	Tungsten	5		Chromite	6
Korea, Rep. of	Talc (etc.)	10	Zaire	Cobalt	54
	Tungsten	6		Copper	5
Jamaica	Bauxite	12		Diamonds	8
Malaysia	Tin	27	Zambia	Cobalt	9
Mexico	Antimony	5		Copper	8
	Fluorspar	20	Zimbabwe	Chromite	7

*Excluded are mineral commodities with total market value in 1978 of less than 100 million dollars.
†Botswana produced less than 5% of the total world value of diamonds in 1978, however, production has rapidly expanded and current production (1984) is about 15% of world production by value.
Source: Callot, 1981.

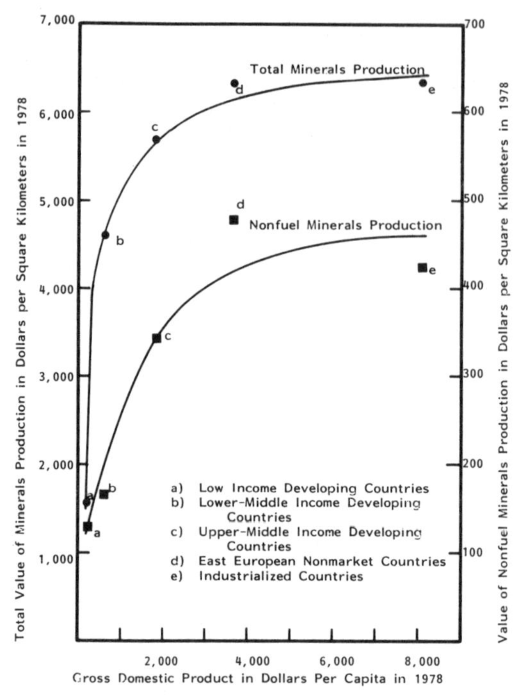

Fig. 1.2.1—Relationship between per capita income and the value of minerals produced per square kilometer for various types of economies (excluding high-income, oil-exporters).

jor anomaly because of its very high per capita income and very high value of oil production per square kilometer. In addition, the East European nonmarket countries produce a higher value of total minerals per square kilometer than the industrialized countries—possibly because of the "closed" nature of their economies. As will be discussed in the following section on "mineral resource potential," the low levels of mineral development in the low-income developing countries is believed to be the result of less investment in mineral exploration and development and not because of lack of mineral resources.

The Role of Mineral Resources in Economic Development

It is generally agreed that the existence of rich mineral resources in a developing country provides a basis for accelerating the overall economic development of a nation. However, in an attempt to place natural resources in a balanced economic perspective, Orris Herfindahl (1969) states:

> No particular type of natural resource is essential to a high level of national income or to economic progress. We can assert with confidence that a country's endowment of natural resources need not exercise a determining influence on the course of its national income over time—if it is able to trade. "All" that is necessary for economic progress is the availability of a substantial quantity of services of capital and labor—with a considerable part of the capital embodied in persons—plus a social system with certain characteristics favorable to systematic improvement of production practices. And the more capital per person the better. The truth of this observation is evident from the economic success of countries with limited natural resources and the success of countries with greatly different natural endowments.

However, Herfindahl further states that:

> From the economic point of view, natural resources are simply a part the capital stock of a country; they are to be regarded as pieces of "equipment" which render productive services. Hence, the more a country has of them and the higher their quality, the better off it will be, other things being equal.

It is important to emphasize that rich mineral resources are but one of the factors important in economic growth. As will be discussed later in this chapter, having rich mineral resources does not insure stable economic growth. Indeed, some have suggested that mineral-led economies are less stable than economies that do not rely heavily on mineral exports (Nankani, 1980).

Rather than attempt to describe all the complexities involved in the role of minerals in the development process, specific areas in which minerals in the developing countries play a more significant role than in the developed countries will be addressed. These areas are: the contribution of minerals to gross domestic product; employment in the mining sector; and mineral commodity exports.

Contribution of Minerals to Gross Domestic Product: It has long been recognized that the role of minerals in the economy declines with increasing industrialization. This factor is illustrated in Table 1.2.4 which shows the contribution of mining and quarrying to gross do-

Table 1.2.4—Contribution of Mining and Quarrying to Gross Domestic Product (GDP) in Selected Economies

Type of Economy	Major Mineral Commodity	Share of Mining in GDP (percent)
Nonfuel Mineral Economies		
Bolivia	Tin	9.9
Botswana	Diamonds	25.9
Chile	Copper	8.7
Jamaica	Bauxite	14.3
Liberia	Iron Ore	15.1
Morocco	Phosphate	4.9
Papua New Guinea	Copper	9.0
Peru	Copper	11.7
Zaire	Copper	8.8
Zambia	Copper	16.1
Petroleum Economies		
Kuwait	Petroleum	67.9
Libya	Petroleum	63.5
Saudi Arabia	Petroleum	53.7
Industrial Economies		
Australia	—	6.5
Canada	—	6.0
Japan	—	0.6
United States	—	3.7

Source: United Nations, 1983.

Table 1.2.5—Countries Having Greater Than 3.5% Employment in the Mining Sector (1970–1980)

Country	Percent
Zambia	15.9
Liberia	11.3
Sierra Leone	9.9
Brunei	9.2
Botswana	6.8
Zimbabwe	6.6
Gabon	6.1
Bolivia	5.6
Jordan	5.5
Trinidad and Tobago	5.5
India	4.0
Turkey	4.0
Swaziland	3.7

Source: International Labor Office, 1976, 1981.

mestic product (GDP) in selected petroleum and nonfuel mineral-producing economies. From this table it is evident that petroleum makes the largest contribution to GDP in selected petroleum-led economies. Nonfuel minerals can typically represent 8% to 20% of GDP in minerals led developing countries whereas in the industrial economies the contribution of nonfuel minerals does not exceed 6.5 percent.

Employment in the Mining Sector: For specific regions within a country, the mining sector can be the largest single employer. At a national level, however, mining employment is relatively small, even in developing countries where minerals are a major part of the economy. Of the 82 countries in the world for which employment data is available, only 13 have greater than 3.5% of their workforce employed in the mining industry. These countries are shown in Table 1.2.5.

The low potential for employment in the mining sector indicates that mineral-led economies must diversify to other more employment-intensive sectors if the chronic unemployment problems of developing countries are to be overcome. Increasingly, mining developments are being examined by governments of developing countries largely from the perspective of their capacity for generating revenue. Nonetheless, employment and localization issues continue to be an important element of most governments' mineral policies.

Mineral Commodity Exports: Eleven of the most important mineral commodities in international trade are petroleum, alumina, bauxite, copper, nickel, tin, lead, zinc, iron ore, manganese, and phosphate (World Bank, 1982). The share of world exports of these commodities from the developing countries (excluding nonmarket economies) increased from about 70% in 1961 to about 90% during the 1974 to 1980 period. With respect to petroleum, developing countries accounted for a fairly constant range of 94% to 97% of world exports during the entire period of 1961 to 1980. Regarding the 10 major nonfuel minerals, the share of world exports from the developing countries averaged 50% with variations of no more than two percentage points in all years except 1970, when the share increased briefly to 54% of the total world exports.

Problems with Nonfuel Mineral-Led Economies[c]

It is generally agreed that mineral-led economies have an opportunity to earn substantial economic rents from mining that can be used to finance other investments, and thereby facilitate economic development. However, the record is mixed; many nations with large reserves of nonfuel minerals have the most dismal records of economic development. Nankani (1980) found that relative to those economies that do not rely on mineral projects, mineral-led economies are more susceptible to a number of economic problems. These problems include instability in export earnings, high unemployment, wage dualism, low marginal savings, high inflation, poor export diversification, and lower growth in the agricultural sector.

Nankani (1980) identifies three major factors responsible for the poor performance of the mineral-led economies. First, mineral resources will eventually be depleted, and this necessitates an effective, long-term strategy for diversification in order to insure sustained economic growth. In addition, even where mineral resources are very large there are limits to the share of the world market that one country can maintain, and expansion opportunities are usually below requirements for sustained economic growth of developing economies. Mineral-led economies tend to have higher exchange rates based on high export earnings from minerals. This results in high wages in the mining sector which increases expectations regarding income and draws people out of the agriculture sector, even though sufficient jobs are usually not available in the mining sector. Both of these factors tend to slow the rate of growth in the labor-intensive agricultural sector.

Second, demand for mineral commodities is subject to short-run instability. This can result in wide fluctuations in mineral commodity export earnings and tends to increase levels of inflation when governments finance budget deficits through borrowing from central banks.

Third, although some mineral deposits produce substantial economic rents, these rents are often difficult to identify because of the complexity of major mining, processing, and marketing operations that can extend to a number of countries and a number of mining companies. The form and level of taxation and the effectiveness of implementation are problems in most developing countries; consequently, taxation of economic rents is quite variable among the developing countries.

The recognition that the minerals sector can produce substantial economic rents, and the concern that these rents are not lost to the foreign investor, has resulted in greater government participation in the minerals sector by most developing countries since about 1960. This has and will continue to result in adjustments in the conditions in which foreign mining companies can operate in developing countries. Although acceptable rates of return on investment will continue to be possible in many developing countries, the opportunities are increasingly becoming limited with respect to capturing a large share of economic rents from the occasional rich mine. This is gradually changing the traditional roles played by governments and foreign mining companies in mineral projects. The major elements of this change are the subject of the following sections of this chapter.

GOVERNMENT OBJECTIVES IN DEVELOPING COUNTRIES

In formulating their development strategies, developing country governments pursue a variety of objectives regarding the extractive sector of the economy. Although it is typical for a government to outline a number of objectives, in reality, policies will focus on only two or three of these objectives. Emphasis on individ-

[c]Much of this section is based on a study by Gobind Nankani, 1980, "Development Problems of Nonfuel Mineral Exporting Countries," *Finance and Development*, Vol. 17, No. 1. The analysis is based on the following 13 countries: Bolivia, Chile, Guinea, Guyana, Jamaica, Liberia, Mauritania, Morocco, Peru, Sierra Leone, Togo, Zaire, and Zambia.

ual objectives will vary from nation to nation and will be influenced by factors such as:

(1) the history of the extractive industry in the nation,
(2) the size of the mining industry in the economy,
(3) government experience with foreign investors in general,
(4) the potential for sustained development in other sectors.

The description that follows is intended to offer a representative, but not exhaustive, sample of mineral objectives pursued by developing countries.

Basic Objectives

In recognizing the diversity of objectives and the differences in emphasis placed on individual objectives by particular nations, it is worth noting certain themes or objectives that are common to the mineral policy statements of nearly all the developing country governments. These objectives are:

(1) that extractive developments should generate revenue for the national treasury,
(2) that the State should exercise some direct ownership and control in extractive projects,
(3) that extractive projects should serve larger goals for development such as regional growth or generating employment opportunities.[d]

Beyond these common objectives, developing nations sometimes advance objectives related to the stabilization of commodity prices or fiscal flows, the maximization of foreign exchange earnings or domestic value added, the development of physical infrastructure, or localization and training programs.

Mineral development can provide a major revenue windfall to the development programs of developing nations. This windfall presents both unique opportunities and potentially disruptive economic forces. That few governments have been able to maximize the development impact of their extractive sectors is more a function of changing expectations and altered investor/host-government relationships than an indictment of the intentions of either party. This is hardly surprising since the notion of "development" itself remains more the product of a dynamic process than of a static state.

Revenue Generation: The establishment of large-scale mining projects represents a source of potential fiscal receipts to developing nations with a limited tax base. The sophistication of the mineral tax schemes used in some of the developing countries reflects (1) the importance these nations place on obtaining fiscal revenues, (2) the large size of these revenues relative to other tax sources, and (3) the depletable nature of mineral deposits. Depletion imposes a finite time frame in which fiscal benefits must be obtained . . . or forever lost.

In pursuing their revenue objectives, host governments use a range of fiscal instruments including royalties, income tax, additional or windfall profits tax schemes, severence or export taxes, dividend remittance taxes, direct equity participation, and user charges for infrastructure. Historically, many of these fiscal instruments have also been used by mineral-producing developed countries including the United States, Canada, Australia, and South Africa. The major difference between the fiscal approaches pursued in the developed countries and those adopted in the developing countries is in attitudes toward (1) concepts of "acceptable" or "normal" profitability, (2) sociopolitical concerns over foreign exploitation, and (3) level of government involvement in the minerals sector.

In structuring fiscal packages, developing nations have only recently started to consider the secondary implications of different fiscal instruments. For example, although the negative impact of a sales-revenue-based royalty on cutoff grade decisions has long been recognized by mineral economists, high royalty rates persist in the fiscal packages of some countries who also have contradictory policies designed to maximize the production of low-grade ores. Similarly, the influence of home country tax laws on the evaluation of overseas projects is of relatively recent concern to taxing authorities in developing countries.

Problems in the fiscal relations between mining investors and host governments are often associated with the transnational nature of mining companies. These problems are often similar to the attitudes of developing countries toward other economic sectors having heavy foreign

[d]See, for example, the Policy Statement of the Sierra Leone Government (1970) or National Development Plan, 1979–85, Republic of Botswana, page 190–194.

involvement. An example of this is the home country tax problem already described. In addition, questions such as the taxation of repatriated dividends, foreign exchange availability for loan or profit disbursements, and tax jurisdictional questions relating to offshore transfer prices can often become sensitive fiscal issues for both mineral and non-mineral investors.

Finally, it should be realized that the revenue objective of developing countries is often at odds with other objectives in the mineral sector. The most prominent of these internal conflicts relates to such issues as environmental expenditures, employment maximization, and regional development. Such objectives often imply an incremental expenditure in areas having low direct productivity which must be met from project cash flow. These expenditures usually result in lower profits to the mining sector, and, consequently, lower tax revenues to the national treasury. As a result, the host government is constantly forced to choose between conflicting objectives.[e] While such choices also face the governments of developed nations, the relative size of the mining sector vis-a-vis other sectors and its accentuated role as a provider of public revenue make these decisions of substantially greater importance to governments of developing countries.

Ownership and Control: Since the early 1960s, the growth of national identity has been a hallmark of the newly independent nations of Africa, Asia, the Caribbean, and Oceania. Nationalism has found expression both in national and in international economic policy.[f] At the domestic level, economic nationalism over mineral resources has most clearly manifested in the principle that the State has title to minerals in the ground and that this right is irrespective of the existence of private claims (Brown and Faber, 1977). State ownership as an expression of national identity can be historically traced for many generations in Latin America.

Internationally, organizations such as the United Nations General Assembly have passed resolutions asserting permanent sovereignty over natural resources. Beginning in 1952 and continuing with subsequent resolutions in 1962 and 1974, the question of permanent sovereignty has repeatedly attracted the attention of the General Assembly. In the General Assembly resolutions 3201 (S-VI) and 3202 (S-VI) of May 1, 1974, the Assembly declared:

> . . . full permanent sovereignty of every State over its natural resources and all economic activities. In order to safeguard these resources, each State is entitled to exercise effective control over them and their exploitation with means suitable to its own situation, including the right to nationalization or transfer of ownership to its nationals, this right being an expression of the full permanent sovereignty of the State. No State may be subjected to economic, political or any other type of coercion to prevent the free and full exercise of this inalienable right.

The notion of resource sovereignty has been adopted by developing nations as a fundamental rationale for assuming increased involvement and control over foreign-owned extractive projects.

Beyond its obvious political and philosophical appeal, the resource sovereignty issue has had several direct policy implications for mineral projects in the developing countries. First, resource ownership implies that royalty payments will accrue to the central government as owner of the resource. In effect, this means that a traditional ownership payment has been converted into a fiscal instrument which is integrated into the nation's overall fiscal package. The logic of resource ownership is easily translated into demands for some form of equity participation. Although equity participation had occurred previously in extractive projects, the resource ownership issue broadened the generally accepted approaches to include various forms of "free" or "earned" or "production sharing" arrangements (Cordes, 1980). Second, the emergence of resource ownership directly challenged the traditional Western notions of sanctity of contracts.[g] Third, it is gradually becoming apparent that resource ownership may become a contentious issue between host governments and financial institutions in establishing a mortgage security for financing new extractive projects (Pintz, 1984).

[e]The problem of conflicting national objectives is specifically acknowledged in Botswana's National Development Plan, 1979–85, Section 8.22, p. 190.

[f]For an excellent overview of the broader ownership and control issues see Walde, T., 1983, "Permanent Sovereignty over Natural Resources," *Natural Resources Forum*, Vol. 7, No. 3, p. 239–251.

[g]For an explanation of the legal theories advanced by developing countries in support of their actions, see Schachter, 1976 "The Evolving International Law of Development," *Columbia Journal of Transnational Law*, p. 1–16.

Developing nations assumed that greater project involvement and control would follow directly from ownership. As stockholders throughout the world have discovered, however, management and ownership are only tangentially related in the modern corporate enterprise. In the case of a highly technical mining project, host governments have found that they seldom possessed either the expertise or the information to effectively attain their ownership prerogatives.

In addition to the equity participation policies which are now almost universally practiced by developing countries, other control devices have been tried with varying degrees of success. During the late 1960s and early 1970s, several developing countries moved to nationalize the holdings of the transnational corporations. The most publicized of these nationalizations involved the mineral industries in Chile and Zambia, but similar courses were followed in Zaire, Peru, and elsewhere. In effect, the nationalized companies became state enterprises. However, a trend seems to be a resurgence of private mining companies in nations where mining had become the more or less exclusive province of state enterprises. Prominent examples of such private-company/state-enterprise systems today exist in such countries as Brazil, Indonesia, Chile, Peru, and Malaysia.

In their relations with private investors, developing nations currently seem interested less in exercising direct, on-going managerial control over projects than in exercising project control through approval of strategic plans relating to aspects of certain projects deemed to be of critical importance to the government. As one commentator notes:

> Indeed, most modern mining agreements provide for government approval of the development plan, of the financing arrangements, of the programs for employment and training of domestic labor and of programs for meeting environmental standards, and for government participation in a variety of activities that were formerly the exclusive domain of foreign enterprise. (Mikesell, 1983)

[h]An interesting analytical framework for evaluating the social benefits and costs to a region is outlined in Gillis M. and Beals R., 1980, *Tax and Investment Policies for Hard Minerals: Public and Multinational Enterprises in Indonesia*, Ballinger, Cambridge, Mass., p. 261–268.

Regional Development Integration: Another objective often advanced by developing countries relates to regional development. There is little doubt that the size and organization of contemporary mining projects can contribute significantly to the development of remote regions. Such contributions may take the form of decreased costs for consumer goods through improved infrastructure, expanded employment and business opportunities for residents, and improved health or education facilities. However, a successful regional development program resulting from the stimulus of a major mine carries with it inherent problems. These problems range from an imbalance between mining districts and adjacent regions, an increased potential for sociocultural friction between those people associated with a project and nonparticipants, and differential standards between national- and project-supplied social services.[h]

The usual expression of sectoral integration in mining projects in developing countries involves pressure for the downstream addition of "value-added" activities such as smelting or refining operations. Many mineral-producing nations pursue the integrated industry notion both as an industrialization strategy and as a means of increasing their tax base while reducing the potential for transfer pricing abuses (Radetzki, 1980). Along with pressure for downstream integration, sectoral objectives may also stress the procurement of local goods and services required for the mine. This stress leads naturally to policies on mine employment/training and to the local procurement of operating supplies or transport services. Such objectives and policies often lead to obligations and expectations beyond those common in the developed world. An investor appreciation of these obligations is essential to successful implementation of projects in developing countries.

Foreign Exchange Earnings and Price Stabilization: With few exceptions, mining projects in developing countries are seen as important in earning foreign exchange. The importance of mineral exports in total export receipts is illustrated for selected countries in Table 1.2.6. Data such as that presented in Table 1.2.6 can easily be misleading if extrapolated to suggest the overall impact of the balance of payments, since extractive projects tend to require substantial imports for continued operation. These "induced" imports can create special policy problems. The

Table 1.2.6—Selected Mineral-Producing Developing Countries: Share of Total Exports by Mineral Commodity

Nation	Commodity	Percent of Total Export Value*
Bolivia	Tin	35.7
Chile	Copper	44.8
Guinea	Bauxite	54.2
Guyana	Bauxite	37.9
Jamaica	Bauxite	17.8
Liberia	Iron Ore	56.0
Morocco	Phosphate	33.3
Peru	Copper	21.2
Papua New Guinea	Copper	35.4
Zaire	Copper	50.1
Zambia	Copper	86.1

*Average 1978–1980.
Source: World Bank, 1982, p. 25.

overall balance between export earnings and induced imports will vary significantly between developing countries. Key determinants of the effect of the overall balance of trade in mineral developments are:

(1) the domestic value added to final exports,

(2) the relative debt position of the aggregate mining sector (since financial flows and debt retirement represent sizable overseas transfers during various stages of a mine's life).

As a minimum, net direct balance-of-payments benefits will be equal to the government's fiscal receipts from mining, but beyond this level it is difficult to generalize about the overall impact of mining on the import-export balance of developing nations.

At a policy level, three aspects of the foreign exchange objectives deserve special comment. First, as illustrated in Table 1.2.6, foreign exchange earnings from mineral sales are of major importance in many countries. Given its importance to the general economy, foreign exchange availability can easily overshadow basic project economics during periods of low mineral prices; uneconomic continuation of mine operations for foreign exchange purposes has been noted by many observers.[i] Such uneconomic continuation of operations may aggravate oversupply conditions in end-markets and accentuate price cycles. Second, governments of developing countries commonly attempt to protect their foreign exchange and revenue earnings against fluctuations through participation in commodity cartels or international price stabilization schemes. Third, depressed foreign exchange earnings sometimes result in the rationing of available foreign reserves among all sectors of the economy. Such rationing can mean that the foreign exchange allocated to the mining sector is inadequate to meet the needs for capital replacement, plant modernization, or exploration.

Conversely, to the degree that a buoyant minerals sector results in a favorable trade balance, the exchange rate of the host country currency may increase relative to other currencies. This upward revaluation of the local currency can have a depressing effect on the export competitiveness of other economic sectors, and result in economic distortions and an increasing vulnerability to international economic fluctuations (Gregory, 1976; Stoeckel, 1979). Thus, for economists in developing countries, optimum foreign exchange earnings from mineral exports must reflect considerations such as stability, sectoral balance, and the maintenance of a modern, adequately capitalized minerals sector.

The cyclical movement of metal prices has created major challenges to planners in developing nations. The degree of disruption associated with price fluctuations must be understood within the context of the domestic minerals industry, and from the perspective of the fiscal policies applied to the industry by the government. With this distinction in mind, governments of developing countries must contend with the dual problem of attempting to protect both the government budgetary revenue and the nation's foreign exchange position.

A convenient way to approach stabilization policy is to distinguish between internal and external stabilization strategies. Internal strategies have been actively pursued by relatively few nations and involve the establishment of some sort of revenue stabilization trust fund by the government. The trust fund normally receives all or a significant fraction of the State's revenues from mining and makes annual transfers to the government's general account. The management of the transfers and the liquidity of the stabilization fund assets largely determine whether the fund can buffer the government from volatile changes in year-to-year revenue receipts. Papua

[i]For an overview of the interaction of declining prices, foreign exchange strategy, and cartel motivations in the copper industry, see *Mining Journal*, 1984, "Copper Conflict Deepens." Vol. 302, No. 7748, pp. 101–103.

New Guinea probably has the best known internal stabilization scheme—which has operated since the mid-1970s.

As an external strategy, stabilization objectives are pursued through producer organizations and through internationally sponsored price or loan schemes. The objective of external stabilization schemes is to moderate the effects of market forces on the domestic minerals industry. Although producer organizations exist for a number of mineral commodities, their success has been limited. Not surprisingly, a factor that often undermines effective producer cooperation is the development strategies and resource endowments of member countries. For example, in the Intergovernmental Council for Copper Exporting Countries (CIPEC), the interests and objectives of a low-cost producer with extensive undeveloped resources, such as Chile, are likely to conflict with the objectives of higher-cost producers such as Zambia or Zaire. Given differing resource and cost structures, cartel members may perceive their national interests in divergent ways. For example, through most of the late 1970s and early 1980s Chile has sought to stabilize its copper revenues by expansion of capacity. On the other hand, the Zambian and Peruvian governments have repeatedly called for production cutbacks as a means of increasing copper prices. Each policy is consistent with the cost structure and resource endowment of the individual country, but the policies are in obvious conflict with each other. Such conflicts are most acute at the very point where harmonization of policy is most needed, i.e., when mineral prices are declining.

Increasingly, mineral-producing developing countries are pursuing stabilization goals in international forums such as United Nations Conference on Trade and Development (UNCTAD), the International Monetary Fund (IMF), and the Lome Convention. In response to pressure from such groups, industrialized countries have developed specialized loan facilities such as the European Economic Community (EEC) Stabilization Scheme for Minerals (SISMIN) or the IMF's Compensatory Financing Facility (CFF) to stabilize export earnings. Although subject both to operational and to conceptual criticism, such schemes have opened channels for consumer support of the stabilization objectives of developing countries.

Finally, by viewing mineral exports within a broader "commodity" framework, producers may someday be able to take advantage of another family of internationally sponsored programs under the United Nations Commonfund Program. Although funding of Commonfund buffer stock arrangements for major mineral commodities has yet to be finalized, copper (as well as iron ore, tin, manganese, phosphates, and bauxite) has been considered for inclusion in the scheme (Rogers, 1979).

Conflicting Objectives

In a discussion of the objectives of developing country governments it is useful to contrast the goals likely to be pursued by transnational corporations with the goals of developing countries. Such a contrast tends to highlight those aspects of developing country/transnational relationships that are most contentious. First, it is reasonable to postulate that the primary interest of foreign investors is the maximization of global profitability available for reuse or for distribution to shareholders (Bosson and Varon, 1977). Such a profit notion suggests potential differences between developing country and foreign investor objectives over such issues as:

(1) repatriation of dividends,

(2) foreign exchange controls,

(3) home-country tax definitions and compatibility,

(4) the prices at which mineral products are transferred across national boundaries.

A second objective pursued by foreign investors involves the assessment and management of risk (United Nations, 1978). In pursuing this risk objective, the multinational investor might seek (1) geographic diversification of his investment portfolio, (2) to minimize equity contributions and maximize limited or nonrecourse debt borrowings, and (3) to minimize nonessential or low-productivity investments in such areas as infrastructure, environment, or social services. Finally, the corporate risk manager usually assigns a higher discount rate with respect to his assessment of project risks in developing countries. The actual perception of risk could be expected to change over the lifetime of a mine.

Another general corporate objective related both to profit maximization and to risk management involves technical decisions about the extent of resource development. Commonly, this resource issue comes down to the real, or perceived, "high grading" of deposits as symbolized by the setting of cutoff grade criteria for defining ore and waste. The investor (in pursuit of his profit motive) is seeking the highest internal rate of return and (in pursuit of his risk minimization motive) the shortest "payback" or investment recovery period. In contrast, host governments tend to consider the extraction of nonrenewable minerals as a depletion of its endowment of natural capital which, if not developed to its fullest extent, will be lost forever. Conceptually, pursuit of such an objective would suggest the maximization of net present value given some social discount rate. To the degree that internal rate of return and social discount calculations vary, different definitions of "optimal" criteria for ore cutoff grade will result.

GOVERNMENT POLICY FORMULATION IN DEVELOPING COUNTRIES

In many respects, governments of developing countries view mineral resource issues from a different perspective than either transnational corporations or the governments of industrial countries. Mineral policies and strategies vary between developing countries, however, in a number of areas, policies and strategies are evolving in a similar direction—particularly among those countries with an active, growing minerals sector. In this section some of the more important elements of policy and strategy trends are examined under the headings of exploration policy, processing policy, taxation policy, financial policy, and environmental policy.

Exploration Policy

The stated objective of most governments of developing countries is to expedite the exploration and development of their mineral resources. The policies regarding who explores for minerals ranges from all exploration being left to private transnational corporations, to all exploration reserved to state corporations. In the majority of the developing countries, the government encourages the involvement of private transnational corporations in mineral exploration. For the most part, exploration is perceived as a risky use of limited government funds. Moreover, many believe that the major transnational corporations may have a comparative advantage in the exploration and evaluation of commercial mineral deposits—a result of the global experience of their exploration geologists, and the availability of advanced technologies as well as proprietary techniques of data analysis.

During the first half of the twentieth century, transnational corporations were often granted vast areas for exploration, as well as long license periods and minimal work commitments. Increasingly, governments have placed restrictions on the terms for exploration in order to insure that exploration programs are accomplished with minimal delay, and to insure against speculative holding of mineral resources. Today, license areas are much smaller, license durations are shorter and specific work commitments are increasingly becoming a part of the license requirements.

Normally, all mineral discoveries must be promptly reported to the government, and mining companies are required to complete feasibility studies for mineral discoveries having possible commercial potential. An area of conflict occurs where a deposit does not appear commercially viable under present economic conditions, yet the company wishes to maintain rights to the deposit (as is common in industrial countries). Governments in developing countries increasingly require ongoing financial commitments as a condition for maintaining rights to undeveloped deposits.

A useful example of modern exploration policies in the mineral-led developing economies is the case of Botswana. The maximum individual license area is 1000 square kilometers, with a maximum initial license period of three years followed by two renewals of two years each. With each renewal, the remaining license area must be reduced by at least 50%, and specific work commitments and minimum expenditure obligations are required in all applications. Data, analysis, and samples must be turned over to the government's geological survey, and, after licenses are relinquished, the government makes the data and analysis available to other interested companies. The reason behind this policy is to promote exploration and to insure that future exploration programs do not repeat exploration already completed.

Public Sector and Aid-Funded Exploration: Most governments have a geological survey responsible for the traditional function of geologic mapping and, to a lesser extent, mineral exploration. In addition, since the later 1950s there has been a relatively rapid expansion of state-owned mining companies which originally were established to manage existing mining operations (Radetzki, 1983). Today, many of these state companies are involved not only in mining but also in mineral exploration and marketing. State companies may also have both important areas and specific mineral resources reserved for their exploration, with limited or no transnational corporation involvement allowed.

Exploration financed through foreign aid and multilateral institutions continues to be a prime factor in many countries. The most important program is operated through the United Nations Development Program which has spent about 200 million dollars for over 200 projects in 75 countries. The United Nation agency primarily involved in exploration is the Department of Technical Cooperation for Development, which operates on a budget of about 20 million dollars per year. This agency is credited with the discovery of a number of commercially viable deposits, including the Mamut copper deposit in Sabah, Malaysia which was subsequently developed by Nippon Mining Company (Lewis, 1983).

A second important agency is the United Nations Revolving Fund for Natural Resources Exploration. This agency handles the exploration of those areas previously defined as geologically favorable for commercial deposits. In the event that a commerical discovery is developed, the government is obligated to pay one to two percent of the gross value of production for a 15 year period. To date, the Revolving Fund program has not been widely used because many governments believe the repayment schedule is too high.

Processing Policy

In the developing countries, domestic processing of most minerals as a share of the world total grew slowly over the period 1960–1980. To a great extent, the increased share of total processing taking place is attributed to two factors: first, the growth in domestic mine production; and second, the emphasis on downstream processing as one of the economic strategies outlined in development plans. The interaction of these two factors cannot easily be separated except on a country-by-country and commodity-by-commodity basis. However, Table 1.2.7 summarizes trends in various processing stages for six major mineral commodities.

Table 1.2.7—Percentage Share of Processing for Six Commodities by Economic Region: 1960–63, 1980*

Commodity	Stage	Developing 1960–63	Developing 1980	Developed 1960–63	Developed 1980	Centrally Planned 1960–63	Centrally Planned 1980
Aluminum	Bauxite	65	54	17	33	19	13
	Alumina	—	23	—	61	—	16
	Primary Al	5	15	74	64	21	20
Copper	Mined	47	47	39	29	15	23
	Smelted	42	38	44	39	15	22
	Refined	20	27	64	48	16	25
Iron	Mined	26	36	48	34	26	30
	Crude Steel	8	17	65	55	27	30
Lead	Mined	37	29	40	42	22	29
	Primary Metal	21	16	48	59	31	24
Oil	Crude	47	54	37	21	16	25
	Refinery Capacity	21	26	66	55	14	19
Zinc	Mined	30	28	51	48	20	24
	Primary Metal	11	17	69	57	20	26

*Percentages derived from data reported in the following sources: W. Gluschke et al. *Copper: The Next Fifteen Years—A United Nations Study*, Reidel, Boston, 1979; *Non-Ferrous Metals Data*, American Bureau of Metal Statistics, 1982; *Price Prospects for Major Primary Commodities*, World Bank, 1982; *Encyclopedia of Petroleum*, 1981, Pennwell; United Nations, *Statistical Yearbook* 1979/1980.

In the past it has often been suggested that developing nations should attempt to increase the "value-added" to their raw material exports; the concept of value-added is generally associated with the concentration, purification, or, less often, the semi-fabrication of the primary product. Many mineral-producing developing countries have embarked on value-added processing activities as part of strategies for industrialization or import substitution. In some cases the character of the primary product is such that in-country processing may be necessary for successful marketing.[j] Moreover, downstream processing activities are sometimes viewed as necessary to control potential transfer pricing abuses by integrated foreign companies.

When embarking on a value-added processing strategy for mineral exports, governments of developing countries face a variety of barriers including reluctance on the part of established producers, marginal economics of projects, and substantial technological risks. In addition, host governments may find that the costs of engaging in low-profit downstream activities are initially quite high. Nevertheless, during much of the 1970s the desire to increase in-country processing was a hallmark of the mineral trade strategy of developing countries. In order to understand this strategy it is necessary to examine the various processing arguments that have been advanced over the years. For convenience, these arguments can be structured into three general classes as follows:

(1) the industrialization strategy,
(2) the trade integration strategy,
(3) economic-environment strategy.

The industrialization strategy has been a primary justification for many processing projects both in large and in small developing countries. This strategy is defended for at least two reasons: (1) exploiting a perceived raw material advantage, and (2) inducing linkage effects in other industrial sectors. The first argument is an extrapolation of the logic of an economic advantage of the basic primary-product industries. That is, it can be argued that the success and economic advantage of the developing countries as primary-product producers should imply a similar advantage in secondary processing. The justifications for this view could include the potential savings in transportation costs, the availability of internationally subsidized capital, and, in some cases, the existence of low-cost energy for processing. For larger, more diversified developing countries (such as Mexico or Venezuela), the development of a local processing industry may be linked to the possibility of inducing or stimulating supply or demand in some other economic sector. For example, the smelting of copper might be encouraged in order to provide a source of sulfuric acid as an input to a fertilizer industry. Such parallel integration strategies would seemingly only be justified in fairly large developing countries where linkage of the processing industry to other industries already exists or can be economically justified.

The trade integration strategy relates to questions of trade dependence and perceived abuses of transfer pricing. Here, the basic motive is to overcome institutional-type threats associated with monopsonistic markets for intermediate products (like copper concentrate or bauxite) while at the same time attempting to insure that economic rents are not transferred to lower-tax countries through artificially set prices. In essence, the trade integration strategy is really a defensive measure wherein the developing country moves into downstream activities in an attempt to gain a greater share of the value of its primary product.

In recent years, increasingly stringent environmental regulation in the industrialized countries has provided a new economic argument for siting processing facilities in developing countries. This argument suggests that not only do developing countries generally require less expenditure on environmental matters, but, as a result of their lower level of industrization, the capacity of their environments to absorb pollution is relatively greater. The potential significance of this "pollution advantage" for copper producers in the developing countries is suggested by a recent U.S. Department of Commerce Report (Haffner, 1979) which states:

> Under the Aggregate Impact scenario, the price at which domestic suppliers could profitably market their copper increases substantially over the price that would be required under Baseline conditions. By 1981, this price level for domestic refined

[j]Examples of this processing-for-marketing strategy include (1) a case in the Philippines where certain high-arsenic copper concentrates had few, if any, smelter-refinery markets; and (2) a case in Indonesia, where the Asahan lateritic nickel ores required cheap, locally available hydropower energy for processing.

Table 1.2.8—Capital Cost of New Alumina Refineries and Aluminum Smelters

	Refineries		Smelters	
	Number	Capital Cost ($ per ton)	Number	Capital Cost ($ per ton)
All projects	11	647	28	4,547
New Plants	6	676	16	4,892
Expansions	5	541	12	3,152
Developing Countries				
New Projects	4	580	9	6,873
Developed Countries				
Expansions	3	543	10	3,220

Source: *Engineering and Mining Journal*, 1981.

copper is projected to be 42% higher than under Baseline conditions.

Baseline conditions assume normal historical growth conditions without the implementation of planned environmental, health, and safety regulations.

Against such seemingly attractive arguments, governments of developing countries considering downstream processing projects must weigh (1) the significantly higher capital costs of projects, (2) considerable technological risks, and (3) the problems of raising developmental capital for such projects. The cost premiums associated with grassroots industrial projects in the developing countries is generally acknowledged. That these premiums often exceed one third of project costs is well understood; a factor more easily overlooked is the differential between "new" projects in the developing world and expansion projects in the industrial nations. Table 1.2.8 compares new and expanded alumina and aluminum facilities in both industrialized and developing nations.

Regarding the technological risks associated with mineral processing projects in developing countries, several examples might be cited where errors in choice of technology seriously jeopardizes both the economics of the processing project and the economics of the associated mining project. An often-discussed example of processing technology that failed to operate as planned is the flash smelter associated with the Seleibe-Phikwe copper-nickel project in Botswana.[k]

In processing, as in other investments in extractive technology, the transnationalism of mining investors can easily come into conflict with the domestic policy orientation of governments of developing countries. Such conflicting perspectives are most acute in cases of mineral commodities such as aluminum where private companies have established integrated operations with mining in one country and processing in a second country. Where the tax rates are different between countries there is a temptation on the part of transnational corporations to set the transfer price of the mineral commodity at a level that maximizes profits to the firm. Although transfer price issues are an important problem in some situations, many of the decisions made by multinationals regarding the location of processing facilities are, in fact, taken for legitimate economic reasons. Examples of this include transportation savings, the availability of inexpensive energy, proximity to supplier or by-product markets, and the requirements of emergent technology, such as continuous cast rod.

Taxation Policy

As previously stated, the central objective of the mineral policies of most developing countries is to maximize government revenues from the minerals sector. The focus of attention is how best to capture the economic rents through a combination of fiscal elements.[l] Throughout most of the world, foreign investors in mining projects are subject to a range of fiscal elements, including: royalties, depreciation, income taxes, withholding taxes, and often provision for gov-

[k]An extended description of the technological problems involved in the Seleibe-Phikwe mine can be found in Mikesell, R., 1983, *Foreign Investment in Mining Projects*, Oelgeschlager, Gunn and Haim, Cambridge, Mass., pp. 51–55.

[l]Economic rent is defined as the value of minerals production minus cost of production which includes the minimum returns to capital necessary to induce investment.

ernment equity. Also, in a number of countries an excess profits tax element is included in the fiscal regime.

From the turn of the century through the 1940s royalties provided a major source of government revenues from mining in many developing countries. Most royalties are based on quantities produced or on gross sales revenue; therefore they are independent of profitability and are an ineffective method of capturing economic rents. As a result, royalties have been gradually replaced by income taxes, which is a far more effective method for taxing profits from mineral projects. For rich deposits, however, income taxes are ineffective in capturing economic rents. Therefore, various forms of taxation have gradually evolved with the highest taxes applying to petroleum, where the largest economic rents occur.

Depreciation has the important effect of reducing taxable income in the early years of a mining project. Therefore, the investor is allowed to recover more rapidly its investment and to increase its internal rate of return. In cases where the government negotiates a mining agreement, flexibility in setting the depreciation rate is an important factor in achieving mutually satisfactory agreements. This flexibility allows the investor to more rapidly achieve its target internal rate of return and allows the government to achieve a higher share of profits after the depreciation period.

Income tax applies to operating profits, usually after deduction of royalties, depreciation, depletion allowances (where these exist), interest on loans, and allowable losses carried forward from preceding years. It is the main source of mining tax revenues in the world today.

Government equity provides the government with an opportunity to share in the financial rewards as well as risks taken by investors of risk capital, and allows the government to exercise greater control over a mineral project. There are a wide range of options for acquiring equity, ranging from free equity, to equity received on concessional terms, to equity purchased at full market value.

Withholding taxes usually apply to dividends transferred out of the country in which the mining operations exist. The purpose is to encourage investors to reinvest a larger share of profits in the country where the profits were earned.

In the mid-1970s an effective taxation feature designed specifically to tax economic rents was first applied in Papua New Guinea. This feature, called the resource rent tax (RRT), was first described by Ross Garnaut and A. Clunies-Ross in 1975. The RRT applies only to those profits above a specified internal rate of return on total funds invested in a project, and can be superimposed on existing income tax regimes. The tax meets the investor's concerns in that it does not apply before the investor has received a reasonable return on total funds invested. It also meets government concerns in that it taxes away a substantial share of the profits above the level necessary to attract investment. The RRT has been applied to two major mines in Papua New Guinea and is the basis on which major exploration programs have been undertaken in that country. The RRT is now fairly widely known in mineral-producing developing countries, and appears likely to become increasingly popular for major new projects. Although the RRT concept is relatively easy to understand and calculate, there is a tendency to become confused by complicated computer tax models that include RRT. As illustrated in Table 1.2.9 the RRT concept is relatively easy to calculate.

The RRT is designed to apply to a project rather than to a company. Taxable profits from a project for each year equal assessable receipts minus assessable expenses defined as follows (Garnaut and Clunies-Ross, 1975):

(1) assessable receipts include gross receipts from a project, including sales revenue and proceeds from the sale of any equipment, but excluding interest revenue;

(2) assessable expenses include:

(i) Operating expenses incurred in earning assessable receipts;

(ii) capital expenses incurred each year, excluding debt servicing;

(iii) exploration expenditures associated with the project;

(iv) all tax and royalty charges levied independently of the RRT; and

(v) any accumulated losses from previous years, including interest on those losses calculated at the hurdle (threshold) rate.

Table 1.2.9 illustrates the application of RRT that applies when an investor has achieved its hurdle internal rate of return (IRR) on investment (defined as 10% in this example). The project has a 15 year life and produces net losses for years one to three, owing to capital and operating costs (columns C and D). In year 11

Table 1.2.9—Illustration of Resource Rent Tax Calculation*

A	B	C	D	E	F	G	H	I
		Deductible Costs						
Year	Gross Receipts	Capital	Operat-ing†	Net Profit	Accumu-lated net losses at 10%	Profits Subject to RRT	RRT payments at 50% rate	Net Profit to investor after RRT
1	—	100	—	-100	-100	—	—	-100
2	—	300	—	-300	-410	—	—	-300
3	50	50	50	-50	-501	—	—	-50
4	200	—	50	150	-401	—	—	150
5	200	—	50	150	-291	—	—	150
6	200	—	50	150	-170	—	—	150
7	200	—	50	150	-37	—	—	150
8	200	—	50	150	109	109	54.5	95.5
9	200	—	50	150	—	150	75	75
10	200	—	50	150	—	150	75	75
11	200	200	50	-50	-50	—	—	-50
12	200	—	50	150	95	95	47.5	102.5
13	200	—	50	150	—	150	75	75
14	200	—	50	150	—	150	75	75
15	200	—	50	150	—	150	75	75

*Derived from R. Garnaut and A. Clunies-Ross (1975).
†Includes other taxes (income tax and royalties).

a capital investment (column D) in expansions or replacement equipment yields a net loss of 50 (column E) and no RRT applies in that year. Losses are accumulated each year plus interest on those losses calculated at the hurdle interest rate of 10% per year in Column F. RRT at a rate of 50% applies when accumulated net losses in column F become positive. To avoid double RRT taxation, accumulated net losses are set to zero after RRT has been applied in any year (whenever a positive number exit in column F). Column G shows profits subject to RRT; column H shows the amount of RRT payment; and column I shows the net profits to the investor.

As might be expected, there are a number of problems encountered in the practical application of IRR criteria to minerals projects. These problems commonly lead to substantial conflicts during negotiations between governments and mining companies over fiscal terms for commercial development of mineral projects.

The first problem is reaching agreement as to whether the IRR calculations are to be made in "current" or "constant" terms. Current terms calculations include inflation for both revenues and costs, and constant terms calculations remove the inflationary effects. A rough approximation can be obtained by subtracting the inflation rate from the IRR in current terms to produce the IRR in constant terms. For example, if a project produces an IRR in current terms of 25% and the inflation rate is 10%, an approximation of constant terms is 25 − 10 = 15% (actual constant IRR is 13.6%). Governments tend to want to negotiate on the basis of an IRR in current terms to the company, whereas the company tends to argue for IRR calculations in constant terms.

A second area of potential conflict is whether the IRR applies to total funds invested by the company (equity plus loans) or only equity. Loans leverage the return to equity, and can readily produce returns to equity in mining projects of more than 50% above the return on the total investment. Most companies consider the IRR on total funds at risk as the most important financial criterion in evaluating investments. Commonly, companies have hurdle or target IRRs that a project must meet before investment will be approved. This target IRR varies significantly among firms and may even be adjusted within a firm for a given project, depending on the perceived risks of a project and the level of the company's interest in a project. A rough approximation is that most companies appear to require a minimum IRR in constant terms of about 10% in industrialized countries and 15% in developing countries.

Financial Policy

Over the past 15 years a revolution has quietly taken place in the ways in which new mining projects are being financed. For mineral-producing developing countries the changes in financial structure have occurred during a period of fundamental change in other policy areas—such as equity participation or the growth of state enterprises. Although most major mineral projects have some unique financing features, certain distinguishing characteristics will inevitably shape policy for financing future extractive projects. These distinguishing characteristics can be broadly grouped under four general headings as follows:

(1) a broadening of the sources of finance,
(2) an increasing concern among financial institutions with risk management and geographic diversification of their loan portfolios,
(3) the relationship between the contingent liabilities associated with equity participation and the ability of the nation to secure debt for general development purposes,
(4) a growing awareness of the impact of project debt on the tax base and availability of foreign exchange.

The contemporary mining project in a developing country will raise finance from a variety of private, public, and multilateral funding agencies. Over the past decade two types of financial institutions have assumed increasing importance to such mineral projects. These are the export credit agencies of industrial nations and international development agencies like the World Bank, the regional development banks, or the European Investment Bank. While the availability, terms, and usefulness of funds from these sources will vary greatly from developing country to developing country, these financial institutions have played increasingly prominent roles in the structuring of mine finance. This growing importance is derived not only from the direct provision of project debt, but also from the implied endorsement of project and political assumptions (World Bank, 1978).

The constraints on funding from such public or quasi-public sources relates to political, industrial, and geographical factors. For institutions like the European Investment Bank, special facilities are available only to signators to the Lome Convention, and loan covenants sometimes require product sales in Europe. Other agencies, like the regional development banks, have a specific geographic mandate for their lending activities, and funding is available only to regional member countries. The lending facilities of Export Credit Agencies are normally available to most developing countries but whether or not such finance is useful to a country depends on its industrial base and mine support sector. In countries such as Mexico or Brazil where substantial mining equipment industries exist, such export credit facilities are less attractive than in nations like Papua New Guinea or Botswana where little or no secondary industry exists. Moreover, although export credit finance may offer attractive terms, it has several inherent limitations (Radetzki and Zorn, 1979).

In addition to increased lending to developing countries, commercial financial institutions have become more sophisticated in the risk management of their debt portfolios. One of the risk management devices widely discussed is the evaluation of a nation's sovereign (i.e., political) risk as part of an assessment of a project's overall creditworthyness.[m]

As well as considering the political stability of the host country, risk assessments may involve criteria relating to the geographic spread of the lender's overall lending portfolio. Such portfolio criteria can result in "country lending limits" designed to minimize the overall exposure to risk. Sometimes such sovereign risk or portfolio diversification criteria can come into conflict with perceptions about the basic economic viability of a project.

Where such conflicts occur, individual lenders may be reluctant to participate in a particular loan syndication and lenders who do participate may impose increasingly stringent terms and conditions as prerequisites for loan approval. Thus, increasing application of risk management techniques to borrowing on the part of developing countries results both in a rationing of interested lenders and in potentially costly loan conditions.

Similarly, developing countries are attempting to manage their risks in the financing of capital-intensive extractive projects. Such risk management is sometimes associated with the

[m]See Valentine, R., 1982, "The Art of Calculated Risk," *Far Eastern Economic Review*, Vol. 121, No. 48, for a critique of sovereign risk assessment techniques.

contingent liabilities that derive from equity participation in new projects. Basically, contingent liabilities are the loan or performance guarantees required by lenders to insure that the debt service capacity of the project is achieved. In practice, shareholders guarantee that if specified performance criteria are not achieved, or if the project fails to generate sufficient revenue to service the debt, then the shareholder will remedy the deficiency (Zorn, 1978). With the current high debt-to-equity ratios, the potential contingent liability exposures of shareholders can easily be several times that of their direct equity outlays. Since, in many cases, developing country government shareholders are dependent on technical evaluations performed by their private joint-venture partners, this contingent financial exposure is of growing importance for policy. Although the interest of the government to define and minimize its financial liability is the same as the interests of the private investors, a key distinction between the two groups exists in their ability to recognize and assume risk.

Another financial issue of relevance to government policy in developing countries is the impact of debt servicing on the taxable base of the mineral project. Interest is normally treated as a tax deduction and is therefore a reduction in potential taxable profit. In contrast to a fully equity funded project, a heavily debt-leveraged project will initially have a smaller taxable income and therefore generate less fiscal revenue to the government. In an attempt to protect their project's taxable income, many host governments have adopted specific policies directed at limiting allowable interest deductions. These policies have taken two forms: (1) policies that specify a minimum equity interest (or maximum debt ratio), and (2) policies designed to insure that the interest transfers is limited to prevailing rates in international markets.

A further aspect of the debt-service ratio of concern to policy makers in developing countries is that of the private and multilateral financial agencies which require automatic availability of foreign exchange for loan servicing. Since both the tax base and the foreign exchange provisions result from a basic debt-equity decision made by investors, governments increasingly want to influence such decisions. In fact, as suggested previously, the growing size and financial requirements of modern extractive projects probably mean that high debt-equity ratios are inevitable and that government regulation in this area is directed simply at insuring that debt is not used or manipulated to indirectly divert or increase repatriated profits.

Environmental Policy

Many developing countries view the environmental issues as secondary to the overall issue of development. Policy conflicts are created to the degree that environmentally disruptive activities, such as mining, conflict with development strategies. These conflicts exist on at least four levels:

(1) the resident versus national problem,
(2) the initial versus retrofit problem,
(3) the legal-regulatory problem,
(4) the baseline study versus project design problem.

Because the environmental question exists in several policy contexts it is often difficult to determine where the true public interest lies, or to get a clear idea of the obligations or responsibility of the investors. When this uncertainty is overlaid by cultural differences between investors and regulators, or by the intervention of third parties, such as lending institutions, questions of environmental policy often can be among the most difficult to anticipate in project planning.

Inevitably, the costs and benefits of a mining project are shared differently by various classes of participants. From an environmental perspective, an important aspect of this differential distribution of costs and benefits is the localized nature of environmental degradation. Clearly, those most affected by pollution or disruptions in land use are the local residents. In contrast, the major beneficiaries from a mining project are likely to be the national treasury and overseas mining investors. When such a differential sharing of project costs and benefits occurs in industrialized societies, the existence of sophisticated communication systems and environmental advocacy groups (e.g., Sierra Club, Friends of the Earth, etc.) facilitates the mobilization of public opinion and political power. However, in many developing nations, communications and institutions for mobilizing political opinion are emerging only slowly. The result is that adjustments to localized environmental costs tend to occur sporadically as a consequence of resident confrontation, rather than gradually as a result of government intervention or regulation. The ultimate cost to the mining project of resident reactions through strikes or riots may be considerably greater than if these issues are resolved at the outset. For example, at the Bougainville copper mine in Papua New Guinea, opposition to the mine's impact by the local people led to a successionist movement at Independence. Since that time, the dumping of mine tailings in the adjacent Jaba River has been a chronic source of resident/company friction.

The often delayed expression of environmental concerns in some developing countries can only be corrected by the retrofitting of pollution abatement equipment to existing production facilities. Such retrofit programs are usually more costly and often les effective than similar facilities incorporated into the original project design. The President of the World Bank, A.W. Clausen (1982), summed up the environmental situation as follows:

> The cost of these environmental and health measures has proved not to place an unacceptable burden on our borrowing countries. And we've learned, as have many private corporations, that the cost tends to be lower the earlier that environmental problems are identified and handled.
>
> We're convinced that its almost always less expensive to incorporate the environmental dimensions into project planning than to ignore them and pay the penalty at some future time.

For many developing countries, environmental regulation is still embryonic. Direct environmental regulation of mining is often the joint responsibility of several authorities operating under various national laws or statutes. Thus, the definition of environmental policy is as much a question of bureaucratic consensus as of regulatory fiat. Such a system contrasts with regulation in industrial countries where a designated agency will have the primary environmental responsibility. The consensus approach may reflect a pragmatic desire to avoid a direct conflict between environmental policy and development strategy. Nevertheless, the environmental regulations that emerge from such consensus are generally less stringent than similar regulations in industrial nations.

A final policy question deserving attention is that of the relationship of mine design to environmental baseline studies. The typical mine feasibility study process is not conducive to the construction of an environmentally sensitive project design. In a typical feasibility evaluation, studies proceed more or less sequentially through the definition of various technical parameters. In turn, these technical assumptions result in a financial evaluation used to establish the viability of the project. Environmental problems with this process are the result of three factors:

(1) investors have an understandable reluctance to commit themselves to a costly environmental study program *before* they know if the project is viable,

(2) design studies occur simultaneously with the baseline studies, not at the conclusion. Hence, environmental data is largely unavailable to designers during the feasibility study,

(3) since the feasibility study is the basic document for obtaining government approvals and raising finance, there is a reluctance on the part of the investors to change the fundamental design or technological assumptions.

Although these arguments obviously pertain to the environmental design interaction of projects both in industrialized and in developing countries, the single major difference between the two situations is the availability of an environmental knowledge base to project planners. Clearly, in an industrialized society there is a considerable body of general environmental information on which project designers can draw. In developing nations this general knowledge base is considerably smaller, with the result that the project's own baseline studies take on an increasing design importance. This problem is complicated because a substantial amount of the environmental experience available from industrialized nations in temperate climates is simply not applicable to conditions in the developing countries of the tropics, nor is environmental information readily catalogued or transferred between the developing countries themselves.

DETERMINANTS OF FUTURE TRENDS

The previous sections of this paper focused on past trends with respect to the role of minerals in developing economies and the evolution of mineral objectives and policies in developing countries. It is believed that these evolving trends provide insight into future patterns. But many other factors will directly influence the pattern of future mineral developments—the most important of which are discussed in the sections following.

Mineral Policies of Developed Countries Regarding Developing Countries

The primary concern of developed countries regarding mineral commodities is to insure that supplies at reasonable prices are available to meet domestic requirements. Preferential treatment is usually given to domestic supplies of any given mineral commodity, rather than importing foreign supplies. Where imports constitute a substantial share of domestic requirements, various incentives are likely to be used in foreign mining exploration and development.

Countries meeting a high percentage of their mineral requirements through domestic sources (such as Australia, Canada, and the United States), are primarily focused on maintaining and expanding their own domestic minerals industry, and are less inclined to develop the same type of resources in developing countries. In the case of Australia and Canada, government assistance in the form of technical aid is given to foreign ministries responsible for minerals and energy. Assistance from the United States tends to be limited to a small number of developing countries.

In those countries with a substantial dependence on imports of raw materials (such as Japan, the United Kingdom, West Germany, and France), mineral policies are more directly designed to encourage increased minerals production in the developing countries.

Historically, the developed countries (excluding those with centrally planned economies) have relied primarily on private enterprise to acquire minerals from foreign sources. An exception to this is in France, where the nationalization of the major domestic mining companies occurred in the early 1980s. Even before the nationalization of the French mining industry, the government-controlled Bureau de Recherches Geologiques et Minieres (BRGM) was involved in active mineral exploration throughout the world.

France, West Germany, and Japan have provided economic incentives to promote foreign mineral exploration: for example, West German companies have been able to obtain subsidies for overseas exploration since 1971. Normally, 50% of the costs of exploration for nonfuel minerals, oil, and gas is refunded when a commercial development occurs, or the project is sold. Subsidies from the French government have applied to copper exploration since 1973 and uranium since 1977, with repayment taking place for successful projects (Crowson, 1979). The Japanese program of incentives for foreign exploration is broadly similar to the French and West German schemes but has a longer history. The United Kingdom does not have similar incentive schemes, but has relied on its London-based transnational corporations to meet the nation's mineral needs (Crowson, 1979).

A number of countries, including West Germany, France, Japan and the United States, have guarantee schemes available to cover noncommercial risks, most notably political risks.

Strategic stockpile programs exist in France, Germany, Japan and the United States. The strategic stockpile managed by the United States is by far the largest and covers the greatest number of mineral commodities. As the United States increases its dependence on mineral imports, there is likely to be more pressure from U.S. based transnational corporations for concessions similar to those offered to competing transnationals from other mineral-importing industrialized countries. However, to date, the response of United States to its increasing dependence on imports has been to establish stockpiles, to spend more on expanding domestic supplies, and to developing substitutes.

Mineral Resource Potential

Developing countries account for 55% of the world's land area excluding Antarctica. In 1978, these countries accounted for about 43% and 49% respectively of the value of world production of nonfuel and fuel mineral commodities. It has long been assumed that because the high-quality mineral resources in the developed countries are being depleted, there will be a shift to the high-quality resources in developing countries.

A combination of economic, technical, and political factors influence the rate of shift in mineral production from the developed countries to the developing countries. However, one of the most important of these factors is the size

Table 1.2.10—Percentage of Total World Mineral Reserves Broken Down by Economic Region for 1970 and 1980*

Commodity	1970				1980			
	Economic Region							
Bauxite	51	38	11	0	76	20	3	0
Chromite	97	1	2	0	98	1	1	0
Cobalt	66	9	25	0	79	4	17	0
Copper	44	31	13	13	56	27	12	5
Gold	71	12	—	8	65	10	24	1
Iron Ore	23	18	43	15	32	31	33	4
Lead	24	65	9	2	26	50	19	5
Manganese	—	—	—	—	64	9	27	0
Mercury	29	29	36	6	59	18	22	1
Molybdenum	19	63	18	0	31	59	10	0
Nickel	42	15	38	5	56	25	19	0
Tin	79	3	16	2	67	7	25	0
Tungsten	13	8	76	4	13	21	65	1
Zinc	15	53	17	14	34	46	12	9
Petroleum	81	10	9	0	81†	9	10	0

*Percentages derived from reserve estimates reported by the U.S. Bureau of Mines, *Mineral Facts and Problems*, Bulletin 650 (1970), 667 (1975), 671 (1980); and *Commodity Data Summaries*, 1970.
†Includes high-income, oil-exporting countries which account for about 70% of world reserves.

Table 1.2.11—Growth of World Reserves for Selected Mineral Commodities (Millions of tonnes near end of decade indicated)

	Copper	Lead	Zinc	Bauxite (gross weight)
1940s	91	30–45	54–70	1,605
1950s	124	45–54	77–86	3,224
1960s	280	86	106	11,600
Reserves: % growth rate/yr 1950s–1970s	7.25	5.0–5.75	4.75–5.25	9.75
Production: % growth rate/yr 1950–1970s	3.75	1.75	2.75	7.0

Source: Crowson, 1982.

and quality of the resources remaining to be mined in the developing countries.

The following two methods provide an indication of the mineral potential of developing countries relative to that of the industrialized countries: (1) a comparison of the size of measured reserves, and (2) a comparison of the value of mineral production for similarly sized areas. Neither method is likely to provide an accurate estimate, however, they do provide a rough semiquantitative basis for estimating the mineral resource potential of developing countries. Table 1.2.10 shows reserve estimates for 15 important mineral commodities broken down by economic region for 1970 and 1980. These commodities account for the majority of the international trade in minerals. From this table it is apparent that the developing countries currently account for over 50% of the reserves of two thirds of the 15 commodities listed.

Reserve estimates for large areas appear to be more a function of the amount of exploration than the size of the mineral resources of an area. As shown in Table 1.2.11, reserve estimates tend to increase over time at least as fast as production—an indication that reserves may not be an accurate indication of the total mineral resource potential.

A second method for estimating the mineral potential of a large, poorly explored region is to assume that its mineral potential per unit area is similar to a large, well-explored region. The larger the regions compared the greater the probability that they will have a similar resource potential per unit area. Since production value can be more readily estimated than resource potential, production value is used here as a proxy for resource potential. The difference in production value per unit area between industrialized countries (heavily explored region) and developing countries (poorly explored region) provides an indication of the remaining potential

Table 1.2.12—Mineral Production and Potential For Developing and Developed Countries

Country Group	Area ('000 km²)	Production of Minerals ($/'000 km²)		Percent of Developed		Additional Production Potential per Year (billion $)	
		Nonfuel	Fuel	Nonfuel	Fuel	Nonfuel	Fuel
Developing*	75,057	352	2,059	55	48	22	171
Developed	31,154	638	4,313	100	100	—	—

*Excludes high-income, oil-exporting nations.
Source: Callot, 1981.

for developing countries. Table 1.2.12 compares the value of minerals production per thousand square kilometers for industrialized and developing countries. As shown in Table 1.2.12, the average value of nonfuel minerals production per unit area is 55% of that of the industrialized countries. For fuel minerals, the average value is 48% of that of industrialized countries. The difference in production value of the two regions multiplied by the total area of the developing countries gives an estimate of the remaining potential in developing countries, that is, 22 billion dollars for nonfuel minerals and about 170 billion dollars for fuel minerals. In both cases the remaining potential is greater than or equal to the value of total present production in the total of all industrial countries. The reader is warned that such figures are only useful to indicate that large potential exists in developing countries. Such comparisons ignore the critical factor of economics which varies greatly with location relative to markets and available infrastructure.

Exploration Levels

The total expenditure on mineral exploration in the developing countries is unknown. Several analyses suggest that the share of exploration expenditures by private companies in developing countries has probably declined (James and Khan, 1978; Mikesell, 1979). However, a recent study by Phillip Crowson (1983) of the trends in nonfuel minerals exploration in the market-economy countries suggests that the share of total exploration expenditures in the developing countries remained at about one third of total expenditures between the early 1970s and the early 1980s. As shown in Table 1.2.13, in the 1970s the total expenditure on mineral exploration (excluding petroleum) in the developing countries increased at the same rate as expenditures in developed countries. Consequently, the share of total expenditure in the developing countries remained in the range of 28% to 35 percent. However, the pattern of exploration appears to have changed substantially within the group of developing countries—with private exploration restricted to fewer countries. Transnational corporations are becoming increasingly selective in assessing technical, economic, and political risks in a given country prior to undertaking exploration. Much of Africa appears to have had a major decrease in nonpetroleum exploration by transnational corporations with only a few exceptions—Botswana and South Africa.

Although the estimates in Table 1.2.13 for

Table 1.2.13—Estimated Expenditure on Non-Petroleum Mineral Exploration*
(Million 1982 US Dollars)

Country Group	Early 1970s		Early 1980s	
	Dollars	**Percent**	**Dollars**	**Percent**
Developed Countries				
Australia	420	18–17	560	17–15
Canada	455	20–18	580	17–16
Europe	200	9–8	410	12–11
USA	500	22–20	800	24–22
Japan	45	2	30	1
Total Developed Countries	1,620	71–65	2,380	71–65
Developing Countries				
Brazil	100	4	180	5
South Africa	180	8–7	250	8–7
by EEC Companies	40	2	55	2
by Japanese Companies	30	1	25	1
on Uranium	50	1	40	1
Other	300–500	13–20	400–700	12–19
Total Developing Countries	680–880	29–35	950–1,250	29–35
Total World	2,300–2,500	100–100	3,330–3,630	100–100

The author notes that these results are rough and should be used with caution.
*Excluding the Centrally Planned Economies.
Source: Crowson, 1983.

nonfuel minerals exploration are incomplete, they indicate an overall rate of growth in exploration expenditures of 2.9% to 4.6% per year during the 1970s. Given the large reserves of almost every mineral commodity, in addition to the relatively slow growth in demand for most mineral commodities between the early 1970s and early 1980s, there appears to be a tendency toward an increasing abundance of mineral reserves. This was illustrated previously in Table 1.2.10 which showed that the growth in world reserves of copper, lead, zinc, and bauxite have increased at a faster rate than the growth in demand. From the perspective of the mineral-producing countries, this suggests that supplies will be more than adequate to meet future demand. Consequently, sustained real price increases for most mineral commodities are unlikely.

Petroleum exploration is estimated to represent greater than 90% of the total expenditure on mineral exploration. Statistical information on worldwide petroleum exploration is much more complete than for nonfuel minerals exploration. Two commonly reported measures of petroleum exploration are the number of wells drilled and total exploration and development expenditures. Such aggregate estimates are useful as an indication of general trends in petroleum exploration. Table 1.2.14 shows the share of petroleum exploration wells drilled in 1972 and 1980 in developed and developing countries. As shown in the table the share of wells drilled in developing countries declined from 10% in 1972 to 6% in 1980. The same trend is indicated by the share of total expenditure on exploration and development which fell from 24% in developing countries in 1972 to 17% in 1980. This trend is the reverse of what would have been expected based on the relative costs of discovering a barrel of petroleum during the 1972 to 1982 period. Even for the higher-cost, petroleum-importing developing countries, exploration costs were only 24% of the cost per barrel in developed countries (Blitzer, et al. 1983). A number of factors probably contribute to the unexpected results shown in Table 1.2.14. However, the main factor may be that investors perceive much higher risk associated with investments in developing countries.

Economics

There is no sharp division in the economics of mineral production in the developing countries as opposed to the industrialized countries. If countries were to be ranked on the basis of mineral production costs, the rankings would show a mix of both developing and industrialized countries (as shown in Table 1.2.15 for copper production). There are so many components of the mineral supply system, most of which vary over time, that broad generalizations can be quite misleading. Indeed, a comparison of averages for developing and industrialized countries tends to mask important changes occurring at the margin (in small groups of nations or mines) that eventually will have widespread implications to the overall economics of minerals supply.

Given that caveat, it is possible to make a number of useful observations about the economics of minerals production in developing versus industrialized countries. These observations are discussed briefly under the following subheadings: ore grade, capital costs, operating costs, energy costs, and transportation costs.

Ore Grade: On the average, the ore grades mined in developing countries are substantially higher than those in the industrialized countries. A major reason for this is that most of the industrialized countries have had an extensive mining history, therefore many of the highest-grade deposits have already been depleted.

Table 1.2.14—Petroleum Exploration Trends*

Country Group	Petroleum Exploration Wells (%)		Investment in Exploration and Production (%)		New Reserves 1972–1982 1980 $/bbl
	1972	1980	1972	1980	
Developed	90	94	76	83	2.56
Developing	10	6	24	17	0.17

*Derived from Blitzer et al., 1983.

Table 1.2.15—Estimated Breakeven Cost of Copper Production For the Major Producing Countries*

Country	Percent Copper "Equivalent" Recovered	Production Cost (cent/lb)	By-Product Credits (cent/lb)	Breakeven Cost for copper (cent/lb)
Canada	2.25	101	84	17
Australia	3.60	100	70	30
Peru	1.02	34	1	33
South Africa/Namibia	1.66	64	26	38
Chile	1.67	42	3	39
Papua New Guinea	0.81	85	33	52
Philippines	0.69	66	11	55
Zambia	2.10	62	4	58
United States	0.79	71	11	60
Zaire	5.23	103	35	68

*Data for 1978. It is important to note that such country comparisons can be highly misleading if applied to specific deposits.
Source: Niarchos, 1982.

Moreover, as the costs of mine developments in more remote areas of developing countries are higher, to be economic the deposits must have higher-grade ore than a comparable mine in the industrialized countries.

An example of the differences in ore grade currently being mined is shown in the case of copper where the weighted average grade of copper ore mined in the industrialized countries is approximately 0.66% as compared to 1.12% in the developing countries (Crowson, 1982). Table 1.2.15 shows the average grade of copper recovered (by-product metal values are converted to "copper equivalent" grades) for the major copper-producing countries. It is clear from Table 1.2.15 that there is no clear-cut division on a country-by-country basis of copper grades between the industrialized and the developing countries. However, the weighted average grade of copper equivalent and the total production of the countries shown in the table indicates an average grade of 1.6% in the industrial countries and 2.0% in the developing countries.

Exceptions to the general rule are Australia and Canada, both of which rival developing countries in the size and high grades of many of their mineral deposits. Australia has some of the largest and highest-grade deposits of iron ore, bauxite, uranium, and diamonds in the world, and Canada has high-grade deposits of potash, nickel, and uranium.

[n]Includes those mineral projects of at least 15 million dollars, and excludes the nonmarket economies.

[o]Some bias may exist in the comparisons because the average size of profits in developing countries may be larger than in developed countries.

Capital Costs: Capital costs for mining and processing facilities in the developing countries tend to be substantially higher than for comparable facilities in the industrialized countries. This is evident when the average capital costs are compared for 141 mineral projects reported to be under development (or are likely to be developed) in both the industrialized and developing countries in 1983.[n] The average capital cost for the 59 projects in the industrialized countries was 255 million dollars, and for the 82 projects in the developing countries the average was about 380 million dollars—or about 60% more than in the developed countries.[o] Total planned investment was 13,252 and 31,236 million dollars respectively for the industrialized and developing countries, an indication that a substantial shift in new investment to the developing countries is occurring.

The basic reason for the high capital costs for mines in the developing countries is that most mines are remote and lack the infrastructure (such as transportation, port facilities, electricity, water systems, hospitals, schools, housing, etc.) necessary to develop and operate a mine. Infrastructure can add 40% or more to the total capital cost of bringing a major mine into production. Similarly, high capital costs occur in remote areas of the developed countries, e.g., Australia, Canada, and the United States (Alaska).

Operating Costs: The developing countries have several characteristics that tend to reduce operating costs. First, labor costs are substantially lower than in the industrialized countries, and local labor usually accounts for 75% to 90%

of the total manpower requirements. Second, as previously mentioned, ore grades tend to be higher thereby reducing per unit production costs. On the negative side, however, the foreign expertise used in projects in developing countries is expensive, and most of the materials used in mining operations are usually imported. Overall operating costs are difficult to assess, but for well-managed mines in developing countries they may be comparable to or lower than those in the industrialized countries.

Energy Costs: The rapid increase in petroleum costs during the 1970s has had an uneven impact on the economics of mining and processing. Many of the most energy-intensive processing activities, such as aluminum smelting, historically have been located near low-cost energy sites (usually hydroelectric facilities), and many of these sites continue to be the lowest-cost sources of electricity. However, energy-intensive industries dependent on imported petroleum, such as those in Japan, have become uneconomic for major expansions in the future. High energy costs are resulting in the development of new, energy-efficient mining and processing technologies. In addition, some processing technologies are being modified to allow substitution of lower cost energy forms for high-cost forms (i.e., coal in place of petroleum). During the 1970s, the rapid increase in petroleum prices widened the gap between low-cost and high-cost electricity supplies from four to one in the early 1970s to ten to one in the early 1980s. This provides a significant comparative advantage to low-cost energy sites in such areas as Australia, Canada, Brazil, Indonesia, Zaire, and the Middle East.

Nickel produced from laterite deposits (most of which are located in the developing countries) is one commodity that has become relatively uncompetitive as a result both of higher energy costs and of the major decrease in the growth rate of consumption. Energy costs account for almost half the sales value of laterite nickel (that is, $1.50 to $1.80 per pound of nickel), whereas nickel from competing sulfide deposits located primarily in the developed countries can be produced for about one third of the energy requirement for lateritic nickel. This results in a savings of about $1.00 per pound for a product that normally sells for about $3.00.

Although energy cost is only one element in the overall economics of mineral supplies, it is the one variable that suffered the greatest change during the 1970s. Nations with low-cost energy supplies will have a significant advantage in attracting energy-intensive processing industries in the later 1980s and 1990s. This factor combined with a stable investment climate, will be a benefit to Australia, Canada, and a few developing countries.

Transportation Costs: The cost of transportation is most critical for the large-tonnage, low unit-value commodities such as iron ore, phosphate, bauxite, and coal, where transport costs may represent half the delivered price of the commodity. Access to low-cost ocean transportation is a key factor; a ton of iron ore can be moved halfway around the world in large bulk carriers for $10 to $15 per tonne. For most metals the transportation element is much less important typically accounting for less than 10% of the delivered price.

Overall Economics: As previously noted, there is no clear division in the economics of minerals production between the developing and the industrialized countries. The most significant cost advantage of mining in the developing countries is the existence of deposits of higher grade than those found in many of the industrialized countries. Also, in some cases there are large supplies of low-cost energy. However, these same two advantages exist in Australia and Canada both of which continue to attract a large share of mineral investments.

Transnational Corporations

Transnational corporations (hereafter referred to as TNCs) have played a highly visible role in many developing countries. For the smaller mineral-led economies, one or two dominant TNCs may provide the most conspicuous business activity, produce the major share of exports, and provide the largest share of government revenues. At times, the business practices of the TNCs (such as transfer pricing) have been in conflict with the interests of the developing countries. This has resulted in a substantial amount of distrust between governments and TNCs, particularly in those economies dominated by a few large mining companies.

The size of TNC operations can be placed in perspective by noting that historically TNCs have controlled directly and indirectly about three quarters of the mineral resources in developing countries (United Nations, 1981). The world mining and petroleum industry is relatively concentrated; 14 TNCs account for about two thirds of the value of nonfuel minerals production, and 12 TNCs account for about two thirds of the value of petroleum production (excluding the centrally planned economies).[p] TNCs based in the United States traditionally have been considered the leaders in the global fuel and nonfuel mineral industries. However, the dominance of traditional U.S. mining corporations is being challenged on a number of fronts, including: (1) expansions of TNCs from Europe, Japan, Australia, and in developing countries, (2) takeovers of traditional mining companies by petroleum companies, and (3) the growth of state-owned corporations.

The role of the Japanese TNCs is a marked departure from the traditional TNCs. Whereas traditional North American and European TNCs are usually vertically integrated and have major equity shareholdings in mines, Japanese TNCs have focused on providing a combination of finance, technology, and long-term sales agreements—often with little direct equity involvement (Widyono, 1980).

As a group, Japanese TNCs have made a concerted effort to develop diversified sources of supply among many different countries and supplying companies. The availability of finance (sometimes on concessional terms) and long-term contracts has probably resulted in mining projects in developing countries that would not otherwise have been developed by the more traditional TNCs. However, the dominant bargaining position with respect to raw material purchases by Japanese TNCs may have resulted in less favorable sales arrangements to some exporting countries than would have otherwise occurred.

Historically, the major petroleum TNCs concentrated on oil and had little direct involvement in the mining industry. Petroleum TNCs account for eight of the 10 largest industrial corporations in the world, and 22 of the top 50 corporations. In contrast, the TNCs producing coal and metal commodities are not among the 50 largest companies (Baxendell, 1982).

The major upheavals in the petroleum industry in the early 1970s and the subsequent politicalization of oil caused the managements of oil companies to search for alternative industries in order to reduce perceived future risks associated with their traditional role in the world oil industry. Some oil companies moved to acquire the major coal resources in the developed countries (Australia, Canada, and the United States) as well as those in a number of developing countries. For the most part, the oil companies viewed the move into coal as a logical extension of the energy business. The coal subsidiaries of major oil TNCs moved aggressively into coal exploration in a number of developing countries—often with exploration commitments substantially in excess of the more traditional coal companies. Shell and EXXON are examples of companies that moved aggressively into coal exploration in developing countries in the 1970s.

Another aggressive move by the oil companies was the acquisition of several major mining companies. The metals industry was seen as being compatible with the oil companies because of the following similarities: both industries use similar exploration techniques and involve extractive technologies, both are international in scope, both consist of large projects with long lead times, large and complex financing requirements, economic and political risks, joint venture agreements, dealing with governments, and international transportation and trade (Baxendell, 1982).

Often overlooked is the role of companies from East European nonmarket countries in developing countries. There are numerous joint-venture arrangement between East European companies and state companies in developing countries. Typically, equity participation in developing appears to vary from 10% to 49% and the major form of participation is in supplying machinery and scientific services with relatively small contributions of hard currency. The goal of the East European arrangements appears to be to obtain long-term supplies of raw material imports. Little is known about the profitability of such joint ventures, however the literature suggests the experience has been mixed (Dobozi, 1983). Although there appears to be a trend toward increasing numbers of joint ventures between East European and developing countries, there is no indication that these joint ventures will displace the more traditional TNC investments in most developing countries.

An important result of the development of a wider range of TNCs in the mining industry is that governments have more flexibility with respect to the terms under which TNCs will op-

[p]The estimate for nonfuel companies is based on the largest 34 companies, and, for petroleum, on the largest 40 companies.

erate in their countries. This is not to suggest that the TNCs are less profit oriented, but that developing countries may have greater opportunities to achieve forms of mineral agreements that are more compatible with national economic and social planning objectives. The entry of petroleum TNC's into the mining industry will provide the obvious advantage of much greater financing capability for major projects. Moreover, with respect to the developing countries, petroleum companies appear to exhibit greater acceptance of new forms of mining arrangements than appears to be the case with some of the more traditional mining companies.

State Enterprises in the Mineral Sector

Following the wave of nationalized enterprises that accompanied political independence in the developing world in the 1960s, governments of developing countries moved to establish state enterprises in the minerals sector. These enterprises were a natural consequence of the economic nationalism accompanying independence, and of the importance of the mining sector to the economics of several mineral-led nations.

Table 1.2.16 presents selected data on the structure of state investment (i.e., in excess of 50% equity ownership) in primary mining and processing of copper and aluminum. Excluding the aluminum smelter capacity based on low-cost natural gas in the Middle East countries, two contrasting trends are evident from the data shown in Table 1.2.16. For aluminum, state investment tends to decline with each subsequent processing stage. While the availability of low-cost electricity plays an important role in aluminum smelters, other factors are also significant. Of these factors, two considerations are worthy of note: (1) the historical (i.e., integrated) structure of private aluminum this companies, and (2) the recognition by the developing

Table 1.2.16—State Ownership in Developing Countries by Stage of Processing (over 50% government equity)

Aluminum*	Mining Capacity ('000 tonnes)	Alumina Capacity ('000 tonnes)	Aluminum Capacity ('000 tonnes)
Africa			
Ghana	300	—	—
Guinea	2,500	—	—
Egypt	—	—	100
Asia			
India	500	250	100
Indonesia	1,800	—	—
Bahrain	—	—	125
Dubai	—	—	135
Iran	—	—	50
Latin America			
Guyana	5,000	350	—
Jamaica	7,300	—	—
Venezuela	—	—	400
Copper†	**Mining ('000 tonnes)**	**Smelting ('000 tonnes)**	**Refining ('000 tonnes)**
Africa			
Zaire	662	469	151
Zambia	704	816	776
Asia			
India	35	48	40
Latin America			
Brazil	34	—	—
Chile	915	960	780
Mexico	76	3	—
Peru	67	91	238

*Radetzki, 1983.
†Data from Intergovernmental Council for Copper Exporting Countries (CIPEC).

countries that most of the economic rents in mineral production exist at the mining rather than processing stages.

In contrast, the pattern of copper processing in developing countries is different. In the four largest copper-producing nations (Zaire, Zambia, Chile, and Peru), there is a high degree of state involvement with downstream processing. Part of the reason for this pattern of public investment is that major copper-producing countries have sought to influence their domestic industry by controlling marketing decisions.

The initial movement toward the establishment of state mining companies has slowed in recent years, and the current trend in new mining projects seems to be toward joint-venture participation rather than toward a nationalized industry. In an insightful study, Radetzki (1983) suggests several possible explanations for this policy shift:

(1) a lessening of the need to rectify unfavorable colonial conditions,
(2) a decline in socialist political ideology,
(3) an increasing sophistication in private sector entrepreneurship and management in the developing countries.

Radetzki goes on to argue that:

> . . . the fast growth of the state-owned share is now over. The tentative conclusion is that in the 1980s and 1990s the expansion of state enterprise in Western World mineral industries will not be much different from the overall growth of the mineral sector. (Radetzki, 1983)

By most measures, the financial performance of state enterprises is below that of privately run mining companies. The reasons for the poorer financial performance are complex but a central reason commonly suggested is that it results from the broader socio-political goals and the bureaucratic character of state corporations. As a product of the state, the mining corporation is expected to serve the broad social objectives of the developing countries. As shown in the preceeding section, social goals such as employment generation or regional development can be costly on the balance sheet and make the management of mineral companies in developing countries less accountable to a measureable performance criteria. In this broadening of corporate objectives, state enterprises in developing countries are reflecting a distinction long recognized by economists as the difference between financial and economic objectives. Mikesell summarizes the different analytical methodologies of state and private mining investors as follows:

> . . . cost calculations for a government enterprise may take into account social opportunity costs and social benefits rather than full monetary costs and money revenues, which would form the sole basis for project evaluation in the private sector, (Mikesell, 1979)

Clearly, one of the most important distinctions between state and private mining concerns is the vastly different attitude of the host government. Although such a statement may seem almost tautological, it is worth examining government attitudes more closely. For the most part, the mildly adversarial tension between government regulator and private mineral firms is replaced by a benevolent attitude toward state enterprises. This benevolence takes many forms and ranges from less severe taxation and dividend policies (Gillis and Beals, 1980) to provision of subsidized finance and infrastructure (Mikesell, 1979).

This different relationship is further reflected in the attitudes of state enterprises toward investment risk. Clearly, a state corporation need not be concerned with the same political risks which concern foreign investors. Similarly, since the government itself backs the investments undertaken by the state corporation, and the available investments are defined wholly within the nation's borders, the ranking and profitability criteria for new projects may be significantly different from those used by private transnational corporations. Finally, the attitude of outside lenders toward contingent liability risks may change. In this latter regard, the distinction between the sort of entreprenurial risk assessments common to (private) project finance and the assessment made for general public debt financing would tend to become blurred. To the degree that state enterprise borrowing tends to preempt general development borrowing, policy makers in developing countries must face difficult choices involving high-risk/high-return investment opportunities in the mineral sector versus investments in social services and infrastructure.

Over the longer term, the maturation and

growth of the state enterprise may provide a vehicle for technical cooperation and exchange between the governments of mineral-producing nations. Such exchange is already widespread among petroleum producers. Increasingly, joint mineral projects have begun to emerge—such as the cooperative arrangements between Brazil and Mozambique, and the proposed arrangement between Zimbabwe and Morocco.

CONCLUSIONS

This chapter has discussed the role of minerals in developing countries with particular reference to the evolution of the relationship between governments and transnational corporations. Developing countries are so diverse with respect to type and size of economy, economic and political stability, minerals potential, and mineral policies and strategies, that few generalizations hold for the entire group. However, as compared to the industrialized countries, mineral-producing developing countries as a group tend to depend on the minerals sector for a larger contribution to (1) export earnings, (2) gross domestic product, and (3) employment.

In some developing countries, the minerals sector is known for its relatively large size, foreign company dominance, and potential for generating substantial economic rents. These factors have led to increased government involvement, particularly in the area of control of mining operations, and efforts to tax a greater share of economic rents. These changes have resulted primarily from an evolution in mineral agreements, many of which have been modified more in form than in substance with respect to actual control and taxing of economic rents.

Probably the best known and possibly the most effective innovation in taxation regimes was established in Papua New Guinea in the mid-1970s. This tax regime contains an "additional profits tax" feature that only applies to profits above a specified internal rate of return to the investor. This taxation feature is a considerable improvement over most existing taxation regimes because it is easy to apply and only taxes economic rents. In the future, this tax regime is likely to be adopted by many developing countries and perhaps by some industrial nations.

The combination of factors such as the entry of new transnational corporations into the mining industry, the acquisition of traditional mining companies by major petroleum companies, increased involvement of East European companies is altering the traditional patterns and arrangements for mining developments in developing countries. Developing countries have an increasing range of options for developing their minerals potential, and this is resulting in an evolution in the traditional forms of agreements and the respective roles of governments and foreign investors.

A period of almost continuous change in the relationship between governments and foreign investors occurred during the 1960s and 1970s. Many developing countries succeeded in gaining greater control of their projects, but failed in greater economic benefits. A smaller group of developing countries were successful both in exercising greater control and in increasing economic benefits to the government. The most successful governments appear to have adopted strategies based on modern financial analysis techniques, on an understanding of the value of their mineral resources in an international context, and on hard negotiations.

The zenith of government activity in taking control of the nonfuel mining industry appears to have been in the 1960s and early 1970s and, in the case of petroleum, the 1970s. However, between the mid-1970s and early 1980s the poor economic performance of most major industrialized nations resulted in a decline in growth rates in consumption and consequently depressed prices for most mineral commodities. For nearly a decade, both governments and transnational corporations have been waiting for a major upswing in prices—an event that has always occurred in the past.

But the past may be a poor gage of the future. Two major trends appear to be offsetting prospects for substantial sustained real price increases for mineral commodities. First, exploration technology has been so successful that the growth rate in reserves of most mineral commodities has exceeded the growth rate in consumption. Second, the current growth rate in consumption appears to be significantly slower than growth in consumption during the 1950s to the mid-1970s. A number of factors appear to be contributing to this overall decline in growth rates including: the saturation of the major markets in industrial nations, reductions in material requirements to produce consumer products, and increased recycling of materials.

In addition, the continued increase in substitution possibilities between materials reduces the comparative advantages of specific materials for individual applications. This will tend to reduce the opportunities for sustained real price increases for specific mineral commodities.

The net result of these trends is a reduction in the opportunity for higher than normal profits in the nonpetroleum minerals sector. The race for resources of the 1960s and 1970s, where developing countries were increasingly pitted against transnational corporations in the struggle to shift greater economic rents to governments may be gradually overshadowed by the urgency to become more competitive in order to produce any economic rents. This may lead to greater flexibility by both governments and transnational corporations in developing more effective arrangements to insure maximum efficiency in the minerals sector, while meeting the objectives of both governments and TNCs.

Present investment trends in new capacity and the large resource potential of developing countries leads to the conclusion that the share of world supplies of mineral commodities from developing countries will continue to increase through the 1980s. However, capturing economic rents may continue to elude most countries and most transnational corporations.

The governments most successful in obtaining benefits from the minerals sector will be those that can establish both efficient government control and effective taxation regimes while maintaining a high degree of efficiency in the mining sector. In many developing countries there will continue to be an important long-term role for those foreign mining companies that have been able both to adapt to the closer involvement of governments and to maintain high levels of operational efficiency.

ACKNOWLEDGEMENTS

The authors would like to recognize the assistance received in the compilation of information and editing of this chapter. We wish to thank Jean Brady for editorial assistance, and Larry Dale and James Dorian for assembling data. In addition, a special thanks to Jim Otto whose help in isolating and categorizing information was invaluable.

References

1. Baxendell, P., 1982, "The Diversification of Oil Companies into Mining," transcript of lecture delivered to the Imperial College of Science and Technology, London, p. 7.
2. Blitzer, C., Cavoulacos, P., Lessard, D., and Paddock, J., 1983, *Overcoming the Oil Exploration Gap in Developing Countries*, final report to the U.S. Agency for International Development, 223 pp.
3. Bosson, R., and Varon, B., 1977, *The Mining Industry and the Developing Countries*, Oxford University Press, New York, p. 133.
4. Brown, R., and Faber, M., 1977, *Some Policy and Legal Issues Affecting Mining Legislation and Agreements in African Commonwealth Countries*, Commonwealth Secretariat, London, p. 3.
5. Callot, F., 1981, *World Production and Consumption of Minerals in 1978*, Mining Journal Books, London.
6. Clausen, A.W., 1982, "Sustainable Development: The Global Imperative," *The Environmentalist*, Vol. 2, No. 1, p. 26.
7. Cordes, J.A., 1980, "Minerals Transnationalism and Economic Development," Ph.D. Dissertation, Colorado State University, Fort Collins, p. 252.
8. Crowson, P.C.F., 1983, "A Perspective on World Wide Exploration for Non Fuel Minerals," presented at the Expert Group Meeting on the Economics of Mineral Exploration, International Institute for Applied Systems Analysis (IIASA), Laxenburg, Austria.
9. Crowson, P.C.F., 1982, "Investment and Future Mineral Production" *Resources Policy*, Vol. 8, No. 1, pp. 3–12.
10. Crowson, P.C.F., 1979, "Approaches Adopted by Other Countries," *Availability of Strategic Minerals*, M.J. Jones, ed., The Institution of Mining and Metallurgy, London, pp. 93–98.
11. Dobozi, I., 1983, "Arrangements for Mineral Development Cooperation between Socialist Countries and Developing Countries," *Natural Resources Forum*, Vol. 7, No. 4, pp. 339–349.
12. *Engineering and Mining Journal*, 1981, "Mining Investment 1981," Vol. 1982, No. 1, pp. 59–81.
13. Garnaut, R. and Clunies-Ross, A., 1975, "Uncertainty Risk Aversion and the Taxing of Natural Resource Projects," *The Economic Journal*, Vol. 85, pp. 272–287.
14. Gillis, M., and Beals, R., 1980, *Tax and Investment Policies for Hard Minerals: Public and Multinational Enterprises in Indonesia*, Ballinger, Cambridge, Massachusetts, pp. 87–89, 261–268.
15. Gregory, R.G., 1976, "Some Implications of the Growth in the Mining Sector," *The Australian Journal of Agricultural Economics*, Vol. 20, No. 2, pp. 71–91.
16. Haffner, B.K., 1979, "The Potential Economic Impact of U.S. Regulations on the U.S. Copper Industry," U.S. Dept. of Commerce, Industry and Trade Administration, p. v-28.
17. Herfindahl, O.C., 1969, *Natural Resource Information for Economic Development*, The Johns Hopkins Press, Baltimore, p. 4–5.
18. International Labor Office, 1976, 1981, *Yearbook of Labor Statistics*, ILO, Geneva.
19. *International Petroleum Encyclopedia*, 1983, Petroleum Publishing, Tulsa.

20. James, C. and Khan, M., 1978, ''Mineral Exploration,'' *Mining Annual Review*, Mining Journal, London, p. 133–134.
21. Lewis, A., 1983, ''The United Nations in Mineral Development,'' *Engineering and Mining Journal*, Vol. 184, No. 1, pp. 68–70.
22. Mikesell, R., 1983, *Foreign Investment in Mining Projects*, Oelgeschlager, Gunn & Hain, Inc., Cambridge, Massachusetts, pp. 51–55, 258.
23. Mikesell, R., 1979, *New Patterns of World Mineral Development*, British North American Committee, Washington, D.C., pp. 38–39.
24. Nankani, G., 1980, ''Development Problems of Nonfuel Mineral Exporting Countries,'' *Finance and Development*, Vol. 17, No. 1, pp. 6–10.
25. Niarchos, M., 1982, ''Competitive Position of Copper Producers in 1978, M.Sc. Thesis, Queen's University, Canada, 89 pp.
26. Pintz, W.S., 1984, *Ok Tedi: Evolution of a Third World Mining Project*, Mining Journal Books, London.
27. Radetzki, M., 1983, ''State Enterprise in International Mineral Markets,'' CP-83-35, International Institute for Applied Systems Analysis (IIASA), Laxenburg, Austria, pp. 59–61, 178–180.
28. Radetzki, M., 1980, ''Mineral Processing in LDCs,'' *Mining for Development in the Third World*, S. Sideri and S. Johns, eds., Pergamon Press, New York, pp. 229–242.
29. Radetzki, M., and Zorn, S., 1979, *Financing Mining Projects in Developing Countries*, Mining Journal Books, London, p. 91.
30. Rogers, C., 1979, ''Non-Fuel Minerals and the Integrated Program for Commodities,'' *Natural Resources Forum*, Vol. 3, No. 4, p. 338.
31. Stoeckel, A., 1979, ''Some General Equilibrium Effects of Mining Growth on the Economy,'' *Australian Journal of Agricultural Economics*, Vol. 23, No. 1, pp. 1–22.
32. United Nations, 1983, ''Monthly Bulletin of Statistics: July,'' United Nations, New York.
33. United Nations, Economic and Social Council, 1981, ''Permanent Sovereignty Over National Resources,'' U.N. E/C.7/119, New York, p. 15.
34. United Nations, 1978, ''Transnational Corporations in World Development: a Reexamination,'' U.N. Centre on Transnational Corporations, New York.
35. Walde, T., 1983, 'Permanent Sovereignty Over Natural Resources: Recent Developments in the Mineral Sector,'' *Natural Resources Forum*, Vol. 7, No. 3, pp. 239–251.
36. Widyono, B., 1980, ''Transnational Corporations in the Mineral Industries of the Asia-Pacific Region,'' presented at the Workshop on Mineral Policies to Achieve Development Objectives, East-West Center, Honolulu, Hawaii.
37. World Bank, 1982, *Commodity Trade and Price Trends*, The Johns Hopkins University Press, Baltimore, p. 25.
38. World Bank, 1980, *World Development Report*, World Bank, Washington, D.C., pp. 110–111.
39. World Bank, 1978, 'Mineral and Fuel Development,'' *Annual Report*, World Bank, Washington, D.C. pp. 20–21.
40. Zorn, S., 1978, ''Conference Review: The United Nations Panel on International Mining Finance,'' *Natural Resource Forum*, Vol. 2, No. 3, p. 294.

1.3

International Trade in Mineral Commodities

Henry N. McCarl* and Gerry Waters†

INTRODUCTION

In the past three decades, world trade has grown from less than $100 billion per year in the years following World War II to slightly more than $300 billion in 1970 and over $2,000 billion in the early 1980s. Adjusted for inflation, the real growth of world trade in 1972 dollars increased from less than $175 billion per year in the early 1950s to slightly more than $342 billion in 1970 and almost $700 billion per year by the early 1980s. Thus trade as a whole has increased a little more than twofold in real terms in the past decade.[a]

Mineral commodities constituted approximately 20% of the value of total world trade in 1970, and significantly more than that proportion in terms of physical volume. By 1980, despite rapidly rising petroleum prices, mineral commodities accounted for roughly 18% of the total value of world trade. Crude oil alone accounted for about 16%, making it the single most important commodity in international trade. The worldwide recession of 1981–1983 brought a dramatic decline in both volume and price of crude oil, shrinking the impact of mineral commodities to roughly 10% of the value of total exports, worldwide, in 1982–1983.

The decline of mineral commodities as a proportion of the value of world trade in the decade 1970–1980 is due largely to the increasing importance of manufactured goods and consumer products such as automobiles and electronic items. During this period, mineral commodity trade expanded roughly sixfold in terms of current dollars to an estimated $350 to 360 billion in 1980. This was equivalent to doubling the real value of mineral commodities in international trade in the 1970–1980 period in constant dollars.

The U.S. Bureau of Mines has expanded the traditional definition of mineral commodities to include semifinished products such as iron and steel (in addition to iron ore), refined metals such as copper, lead, zinc, aluminum, and tin, concentrates, scrap, and petroleum products. Since large quantities of essentially mineral derivatives travel in this form, it is reasonable to use these expanded definitions in stating the role of minerals in world trade. Thus minerals in world trade comprise 30 to 33% of total export trade in all commodities during the period 1976–1980 (see Table 1.3.1).

In addition, the U.S. Bureau of Mines estimates the value of all mineral commodities consumed in the world (at the national level) and compares this value to total worldwide commodity consumption. By this system, mineral commodities range from 30 to 40% of all commodities consumed each year in all countries during the period 1976–1980 (see Table 1.3.2).

The growth in export trade in mineral commodities by commodity group is shown for the period 1976–1980 in Table 1.3.3. These data parallel the swings in worldwide economic activity during the same time period.

By any of the above measures, minerals, especially the fossil fuels, constitute a significant proportion of both total economic activity and

[a]*Economic Report of the President, February, 1983*, Washington, D.C.: U.S. Government Printing Office, Table B-106, p. 282.

*Associate Professor of Economics and Geology, School of Business, The University of Alabama, Birmingham

†Research Associate, School of Business, The University of Alabama, Birmingham.

total international trade. Mineral commodities are an important source of hard currency for many developing nations, and provide the basis for international exchange for many developed economies as well. For economic systems that closely control the circulation of their currency, commodity trade is becoming a more common tool in international exchange agreements. This leads to apparent inconsistencies such as a country both importing and exporting the same commodity. For example, the Soviet Union is both an importer of phosphate rock from the United States and an exporter of phosphate rock to countries in Eastern Europe. Romania imports crude oil and exports petroleum products.

The sheer physical size of some countries also leads to both import and export of the same products. The location of reserves of some mineral commodities sometimes means that it is more economic for a country to export a commodity even when the country is a net importer. Recent developments in the United States concerning the advantages in export of Alaskan crude oil to Japan are an important case in point. There is some economic gain in exporting Alaskan crude to Japan rather than refining it for use in the United States despite a significant need for imported petroleum within the United States. The Trans-Alaskan Pipeline was built along its current route with this export potential in mind.

The significant differences in value, volume, production economics, markets, and end-uses make it necessary to review at least some important minerals as individual commodities to

Table 1.3.1—Value of World Export Trade in Major Mineral Commodities (Million U.S. Dollars)

Commodity Group	1976	1977	1978	1979	1980
Metals:					
All ores, concentrates, scrap	15692	15669	16478	23466	29390
Iron and steel	44720	46703	57117	70628	75949
Nonferrous metals	21546	24235	27729	37182	52573
Subtotal	81958	86607	101324	131276	157912
Nonmetals, crude only	6279	7009	7795	9654	11815
Fossil fuels	199592	222116	222833	333876	477349
Total mineral commodities	287829	315732	331952	474806	647076
All commodities	989261	1124883	1298411	1638302	1993312
Total mineral commodities as a percent of all commodities	29%	28%	26%	29%	33%

Adapted from: *Minerals Yearbook,* Vol. 3, 1981, U.S. Bureau of Mines.

Table 1.3.2—Estimated Value of all Mineral Commodities as a Proportion of Worldwide Commodity Consumption (National and International)

Year	Estimated value of all mineral commodities in millions of U.S. dollars	Change from previous year	Mineral commodities' share of all commodities
1976	$353,200	+13.1	35.7
1977	387,400	+9.7	34.4
1978	407,300	+5.7	31.4
1979	582,600	+43.0	35.6
1980	793,900	+36.3	39.8

Adapted from: *Minerals Yearbook,* Vol. 3, 1981, U.S. Bureau of Mines.

Table 1.3.3—Growth in Value of World Export Trade in Major Mineral Commodity Groups (Percent Change from Previous Year)

Commodity Group	1976	1977	1978	1979	1980
Metals:					
All ores, concentrates, scrap	+9.2	−0.1	+5.2	+42.4	+25.2
Iron and steel	−2.2	+4.4	+22.3	+23.7	+7.5
Nonferrous metals	+15.3	+12.5	+14.4	+34.1	+41.4
All metals	+4.0	+5.7	+17.0	+29.6	+20.3
Nonmetals, crude only	+0.8	+11.6	+11.2	+23.8	+22.4
Fossil fuels	+18.7	+11.3	+0.3	+49.8	+43.0
All major mineral commodity groups	+13.7	+9.7	+5.1	+43.0	+36.3
All commodities	+13.5	+13.7	+15.4	+26.2	+21.7

Adapted from: *Minerals Yearbook*, Vol. 3, 1981, U.S. Bureau of Mines.

put international mineral trade into proper perspective. The following commodity review covers those minerals and mineral groups with both historical and current economic significance in world trade.

WORLD TRADE IN MINERAL COMMODITIES

Petroleum

Crude oil and petroleum products constitute the single most important commodity group in international trade. Not only are they the dominant mineral commodity group, but in both volume and value they are the premier items traveling between the nations of the world. In fact, their dominance has caused international exchange problems, increased trade deficits of many developed economies such as the United States. Fluctuations in petroleum prices have created major international financial crises for countries such as Mexico, Brazil, Argentina, and Nigeria.

In 1980, daily movement of crude oil exceeded 31 million barrels of oil per day on average. The countries of Western Europe, as a group, are the major consumers of crude oil in international markets. The United States and Japan are the two largest individual users of world oil shipped in international trade. The Soviet

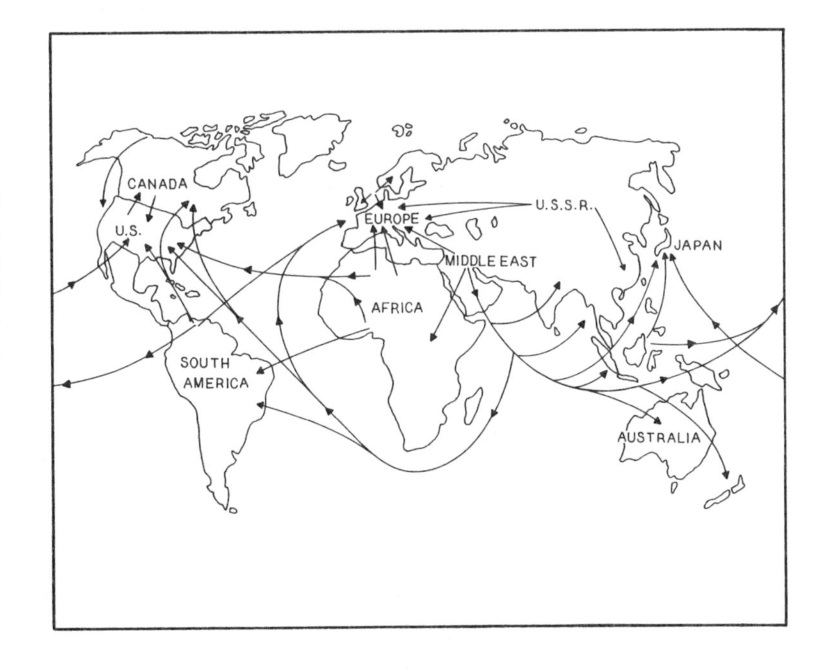

Fig. 1.3.1—Trade patterns in crude oil, 1980 (See Table 4 for quantities). Adapted from: *1982 Annual Energy Review*, U.S. Department of Energy, 1983, p. 38.

Table 1.3.4.—International Trade in Crude Oil, 1980
(Thousand Barrels Per Day)

To/	From Canada	Mexico	U.S.	Ecuador	Trinidad & Tobago	Venezulea	Norway	U.K.	USSR	Middle East	Africa North	Africa West	Far East & Oceania	Other	Total
U.S.	200	507	0	17	115	156	144	173	0	1533	1042	956	376	44	5263
Canada	0	0	84	1	3	166	0	7	0	275	14	3	0	0	553
Caribbean	0	140	198	52	18	501	0	20	100	418	245	167	43	13	1915
Cen. & So. America	0	34	0	18	0	188	0	0	0	704	14	210	30	7	1205
Western Europe	0	147	0	3	0	237	332	596	658	6990	1600	1082	33	85	11763
Eastern Europe	0	0	0	0	0	0	0	0	1550	300	160	25	0	0	2035
Middle East & Africa	0	0	0	0	0	0	0	1	30	1075	39	70	25	0	1240
Japan	0	0	0	21	0	46	0	0	2	3125	45	22	1153	0	4414
Far East & Oceania	0	0	6	0	0	2	0	0	90	2248	59	0	425	30	2860
Total	200	828	288	112	136	1296	476	797	2430	16668	3218	2535	2085	179	31248

Adapted from: *1982 Annual Energy Review*, U.S. Department of Energy, 1980, p. 81.

Union is the single largest consumer and producer of petroleum in the communist bloc, and the major source for petroleum consumed by the nations of Eastern Europe. Details of this trade are presented in Table 1.3.4 and the movements of international petroleum are presented in a general way in Fig. 1.3.1.

Roughly half of all international shipments of crude oil originate in the Middle East. The major producing country is Saudi Arabia, with significant output from Kuwait, the United Arab Emirates, Iraq, and Iran. Petroleum is the single most important export from this area, and has led, in recent years, to a substantial influx of international capital from the major western nations to these countries. While the continuing conflict between Iran and Iraq has disrupted normal trade with these nations, the other countries of the area have had the difficult task of trying to manage rapid economic development. The developing nations of the Third World have also found payments for petroleum an increasing burden on their trade balances.

The necessity to finance national development based on petroleum revenues has caused a major international financial problem for Mexico as it has become increasingly difficult to service a large foreign debt with falling revenues caused by weaknesses in petroleum demand and prices during the early 1980s. Nigeria has faced a similar problem although the magnitude of the crisis has not been as severe.

The discovery and utilization of large petroleum reserves on the north slope of Alaska has only moderated the dependence of the United States on foreign oil. Roughly half of the U.S. consumption of petroleum and petroleum products is imported.

Geography plays an interesting role in the movement of petroleum and other commodities between nations. The United States exports crude oil and petroleum products to Canada and the Caribbean, even though it is heavily dependent on imports of crude oil for domestic consumption. The exported crude oil is refined and exported again as products, some even re-enters the U.S. market.

International exchange is another factor affecting the export of petroleum from net importing nations. Romania, for example, imports crude oil from the Soviet Union and the Middle East, and exports petroleum products to many nations to obtain scarce foreign capital to support its economic needs for other goods and services. The Soviet Union is also involved in international shipments of petroleum and petroleum products for foreign exchange purposes. This type of activity is largely misunderstood by individuals and organizations with easy access to hard currencies and other means to facilitate foreign exchange.

The worldwide recession of the early 1980s has decreased the financial pressures on oil importing nations and brought about declining pe-

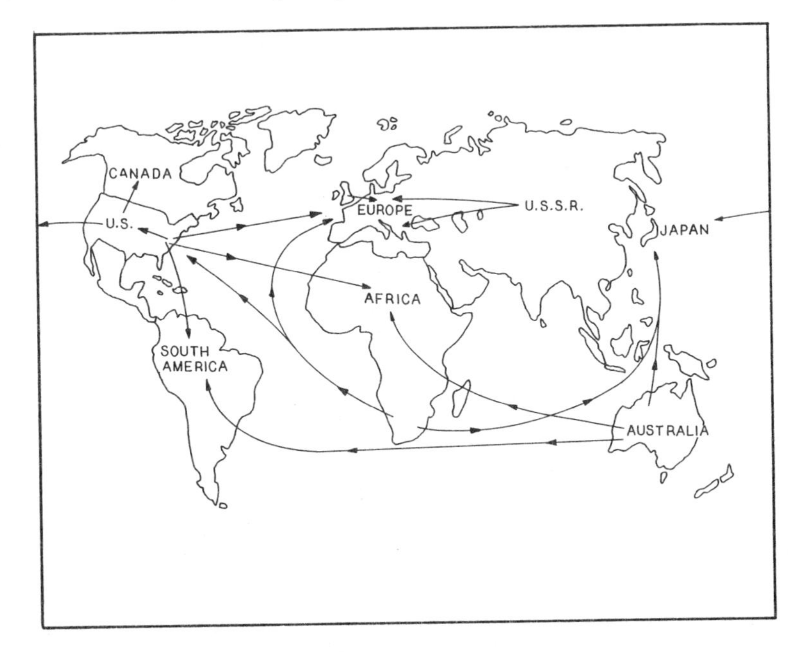

Fig. 1.3.2—Trade patterns in coal, 1980 (See Table 5 for quantities). Adapted from: *1981 International Energy Annual,* U.S. Department of Energy, p. 78.

Table 1.3.5—International Trade in Coal, 1980
(Thousand Metric Tons)

	No. & So. America			W. Europe			E. Europe			Africa	Far East & Oceania				Total
To/From	Canada	US	Other	DFR	UK	Other	Poland	USSR	Other	Africa	Australia	China	Japan	Other	
No. & So. America															
Brazil	734	3845	43	86	43	43	1167	0	0	0	43	0	432	0	6436
Canada	0	20006	0	0	0	302	0	0	0	0	0	0	0	0	20308
Other	475	2852	389	173	43	43	432	173	0	821	173	0	0	821	6395
W. Europe															
Austria	0	0	0	389	43	259	1556	994	1642	0	0	0	0	0	4883
Belgium	43	5272	0	6697	389	907	735	302	43	2463	302	43	0	0	17196
Denmark	302	1172	0	86	778	86	3932	605	0	3846	0	0	0	0	10807
Finland	0	302	0	0	43	216	3932	1642	0	0	0	0	0	0	6135
France	43	8858	0	9938	2031	778	4753	1037	0	10586	1728	43	0	43	39838
DFR	648	3069	0	0	1988	1988	2506	259	1469	1080	864	259	43	0	14173
Italy	86	8253	0	3241	0	216	2895	1210	0	3889	1599	43	0	0	21432
Netherlands	43	5401	43	2463	389	76	1210	0	216	2290	1512	0	0	0	13643
Spain	76	3975	0	76	0	0	1382	130	43	778	864	76	43	0	7443
UK	0	4580	0	216	0	173	778	76	0	173	3154	76	0	0	9226
Yugoslavia	0	1167	0	0	0	0	76	2290	821	0	0	0	0	0	4354
Other	691	3630	0	648	1512	1037	1685	648	259	43	43	0	0	0	10196
E. Europe & USSR															
DDR	0	0	0	0	0	0	3673	7907	1858	0	0	0	0	0	13438
Romania	0	1772	0	0	173	86	519	4494	691	0	389	0	519	0	8643
USSR	0	0	0	0	0	0	8123	0	0	0	0	0	0	0	8123
Other	0	130	0	173	0	346	3154	15037	2809	0	0	0	130	0	21779
Far East & Oceania															
Japan	15900	4230	0	0	0	0	570	3140	0	4100	42100	2900	0	735	73675
So. Korea	1556	1556	0	0	0	0	0	0	0	1667	2938	302	216	2160	10395
Taiwan	216	821	0	0	0	0	0	0	0	1815	2204	0	130	43	5229
Other	346	173	0	0	0	0	130	864	0	76	562	3457	605	76	6289
Middle East & Africa															
Algeria	0	907	0	129	0	0	0	129	76	0	0	0	0	0	1241
Egypt	0	994	0	0	0	0	0	389	0	0	173	0	0	0	1556
Other	0	0	0	0	0	259	0	43	0	432	432	0	0	0	1166
Total	21159	82965	475	24315	7432	6815	43208	41369	9927	34059	59080	7199	2118	3878	343999

Adapted from: *1981 International Energy Annual*, U.S. Department of Energy, 1982, p. 78.

troleum prices. The prospects for economic recovery and renewed economic growth must be balanced by the likely return to fierce competition for available supplies of increasingly scarce petroleum reserves and resurgent prices. The next few decades will almost certainly witness a continuing cycle of price swings and apparent excess supply or demand for petroleum and petroleum products. Oil is likely to continue to be the major commodity moving in world trade during this period.

The ability of OPEC (The Organization of Petroleum Exporting Countries) to control prices and output will vary with the strengths and weaknesses in demand for petroleum in international markets. Oil is unlikely to be displaced as the major source of energy for industrialization and economic growth in the foreseeable future.

Coal

With the advent of rising energy costs in the period 1973–1979, coal became an increasingly attractive alternative energy source in international markets. The international trade in coal is summarized for 1980 in Table 1.3.5 and Fig. 1.3.2 shows the trade patterns that flow from the statistics in Table 1.3.5. The year 1980 is used to illustrate a typical example of the recent movement of coal in international markets. Coal consumption actually reached a peak in 1981, and has declined more recently with the worldwide recession in 1982–1983. Some recovery in coal markets seems to be evident in 1984, but it is unlikely for international demand to return to earlier levels until oil prices stabilize and petroleum supplies again become scarce providing renewed opportunities for coal as an attractive alternative energy source. International growth in demand for iron and steel will support additional trade in metallurgical grade coal.

Historically, coal has been significant in trade among the developed economies of Europe. The demand for coal by Japan, the United States, Europe and the developing economies of Asia, Africa, and South America have provided increased opportunities for export of coal from Australia, South Africa, and the United States. Air pollution regulations have provided markets for low sulfur coal imports in the United States despite a net export of coal largely used for metallurgical purposes. Scarcity of hard currencies for foreign exchange have led to commodity agreements involving coal between Romania and the United States, and Japan has invested heavily in U.S. coal reserves to supply its iron and steel industry.

Some uncertainties continue to exist in the world coal market due to lowered projected demand for steam coal for electric power generation and industrial use in both developed and developing economies. Energy conservation has caused a major slowdown in the growth in demand for electricity, and new technological developments have reduced power requirements in the rapidly expanding markets for consumer electronics.

Of the major coal consuming nations, only Japan imports the majority of its needs. Most of the demand in the United States, the Soviet Union, and the other major coal consuming economies is met with domestic production. Coal is a significant energy source in the world's two most populous economies, India and the People's Republic of China, and China has sufficient reserves to consider export as a means of increasing its position in foreign trade. Large, undeveloped reserves in Colombia also could be a significant factor in future international trade in coal.

Natural Gas

International trade in natural gas has become more feasible with the development of liquefication facilities in countries where a surplus is produced, and the advent of new pipeline networks such as the major Soviet system to provide gas from Siberian fields to Europe. Table 1.3.6 gives the 1980 figures for international shipments of natural gas. Growth in this trade in the near term will decrease dramatically with the completion and operation of the new Soviet pipeline.

Major movements of liquefied natural gas (LNG) are the basis for most exports from Northern Africa and the Middle East. Pipeline networks support the other international exchanges. Higher petroleum prices will provide the impetus for increased trade in natural gas. The major capital investments necessary for LNG production and transportation, and construction of pipeline systems will continue to cloud the outlook for predictable growth in natural gas exports.

Iron ore

International trade in iron ore has been particularly significant in the past three decades. This period has witnessed major shifts in both quantities shipped and sources of supply. World production of iron and steel has varied from roughly 200 million to 500 million metric tons, while ore production (and consumption) to support the industry varied from 250 to over 750 million metric tons. The balance of the iron and steel output was based on scrap recycling.

International shipments of iron ore during the past 30 years have accounted for a modest 16% in the 1950s to roughly 40% of all iron ore currently consumed. In 1980, iron ore moving to international markets was in excess of 250 million metric tons. Iron ore shipments in more recent years have declined somewhat due to the worldwide recession and the smaller requirements of steel in final production such as automobiles and construction.

The countries that provide the bulk of international demand for iron ore are the Soviet Union, Japan, the United States, and the major iron and steel producing countries of Europe. Fig. 1.3.3 indicates the general flows of iron ore from producing countries to consuming countries based on the movements in 1980. This trade is likely to be affected in the future by growing steel industries in Korea and Brazil.

Major export nations for iron ore include Australia, Canada, Sweden, the Soviet Union (also a major producer of iron and steel), Venezuela, Liberia, and India. Smaller, but important shipments of iron ore come from Mauritania, Malaysia, Chile, Brazil, Peru, and South Africa. Canada and Australia will continue to sell large amounts of iron and steel in international markets as well as iron ore.

The Soviet Union probably has the largest reserves of iron ore, but it is closely followed by Australia, with significant, but somewhat smaller quantities of reserves in Brazil, Canada, South Africa, the United States, India, and China. Australia has the largest iron ore reserve base compared to its annual production of iron and steel.

The international demand for iron ore will continue to be volatile for the foreseeable future. It will be subject to significant swings in demand based on the cycles of worldwide economic growth.

Most iron ore moving in international trade travels in oceangoing vessels as concentrates used for sinter feed. Pellets constitute less than 10% of international shipments and lump ore is less than 25% of the total iron ore shipped.

The Japanese iron and steel industry is almost totally dependent on imports of iron ore and the major steel producers in Europe (both EEC and Eastern Europe) obtain 80 to 90% of their ore as imports. Eastern European producers are major consumers of exported iron ore from the Soviet Union. Exports provide the major market for Australian, Canadian, and Swedish iron ore, and for Mauritania and Liberia, iron ore pro-

Table 1.3.6—International Trade In Natural Gas, 1980 (Billion Cubic Meters)

Producing Countries	Importing Countries	Quantity
Canada	United States	22.6
Mexico	United States	2.9
DFR	Austria	0.4
	France	1.1
	Switzerland	1.1
Netherlands	Belgium	10.9
	France	11.2
	DFR	22.6
	Italy	8.1
Norway	Belgium	3.5
	France	2.5
	DFR	9.6
	Netherlands	5.5
	UK	16.1
USSR	Austria	3.1
	Finland	1.1
	France	5.1
	DFR	10.8
	Italy	8.6
	Yugoslavia	1.8
	Bulgaria	5.1
	Czechosovakia	8.6
	Hungary	3.9
	Poland	5.3
	Romania	1.6
	DDR	6.5
UAE	Japan	2.7
Algeria	US	2.4
	France	1.8
	Spain	0.8
	UK	1.2
Libya	Italy	2.1
	Spain	1.6
Total		191

Adapted from: 1981 International Energy Annual, U.S. Department of Energy, 1982, p. 70.

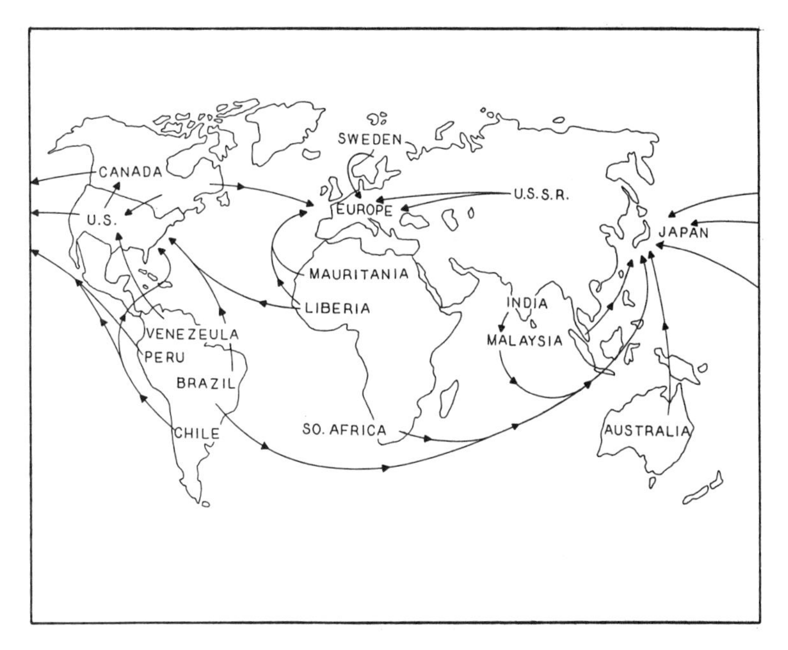

Fig. 1.3.3—Trade patterns in iron ore, 1980. Adapted from: *Mineral Commodity Summaries, 1982,* U.S. Bureau of Mines, p. 77.

duction is a significant part of their total economic activity.

Phosphate and Chemical Fertilizers

In the past 20 years, worldwide consumption of chemical fertilizers has nearly tripled. The largest increases in consumption were realized in the United States and Canada, most of the nations in Europe, and in the Soviet Union. Phosphate rock is a major raw material for the production of phosphatic fertilizer. While some phosphate rock serves as the raw material for elemental phosphorous and phosphorous chemical, over 90% is used in fertilizers. Table 1.3.7 and Fig. 1.3.4 show the international movement of phosphate rock for 1981. Over 41 million metric tons of phosphate rock were shipped to international markets in 1981. Morocco has the largest reserves of phosphate rock, and is the principal source for European consumption. The United States is also a major exporter of phosphate rock and chemical fertilizers, shipping significant quantities to Canada, Europe, and Asia. Important reserves in Jordan and Israel, Tunisia and Algeria, Senegal and Togo, provide major exports of phosphate rock to countries in both Western and Eastern Europe. The Soviet Union contributes about 20% of world production of phosphate rock and is a major exporter to European economies.

Important reserves in South Africa, China, and Vietnam could be developed for export in the 1980s. There is also important production from some Pacific islands for export to Aus-

Table 1.3.7—International Trade in Phosphate Rock, 1981
(Thousand Metric Tons)

Producing Countries	Importing Countries	Quantity
United States	Canada	3200
	Western Europe	3525
	Asia	1486
	Eastern Europe	254
	South America	481
Morocco	Western Europe	10181
	Eastern Europe	2848
	South America	1103
	Asia	859
Algeria & Tunisia	Western Europe	794
	Eastern Europe	807
Israel and Jordan	Western Europe	1431
	Asia	1962
	Eastern Europe	1438
Senegal	Western Europe	826
	Asia	215
Togo	Western Europe	1393
	Eastern Europe	712
USSR	Eastern Europe	4067
	Western Europe	948
Pacific Islands	Australia	1767
	New Zealand	853
	Indonesia, Republic of Korea, Malaysia, Singapore, and Japan	231
Total		41381

Source: *Mineral Commodity Profiles, Phosphate Rock, 1983,* U.S. Bureau of Mines, p. 10.

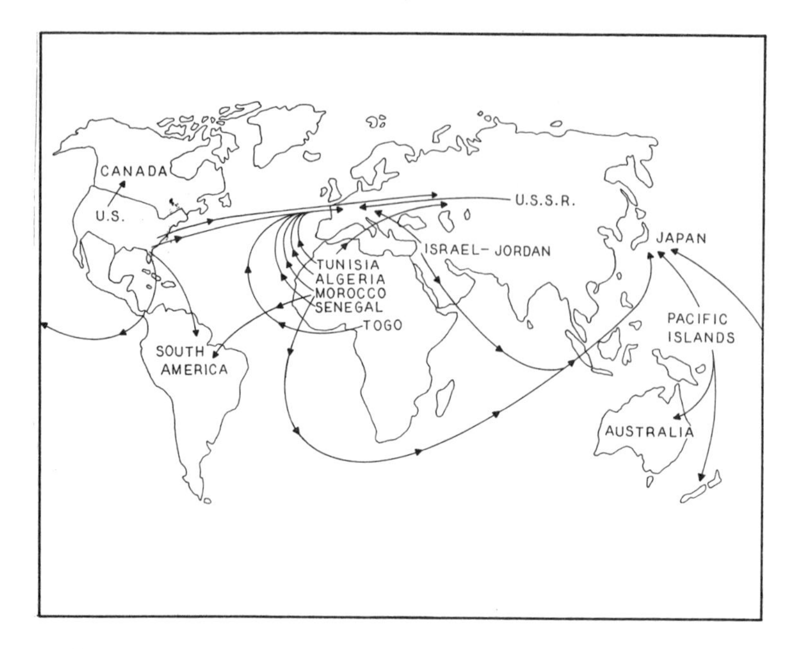

Fig. 1.3.4—Trade patterns in phosphate rock, 1981 (See Table 7 for quantities) Source: *Commodity Profiles, Phosphate Rock, 1983,* U.S. Bureau of Mines, p. 10.

tralia, New Zealand, Japan and other countries in Asia. The reserves on these islands are more limited and could be exhausted by the end of the century.

The Soviet Union has recently concluded a business arrangement with the Occidental Petroleum Corporation to import quantities of phosphate rock from the United States in exchange for ammonia that Occidental will use in nitrate fertilizers. The major components of all fertilizers are nitrogen, phosphorous, and potassium. Canada is the major source of potash for international trade. The Soviet Union, German Democratic Republic, Federal Republic of Germany, France, Jordan, and Israel are important sources of potash for fertilizer. Most nitrogen in fertilizer is derived from the earth's atmosphere.

Cement

While produced largely for domestic consumption, significant quantities of cement are shipped in international trade. Worldwide production capacity is approximately one billion metric tons per year with production between 80 and 90% of this capacity. Between 10 and 20% of annual production is shipped in international trade. Japan is the largest exporter, with various European countries contributing important tonnages, especially for the African and South American markets. Many smaller nations with limited limestone reserves are faced with the necessity to import cement.

Bauxite and Alumina

Aluminum metal is obtained by the processing of alumina in an electric furnace with cryolite made from fluorspar. Alumina is obtained from bauxite by chemical leaching. Most bauxite, alumina, and aluminum metal are therefore linked as a group of related mineral commodities.

Bauxite reserves are normally found in tropical climates where leaching by rainwater has removed silica from weathered rock and left behind aluminum oxides. Australia, Jamaica, Guinea, Surinam, Guyana, Greece, Yugoslavia, Indonesia, the Dominican Republic, Malaysia, and Haiti are important sources of bauxite. These countries export bauxite, and in some cases alumina, to the developed economies of Europe, North America, Japan, and the Soviet Union.

The largest growth in bauxite exports has come from Australia and Guinea in recent years. Brazil is also developing large reserves and began exports to the United States in 1979. Table 1.3.8 and Fig. 1.3.5 present a picture of world trade in bauxite for the year 1980.

Bauxite was an important issue during trade talks between Jamaica and the United States in 1981–82. The United States agreed to purchase 1.6 million tons for the U.S. Strategic Materials Stockpile.

Roughly 45% of all bauxite production in the early 1980s was shipped to international markets. Australia, Guinea, and Jamaica accounted

Table 1.3.8—International Trade in Bauxite, 1980
(Thousand Metric Tons)

To/From	Jamaica	Dom. Rep.	Haiti	Australia	Guinea	Guyana	Surinam	Indonesia	Malaysia	Yugoslavia	Greece	Other	Total
U.S.	11035	1928	997	199	5318	1396	3789	0	0	0	67	67	24796
Canada	0	0	0	0	2327	1662	199	0	0	0	67	1330	5585
Japan	0	0	0	5916	0	67	0	1994	1197	0	0	665	9839
DFR	0	0	0	5584	1597	1330	1330	0	0	665	1330	8642	20478
France	0	0	0	1130	1330	133	67	0	0	0	199	67	2926
Italy	0	0	0	2925	399	67	0	67	0	0	0	266	3724
UK	0	0	0	0	0	0	0	0	0	0	67	465	532
USSR	0	0	0	0	3723	0	0	0	0	1928	1263	133	7047
Other	0	0	0	332	0	598	332	0	133	531	1529	1263	4718
Total	11035	1928	997	16086	14694	5253	5717	2061	1330	3124	4522	12898	79645

Adapted from: Fischman, Leonard, *World Mineral Trends and U.S. Supply Problems*, Washington, D.C., Resources for the Future, 1979, p. 224. *Mineral Commodity Summaries, 1982*, U.S. Bureau of Mines, p. 17.

for 56% of world output and 66% of world bauxite exports. An organization of bauxite producing nations, the International Bauxite Association (IBA) was formed in 1974 to control output and strengthen market prices. Members of the IBA include Australia, Guinea, Ghana, Indonesia, Sierra Leone, and Yugoslavia. Due to the inability to settle disputes over uniform pricing formulas and output quotas, as well as the lack of cooperation of other major exporters, the IBA has not been successful in strengthening the bargaining power of the exporting countries. In the early 1980s, roughly 60% of world production of alumina came from Australia, the United States, USSR, Jamaica, and Japan. Australian exports went mostly to North America (United States and Canada), Japan, United Kingdom (U.K.), Federal Republic of German (F.R.G.), and Norway. Jamaican exports went to the U.S., Canada, Norway, Spain and the U.K. Surinam shipped alumina to the U.S., Canada, F.R.G., U.K., and Norway. Guinea's major markets were the U.S. and Canada, USSR, and the F.R.G. United States exports of alumina went to Canada, Ghana, Norway, Mexico and Venezuela. Approximately 82% of world trade in alumina was supplied as indicated above.

In addition, aluminum metal, from the U.S., USSR, Canada, F.F.G., France and Norway was shipped in international commerce in the early 1980s. The trend in these aluminum-related commodities indicates heavy cross-trading caused to some degree by intracompany shipments within and among multinational corporations.

Copper

Historically, copper has been the most important nonferrous metal in world trade. Significant exports from Chile, Peru, the Philippines, Zambia, and Zaire are shipped to Japan, the United States, and the developed economies of Europe. Canada, Australia, Papua New Guinea, Poland, and South Africa are also copper exporters. The United States and the Soviet Union are important copper producers but consume virtually all they produce. The United States is a net importer of copper. Table 1.3.9 and Fig. 1.3.6 give details on international trade in copper for 1976, a typical recent year for which the data is available.

Much of the past export of copper has been from mines in developing nations that were owned by multinational corporations. The 1970s brought nationalization to most of these operations. Today, copper is exported by operating companies that are owned by the developing nations themselves. The nationalization of copper mining has had a dramatic impact on the international distribution of investment in copper production and on the form of shipment of copper in interna-

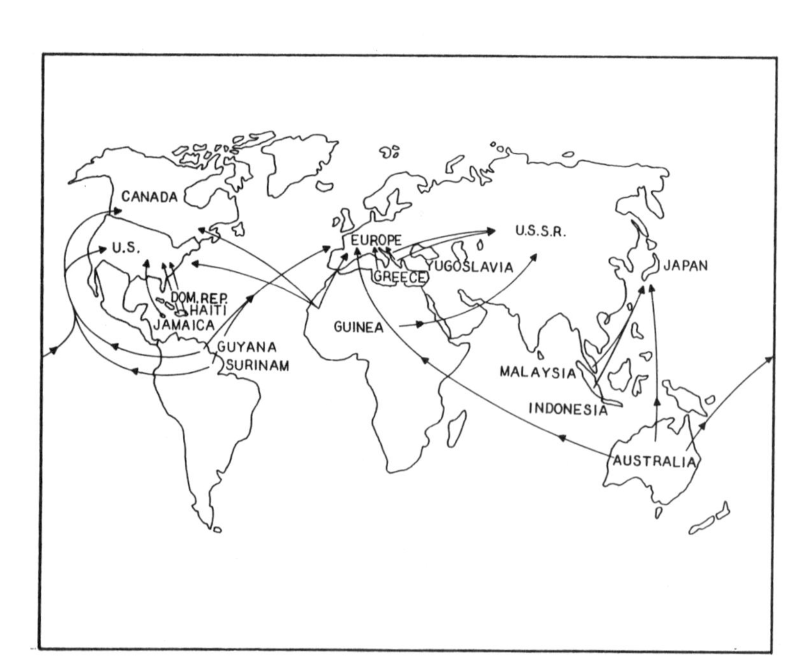

Fig. 1.3.5—Trade patterns in bauxite, 1980 (See Table 8 for quantities) Adapted from: *Mineral Commodity Summaries, 1982*, U.S. Bureau of Mines, p. 17. Fischman, Leonard, *World Mineral Trends and U.S. Supply Problems*, Washington, D.C., Resources for the Future, 1979, p. 224.

Table 1.3.9—International Trade in Copper, 1976
(Thousand Metric Tons)

To/From	Canada	Philippines	Papua N.G.	Chile	Australia	So. Africa	Peru	Zaire	Zambia	Other	Total
U.S.	121	0	1	97	5	6	26	3	123	20	402
Japan	242	201	88	97	49	4	96	203	121	50	1151
Brazil	3	0	0	139	0	0	3	0	0	3	148
DFR	38	0	85	167	139	79	5	0	79	131	723
Spain	0	0	0	43	0	4	1	0	0	23	71
Sweden	17	2	0	11	6	0	0	0	167	12	215
Belgium	14	0	0	21	111	21	2	387	167	40	763
U.K.	86	0	0	85	68	6	34	0	96	58	433
France	22	0	0	56	111	0	0	83	50	36	358
Italy	83	0	0	70	3	0	6	28	72	22	284
Netherlands	3	0	0	6	3	0	6	0	0	14	32
USSR	12	0	0	0	0	0	0	0	0	2	14
Bulgaria	0	0	0	8	0	0	0	0	0	4	12
Yugoslavia	0	0	0	7	0	0	0	0	16	3	26
China	0	0	0	2	0	0	16	0	0	0	18
Other	35	18	5	112	6	8	36	20	126	232	598
Total	676	221	179	921	501	128	231	724	1017	650	5248

Adapted from: Fischman, Leonard, *World Mineral Trends and U.S. Supply Problems*, Washington, D.C., Resources for the Future, 1979, p. 230–233.

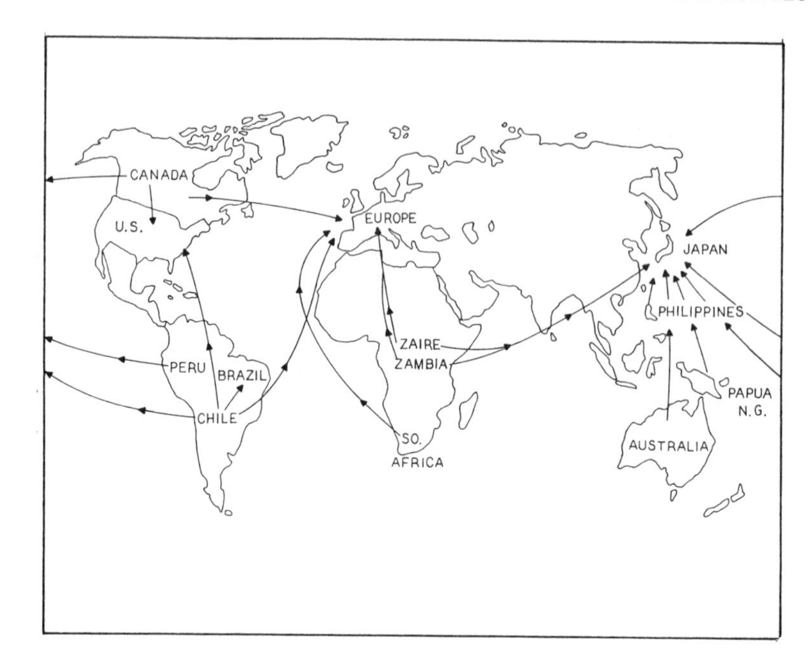

Fig. 1.3.6—Trade patterns in copper, 1976 (See Table 9 for quantities). Adapted from: Fischman, Leonard, *World Mineral Trends and U.S. Supply Problems,* Washington, D.C., Resources for the Future, 1979, p. 230–233.

tional trade. The shipment of copper concentrates is less common than it was in the past. Roughly two thirds of current world shipments of copper are in refined form. Concentrates and blister (smelter) copper each constitute about one sixth of international shipments.

Copper is also one of the most extensively recycled of the common metals. World trade in scrap copper is almost exclusively between the more highly industrialized nations that have the capacity to generate and use scrap copper. The United States is the principal exporter of scrap. Japan is the major scrap importer (mostly from the U.S.) but Belgium, Italy, and the Federal Republic of Germany are also important consumers of scrap copper.

For countries such as Chile, Peru, the Philippines, Zambia, and Zaire, copper production is an important source of employment and a significant source of export revenue. These factors are likely to contribute to a continuing surplus of copper in international markets and depressed world prices.

Air pollution regulation has forced the closing of some copper smelters in the United States. The controversy over acid rain could force further closings in both Canada and the United States. This could lead to stronger demand for refined copper from Third World producers, but the full impact of these actions is difficult to predict at this time.

Tin

World production of tin is on the order of 200,000 to 250,000 tons per year. Since most producers are developing nations such as Bolivia, Malaysia, Indonesia, and Thailand, and consumption is located principally in developed economies in Europe, North America, and Asia (basically Japan), most output enters international trade. The Soviet Union is a major producer of tin, but consumes most of its own output. Australia and South Africa are both significant consumers and exporters of tin.

World tin resources are located principally in Indonesia, China, Malaysia, Thailand, the USSR and Bolivia. Other significant reserves are known in Burma, Brazil, Australia, Nigeria, the United Kingdom, and Zaire. The development of these reserves is unlikely to have a major impact on current trade patterns in tin since most are in countries that are currently major producers. Also, the demand for tin has declined in the past three decades due to the movement away from the more costly tin-plated cans to aluminum, glass, paper, and plastic containers. Substitutes have displaced tin in other major uses such as bearings and solder in addition to the loss of the container market. Due to its value, tin is also widely recycled, with secondary tin accounting for over 25% of final consumption in the United States.

Fluorspar

Total world consumption of fluorspar is on the order of five to six million tons annually valued in excess of $700 million. Of this output, 45% travels in international commerce. Mexico is the world's largest producer with roughly 20% of total world output and 25% of world trade (exports). Other important export countries include Mongolia, South Africa, China, Thailand, Spain, and France. Since fluorspar is used mostly in chemicals, aluminum production (as crYolite), and as a flux in metallurgy, most is consumed in industrialized economies.

The major importers of fluorspar are the United States, Soviet Union, Japan, Federal Republic of Germany, and Canada. The United States and Canada are the major consumers for Mexican fluorspar exports. Virtually all of the Mongolian output is used by the Soviet Union and other centrally planned economic systems. Japan receives most of its imports from China, South Africa, and Thailand.

Lead

International trade in lead has varied substantially from just over one million metric tons per year in the period 1961–1967 to an average of 1.9 million metric tons per year in the period 1976–1981. Government stockpiling, restrictive import quotas, and variable demand have accentuated the swings in year-to-year shipments of lead in international markets. Most of the lead consumed in the United States is produced domestically. From 1979–1983, United States reliance on lead imports has ranged between one and eleven percent. In 1980, the United States actually exported more lead than it imported.

During the period 1976–1981, Canada was the leading exporter of lead concentrates, averaging 144,000 metric tons of contained lead per year. Peru was the second largest exporter of lead during that same period, averaging 90,000 tons of contained lead annually. The Federal Republic of Germany and Japan were the leading importers of concentrates, followed closely by France. Japan's imports were mostly from Canada and Peru. Imports of The Federal Republic of Germany were primarily from Sweden, Canada, Ireland, and Morocco. France imported lead concentrates mostly from Morocco, Ireland, and South Africa.

During the same period (1976–1981), the leading exporters of refined lead were Australia, averaging 156,000 tons per year, Canada averaging 123,000 tons per year, and Mexico, averaging 101,000 tons per year. These three countries accounted for 40% of the world trade total of refined metallic lead. The United States was the world's largest importer of lead metal averaging 159,500 tons per year, and Italy ranked second with 148,000 tons per year. Italy depends primarily on the Federal Republic of Germany, Morocco, Australia, Namibia, and Mexico for refined metal, while the United States depends on Australia, Canada, and Mexico.

A rapidly developing secondary lead (recovery) industry in both the United States and the Federal Republic of Germany should moderate both use and imports of future primary lead and concentrates. Additionally, restrictions are being placed on the use of lead in many consumer applications due to its toxicity, and more attention is being focused on leaded gasoline as a contributor to atmospheric pollution. These factors, and increased substitution by plastics will moderate expanded trade in lead for future international markets.

Zinc

The consumption of zinc normally follows the status of the general economy of the developed nations. Large quantities are used in construction and in the automotive industry. Zinc is mostly consumed in protective coatings (galvanized metals), diecasting, chemicals and pigments. Zinc enters world trade as concentrates and as metallic zinc.

World trade in zinc concentrates was estimated in 1981 to be 1.9 million tons of contained zinc. Concentrates are mainly exported by Canada, Peru, Australia, Sweden, and Ireland; importing countries are mainly those in Western Europe, Japan, and the United States.

During the past decade, the trend has been to smelt and refine the ore in the same country in which it is mined. World trade in slab zinc was estimated in 1981 to be 1.7 million tons or about 30% of world smelter production. The largest exporters were Canada, Australia, Belgium, Netherlands, Finland, and the Federal Republic of Germany. Peru and Mexico have opened smelters since 1980 and slab zinc exports from these countries can be expected to increase. The largest importers of slab zinc were the United States, the Federal Republic of Germany, the United Kingdom, India, and France. Slab zinc consumption has steadily decreased since 1979 due to the introduction of thin-wall diecasting, weight reduction programs in automobiles, and substitution by alternative materials in coatings and pigments.

Chrome

In 1981 world chromite production was approximately 2,554,000 metric tons. The major world producers are South Africa and the USSR which together in 1981 produced two thirds of the world total. About 40% of world production in 1981 was from the centrally planned economies of Albania, Cuba, and the USSR.

World ferrochromium production in 1981 was approximately 1,773,000 tons. The Republic of South Africa and the USSR each accounted for one third of world production.

Chromite has formed the largest percentage of imported chromium. The trend over the past 10 years has been for increasing quantities to be traded as chromium alloys as countries with chromite resources increase their manufacturing capacity. In 1981, chromium ferroalloy and metals trade exceeded that of chromite on a contained chromium basis.

Zimbabwe, which produced approximately six percent of world chromium output in 1981 was ostensibly out of the world market during the past decade. However, the United States imported from Rhodesia during the period from 1972 to early 1977 under the "Byrd Amendment." It is also likely that additional Rhodesian chrome reached the industrial world primarily through South Africa.

Nickel

The United States obtains most of its nickel from Canada. Other sources are New Caledonia, the Dominican Republic, the Republic of South Africa, and Australia. Japan imports most of its nickel from New Caledonia, Indonesia and from scrap originating in the United States. Western Europe obtains nickel from Canada, New Caledonia, Republic of South Africa, and Cuba. Eastern Europe, the USSR and China import nickel from Cuba. The USSR has sold an estimated 30,000 tons of nickel to consumers in Western Europe and the United States since 1975.

MINERAL COMMODITIES AND ECONOMIC DEVELOPMENT

It is difficult to make a general statement about the role of mineral resources and international trade in mineral commodities on the economic development of all economic systems. Suffice it to say that minerals are an especially important raw material input for industrial development and modern consumer products.

Historically, minerals have formed the basis for domestic economic development in most, if not all, of the developed economies in the world. One need only examine the importance of iron and steel to the development of Japan, the United States, and the members of the European Coal and Steel Community to gain some appreciation for the role of mineral commodities in economic development. Perhaps more important is the impact of higher energy prices on the world economy. The major underlying cause of swings in the world economy in the past 10 years has been the rise in oil prices. Since petroleum has become the basis of modern industrial existence, the fortunes of oil have had an interactive impact on the growth of national economies. Nations highly dependent on imported oil have been more dramatically affected than nations, such as the Soviet Union, that are virtually self-sufficient.

At the same time, nations that have depended too heavily on the export of oil have suffered disproportionately when the world economy fails to maintain high levels of demand for petroleum and other energy sources. This has also been true to a varying degree in all developing nations with heavy export dependence on minerals such as copper, tin, iron ore, bauxite, and other natural resources such as rubber and sugar.

Statistics indicate that mineral raw materials contribute only a small proportion of the world GNP, in the range of six to ten percent. But they are far more important to the economies of some nations such as Kuwait, Saudi Arabia, Nigeria, Venezuela, Zaire, Zambia, Chile, Peru, Jamaica, Australia, and South Africa.

While international trade in base metals historically has accounted for less than 10% of total world trade, these same raw materials are used in the machinery and transportation industries that have accounted for roughly one half of international trade in manufactured goods.

Mineral exports have been a vital source of foreign exchange earnings to support the economic growth of many developing nations. Earnings from mineral commodity trade have supported the development of other resources such as land and water, and have made it possible to purchase machinery and other capital goods to support general economic growth. The ability to modernize traditional economic systems with earnings from mineral exports has been most obvious in recent years in the Middle East. While economic change has also brought political and social instability to the nations of this area, the real economic growth in per capita income has been possible because of petroleum export earnings.

Mineral imports are vitally important to maintaining the economic growth of industrial nations. Despite some restrictions on imports such as tariffs and quotas to protect domestic mineral producers in selected nations, most minerals move in world commerce with little or no restrictions. This indicates an international recognition of their basic importance to the total world economy.

THE ROLE OF MINERAL EXPORTS

Mineral exports are normally thought of as most significant for developing economies with a limited industrial base for the export of manufactured goods. However, some minerals are important earners of international exchange for developed economies as well. The Soviet Union is perhaps the largest exporter of mineral commodities among developed nations.

Until the major increases in petroleum prices during the mid-1970s, between 50 and 60% of mineral exports originated from developed industrial economies. Much of this trade comes from the fact that the industrial nations are major processors of raw materials such as mineral ores and concentrates. Japan imports large amounts of iron ore and coal and exports large quantities of iron and raw steel. The United States imports crude oil and exports large amounts of petroleum products and derived petrochemicals. Many Western European nations process metallic ores and concentrates and export refined metals.

Review of Mineral Exports by Industrial Nations

United States: Exports of metals, minerals, and fuels by the United States were valued between \$20 and \$22 billion by the period 1980–1981. They constituted roughly 10% of total exports by the United States. Coal and petroleum products exports were the largest single category valued at approximately \$8 billion in 1980 and just over \$10 billion in 1981.[b] Since total U.S. mineral exports fell from 1980–1981 and the value of mineral fuel exports rose, the proportion of the total represented by mineral fuels rose from 36% in 1980 to about 50% in 1981. Metals, ores and refined metals including iron and steel fell from just over \$12 billion in 1980 to \$8.3 billion in 1981 reflecting a decline in demand due to worldwide recession. Traditionally, metals have been more important than mineral fuels in U.S. exports, but this changed in 1981. If we exclude iron and steel and refined nonferrous metals, mineral fuels are a much larger portion of U.S. mineral exports.

The United States is not a net exporter of mineral fuels. In fact, petroleum is the major mineral import of the U.S., constituting about 50% of annual demand in the past 10 years. Coal exports, however, are a significant portion of U.S. mineral trade. About 10% of U.S. coal production is exported each year, approaching 70 million metric tons of coal exports annually in each of the past five years.

Major U.S. exports in addition to coal, iron and steel, and nonferrous metals include phosphate rock and chemical fertilizers as well as sulfur and scrap metal.

Western Europe: Statistics for European nations are compiled for all developed market economies as well as the European Economic Community and the European Free Trade Association. This discussion utilizes the totals for all of Western Europe.[c]

Exports of metals and minerals from European nations fell from \$141 billion in 1980 to \$133 billion in 1981, reflecting global recession. A significant proportion of these exports is among the nations of Western Europe. The largest single category is mineral fuels, amounting to about half of total exports. The actual figures were \$66.2 billion in 1980 and \$70.5 billion in 1981 for the value of mineral fuel exports. The next largest category was iron and steel, constituting about 30% of international mineral trade originating in Western Europe.

[b]United Nations, *Monthly Bulletin of Statistics*, Vol. 37, No. 2, February, 1983.

[c]United Nations, *Monthly Bulletin of Statistics*, Vol. 37, No. 2, February, 1983.

USSR (Soviet Union): The Soviet Union is a major factor in world mineral trade, exporting close to $44 billion in 1980.[d] Of this total, mineral fuels constituted over 80% or nearly $36 billion. This was divided about 54% to western market economies (mostly Western Europe), 38% to centrally planned economies, and 8% to developing nations with market economies.

The export of natural gas through a major pipeline from Siberian gas fields to Western Europe will increase this total substantially by the mid-1980s. This pipeline was the subject of significant international controversy as the United States attempted to slow its progress in 1982 by imposing an embargo on U.S. pipeline equipment and technology transfer, and attempted to pressure its NATO allies to also stall trade with the USSR due to the political situation in Poland and intervention by the Soviets. This embargo was finally abandoned due to political pressure from U.S. suppliers of pipeline equipment and the potential economic harm to Western Europe, the principal customer for Soviet natural gas. The pipeline should be in full operation by 1985.

Major exports of petroleum and natural gas to both Western Europe and the centrally planned economies of eastern Europe are a significant source of foreign exchange earnings for the USSR. Other Soviet exports include precious metals, coal, phosphate rock, and fertilizer minerals, salt, industrial diamonds, manganese, iron ore, iron and steel, zinc, aluminum and other nonferrous metals and scrap.

Japan: The apparent export of mineral commodities from Japan constitutes 12 to 14% of total Japanese exports. Iron and steel represent about 85% of all mineral exports. The balance is mostly nonferrous metals. This proportion has fallen in the past 20 years from about 17 to 18% in the mid-1960s, mostly due to the increase in export of manufactured goods.

Canada: Mineral trade is a significant contributor to the exports of Canada. With total exports valued at from $63 billion in 1980 to $68.3 billion in 1981, mineral commodities, including fuels and metals were, valued at $19.5 billion in 1980 (31%) and $19.4 billion in 1981 (28%). While mineral fuels constitute roughly one half of all mineral exports, ores and refined nonferrous metals made up about 40 percent. Canada is a significant exporter of nonferrous metals, petroleum, and natural gas (to the United States), potash, precious metals, gypsum, sulfur, salt, and iron ore. The United States is a major market for Canadian iron ore and other nonfuel minerals.

Australia: Mineral exports from Australia in recent years account for between 26 and 29% of total trade originating in Australia and New Zealand. As such, minerals are an important source of foreign exchange for Australia. Mineral fuels, mostly coal, and various nonferrous metals and ores are significant exports from Australia. Australia is also an important source of iron ore. It is the principal supplier to the Japanese iron and steel industry. Important deposits of bauxite, ilmenite (for titanium dioxide paint pigment), corundum, and other nonmetallics have been developed in Australia for export. Minerals have increased in importance as a proportion of Australian exports since the 1950s when agricultural products such as wheat, wool, and beef cattle made the country's balance of payments more sensitive to swings in world agricultural commodity prices.

Historically, both Australia and Canada have been attractive for international investment in mineral production due to their economic stability and government attitudes of cooperation with private developers of mineral resources. Significant tax increases in recent years have dampened the enthusiasm of international investors for new mineral projects in both Australia and Canada.

South Africa: Of the developed economies, South Africa probably has the largest share of mineral exports as a proportion of international trade. South Africa's mineral exports have traditionally exceeded 60% of total exports. Excluding monetary gold, for which South Africa is the primary world producer, minerals still account for more than 40% of total exports. South Africa is an important source of diamonds, gold, chromite, manganese, copper, asbestos, and coal.

Review of Mineral Export by Developing Nations

As indicated in the previous discussion, mineral exports as a proportion of total trade of developed economies varies from 6 to over 50 percent. Countries with diversified manufacturing capabilities tend to have less dependence on

[d] United Nations, *Monthly Bulletin of Statistics*, Vol. 36, No. 5, May, 1982.

mineral exports. As might be expected, minerals often form the major portion of exports by developing economies. For some, such as the Middle Eastern members of OPEC, a single mineral commodity, in this case petroleum, is the overwhelming basis of economic activity and constitutes virtually 100% of the export base.

In 1980, the OPEC nations as a group exported over $289 billion of petroleum and other mineral fuels. Total exports of the OPEC nations as a group were approximately $306 billion, making petroleum and mineral fuels roughly 94% of their total exports. This would be an even larger proportion of total exports if countries such as Venezuela, Indonesia, and Ecuador were not members of OPEC. Export of copper from Peru, Chile, Zaire, and Zambia, bauxite from Jamaica, Guyana, Surinam, Guinea, and Sierra Leone, iron ore from Liberia, and phosphate rock from Morocco, Senegal, Togo, and Jordan are all examples of mineral commodities with dominance in the total exports of the indicated countries.

According to international trade data compiled by the United Nations,[e] international exports by developing nations with market-type economic systems, reached $557.6 billion in 1980. Roughly 62% of this total was mineral fuels, principally petroleum, and 5% other mineral commodities. Excluding petroleum from total exports, 13% of remaining exports were metals and nonfuel minerals.

The outlook for mineral trade from the developing nations is closely tied to the economic growth of the industrial economies as well as continued growth in the developing economies themselves. In 1980, about 70% of all exports from developing market economies went to industrialized countries with developed economic systems. In this same time period, 25% of exports from developing economics were used by other developing nations, and only five percent of these exports went to centrally planned economic systems. Growth in exports from developing nations increased at an annual compound rate of over 21% per year during the period 1975–1980. This growth was not solely dependent on petroleum exports, since oil and mineral fuels exports constituted about 60% of the total in 1975 and grew to about 70% of the total in 1980. Since oil exports increased at an annual rate of about 22% during the 1975–1980 period, all other exports from developing nations grew at roughly 20% per year. By any measures of comparison, the exports by developing nations showed significant growth during this period. This was true for nonfuel mineral commodities other than petroleum as well as petroleum.

While rising prices were a factor during the 1975–1980 period, the growth in exports in real terms exceeded 12% per year after adjustment for changing prices.

During the period 1975–1980, the share of total trade of the developing nations rose from 24% in 1975 to 28% in 1980. This does not indicate a significant change from 1960 when developing nations had roughly 25% of total, and is a decline from the 31.5% share in 1953, when much of the developing world was still part of the colonial empires of Great Britain, France, and other western nations.

This indicates a slowdown in commodity movements from the developing nations since colonial times, but this is not necessarily as negative as it might seem on the surface. In the words of Joseph C. McCaskill, author of the international trade chapter in the 3rd edition of *Economics of the Mineral Industry* published in 1976:

> "For generations many of the poorer nations producing minerals found their capital being mined from the ground, exported to the industrial countries, and their receipts from their exports used to pay the current costs of government and to purchase consumer goods. Over the long run these countries were left with little to show for their efforts and the loss of their resources except holes in the ground while the people remained poor, sick and illiterate."

In the 1960s much of this changed as most former colonies became independent and gained more control over the revenues from their exports. The 1960s was also a period of general prosperity, and exports grew at an annual rate of about six percent. Minerals were responsible for much of this increased trade. The 1970s witnessed the most significant increase in exports from the developing nations, led by oil and related mineral fuels, increasing from a value of $18.3 billion in 1970 to $124.1 billion in

[e] United Nations, *Monthly Bulletin of Statistics*, Vol. 36, No. 5, May, 1982, pp. XXX–LXXXVII.

1975. By 1980, mineral fuels exported from developing market economies were valued at over $345 billion, an increase in nominal value of nearly 19 times, or roughly a tenfold increase in real terms (when adjusted for the purchasing power of the U.S. dollar).

The influx of export earnings has been especially apparent in the nations with significant petroleum exports. While some of these earnings have been invested in military equipment and defense systems, much of the new capital has been invested in *public improvements* such as housing and transportation facilities as well as educational facilities for the citizens of the developing world. A substantial portion of available funds has also been spent for machinery and transportation equipment, an investment in future economic growth.

The ability to export metals such as copper, aluminum, iron and steel, as well as manufactured fertilizers rather than fertilizer raw materials, alumiuna rather than bauxite, and other forms of finished or semifinished products rather than ores and concentrates, speaks well for economic growth in the developing nations.

However, the developing economies are also subject to severe difficulties when the world economy turns downward as it has done in the early part of the 1980s. Dependence on minerals and even products manufactured from minerals has led to severe international payment problems for many of the developing nations as export revenues have failed to keep pace with the foreign debt service of these nations. In addition, the export earnings of developing economies are not evenly distributed, but heavily concentrated in relatively few oil-exporting nations. Less than two percent of the population of the developing world lives in those countries that account for about 55% of all exports (basically oil) from the developing economies.

One of the major issues in international trade policy for the industrial nations is still the design and implementation of policies that would encourage increased imports of finished and semifinished products from the developing nations. These nations have long protested the role of raw material supplier to the developed nations, and significant discussion of these issues continues within the United Nations under the banner of the New International Economic Order.

If the economic gap between the developed and developing economies is to be narrowed in the future, *some program of trade preferences may be necessary*. Not only will the earnings from raw material exports need to be maximized, but new markets must be found for the manufactured products of the developing world, including more highly processed forms of mineral exports. With these changes, more rather than less minerals will be mined and used, and world trade in minerals as well as manufactures will continue to grow.

The Role of Mineral Imports

The ability to obtain raw materials from other nations is an important factor in the ability of most countries to sustain economic growth. For industrial nations, many raw materials are not available domestically, and international trade has been the foundation for the variety and quantity of their economic output. This is especially evident in Europe, where the natural resource base has been well developed for many years, and the energy used to fuel the European economies is mostly petroleum from the Middle East. Even the United States, which has an impressive domestic natural resource base, has become increasingly dependent on mineral imports such as petroleum.

Only the Soviet Union, with the largest natural resource base of any nation in the world, has been able to maintain some semblance of mineral self-sufficiency. This has been possible largely through many years of setting self-sufficiency as a major economic goal in this most advanced centrally planned economic system.

The trade deficit of the developed nations as a group in 1980 was approximately $80 billion according to United Nations data. The developed economies of Europe alone, had a trade deficit of roughly $75 billion in 1980. The dependence on imports of petroleum was the largest single factor in the total trade deficit. Developed economies as a whole imported more than $274 billion of mineral fuels, principally petroleum, on balance in 1980. Europe had a net energy trade deficit of $135 billion in 1980. Japan showed a net importation of over $63 billion in mineral fuels in 1980, but managed to achieve a $6 billion trade surplus in the same time period. South Africa, Canada, and Australia managed to achieve modest trade surpluses in 1980.

The United States has achieved a net trade deficit on the order of $25 to 30 billion annually in 1980 and 1981, with net imports of petroleum valued between $75 and $80 billion.

The above statistics show the significance of imports of mineral fuels to the industrial nations that function as market systems. The correlation between energy consumption and national income is not a simple coincidence. Much of modern economic development rests on an energy base. This is both a benefit and a curse. As long as energy is available, economic progress seems assured. Limitations on energy supplies and/or significant increases in the price of energy have had serious impacts on the worldwide economy. Rapid increases in energy prices such as occurred in the period 1970–1980 have had a major impact on inflation and consequent recession.

The impact has been even more serious on the developing nations outside OPEC. With a much smaller economic base, the non-oil producing countries incurred a total trade deficit of $112 billion in 1980. These countries as a group imported roughly $87 billion of mineral fuels, principally petroleum, in this same period.

The dependence on imports of petroleum has grown as the industrialized and developing nations have prospered. The United States depends on imports of petroleum for roughly half of its average daily consumption. Europe and Japan are almost totally dependent on oil imports.

Mineral imports other than petroleum are also important to the industrial base of the developed economies. The United States imports about 25% of its iron ore. Japan imports more than 90% of its iron ore, 70% of its coking coal, and 25% of its iron and steel scrap. Many of the major European iron and steel producing nations are heavily dependent on imported iron ore as a major raw material input.

The production of primary aluminum in the industrial economies is almost totally dependent on imports of bauxite and alumina. In the United States, imported bauxite supplies roughly 90% of annual consumption, imported alumina accounted for more than 45% of the U.S. supply in 1983.[f] The Federal Republic of Germany, Canada, Japan, and Norway, all major producers of primary aluminum, are almost totally dependent on imports of bauxite and alumina as raw materials. Only the Soviet Union and France have substantial domestic supplies of aluminum raw materials, and even they import bauxite and alumina to supplement their domestic resources. In addition to the import of aluminum raw materials, the United States imports over 20% of its annual consumption of primary aluminum metal. Canada supplies over 60% of U.S. aluminum imports, with Ghana, Venezuela, and Japan accounting for another 25% of metallic aluminum imports.

In addition to the obvious industrial raw materials such as petroleum, iron ore, coal, bauxite, and aluminum, there are other vitally important mineral raw materials that are also essential to advanced economies. Ferroalloys such as chrome, nickel, cobalt, manganese, tungsten, and molybdenum, are major imports to most of the developed market economies with significant iron and steel industries. Commodities such as industrial diamonds are vital abrasives for modern industry and oil drilling. In the past, virtually all industrial diamonds were imported from African nations, with Australia recently becoming a major exporter, and the USSR also a factor in world markets. Major portions of world demand for nonferrous metals such as tin, copper, lead, and zinc are met by imports. Nonmetallics such as sulfur (used for sulfuric acid), potash, and phosphate rock, form the basis for most production of chemical fertilizers, and are important imports to the nations that manufacture and use fertilizers in modern agriculture.

With domestic raw materials being depleted and other limiting factors such as environmental regulation, most industrial economies are likely to increase their dependence on mineral imports. Modern economic growth is measured increasingly in terms of total and per capita consumption of major mineral commodities as well as increasing per capita caloric (food) consumption based on modern agricultural practices that utilize chemical fertilizers.

Minerals in their crude form often become the basis for refined materials that may be worth 10 times their original raw material cost, and these materials then become the basis for products worth 10 times their initial value. It has been estimated by McCaskill[g] and others, that the full impact on Gross National Product may be as much as 1,000 times the initial cost of the mineral raw materials when considering a product such as the automobile.

Trade in mineral commodities ties the nations of the world into a vital interdependence. This

[f]U.S. Bureau of Mines, *Mineral Commodity Summaries, 1984*, p. 17.

[g]McCaskill, J.C. "Minerals in International Trade," chapter in Vogley, W.A. (Ed.) *Economics of the Mineral Industries*, New York: AIME, p. 86.

interdependence is often complex, with many mineral imports becoming the basis for eventual export. Countries such as Great Britain and Japan could not long survive as modern economies without substantial mineral imports. The United States may generate only 10% of its economic transactions in international trade but the materials imported form the basis for products vital to the other 90 percent.

To demonstrate the complexity of minerals in international trade, we need only examine the mineral imports of the United States. Crude oil is purchased from Venezuela, Mexico, Saudi Arabia, Libya, Indonesia, and Canada with many other smaller sources. Iron ore comes from Liberia, Australia, Brazil, Peru, Venezuela, and Canada. Ferroalloys such as chromite come from the Soviet Union, Turkey, South Africa, and the Philippines; manganese arrives from South Africa, Gabon, India, Australia, and Brazil; cobalt is imported from Zaire, Zambia, Canada, Belgium-Luxembourg, Japan, Norway and Finland; nickel is brought in from Canada, Australia, Norway, and Botswana; and tungsten for U.S. markets comes from Canada, Bolivia, and China. Iron and steel from Japan, Canada, and Europe provide 16% or more of domestic demand. Copper arrives from Chile, Peru, Zambia, and Canada. Bauxite and alumina come from Jamaica, Guinea, Surinam, Australia, and Guyana. Malaysia, Indonesia, Bolivia, Thailand, and Nigeria are sources for tin. Lead and zinc originate in Peru, Mexico, Spain, Canada, Australia, and Morocco. Italy and Spain provide mercury. Graphite comes from Sri Lanka, Austria, India, Mexico, China, Brazil, Madagascar, and the Republic of Korea. Gold, diamonds and coal come from South Africa. Antimony is imported from Yugoslavia, Bolivia, China, and Mexico. Arsenic is imported from Sweden, Mexico, France, and Canada. Aluminum metal comes from Ghana, Canada, Venezuela, Japan, and Europe. Other imports include beryl from Rwanda, borates from Turkey and boric acid from France, cadmium from the Republic of Korea, cement from Spain, ferrochromium from Zimbabwe, clays from Great Britain and the Federal Republic of Germany, columbium from Brazil, corundum from South Africa and Zimbabwe, diamonds from Zaire, diatomite from Mexico, feldspar from Canada, Norway, and Sweden, fluorspar from Mexico, South Africa, China, and Italy, gallium from Switzerland, gem stones from Israel, germanium from Belgium-Luxembourg, gold from Canada and the USSR, gypsum from Canada, Mexico, and Spain, titanium raw materials from Australia, India, Canada, and South Africa, magnesium from Canada, Norway, France, and the Netherlands, magnesite from Ireland, Greece, Japan, and India, mica from Canada, India, and Brazil, ammonia from Canada, Trinidad, Tobago, USSR, and Mexico, perlite and pumice from Greece, phosphate rock from Morocco, platinum from South Africa, USSR, and Great Britain, potash from Canada and Israel, quartz crystal from Brazil, rare earth minerals from Australia and Malaysia, rhenium from Chile and the Federal Republic of Germany, salt from Canada, Mexico, the Bahamas, and Chile silver from Canada, Mexico, Peru, and Great Britain, dimension stone from Italy, strontium from Mexico, sulfur from Canada and Mexico, talc and pyrophyllite from Italy, Canada, and France, tantalum from Thailand, Canada, Malaysia, and Brazil, tellurium from Canada, Hong Kong, and Great Britain, thorium from France, the Netherlands, Canada, and Malta, vanadium from South Africa, Canada, and Finland, vermiculite from South Africa and Brazil, and zirconium metal and zircon from France, Japan, Canada, Australia, and South Africa. As the above list indicates, there is hardly a country in the world with mineral production that does not sell some portion of their exports to the United States.

The same kinds of complex trade relationships exist between virtually all developed economies and their mineral suppliers. To a lesser degree, due to the less developed nature of their economies, the nations of the developing world also participate as importers of mineral commodities. While oil is the overwhelming commodity imported by the developing nations that are not oil producers, the other mineral commodities play an important role in filling the needs of the LDC's. This is especially true of fertilizers and construction materials such as cement.

With the growing importance of international sources of minerals, the world has witnessed an increasing interest in capital investment by countries such as Japan in the exploration, development, and production of minerals in the developed as well as developing nations of the world. Japan and Romania have entered into equity participation in metallurgical grade coal production in Alabama and Kentucky, respectively, to increase the capacity of the United

States to export such coal. Japan has also invested in many developing nations such as Chile and Peru, and is substantially involved in Australian iron ore and coal production. Such involvements have important political as well as economic implications for the future, and often offer the opportunity for significant technological exchange.

FACTORS AFFECTING INTERNATIONAL TRADE IN MINERAL COMMODITIES

There are many things that have direct and indirect impact on the actual flows of mineral commodities between nations. These include: government political and economic relations, membership in international economic alliances, monetary exchange rates, relationships with multinational corporations, government purchase and assistance, and military or defense considerations. This section will attempt to deal with the impact of these factors on the flow of mineral commodities as part of general commodity flows between nations.

Barter and Commodity Exchanges

Barter, commonly called countertrade, has staged a major comeback in recent years. The Soviet Union and the members of the Council for Mutual Economic Assistance (COMECON) have used countertrade since World War II due to the tight controls on the exchange of currencies by their respective governments and the related difficulties in establishing a true value of their currencies in international markets. However, the use of countertrade by the rest of the world has increased dramatically since the sharp increases in oil prices during the 1970s.

Countertrade normally develops when the trading nations do not have a sufficient supply of internationally accepted currency (hard currency), but do have an abundant supply of commodities that can be used directly or exchanged with other countries for other commodities or hard currency. In 1976, less than two percent of total trade in international commerce outside the COMECON nations was countertrade. By 1981, countertrade accounted for approximately 20% of world trade outside COMECON.[h]

While barter is the purest form of countertrade, it actually represents only a small part of all commodity exchanges. The term countertrade covers a variety of types of exchange:

(1) Barter—The simplest method of exchanging commodities for other commodities often involves more than two trading partners. In fact, most barter arrangements involve three or more nations and/or international business organizations.

(2) Compensation—Most commonly, this system involves the provision of machinery, technology or production plant construction with payment made in terms of the product or resulting output of the process. An example of this is the agreement between the Occidental Petroleum Corporation and the Soviet Union. The terms of this exchange include the construction of chemical plants in the USSR by Occidental and payments received as ammonia under a 20 year contract.

(3) Counter Purchase—The buyer contracts with the seller to buy some of the contract's value in other goods produced by the buyer. An example of this is the sale of 20 aircraft by McDonnell-Douglas Corporation to Yugoslavia in exchange for partial payment and glass crystal, cutting tools, leather goods, and $3 million in canned hams among other commodities.

(4) Swap—In this arrangement, two sellers, producing the same product, agree to swap buyers to save on transportation or other costs. An example of this might be an American company with a contract to sell phosphate rock or fertilizer to Spain arranging a swap with a French company with a similar agreement in Canada. Another example might be that iron ore from Venezuela to Japan could be swapped for Australian ore being shipped to the United States. Naturally, all trading parties are normally involved in these agreements.

(5) Switch—This type of agreement is designed to facilitate currency exchange and availability. One country ships a product to a second country short of hard currency, the second country ships another product to a third country, and the third country pays the first country in hard currency. An example of this could be an American company selling machinery to a third world nation short of American dollars. The devel-

[h]Truell, Peter, "Barter Accounts for a Growing Portion of World Trade Despite Its Inefficiency," *Wall Street Journal*, August 15, 1983, p. 21.

oping nation might sell a product such as copper, cobalt or iron ore to France or Germany which, in turn, would pay for the commodity by sending the American firm payment for the machinery.

(6) Blocked Currency—This is often used when currency restrictions by a central government prevent the movement of that country's currency out of the country. A firm or country can sell its products for local currency, purchase goods with that currency, and export those products as payment for the original product. This is similar to a barter arrangement but involves a market transaction within the country with the currency restrictions.

The arrangements within the scope of countertrade are as varied as the commodities and countries involved in the exchanges. The safest forms of countertrade normally involve commodities due to the ease of sale or exchange of basic raw materials in world markets. While technology and machinery can be involved, they are normally not taken as payment by the developed nations due to the difficulties in resale. Raw materials are simply easier to sell or consume directly.

One recent countertrade agreement between Italy and the People's Republic of China involves an exchange valued at $500 million. Italy's Technotrade organization will expand Chinese coal mines, modernize a railroad and port facility, and receive payment in coal production at a level of "a couple of million" tons of coal per year for more than 10 years.

Japan has some of the most active trading companies involved in the export of various Japanese products in exchange for South American copper, Australian coal, and various agricultural products.

The United States has also engaged in barter and countertrade with commodities in its supplemental agricultural stockpile. In these arrangements, U.S. agricultural commodities declared surplus have been exchanged for minerals and other commodities held in the government stockpiles.

GOVERNMENT STOCKPILES, MILITARY PROCUREMENT AND NATIONAL DEFENSE CONSIDERATIONS

Strategic and supplemental stockpiles, especially those of the United States, historically have had an important impact on the international trade in mineral commodities. Countries such as the United States, Great Britain, and to a lesser extent, Canada, began to develop strategic materials stockpiles after World War II.

The major purpose of the U.S. government stockpile was to provide strategic materials for a potential three-year wartime emergency. Another purpose has been to permit the government to purchase stockpile materials to support domestic and international prices and to release stockpiled materials in times of shortage or if economic stabilization policy would be served by lower market prices. Government sales of materials from its stockpiles have often brought objections from producers of these materials, therefore, negotiations between government officials and industry representatives normally precede major changes in the stockpile inventory. Industry leaders in several countries believe these stockpiles are a threat to the stability of mineral commodity markets and restrict international trade. These influential individuals have encouraged the disposal of stockpile inventories in all but the most critical materials. There has traditionally been little support for the idea that governments should buy during periods of falling prices and sell when supplies are short.

With relatively little activity by governments to buy or sell stockpiled minerals, the importance of these stockpiles and potential impact on international trade is diminished. The General Services Administration handles purchases and sales of materials in the U.S. government stockpiles. Goals for inventories of the various commodities are determined by the Federal Emergency Management Agency. Specific annual activities of the government stockpile are reviewed by the U.S. Bureau of Mines in its annual report, *Status of The Mineral Industries*. A more detailed discussion of stockpiling, commodity agreements and the impact of government purchases of mineral commodities may be found in McNicol, D.A., *Commodity Agreements and Price Stabilization: A Policy Analysis*, Lexington, Mass.: D.C. Heath and Company, 1977.

INTERNATIONAL TRADE AGREEMENTS

Cartels

In September, 1960, five leading oil producing countries: Iran, Iraq, Kuwait, Saudi Arabia, and Venezuela, founded the Organization of Petroleum Exporting Countries (OPEC) to protect their common economic interests by fixing prices and assigning market shares. The organization is a cartel by definition, since it is a collusive international agreement to avoid price competition and other characteristics of a free market. It was founded to protect export revenues based on a single commodity, crude oil. The organization was largely ineffective in achieving its goals due to political and religious conflicts until 1973.

By 1973, OPEC had expanded its membership to 12 nations with the addition of Qatar, the United Arab Emirates, Algeria, Libya, Nigeria, Indonesia, and Ecuador and added an associate member, Gabon. The oil exports of these nations accounted for over 85% of the world trade in crude petroleum.

By the early 1970s, the world had witnessed an impressive annual growth in the use of energy. Most of this increase had been based on oil exports from the OPEC nations. In 1972–1973 the world market set the stage for increased prices due to apparent shortages since demand had expanded beyond planned production levels. In October, 1973, Israel, in retaliation for being attacked by the armies of its neighbors, bombed the Eastern Mediterranean oil terminal that handled Syrian and Iraq crude oil exports, disrupting a major source of supply in that area. OPEC seized the opportunity to raise crude oil export prices from $3 per barrel to just over $5 per barrel. When the major oil refining companies did not agree with this price hike, OPEC ended the negotiations and raised the price unilaterally. The October, 1973 increases were followed by another price increase in December to the level of $11.65 per barrel. This amounted to a four-fold oil price increase in about three months.

At the same time, OPEC decided to place a total embargo on oil shipments to the United States and The Netherlands due to their economic and political support of Israel, and to cut back oil supply by five percent per month to all other nations in an effort to further pressure countries that had economic and political ties to Israel. Reports that oil was still finding its way to the countries embargoed led OPEC to further cut oil exports by 25 percent.

The embargo was not successful in terminating support for Israel, but it was an economic success in terms of its impact on world oil prices. Transshipment and rerouting undermined the effectiveness of the embargo, but the reduced level of production solidified the price increases.

Oil prices held firm until 1976 when internal disagreement within OPEC led to "split-level" prices effective in January 1977. Saudi Arabia and the United Arab Emirates felt that prices should rise by 5% while the other OPEC nations favored a 10% increase. Failure to agree led to different prices until June, 1977, when Saudi Arabia raised its price by the additional 5% and the OPEC price was restored to $12.70 per barrel for all members.

Throughout 1977 and 1978 it was generally felt that oil supplies were adequate in light of energy conservation and reduced growth in world demand. As a result, inventories were reduced and the stage was set for another round of dramatic price increases in 1979 and 1980.

In December, 1978, OPEC met to discuss oil prices against the background of rising spot prices for crude, and a possible Iranian revolution based on Islamic Fundamentalism. Due to the instability in Iran, OPEC declared that prices would be increased quarterly by "tiers" during 1979 by a total of 14.5 percent. The reasons given included rapid worldwide inflation as the basis for their effort to recover purchasing power. In fact, the oil price increases were as much the cause as the effect of this inflation. Had this announced price increase materialized, OPEC crude oil prices would have been $14.55 per barrel by October, 1979.

On December 26,1978, Iran suspended exports. Other OPEC nations increased their production to offset the Iranian cut. However, the uncertainty of the situation drove the spot price for oil to $44.24 per barrel and set the stage for price increases beyond those announced earlier. Iran restored some of its production and exports in March of 1979, but production did not return to the pre-revolution levels.

Due to the higher spot prices, Saudi Arabia raised its prices by the full 14.5% in April, 1979 and cut back production. Libya increased its prices by even more and had buyers for production near its capacity. The Saudi production cuts caused spot prices to climb once again and other OPEC nations increased prices and maintained near capacity output. Even though Saudi Arabia increased its output to former levels, the other members of OPEC continued to increase prices throughout 1979, 1980, and 1981. By June, 1981, the official OPEC price was $32

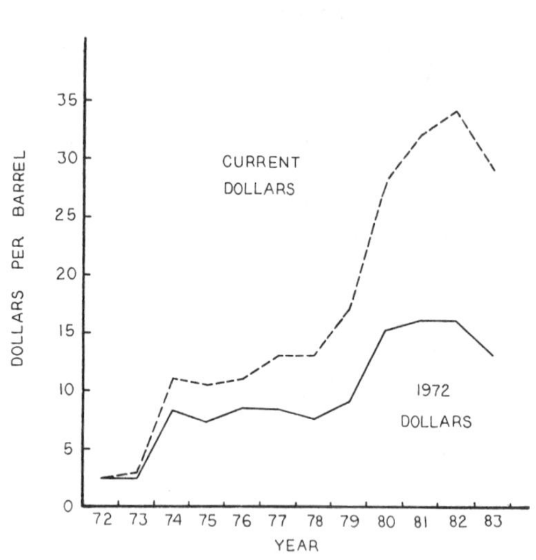

Fig. 1.3.7—Price of OPEC crude 1972–1983. Sources: Hartsharn, J.E., "Two Crises Compared: OPEC Pricing in 1973–1975 and 1978–1980," in *OPEC: Twenty Years and Beyond*, edited by Ragari El Mallakh, Westview Press, Boulder, Colo., 1982, p. 20. Eden, Richard; Posner, Michael; Bending, Richard; Crouch, Edmund; and Stanislaw, Joe, *Energy Economics: Growth, Resources, and Policies*, Cambridge University Press, 1981, p. 252.

per barrel and spot prices were around $30 per barrel. Annual price increases were slowing dramatically in light of the coming recession, brought on to a major degree by the rapid price increases for oil. Although oil prices reached $34 per barrel in late 1981, by 1984, the OPEC price has fallen to $29 per barrel. (Fig. 1.3.7)

Economic theory suggests that higher prices lead to cuts in quantity demanded and provide incentives to increase supply. The rapid increases in oil prices during the 1970s accelerated exploration and development of new oil fields in Mexico, Alaska, the North Sea, and the Soviet Union. These new supply sources weakened OPEC's control over production for world trade, and the higher prices led people to search for methods of energy conservation and alternative energy sources. The combination of these factors as well as the recession of 1981–1983, led to lower levels of demand for energy, and oil prices have varied between $28 and $38 per barrel during this period.

While OPEC demonstrated the power to control prices and output during periods of excess demand, it has shown less ability to control these factors when demand slackens. The continuing instability in the Middle East, especially the Iran-Iraq war, and the situation in Lebanon, create an unclear picture of future economic power in

Table 1.3.10—Percentage of 1981 World Production and Reserves Shared by the Four Largest Countries, Selected Mineral Commodities.

Mineral commodity	Share of world production	Share of world reserves	Countries
Bauxite	64	66	Australia, Jamaica, Surinam, Brazil
Chromium[a]	46	98	South Africa, Zimbabwe, Philippines, Finland
Cobalt	76	52	Zaire, Zambia, Canada, Australia
Copper	53	50	United States, Chile, Canada, Soviet Union
Fluorspar	54	44	Mexico, Mongolia, Soviet Union, South Africa
Gold[a,b]	59	78	South Africa, Canada, United States
Iron Ore	58	71	Soviet Union, Australia, United States, Brazil
Lead[a]	41	51	United States, Australia, Canada, Peru
Manganese	76	90	South Africa, Gabon, Australia, Soviet Union
Mercury	80	58	Spain, Soviet Union, United States, Algeria
Molybdenum[b]	76	61	United States, Canada
Nickel	67	60	Canada, New Caledonia, Australia, Soviet Union
Phosphate rock	79	87	United States, Morocco, Soviet Union, China
Platinum group	99	98	South Africa, Soviet Union, Canada
Potash[a]	50	90	Canada, West Germany, United States, France
Silver[a]	59	61	Peru, Mexico, Canada, United States
Tin	62	44	Malaysia, Soviet Union, Bolivia, Thailand
Vanadium	89	95	South Africa, United States, Soviet Union, China
Zinc[a]	41	59	Canada, Australia, United States, Peru

Source: Calculated from U.S. Bureau of Mines, *Commodity Summaries 1983*, Washington: U.S. Government Printing Office, 1983.
[a] Calculated for noncommunist countries only.
[b] For platinum and gold, figures are for the three largest, for molybdenum, figures are for the two largest.

the hands of OPEC. However, the impressive ability of OPEC to control world prices and market shares during the 1970s has led to renewed interest in creating international cartels in other mineral commodities with important markets in the industrialized nations.

There have been efforts to form cartels in nonfuel minerals such as tin, bauxite, and copper. These have not been effective due to the following:

(1) Ethnic and cultural diversity of proposed members.
(2) Pressure to generate cash to service import purchases and foreign debt.
(3) The desire by most developing nations to move away from commodity exports and engage in industrialization and economic diversity.

There are four characteristics that play an important role in the formation of a cartel. These are:

(1) The production of the commodity must be limited to a small number of members that account for a substantial share of the production. (See Table 1.3.10)
(2) The sensitivity of demand for the commodity to price change is small.
(3) The sensitivity of supply price change expansion potential, especially outside the cartel is relatively small.
(4) Cartel member goals must be compatible with each other. The sources of several important nonfuel mineral commodities are concentrated in a few developing countries. Several of these attempts to organize these cartels will be further discussed. The International Council of Copper Exporting countries (CIPEC) was founded in 1967 by Chile, Peru, Zaire, and Zambia (Table 1.3.11). Since then Indonesia and Australia have joined and Papua New Guinea and Yugoslavia have signed on as associates. If those eight nations comprised the world's copper supply the cartel could work. However 14 other countries including the United States hold large copper reserves. CIPEC faces substantial price elasticity of demand and large increases in supply at sustained high copper prices and unity has eluded CIPEC.

Table 1.3.11—The International Council of Copper Exporting Countries.

Member Countries	Percent of World Production 1979	1981
Chile	17.3	13.2
Peru	6.5	4.0
Zaire	6.5	6.1
Zambia	9.6	7.2
Australia	3.8	2.7
Total	43.7	33.2

Adapted from: Rouig, A. Dan and Richard K. Doran, "Copper A Decade of Change and Its Meaning for the Future," *Mining Congress Journal*, Vol. 32, No. 12, December 1980, p. 32.

Another factor affecting CIPEC is the United States. The United States is both the world's largest consumer and producer. American consumption totals about 1.9 million tons per year, but production including the recycling of copper scrap totals 1.7 million tons per year. This massive concentration of copper makes it difficult for eight countries to control the market.

The Association of Iron Ore Exporting Countries (AIEC) was established in 1975 with an 11-country membership (Table 1.3.12). AIEC accounts for about one quarter of the world's production.

Table 1.3.12—Association of Iron Ore Exporting Countries

Member Countries	Percent of World Production 1973	1981
Algeria	.3	—
Australia	11.0	10.0
Chile	1.2	—
India	4.4	4.8
Mauritania	1.3	—
Peru	1.2	—
Philippines	.3	—
Sierra Leone	.3	—
Sweden	4.3	2.7
Tunisia	.1	—
Venezuela	2.7	1.8
Total	27.1	

Source: U.S. Bureau of Mines, *Minerals in the U.S. Economy: Ten Year Supply–Demand Profiles for Mineral and Fuel Commodities*, 1975.

AIEC has been unable to establish uniform production and pricing policies. Iron ore is usually sold under long term contract and does not have the price volatility that some other commodities have. Also Australia, with huge reserves, maintains an independent role. Finally there is potential supply expansion outside the cartel.

The International Association of Mercury

Producers (IAMP) was formed in 1976 (Table 1.3.13). Although the cartel has potential control over 49% of the world's supply of mercury, slow growth rates will hamper the formation of a strong cartel.

Table 1.3.13—International Association of Mercury Producers

Member Countries	Percent of World Production 1973	1981
Algeria	5.1	12.1
Italy	11.7	—
Peru	1.1	—
Spain	21.8	24.2
Turkey	3.1	—
Yugoslavia	5.7	—
Total	48.5	

Source: U.S. Bureau of Mines, *Minerals in the U.S. Economy: Ten Year Supply–Demand Profiles for Minerals and Fuel Commodities*, 1975.

The Tungsten Producers Association was also formed in 1976. (Table 1.3.14). TPA only accounts for about one fifth of the world's production and one third of world tungsten exports. It is not expected that the TPA will be an effective cartel due to the small share of production and exports and the unpredictable behavior of supplies. Also, no producer holds a large enough market share to provide strong price leadership.

Table 1.3.14—Tungsten Producers Association

Member Countries	Percent of World Production 1973	1981
Australia	3.1	6.8
Bolivia	5.6	5.6
Peru	2.1	—
Portugal	3.9	2.8
Thailand	6.7	2.4
Total	21.4	

Source: U.S. Bureau of Mines, *Minerals in the U.S. Economy: Ten Year Supply–Demand Profiles for Mineral and Fuel Commodities*, 1975.

The International Tin Council (ITC) was established in 1956. Member countries are Australia, Bolivia, Indonesia, Malaysia, Nigeria, Thailand, and Zaire. In 1979, Malaysia, Bolivia, Thailand, Nigeria, and Zaire accounted for 86% of the world's tin production. In addition to the seven producing countries, 22 consuming countries are also members of the ITC.

The major instrument used by the ITC to stabilize prices has been the management of a buffer stock. In practice, this agreement has worked poorly due to the fact the ITC had neither tin nor money to support tin prices.

It is not expected that the ITC will be an effective cartel in the future. Consumption of tin is on the decline. Also, the United States, which consumes about 40% of all tin production has a substantial tin recycling industry.

The International Bauxite Association (IBA) (Table 1.3.15) has control of a substantial amount of the world's bauxite. Members include Australia, the world's largest bauxite producer, Guyana, Jamaica, Guinea, Suriname, and other lesser producers.

Table 1.3.15—The International Bauxite Association

Member Countries	Percent of World Production 1981
Australia	30.0
Guinea	14.1
Jamaica	13.6
Surinam	4.3

Source: U.S. Bureau of Mines, *Commodity Data Summaries, 1983*, Washington: Government Printing Office, 1983.

The IBA has two basic problems. Although bauxite is the most economical source of aluminum it is not the only source. Secondly, Australia has largely acted independently from the IBA.

In general the cartels to date have been ineffective. They have neither full membership of producing countries nor the cooperation of consumers.

What is the future for these cartels?

(1) Most nonenergy, mineral and metal cartels do not pose a major long term economic threat to importing nations.

(2) The world economic climate is conductive to the strengthening of existing cartels.

(3) Cartels are becoming more sophisticated in their analytic capability.

(4) The key to future cartel strength may lie largely with four mineral exporting nations, Australia, Brazil, Canada, and South Africa. There are few commodities where at least one of these nations is not already among the top four exporting nations. In the case of Australia, there has been a move to join cartels, however,

Australia maintains an independence in setting mineral prices.

United Nations' Programs

The United Nations has many programs that are designed to assist the developing nations of the world to utilize their natural resources and encourage favorable terms in international trade and commerce. The following are a few of those programs with specific impact on potential international trade in mineral commodities.

United Nations Revolving Fund for Natural Resource Exploration

The United Nations Revolving Fund for Natural Resource Exploration was approved in 1973. Established as a trust fund, administered on behalf of the Secretary General by the Administrator of the United Nations Development Program, the fund was legally established in 1975 to extend and intensify the activities of the United Nations in the field of natural resources exploration. Currently the Fund's authority covers minerals and geothermal exploration and related feasibility studies. The first mineral exploration project was began in 1976.

The Fund has established a long term financial mechanism which pools high-risk exploration capital to developing countries. Modest repayments or replenishments are only required when production, and thereby government revenue ensue. Currently the Fund is being entirely financed by donor countries principally from the industrialized nations. Any repayment from the developing nations will be channeled back to the developing countries to amortize the debt and to finance future projects.

The Fund is set up to aid in any or all of three stages in production. The three stages include: (1) preliminary activities; (2) reconnaissance work; and (3) detailed exploration and evaluation.

Preliminary activities include a plan to verify and evaluate geological data provided by the applicant's country and limited field checking. *Reconnaissance work* within selected areas include using satellite data, aerial photographic analysis and other types of surveys to define the most promising targets for detailed prospection. *Detailed exploration and evaluation* includes such activities as detailed geochemical and geophysical investigations, trenching, pitting, drilling, and tunneling to determine size and location or ore bodies and geothermal projects.

UNDP

The United Nations Development Program (UNDP) is the principal international agency in the field of technical assistance and preinvestment aid. It is intended to identify, investigate, and present to financial agencies those projects in developing countries that merit investment. Most of the UNDP assistance is actually carried out by specialized agencies and other bodies of the UN.

UNCTAD

The United Nations Conference on Trade and Development (UNCTAD) was founded in 1964 by the developing nations that are members of the United Nations. This organization is headquartered in Geneva. The original purpose of the UNCTAD was to orient the world's economy toward economic development.

Many problems of the developing nations result from their dependence on the exportation of raw materials and agricultural products. Over three fourths of their nonpetroleum export earnings are made up of these commodities. Also, in 30 of these countries over 80% of export earnings come from only three leading commodities; for another 32 the figures lies between 60 and 80 percent.

By the early 1970s the general inability of the industrialized nations to implement new trade policies toward developing nations and the success of OPEC led to a formal statement of demands from developing nations in the *Declaration and Action Programme on the Establishment of a New International Economic Order. (NIEO)* which was adopted by the United Nations in 1974. The NIEO was followed in the same year with the adoption of a *Chapter of Economic Rights and Duties of States* which included such provision as the right of a country to expropriate foreign property and the right to determine how much if any compensation would be paid to foreign owners.

To help manage the NIEO program UNCTAD developed a proposal for an "Integrated Commodity Program," (ICP). It proposed international commodity agreements of the buffer stock variety of each of 10 to 18 primary commodities. The individual agreements would be linked to a "common fund" of $6 billion, made up of contributions by producing and consuming countries.

A buffer stock for each commodity would be operated independently within predetermined upper and lower limits by buying and selling stock. Should the need for additional funds arise, they could borrow from the "common fund."

Although the ICP is the centerpiece of the "New Order," the first reaction of industrialized nations was understandably reluctance and rather negative. Their past experiences with other commodity agreements had been bad and such a large financial commitment could not be justified.

Nevertheless, by 1977 the attitude of rejection was changing, at least on the part of the United States. Policymakers in the industrial nations began to realize in the proposal some ingredients favorable to price stability in their own countries. For example, if the price of tin fluctuates violently, the increase is immediately reflected in the price of goods made of tin. But given price rigidity in a downward direction in many American industries, the reverse does not occur when the price of tin declines. Therefore the net effect of tin price fluctuations is inflationary to the United States, so stabilizing its price would contribute to price stability here, or so that was argued.

What finally emerged in 1980 from lengthy negotiations is a Common Fund of $400 million to finance the buffer stock operation for 18 commodities, and $350 million for market research and export promotion. However, the new program never had a true test. Managers of existing commodity agreements were reluctant to cooperate with the Common Fund and many developing countries became dissatisfied with the small size of the fund. Additionally in 1982, the Reagan Administration requested a reexamination of the entire program to prevent undue interference with the market mechanism.

Economic Alliances and Agreements

This section will review some of the organizations and associations that affect international trade. These will include the European Economic Community (EEC) sometimes referred to as the European Common Market, the European Free Trade Area Association (EFTA), the Organization for Economic Cooperation and Development (OECD), the Latin American Free Trade Association (LAFTA), the Council for Mutual Economic Assistance (COMECON), and the International Tin Agreements. All but the last of these have broader interests in international trade than simply mineral commodities, but they do affect minerals in much the same way that they affect the exchange of other commodities and manufactured products. There are other organizations such as the Organization of African States (OAS) and the North Atlantic Treaty Organization (NATO) that have dealt with issues involving mineral trade, but have not had a primary impact on international trade.

European Economic Community

The European Economic Community (EEC) was established in 1958 and originally included six countries: West Germany, France, Italy, Belgium, the Netherlands, and Luxembourg. On July 1, 1977, three new members, the United Kingdom, Denmark, and Ireland joined the community and Greece joined in 1981. Spain is also in the process of joining the EEC.

The EEC has abolished all tariffs and other trade restrictions among themselves and set up a common and uniform tariff against outsiders. The agreement also ensures free mobility of capital and greater mobility of labor. It also provides for the coordination of fiscal and monetary policies. Harmonization of tax and expenditure programs was deemed desirable to place producers in all countries on an equally competitive basis.

Another feature of the EEC is that a large fund was set up with contributions of member nations to help speed development in the more backward areas of the union. This aspect is known as "regional policy" and has been of special interest to the United Kingdom and southern Italy. In addition, a set of rules was developed to assure competitive behavior in enterprises within the EEC and to prevent the development of cartels. Finally, a common transportation and energy policy was developed and the nations cooperate closely in the development and use of atomic energy.

European Free Trade Association

Great Britain established the European Free Trade Area Association (EFTA) in the late 1950s after it could not successfully negotiate membership in the ECC. Headquartered in Geneva, the organization originally consisted of Great Britain, Austria, Switzerland, Portugal, Denmark, Sweden, and Norway with Finland as an associate member. EFTA is a free-trade area for industrial goods with some special provisions for trade in farm products. Each member of the EFTA retains its own tariff and quota system and follows its own trade policies with respect to third countries.

The EFTA is a much looser organization than the EEC. It does not have many of the EEC's institutional features, has no common economic policies and does not bargain as one unit in the GATT negotiations.

On July 1, 1977 Great Britain and Denmark joined the EEC. Thus the EFTA was reduced in size although it is retained as a formal organization. The EFTA has a preferential trade agreement with the EEC.

The Organization for Economic Cooperation and Development

The Organization for Economic Cooperation and Development (OECD) was established in 1961 and is a Paris-based organization. The goal of OECD is to promote policies to achieve the highest economic growth in member countries and thus to contribute to the development of the world. Member nations include the industrialized nations of Western Europe, North America, Japan, and Australia. Intra-OECD trade amounts to nearly one half of total world trade. The OEDC publishes a large volume of economic statistics for its members and makes forecasts of member's GNP, inflation, etc.

In 1976 the OECD adopted a Declaration of International Investment and Multinational Enterprises. This was a first attempt by the industrialized nations working with the U.N. Commission on Transnational Corporation and Developing Nations to establish a code of fair treatment and protection of foreign investment.

The declaration includes detailed guidelines for Multinational Enterprises that "aim at improving the international investment climate," at strengthening confidence between multinational enterprises and states, "at encouraging the positive contributions of multinational enterprises and states," and "at encouraging the positive contributions of multinational enterprises to economic and social process and minimizing or resolving difficulties that may result from their activities. . . ." Although an intergovernmental consultation procedure has been established, the guidelines are voluntary and the parties to the declaration do not include any of the developing countries.

The Latin American Free Trade Association

The Latin American Free Trade Association was started in 1960 and was the largest integration scheme among developing countries. Member nations include Argentina, Bolivia, Brazil, Chile, Colombia, Ecuador, Mexico, Paraguay, Peru, Uruguay, and Venezuela. These member nations cover a territory twice the size of the United States and have a population exceeding 300 million.

The LAFTA Treaty provided for commitments to negotiate duty reductions between member nations and for integration of single industries through complementarity agreements. Member nations never fully cooperated on trade liberalization due to strong protectionist forces and little progress.

Overall the LAFTA made little contribution to economic development in Latin America. The complementarity agreements fostered a modest degree of specialization but it never did spur investments in any new industrial areas.

In 1981, a stagnating LAFTA was superseded by the Latin American Integration Association (LAIA). The LAIA is less ambitious than the LAFTA; its purpose to create an area of economic preferences rather than a free trade area. LAIA principal mechanism is "partial agreements" that may be negotiated among two or more member nations covering either individual industries or several economic sectors. This method is viewed as the most realistic method to achieve some intergration among Latin American nations.

Council for Mutual Economic Assistance

The Council for Mutual Economic Assistance (COMECON) consists of the Soviet Union, Mongolia, Cuba, Bulgaria, Czechoslovakia, East Germany, Hungary, Poland, Romania, and Vietnam. COMECON has attempted to rationalize trade among its members, but this has been hard to do. The main problem is an internal one because foreign trade is held subservient to the domestic goals of the national plan.

Pricing between members is difficult. Prices in these nations are fixed by an internal planning agency and not by the world market price. Pricing goods traded then becomes a matter of bilateral negotiations. Fixing currency prices among the COMECON countries also is difficult.

COMECON has established the International Bank for Economic Cooperation which grants limited credit to finance intrabloc trade. Politically, intrabloc trade is preferred by COMECON nations rather than trade with the West, however in reality this is not the case.

International Tin Agreements

After a long history of attempting to stablize tin prices, the International Tin Council was established in 1956. The goals of the ITC are to stabilize tin prices, reduce the cyclic nature of shortages and surpluses, and to increase the growth rate in consumption.

The method for controlling tin prices is by managing a buffer stock. In practice the agreement has not worked well, although it accounts for about 86% of the world's production. In reality the buffer stock manager has had neither tin nor money to support ITC policies.

Recent developments in the ITC history include the Fifth International Tin Agreement signed by all members agreeing to a 40,000 ton buffer stock. The pact was undermined by consumer voting on buffer buying.

On July 1, 1982 the Sixth International Tin Agreement went into effect. This agreement provides for a 50,000 ton buffer stock and export controls. Once again there is doubt about the effectiveness of the agreement.

What are the problems with these agreements? One is the fact that the United States has declined to participate. The United States is the world's largest consumer of tin, using about 40% of all tin production. Also the United States has an active tin recycling program capable of supplying one fifth of total U.S. consumption. Finally the U.S. government maintains a strategic stockpile of 204,000 tons of tin, from which the General Service Administration has been authorized to sell all but 33,000 tons.

The demand for tin is declining due to the reduction of use in canning and military applications. Other factors include the entry of China and Brazil in the tin market and economic and political differences between the members.

Barriers to International Trade

Nations have a large variety of methods to restrict and regulate international trade. Most methods are designed to protect portions of their economy that are not competitive with those of other nations. This is often the justification when developing nations impose restrictions on imports in order to protect "infant industries." This argument is generally accepted for some time period to permit new industries and manufacturing facilities to get started and operating at some level of productive efficiency.

However, many older industrialized economies utilize international trade restrictions to protect economic activities that are obsolete or so high cost that they are unlikely to be competitive. When an economy protects a domestic industry with trade barriers, it restricts the flow of products to domestic consumers and raises their costs of purchase. In recent years we have seen the United States impose quotas to restrict the import of automobiles to allow domestic auto manufacturers to charge higher prices without the cutback in sales that would occur if foreign cars were more available. This has also been justified as saving jobs in the auto industry. The United States Steel Industry and the United Steelworkers of America have proposed similar quotas to restrict iron and steel imports. Domestic mining of fluorspar and other minerals have also been protected by restrictions on foreign imports. Countries such as Great Britain have a long history of restrictive legislation to protect domestic producers of tin, China clay, and iron and steel.

Embargoes: Other motives for restriction can be political as well as economic, such as the Arab oil embargo and trade embargoes of South Africa by various African nations. Embargoes basically prohibit trade in specific or general terms. They are the strongest sanctions that can be imposed by one nation toward trade with another. The effectiveness of an embargo is normally related to the importance of the commodities or products to the economy of the embargoed nation and the availability of substitutes or alternative sources of supply.

Embargoes by the Arab nations against trade with Israel have not been successful in destroying Israel's domestic economy due to alternative supplies and sources outside the Arab world. The major commodities in the Arab-Israeli case are crude petroleum and related oil products.

Tariffs: The use of tariffs or trade taxes has been the traditional method by which governments have influenced international trade. The revenue produced from import or export tariffs have often been important sources of government funding. Tariffs have also been used to protect domestic industry and local enterprise by artificially raising the prices of competing producers outside the country imposing the tariff.

It is generally agreed that the imposition of tariffs often causes reprisals in the affected nations, and most industrialized nations have engaged in extensive discussions in recent years directed at lowering rather than raising tariffs and other barriers. The General Agreement on Tariffs and Trade (GATT) is the vehicle most often used to negotiate reductions in trade barriers. It is discussed specifically in a later part of this chapter.

Tariffs can be levied by an importing nation or by an exporting nation. While an import tariff is normally justified on the basis of protecting domestic industry, an export tariff or duty is most normally a revenue measure imposed under the assumption that domestic industry will not be negatively affected in a significant way. At times, duties are imposed to discourage exports of strategic or critical materials. With governments of developing nations taking a more active role in the ownership and control of mining in their countries, the export tariff is becoming less important since the production royalty of license arrangement generally provides revenue directly to the government. The export tariff was most commonly used when mining and other extractive activities were owned and operated by international or multinational corporations.

Quotas: As tariffs have been reduced, the imposition of quantitative restrictions on imports or exports have become a more significant factor in restricting trade. Quotas impose absolute limitations on the amount of a commodity that can be imported or exported. Import quotas are more common than export quotas, and are justified normally to protect domestic production and/or to adjust a balance of payments problem.

Import quotas determine absolute limits in terms of value or quantity on imports of a specific commodity, product, or product group during a specified period of time, usually one year. The quota is generally administered by the issuance of an import license to an individual or corporation.

Export regulations are used to:

(1) Prevent military or strategic products from reaching the hands of unfriendly or potentially dangerous foreign powers.

(2) Assure that all or a certain portion of the total output of an industry is available for domestic consumption.

(3) Permit the regulation of surpluses on a national or international basis to achieve production and/or stability.

Export controls have featured many techniques to improve the market and the price of raw materials including export trade associations, international cartels, and commodity agreements.

In general, quotas are more restrictive than tariffs. Although most industrialized nations have reduced or eliminated most import quotas on manufactured goods, they are still used extensively to protect domestic agriculture and selectively to protect domestic mineral production. Developing nations have a greater propensity to use quotas as part of their international economic policy.

Subsidies: Liberal tax treatment and low cost capital loans are two methods of rendering indirect financial support to mining operations by governments. Financing of research and development, assistance in exploration, construction of transportation facilities, barter arrangements, favorable monetary exchange controls, government export corporations, and other devices are used in varying degrees to encourage the development of domestic mineral deposits and encourage mineral exports. All of these methods affect existing producers and sources by increasing world supply regardless of international demand.

Multinational Corporation

The multinational corporation produces goods and/or conducts its business with operations in two or more independent nations. While there have been corporations for many years doing business in more than one country, the principal reasons today revolve around the advantages to be gained by shifting capital and resources from one country to another to avoid taxes and increase profits. Trade alliances such as the European Common Market give special advantages to corporations within certain nations on matters such as tariffs, quotas, and monetary exchange rates. Corporations which are truly multinational in structure can move their principal offices among the nations in which they operate or locate in a country where there are special tax advantages and/or corporate income is not taxed significantly. Organizations classified as multinational account for more than one fifth of world output and an even higher proportion of international trade.

Multinational corporations tend to be vertically integrated with each manufacturing division or plant producing intermediate goods for resale as well as finished or final goods for consumption or investment. Components are often produced by subsidiaries located in several different countries while assembly into finished products may be carried on in other countries. As components move through the production process, and are transferred from one plant to another, they become part of international trade even though their ownership may not change. The growth of multinational corporations has led to an increasing proportion of international trade as intra-firm transfers. For example, ALCOA might produce bauxite in Jamaica or another country, convert it to alumina in a second country, and ship the alumina to a third country for refining into aluminum metal.

The shifting of components between subsidiaries is an effort to maximize profits for the total corporation rather than seeking maximum profit from each subsidiary. Transfer prices between subsidiaries may vary significantly from world prices for the commodity or product and these transfer prices are normally designed to minimize duties and corporate income taxes. Therefore, if the tax rates differ substantially between nations in which the corporation has its subsidiaries, the corporation will shift its profits to the country with the lowest tax rate. Or if one nation has an export tax on commodities based on value, the commodities will be transfered at cost between subsidiaries. This leads to understatement of values for some trade flows and potential overstatement of values on other exchanges.

Until the 1960s, multinational corporations controlled most of the raw materials markets and were the principal sources of mineral commodities in the majority of the world's nations. This led to less than competitive conditions in most commodity markets based on the high degree of concentration in the mineral industries.

The control of international commodity markets has been changing hands over the past 20 years as more developing nations have nationalized the holdings of multinational corporations and sought to gain a larger share of the commodity value. However, the multinational corporations still control much of the commodity buying in the industrial nations, and are still an important factor in international trade.

Ownership of production facilities from multinational corporations to government owned corporations has shifted significantly in oil, copper, iron ore, and bauxite. This has led to less stability for international commodity prices as the nations have formed cartels and/or commodity associations that have a mixed record of success in controlling markets shares and prices. The most successful economic relationships are those in which the governments share control with the multinationals. This type of arrangement is likely to increase in importance for the foreseeable future.

Monetary Exchange

Relative values of national currencies and restrictions on the circulation of national currencies can have an important impact on international trade. Inflation may significantly alter the purchasing power and relative value of a traditional "hard" or widely accepted standard currency such as the American dollar. Most petroleum trade is conducted in American dollars due to the size of the American economy, the amount of dollars in circulation, and the historical stability of the purchasing power of the dollar as compared with most other "hard" currencies. In fact, it would be virtually impossible to use another national currency due to the size of the world oil trade and limited amount of other currencies available. In fact, the supply of any other currency would likely be dominated by the petroleum market. The U.S. dollar is affected, but not dominated, by the size of the international petroleum market in comparison to the size of the U.S. economy and amount of dollars in circulation.

When engaging in trade, many countries do not have a sufficient supply of western (hard) currencies to finance the purchase of their needs. This has caused a dramatic upswing in countertrade, barter, and other arrangements that avoid monetary transfer and exchange.

The problem is more common between nations with strict controls on monetary flows. Domestic economic policies and activities such as mining are often changed in order to generate foreign exchange or to limit the need for foreign exchange. Countries such as Romania and the Soviet Union often undertake comparatively expensive mining ventures to cut down on international purchases of commodities or to sell commodities on the world market to generate foreign exchange balances for the purchase of other needed products.

Rates of exchange and foreign debt service often affect the policies of nations toward mining and mineral commodity production. Such considerations historically have affected the ability of developing nations to follow international commodity agreements that allocate market shares and attempt to control production in order to stabilize commodity prices.

Commodity Markets

Commodity markets dealing in contracts for future delivery give some stability to market prices for many mineral commodities in international trade. In addition, they provide an opportunity for producers and consumers of mineral commodities to hedge against future price changes. While the London Metal Exchange is perhaps the best known and one of the oldest mineral commodity exchanges, the Commodity Exchange (COMODEX) in New York, New York Mercantile Exchange, and the Chicago Board of Trade also deal in mineral futures. Regular markets are maintained in futures for: crude oil, copper, gold, silver, platinum, palladium, nickel, lead, zinc, aluminum, and tin, among others.

Contracts for current delivery normally follow spot cash prices very closely. Contracts for future delivery reflect the predicted supply and demand conditions for as much as one and a half to two years into the future. A supplier can hedge against future price declines by selling contracts for future delivery, clearing the contract at the time of delivery and selling his commodity at the cash price at the time of delivery. If the expected price falls, the supplier makes money on the futures contract to offset the lower actual price. If the price rises, the seller loses money on the futures contract, but offsets this loss with the unexpected rise in the case price.

A consumer of the commodity hedges against price increases by purchasing a contract for future delivery at a known price. If the price rises, the consumer makes money on the future contract by settling the contract at the higher price to offset the higher purchase price on the spot or cash market. If the price falls, the consumer loses money on the futures contract but buys the commodity for use at the lower cash price.

In either case, both producers and consumers of mineral commodities may reduce uncertainty about future prices by hedging on the commodity markets. While some individuals argue that trading in commodity futures may introduce additional speculation and instability to international market prices, most economists feel that the reduction in uncertainty possible through hedging has the impact of stabilizing rather than destabilizing market prices.

INTERNATIONAL TRANSFER OF TECHNOLOGY

The necessity to transfer technology from the developed nations to the developing nations is of vital importance to economic growth of the international economy, and as such impacts on future international trade. Equally important is the ability of developed nations to engage in technology exchange. Transfer of technology increases the ability of nations to produce all kinds of products, including mineral commodities.

Some barriers to technology transfer do arise as the result of politics or national security. We have seen examples of this in recent years involving technology and equipment for the Soviet Natural Gas Pipeline from Siberia to Europe and concerning well drilling technology for the development of Siberian oil and gas. Other examples include computer technology for industrial process control when the computers could also be used for military applications.

Most observers agree that the United States, Japan, and other developed nations are relatively open in terms of technology transfer. While most governments require export licensing to permit the sale or transfer of technology, this is done mostly to assure that such transfer does not harm the national security of the exporting nation. Normally these export licenses do not significantly impede the international transfer of technology.

Some governments require import licenses before foreign exchange is made available to purchase technology from other nations. This is often done to assure that the purchaser has not neglected possible domestic suppliers of similar products.

Other governments may restrict capital flows for international investment. This is done to encourage domestic investment, but it can have a major impact on technology transfer and international trade. If companies are prevented from investing in foreign resource development, this naturally affects the production and trade of those resources.

Most business favors a free and unhindered transfer of capital and technology. The growth of the multinational corporation has been a significant factor in increased technology and international trade.

Future prospects for continued international transfer of technology are very bright. The ability to produce and productivity of existing economic activities are enhanced by new technology. This naturally leads to increasing trade between nations in all product categories.

Other Organizations with Potential Impact on Mineral Trade

Various trade alliances and United Nations programs that affect the international trade in minerals were discussed in an earlier section of this chapter. There are other organizations that have an impact on international trade and economic development that have not been described. This final section is an effort to present brief descriptions of these agencies and their operation.

GENERAL AGREEMENT ON TARIFFS AND TRADE (GATT)

GATT is an international agreement that sets and regulates the code of conduct for international trade and encourages consultation to avoid contracting problems between nations engaged in trade. It does not deal with difficulties concerning international balance-of-payments problems.

Special provisions are made within its framework for the unique needs of developing nations. Membership in GATT includes all of the industrialized nations, most of the countries in Eastern Europe that belong to COMECON, and at least 50 developing nations. Together, the members of GATT conduct 80% of world trade. The member nations conduct international conferences from time to time to negotiate the lowering of trade barriers and to facilitate international trade in all products.

The Tokyo Round of Trade Agreements, sponsored by GATT, was concluded in April, 1979, after five years of relatively continuous negotiations. The resulting agreement contained provisions for liberalizing nontariff barriers to trade as well as significant tariff reductions to take effect during the period 1980–1988. This was the most recent in a long series of agreements reached by GATT with the goal of more open and unrestricted international trade.

INTERNATIONAL MONETARY FUND (IMF)

The IMF is a specialized agency of the United Nations established to facilitate international monetary exchange and to assist nations in the management of their external financial needs. The IMF has 143 member nations including virtually all market economies and Yugoslavia, Romania, the Peoples Republic of China, Cambodia, Laos, and Vietnam.

The IMF has two major functions: (1) to discourage aggressive behavior among nations with respect to monetary exchange and exchange rates, and (2) to assist member nations to manage their balance-of-payments problems. The IMF monitors international monetary flows and provides technical assistance on other financial matters. The IMF also provides lines of credit to its member nations to assist in purchasing imports and selling exports. As such it performs a critical role in encouraging international trade.

The financial resources of the IMF come from contributions by the member nations and are made available to member nations with balance-of-payments problems.

Since 1973, the financial needs of countries, especially the developing nations, have increased substantially under the influence of higher energy prices, worldwide inflation, and a general recession in the international economy. The IMF has responded to this need by increasing its resources through additional member contributions and borrowing from international financial corporations.

Export-Import Bank (EXIMBANK)

Some of the more developed economies have established banking institutions to facilitate trade with other nations by making credit available for export and import transactions. The Eximbank is an independent agency of the executive branch of the U.S. government. It was established in 1934 and rechartered in 1945.

Eximbank is governed by a five-member board of directors appointed by the President with the advice and concurrence of the U.S. Senate. The bank has $1 billion of nonvoting capital stock paid in by the U.S. Department of the Treasury and may borrow an additional $6 billion from the Treasury as needed.

The functions of Eximbank includes:

(1) To facilitate the financing of U.S. exports through a discounted loan program, various guarantees for commercial loans, an insurance program to reduce the risk of payment default, and other types of direct loans and financial support.

(2) To make loans to foreign purchasers of U.S. products.

(3) To make loans to foreign financial institutions that will, in turn, make loans to purchasers of U.S. products.

(4) To finance the initial development costs for large overseas capital investment projects.

(5) To provide insurance against nationalization of U.S.-owned investments in other nations and to protect the financial liability of U.S. business interests in capital equipment that is leased or used in other countries.

Other national Eximbanks provide similar programs to assist in international business transactions and protect the foreign economic interests of their citizens.

A closely related financial institution that works with the Eximbank is the Private Export Funding Corporation (PEFCO). This organization was established in 1970 by a group of commercial banks and export manufacturers for the purpose of making loans of U.S. funds to foreign purchasers to finance the purchase of U.S. produced goods and services. The impetus for forming PEFCO came from the Banker's Association for Foreign Trade. PEFCO stockholders consist of 54 U.S. commercial banks, one investment banking organization, and seven manufacturing corporations.

Loans made by PEFCO are guaranteed by the Eximbank of the U.S. These loans supplement funds available directly from the Eximbank. PEFCO raises its funds through the sale of foreign repayment obligations that it guarantees, and through the sale of secured notes in U.S. securities markets.

International Bank for Reconstruction and Development (World Bank)

This international financial organization, now called the World Bank, was founded in 1944 to extend loans for reconstruction and development to nations that had been devastated by World War II. Over the years since its founding, the World Bank has developed into the main agency for extending development loans from money raised in the international capital markets. As such, it is largely supported by the world's market economies and available funds flow largely from the developed nations. The World Bank also has a subsidiary known as the International Development Association (IDA) that makes loans primarily to developing nations that do not meet the qualifications for regular loans from the World Bank.

Capital projects are financed for longer periods of time (up to 35 years) and interest rates are determined by market rates of interest. All loans must be guaranteed by the governments of the borrowing nations and must be repaid in hard convertible currencies. Most World Bank loans are used to finance social overhead capital such as transportation systems, electric power development, agriculture, and/or education. While these projects do not directly underwrite foreign trade, their existence makes it possible for the receiving nation to cut the development and production costs of many exports by providing transportation, electric power, and an educated work force.

The IDA was formed to extend credit to high risk projects that would not qualify for regular World Bank financing. IDA loans may be made for up to 50 years duration, repayment in stages after a 10 year grace period with no interest over a relatively modest service charge. IDA does not have its own funds, and depends on World Bank member countries for its funds.

Another related organization, the International Financial Corporation (IFC) does not make loans directly to governments, but instead participates in industrial projects with private investors either as a lending agency, guarantor, or investor in equity. It brings together private investors from the industrial and developing nations. The IFC obtains investment funds from member-country subscriptions to its private stock, from its ability to borrow as a independent corporation, and the sale of equity ownership in its previous investment.

U.S. Agency for International Development (AID)

The Agency for International Development was created in 1961 by the United States Government to establish and administer economic and technical assistance to developing countries. This formalized an existing policy that had created the Development Loan Fund (in 1957) to finance economic assistance projects in Third World nations.

By 1971, the U.S Government-owned Overseas Private Investment Corporation (OPIC) was formed to take over AID's responsibility for investment insurance and loan guarantee program. OPIC is charged with conducting its operations on a self-sustaining basis in support of the foreign investments of U.S. corporations.

The charter of OPIC was amended in 1978 to give preferential treatment to projects in countries with a per capita income of less than $520 (1975 U.S. dollars) per year, and to restrict assistance to projects in countries with a per capita income exceeding $1,000 per year. Another restriction provides an exclusion or prohibition on projects involving the exploration, mining or extraction of copper. No similar restriction exists for other mineral projects.

OPIC operates a program of insuring U.S. private investment in less developed nations and a program of project financing. Project financing consists of technical assistance for U.S. financial institutions and business firms wishing to invest in capital investment projects in developing market economies.

AID sponsors research and other assistance aimed at developing the economic resources of Third World nations. In 1981, the Reagan Administration launched a new foreign aid strategy aimed at strengthening the role of private enterprise in economic development. To support this strategy, a Bureau for Private Enterprise was created within AID to involve individual and corporate investors in the economic development process. This bureau will assist the governments of developing nations to remove barriers to indigenous and foreign private investment. It will support local financial institutions in their efforts to encourage private enterprise.

General Observations

Mineral commodities have been an important part of international trade for most of recorded history. From the salt and precious metals of ancient times to the massive movements of petroleum in the present and foreseeable future, minerals have moved between producing nations and consuming nations.

While petroleum is destined to dominate mineral trade for the next several decades, other minerals are likely to take a shrinking share of total world trade as mineral exporters attempt to capture a larger share of value added in manufacturing by processing minerals into intermediate and finished products.

The process of shifting from mineral exports to products made from minerals will be moderated by the inability to obtain the capital needed for increased industrialization in the developing world. Accelerating capital needs in the developed economies and the investment climate in the developing economies will limit the international flow of investment capital. Despite government-to-government financial assistance, private capital will still be the major source of investment funds in the developing nations. Restrictions on the ability of private investors to recover their investment and secure a reasonable return on their investment will continue to limit private capital investment by individuals and corporations. The more stable a nation's economy and government, the lower the risk to foreign investment, and the more likely it will realize sustained growth in its economy and development of its natural resources.

Emphasis on environmental quality will continue to have an impact on foreign trade in minerals. The more smelters closed by air pollution controls in the developed nations, the more likely it will be to increase international trade in refined metals. Also, developing nations will be able to take advantage of labor cost differentials in the shifting geography of production in many products including those manufactured from minerals.

Prices for mineral commodities are likely to continue to fluctuate depending on the forces of supply and demand in world markets. Efforts to control output and prices will be more successful in periods of growing demand than in periods of economic slowdown. Multinational corporations are likely to dominate the international markets for most mineral commodities and mineral exporters will continue to find a need to develop their markets in the industrialized nations.

There are potential new sources of minerals in the ocean basins and large undeveloped land areas such as the Amazon Basin, Central Africa, Interior Australia, Arctic Canada, and Siberia. The continental shelves are still a potential source of new oil and gas discoveries. The impact of these potential sources will have a short-term destablizing effect on markets for the minerals to be discovered, but in the long-run they cannot help but increase international trade in mineral commodities.

Changing emphasis in the developing world from manufactured products to services will slow and perhaps reverse the trend of ever-increasing per capita consumption of minerals in those nations. But the growing demands of the developing world assure a continuing trend of increasing trade in minerals and products made from minerals for future generations.

Bibliography

1. Adams, F. Gerard and Jere R. Behrman, 1982, *Commodity Exports and Economic Development*, Lexington Books, D.C. Heath, Lexington, Mass., 328 pp.
2. Adams, F. Gerard, Jere R. Behrman, and Manuel Lasage, 1980, "Commodity Exports and NIEO Proposals for Buffer Stocks and Compensatory Finance: Implications for Latin America" in *Export Diversification and the New Protectionism: The Experiences of Latin America*, edited by Werner Baer and Malcolm Gillis, University of Illinois, Urbana, pp. 48–76.
3. Adams, F. Gerard, et al., 1981 "Commodity Exports and NIEO Proposals for Buffer Stocks and Compensatory Finance: Implications for Latin America," *Quarterly Review of Economics and Business*, Vol. 21, No. 2, Summer, pp. 48–82.
4. Adams, F. Gerard and Jere R. Behrman, editors, 1981, "The Linkages Effects of Raw Material Processing in Economic Development: A Survey of Modeling and Other Approaches," *Journal of Policy Modeling*, Vol. 3., pp. 375–397.
5. Adams, F. Gerard and Sonia Klein, editors, 1978, *Stabilizing World Commodity Markets: Analysis, Practice and Policy*, Lexington Books, Lexington, Mass.
6. Agosin, Manuel, 1976, *Preliminary Econometric Models of Sisal, Iron Ore, Rubber, and Copper*, UNCTAD, New York.
7. Assis, C., 1977, *A Mixed Integer Programming Model for the Brazilian Cement Industry*, Ph.D. Thesis, Johns Hopkins University, Baltimore.
8. Banerji, Ranadev, 1977, *The Development Impact of Barter in Developing Countries: The Case of India*, Organization for Economic Cooperation and Development, Paris, 205 pp.
9. Banks, F., 1974, *The World Copper Market: An Economic Analysis*, Ballinger, Cambridge, Mass., 149 pp.
10. Barnet, Richard J. and Ronald E. Muller, 1974, *Global Reach: The Power of the Multinational Corporation*, Simon & Schuster, New York.
11. Barnovin, J.P., 1982, "Trade and Economic Cooperation Among Developing Countries," *Finance and Development*, Vol. 19, No. 6, June, pp. 24–27.
12. Bergsten, C. Fred, Thomas Horst, and Theodore Moran, 1978, *American Multinationals and American Interest*, Brookings Institution, Washington, D.C.
13. Bergsten, C. Fred, 1979, *Managing International Economic Interdependence: Selected Papers of C. Fred Bergsten, 1975–1979*, Lexington Books, D.C. Heath, Lexington, Mass., 317 pp.
14. Bergsten, C. Fred, 1977, "A New OPEC in Bauxite," in *Managing International Economic Interdependence: Selected Papers of C. Fred Bergsten, 1975–1976*, Lexington Books, D.C. Heath, Lexington, Mass., pp. 193–205.
15. Bergsten, C. Fred, 1977, "Reforming the GATT: The Use of Trade Measures for Balance-of-Payment Purposes," *Journal of International Economics*, Vol. 7, No. 1, Jan., pp. 1–18.
16. Blackhurst, Richard, Nicolas Marian, and Jan Tumlir, 1978, *Adjustment, Trade and Growth in Developed and Developing Countries*, Study No. 6, Geneva, GATT Studies in International Trade.
17. Blackhurst, Richard, Nicolas Marian, and Jan Tumlir, 1977, *Trade Liberalization, Protectionism and Independence*, Geneva, GATT Studies in International Trade.
18. Bobrow, Davis B. and Robert T. Kudrle, 1976, "Theory, Policy, and Resource Cartels: The Case of OPEC," *Journal of Conflict Resolution*, Vol. 20, No. 1, Mar., pp. 3–56.
19. Bosson, Rex and Benison Varon, 1977, *The Mining Industry and the Developing Countries*, Oxford University Press, Oxford.
20. Bowers, E.W., 1982, "U.S. Engaged in Another Kind of War—Strategic Minerals," *Iron Age*, Vol. 225, No. 8, Apr. 14, pp. 38–39.
21. British Petroleum, 1980, *British Petroleum Statistical Review of the World Oil Industry*, Britannic House, London.
22. Bronfenbrenner, Martin, 1978, "On Dumping Gold," *South African Journal of Economics*, Vol. 46, No. 12, Dec., pp. 352–359.
23. Callanan, N., 1980, "Why the Metal Cartels Fail to Match OPEC," *Chemical Business*, Nov. 17, pp. 45–46.
24. Callot, Francois G., 1981, "World Mineral Production and Consumption in 1978," *Resources Policy*, Vol. 7, No. 1, Mar., pp. 14–38.
25. Campbell, Colin, 1982, "Who's Trying to Corner Tin?," *New York Times*, Feb. 9, p. A1.
26. Carter, Luther J., 1980, "Phosphate—Debate Over an Essential Resource," *Science*, Vol. 209, No. 4454, July 18, pp. 372–375.
27. Cavendar, D., 1976, "The U.S. and the World Zinc Industry: An Econometric Study of Consumption," *Working Paper*, Iowa State University, Annual Meeting

of the American Economic Association, Atlantic City, N.J.

28. Caves, Richard E., 1979, "International Cartels and Monopolies in International Trade," in *International Economic Policy*, Rudiger Dornbusch and Jacob A. Frenkel, editors, The Johns Hopkins University Press, Baltimore, pp. 39–72.
29. Caves, Richard and Ronald W. Jones, 1977, *World Trade and Payments*, 2nd edition, Little Brown, Boston, 470 pp.
30. Central Intelligence Agency, National Foreign Assessment Center, 1979, *The World Oil Market in the Years Ahead*, NTIS, Aug., Washington, D.C.
31. Chadwick, J.R., et al., 1981, "International Symposium on Small Mine Economics," *World Mining*, Vol. 34, No. 8, Aug., p. 47–51.
32. Chadwick, J. R., 1982, "Malaysia: Tin Market Confused But Expansions Proceed," *World Mining*, Vol. 35, No. 2, Feb., p. 7.
33. Charles River Associates, 1976, *Policy Implications of Producer Country Supply Restrictions: The World Market for Copper*, U.S. Department of Commerce, Washington, D.C.
34. Cline, William R., 1979, *Policy Alternatives for a New International Economic Order*, Praeger Publishers, New York, 392 pp.
35. Corea, Gamani, 1977, "UNCTAD and the New International Economic Order," *International Affairs*, Vol. 53, No. 2, pp. 177–187.
36. Danielsen, Albert L., 1976, "Cartel Rivalry and the World Price of Oil," *Southern Economic Journal*, Vol. 42, No. 3, Jan., pp. 407–415.
37. Danielsen, Albert L., 1982, *The Evolution of OPEC*, Harcourt Brace, Atlanta, 305 pp.
38. Dennis, R.D., 1982, "Countertrade-Factor in China's Modernization Plan," *Columbia Journal of World Business*, Vol. 17, No. 1 Spring, pp. 67–75.
39. Diaz-Alejandro, Carlos, 1976, "International Markets for Exhaustible Resources: Less Developed Countries and Transnational Corporations," *Discussion Paper No. 256*, Yale University, Economic Growth Center, New Haven, Dec.
40. Dorr, Andre, 1975, *International Trade in the Primary Aluminum Industry*, Ph.D. Thesis, Pennsylvania State University.
41. Eckbo, Paul Leo, 1975, "OPEC and the Experience of Previous International Commodity Cartels," *MIT Energy Laboratory Working Paper No. 75-008WP*, MIT, Cambridge, Mass.
42. Edwards, D., 1974, "Foreign Gold Traders: Bullish on U.S. Entry," *Commercial and Financial Chronicle*, Vol. 219, Dec. 30, p. 2.
43. Eiteman, David and Arthur Stonehill, 1982, *Multinational Business Finance*, Addison-Wesley, Reading, Mass., 721 pp.
44. El-Agraa, Ali M., editor, 1982, *International Economic Intergration*, St.Martin's , New York, 287 pp.
45. Ellsworth, P.T. and J.C. Leith, 1975, *The International Economy*, 5th edition, Macmillan, New York.
46. Ember, Lois R., 1981, "Many Forces Shaping Strategic Mineral Policy," *Chemical and Engineering News*, Vol. 59, No. 19, May 11, pp. 20–25.
47. Ethier, Wilfred, 1983, *Modern Internation Economics*, W.W. Norton, New York, pp. 446–447.
48. Eulenstein, Karl H., 1979, *United States Dependence on Imports on Nonfuel Minerals*, U.S. Navy Research and Analysis Office, Washington, D.C., June 21, 88 pp.
49. Fine, D.I., 1980, "Fresh Fears that the Soviets Will Cut Off Critical Minerals,"*Business Week*, No. 2621, Jan. 28, pp. 62, 64.
50. Fine, D. I. 1979, "Growing Anxiety Over Cobalt Supplies," *Business Week*, No. 2581, Apr. 16, pp. 51, 54.
51. Fine, D.I., 1982, "Russia is Using Its Nickle to Raise Cash," *Business Week*, No. 2749, July 26, pp. 38–39.
52. Fink, E., 1980, "Availability of Strategic Materials Debated," *Aviation Week and Space Technology*, Vol. 112, No. 18, May 5, pp. 42–46.
53. Fischman, Leonard L., et al., 1979, *Major Mineral Supply Problems: A Study Prepared by Resources for the Future for the Nonfuel Minerals Policy Review*, Resources for the Future, Washington, D.C., 1108 pp.
54. Force, E.R., 1980, *Is the United States Geologically Dependent on Imported Rutile*?, 4th Industrial Minerals International Congress, Atlanta, 4 pp.
55. Fox, W., 1974, *The Working of a Commodity Agreement*, Mining Journal Books Ltd., London, 418 pp.
56. Fried, E.R., 1976,"International Trade in Raw Materials: Myths and Realities," *Science*, Vol. 191, No. 4228, Feb. 20, pp. 641–646.
57. Garvey, Gerald and Lou Ann Garvey, 1977, *International Resource Flows*, Lexington Books, D.C. Heath, Lexington, Mass., 1979 pp.
58. Gilpin, Robert, 1975, *U.S. Power and the Multinational Corporation*, Basic Books, New York.
59. Gillis, M. and C. McLure, Jr., 1975, "The Incidence of World Taxes on Natural Resources, With Special Reference to Bauxite,"*American Economic Review*, Vol. 65, No. 2, May.
60. Gluschke, W. and J. Shaw, 1977, *Copper: The Next Fifteen Years, A United Nations Study*, D. Reidel, Boston, 177 pp.
61. Goreux, L.M., 1977, "Compensatory Financing: The Cyclical Pattern of Export Shortfalls," *IMF Staff Papers*, No. 24, No. 3, pp. 613–641.
62. Gossling, H.H., 1976, *An Update Summary of the World's Fluorspar Industry, 1975*, Report No. 1814, National Institute for Metallurgy, Johannesburg, 22 pp.
63. Gray, E., 1982, "Shotgun Wedding: How the Canadian Government Engineered the Uranium Cartel," *Canadian Business*, Vol. 55, No. 3, Mar., pp. 96–101.
64. Grichar, James S., et al., 1981, *The Nonfuel Mineral Outlook for the USSR Through 1990*, USBM, Washington, D.C., Dec., 17 pp.
65. Grossman, Gene N., 1982, "Import Competition From Developed and Developing Countries," *The Review of Economics and Statistics*, Vol. 64, No. 2, May, pp. 271–281.
66. Gutierrez, A., 1973, *Chilean Copper: Government Policies and Multinational Corporations, 1955–1973*, Unpublished L.L.D. Thesis, Oxford University.
67. Hall, K. and B. Blake, 1976, "Major Developments in CARICOM 1975," in *The Caribbean Yearbook of International Relations*, Leyden, Sijthoff, pp. 51–54.
68. Hall, Robert B., 1978, *World Nonbauxite Aluminum Resources—Alunite*, USGS Professional Paper 1076-A, Washington, D.C., 40 pp.
69. Handy and Harman, 1979, *The Silver Market*, 64th

Annual Review, 26 pp.

70. Hartshorn, J.E., 1982, "Two Crises Compared: OPEC Pricing in 1973–1975 and 1978–1980," in *OPEC: Twenty Years and Beyond*, edited by Ragaei El Mallakh, Westview Press, Boulder, Colo., pp. 17–22.
71. Harvey, R. E., 1979, "Critical Cobalt Shortages Demand New Strategies," *Iron Age*, Vol. 222, Jan. 15, pp. 75.
72. Heckman, J.S., 1978, "Analysis of the Changing Location of Iron and Steel Production in the Twentieth Century," *American Economic Review*, Vol. 68, No. 1, Mar., pp. 123–133.
73. Hnylicza, E., and R.S. Pindyck, 1976, "Pricing Policies for a Two-Part Exhaustible Resource Cartel: The Case of OPEC," *European Economic Review*, Vol. 8, No. 2, Aug., pp. 139–154.
74. Hogendorn, Jan S. and Wilson B. Brown, 1979, *The New International Economics*, Addison-Wesley, Reading, Mass., 328 pp.
75. Horvitz, Paul M., "The International Monetary Fund," in *Monetary Policy and The Financial System*, 4th Edition, Prentice-Hall, Englewood Cliffs, N.J., pp. 290–294.
76. Hoyt, Eton III, 1980, "The Outlook for Iron Ore," *Mining Congress Journal*, Vol. 32, No. 11, Nov., pp. 1581–1587.
77. Hubbard, David A., 1975, *Nickel In International Trade*, M.S. Thesis, Pennsylvania State University.
78. Hughey, A., 1980, "Growing Boom in Coal Exports," *Forbes*, Vol. 126, No. 4, Aug. 18, pp. 46–47.
79. International Monetary Fund, *Balance of Payments Yearbook*, International Monetary Fund, Washington, D.C., Annually.
80. Jaffe, T., 1981, "Fertilizer After ELF," *Forbes*, Vol. 128, No. 6, Sept. 14, pp. 34–35.
81. Johnson, Charles J., 1976, "Cartels in Mineral and Metal Supply," *Mining Congress Journal*, Vol. 62, No. 1, Jan., p. 30–34.
82. Kessler, Jeffery, 1981, "Cashing in on Strategic Metals," *Financial World*, Vol. 150, No. 6, Mar. 15, pp. 44–48.
83. Kingston, J., 1976, "Export Concentration and Export Performance in Developing Countries, 1954–1967," *Journal of Development Studies*, Vol. 12, No. 4, July, pp. 311–319.
84. Kovisars, L., 1976, *World Production, Consumption, and Trade in Zinc—An LP-Model*, U.S. Bureau of Mines Contract J-0166003, Stanford Research Institute, Palo Alto, California.
85. Krapels, Edward N., 1980, *Oil Crisis Management Strategic Stockpiling for International Security*, Johns Hopkins University Press, Baltimore.
86. Krasner, Stephen, 1977, "The Quest for Stability: Structuring the International Commodities Markets," Chapter 3 in *International Research Flows*, Gerald Garvey and Lou Ann Garvey, editors, Lexington Books, D.C. Heath, Lexington, Mass., pp. 39–58.
87. Kreinin, Mordechai E., 1983, *International Economics*, Harcourt Brace, New York, 432 pp.
88. Kuenne, Robert E., 1978–79, "A Short-Run Demand Analysis of the OPEC Cartel," *Journal of Business Administration*, Vol. 10, Nos. 1–2, pp. 129–164.
89. Labys, W.C., M.I. Nadiri, and J. Nunez del Arco, editors, 1980, *Commodity Markets and Latin American Development: A Modeling Approach*, Ballinger, Cambridge, Mass., 280 pp.
90. Lasaga, Manuel, 1981 *The Copper Industry in the Chilean Economy, An Econometric Analysis*, Lexington Books, D.C. Heath, Lexington, Mass.
91. Law, Alton, D., 1975, *International Commodity Agreements: Setting, Performance, and Prospects*, Lexington Books, Lexington, Mass.
92. Levi, Maurice, 1983, *International Finance*, McGraw Hill, New York, 460 pp.
93. Levy, Walter, 1979, "A Warning to the Oil Importing Nations," *Fortune*, Vol. 99, No. 10, May 21, pp. 48–51.
94. Ley, R., 1977, *Bolivian Tin and Bolivian Development*, Ph.D. Thesis, Washington State University.
95. Lopata, R., 1981, "As Barter Business Booms, Some Firms Miss the Boat," *Iron Age*, Vol. 224, No. 29, Oct. 4, p. 45, 49–50.
96. Losman, Donald L., 1979, *International Economic Sanctions*, University of New Mexico Press, Albuquerque, 156 pp.
97. Major, M., 1981, "Export Coal Prospectus: World Market Trends and Tides," *American Import/Export Bulletin*, Vol. 95, No. 9, Sept., 14–15.
98. Malmgren, Harold B., 1975, "The Raw Material and Commodity Controversy," *Contemporary Issues*, No. 1, Oct.
99. Marer, P., 1980, "Western Multinational Corporations in Eastern Europe and CMEA Integration," in *Partners in East-West Economic Relations*, Z. Fallenbuchl and C. McMillan, editors, Pergamon Press, New York.
100. Martino, Orlando, 1981, *Mineral Industries of Latin America*, USBA, Washington, D.C., Dec., 128 pp.
101. McIntyre, A., 1976, *The Current Situation and Perspectives of CARICOM*, Caribbean Community Secretariat, Georgetown, Guyana.
102. McKern, R.B., 1976, *Multinational Enterprise and Natural Resources*, McGraw-Hill, Sydney, Australia.
103. McNicol, David L., 1977, *Commodity Agreements and Price Stabilization: A Policy Analysis*, Lexington Books, D.C. Heath, Lexington, Mass.
104. Mendershavsen, Horst, 1976, *Coping with the Oil Crisis, French and German Experiences*, Johns Hopkins University Press, Baltimore.
105. Mikdashi, Zuhayr, 1976, "The OPEC Process," in *The Oil Crisis*, edited by Raymond Vernon, W.W. Norton, New York, pp. 203–215.
106. Mikesell, R.F., 1979, *The World Copper Industry*, The Johns Hopkins University Press, Baltimore, 393 pp.
107. Miller, C. George, 1981, "Mineral Demand in the 1980's," *Canadian Mining Journal*, Vol. 102, No. 4, pp. 43–47.
108. Mingst, K., 1976, "Cooperation or Illusion: An Examination of the Intergovernmental Council of Copper Exporting Countries," *International Organization*, Vol. 30, No. 2, Spring, pp. 264–287.
109. Miramon, J. de, 1982, "Countertrade: A Modernized Barter System," *OECD Observer*, No. 114, Jan., pp. 12–15.
110. Moran, Theodore H., 1976, *Multinational Corporations and Politics of Dependence: Copper in Chile*, Princeton University Press, Princeton.
111. NanKani, G.T., 1980, "Development Problems of Nonfuel Mineral Exporting Countries," *Finance and Development*, Vol. 1, No. 3, Mar., pp. 6–10.

112. Niering, F.E. Jr., 1981, "United States: Developing Coal-Export Potential," *Petroleum Economist*, Vol. 48, No. 5, May, pp. 207–208.
113. Noreng, Oystein, 1978, *Oil Politics in the 1980's*, McGraw-Hill, New York, 170 pp.
114. Nziramasanga, M. and C. Obidegwv, 1980, *An Econometric Analysis of the Impact of the World Copper Market on Zambia*, Unpublished report to AID, Wharton Econometric Forecasting Associates, Inc., Philadelphia.
115. Obidegwv, C.F. and M. Nziramasanga, 1981, *Copper and Zambia: An Econometric Analysis*, Lexington Books, D.C. Heath, Lexington, Mass.
116. Osborne, Dale K., 1976, "Cartel Problems," *American Economic Review*, Vol. 66, No. 5, Dec., pp. 835–845.
117. Outters-Jaeger, Ingelies, 1979, *The Development Impact of Barter in Developing Countries*, Development Centre of the Organization for Economic Cooperation and Development, Paris, 130 pp.
118. Palma-Carillo, Pedro, 1976, *A Macro Econometric Model of Venezuela with Oil Price Impact Applications*, Ph.D. Dissertation, University of Pennsylvania.
119. Payer, Cheryl, editor, 1975, *Commodity Trade of the Third World*, Macmillan, London, 192 pp.
120. Pearce, D.W. and J. Rosee, 1975, *The Economics of Natural Resource Depletion*, John Wiley, New York.
121. Pearson, S.R. and J. Cownie, 1975, *Commodity Exports and African Economic Development*, Heath Lexington Books, Lexington, Mass.
122. Poss, J.R., 1981, "Minerals Open the Door to South America," *World Mining*, Vol. 4, No. 6, June, pp. 116–117.
123. Pindyck, Robert S., 1978, "Gains to Producers from the Cartelization of Exhaustible Resources," *The Review of Economics and Statistics*, Vol. 45, No. 2, May, pp. 238–251.
124. Pinto, M.C.W., 1980, "Developing Countries and the Exploitation of the Deep Seabed," *Columbia Journal of World Business*, Vol. 15, No. 4, Winter, pp. 30–41.
125. Radetzki, Marian, 1977, "Where Should Developing Countries' Minerals Be Processed? The Country View Versus the Multinational Company View," *World Development*, Vol. 5, No. 4, Apr., pp. 325–334.
126. Ramsaran, R., 1978, "CARICOM: The Integration Process in Crisis," *Journal of World Trade Law*, Vol. 12, No. 3, May-June, pp. 208–217.
127. Reddy, N.N., 1976, "Japanese Demand for U.S. Coal: A Market Share Model," *Quarterly Review of Economics and Business*, Vol. 16, No. 3, Autumn, pp. 51–60.
128. Regan, R.J., 1978, "Caribbean Bauxite Loses Ground in U.S. Markets, *Iron Age*, Vol. 221, Mar. 6, pp. 30–31.
129. Regan, R. J., 1982, "Caribbean Bauxite Nations Foiled by Aluminum's Slump," *Iron Age*, Vol. 225, No. 9, Mar. 19, pp. 44–45.
130. Regan, R.J., 1979, "Does Zinc Have Sinking Feelings About 1979?," *Iron Age*, Vol. 222, Jan. 22, pp. 27–29.
131. Reid, Stan, 1981, "The Politics of Resource Negotiation: The Transnational Corporation and the Jamaican Bauxite Levy," *National Resources Forum*, Vol. 5, No. 2, April, pp. 115–126.
132. Rippon, S., 1978, "International Symposium on Uranium Supply and Demand," *Energy International*, Vol. 15, No. 9, Sept., pp. 21–22.
133. Risch, Benno W.K., 1978, "The Raw Material Supply of the European Community: The Importance of Secondary Raw Materials," *Resources Policy*, Vol. 4, No. 3, Sept., pp. 181–188.
134. Robok, Stefan, Kenneth Simmonds, and Jack Zwick, 1975, *International Business and Multinational Enterprises*, Richard D. Irwin, Inc., Homewood, Ill., 739 pp.
135. Robinson, Ian M., 1975, "Zinc", in *Commodity Trade of the Third World*, Cheryl Payer, editor, Macmillan, London, pp. 58–73.
136. Roessler, F., 1975, "Access to Supplies: The Role GATT Could Play," *Journal of World Trade Law*, Vol. 9, No. 1, Jan.-Feb., pp. 25–40.
137. Root, Franklin, 1984, *International Trade and Investment*, South-Western Publishing, West Chicago, 543 pp.
138. Rose, S., 1976, "Third World Commodity Power is a Costly Illusion," *Fortune*, Vol. 94, No. 5, Nov., p. 146–150.
139. Rosen, G.R., 1976, "Minerals: Horn of Plenty," *Dunn's Review*, Vol. 108, No. 11, Nov., pp. 68–70.
140. Rosson, R. and B. Varon, 1977, *The Mining Industry and Developing Countries*, Oxford University Press, New York.
141. Rothstein, Robert L., 1978, *Global Bargaining: UNCTAD and the Quest for a New International Economic Order*, Princeton University Press, Princeton, N.J.
142. Rovig, A. Dan and Richard K. Doran, 1980, "Copper: A Decade of Change and Its Meaning for the Future," *Mining Congress Journal*, Vol. 66, No. 12, Dec., pp. 27–29.
143. Rudolph, B., 1983, "Cement is Like Bread," *Forbes*, Vol. 129, No. 11, May 24, pp. 94.
144. Salant, Stephen W., 1976, "Exhaustible Resources and Industrial Structure: A Nash-Cournot Approach to the Oil Market," *Journal of Political Economy*, Vol. 84, No. 10, Oct., pp. 1076–1093.
145. Santos, A., 1976, *International Trade in Iron Ore: An Econometric Analysis*, Ph.D. Thesis, Pennsylvania State University.
146. Segal, Jeffrey and Frank E. Niering, Jr., 1980, "Special Report on World Natural Gas Pricing," *Petroleum Economist*, Vol. 47, No. 9, Sept., pp. 374–378.
147. Schuster, F., 1980, "Barter Arrangement With Money: The Modern Form of Compensations Trading," *Columbia Journal of World Business*, Vol. 15, No. 3, Fall, pp. 61–66.
148. Scott, W. E., 1978, "Australia Faces Future As Major Energy Exporter," *Energy International*, Vol. 15, No. 12, Dec., pp. 23–26.
149. Scott, W. E., 1977, "Australia Unlocks Her Uranium Reserves," *Energy International*, Vol. 14, No. 11, Nov., pp. 25–27.
150. Senghass, Dieter, 1975, "Multinational Corporations and the Third World: On the Problem of the Further Integration of Perpheries into the Given Structure of the International Economic System," *Journal of Peace Research*, Vol. 12, No. 4, pp. 275–292.
151. Shekarchi, Eloraham, et al., 1981, *Zimbabwe (Report on Mineral Perspectives)*, USBM, Washington, D.C., 69 pp.
152. Smith, Gordon W. and George R. Schink, 1976, "The

International Tin Agreement: A Reassessment," *Economic Journal*, Vol. 86, No. 344, Dec., pp. 715–728.

153. Smith, Robert E., 1974, "Private Power and National Sovereignty: Some Comments on the Multinational Corporation," *Journal of Economic Issues*, Vol. 8, No. 2, pp. 417–447.
154. Staloff, S., 1977, *A Stock Flow Analysis for Copper Markets*, Ph.D. Thesis, University of Oregon.
155. Stiglitz, J. E., 1975, "The Efficiency of Market Prices in Long Run Allocations in the Oil Industry," in *Studies in Energy Tax Policy*, G. Brannon, editor, Ballinger Publishing Co., Cambridge, Mass., pp. 55–99.
156. Strauss, Simon D., 1983, "Role of Third World Countries in Mineral Supply," *Mining Congress Journal*, Vol. 68, No. 2, Feb., pp. 33–37.
157. Teleki, Deneb, 1980, "Fertilizer Minerals," *Mining Congress Journal*, Vol. 66, No. 12, Dec., p. 22–26.
158. Thoburn, John, 1981, "Policies for Tin Exporters," *Resources Policy*, Vol. 7, No. 2, June, pp. 74–86.
159. Tilton, John E., 1977, *The Future of Nonfuel Minerals*, Brookings Institution, Washington, D.C., 113 pp.
160. Tilton, John E. and Andre Dorr, 1975, "An Econometric Model of Metal Trade Patterns," in *Mineral Materials Modeling: A State-of-the-Art Review*, William A. Vogely, editor, Resource for the Future, Washington, D.C.
161. Truell, Peter, 1983, "Barter Accounts for a Growing Portion of World Trade Despite Its Inefficiency," *The Wall Street Journal*, Aug. 15, p. 21.
162. Van Alstine, Ralph E. and Paul G. Schruben, 1980, *Fluorspar Resources of Africa*, USGS Report 1487, USGS, Washington, D.C., 25 pp.
163. Van Duryn, Cal, 1975, "Commodity Cartels and the Theory of Derived Demand," *KYKLOS*, Vol. 28, No. 3, Fall, pp. 579–612.
164. Van Rensburg, W. C. Jr., 1981, "Global Competition for Strategic Mineral Supplies," *Resources Policy*, Vol. 7, No. 1, Mar., pp. 4–13.
165. Varon, B. and K. Kakeuchi, 1974, "Developing Countries and Non-Fuel Minerals," *Foreign Affairs*, Vol. 52, No. 3, Apr., pp. 497–510.
166. Velocci, T., 1980, "Minerals: The Resource Gap," *Nation's Business*, Vol. 68, No. 10, Oct. pp. 32–34.
167. Vernon, Raymond, 1977, *Storm Over the Multinationals*, Harvard University Press, Cambridge, Mass.
168. Wargo, Joseph G., 1979, "Cobalt—A Supply Crisis," *Mining Congress Journal*, Vol. 65, No. 4, Apr., pp. 38–40.
169. Warner, Dennis and Mordechai E. Kreinin, 1983, "Determinants of International Trade," *The Review of Economics and Statistics*, Vol. 65, No. 1, Feb., pp. 96–104.
170. Weigand, R. E., 1980, "Barters and Buy-Backs: Let Western Firms Beware!", *Business Horizons*, Vol. 23, No. 6, June, pp. 54–61.
171. Weimer, G. A., 1980, "Iron Ore Falls Prey to Tough Steel Times," *Iron Age*, Vol. 223, Aug. 4, p. MP3.
172. Welt, Leo G. B., 1982, *Countertrade, Business Practices for Today's World Market*, American Management Association, New York, 72 pp.
173. White, D. F., 1981, "Coal-Export Gamble," *Fortune*, Vol. 104, No. 12, Dec. 14, pp. 122–125.
174. Wickham, Gary A., 1980, "Lead/Zinc," *Mining Congress Journal*, Vol. 66, No. 12, Dec., pp. 16–19.
175. Williams, Edward J., 1979, *The Rebirth of the Mexican Petroleum Industry*, Lexington Books, Lexington, Mass.
176. Wilson, Carrol L., 1980, *Coal, Bridge to the Future*, Ballinger Publishing Co., Cambridge, Mass., p. 112.
177. World Bank, *Commodity Trade and Price Trends*, World Bank, Washington, D.C. published annually.
178. Wyant, Frank R., 1977, *The United States, OPEC, and Multinational Oil*, Lexington Books, Lexington, Mass.
179. Zantop, Half, 1981, "Argentina's Potential Copper Potential," *Mining Engineering*, Vol. 33, No. 2, Feb., pp. 137–142.

THE FOLLOWING ENTRIES DID NOT INDICATE AUTHORS

180. "Africa's Mineral Region Offers Opportunity for U.S. Suppliers," *Business America*, Vol. 4, No. 1, Jan. 12, 1981, pp. 11–12.
181. "Algeria: When Barter is Battery," *Economist*, Vol. 281, No. 2205, Oct. 3, 1981, pp. 84, 86.
182. American Iron and Steel Institute, *Economics of International Steel Trade: Policy Implications for the United States*, Putnam, Hays and Bartlett, Inc., Newton, Mass., 1977, 66 pp.
183. *Analysis of Recent Trends in U.S. Countertrade*, U.S. International Trade Commission, Washington D.C., 1982.
184. "Aussie Aluminum Chilled by the Recession's Pall," *Chemical Week*, Vol. 35, No. 131, Sept. 15, 1982, pp. 37–38.
185. "Ban on Chrome Imposed by U.S.: War in Zaire Upset Copper, Cobalt," *Iron Age*, Vol. 219, Mar. 28, 1977, p. 11.
186. "Bankrolling World Resources," *Business Week*, No. 2434, Apr. 26, 1976, pp. 32–33.
187. "Bauxite, Alumia Imports Up in First Half," *Iron Age*, Vol. 233, Oct. 15, 1980, p. 94.
188. "Billion Dollar Play Roils the Tin Market," *Business Week*, No. 2725, Feb. 8 1982, pp. 33, 37.
189. "BP Drive for a Share of US Coal Exports," *Business Week*, No. 2706, Sept. 21, 1981, p. 35.
190. "Burden of Oil Imports on Developing Countries," *Petroleum Economist*, Vol. 49, No. 9, Sept. 1982, p. 366.
191. "Canadian Potash Takeover Leaves IMC Vulnerable," *Commercial and Financial Chronical*, Vol. 220, Dec. 1975.
192. "Caribbean Bauxite, Downstream in Trouble," *Economist*, Vol. 256, No. 6880, July 5, 1975, pp. 123–124.
193. "Cashing in on the Foreign Aid Dollar," *Chemical Business*, May 31, 1982, pp. 49–50.
194. "Cement is Set in a Profitless Recovery," *Business Week*, No. 2795, June 20, 1983, p. 121.
195. "China Opens Aladden's Cave of Rare Strategic Metals," *Economist*, Vol. 279, No. 7183, May 2, 1981, pp. 73–74.
196. *Chinese Coal Industry: Prospects Over the Next Decade*, a research paper, National Foreign Assessment Center, Washington, D.C., 1979, 12 pp.
197. "Coal Boom at Last," *Financial World*, Vol. 150, No. 16, Sept. 1, 1981, pp. 16–20.

198. "Coal: The Global Market Room Forced by OPEC," *Business Week*, No. 2599, Aug. 20, 1979, p. 36.
199. "Cobalt: US is Dependent But Not Vulnerable to Import," *Chemical and Engineering News*, Vol. 59, No. 19, May 11, 1981, pp. 24–25.
200. *Cobalt—World Survey of Production, Consumption and Prices*, 2nd edition, Roskill Information Service, London, 1975, 76 pp.
201. "Congress Bars Imports of Rhodesian Chrome," *Chemical and Engineering News*, Vol. 55, No. 12, Mar. 21, 1977, p. 8.
202. "Copper: CIPEC Scratches," *Economist*, Vol. 253, No. 6848, Nov. 23, 1974, p. 94, 97.
203. "Copper: Even Before Zaire the Outlook was Bleak," *Business Week*, No. 2537, June 5, 1978, p. 69.
204. "Copper Prices and the OPEC that Wasn't," *Citibank Monthly Economic Letter*, Oct. 1978, pp. 11–15.
205. *Copper Statistics*, Report TD/B/IPC/Copper/ACL5, UNCTAD Secretariat, New York, January 28, 1977.
206. "Copper Strike Helped Importers to US," *Iron Age*, Vol. 224, Mar. 2, 1981, pp. 95–96.
207. "Copper: Together at Last," *Economist*, Vol. 258, No. 6917, Mar. 20, 1976, p. 94.
208. *Countertrade Practices in East-West Economic Relations*, Organization for Economic Cooperation and Development, Paris, 1979, 33 pp.
209. *East-West Trade: Recent Developments in Countertrade*, Organization for Economic Cooperation and Development, Paris, 1981, 78 pp.
210. "Economic Developments in the EFTA Countries," *EFTA Bulletin*, Vol. 23, July/Sept. 1983, p. 17.
211. *Economic Report of the President*, U.S. Government Printing Office, Washington D.C., Yearly.
212. "Ending an Export Ban that Hurt Only the US," *Business Week*, No. 2730, Mar. 15, 1982, pp. 39 & 43.
213. "English China Clays: Driving Itself Potty," *Economist*, Vol. 283, No. 7243, June 26, 1982, pp. 80–81.
214. *The European Free Trade Association*, EFTA Secretariat, Geneva, 1980b.
215. "Exploration Activity Unearths Rhodesian Asbestos, Amazon Iron, African Phosphates, and Scottish Coal," *World Mining*, Vol. 30, June 25, 1977, pp. 106–108.
216. "Export Slump Poses Problems for U.S. Phosphate Industry," *European Chemical News*, Vol. 38, Section 2, Jan. 25, 1982, pp. 12–14.
217. "Fertile Prospects at IMC," *Financial World*, Vol. 150, No. 6, Mar. 15, 1981, pp. 62–63.
218. "Four-Fold Rise in World Coal Trade," *Petroleum Economist*, Vol. 46, No. 3, Mar. 1979, pp. 117.
219. *The Future of Nickel and the Law of the Sea*, Mineral Policy Background Paper No. 10, Mineral Resources Branch, Ontario, 1980, 28 pp.
220. *General Statistics Monthly, Overall Trade by Countries*, Organization for Economic Cooperation and Development (OEDC), Monthly.
221. *Handbook of Economic Statistics, 1981*, Nation Foreign Assessment Center, Washington, D.C., Annually.
222. "IMC Agrees Construction Go-Ahead for First Manitoba Potash Mine," *European Chemical News*, Vol. 36, May 25, 1981, p. 6.
223. "IMC and Ashland Link to Buy Hercofina Methanol Plant," *European Chemical News*, Vol. 32, Mar. 31, 1980, p. 4.
224. "IMC: Fertilizers Fine, Chemicals May be Cut," *Chemical Week*, Vol. 126, No. 4, Jan. 23, 1980, pp. 18.
225. "IMC Group Charts Chemical Course," *Chemical Week*, Vol. 117, No. 38, Sept. 17, 1975.
226. *Industrial Minerals*, Proceedings of the Industrial Minerals International Congress 4th Symposium, Atlanta, Ga., May 28–30, 1980, 238 pp.
227. *International Energy Annual, 1981*, U.S. Department of Energy, Washington, D.C., 1982.
228. *International Investments and Multinational Operations*, OEDC, Paris, 1976.
229. *International Monetary Fund*, Annual Report, IMF, Washington, D.C., Annually.
230. *The Iron and Steel Industry in 1978*, Organization for Economic Cooperation and Development, Paris, 1980, 38 pp.
231. "Iron Mines, Shipping Takes a Nosedive," *Industry Week*, Vol. 205, No. 5, June 9, 1980, pp. 28–29.
232. "Is Malaysia Using Loans to Control Tin?," *Business Week*, No. 2745, June 28, 1982, pp. 43–44.
233. "Jamaica Bauxite Buy: Strategy or Politics?," *Iron Age*, Vol. 224, Dec. 28, 1981, p. 56.
234. "Kind Hearts and Cartels," *Economist*, Vol. 285, Vol. 7263, Nov. 13, 1982, pp. 60–61.
235. *The Lead and Zinc Industry in the USSR*, Report ER80–10072, National Foreign Assessment Center, Washington, D.C., 1980, 19 pp.
236. "Malaysia is Forging a New Cartel in Tin," *Business Week*, No. 2730, Mar. 15, 1982, p. 26.
237. *Mineral Commodity Profiles, Aluminum*, U.S. Bureau of Mines, Washington, D.C., 1983, p. 10.
238. *Mineral Commodity Profiles, Chromium*, U.S. Bureau of Mines, Washington, D.C., 1983, pp. 9–11.
239. *Mineral Commodity Profiles: Copper*, U.S. Bureau of Mines, Washington, D.C., 1983, pp. 3, 10.
240. *Mineral Commodity Profiles: Fluorspar*, U.S. Bureau of Mines, Washington, D.C., 1983, p. 8.
241. *Mineral Commodity Profiles: Iron Ore*, U.S. Bureau of Mines, Washington, D.C., 1983, p. 7.
242. *Mineral Commodity Profiles: Lead*, U.S. Bureau of Mines, Washington, D.C., 1983, p. 8.
243. *Mineral Commodity Profiles: Nickel*, U.S. Bureau of Mines, Washington, D.C., 1983, pp. 7–10.
244. *Mineral Commodity Profiles: Phosphate Rock*, U.S. Bureau of Mines, Washington, D.C., 1983, pp. 7–10.
245. *Mineral Commodity Profiles: Tin*, U.S. Bureau of Mines, Washington, D.C., 1978, pp. 8–9.
246. *Mineral Commodity Profiles: Zinc*, U.S. Bureau of Mines, Washington, D.C., 1983, pp. 9–10.
247. *Mineral Commodity Summaries*, U.S. Bureau of Mines, Washington, D.C., Yearly.
248. *Mineral Facts and Problems*, U.S. Bureau of Mines, Washington, D.C., 1975, 1259 pp.
249. *Mineral Trends and Forecasts*, U.S. Bureau of Mines, Washington, D.C., Annually.
250. *Mineral Yearbook 1981, Volume III*, U.S. Bureau of Mines, Washington, D.C., 1983.
251. *Minerals in the U.S. Economy: Ten Year Supply—Demand Profiles for Nonfuel Mineral Commodities*, U.S. Bureau of Mines, Washington, D.C., May 1979, 97 pp.
252. *Mining Annual Review*, Mining Journal, London, Annually.
253. "New International Strategy for U.S. Coal," *Business*

Week, No. 2609, Oct. 29, 1979, p. 91, 95.

254. "New Restrictions on World Trade," *Business Week*, No. 2748, July 19, 1983, pp. 118–122.
255. "Niger's Uranium: Radionative," *Economist*, Vol. 283, No. 7242, June 19, 1982, pp. 85–87.
256. *Outlook for Development in the World Copper Industry*, CIPEC, Paris, 1977.
257. OPEC, *OPEC at a Glance*, OPEC Secretariat, Vienna, 1981.
258. "Outlook for World Nickel Market Through 1985," *World Mining*, Vol. 33, No. 1, Jan. 1980, pp. 76–77.
259. "Oxy and the Soviets are Back on the Track," *Chemical Week*, Vol. 128, No. 18, May 6, 1981, pp. 15–16.
260. "Oxy's Swap with Russia Seems Back on Track," *Business Week*, No. 2687, May 11, 1981, pp. 35.
261. "Phosphates Get Lift as Soviet Embargo Ends," *Chemical and Engineering News*, Vol. 59, No. 18, May 4, 1981, p. 4.
262. "Poland Once Again a Major Supplier of European Coal," *Coal Age*, Vol. 87, No. 10, Oct. 1982, p. 17.
263. "Rapid Increase in Coal Trade," *Petroleum Economist*, Vol. 48, No. 8, Aug. 1981, p. 351.
264. *Restrictive Regional Policy Measures*, Organization for Economic Cooperation and Development, Paris, 1977, 34 pp.
265. "Royal Commission Paves Way for Australian Uranium Exports," *Energy International*, Vol. 14, No. 2, Feb. 1977, pp. 13–16.
266. "Seaga Suggests Bauxite Barter With the United States," *Iron Age*, Vol. 223, Dec. 17, 1980, p. 92.
267. "South Africa: Moving to Dominate the Market in Chrome," *Business Week*, No. 2434, May 31, 1976, pp. 37, 41.
268. "South African Steam Coal Exports Move at Record Levels," *Coal Age*, Vol. 87, No. 8, Aug. 1982, p. 31.
269. "Stabilizing Commodity Prices: Can Rich and Poor Agree?," *Economist*, Vol. 265, No. 7002, Nov. 12, 1977, pp. 68–69.
270. *Status of the Mineral Industries*, U.S. Bureau of Mines, Washington, D.C., Annually.
271. *The Steel Industry in the 1980's*, a symposium, Organization for Economic Cooperation and Development, Paris, 1980, 278 pp.
272. "Survey and Outlook for Mineral Commodities," *Engineering and Mining Journal*, Vol. 182, No. 3, Mar. 1981, pp. 77–169.
273. "Tin: Some Agreement," *Banker*, Vol. 132, No. 7, July 1982, pp. 82–83.
274. "Tin: Swapping Cartels, " *Economists*, Vol. 283, No. 7243, June 26, 1982, pp. 81–82.
275. "Tinpec Comes Into Being as the Tin Market Wallows," *World Mining*, Vol. 10, No. 35, Oct. 1982, p. 57.
276. "Titania: The Largest Producer of Titanium Minerals in Europe," *Mining Magazine*, Vol. 139, No. 4, Oct. 1978, pp. 365–371.
277. *Transnational Corporations in World Development: A Re-Examination*, E1C.10/38, United Nations, New York, Mar. 20, 1978.
278. *Uranium: Resources, Production and Demand*, a joint report by the OECD Nuclear Energy Agency and the International Atomic Energy Agency, Organization for Economic Cooperation and Development, Paris, 1979, 195 pp.
279. "U.S. Heading Toward Entry into Tin Pact," *Commerce America*, Vol. 1, No. 5, Mar. 1, 1976, pp. 19–20.
280. "West Germany: A Rush to Ensure Access to U.S. Coal," *Business Week*, No. 2697, July 20, 1981, p. 70.
281. "What Ails the Market for Metallurgical Coal," *Business Week*, No. 2548, Aug. 21, 1978, p. 27.
282. "What Next: A Cartel for Iron Ore Mining Countries," *Iron Age*, Vol. 214, No. 21, Nov. 18, 1974, p. 42.
283. "Widening Stream of U.S. Coal Exports," *Business Week*, No. 2621, Jan. 28, 1980, p. 35–36.
284. *World Development Report*, The World Bank, Washington, D.C., Yearly.
285. *World Fertilizer Capacity: Ammonia*, TVA, Muscle Schoals, Ala., May 1, 1980.
286. *World Mineral Availability, 1975–2000. Steel, Iron Ore, Coking Coal*, Vol. 3, Stanford Research Institute, Menlo Park, Calif., 1976.
287. *World Mineral Availability, 1975–2000. Ferrous Metals*, Vol. 4, Stanford Research Institute, Menlo Park, Calif., 1976.
288. *World Mineral Availability, 1975–2000. Aluminum, Copper, and Fluorspar*, Vol. 5, Stanford Research Institute, Menlo Park, Calif., 1976.
289. *World Mineral Availability, 1975–2000. Antimony, Lead, Zinc*, Vol. 6, Stanford Research Institute, Menlo Park, Calif., 1976.
290. *World Mineral Availability, 1975–2000. Tin, Mercury, Titanium, Zirconium*, Vol. 7, Stanford Research Institute, Menlo Park, Calif., 1976.
291. *World Mineral Trends and U.S. Supply Problems*, Resources for the Future, Washington, D.C., 1980, 535 pp.

Part 2

PRINCIPLES OF MINERAL ECONOMICS

2.4

The Production of Mineral Commodities

Richard L. Gordon

In this chapter, the principles that determine the procedures selected to transform minerals in the ground into usable commodities are reviewed.[a] I argue here that straightforward modification of the modern general economic theory of production suffices to meet the practical needs of a mineral economist.

Therefore, the chapter is largely a review of the general economic theory of production to which is added discussion of the special problems of the exhaustibility of mineral resources and the economics of finding and developing economically recoverable minerals. In many cases, propositions applicable to all industries are reviewed because they are not treated adequately in prior texts.

In the discussion, use is frequently made of basic concepts of differential calculus widely used in economics. However, this is supplemented by simpler verbal and graphical treatments, so that those either untrained in the calculus or, as is often the case, whose skills have grown rusty, can still follow the discussion. Where the argument depends critically on the mathematics, this is noted and the justification for including the material is provided.

In addition, an appendix outlines the critical mathematics and reviews some of the numerous books published that provide the most readable available introduction to the subject. Furthermore, mineral industries examples are used to suggest the relevance of the theory.

THE RELEVANCE OF THE ECONOMICS OF PRODUCTION

Three basic steps are involved in establishing the relevance of economic analysis to mineral-industries decision making. First, the general theory of production must be defined, and its relationship to the simpler models of production stressed in economics textbooks must be noted. Second, the critical modification—consideration of the problems of discovering and developing mineral deposits—must be developed. It turns out this too involves a special case of the general theory of production. Third, the concept of an economics of exhaustible resources must be explored sufficiently to suggest why its practcal relevance is doubtful (or at least far smaller than the numerous writers on the subject seem to believe).

The present approach is heavily influenced by post-1945 developments in economic analysis. Particularly since World War II, economic theory has drastically changed its scope. First, stress has increased greatly upon viewing the economy as a network of intertwined markets known technically as the general equilibrium system.[b] In this view, not only is the economy viewed as an integrated system, but critical concepts are viewed somewhat differently than was previously fashionable.

Specifically, in general equilibrium models, the concepts of supply and demand are rejected as central concepts in favor of a distinction be-

[a]Among mineral disciplines, no single term fully describes minerals in the ground. Deposits is used here as a shorthand with full recognition that some oil and gas specialists object to description of oil and gas fields as deposits. Meeting their objection seems less critical than simplifying the exposition.

[b]The general equilibrium concepts dates to the 1874 work of Walras (first translated into English in 1954), became more widely known by the publication just before World War II of J. R. Hicks *Value and Capital* (a discussion that attains greater generality than many of its successors) and began to dominate analysis after the war as Hicks became more widely read and various extensions of his work—notably Samuelson's *Foundations of Economic Analysis*—became available.

tween production and exchange. The change is due to recognition that in a general model supply and demand are artificial constructs that imperfectly describe the underlying process. This chapter is heavily influenced by these principles and includes materials further explaining the critical points.

CHAPTER OVERVIEW

Since a central aspect of modern economics is stress that the basis of all economic transactions is trading one commodity for another, the discussion begins with a review of the proposition of the centrality of trade and the associated principle that the key analytic distinction is between production and exchange. This treatment is introduced by review of the basic importance to economic processes of the necessity to make choices. This leads into showing how the terms production and exchange epitomize the available economic choices. This examination of basic concepts is concluded by overviews on the key issues associated with exploration for and exhaustion of minerals.

These sections constitute a longer introduction to the conceptual points critical to understanding of mineral-production economics. Then the remainder of the chapter provides the extensions considered essential for adequate training in mineral-production economics. The last part of this discussion stresses that the exploration process is largely independent of and more important than the question of exhaustion. It also notes that exploration economically is but one element of the more general problem of efficient search for market information.

From these preliminaries, I turn to more extensive development of the theory. The first step is to discuss the economic concept of competition and its importance to analysis. Competition is defined and distinguished from various concepts of monopoly or imperfect competition. Considerable attention is given the difficulties of analyzing imperfectly competitive markets.

I pause to deal with an often discussed aspect of the theory of imperfect competition—namely that the imperfection can give managers considerable discretion in setting corporate goals and how to attain them. I argue that the importance of this flexibility has been overrated. Managers face numerous competitive pressures that limit their discretion. Moreover, the exercise of managerial control is more likely to affect the sharing of profits between managers and stockholders than to lead to material difference in the basic decisions and actual profitability of the firm.

Integrated into this discussion is review of whether the existence of a large number of firms essential for the maintenance of competition is compatible with the existence of advantages of large scale production. I argue that, for most minerals, these advantages are not so great as to limit the number of firms enough to restrict competition. I proceed to contend how in long-lived ventures in which market conditions are variable—key characteristics of mineral ventures—a concept known as present-value analysis is the only appropriate one for profitability analysis. Then, the basic principles of present-value analyses are outlined.

All this enables me to proceed to develop the basic economic rules that guide decision-making by the firm. These are the principles stressed by most economic textbooks, but my presentation attempts to extend the usual approach in several ways. First, the discussion is deliberately delayed until all the necessary background just described is developed. Second, I depart from the dubious tradition of treating a firm presumed to produce only one product to the more realistic case of the multiproduct firm. (However, I periodically revert to the single product case to permit graphical presentation of key points).

Thirdly and least radically, I first present the pre-1940s analyses that assume away questions of the choices of which resources should go into production and what should emerge. (These earlier analyses can deal only with the right level of output and use of those goods worth producing or employing).

Then I go on to present more modern treatments that explicitly treat these questions. I begin discussing the basic "marginal" rules that indicate the most profitable possible combination of inputs and outputs. I proceed to note that the best may not be good enough (or too good). The firm's highest available profit may be more or less than necessary to permit its survival. If too little is earned, the result is ruin, but if too much is earned, the result may be political pressures to capture a portion of the income.

I next note why serious error is produced when profitability analysis of a multiproduct firm is conducted on a product-by-product basis rather than through analyses of the overall situation of the firm. This discussion is provided because of its absence elsewhere. The invalidity of cost allocation rules is frequently asserted. A full presentation of the case, however, does not seem widely available.

I conclude this study of the individual firm by reviewing some critical but often neglected practical questions of applying the principles. In particular, appraisal is given of the best way to organize a firm and establish methods for it to secure the resources needed in its activities.

Firms, of course, do not operate in isolation, and the next step in the discussion is examination of the interaction among firms in the marketplace. The treatment begins with the baseline case of competitive markets and proceeds to cover the complications introduced by imperfections in competition.

The next step in the discussion is dealing with the basic minerals issue of exploration. I argue exploration is best understood as one aspect of the more general problem of the optimal search for information in a world in which knowledge is expensive to obtain. The problem is illustrated by review of a critical case in mineral supply analysis—appraisal of the relative availability of coal, oil, and natural gas.

I then turn to the much discussed problem of the exhaustion of mineral resources. I argue that the problem has been greatly overrated. First, the threat of exhaustion may not be great. Second, the profits available to private investors for saving resources for the future may be a better incentive for solving the problem than government intervention. Third, if private investors err, they can as easily use resources too slowly as to use them too rapidly.

All this then is intended (1) to suggest mineral production problems are not radically different from those in other industries, (2) that modern economic theory fully discusses the critical issues, and (3) unfortunately, many critical points are omitted from standard discussions. These omissions seem due, not to the complexity of the concepts, but to the inability of textbook writers to appreciate their importance.

ECONOMICS AS CHOICE

As Lionel Robbins argued in 1932, the central concepts in economics are the existence of alternative goals and different means to attain them. This point is critical to understanding what follows. The simplest possible case is where there are only two goals and only two ways to secure them. Modern economic theory makes this simplest case the main analytic building block. The critical optimality rules all compare two alternative means to secure two goals.

The ends are the useful things—physical commodities, marketed services, and even the unmarketed intangibles that we may be better able to enjoy because of our participation in the marketplace—that arise from our involvement in an organized economy. The means are various ways that we can alter the ends available to us. The natural starting point is our present situation. By historical forces that are complicated to explain, irreversible, and at best suggestive of how we might proceed in the future, each person ends up with a collection of tangible and intangible assets. The choice concept is meaningful only if these people have ways of altering the *statis quo*.

Many options exist. The usual tradition in advanced studies is to consider first the possibilities of improvement through trades between any one consumer relying upon existing assets and another consumer also employing existing assets. Then, production is introduced. In general, production involves any transformation of existing goods and services into new ones. With production comes further alternatives. Households can engage in production. Separate entities called firms are established to conduct production. Then each household has its own production opportunities and the possibility of trading with firms (and other producing households).

This expositional procedure reflects critical basic aspects of the difference between exchange and production. Exchange is indispensible for the redistribution of resources, existing or produced, to persons better able to use them. Thus, the benefits of production cannot be utilized without exchange, but benefits can be gleaned from exchange even if there is no production. However, the greatest gains from economic processes come from the creation of new goods through production. Thus, exchange is more essential but not necessarily more beneficial than production.

The present chapter stresses the contribution of production. However, because of the need to effect exchanges to distribute the fruits of production more efficiently, exchange issues also are noted.

The argument can be complicated in many ways. Each consumer could engage and, in fact to a limited extent does, engage in some self-production. The most critical point, as noted, is that comparisons between means to choose between two ends proves the basic building block of economic analysis. We proceed from one comparison to appraisal of every possible pair of ends to a pair of means. This suffices to characterize fully the optimization of choice.

Specifically, for each consumer and pair of useful goods, a comparison is made separately between the present endowment, trades with any other consumer, the consequences of self-production, and trades with any producer. Each producer, in turn, makes for every pair of goods comparisons between changes of employment within the firm, deals with each other firm, and deals with each consumer.

MARGINALISM AS THE BASIS FOR CHOICE

Moreover, a single rule that reappears in many guises explains the choice process. Thus, to introduce the specific rules discussed below, the basics are outlined here. I start by stating the key rule, which proves to use several technical terms, and then explain those terms. The rule is stated as indicating that a change is desirable when the marginal added value of making the deal exceeds the marginal cost.

Marginal refers to the valuation attached to any single step made. The concept is borrowed from the differential calculus that stresses that the rigorous way to move to a maximum is to start at some arbitrary point and consider what direction, if any, in which to move to make an improvement. The calculus shows that for well-behaved functions, this process and no other will unfailingly lead to determining the maximum. (See the Appendix to this chapter).

We are better able to judge whether some small moves away from our starting point are improvements than immediately to determine where we should end up. We add improvement upon improvement until no more are possible. By definition, we have maximized.

The benefit is the new goods gained, and the cost is the goods given up. As discussed more fully below, each deal in economic choice is fundamentally the sacrifice of one good to get another. The benefit is the addition of one good; the cost, the loss of the other. A profitable deal is one in which for the initial steps taken, you get what you want for less than you were willing to sacrifice. The actual loss of one good necessary to effect the increase in your supply of the other is less than the sacrifice you were willing to make. (See below for a numerical example).

The analysis is completed by invoking a diminishing net payoff principle that insures that the trading process is self-limiting. Even if some initial beneficial steps exist, a stopping point occurs. Each additional sacrifice becomes less attractive and so even if the deal available is unchanged by your actions, your willingness to proceed further declines after each step. (This is also illustrated below by the numerical example).

The basic presumption is that at each step the lost good becomes more and more dear to you because you have less of it and a gained good becomes less and less attractive for symmetric reasons.

In consumption, this is explained by the proposition that marginal satisfaction declines as use increases. In production, it is similarly suggested that adding more of any one thing (without adding anything else) is less productive or profitable at each step. Thus, in every case, each step is less attractive in the sense that the sacrifice you are willing to make declines.

This impact is reenforced by the opposite pressures at work on your trading partners that lead to an increase in the sacrifice they require of you to make additional deals. As your willingness to sacrifice declines and their requirement of sacrifice increases as more steps are made, the initial situation of a step with your valuation exceeding the required sacrifice is ultimately succeeded by a step with the return matching the sacrifice.

(The term marginal rate of substitution is used to characterize the critical comparative values—of both what you are willing to pay and what you have to pay. The critical equalities, presented below, are among all marginal rates of substitution of all who are involved in production and consumption of any two goods. Earlier generations of economists used a concept of marginal utility—the value of a good considered in isolation. Modern theory indicates that everything of interest is indicated by the marginal rates of substitution because the practical implications of valuation can only be expressed and thus observed in a choice between a specific means to a given end.)

THE CONCEPT OF TRANSACTIONS: A PRELUDE TO PRODUCTION ANALYSIS

Economic transactions, as noted, fundamentally involve exchange of one good for another. As this discussion tries to show, the mechanism used to facilitate trade relies on indirect approaches. Thus, people often lose sight of the basic goals these processes seek to effect. Here, I seek to remind the reader of these basics. We have long recognized that direct barter trade is an exceeding inefficient process and an indirect method of exchange is preferable.

Direct exchange is workable only if deals can be kept simple. In fact, participants in a market economy wish to convert their goods into a diverse range of other goods and have to deal with many others, few of whom want exactly what the others possess. The owner of any resource sells it because of a desire to exchange it for another resource.

However, a two-step process is undertaken. First, the resource we wish to give up is sold for money, and then the money is used to secure what we want. Money is not wanted for itself but rather is the lubricant of trade. Everyone agrees to give or take money in exchange for whatever we really want. Then, we are always assured of a clear-cut way to exchange. (Technically, this is described as using money as a medium of exchange. The details and extension are the subject of numerous texts on money and banking, and the classic advanced treatment is by Patinkin). The use of money makes possible considerably greater trade and far more productivity than could occur in a barter economy (as von Mises often stressed).

A two-stage process is preferable because direct swaps between those wishing to trade one specific good for another specific good are much more difficult to arrange. In practice, those willing to make a deal with us often do not want what we have to offer. They prefer a third good that possibly could be secured by trading what we gave them with someone else who wants to made a deal. Such deals are possible because the *relative* value of any two commodities can, in the absence of exchanges, be radically different among people. Money prices, in turn, provide an epitome of all relative prices. Moreover, trade in a sufficiently vigorously competitive economy manages to insure that all profitable trades will occur. (See Chapter 5.13).

These principles are so basic that review is essential to the subsequent analysis. By historical accident, the principle that relative values are what matter was developed before the general equilibrium model. The development dealt with the special case of trade among nations being motivated by differences in relative values arising from differences in production positions. In this case, the difference is called comparative advantage.

It is rarely explicitly noted that the principle is applicable to differences *from any source* in valuation among any pair of participants or of groups of partcipants. Thus, the arithmetic examples used in most introductory economics texts to explain comparative advantage are, in fact, demonstrations of the *general* case for trade among different parties—be they individuals, firms, or countries. The argument, moreover, is usually presented only for a two commodity world, and generalization is neglected.

TWO COMMODITY EXCHANGE

The critical concept is the marginal rate of substitution or valuation. This is termed a rate of exchange—a measure of the number of units of one good that will be swapped for a unit of another good. As discussed above, the marginal concept indicates this valuation varies as the endowment possessed of the two goods is altered. When economic units operate in isolation from each other, they are likely to adopt different marginal valuations of each pair of goods. Differences in both the amounts owned and how they are evaluated will prevail.

Trade is desirable if there are two people with different marginal valuations of some pair of goods both possess. We can denote the people A and B and the goods 1 and 2. The rates of exchange we can call P_{2A} and P_{2B}—the respective number of units of good 1, A and B are willing to surrender to get a unit of good 2. Three logical possibilities exist:

$$P_{2A} > P_{2B}$$
$$P_{2A} < P_{2B}$$
$$P_{2A} = P_{2B}$$

In the last case, the value is the same, and no

trade is possible. In the other two cases, trade is desirable. Moreover, at least one party gains and the other is no worse off because of the trade. Consider the case in which $P_{2A} > P_{2B}$. This is to be interpreted as indicating that good two is worth more units of good one to *A* than to *B*. Thus a swap arises in which *A* gives good one to *B* in return for some of good two. For example, consider values of 10 and 6, *A* would give up 10 units of one to get a unit of two, but *B* would give only six units of one to get one unit of two (and also only require six units of one to compensate for the loss of one unit of two).

Trade could occur at any price between 6 and 10. At a price above 10, *A* will not buy; at a price below 6, *B* will not sell. At a price of 10, *A* is willing to swap one unit of 2 for the 10 units of 1 that are equal in value to him. *A* is no worse off. *B* has gotten 10 units of one for something worth only 6 units to him—a gain of four units of good one.

Analagously, if the price is 6, *B* is no worse off from the swap and *A* gets his extra unit of two at a cost four units of one less than its value. At an intermediate price, both gain. If the price is 8, *A* gives up two fewer units of good 2 than he is willing to sacrifice. Both gain.

Several key points should be noted. First, a similar argument applies if $P_{2A} < P_{2B}$ except that trade goes in the opposite direction. Thus, *any* difference in valuation will lead to trade. Given the millions of people in the world and the many possible goods to trade and the differences in each person's endowment of goods, it is hardly surprising that many opportunities arise for profitable trade.

Second, money values are a convenient way of developing a general measure of the potential for trade. The measure is trivial in the two good case. The proper money valuation reflects all the marginal rates of substitution. Thus, we can be sure that if money is worth two units of good one to *A*, that money will be worth 20 units of good two to *A*. *A* would not assign a money value to good two different from the money value of the amount of good one (10 × 2) A would surrender for good two. Surrender of good one for money and money for good two is simply another way of getting good two, and no more would be paid going this indirect route than in going directly.

What makes money prices interesting is that they summarize *all* the relevant opportunities. Differences between any two people in their marginal money valuation of a given good always implies that some form of profitable trade can be made. The prior argument generalizes to indicate that a list of money values of all goods to a given individual summarizes his marginal rate of substitution between all pairs of goods. Each will be the ratio of the money prices of the respective goods.

MULTICOMMODITY EXCHANGE

By tedious further extension of the argument to deal with more than two persons and more than two goods, it can be demonstrated that differences in money values of a given good to different people always mean opportunities for advantageous trade. The price differences inevitably mean that there is someone somewhere willing to trade the good with someone else for something else.

More importantly, the differences imply that the deal can be implicitly made through the complex chains of exchange we actually maintain. It is not necessary that the person who wants to sell the good we want also wants what we wish to give up. Both can trade our services for money and turn our money into goods still others provide. (The proof for any actual chain involves considering each step in the process—how much output is produced by our labor, what goods are secured by selling the output and so on until the purchase becomes what we actually want. It will always turn out that if money values differ, the end result of the series of deals is a gain from trade identical in substance from the direct two good exchanges outlined above).

A final pair of key points is that the trading process is such that (1) the result of mutual gain is more likely than a gain to one person and (2) differences in value disappear. This follows from the variability of marginal rates of substitution discussed above. It is presumed that the willingness to sacrifice systematically changes with the amounts possessed. For example, as we get more of good two and less of good one, the valuation of good two in terms of good one declines. This principle suggests that while one trade could occur at the extremes noted, this probably would not be the stopping point.

If an initial deal were made at a price of 10 units of good one, *A* being more satisfied with his supply of good two will reduce his offer for another unit of good two. *B* being less satisfied, however, will raise his value. So long as $P_{2A} > P_{2B}$, more trade would be desirable. The process continues increasing the flow of good two from *A* to *B* until the amount transferred causes P_{2A} to fall and P_{2B} to rise so they are equal and no more trade is profitable.

Moreover, the actual exchange process involves everyone simultaneously agreeing upon prices and quantities. In our two person, two good world, with $P_{2A} > P_{2B}$, *A* and *B* would expand trade, lowering *A*'s value of good two and raising *B*'s value until parity of *P*'s were attained at some price between 10 and 6 with each unit traded at this price.[c] Thus, we might have *A* obtain 2 units of good two at a price of 8 each. This would then give neither *A* nor *B* a gain on the *marginal* trade since both value the second unit of good two at eight units. However, both have gained overall. The first unit of two traded also cost 8 units of one—two less than its worth to *A* and two more than its worth to *B* so both gained on this deal.

More generally, in the economy as a whole, all disparities in valuation will be eliminated. The argument thus far applies to all pairs of goods and persons. Thus, all disparities for all goods and people will be eliminated. In equilibrium, the prices of all goods to all people will be equal. The benefits of trade then are redistributing goods so that differences of marginal values are eliminated. While the last unit gained has a value to the consumer equal to its price, the prior units purchased are worth more than the price because of the diminishing payoff principle discussed above. It is the gains on earlier trades that constitute the benefits of trade.

We have become so used to reliance on money that we simplify our view of economic transactions by looking at the exchange of money for goods. We then define demand as the desire to trade money for goods and supply as the desire to trade goods for money. These conventions are convenient shorthands, but they obscure the underlying processes just outlined.

As noted, money is a critical facilitating device. It was secured by "supplying" and then used in "demanding." On balance, we have a swap in which everyone supplied one good and had their demands for another realized. This can be equivalently expressed as saying we are both suppliers and demanders, or more simply, that we are exchangers. The principles of differential valuation inspiring trades then governs the pattern of exchange.

Exchange could occur simply because everyone started with a collection of given immutable goods but differed among themselves about the value of the goods given amounts owned. As noted, the more interesting world is one in which new goods can be produced.

In contrast to the artificiality of the supply-demand distinction, a significant analytic difference prevails between production and firms on the one hand and individuals and their consumption on the other. In particular, the individual has the ability to express opinions and these opinions alter with the pattern of consumption that is maintained. Money is an unfailing measure of the situation of the firm because it has no attitudes to alter with the amount of money available and the prevailing prices. However, changes in money income and prices do alter the valuation of money by individuals, and the necessity to analyze this impact makes consumer theory more intricate than producer theory. However, this complication is ignored in this chapter (but see Chapter 5.13 for further discussion).

PRODUCTION DEFINED

Economic processes also involve the step that is the concern of this chapter—production. In its broadest sense, production is the securing and use of a variety of goods to produce another set of goods. Thus, we now have a procedure by which one commodity is later used to produce another, and this output is then what is provided in payment for the goods taken in. Again money is used to facilitate matters, and again it is critical to recognize that the money exchange is economically the medium for conducting the more fundamental swap of goods inputted for goods outputted. Production, of course, occurs because the resulting output provides a mix of goods more valuable to society

[c] Those familiar with economic theory will recognize that I employed a standard shortcut. The prevalence of a single price for all transactions is technically called non-discrimination and is insured only when the economy is highly competitive because many people participate. A two-person economy would not be highly competitive, but is a standard convention to assume competition prevails. Some adopt the alternative of a world in which there are two types of people—a million *A*'s and a million *B*'s. See below for further discussion. Note, moreover, that Ps in the prior discussion represents each individual's valuation or offer price, and not the market price.

than the one otherwise available.

Production as just defined is a broader concept than that often used in the mineral industries. In economic analyses, production is a generic term for all activities that use existing resources to produce something new. In this sense, every activity of every firm in the mineral industry or elsewhere is production. The oil-and-gas industry finds it convenient to define production as the actual extraction of oil and gas from the field. In the economic concept of production, exploration, development, transporting, refining, and marketing are also included. Analagous points can be made about the stages for any other mineral. The concern here is with the broader concept of production.

Moreover, production in this sense involves at least four key elements. First, as all analyses make clear, several different goods called inputs are used to effect production. Second, several different goods called outputs are produced. As a matter of historical accident, most discussions continue to emphasize the model in which a single output is produced and introduce, if at all, the general case as an after-thought.[d] Third, the analysis should explicitly recognize that production occurs over time, and interest rate effects should be considered. In this case, an elaborate literature does exist but is largely separate from the economics of production. Finally, the analysis should recognize the existence of transaction costs—the expenses about learning about and reacting to all the available options for conducting production or exchange. This aspect of economic analysis has quite slowly developed, is, however, now well defined, but is often neglected in elementary discussions. This chapter treats all these issues. As noted above, first an introduction is provided to basic questions about the firm competition as its implication and investment analyses. Then the standard multiproduct theory of production is presented. Finally, the concept of transaction costs is introduced and its implication discussed.

In sum, production is any process that transforms resources; normally it involves using several resources to produce several new products; these processes take place over many years. Learning about options, adopting them, and organizing production, moreover, are expensive activities often requiring special skills. These concepts clearly describe all the problems of nonmineral firms and more critically proves to describe all the problems of mineral firms.

THE CONCEPT OF EXPLORATION

The analysis, however, is clarified if we make the exploration problem explicit. It is a variant of the transaction cost question. The information in question is improved knowledge about mineral deposits. Specialists in mineral economics have long stressed the importance of the acquisition and use of such information in industry activity, and their insights are employed here.

A meaningful economic model of the exploration problem and the more general resource availability problem of which it is a part must recognize that a continuous spectrum of possibilities exist throughout. Resources are not simply known or unknown but range from those about which virtually no thought has been given to those whose characteristics have been elaborately analyzed. Similarly, the facilities for exploration can differ markedly in their degree of completion and if completed, in their capabilities. Other aspects of an occurrence about which a range of possibilities exist over both the actual conditions and the accuracy of information on them include the qualities such as ore concentration and impurity content of minerals, the cost of extraction and processing, and the cost of transportation.

In any case, the exploration process usually is defined as those efforts to find minerals. M.

[d] Why this is so is not completely clear. The general case cannot be avoided in analysis of consumer consumption choices. They cannot be meaningfully conducted without consideration of choice among different goods used. However, it is possible to develop a coherent theory of production in which there is only one output so long as we also recognize that there must be one or more inputs. The asymmetry may arise from technical aspects of the mathematical analyses used. In both cases, what is technically known as constrained maximization techniques—Lagrangean multpliers in the conventional calculus or in more modern analyses the Kuhn-Tucker generalization of Lagrange's method—is employed (see below). In the general models of consumption and production, the Lagrangean multipliers relate to an unobservable concept of the value of relaxing a constraint on one resource with nothing else changed. In consumption theory, this unobservable concept happens to have acquired a name—marginal utility of income. However, no such analog has been developed on the production side. In the one product production model, the Lagrangean multiplier has a well-defined meaning—it equals the marginal cost of production. This advantage hardly outweighs the loss of generality. The lack of a clear interpretation of a Lagrangean is hardly problematic since this is far from the only case in economics in which Lagrangean multipliers have no clear meaning but are only building blocks to facilitate development of more meaningful expressions. Alternatively, we might develop an analogy of the marginal utility of income called, say marginal productive utility. We cannot call it marginal productivity because this refers to value in terms of a concrete good instead of some unmeasurable absolutely scaled value.

A. Adelman's effort to develop a generalized analysis of mineral supply has suggested that exploration produces what oil geologists call minerals in place—a stock of "known" but not necessarily currently usable minerals. Some of the minerals in place will not be economic to exploit immediately, and none can be exploited until producing facilities are created. Again following oil industries terminology, the stage of installing produced facilities can be termed development. Finally, comes the actual extraction and whatever further processing is required.

Several points should be noted. Exploration at best provides tentative estimates of the magnitude of known minerals in place and, more critically, of their current economic viability. Even less is known about future economic attractiveness if only because this is so greatly affected by forces such as demand changes and technical progress that occur outside the deposit.

Moreover, the border between exploration and development is another example of the continuum principle. At the point at which new activity takes place at the fringe of an established producing region, it is unclear and of no practical relevance whether this should be considered exploration or development. Finally, the development process normally continues throughout most of the life of the product. At a minimum, equipment fails and must be repaired. Additionally, production tends to deplete a deposit and further development can be undertaken to offset this deterioriation.

Oilfield development involved intricate combinations of equipment replacement and adjustment for deteriorating producing conditions due to the impacts of output on both the reservoir and the well. On the one hand, earth can clog the well; on the other, analysis of reservoir mechanics has established that as production progresses the ability of natural forces to push oil to the surface deteriorates. Increasingly this problem is anticipated early in production by starting immediately to use secondary recovery methods that supplement natural forces. Such methods also can be instituted or augmented at a later stage.

In oil and other forms of extraction, another response to depletion at any one part of a deposit is to move to another location—horizontally to another bench or to a deeper bed in surface mines, or outward or downward from the original shaft in underground mining. These processes can continue through the life of exploitation which in turn can proceed for periods ranging from the few months required to surface mine a small parcel of Applachian coal to the many centuries for which some European mines have managed to persist. These points are better understood after reviewing basic production theory.

THE NATURE OF EXHAUSTION

The need to discover and develop is often associated with the problem of exhaustibility of resources. Exhaustibility does imply exploration and development requirements, but the reverse need not be true. Exploration and development will continue as long as mining persists even if no meaningful exhaustion problem arises. To see this, we must examine the available economics of the impact of exhaustion on private decision making and what public policy implications, if any, arise.

The theory of exhaustion begins with a fundamental principle of investment analysis—investment must be rewarded. In simpler models, stress is on the possibility that delaying output will produce sufficient profit from selling *that unit* to justify the delay. A more general formulation adds recognition of the implications of differences in phyical properties and ease of extraction, production, and processing. The combined effect is differences in the economic attractiveness of different occurrences. Extraction will proceed from use of the most attractive to use of the least attractive.

The delay of output, not only makes more available, but slows down the depletion of better resources. The benefit of this slowdown is another payoff to investment in hoarding of resources for future generations. The theory shows that if exhaustion is a major problem we should observe market developments creating incentives for hoarding. The absence of such evidence suggests reasons to doubt the practical relevance of exhaustibility. This too is best discussed more fully after the basics are presented.

COMPETITION AND PRODUCTION ANALYSIS

Any analyses we undertake must also resolve two further questions—what competitive pressures firms face and what decision rules guide the firm. Conclusive economic analysis often is possible only if special assumptions are made, and this is particularly true of the issue of the vigor of competition. Analysis can readily handle either very strong or very weak competition but has established only that prediction is nearly impossible in intermediate cases.

The most interesting concept of vigorous competition, at least for a mineral industry, is technically referred to as pure competition. It presumes that every firm has so negligible an impact on prices that it acts as if prices were given. Some writers describe this situation as the firms being price takers.

(A more stringent concept of competition can be defined in which besides absence of control over price, all firms are identical, exit and entry are limitless and costless, and knowledge is perfect. While terminology varies, this more restrictive concept is often termed perfect competition. In practice, the perfect information aspect of the definition has a far less clear-cut implication than the identical firm, free entry-exit presumptions. Collectively, the last three assumptions imply that with perfect competition there are never "economic rents"—payments to the owners of superior resources).

(Perfect competition is too restrictive a model to apply to the minerals industries in which economic rents always are important. The critically unrealistic assumption is that all firms are identical. Identical firms are possible only if all resources of a given type are identical. This, of course, is not true in minerals, and for their analysis, pure competition is the relevant case. Therefore, nothing more needs to be said here about perfect competition. Both will have the same impact on efficient resource use. (See Chapter 5.13).

Conversely, the absence of pure competition occurs when the firm does have a perceptible influence on price and takes this into account. Analysis is simple only in the case of pure monopoly—where only one firm influences price. Only its impact has to be considered, and the implications are straightforward. No realistic examples of pure monopoly have every been demonstrated.

At best, the model characterizes the (usually impossible) dream of all participants in economic activities. All would like to form a coalition of like-minded individuals who collectively do possess all the existing power over price. A pure monopoly would then exist. Historically, the coalition represented by the DeBeers diamond group has seen the nearest thing to a pure monopoly actually prevailing. Speculation has emerged that discoveries of diamonds in countries not likely to join with DeBeers may undermine the cartelization.

OLIGOPOLY BEHAVIOR

What usually actually happens is quite different. Often competition is so strong that no coalition can be formed. Alternatively, we may have what is called oligopoly in which a "small" number of firms have control over price. In that case, coalition formation is feasible in principle, but practice is impossible to predict.

Several unresolvable issues arise. First, no clear principles exist about what would make a coalition feasible. Not only is there no evidence about what constitutes a sufficiently small group, but also suspicion exists that the number of firms does not by itself determine fully the feasibility of cooperation. Difference in firm sizes and financial strength of firms also may be influences. If a group of any size consists of a small number (in the range of one to three firms) of very large producers of the commodity and many small firms, the large firms may find market rigging desirable even though the smaller firms fail to cooperate. Well-financed firms are better able to withstand periodic price wars that break out as the cooperative spirit wanes.

Additionally, attitudinal factors can be major influences. Every member of a price rigging scheme becomes simultaneously the potential beneficiary or victim of every other member. The group can move steadily either to ever greater success or to the collapse of accord. If everyone is confident that everyone else will restrict output and put upward pressure on price, they all will restrict output and prices will rise. Conversely, should some members fear that some others will not cooperate and that actually output is heading up and prices down, the optimal strategy is to imitate Nathan Bedford Forrest and seek to get there firstest with the mostest—lead the collapse instead of waiting for someone else to act.

If this were not enough uncertainty, it is also impossible to predict exactly how cooperation works if it does prevail. The ideal for the oligopolists would be to operate exactly as if they were partners in a single monopoly firm. This ideal involves concentrating output in the most efficient facilities, possibly requiring total shutdown by some members.

This, in principle, could be induced by paying the high cost producers to shut down. However, this has rarely if ever been a practically relevant option. Personal pride is a formidable barrier. Further limits are imposed by the difficulties of negotiating and enforcing a compensation agreement. Finally, most cartels require (and may even be stimulated by) governmental aid. Governments must be willing to provide backup to the actions. A probable cost of such assistance is that it becomes necessary to assure the government that employment effects are lessened by continuation of operations.

This leaves us with two major questions. First, how widespread are oligopolistic market conditions and second, where they exist, to what do they lead? The first issue is largely factual. It suffers from the usual problems in economic analysis—available data are inadequate to permit resolution of the question. However, one point to note is that many commentators greatly understate competition by neglecting the competition produced by international trade.

The state of theory, such as it is, has been relegated to the more technical literature. Basic economics texts for some reason sidestep the simple but discouraging points made above that no clear predictions are possible. As noted, the essence of the theory is often suggested. Textbook writers, however, usually present less general analyses of dubious empirical relevance largely because they can be more easily diagrammed.

COMPETITION AND PRICE TAKING

The question of corporate goals proves in practice to be closely related to the question of the vigor of competition. In fact, at least two departures from purely competitive conditions are required to permit a choice of decision-making rules. Vigorous competition implies that firms die if they do not maximize profits—or to be more precise, maximize wealth—the properly weighted value of profits over the lifetime of the firm (see below for a more precise definition).

Competition in the sale of output and in the purchase of resources is the first source of pressure. On one side, the entry of new firms expands supply, lowers prices and puts downward pressures on profits. On the other side, where superior resources exist, competition causes the

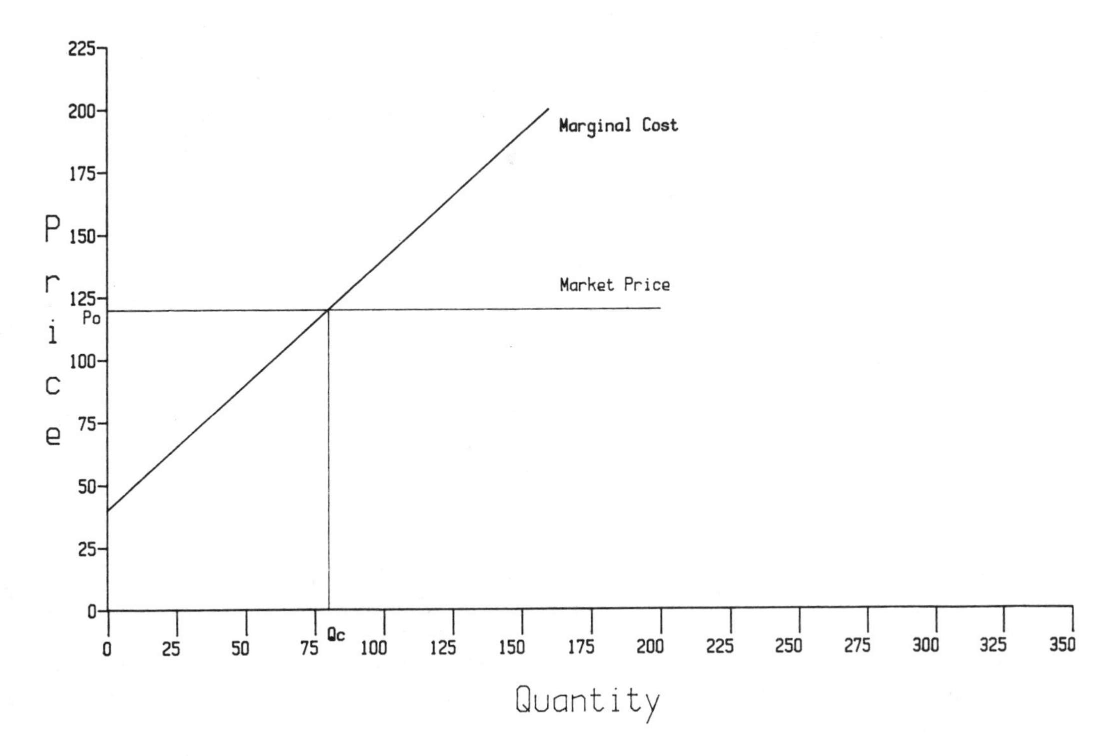

Fig. 2.4.1—The competitive firm.

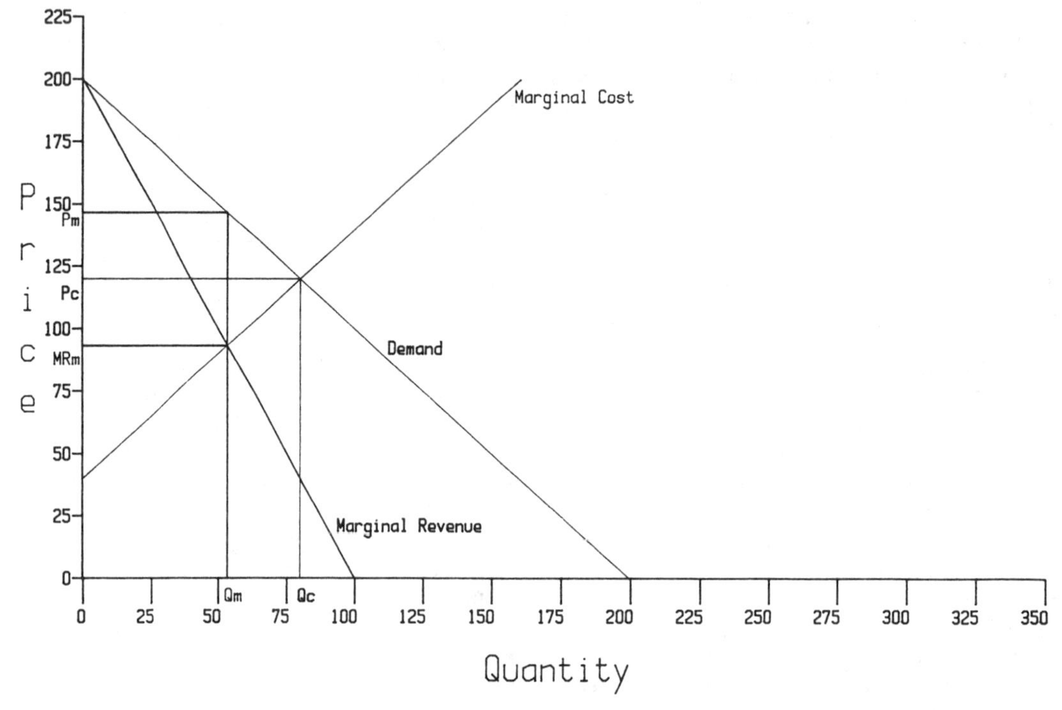

Fig. 2.4.2—Monopoly equilibrium.

prices paid the owners of such resources to rise again tending to depress profits.

Price taking is formally described as indicating that the marginal receipt per unit of output of any competitive firm is simply the going market price. Similarly, the cost of adding a unit of input is the going market price of the input. This is simply a restatement of the point that the firm lacks the power to affect prices notably and so gets or pays the going price.

This concept is diagrammed in Fig. 2.4.1. It shows the demand curve as seen by the individual firm is a horizontal line at the going market price P_o. The firm can sell as much (or little) as it likes at that price. (Here as in the next two curves, I show for future use, the full diagram that shows both sides of the benefit-cost calculation.)

A firm with monopoly power in selling goods must consider that impact. So must a firm with monopsony—power to affect the buying price. For a monopolist, one impact of additional sales of good k is to secure the gain of P_k, the price of good k, also obtained by a competitive firm. The monopolist, in addition, causes prices to fall by an amount dP/dO_k, and given output of Q_k a total loss of $Q_k\ dP/dQ_k$—the product of output and the unit drop in price—occurs.[e]

Thus, by definition, the monopolist faces by itself the market demand curve in Fig. 2.4.2. The marginal revenue concept then indicates that the net payoff of the output change is P_k + $Q_k \frac{dP}{dQ_k}$ with dP/dQ_k negative so the net payoff or marginal revenue is less than P_k. This is shown in the figure by having marginal revenue lie below demand.

Similarly additional buying by monopsonists adds to the portion of costs represented by the input price, P_e, a price rise cost effect $Q_e \frac{\partial P_e}{\partial Q_e}$. This reduction of benefits of an output gain to a monopolist and this added cost of more input use to a monopsonist restrains their buying and selling. A monopolist operating with the same demand and cost conditions as a competitive firm would, as a result, produce less and buy fewer inputs. The optimum limitation is precisely defined only in the case of pure monopoly or of pure monopsony.

As Fig. 2.4.3. shows, in monopsony, the ef-

[e]Note that I implicitly consider continuous variation in price and use the derivative. See Appendix for further discussion.

fect on buying price is an addition to the cost of procurement. A marginal cost curve lies above the supply curve for the good being purchased.

One major impact of the indeterminacy of oligopoly is that the price impacts of the actions of any one firm cannot be unambiguously defined. The effect is a function of both the direct market impacts of the actions of that firm and the difficult to predict response of other firms. Thus, no satisfactory way has been found to express the behavior in equations or diagrams.

COMPETITION ECONOMIES OF SCALE, AND THE GOALS OF THE FIRM

Economic theory usually assumes that a firm maximizes its wealth—the value of its lifetime earnings properly "discounted" to consider the need of money to earn interest, as discussed below. However, some economists have challenged this view. The core of the challenge is the belief that the managers of the firm are different from the stockholders and may have different interests. This, in turn, will inspire pursuit of goals other than wealth maximization. Since resolution of these issues is critical to the choice of models to treat production, the debate is reviewed here.

Several issues are associated with this debate. A central question is whether the nonmaximizing behavior is feasible. This requires lengthy discussion of the limits to managerial discretion that is delayed here until the simpler issues are treated.

Another question is whether the structure of managerial rewards can be altered so that wealth maximization is in the managers' best interests. A third critical concern is definition of the most likely alternative goal that might be adopted. Baumol has suggested that managers prefer higher sales than are justified by wealth maximization. He, therefore, proposed a theory in which sales levels were raised as much above their wealth-maximizing amount as was possible without provoking stockholder revolt. Marris suggested that the preference was for faster growth and that firms would try to grow as fast as possible without provoking reaction.

Williamson postulated that managers wanted to transfer wealth from stockholders to themselves to the extent possible. Such behavior might not alter any of the input or output decisions. The firm would earn the same amount, but managers would appropriate as much of the net as

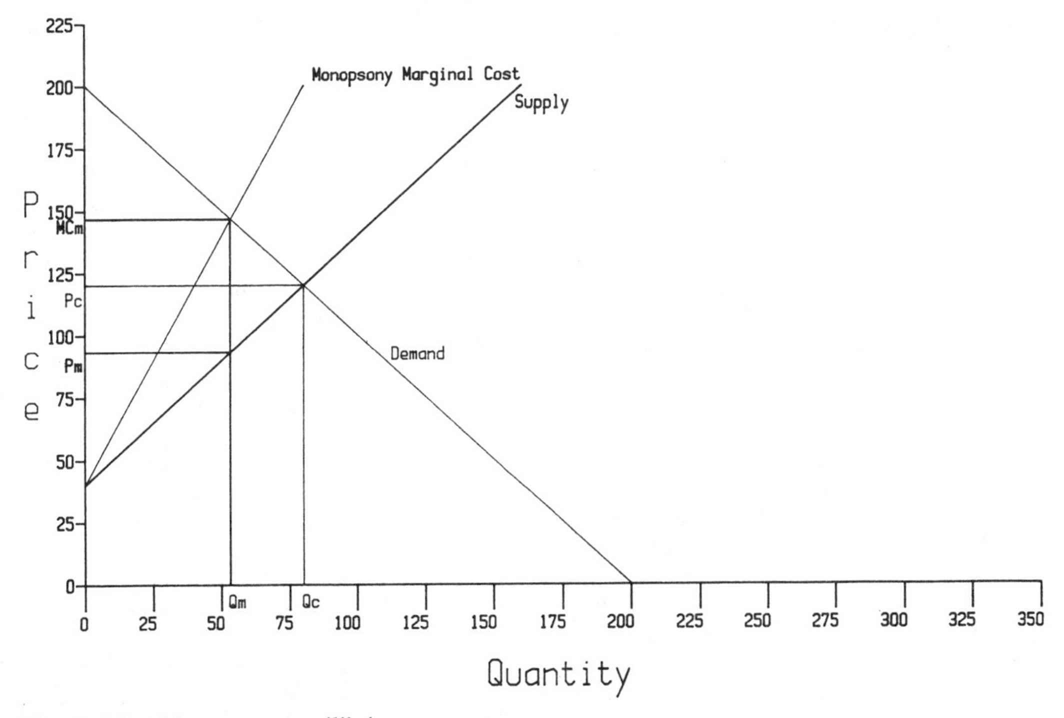

Fig. 2.4.3—Monopsony equilibrium.

possible as higher salaries. Alternatively, the managers could feel that they would be better off to secure nonmonetary rewards in the form, not only of fringe benefits in the usual sense, but of the prestige of having more subordinates. The choice of nonmonetary rewards clearly leads to alteration of input patterns. By definition, the firm uses more inputs. Whether there is an effect on outputs depends on the type of input.

Hiring a personal financial adviser might raise input but not output, but an additional salesperson might lead to higher output. (However, an alternative interpretation of the widely observed tendency to granting perquisites, such as cars, aides, and lodges in resort areas, to management is that they are simply responses to the tax laws. The managers do not get any more before tax income than if there were no managerial discretion but get it in a form that raises after tax income).

The Williamson analysis, as we have just seen, leads naturally to the final key question of whether nonmaximizing behavior had important impacts on company decisions. The Williamson model is the most clear-cut because it moves systematically from a well-defined goal to its implications. Unfortunately, these implications turned out to differ considerably with the circumstances. While less solidly based, the Baumol and Marris models do produce the qualitatively unambiguous implications that the firm uses more than the wealth-maximizing level of inputs to produce a greater than wealth-maximizing level of outputs. In all cases, the key problem is defining the maximum amount of wealth transfer that is feasible.

The theories implicitly suggest that the outcome depends on the circumstances—i.e., how much competitive pressure a given group of managers face. This, in turn, leads us back to the question of whether any ability to avoid wealth maximization can exist.

PROFITS IN COMPETITION

The perfectly competitive model includes enough assumptions to insure that the firm only earns enough income to provide stockholders with returns equal to those available in other industries of equal risk. Free entry and exit implies that when firms outside an industry see high profit, they will enter and low profits will cause exit. The identical cost assumption implies that new firms can produce as well as old firms. Entry can continue until prices fall to the extent to which the entrants are now earning no more than a normal rate-of-return. Since costs are identical among firms, so are profits. All firms end up with only a normal rate-of-return.

When we drop the assumption of identical firms, the conclusion that no excess profits are earned can be maintained only if further assumptions are made. The entrants will settle for and earn only a normal rate of return. Extant lower cost producers could potentially earn more. The usual presumption, however, is that the differences in cost arise from disparities in access to critical productive inputs. For example, variations in costs among mineral-producing firms often are due to differences in access to low-cost deposits.

However, the firm will not normally keep the benefits of this access. When someone else has property rights to the minerals and vigorous competition exists for access to the minerals, the cost saving will be transferred to the landowner through some payment for the right for access. Again, freedom of entry must prevail. With such freedom, the offers for mineral rights will be bid up until no excess profit remains. This tendency of profits associated with use of lower cost resources to end up as income to the owners of such resources had led such income to be characterized as economic rents.

In any case, sufficiently vigorous competition in the marketplace for final output and for inputs prevents excess profits. Therefore, there is nothing for managers to transfer to themselves and no scope for discretionary behavior. The minimum requirement to permit discretion is weak competition in *both* input and output markets.

Additional barriers to managerial discretion, moreover, can arise from the various forms of competition for control of a company. Most obviously, an outside group can threaten a takeover of a company perceived to be transferring wealth to managers. In addition, where individuals compete vigorously for managerial jobs, this can prevent any from securing access to excess profits.

ECONOMIES OF SCALE

Another critical concern is the relationship between economies of scale and the maintenance of competition. Increasing scale may cause costs to rise more or less rapidly than does output. The usual presumption in economic textbooks is that at low levels of operation of costs will rise less rapidly than output because of the ability to employ technologies that are only practical to utilize in large scale operations.

However, as firms become larger, their costs increase faster than sales. When output increases faster than total costs, the firm is said to enjoy economies of scale. When output increases less rapidly than costs, diseconomies are said to prevail. When a firm actually moves from an initial stage of economies of scale to a later stage of diseconomies of scale, the point of transition from economies to diseconomies is termed the point of minimum efficient size. The critical issue is whether the exact output level at which this occurs is a large or small proportion of total industry sales.

The economic theory of competitive firms, for reasons discussed below, indicates firms can operate profitably only if production levels are at or above the minimum efficient size. Thus, the critical question about the possibility of vigorous competition is the number of firms that can operate with outputs at or above the minimum efficient size.

To see whether competition is feasible, it is desirable to examine further the forces creating economies and diseconomies of scale

A major type of potential advantage of large scale arises with the large variety of equipment whose cost is proportional to its surface area and whose output depends on the volume. Examples include pipelines, ships, boilers, furnaces, and all similarly configured devices. Elementary geometry indicates that the volume of any surface increases more rapidly than its area. Thus costs rise less rapidly than capacity.

For example, to double the volume of a cube, it is necessary to increase area by about 59 percent. (Since the volume is of the length of each side cubed, we increase the length of each side in the proportion 1.26 where 1.26 is the cube root of two. The area is six times the square of the length of a side. Multiplying the side by 1.26 multiplies its square by 1.59 or the ⅔rds power of two).

Other advantages of large size include better use of specialized types of employees and the ability to take advantage of scale economies available to suppliers. Coal producers operating on a larger scale can ship by the trainload instead of the carload.

Economies, however, are apparently not unlimited because usually several firms can and do manage to exist in producing any commodity. The standard economic textbook explanations of the limits relate to the attenuation of managerial control occurring when scale becomes too large. Cost control becomes undermined, and expansion causes costs to rise faster than output. Other limits include the technological ones. Attempts to expand size may be precluded or at least made prohibitively expensive by the problems of fabricating or transporting larger pieces of equipment.

Another limit is the ability to fit a large piece of equipment into the production process. All equipment is subject to failure, and a standard way to alleviate the problem is regularly to perform preventive maintenance. The more production is concentrated on one machine, the more its shutdown will cost. Therefore, it may pay to maintain smaller machines because the loss to maintenance is sufficiently less than with larger machines to outweigh the savings in initial costs of the facilities. (This is an argument applicable to all types of industries but is discussed with particular frequency in treatments of the optimal size of units to generate electricity.)

THE INHERENT DISECONOMIES OF MINERAL EXTRACTION

All these forces are at work in various aspects of the mineral industries, but in extraction, a further, in most cases, more critical influence is at work. Generally, deposit size sets a basic limit on scale of operation. It is by no means the sole consideration. No deposit is exploited all at once. Consideration of influences other than deposit size lead to spreading output over an extended period. Nevertheless, the limit to deposit size sets critical bounds on the scale of any mineral-extraction venture.

Save for a few special cases in which low demand coexists with the existence of large deposits, many deposits must be exploited to meet demands and thus a competitive extraction industry is possible. (Another type of exception is where, as in the western United States, very large continuous coal deposits exist, the problems of managing larger units rather than the limits to deposit size may be what restricts the size of the individual mine.)

The net effect of all these considerations is

to make mineral extraction an industry of diseconomies of scale. When substantial increases are involved, it is always best to move on to exploiting more mines than to attempt to expand output by producing more from extant operations. The lower cost mines can undersell the higher cost one and so operate first. Then, as demand increases, higher cost mines will come into operation.

Nevertheless, talk often appears about the existence of scale advantages in mining. The discussion relates to a generally correct proposition that a larger deposit tends to be more profitable to exploit. Some large deposits actually may have disadvantages in long distance from markets, mining conditions, and other factors that outweigh the size effect. More critically, size is an advantage created and *limited* by nature. Mineral producers would like to exploit larger or otherwise more profitable deposits than those presently being exploited.

However, what can be exploited depends upon what is extant, known, and undeveloped. For the reasons just given, the best of the known deposits will be in use. Those starting a new mine must resort to higher cost known deposits or find new ones. Generally, such new deposits will be more costly to exploit than ones already being mined. If a strong likelihood existed that lower cost deposits remained undiscovered, exploration would expand and the deposits would be found. Therefore, it becomes necessary to incur higher costs when expanding output.

WEALTH DEFINED—THE PRESENT VALUE CONCEPT

As noted, a critical consideration in the general theory of production is that alternative investments are available and no particular outlay should be made unless it earns at least as much as these best available alternatives. The competitive process is such that all these alternatives are epitomized by the yields on comparable assets valued in stock markets. Such yields set a lower limit because no one would invest outside the stock market for less than it provided. (Strictly speaking, the argument relates to expected yields.) The competitive processes outlined above insure the stock market yields are also floors. Higher yields produce either entry or bidding for access to superior resources that leads to the elimination of excess profits or their transfer to the owners of the specialized resources.

The yields are normally measured as annual percent rates of return (and used in the actual formulas as decimal fractions). The basic starting point is the compound interest formula—$I = P(1+r)^t$ where I is the final amount, P, the initial amount, r, the rate of return, and t, the time in years involved. This tells how much an initial investment accumulates to when invested for t years at the market rate of interest r.

By simple algebra, this formula becomes $P = I/(1+r)^t = I(1+r)^{-t}$; this is known as the present value formula. It tells us how much must be invested in the stock market to secure I in future income. We would never spend more than P, therefore, to generate I in nonstock market income.

In what follows, I move between explicit and implicit consideration of present value as is appropriate (but usually keep the concept implicit). This is possible by noting that any wage or price at any time can be converted into its present value factor for a given pair of r and t. It is often convenient to ignore this multiplication or introduce a notation that implies the multiplication has been made. Thus, the typical price might be called P_{kot} with the k indicating what commodity was involved, o the time to which the price is discounted, and t the time at which the price is received, $P_{kot} = (1+r)^{-t} P_{ktt}$.

THE BASIC THEORY OF THE FIRM

The standard textbook theory of the firm presumes the existence of a production function that indicates the mix of output possible for a firm given its choice of inputs. The function is defined so that the technically superior options are the only ones included. Technical superiority means that one option differs from another by producing more of one or more goods with no more inputs and no fewer outputs of other goods than another option or the same amount of outputs with less use of some outputs and no increase in the use of other options.

For the purposes of developing the theory, it suffices to agree that such a production function exists and proceed immediately to discuss how its existence affects decision making. However, it can be noted that since World War II considerable effort has been devoted to the use of linear programming to deal simultaneously with the technological and economic choice of techniques. Linear programming allows determination of the profit-maximizing combination of inputs and outputs given a finite number of input output *proportions* allowable. The model also assumes that each pattern can be maintained at any scale and the same input-output proportions will prevail. This is what was defined above as constant returns to scale.

The key difference between linear programming and more traditional methods is that the first assumes that at any given scale, only a finite number of choices exist and the second assumes that infinitely many choices exist. Mathematical principles indicate that where a large number of choices are available either method will produce results with negligibly small errors. The differences between a continuous infinite set of choices and a large finite number of options are trivial. Thus, the choice between the alternative analytic approaches can be made on the basis of analytic convenience. (See the mathematical Appendix to this Chapter).

The finite-choice formulation is desirable in applied studies because computational techniques are available that unfailingly find the optimum, if it exists, for a linear program. In contrast, no comparable techniques exist for solving more complex continuous models. However, general theoretical expositions stress the continuous case. The primary rationale is that the analysis can use a more widely known technique—the optimizing techniques of differential calculus. In practice, a generalization called after its developers the Kuhn-Tucker method of these calculus based techniques that grew out of the mathematical extensions of linear programming techniques is now used to provide a more sophisticated form of the theory.

The basic problem is to determine the wealth-maximizing pattern of input and output. The critical concern is to take account of the limits to profit caused by the costs of securing output. To get more income by producing and selling more of one good, it is necessary either to hire more inputs or reduce the output of other goods. It is these necessities that are measured by the production function.

The resulting analysis gives specific form to the conditions already discussed—that actions are started if their benefits exceed their cost and proceed to the point at which marginal benefits equal marginal costs. To see what is involved, it is desirable first to discuss further the basic benefit-cost calculus that generates the various special conditions derived below.

THE BASIC CONDITIONS OF PROFIT MAXIMIZATION

The conditions, therefore, reflect the basic economic principle that the marginal benefits equal the marginal costs *at the level of production or utilization actually selected for the inputs and outputs involved.* The diminishing payoff principle discussed above applies here. An input or output is introduced in the first place, only if the benefit (in the form of higher sales revenue or lower input cost) of starting the introduction exceeds the cost (in the form of outlays on the additional input or lower sales of an output).

The diminishing payoff principle says that marginal effects decline as the level of activity expands. Any of the substitutions among outputs and inputs, by definition, means expanding the role of one and contracting the role of another. The diminishing payoff indicates that the benefit from expanding declines as expansion continues. It also implies the cost rises. The cost is a contraction of some activity, and at this lower level, the impact of change is higher.

Thus, an input or output is introduced if the benefit of the initial increment of use exceeds the cost. The diminishing payoff principle says that, as more increments are made, benefits will fall and costs rise. The assumptions of a continuous variation in levels implies a similar continuity in the variation of benefits and cost will prevail. Thus, eventually a level of action will be reached at which the excess of benefits over cost will have been replaced by an equality. The equilibrium conditions outlined above will prevail.

The rule then says that the firm should go up to but no further than the point at which the marginal benefits of an action equal its cost. This can be rationalized in at least two ways. First, and most basically, what the firm is concerned about is the difference between benefits and costs. By definition, this difference is the profits that the firm seeks to maximize. Equality of marginal benefits to marginal cost is equivalent to a zero net marginal gain. (If G is gain, B, benefit, and C, cost, $dG/dQ = dB/dQ - dC/dQ$. Then $dB/dQ = dC/dQ$ directly implies $dG/dQ = O$). The rule simply says that an activity should be extended up to the point at which the net gain is zero.

Given the diminishing payoff principle, at lower levels, the net impact is a gain, and further expansive action is needed. Similarly, at higher than optimum levels, each further step produces a loss and should be reversed. The firm should stay at the point of a zero marginal net gain. Those familiar with the principles of optimization in the differential calculus will note that this rule corresponds to the basic optimizing condition that the first derivatives equal zero.

The diminishing payoff principle is related to the concept that the characteristics of the second derivatives determine whether the point at which the first derivatives are zero is a maximum or minimum. (For the two-variable case, the requirement for a maximum is a negative second derivative. For the general case, the requirement is that the Hessian—a matrix of all the second derivatives—have the property known as negative definiteness.)

The alternative approach is to state this argument more implicitly by making comparisons between the gross benefit and gross cost of each step and then deducing that a positive difference implies more should be done, a negative difference means less should be done, and a zero difference implies the optimum has been reached.

The Kuhn-Tucker conditions then add the explicit rule that if there is no way for the benefits of zero production or use of a given product to exceed the costs, no production or use of that product should occur. (Note the condition says that the net benefit, at most, is zero for the zero level of activity. Thus, with diminishing payoff, the net gain would be negative for any positive level of activity. Note also that a comparison between unused activities is unhelpful. It tells which is better but not that the better one is still not good enough. The germane comparisons, made below, are between the unused input or output and ones that it is actually desirable to use or produce).

THE MATHEMATICS OF WEALTH MAXIMIZATION

In the next sections, the specific optimizing conditions are derived and discussed. The first step is to use some basic mathematics to develop building block results. Then these results are manipulated to produce the conditions actually needed. First, the classical calculus case in which all inputs are used and all outputs are produced is considered. Then the Kuhn-Tucker extensions are developed to handle the conditions under which inputs are left unused and outputs unmade. Discussions are provided of the interpretation of each benefit-cost comparison derived. Specifically, the nature of the benefit and cost in each case is explained to simplify discussion, pure competition is assumed and prices are taken as given by each firm. Note is taken of the change produced by removing that assumption.

Mathematically, the production function is called a constraint or side condition on profits. Profits are themselves defined as $W = P_k Q_{kj} - R_e Q_{ej}$ where W is wealth, P_k is the price of output, k, Q_{kj} is the output of good k by firm j, R_e is the cost of input e, and Q_{ej} is the use of input e by firm j.

Were is not for the production function, profits could be made infinitely large by indefinitely expanding output and eliminating inputs. However, the production function implies, as noted, that this is impossible. This phenomenon of the change in one variable necessitating a change in another because of a side condition is treated in advanced differential calculus by the method of Lagrange. This method does not by itself produce particularly interesting results. However, it is the most rigorous way to lead us by further manipulations to the critical rules we need.

An infinite number of ways exist to offset any change in any single variable in this model. Therefore, it would be exceedingly tedious explicitly to consider all possible offsets to a given change. The method of Lagrange allows summarizing all possibilities in a much smaller number of optimizing conditions. Specifically, a new variable λj is introduced for each side condition in the model. It can be shown that all the optimal changes in other variables can be measured by the λj.

Thus, for every output, an optimum condition of the form $P_k + \lambda \partial f / \partial Q_{kj} = O$ prevails in a competitive equilibrium. (λ is what is known as a Lagrangean multiplier and $\partial f / \partial Q_{kj}$ is the derivative of the production function with respect to Q_{kj}). Analagously, $-R_1 + \lambda \partial f / \partial Q_{1j} = O$.

By arrangement, we see that the ratio of any price to any derivative of the production function with respect to the input or output with which the price is associated is equal at least in absolute value to the single Lagrangean multiplier that applies to each constraint. (As the problem was structured, the output ratio equals

$-\lambda$ and the input ratio $+\lambda$.)

As manipulation and economic evaluation of these conditions shows, the constant proportionality results means that we insure that all of the possible desirable shifts in input or output made to offset an initial output or input change are equally costly in equilibrium. We spread the changes evenly enough to insure this result.

These conditions and the production function suffice to determine the most profitable input-output pattern for a competitive firm. The mathematics are completed by noting that these conditions can be derived by establishing the synthetic problem of maximizing the function $L = W+\lambda f(Q_k,Q_1)$ by differentiating with respect to each Q_k, Q_1, and the Lagrangean multiplier.

Kuhn and Tucker generalized this analysis for the case in which all Qs must be nonnegative and it is possible to produce less (or use more inputs) than the production function allows. The modification says that each derivative can be less than equal to zero, the product of the derivative and the variable to which the function is differentiated is equal to zero, and when an inequality prevails, the optimum value of the variable is zero.

Again, rearrangements and economic interpretations show that the additional Kuhn-Tucker conditions indicate that where there is no possibility of profit from any positive level of output or input of a given good, it should not be produced or bought by the firm.

The economic interpretation of the conditions applicable to the relationships among produced outputs and used inputs is developed by a series of manipulations. The substance is the same for all cases. First, the equalities prevailing with the Lagrangean multipliers and any pair of price-derivative ratios is converted into equalities of the ratios. Second, the equalities are rearranged so the ratio of the two derivatives appears on one side of the equation. Third, production functions are examples of the mathematical form called implicit functions.

A basic theorem about implicit functions converts the ratio of two derivatives into a single simpler derivative that measures the underlying trade-offs among inputs and outputs. Fourth, various forms can be developed for the manipulated equations. In every case, one side is a measure of the marginal benefits of the action—i.e., the gain from taking the step contemplated. The other side measures the marginal cost—the sacrifice made to secure the benefit.

THE SPECIFIC CONDITIONS DERIVED—CLASSICAL CALCULUS

The form of the benefit cost comparison is different for each of the three possible types of tradeoffs. (1) Output of one good can be increased by lowering the output of another good. (2) Output of a good can be increased by raising the use of some input. (3) The use of one input can be reduced by using more of another input. Because of the first and third options, the prior notation should be modified to distinguish a second output by the subscript *m* and a second input by the subscript *n*.

Given two inputs and two outputs that are actually used and produced, the following results:

The starting points are the derivatives

$$P_k + \lambda \frac{\partial f}{\partial Q_k} = 0 \qquad (1)$$

$$P_m + \lambda \frac{\partial f}{\partial Q_m} = 0 \qquad (2)$$

$$-R_e + \lambda \frac{\partial f}{\partial Q_e} = 0 \qquad (3)$$

$$-R_n + \lambda \frac{\partial f}{\partial Q_n} = 0 \qquad (4)$$

The respective rearrangements from subtracting the second term from both sides, and dividing by the partial derivatives, are:

$$P_k/\frac{\partial f}{\partial Q_k} = -\lambda \qquad (5)$$

$$P_m/\frac{\partial f}{\partial Q_m} = -\lambda \qquad (6)$$

$$-R_e/\frac{\partial f}{\partial Q_e} = -\lambda \qquad (7)$$

$$-R_n/\frac{\partial f}{\partial Q_n} = -\lambda \qquad (8)$$

Since the four expressions equal minus λ, they equal each other or:

$$P_k/\partial f/\partial Q_k = P_m/\partial f/\partial Q_m = -R_e/\partial f/\partial Q_e = -R_n/\partial f/\partial Q_n \quad (9)$$

To deal with substitution among outputs, we manipulate the first two terms; for substitution among inputs, the last two terms; for raising output by raising inputs, the middle two terms. The critical step is to multiply both sides of each pair of ratios by either of the partial derivatives as follows (with the middle term the first term simplified by cancellation):

$$\text{Output-output } P_k\frac{\frac{\partial f}{\partial Q_k}}{\frac{\partial f}{\partial Q_k}} = P_k = P_m\frac{\frac{\partial f}{\partial Q_k}}{\frac{\partial f}{\partial Q_m}} \quad (10)$$

or

$$P_m\frac{\frac{\partial f}{\partial Q_m}}{\frac{\partial f}{\partial Q_m}} = P_m = P_k\frac{\frac{\partial f}{\partial Q_m}}{\frac{\partial f}{\partial Q_k}} \quad (11)$$

$$\text{Input-output } P_k\frac{\frac{\partial f}{\partial Q_k}}{\frac{\partial f}{\partial Q_k}} = P_k = -R_e\frac{\frac{\partial f}{\partial Q_k}}{\frac{\partial f}{\partial Q_e}} \quad (12)$$

or

$$-R_e\frac{\frac{\partial f}{\partial Q_e}}{\frac{\partial f}{\partial Q_e}} = -R_e = P_k\frac{\frac{\partial f}{\partial Q_e}}{\frac{\partial f}{\partial Q_k}} \quad (13)$$

$$\text{Input-input } -R_e\frac{\frac{\partial f}{\partial Q_e}}{\frac{\partial f}{\partial Q_e}} = -R_e = -R_n\frac{\frac{\partial f}{\partial Q_e}}{\frac{\partial f}{\partial Q_n}} \quad (14)$$

or

$$-R_n\frac{\frac{\partial f}{\partial Q_n}}{\frac{\partial f}{\partial Q_n}} = -R_n = -R_e\frac{\frac{\partial f}{\partial Q_n}}{\frac{\partial f}{\partial Q_e}} \quad (15)$$

The implicit function rule shows that any ratio of a pair of partial derivatives of an implicit function follows the rule

$$\frac{\partial f}{\partial Q_i} / \frac{\partial f}{\partial Q_j} = -\frac{\partial Q_j}{\partial Q_i} \quad (16)$$

so

$$\frac{\partial f}{\partial Q_k} / \frac{\partial f}{\partial Q_m} = -\frac{\partial Q_m}{\partial Q_k}\text{[f]} \quad (17)$$

Substituting this and analagous terms yields

$$P_k = -P_m\frac{\partial Q_m}{\partial Q_k} \quad (18)$$

$$P_m = -P_k\frac{\partial Q_k}{\partial Q_m} \quad (19)$$

$$P_k = R_e\frac{\partial Q_e}{\partial Q_k} \quad (20)$$

$$R_e = P_k\frac{\partial Q_k}{\partial Q_e} \quad (21)$$

$$-R_e = R_n\frac{\partial Q_n}{\partial Q_e};\ R_e = -R_n\frac{\partial Q_n}{\partial Q_e} \quad (22)$$

$$-R_n = R_e\frac{\partial Q_e}{\partial Q_n};\ R_n = -R_e\frac{\partial Q_e}{\partial Q_n} \quad (23)$$

The general term for the derivatives $\partial Q_j / \partial Q_i$ is, as noted, a marginal rate of substitution between i and j (MRS_{ij}). As also noted, it is the

[f]The notation for handling of this type of substitution differs among writers. Many use a dQ_i/dQ_j form. This follows directly from the way a total derivative is written and manipulated to produce the implicit function rule. However, a hybrid sort of derivative is involved in which two variables are changed while others remained constant. Some writers such as Varian (1978) and Nadiri (1982) prefer to use the partial derivative notation to convey the partial nature of the change, and because their approach seems preferable, it is followed here.

amount of good j sacrificed to get more of good i. The derivative, $\frac{\partial Q_k}{\partial Q_e}$, the amount of good k produced when more of input e is used has earned the special name of the marginal productivity of factor e (in producing good k) $-MPP_{ek}$. The differences in signs among the six equations reflect the underlying difference in the sign of its $\partial Q_j / \partial Q_i$ term. These are negative when inputs are traded for inputs or outputs for outputs but positive when input-output changes are made. This simply reiterates the points made before. Greater input use permits greater output. Greater output through adjusting output requires a decline in some other output. Similarly an increase in one input can allow another to decline.

As argued before, the price on the left in equations 18 to 20 is the benefit of the adjustment. The right hand term is the cost. For the equations 21 to 23, the left term is the cost; the right, the benefit. The reversal occurs because inputs are burdens on profits.

In any case, P_k or P_m represents the payoff to a (competitive) firm from increasing output one unit. In the first two cases, the cost is loss of the value of the lost output. The amount of this output lost is given by the marginal rate of substitution. Then the product of the MRS and the unit price of the lost output measures the cost of the loss. In the middle two equations, the benefit again is increased output. In the third equation, we have the expression for the benefits and costs of a unit increase in *output*. The benefit shown on the left is the value of output. The cost is the input price times the number of units of inputs needed to get a unit of output (which is the reciprocal of the marginal productivity) so an alternative formulation is

$$P_k = R_e / \frac{\partial Q_k}{\partial Q_e}. \tag{24}$$

In equation 21, the analysis is in terms of the effects of adding a unit of input. The cost *on the left* is that of one unit of *input*. The benefit *on the right* is the product of the price of one unit of output times the marginal productivity—the number of units of output produced by a unit of input.

Finally, in input-input substitution, the benefit comes from reducing use of one input, the cost, from raising the use of another. The left hand terms give the cost of one unit of output. The right hand gives the benefit—the unit price of the input whose use is reduced times the amount of input reduction made possible by using more of another input. Thus, each equation is a specific form of the principle that an optimum occurs when marginal cost equals marginal benefit.

In production analysis, the generalized concept of marginal cost is often used. This is defined as the cost of increasing output in the optimal way. This cost $\frac{\partial C}{\partial Q_k}$ is equal to *all* the costs of a unit of output derived here.

In equilibrium all the costs equal P_k so we have, in brief

$$P_k = -P_m \frac{\partial Q_m}{\partial Q_k} = R_e \frac{\partial Q_e}{\partial Q_k} = R_n \frac{\partial Q_n}{\partial Q_k} = \frac{\partial C}{\partial Q_k} \tag{25}$$

More generally these qualities prevail for every marginal rate of substitution involving good k and all inputs used and all other outputs produced. The marginal cost of any method of raising output must equal, in equilibrium, the marginal costs of all other methods. This is another consequence of the profit-maximizing assumption. If a cheaper method existed, it would be the one adopted.

The diminishing payoff principle suggests that actually several routes will be pursued. A single-minded reliance on one method of output expansion would raise its costs. At some point, the cost will rise above those of another method and that method would be adopted.

THE GEOMETRY OF THE THEORY

Given that we can only draw in two dimensions, we can only imperfectly graph these relationships. Simple graphs are possible only in dealing with a pair of goods with all others held constant. The respective curves are the production possibility curve for a pair of outputs, marginal productivity curve for input output choice, and the isoproduct curve for inputs.

The traditions with the three types of curves are somewhat different. The productivity curve analysis generally stresses the case of all other inputs and outputs fixed. The isoproduct analysis for inputs invariably involves consideration of output variation. The dominant form of production possibility analysis involves fixed in-

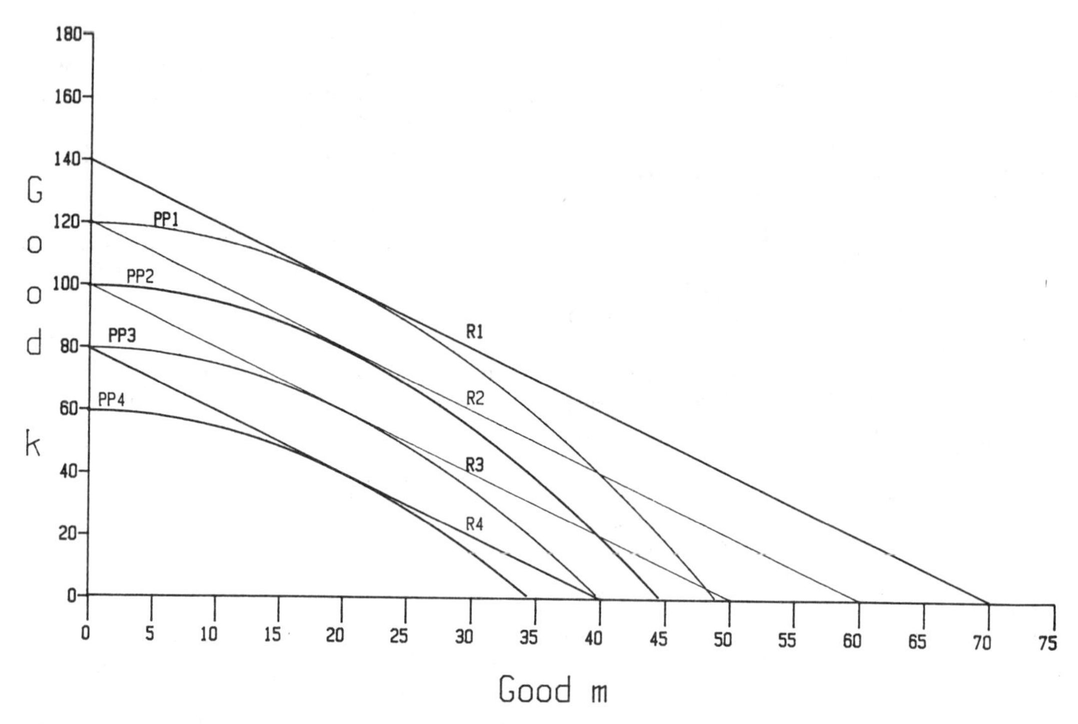

Fig. 2.4.4—Production possibilities.

puts and other outputs, but some work has been done on the case of different input levels.

Fig. 2.4.4 shows the single input level form of production possibility analysis. The production possibility curves (PP_1 to PP_4) measure all combinations of Q_m and Q_k possible with a given set of inputs and all other outputs held constant. The slope is $\frac{\partial Q_m}{\partial Q_k}$ which is negative. The curve is concave from below which can be shown by basic mathematical analysis to indicate that substitution becomes more difficult—each increment of good m requires a greater loss of good k than the prior increment.

Then a set of profit-lines (R_1 to R_4), $R = P_kQ_k + P_mQ_m$ can be defined to show the profitability of different combinations of Q_k and Q_m. The slope of each is the constant $-\frac{P_k}{P_m}$. Rearranging (18) or (19) yields $-\frac{P_k}{P_m} = \frac{\partial Q_m}{\partial Q_k}$ or that the slope of the two lines should be equal. Thus, of the infinitely many lines of the form $R = P_k + P_m$, R_o the one that is tangent (having equal slope) is the preferable one. Curves above R_o are infeasible; curves below are infeasible.

Extension to multiple input levels need not be graphed. A different production possibilities curve exists for each input level and a different profit line is tangent to each curve. Only calculating the profits at each level can determine which of the curves is the best on which to operate. (Higher input levels would be assumed to produce higher output levels). (Discussions of the subject are strangely silent on determining which, if any, of the curves is preferable. It turns out that the answer is a generalization of the principles discussed below for determining the optimum output of the one product firm. Also see below for a discussion of price determination).

Before discussing the geometry of input output relations, we should examine two basic concepts related to a marginal change. First, we have the total—the thing changed. For example, for an output Q_k being changed by varying an input Q_e the total cost is $C(Q_k)$, the total productivity is Q_k itself, the marginal cost is $\frac{dC(Q_k)}{dQ_e}$ and the marginal productivity is $\frac{dQ_k}{dQ_e}$. An average can be computed by dividing the total value by the quantity—$\frac{C(Q_k)}{Q_e}$ for cost and

$\frac{Q_k}{Q_e}$ for productivity.

Moreover, a distinct relationship exists between the average and the marginal value—namely, they are equal when the *average* attains a maximum or minimum. The rigorous proof is found by noting that (1) by definition the total is the product of the quantity and the average value and differentiating this product with respect to quantity. Symbolically $T = AQ$

$$\frac{dT}{dQ} = A + Q\frac{dQ}{dA}$$

By the rules for determining a maximum or minimum $\frac{dQ}{dA}$ is zero at a maximum or minimum so at those points $\frac{dT}{dQ} = A$—the marginal equals the average.

When, as in the productivity case, the presumption is that the average rises and then falls, the vision (see Fig. 2.4.5) is for the marginal to peak and then decline. The result is that the average continues to rise for a while. The decline in the average occurs only when the marginal falls to the average.

This is because the change in the average depends on the relation between the marginal and the average. The relationship between the marginal at one output value and another is by itself not indicative of whether the average rises or falls. Specifically, an addition that is below the average lowers the average and any addition that is above average raises the average.

The algebra of this process is such that simple numerical examples can treat only one of the two principles of the relationship between average and marginal values. This is shown in Table 2.4.1. The first panel shows a series of variations in the average, marginal, and total that illustrates the principle that a marginal value below the prior marginal value but above the average for the prior number of inputs raises the average. Thus, with an increase from 11 to 12 units of inputs the marginal product falls from 15 to 14. However, the average at 11 units was 10, so 14 is an above average increase in productivity and raises the average to 10.33. This example has the average continuing to rise as the input level goes up to about 15 units of input and then declines.

However, we do not show that the actual maximum can be attained where the average

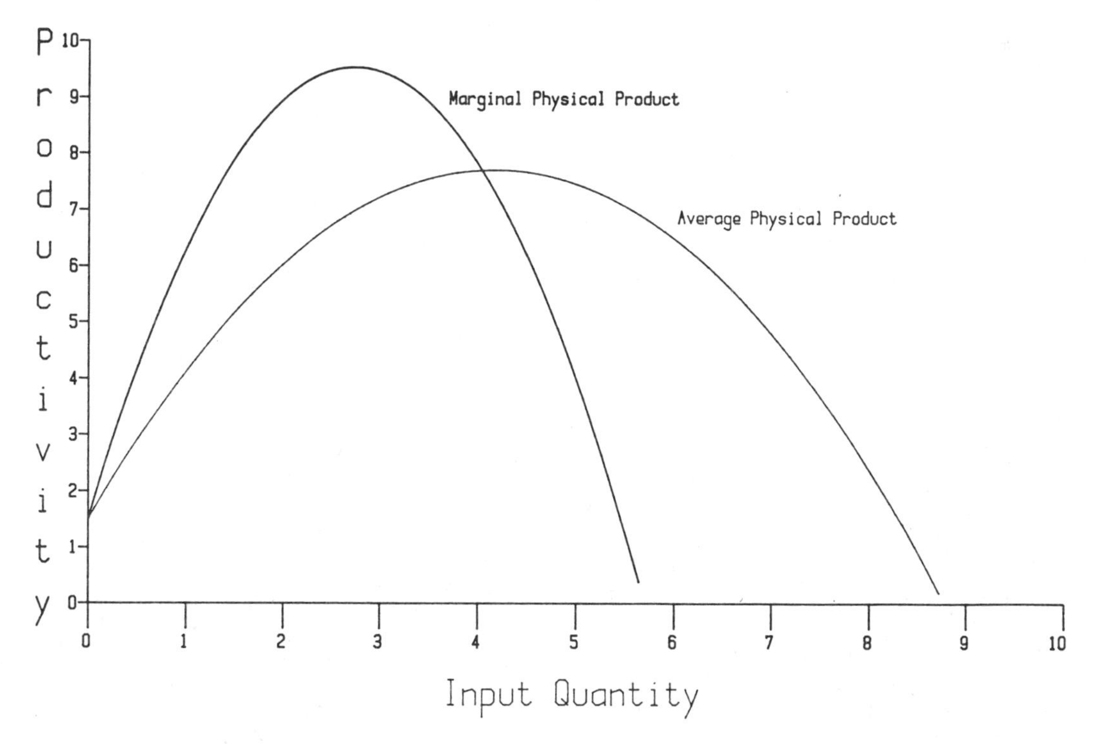

Fig. 2.4.5—Factor productivity.

Table 2.4.1—The Relationship Between Total, Average, and Marginal Values

A. Demonstration That the Rise in Average Depends on Whether the Marginal Is Above or Below the Average

Level of Activity	Total	Marginal	Average
1	5	5	5.0
2	11	6	5.5
3	18	7	6.0
4	26	8	6.5
5	35	9	7.0
6	45	10	7.5
7	56	11	8.0
8	68	12	8.5
9	81	13	9.0
10	95	14	9.5
11	110	15	10.00
12	124	14	10.33
13	137	13	10.54
14	149	12	10.64
15	160	11	10.67
16	170	10	10.63
17	179	9	10.53
18	187	8	10.39
19	194	7	10.21
20	200	6	10.00

B. Demonstration That the Average Peaks When It Equals the Marginal

Level of Activity	Total	Marginal	Average
1	10	10	10
2	24	14	12
3	39	15	13
4	52	13	13
5	60	8	12
6	66	6	11
7	63	−3	9

equals the marginal. This is best shown by another numerical example in which we immediately lower the marginal to the peak average value in the increment after the average is hit.

The same principles apply to situations such as the cost of a firm where marginal costs first fall and then rise. However, then the marginal is regularly below the average and the average is falling both throughout the stage of falling marginal costs and the initial stage in which marginal costs are rising but still below the average. The marginal equals the average at the low point of the average and exceeds the average as the average starts to rise.

In any case, the argument for an initial stage of rising productivity for one input as the given set of other inputs is that at such low levels too few of the variable input are used given the level of the fixed inputs. Thus, a firm might have 10 machines and with five workers, it would be necessary to shift workers between machines. The lost time in shifting would make them less productive on average than if one man were available for each machine. Conversely, once the machines are utilized to capacity, adding workers is likely to be less productive. Some increase might be produced by adding helpers, but this might increase output less than adding the operators.

In any case, we can derive from Fig. 2.4.5, Fig. 2.4.6 that expresses equation 21. We multiply *MPP* by P_k to get the value of an increase of output produced by the input increase—MPP is the number of additional units produced and P_k the input price so their product is the value of the marginal output. This marginal value product is set equal to the going market price P_e of the input. This simply says that the additional income from increased input use exceeds the cost, input use should rise and the stopping point is when the revenue gain just matches the input price.

Analogously, we can define an average value product by multiplying the average productivity by the price. It should be noted that under the assumption of fixed output prices every point on *MVP* and *AVP* is the point on *MPP* and *APP* respectively corresponding to the particular input level scaled up (or down) by the factor P_k. In particular, at the point of maximum average productivity Q_o, we have $MPP(Q_o) = APP(Q_o)$ so $P_k MPP(Q_o) \equiv MVP(Q_o) = P_k APP(Q_o) \equiv AVP(Q_o)$. This says that average value product attains its maximum at the same output as does average productivity.

It should be further noted that only the decreasing average value product portion of the curve is relevant. P_e cuts *MVP* at two input levels Q_2 and Q_1. At Q_2, however, *AVP* is below *MVP* and P_e, so the total revenue, *AVP* times Q_2, is less than the total cost $P_2 Q_2$. However, with decreasing *AVP*, *AVP* exceeds *MVP* and thus P_e so a profit occurs. (Also shown is the equilibrium input Q_o when P_o, the highest input price consistent with profitability, prevails.)

All this is a geometric representation that a maximum occurs when the second derivative is negative. The profit function is $TPP\ P_k - P_e Q_e$ where *TPP* is total physical product and equals $APP\ Q_e$. The first derivative is

$$P_k \left(APP + Q_e \frac{dAPP}{dQ_e} - P_e\right)$$

The second derivative is

$$P_k \left(\frac{dAPP}{dQ_e} + \frac{dAPP}{dQ_e} + Q_e \frac{d^2APP}{dQ_e^2}\right) = P_k \left(\frac{2dAPP}{dQ_e} + Q_e \frac{d^2APP}{dQ_e^2}\right)$$

which will be clearly negative if $\frac{dAPP}{dQ_e}$ is negative—a downward stage and $\frac{d^2APP}{dQ_e^2}$ is not positive by a large amount. The underlying mathematics, in fact, dictates that $\frac{d^2APP}{dQ_e^2}$ will be zero or negative in the decreasing average productivity case or at least not highly positive because then APP will start to rise again.

A simple version of the standard cost curve analysis of economic texts can be derived by shifting from the equation 21 to the equation 20 version of the optimum input output relation-

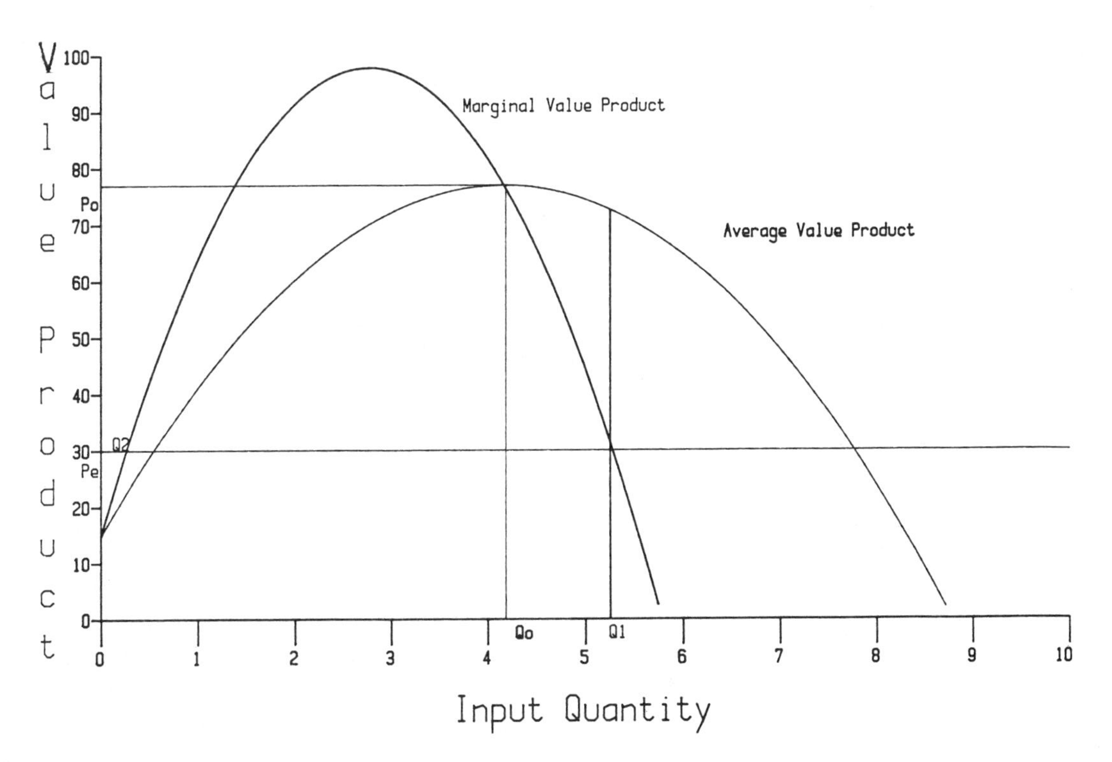

Fig. 2.4.6—Value of factor product.

ship. Here we deal with $\frac{dQ_e}{dQ_k}$ the marginal input needed to produce a unit of Q_k. This is the reciprocal of the MPP. Thus, when MPP is rising, $\frac{dQ_e}{dQ_k}$ is falling, and if *MPP* is falling, $\frac{dQ_e}{dQ_k}$ is rising. A similar relationship applies to average input needs $\frac{Q_e}{Q_k}$. It moves in the opposite direction from *APP*.

Specifically, in the range of outputs from zero to say Q_{kO} associated with a rising *MPP*, $\frac{dQ_e}{dQ_k}$ falls; at the higher outputs associated with *MPP* falls, $\frac{dQ_e}{dQ_k}$ rises. A similar relationship prevails for $\frac{Q_e}{Q_k}$ and *APP*. Combining, so long as $\frac{dQ_e}{dQ_k}$ is less than $\frac{Q_e}{Q_k}$, $\frac{Q_e}{Q_k}$ falls. When $\frac{dQ_e}{dQ_k}$ equals $\frac{Q_e}{Q_k}$ the latter is minimized. Then $\frac{dQ_e}{dQ_k}$ rises above $\frac{Q_e}{Q_k}$ which itself increases with Q_k.

Then marginal and average cost curves can be developed by multiplying $\frac{dQ_e}{dQ_k}$ and $\frac{Q_e}{Q_k}$ by P_e. By reasoning similar to that of the *MVP-AVP* analysis, the point of minimum average cost occurs at Q_o when $APP = MPP$ and $\frac{dQ_e}{dQ_k} = \frac{Q_e}{Q_k}$ (the second equality follows from the first because the terms in the second are reciprocals of the first; if $x = y$, $\frac{1}{x} = \frac{1}{y}$). The same is true for *AC*. The reasoning then extends to having *AC* falls as long as *MC* is below *AC*, hit its minimum when *AC* equals *MC*, and start to rise but lies below *MC* when *MC* begins to exceed *AC*.

Fig. 2.4.7 plots a cost curve and shows that here it is the increasing average cost portion that matters. When *AC* is above *MC*, $MC - P$ implies $AC > P$ and $ACQ > PQ$ cost exceed revenues. When $MC > AC$, we have $P = MC > AC$ so $PQ > ACQ$, the firm at least breaks even. (Again the second-order condition is what is at work). Again the equilibrium for a given P_k is

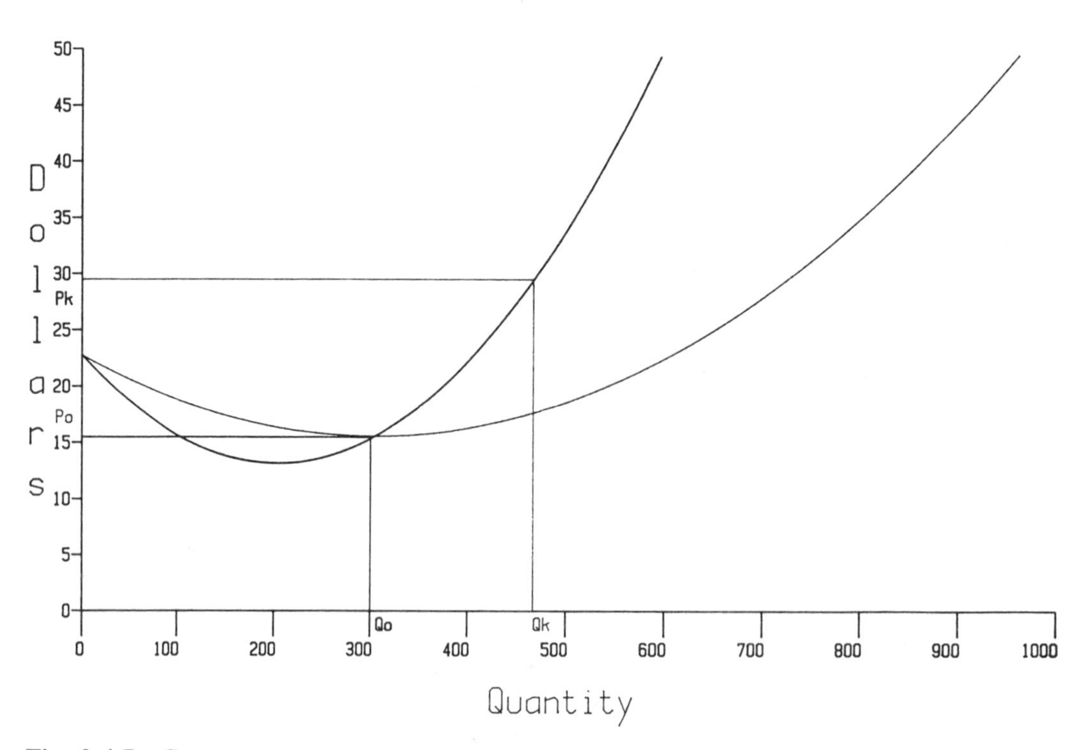

Fig. 2.4.7—Cost curves.

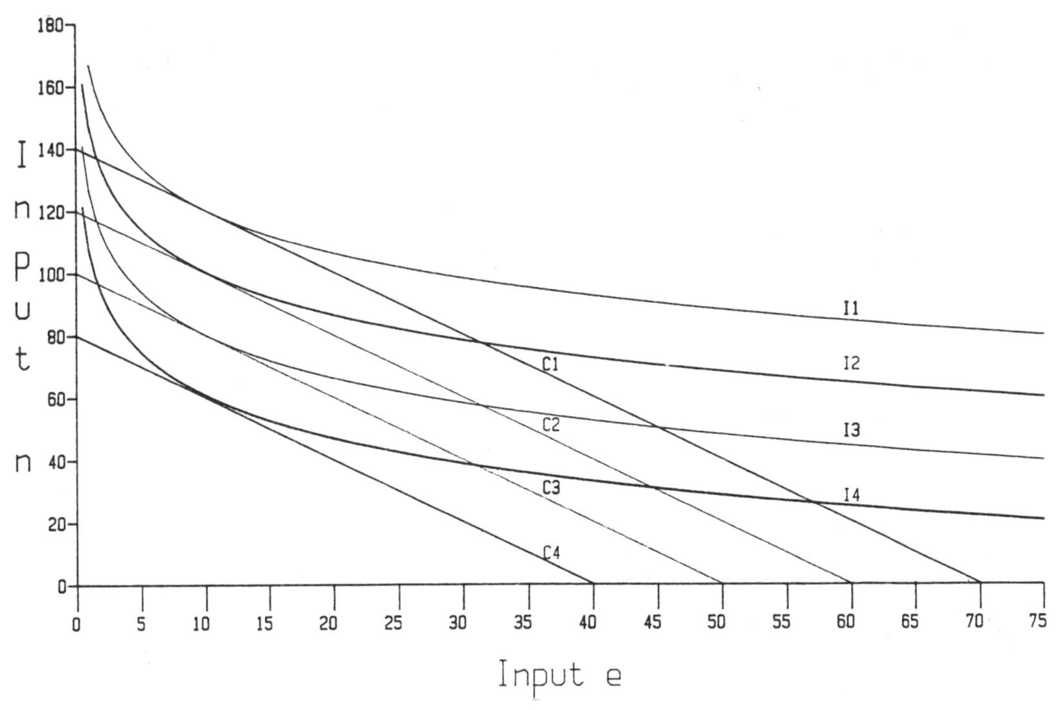

Fig. 2.4.8—Optimal input use.

when $MC = P_k$ and AC is rising.

The treatment of input-input choices is simpler than the input-output case (see Fig. 2.4.8) and similar to the output-output case. We define a curve of the input combinations that produce a given set of output. It has a slope $\frac{dQ_n}{dQ_e}$. This curve has the opposite curvature as the output-output curve. The input-input substitution curve, usually called the isoproduct curve, is convex viewed from below. This says more and more of one input must be added to offset each successive loss of the other input. This is closely related to the prior argument that MPP tends to fall when output rises. This was explained as a rise because a larger number of variable inputs have to share the same amounts of other inputs. In the input-input variation case, the larger number of one input has to work with fewer units of the other and becomes even less productive.

Again, isocost lines of slopes $-\frac{R_e}{R_n}$ can be defined and the optimum for any level of output is expressed by selecting the input pattern where the isoproduct curve is tangent to isocost lines, i.e., $-\frac{R_e}{R_n} = \frac{dQ_e}{dQ_e}$.

This, of course, is only a start of the analysis. The remaining tasks include (1) discussing how all the curves shift when other inputs or outputs vary, (2) how we move from those curves to demand curves for inputs and supply curves for output in competitive models, (3) discussing how these demand and supply curves interact to determine prices, (4) extend the analysis to Kuhn-Tucker cases of nonoutput or nonuse of inputs, and (5) introduce models of imperfect competition.

This discussion makes no pretense fully to develop all this. Economic-theory texts should be consulted for that purpose. Moreover, this section only sketches how the first issue is resolved and treatment of the other four points occurs in later sections.

The discussion of output-output substitution included an indicator of the impact of varying things other than the pairs considered thus far in the diagrams. Briefly, as noted, raising inputs or lowering other outputs will normally shift the production possibility curve outward. Exceptions arise when the inputs become so extensive that they start interfering with each other or when the resources released by reducing other outputs also interfere with production of the two outputs. (Conversely, raising other outputs lowers the output possible of the two goods).

Similarly, raising other inputs or lowering other outputs raises *MPP* and *APP* and lowers *MC* and *AC* for analogous reasons. Raising other outputs lowers *MPP* and *MC*.

A familiar demonstration in economic theory is that as more inputs are profitably added, the lower the *MC* and *AC* curves become. In the limit, it may become possible to increase output without increasing *MC* and thus have *AC* equal *MC* (since $MC = AC + Q \frac{dAC}{dQ}$ for *MC* to be constant $\frac{dAC}{dQ}$ must be zero or $MC = AC$).

Samuelson has called this the case of indeterminacy in purest competition. For a given price, the optimum output is either zero or infinity. If $P > MC$ for any level of output the $P = MC$ rule leads to infinitely increasing output. However, if $P < MC$, production does not pay. The discussion below of how demand sets a price and quantity shows that the market actually sets an output limit (but proves not to provide a limit to any one firm's output).

The ideas that the single product firm in competitive industry never operates in a range of decreasing average costs, has zero or infinite output if costs are constant, and finds its optimum output at an output at or above the point of minimum average costs generalize to the multiproduct firm. The analysis is more complex and best handled with mathematical tools discussed below.

Anticipating, the second point is the simplest to show. The multiproduct analogy to the constant cost or product firm is a firm that gets a percent increase in output exactly equal to the percent increase in inputs. If inputs rise 10%, output rises 10 percent. (This is known as constant returns to scale). Now with given input and output prices, for any arbitrary level of inputs and outputs either a proft or a loss will occur. Changing scale scales up the profit or loss. Thus, if one level of output is profitable, and is, if one level of output is unprofitable, all are.

The problems of failing to cover revenues prove to continue to prevail if output rises more rapidly than inputs and costs thus rise less rapidly. Conversely, a rise in output slower than the rise in the inputs and the resulting tendency of costs to rise faster than revenues are what sets a limit to output for profitable firms.

THE BASIC CONDITIONS IN KUHN-TUCKER ANALYSIS

The prior analysis can be repeated to deal with how the Kuhn-Tucker conditions rule out the use of some inputs or the production of some outputs. If *m* is an unproduced good, at $Q_m = O$, the following holds:

$$P_m + \lambda \frac{\partial f}{\partial Q_m} \leqq O \text{ or } \frac{P_m}{\partial f/\partial O_m} \leqq -\lambda \quad (26)$$

For an unused input *n*, at $Q_h = O$:

$$\frac{-R_n}{\partial f/\partial Q_n} \leqq -\lambda \quad (27)$$

given that output *k* is produced and input *e* is used

$$\frac{P_m}{\partial f/\partial Q_m} \leqq \frac{P_k}{\partial f/\partial Q_e} \quad (28)$$

$$\frac{-R_n}{\partial f/\partial Q_n} \leqq \frac{P_k}{\partial f/\partial Q_k} = \frac{-R_e}{\partial f/\partial Q_e} \quad (29)$$

The manipulations undertaken with the equalities previously presented can be employed here without altering the inequalities. Thus we end up with the following:

$$\text{Output-output } P_m \leqq -P_k \frac{\partial Q_k}{\partial Q_m} \quad (30)$$

$$P_m \frac{\partial Q_m}{\partial Q_k} \leqq -P_k \quad (31)$$

$$\text{Input-output } P_m \leqq R_e \frac{\partial Q_e}{\partial Q_m} \quad (32)$$

$$P_m \frac{\partial Q_m}{\partial Q_e} \leqq R_e \quad (33)$$

$$\text{Input-input } -R_n \leqq R_e \frac{\partial Q_e}{\partial Q_n}; R_n \geqq -R_e \frac{\partial Q_e}{\partial Q_n} \quad (34)$$

$$-R_n \frac{\partial Q_n}{\partial Q_e} \leqq R_e \; ; \; R_n \frac{\partial Q_n}{\partial Q_e} \geqq -R_e \quad (35)$$

Where a strict inequality applies (i.e., the left is less than rather than equal to the right), the interpretation is that even at zero levels, the unmade output is less valuable than its costs. Similarly, unused inputs cost more than their benefits. An equality at zero output or use does not change the outcome. As previously noted, the diminishing payoff concepts implies that any nonzero level of input or output will have a lower (than zero) net payoff and be undesirable. This undesirability is clearly worse if the diminution occurs from an initially negative level. Use or output occurs when an equality of costs to benefits occurs at positive levels of input and output.

The prior geometry can be modified to illustrate the Kuhn-Tucker results. Here the nature of the geometry is sketched but not shown in all cases. In the input-input case of Figure 2.4.4 and the output case of Figure 2.4.8, the geometry involves what are called corner solutions. For example, if it is optimum not to produce good m, it is because from 30, $\frac{P_m}{P_k} \leqq \frac{-\partial Q_k}{\partial Q_m}$ or $\frac{-P_m}{P_k} \geqq \frac{\partial Q_k}{\partial Q_m}$—algebraically the slope of the isoprice line (R_o^1) in 2.4.4 is larger—i.e., less steeply declining—than the production possibility curve at $Q_m = O$. Thus, it pays to stay at the corner because R_o is the highest R curve consistent with the production outputs of Q_m.

The nonuse of input Q_n in 2.4.8 would involve an isocost line more steeply falling than the isoproduct line at $Q_n = O$, i.e.,

$$\frac{-R_n}{R_e} \leqq \frac{\partial Q_e}{\partial Q_n}$$

(which is equations 34 and 35 rearranged).

The nonuse of an input to produce an output occurs when there is no marginal value product above input price. The nonoutput arises when no marginal cost is below price.

EXTENSIONS OF THE THEORY

We would expect that the output of goods would rise if their prices rose and that the purchase of inputs would fall if input prices rise. Unfortunately, it turns out these intuitively appealing propositions can only be proved by mathematical manipulation. This section presents the proofs which are somewhat complicated. Thus, those who are unfamiliar with the underlying mathematics must be content to rely on the basic reasonability of the propositions.

Modern discussions of these propositions rely on two lemmas (preliminary propositions that are considered uninteresting except as the basis for more important further analysis). One key lemma is credited to Hotelling. It holds that the firm's supply function

$$Q_k(P_k, R_e) = \frac{\partial W}{\partial P_k} \quad (36)$$

and the demand function for an input

$$Q_e(P_k, R_e) = -\frac{\partial W}{\partial R_e} \quad (37)$$

Varian suggests the following proof: define the synthetic function g such that $g\,(P_k,R_e) = W\,(P_k,R_e) - (P_kQ_k^* - R_eQ_e^*)$ where W is actual net profit, the two terms of the last expression are vectors of output and input prices, and the stars denote the profit maximizing values at some set of prices P_k^* and R_e^*.

By the definition of profit maximization, the actual profit maximizing output-input pattern at any arbitrary set of prices will be at least as profitable as the profits at those prices as the pattern (Q_k^*, Q_e^*). At P_k^*, R_e^*, however, the profits will be equal to $(P_k^*Q_k^* - R_e^*Q_e^*)$. Thus, g reaches a minimum at (P_k^*, R_k^*). Since the derivatives of a function at a minimum equal zero:

$$\frac{\partial g}{\partial P_k} = \frac{\partial W}{\partial P_k} - Q_k = 0 \quad (38)$$

$$\frac{\partial g}{\partial R_e} = \frac{\partial W}{\partial R_e} + Q_k = 0 \quad (39)$$

Rearranging, we get as postulated

$$Q_k = \frac{\partial W}{\partial P_k} \quad (40)$$

$$Q_e = -\frac{\partial W}{\partial R_e} \quad (41)$$

This proof applies for any set of output and input prices and so proves Hotelling's lemma. Then the slope of the output supply function is found by differentiating the first of the two results of Hotelling's lemma (i.e., equation 40).

$$\frac{\partial Q_k}{\partial P_k} = \frac{\partial}{\partial P_k}\frac{\partial W}{\partial P_k} = \frac{\partial^2 W}{\partial P_k^2} \qquad (42)$$

which should be positive if higher prices raise outputs.

Varian's proof of this last point relies on establishing that a profit function has the properties associated with a positive second derivative. The key is that with such a derivative, the curve lies on or below a plane connecting any two points on the curve. The input and output prices (P_k'' and R_e'') between two arbitrary levels will be weighted average of the prices at the end points—(P_k, R_e) and (P_k', R_e') respectively; i.e.,

$$P_k'' = tP_k + (1-t)P_k' \qquad (43)$$

$$R_e'' = tR_e + (1-t)R_e' \qquad (44)$$

where $O \leqq t \leqq 1$.

By definition

$$\begin{aligned} W(P_k'',R_e'') &= P_k''\,Q_k'' - R_e''\,Q_e'' \\ &= t\,(P_k\,Q_k'' - R_e\,Q_e'') \\ &\quad + (1-t)\,(P_k'\,Q_k'' - R_e'\,Q_e'') \end{aligned} \qquad (45)$$

The profits of the optimal plan at prices (P_k'', R_e'') can be written as weighted averages of some higher and lower prices times the optimal qualities for (P_k'', R_e''). By arrangement, these become the weighted sum of the profitability of the (Q_k'', Q_e'') pattern at the lower and at the higher prices. However, by the assumption of profit maximization, the actual profits from the optimal output and input at prices (P_k, R_e) and (P_k', R_e') can involve patterns different from (Q_k'', Q_e'') and possibly higher profits. Thus,

$$W(P_k, R_e) \geqq P_kQ_k'' - R_eQ_e'' \qquad (46)$$

$$t\,W(P_k, R_e) \geqq t\,(P_kQ_k'' - R_eQ_e'') \qquad (47)$$

Similarly

$$(1-t)\,W(P_k', R_e') \geqslant (1-t)\,(P_k'Q_k'' - R_e\,'Q_k'') \qquad (48)$$

Combining

$$\begin{aligned} t\,W(P_k', R_e) &+ (1-t)\,W(P_k', R_e') \\ &\geqq t\,(P_kQ_k'' - R_eQ_e'') \\ &+ (1-t)\,(P_k'Q_k'' - R_e'Q_k'') \\ &= W(P_k'', Q_e'') \end{aligned} \qquad (49)$$

The first expression gives the value of any point on the plane connecting the two price points, and the combination shows that such a point lies on or below the profit curve between the two points. Thus, a positive second derivative prevails.

(A second key lemma is due to Shephard. He shows that the cost minimizing choice of inputs for a given production pattern is equal to the derivative of costs with respect to input price—$\frac{\partial C}{\partial R_e}$. Again a synthetic function (R_e) is defined as $C(R_e, \overline{Q}_k) - R_e\,Q_e^*$ where $\overline{Q}_k$ is a set of given outputs. The actual cost at arbitrary prices will equal or exceed the cost that is minimizing for any one set of input prices. It will equal those costs when the prices assumed in the right hand expression hold. Thus, the maximum of g occurs at $\frac{\partial C}{\partial R_e} - Q_e^* = O$ which leads to the lemma).

THE IMPACT OF HISTORICAL EXPERIENCE

The theory of production lays great stress on the influence of past decisions upon what can be done at any moment. Unfortunately, this is one of the many critical economic issues that are normally treated in an oversimplified fashion. The usual concept is of something called fixed costs normally conceived of as consisting of the expenses of holding capital equipment or possibly certain types of managerial labor.

It is suggested that at any point, the firm may have more or less of the equipment than would be consistent with satisfaction of the equilibrium conditions presented above. In either case, costs *in toto* would be higher than the profit maximizing level determined by the equilibrium conditions. However, part of that cost would be the uncontrollable expenditures and fixed equipment, and only the remaining variable costs of those inputs that can vary affect the operating decisions of the firm. The analysis proceeds to point out that with an excess, less is used of other inputs than would have been had the firm been free to adjust all inputs. Conversely, with a deficiency, more of other inputs are used than would be the case if all inputs could be varied.

This approach has several limitations as an adequate explanation of the implications of past commitments on current decision making. The identification of fixed costs with capital or management labor is a gross oversimplification. There can be fixed commitments to buy any input. Long-term contracts can be made for labor services other than those needed to continue the existence of the firm. Key materials such as fuel for electric power plants are often purchased on contracts.

Conversely, not all capital resources are fixed, and, more critically, not all commitments are as irreversible as assumed in fixed cost analyses. Numerous examples exist in which it is possible to transfer certain types of equipment into or out of any operation. Office space can, within limits, be expanded by additional leasing or contracted by subletting or sale. Many types of equipment such as that for earth moving can be, and often are, transferred, to different jobs sometimes in different industries. Shifts can be made between coal mining and road building, for example.

More subtlely, those with excess resources can arrange to trade with those with deficient resources. Such trading has been highly developed by the electric-power industry. Companies long on capacity regularly sell to capacity-short utilities. Arrangements can range from a deal to handle an emergency starting and ending in a few moments to agreements to sell power steadily for several years.

In addition, only some of the commitments made are unbreakable at any cost. The only clear-cut examples are investments in extremely specialized, immobile facilities such as mine shafts and petroleum wellholes. At a price, most contracts can be broken.

Another neglected consideration is the evolution over time of the undesirable level of fixed costs. A deficiency can be eliminated as quickly as it takes to recognize the need and undertake whatever is required to put the addition in place. An excess can be removed as fast as the commitment can be terminated. Which process is more rapid depends upon the nature of the input. One that can be built rapidly would be easier to add than drop. An office building in a leading city, however, might be more quickly sold than added. In any case, the fixed-cost problem should correct itself over time as adjustments are made.

Finally, the usual analysis only treats the adequacy of inputs at any one moment. The apparent implicit assumption is that the problem will persist until the use of the input can be altered. The treatment of the procurement of rigid inputs for a firm facing demand growth, fluctuation, or both is largely neglected. However, the prior optimizing conditions implicitly handle this problem. The conditions can be interpreted to show that an inflexible input will be procured in the quantity that produces the greatest amount of wealth for the firm. The firm will endure whatever temporary deficiencies or excesses of availability that might occur so long as such an action is cheaper than trying to vary procurement with immediate needs.

This last point is a critical one for evaluation of mineral production and particularly metal mining. When, as often occurs, the industry is in one extreme or other of the swings to which it is prone, pronouncements become widespread about the allegedly severe errors made in investment decisions. This may be the case in some instances.

However, in others, it is the criticism that is erroneous. For the reasons just outlined, and industry facing wide swings in demand is best advised to adopt a level of capacity somewhere between that suited for the high and the low level of demand. It is the long-run average (appropriately weighted by present value factors) that matters.

BREAKEVEN—THE BEST MUST BE GOOD ENOUGH

It was previously noted that firms facing vigorous competition cannot earn more than enough to yield a normal rate of return to investors. However, if the firm earns any less, it will go out of business. It is not necessary here to explore how firms survive this balancing process. Observation suggests the task is far less difficult in practice than it might appear from observing the apparently severe conditions imposed by theory. Firms have survived for many generations in such allegedly perilous industries as nonferrous metal mining (and die frequently in the supposedly stable realm of retailing). This suggests that many managements do learn how to adapt. Instability is less important than unpredictability. Thus, retailing proves more problematic than mining because it involves more unforeseeable changes.

What is critical is discussing the implications of the requirement that firms must break even. The essence is that in addition to the requirement that all the marginal conditions presented above are satisfied, it must be true that the resulting production pattern is more profitable than not producing any more.

Thus, the firm in general regularly must implicitly compare the situation if it liquidated (now or at some later date) to that of continued operations. The prior discussion of the varied nature of existing commitments and their duration should suggest that the actual calculation can prove quite complex. The ambiguity of the concept of variable cost implies comparable problems in applying the rule that the present value of revenues exceed the present value of variable cost.

The desirability of continued existence can be over or understated by different errors in appraising costs. On the one hand, the existence of enforceable commitments to compensate various employees and suppliers for the termination may lessen the attractiveness of such a move. Conversely, as the decay of commitments noted above proceeds, fewer costs are fixed, and termination becomes increasingly attractive. A chronic source of criticism of government policies is the tendency to overlook such natural transitions. Governments tend to fight the decline of established industries. The aid often tends to go beyond protecting owners of existing resources and encourages replacement that otherwise would not have occurred.

THE GENERAL MATHEMATICS OF BREAKEVEN

A basic consideration in economic theory is the relationship between breakeven for a competitive firm and economies and diseconomies of scale. The prior analysis of the single product firm showed that the critical requirement for breakeven of such a firm was operating at or above the level of minimum average cost. In this section, a more general approach is developed but again it is a purely mathematic argument not readily reduced to verbal explanations. The basic points are that the mathematics shows that the one product case does generalize, but the concept of average costs disappears. A general argument is derived using Euler's theorem for homogeneous functions.

The argument is simplified without loss of generality if we first compute the case of one output firm. A homogeneous function is one in which when all the dependent variables are increased in the proportion a, the value of the function $f(x_i)$ increase in the proportion a^m—where m is the measure of the degree of homogeneity. Euler's theory shows that for such a function

$$mf = \sum_{i=1}^{n} \frac{\partial f}{\partial x_i} x_i$$

Thus, the sum of the products of all the partial derivatives of a function with the value of the variable with which the function is differentiated equals m times the value of the function.

To see what is involved economically, we multiply both sides of the equation by P_1, the price of the single good produced.

In the one-product-firm case, $f(x_i)$ is the output of that one product. $P_1 f$ is the revenue from that product and so $m\ Pf$ is an amount whose relationship to revenue depends upon the size of m. Each term in the sum on the right-hand side has the form

$$P_1 \frac{\partial f}{\partial Q_i} Q_i$$

Moreover, since $Q_1 = f(x_i)$ in this case, this can be rewritten

$$mP_1Q_1 = \Sigma\ P_1 \frac{\partial Q_1}{\partial Q_i} Q_i$$

From equation 21 above, the right-hand expression transforms into the cost of production if competitive firms are involved. The right-hand side has become the sum of payments to input suppliers.

The right-hand side measures total cost. Thus, Euler's theorem and the prior analysis combine to indicate that $mP_1Q_1 = TC$ —that m times the revenue equals the total cost. This can be rearranged into $P_1Q_1 = TC/m$. From this it should be clear that if m equals one, revenues equal costs, that if m exceeds 1 revenues fall short of costs and that if m is less than 1 revenues exceed costs.

The critical point to note is that the degree of homogeneity is the measure of the effect of scale. Constant returns of scale occur when $m = 1$—an equal increase in inputs increases output in the same proportion. Economies of scale occur with m greater than one—an increase in inputs in the same proportion raises output by a greater proportion. Diseconomies of scale occur when m is less than one—equal increases in inputs produce an increase in output less than the increase rate for inputs. Thus, the theorem shows that with constant returns to scale a competitive firm would lose money and with diseconomies, profits are earned. This argument can be extended to the multiproduct firm where f is reinterpreted as a fixed optimally configured pattern of outputs.

The two cases where exact breakeven fails to occur lead to opposite results. The economies-of-scale case with its problem of inadequate profits leads to consideration of the need to tolerate monopolies and the problems of ameliorating the impacts of their existence. The profitability of industries with diseconomies of scale leads to discussion of the source of diseconomies and the implications of these sources. In particular, it is argued that the difference usually is attributable to differences in the quality of inputs and that competition for access to the higher quality inputs will bid up their prices and transfer the difference in cost to the owners of the resource. The transferred incomes are then termed economic rents to specialized resources.

THE FALLACY OF COST ALLOCATION

Another critical point to note is that the criterion applies to the firm as a whole and cannot usefully be transformed into a rule relating to individual products. More precisely, the marginal conditions provide all the information needed about individual inputs and outputs in the optimal plan, and the breakeven rule tells whether optimum operation is better than closing down.

This point is important because cost accountants and policy observers regularly attempt to supplement the marginal requirements by individual product counterparts of the breakeven rule. Such efforts can at best lead to wasted time, and any effects using the allocations of costs to individual products can only be bad.

The critical argument here is that all the interesting questions are answered by the marginal efficiency and global breakeven conditions. The latter appraises the basic question of whether the firm should survive; the former tells the best pattern of survival. Nothing is left to appraise.

However, the persistence of allocation schemes and the absence of a readily available discussion of the shortcomings of such allocation schemes justify consideration of their inherent defects. The basic conclusion has been stated, so let me turn to its rationale. The first need is to make explicit the probable use of allocation of costs to separate products. The most likely application is to provide the single product counterpart of the overall breakeven rule. The natural choice is requiring that each product would be worth producing only if its price equalled or exceeded the imputed average cost.

Whatever workability such rules will have differs with the overall profitability of the firm. Three possibilities exist: (1) the firm will breakeven overall, (2) the firm will earn more than enough to breakeven, or (3) the firm will lose money. The first case is *easiest* to treat. The combination of the rule that marginal costs of each product equal price and Euler's theorem provide the basis for the analysis of allocation of average costs to individual products. Specifically the only workable cost allocation is the redundant one that average cost equal marginal cost.

Euler's theorem and the equality of marginal cost to price indicates that the sum of the products of the marginal costs of each product times its output will equal total costs. Therefore, average costs equal to marginal costs will add up to the total costs of the firm and fully allocate the costs. The identity of marginal and average cost insures that products profitable under marginal cost rules are profitable under average cost rules since the two costs are the same.

Any alternative allocation will inevitably lead to incorrect decision. If any product is assigned an average cost less than its marginal cost, it will appear to have made money. More critically, less of the total cost will have been allocated to the product, and some other product or products would have to bear this cost. This would raise their average cost above price and incorrectly produce the conclusion the other product was unprofitable. Thus, in the case of a competitive firm that breaks even, the only cost allocation that does not produce incorrect decisions is the redundant one of calling marginal costs average costs.

When the competitive firm has revenues greater than cost, it is no longer possible to divide up the total cost among firms by using the rule that imputed average costs equal the marginal costs of optimal output. The sum of the products of marginal costs times output would again equal revenue but exceed costs (since the latter by definition are less than revenues.

Nevertheless, marginal costs remain critical as the upper limit to the allocation of average costs to any product. No profitable-to-produce product should be assigned an average cost higher than the marginal cost of optimal output. With an imputed average cost above the critical level, the product would incorrectly appear unprofitable.

For a competitive firm with income in excess of the minimum needed, proper costing rules require allocating so that the average cost of all profitable products is no more than the marginal cost of optimal output. To limit allocated costs so their sum does not exceed total costs, some products must have an imputed average cost below the marginal cost of optimal output.

Finally, where losses occur, allocation of costs to products will always ultimately produce the correct conclusion that the firm should go out of business. However, the evaluation process will be an awkward and probably inordinately slowly working one. We now have a situation in which total costs exceed total revenue and thus if average costs are set equal to marginal costs whose product with output sums to revenue, imputed costs will fall short of actual total costs. To restore equality of the sum of the product of total costs and output to total costs, some imputed costs must be raised. The neatest choice is to set all accounting costs above marginal costs; this immediately leads to shutdown. All costs exceed all prices; all products lose money; shutdown is the inevitable choice.

Any other allocation will produce the same results more tediously. The alternative to all products making losses is that only some do. This will lead to the decision to drop these products. Since this means moving away from the wealth-maximizing pattern, costs will fall less than revenues. The resulting required reallocation of costs will make some of the remaining products unprofitable. They then will be dropped. This process must continue until all products are eliminated.

The existence of imperfect competition and an excess of price over marginal cost does nothing to restore the utility of allocating costs. Where the firm only breaks even, even with control over price, total costs exceed the sum of the products of the marginal costs times the outputs. (This can be seen by using Euler's theorem and the proposition that marginal prices exceed costs in imperfect competition).

Thus, some products must be assigned an average cost in excess of the marginal cost to insure that the costs are fully allocated. However, care must be taken that no cost imputation is too great. The only way this can be done is to insure that no cost is greater than the price realized on optimal sales levels. The only way this can be insured is to conduct the marginal evaluations.

Cost allocation is no more useful when profits are possible. Just where the average costs estimates should be set is not clear. However, again, we must use the marginal rules to insure no undesirably high average cost allocations are made. Finally, the argument for a money losing imperfect competitor is basically the same as that for a purely competitive firm.

In every possible case then, satisfactory average cost calculation rules only can be developed after the marginal cost of optimal production and the overall profitability of the firm are known. Satisfactory cost allocations merely are redundant restatements of the marginal rules and the overall profitability rule. The best possible result of assigning average costs to individual product is wasted effort. Rules devised by these methods can lead to serious error. Thus, the best a cost allocation can be is an innocuous waste. Most formulas used fail to observe the critical rules and are harmful.

THE PROBLEM OF OPTIMAL ORGANIZATION OF THE FIRM

Another important, but often-neglected, aspect of transaction-cost economies is the optimal organization of the firm. Coase (1937) pointed out that an important trade-off is associated with the procurement of any input by the firm. Every method of arranging for the supplies imposes different transaction costs. Differences also can prevail in the production costs arising from alternative approaches.

The primary productive effect is that increasing the responsibilities of management may lead to increasing marginal costs. The transaction costs include those of learning about input availability, those of arranging the agreement to secure the needed inputs, and those of monitoring the compliance with the accord.

Coase concentrated on a simple distinction between undertaking production of inputs—vertical integration—and buying these inputs. He was well aware that the analysis could be extended. A wide range of procurement options can and do prevail. Different options can be adopted for any input—natural or produced.

There are many different ways in which to buy ranging from casual short-term purchases in amounts to meet the needs of a few days or even hours to contract commitments lasting several decades. Arrangements arise that combine aspects of integration *and* purchase. Financial involvements of various types can be undertaken with outside firms. These can be equal partnerships, deals in which the outsider provides a limited portion of the inputs such as just the management or the management and the mining equipment, or simple guarantees of the investment. (All these approaches have been used in electric-utility coal and uranium procurement.)

While some of these options do not exist with labor inputs, it is still possible to adopt many different forms of relationships in this area too. We can note the existence at one extreme of suppliers of fill-in labor on a daily basis and at the other of formal contracts guaranteeing employment income over several years.

Coase suggested that in the integration versus purchase alternative, the former brought lower transaction costs but produced a greater tendency to increased marginal production costs than the latter. Integration should be extended so long as the marginal benefit in lower transaction costs was less than the marginal cost of higher production costs. Again, the process is self-limiting. As integration increases, the greatest transaction cost savings are exhausted, and the strain on production rises. Thus, benefits fall and costs rise until parity is attained.

When we turn to the multiplicity of alternatives available, the principles generalize to suggest that the greater the degree of direct involvement of the firm in input procurement the higher the saving on transaction costs but the greater the strain on management. Contracts, or more precisely the recognition of the desirability of continued relations that makes it possible to reach an accord, are a greater commitment than short-term purchases; participation, a greater involvement than contracts; integration, the greatest possible dedication. This generalization probably at best expresses the average tendency. In any particular case, exceptions may arise. Often a stable tacit relationship with suppliers can be cheaper to administer than a formal contract.

Still another aspect of optimal organization is the number of different outputs produced. Here the economics are much simpler. Outputs can be added if their production is at least as cheaply conducted jointly as when undertaken as a separate venture. Outputs should be added if joint production leads to lower total costs than separate production.

ORGANIZATION IN PRACTICE

The actual organization of mineral production differs considerably. In petroleum and natural gas, the dominant tradition is for combining the extraction of the two because it is clearly much cheaper to integrate than to separate the activities. To a lesser but still substantial degree, the refiners of crude oil also are engaged in both extraction and in a considerable part of the marketing of products—particularly gasoline and lubricants. The only thing that seems clear about these patterns is that total integration has not been found desirable. Not only do oil companies not try to insure equality between corporate crude-oil production and refinery consumption of crude, but they often choose simultaneously to sell much of their crude to other companies and buy to meet refinery needs.

Whether the degree of integration is close to optimum, and if not, in what direction the correction should go is far less clear. It is difficult to produce convincing evidence in either direction, and, therefore, no good reason to force an expensive industry restructuring.

Coal is quite a different situation. Until the late 1960s, coal production was separate from the production of other fuels. Only one customer group—the steel industry—maintained substantial vertical integration into coal production. In western Europe, government ownership of coal was widespread. Great Britain and France resolved long-standing debates over the future of their coal industries by nationalization right after World War II. During the coal crises of the 1960s, West Germany centralized the management of most Ruhr mines in a government-sponsored corporation.

Starting in the 1960s, the organization of the U.S. coal industry changed considerably. Several of the leading established producers were acquired by oil companies or other outsiders including metal-mining companies. Coal production began to grow significantly west of the Mississippi. New companies tended to dominate this western production, but these new companies often were subsidiaries of established firms in other industries—notably petroleum, electric utilities, and construction. Again the evidence does not suggest major problems or benefits from the pattern.

Metals industry structure is even more varied. First we have the steel industry that traditionally has limited final output to products of iron and steel, had heavy integration among iron ore and coal mining, iron ore processing, coke manufacture, iron making, and the fabrication of steel shapes, and operates the last three stages on a largely national basis. Aside from a few European companies with plants in two or three countries, the typical steel company operates mills in only one country. Ventures abroad have largely been in iron-ore mining. In short, there was extensive integration within steel related areas but little activity in other areas.

The producers of many nonferrous metals—notably copper, lead, and zinc—tend, not only to be integrated into several stages of production of a given metal, but also involved with several different metals with extensive international operations. Aluminum producers tend, however, to differ from both steel and other nonferrous metal producers. The industry was originated by companies specializing in aluminum. Many were centered in a single country. When the developers of the U.S. aluminum industry started operating in Canada, they established a separate corporation. Later entrants to the U.S. industry included companies with multimetal or multicountry involvement. Some European companies, however, began international operations at an earlier stage. The established companies have participated in developing bauxite mines abroad.

Some of these characteristics are quite natural. Economies do arise in combining aspects of processing. Since various metals occur together in nature, simultaneously mining and processing these metals is more efficient. Skilled, interested, well-financed firms were needed to develop the minerals in less developed countries. The established processing companies were the most natural ones to invite to undertake the developments.

Again the evidence does not suffice to explain all the patterns, but no reason exists to believe radical changes are desirable.

A NOTE ON THE OPTIMUM DURATION OF EXTRACTION

Exploitation of mineral deposits involves the special problem of depletion. Even if there are none of the exhaustion problems discussed below, optimum life problems still arise in most cases. The critical problem is the existence of immobile facilities in the mine. Tradeoffs have to be made between the benefits and costs of speeding depletion using more immobile capital.

The main benefit is that production occurs sooner and thus in a competitive industry has a higher gross present value. Several possible costs can occur. The most clear-cut is that to increase the rate of annual production more must be invested in fixed facilities. At some point, the cost of this investment increase will outweigh the benefit of the earlier sales revenues. This will certainly be true if the capacity costs rise at the same rate or higher rate than the output increase. Even if the rise in costs is slower than the rise in exploitation rates, it is likely that the firm will still decide to spread exploitation over several years.

Further problems arise with the impact of changes in the rate of exploitation on subsequent costs and on the recovery of minerals. Oil fields, for example, can have characteristics that cause total recovery to decline if the rate of extraction is increased. More generally, operation costs may be higher or lower in larger facilities.

Thus, for minerals more than for most commodities, decision-making must consider the optimum life of the facility and then the annual rates of production during each year of operation.

INTERACTIONS AMONG FIRMS, INPUT SUPPLIERS, AND OUTPUT BUYERS—MARKET EQUILIBRIUM—THE PARTIAL EQUILIBRIUM COMPETITIVE CASE

Thus far the discussion has been limited to treatment of the individual firm responding to given prices. To see how these prices are determined, it is necessary to examine market behavior. Two levels of generality are possible—that of the single market or that of all markets simultaneously. Alternative assumptions can be made about the vigor of competition. The market may be highly competitive, totally monopolized or monopsonized, or subject to oligopoly or oligopsony. The analysis here stresses the single market competitive case but sketches the nature of the other possibilities. A general discussion is followed by presentation and discussion of the relevant diagrams.

A key implication of the price taking assumption is that demand and supply functions are derived from manipulation of the optimum conditions outlined above. These conditions provide a set of equations that can be solved to determine the equilibrium quantities associated with the given set of prices. The firm's supply of a given output most easily can be traced in what is termed partial equilibrium analysis. In this approach, we examine how output changes when only the price of the given good changes. A similar partial equilibrium input demand function is derived by tracing the change in quantity demanded as the price of only one input changes.

What is termed a general equilibrium demand-supply function indicates the output of all goods produced and the inputs used as a function of all prices. In particular, the general formulation shows the effect of changing each price on the use of other inputs and the production of other than that whose price has changed. The first type of analysis is termed partial because, not only does it consider only one change at a time, but it fails fully to analyze the repercussions of the change. Thus, the input-output changes in the sector being studied require input-output changes elsewhere that are not treated in partial analysis.

As was shown above, the supply of a good is an increasing function of its price, but the demand for an input is a decreasing function of its price. The argument does not transfer completely to household demand and supply of goods. For technical reasons that need not be discussed here, it is possible under special circumstances for demand to rise and supply to fall with price rises. However, the usual presumption is that household demand falls as prices rise and supply rises.

Thus, for any firm or household and for any good, there can be defined functions relating decisions to price. If the firm or household tends to be an acquirer of the good, it has a demand function that declines with price. If the firm or household tends to dispose of the good, a supply curve results and increases with price.

Finally, a market is in equilibrium when the desire to buy at prevailing prices equals the desire to sell. Equilibrium means a set of prevailing prices such that there are no incentives to change them. This is true only of prices such that the amount offered at the price equals what people wish to buy at that price. The usual way this is shown is to point out that outside forces arise that push prices to the required levels. If prices are above the equilibrium level, the amount people wish to buy falls from the equilibrium level while the amount supplied rises. The offer exceeds what can be absorbed. This inspires direct price falls or output declines that indirectly produce price rises. The process continues until equilibrium returns. Conversely, at below equilibrium prices the desire to buy exceeds the willingness to sell. This tends to bid up prices towards equilibrium.

COMPETITIVE EQUILIBRIUM

The diagramatics of supply and demand for a single good arise from extensions of the prior analysis. The discussions of Fig. 2.4.6 to 2.4.8 provide the starting point for supply and demand analysis. 2.4.6 and 2.4.7 are the simplest to use. 2.4.6 showed how to determine the firm's demand for an input at a given output price. Thus, at the price P_e, we get a demand Q_1. Generally, the quantity demanded is that at which the MVP equals the price. However, the requirement to operate in the range of decreasing average value product means that input demand only starts with $P_o = MVP_o$ at the input level Q_o and rises as P_e declines.

Similarly, the supply of output at a given price is the quantity at which $MC = P$. The increasing average cost criterion means the curve only starts at the level Q_o at a price P_o equal to AC_o, the minimum level of average costs.

Analogous supply and demand curves exist for consumers. They too tend to reduce demand and raise supply as prices rises.

Then total demand is found by summing the demands of all consumers and producers at each price. Total supply is calculated by summing all supplies at each price. This summation produces market supply and demand curves such as S and D in Fig. 2.4.9. Market clearing is represented by the intersection point with price P_o and quantity Q_o.

Where the costs of each supplier are constant and equal, a flat supply curve applies the equilibrium price is the average cost, and the equi-

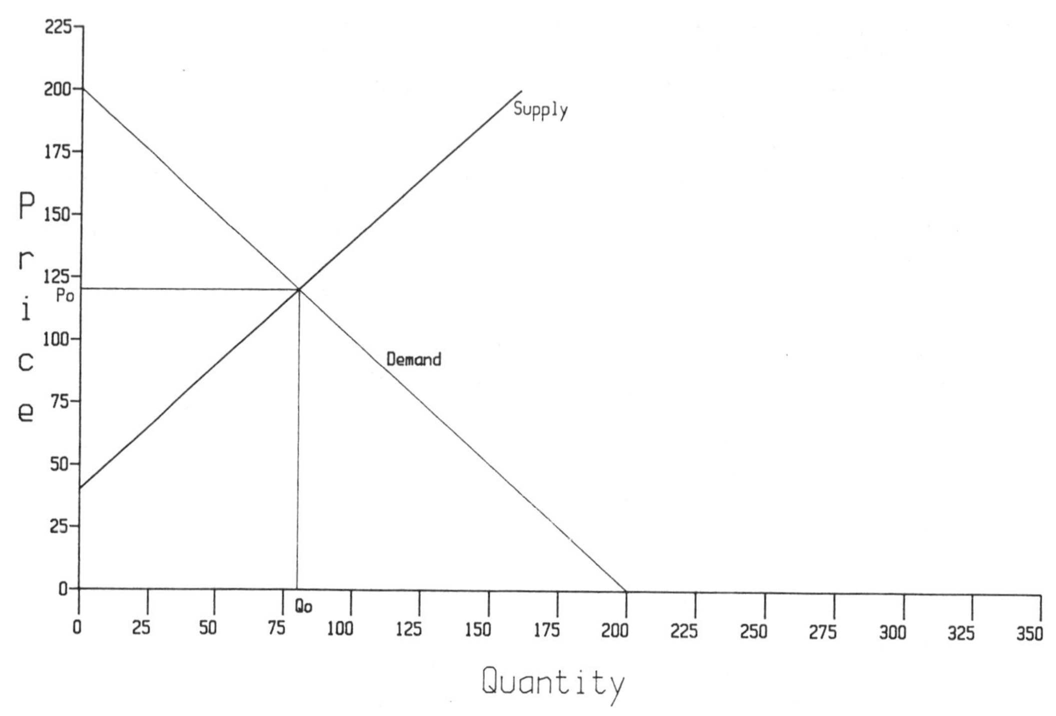

Fig. 2.4.9—Competitive equilibrium.

librium output is the demand at the cost. However, since each firm can produce any amount at the cost, an infinite combination of output at all locations among firms is possible. However, the practical significance of this is limited. It is unlikely that, in practice, firms can increase output indefinitely without increasing marginal costs.

No good geometry exists to handle multimarket equilibrium. Each single market curve is defined for equilibrium quantities and prices of all but two goods—the one being demanded and the standard of value used to price it. For reasons discussed in the texts (see especially Hicks), the impacts of changes in one market or another are too complex to treat here.

Suffice it to say, much analysis has been devoted to showing that the pressures toward equilibrium that exist in a one-good market also arise in a multimarket economy.

MONOPOLY AND MONOPSONY MARKET BEHAVIOR

As noted, monopoly and monopsony are the only cases of imperfect competition in which clear-cut conclusions are possible. Such definite results, however, depend on the further assumption that the monopsonist or monopolist deal with purely competitive suppliers or buyers. These assumptions combine to define precisely the behavior of all the participants.

Thus, in the monopoly case with competitive buyers, the same demand curve used in competitive analysis is also the proper one to model *buyer* behavior. Buyers continue passively to accept prices as given and purchase the quantity whose marginal value equals the prevailing price.

The concept of supply is inapplicable to monopoly markets. Supply relates output only to the cost of production. Output decisions by monopolists, as was pointed out above, are influenced by costs *and* the effects of output on market price. These effects depend upon the nature of the demand curve.

Nevertheless, the starting points of monopoly analysis are similar to the starting point of competitive analysis. We begin with the marginal cost curve of the producer and the market demand curve. The next step is quite different. A new relationship—the marginal revenue curve—is derived from the demand curve. The marginal revenue measures the net impact of increased sales (assuming for the moment that all customers pay the same price). This influence is the price received for additional output less the loss of income from extending to all customers the price cut necessary to secure the extra sale.

As noted above, the mathematical formula for marginal revenue is $dR/dQ = P + Q\ (dP/dQ)$ where R is revenue, Q output and sales, and P price. This formula is rigorously derived by differentiating with respect to Q the definition $R = PQ$. Marginal revenue, like price, decreases with output (d^2R/dQ^2 is negative). The P term clearly declines with Q; the loss tends to rise because Q rises. The degree of competitiveness of an industry is indicated by the magnitude of dP/dQ for *the individual firm*. Where dP/dQ is very small, we have a highly competitive industry; large values imply highly imperfect competition. I return to this point later, but first the nature of monopoly decision making must be explained.

The marginal revenue function measures the net benefit the firm receives from increased sales—i.e., the value of sales gain less the cost of price reduction. Marginal revenue, thus, is below the price at which an output is sold. (See Fig. 2.4.2). This is a critical point because in the complete analysis, we must consider both the optimal marginal revenue and the price associated with it. The firm charges the price associated with the optimal marginal revenue.

The marginal cost curve in Fig. 2.4.2 measures the increased production cost associated with a sales and output rise. The general optimizing principle that marginal benefits be equated to marginal costs here translates into the conclusion that the optimum sales level is one at which marginal revenues—the net benefit from sales—equal marginal costs—the production costs. This is equivalent to, but much simpler to explain and diagram, than a comparison of gross benefits (i.e., the price) to both costs (i.e., marginal costs of production and the marginal price cut impact). The second comparison involves a rearrangement of the benefit cost equations that merely adds the same term to both sides. This extra step achieves purity at the cost of undesirable complexity. We get an equilibrium in Fig. 2.4.2 at an output Q_m at which the marginal revenue curve cuts the marginal cost curve at the level MR_m and the price is P_m as given by the demand curve.

The resulting output is necessarily less than that of a competitive industry with an aggregated marginal cost curve identical to that of the monopolist. Thus, if the marginal cost curve were the supply curve for a competitive industry, it would intersect the demand curve at Qc leading to the price Pc. Such an industry would ignore the impact of individual firm output on price, operate on a supply curve represented by the aggregated marginal cost curve reflecting the firms' decision to equate marginal costs to the given market price, and produce where the quantity supplied equaled the quantity demanded. Given the nature of firm decisions in the competitive case, this price would equal marginal cost.

The monopolist, however, equates marginal cost to marginal revenues—a lower amount for any given output. Thus, the marginal revenue for the output a competitive industry with identical costs would choose is below marginal costs. (The marginal revenue of that output is less than the price which equals the marginal cost and so is also less than the marginal cost.) Reducing output raises marginal revenues and for a viable monopoly industry, leads to the optimum.

FURTHER ANALYSIS OF MONOPOLY

This result is most clear cut when marginal costs are constant or increasing with output. Then, output reduction both raises marginal revenues and lowers or leaves unchanged the marginal costs. When marginal costs fall with output, the optimality of production lower than the output at which marginal cost equals price still prevails, but the proof is more complex.

First, as shown above, decreasing marginal costs are inconsistent with the persistence of competition. Prices only equal to marginal costs imply losses. This is proved more simply in the single product case where marginal costs can by use of an analogy to the prior analysis of marginal revenues be shown to be $AC + Q\ dAC/dQ$. So long as dAC/dQ is negative, marginal costs fall short of average costs, prices equal to marginal costs, therefore, fall short of average costs, and revenues (price times output) fall short of total costs (average costs times output).

Industries where firms enjoy economics of scale cannot survive with a uniform price for each unit sold of each commodity equal to the marginal cost of production. It no longer makes sense to compare monopoly output to competitive output, but as discussed later, a comparison to the output at which price is equal to marginal costs is important.

The second consideration about decreasing marginal costs is that it is conceivable that reducing output could raise marginal costs more than it raised marginal revenue. However, in that case, producing that commodity would be unprofitable, at least in the range of output in which marginal costs were more steeply falling than marginal revenues.

Consider, for example, the case in which marginal costs always have a steeper decline than marginal revenues. If there is any intersection of the curves at a positive price, it will be a point of profit minimization. Drawing the curves shows that every output from zero to that at which marginal costs equal marginal revenue will have a marginal cost in excess of marginal revenue. Thus, the total situation measured by summing marginal costs and marginal revenues will be one of losses since each marginal cost exceeds each marginal revenue and so the sum of costs exceeds the sum of revenues.

Again, the calculus can be used to derive a more rigorous basis for the argument. The critical second order condition requires that d^2W/dQ^2 be negative. This can be restated as requiring that $d^2R/dQ^2 - d^2C/dQ^2$ be negative or equivalently d^2R/dQ^2 be less than d^2C/dQ^2. Since d^2R/dQ^2 is always negative, this rule is readily satisfied when d^2C/dQ^2 is zero or positive. When d^2C/dQ^2 is negative, the satisfaction of the rule requires that it be larger algebraically than d^2R/dQ^2. This in turn means a lower absolute value or less steep slope. Thus, generally a steeper slope for marginal costs than for marginal revenues is inconsistent with maximization and, in fact, implies minimization.

Fig. 2.4.10 illustrates this case. It shows because of the price—*MR* and *MC-AC* relations discussed above, when *MC* cuts *MR* from above *AC* will exceed *P* and so the firm is unprofitable as when $MR = MC$.

The analysis of monopsony follows basically similar steps appropriately adjusted to account for the difference between buying and selling. Related to the competitive supply curve (analogously to the relationship between demand and marginal revenue) is a marginal procurement cost curve. The latter is derived by adding to the price of an additional unit of purchase the additional cost necessitated by having to pay more for inputs. Equilibrium occurs when marginal benefits equal total marginal costs—namely, where the marginal valuation cost curve inter-

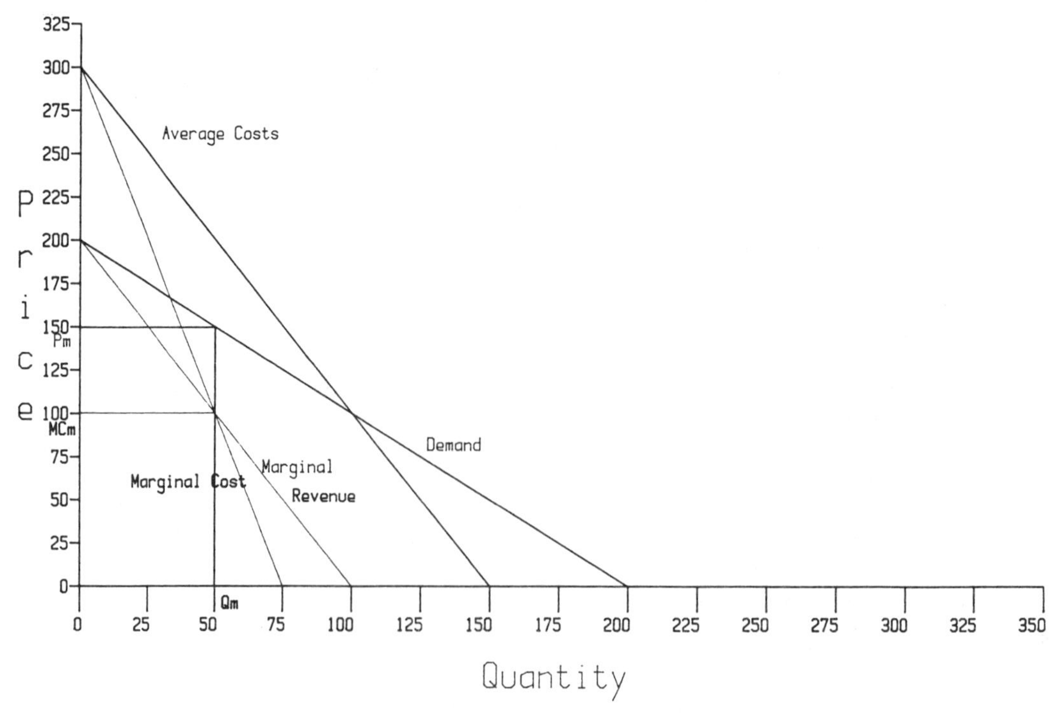

Fig. 2.4.10—Unstable monopoly.

sects the marginal procurements cost curve. Then the marginal gross benefit is equated to the marginal procurement cost. Less is procured than under competition.

Figure 2.4.3 show this procedure. The monopsonist sets purchases where marginal cost hits demand at a value of MC_m and a quantity of Q_1 for which input supplies receive a price of P_m. If the supply were bought competitively Q_c would be bought at a price P_c.

THE PROBLEMS OF MONOPOLY

Three critical further issues arise with monopoly and monopsony—their undesirability, the further questions of measuring impacts, and the benefits to the monopolists, monopsonists, and in some cases, society of varying the price charged or paid with the identity of the user or the amount used. A basic tenet of economic theory is that monopoly and monopsony are inefficient—they unambigously lead to less production and consumption than is possible. The gains extracted by monopolists and monopsonists are shown to be less than the losses inflicted on those with whom they deal.

The key point was suggested early in the chapter. The discussion of exchange showed that an offer price measured the amount of resources the buyer was willing to trade to get more of a good. The treatment of cost minimization suggested that the marginal cost of production measured the amount of resources that must be input to produce a good. Combining, an offer price in excess of marginal costs implies the potential gain to society of the difference between the sacrifice the buyer is willing to make and the actual resource cost. Preventing trade to continue until marginal cost equals price precludes these gains.

Monopolists and monopsonists worry, not only about these gains from trade, but on the way the gains are split. Transactions that are socially beneficial are thwarted because the monopolist or monopsonist is better off depriving others to prevent a transfer of wealth.

The point is that the monopolist's gain is less than its victims' loss. The crux is that the victims both pay a higher price on what they buy from the monopolist *and* lose output. Only the first part of that loss goes to the monopolist. The loss of extra output to consumers is a cost to them not producing a gain to the monopolist. This is illustrated by Fig. 2.4.11 for the case of

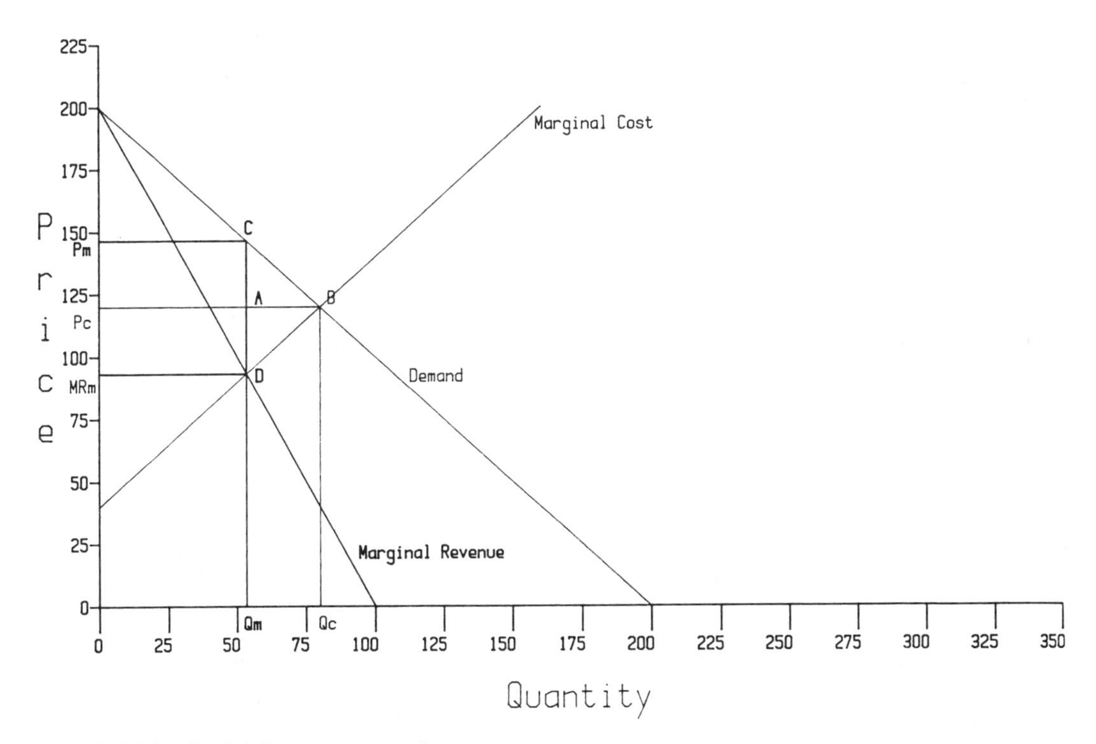

Fig. 2.4.11—Social loss to monopoly.

constant and increasing costs. The competitive equilibrium for an industry with costs *PC* is Q_c at a price P_c. Monopoly lowers output to Q_m and raises price to P_m. The gain to the monopolist is the extra profit $(P_m - P_c) Q_m$. This is a transfer from consumers.

Additionally, an output decline of $Q_m - Q_c$ occurs. This is necessary to produce the profit rise so all its impacts on producers have already been measured. However, an additional impact on consumers has been ignored—namely the payoff from expanding consumption from Q_m to Q_c.

This payoff is traditionally represented by the triangle ABC. The point is best rationalized when the demands are input users. At each point, the demand price is the marginal value product of the input. Again, on marginal use, a difference between *MVP* and P_c prevails. Output cutback causes the loss of all these gains, represented by ABC. It is this extra loss that makes the benefits to monopolist less than the loss of their victims.

(The case is more complex when the victims are consumers because of the problem noted above that the value of money is not a constant to consumers. Nevertheless, the qualitative point remains valid—the opportunity to buy extra output whose benefit exceeds its cost is lost). With increasing costs, the output restriction also leads to a loss of economic rents of ABD.

It is worth noting parenthetically that a version of this argument is often used to justify government efforts to help restrict output in individual markets. It is argued that the sellers are worthy people deserving of protection from income transfers.

Economic analysis argues that a better route is to let the markets operate without control and to use direct aid to offset the income-distribution impacts of the market outcomes. Observation of practice suggests that the beneficiaries of government-sponsored output restrictions are not noticeably deserving of aid. In fact, all too often they are merely the politically powerful trying to secure by political means the market power that is otherwise unattainable. Perversely, what when practiced privately is condemned as undesirable monopolization is excused as producing equity when practiced by government.

A second concern is over measuring monopoly power. The preferred measures are those derived from the existence of a nonzero dP/dQ for a firm. Two are available—the elasticity of demand facing the firm and Abba Lerner's index of monopoly and monopsony power. Generally an elasticity is the ratio of the relative change in one variable to the relative change in another variable presumed to have caused the other change. For example, the elasticity of demand with respect to the price of the commodity itself is $(dQ/Q)/(dP/P) = (P/Q)(dQ/dP) = dln\ Q/dln\ P$. dQ/Q is the percent *change* in quantity; dP/P the percent change in price. Thus, elasticity gives a comparison of whether the percent quantity change from a given percent price change is greater or less than the percent price change.

Since price and quantity move in opposite directions, this elasticity is a negative number. Economic practice tends to cause minor confusion by talking about the absolute value in verbal discussions but using the algebraic value in formulas.

Very elementary algebra allows us to relate elasticity to marginal revenue. From the prior rule, $MR = P + Q\ dP/dQ$, we get by multiplying the right-hand expression by P/P and factoring:

$$\begin{aligned} MC &= P + (P/P)\ Q\ (dP/dQ) \\ &= P[1 + (Q/P)\ (dP/dQ)] = P[1 + (1/E)] \end{aligned}$$

where E is the elasticity of demand. This indicates that the more responsive price is to output changes, the lower is (the absolute values of) the elasticity of demand. The relationship also illustrates the inconsistency between profit maximization and the observation that demand has a low responsivity to price. Profit maximization requires that the absolute value of elasticity exceed one or the algebraic value be less than minus one.

This follows from the condition that the optimal marginal revenue equals a positive marginal cost so itself must be positive so $1 + (1/E)$ must be positive or $-(1/E)$ must be less than one which requires $-E$ greater than 1.

Lerner's monopoly index is $(P - MC)/P$—the percent departure from efficiency. This ratio is zero when marginal cost equals price and rises with the gap created by output-restricting actions. However, the ratio can never reach unity since any production will have nonzero costs. When $MC = MR$, we get the index equals $(P - MR)/P$. Substituting the prior formula for MR leads to a value of $\{P - [P(1 + 1/E)]\}/P = 1 - 1 - (1/E) = -(1/E)$. For a profit maximizing monopolist, the Lerner index is the reciprocal of the absolute value of the elasticity of demand.

While these are the preferred measures of the vigor of competition, others have been proposed. The most important are those that attempt to relate market power to shares of the market. This approach has severe drawbacks. First, it postulates an invalidly absolute view of the nature of competition. All or no decisions are the only ones possible in making market-share measurements. A given competitor can only be totally included or totally excluded. No way exists to weight, as do elasticity measures, the actually-prevailing differences in the strength of various participants. Sterile arguments thus arise over the unanswerable question of what constitutes the appropriate criterion of significant influence.

Thus, what is being shared is difficult to define. A further problem is that no principles exist to determine what would constitute a dangerously uncompetitive pattern of market shares. Therefore, more elaborate market analyses that consider and weight all the influences are needed to appraise the vigor of competition.

PRICE DISCRIMINATION

A final issue is that of price discrimination—the effort to charge different prices depending upon the identity of customers, the amount bought, or both. The desirability of such discrimination to a monopolist and comparable discrimination in buying prices by monopsonists follows directly from the prior analysis. A major restraint on monopoly selling is the price fall induced. This disadvantage can be reduced if the degree of price cutting can be limited.

Various devices can be designed to prevent the sharing of price cuts with all consumers of all outputs of a product. Two important cases are differentiation of prices among consumer classes—most notably selling at different prices in different regions—and varying rates with the amount bought as with the traditional methods of pricing electricity.

The critical requirement to permit discrimination is the ability to distinguish and control the flow of goods to different customers. A price that varies with the amount of consumption can be maintained only if the seller has knowledge of the amounts used as is the case with metered service by gas or electric utilities. Different customers can be charged different prices only if some barriers make it prohibitively expensive for the favored customers to resell to the disfavored ones. Transportation costs are an important source of such barriers. Discrimination in the form of a lower f.o.b. price is feasible if it can be limited to sales made and *shipped* to a distant customer. So long as the discrimination is sufficiently modest, it will not pay the favored customer to resell. The cost of having had the goods shipped to it and then of returning the goods can be prohibitively expensive unless an extremely attractive f.o.b. price was made available.

The benefits of discrimination to monopolists and monopsonists are clear, but ambiguities arise about the impact on customers. There is one realm in which discrimination is considered an appropriate solution—the decreasing cost industry. Different prices for different amounts of production could allow efficient output without losses. Low levels of consumption would be charged higher prices than each increment. Ideally, each consumer would face a rate schedule such that its demand at the socially efficient marginal price would be within the range available to it at a marginal rate equal to that price. The higher prices on inframarginal rates of consumption would transfer income to suppliers enough to eliminate losses. (See Chapter 5.13).

Other forms of discrimination such as geographic price discrimination tend to alter burdens compared to those under uniform pricing. The essence is that the group paying higher prices loses while the group paying lower prices gains. Discrimination of this sort is giving the favored group more at a lowered price by lessening the price cuts made to the less favored group. Ironically, international-trade laws regularly presumes that it is worse to benefit from discrimination than be subject to uniform prices. Such benefits are characterized as dumping and regularly used to justify restrictions on international trade.

Attacks on these and other forms of low prices are often rationalized by the concept of predation. It is contended that the price cuts are only temporary ones to drive competitors out of business. Once they have gone, prices can be raised to much higher levels. However, predation is considered by the economists who have analyzed it to be exceedingly rare if not nonexistent. The unlikelihood of predation is explained by the point that it is not enough to force competitors out. They must also be unable to return when prices rise. This last is most unlikely.

The real reason dumping has been disfavored seems to be because of the influence on legislation of producers and their workers. Antidumping legislation is a concession to these protectionist pressures.

A final problem is distinguishing discrimination from price differences arising from cost differences. The marginal cost principle says the marginal costs of actually serving a user should be equated to the price that user pays. Thus, if the cost of service differs, so should the price. Considerable difficulties can arise about distinguishing cost-based price differences from discrimination.

ON THE PROBLEMS OF EXTENSIONS TO OLIGOPOLY, OLIGOPSONY, AND BILATERAL MONOPOLY

The prior analysis attained definite results by ignoring totally the reality that, to the extent market power exists, it is possessed by more than one firm. Thus, we have the cases of oligopoly—few sellers, oligopsony—few buyers, and bilateral monopoly—a small group of sellers dealing with a small group of buyers. In all cases, the analysis is rendered inconclusive by the problems of dealing with the simultaneous behavior of all the participants. A particular problem is that each participant's decision depends upon what he *thinks* all the other participants will do. The actual outcome will differ from the expected one if the others' actions are incorrectly anticipated.

The best that the theorists have been able to do is suggest a range of outcomes and deal with the difficulties of different ways of incorporating rivals' behavior into decision making. The monopoly models previously discussed represent a critical starting place for more general models. It is this contribution that justifies extensive development of a case that by itself has no apparent empirical use.

Behaving like a monopolist or monopsonist represents the goal that oligopolists or oligopsonists wish to attain. This follows from the nature of the optimum—it produces the greatest possible benefit to the sellers or buyers, respectively. In monopoly, the cost of any given output is minimized, and the actual output is set at the level which yields the greatest gap between revenues and cost. This is precisely what would make oligopolists *as a group* best off.

The existence of economies or diseconomies of scale at the firm level affect the difference between a single firm monopoly and an oligopoly. As already noted, the usual economic assumption is that up to some critical scale, the firm enjoys economies of scale, and expansion beyond that size produces diseconomies. Then in the case of identical firms, all firms will operate at the scale at which the shift from economies to diseconomies occur. The market will accommodate as many firms as are needed to produce the quantities demanded at the prices necessary to keep firms operating at the optimum scale.

PROBLEMS OF OLIGOPOLY

The greatest problems for oligopolies arise when there are more firms operating than are necessary to produce the collective profit-maximizing outputs. The most profitable solution would be for some firms to shut down and be compensated for doing so by an equal share in the profits. This is rarely a practical option because of reluctance of any firm to shut down and numerous legal barriers to such accords. The actual solutions are to sacrifice some profits by allowing all firms to stay in business. The opposite case of enough or too few firms has expectedly opposite implications. Several firms of a more efficient scale cooperating is better for society than a monopoly because production costs are lower. A potential exists to make the situation even better by allowing the number to rise to that needed to produce the optimum output with each firm operating at the optimum scale.

When firms differ, entry of a higher cost firm is desirable only if otherwise the lower cost firms operate at a level so much above the minimum efficient size that costs have risen above the costs the entrant would incur. This complicates the description of the process to encompass determination of how much an increase in scale past the point of minimum efficient size is desirable for lower-cost firms. However, the essentials are unchanged. There is still a right number of firms; cost minimization is still retarded if the number of actual firms exceeds the optimum; more firms are still desirable if the existing number is below the optimum.

The number of firms an oligopoly can accommodate depends upon how large a percent of the optimum output is accounted for by the average-cost minimizing output of each firm. If this output were half the monopoly optimum, two firms could operate. If a third, three firms; if a tenth, ten firms. Thus, the number of firms possible increases as the ratio of average-cost minimizing output for the firm to total industry output decreases.

These complications about scale are warnings that differences do exist between single firm monopolies and efforts of a small group to rig prices. The latter can be somewhat more or less profitable than a single firm monopoly depending upon the number of firms operating. Thus, the qualified description of oligopoly goals is to come as close to monopoly as possible. This requires that everyone know what the ideal total is and be willing to set an amount such that at the end of the day the collective decisions would have cumulated to exactly total the group ideal.

This is the great barrier to success. To effect and maintain the outcome that the oligopolists prefer, several actions must be taken. The best total must be calculated. A consensus must be reached about the shares to be allotted to each participant. Each must faithfully adhere to its allotment. Accord is difficult to establish, and even more difficult to maintain. It has been suggested that each member of the group can victimize every other member. The process will ultimately be self-defeating, but fears of its development can, nevertheless, produce the undesired outcome.

The problem may be called the Cournot paradox in honor of the economist who provided the basis for the argument. He showed that necessarily as long as other outputs were unchanged, a given firm in an oligopoly gained at the expense of the others from raising output so that the group total exceeded the level that maximized group profits. By construction, all the benefits—the increased sales—go to the output-increasing firm. However, the cost of the price cut induced is shared with the other firms. The net effect is a short term gain for the firm.

At the monopoly optimum, *industry* marginal cost equals marginal revenue. The marginal revenue for one firm when it changes output and the others do not is higher—the respective amounts being $P + Q_i\ dp/dQ$ and $P + Q_T\ dp/dQ$ with Q_i the firm's output and Q_T, the industry's. Since Q_i is less than Q_T, the firm's *MR* has a less negative second term so its *MR* is larger than the industry's.

All that is involved here is that, by construction, the undetected output raiser gets all the benefits of the sales expansion but the costs are spread over the whole industry. Thus, undetected output expansion from the group optimum is initially quite profitable because of the difference between the industry's $Q\ dP/dQ$ and the firm's absolutely smaller $Q_i\ dP/dQ$ (since the same dP/dQ is involved but, by definition, $Q_i < Q$).

Cournot's proceeded to note that actually the output change would provoke reaction by others that would offset the gain. In fact, his theory of oligopoly was one in which this process of undercutting proceeded to the limit (which turns out to be an output between the monopoly and competitive level). This theory has been widely criticized for its clear neglect of the ability of oligopolists to recognize that retaliation will wipe out the initial gain. The criticism may have gone too far because recognition of the danger of retaliation does not invariably suffice to prevent output expansion.

Firms are also aware that every other firm might try to increase output in search of at least temporary advantage. In fact, experience with price rigging efforts is that usually someone does yield to temptation and raises outputs and undermines prices. As fear of such actions mount, firms will be more interested in beating the rush than in preserving what seems to be a lost cause.

Thus, the ideal is hard to attain. The difficulties are such that many economists expected that the run up in world oil prices in the 1970s would not last. However, despite considerable disarray, the price has remained high. The main consideration seems to have been that a few countries with both the appreciation of the benefits of restraint and the ability to tolerate limited erosion of their market share have been able to lead the process. (Whether this will last remains controversial and the subject of another chapter in this book).

FINDING MINERALS—THE ROLE OF EXPLORATION IN PRODUCTION ECONOMICS

It is frequently argued that the finiteness of mineral resources is a vital influence on supply. The nature of the natural stock of minerals poses two basic problems. First, information about the economically relevant characteristics of the stock can only be learned through expensive efforts. Second, the limitations on the supply can lead to restricting immediately production to preserve supplies for future needs. The second problem has received far more attention than the first and during the 1970s was a popular area of research in economic theory. However, the exploration problem appears to be a far more relevant one. The search for information on resource characteristics is a vital part of obvious mineral-industry practice. However, the practical relevance of supply limits—exhaustibility as it is usually termed—appears almost nonexistent.

In many senses, the interesting thing about the need to secure information about mineral resources is its similarities to information searches in other industries. Adelman (1974) noted that in French the same word is used to encompass both research (as it is usually defined in English) and exploration for minerals. He suggests this is an apt choice. From an economic standpoint, the problems of getting ideas about what a manufacturing firm should produce are similar to those of searching for minerals. In both cases, market pressures require an inflow of new prospects. The same basic pressures, in fact, motivate the search. Every firm knows others are at work finding better alternatives and are thus under pressures to match them or be overwhelmed.

The main difference between extraction and manufacturing is that the former always means steadily rising costs if new opportunities are not found. If present deposits continue to be exploited *with present technologies,* costs inevitably rise as the best part of the deposit is extracted. This problem does not necessarily constitute a threat to the firm. Competitive position involves relationships to other firms. If all firms endure cost increases and industry demand is highly inelastic, most of the cost increase can be passed on to customers. A manufacturing firm with steadily falling costs can go bankrupt because new products from rival firms take over the market.

This suggests that the theory of securing information about mineral resource characteristics is a special case of the general problem of information search. Investment in finding new deposits is an alternative to developing and implementing new technologies to extract, process, and use the mineral in existing deposits. If any fundamental difference exists, it probably is in perception of the problem. Analysts of new technologies recognize that uncertainties abound. Too many observers of mineral availability believe all the essential information is known. They confuse the fact that the minerals exist with availability of accurate information about actual occurrences.

Whatever the realities, several basic problems still hinder elaboration of the theory. First, the specifics of the information search process are so diverse and imperfectly known that only broad statements are possible. Once we note that again marginal benefits must be compared to marginal costs, there is little more we can fruitfully add in the way of generalization. It is possible to suggest how the concept relates to the various aspects of information gathering in the minerals industries. Then the arguments can be used to appraise popular concepts of mineral availability.

Information search is a broader task than the creation of new knowledge that has been the concern thus far in this section. In particular, knowledge of existing technologies, business regulations, the state of the market, and numerous other details critical to operating a business takes effort to secure. Firms are regularly weighing the costs and benefits of efforts to improve knowledge. It is a recognizably difficult process because the payoffs are always so hard to predict.

The shorthands used in this area, as in most areas of economics, tend to suggest that the processes can be more neatly distinguished than is possible in practice. For example, in mineral exploitation, the stages of exploration, development, production, processing, and marketing are usually delineated. At least among the first three, the borders can be quite unclear. We cannot (and need not) always precisely decide when oil or gas well drilling constitutes exploration of a new region or development of an old one. Similarly, the activities to refurbish mines and wells cannot be neatly divided between production-support and development activities. One state leads gradually into another, and it seems better to recognize this fuzziness than to waste time worrying about the operationally-irrelevant question of refining the terminology.

Even more important points are (1) that knowledge about the economic characteristics of minerals must be secured by mineral firms, about new technology and (2) that the knowledge is obtained in all phases of operation. The occurrence of minerals is but one dimension of their economic attractiveness. Location, ease of extraction, and chemical composition and physical state with their impacts on the ease of processing and use all must be considered. Each of these concepts, moreover, encompasses numerous subareas. Location, for example, involves proximity to markets in terms of both the actual distance and the ease of transportation, terrain, and the availability of critical productive inputs.

Information on these matters always properly remains incomplete even after the minerals are produced and consumed. At no stage does it pay to discover every possible relevant datum. The relevance of information, moreover, differs at different points. Investigation should be, and is, timed to coincide with needs. The magnitude of fact gathering is tailored to the extent of the investment involved. Less effort is made to justify optioning a property than to undertake basic reconnaissance. As the firm moves on to even more costly activities such as extensive drilling and then development for production, increasing amounts of information are required. Even so, in every case, the effort will be incomplete. Information not immediately needed will not be acquired (at least unless a sufficient cost saving accrues to adding the effort on to the present program rather than undertaking a potentially much more expensive later effort).

M.A. Adelman's efforts to characterize the process of mineral supply evolution has properly concentrated on the tradeoffs made among the activities directly related to the mineral occurrences themselves. He starts with the vision, described earlier in this section, of the natural deterioration of an extractive property due to decay of the equipment or depletion of the best resources. He presents this as a generalization of the observations of petroleum engineers about the decline of oilfield production. He might alternatively have presented the process as a special case of the deterioration inevitable in all things.

He proceeds to note that in broad terms, a choice exists among allowing the cost rise to proceed, acting to retard it by refurbishing the existing properties, developing new facilities in properties already known to contain minerals, or finding new properties. He is well aware of my prior points that many options are included in each area and the borders are imprecise. He is similarly aware that technical advance is a critical alternative to finding and developing new properties.

The principles of choice here are applications of the basic marginal principle with consideration of the peculiarities of its implementation in different phases. A critical basic point is that while more *types* of expenditures arise as one moves from refurbishing an existing property, it most clearly is not true that, what matters, the total cost will be higher. Quite the contrary, exploration and new field development occur because of an anticipated cost advantage. For some existing fields, the single step of adding new producing facilities can be far more expensive, then the finding and developing of a new field. The old facility may be so depleted and the new so rich that the latter involves a much lower cost than the former.

Conversely, the uncertainties about what will occur increase as we move away from known areas to new ones. In particular, the outside analyst is better able to observe the economics of existing properties. The information from operating and development practice provides conservative estimates of the upper limits to future prices. The data can only show what would happen without further discoveries. This is always an upper limit and is independent of the skills of the explorationists.

The worst thing that can occur in exploration is that nothing worth exploiting is found. Then nothing new will operate, and costs actually match those estimated for the zero-exploration case. Success by definition is discovery of something better than exploiting known resources. This necessarily leads to costs below those in the worst-case estimate. The more successful the effort, the lower cost will be. Successful exploration is a major element in preventing the cost rise that would otherwise occur.

The decision-making process in minerals that emerges from the Adelman analysis is one in which these considerations are used to allocate the optimal amount of resources among the specific opportunities available in production, development, exploration, and technological improvements. The only thing that might be added about this has already been suggested—other options arise in technology development and in other areas of information acquisition and must receive their proper share of resources. Thus, the property development process is only one element of the broader process of seeking various ways to overcome depletion, lower costs, and raise profits.

THE MCKELVEY DIAGRAM

This process is not readily reducible to simple terms. Therefore, efforts to summarize resource development by such devices as the McKelvey diagram used by the U.S. Department of the Interior are at best suggestive. The diagram is based on the contention that there are two critical dimensions of mineral availability—the economic attractiveness and the extent of knowledge. It is further presumed that distinct borders exist between categories of both economic attractiveness.

This provides a useful first start, but several critical drawbacks should be noted. First, while cost of supply to final consumers constitutes a valid index of economic attractiveness, no counterpart exists on the degree of knowledge. I argued earlier much information is relevant and that no one ever learns everything. The McKelvey diagram can be interpreted as imagining that a measure of the quality of information is available. An alternative interpretation, namely, that information about the existence of resources is all that matters, is unsatisfactory. It ignores the possibility that certain mineral occurrences whose existence is well-known may require more expensive study before exploitation than some other less well delineated ones.

Recognition of this point leads to consideration of another point neglected in the McKelvey analysis but already handled here. We need a method for deciding between exploiting "lower-cost" less known resources and continuing to use known ones. The prior analysis provided such a basis. However, that basis implies a fundamental flaw in the McKelvey dichotomy. Costs do not summarize part of the process; properly defined, they encapsulate *everything*.

Thus, one can view McKelvey's effort as either an overly complex or overly simple approach. It is too complex in the sense that uncertainty is something we can overcome at a cost. Thus, uncertainty is part of cost and a measure of probable costs is all we need. Alternatively, I have already shown that cost has many elements and this is partially true for the uncertainty reduction elements of costs.

McKelvey leaves the impression that the only uncertainty is whether something exists. Sad experience shows being there is not enough. We must also know how costly the material is to extract, or burn.

The critical aspect of knowledge is that it is costly to obtain. We can thus make all resources comparable by a single index by computing their total costs including the expected cost of required information. Obviously such rankings suffer, as do the presently used ones. from the uncertainties involved. Even so, problems remain. Neither rankings nor borders can be expected to remain fixed perpetually. Technical advance in the extraction, processing, and utilization related fields, changing public policies, and other forces can alter the ranking of different resources. In an uncertain world, we cannot be sure that we always explore for the most attractive resources. Our errors add to the store of unexploited known resources and obviate future expenditures on finding them. This improves the position of such discoveries in the cost hierarchy; they are more attractive than if they had never been found and finding costs remained to be incurred.

Similarly, the limit to economic viability is ever changing—and not necessarily increasing. Technical progress by easing production, processing, and distribution and by lessening demand growth can cause prices to fall over time.

In sum, the information gathering process in minerals and other industries is a major influence. Firms are perennially making tradeoffs between investments in the known and the *unknown*. These are further split between efforts to apply existing technology to find and develop new producing facilities and efforts to improve technology of finding, developing, extracting, processing, transporting, and use. The former are easier, but it is the latter that are essential to preserving the progressive industrial society on which we have become reliant.

COAL IN THE ENERGY MARKETS—A CASE STUDY IN SUPPLY ANALYSIS

To illustrate these points, the case of coal supply predictions can be examined as a particular apt case study. A chronic tendency in popular discussions of energy is to overrate the rapidity with which it will be necessary to move from oil and gas back to coal. The problem is discussed here because it represents neglect of the understanding of the economic principles of mineral supply outlined in the prior section. The prognosticators proceed by ignoring principles so basic that their firms would go bankrupt if they failed to apply them to their actual decision making.

The critical question about coal is where is it likely to lie in the cost distribution of mineral supplies. Specifically, the natural endowments of each mineral differ in their costs, or more precisely, in their economic attractiveness. The first distinction here is that cost can be defined in several ways:

(1.) cost after extraction,
(2.) cost after processing,
(3.) cost after delivery, or
(4.) total cost to the user.

Each successive cost adds to the prior costs of the additional action that moves you to the next stage. Extraction cost is everything spent on getting the material out of the ground. Then you add processing costs to extraction costs to get the cost after processing. The cost to users is then the sum of extraction, processing, distribution, and other use costs such as those for buying the facility in which the material is consumed and paying for those who operate it.

It is this cost to users that matters. Choice depends on total utilization costs—which is a major problem for coal. While it is cheaper than oil and gas at the point of extraction, higher costs at later stages often offset the extraction cost advantage.

A second key point is that for each mineral, there are differences in the utilization costs of different occurrences. Thus, for each mineral there is a unique cost distribution. In the fuel case, we thus have separate distributions for each fuel be it coal, oil, gas, uranium, or solar.

The cost-of-final-use concept allows us to add up these distributions into an overall distribution of total economic attractiveness. The basic economic rule is that we should move from the most attractive to the least attractive resources in precisely their order of economic attractiveness. It is more efficient to delay the agony of higher costs. The major rationale lies in the time value of money—the fact that money earns interest. When confronted with the choice between a cheaper and a dearer resource, the cheaper is chosen because the saving can be invested and earn interest. The investments made ease the pain of later using the dearer resources.

This analysis, of course, is oversimplified. The cost ranking complicated by the differences among users. Use cost differs because for any given energy use by a given entity, post-extraction costs differ markedly. For coal, we can go from the lowest cost case of meeting the electricity needs of customers near a coal mine and proceed all the way on to the cost of turning coal into synthetic gasoline. Thus, we have at any point in time a distribution of users with different needs and an optimal mix of fuels that best supply this mix of needs.

This complicates the calculation of the cost hierarchy. Specific deposits can become more or less attractive in a world of diverse customer needs than they would have been in a world of uniform needs. The position change depends upon how poorly or well the fuel meets specific needs and the strength of demands of those with those needs. Nevertheless, an attractiveness distribution still exists, and we still go from using best to worst resources.

At least two further points must be made. By introducing needs, I have made it necessary to consider demands as well as supplies in establishing the attractiveness hierarchy. Second, the actual distribution cannot and should not be fully known in advance. Part of the ranking depends on unknowable future events. We cannot guess how new technology will alter relative resource attractiveness.

Even that which we can know can only be learned at a cost. It is never economic at any moment to acquire more information than we can use profitably at that moment. Therefore, we deliberately limit our knowledge search to that necessary to determine the investments we are currently making. Thus, we move along an imperfectly known cost distribution trying as well as we can to pick the next most attractive resource. This all has many implications.

In the United States, we have had numerous recent cases where coal producers ran afoul of inadequacy of their knowledge of characteristics other than physical availability (which may be considered a misunderstanding of where coal lies in the McKelvey box for energy). Several acrimonious disputes have arisen over mines that proved far more expensive to mine than anticipated. Another case arose in which the coal contained impurities that unexpectedly proved to hinder burning. In these cases, occurrence was well known, but other critical characteristics were not.

In formal terms, this argument explains that coal attractiveness is overrated because of a tendency to be overly optimistic about where coal lies in the attractiveness distribution. The interesting question, of course, is why this is the case. The primary reason historically has been concentration on the single dimension of availability of known coal resources to the exclusion of many other considerations. These last include that oil and gas are more expensive to locate than coal and have far lower post-extraction costs than coal. As I have suggested, this is aggravated by a tendency to neglect the other dimensions of information such as mineability and burnability.

The grossest form of the error occurs when the generous estimates by geologists of the volume of coal occurrences are compared to proved reserves of oil and gas. These last consist only of discovered oil and gas known to be economically recoverable and usually close enough to development that the technical ability to recover them is already assured. This comparison excludes not just undiscovered oil and gas but also the great amounts of oil and gas known to exist but currently uneconomic to exploit. A shift of concepts to estimates of what can be recovered of all fuels gives a more valid comparison but one that forces us back to using my attractiveness ranking concept. All estimates of ultimate discoveries suggest that it is physically possible to continue oil and gas use at current rates for many decades.

A new argument is that oil and gas are so much better than coal at serving certain tasks—petrochemicals manufacture or transportation fuel—that their current use for other tasks should be curtailed. More oil and gas then would be available for superior uses later on. In particular, we are supposed to save oil and gas as petrochemical feedstocks and motor fuels and reduce their use as boiler fuel.

The defects in this method of ranking are myriad. First, inspection of almost any room immediately suggests that it is nonsense to categorize all petrochemical use as unambiguously better than boiler fuel use. Both range wildly in urgency. A vast amount of petrochemical use goes into activities that once were served by other materials or deemed unnecessary—consider all the plastic wraps with which we have become provided. Some boiler fuel is used to maintain life itself, and a vast amount is used to keep us from being injured by cold. Thus, the worship of petrochemicals over boiler fuel proves perverse under scrutiny. It favors better wrapping paper over basic human needs.

However, the key point is that, as South Africa's Sasol process proves, coal can, at a cost, meet these petrochemical and nonfuel needs. It can be more efficient presently to use oil and gas for boiler fuel and later to use synfuels from coal to meet motor fuel and petrochemical needs. South Africa's pattern has not been widely imitated because other countries do not face the same oil supply disruption problems South Africa faces.

In sum, the errors about coal are prime examples of the perils of neglecting basic economic concepts. When the error relates to immediate decisions, it can lead to bankruptcy. No businessman could afford to use analysis of this sort to guide all decisions so very little has been invested in following up on these erroneous resource-availability conjectures.

So long as nothing is at stake, observers can leave aside their hardheadedness and indulge in flights of fancy without penalty. This is not always the case. Western European governments have wasted billions upon billions on the basis of erroneous estimates of the competitive position of coal.

THE ECONOMICS OF EXHAUSTIBLE RESOURCES

Since at least the early twentieth century, economic analysis has noted the implications of the exhaustibility of mineral resources. The standard citation of pioneering work published in 1914 is that of an agricultural economist named L.C. Grey. He showed that in anticipation of exhaustion, firms would withhold production in early years to meet the needs of the future. A more formal, more comprehensive analysis was published by Harold Hotelling in 1931. Hotelling, a pioneer in the use of mathematics in economic analysis, was at the time interested in the application of a tool known as the calculus of variations—a method for selecting a function that maximizes the value of a specific integral. The technique thus is related to the notoriously difficult area of differential equations.

In any event, Hotelling chose to use the tool to deal with exhaustion. The decision in retrospect appears unfortunate. More familiar techniques more readily produce all the critical results. In fact, Hotelling himself was so imprisoned by his methodology that he failed to develop a comprehensible interpretation of his most general and most useful results. Such an interpretation was not to emerge until a 1969 article by Cummings.

Very little work on exhaustion theory occurred between 1931 and the 1970s. Much was done on the analysis of resource scarcity in practice. The U.S. government conducted the Paley Commission study in the 1950s. Those connected with the Commission inspired the creation of Resources for the Future which has done extensive work on resource availability including particularly the Potter and Christie compilation of historic price data and the use of the data by Barnett and Morse to analyze the prevalence of scarcity.

Work on the theory was limited to Orris C. Herfindahl of the Resources for the Future staff and Anthony Scott of the University of British Columbia. A boomlet of work was inspired in the middle sixties when Herfindahl and Scott prepared papers, for a symposium on minerals taxation, relating to exhaustion theory and taxation. This, inspired me and several others to proceed further. However, it was not until the early seventies that concern over soaring prices of oil and price rises for many other minerals created extensive interest in the subject. The leading journals devoted to economic theory became filled with articles on exhaustible resource economics.

The review here attempts to suggest the critical aspects of exhaustion theory in as simple fashion as possible. (An alternative discussion treating some issues in more detail appears in Gordon, 1981). The formal analysis of exhaustion expresses the problem as one of a limit to the *cumulative total* output of an industry during its lifetime. This limit is expressed as a side-condition of the type discussed above in the treatment of conventional production economics.

The basic conclusion is that so long as demand for the product endures long enough to deplete the supply of a commodity and its close substitutes market forces will arise to alter the pattern of production from that which would occur in the absence of supply limits.

The principles of present value analysis govern the need for incentives. Income now can earn interest. To sacrifice income now for income later, that later income must include the interest lost by delay. Again a marginal concept is involved. You hoard for the future if nonzero hoarding produces a marginal gain greater in present value than current production. You cease hoarding at the point at which the marginal benefit (increased future income) equals the marginal cost (lower current income).

Several incentives can induce saving output for a future generation. First, a sufficiently high price can be offered in later years than in earlier years. This can occur because, as Hotelling stressed, the process of anticipating exhaustion reduces production over time or because rising demand increases the willingness to pay.

Second, where marginal costs increase with output per period, reducing output can raise unit profits by lowering marginal costs—the effect analyzed by Grey. Third, technological progress can lower the marginal costs of any given output at any future time. Fourth, slowing depletion leaves, not only more resources, but higher quality ones since they are mined first. This improvement in quality is another benefit of delayed output.

The first impact was demonstrated as an introductory exercise in Hotelling's article (and a distressingly large number of the writers in the 1970s failed to realize that there was more to the theory). Grey's earlier analysis stressed the effect of the rate of production per period on costs. Hotelling's analysis incorporated this effect of resource depletion but made little note of it. The implication of the preservation of lower-cost resources was the inadequately-developed point on which I previously criticized Hotelling.

Two critical further implications of the theory relate to the efficiency of market processes. As Chapter 5.13 discusses, economic theory distinguishes a variety of conditions such as monopoly, imperfect institutions for financing risky investments, and taxes that distort economic choices produce inefficient use of resources. Such defects will similarly lead to inefficient use of exhaustible resources. However, no additional special sources of inefficiency are associated with the existence of exhaustible resources. Moreover, as shown below, the implications of these market defects are as ambiguous for exhaustible resources as for inexhaustible ones. In particular, the "conservationist" presumption that market imperfection leads to overly rapid exploitation is not necessarily true. In particular, the inability to finance risky investments might be thought surely to retard the saving of minerals for future generations. However, this is not always the case.

MODELS OF EXHAUSTION

As noted, the crux of models of exhaustion is demonstration of the numerous ways the market can provide incentives for altering production and sales patterns to account for exhaustion. It was further noted that the critical step is to introduce limits on cumulative production into the decision function for the firm. The full model then includes market demand equations and exhaustion constrained wealth functions for each demand equations and exhaustion constrained wealth functions for each firm producing minerals. At the firm level, the requirement is to produce a wealth-maximizing output pattern, given the market prices. These prices must satisfy the usual requirement of market clearing; what firms offer in each time period at the price in that and other time periods must equal the quantity demanded at the prevailing price. Again, I begin wth a mathematical presentation and then discuss its meaning.

The best available statement of the constrained maximization problem is due to Modiano and Shapiro (1980). They provide a discrete time analysis of the problem where both the amount of output per period—q^t and cumulative output to date $\sum_{j=0}^{t-1} q^t$—affect profits. The version here differs slightly from theirs which was designed to deal with a planning horizon that might be less than the actual life of the firm. Thus, they introduce a salvage value term to handle any remaining resources. The goal here is to provide the equivalent condition over the actual lifetime, i.e., when all the economically desirable resources have been exploited. In that case, no salvage term is needed. However, time is added as a variable to allow inclusion of the impacts of demand shifts, technical progress, and government regulation in altering profitability.

Using the notation, R is revenue, C is cost, q is output, r is the relevant discount rate; t and j are time indexes. Thus we maximize the Lagrangian:

$$L = \sum_{t=0}^{T} (1 + r)^{-t} \quad C^t \left(\sum_{j=0}^{t-1} q^j, q^t\right)$$

subject to

$$\sum_{t=0}^{T} q^t \leqq K$$

The critical optimizing condition is then

$$(1 + r)^{-t} \left[\frac{dR^t}{dq^t} - \frac{\partial C^t}{\partial q^t}\right] - \sum_{j=t+1}^{T} (1 + r)^{-j} \frac{\partial C^j}{\partial q^t} - \lambda \leqq 0,$$

or

$$\frac{dR^t}{dq^t} - \frac{\partial C^t}{\partial q^t} - \sum_{j=t+1}^{T} (1 + r)^{t-j} \frac{\partial C^j}{\partial q^t} \leqq \lambda (1 + r)^t$$

To simplify further discussion, the three terms on the left of the last inequality subsequently are designated respectively as *MR, MC,* and *CCS* for marginal revenue, marginal cost, and cumulative discounted cost saving.

The final equation encompasses most of the critical cases in exhaustible resource economics. It is more convenient separately to discuss the cases of nonbinding and binding constraints. Kuhn-Tucker theory shows that when the constraint is nonbinding, the Lagrangean is zero. The expression reduces to $MR - MC = CCS$. The point is simply that exhaustible resource industries incur an additional cost to that of current production—namely, the higher future costs due to depletion. Marginal revenues are equated to the sum of marginal production costs and the present value of marginal cost savings. This can alternatively be interpreted that marginal revenues exceed marginal production costs by the amount of the additional discounted saving in future cost. This extra cost thus causes firms to operate at lesser scales than if exhaustion were not a problem.

This case is probably the most interesting and thus should be examined particularly closely. The basic situation is that if anything matters it is the steady deterioration of the quality of remaining resources. We never actually physically deplete the stock but simply use up the best portion. The industry peters out at that point. The process is a regular rise in prices but fall in the gap between prices and production cost as we move to extinction. In early years, there are more high quality reserves left, a greater benefit to preservation, greater restraints to current production, but a greater ability to produce. By the end of the process, nothing is left to save, production is thus at the point where marginal costs are equated to marginal revenues, but depletion is so advanced, that these costs are quite high and lead to high prices and low output.

While prices rise continually, nothing can be said about the pattern. The rule does not impose any restrictions on the rate of increase. All we can say is that some increase prevails. This vindcates the intuitive decision by Barnett and Morse to argue that the presence or absence of price increases is the critical indicator of the extent of exhaustion pressures. Conversely, their argument that technical progress is more important than depletion as an influence on mineral markets directly transfers to exhaustible resource models. It can be interpreted as indicating that the impacts of depletion are swamped by other forces and prices do not rise perceptibly because of exhaustion. Some combination of technical progress in production, processing, and use and of greater than realized availability of minerals has prevented cost and price run-ups.

PRICE BEHAVIOR WITH EXHAUSTION

The case of a binding exhaustion constraint only superficially alters the argument. Here we have an equality of the present values of *MR-MC-CCS* in all time periods or alternatively a growth in the value of *r* percent per year. Thus, we do have something growing smoothly but in any interesting case, that something is not observable.

At best, we can show that what we can observe—the price of produced minerals will grow far less rapidly than *r* percent. Consider, for example, the extreme case that was Hotelling's simple example—where the price increase produces the entire rise in *MR-MC-CCS* because *MC* is a constant and resources are homogeneous so that *CCS* is zero. Then, in a competitive model, the key expression reduces to *P-MC* where *MC* is the constant level of marginal cost. By the nature of a percentage, a *r* percent rise in *P-MC* involves *P* rising by less than *r* percent. *P-MC* is necessarily less than *P*; *r* percent of *P-MC* thus is also less than *r* percent of *P*.

Every complication leads to lesser rates of price rise. Thus, if costs fall because current output is being reduced, this contributes to the required *r* percent rise in *MR-MC-CCS*, and prices need rise less. The existence of quality deterioration adds the contribution of saving of future costs and lessens the need to raise current prices or lower current marginal costs. Thus, with depletion effects on resource quality, not only do we not have an *r* percent rise in prices but even the gap between marginal revenue and marginal production cost grows less rapidly than *r* percent. In general, the *r* percent rule applies to the sum of observed historical revenues and costs and the cumulative percent worth of expected cost savings which are even harder to observe than current costs.

Thus, the much overcelebrated *r* percent rule never in practice refers to prices and, even when the rule does hold, does not apply to an easily measured set of variables. As already noted, the *r* percent rule need not apply at all.

Among the important further points to note is that the possibilities become even more complex in a technically progressive world. When the world is progressive, the possibility exists that at least for much of the life of an extractive industry, output will rise steadily. This is not possible with an unprogressive (or even slowly growing) world. The need for an *r* percent growth in *MR-MC-CCS* requires in slow growth cases declines in output to raise *MR* and lower *MC* sufficiently to have the required increase. Technical progress and economic expansion can sufficiently raise the demand price and lower the marginal production cost of a *given* output that if output were kept constant over time, *MR-MC-CCS* would grow faster than *r* percent. To reduce the growth down to the *r* percent level, output would have to rise.

While any combination of demand rise and cost fall of sufficient magnitude could produce the desired result, a case of particular interest is where the main changes are on the cost side. A sufficient downward shift in the cost function could allow the satisfaction of the *r* percent rule with constant or even falling prices.

INEFFICIENT EXHAUSTION

The enormous literature on exhaustible resources has further established that the efficiency of market processes to save resources will depend upon the same considerations as the efficiency of markets in the absence of exhaustion. The standard list of concerns includes inadequate competition in the input or output markets, imperfect competition in capital markets, the inability to treat environmental side effects, and the distortionary effects of taxes.

Each of these defects has a different nature and implication. The nature of the impacts of imperfect competition was sketched above. Productive output and input use tends to be restricted below the socially desirable level. This result carries over into exhaustible resources except under very special circumstances in which monopoly makes no difference. Thus, monopoly still leads to inefficiency but always in the form of too little use. Thus, as is often noted, monopoly is the friend of conservation if the latter is taken to mean lesser use.

The failure to recognize environmental impacts, however, leads to overproduction in both the exhaustion and nonexhaustion cases. Thus, environmentalism does imply lesser use. However, contrary to suggestions that sometimes emerge (see e.g., the Ford Foundation Energy Polity Project report), lesser use does not similarly imply better environmental quality. We can guarantee that programs that directly make consumers and producers aware of the environmental impacts of their actions will induce some restrictions of consumption and production.

However, we cannot be sure that by reducing consumption and production we will reduce pollution at all, let alone in the best possible way. First, production and consumption controls designed without explicit regard to pollution damages may not control the right things, and even if they do, may not produce the right levels of output restrictions. More critically, reduced output is but one of the optimal ways to reduce pollution. Efforts to reduce pollution per unit of input or output can also be adopted. Controls of pollution leave open the choice between lowering input or output or reducing per unit pollution. Direct input or output controls foreclose the second type of response.

Finally, the imperfection of capital markets is the trickiest issue. Difficulties are associated with both defining the problem and with analyzing its implications. The various expressions of the argument all lead to the same point that the absence of sufficient devices to reduce risk lead to excessive investor timidity and excessively high interest rates.

Severe doubts can be raised about the empirical relevance of the argument. We observe the existence of an impressive array of devices to facilitate the flow of capital. We similarly recognize that these devices are costly to maintain and thus are created only when clear benefits arise. Thus, we should not expect, as do theorists thinking of a world of zero transactions costs, that every imaginable risk reducing alternative is worth providing. We could go further and point out that the faith that governments could wisely provide the needed extra institutions is highly questionable. The government behavior has been that of being the primary barrier to creation of imaginative new financial devices and of devoting funds to protecting dying industries instead of encouraging emerging ones.

A more critical concern is that even if interest rates are too high, the implications of lowering them is unclear. The result can be faster or slower depletion. Higher interest rates do make the future less attractive and thus directly lower, among other things, the desirability of holding minerals for later generations. However, higher interest rates also raise costs and this tends to retard present as well as future activities.

The net outcome is indeterminant. In a simple analysis of the mining effects, I was able to show (see Gordon, 1966) that at low interest rates costs would be very low relative to the prices consumers were willing to pay but that as interest rates rose costs rose and at some point reached levels that precluded sales. As interest rates reached the point at which costs were getting near to prohibitive levels, the tendency was for exploitation to be extended as interest rates rose. In contrast, with a big gap between price and marginal costs, the main effect of a higher interest rate was to increase the present value of delayed output and future cost savings and thus tended to encourage current production. Thus, at lower rates of interest, a rise would speed depletion.

It was also pointed out that certain visions of accelerating investment clearly led to faster depletion. One much discussed view is that there exists a finite store of profitable investments and that eventually all these investments will have been undertaken and economic growth will cease. Extraction of minerals is one of these investments so it too will cease eventually. Lowering interest rates speed up the investment process and thus the end of investment and thus the end of mining.

Thus, we cannot be sure that imperfect capital markets speed or retard exhaustion but have reason to suspect that in the long-run the net effect would be to retard exhaustion by generally slowing growth and the drain it places on mineral resources.

SUMMARY AND CONCLUSIONS

This chapter has sought to demonstrate that modern economic theory suffices to provide principles for effective analyses of mineral production decisions. However, satisfactory analyses depends on taking a broader view than most textbooks provide. We must start from the basic consideration that economics is about choices that arise in by reliance on exchange and production. We must be cognizant of such complications as the possible existence of many different degrees of competitiveness in an economy and particularly of how difficult it is to determine the vigor of competition or how weaknesses in competition will affect decision making.

We must recognize that we live in a world of long-lived equipment and the need to use present value analysis to treat such investments. We also must recognize that information of which data on mineral resources is a key element is expensive to obtain and these costs must be included in our analysis. Once these points are recognized, the basic marginal principles of traditional economics can be used to guide appraisal of mineral supply problems. This appraisal, moreover, suggests exhaustion of minerals is a much less serious problem than popularly believed.

APPENDIX

THE USE OF MATHEMATICS AND THE LITERATURE OF PRODUCTION ECONOMICS

A vast literature exists dealing with economic theory and the use of mathematics in this theory. The basic literature includes the many texts designed for introductory courses and for a more advanced undergraduate theory course (conventionally termed intermediate theory because it lies between introductory and graduate level courses).

Shortly after World War II, Samuelson wrote the first edition of what was long the most popular introductory text and greatly influenced the content of subsequent texts. Among those of interest are those of Lipsey and Steiner and of Miller.

George Stigler was the model for most intermediate theory courses. Although his book has not been updated for several years, it is still worth reading. The most interesting more recent texts with which I am familiar are by Hirshleifer and by Mansfield. A particularly thorough, if not particularly imaginative, text was prepared by Sher and Pinola.

At the more advanced level, a wide profusion of books exist. At least three basic types of approaches can be identified—books that stress mathematical tools for economists, those that are largely devoted to applications of mathematics to economic theory, and books that treat both. (Many books of the middle type have brief mathematical appendixes).

The classic guide to mathematics for economists is by R.G.D. Allen in a pre-World War II text that stresses the calculus. A second Allen book in the 1950s added some other topics including linear programming, but not the Kuhn-Tucker theories, and also discussed applications. Chiang seems the best of the more recent simpler introductions to the mathematics.

Nikaido has prepared both a very advanced treatment (1968) of the mathematics and a slightly simpler but still difficult book (1970) treating mathematics and its use. Other useful combined mathematics and economics text include those by Intrilligator, Lancaster, and Takayama. The first is one of the most lucid; the last, one of the most comprehensive. Baumol (1972) has a useful, less rigorous text of this type.

Samuelson also wrote one of the basic treatises applying mathematics to economics and in 1983 added a long appendix to the revised version. Henderson and Quandt prepared a somewhat simpler treatment that they regularly revise. Other interesting texts include Malinvaud, Shone, Varian (the last misleadingly disguised as an intermediate theory course), and Arrow and Hahn. The middle volume of an encyclopedia of mathematical economics edited by Arrow and Intrilligator contains useful surveys with excellent bibliographies of the critical aspects of consumption and production theory and related issues.

Investment theory is generally only sketched in most texts but many special texts are available on the subject. Fisher (1930) remains a useful guide, and the best modern introduction is by Hirshleifer. More applied books are available by such writers as Bierman and Smidt or Merrett and Sykes.

A vast range of mathematical tools are used in economics. One that has become critical because of its wide use in enabling complex expressions to be presented more compactly is matrix algebra. A matrix is simply a rectangular collection of elements such as $\begin{pmatrix} 1 & 2 & 3 \\ 4 & 5 & 6 \end{pmatrix}$.

The length and width of a matrix is unlimited. However, the rules are such that operations are possible only between matrices with proper relationships among each other.

Addition, for example, involves adding each element of one matrix to the corresponding element of another matrix. This is possible only if the matrices each have the same length and width. The same is true for subtraction.

Even more complex rules apply to multiplication. Each element of a product matrix is a sum of the product of specified elements of a row of one matrix by the corresponding elements of a column of another matrix. Thus, multiplication is possible only if the number of columns (the number of elements in a row) of the first matrix equals the number of rows (the number of elements in a column) in the second matrix. The resulting matrix ends up having the same number of rows as the first and the same number of columns of the second since its elements include, in a carefully specified order, the sums of products of the elements of each row in the first matrix with the elements in every column in the second. It should be noted that the only matrices of identical size that can be multiplied are square ones—ones with equal numbers of rows and columns.

The matrix analog to division is a generalization of the equivalent notion that division by X is multiplication by its inverse $1/X$. Inverses can be defined as a matrix X^{-1} such that $X X^{-1} = X^{-1} X = I$ where I is a square matrix which has one on its diagonal elements and zeros elsewhere. The principles of matrix algebra prove to produce the result that, for the matrix I that can be multiplied with a matrix B, $IB = B$ or $BI = I$ (note a different sized I may be involved in IB than in BI). Therefore, I plays a role similar to one in ordinary multiplication. The virtue of all this is that a system of equations can be compactly written as a matrix relationship such as $Y = A\ X$ and a solution $Y = A^{-1}\ X$ can possibly be found. However, no matrices except square ones can ever have an inverse and many square ones may not have an inverse.

(More details are available in numerous places including books, such as Hadley and Kemp, devoted primarily to this type of mathematics, all the texts in mathematics for economics, and innumerable appendixes to price theory and econometric texts).

The great virtue of a matrix is that it can stand for many coefficients or variables, and numerous proofs for two variable cases are generalized by replacing the single variables with matrices and showing that the same type of relationships prevail. One important simple set of applications is simply to derive certain basic results such as the Hotelling and Shepherd lemmas discussed in the test by simple algebraic manipulations of matrices. Complex optimization models similarly can be most simply shown in matrix form (see Intrilligator or Chiang for some of the best presentations of this point).

The historical starting point for optimization in economic theory, as represented by Samuelson's unified optimization approach in his *Foundations*, was the differential calculus. The basic concept in the calculus is the function—a relationship among variables. The differential calculus deals with measuring the impact of changing one or more variables on the other variables. The calculus, moreover, simplifies the analysis by dealing with approximate estimates of the change. The actual impact is usually a function of the size of the change. Consider, for example, the equation $y = x^2$. If x increases to $(x + \Delta x)$, y rises to $(x + \Delta x)^2 = x^2 - 2x\Delta x + (\Delta x)^2$ so the net increase in the value of the function is

$$\begin{aligned} \Delta y \equiv x^2 + 2x\Delta x + (\Delta x)^2 - x^2 \\ = 2x\Delta x + (\Delta x)^2. \end{aligned}$$

The usual approach is to calculate the relative change $\dfrac{\Delta y}{\Delta x} = 2x + \Delta x$.

The practice then is to approximate the change by the derivative $\dfrac{dy}{dx} = 2x$. Generally derivatives are always calculated in this fashion—determining the relative change and dropping terms involving Δx from the expression. The rationale is that for sufficiently small changes, Δx terms will have minor effects.

More precisely, the validity of dropping Δx terms is established by showing that for whatever definition of a negligible error you adopt, a small enough Δx can be selected so that the derivative estimates the change with the desired accuracy.

The efforts to explain this are best stated in a form that the variation in x is continuous. The functions are defined for any value of x and Δx can be made as small as desired. Thus, the derivative is a measure of change near a point on the function with the size of the neighborhood near the point to which the derivative applies being a function of the definition of acceptable error. In practice, to keep errors low, we assume that the change also is quite small and moves us an imperceptible distance from x.

For complex enough functions, numerous higher orders of derivatives can be defined. For example, a second derivative is the change in the derivative when a variable is changed and the nth derivative is the change in the $n - 1$ derivative when the variable changes.

Optimization in classical calculus involves first calculating stationary points in a function—points at which change of any variable leaves the value of the function unchanged. However, this only tells us that an extreme may have been reached. Further checks are needed to see whether we have reached a maximum or minimum, or neither.

Second derivatives are used for this purpose. A negative second derivative says any further change lowers the value so we are at the maximum or more strictly speaking at a "local" maximum—the highest value near this point is defined. An involved function can rise and fall and have several local maximums and minimums and the further step of calculating the value of the function at each point is needed to determine which of the local highs is the highest or global maximum. Similarly, a positive second derivative denotes a minimum.

Matrix algebra and associated tools come into play here to provide expressions, technically known as quadratic forms, that calculate the relevant weighted value of derivatives needed to identify if a maximum or minimum occurs when the first derivatives are zero. (Note that, moreover, some situations can occur when the first derivative is zero and neither a maximum or minimum occurs or conversely cases can arise when the value of the first derivative cannot be unambiguously defined—e.g., in a function with a discontinuity—but the behavior of second derivatives allows identification of a maximum or minimum).

A class of optimization problems critical to economics is that of maximization subject to which are termed constraints or side conditions. Each constraint defines a degree of interdependence among the variables in the original equation such that one cannot change one variable in the original equation without making a compensating change in other variables. In principle, it then becomes necessary to consider all possible offsetting changes—be they confined to one or many of the other variables. The method of Lagrange shows that it is unnecessary actually to consider explicitly all possible changes in other variables. If we introduce a new variable λ_j for each constraint g_j, the generalized effect of changing other variables to compensate for changing an x and still satisfying the additional equations (constraints) is

$$\frac{\partial y}{\partial x_i} + \Sigma\lambda_j \frac{\partial g_j}{\partial x_i}$$

where $\frac{\partial y}{\partial x_i}$ is the change in the basic function and the second term is the sum of the product of the derivative of each side condition j with respect to x_i multiplied by the λ_j for that equation. This can be shown to epitomize every possible change in other variables. Additionally, the system is completed by adding in the side conditions as further elements in the solution.

Moreover, these critical conditions can be derived by establishing a synthetic function $L = f(x) + \Sigma\lambda_j g_j(x)$ and differentiating with respect to each x and λ. A concept of examining a weighted sum of second derivatives to determine whether a maximum or minimum has been obtained can then be developed. The sum includes both second derivatives of the main function and first derivatives of the side condition.

These tools are excellent for deriving basic rules such as shown in my text, but numerical solutions often are difficult to derive. For that reason, during World War II, George Danzig devised the concept of linear programming. The term denotes the proposition that both the expression to be optimized (termed the objective function) and the constraints are linear—i.e., have a form such as $Z = a_1x_1 + a_2x_2 + a_3 x_3 . . .$

Moreover, Danzig added the characteristic that the constraints were inequalities rather than equations thus allowing doing more or less than required as appropriate. For example, instead of restricting a machine to a given output level, you could set a ceiling, a floor, or both to output of that machine and allow the level to lie between as well as at the limits. He also imposed further conditions excluding negative values for the variables.

This approach involves considering discrete changes in the value of the function. Specifically, a brute force solution process is followed by which you start at some point where all the constraints are satisfied, calculate the value of the objective function (which can be shown to involve no more variables at positive levels than there are constraints).

The next step is a determination of what change in the set of included variables produces the greatest local increase in the value of the objective function. (Note we change each variable by a non-negligible amount).

The optimum of the model is found by taking one improvement after another until no further improvement is possible. Danzig was able to show that the process always found the optimum, if it existed. The process inevitably moved you from any starting point to the actual best point because it guaranteed that no better point could be inadvertently skipped.

The computational ease of linear programming made it a popular empirical tool, and numerous linear program models have been used in minerals. In 1974 the Federal Energy Administration devised a complex linear program representation of the interaction of demand and supplies of energy fuels in the United States. It was initially named PIES—Project Independence Evaluation System after the Project Independence study for which it was originally designed. As Project Independence became an embarassing memory, the name was shifted to MEFS—the Midterm Energy Forecasting System. (Contrary to most concepts of midterm—say two to five years, the time frame was to 1995 and long term was defined as 2000 or later). Similarly, many models of the coal market were developed. (For an overview of energy models see the articles in Lev; on coal models, see my various writings and the references in them).

Early writers described linear programming as a more realistic description of actual practice than the calculus. This view periodically reappears. However, it was long ago given the *coup de grace* in Dorfman, Samuelson, and Solow's classic introduction to the subject (DOSSO). They pointed out that linear programming was another approximation in which it was assumed that changes were "large," but it was possible to extend the model to reduce the size of the changes that could be considered. Thus, classical calculus results and linear programming results would differ negligibly if the former dealt with sufficiently high levels of errors in the derivative and the latter considered small enough changes.

Kuhn and Tucker's effort to generalize linear programming to nonlinear models completed the process by showing that the generalization was a modification of the method of Lagrange that handled the additional considerations of insuring nonnegative values for variables and allowed for inequality constraints. Nonnegativity is handled simply by keeping to zero all variables whose optimal value would be negative in classical calculus. The idea is that going towards zero heads you in the direction you would go in the classical calculus. The stopping point then takes account of the nonnegativity constraint. Nonsatisfaction of a constraint is shown to be represented by a zero Lagrangean (on all this Chiang, as usual, is the best introduction, and Intrilligator gives a more rigorous but still clear account).

The other branch of the calculus—integral calculus—also has a role in optimization and other forms of economic analysis. An integral provides a measure of the area under a curve. Crudely, the area under any curve can be approximated by dividing the curve with many segments and contructing rectangles in each segment. Two such rectangles can be constructed for each segment—one with a height equal to the highest point on the curve and the other with a height equal to the lowest point on the curve. The true area then lies between the areas of the rectangles. As the size of the segments decreases, the gap between the two estimates closes and approaches the true area. Thus, as the size of the segment gets closer to zero, we get adequately close to the true value of the integral.

A further point to note is that, in a complex sense, derivatives and integrals are the reverse of each other. The simpler part is that if you solve an integral and then differentiate it, the result is the function originally integrated. Going from an integral to the derivative that could have produced it is more complex, primarily because the derivative of any constant is zero. Thus when integrating, some method must be found to determine what that constant if any might have been in the equation differentiated to produce the function being integrated.

This is unknowable for an integral of unspecified length, but if the integral has an upper limit a_2 and a lower limit a_1 the solution is $f(a_2) - f(a_1)$—where f is the value of the terms other than the constant in the integral and $f(a_2)$ is the value of the function at the upper limit and $f(a_1)$ is the value at the lower limit.

At least three major uses of integral calculus arise in economics. The simplest is that present value analysis can be conducted on a continuous time basis. An integral of $F(t)e^{-rt}dt$, where $F(t)$ is net income, (t) is time, e is a basic mathematical constant and r is the market rate of interest, can represent present worth. Where $F(t)$ takes a simple form, the integral can be solved.

A more complex use is in the calculus of variations. This is a form of maximization in which one selects a function that will maximize the value of some integral. This technique is widely used in among other areas exhaustible resource economics.

A final use is differential equations—equations involving one or more terms that are derivatives. The principle that integration reverses differentiation is used to work towards reducing these equations to more manageable forms (the calculus of variations is actually an offshoot of differential equations but was introduced first here because its use is more critical to this chapter).

In practice, most differential equations do not solve neatly if at all; in many cases, the best we can do is make computer approximations of the values. In any case, the motions of a physical or economic system can be represented by a differential equation. Study of the behavior of these equations then can be used to determine the stability of the system.

Specifically, we start with a set of basic equations describing a system and a primary step with analysis to determine the solution points (i.e., the points at which all conditions are satisfied). Stability analysis then describes how the system will move if we start from any arbitrary point away from the solution. The possibilities include situations in which the system always moves to a solution, systems in which it never does, and systems that do so in some but not all cases.

Other areas in mathematics in use in economics include various aspects of point set topology. The materials on the Hotelling and Shephard lemmas are simplifications of some of the simpler uses. More abstract concepts called fixed point theorems have been used to deal with the technical issue of the existence of competitive equilibrium.

At issue is under what circumstances the idealized textbook model of a competitive general equilibrium economy has a solution. The key concepts used are fixed point theorems. Roughly such theorems deal with functions that take a set of variables and generate a new set of variables drawn from the same universe as the first. For example, in simple applications in economics, the functions start from an arbitrary set of prices for all goods and define the sort of price changes that should occur if markets do not clear. A fixed point is a value of the function such that the initial variables produce themselves as the dependent variables. A fixed point set of prices would be ones that lead to no price changes—the key property of equilibrium prices. The critical concern then is to insure that a fixed point is associated with the equilibrium prices of the competitive model. This implies equilibrium prices exist.

The concern arises because the model is so great an idealization of reality that one cannot be sure whether it contains a solution. Not surprisingly it proved necessary explicitly to add to the model assumptions to guarantee existence. Fortunately, however, these required assumptions proved reasonably realistic (Debreu wrote the classic book on the subject, but it is very difficult. Arrow and Hahn are a bit better particularly in their discussion of the simpler cases. DOSSO provides a somewhat dated but clear intuitive view of the issues, and a more difficult but still clear discussion appears in Henderson and Quandt).

References

1. Adelman, M.A., 1970, "Economics of Exploration for Petroleum and Other Minerals," *Geoexploration* 8, pp. 131–150.

2. Adelman, M.A., 1972, *The World Petroleum Market,* Baltimore: The Johns Hopkins University Press for Resources for the Future.
3. Allen, R.G.D., 1938, *Mathematical Analysis for Economics,* London: Macmillan and Co.
4. Allen, R.G.D., 1956, *Mathematical Economics,* London: Macmillan and Co., and New York: St. Martins Press.
5. Arrow, Kenneth J., and Frank Hahn, 1971, *General Competitive Analysis,* San Francisco: Holden Day.
6. Arrow, Kenneth J. and Michael D. Intrilligator, eds. 1982, *Handbook of Mathematical Economics,* V. II, Amsterdam: North-Holland Publishing Co.
7. Barnett, Harold J., and Chandler Morse, 1963, *Scarcity and Growth: The Economics of Natural Resource Availability,* Baltimore: The Johns Hopkins University Press for Resources for the Future.
8. Baumol, William J., 1967, *Business Behavior, Value and Growth,* revised edition, New York: Harcourt, Brace & World, Inc.
9. Baumol, William J., 1977, *Economic Theory and Operations Analysis,* 4th ed., Englewood Cliffs, N.J.: Prentice-Hall.
10. Bierman, Harold, Jr., and Seymour Smidt, 1975, *The Capital Budgeting Decision: Economic Analysis and Financing of Investment Projects,* 4th ed., New York: Macmillan Publishing Company.
11. Chiang, Alpha C., 1974, *Fundamental Methods of Mathematical Economics,* 2d ed., New York: McGraw-Hill Book Company.
12. Coase, Ronald H., 1937, "The Nature of the Firm," *Economica,* New Series 4, pp. 386–405. Reprinted in Kenneth E. Boulding, and George J. Stigler, eds., *Readings in Price Theory,* pp. 331–351. Homewood, Ill.: Richard D. Irwin, 1952.
13. Cournot, August, 1897, *Researches into the Mathematical Principles of the Theory of Wealth,* (first English translation of book first published in 1838). A revised edition with notes by Irving Fisher appeared in 1927 (New York: Macmillan and Co.) and in a reprinted edition in 1963 by Richard D. Irwin.
14. Cummings, Ronald, G., 1969, "Some Extensions of the Economic Theory of Exhaustible Resources," *Western Economic Journal* 7:3, pp. 201–210.
15. Danzig, George B., 1963, *Linear Programming and Extensions,* Princeton: Princeton University Press.
16. Dasgupta, P.S., and G.M. Heal, 1979, *Economics Theory and Exhaustible Resources,* Cambridge, England: Cambridge University Press.
17. Debreu, Gerard, 1959, *Theory of Value: An Axiomatic Analysis of Economic Equilibrium,* New York: John Wiley and Sons, Inc.
18. Dorfman, Robert, Paul A. Samuelson, and Robert M. Solow, 1958, *Linear Programming and Economic Analysis,* New York: McGraw-Hill Book Company.
19. Fisher, Irving, 1930, *The Theory of Interest,* New York: Macmillan Publishing Company.
20. Ford Foundation Energy Policy Project, 1974, *A Time to Choose: America's Energy Future,* Cambridge, Mass.: Ballinger Publishing Company.
21. Furubotn, Eirik G., and Svetozar Pejovich, eds., 1974, *The Economics of Property Rights,* Cambridge, Mass.: Ballinger Publishing Company.
22. Gaffney, Mason, ed., 1967, *Extractive Resources and Taxation,* Madison: University of Wisconsin Press.
23. Gordon, Richard L., 1966, "Conservation and the Theory of Exhaustible Resources," *Canadian Journal of Economics and Political Science* 32:3 (August), pp. 319–326.
24. Gordon, Richard L., 1967, "A Reinterpretation of the Pure Theory of Exhaustion," *Journal of Political Economy* 75:3 (June), pp. 274–286.
25. Gordon, Richard L., 1981, *An Economic Analysis of World Energy Problems,* Cambridge, Mass.: The MIT Press.
26. Gray, Lewis C., 1914, "Rent Under the Assumption of Exhaustibility," *Quarterly Journal of Economics* 28 (May), pp. 466–489. Reprinted in Mason Gaffney, ed., *Extractive Resources and Taxation,* pp. 423–446, Madison, Wisc.: University of Wisconsin Press, 1967.
27. Hadley, George, and M.C. Kemp, 1972, *Finite Mathematics in Business and Economics,* Amsterdam: North Holland Publishing Company, and New York: American Elsevier Publishing Company.
28. Henderson, James M., and Richard E. Quandt, 1980, *Microeconomics Theory: A Mathematical Approach,* 3d ed., New York: McGraw-Hill Book Company.
29. Herfindahl, Orris C., 1974, *Resource Economics: Selected Works,* David B. Brooks, ed., Baltimore: The Johns Hopkins University Press for Resources for the Future.
30. Hicks, J.R., 1946, *Value and Capital,* 2d ed., Oxford: Oxford University Press.
31. Hirshleifer, Jack, 1970, *Investment, Interest, and Capital,* Englewood Cliffs, N.J.: Prentice-Hall.
32. Hirshleifer, Jack, 1980, *Price Theory and Applications,* Englewood Cliffs, N.J.: Prentice-Hall.
33. Hotelling, Harold, 1931, "The Economics of Exhaustible Resources," *Journal of Political Economy* 39 (April), pp. 137–175.
34. Intrilligator, Michael D., 1971, *Mathematical Optimization and Economic Theory,* Englewood Cliffs, N.J.: Prentice-Hall.
35. Kuhn, H.W., and A.W. Tucker, 1951, "Nonlinear Programming," in J. Neyman, ed., *Proceedings of the Second Berkeley Symposium on Mathematical Statistics and Probability,* Berkeley: University of California Press, reprinted in Peter Newman, ed., *Readings in Mathematical Economics,* Baltimore: Johns Hopkins Press, 1968.
36. Lancaster, Kelvin, 1968, *Mathematical Economics,* New York: The Macmillan Company.
37. Lerner, Abba P., 1934, "The Concept of Monopoly and the Measurement of Monopoly Power," *Review of Economic Studies,* 1 (June), pp. 57–75.
38. Lev, Benjamin, ed., *Energy Models and Studies,* Amsterdam: North-Holland Publishing Company.
39. Lipsey, Richard G., and Peter D. Stenner, 1983, *Economics,* 7th ed., New York: Harper and Row.
40. Logistics Management Institute, 1979, *The Integrating Model of the Project Independence Evaluation System,* 6 V., Washington, D.C.: U.S. Department of Energy.
41. Malinvaud, E., 1972, *Lectures on Microeconomic Theory,* Amsterdam: North-Holland Publishing Company.
42. Mansfield, Edwin, 1979, *Microeconomics: Theory and Applications,* 3d ed., New York: W.W. Norton & Company.
43. Marris, Robin, 1964, *The Economic Theory of "Managerial" Capitalism,* New York: Basic Books Inc.

44. Marris, Robin, and Adrian Wood, eds., 1971, *The Corporate Economy, Growth Competition and Innovative Potential,* Cambridge, Mass.: Harvard University Press.
45. Merrett, A.J., and Allen Sykes, 1973, *The Finance and Analyses of Capital Projects,* 2d ed., New York: John Wiley & Sons.
46. Miller, Roger Leroy, 1982, *Economics Today,* 4th ed., New York: Harper and Row.
47. Modiano, Eduardo M., and Jeremy F. Shapiro, 1980, "A Dynamic Optimization Model of Depletable Resources, *The Bell Journal of Economics,* 11:1 (Spring), pp. 212–236.
48. Nadiri, M. Ishaq, 1982, "Producers Theory," in Kenneth J. Arrow and Michael D. Intrilligator, eds., *Handbook of Mathematical Economics,* Vol. II, Amsterdam: North-Holland Publishing Company, pp. 431–490.
49. Newman, P., 1965, *The Theory of Exchange,* Englewood Cliffs, N.J.: Prentice-Hall.
50. Nikaido, Hukukone, 1968, *Convex Structures and Economic Theory,* New York: Academic Press.
51. Nikaido, Hukukone, 1970, *Introduction to Sets and Mappings in Modern Economics,* Amsterdam: North-Holland Publishing Company, and New York: American Elsevier Publishing Company.
52. Patinkin, Don, 1965, *Money Interest and Prices,* 2d ed., New York: Harper and Row.
53. Potter, Neal, and Francis T. Christy, Jr., 1962, *Trends in National Resource Commodities,* Baltimore: The Johns Hopkins Press for Resources for the Future.
54. Robbins, Lionel, 1932, *An Essay on the Nature and Significance of Economic Science,* London: Macmillan and Company Ltd.
55. Samuelson, Paul A., 1980, *Economics,* 11th ed., New York: McGraw-Hill.
56. Samuelson, Paul A., 1983, *Foundations of Economic Analysis,* expanded edition, Cambridge, Mass.: Harvard University Press. (First edition 1947).
57. Scott, Anthony, 1967, "The Theory of the Mine Under Conditions of Certainty," in Mason Gaffney, ed., *Extraction Resources and Taxation,* Madison, Wisc.: University of Wisconsin Press, pp. 25–62.
58. Shephard, Ronald W., 1970, *Theory of Cost and Production Functions,* Princeton: Princeton University Press.
59. Sher, William, and Rudy Pinola, 1981, *Microeconomic Theory: A Synthesis of Classical Theory and the Modern Approach,* New York: North-Holland Publishing Company.
60. Shone, R., 1976, *Microeconomics: A Modern Treatment,* New York: Academic Press.
61. Smith, V. Kerry, ed., 1979, *Scarcity and Growth Reconsidered,* Baltimore: The Johns Hopkins University Press for Resources for the Future.
62. Stigler, George J., 1966, *The Theory of Price,* 3d ed., New York: Macmillan Company.
63. Takayama, Akira, 1974, *Mathematical Economics,* Hinsdale: The Dryden Press.
64. U.S. Federal Energy Administration, 1974, *Project Independence Report,* Washington, D.C.: U.S. Government Printing Office.
65. U.S. President's Materials Policy Commission (The Paley Commission), 1952, *Resources for Freedom,* 5 V., Washington, D.C.: U.S. Government Printing Office.
66. Varian, Hal R., 1978, *Microeconomic Analysis,* New York: W.W. Norton & Company.
67. Walras, Leon, 1954, *Elements of Pure Economics or the Theory of Social Wealth,* (translation by William Jaffe of 1926 edition of book first published in 1874), London: George Allen and Unwin.
68. Williamson, Oliver E., 1967, *The Economics of Discretionary Behavior: Managerial Objectives in the Theory of the Firm,* Chicago, Ill.: Markham Publishing Company.
69. Willig, R., 1976, "Consumer's Surplus without Apology," *American Economic Review,* pp. 589–597.

2.5

Theory of Mineral Demand

Gary A. Campbell*†

INTRODUCTION

The theory of mineral demand often is slighted in the study of mineral economics. The supply side of the mineral industries usually commands a great deal more attention and effort by mineral economists than the demand side does because the study of mineral supply includes topics such as exploration, resource availability, and exhaustion theory that are generally considered unique to the mineral industries. Mineral demand, on the other hand, is thought of as being more general in nature and therefore of less concern to mineral industries specialists since it is similar to the demand for other inputs of production.

This similarity with other inputs of production, however, does not negate the importance of a thorough understanding of mineral demand. It is the interaction of supply and demand that determines market behavior and not supply alone. Mineral demand is a key component among the factors that influence mineral production; therefore, a good understanding of mineral demand can give one better insights into the industry behavior and trends than one would otherwise have. A strong working knowledge of the demand side of the mineral industries is a powerful tool for analyzing and predicting mineral market behavior.

*School of Business, Michigan Technological University, Houghton, Michigan 49931 USA

†The chapter was improved by the thoughtful comments of several persons. My thanks go to B. Patrick Joyce, Peter A. Gaines, the AIME reviewer, and especially Betty C. Heian for her very thorough reading and suggestions on the economic portions of this chapter. The errors and omissions remain mine.

The material presented in this chapter does not depend on any special background in economics and should be accessible to all interested readers. The chapter is organized so as to permit an orderly progression of the basic topics of the theory of mineral demand while allowing digressions on related issues of special interest. The first section of the chapter is on the determinates of mineral demand. The discussion begins with the introduction of consumer demand and some of the necessary terminology of economics and proceeds to the particular case of the derived demand for an input of production such as minerals and the characteristics of such demand. The second section of the chapter is on the response of mineral demand to changes in economic variables like price and aggregate income. This second section of the chapter begins with a general presentation of elasticity of demand and continues with the particular case of the price responsiveness of mineral demand and its nature in the short and long-runs. A brief look at the response of mineral demand to price changes for other products is also included. The chapter concludes with a discussion on the effect of aggregate income changes on mineral demand and its nature in the short and long-runs.

CONSUMER DEMAND

Some Basic Definitions

The ultimate goal of any economic system is the allocation of scarce resources among competing factions. To accomplish this goal, an economic system must explicitly deal with the supply and demand of goods and services as well as the interaction between the two. The material in this section of the chapter is centered around the way material demand is involved in this process.

The typical role of minerals is as inputs of production used to produce desired final goods and services for consumers. Minerals are rarely demanded as a final product by consumer. This type of indirect demand is referred to as a derived demand. But before the topic of the derived demand for minerals can be addressed, the topic of consumer demand and its characteristics must be developed because this final product demand is what drives the forces that shape the demand for minerals.

To introduce the fundamentals of consumer demand, it is advantageous to use simple economic models that avoid much of the complications observed in the actual marketplace while retaining the essential characteristics being sought for study. Such economic models have been criticized by some individuals for being too abstracted from reality and therefore of little practical value. Closeness to reality, however, is not the purpose of such models. The test, as in the physical sciences, is how well does the model predict basic behavior. The following types of models have done well in that regard.

The most general type of model is called a general equilibrium model and considers all markets and their interactions within a competitive situation. A market is defined as all the buyers and sellers of a particular good and service—not just a physical location. A competitive situation exists in a market(s) when there are numerous small buyers and sellers who individually have no effect on the market and its price. All such markets and their interactions within an economy can then be studied together as an unit. A general equilibrium occurs when the quantity supplied by sellers equals the quantity demanded by buyers in each market. This condition is reached through adjustments in price and quantities of the goods and services. An existing equilibrium condition is characterized by a tendency not to change unless affected by external forces. A stable system not in equilibrium will converge to an equilibrium position.

A general equilibrium model is difficult to use for studying activity in a particular market because of its broad nature and general conclusions. This type of model simply is not appropriate for dealing with isolated markets. Instead, a partial equilibrium model is used. This kind of model is constructed around one particular market and usually does not directly consider interactions with other markets. Equilibrium in this analysis occurs when supply and demand for this one market are equal. This type of analysis while being more restrictive than the general equilibrium model is more particular and can be very informative. Actual markets typically do not achieve equilibrium positions because external factors that affect the markets usually are changing. This fact, however, does not detract from the useful results derived from this type of model.

Analyses can be conducted using the partial equilibrium model by taking a market initially in equilibrium and introducing one or more changes to it and observing the new equilibrium position. The study of what determines equilibrium values and what the changes are is called comparative statics. The study of how the market moves from one equilibrium position to another is called dynamics.

Another key consideration in the workings of the model is the nature of consumption of the goods and services by consumers. Minerals are considered private goods. Private goods are products that users can be excluded from (can be forced to pay), and an individual's use of the product prevents other users from consuming it (the product is used up). Most goods and services are of this nature. Another type is a public good. For this type of product, consumers can not be reasonably excluded from its use once it becomes available, and one's use of it does not preclude others from using it as well. One example is radio and television signals. The signal can be used by anyone with a receiver, and one's use does not prevent others from also using the same signal.

Introduction to the Demand Curve

So far, the discussion has revolved around terminology and assumptions without much comment on the economic theory and analysis needed to operate the models. Now that some of the basic simplifications and definitions have been introduced to the reader, the discussion can proceed to this next stage.

A basic analytical tool of partial equilibrium analysis and economics in general is the supply-demand graph. To present a particular market, one places quantity of the good or service on the horizontal axis and the product price on the vertical axis. Only the first quadrant is shown since for economic purposes only zero and positive prices and quantities are meaningful. The representation is completed by curves on the graph showing the behavior of buyers and sellers. For demand (buyers), this curve is called the demand curve, and it is derived from the demand schedule. The demand schedule is defined as the quantity of a product desired by consumers at any given price at a particular time. An equivalent schedule is defined for supply (sellers). Fig. 2.5.1 shows a typical graph with a demand curve that comes from the demand schedule shown in Table 2.5.1.

Table 2.5.1—A Typical Demand Schedule

Price (per unit)	Quantity Demanded (units/year)
$ 1	6
2	5
3	4
4	3
5	2
6	1

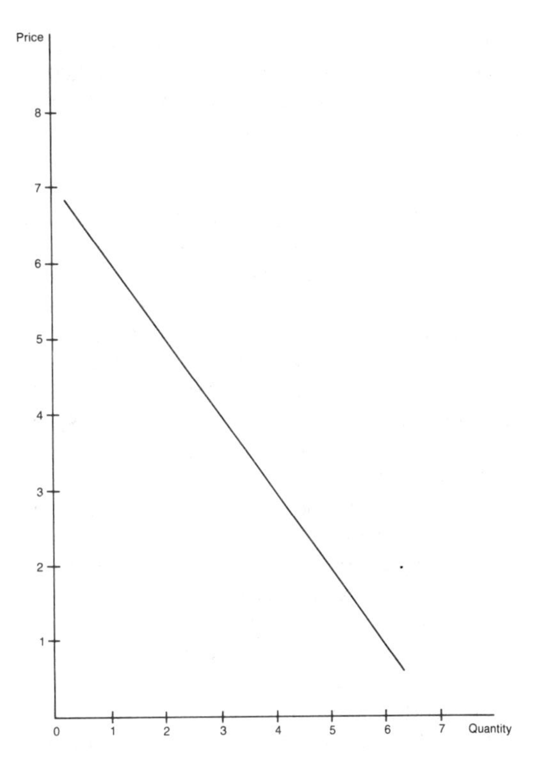

Fig. 2.5.1—The demand curve for Table 2.5.1.

The downward slope of the demand curve shown in Fig. 2.5.1 is assumed because of the law of demand. This law states that the quantity of a commodity demanded by consumers is inversely related to its price if all other influencing factors are held constant. In other words, as the price goes up for a product, consumers buy less, and as price goes down, consumers buy more. This behavior is easily seen in daily life. A store that is overstocked in certain items has a sale. Tickets to a very popular, sold-out athletic event are ''scalped'' at much more than face value.

The law of demand is valid because of the substitutions and income effects. The income effect is based on the amount of money consumers have to spend. In general, the more income consumers have the more of the product consumers are likely to buy. The opposite also holds. The less income consumers have the less of the product they are likely to buy. This effect occurs for most goods and services (normal). The income effect is important to the shape of the demand curve because if the price of the product rises it reduces the consumer's income (which is assumed fixed) and he will demand less. If the price drops it increases the consumer's income and he will demand more. The substitution effect is similar. If the price of a product rises, consumers are likely to substitute other products for it. When the price of beef increases, consumers tend to buy more chicken. The opposite is also true. If the price of a product drops, consumers generally will expand their use of the product. The combined result of these two effects is the inverse relationship between a product's price and the quantity demanded by consumers known as the law of demand.

Formal Development of Consumer Demand

The first requirement in the development of the theory of consumer demand is determining the general conditions that influence the consumer's desire for a particular quantity of a good or service. In actuality, there are many relevant factors and they will vary greatly across individuals. Fortunately, there are several important ones that are common to all consumers and will be sufficient for analytical purposes. The first of these conditions is the price of the product itself. This factor has been discussed earlier as the law of demand and is the variable used in developing the demand schedule. Often this is the only variable that is considered in simple models. All else is just ignored (assumed constant). The second condition is the price of complements and substitutes. Complements are other goods and services that are used concurrently with the product of interest. For example, the normal use of a car requires gasoline (or oil in some form). Accordingly, the price of gasoline can influence consumer car buying decisions. This effect was demonstrated during the late 1970s when buyers moved from large, traditional cars to smaller, fuel-efficient ones due to the large jump in fuel prices. Substitutes are goods and services used in place of the product of interest. If the price of a substitute rises, the product appears more desirable to use. If the price of a substitute drops, the product looks less appealing to the user. The third condition is the individual's income level. How much an individual demands of a product at a given price is directly and positively related to the amount of income the individual has to spend. This condition is broader in scope than the income effect used to explain the law of demand. For the law of demand, changes in income are brought about by changes in the product's price. Here, the consumer's total income level is involved. The final factor is one that is important but very difficult in practice to handle in a model. This is the psychological component of a consumer's demand. It is based on his tastes, preferences, and expectations and is unique to each individual. Economists, in general, do not try to estimate individual weights for this final component but do try to draw conclusions that are general enough to incorporate the wide variations seen in individual behavior. The derivation of such general behavior is the task of the next section.

Indifferences Curve Analysis: Consumer decisions play an important role in the demand model and must be incorporated into the analytical framework. As previously explained, it is not feasible to model the particular psychological behavior of each individual. However, general postulates on consumer behavior are possible. First, the consumer's goal is to get the most satisfaction possible out of his resources. Utility is the term economist's use to describe the satisfaction an individual receives from the consumption of a good or service. There is no established criteria to measure explicitly the level of satisfaction a particular good or service will give an individual. Consumers simply are described as making choices among alternatives such that they get as much satisfaction as they can. Second, to maintain the logical consistency of the model, consumers are assumed to exhibit rational behavior. Rational economic behavior means that a consumer will always make the choice that maximizes his utility to the best of his knowledge, and will not knowingly choose an option that provides less utility over another that would provide more. If rational behavior is not postulated, it is impossible to draw conclusions about consumer behavior from this type of economic model because consumer behavior would follow no coherent pattern. Finally, it is assumed that consumers normally prefer more to less of a particular good or service, all other factors being constant (nonsatiation). If a consumer is able to get more of a good without giving up some of another, he will do so.

These three postulates of utility maximization, rational behavior, and nonsatiation are used in the development of indifference curve analysis. An indifference curve links points (bundles) that give equal utility to a consumer. Each point represents different combinations of amounts of available goods and services. An individual is indifferent (neutral) about which of the bundles he consumes on a particular curve because each bundle gives him equal satisfac-

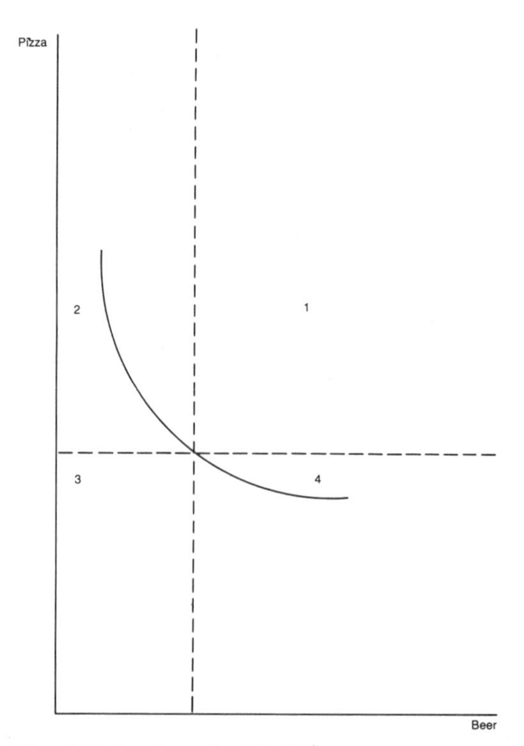

Fig. 2.5.2—A typical indifference curve.

tion. Consider Fig. 2.5.2. A consumer has a choice between combinations of two goods, beer and pizza. To permit graphical analysis, a two-good world is assumed. This restriction causes no later problems for the general case since the results can be easily extended mathematically if not graphically. For illustration purposes, the consumer presently has bundle A. From the postulate of nonsatiation, it is clear that all bundles in region 1 are ''better'' than A because the consumer gets more of both goods. By the same reasoning, bundles in region 3 must be ''worse'' than bundle A because the consumer gets less of each good. What about regions 2 and 4? No obvious answer is available because the bundles contain less of one good but more of the other. Therefore, it is necessary to know something about the trade-off in satisfaction between consuming goods beer and pizza before the ambiguity in regions 2 and 4 can be resolved.

To deal with the problem of regions 2 and 4, the assumption of substitutability is made. This assumption states that a subtraction of a small amount of one good can be exactly offset by the addition of a certain amount of the other good. This required trade-off is called the marginal rate of substitution (MRS). Marginal refers to the changes brought about by the introduction of one more unit. For the MRS between beer and pizza for a particular indifference curve, the question is how much beer must be given up to offset the addition of one more unit of pizza. Geometrically, the MRS is the slope of the indifference curve.

The indifference curve can now be drawn (Fig. 2.5.2). Notice that the curve slopes downward to the right and is convex to the origin. This shape is based on a psychological assumption called the law of diminishing marginal utility. This law asserts that the more you have of one particular good the smaller the increase in the amount of the other good needed to offset a unit decrease in the first good. To put it differently, the less beer you have, the greater its value relative to pizza. For another example, an individual likes apples and oranges equally well. If he has a bushel of apples and just a couple of oranges, he is probably willing to give up several apples for an additional orange. However, if the reverse is true, he would probably require many oranges before he would give up an apple in trade. Graphically, if one stays on a particular indifference curve, this law results in a convex to the origin curve as the trade-off between beer and pizza varys depending on the consumer's holdings of beer and pizza. There is an indifference curve associated with each different level of utility. The utility increases as the curves move rightward due to the postulate of nonsatiation because there is an increase in the amount of both goods in the bundles.

It has been pointed out that consumer demand is positively related to the level of income. For this simple two-good case, it is assumed that all the consumer's income (I) must be spent for some combination of X (quantity of beer) and Y (quantity of pizza) whose prices are P_{beer} and P_{pizza}. Algebraically, this relationship can be expressed as

$$P_{\text{beer}} X + P_{\text{pizza}} Y = I \qquad (1)$$

This equation is an expression of a straight line and is referred to as the budget constraint. It shows the feasible consumption possibilities for a particular income level. The budget constraint for the consumer is now placed on the same graph with his indifference curves (Fig. 2.5.3).

It is now possible to model the consumer's utility maximizing decision. A rational individual wants to get on the highest (rightmost) in-

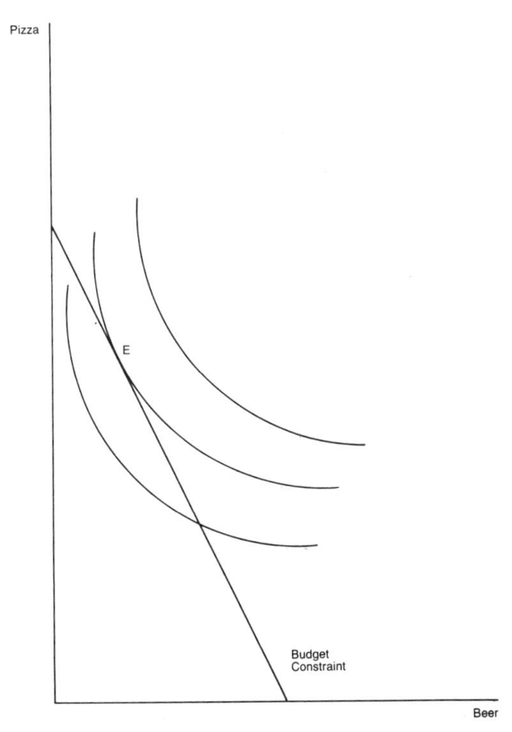

Fig. 2.5.3—Indifference curve analysis.

difference curve he can with his given income level. Geometrically, this position is reached when the budget constraint is just tangent (has the same slope) to an indifference curve. This is shown as point E on Fig. 2.5.3. With income I, it is not possible for the consumer to get more satisfaction with this particular utility ranking. Algebraically, this condition is when the slope of the budget constraint (shown by simple algebraic manipulation to be the price ratio) equals the slope of the indifference curve, the marginal rate of substitution. With more than two goods and services, it is difficult geometrically to establish the optimal point, but the problem remains simple algebraically. The price ratio and the MRS for any two goods must be equal to each other.

The tools have not been developed that are necessary to derive the consumer demand for a particular good or service. Remember that a demand curve shows the quantity demanded at a given price. Fig. 2.5.4 shows the process used to derive an individual's demand curve for good *X* with indifference curve analysis. One axis of the graph is the quantity of *X* and the other axis is the money not spent on buying *X*. The process starts with an initial income and price of *X* (line A). Then, the price of *X* is allowed to vary. If the price of *X* decreases, more *X* can be bought causing an outward rotation of the budget constraint (line B). An increase in price leads to a decrease in the quantity of good *X* that can be purchased and results in an inward rotation of the budget constraint (line C). Each of these representative budget constraint lines is tangent to an indifference curve showing the quantity of *X* demanded and the money balances held for each case. It is a simple matter to determine the quantity of *X* demanded at any given price for a set income level. One takes the quantity of *X* from the horizontal axis, and price can be determined by taking the horizontal intercept of the budget constraint and dividing it into the dollar amount of the vertical intercept. Since price and quantity are now known, the information needed for a demand schedule is available, and a demand curve can be drawn. Table 2.5.2 shows a numerical example and Fig. 2.5.4b. is the result.

Table 2.5.2—Deriving an Individual's Demand Curve

Quantity of *X*	**Money Not Spent On *X***	**Price of *X*** $\left(\frac{\$12\text{—column 2}}{\text{column 1}}\right)$
1	$ 6	$ 6
2	4	4
4	3	2.25

It is now possible to get the demand curve for a particular individual and this same process can be followed for other consumers. The next step is to derive the market demand curve from these individual cases. For a private good, it is a simple matter. One merely sums the individual demand curves horizontally. Fig. 2.5.5 presents a graphic example of a two-user case. An individual's consumption of a private good excludes others from its use; therefore, the market

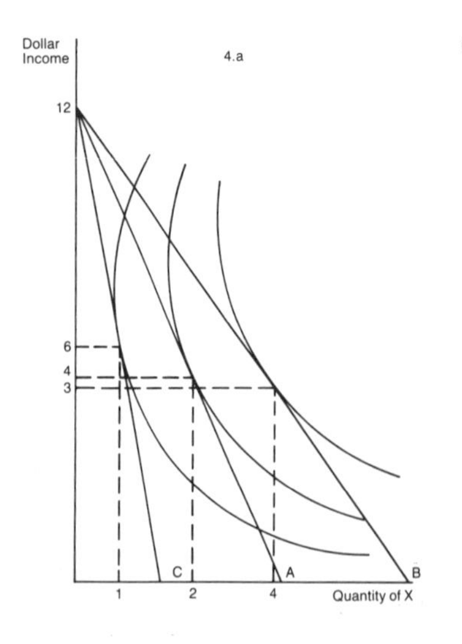

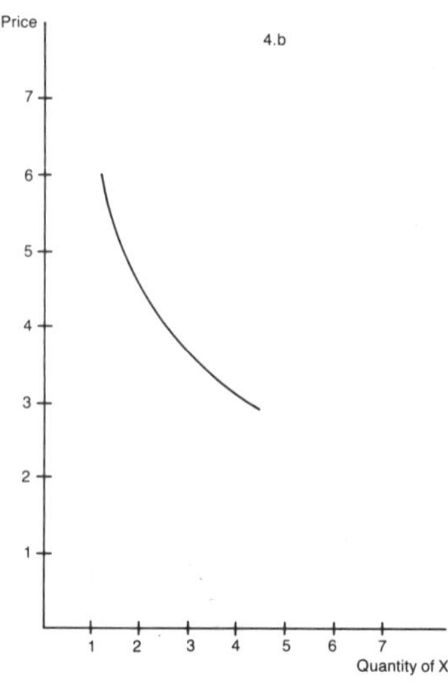

Fig. 2.5.4—Deriving an individual's demand curve.

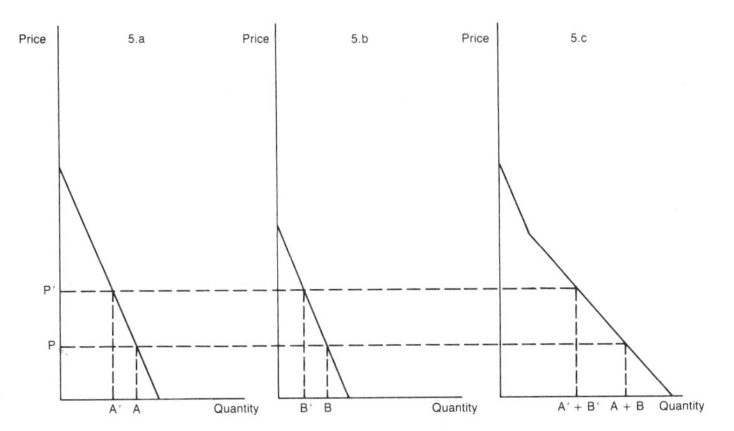

Fig. 2.5.5—Summing individual demand to get the market demand.

demand is simply the summation of all the individual demands.

Shifts of the Demand Curve: Care must be taken when manipulating a demand curve. The demand curve is constructed on the assumption that price is variable, and all other factors are constant. Accordingly, a change in the good's price and the associated change in the quantity of the good demanded is a movement along the existing curve. There is no change in the position of the demand curve. Change in the other determinants of demand cause a shift in position of the demand curve because there is now a new quantity associated with each price. This type of change is referred to as a shift in demand. The shift can be rightward, representing an increase in quantity for a given price or leftward, showing a decrease in quantity for a given price.

A rightward shift in an individual's (or market) demand curve can be caused by an increase in income, an increase in a substitute's price, a decrease in a complement's price, a favorable change in tastes and preferences, and other such reasons. A leftward shift can be caused by the opposite of all the reasons listed for a rightward shift. Fig. 2.5.6 illustrates what happens to the demand for copper if there is a discovery that copper bracelets actually reduce arthritis pains (*A*) and if the U.S. government reduces the amount of copper in pennies (*B*). In case *A*, the demand for copper increases due to the increased use of copper for medical purposes. In case *B*, the U.S. government's preference for copper in pennies has changed negatively and the new demand for copper reflects this. Joined with a supply curve, this type of graphical analysis can be very informative about market behavior.

DERIVED DEMAND

Introduction

Up to now, the consumer demand for final goods and services has been the focal point of the discussion. However, minerals frequently do not fit into this category. Consumers do demand precious metals and gems for wealth and jewelry purposes and certain minerals for health-

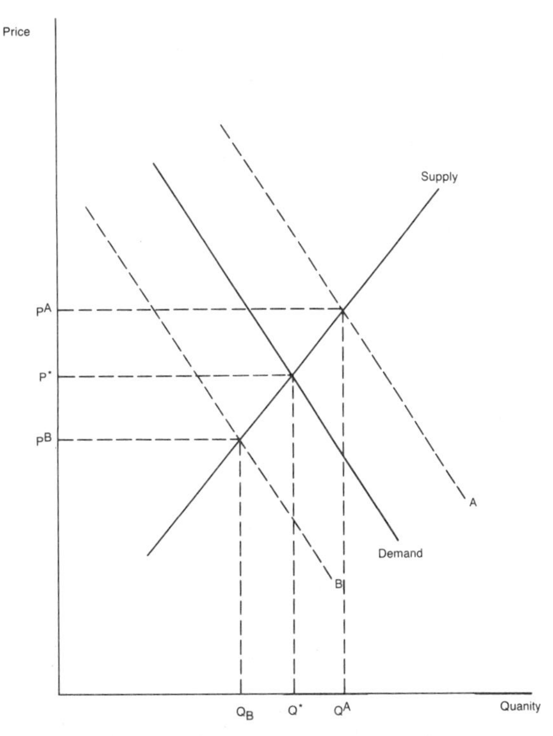

Fig. 2.5.6—Shifts of the demand curve.

related uses, but these uses are a small fraction of total mineral use. Minerals usually are employed as inputs for the production of final goods and services that are desired by consumers. The demand for this type of good is known as a derived demand. A formal definition of the condition of derived demand is that the demand for any productive factor ultimately depends upon, or derives from, the demand for the final product or products that the factor is used to produce (Waud, 1983, p. 639).

There is a clear relationship between the consumer demand for final goods and services and the derived demand by producers for inputs to produce these final goods and services. Yet, more than a clear understanding of consumer demand is needed to explain the derived demand for an input. There is little demand for a particular input of production—only for the services and properties it can provide with existing technologies. If another type of input is found that can do the job better, there is no hesitancy by producers about using it in place of the present input. The consumers of the final product are not concerned about which input factors are used as long as the final product meets their specification. Therefore, the derived demand for a particular input is also strongly dependent on existing technologies and the availability of competing and complementary input factors of production as well as the factors associated with consumer demand.

Since the demand for minerals normally is derived, the demand for an input by a producer is a key consideration in understanding the theory of mineral demand. The previous discussion of consumer demand for final goods and services has presented the background on the forces that ultimately create the demand for inputs of production, but consumer demand is less informative in explaining how the particular demand for an input is determined. This latter question is the one of immediate concern to the minerals industries.

Deriving the Input Demand Curve[a]

The derived demand for minerals is linked closely to the production decision and must be considered in that context. The relationship between the inputs of production and the resulting final goods or services is called the production function. The production function shows the most output that can be obtain from each possible combination of input factors for a particular level of technology. The information available from a production function can be presented in many useful ways. Fig. 2.5.7a. shows the production function plotted from the data in Table 2.5.3. The data described one producer in a competitive market who has only one variable input factor of production—minerals. All other input factors are held constant. The vertical axis shows total product, and the horizontal axis is for units of minerals. Fig. 2.5.7b. shows the marginal physical product (MPP) for the use of minerals associated with Table 2.5.3 and Fig. 2.5.7a.. The vertical axis is for the MPP per unit of minerals, and the horizontal axis remains the same. Marginal physical product of an input is the extra output of the final product that results from the use of one more unit of that input while other inputs are held constant.

The shape of the curves in Fig. 2.5.7 is due to the law of diminishing returns. This law, which is really only an observation, states that

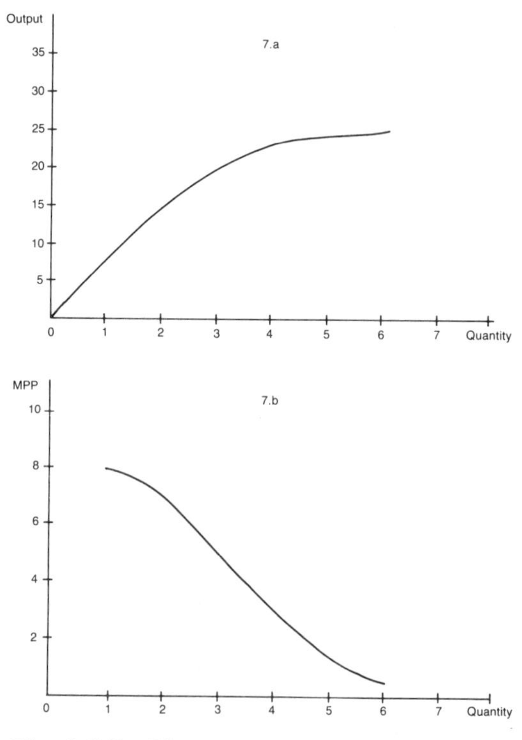

Fig. 2.5.7—The production function and MPP curve for Table 2.5.3.

[a]An excellent and fuller discussion can be found in Waud, 1983, pp. 635–655.

Table 2.5.3—Deriving the Derived Demand Curve For An Input of Production

Quantity of the Input (000 tons/year)	Output (units/year)	Marginal Physical Product (units/year)	Price (per unit)	Value of the Marginal Physical Product
0	0			
1	8	8	$ 2	$ 16
2	15	7	2	14
3	20	5	2	10
4	23	3	2	6
5	24	1	2	2
6	24.5	0.5	2	1

the MPP of any input factor tends to decline as more of it is employed in proportion to the other input factors used with it. Consider digging a hole. The amount of labor (you) is fixed, but the number of shovels is allowed to vary. Productivity is measured in dirt moved per time period. The MPP of the first shovel when one has had only his bare hands is very high. Dirt can be moved much faster and easier. The second shovel is useful as a spare in case the first one is broken but the gain from having it is not nearly as high as for the first. The third shovel is another spare and so forth. It should be clear that the addition of each shovel has a positive but declining MPP. This condition is shown in Fig. 2.5.7a. as a decrease in the rise of the curve as more units of minerals are used. Correspondingly in Fig. 2.5.7b., the curve slopes downward showing that each additional unit of minerals contributes less to total production.

The next step is to determine how much more value is obtained with the use of one more unit of minerals. The procedure is straightforward. The *MPP* for each unit of minerals is shown in Fig. 2.5.7a. By definition, the *MPP* shows the gain to output by the use of one more unit of the input, all else constant. The MPP is multiplied by the selling price of the final product. The result of this multiplication is the value of the marginal physical product (*VMP*)—the increase in total revenue due to the one unit increase in the variable input factor of production. Fig. 2.5.8 presents the VMP curve for the example in Table 2.5.3. Remember that the final product price is constant for a film in a competitive market no matter how it changes production.

How much of the variable input (minerals) should a producer use to maximize his net revenue? Theoretically, the answer is easily determined. He uses minerals until the value gained from the last unit used is offset by the cost of the additional unit. In more precise terminology, the producer should employ the variable input up to the point at which the marginal cost of the factor is just equal to the marginal revenue gained from its use. If the cost for an unit of minerals is less than the value gained from its use, why not use it? The user will gain net revenue on each unit employed. On the other hand, if the cost of the input is greater than the value gained from its use, why reduce net revenue by using it? The equilibrium point is reached when the cost of the input is equal to the revenue gained

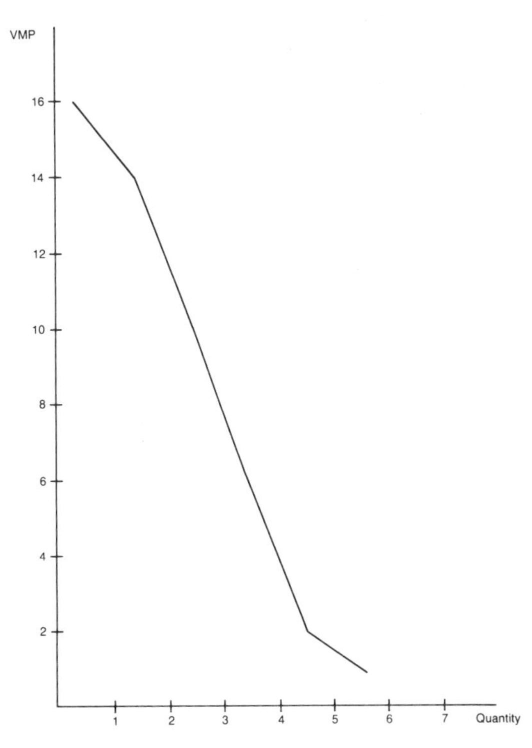

Fig. 2.5.8—Derived demand curve for Table 2.5.3.

from its use. The relationship can be expressed as:

$$MC_{\text{minerals}} = P_{\text{final good}} \cdot MPP_{\text{minerals}} \quad (2)$$

The marginal costs for minerals in a competitive market is equal to the price of minerals since the additional unit of minerals can be bought at the constant market price. With this substitution, the new expression is:

$$P_{\text{minerals}} = P_{\text{final good}} \cdot MPP_{\text{minerals}} \quad (3)$$

Since the VMP curve tells one how much of the input factor a firm will want to use at each price, the VMP curve is also the producer's demand curve for the input.

So far, the discussion about the demand for an input of production has been limited to competitive markets for the inputs and outputs. In a competitive output market, a single producer faces a horizonal demand curve for the final product he produces because he is so small relative to the overall market that his production decision does not affect in any way the market. However, for some mineral markets this is not the case. Individual producers control a large enough portion of the total market for their output to have a noticeable effect on its behavior when they make production changes. The U.S. steel industry is an example. In this case, the producer faces a downward sloping demand curve for the final product and the competitive assumptions no longer hold.

Some modifications of the derived demand model are needed to adjust the input demand curve for this ''imperfect'' competition. Under competition, the downward slope of the *VMP* curve is due only to falling marginal physical product (law of diminishing returns) because the output market price is constant (a horizonal demand curve for the final product). There is an added complication in the imperfect competition case. Both the MPP and the market price for the final product fall. The market price falls because the producer faces a downward sloping demand curve for the final product (law of demand again). The result is that the demand curve for the input under imperfect competition will fall more quickly than under competition causing the imperfect competition curve to be to the left and to have a steeper slope than the competitive curve.

A similar type of complication is present when one tries to determine the market demand for an input under competitive conditions. Since the consumption of minerals as an input is a private good, one would think that the demand by individual users could be summed horizonally to get the market demand for an input as in the consumer demand case. This is not so because of the derived nature of the demand for an input. Changes in the market price for the input bring about changes in the cost of producing the final product for a producer. Over the whole market, this causes changes in the quantity and price of the final product. The price of the final product must be constant for the horizonal summing to be appropriate. Instead, if the price of the input falls, the price of the final product falls and vice-versa. The result is a market-wide change in the quantity demanded of the final product. Accordingly, the input demand curve for each producer will be steeper when this effect is taken into account. These modified input factor demand curves then can be summed horizonally to get the correct market demand for the input. This market demand curve will have a steeper slope than one obtained from the unmodified individual factor demand curves.

Shifts of the Input Demand Curve

The section on consumer demand presents important factors that cause shifts in the demand curve. The same needs to be done for the input demand curve because some of the influencing factors are different for input demand. These different factors reflect the derived nature of the demand in addition to consumer desires and behavior.

The first of these factors is the change in consumer demand for the final product. Obviously, if the demand for the input factor ultimately is derived from the demand for the final product, there should be a direct and positive relationship between the consumer demand and the associated derived demand. An increase in demand for the final product will increase demand for the input factor to produce it. On a graph this translates to a rightward shift in the input factor demand curve. A decrease in demand for the final product will cause a leftward shift in the input factor demand curve. Under imperfect competition, the tendency is the same but the final outcome is complicated by price changes of the input factors and final product.

The second factor is technological change. The production function and MPP curve are developed on the basis of a fixed technology. Changes in technology will directly influence the production function and the corresponding *MPP* curve that are used in deriving the input demand curve. A technological change that improves the productivity of the input factor or the productivity of a complementary input factor will shift the input factor demand curve to the right. A change that improves the productivity of a substitute input factor or technique will cause a leftward shift in the input factor demand curve.

The third factor is the change in the mix of the input factors used to produce a final product. The derivation of the input demand curve is based on one variable input factor with all other input factors being held at some constant level. However, if the other factors are allowed to vary as well, the input demand curve will shift. The resulting shift in the input demand curve depends on the relative proportion of the new input factor mix. The shift will be rightward if less of the particular input is needed in the mix and leftward if more of the input factors is used.

The fourth factor is the change in the prices of other input factors. This factor is similar to the case of substitutes and complements for consumer demand, with an additional twist. There is no difference in the effect of changes in the price of a complement. A decrease in the price of a complementary input factor causes a rightward shift in the input factor demand curve and an increase causes a leftward shift. Price changes for substitute input factors are more complicated to consider. There is the typical substitution effect. A rise in the price of a substitute input factor causes a rightward shift in the input factor demand curve, and a price decrease for a substitute input factor causes a leftward shift of the input demand curve. However, there is also an output effect. A decrease in the price of a substitute input factor will reduce the cost of producing the final product which leads to higher production. Higher production requires more of both the input and the substitute. The opposite situation also holds. The substitution and output effects are in opposite directions. The final result depends on which effect is larger and varies from product-to-product.

ELASTICITY OF MINERAL DEMAND

Introduction

The previous sections deal with the general characteristics of mineral demand from the consumer's demand for a final product through the derived demand for an intermediate good of production. The remainder of the chapter is devoted to the study of changes in the derived demand for minerals in response to changes in mineral prices and to changes in consumer incomes. A basic concept used in this type of analysis is the elasticity of demand. Elasticity is the degree of change in the quantity demanded of a product due to changes in factors such as price or income. Using own price as an example, an elasticity coefficient (*Ed*) is defined as the percent change in the quantity demanded (*Qd*) of the mineral divided by the percent change in its price (*P*):

$$Ed = \frac{\text{percent change in } Qd}{\text{percent change in } P}\,.$$

The result is a ratio that shows the percent change in the quantity demanded of minerals for an one percent change in price. The coefficient can be positive or negative, depending on the relationship between demand and the factor. For the case of own price, the sign is negative because of the law of demand.

The estimation of the elasticity coefficient can be an important and useful means of moving from a general consideration of demand theory to the case of a particular market and industry. From the definitions for the price elasticity of demand and the demand schedule, it can be shown that the slope of the demand curve is related to the price elasticity of demand. In fact, it is an inverse relationship. Knowing something about the price elasticity of demand allows one to estimate the nature of the demand curve and to make more powerful statements about behavior in that market. It is the study of elasticity of demand that allows one to incorporate some of the characteristics of the mineral industries into the general theory of demand to develop the theory of mineral demand.

First, it is necessary to present more background information on the use of this important tool, elasticity. When discussing the numerical value of the elasticity coefficient, it is standard practice for economists to place it into one of

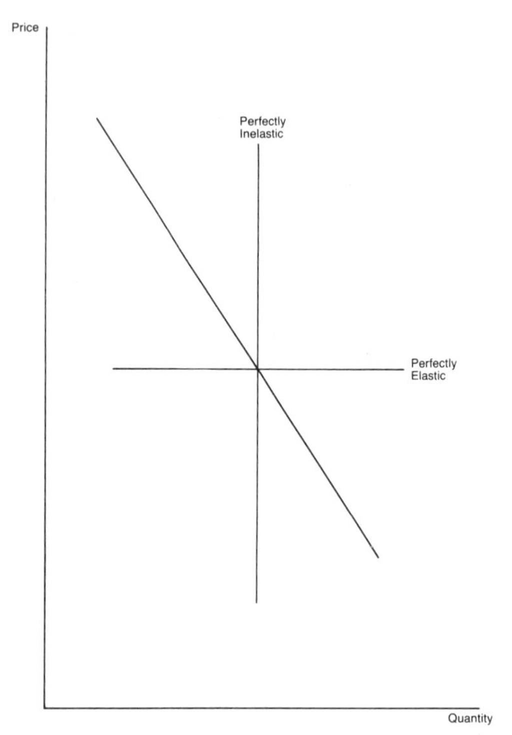

Fig. 2.5.9—Different levels of price elasticity of demand.

three categories. If the absolute value (it can be positive or negative) is between zero and one, it is known as inelastic—little change in the quantity demanded in response to changes in the influencing factor. The extreme case is when $Ed = 0$. This condition is called perfectly inelastic demand. An one-to-one change is called unitary elasticity. An absolute value greater than one is known as elastic—a large response in the quantity demanded due to a change in the relevant factor. If $Ed = \infty$ demand is perfectly elastic. Fig. 2.5.9 shows examples of demand curves with different price elasticities.

Elasticity is also a function of the time period involved. For prices and income, the distinction is often made between the short-run and the long-run. The short-run is defined as the time period over which some of the market conditions are considered fixed. This would include things like a lease, existing production capacity, an agreed upon labor contract, and so forth. The long-run has no such fixed costs and conditions. Everything is free to change. Notice that a definite time period is not part of the definition. The definitions are based on the degree of freedom to change that is available to the user. As such, the relevant time period will vary from industry to industry. It should be pointed out that the long-run can be interpreted differently. In a static sense, the long-run has a fixed technology and only economic factors are considered variable. The long-run is considered in a more dynamic sense for purposes of this chapter. In this case, all factors, both economic and technical, are assumed variable. The distinction between the short-run and the long-run plays an important role in the development of the theory of mineral demand.

Price Elasticity of Intermediate Product Demand

Since the demand for minerals is usually derived, a key consideration is the price elasticity of intermediate product demand for minerals or how the demand for minerals changes with mineral prices. The concept is very similar to that of price elasticity of demand for a final product, but several of the key factors that influence the behavior of intermediate demand are different. Some of the important influencing factors for the mineral industries are the percent of total cost due to mineral input cost, the price elasticity of demand for the final product, and the degree of input substitution available.

The key to the first factor is the relative importance of the cost of the input used to the total cost of the final product. If the portion of the final cost associated with the input is large, the demand for the input will tend to be price elastic. This is because if the input cost is a large part of the total production costs, any input price change will have a major impact on total costs and reactions to it will be substantial. Conversely, if the input cost is a small part of total costs, an input price change will have a small impact on total costs and the producer reaction to the price change will usually be minimal (inelastic).

The second factor is directly related to the first. If the cost of the input is an important part of the final product cost, there is a close relationship between the price elasticities of demand for the input and the final product because an input price change will have a direct and major effect on the final product price. In turn, a change in the final product price affects the quantity demanded of the final product which changes the quantity demanded of the input. Therefore, the more price elastic the demand for the final product is the more price elastic the demand for the input. The equivalent condition is true for inelastic demand too. However, if the input's share of the final product cost is small, the link between the price elasticities of demand for the input and the final product is slight because an input price change has very little impact on the final product price and there is little change in the quantity demanded of the final product. The importance of this second factor is dependent on the degree of the relationship between input cost and final product cost.

The third important factor is the availability of substitutes for the input. The consideration here is the degree of freedom the producer has to make changes in his inputs. The more substitution possibilities there are for a particular production need the more freedom a producer has to change his usage of an input (elastic). On the other hand, if the available substitution options are limited, the likelihood of a change in the use of an input by a producer is also limited (inelastic). The number of substitution possibilities a producer has for a particular input depends on many factors like technology, the available time period, physical availability, costs, and government regulations.

The Short Run: The next step for the readers is to consider how the price elasticity for the derived demand for minerals can be categorized by using the three factors presented in the previous section and the type of market behavior that is likely to be associated with such conditions. Although, each mineral market has its own unique characteristics, some important common trends to mineral demand can be developed and discussed. An important distinction should be made between the short-run and the long-run price elasticities because each has its own issues and properties. The analysis begins with the short-run price elasticity of demand. For the mineral industries, the short-run is roughly up to two years (Tilton, 1981, p. 247).

In the short-run, the price elasticity of mineral demand is determined by two factors: the mineral cost as related to the total cost of the final product and substitution possibilities. First, the cost of minerals are often a small part of the total cost of a final product. There are many examples including cobalt in jet engines and copper in electrical motors. Any change in the mineral price will have a minor impact on total costs and the production decision because it is such a small percent of the total cost. Accordingly, a price change has little effect on mineral demand (in the short-run), and the demand is price inelastic. Second, a common characteristic of mineral usage is the difficulty of substituting away from it in the short-run. Normally, the use of minerals as an input ties it into a system of processes, materials, and labor skills operating under a particular technology. Often changes in the use of minerals necessitate important changes throughout the system that is already in operation. It is likely that producers will be slow to make any such adjustments until they are sure that economic conditions warrant it. This type of situation severely limits any substitution possibilities in the short-run. The result is again a tendency for mineral usage to be price inelastic in the short-run.

So as an answer to the first part of the question posed in the beginning of this section, the derived demand for minerals is in general price inelastic in the short-run. As a result, the physical demand for minerals is not likely to change much in the short-run in response to a change in price, all else constant. The reasons for this are the small portion of total costs of a final product that is associated with mineral's usage and the difficulty of substituting for a mineral in industrial processes in the short-run.

The Long Run: To complete the discussion on the price elasticity of mineral demand it is important to examine the long-run where all relevant factors are free to vary. Since the absolute magnitude of the price elasticity of mineral demand is dependent on the available freedom to change, one would expect that mineral demand is more price elastic in the long-run than the short-run because of the increased flexibility. Users are no longer constrained by existing facilities, agreements, and technologies. This obvious conclusion is not what is important about the long-run price elasticity of mineral demand, though. The importance lies in studying how adjustments in mineral demand occur over time and if this adjustment is adequate to meet the needs of a society that presently depends on nonrenewable minerals. This is a question well worth further study. To do so, the reader must develop an appreciation of the issues involved with conservation and its need, as well as an understanding of how mineral demand adjusts to long-run pricing trends.

The first problem in considering the issue of conservation is determining exactly what is meant by the term conservation (as related to mineral usage). A partial list of suggested definitions include: an absolute reduction in mineral use from present levels, the economically efficient use of minerals, a perpetual steady state use of minerals, and the least physically wasteful use of minerals. Most individuals probably think of the first definition in the list when the term conservation is used. Is this really the most appropriate one? Remember that the demand for minerals is derived. Minerals have value because they provide certain qualities that consumers desire in the final products they buy. So from an economic standpoint, society would like to get the maximum value it can from the mineral resources it is endowed with. A reduction in mineral usage for its own sake is not the appropriate goal. The same comment can be made about the steady state use and least physically wasteful use definitions. These goals are based on inflexible rules that do not change with the changing needs and economic conditions of society. The emphasis should be on the economic gains from the use or non-use of minerals—not the physical availability of minerals. One needs only to consider the volume of minerals present in the earth's crust and the oceans, in context with the law of conservation of matter (matter can not be destroyed) to realize that the long-run issue is the cost of mineral usage and not the physical availability of a particular mineral. From an economic viewpoint, then, the best definition of conservation is the efficient use of minerals.

This working definition of conservation of course leads to a second question. What is meant by efficient? In economic terms, efficient is getting the most discounted net value (benefits minus costs as adjusted for the time value of money) possible out of society's mineral resources as they are or are not consumed over time. A common misconception of this rule is that it means one should exploit mineral reserves as much as possible today. That is not necessarily the case. What it says is that society should use minerals over the economic lifetime of the resources such that the maximum discounted net benefit is received from them. When this situation exists, there are no further economic gains available by changing the amount of minerals used in a particular time period. This condition is known as economic efficiency. The determination of the efficient rate of mineral usage is a difficult problem and is an important part of the conservation issue.

How can an economically efficient use of minerals be achieved? One possibility is through market adjustments. The belief is that through free market interactions the maximum benefit from mineral usage over time will be obtained. An important assumption is that market prices accurately reflect the actual conditions of supply and demand. If so, material substitution and technological change will occur in response to long-run pricing trends in order to adjust mineral demand to the realities of changing costs of supply. It is useful to take a close look at how material substitution and technological change affects mineral demand.

In the short-run it is very difficult to adjust mineral usage because of the nature of its demand. In the long-run mineral demand is better able to respond to changes in price. Materials substitution and technological change have been long recognized as major means of avoiding the depletion of minerals. When the long-run price for a mineral is rising (reflecting greater demand than supply), the user of the mineral looks for cheaper substitutes to replace the mineral in order to keep production costs down. This substitution can come in several forms: substituting one material for another in the existing product, replacing the product with a different product that uses other materials, and reducing the amount of the mineral used in the existing product (Tilton and Vogely, 1981, pp 3–4). An example of the first type is the substitution of aluminum for copper in high-power transmission lines because of the cost advantages of aluminum over copper. The second type is illustrated by the replacement of long distance telephone lines with communication satellites. An example of the third type is the sizeable reduction in the amount of cobalt used in permanent magnets when adjustments in the alloy mixture were made in response to rising cobalt prices.

In all of the above cases, the demand for the particular mineral was reduced. This reduction permits more of the mineral to be available for uses where substitution is less feasible economically. If the price signals are accurate, market adjustments will result in an efficient use of the mineral because users will substitute for the mineral as the price rises, leaving only the high-value needs. Eventually, the price becomes so high that nobody wants to use it, just as economic sources are depleted. This condition is known as economic exhaustion. Due to the difficulties of making changes in the production process, material substitution often occurs in discrete events which are not likely to be reversed soon. A change over takes too much effort, expense, and risk. Therefore, mineral suppliers are very concerned about substitution trends because they can have a sudden and dramatic effect on the market for their mineral.

The above scenario is based on the belief that free market interactions are all that are necessary to bring about the efficient use of minerals. In this case, conservation is automatically obtained and running-out (economically) is not a concern. If society runs out of a particular mineral, it is because it is time to run out of the mineral for the maximum benefit to society and alternatives are available. However, many individuals feel that the competitive market place alone, will not insure the efficient use of minerals because of the failure of the correct pricing signals to be made and received in the market. The fear is that prices will be too low causing the over use of minerals. Here, an explicit policy of reducing mineral demand must be imposed on the market to assure the efficient use of minerals. Appropriate policies would include taxes, physical use restrictions, and changes in lifestyle. Such market failures can occur either in the use of minerals at any given time or in the use of minerals allocated between generations.

The failure of a market to allocate efficiently the use of minerals for a given time period is caused by conditions that distort the information about costs and benefits provided by the pricing system. There are several ways for this distortion to happen (Page, 1977, p. 5). One way is to have a price that is not directly responsive to market conditions. An example of this problem is a regulated pricing system like that presently used for freight train rates and utility rates. Experience has shown that the regulators who set the prices are often unable to act as good proxies for the market. The result is an inappropriate price that distorts resource usage and availability.

Another cause of distortion is when market prices do not accurately reflect actual production costs and user benefits. This situation is probably the most common distortion problem that the mineral industries face. A classic example of this problem is the inability to "internalize" all costs and benefits into the market price. This is illustrated vividly by the case of pollution and environmental protection. The cost of supplying minerals should include the effects of pollution and environment damage because they are real costs that must be borne by someone. Often though, that someone is not directly involved in the market for the final product being produced. Since the costs of the pollution and the benefits of its reduction do not occur to individuals in the market, there is no incentive to include it in the market prices. Therefore, the actual marginal costs of production to society are understated by the market price and society's demand for minerals will be correspondingly too high (law of demand). The U.S. government tries to correct this problem through a complex system of regulations, subsidies, and taxes. The result is often a confused market situation that obtains no goals.

A similar type of distortion occurs when the suppliers or the buyers of a mineral have enough market power to influence the market price in accordance with some strategy. Such market power is sometimes present in the supply side of the mineral industries. This situation allows suppliers to charge prices that might reflect other factors beside marginal costs. Both suppliers and buyers will still make the correct decisions given the available information, but society as a whole will be worse off because of the price distortion that leads to less than optimal mineral usage. An example of this type of behavior is illustrated by U.S. copper producers in the 1950s and 1960s. The copper producers' concern was that aluminum was taking over many traditional copper markets. Recalling the previous discussion of material substitution, the losses to copper producers could have been large and permanent. To combat this encroachment of aluminum, copper producers maintained a posted price to encourage the use of copper. The price of copper was not an accurate reflection of the costs and benefits of copper usage. On the other side, aluminum was priced in much the same manner to encourage its movement into new markets. In both cases, physical rationing had to be used periodically to deal with excess demand. It is hard to assume efficient mineral use in these kinds of markets.

The second type of market failure is the inability of markets to allocate efficiently the use of minerals between generations. The argument is that no matter how efficient markets are in allocating minerals today they are unable to obtain the efficient allocation of minerals over the years. A suggested reason for this is the lack of knowledge about future supply and demand conditions, technological changes, social-economic changes, and other such variables. Expert opinion is split over whether or not there is such a problem. Many of the arguments are either highly theoretical or philosophical in nature and are not included here.

However, there is another part of the conservation issue not covered by economic efficiency. Economic efficiency does not address the question of the fairness of mineral usage between generations. Does the present generation put too high of a value on its consumption of minerals and not enough value on the consumption of minerals by future generations? This is a difficult question to answer and much debate centers around it. One side points out that each new generation has been wealthier than the previous one due to improved technology and society should be concerned with its present problems. The other side argues that there is a limit to technological improvement and mineral reserves, and we are the only guardians the future has. The well-being of the future depends on us today. The debate continues with no side gaining an advantage.

As the reader has learned, conservation is a complex issue with many questions and few answers. A good working definition of conservation of minerals is the economically efficient use of minerals over time—the largest possible discounted net gain from their use. In the long-run, if market signals are able to reflect the true market conditions then adjustments through material substitution and technological change will move society through the efficient use of minerals, and conservation is automatically achieved. However, if the market does not function properly due to price distortions, government interventions, costs and benefits that do not figure in the market price, and other such similar factors, an explicit policy of adjusting mineral usage is needed to reach the efficient usage. The position one holds on the conservation issue depends on one's belief about general market behavior—do mineral markets function efficiently or is there market failure?

Cross Price Elasticity of Demand

So far, the discussion of price elasticity of mineral demand has been limited to the effect of changes in the price of a mineral on the demand for that mineral. There is also the effect of changes in the prices of other goods and products on the demand for the mineral to be considered. This effect is known as the cross price elasticity of demand and is defined as

$$E_{x/d} = \frac{\text{percent change in the quantity demanded of a mineral}}{\text{percent change in the price of another product}}$$

A negative coefficient of elasticity denotes a complement to the mineral. The demand for a complement moves the same way the demand for the mineral does so the sign of the coefficient should be the same as in the own price case. A positive coefficient shows a substitute for the mineral. A substitute is used in the place of the mineral so a decrease in the substitute's price brings about a decrease in the quantity demanded of the mineral and vice-versa (remember the law of demand). As before, the more responsive the quantity demanded to changes in the cross product prices the higher the absolute value of the coefficient of elasticity.

The cross price elasticity of demand is not widely used in analyzing the demand in a market, but it can provide some very important information for the mineral industries. This is particularly true in markets where there are close substitutes or important complements. Copper producers are very concerned about the effects of a reduction in aluminum prices on copper sales. The tin industry is greatly interested in the price of lead and how it affects solder use. There are many more examples. Therefore, any discussion of the price elasticity of mineral demand should include the effects of price changes of substitutes and complements to be complete.

Income Elasticity of Mineral Demand

Another important factor that influences the demand for minerals is the changing level of economic activity associated with consumer spending. This activity is also referred to as the business cycle. Usually, economic activity is measured by the Gross National Product (GNP) or some similar indicator statistic. The response of mineral demand to changes in economic activity is known as the income elasticity of mineral demand. The term income is used because consumer spending is the driving force of the business cycle and consumer spending is in turn based on income levels. The coefficient of elasticity is defined as:

$$E_{y/d} = \frac{\text{percent change in the quantity demanded of a mineral}}{\text{percent change in the level of economic activity (income)}}$$

The sign of the coefficient will be positive for any normal good. Demand and income generally move in the same direction. The important information is the value of the coefficient. Is mineral demand elastic or inelastic in response to changes in economic activity? An elastic response would indicate a market that shows large fluctuations in demand during the business cycle while an inelastic response would imply a market has a relatively stable demand throughout the business cycle. This result requires that all other economic factors be constant as usual. The difference between elastic and inelastic is important because each of these types of markets has its own characteristics and behavior. Knowing the degree of elasticity, therefore, would be useful in explaining mineral market behavior. As in the price elasticity case, the short-run and the long-run are considered separately.

The Short Run: The factor that dominates the degree of response of mineral demand to changes in economic activity in the short-run is the type of industries that use minerals. Mineral demand in the short-run is tied to certain industries and products. Accordingly, as these industries alter production in response to changes in the level of consumer activity, it has a direct effect on mineral demand as an intermediate good of production. The industry sectors that use the largest share of minerals, particularly metals, are transportation, construction, consumer durables, and capital equipment. A common characteristic of all these sectors is a relatively high income elasticity of demand. The reason for this characteristic is the nature of the sectors' final products. These products (automobiles, houses, refrigerators, and so forth) are items that many consumers can put off buying during economic downturns. Buyers can make-do with their existing goods and spend their income on other more urgent needs such as food and clothing. Purchases of durable goods and capital equipment, on the other hand, usually pick up quickly when the economy turns upward, and consumers are once more able to spend income on wants as well as their daily needs. The same cyclical effect is passed onto the minerals demanded in these sectors, creating a high income elasticity of mineral demand. Of course this effect is not universal to the mineral industries. Some minerals, particularly industrial minerals, are used in industries that are less cyclical in nature. A classic example is fertilizer for agriculture. Food demand is less cyclical than consumer durables demand.

In conclusion, the mineral industries, particularly metals, will be more extreme in their variation of activity in the short-run than the general economy. During economic downturns, mineral demand and output will be very low as the demand and production of consumer durables and capital goods is greatly reduced. Conversely, mineral demand and output will be very high during economic upswings as the production and consumption of consumer durables and capital goods surges. This type of behavior raises concern about the stability of the mineral industries and the effect such problems have on the market participants. An excellent discussion of this phenomenon for the metal industries can be found elsewhere in this volume.

The Long Run: The nature of the income elasticity of mineral demand over the long-run has been a topic of a great deal of interest and speculation in recent years. One of the reasons for that interest is a desire to make long range forecasts about the demand for minerals. There is concern about the future adequacy of mineral supplies to match society's growing need for minerals. Forecasts are needed to see if there is a developing problem. The fear is that the lack of minerals could prevent desired economic growth. A major task in these kinds of forecast is finding the appropriate information upon which to base the forecast. One widely used source of information is the GNP—value of the final goods and services an economy produces. This measure is intensely studied and readily available for many countries. Since the measure is also an aggregate one, its projection into the future can be adjusted to the target goals of a government without knowledge of the details needed to achieve those goals. Mineral demand forecasters would like to use projections of the GNP in their efforts to describe the future for these reasons. However, it is necessary to determine the long-run relationships between income and mineral demand before the GNP measure can be properly used. Sometimes this proposed relationship is referred to as the law of demand for minerals (for income not prices).

At first, the simple assumption was made that

mineral demand grew proportionally to income. This kind of assumption is a common one for the type of natural resources availability study examplified by *Limits to Growth* (1972). In the *Limits to Growth* study, mineral demand was assumed to grow exponentially with income worldwide. The exponential growth of mineral demand coupled with an obviously finite supply of minerals meant that the physical exhaustion of mineral resources was a real and growing threat. An immediate program of explicit conservation practices (in the sense of reduced use) was needed to avoid this catastrophic event. However, this type of study does not incorporate prices and market adjustments into the analysis. Substitution possibilities are also ignored.

A more sophisticated approach has been suggested by W. Malenbaum (1975). He has developed an intensity-of-use hypothesis to explain the behavior of long-run mineral demand in response to changes in income. He argues that the amount of minerals that society demands follows an inverted-U pattern when plotted against increasing per capita income. Fig. 2.5.10 shows a representative curve. The vertical axis is intensity-of-use of minerals in some measurable

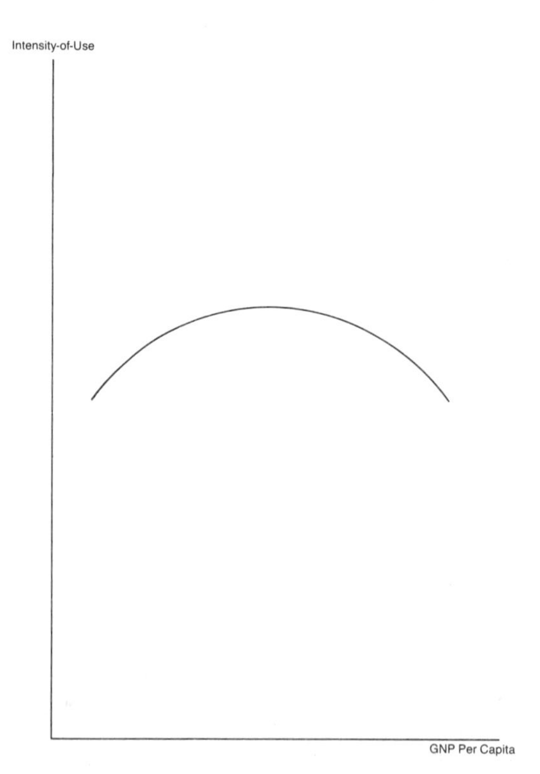

Fig. 2.5.10—Intensity-of-Use hypothesis.

form like pounds per year per capita. The horizonal axis is GNP per capita. The GNP is divided by population in order to avoid demand increases or decreases due simply to population changes. The result is a curve that initially rises with income (positive elasticity coefficient) then flattens out (zero elasticity coefficient) and finally decreases (negative elasticity coefficient)—an inverted-U shape.

Theoretically, the curve shape is postulated on the basis of three causes. One cause is the changes in the nature of a nation's economy as per capita income rises. An undeveloped economy moves from agriculture to manufacturing as it modernizes. This change-over is very mineral intensive and is represented on the graph by the rising portion of the curve. At some point, industrial growth slows and mineral demand remains fairly constant. This condition is the flat portion of the curve. Eventually, the economy shifts from manufacturing goods to more service-oriented activities as the population becomes wealthier. Services are products like education, information, entertainment, government activities, and the like. Malenbaum asserts that a services economy requires less minerals per capita than a growing, manufacturing economy so the curve will slope downward. The second cause is the substitution for minerals of other materials in accordance with changes in relative price differentials. The third cause is technological changes in the production of goods and services that alter the use of minerals. These causes are seen by Malenbaum as reenforcing the inverted-U shape of the curve. Some empirical work based on the degree of use of a particular mineral over different income levels supports the hypothesis (Malenbaum, 1975).

The impact of this hypothesis on forecasting long-run mineral demand and possible mineral exhaustion is important. If correct, the hypothesis implies that forecasts based on the proportional growth of mineral demand with income will overstate demand and increasingly so. The hypothesis also suggests that mineral "conservation" automatically comes about as the population becomes wealthier and passes its initial industrializing phase, a point directly opposite to that of the "Doomsday" school. The intensity-of-use hypothesis is certainly a more optimistic law of demand than previous ones, but is it any more correct?

A basic issue with the intensity-of-use hypothesis is that one has to accept the assertion that a service-oriented economy uses less minerals per capita than a manufacturing one does. On the surface, for example, it would appear that college classes do not require as much in the way of minerals as auto production does. Yet, there is still the demand for furniture, buildings, electricity, heating, transportation and much more that requires minerals. It is quite possible that the service sector of the economy requires just as much minerals per capita as the manufacturing sector does. As a secondary point, it is not clear exactly what the relationship between mineral prices, technological change, consumer tastes, and the intensity-of-use of minerals is. The relationship is probably more complex than the inverted-U assumption.

If there are doubts about the appropriateness of the intensity-of-use hypothesis, how does one explain the supportive empirical evidence? First, the empirical work is not totally in support of the hypothesis as shown by Vogely (1977). Second, a study of the behavior of per capita income over the modeling period shows that it has moved much like time (positive and increasing arithmatically). Therefore, the income variable could be simply measuring some movement over time that is occurring in mineral demand. This possibility suggests numerous alternative hypotheses. One hypothesis is that time is a proxy for technological change that is presently mineral-saving in nature. Another hypothesis is that a particular mineral demand has a life-cycle in which it shows rapid growth into new markets when first introduced, stabilizes when mineral usage is mature, and declines as new materials move into its markets. The curve can also shift in response to new demands and technologies and start the whole process over again. Hypotheses of these types can be theoretically justified equally as well as the intensity-of-use hypothesis and are supported by the same empirical evidence.

The debate continues but no clear decision has emerged. While it is very desirable to have a "law of demand," no one has been able to show that one exists. Certainly, there does not appear to be a simple relationship between long-run mineral demand and income that can be used as there is between demand and price. The true relationship is probably a complex interaction between price differences between materials, technological change, material life-cycles, consumer tastes, and other such factors. It may require a great deal more effort and time before the link between long-run mineral demand and income is truly understood.

References

1. Barnett, H.J. and Morse, C., 1963, *Scarcity and Growth,* John Hopkins, Baltimore.
2. Brooks, D.B., 1976, "Mineral Supply as a Stock," *Economics of the Mineral Industries,* W.A. Vogely, ed., 3rd ed., AIME, New York, pp. 127–207.
3. Cammarota, A., Jr., Mo, W.J., and Klein, B.W., 1980, "Projections of Forecasts of U.S. Mineral Demand by the U.S. Bureau of Mines," *Proceedings of the Council of Economics,* 109th Annual Meeting of the AIME, Feb. 24–29, pp. 69–72.
4. Ciriacy-Wantrup, S., 1952, *Resource Conservation,* University of California Press Berkeley.
5. Dorfmar, R., 1978, *Prices and Markets,* 3rd ed., Prentice-Hall, Englewood Cliffs, N.J.
6. Labys, W.C., 1980, *Market Structure, Bargaining Power, and Resource Price Formation,* D.C. Health, Lexington, Mass.
7. Lovejoy, W.F., 1976, "Conservtion," *Economics of the Mineral Industries,* W.A. Vogely, ed, 3rd. ed., AIME, New York, pp. 684–692.
8. Malenbaum, W., 1975, "Laws of Demand for Minerals," *Proceedings of the Council of Economics,* 104th Annual Meeting of the AIME, Feb. 16–20, pp. 147–155.
9. Meadows, D.H. et al., 1972, *Limits to Growth,* Universe Books, New York.
10. Miller, R.L., 1982, *Economics Today,* 4th ed., Harper and Row, New York.
11. Newcomb, R.T., 1976, "Mineral Industry Demand and General Market Equilibrium," *Economics of the Mineral Industries,* W.A. Vogely, ed., 3rd. ed., AIME, New York, pp. 271–316.
12. Page, T., 1977, *Conservation and Economic Efficiency,* John Hopkins, Baltimore.
13. Samuelson, P.A., 1976, *Economics,* 10th ed., McGraw-Hill, New York.
14. Tilton, J.E., 1977, *The Future of Nonfuel Minerals,* The Brookings Institution, Washington, D.C.
15. Tilton, J.E., ed., 1980, "Material Substitution: The Experience of Tin-Using Industries," Workshop on Material Substitution sponsored by Resources for the Future, May, National Science Foundation.
16. Tilton, J.E. and Vogely, W.A., eds., 1981, "Market Instability in the Metal Industries," *Materials and Society,* Vol. 5, No. 3.
17. Tilton, J.E., 1981, "Prices, Innovation, and the Demand for Minerals," *Proceedings of the Council of Economics,* 110th Annual Meeting of the AIME, Feb. 22–26, pp. 41–45.
18. Vogely, W.A., 1976, "Is There a Law of Demand for Minerals?" *Earth and Mineral Sciences,* The Pennsylvania State University, Vol. 45, No. 7, April.
19. Vogely, W.A. and Bonczar, E.S., 1977, "The Demand for Natural Resources Revisited," *Proceedings of the Council of Economics,* 106th Annual Meeting of the AIME, Mar. 6–10, pp. 165–170.
20. Waud, R.N., 1983, *Economics,* 2nd ed., Harper and Row, New York.

2.6

Mineral Resource Information, Supply, and Policy Analysis

DeVerle P. Harris*

INTRODUCTION

Perspective

The phrase "mineral[a] resource appraisal" is widely and loosely used to refer to any quantitative description of a naturally occurring source of fuels, metals, or nonmetals. This leads to confusion when mineral resources are so conceived, for the term refers generically to information on stocks—mineral endowment, resources, reserves, and potential supply—that differ in ways that are important in economic analysis.

Mineral resource and other stock measures—these are not flows, e.g., the supply of minerals—have precise technical meanings, which are presented in this paper. Quotation marks are used to separate the nontechnical use of mineral resource from the technical.

The term appraisal also leads to some confusion because it too is loosely used. When an appraisal employs probability and statistical methods, the appropriate term is estimation[b], not appraisal or assessment.

This paper presents a conceptual framework for mineral resource information which renders technical, or well defined, relationships of mineral endowment to physical attributes of the earth's crust and earth processes, as distinct from economic values, and identifies estimation approaches that lead to different economic measures of stocks and flows. An important result of this approach is the distinction between the relevance of mineral resource information to economic and policy issues about potential supply (stocks) and dynamic supply (flows across time). With regard to policy analysis, this paper recognizes that mineral resource information has a cost as well as an imputed value. Such a treatment implies that there is an optimum state of resource ignorance, or enlightenment, for a given policy issue and that the relevant economic issue is the selection of that information strategy and appraisal methodology that is indicated by the optimum state.

Until the initiation of the OPEC oil embargo, the demand for information on mineral resources and potential supply was limited to occasional studies commissioned by Congress or the President of the United States. In these, mineral and energy resources remained loosely described. When quantitative estimates were supplied, they were not rigorously made, and the estimates and the methodology accordingly escaped close scrutiny. Usually, neither the geological nor economic bases for estimation were carefully described or critically examined.

The imposition by OPEC nations of the embargo on oil caused a sudden change in the public and scientific demands for specific 'resource" appraisals, especially of oil and of uranium. These

*Director of Mineral Economics, Department of Mining and Geological Engineering, College of Mines, The University of Arizona, Tucson

[a]Mineral will be used throughout this paper to include not just natural occurring metallic and nonmetallic chemical compounds having crystalline structure but also noncrystalline hydrocarbon chemical compounds (coal, petroleum, and natural gas), which technically are not minerals but mineraloids.

[b]"In general it [estimate] lacks the definitiveness of other terms, especially appraise, which stresses expert judgment. Assess implies authoritative judgment; it involves setting a monetary value on something as a basis for taxation." p. 449, *The American Heritage Dictionary*.

demands require levels of information today technically quite different from those of the past. Most users of appraisals desire estimates of potential supply, i.e., the stocks of material that could be both discovered and produced for a specified price. Usually probabilistic estimates are required, as well as some assurance of the credibility of the estimates. This assurance may take different forms, but it includes a documentation of methodology and data.

In the post-embargo period resource estimates and estimation methods received intense scrutiny. Unsurprisingly, inadequacies were found, and inquiries were made of resource concepts, resource and geological data, and methods of analysis. As a result, the last decade has witnessed unparalleled research expenditure on improvements in data collection and methods of "resource appraisal." In spite of this progress, mineral and energy resource science and methods for appraisal of resources and potential supply are in their infancy.

Motivations for Appraisal of Mineral Resources

The most apparent motivation for appraisal is the examination of future resource adequacy. It was for this loosely understood concept that the nation's mineral resources were appraised in 1952 by the Paly Commission [49], that the nation's oil and gas resources were appraised in 1975 by the U.S. Geological Survey, and that the nation's uranium resources were appraised through the NURE (National Uranium Resource Evaluation) Program in 1980 [67]. In reality, the concept of future adequacy is complex and undefined without a clearly identified structure for resource information and correspondingly well specified future economic circumstances.

Complexity stems from several sources, one of which is the determination of an optimum level of imports, a determination which requires the consideration of political economic issues, e.g., security. Another complexity stems from the essentially dynamic process of resource exploitation over long periods of changing materials technology and uses. The states of potential and dynamic supply as information are in part a function of such changes. To be useful, resource appraisal, like technical change, shares the problematic character of the discovery of scientific information and its uncertain costs and benefits.

In spite of the complexity of the concept of resource adequacy, the practice often has been simply to compare projected requirements to the magnitude of known or estimated resources. This simple exercise was repeated often in the aftermath of the OPEC embargo, during the period of the energy scare and the subsequent rises in petroleum price.

A motivation for the NURE program, which is the most costly and comprehensive resource appraisal in history, was the "breeder question," which at the time of the initiation of NURE was perceived to be the following: are our uranium resources large enough to meet the feed requirements of light water reactors if they were to meet projected future requirements for nuclear generated electricity? Or, are these resources so small that as a nation we should commit heavily to the development of the breeder reactor? While these questions do not seem pressing at this time, their great urgency as perceived during the 1970s is indicated by the massive effort initiated to appraise the nation's uranium resources.

Probably, future demands for mineral resource information will derive more from efforts to support management of the nation's lands and nonmineral resources than from the desire to examine issues of resource adequacy. Management decisions at two different levels require, in principle, the consideration, along with other factors, of the "mineral potential" of federal lands. At one level, Congress and the President make decisions on withdrawal of public domain as wilderness areas. At another level, federal agencies, e.g, U.S. Bureau of Land Management, have been given the responsibility to manage the many parcels of public land and their mineral and nonmineral resources in a way that maximizes society's welfare. If decisions were actually made by a comprehensive determination and comparison of values of alternative and multiple uses, more costly resource "appraisals" would be required. One example of an appraisal to support land management is that mandated by the U.S. Geological Survey in 1978 of the mineral potential of Alaska [19,21,43,46,50].

A mineral resources appraisal also may be performed to support the selecting of optimum development programs. Witness the resource appraisal in 1969 of the Province of British Columbia and the Yukon Territory [2,40] for the selection of the optimum route for the construction of a railroad.

Finally, a new demand for mineral resources appraisal may emerge in the future as appraisals

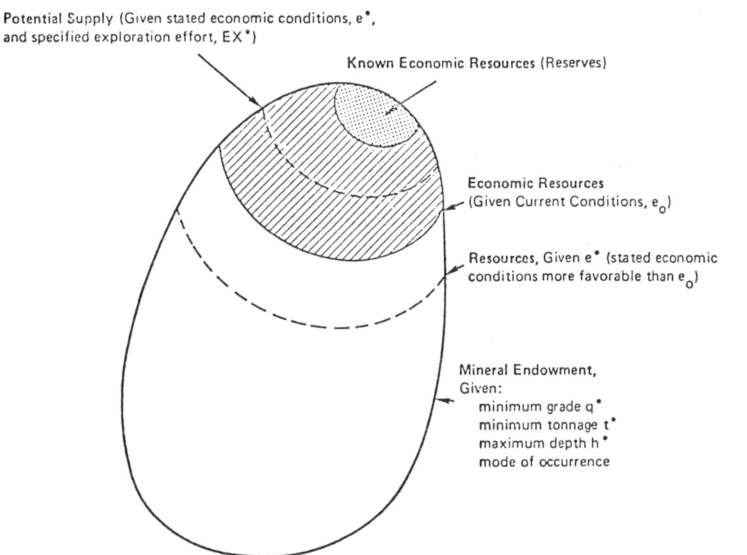

Fig. 2.6.1—Resource terminology and relations. Source: Harris, 1978 [25].

improve and as research on modeling of dynamic supply, i.e., supply across time in response to depletion and new discoveries, continues. Dynamic supply models based upon explicit description of mineral resources, exploration, and production may be better able to explore wide ranges of price, costs, and technology than can traditional aggregate econometric models based on time series data.

STOCK CONCEPTS AND MEASURES

Traditional Resource Information

The terms most commonly used to convey mineral resource information are reserves and resources. Communications between professionals of different disciplines and between analysts and either the policymaker or public often break down because of confusion over these two terms. Communications would be more successful if the term reserves were replaced by its equivalent, information on known resources at current costs and prices. The term reserve is commonly used by mining and geological engineers working with a specific mineral deposit. To them it is understood that a quantity known as an ore reserve is the size of the current working inventory of material that is known to exist and can be produced given the costs and prices at the time. When it is desired to specify the inventory by degree of certainty of the quantity having a specific average grade, a modifier is used:[c] proven, probable, possible. However qualified, the critical features of a reserve are that it is known to exist, and it is economically producible at current prices and costs. Since delineating reserves requires investment in drilling, stated reserves of a mine may be approximately the same over many years. This reflects the planning function of management (mine or oil production management). Typically, management will desire to have enough reserves to support production capacity for its planning horizon, e.g., 5 to 10 years. Thus as proven reserves are produced, more may be added by additional drilling. For that reason, it is useful to consider quoted reserves as the *current working inventory*.

Contrary to some perceptions a change in product price or factor cost will alter the magnitude of a reserve; this reflects the definition of a reserve by current economics. Reserves, once delineated, do not necessarily exist forever simply because they are known, for if product price falls below break-even costs, reserves of a mine cease to exist. Furthermore, contrary to

[c]An alternative classification of ore reserves is measured, indicated, and inferred. Measured and indicated are considered to describe the same material as proven and probable, but inferred includes probable and some material for which grade and quantity are less well known than the material referred to as probable ore reserves.

some perceptions, *the non-ore quantity is not a resource at the same low price*; it is properly referred to as a resource only by specifying economic conditions that make it profitable to produce. An ore reserve is simply a known quantity of mineral resource that is economically producible under *existing* economic conditions. If the deposit currently is not producible, it is properly referred to as a *known resource at the higher price* required to meet production costs (see Fig. 2.6.1).

Resources exist only for an economic and technologic reference. In this sense, a resource is like a reserve; however, when the term resource is qualified only by a specified price, it refers to the sum of the element or compound in all (known plus unknown) accumulations producible at that price, (ignoring discovery costs), given current and near feasible technology. If the set of accumulations that comprises resources be restricted to only those that are also discoverable, the sum of the element or compound in these accumulations constitutes the stock known as potential supply [30]. These relations and definitions are shown schematically in Fig. 2.6.1 and 2.

Additional Measures

Two additional measures are required for a complete description of mineral stocks: resource base and endowment. These two stocks differ from reserves, resources, and potential supply in that they are not defined by economics or technology. Resource base is the total amount of an element or compound that is present in the earth's crust within the region of interest. For example, the resource base for copper includes not just copper in ore minerals but copper that substitutes for other ions in common silicate

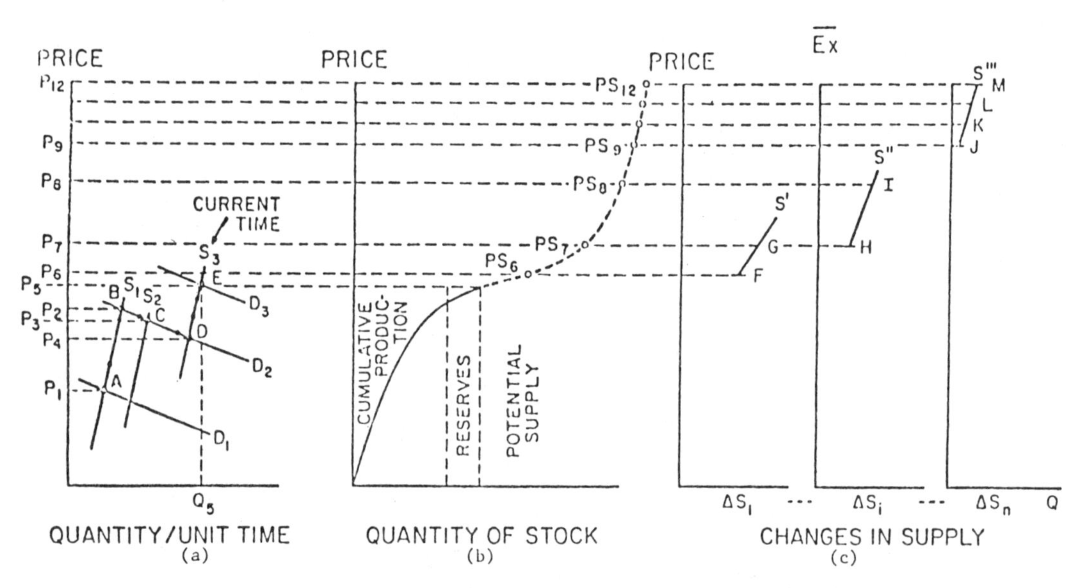

Fig. 2.6.2—Industry dynamics and potential supply. Source: Chavez-Martinez, 1983 [12]. Fig. 2(a)—depicts the historical shifting of supply (S) and demand (D) schedules as their determinants change, leading to the current market represented by S_3, D_3, and P_5. Fig. 2(b) shows that the cumulative effect over time of previous market transactions is the depletion of the original stock by the amount labeled cumulative production for a price slightly below P_5. This material has been produced and is no longer available. At current time, given fixed technology and factor prices, mineral material will continue to be produced according to S_3 until the currently available stock, labeled reserves, is exhausted. Additional production from new deposits will require a higher price than P_5 and capital investment in exploration and development to convert part of the potential supply stock to new reserves and a flow according to a new supply function, S', as indicated in Fig. 2(c). Of course, the higher prices required to elicit incremental new supply will also create new reserves in old deposits to the extent that the old deposits contain low grade material not previously producible economically at P_5.

minerals. Thus, for resource base, the basis for measurement is molecular.

Mineral endowment can be considered a component of resource base. If from the earth's crust those occurrences of the element or compound are selected that meet specified minima of concentration and size and a maximum depth within the crust, the sum of the amount of the element or compound in all accumulations is known as mineral endowment [30].

A Formal Statement of Stock Measures and Determinants[d]

Consider, for convenience, at a given point in time, a single metallic element in a single region of the earth, so that notationally we may ignore both time, metal varieties, and places. Our experience has shown that the ultimate deposit of this metal in that region of the earth consists of many smaller deposits occurring in varied geologic environments and possessing various characteristics of grade, size , shape, mineralogy, depth, host rock, etc.

Suppose that there are NM of these deposits and that they constitute a set RB:

$$RB = \{r_1, \ldots, r_{NM}\}$$

Let us represent our knowledge about the ith member of set RB by a set, Z_i, of NC characteristics:

$$Z_i = \{k_{i1}, \ldots, k_{i,NC}\}$$

The set of NC characteristics includes all physical properties of the NM metal deposits.

Suppose that RB were partitioned solely on the basis of only NCM of the NC characteristics, $NCM \leq NC$, with no thought given to economics or technology. The set of metallizations so formed is D, and the quantity of metal in D is referred to as m, metal endowment:

$$m = \sum_{D} \gamma(r_i)$$

where[e]

D is a subset of RB, $RB = D \cup \overline{D}$, such that for $r_i \varepsilon D$ requires that $k_{i,j} \geq k'_j$, $j = 1, 2, \ldots, NCM$. $NCM \leq NC$.
Otherwise, $r_i \varepsilon \overline{D}$.

For future reference, let us refer to the level of the NCM conditions used to define D as Z'. Thus, for every $r_i \varepsilon D$, $Z_i \geq Z'$.

Suppose that a function, f, is known, which for specified economic and technological conditions describes the present value, v_i, for each $r_i \varepsilon RB$.

$$v_i = f(k_{i1}, k_{i2}, \ldots, k_{i,NC}; e_1, e_2, \ldots, e_{NE}),$$

or in vector notation

$$v_i = f(z_i, E)$$

where

Z_i is the set of NC characteristics of the ith deposit, as previously described.
E is the set of NE economic factors:

$$E = \{e_1, e_2, \ldots, e_{NC}\}$$

The set E includes operating costs, capital costs, prices, rate of return, etc. Naturally some of these factors reflect the state of technology.

Suppose that levels of the NE economic conditions for currently feasible and near-feasible technology are specified. While these levels must reflect currently feasible and near-feasible technology, some of them, such as product price, need not be those that currently prevail. Then, given RB and the function, f, the present value for each $r_i \varepsilon RB$ can be computed, giving rise to the set, V^R.

$$V^R = \{v_1, v_2, \ldots, v_{NM}\}$$

V can be employed to partition RB into two subsets, R and $\overline{R}$:

$$RB = R \cup \overline{R}$$
$$R = \{r_i, \ldots, r_{NR}\},$$

where

$r_i \varepsilon R \rightarrow v_i \geq 0$
$r_i \varepsilon \overline{R}$, otherwise
$NR \leq NM$

That is, R contains all deposits in RB that could be produced economically given the specified economic conditions and given currently feasible or near-feasible technology. Let us des-

[d]The definitions of this section are reproduced from the author's book, *Mineral Resources Appraisal—Mineral Endowment, Resources, and Potential Supply: Concepts, Methods, and Cases*, Oxford University Press, 1984.

[e]The function γ maps the characteristics of the ith deposit, r_i, into m_i, which is quantity of metal in deposit r_i. $\overline{D}$ means "not D". $D \cup \overline{D}$ means union sets of D and $\overline{D}$. $r_i \varepsilon D$ means r_i belongs to set $\overline{D}$.

ignate rs as the quantity of metal in R:

$$rs = \sum_{R} \gamma(r_i)$$

The quantity, rs, measures the magnitude, usually by weight, of the metal resource.

The set R can be considered to be a subset of D if Z', the required levels of the NCM characteristics, is specified so that none of the metallizations excluded from D would be economic to produce given the conditions for R. For example, if grade, one of the elements of Z', were set at an order of magnitude lower than that which would allow profitable mining for conditions specified for R, then $R \subset D$; $D = R \cup \overline{R}$. For such a circumstance, $rs \leq m$. For one class of resource models, that which is based upon estimating metal endowment, this is a useful perspective and is the view of metal endowment as it is employed here. While there is nothing in the definition of metal endowment, as previously given, that stipulates this relationship between R and D, the primary need for the term arises when an inventory of deposits having specified physical characteristics is estimated as an initial step to appraising resources and potential supply. Given this motivation, it makes little sense to set Z' such that resources for the economic and technologic conditions of interest exceed the physical inventory, the metal in deposits belonging to D. It will be assumed that the term metal endowment is a more inclusive term than resources in the sense of the metal occurrences that are implied by the term. Thus, given an estimate of D and m, resources (rs) can be determined by using the value function, f, and the economic and technical conditions, E, to compute a present value, v_i, for all $r_i \varepsilon D$, creating a set V^D which contains only NMM elements:

$$V^D = \{v_1, v_2, \ldots, v_{NMM}\},$$

where

NMM are the number of metallizations in D, $NMM \leq NM$.

By selecting all v_i in V^D which are greater than or equal to zero, the set of NR metallizations that constitute R (the same set as was formed from RB) is formed from D:

$$R = \{r_1, r_2, \ldots, r_{NR}\},$$

where

$r_i \varepsilon R \rightarrow v_i \geq 0$
$r_i \varepsilon \overline{R}$, otherwise
$NR \leq NMM \leq NM$

Thus, given the perspective described, which is that employed here, $R \subset D \subset RB$; and $rs \leq m \leq$ resource base.

Suppose that a set $V^{R'}$ is formed by specifying fully the currently prevailing economic conditions and technology. Then a new set, R', can be formed from R by selecting all $r_i \varepsilon R$ for which $v_i \geq 0$ for the current status of economics and technology:

$$R' = \{r_1, r_2, \ldots, r_{NER}\},$$

where

$R = R' \cup \overline{R}'$
$r_i \varepsilon R \rightarrow v_i \geq 0$
$r_i \varepsilon \overline{R}'$, otherwise
$NER \leq NR \leq NMM \leq NM$

Let us designate rs' as the quantity of metal in R' and rs'' as the quantity of metal in $\overline{R}'$. Then

$$rs' = \sum_{R'} \gamma(r_i)$$

Similarly,

$$rs'' = \sum_{\overline{R}'} \gamma(r_i).$$

The quantity rs' is economic resources, while rs'' is subeconomic resources. Thus, $R' \subset R \subset D \subset RB$; $rs' \leq rs \leq m \leq$ resource base; and $rs = rs' + rs''$.

Potential supply, ps, can be formed directly from R or indirectly from D. Consider set R, which contains those metal occurrences which would be economic to produce for specified economic conditions and currently feasible or near-feasible technology, if the occurrences were known. Suppose that R were partitioned to R^d and $\overline{R}^d$ such that R^d contains those metal occurrences of R that would be discovered by an optimum exploration effort, EX^*.

$$R = R^d \cup \overline{R}^d$$

where

$r_i \varepsilon R^d \rightarrow v_i - c_i \geq 0$ and discovery
$r_i \varepsilon \overline{R}^d$, otherwise.

Then,

c_i = the share of EX^* for the ith metal occurrence that was discovered by EX^*

$$ps = \sum_{R^d} \gamma(r_i)$$

Thus, $R^d \subset R \subset RB$ and $ps \leq rs \leq$ resource base.

Suppose that D were formed from RB by specifying Z' such that $R \subset D$. It has already been shown that by consideration of exploration, R^d, the subset of metal occurrences of R that would be discovered, can be formed from R. Then $R^d \subset R \subset D$. Since ps is the sum of metal in R^d, the relationship between ps and metal endowment, m, is obvious: $R^d \subset R \subset D \subset RB$ and $ps \leq rs \leq m \leq$ resource base.

Let us examine further the concept of an optimum level of EX. To contribute to potential supply, deposits must be of a quality such that their exploitation covers production costs and the costs of discovering them. Allocating to each deposit discovered its share of the exploration effort, EX, gives a net present value (net of exploration and production costs). Naturally, increasing EX to a higher level than EX^* discovers more deposits, but since EX is charged against only those deposits discovered, increasing EX beyond EX^* loses more economic and discoverable resources than are gained. This is because at any progression in the optimizing path of exploration, the deposits which remain to be discovered require a greater expenditure per unit of resources than those already discovered, and because when interacting exploration and exploitation with endowment in a "one-shot" or "single contract" kind of optimization in which the sequential timing of incremental exploration expenditures are suppressed, there are no "sunk" exploration costs. Therefore, diminishing returns to exploration, due to the greater difficulty of discovery of progressively larger fractions of the endowment, means that exploration costs allocated to the deposits discovered increase while at the same time additional deposits are discovered by a greater expenditure. Since the set R^d consists of only those occurrences that would be discovered by EX *and* would be economic to produce when all exploration and exploitation costs are considered, then there is an optimum level of EX, EX^*.

If we now relax the assumption of unlimited markets but invoke the assumptions that the single large firm would seek to maintain prices and that there is no technological change across time, then exploration would be spread across time as warranted by demand and depletion. Thus, at any point in time, EX_t would be less than or equal to EX^*; consequently, the set of known and exploited deposits, $\tilde{R}_t^d$, would be some subset of R^d.

$$R^d = \tilde{R}_t^d \cup \bar{\tilde{R}}_t^d$$

Of course, the sum of metal in deposits of $\tilde{R}_t^d$ must be less than or equal to the sum of metal in deposits belonging to R^d.

$$\sum_{\tilde{R}_t^d} \gamma(r_i) \leq \sum_{R^d} \gamma(r_i)$$

Let us designate this sum as a stock measure of supply, s_t:

$$s_t = \sum_{\tilde{R}_t^d} \gamma(r_i)$$

Then, at any point in time in our simplified and hypothetical world, supply is less than or equal to potential supply:

$$s_t \leq ps$$

Obviously,

$$\lim_{t \to \infty} (s_t) = ps$$

In terms of an individual area and a single metallic ore type hypothesized here, the more intensely the area has been explored the more closely s_t will approach ps. The relationships of the stock terms from resource base to supply can be summarized as follows:

$$s_t \leq ps \leq rs \leq m \leq \text{resource base} \Rightarrow \bar{R}_t^d \subset R^d \subset R \subset D \subset RB$$

Suppose that we define $b(t)$ as reserve addition at t, then in this highly constrained model of economic activity supply as a flow is related to supply as a stock in the following way:

$$Q_s(t) = \frac{ds(t)}{dt} - b(t)$$

Alternatively,

$$ds(t) = Q_s(t)dt + b(t)dt$$

And,

$$s(t) = \int_0^t Q_s(\tau)d\tau + \int_0^t b(\tau)d\tau$$

Of course, this simple relation is valid only for the foregoing assumptions of maintained product price and fixed factor prices. In actuality, these prices, which are determinants of the sets R' and $\tilde{R}_t'^d$, vary across time in response to economic dynamics. It is the behavior of $Q_s(t) = \phi[e_1(t), e_2(t), \ldots, e_m(t)]$ that is a subject of interest to economists. Variation across time of economic factors makes realized $s(t)$ an aggregate effect of historical variations:

$$ds(t) = \phi[e_1(t), e_2(t), \ldots, e_m(t)]dt + b[e_1(t), e_2(t), \ldots, e_m(t)]dt$$

And,

$$s(t) = \int_0^t \phi[e_1(t), e_2(t), \ldots, e_m(t)]dt + \int_0^t b[e_1(t), e_2(t), \ldots, e_m(t)]dt$$

Such a condition confounds to some degree a simple use of realized $s(t)$, or characteristics of the deposits that have contributed to $s(t)$, to estimate undiscovered stocks. Because of the importance of the contamination of stock measures by economics to the estimation of potential and dynamic supply from undiscovered deposits, economic contamination is considered in greater depth in a subsequent section.

NOTIONS OF SUPPLY

Stock, Flow, and Economics

The introduction to this paper presented the idea that the magnitude of mineral resources seldom is the real objective of an appraisal; rather the appraisal of mineral resources is a means to an end, which is the estimation of a supply measure, either potential supply or dynamic supply. These supply concepts are not equivalent to either the short or long run supply of traditional economic theory. As commonly understood, a supply function describes *flow* per unit of time for various prices, given fixed determinants such as capacities, number of suppliers, and factor prices. The long run allows changes in these determinants.

Although in graphical or functional form a potential supply curve appears like the economist's supply curve, the two curves describe very different things. The potential supply curve describes the size of the *stock* of material by cost of production.

Fig. 2.6.2 depicts market dynamics—the shifting of supply and demand across time—and how price resolution and market flows are related to stock measures—cumulative production, reserves, and potential supply from undiscovered deposits. Specifically, at a moment in time the market clearing quantity reflects not only current consumer preferences, producer technologies, and growth in the economy, but also the cumulative effect of past economic activities in depleting the original stock of producible material (initial potential supply). The position of cumulative production on the potential supply curve indicates the extent of this depletion. At the current price level (P_5) supply will continue to flow until the block of material labeled reserves has been produced; thereafter, ceteris paribus, production would cease unless net price increases. For current technology, the magnitudes of the stock of material that are estimated to be producible at prices above P_5 are described by the potential supply curve. *It is incorrect to draw a series of shifting (future) demands on the potential supply curve and read on the price axis the market clearing price for each successive time shift of demand.* This is because a potential supply function describes a stock, and the quantity consumed in one period is *unavailable* for the next period. A potential supply curve implies nothing as to when quantities of the metal or compound will be produced, except that they *will not* be *produced now. Current delivèrability is not implied* by potential supply.

Even though the distinctions between potential supply, reserves, and supply conceptually are easily made, the terms often are incorrectly used or interpreted in communications. For example, the sometimes sharp criticisms by industry of the quantities of potential supply estimated by an agency of the government, e.g., uranium by U.S. DOE, seem to reflect skepticism about *deliverability* more than about the postulated magnitude of the stock for the stated price. Of course, by definition potential supply is not intended to imply a supply schedule (deliverability). If it is incorrectly interpreted to be supply, disagreement over estimates is inevitable in most cases because of the very different magnitudes involved.

Dynamic Supply

Dynamic supply, as the term is used here, refers to supply in response to price across time as deposits are depleted (some exhausted) and replaced by new ones which are of lower quality either in terms of production cost per unit or cost of discovery. While dynamic supply is only relevant in the very long run, as defined, this supply is not equivalent to long run supply because of depletion of an exhaustible resource and the consequent dynamics of new supply. Given that a resource appraisal provides estimates of the number of undiscovered deposits and estimates of deposit characteristics (size, grade, depth), it provides a means for modeling dynamic supply.

GEOLOGY OF MINERAL OCCURRENCE

Crustal Abundance

Geological investigations find the earth to be nonhomogeneous. This nonhomogeneity is expressed by the recognition of the following zones, beginning at the center and proceeding outward:

> inner core (solid), outer core (liquid), mantle, upper mantle, asthenosphere, and lithosphere

The lithosphere is the rigid, relatively thin, outer part of the earth and consists of solid rock. The upper portion (5 to 60 km) of the lithosphere contains the oceanic and continental crusts. It is from the continental crust that man has obtained the bulk of his mineral supply. While this is expected to continue for some time, mineral supply may one day also derive from oceanic crust. For example, the polymetallic massive sulfide deposit in the Atlantis II Deep of the Red Sea has received considerable attention not only for its scientific value but also for what it may portend as a future source of metals [56]. The oceanic and continental crusts differ slightly in bulk composition; throughout the remainder of this paper, any reference to the earth's crust is of the continental crust.

In view of the fact that man's mineral supply has derived from the continental crust, it is informative to examine the composition of this fundamental feature of the earth and ultimate source of mineral substances. The following nine elements account for 99% by weight of this crust: oxygen (45.2), silicon (27.2), aluminum (8.0), iron (5.8), calcium (5.1), magnesium (2.8), sodium (2.3), potassium (1.7), and titanium (0.9) [39]. Noteworthy is the fact that all remaining elements, including some that are commonly used and some that are highly prized—copper, lead, zinc, silver, gold, and carbon—account for only one percent of the crust. Given the high concentration (abundance) of oxygen and its tendency to form complex anions with silicon, it follows that the bulk of the earth's crust consists of oxide and silicate compounds. The point to be made here is that while the ultimate source of mineral substances is the earth's crust, most of the mineral substances sought by man to support his economic activities comprise a very small part of crustal material. In essence, most of these substances are present only in "trace" amounts. In spite of this fact, the resource bases of these substances are very large, a fact that at times is misinterpreted to indicate a state of cornucopia vis-a-vis availability of materials. Such interpretations often ignore the fact that much of the substance indicated by the resource base is recoverable by man only at high cost because of its low concentration or its chemical form, e.g., as ion substitutes within silicate crystal lattices. U.S. uranium is used in the following example to explore these ideas.

Let us refer to the concentration of an element, expressed as a decimal fraction or as parts per million (*PPM*), within the earth's crust as crustal abundance, A. Then the stock measure known as resource base (*RB*) is the product of A and W, the weight of the crust within the region of interest:

$$RB = A \cdot W$$

For example, consider uranium in the U.S. crust. Let us specify the weight of the U.S. crust to a depth of 600 feet (a depth that includes most U.S. uranium deposits) to be 4.06×10^{15} mt. One estimate of A in terms of U_3O_8 is 3 *PPM*, which is equivalent to the fraction 0.000003. Thus

$$\begin{aligned} RB &= (3 \times 10^{-6})(4.06 \times 10^{15}) \\ &\approx 12.18 \times 10^{9} \text{ mt of } U_3O_8 \end{aligned}$$

This is a staggering quantity in view of 1977 annual consumption[f] of 15,900 st [65]. However, if we had to extract this annual requirement from ore having a grade of only 0.0003%

[f] Receipts at buying stations and mills.

U_3O_8 (average crustal grade), the cost per lb of U_3O_8 also would be staggering. The average grade of uranium ore produced in the U.S. during 1977 was about 0.154% [65]. Thus, those deposits upon which uranium supply is based represent accumulations of uranium that had been enriched during their formation to approximately 500 times that concentration of the common crust. As rock material of the crust, these deposits differ considerably in concentration of U_3O_8 from typical material. Let's consider this further by drawing upon the work of Deffeyes and McGregor [15], who, after considering the distribution of U_3O_8 in a number of mining districts and also in the various rocks that comprise the crust, concluded that the distributions of uranium concentrations (grade) are consistent with a lognormal crustal abundance model. Let us consider the crustal abundance of uranium as a percentage (0.0003%) to be the mean of this distribution. Then, let us postulate the following model:

$$\ln q \sim N(\mu,\sigma^2)$$

where

$$A = e^{\mu+\sigma^2/2}$$

The two parameters (μ and σ^2) were estimated using DOE's preproduction inventory [36]: $\mu = -9.144$ and $\sigma^2 = 2.065$. DOE's production inventory (reserves plus all produced material plus subeconomic resources present in known deposits) shows, for a cutoff grade of 0.06% U_3O_8, 677×10^6 mt of mineralized material having an average grade of 0.15% U_3O_8, implying 1.016×10^6 mt of U_3O_8. The lognormal endowment model indicates the following for the same cutoff grade:

mineralized material $= 9.215 \times 10^6$ mt;
average grade $= 0.108\%$ U_3O_8;
amount of $U_3O_8 = 9.95 \times 10^6$ mt

According to the model, approximately nine times the known uranium occurring in accumulations which have grades of at least 0.06% U_3O_8 remains to be discovered. Thus, if uranium is lognormally distributed and these estimates of parameters of the distribution are reasonable, one must consider the possibility that large quantities exist. Since 0.06% U_3O_8 is near cutoff grades of some actual operations, we have no compelling reason to expect this uranium to be nonavailable because of its chemical form.

For the lowest cutoff grade of DOE's preproduction inventory, 0.01% U_3O_8 [66], the lognormal crustal abundance model indicates 2.206×10^{12} mt of mineralized material having an average grade of 0.018% U_3O_8, giving 397.1 $\times 10^6$ mt of U_3O_8, a very large quantity (see Table 2.6.1). In 1977 known endowment amounted to only 1.564×10^6 mt of U_3O_8. Thus, the model indicates that the quantity of U_3O_8 in material having concentrations of at least 0.01% U_3O_8 is approximately 254 times known endowment. Even if this uranium were chemically available, estimating the amount of resources and potential supply for specified economics would be very difficult, giving highly speculative results, for we are uncertain as to how this uranium occurs. Costs of production and efficiency of exploration are strongly influ-

Table 2.6.1—Crustal Abundance Model Estimates

		lnQ ~ N(−9.144,2.056) Model Estimates				
q′ (%)	ln(q′)	Probability	Mineralized Material (mt)	Average Grade (%)	Amount of U_3O_8 (mt)	DOE's Preproduction Inventory of U_3O_8 (mt)
→ 0	− ∞	$P(\ln Q \geq -\infty) = 1.0$	4.06×10^{15}	0.0003	12.18×10^9	—
0.01	−4.605	$P(\ln Q \geq -4.605) = 5.433 \times 10^{-4}$	2.206×10^{12}	0.018	3.971×10^8	1.564×10^6
0.04	−3.2189	$P(\ln Q \geq -3.2189) = 8.867 \times 10^{-6}$	3.60×10^{10}	0.07	2.52×10^7	1.176×10^6
0.06	−2.8134	$P(\ln Q \geq -2.8134) = 2.2697 \times 10^{-6}$	9.215×10^9	0.108	9.95×10^6	1.016×10^6
0.10	−2.3026	$P(\ln Q \geq -2.3026) = 3.4877 \times 10^{-7}$	1.416×10^9	0.182	2.58×10^6	0.741×10^6

Source: Harris and Chavez, 1984 [36]

enced by the size of the deposit and the distribution of its grades within the deposit.

Geologic Processes

While the notion of crustal abundance is useful as an ultimate determinant of availability of a mineral substance, it and its companion stock measure, resource base, can lead to erroneous perceptions, because they treat the crust as a homogeneous mass having the chemical composition indicated by the crustal abundance measures. In reality, that part of the crust that is exploited by man, the outer two miles of the continental crust, is heterogeneous, and if crustal abundance measures are an accurate description of crustal bulk composition, they are so only for large crustal blocks which average out the effects of the heterogeneity.

The last two decades have witnessed a great unification of observation and theory in geology via plate tectonics. Plate tectonics is based upon the established fact that the crust of the earth consists of crustal blocks that are moving. While the precise energy mechanism driving these movements is not well understood and the subject of controversy, its effect is well documented. Very simply stated, the energy system of the earth creates a recycling of crustal material: ocean floors are spreading; as new crust is formed by upward movement of molten material, crustal blocks move away from the spreading zones. This outward movement causes some crustal plates to collide. At the collision boundary, oceanic sediment may accrete on continental plates and one plate may subduct. The subducted plate may partially melt in response to high temperatures accompanying deep subduction. This melting may give rise to volcanic activity and magmatic intrusions, which may under certain conditions result in the formation of new mineral deposits. Once formed, the metal in these new deposits may be preserved or it may be remobilized by other earth processes (e.g., oxidation, weathering, erosion) and reconcentrated.

As with metals, plate tectonics is useful in explaining the macro features of petroleum and gas deposits. For example, fore-arc basins formed on the continental margin as a result of the collision of two plates and subduction of the oceanward plate, may be shallow with irregular basin profiles. Typically, sedimentary rocks include shales, carbonates, and interbedded volcanics. Due to low heat flow, poor reservoir rocks, and a deficit of trapping mechanisms, fore-arc basins generally offer poor potential for occurrence of productive petroleum accumulations [44]. Classifications of basins [54] have been related, where appropriate, to plate tectonics. Plate tectonics explains in a fundamental way the existence and formation of geosynclines and basins. Since the different kinds of basins have been observed to vary somewhat systematically in major structural and stratigraphic ways and in petroleum resources, the tectonics associated with plate movements are determinants of petroleum endowments. Of course, petroleum deposits, like metal deposits may be remobilized, retrapped, and upgraded by subsequent geologic processes.

For a paper dealing with mineral resource appraisal and supply, a focus for the foregoing discussion may be helpful. Specifically, the major point to be made is that mineral substances of the crust are acted upon by both macro and micro earth processes and often by more than one cycle of them. In a broad and general sense, the result of the sequential operation of these processes is the depletion in an element or compound of some crustal material for the enrichment of other material. Thus, nonhomogeneity of the crust in element concentration is accompanied by nonhomogeneity of geology. Identifying the combinations and sequences of earth processes involved in creating nonhomogeneities is a task of geoscience. One result of these efforts is the identification of geologic environments for various phenomena, e.g., stratigraphic type oil deposits, porphyry copper deposits, sandstone roll-type uranium deposits, etc. Usually, for a given element or compound, there are a number of environments, giving rise to different kinds of deposits. For example, environments for copper include porphyry systems, volcanogenic massive sulfide, sedimentary, replacement, and strataform. For the resource analyst, the importance of this structure and taxonomy is that the different processes that were operative in these environments embodied the enriched crustal material in different chemical and physical forms. The following four characteristics are especially relevant to supply because of their effects on costs:

mineralogical form
deposit size
average grade
intradeposit grade variation

Although there is much variation in these characteristics within a given deposit type (mode of occurrence), there also are some general patterns across modes. Look, for example, at the average size ($\bar{t}$) and grade ($\bar{q}$) for the three major modes of copper occurrence [58].

porphyry	$\bar{t} = 548 \times 10^6\ st;$	$\bar{q} = 0.63\%$
massive sulfide	$\bar{t} = 10.3 \times 10^6\ st;$	$\bar{q} = 2.92\%$
strataform	$\bar{t} = 91 \times 10^6\ st;$	$\bar{q} = 3.78\%$

Important recent contributions to resource appraisal include (1) the construction of empirical grade and tonnage models, and (2) identifying genetic and occurrence models for different modes of occurrence. The first of these is useful to (1) geologists who by interpreting the geology estimate numbers of deposits within a region, and (2) resource and supply analysts who model the economics of discovery, development, and exploitation.

One characteristic of deposits that has been neglected and is very important in supply analysis is the intradeposit grade variance. Although a deposit may be characterized by its total amount of enriched material and the average concentration of the element within this material, most deposits exhibit a variation in quality (grade) within the mineralized material. Typically, all mineralized material is not produced, because profit maximization leads to identifying an optimum cutoff grade which either leaves material in the ground or in the waste pile. To not account for this effect in supply analysis ignores an important economic effect. Incorporating this effect requires an intradeposit grade variance and a grade distribution.

Although the foregoing comments use the language of metal deposits, the same ideas apply to petroleum. Macro processes create different kinds of basins, and there are patterns in (1) the amount of oil entrapped, (2) field size distribution, and (3) recovery across these basin types. In the case of petroleum, recovery is a quality characteristic, much like grade, and this characteristic varies considerably across deposits of different major types, reflecting the sediment type, degree of cementation, grain size, sorting, and lithification caused by the processes that formed the basin and caused the source, migration, and entrapment of the liquid and gaseous hydrocarbons.

GEOLOGIC AND TECHNOLOGIC DETERMINANTS OF COST AND PRICE

Crustal Abundance and Relations Across Elements or Compounds

Suppose, as a gross generalization, that a deposit of X tons of an element or compound having low abundance within the earth's crust represents the outcome of a greater number of sequential depletions and enrichments than does a deposit of X tons of an element of high abundance. Then, a deposit of the scarce element that is both large in size and of high grade would be in a relative sense a very rare event. Deposits of elements having low abundance in general would tend to be of smaller size than those of high crustal abundance. If crustal elements under consideration also were of similar chemical reactivity, then the numbers of deposits of the various elements and their average deposit sizes and grades would be, ceteris paribus, proportional to crustal abundances.

In reality, elements differ considerably in ionic size, valence, electron potential, and in their geochemical cycle. Consequently, the simple postulated proportionality is deficient as a means of making generalizations about accumulations. Suppose, however, that an additional characteristic, d, referred to as the coefficient of mineralization [5, 6] is identified. Consider d to represent the tendency of an element or compound to form accumulations of high concentration in the processible compounds of the element. Then, together, crustal abundance (A) and the coefficient of mineralizability (d) should be related generally to the quality and quantity dimensions of a mineral substance.

Suppose further that the crust of a particular region supports a complex economy producing many varied goods and giving rise to derived demands for many materials. Suppose also that demands are either constant over time or shift in unison in a regular way. Suppose further that there is a limited degree of substitutability among the materials factors for at least some final goods and that both final goods and factors are priced competitively. Given these suppositions, there should be, in a broad sense, a relationship between prices of the materials (production factors) and the two characteristics of their occurrence, A and d:

$$p = \gamma_0 A^{-1} e^{-\gamma_1 d}$$

Table 2.6.2—Long Term Price Predictions by Crustal Abundance Relations

Metal	Crustal Abundance A	Coefficient of Mineraliza-bility d	1971 Actual Price[4] ($1971/lb)	1971 Predicted Price ($1971/lb)	1979–1982 Actual Prices[2,5] ($1982/lb) Low	Average[6]	High	Predicted[5] price ($1982/lb)
Copper	70	0.1981	0.514	0.88	0.54	0.98	1.78	2.67
Lead	16	0.2793	0.139	0.176	0.20	0.46	0.71	0.53
Zinc	8	0.2126	0.169	0.194	0.27	0.45	0.52	0.59
Uranium[1,3]	3	0.2003	6.067[7]	8.35	24.86	35.00	57.64	25.30
Gold	10^{-3}	0.3547	446.52	409.00	1353	6402	11865	1353

[1]Units of U_3O_8.
[2]*Metal Statistics*, 1983, unless otherwise noted, e.g., uranium.
[3]Estimated from various reported prices.
[4]Producers price FOB refinery.
[5]Inflated to $1982 by Producer Price Index for Industrial Commodities.
[6]Average of annual averages, where an annual average is simply the average of the low and high for that year.
[7]Average of NUEXCO monthly spot prices.

where

p = $ per lb
A = crustal abundance in *PPM*
d = coefficient of mineralizability
γ_0 and γ_1 are parameters.

Brinck [6] estimated γ_0 and γ_1 for several metals and characterized the equation as describing long term metal prices:

$$p = 3558.18A^{-1}e^{-25.5688d}$$

where p is in 1971 dollars.

Table 2.6.2 shows A, d, and the average annual and estimated prices for five metals for 1971 and 1983.

Phillips and Edwards [52] offer the observation that "relative prices of metals are not arbitrary and are related in some way to their scarcity." Acknowledging the possibility that for a given metal noncompetitive factors, e.g., cartels, barriers to entry, may influence prices, they, nevertheless offer the tentative hypothesis that relative prices are dictated by costs on the rationale that few metals are free from competition from substitutes. Cost of metal production is specified to consist of two components. The first component consists of the costs of discovery, development, mining, and milling and is considered to be a function of crustal abundance of the metal. The second component consists of smelting and refining (including hydrometallurgical recovery), and is denominated in the energy required to liberate the element from its chemical bonding. This energy is represented by the change in Gibbs free energy, a measure known as ΔG, which can be obtained from a handbook of chemistry and physics. Specifically, G, Gibbs free energy, is related through a law of thermodynamics to entropy, S, absolute temperature, T, volume, V, pressure, P, chemical potential of species k, and number of gram molecules present of species k:

$$dG = -S \cdot dT + V \cdot dP + \sum_k \mu_k \cdot dn_k$$

At constant temperature and pressure $dG = \sum \mu_k \cdot dn_k$, the energy involved in the chemical reactions. This measure was taken as a proxy for the energy required to refine the metal, allowing that this is a conservative estimate because of thermal losses in refining [53]. In their investigation, Phillips and Edwards found that the median grade of ores being mined was a better determinant than crustal abundance; consequently, their model for metal prices was accordingly modified, as follows:

$$p = \gamma_1\left(\frac{1}{q}\right) + \gamma_2(\Delta G)$$

Statistical analysis of prices, grades, and ΔG's for 23 metals produced the following equation:

$$p = 0.9042 \times 10^{-2}\left(\frac{1}{q}\right) + 1.0018 \times 10^{-2}(\Delta G)$$

where

q = median grade of ore produced expressed as a proportion
ΔG = change in Gibbs free energy times a (-1), expressed as MJ/kg
p = price in 1968 dollars per lb of metal.

For example, the median grade of all copper ore produced in the world was taken as 0.01305 and a representative value of the change in Gibbs free energy was estimated to be 0.678; these values of the explanatory variables produce an estimated price for copper in 1968 of \$0.70/lb:

$$p_{cu} = 0.9042 \times 10^{-2}\left(\frac{1}{0.01305}\right) + 1.0018 \times 10^{-2}(0.678)$$
$$p_{cu} = 0.6929 + 0.0068$$
$$= 0.6997 \approx \$0.70/\text{lb}.$$

Noteworthy is the fact that for copper, metal price essentially is determined by the first component; this is supported by engineering cost analyses, which show the cost of milling to dominate copper costs. Contrast this with aluminum, for which the above equation estimates a price in 1968 dollars of approximately \$0.31/lb, 93% of which, according to this model, arises from refining costs.

Using a producer's price index for metals, this 1968 price for copper of \$0.70/lb would imply a price in 1983 dollars of \$1.73/lb. This is approximately twice the current depressed price, but quite close to estimates of the cost on the margin for new copper capacity, which is estimated to be in the range of \$1.50 to \$1.60/lb.[g] Fig. 2.6.3 shows the plot of observed metal prices with those estimated by the above equation by Phillips and Edwards. With exception of cadmium, cobalt, and antimony, the relationship is remarkable in its description of prices across this large set of varied metals which occur in diverse chemical compounds and geological environments, thus exhibiting wide ranges in deposit size, average grade, and deposit morphology.

The foregoing relations are intriguing as descriptions of cost determinants, and it is as such that they are here presented. While Brinck [6] and Phillips and Edwards [52] use them to comment upon price, as price relations they tax economic sense considerably because they do not explicitly consider demand factors. Appealing to long run equilibrium, to equate cost and price is not inconsequential when the relation of cost is across elements among which there is some substitutability. Strong assumptions about product demands, production technologies, and particulars of mineral occurrences (size, depth, grade, grade distribution) are required to consider these relations as describing long term price.

[g]If one considers disequilibrium in current exchange rates of the dollar to the currencies of major copper producing countries, such as Chile, Mexico, and Peru, the current value of copper sold in the United States from these countries may be close to \$1.30/lb.

Cost Determinants for a Specific Element or Compound

General: Even if man had knowledge of a general law of cost *across* a group of elements or compounds, both the business of generating new supply and the activity of estimating supply would need to consider for a given element or compound the great variation in the characteristics of the natural occurrences upon which supply must be based. Consider, for example, the 95% confidence intervals for average grade and tonnage of porphyry copper deposits in the United States and Mexico:

$$19{,}900{,}000 \leq \text{deposit size} \leq 5{,}779{,}000{,}000 \text{ (tons mineralized material)}$$
$$0.36 \leq \text{deposit average grade} \leq 0.90 \text{ (\%)}$$

Clearly, such ranges in size and grade alone represent important variations in cost.

The earth processes which create accumula-

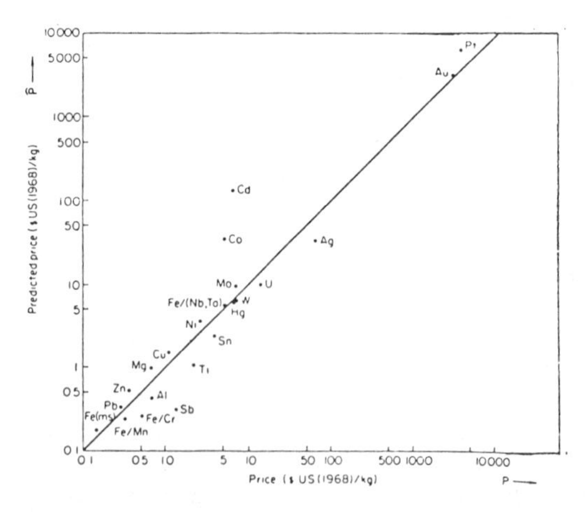

Fig. 2.6.3—Predicted prices from Trial B plotted against actual price. Source: Phillips and Edwards, 1976 [52].

tions of an element or compound operate at differing magnitudes, intensities, and combinations; consequently, the deposits of the element or compound occur in many combinations of mineral form, morphology of deposit, amount of mineralized rock, and quality (grade or concentration). For a given mode of occurrence of an element, e.g., strataform chromite, the most important determinants are size, grade, depth to deposit, and quality variation (intradeposit grade variance).

Some Examples: Some recent [11,36] modeling efforts for the estimation of potential or dynamic supply have employed an index of discoverability as a means of ordering discoveries as they would be made by industry. Tonnage of mineralized material, depth, and grade are all important in this ordering. For example, consider the discoverability equation for tabular sandstone uranium deposits [36]:

$$I = 1000 - 0.23h^{0.64}q^{-0.48}t^{-0.80}$$

Discoverability index (I) varies inversely with depth (h) and directly with grade (q) and tonnage of mineralized material (t), being most elastic to tonnage. This equation describes discoverability of a deposit in isolation, i.e., not adjacent to an already discovered deposit. Once a discovery is made, the discoverability of associated deposits increases greatly until the value of this information gain is offset by discovery depletion. This tradeoff is described by the following equation which shows the fraction, A, of the effort required for the initial discovery that is needed for the next discovery, given that N deposits are present and n already have been found:

$$A = 2.074n^{0.43}N^{-1.66}, \quad 1 < n < N$$

Unsurprisingly, this is a direct function of n but an inverse function of endowment, N. Even for a small N, say 3, the value of information from the initial discovery is high, for it reduces the relative effort to discover the second and third deposits to 0.44 and 0.59, respectively. Thus, if both deposits were of the same size and grade, discovery cost, collectively, would be approximately equal to that for the initial discovery. If N is large, say 10, and $n = 1$, the proportion of initial discovery cost required for the second deposit, given it is of the same size, is quite small, 0.06. Of course, usually the largest and richest deposits are discovered first, meaning that in actuality these costs would be larger proportions of initial discovery cost because of de-

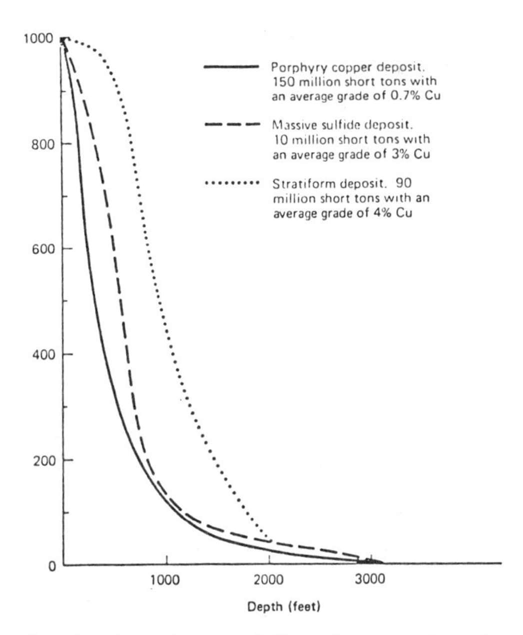

Fig. 2.6.4—Discoverability of copper deposits using current technology. A discoverability of 1000 implies certainty of discovery given best-practice technology. Source: Harris and Skinner, 1982 [39].

creases in deposit sizes and grades of subsequent discoveries.

Fig. 2.6.4, which shows the effect of depth on discoverability of a copper deposit, was developed from discoverability equations similar to those shown for uranium [39]. Especially noteworthy is the rapid decrease in discoverability for all modes of occurrence as depth of cover increases. Consider a porphyry copper deposit of 150×10^6 st of ore at an average grade of 0.7% Cu. When some part of this deposit is exposed, the explorationist is certain that current exploration practice would discover it, giving it the maximum relative discoverability of 1000.0. With 1000 feet of cover, relative discoverability is reduced to approximately 100. The implication of this figure is twofold. First, considering what has been discovered in light of the rapid decay in discoverability with depth, there probably are quite a few copper deposits remaining to be discovered; second, costs for the discovery of remaining deposits will increase unless there are some noteworthy improvements in sensing technologies or reductions in drilling costs.

These same determinants—deposit size, grade, and depth—also affect production costs. Consider, for example, the equation [14] for average total (mining + milling + smelting) costs (c_m) ($1978) per pound of copper:

$$c_m = 0.893 + \frac{0.146}{q} - 0.000271t^{1/2} - 0.82x_1 + 0.0627x_2 + 0.169x_3 - 0.40x_4$$

where

q = grade in percent copper
t = ore reserves (1000 mt)
x_1 = by-product value—gold per pound of copper multiplied by the price of gold plus silver per pound of copper multiplied by price of silver ($1978)
x_2 = ratio of ore reserves to ore plus waste rock
x_3 = 1 if mine is in Alaska; $x_3 = 0$, otherwise
x_4 = 1 if mine is operating or has operated; $x_4 = 0$, otherwise.

As expected, c_m is inversely related to grade and tonnage. For a tonnage of 300×10^6 mt of ore at an average grade of 0.6% Cu, which is approximately the average for U.S. mines in recent years [60]; average total cost is $1.02/lb Cu, of which 89% is *fixed*, i.e., does not vary with cost determinants.

As another example, consider Fig. 2.6.5, which for U_3O_8 is the $15/lb ($1975) isocost of deposit grade and tonnage [30]. This isocost represents comparable combinations of grade and tonnage of ore of U_3O_8 on average, when frequencies of depths of the population of known uranium deposits in the San Juan Basin of New Mexico are used as weights in averaging across depth. Suppose that this isocost is roughly approximated by two segments: for $t \leq 100{,}000$, $q = 0.31 - 0.0174 \ln t$, and for $t > 100{,}000$, $q = 0.11$. Then, if the population of uranium depostis in the San Juan Basin were represented by an independent bivariate lognormal probability density, $f(\text{lnt},\text{lnq})$, the percentage of the population for which production costs are greater than $15/lb U_3O_8 is approximately 23%:

$$0.23 \approx \int_{-\infty}^{\ln(100{,}000)} \int_{-\infty}^{\ln[0.31-0.0174\ln t]} f(\text{lnt},\text{lnq})\, \text{dlnq dlnt} + \int_{\ln(100{,}000)}^{\infty} \int_{-\infty}^{\ln(0.11)} f(\text{lnt},\text{lnq})\, \text{dlnq dlnt}$$

where

$$\ln Q \sim N(-1.747, 0.0968)$$
$$\ln T \sim N(10.7034, 1.805)$$

A final example of the use of deposit parameters in resource and supply analysis is U.S. coal [72]. An estimation of U.S. deep coal resources was based upon production (Q) and cost (TC) functions involving a deposit parameter (thickness, Th) [73]:

$$Q = 0.7568Th^{1.1071}S^{0.7915}O_p^{0.0283}$$
$$TC = 3{,}914{,}764 + 2{,}122{,}480 \cdot [Q^{1.2796}(1597.6Th^{1.1071}\varepsilon)],$$

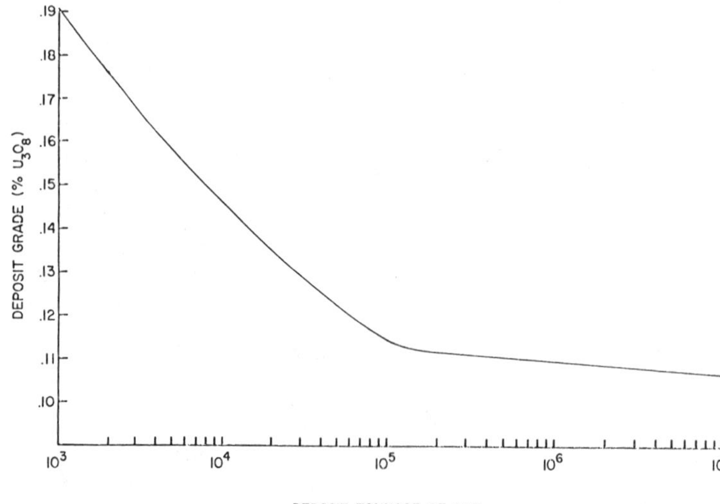

Fig. 2.6.5—The $15/lb U_3O_8 isocost (1975 dollars forward cost) for uranium deposits of San Juan Basin, New Mexico. Source Harris, 1977 [23].

where

Th = seam thickness
S = number of sections in mine
O_p = number of openings in seam
Q = tons of coal output
TC = total annualized cost of mining.

By converting TC to average cost, AC, and differentiating with respect to output Q and setting equal to zero, the output rate that minimizes average cost can be determined. By substituting this quantity into the average cost equation, an estimate of minimum average total cost AC^* can be made; for deep mining, this produced the following relationship [74]:

$$AC^* = \frac{2567}{Th^{1.1071}\varepsilon}$$

where ε is a lognormally distributed error term representing those elements of cost not included in the analysis.

Given the assumption that ε is lognormally distributed, it has been shown [75] that the distribution of deep coal by cost is also lognormally distributed:

$$\ln c \sim N(\mu_c, \sigma_c^2),$$

where

$$\mu_c = \ln 2567 - 1.1071 \ln (Th)$$
$$\sigma_c^2 = (1.1071)^2 \sigma_{\ln Th}^2 + \sigma_{\ln \varepsilon}^2$$

Of course, before the distribution of coal resources by cost can be computed, one must have estimates of additional parameters: E, initial coal endowment; $\mu_{\ln Th}$ and $\sigma_{\ln Th}^2$, the parameters of the lognormal distribution for thickness; and $\sigma_{\ln \varepsilon}^2$, the variance of the error term representing geologic factors not accounted for.

Inspection of a plot [76] of the logarithm of coal seam thickness in Pike County, Kentucky, gives the following:

$$\ln Th \sim N(3.611, 0.0406)$$

The regression analysis performed to estimate the parameters of the production function for coal provided a standard error of the regression estimate of 0.91 [77]. This serves as an estimate of σ_ε. By putting these estimates together, we have an estimate of σ_c^2:

$$\hat{\sigma}_c^2 = (1.1071)^2(0.0406) + (0.91)^2 \approx 0.88$$

Thus, we have the following:

$$\ln Th \sim N(3.611, 0.0406)$$
$$\ln c \sim N(\mu_c, 1.01),$$

where

$$\mu_c = 7.8505 - 1.1071 \ln Th = \lambda(\ln Th)$$

Production records indicate cumulative production of 4.12×10^9 tons (cumulative production through 1974 for Kentucky) [57]. Suppose, for demonstration purposes, that this production were in response to a stable price of approximately \$25.00. Then, we could estimate the initial coal endowment (E) from the basic relations of the model:

$$4.12 \times 10^9 = E \cdot \int_{-\infty}^{\infty} \int_{-\infty}^{\ln(25)} \cdot \phi[\ln c; \lambda(\ln Th), 0.88] \cdot f(\ln Th; 3.611, 0.0406)\, d\ln c\, d\ln Th$$

E could be roughly approximated by discretizing the thickness variable into 100 thickness intervals up to a maximum thickness of 90 inches; then, evaluating the right hand side for each interval and summing the products, gives $4.12 \times 10^9 = 0.255\hat{E}$, which implies $\hat{E} = 16.16 \times 10^9$ tons. Now, suppose that an estimate were desired of the *remaining* resources at \$50.00/ton [$R(50) - R(25)$]:

$$\hat{R}(50) - \hat{R}(25) = 16.16 \times 10^9 \int_{-\infty}^{\infty} \int_{\ln(25}^{\ln(50)} \cdot \phi[\ln c; \lambda(\ln Th), 0.88] \cdot f(\ln Th; 3.611, 0.0406)\, d\ln c\, d\ln Th$$

Using the same class intervals, this estimate is approximated: $\hat{R}(50) - \hat{R}(25) \approx (16.16 \times 10^9)(0.269) \approx 4.35 \times 10^9$ tons. Thus, at \$50/ton, only about 50% of the initial coal endowment would constitute resources, and most of that is already produced; the remainder of the endowment (1.07×10^9 tons) would be a nonresource at the price of \$50/ton.

Petroleum is well known for the wide range in its production costs, reflecting quality (gravity) of the hydrocarbon compound, the porosity and permeability of the rock containing the accumulation (deposit), and the size of the deposit. Crude petroleum is produced from the exceptionally large and high quality deposits in Saudi Arabia for less than a dollar per barrel, giving large resource rents even after maximal transportation costs are deducted from price. On the other hand, the same price that creates the large resource rents in the Saudi Arabian deposits just covers exploration and production for many of the small deposits recently discovered in the United States or for the very deep deposits of the Overthrust Belt. The same is true for some deposits of the North Sea.

Oil deposits, like many of the mineral deposits, have been considered to be lognormally distributed, with the bulk of the endowment of a region contained in a few very large deposits. As with metals, the grade, size, depth, and mode of occurrence (stratigraphic trap, anticlinal trap, etc.) are all determinants of exploration and production costs and of supply.

ESTIMATION ISSUES, APPROACHES, AND DIFFICULTIES

Basic Aproaches—Geologic and Economic

The perspective here is that the estimation of potential or dynamic supply is the ultimate objective. Consider the potential supply from undiscovered deposits. If estimation of this potential supply were to follow classical approaches, it would gather sample information as a means of making inference to the population. But, since new sample information on the presence or absence of mineral deposits is particularly costly, approaches to estimation of potential supply have relied upon other kinds of information. Generally, there are two main, very different, approaches: (1) estimation by projecting economic relations [8,30,42,45,55,63], e.g., life cycles of discovery and production or discovery rates; and (2) estimation based upon inference from related geological information [1,19,21,22, 30,41,43,46,48,50].

Disaggregation of the Estimation Process

The Economic Motivation: The simple taxonomy of the estimation of potential supply as geologic and economic seems to belie the possible combinations of methods and supportive information strategy that confront the analyst when he surveys relevant literature. In part, this impression is reflective of the wide variation in objectives of the various applied studies, which include besides "resource appraisal" statistical methods and models of exploration decision-making and optimization and highly specific studies on statistical and operations research methodologies. Restriction of objective to the estimation of potential supply narrows considerably the range of appropriate methods. Even so, there is a need to further structure the taxonomy to include the choice of estimating potential supply either directly or indirectly, i.e., by first estimating endowment and then subjecting the endowment to the economics of exploration and exploitation [30,34,37,38,40]. This latter approach requires that the estimation of potential supply be viewed as a systems problem, which, appropriately, requires a potential supply system.

The main components of a potential supply system, shown schematically in Fig. 2.6.6, include mineral endowment, exploration, and exploitation models. In practice, such systems range from models which are basically mathematical, e.g., crustal abundance models [7,18,30], to models that include the computer simulation of exploration, disaggregated cost analysis, and mine optimization [34,38].

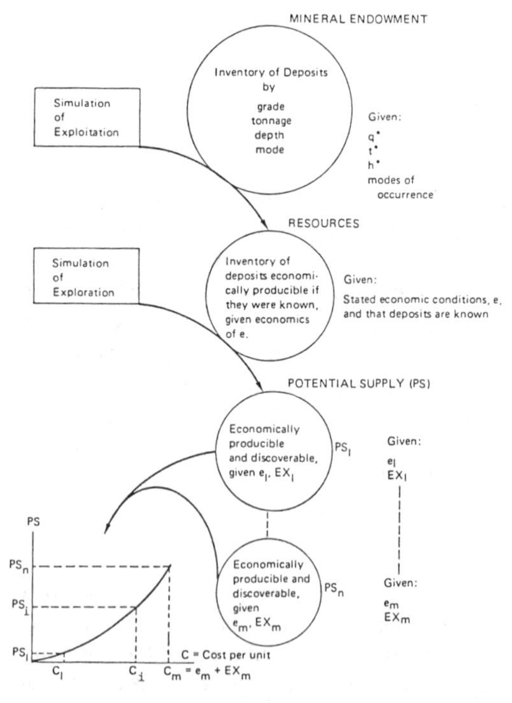

Fig. 2.6.6—A potential supply system (model). Source: Harris, 1978 [25].

The move to estimation of potential supply by a system, which takes as input estimated endowment, reflects the keen interest during the aftermath of the OPEC oil embargo in assessing the impact of high prices, new technology, or policy instruments, in fostering greater energy from fossil energy sources (coal, oil, gas, uranium, and oil shale). The desire to estimate potential supply for prices that were much higher than those upon which previous resource development was predicated, led nautrally to a potential supply system in which "resources" are described in physical, not economic terms. When a "resource appraisal" has produced a quantity that is by its nature an estimate of resources or potential supply, it already has built in the economic dimension, making it of limited use in the examination of other possible economic scenarios. For example, consider the widely heralded estimate by M. King Hubbert [42] of "ultimately recoverable oil" of the United States. By definition, and implicit to his methods, this estimate includes economics and technology not of the present but of the future; however, since these factors are only implied by embodied time trends, it is impossible to unravel the mix of implied endowment, economics, and technology so that other, specific scenarios can be examined. Clearly, economic analysis of potential supply is best supported, ceteris paribus, by a probabilistic description of mineral endowment.

The Inferential Motivation: The pressing economic and policy issues of the post-embargo period were manifest in two resource appraisals of national scope: NURE for uranium [67] and the oil resource appraisal resulting in USGS Circular 725 [48]. Both of these appraisals employed (1) subjective geological analysis as a means to estimating the magnitude of undiscovered deposits, and (2) the description of quantitative estimates by subjective opinion. Because of the urgency of the policy issues that were examined with regard to estimates of the quantity and quality of undiscovered deposits, methodologies of estimation were scrutinized more carefully than ever before. In particular, critical examination was given to appraisal by expert opinion and the heuristics employed by an estimator in providing subjective probabilities. The principal heuristics, identified by psychometricians in controlled experiments, included anchoring and adjustment, representativeness, and availability. Tversky and Kahneman [61,62] found that these heuristics lead to predictable biases. In general, it was found that man behaves as though he knows more than he actually does, tending to provide subjective probability distributions that are narrower than they should be by approximately 50% [59]. This tendency results from the inability of man to imagine the combinations that result in extreme values of the random variable; heuristics used to approximate actual calculations understate probabilities for events in the tails of the distribution.

There is logical appeal in the proposition [23,30,33] that when the uncertain event is a compound and complex event, such as is the magnitude of mineral endowment as inferred from geologic evidence, heuristic bias may be decreased by decomposing the estimation task to its components and then subsequently recombining them by computer and mathematical analyses. The point to be made here is that the motivation from economics for predicating potential supply analysis on an endowment appraisal was coupled with a motivation to improve on the nature of subjective probabilities. Both motivations led to having the geologist estimate mineral endowment, which may be decomposed to its components, e.g., number of deposits, size of deposits, grade of deposits, and intradeposit grade variance.

Decomposition of the task of endowment estimation was pursued to great lengths in an experiment on the subjective appraisal of uranium endowment [33]. For each of five expert geologists, the geoscience of uranium deposit formation was formalized as an inference net which served as the structure for a computerized geologic decision model, a submodel in an endowment appraisal system. Once the geologic model was calibrated on known test areas, the geologist (creator of the model) employed it to make endowment estimates of less well known areas by answering questions posed by the model about the geology and states of relevant earth processes for each area. The endowment appraisal system then computed from these inputs a probability distribution for each evaluation area. Fig. 2.6.7 shows probability distributions for the uranium endowment of the entire San Juan Basin of New Mexico that resulted from the disaggregation of estimation and the formalization of geoscience (System) and from unstructured subjective estimation (Implicit 1 and Implicit 2). Implicit 2 estimates are Implicit 1 estimates subsequently modified by the geologist to conform

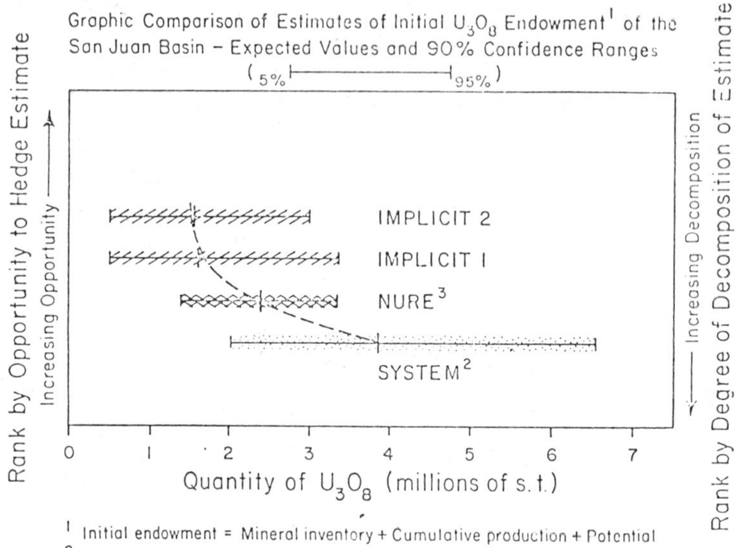

Fig. 2.6.7—Graphic comparison of estimates of initial U_3O_8 endowment of the San Juan Basin—expected values and 90% confidence ranges. Source: Harris and Carrigan, 1980 [32].

as closely as possible to his intuitive feelings. The NURE estimate is not strictly comparable to the other three, because it was not made by the five geologists involved in this experiment; it is presented here solely for comparison with the results of the experiment.

Basically, this study found that use by geologists of formal and explicitly stated geoscience results in estimates of uranium endowment that are considerably greater than those made by NURE and by Implicit assessment. However, this study also found that there exists much greater uncertainty about the magnitudes of the uranium endowment of this region than had previously been expressed. This experiment verified the heuristic bias reported by psychometricians in that the Implicit distributions are approximately one half as broad as the System distribution. In addition to this heuristic bias, the experiment revealed a pronounced motivational bias, exhibited by the shift to the left of expected values and percentile values of the Implicit distributions relative to the System distribution. While possible motivations are numerous, chief among them are (1) the belief in the virtue of conservatism—it is better to be always pleasantly surprised than occasionally disappointed, (2) a mistrust of the use of estimates (protect the public from policymakers), and (3) the desire to conform to previously made estimates, either his own or others, e.g., those of prestigious scientists or of the company or institution of his affiliation [33].

Considerations in Application

As indicated, a useful foundation for *economic analyses of potential supply* is a mineral endowment inventory. Clearly, this inventory includes producing deposits, prospects, and unknown but postulated deposits. In many cases the latter component, the unknown deposit, makes up by far the greatest part of the endowment. Consequently, there is great uncertainty attendant on the estimates of the endowment. Since this part of the inventory often must be estimated without any *direct* (sample) information about the presence of a deposit, estimation must be based upon some kind of model. Therefore, the preferred endowment inventory is probabilistic.

A useful probabilistic description of the mineral endowment inventory would deal with each kind (mode) of deposit separately, for example, porphyry copper, and would consist of four probability distributions [39]:

(1) Probability distribution for number of deposits or for fraction of the earth's crust which consists of mineralized rock.

(2) A bivariate dependent distribution of deposit tonnage and of the deposit average grade.

(3) A probability distribution for grade, conditional upon deposit tonnage and deposit average grade.

(4) A probability distribution for depth to the deposit.

The first three of these distributions would be conditional upon the cutoff grade and minimum size specified for the description of the endowment.

With respect to the methodology that would support in-depth economic analysis, most resource estimates made to date have been inadequate. Clearly, *a resource appraisal method that directly yields potential supply cannot support a long-term supply model which is specified in the comprehensive manner just described*, for it denies the description of the physical features of mineral deposits and the interaction of technology and economics with these features. Thus, resource estimates made by geological analogy as traditionally applied are not useful. Estimates by life cycle or discovery rate models [42,45] are even less useful, for not only are the physical features of the endowment suppressed, but economic conditions and technology are prescribed by unstated trends in the response variable.

Estimation of potential supply by a system leaves open the methodology for estimating mineral endowment. In certain circumstances endowment can be estimated by exploration process models [16,17,68], as demonstrated in a subsequent section. Other circumstances will require that estimation be based upon geological information. Even so, such estimation can be variously made.

The use of geological information to estimate the magnitude of mineral endowment requires a model which is linked to either number of deposits or total amount of mineral. This linkage in practice varies from basic statistical relations, in which quantitative measurements of geodata are correlated with magnitude of endowment, to genetic models which may be used (1) only to guide the geologists' thinking and subjective estimation, (2) as a theoretical structure for the construction of an expert system, or (3) as a basis for structuring information for the construction of a multivariate statistical model. In practice, each of these approaches has its advantages and disadvantages. For example, unconstrained subjective estimation is intuitive and quickly done, but subjective estimates may have undesirable motivational and cognitive biases. In spite of this shortcoming, subjective analysis may be selected because time permits only that approach. Multivariate techniques are desirable because they are objective, but some geological information is either qualitative or suffers considerable information loss when quantified. Expert systems formalize subjective analysis but are expensive and time consuming [9,20,32]. As will be discussed subsequently, the selection of the appropriate methodology must be a result of careful consideration of the use of the estimate, the time and resources available, and the value of additional information.

Determinants of Methodology

Information Strategy and Use: Since methodology is not independent of kinds and levels of information, the pursuit of the "*best methodology*" for *all circumstances* is a fruitless one and should be abandoned. First of all, it reflects a far too simple-minded perception of the highly varied circumstances and objectives of appraisal. Second, it ignores real constraints on methodology that are imposed by time, data, and expertise. In its place there should be an analytical structure for the selection of the methodology which best supports the optimum strategy for information. The important point here is that information and methodology are intimately linked, and the selection of both should result from the broad evaluation of benefits and costs.

When potential supply is estimated as a means to examining issues of resource scarcity and economic growth, the implied time frame is very long; consequently, allowance must be made for potential supply from currently undiscovered deposits. Of course, when location within the region is not important, estimation of potential supply need not be limited to geology-based methods, provided that alternative methods provide the flexibility that is needed for the examination of the economic scenarios of interest.

The position taken by some analysts is that until direct sample evidence of the existence of a deposit has been obtained, there is no justification for making estimates of potential supply, because these are only "guesstimates." Certainly, it is true that estimates based upon direct information are *much more credible* than estimates based upon indirect evidence, e.g., geology, geophysics, and geochemistry. Clearly, the means for obtaining direct sample information on endowment have been available:

drilling directed at deposit recognition. Unfortunately, while potential supply estimates made as a result of such a program would be highly credible, they could be made only at great cost. So far, circumstances have not warranted such expenditure. Of course, it would be premature to conclude that circumstances could never develop for which the value to society of the highly credible estimates that would result from drilling for target detection would justify this high cost. More importantly, selection of the best strategy for information acquisition should be a result of a comprehensive evaluation of benefits and costs of information vis-a-vis the decision variable.

The Time Dimension: The relevant time dimension is that time frame of the central motivation for estimation, e.g., some policy issue. Consider the case in which the motivation is one of a short time frame, say two years. For this case, the time frame is not long enough for the initiation of completely new production facilities, which in mining typically take from two to ten years to complete after the deposit has been discovered. Nor, is it long enough for the realization of production from significant expansion in reserves of known deposits. Here, if stocks are the central issue, then the relevant stock measure is current reserves, perhaps amplified by a model of conversion of probable and possible reserves to proven reserves. If flows are the issue, then they can be best estimated by conventional econometric modeling or by time series forecasting methods.

A time frame of two to ten years is an intermediate case, for it is not long enough for significant enlargement of stock measures through new exploration. But, it is long enough for the expansion of reserves of existing mines and for the development of new production capacity from the inventory of known (shelf) but undeveloped deposits. It is for this time frame that for some minerals we have considerable institutional capability for estimating potential supply for various economic scenarios. The Minerals Availability System (MAS) of the U.S. Bureau of Mines is one example of impressive capability [3,51,64]. For some mineral commodities, such as copper, this system documents the physical characteristics of each known deposit, including those not yet producing. Furthermore, the system is replete with extensive engineering costing models for mining, milling, smelting, and refining. For specified economic scenarios, initiation of this system for a commodity produces what is called an availability curve, which describes potential supply from known deposits. Similar capability exists in DOE for the estimation of potential supply for uranium and in the U.S. Geological Survey for petroleum from known deposits. These capabilities are generally very impressive, a fact that was overlooked during the mid 1970s and the great concern over future energy supplies. There is much irony in the harsh criticism [13] made of DOE for its methodology for the estimation of potential supply of uranium from known deposits, when, in fact, never before had there been made by an institution the reserve and production accounting, engineering cost analysis, and statistical inference (based upon geostatistical techniques) as comprehensive and accurate as that made by DOE's ore reserves division. In retrospect, some of this criticism seems to have reflected dissatisfaction not with reserves data and computation methods, but with the "forward cost" economic reference.

In some cases, the real economic issue was not properly identified. For example, the sometimes sharp criticisms by industry of the quantities of potential supply estimated by an agency of the government, e.g., uranium by U.S. DOE, reflect skepticism about *deliverability* more than about the postulated magnitude of the stock for the stated prices. Of course, since potential supply is a stock, it implies nothing about deliverability. Indeed, deliverability of a given quantity by a stated time at the specified price may be an impossibility. If deliverability is the primary issue, then dynamic supply, not potential supply is the appropriate concept and estimate. Such estimation is more demanding because it must deal with the time dimension and with flows.

Given the long lead times in exploration and development, unless the relevant time reference is *at least 10 years* distant, economic questions are answerable by consideration of ore reserves and on-the-shelf deposits (deposits held in inventory by mining companies but not yet developed). Our capability to provide *credible* estimates of future supply from this *known* stock for specified economic or policy issues is considerable. For Canada, the work described by Williams [71] and by Zwartendyk [78] are further examples of the analytical frameworks, the data systems, and the computer systems that have been developed and are being refined for analysis of these intermediate term issues.

It is for a time frame of greater than 10 years

that the contribution of potential and dynamic supply from currently unknown deposits must be considered. The longer this time frame, the larger the region, and the greater the change in economics, the greater the possible change in potential supply. Thus, in an *a priori* sense, the potential of unexplored regions becomes very important in this time frame. Often, it is the relatively unexplored regions that could contribute most to potential supply for this time frame, and, ironically, it is for these regions that information often is very limited, that estimation is the most difficult, and that controversy over the best methodology and the usefulness of estimates is the greatest.

Degree of Advancement of Exploration: Preference for a methodology may arise as a result of the degree of advancement of exploration. Consider the case in which exploration of a sedimentary basin is approaching maturity and the objective is an appraisal of the number of undiscovered petroleum deposits of specified size. In this case the appraisal could be made by geological analysis. And, if this analysis were supported by maps of the probability that deposits of various sizes could be present and could have escaped detection by the current drilling density, the geologist's estimated probability distribution for the number of deposits that actually remain may be highly credible. The point being made here is that for an estimate of remaining endowment to be highly credible, it must consider the very important information present in previous exploration results, for, such information, is about both economics of exploration and the petroleum endowment of the region.

An alternative to the foregoing approach is the estimation of endowment through an exploration process model. Where considerable exploration has taken place, this approach can yield highly credible estimates. In a particularly impressive demonstration, the U.S. Geological Survey [68] used drilling and discovery data only through 1961 to estimate an exploration process model for deposits of size class 10 (1.52×10^6 to 3.04×10^6 bbls of oil equivalent):

$$F_{10}(w) = F_{10}(\infty)(1 - e^{-CAw/B})$$

where

$F(\infty)$ is the desired endowment estimate, i.e., the number discovered when there has been an infinite amount of drilling

B = basin area

A = average areal extent of the fields in the given size class and depth interval

w = cumulative number of wells for the depth interval

C = the efficiency of exploration. For random drilling, $C = 1$; if exploration is twice as effective as random, $C = 2$.

For deposits of the tenth size class in the Permian Basis, there were by 1961 14,243 net wells, which yielded 59 discoveries. Given this amount of exploration and the following basin parameters ($B = 100{,}000$ mi^2, $A = 2.2$ mi^2, and $C = 2.0$), we have an initial endowment estimate of 126.7 size-class-10 oil fields:

$$59 = F_{10}(\infty)(1 - e^{-(2.0)(2.2)(14{,}243)/100{,}000})$$
$$F_{10}(\infty) = 126.7$$

Therefore, this model indicates that 68 deposits of size class 10 remain to be discovered. The model then was tested by predicting discoveries for the interval of 1961–1974. Cumulative drilling by 1974 amounted to 25,055 net wells for size class 10; thus $F_{10}(25{,}055) = 84.5$, meaning that an additional 25.5 fields of size class 10 should have been discovered by 1974. In fact, during this period 25 discoveries were made, testifying to the accuracy of the model.

Consider the case of a frontier or lightly explored region which has no known deposits but for which considerable geologic, stratigraphic, seismic, and gravity data are available. For this circumstance, the only reasonable approach to the estimation of its petroleum endowment is through geological analysis. Although this fact is obvious, there are a number of decisions to be made as to how this geologic estimation should be made. Options vary all the way from unconstrained subjective analysis by an expert geologist to computerized expert systems and to one or more strictly quantitative multivariate models which have been constructed and estimated on one or more well explored areas.

The Past as a Statement on the Present and Future

Historical resource estimates should *not be used* as a measure of the current credibility and capability of geologists to estimate mineral or energy endowment and resources. Moreover, it is not appropriate to judge the potential of geoscience in resource analysis by the widely differing past estimates made by various individuals, mostly by U.S. Geological Survey geologists, of petroleum resources, no matter how prestigious the scientists. *Even the prestigious scientist of a decade ago was not pressed to describe sharply the object of estimation nor to defend his methodology of appraisal.*

Critics must be aware of current and emerging developments in geoscience and resource appraisal methodologies. Only recently have geologists become formally engaged in the use of geoscience to estimate mineral endowment, resources, or potential supply. Both geologist and nongeologist have assumed that since mineral deposits are geological phenomena, geologists have been prepared by virtue of their science to estimate mineral endowment or resources. This simply has not been so! *Until recently, most geologists did not consider resource appraisal to be a geologist's professional function.* In the past, neither academic nor professional experience prepared a geologist for this task. Therefore, it may be premature to pass judgment on how useful geoscience now is and can be as a means of estimating mineral endowment, resources, or potential supply.

Contamination of Geologic Data by Economics

There can be little argument with the proposition that data on deposit characteristics are important to the estimation of potential supply. For this reason, many models employ distributions of deposit size and grade (quality). Only in the past few years has the information content of such data been critically examined. As a result, these data have been found to be both truncated and translated [11,25,30] (see Fig. 2.6.8), and use of them as they are routinely reported poses some difficulties, particularly if economic scenarios very different from the past are to be considered for their impact upon potential supply. From Fig. 2.6.8, it is apparent that deposit size and grade data, as they are reported, are a mix of the physical attributes of the deposit and of the economics of exploration and profit maximization in the exploitation of the deposit.

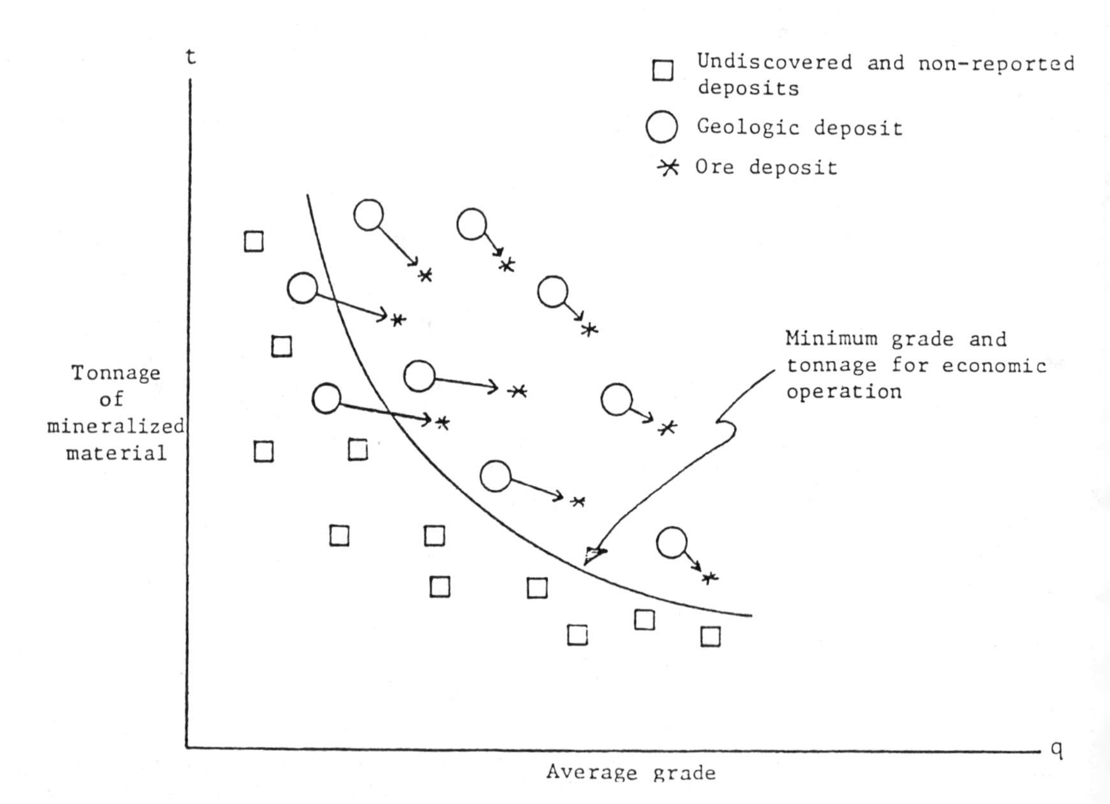

Fig. 2.6.8—Schematic illustration of truncation and translation of ore deposit to reserves due to economic and technologic effects. Source: Harris and Agterberg, 1981 [31].

Modeling endowment by probability distributions constructed from these data tends to overstate the relative frequency of occurrences of high grades and large sizes of deposits. Such distortions cause distortions of exploration outcomes and of reserve expansion in response to improved price. Furthermore, economic contamination interferes considerably with the determination of the presence of dependency between size and grade, an issue of considerable importance in modeling of potential supply and the investigation of resource scarcity.

Consider Fig. 2.6.9, which portrays for fields of the Western Gulf of Mexico area the truncation effect of economics [16]. In fact, because of this contamination effect, it has been proposed [31] that the lognormal distribution, which long ago was accepted as the cornerstone of petroleum process modeling, should be replaced by a log geometric distribution because the lognormal model does not explain well the very large numbers of oil deposits of small sizes: define the random variable X as field size class; then, $P(X = x+1) = P(X = x) \cdot 2^{-\theta}$, where $P(X = x) = c2^{-x\theta}$; c and θ are parameters. Basically, the proposition that the log geometric model is the proper model implies that the previous selection of the lognormal was predicated in part upon data that exhibited a truncation or censoring of the full population by economics of exploration and production. The distortion of the petroleum field size data by economics is even more serious than that indicated if these data are to serve as bases for modeling petroleum endowment, for the sizes of those deposits in the untruncated part are of reserves, which represent recovery of initial endowment for current economics. For endowment representation, these sizes should be total petroleum in place.

A similar phenomenon recently became recognized for size and grade distributions of uranium deposits of the San Juan Basin, New Mexico. Investigations by the U.S. Geological Survey suggest that from a geological point of view, many of the ore deposits of the Basin represent high grade portions of a single deposit. Such a perception contrasts sharply with previous views. More importantly, such a finding carries significant implications to exploration economics and efficiency and to the impact of improved economics on the potential supply from known deposits or mining districts.

Another consequence of economic contamination is its impact on the geologist's estimate

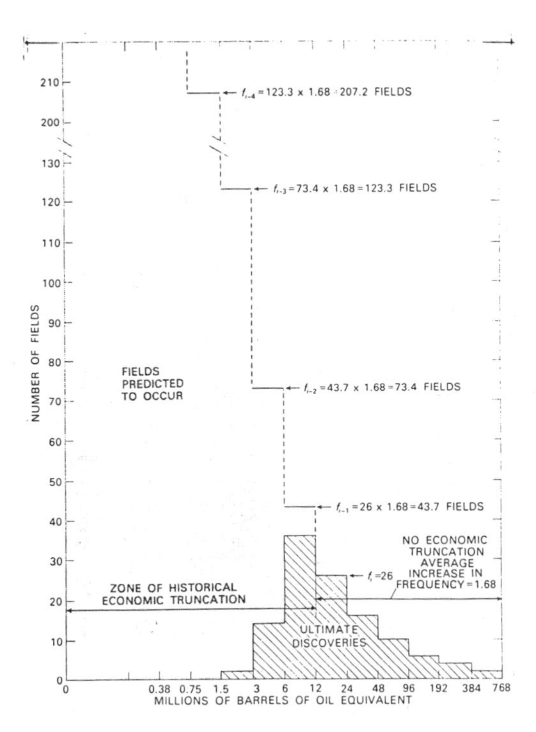

Fig. 2.6.9—Schematic diagram of the method used to estimate the part of the field-size distribution below the historical level of economic truncation. Source: Drew, et al, 1982 [16].

of number of undiscovered deposits. Clearly, the translation of geology into numbers of deposits deals with physical and geologic process relations. This activity is best supported by information on sizes, average grades, and grade distributions of the deposit as a geologic phenomenon, not one of economics and geology. Economics can be introduced later through proper analytical models of exploration and exploitation. In the case of uranium, there is hope for improvement in the future, for the U.S. Geological Survey is conducting a pilot study on defining deposits as geological phenomena.

POTENTIAL SUPPLY FROM UNDISCOVERED ENDOWMENT AS A SUPPORT TO POLICY ANALYSIS

Perspective

When the subject of policy is an uncertain quantity, such as is the magnitude of potential supply from undiscovered mineral resources or future mineral supply, policy options may include delaying commitment to a specific program, e.g., industry subsidization or stockpiling, until after the gathering of additional geologic or resource information [28]. In the language of statistical decision theory, such an option (strategety) may be feasible when the expected value of uncertainty reduction by additional information outweighs the cost of gathering the information. Such an option is considered to be *optimum* when the expected loss of commitment now to identified programs, without gathering additional data, exceeds the direct cost of gathering data *plus* the opportunity loss of deferring program commitment until after analysis of data [28].

Geoscience Information—A Complication

The making of a rational decision to defer commitment to a policy option until after the acquisition of additional information presupposes that the value of the cost of information on the subject of policy can be measured. In the case of undiscovered mineral and energy resources, this measurement at present is problematic. For, except for very rare cases, direct information on the magnitude of the undiscovered resources is meager and costly to acquire. Usually, undiscovered mineral and energy resources are inferred from geoscience data, e.g., the NURE geologic and hydrogeochemical program for uranium resource appraisal. Such inference usually is attended by considerable uncertainty.

It is both appropriate and useful to consider the magnitude of a mineral or energy resource to be a conditional random variable. In other words, given all available geodata and information at the time of policy analysis, the magnitude of the resource is described by a probability density function (pdf). A rational decision to postpone commitment to any policy option until more geodata are obtained requires a determination of the expected value to society of additional information.

The expected effect of acquiring and analyzing additional geoscience data can be viewed simply as a modification of the shape of this pdf. However, evaluation of the benefits of a governmental program to generate geoscience data cannot be made by examining solely the probabilities before and after additional geodata. This evaluation of benefits and costs must *combine conditional losses to society of possible resource states with the resource probabilities before and separately with those expected as a consequence of additional data acquisition*, with allowance made for the expected cost of acquiring and analyzing the additional geodata. The lack of an institutional capability to estimate the conditional losses for each of the alternative energy sources was camouflaged during the post-embargo period by the intensive focus of attention on resource estimates. While welfare theory and the theory of risk-bearing provide theoretical foundations for the estimation of conditional losses, given the interdependency between some minerals and energy sources, the availability of imports, and specific policy alternatives, e.g., stockpiling, our capability of making a quantitative determination of conditional losses vis-a-vis alternative policy options is probably no better than the capability of geologists to estimate the probability distributions for mineral endowment or resources.

Resource Ignorance and Policy

An increment of information is achieved only through an increment of investment; consequently, there is an optimum state of ignorance (knowledge), a state for which marginal benefits of information equal marginal costs. *Commonly, this information state is not the total absence of ignorance*. Ignorance about the mineral endowment of a region can be reduced to zero simply by drilling on a grid such that no deposit of interest could be missed by the drill. So far, and probably in the future, the benefits of zero ignorance are less than its costs.

For the sake of argument only, let's examine the hypothetical state of "ignorance." Even when a geologist claims ignorance, suggesting a rectangular probability distribution for mineral or energy endowment, he seldom, if ever, is totally indifferent about the upper limit of that distribution, the maximum endowment of the region; often, neither is the geologist, even in his professed ignorance, indifferent to the minimum endowment. Since the parameters of a rectangular distribution are the minimum and maximum values of the random variable, this is tantamount to being not indifferent about the parameters of the distribution, and to being not indifferent to the expectation for mineral endowment [28]. That being the case, are there any regions about which we are *totally ignorant*? On the contrary, geological information invariably bounds the values of many factors. Furthermore, a large region, like the United States, is not homogeneous with respect to the circumstances attendant to the geologic estimation of undiscovered mineral endowment. These circumstances can be, and usually are, highly varied with respect to available geodata and data on already discovered mineral resources. There are subregions within the large region for which our level of geoscience and resource information is high, and for these subregions estimates of the magnitude of undiscovered mineral endowments are highly credible. While we are nearer a state of total ignorance for the entire region (collection of subregions) than for the better known subregions, an hypothesis of ignorance for the region ignores the high-information subregions; consequently, such an hypothesis cannot generally lead to optimum mineral policy.

An ever present question, particularly for appraisals made by elicitation of judgments of experts, e.g., NURE, is the credibility (information content) of an appraisal [70]. Usually, about all that can be offered by way of assurance is documentation of a well conceived and executed methodology, possibly supplemented by a documentation of performance on control or validation areas. For this reason, great emphasis should be given to methodology. The following two references are attempts to go beyond methodology and use exploration results, albeit in a weak sense, to comment upon the credibility issue. Consider the judgment recently offered by R. J. Cathro (July, 1983) at a symposium on predictive metallogeny [10], a judgment which contrasts quite sharply with a state of resource ignorance. Cathro was speaking as one of three invited discussants for the Predictive Metallogeny Symposium, sponsored by the Geological Association of Canada and the Mineralogical Association of Canada (GAC-MAC). In discussing predictive metallogeny and the usefulness of resource appraisal, he made the following reference to an appraisal made in 1969 of the mineral resources of the Yukon Territory and Province of British Columbia of Canada[h] (for a description of that appraisal, see Harris et al., 1971 [40]; Barry and Freyman, 1970 [2]):

> The results were synthesized [see maps of Fig. 2.6.10 and 11] and appear to have been remarkably perceptive. I am sure none of us who took part in that project 15 years ago would have guessed that it would turn out so well (*Geoscience*, June 1983, p. 100).

This judgment is particularly noteworthy because it reflects at least a partial test by time (exploration results).

As a final commentary consider Fig. 2.6.12, which shows two maps of contours of the probability for the occurrence of a porphyry copper deposit in southeastern Arizona and southwestern New Mexico [27]. In the first of these maps, the probability contours represent the potential as discerned from surface geology only, which for many subdivisions of the region included alluvial cover. The second of the maps shows estimated probability for occurrence if the al-

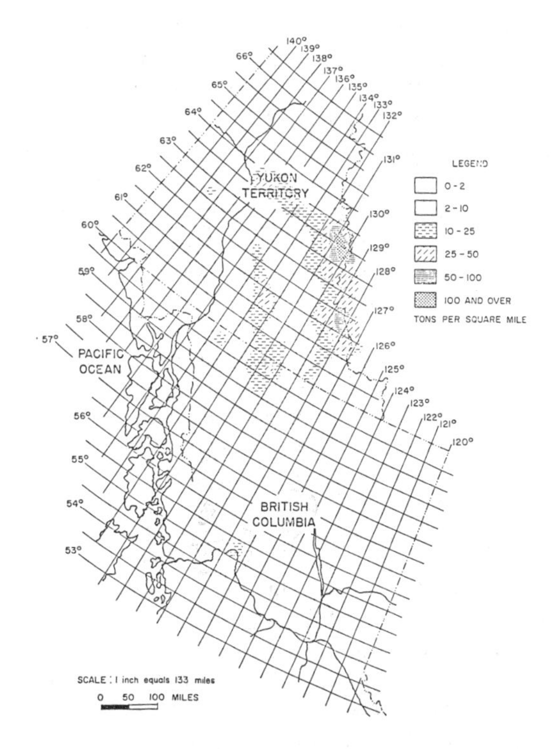

Fig. 2.6.10—Geographic distribution of the endowment of tungsten in undiscovered but expected deposits—Canadian Northwest. After Barry and Freyman (1970) [2].

[h]Each of 313 separate areas, defined by superimposing a grid on a base map of the region, were appraised by obtaining subjective probability distributions for each area from each of 30 active explorationists on number of deposits, size, and grade for each of 11 mineral commodities.

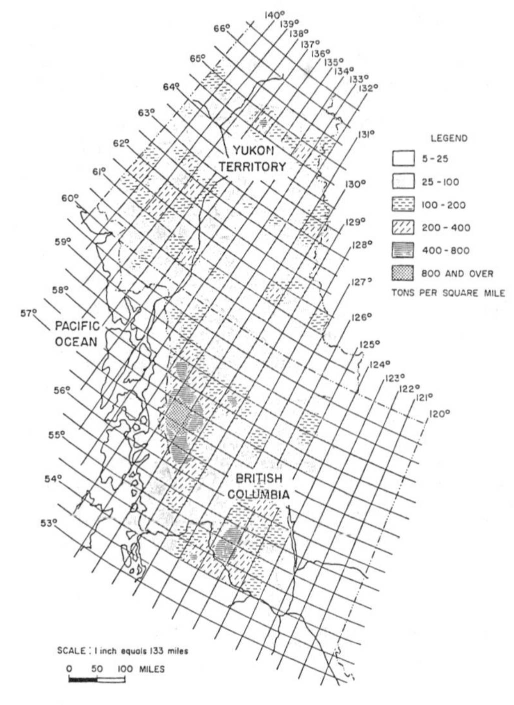

Fig. 2.6.11—Geographic distribution of the endowment of copper in undiscovered but expected deposits—Canadian Northwest. After Barry and Freyman (1970) [2].

luvium were to be stripped away, exposing estimated geology of the underlying rocks. The contours were based upon quantified reconnaissance geological data and discoveries of 1964. These data were analyzed by a combination of multivariate statistical techniques[i] (factor analysis, trend surfaces, multiple regression, and multiple discriminant analysis combined with Bayesian classification). The second map also shows the location of actual major discoveries made from 1964 to 1981. Thus, the locations of actual discoveries can be compared to the region that in 1964 would have been estimated to have high probability for occurrence using pre-1965 discovery data and reconnaissance geology. As indicated in Fig. 2.6.12, exploration results basically confirm inference that could have been made from information that was available two decades earlier.

[i]This figure, as well as a description of the geostatistical model and analyses, were presented at the Tenth Geochautauqua on Computer Applications on the Earth Sciences, organized by Dr. Frits Agterberg, Head of the Department of Geomathematics of the Canadian Geological Survey.

The Need for a Broader Perspective

Focusing attention on the polar case of ignorance may inhibit the perception of the more complete, comprehensive, and useful idea:

> If policy were evaluated according to the decision-theoretic approach generally referred to in this paper, total ignorance about the mineral endowment of a region would be only a special circumstance for the *prior* probability distribution. *Best policy will be made only by considering the prior probabilities (no matter how diffuse), conditional losses, and the expected value of information for various information strategies. The optimum action, given a comprehensive analysis, could be to* postpone *an ultimate selection of a policy option until the acquisition of additional data.*

Since a state of total ignorance about mineral endowment could be due to lack of information, not deficiency of geoscience, the relevant question is the policy action taken given this state, and this could include policy deferment and an information strategy with a conformable endowment appraisal [28]. The decision to keep an area open for exploration, or to close it, should be made *only after* analyzing the information available at that time, *even if the state of geological knowledge were that of "total ignorance,"* for exploration, whether done by government or industry, is the acquisition of information, and this information has a cost as well as an imputed value. *No matter how poor the prior information is on mineral endowment, an optimum decision on land use would at least reflect the expected value and costs of additional information, given the priors.*

Some Examples of "Resource Appraisals" for Policy Decisions

NURE: NURE (National Uranium Resource Evaluation), the largest and most comprehensive of all "resource appraisals" to date, initially was motivated by questions as to policy vis-a-vis development of the breeder reactor and the processing of plutonium. Political issues ranging from the general capability of meeting future energy requirements and of energy supply independence to safety and proliferation of nuclear weapons were involved. The desirability of at least some policy options initially was per-

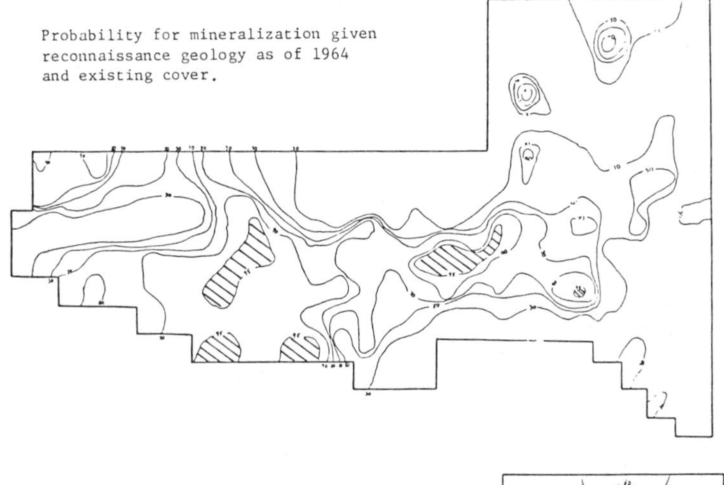

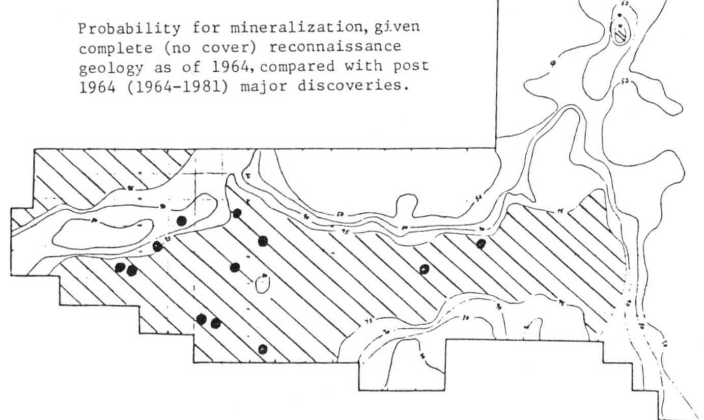

Fig. 2.6.12—A comparison of actual porphyry discoveries (1964–1981) with areas of high estimated probability, based upon 1964 reconnaissance geology and multivariate statistical inference. Source: Harris, 1981 [27].

ceived to be highly dependent upon the magnitude of the nation's resources of U_3O_8; consequently, NURE was created to provide enlightment on the existing policy controversies.

The approach employed in NURE combined principles of geological analogy, as evidenced by the effort to describe thoroughly the geology and resource parameters on control areas, with subjective probability [26,67]. Important modifications over previous practice include the following: (1) estimation by the geologist of U_3O_8 endowment instead of resources or potential supply, (2) decomposition of endowment to its components (A, F, T, G, and P_0—see Fig. 2.6.13), (3) elicitation of subjective probabilities on the components, and (4) the recombining through mathematical analysis of these components to give a probability distribution for E (U_3O_8 endowment) by the following relationship [67]:

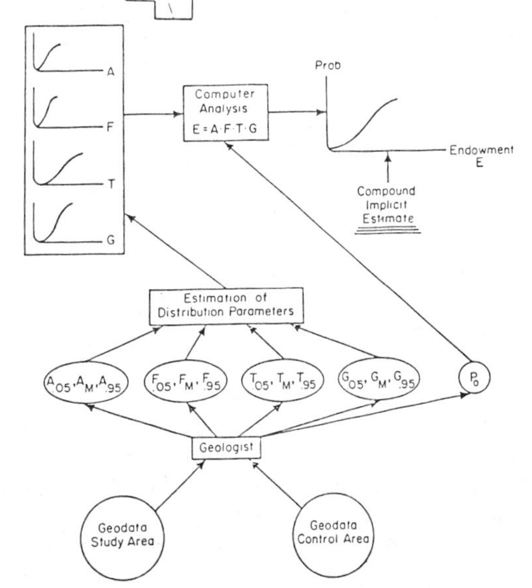

Fig. 2.6.13—Implicit methodology of U.S. Department of Energy—NURE. Source: Harris, 1983b [29].

$$E = A \cdot F \cdot T \cdot G$$

where

A = size of favorable area
F = fraction of favorable area underlain by mineralization
T = tonnage of mineralized material per square mile of favorable area
G = average grade of mineralized material
P_0 = probability for at least one deposit having at least 10 tons of U_3O_8 given a cutoff grade of 0.01% uranium.

Subsequent to their geological investigations, geologists provided most likely plus 5th and 95th percentile estimates for each of A, F, T, and G. They also provided a single point estimate of P_0. Distributions were fitted to A, F, T, and G, and these distributions and P_0 were combined appropriately to yield a probability distribution for uranium endowment (see Fig. 2.6.14). Although the original design of the appraisal methodology, as shown in Fig. 2.6.13, treated A as a random variable, as the methodology was applied, only the most likely estimate of A was employed.

Potential supply of U_3O_8 was estimated by analyzing the cost of finding and producing the endowment. The results of the potential supply analysis are presented in Fig. 2.6.14; the right most curve of case d is the probability distribution for E; the remaining curves are potential supply for forward costs of \$30, \$50, and \$100/ lb U_3O_8.

Rail Construction Cost/Benefit Analysis—Canada: A previous section (Resource Ignorance) referred to an appraisal of the resources of the Yukon Territory and British Columbia Province of Canada. The reason this appraisal was made is that in 1969 the Canadian Government was considering the merits of constructing a railroad extending north from Vancouver through British Columbia and across the Yukon Territory. Of the several routes that had been proposed, the government wished to identify the one which was optimum in the sense that it created the greatest benefit/cost ratio when benefits are measured comprehensively, in accordance with regional accounting and economic analysis. The basic approach was to simulate both the construction of the railroad over a period of 50 years and the commerce that would develop as a consequence of reduced transportation rates and of economic growth. The chief benefits were considered to derive ultimately from three basic sectors: agriculture, forestry, and mining. Thus, one input required for this analysis was a description of potential supply in terms of the media of commerce, e.g., ore concentrates, bulk ore, etc. Furthermore, in this case geographic location of this potential supply was critical.

Accommodating the needs of this economic analysis required dividing the two major regions into 313 geographic subdivisions referred to as cells, and obtaining for each cell a probabilistic description of its endowment in each of 11 mineral commodities (copper, lead-zinc, molybdenum, nickel, tungsten, asbestos, coal, iron, uranium, mercury, and silver-gold) [40]. Endowment distributions for a cell were described by averaging the subjective probability distributions estimated by 20 explorationists, resulting in one average univariate probability distribution for number of deposits and one average joint probability distribution for size and grade per deposit for each commodity (see Harris, et al, 1971 [40]). These endowment descriptions served as inputs to a potential supply system that simulated the exploration for and development of mineral deposits, given projected regional demands and prices and scenarios of transportation rates.

For demonstration purposes, a set of 36 cells (2 rows of cells) on the Yukon-British Columbia border were analyzed by the potential supply system for selected scenarios. Employing the input to and output of the potential supply system, response functions were estimated for tonnage (T) of the appropriate medium of commerce and number of deposits (N) as functions of price (P) of product and multiple (K) of basic transportation rates. The following equations for copper and asbestos are representative of the analysis [40]:

Asbestos[j]

$$T = 13{,}826{,}110\ P^{0.2246}\ K^{-1.0274}$$
$$N = 0.966P^{0.5116}\ K^{-1.1889}$$

Copper[j]

$$T = 84{,}255{,}838P^{0.8255}\ K^{-0.7606}$$
$$N = 6.193P^{0.6112}\ K^{-0.801}$$

[j] T is tons of ore.

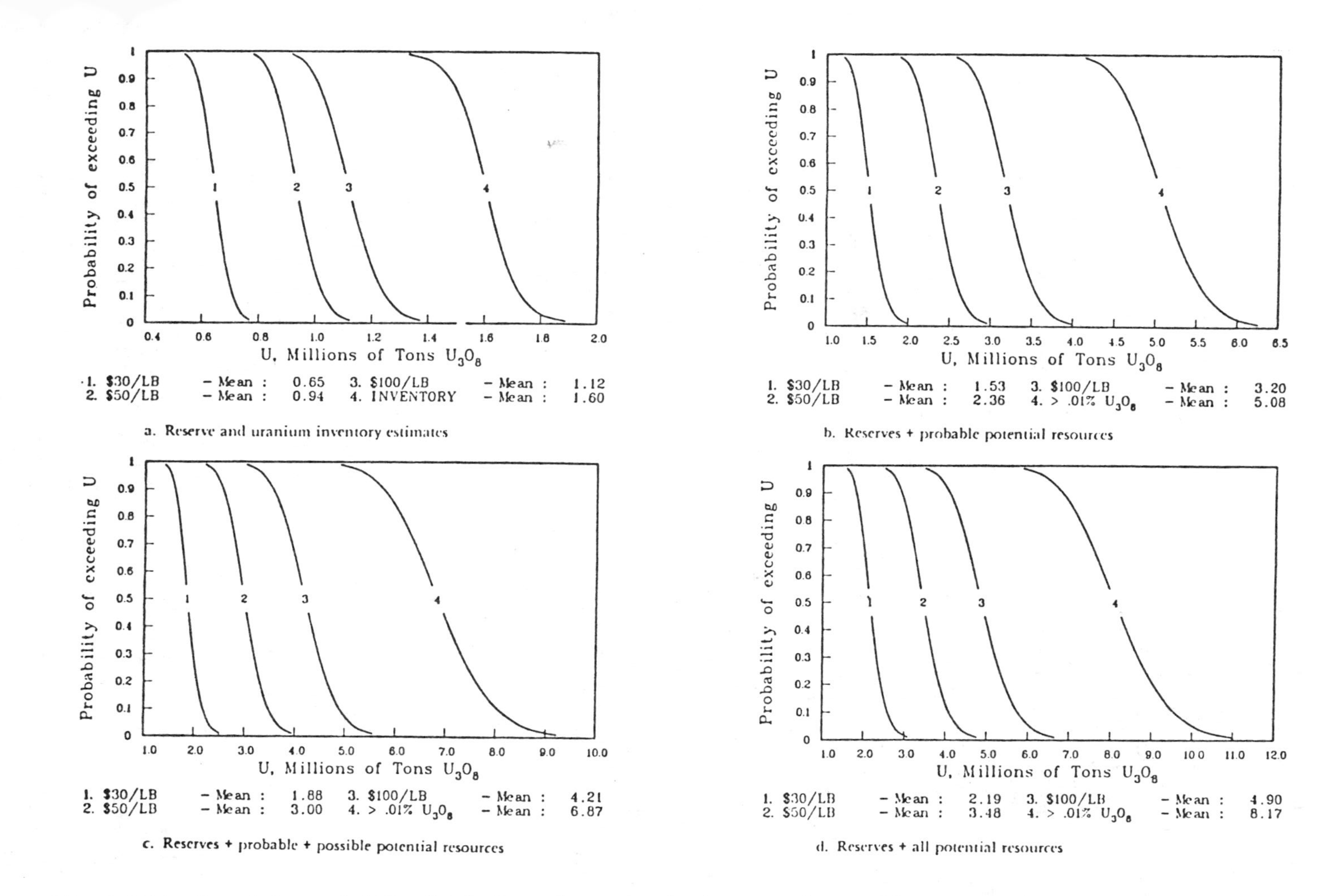

Fig. 2.6.14—Cumulative probability distributions of reserves plus incremental potential resources of the United States. Source: U.S. DOE, 1980 [67].

Infrastructure and Potential Supply of Base and Precious Metals—Sonora, Mexico: The overall approach in this study was similar to that just briefly described for Canadian rail construction in that one objective was to compute the cost-benefits of constructing a road from northern Sonora south to Hermosillo. The analysis was based upon the simulation of occurrence, exploration, and production of deposits of base and precious metals. Endowment of each of 64 subdivisions (cells) of the region in base and precious metals was described by endowment probability distributions (one for number of deposits and one for tonnage and grade per deposit) synthesized from subjective probabilities estimated by each of nine explorationists representing the most active major mining firms in the region.

One important feature of this study was the introduction of a dynamic programming solution for the optimum location of infrastructure linking cells to the proposed road. This was accomplished by representing the topography of the region as a matrix of elevations. Analysis found that benefits of increased potential supply due to reduced costs of transporting concentrates to Hermosillo outweighed considerably the costs of road construction.

Suppose that the potential supply system were executed for each of a number of combinations of transportation parameters; then, by statistical analysis of the response of the system and the parameters, potential supply could be described parametrically. The following equation for cell 39 was determined in such a fashion [37]:[k]

$$\hat{ps}_j = (7 \times 10^8) \cdot \left\{ (TR)^{-\frac{TD^{2.59}}{7.12}} \right\} \cdot \left\{ (CC)^{-\frac{CD^{0.41}}{2.14}} \right\} \cdot (h_j/r_j)$$

where

$\hat{ps}_j$ = the estimated potential supply of the jth metal; TD and CD are in units of 100 miles.

h_j = value proportion for the jth metal of the five-metal aggregated value, i.e., that fraction of the total value of a ton of ore that is due to the value of the jth metal

r_j = price of the jth metal

TR = transportation rate

CC = transportation capital cost

CD = construction distance

TD = transportation distance.

Suppose that CD were zero and TD were 400 miles, then from the above equation, we have

$$\hat{ps}_j = (7 \times 10^8)\, TR^{-\frac{36.2}{7.12}} (h_j/r_j)$$

The elasticity of $\hat{ps}_j$ with respect to TR, the multiple of the basic carrying rate, is approximately $-5 \approx -36.2/7.12$. For $CD = 0$ and $TD = 200$, we have

$$\hat{ps}_j = (7 \times 10^8)\, TR^{-\frac{6.03}{7.12}} (h_j/r_j),$$

indicating an elasticity of approximately -0.85, a much reduced elasticity of $\hat{ps}_j$ with respect to TR.

Suppose the following: $TR = 1.5$, $CC = 1.5$, $TD = 2.0$, $CD = 2.0$, $h_1 = 1.0$, and $h_2 = h_3 = h_4 = h_5 = 0$; and $r_1 = \$0.56/\text{lb}$ (the basic price). The implication of setting $h_2 = h_3 = h_4 = h_5 = 0$ and $h_1 = 1.0$ is that the resource appraisal will be in terms of copper equivalent. Then,

$$\hat{ps}_j = \frac{(7 \times 10^8)(1.5)^{-0.85}(1.5)^{-0.62}}{0.56}$$
$$= 690{,}600{,}000 \text{ lbs copper equivalent}$$

This quantity is approximately 14% of the metal expected to occur in the cell (metal endowment) and about 58% of potential supply when transportation and carrying distances are near zero.

The expected metal endowment of the 64-cell (400 sq. miles each) area was estimated to be 66×10^6 tons of copper. Of this quantity, under economic conditions and the infrastructure at that time, 17×10^6 tons of copper were estimated to constitute potential supply.

Public Land Withdrawal—Alaska: During the latter part of the 1970s, the U.S. Congress was struggling with difficult policy decisions on the withdrawal of Alaskan land from mineral exploration and exploitation. In principle, these decisions weigh the present value of expected net benefits from exploitation of the mineral resources against those for other nonmineral uses. While land management decisions in practice usually do not honor completely these principles, a consideration is given to the so-called

[k]The presence of r_j, price of the jth metal, in the denominator seems to contradict economic theory. However, a change in r_j is accompanied by a change in h_j; consequently, the overall effect of a price change must be described by h_j/r_j.

Table 2.6.3—Grade and Tonnage Models

Deposit type	Tonnage and grade variables (units in parentheses)	Number of deposits used in developing model	Correlation coefficient of listed variable with variable on line with it in column 2	90% of deposits have at least	50% of deposits have at least	10% of deposits have at least
Porphyry copper	Tonnage of ore (millions of tons)	41		20.0	100.0	430.0
	Average copper grade (percent)	41	with tonnage of ore = −0.07 NS	0.1	0.3	0.55
	Average molybdenum grade (percent Mo)	41		0.0	0.008	0.031
Island-arc porphyry copper	Tonnage of ore (millions of tons)	41		20.0	100.0	430.0
	Average copper grade (percent)	41	with tonnage of ore = −0.07 NS	0.1	0.3	0.55
	Average molybdenum grade (percent Mo)	41		0.0	0.008	0.031
	Average gold grade—locally significant but not determined					
Porphyry molybdenum	Tonnage of ore (millions of tons)	31		1.6	24.0	340.0
	Average molybdenum grade (percent Mo)	31	with tonnage of ore = −0.05 NS	0.065	0.13	0.26
Podiform chromite	Tonnage of Cr_2O_3 (tons)	268		15.0	200.0	2,700.0
Copper skarn	Tonnage of ore (millions of tons)	38		0.08	1.4	24.0
	Average copper grade (percent)	38	with tonnage of ore = −0.44**	0.86	1.7	4.5
	Average gold grade—locally significant but not determined					
Mafic volcanogenic	Tonnage of ore (millions of tons)	37		0.24	2.3	22.0
	Average copper grade (percent)	37	with tonnage of ore = −0.13 NS	1.1	2.2	4.1
	Average zinc grade excluding deposits without reported grades (percent)	19	with tonnage of ore = 0.03 NS	0.3	1.3	5.5
	Average gold grade—locally significant but not determined					
Felsic and intermediate volcanogenic massive sulfide	Tonnage of ore (millions of tons)	89		0.19	1.9	18.0
	Average copper grade (percent)	89	with tonnage of ore = −0.41**	0.54	1.70	5.40
	Average zinc grade excluding deposits without reported grades (percent)	41	with tonnage of ore = 0.25 NS	1.40	3.80	10.0
	Average lead grade excluding deposits without reported grades (percent)	14	with tonnage of ore = −0.02 NS	0.20	0.95	4.80
	Tonnage of contained gold excluding deposits without reported gold (tons)	38	with tonnage of ore = 0.78**	0.27	2.90	32.0
	Tonnage of contained silver excluding deposits without reported silver (tons)	46	with tonnage of ore = 0.82**	5.0	80.00	1,300.0
Nickel sulfide	tonnage of ore (millions of tons)	48		0.23	1.20	5.90
	Average nickel grade (percent)	48	with tonnage of ore = −0.03 NS	0.32	0.61	1.20
	Average copper grade (percent)	48	with tonnage of ore = 0.03 NS with nickel grade = 0.04 NS	0.18	0.47	1.20
Mercury	Tonnage of contained mercury (tons)	165		0.09	3.10	120.0
Vein gold	Tonnage of contained gold (tons)	43		0.29	3.30	38.0
Skarn/tactite tungsten	Tonnage of ore (millions of tons)	31		0.024	0.63	17.0
	Average tungsten grade (percent W)	31	with tonnage of ore = −0.34 NS	0.24	0.51	1.10

Notes: Related data occur on line from column to column. All data in metric units. NS = not significant; * = significant at 5% level; ** = significant at 1% level.
Source: Hudson and DeYoung, Jr., 1978 [43].

"mineral potential" of the lands. It was for such a consideration that the U.S. Geological Survey was mandated to appraise the mineral resources of Alaska. As is often the case, the appraisal had to be made quickly, using data, information, and expertise available at that time. Unsurprisingly, the methodology selected was the quantification of judgment of knowledgeable geologists. Specifically, the geologists delineated favorable areas on the basis of either (1) mineral "shows" or deposits, or (2) the presence of geological conditions favorable for the formation of a particular mineral deposit. Then, for each favorable area, the geologist provided subjective estimates of the 5, 50, and 95 percentile values for number of deposits that occur. Besides general geological information, the geologist's analysis was supported by tonnage grade models, such as those presented in Table 2.6.3. These models, which were developed from data on existing deposits, aided the geologist in converting geological conditions to implied deposit occurrence; furthermore, they are needed in economic analyses required to estimate mineral resources or potential supply from the endowment appraisal.

In a subsequent analysis, Poisson and negative binomial models were fitted to the percentile estimates of number of deposits of copper for each favorable area and the means of the best fitting distributions were added across areas [11]. Table 2.6.4 shows the results by mode of occurrence of copper. If each of the 136 porphyry deposits expected to occur had the expected ore tonnage and grade indicated by the tonnage and grade models, these deposits would constitute 81,000,000 tons of copper [11].

Table 2.6.4—Expected Number of Copper Deposits in Alaska

Mode of occurrence	Expected number of deposits
Porphyry	136
Mafic volcanogenic massive sulfide	8
Felsic and intermediate volcanogenic massive sulfide	68
Skarn	16
Contact metamorphic	5

Source: Charles River Associates, Inc., 1978 [11]

Potential Supply of Petroleum—Permian Basin: There have been numerous economic and resource analyses made of domestic oil. Some comments already have been provided regarding the most famous of the "resource estimates," that of M. King Hubbert [42], in which models of production and discovery trends and of discovery rates were used to estimate "ultimate recoverable oil." Another widely cited oil resource appraisal is that performed by the U.S. Geological Survey [48], in which an assortment of methods, including geological analogy, were combined with subjective probability and the lognormal model to estimate oil and gas "resources." Actually, both of these studies provide estimates of potential supply [23,24,30]. Rather than commenting at greater length here on these methods and estimates, a brief description is presented of the estimation of the marginal cost of future oil supply from the Permian Basin made by an interagency (DOE and USGS) analysis [68]. This particular study is relevant to policy analysis not so much for its policy motivation but for the utility of this kind of analysis as a support to policy evaluation.

Detailed analysis of discovery and drilling data on the Permian basin led to the estimation of the parameters of separate exploration process models for each combination of depth interval and field size class [68]:

$$F_{ij}(w) = F_{ij}(\infty)\,[1 - e^{-CAw/B}]$$

where

C = the efficiency of exploration
B = basin area
A = average areal extent of the fields in the given size class and depth interval
w = cumulative number of net wells for the ith size class and j^{th} depth interval
$F_{ij}(\infty)$ = endowment ($w = \infty$) of deposits for the ith size class in the jth depth interval of the basin
$F_{ij}(w)$ = cumulative discoveries of the ith size class in the jth depth interval given cumulative drilling of w.

$$F_{ij}(w) = F_{ij}(\infty)\,[1 - e^{-(2.0)(2.2)\cdot(w/100{,}000)}]$$

Clearly, this model provides conditional discoveries, conditional upon cumulative drilling. But, an increment of drilling occurs only when its expected benefits outweigh expected costs. Thus, use of this model to estimate potential supply requires (1) explicit consideration of the costs of drilling, developing, and producing deposits of each size class and depth, (2) an estimation of surplus net present value, and (3) a relationship of surplus net present value to drilling effort.

The approach in this study was to develop engineering relations which either (1) ascribed physical parameters, e.g, oil gas ratio, important to costing, or (2) estimates for each of a number of cost categories. Table 2.6.5 lists each of the engineering or economic factors and the physical variables by which each factor was estimated. Thus, by employing these relations, each discovery was given values for the physical parameters and costs required for economic analysis. Potential supply from undiscovered endowment was estimated by finding that level of drilling for which the positive surplus of net present value from the discovered fields just offset the cost of exploration [69]. The results of this analysis were summarized by graphs of marginal cost schedules for each of three rates of return (5%, 15%, and 20%). By considering rate of return (r) as a parameter (shift variable in marginal cost), the results can be generalized to the following model of marginal cost of potential supply:

$$\hat{mc}(ps;\, r) = 3.73\, r^{0.339} ps^{0.202} e^{0.0665\ ps^2}$$

where

r = rate of return on capital as a percentage

ps = quantity of petroleum (oil + gas) in billions of barrels

$mc(ps;\, r)$ = marginal cost in dollars per bbl, given a required return of r and a quantity of potential supply of ps.

For example, consider a rate of return of 15% and a potential supply of 3×10^9 bbls:

$$\hat{mc}(3 \times 10^9;\, 15) = 3.73(15)^{0.339} 3^{0.202}\, e^{0.0665(9)} = \$21.21/\text{bbl}$$

Marginal cost increases rapidly as the size of the stock increases; for example, if rate of return remains at 15% but ps is increased from 3 to 5 (billions of barrels), marginal cost triples:

$$\hat{mc}(5 \times 10^9;\, 15) = 3.73(15)^{0.339}\, 5^{0.202}\, e^{0.0665(25)} = \$68.17/\text{bbl}$$

By recasting the data, they can be statistically analyzed to estimate potential supply (ps), in which potential supply is a function of the marginal cost for each of three rates of return:

Table 2.6.5—Engineering and Cost Relations Used to Compute Potential Supply

	Economic or engineering factor	Explanatory variable
1	Average exploratory well depth	Cumulative exploratory wells
2	Exploratory drilling and equipment costs	Well depth
3	Total exploration cost	Drilling and equipment costs
4	Ration of oil fields to total fields	depth, size class
5	Ultimate oil recovery per field	depth, size class
6	Ultimate oil recovery per well from primary oil fields	depth, size class
7	Ultimate associated-dissolved gas recovery per oil well from primary fields	depth, size class
8	Ultimate oil recovery per well from secondary and pressure maintenance fields	depth, size class
9	Ultimate gas recovery per oil well from secondary and pressure maintenance fields	depth, size class
10	Ultimate nonassociated gas recovery per field	depth, size class
11	Ultimate nonassociated gas recovery per well	depth, size class
12	Ratio of primary oil fields to total oil fields	depth, size class
13	Cost of lease equipment per nonassociated gas development well	depth, size class
14	Development drilling and equipment costs per well	well depth
15	Cost of development of dry hole	well depth
16	Exponential oil well decline rates per year	depth, size class
17	Annual direct operating expenses for nonassociated gas wells	depth, size class
18	Lease equipment cost per well for primary oil production	well depth
19	Oil production decline curves, by size class and depth	time (years)
20	Per cent of expected ultimate gas recovery	per cent of expected ultimate oil recovery
21	Annual gas production, by size class and depth	time
22	Annual direct operating costs per producing oil well for primary recovery	well depth
23	Annual direct operating costs per producing oil well for secondary recovery	well depth

Interpreted from USGS Circular 828, 1980 [68]

$$\hat{ps}(mc; r) = 1.0021 \times 10^9 \, e^{0.08749mc_5 - 0.0012415mc_5^2 + 0.058501mc_{15} - -0.00052982mc_{15}^2 + 0.00107155mc_{20}^2}$$

where

$ps(mc; r)$ = potential supply of oil equivalent, given marginal cost of mc and rate of return r

mc_5, mc_{15}, mc_{20} = marginal costs for 5, 15, and 20% rates of return, respectively.

For example, suppose we wish an estimate of the potential supply of petroleum from undiscovered deposits of the Permian Basin for a marginal cost of $30/bbl, given a required return on capital of 15 percent. For this circumstance,

$\mathrm{mc}_5 = mc_5^2 = mc_{20}^2 = 0.0$; $mc_{15} = 30$; $mc_{15}^2 = 900$; $r = 15$; and $\hat{ps}(30; 15) \approx 3.6 \times 10^9$:

$$\hat{ps}(30;15) = 10021 \times 10^9 \, e^{0.058501(30) - 0.00052982(900)} \approx 3.6 \times 10^9$$

RESOURCES AND THE MODELING OF DYNAMIC SUPPLY

Perspective

Dynamic supply, unlike potential supply, is a flow. Thus, in this sense it is like the supply of economic theory; however, as the term is used here, dynamic supply is not identical to conventional short or long run supply concepts, for it is an across-time notion of supply. Specifically, dynamic supply refers to supply in response to price and resource depletion. As known deposits are depleted, they are replaced by new ones which, ceteris paribus, are of lower quality, either in discoverability or in costs of mining and processing. It should be noted that the foregoing definition does not imply that all later-found deposits are of lower grade or smaller size than those found at an earlier date, for supply is dynamic with regard to depletion and resource information. Thus, a high grade deposit may be found later because of the gain in resource and geoscience information as exploration proceeds.

Motivations

In the aftermath of the OPEC oil embargo, policymakers turned to econometric models of various kinds for policy guidance. One particular kind of prediction that was sought was that of domestic supply for a wide range of economic scenarios. Econometric models constructed at that time have received some harsh criticism [47]. Of course, given the lack of an economic history for such high oil prices, poor performance of the econometric models at that time is not too surprising. Basically, the domestic and world economies experienced an economic perturbation of major dimension. *A priori*, responses of economic systems ex post this perturbation would not be well explained by economic relations implied by historical economic data and conventional models.

Search for improvements or alternative approaches led in different directions, one of which is the construction of a resources-based dynamic supply model. The foundation of such a system on an inventory of endowment and upon engineering cost models seems to offer a means to examining variations in factor and product prices greater than those of our experience. Motivations other than the foregoing for the construction of a resource-based dynamic supply system may include (1) the identification of the impact of a new technology on future supply and prices, and (2) the examination of depletion and the early recognition of scarcity effects. The central idea is that a dynamic supply system that is built upon a description of mineral endowment provides both flexibility and constraints through an explicit modeling of endowment and engineering processes. Of course, achievement of such capability is attended by various costs, chief among them being the complexities of highly disaggregated models and the difficulty of economic coordination, the linking of activities, e.g., exploration, to more aggregate economic relations, e.g., supply.

Approach

Previous sections established the concept of a potential supply system, which consists of endowment, exploration, and exploitation components, as a means of estimating potential supply. Given the completion of such a system, interest flows naturally to its modification for the purpose of estimating future flows of mineral supply, for in concept, future flows must at some point derive from the deposits that make up endowment. Furthermore, future flows will occur in the short run only if production is economic and in the long run only if *exploration and production* are economic. Since the potential supply system contains submodels for exploration and exploitation, it appears to provide the essential components for the analysis of dynamic supply. However, there is much more involved than the system's integration of these components in estimating dynamic supply. In particular, dynamic supply, being a flow, cannot be estimated without some treatment of future demands. Consider Fig. 2.6.15 and 16, which are stylized, simplistic descriptions of potential and dynamic supply models.

As indicated in Fig. 2.6.15, in potential supply estimation, the size of the stock for each of k input prices is estimated. Notice that for such analysis, there are no sunk costs and *all* decisions made for the previous specified price are *reversible*, i.e., they are remade at the higher price. Such provisions of a potential supply system contrast sharply with those of a dynamic supply system in which production decisions made at one price are not reversible, i.e., remade at a higher price. Furthermore, in dynamic supply, price is determined endogenously. As shown in Fig. 2.6.16, which is a simplified and stylized description of a dynamic supply system, prices of future periods are resolutions of supplies and demands, giving rise to time paths of future price and future market clearing quantities. Clearly, modeling of dynamic supply requires as input projected demand functions for the relevant future time periods and the wedding of market interactions with exploration and mining activities at the deposit level.

A complete potential supply system consists of endowment, exploration, and exploitation submodels and of an economic framework that provides for exploration optimization and the use of profit maximization, given an exogenous price, in mine development. Thus, it is replete with cost functions parameterized on deposit characteristics and with a cost accounting and discounted cash flow model. Even so, the step from estimation of potential supply to estimation of dynamic supply is a very great one in terms of modeling difficulty, even when the model is purposefully simplified.

Fig. 2.6.16 shows in very general and stylized terms the major components of a dynamic supply system. Those components of the system that are required for dynamic supply but not for potential supply are marked with an asterisk.

POTENTIAL SUPPLY

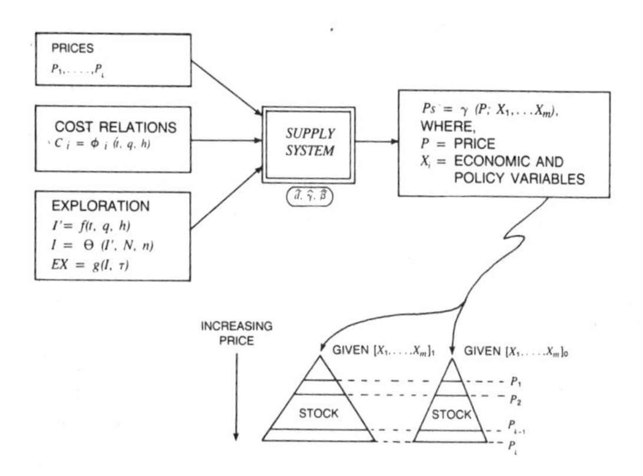

Fig. 2.6.15—Stylized description of use of supply system to estimate potential supply (stock). Source: Harris and Chavez, 1983 [35].

x_j, $j = 1, \ldots, m$ = economic and policy variables other than product price;
P_i, $i = 1, \ldots, k$ = k prices of U_3O_8;
t, q, h = tonnage and average grade of deposit, and depth to deposit, respectively;
I', I = discoverability index of deposit in isolation and in association with $N - n$ other deposits after the discovery of n deposits, respectively;
EX = exploration expenditure.

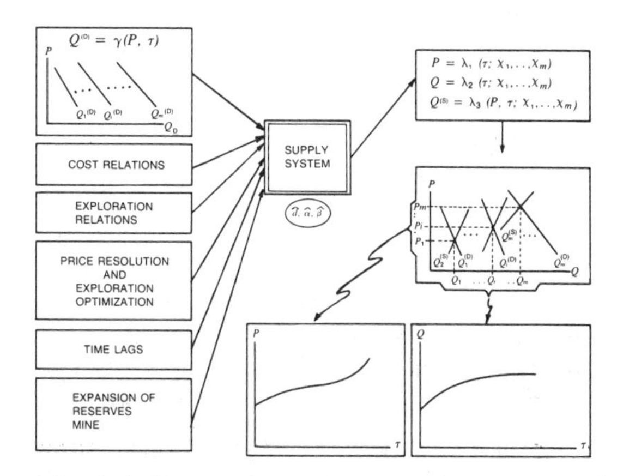

Fig. 2.6.16—Stylized description of a dynamic supply system, one that estimated time series of future price and of market clearing quantities. Source: Harris and Chavez. 1983 [35].

The identity of these components is hardly surprising, for they are indicated by the economics of mineral deposit discovery and exploitation. Even so, constructing a computer system that integrates these components within an economic framework is a challenging task.

Consider, for example, the economics of a shift of the demand curve from one period for which the system is at equilibrium to higher levels for future years. There are three kinds of supply responses. One is an increase in production levels of existing mines as producers maximize profits. While this is straightforward, for a dynamic supply system to have such capability requires that each producing mine be represented by a marginal cost function, allowing the identification for each mine of its profit-maximizing production level. In actual practice, these marginal cost schedules (functions) usually are not available. As a consequence, either the effect of the price increase on short run supply would be ignored, i.e., mines produce at minimum average cost provided price exceeds average variable costs, or it would be approximated. One means of approximation is the construction of production functions, one for each of several classes of deposit sizes, grades, and depths. Marginal costs could be estimated from these production functions and described parametrically by a response function which could be used to compute a percentage increase in production for mines of each class. The dynamic supply model would increase the production of each mine as indicated by the depth, size, and grade of its ore body. While such an approach could never be more than a rough approximation, if information permitted the construction of credible production functions, such an approach would be better than ignoring this contribution to supply.

A second supply response is the increase in ore reserves induced, ceteris paribus, by increased price. Such an increase may be accompanied by an expansion of existing production capacity if long term expectations are for the persistence of higher prices. Modeling this supply response in a credible, but simple, fashion requires at the very minimum the following:

(1) that the deposit of each producing or "shut down" mine be defined by its total mineralized material, its average grade, and by its intra-deposit grade variance,
(2) the cumulative tonnage of produced ore,
(3) the production capacity of existing plant,
(4) an expectational relationship for price and levels of demand, and
(5) time lags in the creation of new capacity.

Expansion of production capacity requires, of course, that these components be integrated appropriately to represent the across-time allocation of capital and maximization of resource rents. For depletable mineral resources, such optimization in principle requires the consideration of user cost. Of course, to the extent that (1) technologic change is provided for, (2) capacity expansion is based upon price expectations, and (3) magnitude and quality of remaining endowment is uncertain, user cost may be very small.

Finally, if long term expectations are for higher prices relative to costs even when allowance has been made for reserve and capacity expansions, supply response may include the finding of a deposit and the development of a totally new production facility. Representing this response credibly requires honoring (1) the time lags, risk, cost, and efficiency of exploration, and (2) the "shelf deposits." Typically, the conduct of exploration produces some discoveries that are not economic for the current and expected prices. Some of these deposits are held in inventory (shelf deposits) until economic conditions improve, making them profitable to develop and produce. Sometimes the long term response to expectations for improved price is dominated by the development of "shelf deposits." Consider the following statement made in 1978 about uranium supply by a recognized exploration authority [4]:

> The major effort of the U.S. uranium exploration industry over the past few years, because of the recent change in uranium prices, has been directed toward inventorying existing deposits and discoveries. This has resulted in very few new uranium "finds" and the mistaken impression of inactivity on the part of the explorationist. The fact that is sometimes overlooked is that these same increased prices have prompted the initiation of many conceptual exploration programs that, aided with recent advances in exploratory procedure, should begin showing substantial additions to the total uranium resources of the U.S. in the near future.

The suggestion here is that the first response of a firm considering the creation of new production capacity is examination of the shelf deposits for one or more deposits that could be profitably exploited, given demand expectations. If there were no such deposit in inventory, then new exploration would be initiated. Of course, one outcome of exploration may be another subeconomic deposit to put on the shelf; in this case both capital and time would have been expended without achieving the objective. Exploration would be continued until either new capacity expectation from discoveries are sufficient or until the expectations for size, grade, and depth are such that expected rent does not justify the next increment of exploration effort. Clearly, there are two dimensions of expectations involved: one is for future demand and the other is for characteristics of remaining endowment.

The point to be made here is that turning to highly disaggregated, endowment-based systems to estimate dynamic supply carries the modeler into relatively uncharted territory, one dealing with uncertain endowment characteristics, expectational relations, across-time allocation, and optimization of activities. Representing these credibly in a dynamic supply system is a challenging task. While a dynamic supply system may provide quantitative economic descriptions not otherwise credible, this modeling form poses some major challenges to systems design and to data to support the specification and estimation of all required relations.

Some Selected Comments on CRUSS—An Unfinished Example

Relevant Circumstances: CRUSS is an acronym for crustal abundance uranium supply system. This system initially was designed as a potential supply system [12,34,35,36]. Subsequent interests motivated its extension to estimate dynamic supply. This work is still in progress. Basing dynamic supply upon crustal abundance raises a number of modeling problems [35,36]. Some of these are generic with regard to dynamic supply; others are unique to CRUSS. A thorough description of the system here is not appropriate; for such see Harris et al [34]. What follows are some selected comments and descriptions.

Criticisms of U.S. DOE estimates of potential uranium supply, made during the execution of NURE, included concern that U.S. policy on nuclear energy could be too heavily influenced by the few DOE geologists who estimated the potential for undiscovered deposits in the primary resource regions: New Mexico, Wyoming, and Texas. In response to such concern, DOE funded research on the construction of a crustal abundance supply system (CRUSS) to generate a "second look," one made by a different approach.

System's Structure: Of the alternative approaches considered at that time, a system based upon a geostatistical crustal abundance model was especially attractive because of the popularity of the crustal abundance approach in Europe, especially the use of Brinck's [7] model by EURATOM, the European equivalent to the previous U.S. Atomic Energy Commission. Some of the concepts of Brinck's approach are used to describe uranium endowment, as shown in Fig. 2.6.17, which identifies the major components of CRUSS. What isn't revealed by this figure is the communication of these components through an economic structure and the

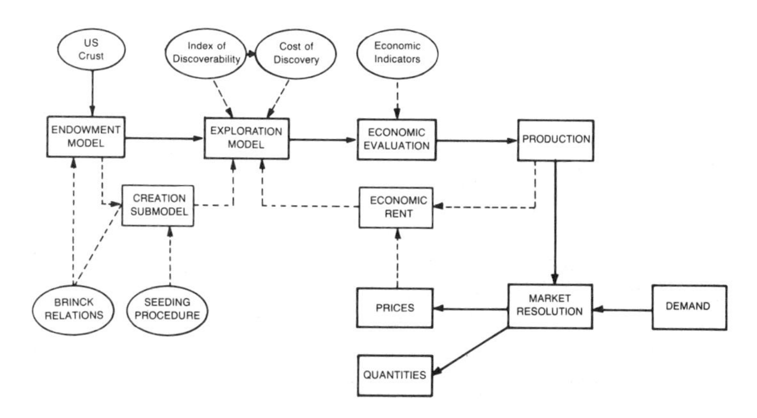

Fig. 2.6.17—Simplified schematic diagram of a supply system based on a crustal abundance endowment model (CRUSS).

rather extensive system which accounts for production by mine and in aggregate as supply flows.

While predicating dynamic supply upon crustal abundance is not new in concept, the construction of such a system does present new modeling problems, the solutions of which require innovative measures. One such problem is extension of the crustal abundance notion to include deposit morphology. Brinck's approach to crustal abundance modeling employs a binomial model to describe frequencies of grade for a given deposit size. Unfortunately, size, a deposit parameter that is an important cost determination, is not intrinsic to the model. Consequently, the endowment component of CRUSS augments Brinck's binomial model to give deposits an intrinsic size, grade, depth, and intradeposit grade variance. Specifically, the endowment submodel of CRUSS employs Brinck's binomial crustal abundance relations to partition the crust of the United States into exploration regions and into geologic deposits. Consider W to be the weight of the earth's crust

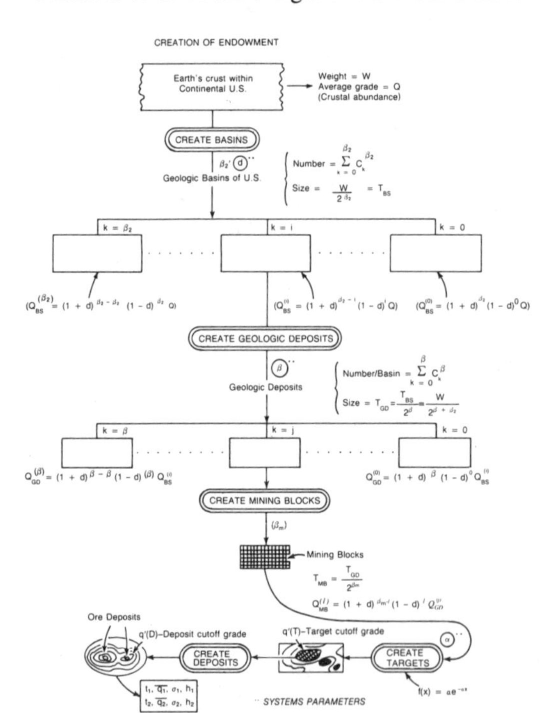

Fig. 2.6.18—Schematic representation of the use of a binomial crustal abundance (Q) model (parameters: W, Q, β, and d) and a clustering function (parameter: α) to create exploration targets and ore deposits that are functions of cutoff grade. Source: Harris and Chavez, 1984 [36].

to a specified depth (h): $W = A \cdot h/F$, where A is area, h is depth, and F is the number of tons per cubic foot of crustal material. Then, W can be partitioned into 2^β smaller regions, each of size T_{BS}: $T_{BS} = \frac{W}{2^\beta}$. Given that the crustal abundance for material comprising W is known to be $\overline{Q}$, the grade, q, of each region is described by the binomial law [5, 7]:

$$q = \overline{Q}(1 + d)^{\beta - k}(1 - d)^k$$

where d is an enrichment factor and constitutes a parameter that can be estimated from sample data that describe grade variation.

For given values of parameters $\overline{Q}$, d, and β, grade (q) is a function of k. Consequently, by allowing k to take on all integer values from 0 to β, a distribution of grades is defined, the richest and poorest being $\overline{Q}(1 + d)^\beta$ and $\overline{Q}(1 - d)^\beta$, respectively. Since each grade is associated with a specific k, the number of regions having a grade is simply the binomial coefficient $C_k^\beta = \frac{\beta!}{(\beta - k)!\,k!}$. Of course, $\sum_{k=0}^{\beta} C_k^\beta = 2^\beta$, the total number of regions.

Fig. 2.6.18 shows how relations of the binomial crustal abundance model are used first to create regions and then a second time to partition the regions into smaller volumes referred to as geologic deposits. The final act in creating the mineral endowment of a region is the subdivision of the geologic block into mining blocks, which are then seeded (rearranged) to form mineral deposits, each being given a size, t, an average grade, $\overline{q}$, a depth, and an intradeposit grade variance.

Parameter Estimation and System Design: As noted in Fig. 2.6.18, the endowment submodal requires four parameters: α, Q, d, and β (β_2 and β_m are defined by specifying the desired region and mining block sizes). Of the four parameters, one, d, has been estimated [30] exterior to the system, and one of the available measures of crustal abundance can be taken as Q. The remaining parameters (α and β) are appropriately referred to as systems parameters, for they are estimated by using the dynamic supply system as a search algorithm; specifically, numerical values for α and β are searched for which make the output of the system (ura-

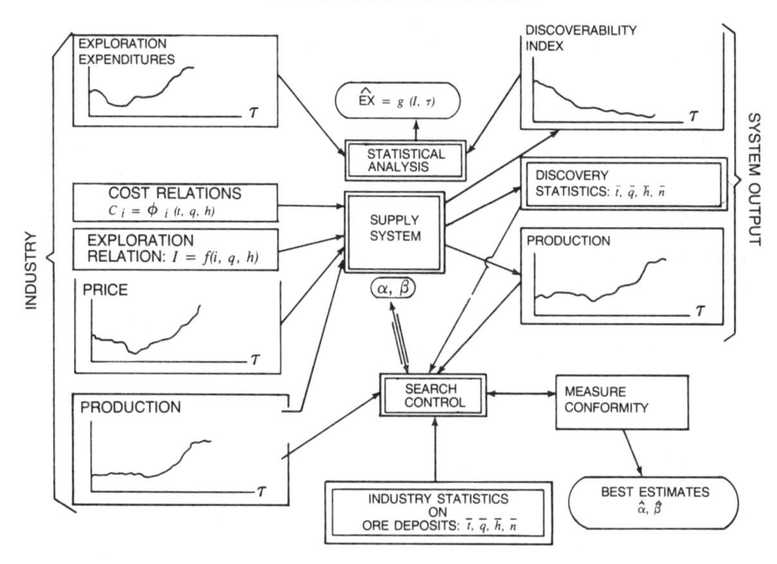

Fig. 2.6.19—Stylized Use of Supply System and Time Series Data to Estimate Parameters (α, β) of the System—asterisks denote those items that either control the search or are products of the search. Sources: Harris and Chavez, 1983 [35].

nium production) match closely historical time series production data.

This approach to parameter estimation reflects the perception that the parameters for which uncertainty is the greatest and available data the least useful are those that (1) define the size and average grade of the geologic deposit (not ore deposit) and the grade variation within the deposit, and (2) the relationship of exploration effort to deposit characteristics, discovery depletion, and information gain. Available statistical data are for the ore reserves of mining properties. As indicated in a previous section (Contamination of Geologic Data by Economics), these data do not describe just the characteristics of a physical-chemical phenomenon, referred to as geologic deposit, but rather a mix of effects, such as the geologic deposit, economics of mining, and competition among firms for land positions during exploration. The statistical data on ore reserves—along with the time series of production data—also were used in the estimation of systems parameters (α, β). In principle, the sizes of the ore bodies developed from geologic deposits by the system were compared to statistical data on ore reserves. In this way, geologic deposits of the model have those physical characteristics which through exploration and exploitation economics yield ore reserves similar to those reported by industry. Thus, the estimation of systems parameters (α and β) was based upon comparing time series and statistical data of the economy with those of the dynamic supply system. Fig. 2.6.19 is a schematic representation of parameter estimation.

Other Components: Estimation of systems parameter in the manner described places great importance on the modeling of exploration and exploitation. Accordingly, the exploitation submodel of CRUSS contains detailed cost relations and grade-tonnage relations coupled with an algorithm that imitates mine design and profit maximization. Discoveries are ordered by a discoverability index that is a function of (1) deposit size, grade, and depth, (2) information gain, and (3) discovery depletion—see section on Geologic and Technologic Determinants of Long Term Relative Cost/Price for a brief description. This exploration subprogram imitates the allocation of effort both within a region and across regions by readjusting discoverability indexes of deposits in response to progress of exploration and attendant discovery depletion and information gain. While the exploration submodel provides a useful ordering of discoveries, it does not of itself define the level of exploration expenditure. This relationship is defined by the system once parameters α and β have been estimated, as indicated in Fig. 2.6.18.

Use of the System: Once the systems parameters have been estimated, the system can be used to estimate dynamic supply. In the process of estimating systems parameters, the system creates a state of supply components and activities for the most recent time period for which data were available. This is the model's best representation of the uranium endowment and the industry's current supply activities. The system contains producing mines at various stages of depletion. Cumulative production, remaining life, costs, and current flow for each mine are documented by the system. The system has a current inventory of "shelf deposits," previous discoveries that were subeconomic. Finally, the system defines the uranium deposits of each region that have not yet been discovered. Consequently, given future demand schedules or functions, CRUSS proceeds from this current, base state and simultaneously simulates supply activities and market resolution to generate across future time an estimated time profile for price and quantity of U_3O_8.

Important Unfinished Modeling Tasks: The foregoing is a description of design objective. At present, this objective has not yet been fully achieved. Besides a number of minor technical modeling problems to overcome, the major substantive features of the system that remain to be defined are the expectational relationships for price and demand and their coordination with the timing of exploration, mine expansion, and new mine development. Although unfinished, this system serves to identify issues and problems in the design and use of an endowment-based system to estimate dynamic supply. While the existence of problems is apparent, this kind of supply modeling is new, and given continued improvements in methods of "resource appraisal," and in the modeling of exploration and exploitation, it may provide a useful approach to the examination of some issues in mineral supply.

References

1. Agterberg, F. P., and Divi, S. R., 1978, "A Statistical Model for the Distribution of Copper, Lead, and Zinc in the Canadian Appalachian Region," *Econ. Geol.*, Vol 73, pp. 230–45.
2. Barry, G. S., and Freyman, A. J., 1970, "Mineral Endowment of the Canadian Northwest," *Mineral Information Bulletin*, MR 105, Mineral Resources Branch, Dept. of Energy, Mines and Resources Canada, Ottawa, Canada, pp. 57–95.
3. Bennett, H. J., Moore, L., Welborn, L. E., and Toland, J. E., 1973, *An Economic Appraisal of the Supply of Copper from Primary Domestic Sources*, Information Circular 8598. U.S. Bureau of Mines, Washington, D.C.
4. Bonner, J., 1978, *The Changing U.S. Uranium Exploration Industry*. Topical Symposium on Uranium Resources sponsored by American Nuclear Society and U.S. Dept. of Energy, Sept. 10–13, 1978, Las Vegas, Nevada.
5. Brinck, J. W., 1971, "MIMIC, the Prediction of Mineral Resources and Long-term Price Trends in the Nonferrous Metal Mining Industry is No Longer Utopian," *Eurospectra*, Vol. 10, pp. 46–56.
6. Brinck, J. W., 1972, "Prediction of Mineral Resources and Long-term Price Trends in the Nonferrous Metal Mining Industry," in *Section 4-Mineral Deposits, Twenty-Fourth Session International Geological Congress, Montreal*, pp. 3–15, 24th International Geological Congress, Ottawa, Canada.
7. Brinck, J. W., 1974, *The Geochemical Distribution of Uranium as a Primary Criterion for the Formation of Ore Deposits*. Symposium on the Formation of Uranium Deposits, International Atomic Energy Agency.
8. Cargill, S. M., Root, D. H., and Bailey, E. H., 1980, "Resource Estimation from Historical Data: Mercury, a Test Case," *J. Intl. Assn. Math. Geol.*, Vol. 12, No. 5, pp. 489–522.
9. Carrigan, F. J., 1983, *Computer-Assisted Decision Aid for the Estimation of Mineral Endowment: Uranium in the San Juan Basin, New Mexico, A Case Study*, Ph.D. Dissertation, Univeristy of Arizona, Tucson.
10. Cathro, R. J., 1983, Invited discussions (Predictive Metallogeny Symposium). *Geoscience Canada*, Vol. 10, No. 2, June 1983, p. 100.
11. Charles River Associates, Inc., 1978, *The Economics and Geology of Mineral Supply: An Integrated Framework for Long-Run Policy Analysis*, CRA Report No. 327, November 1978.
12. Chavez-Martinez, M. L., 1983, *A Potential Supply System for Uranium Based Upon a Crustal Abundance Model*, Ph.D. Dissertation, University of Arizona, Tucson.
13. Committee on Nuclear and Alternative Energy Systems (CONAES), National Research Council, 1978, *Problems of U.S. Uranium Resources and Supply to the Year 2010*, National Academy of Sciences, Washington, D.C.
14. Dale, L., 1983, "The Development of Copper Deposits in the United States," *Materials and Society*, Vol. 7, No. 2, pp. 183–199.
15. Deffeyes, K. S., and MacGregor, I.D., 1980, "World Uranium Resources." *Sci. Am.*, Vol. 242, No. 1, pp. 66–76.
16. Drew, L. J., Schuenemeyer, J. H., and Bawiec, W. J., 1982, *Estimation of the Future Rates of Oil and Gas Discoveries in the Western Gulf of Mexico*. U.S. Geological Survey Professional Paper 1252.
17. Drew, L. J., Schuenemeyer, J. H., and Root, D. H., 1980, *Petroleum-Resource Appraisal and Discovery Rate Forecasting in Partially Explored Regions—An Application to the Denver Basin*. U.S. Geological Survey Professional Paper 1138-A.
18. Drew, M. W., 1977, "U.S. Uranium Deposits: a Geostatistical Model," *Resources Policy*, Vol. 3, No. 1, pp. 60–70.
19. Eberlein, G. D., and Menzie, W. D., 1978, "Maps and Tables Describing Areas of Metalliferous Mineral Resource Potential of Central Alaska," USGS Open File Report 78-1-D, Menlo Park, Calif.
20. Gaschnig, J., 1980, *Development of Uranium Exploration Models for the Prospector Consultant System*, Final Report, SRI Project 7856 (March 1980), Stanford Research Institute International, Menlo Park, Calif.

21. Grybeck, D., and DeYoung, J. H., Jr., 1978, "Map and Tables Describing Mineral Resource Potential of the Brooks Range, Alaska," USGS Open File Report 78-1-B, Menlo Park, Calif.
22. Harris, D. P., 1968, "Alaska's Base and Precious Metals' Resources: a Probabilistic Regional Appraisal, *Mineral Resources of Northern Alaska*, MIRL Report No. 16, Mineral Industries Research Laboratory, University of Alaska, pp. 189–224.
23. Harris, D. P., 1977a, *Mineral Endowment, Resources, and Potential Supply: Theory, Methods for Appraisal, and Case Studies*, MINRESCO, Tucson, Ariz., January 1, 1977.
24. Harris, D. P., 1977b, "Conventional Crude Oil Resources of the United States: Recent Estimates, Methods for Estimation and Policy Considerations," *Materials and Society*, Vol. 1, pp. 263–86.
25. Harris, D. P., 1978, "Undiscovered Uranium Resources and Potential Supply," In *Workshop on Concepts of Uranium Resources and Producibility*, National Research Council, National Academy of Sciences, pp. 51–81.
26. Harris, D. P., 1980, *Critique of Resource Estimation Methodology*, Report prepared for U.S. Department of Energy under Subcontract No. 80-469-S, Grand Junction Office, Colorado, July 3, 1980.
27. Harris, D. P., 1981, *Multivariate Geostatistical Analysis for the Prediction of Mineral Occurrence*, Paper presented at the Workshop on Interactive Graphic Computer Programs Preceding Tenth Geochautauqua on Computer Applications in the Earth Sciences, Geological Survey of Canada, Ottawa, Canada, October 20–22, 1981. Unpublished.
28. Harris, D. P., 1983a, *Mineral Resources Appraisal and Policy—Controversies, Issues, and the Future*, Prepared for Conference on the Role of Earth Sciences Information in the Mineral Policymaking Process, Carnegie Institution of Washington, May 1983.
29. Harris, D. P., 1983b, "An Investigation of the Estimation Process of Predictive Metallogeny (Predictive Metallogeny Symposium), *Geoscience Canada*, Vol. 10, No. 2, June 1983, pp. 82–96.
30. Harris, D. P., 1984, *Mineral Resources Appraisal—Mineral Endowment, Resources, and Potential Supply: Concepts, Methods, and Cases*, Oxford University Press.
31. Harris, D. P., and Agterberg, F. P., 1981, "The Appraisal of Mineral Resources," *Econ. Geol. 75th Anniversary Volume*, pp. 897–938.
32. Harris, D. P., and Carrigan, F. J., 1980, *A Probabilistic Endowment Appraisal System Based Upon the Formalization of Geological Decisions—Final Report: Demonstration and Comparative Analysis of Estimates and Methods*, Open File Report No. GJBX-383(81). Prepared for U.S. Department of Energy, Grand Junction Office, Colorado, Dec. 1980.
33. Harris, D. P. and Carrigan, F. J., 1981, Estimation of Uranium Endowment by Subjective Geological Analysis—a Comparison of Methods and Estimates for the San Juan Basin, New Mexico, *Econ. Geol.*, Vol. 76, pp. 1032–55.
34. Harris, D. P., and Chavez, M. L., 1981, "Crustal Abundance and a Potential Supply System," *Part II, Systems and Economics for the Estimation of Uranium Potential Supply*. Report prepared for U.S. Department of Energy, Grand Junction Office, Colorado, Subcontract No. 78-238-E, July 1981, p. 385–506.
35. Harris, D. P. and Chavez, M. L., 1983, *Modeling Dynamic Supply of Uranium—An Experiment in the Integration of Economics, Geology, and Engineering*. Paper presented at 1983 Convention of Allied Social Science Associations (ASSA), Joint AEA/AERE Session on Modeling Resource Supply, San Francisco, Calif., Dec. 27–30, 1983.
36. Harris, D. P., and Chavez, M. L., 1984, Modelling Dynamic Supply of Uranium—an Experiment in the Integration of Economics, Geology, and Engineering," *18th International Symposium on Application of Computers and Mathematics in the Mineral Industries* (March 26–30, 1984), Institution of Mining and Metallurgy, London, pp. 817–92.
37. Harris, D. P., and Euresty, D. E., 1973, "The Impact of Transportation Network upon the Potential Supply of Base and Precious Metals from Sonora, Mexico," *Proc. 10th International Symposium on Application of Computer Methods in the Mineral Industry*, The South African Institute of Mining and Metallurgy, Johannesburg, pp. 99–108.
38. Harris, D. P., and Ortiz-Vértiz, S. R., 1981. "Potential Supply Systems Based upon the Simulation of Sequential Exploration and Economic Decisions—Systems Designed for the Analysis of NURE endowment," *Part I, Systems and Economics for the Estimation of Uranium Potential Supply*. Report prepared for U.S. Department of Energy, Grand Junction Office, Colorado, Subcontract No. 78-238-E, July 1981, p. 9–384.
39. Harris, D. P., and Skinner, B. J., 1982, "The Assessment of Long-Term Supplies of Minerals," In (V. K. Smith and J. V. Krutilla, eds.) *Explorations in Natural Resource Economics*, The Johns Hopkins University Press, Baltimore, MD (For Resources for the Future, Inc.), pp. 247–326.
40. Harris, D. P., Freyman, A. J., and Barry, G. S., 1971, "A Mineral Resource Appraisal of the Canadian Northwest Using Subjective Probabilities and Geological Opinion," *Proceedings of 9th International Symposium on Techniques for Decision-Making in the Mineral Industry* (June 14–19, 1970), Special Volume 12, Canadian Institute of Mining and Metallurgy, Montreal, pp. 100–16.
41. Hetland, D. L., 1979, "Estimation of Undiscovered Uranium Resources by U.S. ERDA," In *Evaluation of Uranium Resources*, Proceedings of an Advisory Group Meeting, International Atomic Energy Agency, Vienna, pp. 231–50.
42. Hubbert, M. K., 1969, "Energy Resources," In *Resources and Man* (A study and recommendations by the Committee on Resources and Man, National Academy of Sciences-National Research Council), Chapter 8, pp. 157–242, W. H. Freeman & Company, San Francisco (For National Academy of Sciences).
43. Hudson, T., and DeYoung, J. H., Jr., 1978, "Map and Tables Describing Areas of Mineral Resource Potential, Seward Peninsula, Alaska," USGS Open File Report 78-1-C, Menlo Park, Calif.
44. Klemme, H. D., 1977, "World Oil and Gas Reserves from Giant Fields and Petroleum Basins (Provinces)," In (R. F. Meyer, ed) *The Future Supply of Nature-Made Petroleum and Gas*, Pergamon Press, New York, pp. 217–70.
45. Lieberman, M. A., 1976, "United States Uranium Resources—an Analysis of Historical Data," *Science*, Vol. 192, No. 4238, pp. 431–36.

46. MacKevett, E. M., Jr., Singer, D. A., and Holloway, C. D., 1978, "Map and Tables Describing Metalliferous Mineral Resource Potential of Southern Alaska," USGS Open File Report 78-1-E, Menlo Park, Calif.
47. Mayer, L. S., 1977, *The Value of the Econometric Approach to Forecasting Our Energy Future*, Paper presented at the International Conference on Energy Management, Tucson, Ariz., October 1977.
48. Miller, B. M., Thomsen, H. L., Dolton, G. L., Coury, A. B., Hendricks, T. A., Lennartz, F. E., Powers, R. B., Sable, E. G., and Varnes, K. L., 1975, *Geological Estimates of Undiscovered Recoverable Oil and Gas Resources in the United States*, U.S. Geological Survey Circular 725.
49. Paley, W. S., 1952, *Resources for Freedom*, A Report to The President by The President's Materials Policy Commission, William S. Paley, Chairman, June, 1952.
50. Patton, W. W., Jr., 1978, "Maps and Table Describing Areas of Interest for Oil and Gas in Central Alaska," USGS Open File Report 78-1-F, Menlo Park, Calif.
51. Peterson, G., Davidoff, R., Bleiwas, D., and Fantel, R., 1981, *Alumina Availability—Domestic: A Minerals Availability System Appraisal*. U.S. Bureau of Mines, Washington, D.C., NTIS PB82-135468.
52. Phillips, W. G. B., and Edwards, D. P., 1976, "Metal Prices as a Function of Ore Grade," *Resources Policy*, September 1976, pp. 167–178.
53. *Ibid*, p. 172.
54. Riva, J. P., Jr., 1983, *Worldwide Petroleum Resources and Reserves*, Westview Press, Boulder, Colo.
55. Roberts, F., and Torrens, I., 1974, "Analysis of the Life Cycle of Nonferrous Minerals," *Resources Policy*, Vol. 1, No. 1, (September 1974), pp. 14–28.
56. Rona, P. A., 1983, "Potential Mineral and Energy Resources at Submerged Plate Boundaries," *Mineral Resources Forum*, Vol. 7, No. 4, October 1983, pp. 329–38.
57. Schmidt, R. A., 1979, *Coal In America: An Encyclopedia of Reserves, Production and Use*. McGraw-Hill, New York, NY
58. Singer, D. A., Cox, D. P., and Drew, L. J., 1975, *Grade and Tonnage Relationships Among Copper Deposits*, USGS Professional Paper 907-A, p. A1–11.
59. Slovic, P., 1972, From Shakespeare to Simon: Speculations—and Some Evidence—about Man's Ability to Process Information," *Oregon Research Institute Monograph*, Vol. 12, No. 12, April 1972.
60. Sousa, L. J., 1981, *The U.S. Copper Industry: Problems, Issues, and Outlook*, U.S. Bureau of Mines, October 1981.
61. Tversky, A., and Kahneman, D., 1972, "Anchoring and Calibration in the Assessment of Uncertain Quantities," *Oregon Research Institute Research Bulletin*, Vol. 12, No. 5.
62. Tversky, A. and Kahneman, D., 1974, "Judgment under Uncertainty: Heuristics and Biases," *Science*, Vol. 185, No. 4157, pp. 1124–31, 1974.
63. Uri, N. D., 1980, "Crude Oil Resource Appraisal in the United States," *The Energy Journal* (IAEE), Vol. 1, No. 3, pp. 65–74.
64. U.S. Bureau of Mines, 1978, *Minerals Availability System Deposit Information Manual*, July 1978. Unpublished.
65. U.S. Department of Energy, 1978, *Statistical Data of the Uranium Industry*, GJO-100(78), Grand Junction Office, Colo., pp. 8.
66. *Ibid*, p. 18.
67. U.S. Department of Energy, 1980, *An Assessment Report on Uranium in the United States of America*, Report No. GJO-111(80), Grand Junction Office, Colo.
68. U.S. Geological Survey, 1980, *Future Supply of Oil and Gas from the Permian Basin of West Texas and Southeastern New Mexico*, U.S. Geological Survey Circular 828.
69. *Ibid*, p. 40.
70. Vogely, W. A., 1983, "Estimation of Potential Mineral Reserves, and Public Policy," *Earth and Mineral Sciences*, Vol. 52, No. 2, The Pennsylvania State University, College of Earth and Mineral Sciences, University Park, Pa.
71. Williams, R., 1982, *Uranium Production Capability—Concepts and Procedures*, A Seminar Presentation to Uranium Resource Appraisal Group, Energy, Mines and Resources Canada, Ottawa, Canada, April 20, 1982.
72. Zimmerman, M. B., 1981. *The U.S. Coal Industry: The Economics of Public Choice*, The MIT Press, Cambridge, MA.
73. *Ibid*, p. 29 and 31.
74. *Ibid*, p. 35.
75. *Ibid*, p. 40.
76. *Ibid*, p. 38.
77. *Ibid*, p. 29.
78. Zwartendyk, J., 1982, "Monitoring and Assessing the Mineral Supply Process from Mineral Policy Formulation: the Role of Scientific and Technical Knowledge," *Proceedings of the Tenth CRS Policy Discussion Seminar*, June 22–24, 1982, Centre for Resource Studies, Kingston, Ont., Canada.

Part 3

MINERAL INDUSTRY ANALYSIS

3.7

Mineral Investment and Finance

Alfred Petrick, Jr.*

INTRODUCTION

The scope of mineral finance extends beyond the economic aspects of property evaluation to all of those activities involving acquisition and utilization of funds with an objective to optimize the value of the firm. It must, therefore, encompass consideration of the overall financial goals of the firm, decisions involving external versus internal growth and domestic versus foreign investments, management of working capital, efforts to minimize the cost of capital, and decisions with regards to the instruments of long-term debt and equity financing. The scope of this section is limited to considerations of special interest to the mineral or petroleum engineer, which include the following aspects of mineral finance:

(1) The technical aspects of mineral or coal property or oil and gas reservoir evaluation which provide the basis for estimation and forecasting of revenues and costs.

(2) Estimation and forecasting of costs and prices in a inflationary economic environment.

(3) Calculation of cash flows which model the investment in minerals, coal or oil, and gas. Such models require assumptions as to the timing of cash flows, escalation of capital and operating costs, base prices, and escalation, the tax regime, royalties, division of profits and the economic aspects of loans.

(4) The role of discount rate in the economic analysis of an investment which requires consideration of the opportunity costs of the firm, cost of capital, risk premiums.

(5) Risk analysis including application of the concepts of conditional net present value, expected value, and risk preference as well as the application of the techniques of sensitivity and simulation.

(6) Selection of application criteria for investment decision making.

(7) Capital structure with particular reference to the effect of debt financing.

(8) Consideration of sources of funds utilized by the minerals industries.

(9) Long-range planning including portfolio analysis.

This section is limited to principles, methodologies, and trends in mineral investment, finance, and long-range planning in the United States with a perspective from the year 1984. Consideration of the key concepts is supplemented by case examples of investment risk analysis and financial structure.

Important factors affecting mineral investment evaluation during the 1980s include volatile metals prices, cost inflation, and the introduction of techniques requiring the computer and its capacity for improved technical and economic analysis into the hands of those who know most about the industry. The personal computer is replacing the calculator on the desk of the individual analyst and is rapidly becoming a required tool for the study of engineering and economics. Examples of change brought about by the micro-computer include greatly improved techniques of cost estimation, wider application of the computer for project analysis, computer graphics, reserve estimation, geostatistical analysis, mine design and other techniques that capitalize on the ability of the computer to handle

*Director, Petrick Associates, Evergreen, Colorado.

large volumes of data in a short period of time. The micro-computer also allows new techniques which consider investment criteria beyond net present value and rate of return; these techniques often require introduction of probabilities, utilization of ranking techniques, and consideration of subjective variables not quantified in the past.

Accepted industry standards for appraisal of mineral, coal, and oil and gas properties are divided into three major components:

(1) A technical evaluation includes development of an extraction and processing plan and a transport concept for the product. The extraction plans and processing flow sheets provide a basis for estimates of the equipment, manpower, and materials required for development and production. Technical evaluation of geological conditions and physical characteristics of the reservoir or deposit provide information for the calculation of reserves, the quality of the product, and identification of potential extractive or processing problems. Reserves must be sufficient to support production at an assumed production rate over the life of the project.

(2) An economic evaluation converts the physical data of the technical evaluation into a pattern of cash flows for the project. These cash flows and choice of an appropriate discount rate lead to calculation of net present value, rate of return, and other investment decision criteria. The "base case" analysis usually models the most likely situation expected by the investor. A series of "sensitivity analyses" are prepared and influence the investment decision by quantifying the impact of possible alternative outcomes in the future. "Simulation" may also provide additional quantification of risks by calculating project economics using a range for each uncertain variable in contrast to the point estimate of a sensitivity analysis.

(3) Other relevant factors. There may be socio-economic, legal, regulatory, environmental, or financing factors which are not adequately quantified by the economic evaluation.

(4) Implicit in the preceding concentration on project analysis, is the assumption that it has been *preceded* by extensive long-range planning. Such planning includes analysis of the prospects of alternative business sectors of interest to the firm as well as analysis of the company position in the sector. The objective is to identify attractive sector prospects which in turn lead to consideration of the technical and economic aspects of specific projects.

TECHNICAL ASPECTS OF MINERAL PROPERTY EVALUATION

In their 1957 edition of "Examination and Evaluation of Mineral Property," Charles H. Baxter and Roland D. Parks (Parks, 1957) described a sequence for examination of mineral and petroleum properties as well as techniques of financial evaluation. There have been major changes in approach during the 1960s and 1970s, many due to the development of the computer which allows for more detailed analysis of the variables controlling the value of a discovery. Ability to do extensive sensitivity analysis and simulation as well as geostatistical reserve estimation has been enhanced by the development of computer techniques. The acceptance of these new analytical techniques have also had an impact on the techniques used in sampling and the approach to mine valuation described in the Parks book.

Ore Deposit Definition and Ore Reserve Estimates

Geological Investigation and Sampling: The basic problem is delineation of an orebody using structural geology aided by sampling. the geologist must rely on theories of mode of occurrence, mineral association, and alteration in order to aid in his description as to the probable size and shape and physical characteristics of the deposit. Knowledge of the regional geology is a prerequisite to understanding the specific ore body and for this the geologist turns to sources of data including state and national geological surveys, company reports, maps of adjoining mines, drilling, and geophysical records. This information is supplemented by field investigations including test pitting and trenching, drilling, core drilling, geophysical surveys, laboratory analysis of sampling, mapping, and interpretation of geology. Major changes have taken place in sampling because of the demands of geostatistics for proper location of samples and proper weighting of assays. The problems of intentional or unintentional distortion of sample results are still important to the investigator who must advise the investors as to the economic viability of the mineral occurence.

Conventional Reserve Estimates: Conventional techniques of reserve estimation work with cross sections and plans to plot sampling results, then to weight these samples correctly to estimate ore grades between sampling points. A detailed discussion of the techniques is beyond the scope of the present chapter; they are described in Parks (Parks, 1957), Koch (Koch, 1971), Peters (Peters, 1978), and David (David, 1977). Reserve calculations techniques rely on assigning grades to blocks of material between samples. For example, the cross section method divides the deposit into several blocks utilizing geologic cross sections. Sample data are plotted on the cross sections which may represent results from a line of drill holes or are controlled by existing mine workings or test pits. One approach assumes the grade or ore varies gradually from one cross section to the next; another defines a block or ore by one cross section and assigns the ore grade to one-half the distance to adjacent cross sections. The area of a cross section may be calculated by planimeter, then block volumes calculated by multiplying cross section areas by the distance between sections. The grade for a cross section is determined by weighted averaging of grades of assays from drill holes or other samples adjacent to the sections. Volumes are calculated from the dimensions of cross sections and distances between them; this volume is converted to tonnage of individual blocks by applying a tonnage/volume factor for the type of material involved.

Most of the "conventional" techniques for ore reserve estimate were derived prior to 1930 and include cross section, isoline, polygonal, and triangular methods; each technique takes its name from the arrangement of samples and an assumption as to how ore grade changes between samples. For example, isoline techniques join points of equal value assuming continuous change between sample points.

Geostatistical Reserve Estimates: Geostatistics refers to the use of statistical methods for analysis of ore deposits and determination of ore reserves. Early work by Krige (Krige, 1960) and Matheron (Matheron, 1963) was followed by a long list of contributors bringing this technique to its present level of development. Geostatistical techniques are relatively new and depend on the computer for analysis of a large quantity of data and calculation of the statistical relationships. Use of the techniques requires a knowledge of basic statistics and "the theory of regionalized variables." The theory of regionalized variables was developed by G. Matheron and is the mathematical basis of geostatistics. A key assumption is that the geological process active in the formation of ore deposits is a random process; the grade at any point in the deposit is considered an outcome of a random process. This *does not assume* random sampling as in conventional statistical analysis. The term "regionalized" is used to indicate that such variables are spatially correlated to some degree, with the variogram being the tool to capture the structural aspects of regionalized variables by measuring the degree of similarity of two samples taken some distance apart in some direction. Examples of regionalized variables include grade of ore, thickness of formation or elevation of the surface of the earth. The fundamental difference between geostatistics and classical statistics is that geostatistics *does not demand that samples be independent*, rather it assumes that adjoining samples are correlated spatially and that the correlation between samples can be captured in a function called a "variogram" which quantifies the correlation between samples. Geostatistics makes an important contribution to mineral industry financial analysis by quantifying the uncertainty with respect to ore tonnage and grade of the ore deposit under consideration. Lack of such information is a principle shortcoming of conventional ore estimation techniques. Financial analysis requires maximum quantification of the variables that influence the decision to invest or not to invest; the contribution of geostatistics is therefore a major improvement in the decision making process.

The tools of geostatistics include:

(1) The variogram
(2) Block variance
(3) Estimation variance
(4) Kriging

The variogram measures the continuity of mineralization, zones of influence for individual samples, and whether the zones of influence vary in different directions. It measures similarity or dissimilarity between ore grades for some distance (h) apart. The criteria, designated by gamma, is calculated by squaring the differences between two grades (h) meters apart. The calculation is repeated for all the samples that are (h) meters apart and the average squared differences obtained. If this calculation is repeated for increasing distances (h), we find that the values of gamma increase with increasing distance. This result is intuitively logical because as the distance between one ore grade sample and another increases, we expect that at some point there will be no correlation between the two. The distance (h) at which the variogram levels off is called the range. Beyond that point samples are no longer correlated; they are independent. The value of the variogram where it levels off is called the sill value which is also the variance of all the samples used in variogram development. Fig. 3.7.1 illustrates the variogram and its key characteristics.

If one considers a gold placer deposit it seems logical that a gold nugget found in one sample could have apparent zero correlation with gold in an adjacent barren sample. This gives rise to the concept of "nugget value" which is the value of a variogram when distance (h) equals zero. A positive nugget value, usually expected to be small, may have real physical meaning as in the case of the gold placer or it may be due to errors in sampling assaying. For example, in uranium deposits high nugget values are often found because of the difficulty in obtaining accurate assays of the deposits. Sometimes the variograms indicate that mineralization is more continuous in one direction than another. This is called "anisotrophy" which becomes important in comparing the range of influence for samples in different directions. Though it is only one component of a complete one reserve estimate, the variogram alone can be extremely helpful in understanding variations in the mineralization within a deposit.

Block variance, the second major geostatistical tool, is important because if the mineral deposit is split up into blocks of a certain

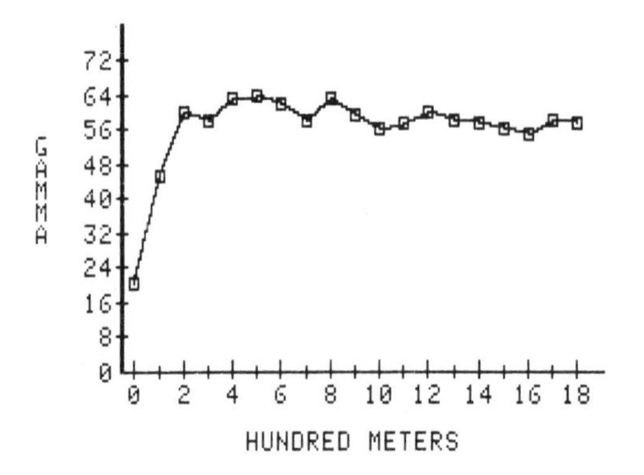

Fig. 3.7.1—Variogram.

size,shape, and orientation the variance of ore grade of the blocks is likely to be very different than the variance of the sampling unit used on the blocks. This difference in variance has a large effect on the calculation of tonnages at different cutoff grades.

The third geostatistical tool is estimation variance which measures the variance of error between the true grade z and the estimated grade z^* of a block. The magnitude of the estimation variance depends on the characteristics of mineralization, variance of the samples, sampling grid and estimation method. As mentioned before, a strong point of geostatistics is the ability to calculate estimation variance and thus to quantify confidence limits for a range of ore reserve grade and tonnage. This information is extremely valuable in the preliminary economic analysis of the potential orebody. Geostatistical techniques utilize the variogram to derive estimation variance for any drilling grid. Thus the concepts of the variogram, block variance and estimation variance combine to provide a confidence interval for the drilling grid. If the calculated confidence interval is greater than acceptable as indicated by the economic evaluation, then the drilling grid can be modified to an acceptable level. For example, the range of grade or tonnage for drilling grid "A" may indicate a range of uncertainty that produces a large number of possible future outcomes which are uneconomic. The evaluation team may propose a tighter drilling grid with the expectation that the range will be reduced and thus bring the return on investment within acceptable limits.

The fourth and final geostatistical tool is kriging, used to estimate the grade of a block as a linear combination of available samples in or near the block. The technique finds a set of weights that minimize the estimation variance for the block including consideration of the geometry of the problem and character of the mineralization. Low weights are assigned to distant relative to near samples. The technique also takes account of the position of the sample relative to the block under consideration. The technique is attractive because it produces an unbiased minimum variance estimate for each block kriged.

In summary, the geostatistical techniques are an important advance in improved technical evaluation of a mineral property. The major stages in utilization of the techniques for ore reserve estimation include:

(1) Data preparation
(2) Variogram computation
(3) Kriging of individual blocks and summing the blocks to obtain the ore reserve estimate.

The details of the art and science of geostatistical ore reserve estimation are beyond the scope of the present chapter and the reader is referred to publications listed in the bibliography, for example, publications by David (David, 1977), or Knudsen and Kim (Knudsen, 1978).

Engineering

At the prefeasibility stage the technical analysis may be simply a sketch of a mine plan or flow sheet indicating the major physical components of a project. The plan or flow sheet may have been adapted from an existing mining property with modifications to fit the new situation. Before an economic analysis can proceed, the engineering analysis must supply the following required information:

(1) A mine plan and/or process flow sheet in the form of a block diagram or rough sketch
(2) A heat and materials balance for the plant including quantities of feed, effluents, products, recoveries, capacity, and conversion requirements
(3) An equipment list for mine and plant
(4) A utility balance which may be calculated as a ratio based on process requirements
(5) A plot plan showing location of the mine and plant and indicating any special site requirements
(6) Projected engineering costs estimated from similar jobs

The preparation of detailed cost estimates requires detailed mine plans and process flow sheets. These detailed mine plans and flow sheets allow preparations of lists of equipment and precise specifications for structures, foundations, piping, electric installations, insulation, and equipment arrangement. A detailed cost estimation is based on firm quotations and contractor bids which in turn must be based on completion of a large portion of the engineering work.

The high cost of the detailed estimates is a result of the requirement for detailed engineering work to provide specifications for high accuracy cost estimation. Resolution of process scale-up questions is likely to require detailed laboratory or pilot plant work which is expensive. Similarly, the detailed mine plan may require additional drilling, sampling, and surveying which also is expensive.

TECHNICAL ASPECTS OF COAL PROPERTY EVALUATION

A hypothetical coal property provides a specific example of the components of a technical evaluation. The technical evaluation of a coal seam is unique because of the geologic, marketing, and engineering characteristics of the commodity. The prefeasiblity technical evaluation also will be based on existing geologic and engineering information on a deposit.

Coal Seam Definition and Reserve Estimates

A geologic evaluation will establish the following:

(1) Tonnage, volume, location, and geometry of coal beds within the property boundaries to a specified depth. These data describe the coal resource base of the property and when coupled with technical and economic data, provide a base for calculation of recoverable coal reserves.
(2) Classification of the coal as suitable for metallurgical or power generation, or other uses.
(3) Establishment of the rank of the coal.
(4) Establishment of the quality of the coal in terms of Btu content, moisture, volatile matter, carbon, and ash. The quality evaluation will lead to definition of the market and help to estimate a price for the product. It will also indicate whether the coal must be washed prior to use in order to meet market requirements.

Definition of the geological characteristics of the coal seam coupled with engineering and economic data leads to an estimate of recoverable reserves. These will be classified first into possible, probable, and proven categories and second in terms of amount of overburden stripping necessary to expose strip mine coal. Reserves often serve as collateral for debt financing of the investment in mine and plant and thus are critical input for the economic analysis to follow; a reasonable approximation is that reserves must be twice that needed to amortize the investment. Other geological evaluations will characterize the coal seam in terms of its depth, dip, amount of overburden, thickness, and quality. This characterization of the coal is necessary for the economic evaluation to determine whether the product is acceptable and if acceptable what price it will command in the market.

Title Check and Leasehold Evaluation

A check of land title and other potential land problems such as environmental constraints will be part of the initial effort prior to economic evaluation. In addition, a minimum-sized coal property must have sufficient reserves to amortize and pay the required return on the investment. For example, an underground mine designed to produce 200,000 tons per year is likely to require at least 10 years to amortize and pay a competitive return on the investment. Thus the investor will look for two million tons of recoverable coal. In coal beds 4 to 12 feet in thickness, utilization of underground techniques allows recovery of only 33% to 50% of the coal in place. Using the lower recovery limit, the in place coal should be about three times production or at least six million tons; thus, for bituminous coal at 24.6 cubic feet per ton and four feet thick there will be 7,084 tons per acre so that the required land holding will be 847 acres plus requirements for loading and washing facilities to support a mine. Another example is an opencast mine designed to produce half a million to ten million tons per year that would also require at least ten years to amortize and pay a return on the investment. This mine requires production of 50 to 100 million tons of coal over 10 years. In this case with a recovery of 80% the investor needs 125 million tons of coal which at 24.6 cubic feet per ton and 10 foot thickness requires 7,058 acres of land plus requirements for loading and washing.

Economic evaluations will help guide selection of a mining rate and other variables that will optimize the net present value of the property given the physical characteristics of the coal beds, size of the reserves, and market constraints. Coal reserves that expand an existing mining unit already in production will have the higher value than similar properties in an undeveloped area which probably will be more difficult to develop and bring into production.

Engineering

The type of mining employed for a specific coal bed varies with the physical conditions and production requirements of the property. Methods are either surface or underground and if underground are often characterized by the type of access or method of extraction. The physical characteristics affecting choice of mining method include:

(1) overburden—nature and amount;
(2) coal seam—thickness, pitch, cleavage, hardness, structural strength, and presence and amount of explosive gases;
(3) water—amount likely to be encountered;
(4) roof—strength above coal seam;
(5) floor—strength below coal seam; and
(6) previous mining activity above or below the seams.

The choice of mining and processing technology will have a significant impact on the economic evaluation. For example, surface mines generally are less costly to operate than are underground mines. Coal recovery is higher (85% average compared with 50% average underground), and accident rates are lower. Surface mining is limited primarily by stripping economics. The amount of overburden to be removed to recover a ton of coal is limited by the price of the coal and other operating costs. The boundaries of a surface mine are often determined by the break-even cost of recovery where the cost of stripping overburden is the major variable.

The product of the technical evaluation includes a mining plan based on a concept of surface or underground mining methods, a flow sheet of the preparation plant with energy and material balances, a description of the transportation system, and details of plant capacity and product quality specifications as they relate to the consumer. The mine plans and flow sheets allow estimators to prepare manning tables, equipment and other input requirement lists. When coupled with information on labor rates, materials prices, and equipment costs, the lists lead to the estimates of capital and operating costs, for the project. The electric utility consumer will specify heat (Btu) content, ash, and sulfur and so define acceptability of the coal for steam generation. Plant capacity will control the rate of production which, in turn, governs the flow of profits and the time needed to amortize and achieve a return on the investment.

The technical risks facing the investor are caused by imperfections in the available coal seam data including quantity, quality, and geological characteristics. Errors in the data contribute to errors in the estimate of capital and operating costs of the venture. There is also a risk related to the failure of human judgment through choice of an incorrect mining, preparation, or transport concept for development of the property.

TECHNICAL ASPECTS OF OIL AND GAS RESERVOIR EVALUATION

A second example of the problems involved in technical evaluation is this hypothetical wildcat oil/gas well. Like the coal property, the technical evaluation is unique because of the unique geologic, marketing, and engineering characteristics of the commodities. Analysis will require preliminary estimation of the reservoir characteristics and potential reserves as well as estimates of the engineering required to explore and if successful to develop the potential producer.

Reservoir Characteristics and Reserve Estimation

Initial geological work will indicate the shape and size of an expected oil trap. These come in a large variety of shapes and types but the anticline is representative of a large number of oil fields. In this case assume that the reservoir is a layered formation of marine sedimentary beds consisting of porous sandstone covered with a cap of impermeable rock. Gas, oil, and water are segregated by their densities with the gas at the top underlain by oil and then water.

The final step in exploration of a potential oil structure is the drilling of a wildcat well. Prior to that final step, investment in exploration proceeds primarily by geophysical methods. These geophysical investigations provide engineering and geologic data critical to the economic analysis of the investment in the wildcat well including:

(1) Data on the size and shape of the expected oil trap.

(2) Information on the probable producing formation and its geologic age and its physical characteristics.

(3) Depth to the potentially producing formation.

(4) Thickness of the "net pay" or expected portion of the formation expected to be porous and contain oil.

(5) The expected porosity or pore space in the rock expected to contain oil, gas, and water.

(6) The expected water saturation or the portion of pore space occupied by water.

(7) The average formation volume factor which relates the expected volume of hydrocarbons at the surface to the volume underground.

(8) The initial gas-oil ratio which measures the amount of gas associated with each barrel of oil in the formation.

The above information allows an estimate of the expected average oil and gas in place and technically recoverable reserve. This is the basis for an engineering and economic analysis that leads to an estimated economically recoverable reserve in the field.

Engineering

Prior to the drilling of the first well, lease acquisition and data acquisition costs will be the principle expenditures for the petroleum project. Geophysical and geological data may indicate the presence of a formation with a potential for a commercial accumulation of hydrocarbons. This might be a recognizable geologic structure such as an anticline or a stratigraphic trap or reef at great depth that can be examined only with an expensive investment in seismic geophysical testing. The objective is to purchase additional information on the formation at depth utilizing the most efficient exploration techniques. There is much uncertainty because no reservoir analysis short of drilling a hole provides the answer as to whether commercial amounts of petroleum are present. The petroleum geologist or engineer will concentrate on the three critical preconditions for the existence of a reservoir:

(1) There must be a reservoir (rock bed capable of holding liquid petroleum).

(2) There must be a seal on the reservoir (a trap to hold in the petroleum).

(3) There must be an identifiable source of the hydrocarbons.

A firm may spend from $5,000 to $10,000 per mile for seismic work in connection with a prospect. Subsequently drilling may go from 9,000 to 25,000 feet which means an investment in drilling of $5 to $10 million per well. It is clear that deep wells will quickly absorb the total budget of the company hence efficient use of geophysical techniques as well as efficiency in drilling are critical to the economics of petroleum exploration.

When one or more exploratory wells indicate the presence of a potential commercial field, engineering analysis of cuttings, penetration rates, and logging of the drilling fluid will provide additional engineering information to confirm the presence and recoverability of commercial amounts of oil or gas. Logging tools measure factors such as effective porosity, water saturation, and the presence of hydrocarbon. Formation pressures, bottom-hole fluid samples and production rate pressure tests allow forecast of production rates and potential cumulative production. At this stage, well costs constitute the major percentage of investment in the drilling program so that efficiency in obtaining information is paramount. Engineering will be directed toward estimates of the total investment in drilling as well as developing a list of equipment necessary for exploitation of the field. Pumping equipment, electric motors, storage tanks, power lines, transformers, and pipe line costs are needed to estimate the total investment necessary to develop the field.

ESTIMATION AND FORECASTING OF COSTS

The accepted method of valuation of mineral projects requires us to forecast and capitalize future net cash flows. Net cash flow is the difference between cash inflows, usually consisting of net sales revenues, and cash outflows, usually consisting of operating and administrative costs and taxes. The objective is a set of conceivable values for each. Since the revenue and cost forecasts depend entirely upon assumptions about the future, making "reasonable" or "conservative" assumptions is the most important part of the evaluation procedure.

Cost estimation involves forecasting capital and operating costs for the project. The accuracy of a "best" estimate, varies with the information, money, and time available for the estimating process. Cost can vary from a few hundred dollars for a quick "order-of-magnitude" estimate to several million dollars for detailed estimates. This cost is balanced against the need for accuracy which, in turn, depends on the proposed use of the estimate. For justification of additional drilling, an order of magnitude estimate may be sufficient. For financing or marketing agreements, accuracy must be much higher and the higher cost is justified. The error of cost estimates may range from as high as plus or minus 40% in order of magnitude estimates to as low as plus or minus three percent on detailed cost estimates.

The uncertainty in cost estimates varies widely. For an operating property we have some past data and by looking at known changes in cost components such as labor, supplies, power or administration it is possible to make a fairly accurate prediction of costs expected in the near future. The uncertainty of expected future costs increases with the length of the forecast.

A cost forecast requires an estimate of all geologic, mining, mineralogic, and metallurgical variables, some unknown, others uncertain in the venture. Thus there is a demand for a thorough technical analysis of the project before estimating can begin. A good cost estimate starts with a careful analysis of existing physical data, then develops a concept of how the orebody could be explored, mined, and processed, as a basis to estimate costs for the system; it is the essence of creative engineering.

Cost studies evaluate engineering designs in economic terms. Most likely the design is a new combination or pre-existing technology, while the cost estimate itself is the numerical forecasting of probable economic results. Because of its multi-disciplinary nature, cost estimation requires a grasp of the technical aspects of geology, geophysics, mining, and metallurgy as well as a knowledge of economics, and mathematical techniques; typically it relies on a team of experts rather than a single individual. The leader of the team should be an experienced engineer able to integrate the views of the individual specialists into a consistent and accurate forecast of costs.

Actual costs deviate from estimates because no individual or technique can foresee the future and there are always limitations in time, money, and staff available to develop the estimate. The deviation of the actual cost from the estimate will be a function of four major factors:

(1) *Risk*. Risk may be subjectively or objectively estimated. When statistical data allows calculations of probabilities, a variable may take on several possible values according to the known probability distribution. An example might be the climatic conditions expected during the life of the project, the risk of accident or injury in

operations, or the risk of delays affecting the transportation system.

(2) *Uncertainty*. Uncertainty applies to all of the risk probabilities because they relate to future rather than past conditions. Estimates of inflation, variation in operating conditions, ore grade or mineralogy involve uncertainty which often increases with the length of the forecast. There are many examples where the cost estimate, presumably accurate within 5 or 10% of the total cost turns out to be different by 50 or 100% from the final expenditure.

The accuracy of the estimate will be directly proportional to the amount and quality of the physical data available on the mineral deposit thus it will change continuously drillhole by drillhole with results of new testing and research.

(3) *Mistakes*. Mistakes may be errors in calculation or errors in technique. The development of cost estimating systems based on concepts of standard crews, standard equipment and materials list, and an organized approach to costing has led to major improvements in the quality of cost estimates. More recently, a computer and micro-computer cost estimating programs have further promoted a standard approach which is helping to eliminate errors in calculation and technique.

(4) *Errors in Concept*. Every cost estimate must begin with a plan or a design concept. If this concept is in error, the cost will be in error. If the engineer fails to recognize that "heavy ground" may require additional support, or that metallurgical recoveries will be low, the estimate will have fatal errors and lead to erroneous investment decisions.

Types of Cost Estimates

Four fundamental types of cost estimates can be defined. The type used and the accuracy of the result depend on the amount of data available and the time allotted for preparation of the study. The types include (1) order of magnitude studies, (2) factored estimates, (3) definitive estimates, and (4) detailed estimates. The use of this terminology varies from one organization to another, but the idea of increasing time, cost, and money and required informational detail is consistent with the progression from type one to type four. The benefit from increased expenditures of time and money is an estimate of higher accuracy that can be used to justify increasing larger expenditures by investors.

The differences between types one and four is becoming less and less sharp as computer based systems allow specification of great detail in the final estimate even when project data is very preliminary. A prime example of this is the ICARUS model, described later, which provides great detail in the estimated cost, therefore, a basis for questioning assumptions and refining the estimate.

Purpose: The use of cost estimate varies widely. An order of magnitude estimate may be an exploration target model or for management decisions on approval of funds for further exploration of a potential orebody. As additional information on geology, mineralogy, structure, and other physical characteristics on the deposit increases and the engineering work proceeds from a preliminary concept to a more detailed plan, the "factored" estimate may be justified. Usually this is the first estimate of a specific project or as a followup on an order of magnitude estimate. Computer assisted estimating methods are replacing the factored estimate as the first estimate of a project because the capability of the computer to provide detail helps to quantify assumptions which may be implicit in the factored estimate. Such quantification allows the estimator to identify potential sources of error and to seek additional data as necessary.

As the amount of information increases the estimator may proceed to a "definitive" estimate for appropriation of funds to establish contract prices or establish a budget and basis for project cost status reports. It also establishes a format for the final detailed cost reports. The final type of cost estimate is the "detailed" estimate which is used to establish contractor price, obtain financing and negotiate contracts for projects. It requires the maximum amount of detailed information on all physical aspects of the project.

Accuracy Relative to Cost and Time Requirements: Order of magnitude estimates may be made in a matter of hours to a week at a cost of one half to five percent of the cost of a detailed estimate. As a result, accuracy is highly variable or even undefinable as in the case of an exploration model representing a possible discovery. In the case of an order of magnitude feasibility study expected accuracy may be of the order or plus or minus 20% to plus or minus 40% at best.

Factored estimates may be prepared in two or three weeks to three or four months at a cost of 20 to 30% of the detailed estimate. Again, accuracy is highly variable from plus or minus 10% to plus or minus 40% depending on the scope of the estimate and the reliability of the data. Computer assisted methods allow preparation of factored estimates within hours to three or four months at a cost of 5 to 15% of the cost of a detailed estimate. Because of the greater consideration of the detailed accuracy may be plus or minus 10% to plus or minus 20% depending on the available information.

Definitive estimates are likely to require three to four months and require expenditures up to 90% of a detailed estimate. As a result of the increased engineering and geologic detail, estimates may provide an accuracy of plus or minus 5% to plus or minus 15% of true cost.

The final stage in the cost estimating process is the detailed estimate requiring 6 to 18 months for completion because of the great amount of engineering detail required. The final product is expected to be plus or minus three percent to plus or minus five percent from true cost. It is critical to recognize that the three to five percent only partially covers risks of projecting todays costs into the future. The assumption is that current cost data can be used to forecast future costs which in the case of a mining project may be five or even ten years in the future. It is not surprising then, that actual costs for mining projects sometimes turn out to be orders of magnitude different from the detailed cost estimate. Fig. 3.7.2 summarizes relationships between types of cost estimate, accuracy and the percentage of engineering necessary for each type of estimate.

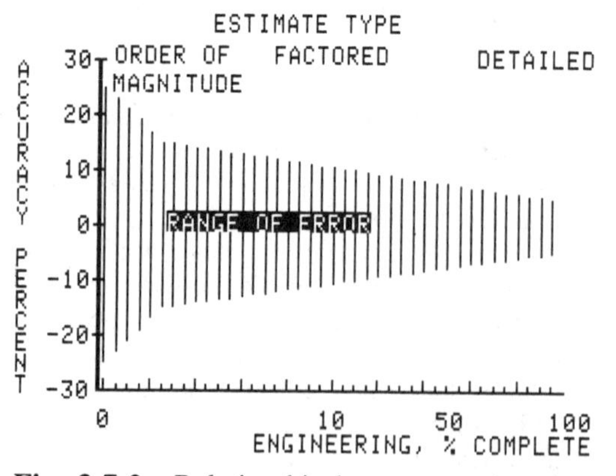

Fig. 3.7.2—Relationship between engineering work completed, type of estimate and accuracy.

Required Information and Methodology: Table 3.7.1 shows approximate relationships between required information, types of estimates, and expected accuracy. The requirements apply equally to conventional or computer assisted estimates.

As the amount of information required increases the analyst moves from the order of magnitude estimate, to the factored estimate and then to the detailed estimate. The initial order of magnitude estimate may rely on a very rough idea of the mine plan and a process flow sheet based on similar installations. By the time the detailed estimate is made, the engineering for mine development is much more complete and process flow sheets show detailed information sufficient to negotiate subcontracts for electrical, structure, and foundation work, and obtain quotations on individual pieces of equipment.

Methods also vary with type of estimate. The order of magnitude estimate is likely to rely on "parametric" estimating where a capital or an operating cost is related to major variables that cause changes in the cost.

One example of parametric estimating is a major study made by STRAAM Engineering (formerly A.A. Mathews, Inc.) for the U.S. Department of the Interior, Bureau of Mines completed in December of 1977. The product of the study is a *Capital and Operating Cost Estimating Handbook for Metallic and Nonmetallic Minerals* (STRAAM, 1977). The STRAAM handbook was prepared for use in the Minerals Availability System (MAS) program of the U.S. Bureau of Mines. The MAS program requires costing of mineral occurrences using an order of magnitude estimating system based on the functional methods of estimating capital and operating costs. The objective is a preliminary financial analysis to evaluate the economics of mineral resources particularly the market price that would make a particular group of orebodies economically viable.

The USBM handbook was developed for a user with knowledge and experience in mining and estimating procedures. It is to be used for rough estimates of capital and operating cost of entire mining and beneficiation systems, not for use to determine the cost of a particular component of the system because of its reliance on averages and approximation. Resulting estimates are expected to be within 25% of expected actual cost.

A second example of parametric estimating

Table 3.7.1—Relationships between Type of Estimate, Required Information and Accuracy (Process Plant Example)

X = Not required D = Desirable R = Required

	Information	Order of Magnititude	Factored	Detailed
I.	**General Plant Information:**			
	Type Plant & Process	R	R	R
	Plant Location	D	R	R
	Time Frame for Cost Escalation	R	R	R
	Milepost Schedule	D	R	R
	Plot Plan	D	D	R
	Equipment Layouts & Elevations	X	D	R
	Piperack Sizes & Number of Levels	X	D	R
	Building Sizes & Usage	D	R	R
	Building Details	X	D	R
	Structure Sizes & Floor Landings	D	R	R
	Craft Wage Rates & Benefits	X	R	R
	Engineering Mark-up	X	D	R
	Engineering Preliminary Def. Est.	X	D	R
	Shipping Restrictions	X	D	R
	Freight Costs	X	X	D
	Wind Loading (MPH)	X	D	R
	Major Sub-Contracts	X	D	R
	Sales Taxes	D	R	R
	Royalty or Licensing Fees	D	R	R
	Contractors' Fee	X	D	R
	Mechanical Flow Diagrams	X	D	R
II.	**Civil Information:**			
	Soil Type, Density & Loading	X	D	R
	Special Soil Conditions	X	D	R
	Ground Water or Drainage Problems	X	D	R
	Depth of Footing	D	R	R
	Site Topography	D	R	R
	Type Fdn. (Piling, Pad, Spr'd., etc.)	X	R	R
	Ready-Mix Concrete Cost	X	D	R
	Fire Resistance Rating	X	D	R
III.	**Process Equipment Information**			
	Major Parameters	R	R	R
	Materials of Construction	R	R	R
	Data Sheets	X	D	R
	Vendor Prel. Quotes for Costly Items	X	D	R
	Spare Parts List	X	D	R
IV.	**Piping Information:**			
	Type Utility Headers	X	D	R
	Offsite & Inter-Area Piping	D	R	R
	Process Pipe In Piperacks	X	D	R
	All Piping in Main Piperacks	X	D	R
V.	**Electrical Information:**			
	Main Feeder Voltage	D	R	R
	Main Feeder Tie-In Point	X	R	R
	Class & Division Classification	X	R	R
	Buried or AG Cable	X	D	R
	Conduit or Armored Cable	X	D	R
	MCC Location	X	R	R
	Substation Type	X	D	R
	Radial or Spot Distribution System	X	D	R
	Demand Diversity Factor	X	D	R

Table 3.7.1—Relationships between Type of Estimate, Required Information and Accuracy (Process Plant Example)

X = Not required D = Desirable R = Required

	Information	Order of Magnititude	Factored	Detailed
	Power Factor	X	D	R
	Percent Excess Transformer Capacity	X	D	R
	Transformer Capacity (KVA)	X	R	R
VI.	**Instrumentation Information:**			
	Density—Standard or Complex	X	R	R
	Pneumatic or Electrical	X	D	R
	Control Room Location	X	R	R
	Instrument Panel Display Type	X	D	R
	Type Signal Wire	X	D	R
	Type Thermocouple	X	D	R
	Type Thermocouple Wire	X	D	R

is a system developed by T. Alan O'Hara published in an article in the Canadian Institute of Mining and Metallurgy (O'Hara, 1978) which summarizes capital and operating costs in the form of a series of cost curves for open-pit and underground mining and beneficiation. The curves are less detailed than found in the STRAAM Handbook but more detailed than the crude guides that relate project costs only to mine and plant size. The concept behind the O'Hara curves is that a major variable affecting all items of capital costs, revenues, and operating costs is the daily tonnage of ore treated by the process plant. O'Hara postulates that an open-pit estimate can be better prepared if items such as site preparation, preproduction stripping, and equipment are handled separately and modified for other variables such as the type of overburden, the characteristics of the topography or relationships between the truck fleet and shovel size. In the case of underground mines a similar approach calculates the costs of shafts not only on the basis of tons of ore hoisted per day but also on the basis of shape of the shaft, type of lining and consideration separately of the characteristics of the hoist and head frame. In the case of the plant site clearing, concentrator building, foundations, equipment, storage, and conveyors are considered separately in order to allow modifications for additional variables.

The O'Hara system is readily adapted to a micro-computer and provides a quick method for conceptual capital and operating cost estimation. The system has been updated and included in the Canadian Institute of Mining and Metallurgy volume *Mining and Mineral Processing Equipment Costs and Preliminary Capital Cost Estimations* (Mular, 1982).

An example of factored estimating is also included in the volume edited by Professor Mular (Mular, 1982). The volume, an update of the previous volume 18 dated 1978, consists of a section describing methods of factored capital cost estimation and a second section that includes all of the cost curves necessary to prepare an estimate. Use of the Mular volume requires that specific items of equipment be specified, assuming the availability of at least a preliminary flow sheet of the processing system. The estimator then refers to the cost curves in the book for estimates of capital costs which vary by capacity or size of the individual piece of equipment with corrections for physical variables such as material of construction, type of drive, or characteristics of the feed material. The cost curves result in an estimate of purchased equipment costs which are the first step in a factored capital cost estimate. Factors are then used to convert the initial estimate of equipment into and estimate of the fixed capital cost of the entire plant. As indicated, this volume of the CIMM publication also includes a summary of the O'Hara system for mining capital and operating cost estimation.

Other systems which rely on the factored approach include those developed by R.S. Means and Richardson. There are a number of other reference manual systems available for estimation of building construction and process plant construction. The Cost Information Systems Division of McGraw-Hill provides data for estimation of systems costs for building construction and heavy construction costs to estimators (McGraw-Hill, 1983). The R.S. Means Co., Inc. (Means, 1983) provides cost estimation manuals for building construction and site work useful to the mineral cost estimator. The company has also developed a computerized general estimating program which represents the latest systematic way to make construction estimates rapidly. Quantity surveys lead to estimates of the detailed components of the project which then are summarized and brought forward to a total estimate for a project.

MARKET SURVEYS AND PRICE FORECASTING

The mineral or petroleum market survey seeks to answer several questions that are of critical importance to the investment decision. These include:

(1) Size of the market and its rate of growth.

(2) The potential sales of a new supplier in the market.

(3) The pattern of costs and prices in the past and projection of future trends.

(4) Geographic location of markets for the product.

The evaluator can turn to a large number of books, periodicals, government publications, surveys, censuses, and special public and private surveys that summarize the market outlook for a particular commodity. An analyst beginning a survey involving a new commodity can usually benefit from one or more of the publications of the U.S. Department of the Interior, Bureau of Mines, McGraw-Hill Mining Informational Services, Roskill Information Service Ltd., the National Technical Information Service, Resources for the Future, the Society of Petroleum Engineers and the American Petroleum Institute, and the American Institute of Mining, Metallurgical and Petroleum Engineers including *Proceedings of the Council of Economics*. In some cases, special commodity studies have been made by the preceding organizations or others. For example, McGraw-Hill published *Alumimum: Profile of the Industry* in 1982 (Berk, 1982).

An important output of the market study and a key input into the financial analysis is the assumption with regard to future prices for the product. Three approaches can be identified as the basis for the price assumption:

(1) A review of historical price trends and cycles in relations to past supply, demand, and inventory situations. This combined with a judgment regarding future economic conditions provides a basis for investment analysis

(2) Cost of competitive producers are reviewed in relations to expected supply/demand conditions to forecast price levels that seem reasonable to bring future supplies into production

(3) Econometric models quantify relationships between supply, demand, inventories, technical factors, and expected price trends in the future

Historical Prices and Judgment

The price history of copper on the COMEX market shown in Fig. 3.7.3 is best understood in terms of economic conditions during the period shown as indicated in Fig. 3.7.4. Basic data

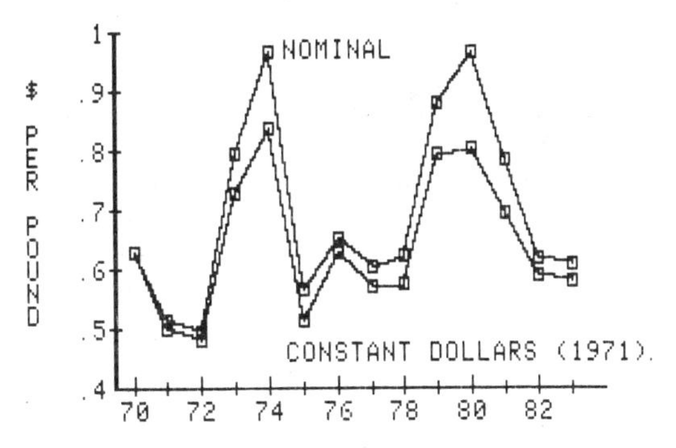

Fig. 3.7.3—Price history of copper on the COMEX Market.

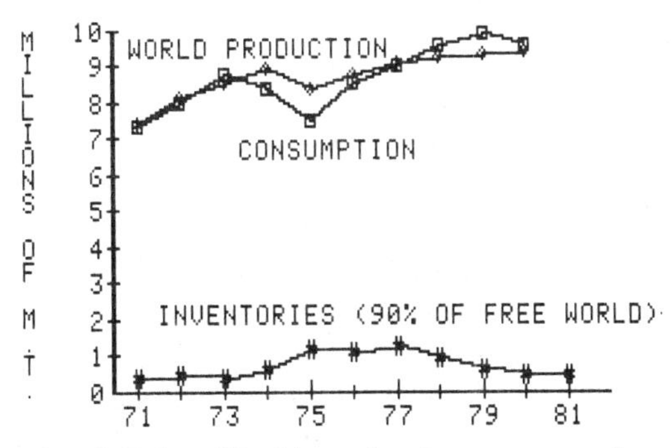

Fig. 3.7.4.—World production, consumption and inventories of refined copper.

for these charts comes from American Metal Market *Metals Statistics* (American Metal Market, 1982, p. 57).

During 1971, prices were depressed due to an excess of capacity relative to demand. The industry operated under price controls imposed by the Nixon administration but the demand for copper was insufficient to permit an increase in prices to the limit of \$0.65 per pound permitted by the government.

During 1973, the price of copper began to rise rapidly as a result of increasing demand. Once price controls were removed copper prices rose to extremely high levels, partially in anticipation of a strike anticipated by the industry. When the strike did not materialize and demand fell drastically due to a decline in the world economy the price of copper collapsed. This led to accumulation of enormous stocks of copper in 1974 and 1975 which continued to depress prices sharply through 1978.

The copper price recovery after 1978 was shortlived because of the economic depression. Even through 1984, uncertainty in the world economy has restricted the demand for copper and kept prices below that necessary for economic operation of many of the world's copper producers. During 1978 and 1979 major foreign producers of copper such as Chile produced more rather than less copper at the low prices. Chile depends heavily on foreign exchange generated by metals exports; in addition they have some of the lowest cost copper in the world. Since Chile was very dependent on the export of metals for foreign exchange, they were required to produce larger quantities during the period of low prices in order to secure foreign exchange necessary for imports.

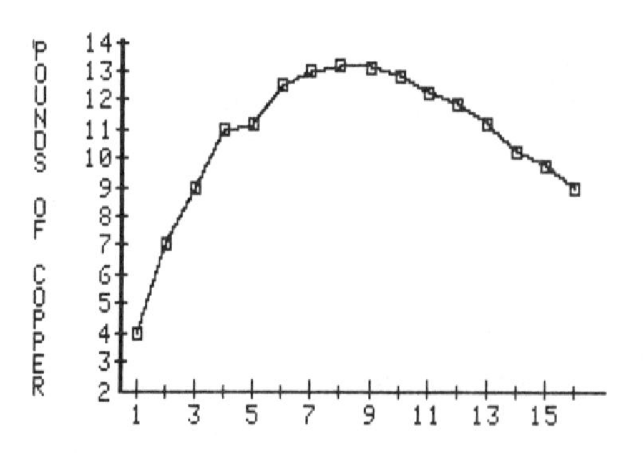

Fig. 3.7.5—Intensity of use.

The decline of stocks of copper after 1977 resulted in higher prices in 1979 and 1980. The decline in stocks was helped by the interruption in Africa and South America, voluntary reductions in the United States, and strikes in Canada. After 1980, prices went down again as a result of another world economic recession which resulted in reduced demand especially in the sectors of transport and construction. In 1984 the industry still has not recovered to expected price levels.

An analysis of historical price fluctuations and trends identifies speculative highs and depressed lows in terms of production, consumption, and inventories. Usually such analysis helps in making a judgment about projected price for use in the investment analysis. Price projections require judgment about future levels of supply, demand, and inventories.

Long-term demand projections often require international data which is often much less detailed than the U.S. data base. The "intensity of use" approach used by Malenbaum (Malenbaum, 1978) is useful for world supply/demand analysis. In this approach economic data available on a worldwide basis is used to forecast commodity demand. "Intensity of use" is defined as the amount of material used in a particular year per unit of gross domestic product (GDP) of the country involved. But, the composition of gross domestic product changes as the welfare of the individuals within the country changes. A measure of this welfare is GDP per capita which is gross domestic product divided by population. Malenbaum's study indicates that as the gross domestic product per capita increases, there is a large increase in the use of materials initially but that this intensity of use levels off and even declines as gross domestic product per capita reaches high levels. The reasons for the changes are complex but logically would include the shift of the composition of gross domestic product from final goods to services, the development of new technologies that make for more efficient use of materials and the possibilities of substitution among materials as technology, demand and supply alter relative materials prices. Following the Malenbaum approach, Fig. 3.7.5 indicates that for many commodities "intensity of use" increases at a decreasing rate as the economy reaches a high level of development as measured by gross domestic product per capita.

Combining "intensity of use" projections with

projections of population and gross domestic product allow forecasts for commodity demand on a worldwide basis. The technique has the advantage of reliance on statistical data that is generally available on a worldwide basis.

Alternate Source Cost

A second concept useful in forecasting future commodity prices is the expected cost from alternate sources. In January of 1978, the U.S. Bureau of Mines published a chart showing present and potential domestic production of copper through the year 1990. The chart shown in Fig. 3.7.6 is derived from the curve published by the U.S. Bureau of Mines but has been updated to 1982.

The chart shows the amount of recoverable copper in the United States available at alternate prices sufficient to pay all costs amortize and pay a 15% return on unamortized investment over project life. The price indicated on the chart is for November 1982 and it indicates a very small portion of the industry that could be considered economic under price levels of that date. The logic of analysis suggests that if future demands for copper put pressure on supply, that prices would have to rise for such additional supply to appear in the market. There are implications, therefore, for the price of the commodity.

The copper market is international in nature, so a second chart is needed. Fig. 3.7.7 is an approximate plot of the international cost of copper relative to world capacity; it indicates that a large portion of the world capacity would require about $1.00 per pound just to break even on total cost. To interest investors in new projects, the price must cover not only total cost but must be sufficient to amortize very large investments and pay a competitive rate of return on these investments. Internationally, prices between $1.00 and $3.00 per pound are required to justify development of new properties.

Following the logic indicated by these supply curves, a study by the U.S. Bureau of Mines (Tomimatsu, 1980, p. 10) indicated that the copper industry would require a price of $0.82 per pound in 1978 to generate a rate of return of 15% on past copper mine investments in the United States. Assuming an increase production costs of 30 to 40% since 1978 leads to a required price of approximately $1.00 per pound for properties in operation. Investment in new properties, according to the Bureau of Mines study, would require $2.50 to $4.50 per pound of copper to justify and pay a competitive return on investment. On the assumption that costs in the United States are approximately 10% higher per pound than the world average, a world price of $2 to $4 per pound is required to justify new investments. To utilize this information for price forecasting, the analyst must decide when to expect growing demand sufficient to utilize existing world capacity and require additional investment in new capacity. The decision is difficult and uncertain especially when it involves governments in several countries whose objectives may be very different than a competitive return on invested capital. Nevertheless, the cost based required price approach is helpful in making a reasonable forecast of expected future prices.

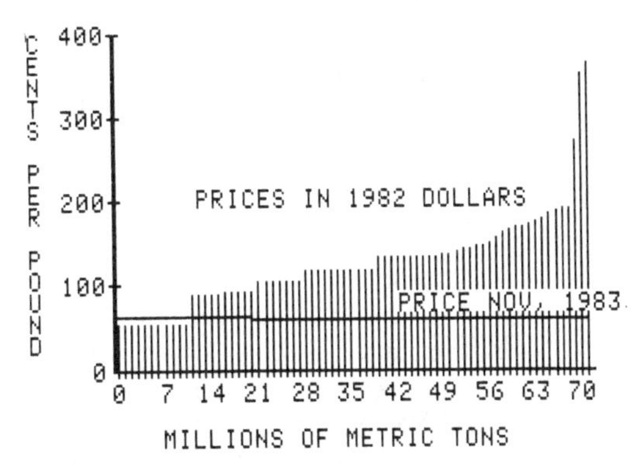

Fig. 3.7.6—Availability of U.S. recoverable copper at alternate prices.

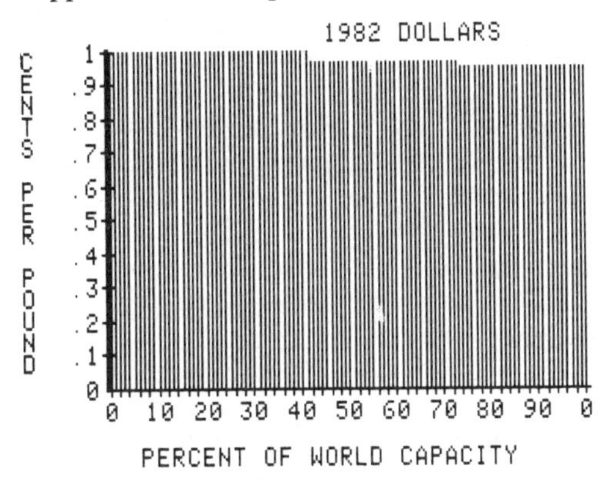

Fig. 3.7.7—International cost of copper.

Econometric Models

Econometric models attempt to quantify relationships between key demand, supply, inventory variables, and price. Several firms utilizing econometric models provide forecasts for use in economic analysis of mineral and fuels commodities. For example, Chase Econometric provides forecasts of more than 350 industrial commodity prices based on multiple factors domestic and international. Long-term forecasts are issued quarterly and extended for a horizon of 10 years. Each price in the forecast is based on econometric equations developed by analyzing relationships between prices and historical supply/demand factors and cost trends. The most important variables for a particular commodity are combined into an equation.

The econometric model is attractive because it forces explicitly stated assumptions which might be hidden under the "subjective judgment" approach described in previous sections. The limitation of such models is that they rely on historical data and historical relationships which may be inadequate to forecast the future. In addition, final equations in an econometric model also require subjective political and technological forecasting that may introduce errors into the final price forecast.

Econometric models can usually be accessed by portable or desk top micro-computer terminals so that clients can simulate the model and generate alternative forecasts based on their own assumptions.

CALCULATION OF CASH FLOWS

Estimation and forecasting of costs, development of investment schedules, and the supply/demand/price forecasts allow expression of the economics of the mineral, or energy project in terms of cash flows over the period of investment and production. The final step in analysis is the discounting of a pattern of cash flows which are an economic representation of the technical and economic forecasts of a large number of variables.

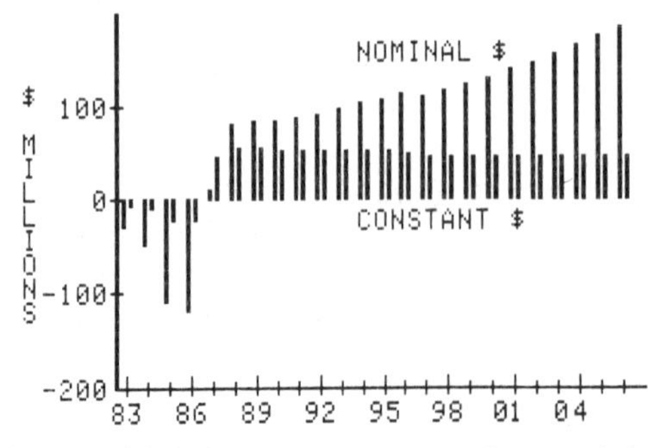

Fig. 3.7.8—Cash flow pattern for a mining property.

Three Most Common Approaches to Evaluation

The longer bars in Fig. 3.7.8 represent the nominal dollar cash flows of a mining project with an investment period of five years and a production life of 20 years beginning with the last year of preproduction investment. Development of this cash flow pattern requires assumptions of the following key variables:

(1) The first year and the last year of preproduction development expenditures;

(2) The first and last years of production;

(3) Product price at the beginning of production and escalation of this price over the life of the property;

(4) Expected level of operating cost for the first year of production and escalation of these costs over the life of the property;

(5) The rate of production for all future years;

(6) Preproduction capital investment and replacement investment and escalation of these costs over the life of the property;

(7) Royalty rates, depreciation rates, and tax rates over the life of the property.

The three most common approaches in cash flow analysis are the nominal dollar evaluation, the constant dollar evaluation, and the constant cash flow approach. First, consider the nominal dollar situation; timing, royalties, and the tax regime deserve specific consideration.

Nominal Dollar Analyses

The nominal dollar value of the cash flows shown in Fig. 3.7.8, when discounted at 15%, is $110 million. If the project is delayed for a period of 10 years, whether due to problems in marketing, development, startup, governmental regulations or other reasons, the net present value discounted at 15% declines to minus $35 million, which means it would be rejected for more attractive investment opportunities. Delays in project implementation also increase uncertainties involved in forecasting future economic conditions. This uncertainty increases with time therefore increasing the risk of the project.

Royalty payments are not included in the cash flows shown in Fig. 3.7.8. If there is a deduction for royalty payments, net cash flows and net present value (NPV) will decline. An alternative analysis introduces a federal royalty payment equal to 12.5% of revenues, and assumes that the surface owners will ask for an overriding royalty equal to 50% of the federal royalty. The total royalty of 18.75% reduces the net cash flows and the corresponding NPV at the 15% discount rate. The new NPV is $24 million or 22% of the $110 million NPV without royalty. Thus the valuation of the project is extremely sensitive to the royalty assumption.

Tax assumptions also have a strong impact on net present value. In the first analysis, the assumed effective federal, state and local tax rate on net taxable income is 40%. If this tax rate is increased to 50%, the nominal dollar cash flows decline, and the NPV is reduced to $63 million. If the increase in royalties from zero to 18.75% is combined with the increase in the effective income tax rate from 40% to 50%, the combined effect on the project is to reduce the present value to minus $9 million; the property could not meet the minimum acceptable return criterion of 15% after taxes.

Nominal dollar economic evaluation is based on forecasted costs and revenues over the life of the project. Each future year is a forecast of the level of revenue and cost expected in that year. The procedure for a 1984 evaluation might proceed as follows:

(1) Estimate preproduction capital investment in terms of 1984 dollars. Preparation of the cost estimate also yields a preproduction schedule which provides a time pattern for the investment expenditures. The nominal dollars are calculated by multiplying the base year (1984) estimate by $(1 + r)^n$ where r is the average rate of escalation per period (decimal percent) expected between 1984 and the year of the expenditure. The exponent "n" is the number of periods in the future that the expenditure is scheduled. Some companies may use continuous compounding, but the concept is the same. The objective is to predict the actual dollar expenditure for each year of the preproduction period.

(2) Escalate prices and costs. Prices and production costs over the life are not likely to be constant from year to year and are therefore escalated or de-escalated according to expectations. Rates are chosen based on historical data, forecasts of supply and demand and contract arrangements. They are the best estimate given available data.

(3) Calculate cash flows for each year of the project. These cash flows will depend on some unique characteristics of the project and the tax system involved.

During the preproduction investment phase of the project the investment expenditures will be modeled by negative cash flows which usually represent the actual expenditures less any tax credits which may reduce net cash flows. For example the investment tax credit or development costs which are expensed may reduce company taxes on income from other sources; the net cash flow for the new project is effectively reduced because of these tax benefits.

An example of production cash flows, assuming the U.S. tax system, follows:

$$CF = RV - AR - OC - IT - AD$$
$$DP - TX + AD + DP - AM - CE$$

Where:

CF = Net cash flow

RV = Annual revenues from sales of metals or concentrates

AR = Annual royalty assumed as a percent of gross revenues. Royalties in connection with minerals projects may be paid to private parties, the federal government or to Indian tribes. Rates vary widely from 5 to 20%.

Severence taxes imposed by some states on gross revenue are similar to royalties. The state of Colorado, for, example, has a severance tax on metallic minerals of 2.25% with the first $11,000,000 of gross income exempt. Other severance taxes may be on a per ton basis, for example, $.15 per ton on molybdenum ore with no exemptions or credits allowed against other taxes. Severance taxes may be imposed on gross proceeds or net proceeds of mining operations; the gross proceeds tax is easy to administer but the net proceeds approach has an advantage of not taxing a marginal producer out of business.

OC = Annual total cash costs including operations, general and administrative, and other overhead

IT = Annual interest charges on loans

AD = Annual depreciation calculated under the

Accelerated Cost Recovery System.

Depreciation on the property placed in service after 1980 in the United States is covered by the Accelerated Cost Recovery System (ACRS). This is a simplified form of accelerated depreciation over a three year, five year, 10 year, or 15 year recovery period depending on the type of property. Three year property includes property with a short useful life like vehicles, tractors, trucks. Five year property includes most equipment, office furniture, fixtures. Ten year property includes mobile homes and manufactured homes. Fifteen year property includes buildings other than designated as 10 year property. To calculate ACRS deductions, the investment must be divided into the various life categories and depreciated using the following schedule:

3-year property	**Cost Recovery Rate**
1st year	25%
2nd year	38%
3rd year	37%
5-year property	
1st year	15%
2nd year	22%
3rd through 5th year	21%
10-year Property	
1st year	8%
2nd year	14%
3rd year	12%
4th through 6th year	10%
7th through 10th year	9%

The percentage for 15-year real property depends on when you place the property in service during your tax year. The table that follows shows the percentages.

For the purposes of analysis, the investments are divided into three year, five year, and 15 year categories and the resulting depreciation schedule calculated.

DP = Depletion calculated as a percent of revenues minus royalties but limited to 50% of the net income after depreciation but before depletion

The depletion allowance is an important characteristic of the U.S. income tax law. The concept behind depletion is that the mining company has purchased and owns an orebody either through direct expenditures and land acquisition or through exploration expenditures. Cost depletion is usually figured by dividing the adjusted basis of the mineral property by the total number of recoverable units in the deposit then multiplying the resulting rate by the number of units sold during the tax year. The adjusted basis for the property is the original cost plus capitalized cost minus previous depletion allowed.

Percentage depletion is a certain percentage specified for each commodity multiplied by the gross income after royalties from the property during the tax year. The deduction is limited to 50% of the taxable income after all deductions except depletion and loss carry forward deductions but before the depletion calculation. The percentage depletion allowance varies by commodity as follows:

Commodity	*Percentage Depletion (%)*
Oil and Gas	15
Sulfur and uranium; and if from deposits in the United States: asbestos, mica, lead, zinc, nickel, molybdenum, tin, tungsten, mercury, vanadium, and certain other ores and min-	

Month Placed in Service

Year	1	2	3	4	5	6	7	8	9	10	11	12
1st	12%	11%	10%	9%	8%	7%	6%	5%	4%	3%	2%	1%
2nd	10%	10%	11%	11%	11%	11%	11%	11%	11%	11%	11%	12%
3rd	9%	9%	9%	9%	10%	10%	10%	10%	10%	10%	10%	10%
4th	8%	8%	8%	8%	8%	8%	9%	9%	9%	9%	9%	9%

erals including bauxite	22
If from deposits in the United States, gold, silver, copper, iron ore and oil shale	15
Coal, lignite and sodium chloride	10
Clay and shale used in making sewer pipe, bricks or used as sintered or burned lightweight aggregates	7½
Gravel, sand, stone	5
Most other minerals and metallic ores	14

The tax treatment of depletable resources is complicated by Internal Revenue Service regulations relating the preproduction investments to allow depreciation and depletion. For example:

(1) Acquisition or lease bonus costs for mineral properties or petroleum properties must be capitalized and recovered through the depletion allowance or abandonment loss tax deductions (Stermole, 1980, pp. 345–347).

(2) For minerals properties, prospecting (defined as any activity designed to localize an area of interest for exploration) must be capitalized into a cost depletion basis.

(3) For a mineral property, legal costs, recording fees, and assessment work costs are capitalized into the cost depletion basis.

(4) Exploration (defined as the activity of delineation of an ore body) is capitalized into a cost depletion basis or treated as expenses with no dollar limit and deducted in the year incurred.

(5) In the case of petroleum properties, intangible drilling costs are distinquished from tangible drilling costs for purposes of taxation. Intangible drilling costs include site preparation, rig placement, labor, fuel, repairs, hauling, supplies, and fracturing. These costs do not create depreciable assets so they are expensed as operating costs in the year incurred or capitalized for cost depletion. In contrast, other investment expenditures create tangible depreciable assets which are recovered through the accelerated cost recovery system.

TX = Effective federal/state/local tax rate calculated as a percent of the taxable income.

The corporate tax in the United States is assessed on taxable income defined as gross income less deductions. Gross income includes gross receipts less the cost of goods sold from inventories. Deductions include business expenses including total operating costs, interest, depreciation, and depletion. The current federal corporate tax rate schedule as follows:

Taxable Income	Tax Rate
$0 - $25,000	16%
$25,000 - $50,000	19%
$50,000 - $75,000	30%
$75,000 - $100,000	40%
Over $100,000	46%

State taxes are likely to be of the order of 10 to 15% of the federal tax. In preliminary analyses a single combined federal-state income tax rate is sometimes used combining the effect of both.

AM = Annual amortization of loans

CE = Annual capital expenditures recovered subsequently through depreciation deductions. Under certain conditions small capital expenditures may be expensed and charged against operating cost providing "instant depreciation" in the year of the expenditure. Capital expenditures during the preproduction phase may be reduced because of the ability of a producing company to expense development of the mine against current income from other properties. Capital expenditures may also be reduced from special incentives such as the investment tax credit.

(4) Discount the cash flows to calculate net

present value and the rate of return on the project; these are the two investment criteria most widely accepted in the United States. If the investor or the host country is represented by a foreign government, other criteria may be relevant, for example, "retained value" which sums the total of taxes, wages and salaries, local expenditures and similar measures that accrue to the country in the form of foreign exchange. The discount rate is usually described as the rate of interest that the investor could achieve on alternate investments, his "opportunity cost."

Constant Dollar Analyses

The procedure for converting nominal dollar cash flows to constant dollar cash flows requires discounting with the factor $1/(1 + f)^n$ where f is the average expected rate of general inflation over the life of the project, and n is the number of years in the future that a cash flow is generated. The product of this factor and the nominal dollar cash flow is the constant dollar cash flow where the element of general inflation has been removed by the discounting process.

In the cash flow pattern of Fig 3.7.8, the shorter bars are the constant dollar cash flows. To calculate the constant dollar cash flow, a factor of general inflation is used to discount all nominal dollar cash flows. In the example, the assumed general inflation rate is 6% and nominal dollar forecasts are 6% for costs, 6% for revenues and 10% for investment expenditures. Even though the percentage escalation applied to revenues and costs are equal, the cash flows grow in absolute terms because the increases in revenues are greater than the increases in costs. When these same cash flows are discounted by the factor of general inflation also equal to six percent, the resulting constant dollar cash flows decline over the life of the project because some components of the cash flows are escalated, and others like depreciation do not escalate. When items, like depreciation, are not escalated under conditions of high inflation, the investor is likely to find himself paying higher and higher taxes and his net cash flows in real terms declining year by year.

Constant Cash Flow

There is a third type of analysis holds the difference between revenues and costs constant over the production life. Often the investment is estimated in nominal dollars; then the cash flow of the first year of production is estimated and held constant over the life of the project except for changes in production rate or technical changes that affect production cost. This becomes a mixture of nominal dollars and constant cash flows (not constant dollars) that can not be described as nominal or constant dollar analysis. This approach has been made obsolete by the availability of the computer which easily allows for modeling of individual year cash flows whether expressed in nominal or constant dollars.

Other Relevant Factors

There are socio-economic, legal, regulatory, environmental, and financing factors that are not fully quantified in the preceding decription of NPV and rate of return analysis. It is possible to model potential delays in the project because of environmental blocks to development or to other legal, regulatory, or socio-economic problems. The problem is in assigning probabilities of occurrence to these factors. They usually influence the investment decision via subjective judgment.

SELECTION OF THE DISCOUNT RATE

A common approach used by most mineral firms is to discount all cash flow streams at the same discount rate and then to evaluate the results including an adjustment for the differing risks of the projects. This adjustment, usually subjective, is a practical solution to avoid the confusion of comparing the value of several projects, each discounted at a different rate.

The concept of a "risk adjusted" value is an extension of the Capital Asset Pricing Model (CAPM). The Capital Asset Pricing Model states that the required return on any asset is composed of the interest rate on riskless debt plus a risk premium which is a function of the asset's systematic risk. In terms of an equation the following applies:

$$ER_j = R_f + (R_m - R_f) \times B_j$$

where: ER_j = the expected return required to make asset J competitive with other investments

R_f = a risk-free return usually measured by the return on government bonds

$(R_m - R_f)$ = the market risk premium which is a measure of market returns where unsystematic risks are eliminated through diversification

B_j = a factor to adjust for the riskiness of the individual asset J

The equation is the security market line (SML) which measures the relationship between expected returns and increasing risk. In the securities business the risk premium is a statistical calculation based on the concept of the expected volatility of the individual investment relative to market and to safe rate returns. The model is theoretical and based on portfolio analysis of securities. Its advantage is that it provides a concept for describing risk. Its disadvantage is that it is based on a series of very restrictive assumptions if it is accepted literally.

An important concept behind the CAPM is that the riskiness of an investment can be measured by the dispersion of probable outcomes as illustrated in Fig. 3.7.9. Thus the investment can be evaluated under alternative possible outcomes and the dispersion measured. The coefficient of variation of returns is a measure of risk.

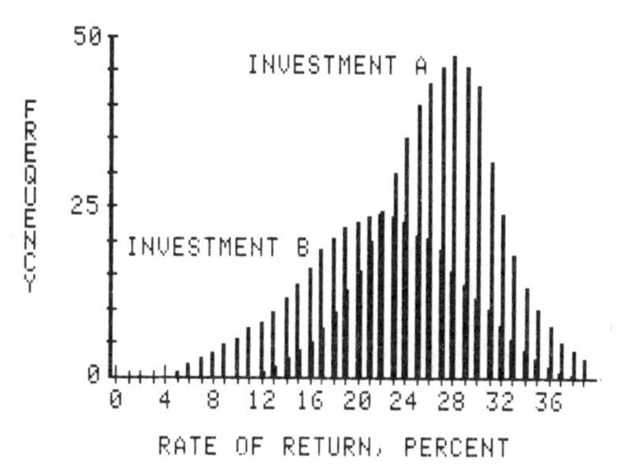

Fig. 3.7.9—Dispersion of cash flows as a measure of risk.

Safe Rate Plus Premium for Risk

Lessard and Graham describe the project risk of extractive ventures in terms of the following risk elements (Lessard and Graham, 1976). Some elements of risk are described as "systematic" which is that proportion of risks common to all of the risky assets available to investors. It is this systematic risk that can not be reduced or eliminated by diversification. Other risks are described as "unsystematic" which can be reduced or eliminated by diversifying the portfolio of investments. The concepts are extended to extractive ventures as follows:

(1) *Mineral reserve risks* which reflect the uncertainties associated with various orebodies, independent from each other and from general economic behavior. This element therefore contributes little or nothing to systematic risk of the project.

(2) *Development risks* which are subject to the uncertainties of cost and timing including delays in design, installation, and phasing-in of different parts of the project. Because these cost overruns have a profound affect on present value, they are critical in economic analysis. They are largely unique to the project and are considered mostly unsystematic.

(3) *Operating risks* include those risk elements that lead to changes in the cost of production or disruptions in the flow of minerals. These include strikes and slowdowns by labor and unanticipated failure of equipment. Some operating risks, such as equipment breakdown, overlap development risks which are unique to projects, therefore, diversifiable and unsystematic. Some labor related risks are considered systematic because of their correlation with general economic conditions.

(4) *Market risks* include the risk of not being able to sell full output or encountering unexpected changes in price. These risks are complex and depend on how much control the producer has on selling price as well as the cost structure of the producer. For example, a high fixed cost operator will show much greater fluctuation in profitability for a given change in price than will a low cost producer. Because of the close relationship between market risks and economic conditions, they are considered "systematic" therefore are handled by a discount rate that increases with the risk.

(5) *Political risks* include the range of government actions which affect project cash flows. It is often the least predictable of the five categories and it is difficult to decide whether the risks are systematic or not. In general it is assumed that they are unique to the project and can be reduced through diversification.

The preceding suggests that the discount rate

for a specific project should vary in proportion to those ''systematic risks'' which can not be reduced or eliminated by diversification. In addition, it is possible to identify the risk elements which vary from one stage of a project to another and further complicate the calculation of the present value of a project. The practical response to these complications at the present time is to settle on a single overall cost of capital for a firm to be used on all projects and then to make subjective adjustments of the value as specific elements are considered.

Cost of Capital

The following illustrates calculation of a weighted average of the firms cost of debt and equity. The mining company's cost of equity is 18%; before tax cost of debt is 12% and its tax rate is 40%. The following balance sheet data allow calculation of an after-tax weighted average cost of capital.

Assets

	($000)
Cash	$ 100
Accounts receivable	200
Inventories	300
Plant and equipment, net	1,800
Total asets	$2,400

Liabilities

	($000)
Accounts Payable	$ 200
Accrued taxes due	200
Long-term debt	400
Equity	1,600
Total Liabilities	$2,400

The sum of long-term debt plus equity for the firm is $1,600 + $400 = $2,000. The proportion of long-term financing that is debt is 400/2,000 or 20%. The proportion that is equity is 80%. Interest is deducted before taxes therefore its after tax cost is $(1 - .40) \times 12\% = 7.2\%$. The weighted average cost of capital therefore is $.2 \times 7.2 + .8 \times 18 = 15.8\%$. The figure represents a historical weighted average cost of capital but it is not the relevant discount rate for investment decisions because the cost of new financing (the marginal cost of capital), not the weighted average historical cost is the relevant cost to be used. To calculate the marginal cost the cost of additonal debt and equity financing must be calculated as follows

Cost of additional debt at 12% rate

$$.12 \times (1 - .4) = 0.072 \text{ or } 7.2\%$$

Cost of additional equity (internal) where:

(1) the dividend at the end of period one (D_1) is $2.00 and is growing at 5% per period,

(2) the price of the stock at time zero (P_0) is $50 per share, and

(3) the growth rate over period one (G) for stock price is 5% $(D_1/P_0) + G = (\$2.00 \times 1.05/50) + .05 = .092$ or 9.2%

The weighted cost of capital where additional capital will consist of 30% debt and 70% equity

$$0.30 \times 7.20 + .70 \times 9.20 = 8.6\%$$

The preceding calculations are based on historical data that include assumptions of diversification by the firm to minimize or eliminate unsystematic risks. There will be a risk adjustment for the specific project.

Constant Dollar Discount Rate

The discount rate for calculating net present value has been described as a minimum acceptable rate that represents the minimum expected return if the capital was invested in alternative projects of equal risks.

Cash flows may be expressed both in nominal and constant dollars. The meaning of net present value or rate of return varies under the nominal versus constant dollar assumptions. The nominal dollar cash flows include specific assumptions with regard to price and cost changes over the life of the project; this increase can be thought of as general inflation with a plus or minus correction for each component. This correction may reflect future expectations for supply, demand or special cost impacts such as increasing energy costs above and beyond the general inflation rate. For example, a specific price may rise faster than general inflation if the demand/supply conditions indicate shortages; a specific cost may rise slower than general inflation if an excess supply is expected over the life of the project.

The appropriate discount rate for net present value or the appropriate minimum acceptable rate of return must vary with the nominal and constant dollar analyses. For a specific pattern of nominal dollar cash flows and the assumption of a general inflation rate, the relationship between the appropriate rate for the nominal dollars and the appropriate discount rate for the constant dollars is mathematically determinable (Stermole, p. 169).

$$\frac{1}{(1+i)^n} = \frac{1}{(1+f)^n} \times \frac{1}{(1+i')^n}$$

$$(1+i) = (1+f) \times (1+i')$$

$$i' = \frac{1+i}{1+f} - 1$$

where: i = current dollar rate of return
i' = constant dollar rate of return
f = assumed rate of general inflation

Thus for a nominal dollar discount rate of 15% and an assumed general inflation rate of 6%, the appropriate discount rate for constant dollars is:

$$i' = \frac{1.15}{1.06} - 1 = .085 \text{ or } 8.5\%$$

The computer programs used for the case study examples in this chapter use a specified discount rate for nominal, then reduced the discount rate for constant dollars using the preceding formula. When this is done the NPV of the nominal dollar cash flows discounted at the current dollar rate is exactly equal to the NPV of the constant dollar cash flows discounted by the constant dollar discount rate.

The preceding description is over simplified. Each component of price or cost in the nominal dollar analysis has its own escalation rate and its own relationship to general inflation. The average escalation for the final cash flows is implicit in the assumptions for all of the components with corrections for taxes. In some cases it may be logical to vary escalation rates for components or the general inflation rate over the life of the project, therefore the analysis is not only a composite but also varies with time. The micro-computer makes all of this as easy as specifying the assumptions for each component of cash flow and the assumption for general inflation; first the nominal dollar net present value and rate of return are calculated; then the constant dollar and net present value and rate of return are calculated; then the constant dollar and net present rate of return are calculated using an appropriately corrected discount rate. Some evaluators prefer ranking projects on constant dollar rate of return rather than nominal dollar rate of return because the constant dollar approach reduces error due to errors in assumed escalation rates.

CRITERIA FOR DECISION MAKING

Current practice in economic evaluation combines a list of criteria for decision making with sensitivity and simulation analysis to provide a basis for economic evaluation of investment alternatives. The fundamental objective underlying all quantitative techniques is to maximize the net present value of the firm or the after tax earnings per share for stockholders. Variations on this central theme include maximization of rate of return where the investor expects to amortize his investment and obtain a rate of return or interest on the amortized capital over project life which are optimum. Simple definitions of financial viability are made complex by a long list of variables which must be forecasted to determine what is "optimum" or competitive. Some characteristics of an investment opportunity may be extremely difficult to quantify and require subjective estimates.

Other criteria may be required when the investor or host is a government. Concepts such as cost/benefit ratio and retained value are designed to reflect governmental objectives to maximize foreign exchange or to measure benefits and costs not included in the usual investment for private capital. In some cases employment, the acquisition of new technology or management expertise, market connections, or new infrastructure may be important objectives. Then, the criteria selected for evaluation will include measures beyond those accepted for private capital decision making.

Net Present Value

The present value of a future sum is that cash flow which when invested at the given interest rate compounded over the given period of time will equal the future cash flow. In a sense, investors are buying future cash flows and they are interested in the purchase price for these future cash flows which is competitive with other investment opportunities. This purchase "price" may be the cumulative expenditures in land acquisition and exploration, or it may be cash in the form of a bonus payment, royalty agreement or cash price.

A mining project usually consists of a number of negative cash flows modeling after tax investment and a number of positive cash flows representing sales less all cash costs. In an analysis the positive and negative cash flows will be discounted and the net present value calculated as indicated in the following formula:

$$\text{Net Present Value} = \text{Sum}\,\frac{1}{(1+i)_n} \times CF_{Inv} + \frac{1}{(1+i)_n} \times CF_{op}$$

In the formula the CF_{Inv} represents the negative after tax investment cash flows and CF_{op} represents the after tax operating cash flows of the investment. The symbol i is the discount rate.

Discounted Cash Flow Rate of Return

Discounted cash flow rate of return is the interest paid on unamortized capital over the life of the project. The firm or organization will think of terms of a ''hurdle rate'' which is competitive with the other investment opportunities available. When investment and operating cash flows are discounted at the rate of return, the net present value is zero.

The Reinvestment Question and Growth Rate Analyses

The discounted cash flow rate of return measures the average rate of interest paid on the unamortized balance of the original capital investment. Because of this fact, two rate of return calculations may be identical yet have a different total cash flow. Which investment is preferred will depend on the opportunity that the investor has for his return capital. If the opportunities are likely to be higher than the project under consideration, the investor will probably prefer rapid amortization. If future opportunities are likely to be lower returns, the investor probably will prefer the slower amortized project. For purposes of analysis, it may be desirable to put alternative analyses with different amortization rates on the same basis by introducing an assumed rate of reinvestment of amortized capital.

Growth rate of return is, by definition, the compound interest rate expressed as a percent at which investment dollars grow, for example, the interest paid on bank deposits which compound periodically without amortization (Stermole, p. 54). In contrast to the unamortized bank deposit, a discounted cash flow rate of return is a percent rate on the unamortized capital. If the analyst fails to distinguish between the growth rate for capital and the discounted cash flow rate of return, there will be a built in assumption that the amortized capital is reinvested at the rate of return of the project. Comparison of these two entirely different concepts is responsible for much confusion in the literature sometimes leading to the erroneous statement that the discounted cash flow rate of return implicitly assumes reinvestment of cash flows at the calculated rate of return (Stermole, p. 57).

One logical approach to the reinvestment problem is to make some specific forecasts of expected future investment opportunities in order to decide whether a rapid amortization or a slow amortization is desired. In project analysis, the existence of any other variables contribute to the investment decision; thus it is unlikely to be necessary to make quantitative corrections for reinvestment especially if the reinvestment rate is assumed as the company minimum acceptable return rather than perceived variations in future investment opportunities.

When a growth rate of return is desirable, the simplest approach is to assume reinvestment at a given rate, calculate a future value at the end of the project including reinvestment, and then discount the future value to time zero. The results introduce the impact of rapid or slow amorization and put all alternatives on a growth rate basis.

Benefit-Cost Ratios

Benefit-cost ratios often are used on governmental projects where the objective is to quantify as best possible both tangible benefits such as operating cash flows and intangible benefits measured in a different manner. The cost may also be difficult to quantify and include both tangible and intangible components. Usually both the benefits and the costs are discounted to calculate a present value of benefits and a present value of costs. The discount rate selected for governmental investments is often lower than the discount used by private capital giving weight to governmental objectives over the longer term. The benefit-cost ratio can be expressed as follows:

$$\text{Benefit-Cost Ratio} = \frac{\text{Present Value of Benefits}}{\text{Present Value of Costs}}$$

Retained Value

Objectives of host governments in developing countries often seek to measure foreign exchange and other resource impacts resulting from mineral industry investments; special criteria for evaluation are required. Mikesell (Mikesell, 1971, pp. 23–25) discusses a series of indicators designed to reflect economic effects of investment in host countries. Some of the impacts are fairly easy to express in quantitative terms and others tend to defy quantification. The concept of ''retained value'' is used in several of the case studies in the book to quantify directly measurable impacts of a mineral investment. Retained value or net foreign exchange effect is expressed by the following equation:

$$R_{es} = W + L_p + T_a + T_r$$

where: R_{es} = Retained value
W = is wages and salaries less remittance abroad of nonnationals
L_p = local purchases
T_a = tax payments to government
T_r = other transfers

These sources of foreign exchange may be calculated by the formula

$$R_{es}/RV = \%\text{ of project revenues}$$

where: RV = annual revenues generated by the project

These impacts do not take into account certain indirect effects such as (1) the potential of the investment to stimulate other local projects, (2) the potential for improved labor, technical, or managerial productivity, (3) forward and backward linkages which may result in development of local supplier industries or downstream processing. This measure and others like it are not likely to be part of a U.S. domestic evaluation but may be extremely important to

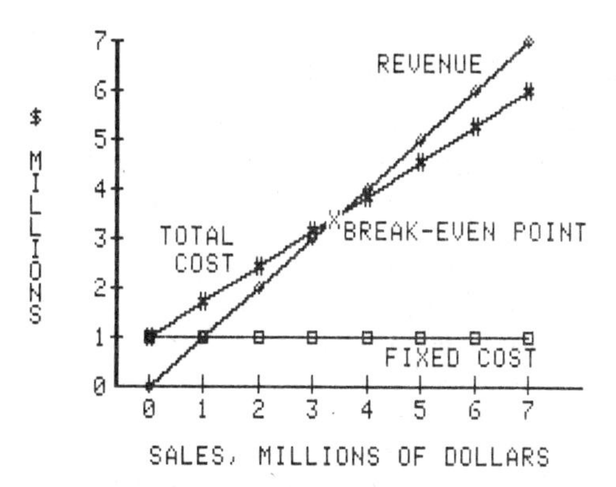

Fig. 3.7.10—Breakeven chart.

those considering foreign investment in developing countries.

Breakeven Analysis

The term ''breakeven analysis'' is used in two different contexts, the first expressing the relationship of income to total cost as a function of operating rate or level of sales (Weston, pp. 224–235). As indicated in Fig. 3.7.10 the breakeven point is that where revenue is equal to total cost. Above the breakeven point operations are profitable and below the breakeven point there is a loss.

The equation relates revenue and total cost as follows:

$$P \times Q = v \times Q + F$$

Where: P = unit price
Q = units produced and sold
v = variable cost per unit
F = level of fixed cost

The breakeven technique may be used in a number of ways for investment analysis:

(1) Analysis of required sales volume for profitable operation.

(2) Analysis of the risk of operating at a loss due to changes in price, volume or cost.

(3) Analysis of the risk of ''leverage''; the risk associated with increased fixed cost due to interest payments resulting from debt financing.

(4) Analysis of the financial effects of a potential expansion or shut down of operations.

(5) Analysis of a change in technology for a given project that would shift variable costs to fixed costs or vice versa.

(6) Analysis of pricing policy which may af-

fect volume of sales and is related to the risk of operating below breakeven point.

A second use of the term "breakeven" is usually in connection with rate of return analysis of cash flows of a project where the dependent variable is changed in order to aid in an investment decision (Stermole pp. 225–227). An initial question may be stated as follows:
"What is the rate of return of this project given data on production rate, prices, costs, taxes, and timing over the life of the project?

The question may be rephrased as follows:

What price is necessary to yield a minimum acceptable rate of return of 15% given data on production rate, prices, costs, and taxes and over project life?

Sometimes it is valuable to identify the breakeven price in order to compare this price with what seems reasonable for the product.

An example of the required breakeven price for uranium oxide follows:

Given the amount and timing of investment and production cost data, the derivation of required U_3O_8 price is a four step process.

(1) Estimation and updating of capital and operating costs based on the available information.

(2) Calculation of the required net operating cash flow necessary to amortize and pay a 12% return on the capital investment.

(3) Determination of the required revenue necessary to meet the cash flow requirements calculated in step two. This requires knowledge of production costs and the depreciation, depletion, tax and royalty rates.

(4) Division of the total revenue requirement determined in step three by the expected annual production of U_3O_8. This requires knowledge of the tonnage operating rate, metallurgical recovery, and ore grade of the deposit. The result of the calculation is the U_3O_8 price.

When the discounted cash flow method of analysis is used on an individual project and the discount rate is the minimum acceptable rate the result indicates the minimum price required to make the project look attractive. The price calculated must be sufficient to recover all investment, pay all cash costs, and yield a sufficient margin to pay the minimum acceptable rate of return on invested capital.

EXAMPLES OF INVESTMENT RISK ANALYSIS

Minerals

The example is a hypothetical underground molybdenum mine similar to the Henderson mine in the state of Colorado. The analysis was made using a computer program in Applesoft BASIC and an Apple II Plus computer with 64 kilobytes of memory.

We assume that the technical analysis of the mineral property indicates proved and probable ore reserves of 303 million short tons of recoverable ore capable of producing almost seven million pounds of molybdenum sulfide concentrate during the first year of production rising to almost 54 million pounds at full capacity during the fourth year of production. The deposit is a stockwork with molybdenite distributed throughout or near contacts of associated siliceous intrusives. Molybdenite occurs as veinlets associated with quartz, and lesser amounts of other sulfides, oxides, and gangue, fissure veins, paint in fractures, breccia filling, and rarely in disseminated grains.

The geological evaluation describes an orebody contained within an Oligocene intrusive complex at a depth of 3000 to 4000 feet below the surface. Multiple intrusions of rhyolite porphyry form a subvolcanic stock that enlarges at depth in Precambrian granite country rock. Granite porphyry intrusions at depth host the quartz-molybdenite stockwork that forms the orebody. The orebody is elliptical in plan (3000 ft × 2200 ft) and elongate in a N 30° E direction. Ore thickness varies from 400 to 800 feet with an arcuate section concave downward. Alteration and mineralization halos outward and upward from ore are similar to those found at typical porphyry-type deposits in the western United States.

The mining concept for the orebody is a panel caving system with rock classified as "fairly-difficult-to-cave." Three shafts handle waste rock and transport men and materials underground and provide ventilation outlets. Mining equipment includes load-haul machines for ore transfer, drilling machines for development activities and associated equipment for service, compressed air and pumping. Mine development openings comprised a substantial portion of the total preproduction investment.

The mining rate at full capacity is 30,000 short tons per day mill feed for 10.5 million tons per year assuming 350 operating days. The stoping method is a high output, mechanized panel caving system in an orebody defined as suitable for a caving but with rock competent enough to support itself in relatively large openings necessary for operation of the load-haul-dump (LHD) units used for production. The high capacity, rubber-tired units deliver ore from drawpoints to loading chutes above the main haulage level for delivery to the unit trains carrying the ore to the mill through the main haulage tunnel.

The flow sheet of the mill is the basis for capital and operating cost estimates. A large primary gyratory crusher reduces mine-run ore to minus nine inches. Coarse ore is delivered to a semiautogeneous grinding system in closed circuit with cyclone classifiers. Mill undersize is delivered to rougher flotation cells, followed by four cleaner stages, interspersed with three regrind stages. The final product is a 90 to 95% molybdenum sulfide concentrate which is filtered, dried, and packed for shipping. The product of a technical analysis is a flow sheet of the plant with sufficient detail for take-offs of equipment and estimation of the capital and operating costs.

Price and volume assumptions for the economic evaluation are based on a market study. Most molybdenum contained in molybdenum sulfide concentrates is priced based on metal content f.o.b. Climax, Colorado. The world price is based on a minimum 85% molybdenum sulfide concentrate and the U.S. price is based on a 95% molybdenum sulfide concentrate f.o.b. shipping point.

Molybdenite produced as a byproduct of copper output is sold at a discount from the Climax price depending on copper, lead, and impurity content. U.S. molybdenum prices rose from $1.72 per pound contained molybdenum in 1969 to $3.45 per pound at the end of 1976. Subsequent price increases were even more spectacular. By December of 1979, Climax concentrate for export was selling at $8.84 per pound of molybdenum contained in molybdenum disulfide. Some byproduct molybdenum from the copper mines was selling at $20 to $23 per pound rather than at a discount of the Climax price. Price "trends" are a dangerous basis for price forecasting. This is made very clear by the collapse of molybdenum price after 1980. The price of molybdenum in Climax concentrate in November of 1983 was $3.60 per pound of molybdenum metal contained in sulfide concentrates.

The project is evaluated on a 100% equity basis. Capital expenditures are assumed met from retained earnings of the company supplemented by loan and debenture funds secured by assets of the company rather than by the project itself. Thus while there is no debt directly connected to the project, the investor will look for a "higher leveraged" return on equity capital because of the debt risks assumed by the company.

The micro-computer program used for the financial analysis requires the input data shown in Table 3.7.2.

The project investment excludes sunk exploration costs prior to the decision point for development of the new mine; these costs are assumed previously expensed or recovered through future percentage depletion. Accelerated depreciation of mine and mill facilities is calculated based on the Accelerated Cost Recovery System (ACRS). Percentage depletion for molybdenum is based on an allowance of 22% of the gross revenue limited to 50% of the net revenue after depreciation but before depletion.

Cash Flow Pattern: During the preproduction investment phase of this minerals project the cash flows are calculated as follows:

$$CF = INV - DTC - ITC$$

where: CF = nominal dollar cash flows during preproduction investment

INV = nominal (escalated) dollars invested in property, plant and equipment prior to tax deductions

DTC = development tax credits. Certain development costs can be expensed against other income of the company provided a deduction in the year incurred rather than capitalizing the expenses for recovery by depreciation. In the example, 28% of the escalated investment costs are deductible development costs and the tax savings are the company marginal tax rate of 46%.

ITC = investment tax credit which is a direct credit against taxes paid on other income. This is a one time tax deduction to provide incentive for investment. It reduces tax liability by up to 10% if the investment qualifies. To qualify it must be depreciable, tangible, property with the useful life of at least three years and must be placed in service during the year of the deduction. The credit does not reduce the basis for future depreciation. In the example, 57% of the total escalated investment is assumed subject to the investment tax credit at an average rate of 9.5 percent.

Cash flows resulting from production may be summarized as follows:

$$CF = RV - AR - OP - IT - AD - DP - TX - AM + AD + DP - RI$$

where: CF = nominal dollar cash flows resulting from production

RV = revenues calculated as the product of annual tonnage times ore grade times metallurical recovery times price per unit

AR = annual royalty if applicable calculated as a percent of the gross revenue

OP = annual total cash costs of production exclusive of taxes

IT = annual interest paid on loans

AD = annual depreciation calculated under the accelerated cost recovery system (ACRS)

DP = annual depletion deduction

TX = annual income taxes on the taxable base which is

$$RV - AR - OP - IT - AD - DP$$

AM = annual amortization of loans

RI = replacement investment capitalized and recovered later through accelerated cost recovery system

The pattern of cash flows for the project are shown in Fig. 3.7.11.

Table 3.7.2—Data and Assumptions for Analysis of a Molybdenum Project

First year of investment cash flows	year 1
Last year of investment cash flows	year 12
Discount rate for calculation of net present value	10%
Total investment prior to escalation in $ of the first year	300,500,000
Investment tax credits	average 9.5% on 57% of total investment
Tax for expensing of development costs	average 13.2% of investment prior to escalation
Distribution of investment cash flows	distributed over years one through 12
General inflation rate	7%
Ore reserves	303,000,000 short tons
Escalation rate for investment expenditures	7.6%
Assumed price escalation over project life	6%
Assumed cost escalation over project life	6%
First year of production	year 9
Last year of production life	year 20
Assumed mine life	year 28
Price during first year of production	$3.60 per pound molybdenum in molybdenum sulfide concentrates
Loan data	none
Rate of production	10,500,000 short tons annually at full production
Startup production during first year	13.3% of full capacity and rises to full production after four years
Cost per ton first year of production	$9.93 per ton
Royalties	none
Overall tax rate, federal, state, local	50%
Average ore grade to mill including dilution	0.49% MoS_2
Depletion rate for tax purposes	22%
Metallurgical recovery	85% during first year of startup increasing to 87% by the third year and thereafter
Productivity	12.1 tons per man shift of total employment

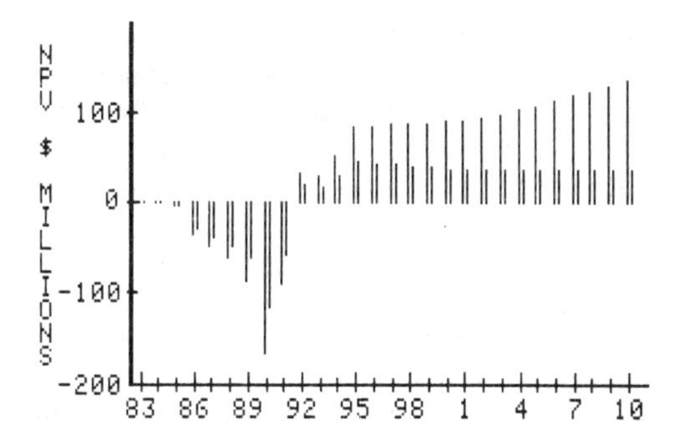

Fig. 3.7.11—Nominal dollar and constant dollar cash flows of a molybdenum project.

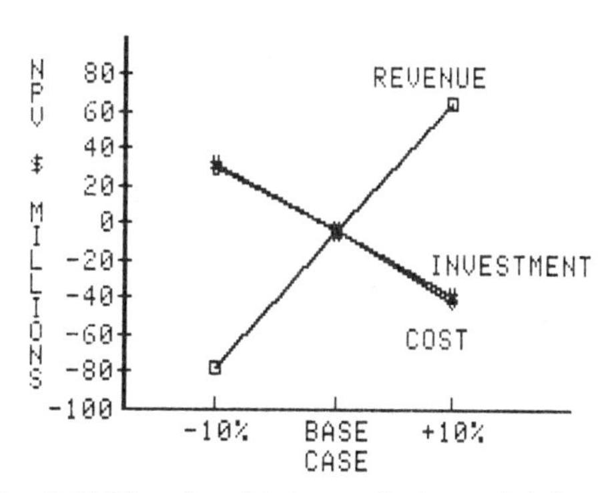

Fig. 3.7.12—Sensitivity analysis: molybdenum project.

Sensitivity Analyses: The first analysis of the investment will probably be a "most likely" or "base case" evaluation. But the single point estimate of such an evaluation must rely on precise assumptions for a large number of variables listed in Table 3.7.2 that are forecast only with uncertainty. Additional or "sensitivity" analyses use alternative possible outcomes which provide a basis for assessing the risks of the project. Risk usually refers to situations where probabilities can be estimated and uncertainty to situations where the probabilities of various outcomes are unknown. The variables in an analysis consist of a mixture of risk and uncertainty as well as the risk due to uncertainty. In this example we assume 10% charges in revenue items, cost items, and investment items. The impact on base case results of these revised assumptions is summarized in Fig. 3.7.12. Fig. 3.7.12 indicates that the value of the discovery is most sensitive to changes in revenue, followed by changes in operating cost and investment. Fig. 3.7.12 is for illustrative purposes only. A complete sensitivity analysis must consider each component of cash flow separately from the point of view of the probability of a specific change and the likely magnitude of a change. Unique characteristics of each investment are evaluated using sensitivity analysis which in turn may indicate the need for development of more precise estimates through additional market surveys, contract negotiation, pilot plant or pilot mine operations or any other means of reducing uncertainty surrounding specific variables.

Risk Analysis Using Simulation: Simulation allows the investment evaluator to describe any or all values in terms of distributions of possible values, for example a distribution of possible price, costs, investment, recovery, or ore grade. Expression of each parameter in terms of a probability distribution makes it possible to select a single value for each of a series of calculations of present value or rate of return. As the process is repeated a distribution of results reflects the distributions of the input values. The simulation method sometimes described as "Monte Carlo simulation" has the advantage over point estimate sensitivity analysis in that all parameters can assume values over a range according to the probability distribution chosen for the parameter. The limitation is that very often the evaluator must subjectively estimate the probability distribution or he may have no idea at all other than some distribution exists because of the uncertainty involved. The reader is referred to discussions of simulation by Newendorp (Newendorp, 1975, pp. 369–490), Stermole (Stermole, 1980, pp. 157–164) and McCray (McCray, 1975, pp. 190–221). The example here is one of a large number of possible approaches to simulation. The example uses a probability density function in the form of a triangle. In a specific case other distributions may be more representative for a parameter. They may be discreet or continuous, described by a histogram of existing data or determined by judgment; they may be triangular indicating that a most likely value is identifiable or rectangular indicating that evaluator has no idea of the most likely value, only an idea of the range. On each pass of the simulation procedure one value will be selected for each of the parameters and the decision criteria calculated.

Assume that a specific parameter for investment analysis is modeled by a triangular distri-

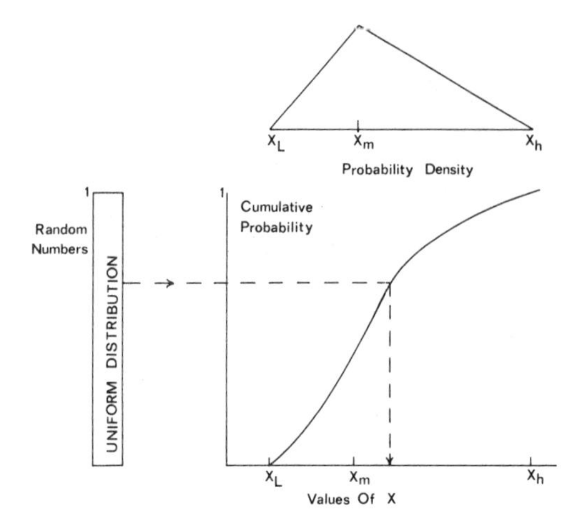

Fig. 3.7.13—Selecting random values from a triangular distribution.

bution such as the one shown in Fig. 3.7.13. If the parameter is ore grade, a range may be a much better description of our present knowledge than a specific point estimate. The value of X_m is the most likely expected ore grade while X_L is the lowest possible average grade to the mill and X_h is the highest possible grade to the mill.

In our example, assume the value of X_L is .42% molybdenum disulfide, the value of X_m is .49% molybdenum disulfide, and the value of X_h is .60% molybdenum disulfide. These numbers correspond to a range of ore grade for a hypothetical ore body only partially explored. Note that the use of a low estimate (X_L), a most likely estimate (X_m), and a high estimate (X_h), allows us to objectively or subjectively express our uncertainty about the grade of the new deposit without specifically assigning probabilities to their occurrence.

The probability density function tells us that there is no chance at all of the grade being below X_L or above X_h. The probabilities of each of these values is 0. Even though we have not specified the probabilities of occurrence between X_L and X_h they are implicit because the area enclosed by the triangle equals one and between X_L and X_h we say that there is a 100% probability that the actual average grade has been bracketed.

Given the triangular distribution it is a matter of simple mathematics to calculate the cumulative probability function shown in Fig. 3.7.13. With the cumulative distribution we can use a random number generator in the form of an equation in our computer program. A uniform distribution generates random numbers. Each time a random number is generated the corresponding value of the ore grade is determined from the cumulative probability function.

The chart below shows the results of the first simulation and the two hundredth simulation. The random number has meaning in terms of the triangular distribution. In this case we assume that the lowest conceivable grade for the orebody is 0.40% MoS_2, that the most likely grade is 0.45% MoS_2, and that the highest conceivable ore grade is 0.60% MoS_2. Note that the ore grade is not necessarily the average grade of the deposit because it includes dilution during mining.

The random numbers come from a distribution uniformly distributed between zero and one. The random number 0.904 used in simulation number one can be interpreted as meaning that 90.4% of the ore grades calculated will be below the 0.55% MoS_2 using the probability distribution selected. Likewise, in the two hundredth simulation 34.3% of the ore grades generated can be expected to be below 0.46% MoS_2. For this example we assume a perfect correlation between ore grade and metallurgical recovery. This may or may not be true for a specific deposit but it is important to recognize correlation between parameters that are not independent. In this case the ore grade used to calculate the metallurgical recovery assumes that a constant

Project Economics Simulation

Simulation Number	Random Number	Ore Grade % MoS_2	Metallurgical Recovery %	NPV 12% ($ million)	DCFROR (%)
1	0.904	0.55	91.8	53.1	15.2
..					
..					
..					
200	0.343	0.46	90.2	2.4	12.2

.9 pounds of molybdenum sufide will be lost in the milling process, therefore, metallurgical recovery varies with ore grade.

The results of two hundred simulations are summarized in Fig. 3.7.14. The most frequently calculated rate of return is between 12.1% and 12.3% and 66% of the simulations are between 11.2% and 14.1% discounted cash flow rate of return. Stating the results in another way, we can relate to the probability of being at or below our minimum acceptable return. For example, 27% of the simulations result in a rate of return of 12% or less. Fifty percent of the simulations result in a rate of return below 13%. Eighty-six percent of the simulations result in a rate of return of 15% or less.

Great caution must be used in interpreting the results of the simulation. In this case it is simply an indicator of our uncertainty in grade. The actual outcome of the project will either be profitable, marginal, or uneconomic. The miner, unlike the petroleum explorationist, does not have the luxury of playing the game a sufficient number of times to balance losses against gains. There is one ore body and essentially one decision to be made, to go ahead with development or to delay it. Nevertheless the simulation approach does seem helpful in quantifying risk.

Of course any of the other variables in the analysis can also be modeled in the same way if a subjective or objective probability density function can be estimated. In this case all of the uncertainty in the outcome is due to the relationship of project economics to grade and metallurgical recovery.

The question of dependency between parameters or variables is important enough to mention a second time. In the example analysis, a base year price is selected using the simulation technique but the price is escalated at a given rate indicating perfect correlation from one year to the next. This may introduce an error because it is reasonable to expect that this correlation may not be perfect but subject to a variation requiring another use of simulation. The amount of such detail built into a computer program will depend on the value of the additional "realism" to the evaluator. At some point, refinement of the technique will not seem consistent with the crudeness of the data supplied.

Oil and Gas

This section summarizes an economic evaluation of a hypothetical wildcat oil well in Colorado. After presentation of the basic technical and economic considerations in the evaluation, sensitivity analysis and expected value are utilized to evaluate risks under alternative probabilities of success and alternative development agreements. There are seven steps in defining the expected profitability of an oil producing project:

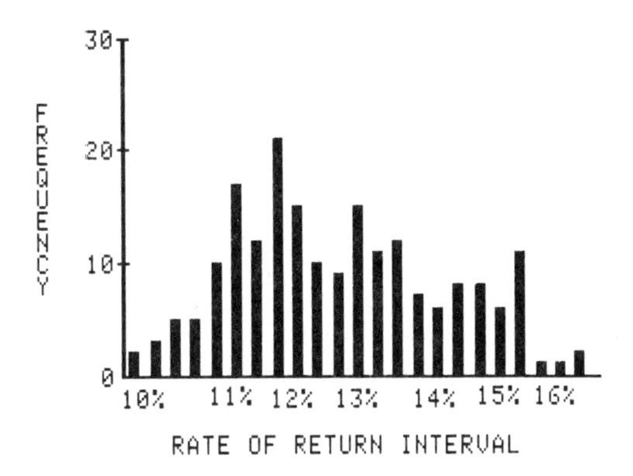

Fig. 3.7.14—Simulation results.

(1) Estimate the oil reserves in place. This is a volumetric analysis designed to estimate the porosity of the rock and the space occupied by oil, water and gas in the pore space.

(2) Determine the rate of recovery from the formation which will depend on reserves and the energy in the formation to drive the oil.

(3) Estimate the capital and operating costs of the project assuming if successful and if it results in failure—a "dry hole." This requires study of the physical characteristics of the oil field in order to make engineering assumptions for a technical analysis, then utilization of the engineering assumptions to prepare the financial evaluation.

(4) Estimate the cash flow stream expected from the prospect if it is successful. This requires forecasts of future prices and costs, taxes, and government regulations in order to develop the operating cash flows of the project. The specific development agreement specifies how the investment and operating cash flows are to be divided among the joint venture partners.

(5) Define the *technical* risks involved in the investment. This often requires subjective estimates of the probabilities involved.

(6) Define the *economic* risks involved in the investment. This often requires subjective es-

timates of future prices, costs, tax regimes, and political or legal variables that affect timing or cash flows.

(7) Calculate the present value and rate of return of the investment. Utilize "risk adjusted return" in the form of an expected monetary value as a basis for recommending action.

This wildcat oil prospect is assumed in the state of Colorado where 640 acres are involved in the oil play. The initial proposal is to drill three test wells at the high point on the anticline structure and to run 10 miles of seismic line to gain further information on the anticlinal stratigraphic trap. A step by step listing of basic data, well profitability, project profitability, and risk analysis follows.

Basic Data

(1) Type of trap: Anticline.

Previous data indicates the shape and size of the expected oil trap. These come in a large variety of shapes and types but the anticline is representative of a large number of oil fields.

(2) Objective geologic formation: Miocene.

Geologic cross sections indicate the probable producing formations and their age. In addition physical characteristics of this formation are available. It is a muddy sandstone with certain physical characteristics that affect production.

(3) Depth: 7000 ft. total depth to producing formation.

(4) Average net pay: 7 ft.

This is the expected portion of the formation expected to be porous and to contain oil.

(5) Average porosity: 20%.

This is the pore space in the rock which may contain oil, gas or water.

(6) Average water saturation: (30%).

This is the expected water occupying pore space in the rock.

(7) Average formation volume factor: 1.20.

This factor relates the expected volume of the hydrocarbons at the surface to the volume underground. Oil at depth is subjected to approximately 0.434 pounds per square inch per foot of depth. Thus at a depth of 10,000 ft the rock would be subjected to 4340 pounds per square inch hydrostatic head. A barrel of oil coming to the surface may come from a temperature of 160° F and a pressure of 3000 pounds per square inch to atmospheric temperature of 60°F and a pressure of 15 pounds per square inch. When the oil comes to the surface the gas comes out of solution and the barrel shrinks as a result. The formation volume factor measures this difference. A volume factor of 1.2 indicates that the volume at the surface will be 1/1.2 or 83% of the oil underground. The smaller the formation volume factor, the less energy to drive the oil to the surface but the more oil you have underground.

(8) Average initial or solution Gas Oil Ratio (GOR): 354 standard cubic feet per barrel (scf/bbls.)

This ratio measures the amount of gas associated with each barrel of oil in the formation. Since oil and gas recoveries from the formation are quite different (e.g. 30% of the oil and 60% of the gas) this ratio does *not* indicate the amount of gas that will be produced per barrel of oil produced. The ratio is used to estimate the gas reserve in the formation and the ratio of oil recovered to gas recovered is used to estimate the average gas production per barrel of oil.

(9) Average oil in place:

There are 43,560 square feet in an acre or 43,560 cubic feet in a volume one acre in area and one foot thick. There are 5.61 cubic feet in a (42-gallon) barrel of oil; therefore, there are 7758 barrels per acre per foot of formation thickness.

(a) $7758\,(0.2)(1-0.3) \times (1/1.2) = 905.10$ B/NAF

(b) $7 \times 905.10 = 6335.70$ B/Acre

Calculation (a) is the product of the barrels per acre-foot factor times the porosity (0.2) times the pore space not occupied by water $(1-0.3)$ times the reciprocal of the volume factor (1.2). The result is the estimated barrels per net acre foot (B/NAF) of the formation.

Calculation (b) uses the estimated formation thickness (7 ft) and the expected B/NAF to calculate the estimated barrels per acre (B/Acre) on the prospect. The result of these calculations is the volume of oil per acre of reservoir under the well. These barrels are *potentially* recoverable and are measured at the surface.

(10) Average well spacing: 40 Acres.

This is an engineering calculation based on an assumed spacing for maximum recovery from the field.

(11) Average oil in place per well: $40 \times 6335.70 = 253{,}428$ barrels per well.

(12) Estimated recovery factors:

Oil—30%

Gas—70%

Recovery will depend on the type of trap and the type of drive pushing the oil. Gravity drainage is most effective but slow, dissolved gas drive is least effective but rapid. The third drive is water, which usually falls in between. A weak water and gas drive results in a range of oil recovery between 18% and 38%. The gas comes out of solution as the oil comes to the surface.

(13) Total estimated economically recoverable reserves per well:

Oil recovery is the product of the expected barrels of oil under each 40 acre well multiplied by expected recovery. Gas reserves are the product of the total oil contained in the formation under a 40 acre well times the gas oil ratio (GOR) times the expected gas recovery.

(a) Oil: 253,428 × 0.3 = 76,028 barrels

(b) Gas: 253,428 × 354 × 0.7 = 62,799,458 standard cubic feet (62.8 MMSCF)

(14) Average initial producing rate: 30 barrels of oil per day per well (B/D/W) and 826 standard cubic feet of gas per day per well (826 SCF/D/W).

The initial producing rate is the rate of production assumed during the flat life of the project or as the beginning of declining production. This rate is an assumption based on performance of similar wells. The producing rate may also be controlled by maximum allowable production rates; in this analysis it is assumed that 100% of the 30 barrels per day is produced.

(15) Flat Life = 1 year; Flat Life Reserves: 10,950 barrels

In some cases there will be a flat life where the production rate maintains a constant annual rate. Thereafter production will decline as the drive pressure gets lower and the oil and gas reserves are depleted.

For this project, the flat life is one year. The annual rate of production declines each year after the first year of 1 × 365 × 30 = 10,950 barrels

(16) Economic limit: one barrel per day.

It is assumed that the well will flow down to a one barrel per day ultimate rate of production. There is, however, an economic limit which depends on oil and gas prices and total costs of production. This economic limit will vary with cost and price levels and thus will be evident when the cash flows become negative in the economic analysis of the project.

(17) Years on decline = 20.7.

For purposes of this analysis, it is assumed that exponential decline will follow the flat life production. In order to calculate annual production during the years of declining output, the analyst must apply some of the petroleum engineer's models of decline curves. Sometimes the curves can be extrapolated from known production histories but in the case of an entirely new region, the engineer may have to resort to theoretical relationships. Most of the decline curves from the hyperbolic family and include exponential decline and harmonic decline models. Following the example in McCray (McCray, 1975 pp. 323–324) we can calculate the expected period of decline with the following formula which assumes an exponential relationship.

$$T = \frac{N_{pd}}{q_o - q_f} \ln \frac{q_o}{q_f}$$

where: N_{pd} = oil production on decline, bbls
q_o = initial annual production rate, bbls/yr
q_f = final annual production rate, bbls/yr

therefore:

$$T = \frac{76028 - (1 \times 30 \times 365)}{365\,(30 - 1)}\, ln \frac{30}{1} = 20.7 \text{ years}$$

(18) Annual Production during decline

Utilizing the calculated decline life, the annual production for each year during decline can be calculated using additional based on exponential relationships

$$R = \frac{q_f^{\,1/t}}{q_o} = \frac{1^{1/20.7}}{30} = .848$$

$$Y_1 = \frac{1 - R}{ln\,\frac{1}{R}} \times Q_0 = \frac{1 - .848}{ln\,\frac{1}{.848}} \times (365 \times 30) = 10095 \text{ bbls, first year of decline}$$

$$Y_2 = R \times Y_1 = .848 \times 10095 = 9308 \text{ bbls, second year of decline}$$

In similar fashion the production rates for all the years declining production are calculated as a product of the decline factor R and the previous year production.

(19) Operating Cost.

The operating cost used in the analysis is $2.18 per barrel which includes the $1.54 of direct cost, $0.15 of overhead and a 4% production tax.

Total Project Profitability

The evaluation to this point is in terms of an average development well. It remains to calculate the number of development wells on the productive acreage of the prospect and the investment required to develop the project. In this example a land position is secured through leasing and rentals prior to initiating geological and geophysical reconnaisance over a period of two years.

Success in initial reconnaisance narrows down the exploration target for detailed seismic exploration and geologic interpretation. A total of three exploratory test wells determine whether the 40 acre prospect is dry or productive. If the prospect is dry it is abandoned and the loss is absorbed by the company. If the prospect is productive, investment continues to complete three additional test wells, 16 development wells, and two dry development holes. Thus the total investment for a productive field is larger and must be justified by future cash flows. After two years of production, pressures have declined sufficiently to install pumping equipment for artificial lift of the petroleum.

The total investment for both a dry hole and a producing venture can be summarized in terms of before tax unescalated dollars. Escalation calculated in the computer program will increase the total dollars required while the tax credits such as the investment tax credit will reduce the total required dollars.

In the analysis that follows, the total investment is measured in base year dollars and must be escalated to calculate the nominal dollar cash flows for the project. It is this *escalated* investment that is used in the calculations of net present value, rate of return and expected monetary value. In the analysis, the estimated investment is reduced as a result of the investment tax credit.

(30) Estimated probability of success: 20%

Probabilities are subjective estimates based on previous experience. In the United States onshore drilling success averaged one out of 14 or 7% while offshore exploration success was one in three or 33%. In this analysis the subjective estimate is 20%, that is, under similar conditions one well out of five is expected to be a success.

(31) Potential investment = $3,865,000

The potential investment when the probability of success is 20% is calculated as four dry holes plus one investment in a productive prospect which in this case totals to $3,865,000.

(32) Expected value:

The expected value is equal to the probability of success (0.20) times the present value of success ($1,104,000) minus the probability of failure (0.80) times the cost of a dry hole ($363,000) thus

Cost Estimates:	**Prospect** Dry	**Prospect** Productive
(20) Leasehold costs	$ 35,000	
(21) Rentals	$ 72,000	
(22) Geological and geophysical expense	$ 16,000	
(23) Cost of three exploratory test wells. The three exploratory test wells would be expanded to five if gas is to be sold from the property.	$240,000	
(24 Total investment if prospect tests dry	$363,000	$ 363,000
(25) Completion cost of three test wells		$ 60,000
(26) 15 subsequent development wells		$1,650,000
(27) Cost of artificial lift or special facilities (2 yrs. @ $15,000/well)		$ 240,000
(28) Cost of two development dry holes:	$100,000	
(29) Total investment if prospect is productive		$2,413,000

$$0.20 \times \$1{,}104{,}000 - 0.80 \times \$363{,}000 = -\$69{,}600$$

Table 3.7.3 summarizes the inputs required for the financial evaluation of the wildcat petroleum prospect.

Table 3.7.4 summarizes the cash flows for the wildcat petroleum prospect.

The preceding analysis represents a wildcat effort in a very small field. The combination of a low conditional net present value and a low probability of success makes the expected monetary value negative indicating that the long-run results of such activities are likely to be uneconomic. Such fields can also easily be made uneconomic by over drilling of development holes. To avoid such errors some producers seek to limit development drilling to three or four holes at $300,000 to $500,000 each.

In 1982, a typical budget for a medium sized petroleum exploration firm might be of the order of $43,000,000. This might be distributed among geological regions as follows:

17 million	Overthrust Belt
14 million	Williston Basin
6 million	Frontier
4 million	Powder River Basin
2 million	Operations

The basic objectives in such an exploration program will include the three basic objectives indicated previously:

(1) There must be a reservoir (a rock bed capable of holding liquid petroleum).

(2) There must be a seal on the reservoir (a trap to hold in the petroleum).

(3) There must be an identifiable source of the hydrocarbons.

A firm working in the Colorado area will expect to spend $3000 to $4000 per mile for seismic work on a quarter mile grid in connection with each of the prospects. Drilling will go

Table 3.7.3—Technical and Financial Data for Evaluation

First year of investment	Year one
Last year of investment	Year two
Discount rate for calculation of present value	15%
Total investment for productive prospect (before escalation)	$2,413,000
Investment tax credit	9.5% of total investment
General inflation rate	6%
Thickness of net pay formation	7 ft
Porosity	20%
Water saturation	30%
Volume factor	1.2
Gas/oil ratio in place	354 standard cu ft. per bbl
Land area per well	40 acres
Oil recovery	30%
Gas recovery	70%
Initial oil production rate	30 bbls per day
Final oil production rate	one bbl per day
Size of oil field	640 acres
Distribution of investment	1/3 first year, 2/3 second year
Type depletion	percentage
Escalation rate for investment expenditures	8.56%
Investment in dry hole	$363,000
Probability of success	20%
Depletion rate assumed	22%
First year of production	year one
Last year of production	year 21
Escalation rate for replacement expenditures	6%
Overall average tax rate	50%
Escalation rate for oil prices	6%
First year oil price	$11.50 per bbl
Escalation rates for gas prices	short term 15% annually longer term 6% annually
Escalation rate for costs	6%
Base year cost	$2.18 per bbl
Royalty rate	25%

Table 3.7.4—Cash Flows for the Wildcat Petroleum Prospect

	PHASE 1	PHASE 2					PHASE 3		
YEAR	1977	1978	1979	1980	1981	1982	1983	1984	1985
Price of Oil ($/BBL)	11.5	12.19	12.92	13.7	14.52	15.39	16.31	17.29	18.33
Price of Gas ($/Thousand Std. Cu.Ft.)	.8	.92	1.06	1.22	1.4	1.61	1.71	1.81	1.92
Annual Production Rate of Oil (000 BBL)	175	162	137	117	99	84	72	61	52
Annual Production Rate of Gas (000 SCF)	144715	133559	113510	96471	81990	69682	59222	50332	42777
Average Total Cost ($/BBL Oil)	9.71	10.34	11.05	11.8	12.62	13.51	13.79	14.63	15.52
Average Operating Cost ($/BBL Oil)	2.54	2.69	2.86	3.02	3.22	3.42	3.59	3.82	4.03
Revenue ($000 U.S.)	2131	2094	1896	1717	1556	1410	1271	1145	1031
Less Annual Royalty ($000)	533	523	474	429	389	353	318	286	258
Less Annual Operating Cost ($000)	405	396	357	321	290	261	235	212	191
Equals Gross Income ($000)	1193	1174	1065	966	877	797	718	647	583
Less Annual Depreciation and Amorization ($000)	96	96	96	96	96	96	12	12	12
Less Depletion ($000)	240	236	213	193	175	159	143	129	116
Taxable Income ($000)	857	843	756	677	606	542	563	506	455
Effective Income Tax Rate (%)	50	50	50	50	50	50	50	50	50
Less Tax ($000)	429	421	378	339	303	271	282	253	228
Equals Net Income After Tax ($000)	429	421	378	339	303	271	282	253	228
Plus Non Cash Flows ($000)									
Depreciation and Amorization ($000)	96	96	96	96	96	96	12	12	12
and Depletion ($000)	240	236	213	193	175	159	143	129	116
Less Replacement Investment ($000)	34	37	40	43	47	51	55	60	65
Equals Nominal $ Cash Flow ($000)	731	716	647	585	527	475	381	334	290
Constant $ Cash Flow ($000)	689	638	544	463	394	335	253	209	172

to depths of 9000 to 10,000 feet in the Overthrust Belt with costs of a 12 to 13,000 foot well being $5,000,000 to $6,000,000. Such deep wells quickly absorb the total budget of the company with 10 to 15 opportunities. It is the responsibility of the exploration manager to sort through 100 to 150 opportunities made available through reconnaisance to decide which 10 to 15 his company will gamble on over the next year.

Expected Monetary Value Concept

The preceding pattern of cash flows, when discounted, results in a conditional net present value which is dependent on a successful outcome of the drilling project. Petroleum explorationists, and to a lesser extent, mineral explorationists use expected monetary value (EMV) to rank investment alternatives. The goal of the investor is to maximize the value of his firm over time. The decision rule regarding expected monetary value is consistent with this objective; it is to select the alternative with the highest positive expected monetary value (EMV). There may also be a risk preference involved; the explorationist may choose a lower than optimum EMV because potential losses on the higher EMV are unacceptably high. This might be considered in terms of the assets of the firm and the possibility of gambler's ruin because of large potential losses. See Newendorp (Newendorp, 1975, p. 137) for a discussion of preference theory. If the decisions are made entirely on the basis of EMV, the firm is indifferent to the size of potential gains or losses.

EMV includes all of the risk and uncertainty of exploration into one estimate of the probability of success for a particular outcome. If all possible outcomes are included in the analysis, the sum of the probabilities must be equal to 1.0; any number of alternatives or outcomes can be considered. Although there are various ways to express the "value" of success or failure, we will use net present value of future cashflows in the analyses that follow.

Consider an oil well drilling venture where the net present value of success includes drilling the initial wildcat well plus developing and equipping production wells. The net present value of a successful venture is estimated at 10 million dollars. If only the wildcat well is drilled and it is a dry hole the cost is 500,000 dollars. Based on geological data and success ratios in the region the probability of success is esimated at 0.3 and the only alternative considered is a failure in the form of a dry hole with a probability of 0.7. In this case the expected monetary value of the venture (EMV) is calculated as follows:

Table 3.7.4 (cont.)

PHASE 3		PHASE 4					PHASE 5				
1986	**1987**	**1988**	**1989**	**1990**	**1991**	**1992**	**1993**	**1994**	**1995**	**1996**	**1997**
19.43	20.59	21.83	23.14	24.53	26	27.56	29.21	30.97	32.82	34.79	36.88
2.03	2.15	2.28	2.42	2.56	2.72	2.88	3.05	3.24	3.43	3.64	3.86
44	37	32	27	23	20	17	14	12	10	9	7
36355	30898	26260	22318	18968	16121	13701	11644	9896	8411	7148	6075
16.46	17.47	18.53	19.67	20.87	22.15	23.52	24.97	26.52	28.16	29.92	31.79
4.3	4.6	4.79	5.12	5.41	5.61	5.94	6.5	6.83	7.39	7.39	8.57
929	837	754	679	612	551	497	447	403	363	327	295
232	209	188	170	153	138	124	112	101	91	82	74
172	155	139	126	113	102	92	83	75	67	61	55
525	473	426	384	346	311	281	253	228	205	185	167
12	12	12	12	12	12	12	12	12	12	12	12
105	94	85	76	69	62	56	50	45	41	37	33
409	367	330	296	265	238	213	191	171	153	136	122
50	50	50	50	50	50	50	50	50	50	50	50
204	184	165	148	133	119	107	95	85	76	68	61
204	184	165	148	133	119	107	95	85	76	68	61
12	12	12	12	12	12	12	12	12	12	12	12
105	94	85	76	69	62	56	50	45	41	37	33
71	77	83	91	98	107	116	126	136	148	161	175
250	213	178	145	115	86	58	32	6	−19	−44	−69
139	112	88	68	51	36	23	12	2	−6	−14	−20

$$
\begin{aligned}
\text{EMV} &= 0.3 \times \$10{,}000{,}000. \\
&\quad + 0.7 \times (-\$500{,}000) \\
&= 3{,}000{,}000 - 350{,}000 \\
&= \$2{,}650{,}000.
\end{aligned}
$$

Expected Value of Petroleum Prospect: Refer to the cash flow model of the Colorado Wildcat oil well prospect. The prospect was located in the state of Colorado where 640 acres were involved in an oil play. The proposal was to run 10 miles of seismic line to gain information on a stratigraphic trap, then, given favorable results, to drill three test wells. How can we use EMV to evaluate this opportunity? In this section we follow an approach outlined by Newendorp (Newendorp, 1975 p. 77). Analysis of the prospect indicates that if the project is successful it will require an investment of \$2,413,000, unescalated and before tax credits, to bring the field into production. If the project results in a dry hole, the total investment is \$363,000. Table 3.7.4 indicates that a successful project is expected to develop 1.2 million barrels of oil and 1004 million standard cubic feet of gas. If the project is successful the conditional net present value of the project is \$1,104,000 and this is expected to occur only one out of five times indicating as probability of success of 20% on this type of situation. The expected value is calculated as follows:

$$
\text{EMV} = 0.2 \times \$1{,}104{,}000 - 0.8 \times \$363{,}000 = -\$69{,}600
$$

The reason for the negative expected value is a combination of low probability of success (20%) and the small size of reserves, hence low conditional net present value in relation to exploration costs. In contrast, we might assume that the Colorado wildcat prospect has a potential for reserves greater than the 1.2 million barrels and that we are able to estimate the probabilities of these reserves as follows:

Possible Outcomes	**Probability Outcome Will Occur**
Dry Hole	0.80
1.2 million barrels with associated gas	0.08
3 million barrels with associated gas	0.08
5 million barrels with associated gas	0.03
7 million barrels with associated gas	0.02

These reserve estimates may be based on study of the physical characteristics and the historical record of past discoveries under similar geological conditions or they may be highly sub-

Table 3.7.4A—Basic Assumptions for the Wildcat Petroleum Prospect

First Year of Investment Cash Flows			1976
Last Year of Investment Cash Flows			1977
Minimum Acceptable Discount Rate (%)			15
Exchange Rate Used in this Analysis (Per/$US)			1
Total Estimated Investment (Base Year ($000))			2413
Less Tax Deductions from Investment Tax Credit ($000)			229.235
Equals Net Equity Investment ($000)			2184
Projected Rate of Escalation for Investment Expenditures			9
Summary of Investment Cash Flows ($000)			
Year	Cash Flow Base Year $	Cash Flow Escalation $	
1976	728	728	
1977	1456	1580	
Net Equity Investment ($000)			
Total	2184	2308	
Depreciable Investment ($000)			577
Intangible Investment Recovered via Depletion			1731
Expected Investment ($000)			3760
Projected Rate of Oil Price Escalation			6
Projected Rates of Gas Price Escalation (%)			15–6
Projected Rate of Production Cost Escalation			6
Projected Rate of Replacement Inventory Escalation			7
Average Net Pay (Feet)			7
Average Porosity (%)			20
Average Water Saturation (%)			30
Average Formation Volume Factor			1.2
Average Initial Gas/Oil Ratio (Scf/Bbl)			354
Average Oil in Place (Bbl/Net Acre Ft.)			905.1
Average Spacing (Acres)			40
Estimated Oil Recovery (%)			30
Estimated Gas Recovery (%)			70
Total Recoverable Oil Reserves per Well (000 Bbl)			76
Total Recoverable Gas Reserves per Well (000 Scf)			62799
Average Initial Oil Production Rate (Bbl/Day/Well)			30
Average Initial Gas Production Rate (000 Scf/Day/Well)			24780
Flat Life (Years)			1
Economic Producing Limit (BBl/Day)			1
Annual Decline Rate (%)			15
Declining Life (Years)			21
Total Life (Years)			22
Productive Acreage Expected (Acres)			640
Number of Dev. Wells			16
Total Oil Reserves (000 Bbl)			1216.45
Associated Gas Reserves (000 Scf)			1004791.33
Estimated Probability of Success (%)			20
Investment for Dry Prospect ($000)			363

Table 3.7.4B—Summary of Cash Flows for the Wildcat Petroleum Prospect

The Nominal $ Net Present Value at 15 Percent Discount Rate is ($000)	1104
Nominal $ Expected Value of this Project ($000)	−69.6
Gamblers Ruin Investment ($000) is	1815
The Nominal $ DCFROR Is 64.6 Percent	
Assumed Rate of Inflation Is 6 Percent	
The Constant $ NPV at 8 Percent Discount Rate is ($000)	1104
The Constant $ DCFROR Is 55 Percent	

jective guesses based on limited information.

Effect of Alternative Development Agreements on EMV: Three possible alternatives are identified for development of the prospect:

1. Drill the well and, if productive, continue to hold a 100% working interest during development and production.
2. Farm out the acreage and retain a 10% overriding royalty on the 640 net acres.
3. Participate in a "back-in" agreement. We

assume that we do not choose to participate in the risk of drilling the well. Another company will do the drilling and be allowed to recover say 150% of their investment before we "back-in" and share in additional revenues. In this case we share no risk or revenues until the well has recovered the previously determined amount of revenue.

Our engineering appraisal of the venture required estimates of production, costs, and investments of the prospect under assumption of different possible outcomes. Each engineering analysis led to an economic evaluation similar to the first which resulted in an expected monetary value. We need to select the agreement which maximizes the expected monetary value given the uncertainties of future outcomes. The results are summarized in Table 3.2.5.

The key questions in connection with the leasehold area include:

(1) How much can the purchaser afford to pay for leasehold rights?

(2) Which of the decision alternatives will maximize the expected monetary value?

The answers as indicated in Table 3.7.5 are the following:

(1) Under the penalty with back-in alternative, we can afford to pay up to $426 per acre for the 640 acre tract.

(2) The decision alternative which maximizes expected monetary value is the penalty with back-in alternative.

The preceding analysis recognizes uncertainty in the size of field but ignores uncertainty in the probability of discovery of hydrocarbons. In the example, the assumption is that the probability of a dry hole is 0.80, therefore the probability of finding hydrocarbons is 0.20. If there is a possibilility of higher or lower ratios of success, a different exploration agreement may be the optimum.

The Effect of a Change in the Probability of Discovery: The assumptions of Table 3.7.5 can be modified to consider the probability of a dry hole equal to 0.20 rather than the 0.80. The recalculation changes all of the expected monetary values as indicated in Table 3.7.6.

The change in probability of a dry hole has a drastic effect on the choice of exploration joint venture agreements. In the latter case the drill with 100% working interest becomes the best agreement and its expected value is $3187 per acre as compared to $1702 per acre under the back-in agreement.

Fig. 3.7.15 presents the results of calculation of expected monetary values for a range of probability of a dry hole from zero to one. The technique is a powerful tool for the determination of the best decision alternative under conditions of uncertainty and risk.

Decision Trees: Decision trees are an extension of the expected monetary value concept. They are a graphical approach to long range planning wherever there is a chain of decisions to anticipate. The technique is most widely used in the evaluation of oil and gas drilling ventures but it has application in the minerals area as well.

Fig. 3.7.16 is a partial decision tree for the Colorado wildcat oil/gas prospect. The data for the alternative to drill with a 100% working

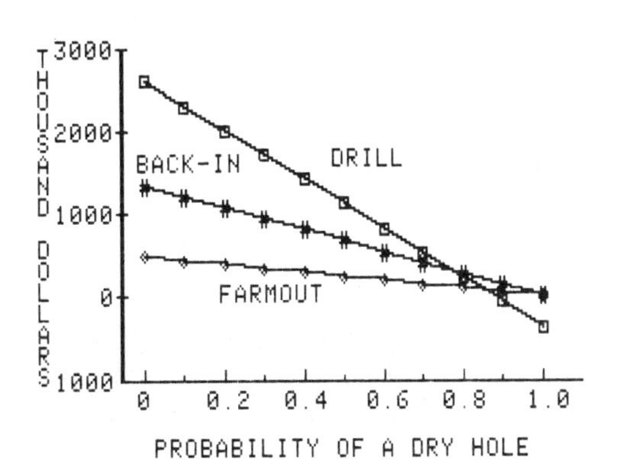

Fig. 3.7.15—Expected monetary value under and the probability of a dry hole.

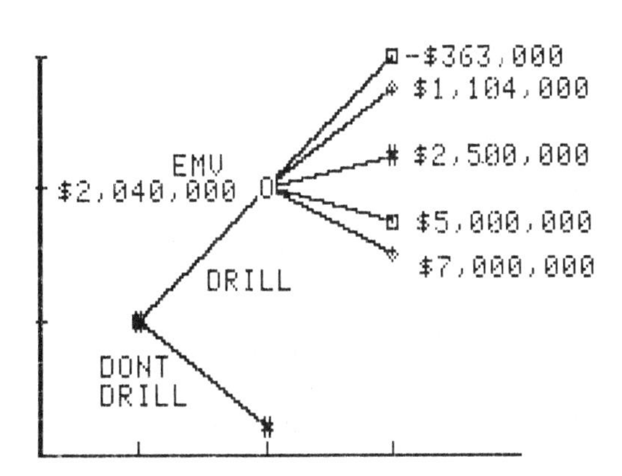

Fig. 3.7.16—Partial decision tree for the Colorado Wildcat Prospect.

Table 3.7.5—Expected Monetary Value of Alternative Outcomes; Colorado Wildcat Prospect

Possible Outcomes	Probability Outcomes Will Occur	Drill with 100% Working Interest		Farm Out with Overriding Royalty		Penalty with Back-in	
		Conditional NPV ($000)	Expected NPV ($000)	Conditional NPV ($000)	Expected NPV ($000)	Conditional NPV ($000)	Expected NPV ($000)
Dry Hole	0.80	−363	−290	0	0	0	0
1.1 MM BBL	0.08	1104	88	352	28	352	28
3 MM BBL	0.08	2500	200	476	38	1190	95
5 MM BBL	0.02	5000	100	859	17	2891	58
7 MM BBL	0.02	7000	140	1058	21	4558	91
Sum	1.00		238		105		272
Acres			640		640		640
EMV ($/Acre)			371		163		426

Table 3.7.6—Expected Monetary Value of Alternative Outcomes; Colorado Wildcat Prospect

Possible Outcomes	Probability Outcomes Will Occur	Drill with 100% Working Interest		Farm Out with Overriding Royalty		Penalty with Back-in	
		Conditional NPV ($000)	Expected NPV ($000)	Conditional NPV ($000)	Expected NPV ($000)	Conditional NPV ($000)	Expected NPV ($000)
Dry Hole	0.20	−363	−73	0	0	0	0
1.1 MM BBL	0.32	1104	353	352	113	352	113
3 MM BBL	0.32	2500	800	476	152	1190	381
5 MM BBL	0.08	5000	400	859	69	2891	231
7 MM BBL	0.08	7000	560	1058	85	4558	365
Sum	1.00		2040		418		1089
Acres			640		640		640
EMV (%/Acre)			3187		654		1702

interest is displayed as a decision partial tree; the convention and decision tree analysis is to label decision node with a box and chance nodes with a circle. The tree is constructed from left to right showing the alternatives available and labeling the branches of the chance nodes with the expected probabilities. Conditional net present values based on economic analysis of cash flow models are assigned to the possible outcomes.

Decision trees are solved in reverse, from the end point working backwards toward the first decision node. In our example, the alternative to drill leads us to a chance node with several branches each of which leads to an outcome and a value. Multiplying the probabilities and the net present values, the EMV at the chance node is $2,040,000. Since no other alternatives are considered in this example, the decision is between (1) drill with working interest or (2) don't drill. If our objective is to maximize EMV we will select the drilling alternative. The assignment of zero to the don't drill alternative may be misleading; there are probably opportunities for the capital invested in drilling that have positive EMV's.

The following is the step by step sequence for constructing and solving a decision tree.

(1) Construct the tree showing graphically the alternatives for action available to the decision maker.

(2) Identify decision nodes and chance nodes.

(3) Estimate probabilities to all branches from the chance node. The sum of the probabilities must equal 1.0.

(4) Estimate conditional net present values for final outcomes.

(5) Solve the tree. Begin with the net present values, then multiply by probabilities to calculate expected monetary value at the chance node.

(6) Continue until the current decision node is reached.

THE FEASIBILITY STUDY

The feasibility study is the final comprehensive report that brings together both the technical and the economic aspects of a minerals or fuels project. The feasibility study of a mining project will summarize geology, metallurgy, mining, marketing, capital, and operating costs, and cash flows as well as the inferences drawn from the criteria of evaluation such as net present value or rate of return. In the early stages of project analysis, preliminary feasibility studies justified the next step in evaluation of a potential investment. This may include an additional investment in geological reconnaissance or diamond drilling, metallurgical studies involving pilot plants, small scale mining of the ore body to test mining conditions, or processing of the ore in existing plants. When engineering and economic data are available, a final feasibility study of a project is the one used to secure project financing and to negotiate contracts for sales. It requires 40 to 60% of engineering complete in order to prepare the detailed estimates of capital and operating costs as well as a project schedule for construction.

Key components of the feasibility study will vary with the type and available data but generally include:

(1) A summary of the proposal.

(2) Location and plant site requirements.

(3) Mine and processing plant equipment requirements.

(4) Requirements for supplies, labor, power, water, and other production inputs.

(5) A summary of the total investment and its timing.

(6) A summary of the detailed cash flows of the project over its life.

(7) A proforma balance sheet and income statement for the company indicating the impact of the project.

(8) Summary of the criteria used for evaluation including net present value, rate of return.

(9) An appendix of material to support the analysis including maps, plans, specifications of the mine and plant, economic data, proposed contracts, and engineering data on development and production.

FINANCIAL AND CAPITAL STRUCTURE

Financial structure is the entire right hand side of a company balance sheet including current liabilities, long-term debt, other liabilities, as well as shareholder equity in the form of capital stock or reinvested earnings. Capital structure is long-term debt only, including preferred stock, common stock, and retained earnings. This long-term debt and equity is the focus of the present section. The right hand side (liabilities) of the consolidated balance sheet for Exxon Corporation in 1980, shown in Table 3.7.7, is illustrative of a mixture of short-term, long-term debt and equity necessary for operations and growth.

In 1980, 54% of Exxon financing came from current liabilities including short-term notes and loans, accounts and income taxes payable and other accrued liabilities. A major portion of the financial manager's effort will be in managing these short-term accounts. Fifteen percent of Exxon Corporation financing in 1980, came from long-term debt which was primarily in the form of debentures, long-term bonds not secured by the pledge of any specific property. The balance of the long-term debt was in the form of bonds which are simply long-term promissory notes. Weighted average interest rates on the debentures and revenue bonds were in the range of 5.8 to 6.5% annually. Shareholder equity in the firm was 93% in the form of reinvested earnings and the balance was 453 million shares of authorized capital stock that had been issued.

Benefits and Risks of Financial Leverage

The Exxon balance sheet data indicates a relatively low financing by long-term debt. In contrast some international mining projects are financed with extremely high proportion of debt. As a result the investors are able to "leverage" their rate of return on equity but must also assume the risk associated with the high interest payments on debt portion of financing.

Financial leverage measures the extent to which assets are financed with debt. This debt shows up as a fixed annual interest expense which may cause operating income to be more sensitive to changes in prices and the volume of sales. This variability is a measure of the increased risk due to debt financing. The objective is to balance the additional risk with a higher return on equity capital.

The effects of leverage on rate of return and on risk are best illustrated with an example. In 1970, the long-term financing of the Freeport Indonesia project was pictured as shown in Table 3.7.8. The expected total cost of the project in 1970 was $120 million of which $100 million or 83% would be debt financing. Assuming a U.S. prime rate of 8.5%, and taking into account all guarantee fees and compensating balances, the effective rate of interest on the $100 million debt was 9.7%. Thus the project was heavily leveraged and assumed both the benefits and the risks of heavy debt financing. An evaluation of the project under 100% equity financing is summarized in Table 3.7.9; the effects of the 83% debt on the project are summarized in the cash flow evaluation of Table 3.7.10; Table 3.7.11 assumes thhe 83% debt but with changes in the economic assumptions assuming slower price escalation, and more rapid capital and op-

Table 3.7.7—Consolidated Balance Sheet, Exxon Corporation: Liabilities (Exxon Corp., 1980, p. 25)

Liabilities	**$ million**
Current liabilities	
Notes and loans payable	1,537.4
Accounts payable and accrued liabilities	12,481.9
Income taxes payable	2,865.0
Total current liabilities	16,884.2
Long-term debt	4,717.1
Annuity and other resources	1,891.9
Deferred income tax credits	6,218.0
Deferred income	138.9
Equity of minority shareholders in affiliated companies	1,313.7
Total liabilities	**31,163.9**
Shareholders' equity	
Capital stock	1,695.0
Earnings reinvested	23,717.7
Total shareholder's equity	**25,412.6**
Total liabilities and shareholder's equity	**$56,576.6**

Table 3.7.8—Freeport, Indonesia Financial Structure, 1970 (Petrick, 1980, p. 36)

	Basic Estimates $ million
Senior Loans	
U.S. Insurance companies (AID extended risk guarantee)	40.0
U.S. Commercial banks (Eximbank guarantee)	18.0
Kreditanstalt fur Wiederaufbau (KFW)	22.0
	80.0
Subortinated Loan	
Japan (Copper consumers and trading companies)	20.0
Equity	
Freeport Minerals (87%)	16.9
Norddeutsche Affinerie (5%)	2.5
Zuid-Pacific Koper (5%, carried)	–
Julius Tahija (3%)	0.6
	20.0
Total	120.0

erating cost escalation.

Under the 100% equity assumption the net present value of the project is estimated as $129.4 million. The net present value is the same under nominal and constant dollar analyses because the discount rate is adjusted for inflation when applied to the constant dollar cash flows. Probably the most relevant rate of return is the 19% return based on constant dollar analysis. The constant dollar analysis removes much of the effect of the assumed escalation rates for prices and costs. Measured from the point of view of the country, the retained value for the project is 30% of revenues during the first year of production rising to 42% of revenues in the last year of production.

The second analysis, assuming 83% debt at 9.7% interest, increases the present value of the project to $145.1 million. The constant dollar rate of return is increased to 46%; the after tax cost of debt is much lower than the overall return on funds invested in the project thus the return on the equity portion is increased dramatically. Retained value from the point of view of the country is decreased during the early years of the project because the interest on loans is not taxed. After the loans are paid off in 1982, retained value is equal to the 100% equity situation.

The third cash flow summary can be described as a situation where prices rise slower than expected and operating and investment rise faster than expected. The 5% escalation for the price of copper in the first two analyses is reduced to 2%, the 5% escalation for operating cost is increased to 10% and the 5% escalation for investment costs is increased to 15%. Actual data on the project indicates that the price escalation is too low but that the cost and investment corrections were insufficient to equal actual costs. For example the cost of the investment exceeded $200 million. The net result of the "poor economic conditions" reduces the present value of the project to $49.5 million, only 31% of the estimate under leveraged conditions. The constant dollar rate of return is reduced from 46% to 28% a decline of 61%. Thus the risks of high debt financing are evident; fixed interest costs are higher under the "poor economic conditions" because of the necessity to finance the larger investment.

SOURCES OF FUNDS

The liabilities side of the Exxon balance sheet in Table 3.7.7 indicates that accounts payable by the company are more than 12 times the financing through notes and loans and more than two and a half times the money raised through long-term debt. The size of accounts payable supports the idea that the financial manager will spend a large portion of his time managing working capital. the financial manager may also turn to the commercial bank for short-term loans on inventories or accounts receivable and if the firm is large they may issue "commercial paper" in the form of unsecured, short-term promissory notes. He needs a supply of short-term credit to meet the requirements for working capital as they fluctuate from month to month. Sources of long-term credit are matched to the long-term growth pattern of the firm.

A focus on the long-term financing needs of the company highlights the importance of equity, especially in the financing minerals and petroleum development. The Exxon balance sheet shows 45% of total liabilities and stockholder equity coming from shareholder equity primarily in the form of reinvestment earnings. Thus the short-term financing of the firm and shareholder equity accounts for 75% of the financing needs of the company.

The logic behind specific packages of debt and equity varies with differing objectives and financing conditions unique to the firm, general

Table 3.7.9—Gunung Bijih Cash Flow Evaluation 100% Equity, 1969–1985

YEAR	PHASE 1 1973	1974	1975
Ore Grades			
Copper (%)	2.5	2.5	2.5
Gold (Ounces Per Metric Ton Ore)	.025	.025	.025
Silver (Ounces Per Metric Ton Ore)	.265	.265	.265
Price Per Pound of Copper Net of Smelting and Ref. Charges ($U.S.)	.45	.47	.5
Price Per Ounce Silver ($U.S.)	1.6	1.68	1.764
Price Per Ounce of Gold ($US)	35	36.75	38.59
Add Byproduct Credit for Gold and Silver ($/Lb. Copper)	.025	.026	.028
Annual Copper Production Rate (000) Pounds of Copper ($000)	97159	129546	129546
Metallurgical Recovery Percent	94	94	94
Avg. Total Cost ($/Lb. Cu.)	.25	.25	.26
Avg. Operating Cost ($/Lb. Cu.)	.05	.05	.05
Revenue ($000 U.S.)	46184	64657	67890
Less Annual Royalty ($000)	0	0	0
Less Annual Operating Cost ($000)	4706	6589	6918
Equals Gross Income ($000)	41477	58068	60972
Less Annual Depreciation and Amortization ($000)	8520	8520	8520
Equals Taxable Income ($000)	32957	49548	52452
Effective Income Tax Rate (%)	35	35	35
Less Tax ($000)	11535	17342	18358
Equals Net Income After Tax ($000)	21422	32206	34094
Plus Non Cash Flows			
Depreciation and Amortization ($000)	8520	8520	8520
Less Replacement Inv. ($000)	0	0	0
Equals Nominal $ Cash Flow ($000)	29942	40726	42614
Constant $ Cash Flow ($000)	22375	28711	28341
Wages and Salaries of Nationals Including Non-Monetary Benefits ($000)	918	1285	1349
Incomes Taxes	11535	17342	18358
Domestic Purchases of Materials and Services	871	1219	1280
Other (Import Duties Other Taxes Wages and Salaries Not Transferred Abroad)	348	488	512
Total Retained Value	13672	20333	21499
Retained Value as a Percent of Revenue	30	31	32
Freeport's Portion of Cashflow ($000)	29942	40726	42614

Table 3.7.9A—Gunung Bijih Basic Assumptions for Cash Flow Evaluation 100% Equity, 1969–1985

First Year of Investment Cash Flows	1969
Last Year of Investment Cash Flows	1972
Minimum Acceptable Discount Rate (%)	10
Projected Rate of Escalation for Investment Expenditures (%)	5

Summary of Investment Cash Flows ($000)

Year	Cash Flow Base Yr. $	Equity CF Esc. $	Debt Esc. $	Total Esc. $
1969	7247	7247	0	7247
1970	17718	18604	0	18604
1971	39396	43434	0	43434
1972	43809	50714	0	50714
Total	108170	119999	0	119999

Depreciable Inv. ($000)	85200
Mineable Ore Reserve (000 M.T.)	32700
Mine Life (Yrs.) = Ore Reserves/An. Prod. Rate	13
Total Production Employment	338
Assumed General Inflation Rate (%)	6

Table 3.7.9B—Summary of Gunung Bijih Cash Flow Evaluation 100% Equity, 1969–1985

The Nominal $ Net Present Value at 10 Percent Discount Rate is ($000)	129448
The Nominal $ DCFROR Is 26.6 Percent	
Assumed Rate of Inflation is 6 Percent	
The Constant $ NPV at 4 Discount Rate is ($000)	129448
The Constant $ DCFROR Is 19 Percent	

economic conditions and the timing of financing needs. The packages may be explained in a several different ways:

(1) The firm will try to minimize the cost of capital. Usually debt is cheaper than equity and short-term loans are cheaper than long-term loans but the mixture must be related to the financial risk imposed by the increased fixed interest cost of debt. Minimization of the cost of capital may

Table 3.7.9 (cont.)

PHASE 1		PHASE 2					PHASE 3		
1976	**1977**	**1978**	**1979**	**1980**	**1981**	**1982**	**1983**	**1984**	**1985**
2.5	2.5	2.5	2.5	2.5	2.5	2.5	2.5	2.5	2.5
.025	.025	.025	.025	.025	.025	.025	.025	.025	.025
.265	.265	.265	.265	.265	.265	.265	.265	.265	.265
.52	.55	.57	.6	.63	.67	.7	.73	.77	.81
1.852	1.945	2.042	2.144	2.251	2.364	2.482	2.606	2.737	2.873
40.52	42.54	44.67	46.9	49.25	51.71	54.3	57.01	59.86	62.85
.029	.03	.032	.034	.035	.037	.039	.041	.043	.045
129546	129546	129546	129546	129546	129546	129546	129546	129546	129546
94	94	94	94	94	94	94	94	94	94
.27	.28	.3	.31	.32	.34	.35	.37	.39	.41
.06	.06	.06	.06	.07	.07	.08	.08	.08	.09
71285	74849	78591	82521	86647	90979	95528	100305	105320	110586
0	0	0	0	0	0	0	0	0	0
7264	7627	8009	8409	8830	9271	9735	10221	10733	11269
64020	67221	70583	74112	77817	81708	85793	90083	94587	99317
8520	8520	8520	8520	8520	8520	8520	852	852	852
55500	58701	62063	65592	69297	73188	77274	89231	93735	98465
35	35	35	35	35	35	35	42	42	42
19425	20546	21722	22957	24254	25616	27046	37254	39134	41109
36075	38156	40341	42635	45043	47572	50228	51977	54601	57356
8520	8520	8520	8520	8520	8520	8520	852	852	852
1800	1908	2022	2144	2272	2409	2553	2707	2869	3041
42795	44768	46838	49011	51291	53683	56194	50123	52584	55167
26850	26498	26154	25818	25490	25169	24855	20914	20699	20487
1417	1487	1562	1640	1722	1808	1898	1993	2093	2197
19425	20546	21722	22957	24254	25616	27046	37254	39134	41109
1344	1411	1482	1556	1633	1715	1801	1891	1986	2085
538	564	593	622	653	686	720	756	794	834
22723	24008	25358	26775	28263	29825	31465	41895	44007	46225
32	32	32	32	33	33	33	42	42	42
42795	44768	46838	49011	51291	53683	56194	50123	52584	55167

be an objective but it may be balanced by a realistic appraisal of the financial risks involved.

(2) There are tax benefits accruing with alternative mixtures of debt and equity. The objective often is to maximize these tax benefits given the requirements of a project.

(3) There may be a credit impact on the company as a result of increasing the debt burden that results in higher capital cost and should be avoided.

(4) Availability of guarantees may dictate certain sources of funding as preferred to others.

(5) Legal requirements may dictate the financing package.

(6) In some cases there is great value in sharing risk through joint venture or concentrating the financial resources from a wide variety of sources.

(7) Project versus non-project financing will control the sources available for a specific package.

The financial manager of the minerals or energy company supplies the link to financing sources of many different types, which may be classified as debt, equity or other forms as follows:

Debt Financing Sources

(1) Commercial Banks (term loans, promissory notes, other private placement)
(2) Savings and Loan Associations
(3) Finance Companies
(4) Insurance Companies (term loans)
(5) Pension Funds
(6) Equipment Dealers
(7) Holding Companies
(8) Governments
(9) International Agencies
(10) Customers
(11) Development Banks
(12) Companies

Table 3.7.10—Gunung Bijih Cash Flow Evaluation 83% Debt, 1969–1985

YEAR	PHASE 1		
	1973	1974	1975
Ore Grades			
Copper (%)	2.5	2.5	2.5
Gold (Ounces Per Metric Ton Ore)	.025	.025	.025
Silver (Ounces Per Metric Ton Ore)	.265	.265	.265
Price Per Pound of Copper Net of Smelting and Ref. Charges ($U.S.)	.45	.47	.5
Price Per Ounce Silver ($U.S.)	1.6	1.68	1.764
Price Per Ounce of Gold ($U.S.)	35	36.75	38.59
Add Byproduct Credit for Gold and Silver ($/Lb. Copper)	.025	.026	.028
Annual Copper Production Rate (000) Pounds of Copper ($000)	97159	129546	129546
Metallurgical Recovery Percent	94	94	94
Avg. Total Cost ($/Lb. Cu.)	.32	.3	.3
Avg. Operating Cost ($/Lb. Cu.)	.05	.05	.05
Revenue ($000 U.S.)	46184	64657	67890
Less Annual Royalty ($000)	0	0	0
Less Annual Operating Cost ($000)	4706	6589	6918
Equals Gross Income ($000)	31816	49022	52600
Less Interest On Loans ($000)	9661	9046	8372
Less Annual Depreciation and Amortization ($000)	8520	8520	8520
Equals Taxable Income ($000)	23296	40502	44080
Effective Income Tax Rate (%)	35	35	35
Less Tax ($000)	8154	14176	15428
Equals Net Income After Tax ($000)	15143	26326	28652
Less Amortization of Loan ($000)	6340	6955	7629
Plus Non Cash Flows			
Depreciation and Amortization ($000)	8520	8520	8520
Less Replacement Inv. ($000)	0	0	0
Equals Nominal $ Cash Flow ($000)	17323	27892	29543
Constant $ Cash Flow ($000)	12945	19662	19648
Wages and Salaries of Nationals Including Non-Monetary Benefits ($000)	918	1285	1349
Income Taxes	8154	14176	15428
Domestic Purchases of Materials and Services	871	1219	1280
Other (Import Duties Other Taxes Wages and Salaries Not Transferred Abroad)	348	488	512
Total Retained Value	10290	17167	18569
Retained Value As a Percent of Revenue	22	27	27
Freeport's Portion of Cashflow ($000)	17323	27892	29543

Table 3.7.10A—Gunung Bijih Basic Assumptions for Cash Flow Evaluation 83% Debt, 1969–1985

First Year of Investment Cash Flows	1969
Last Year of Investment Cash Flows	1972
Minimum Acceptable Discount Rate (%)	10
Projected Rate of Escalation for Investment Expenditures (%)	5

Summary of Investment Cash Flows ($000)

Year	Cash Flow Base Yr. $	Equity CF Esc. $	Debt Esc. $	Total Esc. $
1969	7247	7247	0	7247
1970	17718	3664	14940	18604
1971	39396	3594	39840	43434
1972	43809	5894	44820	50714
Total	108170	20400	99599	119999

Depreciable Inv. ($000)	85200
Mineable Ore Reserve (000 M.T.)	32700
Mine Life (Yrs.) = Ore Reserves/An. Prod. Rate	13
Total Production Employment	338
The Loan Is ($000)	1

Table 3.7.10A (cont.)

The Period of the Loan Is	10
Loan Repayment Is Scheduled During Phases 1 and 2 Over 10 Years	
The An. Payment ($000) Is 16001.0376	
Assumed General Inflation Rate (%)	6

Table 3.7.10B—Summary of Gunung Bijih Cash Flow Evaluation 83% Debt, 1969–1985

The Nominal $ Net Present Value at 10 Percent Discount Rate is ($000)	145080
The Nominal $ DCFROR Is 54.6 Percent	
Assumed Rate of Inflation is 6 Percent	
The Constant $ NPV at 4 Discount Rate Is ($000)	145080
The Constant $ DCFROR Is 46 Percent	

Table 3.7.10 (cont.)

PHASE 1		PHASE 2					PHASE 3		
1976	**1977**	**1978**	**1979**	**1980**	**1981**	**1982**	**1983**	**1984**	**1985**
2.5	2.5	2.5	2.5	2.5	2.5	2.5	2.5	2.5	2.5
.025	.025	.025	.025	.025	.025	.025	.025	.025	.025
.265	.265	.265	.265	.265	.265	.265	.265	.265	.265
.52	.55	.57	.6	.63	.67	.7	.73	.77	.81
1.852	1.945	2.042	2.144	2.251	2.364	2.482	2.606	2.737	2.873
40.52	42.54	44.67	46.9	49.25	51.71	54.3	57.01	59.86	62.85
.029	.03	.032	.034	.035	.037	.039	.041	.043	.045
129546	129546	129546	129546	129546	129546	129546	129546	129546	129546
94	94	94	94	94	94	94	94	94	94
.31	.32	.33	.33	.34	.35	.36	.37	.39	.41
.06	.06	.06	.06	.07	.07	.08	.08	.08	.09
71285	74849	78591	82521	86647	90979	95528	100305	105320	110586
0	0	0	0	0	0	0	0	0	0
7264	7627	8009	8409	8830	9271	9735	10221	10733	11269
56389	60402	64653	69160	73937	79003	84379	90083	94587	99317
7631	6820	5929	4952	3880	2705	1415	0	0	0
8520	8520	8520	8520	8520	8520	8520	852	852	852
47869	51882	56133	60640	65417	70484	75859	89231	93735	98465
35	35	35	35	35	35	35	42	42	42
16754	18159	19647	21224	22896	24669	26551	37254	39134	41109
31115	33723	36487	39416	42521	45814	49308	51977	54601	57356
8370	9181	10072	11049	12121	13296	14586	0	0	0
8520	8520	8520	8520	8520	8520	8520	852	852	852
1800	1908	2022	2144	2272	2409	2553	2707	2869	3041
29465	31154	32912	34743	36648	38629	40689	50123	52584	55167
18487	18440	18378	18302	18213	18111	17997	20914	20699	20487
1417	1487	1562	1640	1722	1808	1898	1993	2093	2197
16754	18159	19647	21224	22896	24669	26551	37254	39134	41109
1344	1411	1482	1556	1633	1715	1801	1891	1986	2085
538	564	593	622	653	686	720	756	794	834
20052	21621	23283	25042	26905	28878	30970	41895	44007	46225
28	29	30	30	31	32	32	42	42	42
29465	31154	32912	34743	36648	38629	40689	50123	52584	55167

Equity Financing

(1) Investment or Drilling Funds
(2) Other Companies
(3) Governments (equity participation in infrastructure)
(4) Individuals
(5) Sales of Common Stock (direct or through investment bankers)
(6) Retained earnings
(7) Investment Bankers (Purchasing and distributing of new stocks or bonds, underwriting)

Other Financing Sources

(1) Leasing company (sales leaseback agreements, operating leases, financial leases)
(2) Mergers and acquisitions

It is not possible to define an ''optimum'' financial plan; each financial package must be related to a unique situation, therefore the subject is most easily defined in terms of specific times, places and economic conditions. Some examples follow:

Freeport Indonesia: Gunung Bijih

The example of financial leverage summarized in Table 3.7.8 is based on estimates of the financial structure of Freeport Indonesia, a subsidiary of Freeport Minerals Company (now Freeport-McMoRan). Freeport Indonesia was created to develop and produce the copper ores from Gunung Bijih, formerly the Ertsberg, in Irian Java. When the Ertsberg came into production in late 1972, Freeport Minerals was engaged exclusively in the business of mining and processing sulfur, phosphoric and sulfuric acid, potash, kaolin, oil and gas. Freeport Indonesia was established as an 80% owned unconsoli-

Table 3.7.11—Gunung Bijih Cash Flow Evaluation 83% Debt, with Changes in Economic Assumptions 1969–1985

YEAR	PHASE 1 1973	1974	1975
Ore Grades			
Copper (%)	2.5	2.5	2.5
Gold (Ounces Per Metric Ton Ore)	.025	.025	.025
Silver (Ounces Per Metric Ton Ore)	.265	.265	.265
Price Per Pound of Copper Net of Smelting and Ref. Charges ($U.S.)	.45	.45	.46
Price Per Ounce Silver ($U.S.)	1.6	1.68	1.764
Price Per Ounce of Gold ($U.S.)	35	36.75	38.59
Add Byproduct Credit for Gold and Silver ($/Lb. Copper)	.025	.026	.028
Annual Copper Production Rate (000) Pounds of Copper ($000)	97159	129546	129546
Metallurgical Recovery Percent	94	94	94
Avg. Total Cost ($/Lb. Cu.)	.39	.36	.37
Avg. Operating Cost ($/Lb. Cu.)	.05	.05	.06
Revenue ($000 U.S.)	46184	62324	63084
Less Annual Royalty ($000)	0	0	0
Less Annual Operating Cost ($000)	4706	6903	7593
Equals Gross Income ($000)	29695	44389	45281
Less Interest On Loans ($000)	11783	11033	10210
Less Annual Depreciation and Amortization ($000)	10391	10391	10391
Equals Taxable Income ($000)	19304	33998	34890
Effective Income Tax Rate (%)	55	55	55
Less Tax ($000)	10617	18699	19189
Equals Net Income After Tax ($000)	8687	15299	15700
Less Amortization of Loan ($000)	7732	8482	9305
Plus Non Cash Flows			
Depreciation and Amortization ($000)	10391	10391	10391
Less Replacement Inv. ($000)	0	0	0
Equals Nominal $ Cash Flow ($000)	11345	17208	16786
Constant $ Cash Flow ($000)	8478	12131	11164
Wages and Salaries of Nationals Including Non-Monetary Benefits ($000)	918	1346	1481
Income Taxes	10617	18699	19189
Domestic Purchases of Materials and Services	871	1277	1405
Other (Import Duties Other Taxes Wages and Salaries Not Transferred Abroad)	348	511	562
Total Retained Value	12754	21833	22637
Retained Value As a Percent of Revenue	28	35	36
Freeport's Portion of Cashflow ($000)	11345	17208	16786

Table 3.7.11A—Gunung Bijih Basic Assumptions for Cash Flow Evaluation 83% Debt, with Changes in Economic Assumptions 1969–1985

First Year of Investment Cash Flows	1969
Last Year of Investment Cash Flows	1972
Minimum Acceptable Discount Rate (%)	10
Projected Rate of Escalation for Investment Expenditures (%)	15

Summary of Investment Cash Flows ($000)

Year	Cash Flow Base Yr. $	Equity of Esc. $	Debt Esc. $	Total Esc. $
1969	7247	7247	0	7247
1970	17718	2155	18221	20376
1971	39396	3512	48589	52101
1972	43809	11965	54662	66627
Total	108170	24880	121472	146352

Depreciable Inv. ($000)	103910
Mineable Ore Reserve (000 M.T.)	32700
Mine Life (Yrs.) = Ore Reserves/An. Prod. Rate	13
Total Production Employment	338

Table 3.7.11A (cont.)

The Loan Is ($000)	1
The Period of the Loan Is	10
Loan Repayment Is Scheduled During Phases 1 and 2 Over 10 Years	
The An. Payment ($000) Is 19514.9389	
Assumed General Inflation Rate (%)	6

Table 3.7.11B—Summary of Gunung Bijih Cash Flow Evaluation 83% Debt, with Changes in Economic Assumptions 1969–1985

The Nominal $ Net Present Value at 10 Percent Discount Rate Is ($000)	49457
The Nominal $ DCFROR Is 35.5 Percent	
Assumed Rate of Inflation Is 6 Percent	
The Constant $ NPV at 4 Discount Rate Is ($000)	49457
The Constant $ DCFROR Is 28 Percent	

Table 3.7.11 (cont.)

PHASE 1		PHASE 2					PHASE 3		
1976	1977	1978	1979	1980	1981	1982	1983	1984	1985
2.5	2.5	2.5	2.5	2.5	2.5	2.5	2.5	2.5	2.5
.025	.025	.025	.025	.025	.025	.025	.025	.025	.025
.265	.265	.265	.265	.265	.265	.265	.265	.265	.265
.46	.47	.47	.48	.48	.49	.49	.5	.5	.51
1.852	1.945	2.042	2.144	2.251	2.364	2.482	2.606	2.737	2.873
40.52	42.54	44.67	46.9	49.25	51.71	54.3	57.01	59.86	62.85
.029	.03	.032	.034	.035	.037	.039	.041	.043	.045
129546	129546	129546	129546	129546	129546	129546	129546	129546	129546
94	94	94	94	94	94	94	94	94	94
.37	.37	.37	.38	.38	.38	.39	.36	.37	.38
.06	.07	.08	.09	.09	.1	.11	.13	.14	.15
63858	64647	65451	66271	67108	67962	68834	69723	70632	71561
0	0	0	0	0	0	0	0	0	0
8352	9187	10106	11117	12228	13451	14796	16276	17904	19694
46198	47142	48114	49115	50147	51212	52312	53447	52729	51867
9307	8317	7231	6040	4732	3299	1726	0	0	0
10391	10391	10391	10391	10391	10391	10391	1039	1039	1039
35807	36751	37723	38724	39756	40821	41921	52408	51690	50828
55	55	55	55	55	55	55	55	55	55
19694	20213	20748	21298	21866	22452	23056	28825	28429	27955
16113	16538	16975	17426	17890	18370	18864	23584	23260	22872
10208	11198	12284	13475	14782	16216	17789	0	0	0
10391	10391	10391	10391	10391	10391	10391	1039	1039	1039
2195	2327	2467	2615	2771	2938	3114	3301	3499	3709
14101	13404	12616	11727	10727	9606	8352	21322	20800	20203
8847	7934	7045	6178	5331	4504	3694	8897	8188	7503
1629	1792	1971	2168	2385	2623	2885	3174	3491	3840
19694	20213	20748	21298	21866	22452	23056	28825	28429	27955
1545	1700	1870	2057	2262	2488	2737	3011	3312	3643
618	680	748	823	905	995	1095	1204	1325	1457
23486	24384	25336	26345	27418	28559	29774	36214	36557	36896
37	38	39	40	41	42	43	52	52	52
14101	13404	12616	11727	10727	9606	8352	21322	20800	20203

dated subsidiary exclusively concerned with the production of copper concentrates from the Gunung Bijih orebody. Because of this relationship, the annual reports and 10-K made by Freeport Minerals to the Securities and Exchange Commission provide a clear picture of the financing of this unique project.

Various start-up problems were experienced by the project in 1973 and 1974. Expenditures incident to these changes and additions increases the original estimated capital investment in the project, including working capital to approximately \$200 million. The increased project costs, amounting to approximately \$80 million were financed from the cash flows of Freeport Indonesia.

By 1975, Freeport Indonesia was contributing 20% of Freeport Minerals total sales and the right hand side of the Freeport Indonesia balance sheet reveals the financing pattern shown in Table 3.7.12.

At the end of 1975, the total investment in property, plant, equipment, exploration, and feasibility studies for the project totaled \$195.8 million; there was an additional investment in materials and supplies of \$22.3 million, and cash on hand totaled \$2.8 million. Of this over \$200 million investment, 37% was stockholder equity, including retained earnings and 50% came from long-term debt.

The retained earnings were critical sources of financing during the difficult startup period. Agreements signed with five U.S. insurance companies, seven U.S. banks, the German development bank (Kredietanstalt fur Wiederaufbau), and eight Japanese copper smelters provided for a 20% overrun contingency but not for overruns to the extent actually experienced.

Table 3.7.12—Freeport Indonesia Financial Structure, 1975 (Freeport Minerals, 1976)

LIABILITIES:	1975
Current liabilities:	
Accounts payable, trade creditors	$ 1,917,000
Accrued liabilities	1,197,000
Accrued income taxes	1,030,000
Long-term debt due within one year	15,128,000
Total current liabilities	19,272,000
Long-term debt, less portion included in current liabilities	86,681,000
Reserve for future income taxes	21,873,000
STOCKHOLDERS' EQUITY:	
Common stock, par value $100; authorized, issued and outstanding 224, 210 shares	22,421,000
Paid-in additional capital; excess of amount paid in over par value of common stock	1,579,000
Retained earnings, as annexed	51,652,000
Prepaid and deferred items chargeable to future operations	75,652,000
Total Liabilities and Stockholders' Equity	$203,478,000

U.S. Insurance Companies: Five U.S. insurance companies committed $40 million of senior debt plus an additional 20% in case of cost overrun. These funds can be described as supplied by term loans to be repaid from the third through the tenth year of operation. Usually such loans are amortized over the repayment period in the form of an annuity which during the earliest years consists of a high interest portion and a low amortization portion and during the last years a high amortization and low interest. The repayment schedule has an important impact on economic analysis because of the deductibility of interest and its impact on lowering taxes during the early years of the project.

For Gunung Bijih project, the loans and interest were covered by a comprehensive guarantee by the Agency for International Development (AID) against all risks under the AID extended risk program. The AID program, now known as the Overseas Private Investment Corporation (OPIC), insures investments against expropriation, war or insurrection, and restrictions on foreign exchange. Such risks are likely to prevent insurance companies financing of the project. Typically these risks are uninsurable by commercial insurance companies. As of 1983, 24 countries sponsored major investment guarantee programs including Australia, Austria, Belgium, Canada, Denmark, Finland, France, West Germany, India, Italy, Japan, Netherlands, New Zealand, Norway, Republic of Korea, South Africa, Sweden, Switzerland, United Kingdom, and the United States. OPIC insurance is described in detail in a Colorado School of Mines MS thesis by James Otto (Otto, 1984).

U.S. Commercial Banks: A second major source of financing for the Gunung Bijih project was the seven U.S. commercial banks that provided $18 million of senior debt financing with an additional 20% in case of cost overrun. These were term loans scheduled for repayment during the second through fourth year of operation. The commercial bank loans were guaranteed by the Export-Import Bank of the United States (EXIMBANK). EXIMBANK was established in 1934 with the objective to increase U.S. exports of goods and services. The institution has three major areas of activity (1) direct loans associated with long-term lending for the purchase of U.S. goods and services, (2) financial guarantees backed by the full faith and credit of the United States, and (3) export credit insurance covering credits extended by U.S. suppliers to overseas purchasers. In the case of the Freeport Indonesia loan, insurance was provided against political risks such as expropriation, inconvertibility, and war and commercial risks such as default by the purchaser. The program, operated in collaboration with 50 private insurance companies, makes every effort to match the competitive terms quoted by foreign suppliers. EXIMBANK strives for "mutuality of benefits" with the U.S. supplier of capital gaining a customer and the developing country gaining a low cost source of financing.

German Development Bank: The third source of senior loan capital for the Gunung Bijih project was Kredietanstalt fur Wiederaufbau (KFW), the German Reconstruction Loan Corporation which provided $22 million plus 20% in case of overrun. This also was a term loan scheduled for repayment from the second through the tenth year of operation. German government guarantees protected the KFW loan against political and economic risk while the U.S. EXIMBANK agreed to guarantee a portion of the KFW loan not guaranteed by the German government.

Japanese Financing Package: The Japanese provided $20 million in the form of a subordinated loan for the Gunung Bijih project plus 20% in case of overrun. Here the money was provided by the Japanese export-import bank and a group of Japanese commercial banks. Guarantee were provided by the Japanese copper smelting industry and five Japanese trading companies that were interested in the copper concentrates. The term loan was scheduled for repayment from the third through the eighth operating year of the project with loan repayments deducted from payments for copper concentrates. This scheme might be described as financing by production payments where loan repayment comes in the form of production from a proven mineral reserve (Nevitt, 1979, p. 128).

Southern Peru Copper Corporation: Cuajone

The Cuajone project of Southern Peru Copper Corporation provides another example of a major international mineral financing scheme. The basic financing for the Cuajone Project consisted of $216 million in equity contributions and $406.7 million in project loans supplied by 54 banks in nine countries and included loan guarantees from the Export-Import Bank ($75 million), the International Finance Corporation of the World Bank ($15 million), the Export Credits Guarantee Department of the United Kingdom (approximately $75 million of guarantees), and the government of Japan (approximately $50 million in guarantee commitments). All the loan agreements are related through an inter-creditor agreement which provides that a default against any one lender is a default against all. The basic security for the loans is provided by assignment for the benefit of the lenders of the proceeds of sales of its production under term contracts with 13 purchasers of copper in the United States, Europe, and Japan.

In mid-1976 it was necessary for Southern Peru Copper Corporation to acquire an additional $106 million in loans to complete the project. This consisted of a $36 million overrun on the $620 million project cost, including interest during construction, and $70 million for permanent working capital, required to finance warehouse stocks and the product pipeline to market.

The plan to provide the required additional long-term credit was worked out with certain major lenders under which stockholders of Southern Peru and others provided $48 million in additional equity funds and the balance of the $106 million to be provided by banks which participated in the basic financing.

Table 3.7.13—Southern Peru Copper Corporation Long-Term Debt Commitments

	Thousands of Dollars
Chase Manhattan Bank Consortium (29 banks)	$ 200,000
Equipment Financing	
Export-Import Bank of US/Wells Fargo	64,684
J Henry Schroder Banking Corp. Syndicate	10,316
Wells Fargo	5,650
Manufactures, Irving Trust, Marine Midland (Guaranteed by Banco de Credito)	15,000
Lazard Brothers and Baring Brothers	44,613
Cobrasma, S A (Brazil)	2,854
Financing Arranged by Copper Purchasers Holding	
15 Year Copper Contracts	
Lloyds Bank International	23,500
Orion Bank Ltd	6,375
Bank of Toyko, Ltd	23,750
International Finance Corporation (Additional Commitment of $5 million subject to approval of other lenders)	10,000
Total	$ 406,742

A summary of the long-term debt commitments prior to the 1976 financing efforts follows in Table 3.7.13:

The Chase Manhattan Consortium: The Cuajone financing again indicates the importance of the commercial bank as a financier of large mining projects. These banks hold the position of debtor and public trustee of funds deposited in the banks. A return of loan principle and interest earnings at the scheduled time must be assured. Thus the project must have a well defined earnings with known mineral values and rates of recovery. Such financing is usually provided after the private investor has contributed substantial equity or risk capital, a requirement that often disqualifies properties in new areas where there is insufficient information to support reserve estimates or to define operating hazards.

There may also be questions as to market ability of the product at a price sufficient to pay costs including interest and amortization of loans. In the case of the Cuajone project a marketing agreement for the product was a key factor in obtaining financing. Long-term sales contracts for planned output were signed with 13 purchasers in the United States, Europe, and Japan with the approval of the Peruvian government. A major surprise was the decline in continued low price levels for copper after 1975 after most of the money had been committed to development. In addition, as a result of escalating costs and the needs to provide additional working capital, the financial arrangements made in 1974 were not adequate to complete the project. Revised project costs totaled $726 million in mid-1976 leading to discussion with the various lenders and others with a view to arranging waiver of debt limitations and providing necessary additional financing. The partners also had to increase their investment.

The period between 1977 and 1984 has been an extremely difficult one for the copper industry. Prices of $0.60 to 0.80 per pound were completely inadequate for most of the copper industry to justify expansion or even continued production. The Southern Peru Copper Corporation was required to renegotiate loans in order to extend debt payments. During the same period the effective Peruvian tax rate increased to 68.5% as compared to the expected effective rate of 48%.

Equipment Financing: A second major source of long-term financing for Southern Peru Copper Corporation were the banks involved in supplying funds for purchase of equipment. This included the U.S. Export-Import Bank and commercial banks in the United States, Brazil and Peru.

The Export-Import Bank of the United States, formerly EXIMBANK, is a source of equipment financing for international projects. The bank recognizes that credit availability is an important competitive tool and seeks to provide American exporters with facilities at least as good as those made available to exporters in other countries by their governments. They provide direct credits, guarantees, and insurance for U.S. exports with funds derived from a number of sources including capital stock in accumulated reserves, certificates of participation in its portfolio, debentures, and short-term promissory notes.

Copper Purchasers: A third major source of funds for the Cuajone project were those interested in its copper product. Cuajone was expected to market 100,000 short tons of blister copper annually with 30,000 tons going to the United Kingdom, 30,000 tons to Japan, and 40,000 tons to Holland. Thus with long-term sales contracts in hand this source of financing became available for the project. The details of the Cuajone arrangement are not known but this is the type of financing described by Nevitt that sometimes takes the form of production payments or carveouts or advanced payments for the product (Nevitt, 1979, p. 127–138).

International Finance Corporation: The fourth major financier for the Southern Peru Copper Corporation Caujone Project was the International Finance Corporation (IFC). The IFC is an affiliate of the International Bank for Reconstruction and Development (World Bank) which has been a major supplier of international capital. After World War II, meetings of 44 allied nations convened at the United Nations Monetary and Financial Conference at Bretton Woods, New Hampshire, USA. The bank was created to make or guarantee loans for reconstruction and development projects and were to be made from the bank's own capital or through mobilization of private capital. Risk is shared by all member governments roughly in proportion to economic strength.

The bulk of the financial resources for the organization come from world capital markets. The role of the international finance corporation is to support private enterprise in developing countries where there is a good prospect of profitable operation and where the project is a benefit to the economy. In the case of Cuajone, the participation of the IFC is relatively small amounting to only 2% of the total long-term debt commitments. The participation of the organization is generally less than 25% and the organization has the objective to sell its investment once the project is successful (Ulatowski, 1979).

Rosario/Simplot: Pueblo Viejo

The Pueblo Viejo gold-silver project in the Dominican Republic began as a joint venture with three partners—Campbell, Chogougam au Mining Co., Toronto; Simplot Industries, Inc., Boise, and Toronto; Rosario Resources Corporation formerly New York and Honduras Rosario Mining Company. After six months, Campbell dropped out of the project leaving it a joint venture between Rosario and J.R. Simplot Company, with Rosario as operator. Rosario is a natural resources company with a solid organization of capable people dedicated to exploration, development and production of mineral wealth. The company is a diversified producer of silver, lead, zinc, gold, and cadmium, oil and gas, granite aggregates, limerock, chemical lime, agriculture limestone, and industrial minerals.

Initially, the objective for development of the Pueblo Viejo Gold Mine was for $18 million of debt financing and $6 million of equity with the equity split 50-50 between Rosario and Simplot. The complete financing package as it was invisioned in February of 1973 is shown in Table 3.7.14.

Table 3.7.14—Financing Plan for Pueblo Viejo, Feb. 1973

Participant	Financing
Rosario and Simplot	$ 6,000,000
Export-Import Loan	3,000,000
Chemical Bank (with EXIMBANK guarantee)	5,933,000
Chemical Bank (Eurodollars)	6,167,000
Chemical Bank (OPIC insurance)	3,000,000
	$24,000,000

The financing plan never materialized because the Export-Import Bank (EXIMBANK) and the Overseas Private Investment Corporation (OPIC) could not agree on the separation of insurance responsibilities. The contracts written by these two organizations differ with the OPIC contract being more inclusive. In particular, EXIMBANK would not provide any insurance for "creeping nationalization" while OPIC policies did cover this type of threat. Unable to get agreement between insurer and guarantor, Rosario went to the Royal Bank of Canada in Montreal and secured a $19 million loan, including insurance and a completion guarantee but the financing scheme was about to take a 180° turn.

In May of 1973, the Central Bank of the Dominican Republic decided that they wanted equity in the project. The Central Bankers got interested in equity in the project because of the rising price of gold and because high sugar prices created a very favorable balance of payments on the island. During the fall of 1973 gold and foreign exchange holdings of the Dominican Central Bank were high—$35 to $40 million. Dominican refinancing would eliminate growth in the money supply of the island, whereby controlling prices. It appears that this strategy was successful over the 1973–1976 period; the rate of inflation in the country was low. The agreement to borrow the $18 million from Dominican banks was consummated on August 30, 1973.

This preliminary financing plan provided equity from the profits of existing operations of the joint venture partners coupled with commercial bank and the Export-Import Bank debt financing. One new source of funds indicated is the Eurodollars supplied by the Chemical Bank. A Eurocurrency deposit is created when a banking office in one country accepts a deposit denominated in the currency of another country. The Eurocurrency market denominated in U.S. dollars is referred to as the Eurodollar market which includes loans of various maturities.

Many U.S. firms utilize the Euromarket as borrowers because rates on long-term borrowing are sometimes below rates on comparable securities in the United States. Relative rates depend on supply and demand for dollars but during the 1970's competition for a loan business reduced the cost of credit making the market an attractive source of capital. Eurodollar loans typically are in multiples of $1 million and have maturities from 30 days to five to seven years (Weston, 1981, p. 1037).

The origin of the Eurodollar system in the early 1950s was the result of banks accepting interest bearing deposits in currencies other than their own. Most of the activity was in Europe and predominantly was the U.S. dollar. As the system developed it included currencies of many different countries and was designated the Eurocurrencies system.

The interest of the Central Bank of the Dominican Republic as an equity participant and principle source of loans for the Pueblo Viejo mine created a unique financing structure for Rosario Dominicana S.A. When the project came into production in 1975, Rosario and Simplot Industries Inc. each had a 40% equity interest in the project and the Central Bank of the Dominican Republic owned a 20% equity interest. Table 3.7.15 summarizes the liabilities and stockholder equity for Rosario Dominicana, S.A. at the end of 1974 prior to the generation of additional retained earnings due to production. The table shows short-term financing from accounts payable, accrued expenses, taxes, and other liabilities and the major long-term debt financing for the project.

The long-term debt financing consists of $32.2 million of bank financing in the form of two loans bearing interest at 9.5%. These were se-

Table 2.7.15—Rosario Dominicana S. A. Liabilities and Stockholders' Equity, December 31, 1974 (Rosario Resources Corporation, 1975)

Liabilities and Stockholders' Equity	1974 US $
Current liabilities:	
Bank Overdraft	462,524
Current maturities of debt	—
Accounts payable:	
Contractors	727,971
Trade	133,394
Total accounts payable	861,365
Due to affiliates	225,665
Accrued expenses and other liabilities	375,168
Income taxes	—
Total current liabilities	1,924,722
Long-term debt, excluding current maturities	
Stockholders' Equity:	
Common stock-authorized 120,000 shares of RD $100 par value each issued 100,000	10,000,000
Legal reserve	—
Retained earnings	—
Total stockholders' equity	10,000,000
Total liabilities and stockholder's equity	$42,524,722

cured by mortgages and liens on substantially all property, plant and equipment, and inventory. Insurance policies endorsed to the lenders cover the values of assets acquired with the loans. Repayment was scheduled in 10 equal semiannual installments. In addition the company was required to maintain compensating checking account balances with the banks equivalent to 10% of the outstanding loans or alternatively to make payment equivalent to interest at 10% on the amount of such balances used (Rosario Resource Corporation, 1975). The source of these loans were banks in the Dominican Republic subject to the request by The Central Bank that Rosario Dominicana borrow all of the money in pesos from banks within the country. It was an unusual decision but the government felt the country could absorb the drain on local currency, that the interest would be paid in pesos within the country and that given the profitability of the venture, the loan would be repaid very quickly. The structure of the financing highlights the uniqueness of each financing scheme adapting to specific economic conditions and specific characteristics of the interested parties.

AMAX: Henderson

The financing of the Henderson project in Colorado by AMAX Inc., is an excellent example of non-project financing; lenders look to the cash flows and earnings of the company rather than the new economic unit as a source of funds from which the loan will be repaid. The massive capital expenditures for the $500 million project in Colorado were met from retained earnings of AMAX supplemented by loan and debenture funds raised by AMAX against other company assets and earnings during the project development.

Early expenditures on the Henderson project began after discovery of the orebody during 1965 and continued until full production capacity was reached in 1979. Heaviest investment occurred during the major construction phase between 1971 and 1976 as the mine approached initial production.

Table 3.7.16—AMAX sources and Uses of Funds, 1971–1975

Sources	
Net earnings	24%
Depreciation and depletion	9
Proceeds from sale of future production	6
Long-term debt securities	19
Sale of common stock	15
Increase in Accounts Payable	6
Sale of ALUMAX, Inc.	6
All other	15
	100
Uses	
Property, Plant, Equipment	64%
Investment in other companies	5
Dividends	11
Repayment of long-term debt	6
All other	14
	100

Table 3.7.16, calculated from data taken from AMAX form 10K annual reports, shows the sources and uses of funds during the period of the Henderson financing, 1971–1975. AMAX "cash flow" or net earnings plus depreciation plus depletion from producing operations provided 31% of the funds during the period. Another 44% of funds came from sale of common stock, long-term debt, and the sale of existing

assets in ALUMAX Inc. Two thirds of these sources were used for investment in property, plant, and equipment of which a large part must have been the Henderson investment.

AMAX is engaged, directly or through ventures in which it holds substantial interests, in the exploration for and mining of ores and minerals and the smelting and refining and other treatment of minerals and metals. Principle products include molybdenum, coal, iron ore, copper, lead, zinc, petroleum and natural gas, potash, nickel, and forest products. The consolidated balance sheet covers all activities of the company, making it impossible to identify funds specifically directed to the Henderson project, however, liabilities of the 1976 balance sheet reflect the structure of financing at that point which included $414 million expended on the Henderson project as of December 31, 1976. In addition, the company was projecting addi-

Table 3.7.17—AMAX Inc. and Consolidated Subsidiaries, Consolidated Statement of Financial Position, Liabilities and Deferred Credits, December 31, 1976 (Amax, 1977, p. F-4)

	1976	
Liabilities	**$ millions**	
Current Liabilities		
Accounts and drafts payable, trade		87.6
Accrued liabilities		
Payrolls and vacation pay	11.1	
Taxes	54.0	
Interest	18.3	
Other	31.4	114.7
Short-term borrowings		43.0
Current maturities of long-term debt		17.1
Dividends payable		3.3
Other current liabilities		12.7
Total current liabilities		278.5
Long-term debt		630.1
Equipment lease contracts		7.3
Other non-current liabilities		18.4
Reserves		34.3
Proceeds from sale of future production		191.8
Deferred income taxes		70.4
Minority interest in consolidated subsidiaries		
Capital stock	1.0	
Retained earnings	1.3	2.2
		1,232.9
Shareholders' Equity		
Capital stock, statement annexe		
Preferred Stock, $1 par value, authorized 10,000,000 shares		
Series A Convertible, issued and outstanding, 1976, 1,328,001 shares; 1977, 1,227,566 shares; preference in liquidation, 1976, $132,800; 1977, $122,760	1.3	
Series B Convertible, issued and outstanding 2,000,000 shares; preference in liquidation, $100,000	2.0	
Series C, issued and outstanding, $1,500,000 shares; preference in liquidation, $150,000	—	
Common Stock, $1 par value, authorized 1976, 50,000,000 shares; 1977, 75,000,000 shares; issued: 1976, 32,112,175 shares; 1977, 32,456,012 shares	32.1	
Paid-in capital, statement annexed	709.2	
Retained earnings, statement annexed	821.9	
	1,566.6	
Deduct cost of Common Stock in treasury 1976 13,824 shares 1977, 12,076 shares	(500)	1,566.1
		$2,799.0

Table 3.7.18—AMAX Long-term Debt Excluding Current Maturities, December 31, 1976 (AMAX, 1977, p. F-12)

Senior Debt	1976
6¼% to 9⅜% sinking fund debentures payable 1979 to 2001	$210,100
4½% to 9¾% notes payable 1979 to 1996	238,180
8¾ sinking fund notes payable 1983 to 1997	—
6% to 7% short-term borrowings to be refinanced	—
7% mortgage payable 1979 to 1996	23,630
Other 3.8% to 10¼% mortgages, notes and bonds payable 1979 to 2007	47,820
	519,730
Subordinated Debt	
8¼% senior subordinated notes payable 1979 to 1997	50,000
8% subordinated sinking fund debentures payable 1981 to 1986	74,160
	124,160
Total	643,890
Less unamortized discount	(13,820)
Long-term debt, net of discount	$630,070

tional mine development, capital replacement, and possible capacity expansion of the mine requiring an additional $140 million between 1976 and 1985.

The table indicates that AMAX made extensive use of short-term financing through the use of a accounts payable, accrued liabilities, short-term borrowings, dividends payable, and other current liabilities. These current liabilities totaled 10% of total liabilities plus stockholders' equity at the end of 1976.

Long-term Debt: AMAX long-term debt is 23% of total liabilities plus shareholders equity. This debt structure, as indicated in Table 3.7.18, includes both senior and subordinated debt in the form of sinking fund debentures, notes payable, sinking fund notes payable, and mortgages.

The senior/subordinated classification separates liabilities which have a senior claim on assets from those having a claim only after the specified senior claims hae been paid off. The sinking fund classification indicates an agreement under which periodic payments are made at the applicable interest rate with the objective to accumulate to a target amount necessary to pay off the obligation. Debentures are long-term debt instruments not secured by a mortgage on specific property; a promissory note establishes the relationship between borrower and lender including the specified rate of interest and the period of the loan. In contrast to a debenture, a mortgage represents a pledge of specific property for a loan.

Common and Preferred Stock and Retained Earnings: AMAX shareholders' equity in Table 3.7.17 consists of preferred and common stock capitalized at par value. Common stock stockholders theoretically have legal control of the corporation and carry both rights to income and risks of loss; the common stockholder as receiver of residual income paid in the form of dividends may or may not receive a return on their investment in any specific year.

Carrying a fixed commitment for periodic payments, preferred stock is a hybrid between debt and common stock. Liquidation of claims of preferred stockholders take precedence over those of common stockholders, however, preferred dividend payments do not result in bankruptcy as does nonpayment of interest on bonds. In the example, the preferred stock is "convertible" which means that it is convertible into common stock at some ratio of preferred stock to common stock. The advantage of the conversion agreement to the company is a potential future sale of common stock at a price above the market on the date that the preferred is issued. Characteristically, convertibles have a provision that allows the firm to "call" the convertible at a specified price. If the company calls the preferred stock, the stockholder has the option to either convert to common stock or to redeem the preferred stock at the call price. Usually the preferred stockholder would choose the conversion which gives the company a means of forcing conversion. The convertible stock may be viewed as a temporary financing device which provides less expensive capital than debt and converts from hybrid debt to common stock at a later date. The disadvantage of a convertible, of course, is that if the common stock increases greatly in price, the company would have been better off if it had simply waited and sold common stock at a future date.

The paid-in capital indicated in Table 3.7.17 is an accounting method of adding the excess of sales price over the par value assigned to capital stock for shares which are sold at prices higher than par value. These may represent capital stock issued in exercise of subscriptions or options under a stock ownership plan or shares of common stock charged to earnings retained and invested in the business.

The retained earnings portion of shareholder equity is that portion of earnings not paid out in dividends. The figure that appears in Table 3.7.17 is the sum of retained earnings over the life of the company.

AMAX, Inc. Capital Structure: Study of the balance sheets of AMAX over the 1971–1977 period when the Henderson project was under construction, indicates a changing mixture of debt and equity. The increasing cost of debt brought about by rapid inflation after the mid-1970s had an impact on capital structure and the company objective to minimize the weighted cost of capital from debt and equity sources. The result of these changes appears in Fig. 3.7.17 indicates as a shift of long-term debt from 43% of total assets in 1971 to 29% of total assets in 1977.

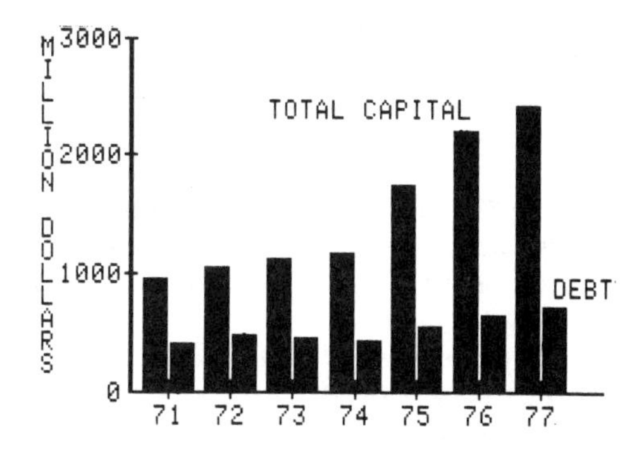

Fig. 3.7.17—AMAX capital structure.

Other Sources of Financing

The preceding examples cover debt financing by commercial banks, insurance companies, equipment dealers, governments, international agencies, customers, development banks, and joint ventures including two or more companies. The example also covers equity financing involving several companies, governments, sales of common stock, retained earnings, and the role of investment bankers in financing through the sale of new stocks and bonds. All of the preceding were part of the more comprehensive list of financing sources for minerals or petroleum projects. The scope of the present chapter does not include a comprehensive survey of all sources of financing but a brief review of some additional sources is in order.

Leasing: Lease financing especially leases of capital equipment are an important source of financing for the minerals industries. For example, during 1976, AMAX concluded arrangements for lease financing over a two year period of approximately $200 million of coal mining draglines. The two types of lessors involved in mineral project financing include (1) third party leasing companies and (2) sponsors or parties interested in the completion of the project. Leases may be leveraged with loan capital or nonleveraged. The reader is referred to a number of publications that discuss in detail the characteristics of alternative types of leases (Nevitt, 1979, pp. 25–53), (Weston, 1977, pp. 850–874), (Woolley, 1975), (Jenkins, 1970), (Hamel, 1968). Under a true lease, the leasing company owns the equipment during and at the end of the lease term, and thus can claim accelerated cost recovery (ACRS) deductions. The lessee may deduct the full lease payments for tax purposes which makes the efficient use of tax benefits one of the principal advantages of the leasing arrangement. Several advantages of using a lease from a leasing company to finance a project (Nevitt, 1979, p. 25–37) include:

(1) Relatively low cost under conditions where the lessee is unable to make efficient use of tax benefits. This is especially true when the lessee can not take advantage of tax benefits such as depreciation, interest, and the investment tax credits; the lessor can take the benefits and pass most of them on to the lessee and still make a profit.

(2) Improved cash flow for the lessee as compared to committing to loan payments may increase the lessee's present value of the future cash flows.

(3) Useful in joint venture partnerships where tax benefits are not available to one or more of the joint venture partners because of the way the joint venture is structured or a particular tax situation.

(4) Useful in project financing through subsidiaries not consolidated for tax purposes and not in a position to claim and use tax benefits from equipment acquisitions.

(5) On or off balance sheet financing which may be consistent with accounting objectives of the firm or supply an alternate source of financing when existing loan and note agreements restrict alternatives.

(6) A properly structured lease may reduce impact on book earnings.

(7) Offers predictable future fixed rate lease payments through pre-tax rentals rather than depreciation and after-tax cash flow.

(8) As a hedge against inflation where lease rates are determined by current price levels and paid in inflated dollars.

(9) Leases offer convenience in the form of a payment schedule designed to coincide with earnings, leasing terms often longer than loan terms, and level payments permitting the matching of rental expense to cash generated. In addition, the company avoid budget limitations and dilution of company ownership.

(10) Other advantages may include the fact that no public disclosures are required for a lease transaction as they are for public offering of debt and equity; the lease will provide 100% financing while a loan would require an initial down payment; capital is retained and can be utilized elsewhere; and provides for financing charges for delivery, interest, sale and use taxes not usually financed under alternative methods.

There are also some disadvantages to leasing including loss of residual value of the equipment, higher overall costs, timing which may not be ideal for the lessee, loss of the prestige of ownership, flexibility in use of the equipment, and creation of a long-term senior fixed obligation against the project. In general the leasing alternative is not as attractive for equipment located outside of the United States because of the lack of the investment tax credit and lower depreciation rates under other tax systems.

Mergers and Acquisitions: Merger activity has been an important source of financing external growth for U.S. firms especially those in the minerals and petroleum industry. Merger activity for U.S. industry resumed in 1977 and continued through the early 1980s. The year 1981 was described by *Fortune* as a fabulous one for big deals when sales of 50 companies

Table 3.7.19—Merger Deals of 1981: Oil and Gas, Mining, and Other Natural Resources

Rank	Company	Value ($ 000)	Transaction	% of Book Value	Industry
1	DU PONT CONOCO	7,214,858	Acquisition for cash and common stock, closed 9/30	156	Chemicals, fibers Oil, gas and coal
2	ELF AQUITAINE TEXASGULF	2,741,670	Acquisition of 65% for cash, 9/28	212	Oil and Gas (France) Natural resources
3	SEAGRAM CONOCO	2,552,275	Cash tender for 32%, 8/7	171	Distilling (Canada) Oil, gas, and coal
4	KUWAIT PETROLEUM SANTA FE INTERNATIONAL	2,500,774	Acquisition for cash, 12/4	376	Oil and gas (Kuwait) Contract drilling, engineering and construction
5	FLUOR St. Joe Minerals	2,342,945	Acquisition for cash and common stock, 8/3	190%	Engineering and construction Mining

6	STANDARD OIL OF OHIO KENNECOTT	1,767,592	Acquisition for cash, 6/4	123%	Oil and gas Copper, metal products
7	DOME PETROLEUM	1,675,000	Acquisition of 53% from Conoco for cash and Conoco common stock, 6/10	531	Oil and gas (Canada)
	HUDSON BAY OIL AND GAS				Oil and Gas (Canada)
8	DOME PETROLEUM CONOCO	1,430,000	Cash tender for 20%, 6/3	147	Oil and gas (Canada) Oil, gas and coal
19	OCCIDENTAL PETROLEUM IOWA BEEF PROCESSORS	770,330	Acquisition by exchange of common stock 8/12	240%	Oil and gas Meat processing
27	HOME PETROLEUM (Hiram Walker Resources	630,000	Acquisition for cash, 3/11	N.A.	Oil and gas (Canada)
	OIL AND GAS PROPERTIES OF DAVIS OIL (owned by Marvin Davis)				Oil and gas
31	STANDARD OIL OF OHIO COAL PROPERTIES OF U.S. STEEL	600,000	Acquisition for cash, 9/30	N.A.	Oil and gas Coal mining
32	CSR DELHI INTERNATIONAL OIL	591,130	Acquisition for cash, 11/20	1605	Diversified Resources (Australia) Oil and gas in Australia
34	ONTARIO ENERGY RESOURCES (Ontario Energy Corporation) SUNCOR (Sun)	543,400	Acquisition of 25% for cash and notes, 12/23	N.A.	Oil and gas (Canada) Oil and gas (Canada)
41	APEX OIL CLARK OIL AND REFINING	478,115	Acquisition of 90% for cash, 10/30	232	Petroleum trading Petroleum refining
43	SULPETRO CANDEL OIL (St. Joe Minerals)	460,166	Acquisition of 92% for cash, 4/3		Oil and gas (Canada) Oil and gas (Canada)
45	FREEPORT MINERALS McMORAN OIL AND GAS	456,623	Merger by exchange of common stock to form Freeport-McMoran, 4/7	1056	Mining Oil and Gas
49	TENNECO HOUSTON OIL AND MINERALS	417,000	Acquisition by exchange of common stock, 4/24	133	Oil and gas Oil, gas and minerals
50	NATOMAS MAGMA	415,947	Acquisition of 92.5% for cash	831	Oil and gas Geothermal Resources

totaled nearly $15 billion and 18 deals involved oil and gas or other natural resource mergers (*Fortune*, Jan. 25, 1982, p. 36). The 18 deals are listed in Table 3.7.19. The first eight deals in the *Fortune* list involve natural resources including the largest of all, the $7.2 billion purchase of Conoco by Du Pont, the biggest corporate deal in U.S. history.

Reinhardt defines "external" expansion as growth and diversification resulting from the acquisition of other ongoing business firms (Reinhardt, 1972). Most financing is discussed in terms of internal growth, however, external rather than internal expansion may be the logical choice at a given point in time. Reinhardt provides the following list of reasons for choosing a business combination over a new investment (Reinhardt, 1972, p. 2):

(1) The possibility of obtaining facilities at prices below their true economic value.
(2) The advantage of almost instantaneous growth.
(3) The desire to obtain unique resources (e.g., managerial talent or research and development capability).
(4) Access to otherwise unavailable financing.
(5) Increase in market power.
(6) Tax advantages.
(7) Synergism.
(8) Personal reasons (e.g., an executive's desire or vision for the company's future).

Another related source of financing was holding companies which are operated for the purpose of owning the common stocks of other corporations. The other corporation may be a minerals firm in need of financing for a new project so that the stock sale is a source of funds. From the point of view of the holding company, the fractional ownership may give the holding company working control or influence over the operations of a much larger company; it may be an opportunity for an operating company in a declining industry to begin a transformation by holding assets in industries having more favorable growth potential because operating companies in a holding company system are separate legal entities, catastrophic losses incurred by one are not transmitted as claims on the assets of others. The holding company may also be used for "pyramiding" where one holding company with a low investment controls the total assets of an operating company and a second holding company with an even smaller investment controls the assets of holding company number one. Such leveraging can result in high profits but also introduces the risk of collapse if sales of the operating company decline (Weston, 1981, p. 939).

A detailed discussion of mergers and acquisitions is beyond the scope of the present chapter; the reader is referred to the periodicals covering mergers and acquisitions for example, *Mergers and Acquisitions*, bimonthly, *The Journal of Business*, monthly, *The Economist*, monthly, *The Harvard Business Review*, bimonthly, *Financial Analysts Journal*, bimonthly, *The Wall Street Journal*, daily, and the *American Economic Review*, monthly. Most of the texts on finance also include coverage of external growth for example Weston (Weston, 1981, pp. 916–951.

Drilling Funds: Limited partnership drilling funds as a source for capital for petroleum exploration and development is discussed in Nevitt's book on project financing (Nevitt, 1979, pp. 143–149). The parties to a drilling fund partnership are (1) The sponsor who acts as general partner and pays all capital costs and (2) The individual investors as limited partners who pay all non-capital costs which can be deducted for tax purposes immediately. Much of the risk of success of the venture is on the limited partners because the general partner will not incur major capital expenses until the drilling is completed and tests indicate the likelihood of a producing formation. This financing approach shares risks, provides financing, and has income tax benefits, for example, the limited partners have immediate deductions and the general partner can claim the investment tax credit and depreciation. The partnership is also a means for bringing individual's capital together in amounts sufficient for a major exploration venture.

PORTFOLIO ANALYSIS AND THE DIRECTIONAL POLICY MATRIX

Long-Range Planning with the Directional Policy Matrix

The preceding emphasis on project analysis and finance would be misleading without reference to the problems of long-range financial planning that require consideration of attractive market sectors and strengths of the firm within those sectors. There are many approaches to the objective of a balanced portfolio of investments. This section focuses on a specific program of long-range planning to achieve that objective for minerals firms.

For more than 10 years the Shell Chemical Companies have been analyzing their company strength and prospects with a technique known as the "directional policy matrix." It is a technique to supplement the conventional rate of return, present value and profitability yardsticks used to guide diversifications of a major corporation. This section is based on an article by Shell corporate planners (Robinson, S.J.Q., 1978) with adaptations to the mineral sector (Lewis, et al, 1982).

In a stable economic environment, rate of return or net present value are the conventional criteria used for deciding between investment alternatives. However, these conventional techniques are not sufficient in themselves for long range planning for several reasons:

(1) They fail to identify the reasons for strength or weakness in particular business sector and they fail to explain why a particular company is strong or weak in the business sector.

(2) Profitability and rate of return alone fail to explain how the growth and dynamics of business sectors will contribute to a balance for the company that leads to stable and prolonged growth.

(3) When investment in a new sector is being considered there is lack of business by definition. Even if entry is by acquisition of an operating company the performance of the company in the recent past may not be a good indicator of the future.

(4) World wide inflation especially after 1973, severely weakened the credibility of the conventional financial forecasting techniques.

Normally there are a number of investment alternatives facing the company. The factors to be considered in developing a long range plan include the following:

(1) Market growth.
(2) Industry supply/demand balance.
(3) Prices.
(4) Costs.
(5) Future market shares.
(6) Ability to compete.
(7) Research, development, and exploration strength.
(8) Activities of competitors.
(9) Future business environment.

The directional policy matrix approach identifies:

(1) The main criteria for which the prospects for the *business sector* may be judged favorable or unfavorable. Favorable in this context means a sector with high profit and growth potential for the industry generally.

(2) Those factors by which a *company's position in the sector* may be judged strong or weak.

Criteria are used to construct ratings of sector prospects and of the company's competitive capability and the ratings are plotted on a matrix as indicated in Fig. 3.7.18.

It should be emphasized that the zones shown in Fig. 3.7.18 are not as precise as the rectangular division might suggest. The zones are sometimes overlapping and often shade into one another but are designed to suggest alternative strategies for the company.

Details of the Technique

Scope of Analysis: The analyst must first define the business sector and second the geographical area for study. In the minerals context there are two problems if exploration is considered, first the regions of probably discoveries and second the regions which would be the market for the product.

A time horizon for analysis must be selected; usually 10 to 20 years is appropriate for minerals or petroleum investments.

Analysis of the Business Sector: The four

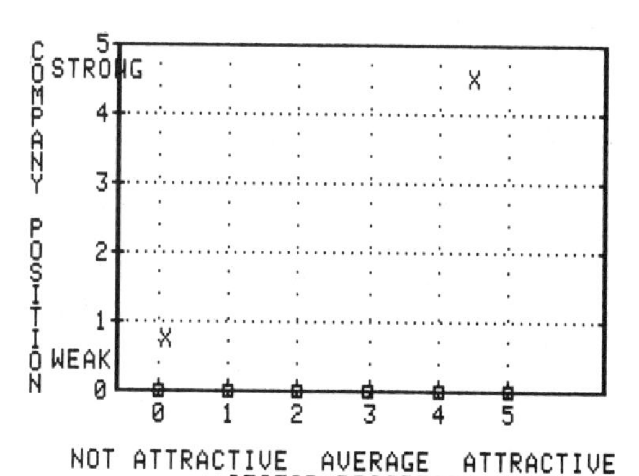

Fig. 3.7.18—The directional policy matrix.

main criteria for analysis of profitability of a particular sector are:

(1) Market growth rate.
(2) Market quality.
(3) Prospect for discovery and development.
(4) Environmental aspects.

Market Growth Rate: Market growth is a necessary condition for profitability although sectors with high market growth are not always those with the greatest profits. Some sort of an index must be developed for ranking growth rates measured in percentages. The Shell Corporation approach assigns one to five stars or points in order to rate sector prospects, for example:

Percent	Ranking	
0-3	*	minimum
3-5	**	
5-7	***	average
7-10	****	
10 and over	*****	maximum

When the weighting system is applied to different industries, a scale must be developed for each with a center point appropriate for the average growth rate for the particular sector.

Market quality is interpreted in terms of profitability. Some markets have very wide swings in profits with changes in market conditions; this is particularly true of many of the commodity type mineral markets. Also some sectors may have poor profitability because of market domination by a small group or powerful customers able to keep prices down.

Some sectors remain profitable even in depressed conditions because well established producers are content to let sales fall when demand goes down rather reduce prices. Other factors such as a high technical content for the product or careful tailoring of the product to the needs of the consumer lend stability to markets.

Market Quality: Market quality is extremely difficult to quantify and there is a requirement for some subjectivity in analysis. Shell lists the following important questions before assigning an index number:

a. Does the sector have profitability?
b. Are margins usually maintained when capacity exceeds demand?
c. Is the product resistant to volatile changes in prices?
d. Is the technology freely available?
e. Is the market free from domination by a small group of producers or customers?
f. Does the product have high added value when converted by the customer?
g. Is the market destined to remain small enough not to attract too many producers?
h. Is the product one where the customer has to change his process or machinery if he changes supplier?
i. Is the product free from risk of substitution?

The sector which answers most of the questions yes would attract a four or five star market quality rating.

Prospects for Discovery and Development: Expansion of productive capacity in the chemical industry may be limited by feedstock supply. Prices for feedstocks may be high because of competition from an alternative use or there may be difficulty in obtaining large quantities of the materials.

If the feedstock is a by-product of another process its price may be forced up or down because of its by-product nature. By definition a by-product is incidental to the main product and its production will vary with demand and price for the main product. This may create a situation of over supply with consequent price cutting efforts to stimulate consumption or conversely if there is a strong demand for the by-product in amounts in excess of that indicated by the main product its price may go up because of the lack of increased production as demand increases.

Environmental Aspects: Environmental regulation may also be critical if restrictions or the manufacture, transportation or marketing of the product was involved. Environmental regulations may add significantly to the required investment or to operating costs for the product. The effect on economics may be to discourage additional investment.

Analysis of the Competitive Position of the Company

The three main criteria for analysis of company position are:

a. Market position.
b. Production capability.
c. Exploration, research and development.

The analysis is done in terms of the company position at point of analysis. Other points may be based on assumptions of possible future positions.

Market Position: Market position is measured in terms of percentage share of the total market and a measure of the degree to which the share is secured. Star ratings are given again as follows:

- ***** Leader. The company holds a significant portion of the market and has a well established marketing organization.
- **** Major producer. There may be no leaders in the industry but the company is one of several major producers.
- *** Strong stake but below top level.
- ** Minor market share. Market share is less than adequate to support research, development or exploration.
- * Current position negligible.

Production Capability: Production capability is measured by a combination of criteria based on process economics, capacity, location, number of plants, and ore reserves. Assignment of ratings depends on answers to the following questions:

(a) Does the producer employ a modern economic production process?
(b) Does the company have patents or licenses for process?
(c) Can the company keep up with advances in process technology?
(d) In the area of plant and equipment, is current capacity sufficient to meet market demands and can existing facilities be produce at competitive cost?
(e) Does the company have a favorable land position for exploration and adequate ore reserve for production?

Exploration Research and Development: In the area of exploration, research, and development the key questions are:

(a) Does the company have a record of competence in the research and development area for the particular industry?
(b) Does the company have a exploration team adequate to the task of finding and development of new deposits?
(c) Does the company have sufficient income to finance the team?

Assignment of Ratings

Specialists working with nonspecialists should develop the ratings for the components of the matrix. The final result is a consensus where differences are resolved by discussion or in some cases averaged. In other cases more sophisticated methods of sampling opinion using computer techniques may be useful but Shell's experience showed that the group discussion method was preferred.

In a simplified form of the technique each of the main criteria is given equal weighting and a final index developed for the business sector and for the company competitive capabilities. The system of equal weighting gave good results for Shell but it would very likely be replaced by a system of unequal weights when applied to the very diverse investment opportunities in the minerals sector.

Example of Shell's Simplified Weighting System

Tables 3.7.20 and 3.7.21 summarize an analysis made by Shell Chemical Company considering a thermoplastic product suitable for engineering industry application. One table summarizes sector prospect and the other company capabilities.

Weighting System

In many businesses it is unrealistic to assume that all factors have equal weight, therefore a weighting system must be developed. An example of a weighting system is shown in Table 3.7.22

The Second Order Matrix

The objective of the first order matrix is to define a product strategy. The second order matrix is designed to look at a specific *investment* decision. One type of second order matrix described by the Shell group relates the product strategy to nonproduct strategy.

Non-Product Strategy Options

Table 3.7.23 is a summary ranking of business sectors in order of priority. These are the result of the first order matrix analysis where the criteria for ranking is:

(1) Matrix position.
(2) Profit record.

Table 3.7.20—Sector Prospects for Shell Thermoplastic

Criteria	Prospects	Stars	Points
Market Growth	15–20% per year forecast	*****	4
Market Quality			
Sector profitability record?	Above average.		
Margins maintained in over-capacity?	Some price-cutting has taken place but product has not reached commodity status.		
Customer to producer ratio?	Favorable. Numerous customers only two producers so far.		
High added value to customers?	Yes. The product is used in small scale, high value, engineering applications.		
Ultimate market limited to size?	Yes. Unlikely to be large enough to support more than three or four producers.		
Substitutability by other products?	Very limited. Product has unique properties.		
Technology of production restricted?	Moderately. Process is available under license from Eastern Europe.		
Overall market quality rating.	Above average.	****	3
Industry feedstock	Product is manufactured from an intermediate which itself requires sophisticated technology and has no other outlets.	****	3
Environmental aspects	Not rated separately	—	—
Overall sector prospects rating			10

Table 3.7.21—Company's Competitive Capabilities

	Competitors A	Competitors B	Competitors C	Ratings A	Ratings B	Ratings C
Market position Market share	65%	25%	10%	*****	***	***
Production capability Feedstock	Manufacturers' feedstock by slightly out dated process	Has own precursors. Feedstock manufactured by third party	Basic position in precursors. Has own second process for feedstock			
Process economics	Both A and B have own "first generation" process supported by moderate process R & D capacity		C is licensing "second generation" process from Eastern Europe			
Hardward	A and B each have one plant sufficient to sustain their respective market shares		None as yet. Market product imported from Eastern Europe			
Overall production capability ratings				****	***	**(*)
Product R & D (in relation to market position)	Marginally weaker	Comparable	Stronger	****	***	**(*)
Overall competitors' ratings				13	9	9

Table 3.7.22—Examples of Weighting on Company's Competitive Capabilities Axis

Criteria	Businesses W	X	Y	Z
Selling and distribution	2	3	6	3
Problem solving	2	4	3	1
Innovative R & D	4	1	0	1
Manufacturing	2	2	1	5
	10	10	10	10

Table 3.7.23—Classification of Business Sectors in Order of Priority

Category	Characteristics
5	Hard core of good quality business consistently generating good profits. Example: Engineering thermoplastic
4	Strong company position. Reasonable to good sector prospects. Variable profit record. Examples: Dyestuffs. Chlorinated Solvents
3	Promising product sectors new to company. Example: New Chemical Business
2	Reasonable to modest sector prospects in which the company is a minor factor. Variable profit record. Example: Chemical Solvents
1	Business with unfavorable prospects in which the company has a significant stake. Example: Detergent Alkylate

The second order matrix will combine the results of the five categories from the first order matrix with the following non-products strategic options:

Category	Strategy
5	Joint venture to make olefins with petroleum company having secure oil feedstocks.
4	Make maximum use of land and infrastructure at existing sites.
3	Develop new major coastal manufacturing site in the European Economic Community.
2	Develop a foothold in the U.S. market
1	Reduce dependence upon investment in Europe.

(3) Other product related criteria.

(4) Judgement.

These two variables are combined in a second order matrix as is shown in Fig. 3.7.19.

The second order matrix provides another screen for ranking feasible alternatives. It is another powerful tool for keeping track of all important factors some of which may be subjective but which bring together combinations of easily quantifiable factors and judgement.

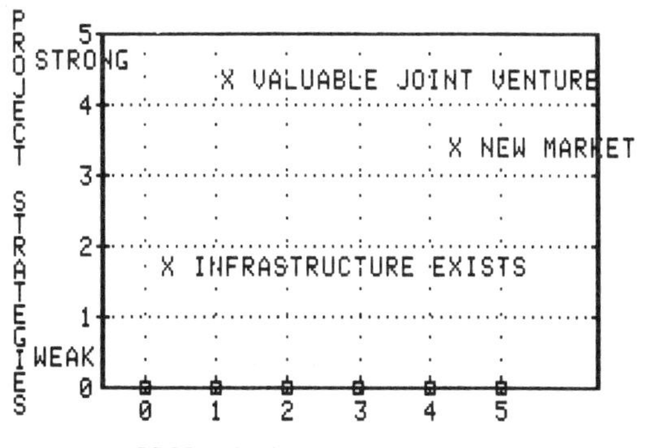

Fig. 3.7.19—Second order matrix.

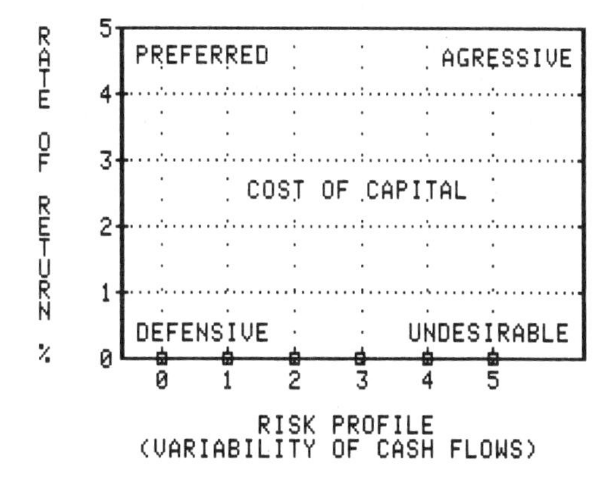

Fig. 3.7.20—Alternative second order matrix.

An Alternative Second Order Matrix

Fig. 3.7.20 is another second order matrix designed in this case to compare investment rate of return directly with an index designed to measure risks of a specific investment. In this the vertical axis is easily quantifiable as rate of return, present value or return on investment criteria. The horizontal axis requires a ranking of investment risks which has to be developed in manners similar to those described for product strategy. The risk of a particular investment may be described in the form of a weighted index including such factors as expected price stability or cash flow stability, geographical risks, political risks, technical risks, and similar measures.

Extension of the Directional Policy Matrix to Minerals

The directional policy matrix, developed by Robinson, Hichens, and Wade for strategic planning by the Shell Chemical Company, can also be used to set the direction of exploration effort, new mining project development, or mergers and acquisitions in the minerals industry.

For purposes of description, we define a hypothetical minerals firm, its strengths, and the advantages to the firm of entering new market sectors. In this section we wish to determine whether the hypothetical company position in each of proposed target sectors is strong or weak. The target sectors to be analyzed are copper, molybdenum, and tin. The relevant time horizon is limited to entry within the next 10 years. Geographic limitations are based primarily on political climate, and not location per se. The company is interested primarily in investments in areas where long term experience has proved the region to be politically stable. The firm is interested in investments within the United States, Canada, Australia, Western Europe and selected Third World countries. In any of these areas, the company is interested in the possibility of a grass roots exploration program, and acquisition of reserves through lease purchase or option agreements, a joint venture, or the merger with or acquisition of an existing firm in the sector. The best method of entry into a particular sector will be determined and recommendations made, after the optimum target sector has been identified, using the directional policy matrix.

Assume that the company is a major industrial corporation engaged primarily in nonferrous metal mining, metal processing, and the manufacture of metal products. It is a wholly owned subsidiary of a major industrial corporation engaged primarily in the production, refining and sale of petroleum and petroleum based products. The company is involved in the production and sale of primary copper, primary and fabricated aluminum products, molybdenum, and uranium oxide concentrates. the company has also been involved in the production of small amounts of silver, gold, nickel, lead, and zinc in the United States and tin in South East Asia. The company employs personnel with expertise in exploration, development, mining, processing and marketing of these products. The financial strength of our parent corporation will enable us to take advantage of whatever opportunities meet our financial evaluation criteria, set out in a separate section below.

Company Objectives: Assume that the company's overall objective is to remain a major profitable producer of nonferrous metals. As such, we must continue to search for new investment opportunities and develop new ore reserves to replace the reserves which are depleted. The firm is primarily interested in developing new resources within areas where we have existing expertise, but because of the large reserves of copper and aluminum which we now hold, we are primarily interested in high grade, low cost sources of these minerals. An expansion program will emphasize development in other mineral sectors, in order to diversify our resource base and reduce our dependence on these two traditional sectors.

General Premises: Management of the firm assumes that there will be no radical changes in the structure of the world business environment and that traditionally stable political environments will remain stable. Interest rates will remain at relatively high levels. Labor and capital costs will continue to escalate at rates which can be predicted from historical trends. Likewise, the supply and demand picture for minerals, as well as the future price of commodities, can be reasonably predicted from past performance and information about current developments. There

Table 3.7.24—Market Position of Commodities Produced

Mineral	Percent of Domestic Production	Percent of Non-Socialist World Production
Copper	11.7	1.6
Molybdenum	8.7	5.4
Alumininum	6.3	1.5
Uranium	8.6	2.1
Silver	9.7	4.2
Gold	3.5	—
Nickel, Lead and Zinc	less than 1% each	—
Tin	—	3.5

will not be any radical changes in the structure of the world commodity markets and the legal environment as it relates to minerals development. There will be no major technological breakthroughs in the mining or beneficiation technology.

Financial Premises and Criteria: The analysis of each business sector focuses on the impact of this sector on profitability of the company. Sector evaluations are based initially on hypothetical investments whose parameters are drawn from collective past experience. Prices and market supply and demand characteristics are forecast based on the best available historical data. Assume that a 15% discounted cash flow rate of return (DCFROR) is the minimum acceptable rate for any project in any target sector. Only relatively large investment projects with annual cash flows over $1 million are acceptable for in-house development. If a resource is acquired through merger, any form of stock for stock or stock plus cash transfer is acceptable, so long as the fair market value for the price paid reflects a minimum 15% DCFROR, conservatively calculated from projected cash flows. In no case will we pay more for a firm that the fair market value of the physical assets acquired. All stock transfers will involve the stock of our parent corporation. We will remain a wholly owned subsidiary. In addition, all cash needs will be financed through our parent corporation.

Strategies: All relevant target sectors must be evaluated fully, and new opportunities continuously investigated and periodically updated. The most promising target sectors which meet our minimum financial criteria will be investigated further to identify specific investment opportunities. Sectors which do not promise to meet our minimum criteria will be rejected. The Directional Policy Matrix technique will be used to identify the sectors to be evaluated more fully.

Matrix Analysis of Company's Competitive Capabilities: The company's competitive capability in each of three target sectors (copper, molybdenum, and tin) are selected for the initial investigation defined by the following factors:

(1) Expertise in the general mining and production techniques.
(2) Current market position and potential position after and acquisition.
(3) Quality of marketing organization.
(4) Ability to meet financial commitments.
(5) Potential legal problems (antitrust, environmental, etc.).
(6) Geographic limitations.
(7) Exploration abilities for target mineral.
(8) Ability to acquire existing resources in the sector.
(9) Applicability of sector to corporate long term plans.
(10) Are the sector returns adequate to meet the company's minimum financial criteria?

The company is ranked using the five point system in each of the 10 analysis areas. The results of this evaluation provide the data needed for the vertical axis of the Directional Policy Matrix. A score of five indicates great strength in an area, a score of one indicates weakness from the firm's point of view.

Analysis of Business Sector Prospects: The prospects for development within a particular business sector will depend to a large extent on the strength of that sector. The overall market outlook for each target sector is reviewed with the results summarized in Tables 3.7.28 to 3.7.30. Several factors were considered in evaluating the growth potential of each sector. Each sector was evaluated based upon the following criteria.

(1) Market growth rate; prospects for increased demand.
(2) Production characteristics and existence of excess capacity.
(3) Price history, and future price prospects; stability of profits.
(4) Industrial organization and concentration.
(5) Market size.
(6) Magnitude of market cycles.
(7) Governmental interference in the market sector.
(8) Geological occurrence and quality of mineral.
(9) Political-geographic location and risk.
(10) Transportation and delivery.
(11) Industry cost structure.

Each sector was evaluated using a five point system as required by a directional policy matrix. A score of five indicates a favorable sector for development or entry by a new firm. A score of one indicates that entry into the sector is not likely to be profitable.

Table 3.7.25—Analysis of Company's Capabilities in Copper Sector

Evaluation Criteria	Company's Ranking: Rank Weight	Weighted Rank
Expertise (Engineering)	5 × 20 =	100
Current Market Position	4 × 12 =	48
Sales Organization	5 × 15 =	75
Financial Commitment Needed	4 × 8 =	32
Legal Aspects	3 × 7 =	21
Geographic Limits	5 × 5 =	25
Exploration Ability	5 × 6 =	30
Ability to Acquire Existing Resources	4 × 10 =	40
Company's Objectives	3 × 7 =	21
Financial Returns Needed	2 × 10 =	20
Total		412

Comments: Total Weight = 100.0
Overall Rank = Total Weighted Rank ÷ Total weight
Overall Ranking of Company's Capabilities in the sector: 4.12

Table 3.7.26—Analysis of Company's Capabilities in Molybdenum Sector

Evaluation Criteria	Company's Ranking: Rank Weight	Weighted Rank
Expertise (Engineering)	4 × 20 =	80
Current Market Position	1 × 14 =	14
Sales Organization	3 × 16 =	48
Financial Commitment Needed	3 × 5 =	15
Legal Aspects	4 × 6 =	24
Geographic Limits	4 × 5 =	20
Exploration Ability	4 × 7 =	28
Ability to Acquire Existing Resources	4 × 10 =	40
Company's Objectives	2 × 7 =	14
Financial Returns Needed	1 × 10 =	10
Total		293

Comments: Total Weight = 100.0
Overall Rank = Total Weighted Rank ÷ Total Weight
Overall Ranking of Company's Capabilities in the sector: 2.93

Table 3.7.27—Analysis of Company's Capabilities in Tin Sector

Evaluation Criteria	Company's Ranking: Rank Weight	Weighted Rank
Expertise (Engineering)	3 × 18 =	54
Current Market Position	2 × 14 =	28
Sales Organization	4 × 16 =	64
Financial Commitment Needed	5 × 5 =	25
Legal Aspects	5 × 7 =	35
Geographic Limits	3 × 5 =	15
Exploration Ability	5 × 6 =	30
Ability to Acquire Existing Resources	3 × 11 =	33
Company's Objectives	5 × 7 =	35
Financial Returns Needed	5 × 11 =	55
Total		374

Comments: Total Weight = 100.0
Overall Rank = Total Weighted Rank ÷ Total Weight
Overall Ranking of Company's Capabilities in the sector: 3.74

Table 3.7.28—Analysis of Business Sector Prospects in Copper

Evaluation Criteria	Company's Ranking: Rank Weight	Weighted Rank
Market Growth	3 × 5 =	15
Production Characteristics	5 × 11 =	55
Prices/Profits	3 × 15 =	45
Industrial Concentration	4 × 7 =	28
Market Size	5 × 8 =	40
Market Cycles	3 × 5 =	15
Government Interference	4 × 8 =	32
Geological	5 × 12 =	60
Political/Geographic	5 × 10 =	50
Transportation	4 × 11 =	44
Industry Cost Structure	2 × 8 =	16
Total		400

Comments: Total Weight = 100.0
Overall Rank = Total Weighted Rank ÷ Total Weight
Overall Ranking of Business Sector Prospects: 4.00

Table 3.7.29—Analysis of Business Sector Prospects in Molybdenum

	Company's Ranking	
Evaluation Criteria	**Rank Weight**	**Weighted Rank**
Market Growth	3 × 6 =	18
Production Characteristics	1 × 11 =	11
Prices/Profits	1 × 15 =	15
Industrial Concentration	1 × 7 =	7
Market Size	2 × 8 =	16
Market Cycles	1 × 5 =	5
Government Interference	4 × 8 =	32
Geological	3 × 12 =	36
Political/Geographic	4 × 10 =	40
Transportation	4 × 8 =	32
Industry Cost Structure	2 × 10 =	20
Total		232

Comments: Total Weight = 100.0
Overall Rank = Total Weighted Rank ÷ Total Weight
Overall Ranking of Business Sector Prospects: 2.32

Table 3.7.30—Analysis of Business Sector Prospects in Tin

	Company's Ranking	
Evaluation Criteria	**Rank Weight**	**Weighted Rank**
Market Growth	3 × 5 =	15
Production Characteristics	5 × 11 =	55
Prices/Profits	3 × 16 =	48
Industrial Concentration	4 × 7 =	28
Market Size	4 × 8 =	32
Market Cycles	3 × 5 =	15
Government Interference	5 × 8 =	40
Geological	5 × 10 =	50
Political/Geographic	2 × 10 =	20
Transportation	3 × 10 =	30
Industry Cost Structure	3 × 10 =	30
Total		363

Comments: Total Weight = 100.0
Overall Rank = Total Weighted Rank ÷ Total Weight
Overall Ranking of Business Sector Prospects: 3.63

Examination of the directional policy matrix suggests the following course of action for each commodity:

Copper: expansion at producing mines and increased exploration;
Molybdenum: Phased withdrawal from primary production;
Tin: increase production, reduce costs and market more aggressively.

Presented below is the rating system used for each commodity:

Evaluation Criteria	Sector Prospect	Point Award
1. Market Growth	**Growth Rate Per Year**	
	0–2%	1
	2–4%	2
	4–6%	3
	6–8%	4
	8% and over	5
2. Product Characteristics	**Engineering Considerations**	
	Underground—access through shaft	1
	Underground—access through decline	2
	Open Pit and Underground/shaft	3
	Open Pit and Underground/decline	4
	Open Pit	5
3. Prices/Profits	**Annual % Variation of E.P.S.**	
	Greater than 20%	1
	12–20%	2
	8–12%	3
	5–8%	4
	Less than 5%	5

4. Industrial Concentration	**Number of Firms**	
	Dominated by 1 or 2 firms	1
	Few Large Firms	2
	Several Large Firms	3
	Many Small Firms; Few Large Firms	4
	Many Small Firms; No Large Firms	5
5. Market Size	**Produce Mix and Geographic Distribution**	
	Small, Diversified Domestic Market	1
	Small, Homogenous World Market	2
	Small, Diversified World Market	3
	Large, Homogenous World Market	4
	Large, Diversified World Market	5
6. Market Cycles	**Annual % Variation of Market Prices**	
	20% and over	1
	15–20%	2
	10–15%	3
	5–10%	4
	Less than 5%	5
7. Government Interference	**E.P.S Measured With and Without Regulation**	
	Mandated Suspension of Operations	1
	Development Obstructed	2
	Curtailed Operations	3
	Socially Acceptable Regulation	4
	Regulations Subject to Negotiation	5
8. Geological	**Ore Deposit Characteristics**	
	Surface/Underground; Poor Recovery	1
	Surface; Poor Recovery	2
	Underground; Good Recovery	3
	Surface/Underground/Good Recovery	4
	Surface; Good Recovery	5
9. Political and Geographic	**Location and Political Stability**	
	Foreign; Unstable	1
	Foreign; Unstable; Military	2
	Foreign; Stable; Democratic	3
	Domestic; Anti-Mining State	4
	Domestic; Pro-Mining State	5
10. Transportation	**Degree of Infrastructure Development**	
	Foreign; No Infrastructure	1
	Foreign; With Infrastructure	2
	Foreign; Near Navigable Water	3
	Domestic; No Infrastructure	4
	Domestic; With Infrastructure	5
11. Industry Cost Structure	**Cost Structure Relative to Competition**	
	Bottom Third of World Industry	1
	Bottom Third of Domestic Industry	2
	Average	3
	Domestic Industry Leader	4
	World Industry Leader	5

Resulting Directional Policy Matrix: The ranking of business prospects in each sector provides the data needed for the horizontal axis of the directional policy matrix. The matrix is the result of plotting the combined scores of company strength and market attractive. Fig. 3.7.21 is the resulting matrix, defining the company's competitive capabilities and the prospects of each business sector.

The Second Order Matrix for Analysis of Non-product Characteristics: The directional policy matrix can be extended to the analysis of specific projects within acceptable business sectors. The mix of important considerations will vary with the objectives of the company, sector characteristics and the mix of projects under study. Table 3.7.31 is a list of possible considerations, assuming the same one to five point ranking system described. Intermediate ranking (point awards two through four) are left to the reader's discretion.

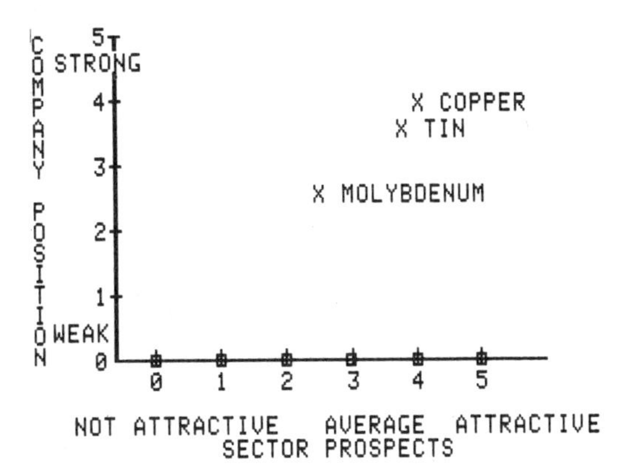

Fig. 3.7.21—Directional policy matrix for a mineral firm.

Table 3.7.31—Evaluation Criteria and Point Awards in a Second Order Matrix

Evaluation Criteria	Project Characteristic	Point Award
1. Volume of Sales	Potentially highly variable and unpredictable	1
	Secured by long-term contract, stable and predictable	5
2. Price determiniation	Determined by highly variable market forces. Subject to wide cyclical swings	1
	Guaranteed by long-term contract, cartel or monoply	5
3. Deductions (eg. Smelter charges)	Uncertain and likely to be unfavorable to project economics	1
	Predictable and favorable to project economics	5
4. Sales contact	Critical aspects subject to change which may be detrimental to project economics	1
	Contract firm and unlikely to change in a matter detrimental to project economics	5
5. Market cycle	Current outlook includes high probability of over-supply or depressed demand condition at startup	1
	Supply/Demand outlook is for stability or rising cycle of demand relative to supply	5
6. Risk of technical problems in mining	Unproven. System and ore deposit characteristics indicate high probability of problems	1
	Proven technology in "textbook case" application	5
7. Risk of technical problems in processing	New technology. Possible scale-up and other unforseen technical problems	1
	Proven technology on specific application	5
8. Infrastructure	Non-existent	1
	Virtually no additional infrastructure required	5
9. Location	No special advantages	1
	Excellent site consistent with future exploration and development plans	5
10. Operating cost estimates	Forecasts subject to wide and unpredictable variations. Range includes costs which would make the project marginal or uneconomic	1
	Forecast probably very reliable predictors of actual costs. Entire range of expected cost yields profitable results	5

Table 3.7.31 (cont.)

Evaluation Criteria	Project Characteristic	Point Award
11. Level of operating costs	Upper 20% relative to competing sources	1
	Bottom 20% relative to competing sources	5
12. Investment costs	Probability of large cost overrun high	1
	Probability of cost overrun next to zero	5
13. Capital exposure	Large relative to company assets and of long duration	1
	Small relative to company assets. Failure would not threaten company survival	5
14. Labor	Inadequate and unskilled. Likely high turnover	1
	Adequate skilled, stable. Low expected turnover	5
15. Weather	Likely to severely affect construction and operations	1
	Weather problems very unlikely	5
16. Environmental	Probable conflicts which could severely impact construction or operations	1
	None	5
17. Profitability	Marginal present value and rate of return	1
	Present value and rate of return high and remain acceptable under wide range of expected future conditions	5
18. Reserves	Likely to be less than estimated. Uncertain. Low grade	1
	Certain. High grade. Geology known simple, predictable	5
19. Competition	Other new large discoveries likely during life of this project	1
	No other large discoveries are likely	5
20. Construction schedule	Delays probable. Budget overrun impact large	1
	Delays or budget overrun improbable	5
21. Government contracts	Not very attractive and probably will change for the worst	1
	Government and investor objectives coincide. Contract attractive and stable	5
22. Economy	Economy of host country weak, e.g. high inflation-high unemployment, political unrest	1
	Economy stable and strong	5
23. Political	Project likely to be the victim of political change prior to pay-out	1
	Project coincides with political objectives. Has value to government beyond economics	5
24. Taxation	High and likely to get higher. Possible severe impact on project economics	1
	Low and stable	5
25. Exploration	Project stands alone, will not lead to other opportunities	1
	Project likely to lead to additional discoveries and investment opportunities	5
26. Social problems	Likely high turnover and low productivity due to social problems	1
	None	5
27. Culture	Cultural conflicts likely to lead to delays and cost increases	1
	Culture understood. Project designed to minimize conflicts. Unlikely that there will be problems	5
28. Joint Venture	Very probable to lead to problems leading to need for renegotiation opposed by one or more participants	1
	Likely to avoid need to renegotiate	5
29. Land status	Confused and/or likely to lead to conflict	1
	Clear and without problems	5
30. Financing	Financial leverage results in excess risk given project characteristics	1
	Financial leverage improves return without undue risk to company	5

Bibliography

1. Abraham, C.T., Prausand, R., and Ghosh, M., 1968, "A Probabilistic Approach to Cost Estimation," 68-10-001, IBM Corp., Armonk, NY.
2. AMAX Inc., 1977, *Securities and Exchange Commission Form 10-K for Fiscal Year Ending December 31*, Commission file 1-229-2, AMAX Inc., New York, NY.
3. American Association of Cost Engineers, 1979–1983, *AACE Transactions*, annual, AACE, Morgantown, W. Va.
4. American Association of Cost Engineers, 1983, *Cost Engineers' Notebook*, annual update, AACE, Morgantown, W. Va.
5. American Metal Market, 1982, *Metal Statistics*, Fairchild Publications, New York, NY.
6. Berk, R., et al., 1982, *Aluminum: Profile of the Industry*, McGraw-Hill, New York, NY.
7. Borgman, L. and Frahme, R., 1976, "A Case Study: Multivariate Properties of Bentonite in Northeastern Wyoming," *Advanced Geostatistics in the Mining Industry*, M. Guarascio, M. David and C. Huijbregts, eds., Reidel, Dordrecht, Netherlands, pp. 381–390.
8. Brentz, J.N., 1980, *Fundamentals of Product Analysis for Off-Highway Trucks*, Wabco Construction and Mining Equipment, Peoria, IL.
9. Brentz, J.N., 1980, *Application of Off-Highway Trucks for Mining, Quarry, and Construction Projects*, Wabco Construction and Mining Equipment, Peoria, IL.
10. Bubenicek, L. and Haas, A., 1968, "Methods of Calculation of the Iron Ore Reserves in the Lorraine Deposit," *A Decade of Digital Computing in the Mineral Industries*, A. Weiss, ed. AIME, New York, pp. 179–210.
11. Castle, G.R., 1975, "Project Financing Guidelines for the Commercial Banker," *Journal of Commercial Bank Lending*, Robert Morris Associates, Philadelphia.
12. Chase Econometrics/Interactive Data Corporation, 1983, "The Inflation Planner Service," Advertising Brochure.
13. Clark, F.D., and Lorenzoni, A.B., 1978, *Applied Cost Engineering*, Marcel Dekker, New York.
14. Clark, I., 1979, "Does Geostatistics Work," *Proceedings*, SME/AIME, 16th APCOM, New York, p. 13.
15. Clement, G.K., et al., 1979, *Capital and Operating Cost Estimating System Handbook Mining and Beneficiation of Metallic and Non-metallic Minerals Except Fossil Fuels in the United States and Canada*, STRAAM, U.S. Dept. of Interior, Bur. of Mines, Denver.
16. Cortez, L.P., Muge, F.O. and Pereira, G.G., 1974, "Ore Reserve Estimation of a Gold Orebody with Imbricated Structures," *Proceedings*, SME/AIME, 12th APCOM Symp., Golden, Colo., pp. F30–F49.
17. David, M., 1977, *Geostatistical Ore Reserve Estimation*, Elsevier Scientific, Amsterdam.
18. David, M., 1971, "Geostatistical Ore Reserve Calculations, a Step by Step Case Study," *Decision Making in the Mineral Industry*, C.I.M.M., pp. 185–191.
19. David, M., Dowd, L. and Korobov, S., 1974, "Forecasting Departure from Planning in Open Pit Design and the Grade Control," *Proceedings*, SME/AIME, 12th APCOM Symp., Golden, Colo., pp. F-131–F-194.
20. Dieneman, P.R., 1966, "Estimating Cost Uncertainty Using Monte Carlo Techniques," RM4854PR, The Rand Corp., Santa Monica, Calif.
21. Exxon Corporation, 1980, *Securities and Exchange Commission Form 10K*, Exxon Corporation, New York, NY.
22. Fisher, I., 1922, *The Making of Index Numbers*, Houghton Mifflin, Boston.
23. Freeman, H.E., Rossi, P.H., Wright, S.R., 1980, *Evaluating Social Projects in Developing Countries*, OECD, Paris.
24. Freeport Minerals Company, 1976, *Securities and Exchange Commission Form 10-K for Fiscal Year Ending December 31, 1976*, Commission File 1-605, Freeport Minerals Company, New York, NY.
25. Gonzalez, R.J., 1976, "Oil and Gas," *Economics of the Mineral Industries*, Vogely, W.A., et al., eds., 3rd ed., AIME, New York, pp. 486–497.
26. Guarascio, M. and Raspa, G., 1974, "Valuation and Production Optimization of a Metal Mine," *Proceedings*, SME/AIME, 12th APCOM Symp., Golden, Colo., pp. F50–F64.
27. Guarascio, M. and Turchi, A., 1976, "Ore Reserve Estimation and Grade Control at AGIP Uranium Branch," *Proceedings*, SME/AIME, 14th APCOM Symp.
28. Hamel, H.G., 1968, *Leasing in Industry*, Oak Brook: 1968 National Industrial Conference Board, Inc., 1968.
29. Harris, C.C., 1978, *The Break-Even Handbook: Techniques for Profit Planning and Control*, Prentice-Hall, Inc., Englewood Cliffs, NJ.
30. Hoskin, J.R., compiler, 1982, *Mineral Industry Costs*, Northwest Mining Association, Spokane, Wash.
31. Huijbregts, Ch., 1975, "Estimation of a Mass Proved by Random Diamond Drill Holes," *Proceedings*, SME/AIME, 13th APCOM Symp., Clausthal, pp. A1-1–A1-17.
32. Huijbregts, Ch. and Segovia, R., 1973, "Geostatistics for the Valuation of a Copper Deposit," *Proceedings*, SME/AIME, 11th APCOM Symp., University of Arizona, Tucson, pp. D-24–D-43.
33. ICARUS Corp., 1983, *Cost System User's Manual*, annual update, 7th ed., Vol 1 & 2, ICARUS, Rockville, MD.
34. Imboden, N., 1978, *A Management Approach to Project Appraisal and Evaluation*, OECD, Paris.
35. International Monetary Fund, 1982, *World Economic Outlook*, International Monetary Fund, Washington, D.C.
36. Jelen, F.C., ed., 1979, *Project and Cost Engineers' Handbook*, American Association of Cost Engineers, Morgantown, W. Va.
37. Jelen, F.C. and Black, J.H., eds., 1983, *Cost and Optimization Engineering*, 2nd ed., McGraw-Hill, New York.
38. Journel, A., 1973, "Geostatistics and Sequential Exploration," *Mining Engineering*, Vol. 25, No. 10, pp. 44–48.
39. Journel, A., 1974, "Grade Fluctuation at Various Scales of a Mine Output," *Proceedings*, SME/AIME, 12th APCOM Symp., Golden, Colo., pp. F78–F94.
40. Journel, A., 1980, "The Lognormal Approach to Predicting Local Distributions of Selective Mining Unit

Grades," *Mathematical Geology*, Vol. 12, No. 4, pp. 285–303.
41. Journel, A. and Huijbregts, Ch., 1972, "Estimation of Lateritic-type Deposits," *Proceedings*, SME/AIME, 10th APCOM Symp., Johannesburg, South Africa, pp. 207–212.
42. Jenkins, D.O., et al., 1970, "Leasing and the Financial Executive," *Financial Executive*.
43. Kim, Y.C., Myers, D.E., and Knudsen, H.P., 1977, "Advanced Geostatistics in Ore Reserve Estimation and Mine Planning (Practitioner's Guide)," GJBX-65(77), Oct., U.S. Department of Energy, Grand Junction, Colo.
44. Knudsen, H.P. and Kim, Y.C., 1978, "A Short Course on Geostatistical Ore Reserve Estimation," University of Arizona, Tucson, May.
45. Knudsen, H.P. and Kim, Y.C., 1977, "Geostatistical Ore Reserve Estimation for a Roll-Front Type Uranium Deposit (Practitioner's Guide)," GJBX-3(77), Jan., U.S. Energy Research and Development Administration, Grand Junction, Colo.
46. Knudsen, H.P., Kim, Y.C., Mueller, E., 1977, "A Comparative Study of the Geostatistical Orc Reserve Estimation Method Over the Conventional Method," *Trans. SME-AIME*, Dec.
47. Koch, G.S., Jr., and Link, R.F., 1971, *Statistical Analysis of Geological Data*, Vol. II, John Wiley & Sons, Inc., New York.
48. Krige, D.G., 1973, "Computer Applications in Investment Analysis, Ore Valuation and Planning for the Prieska Copper Mine," *Proceedings*, SME/AIME, 11th APCOM Symp., Tucson, pp. G31–G47.
49. Krige, D.G., 1976, "A Review of Development of Geostatistics in South Africa," *Advanced Geostatistics in the Mining Industry*, M. Guarascio, M. David and C. Huijbregt, eds., Reidel, Dordrecht, Netherlands, pp. 279–294.
50. Krige, D.G. and Rendu, J.M., 1975, "The Fitting of Contour Surfaces to Hanging and Footwall Data for Irregular Orebody," *Proceedings*, 13th APCOM Symp., Clausthal, pp. CV-1–CV-12.
51. Kuestermeyer, A.L., 1982, "Capital and Operating Cost Estimation Handbook for the Milling of Uranium Ores in the United States," M. Thesis, Colorado School of Mines, Golden.
52. Lane, K.F., 1963, "Choosing the Optimum Cut-off Grades. Q," M.S. Thesis, Colorado School of Mines, Golden, pp. 811–829.
53. Lasky, S.G., 1950, "How Tonnage and Grade Relations Help Predict Ore Reserves," *Engineering Mining Journal*, pp. 81–85.
54. Lessard, D.R., 1980, "Evaluating International Projects: An Adjusted Present Value Approach," *Capital Budgeting Under Conditions of Uncertainty*, Crum and Derkindeven, eds., Martinus Wijhoff, the Hague.
55. Lessard, D.R and Graham, E.M., 1976, "Discount Rates for Foreign Mining Ventures," Massachusetts Institute of Technology, Alfred P. Sloan School of Management, Boston.
56. Levi, M., 1983, *International Finance*, McGraw-Hill, New York.
57. Lewis R., W. Balaz, F. Serrano, Unpublished paper Mineral Economics Department, Colorado School of Mines, April 1982.
58. Lindley, A.L. et al., 1976, "Mineral Financing," *Economics of the Mineral Industries*, Vogely, W.A., et al., eds., 3rd ed., AIME, New York.
59. Malenbaum, W., 1978, *World Demand for Raw Materials in 1985 and 2000*, McGraw-Hill, New York.
60. Marino, J.M. and Slama, J.P., 1972, Ore Reserve Evaluation and Open Pit Planning," *Application of Computer Methods in the Mineral Industry*, Salomon (Editor), S.A.I.M.M., pp. 139–144.
61. Matheron, G., 1961, "Precision of Exploring a Stratified Formation by Boreholes with Rigid Spacing—Application to a Bauxite Deposit," *Proceedings*, Int. Symp. on Mining Research, Univ. of Missouri, pp. 407–423.
62. Matheron, G., 1963, "Principles of Geostatistics," *Economic Geology*, Vol. 58, pp. 1246–1266.
63. McCray, A.W., 1975, *Petroleum Evaluations and Economic Decisions*, Prentice-Hall, Inc., Englewood Cliffs, NJ.
64. McGraw-Hill Cost Information Systems, 1983, *Data from Dodge*, Vol. 1, *1983 Dodge Systems Costs for Building Construction*, Vol. 2, *1983 Dodge Manual of Pricing and Scheduling*, Vol. 3, *1983 Dodge Digest of Building Cost and Specifications*, Vol. 4, *1983 Dodge Guide to Heavy Construction Costs*, McGraw-Hill, New York.
65. Means, R.S. Co., Inc., 1983, Series of "Building Manuals and Micro-computer Programs," R.S. Means Co., Inc., Kingston, Mass.
66. Mikesell, R.F., 1971, *Foreign Investment in the Petroleum and Mineral Industries*, Resources for the Future, Johns Hopkins Press, Baltimore, MD.
67. Mular, A.L., 1982, *Mineral Processing Equipment and Preliminary Capital Cost Estimation*, Spec. Vol. 25 (update of Vol. 18), The Canadian Institute of Mining and Metallurgy, Montreal.
68. Murelius, O., 1981, *An Institutional Approach To Project Analysis in Developing Countries*, OECD, Paris.
69. Nevitt, P.K., 1979, *Project Financing*. Euromoney Publications, London.
70. Newendorp, P.D., 1975, *Decision Analysis for Petroleum Exploration*, Pennwell Publishing Company, Tulsa, Okla.
71. Newton, H.S., 1973, "The Application of Geostatistics to Mine Sampling Patterns," *Proceedings*, 11th APCOM Symp., Univ. of Arizona, Tucson, pp. D44–D58.
72. O'Hara, T.A., 1980, "Quick Guides To The Evaluation of Orebodies," *Canadian Mining and Metallurgical Bulletin*, Feb., Canada.
73. O'Hara, T.A., 1981, "Analysis of Risk in Mining Projects," *83rd CIM Annual Meeting*. HBMS, Toronto.
74. OECD, Development Assistance Committee, 1982, *Investing in Developing Countries*, 5th ed., OECD, Paris.
75. OECD, Development Centre Studies, *Manual of Industrial Project Analysis in Developing Countries*, OECD, Paris.
76. OECD, International Energy Agency, 1982, *World Energy Outlook*, OECD, Paris.
77. Ostwald, P.F., 1974, *Cost Estimating for Engineering and Management*, Prentice-Hall, Englewood Cliffs, N.J.

78. Otto, J., 1984, "The Management of Risk Associated with Natural Resource Projects Located in the Developing Countries," Colorado School of Mines Master of Science Thesis T-2783, Golden, Colo.
79. Parker, H., Journel, A. and Dixon, W., 1979, "The Use of the Conditional Lognormal Probability Distribution for the Estimation of Open Pit Ore Reserves in Stratabound Uranium Deposits," *Proceedings*, SME/AIME, 16th APCOM, p. 133.
80. Parks, R.D., 1957, *Examination and Valuation of Mineral Property*, Addison-Wesley, Reading, MA.
81. Patterson, J.A., 1959, "Estimating Ore Reserves Following Logical Steps," *Engineering Mining Journal*, Vol. 160, No. 9, pp. 111–115.
82. Peters, W.C., 1978, *Exploration and Mining Geology*, John Wiley & Sons, New York.
83. Petrick, A., 1980, *The Economics of Minerals and Energy Projects*, 2 vols., Petrick Associates, Evergreen, Colo.
84. Powers, T.A., ed., 1981, *Estimating Accounting Prices for Project Appraisal*, Inter-American Development Bank, Washington, D.C.
85. R.S. Means Co., Inc., 1983 "New Ideas for Building," Advertising Brochure.
86. Raymond, L.C., 1976, "Evaluation of Mineral Property," *Economics of the Mineral Industries*, Vogely, W.A., et al., eds., 3rd ed., AIME, New York.
87. Reinhardt, U.E., 1972, *Mergers and Consolidations: A Corporate-Finance Approach*, General Learning Press, Morristown, NJ.
88. Rendu, J.M., 1979, "Kriging, Lognormal Kriging and Conditional Expectation: Comparison of Theory with Actual Results," *Proceedings*, SME/AIME, 16th APCOM, p. 155.
89. Rendu, J.M., 1979, "Normal and Lognormal Estimation," *Mathematical Geology*, Vol. 11, No. 4, pp. 407–422.
90. Richardson Engineering Services, Inc., 1984, *Process Plant Construction Estimating Standards*, annual update, Vols 1–4, Richardson Engineering Service, San Marcos, Calif.
91. Richardson Engineering Services, Inc., 1983, *Instruction Manual for Construction Estimating*, Richardson Engineering Service, San Marcos, Calif.
92. Richardson Engineering Services, Inc., monthly, *Richardson Construction Cost Trend Reporter*, Richardson Engineering Service, San Marcos, Calif.
93. Robinson, H.J., 1971, *Prospectus Preparation for International Private Investment*, Praeger, New York.
94. Robinson, S.J.Q., R.E. Hichens and D.P. Wade, "The Directional Policy Matrix: Tool for Strategic Planning," Long Range Planning Vol. II, June 1978.
95. Roghani, F., 1980, "Conventional and Geostatistical Methods of Ore Reserve Estimation and Their Application to the Economics of Mineral Exploration, with a Case Study," PhD Dissertation T-2369, Colorado School of Mines, Golden.
96. Rosario Resources Corporation, 1975, *Securities and Exchange Commission Form 10-K for Fiscal Year Ending December 31, 1975*, Commission File 1-1189, Rosario Resources Corporation, New York, NY.
97. Screiber, H.W., "The Role of the Independent Consulting Firm in Project Financing," Unpublished Paper, Behre Dolbear.
98. Stanley, B.T., 1976, "From Drill Hole to Total Estimate, a Workable Geostatistical Case Study," *Proceedings*, SME/AIME 14th APCOM Symp., Pennsylvania State University.
99. Stermole, F.J., 1980, *Economic Evaluation and Investment Decision Methods*, Investment Evaluations Corporation, Golden, Colo.
100. Stewart, R.D., 1982, *Cost Estimating*, John Wiley & Son, New York.
101. STRAAM Engineers Inc., 1977, *Capital and Operating Cost Estimating System Handbook Mining and Beneficiating of Metallic and Nonmetallic Minerals Except Fossil Fuels in the United States and Canada*, prepared for the U.S. Department of the Interior, Bureau of Mines, Contract No. J0255026, Denver, Colo.
102. Suboleski, S.C., 1981, "Engineering Analysis of Mine Financing Projects," Unpublished Paper, Illinois National Bank and Trust.
103. Synergic Resources Corporation and McLean Research Center, Inc., 1981, "Development of Underground-Mine Cost-Estimation Equations," DOE/EIA-0331, March 6, U.S. Dept. of Energy.
104. Taylor, H., 1982, "Modeling and Orebody Having Lognormal Grade Distribution," *Proceedings*, SME/AIME 12th APCOM, pp. 784–788.
105. Tomimatsu, T.T., 1980, "The U.S. Copper Mining Industry. A Perspective on Financial Health," U.S. Bureau of Mines IC 8836, Department of the Interior, U.S. Government Printing Office, Washington, DC.
106. Ulatowski, T., 1979, "Structuring of the Credit," *SME-AIME*.
107. Ulatowski, T., 1979, "Sources of Funding for Mineral Projects," *SME-AIME*.
108. United Bank of Denver, 1975, "Energy and Natural Resources Financing," United Bank of Denver, Denver, Colo.
109. VonBauer, E., 1981, "Meaningful Risk and Return Criteria for Strategic Investment Decisions," *Mergers and Acquisitions*.
110. The World Bank, 1978, *World Development Report 1983*, Oxford University Press, New York.
111. Wanless, R.M., FCA, 1983, *Finance for Mine Management*, Methuen, Ontario, Canada.
112. Williamson, D.R. and Mueller, E., 1976, "Ore Estimation at Cyprus Pima Mine," AIME, Annual Meet., Las Vegas.
113. Weston, J.F. and Sorge, B.W., 1977, *Guide to International Financial Management*, McGraw-Hill, New York.
114. Weston, J.F., and Brigham, E.F., 1981, *Managerial Finance*, 7th Edition. The Dryden Press, Hinsdale, Ill.
115. Woolley, M.B., 1975, "A Primer on the Economic Evaluation of Equipment Lease Transactions," Colorado School of Mines, MS Thesis, T1780, Golden, Colo.

3.8

Energy Modeling

Walter C. Labys* and David O. Wood†

INTRODUCTION

The past decade has witnessed a literal explosion in formal modeling of energy systems and markets. This rapid development was stimulated primarily by policy issues arising from the energy price shocks of 1973–1974 and 1979–1980. The suddenness of this interest is illustrated by comparing the current and earlier editions of *Mineral Economics*. The third edition, published in 1976, includes an addendum to the section of projection and forecasting methods because,

> Since the foregoing chapter was written, major events in petroleum have caused an explosion in formal modeling efforts in the energy field.[a]

In contrast, this fourth edition includes the present chapter on energy modeling, and a separate chapter on nonfuel minerals modeling.

The considerable interest in formal energy models is based on the complexity of the system, in particular the interaction between technical and engineering data and designs, and economic behavior of energy producers and users. Energy is a vital component in the economic and social well-being of nations, and must increasingly be explicitly considered in developing economic growth and welfare policies. Energy models can contribute to policy development and analysis, both by improved understanding of the energy system, and of the interaction between energy markets and the economy, and as an explicit means of analyzing and comparing the implications of alternative policy actions. At their best, formal models can clarify and illuminate policy options.

In this chapter we will survey the objectives, methods, and potential contributions and limitations of energy system modeling. Section 2 considers the objectives, scope, and approaches to energy system and market modeling. Methodological approaches are then discussed more formally in section 3. Our approach is to introduce each model type and method via relatively simple generic models, illustrating their use by a selective review of the applications literature. Section 4 shifts the focus from modeling methods to procedures for model evaluation and improving model credibility especially in policy research and analysis. We conclude with some brief remarks on directions and opportunities for future modeling research.

ENERGY MODELS: OBJECTIVES, SCOPE AND APPROACHES

The concept of a model usually evokes an image of a complex, computerized system of mathematical or econometric equations providing detailed information concerning the operation of the process being modeled. In fact, models may be simple or complex, formal or mental, depending upon the purposes for which the model is intended. Simple judgmental models may be most appropriate when monitoring the overall performance of a process. When more detailed information is required and/or when the model

*Department of Mineral and Energy Resource Economics West Virginia University
†Energy Laboratory Massachusetts Institute of Technology

[a]See Vogely (1976), pp. 373–376. The addendum was to a section by Morrison titled "Projections and Forecasting Methods."

is used for planning of complex decision steps, such as the choice of an optimal generation mix for an electric utility, then more complicated models are appropriate. The choice of theoretical structure, implementation methods, and the level of detail represent the art as distinct from the science of modeling. The first order of business in considering any model, then, is to determine the appropriateness of the detail, theory, and implementation methods in relation to the purposes for which the model is intended.

The scope of energy system modeling ranges from engineering models of energy conversion processes (e.g., nuclear reactors) or components of such processes to comprehensive system models of the nation's economy in which the energy system is identified as a sector. The models considered in this chapter are characterized by the coverage of various fuel supplies, and demands, and by the methodology employed. Thus, the scope of the models reviewed includes addressing the supply and/or demand for specific energy forms such as natural gas and electricity, analysis of interfuel substitution and competition in a more complete energy system framework, and analysis of the interrelationships between energy, the economy, and the environment.

Energy models are employed for both normative or descriptive analyses and predictive purposes. In normative analysis, the primary objective is to measure the impact on the system of changing some element or process that is an exogenous, or independent, event in the model. Predictive models are used to forecast energy supply and/or demand and attendant effects over a particular time horizon. Most models have both normative and predictive capability, and a partition of models into these classes can be misleading. Whenever such a classification is used here, it is intended only to identify the primary objective of the model.

Geographical detail appropriate for a given model again depends upon the purposes for which the model is designed. A model of energy flow in a particular production process is specifically related to the plants in which that process operates. Such a model has no geographical dimension. However, a model of utility electricity distribution has a very explicit regional dimension defined by the utility service area being modeled.

Treatment of uncertainty in a model is an important distinguishing characteristic. Uncertainty may arise because certain elements of the process to be modeled arc characterized by randomness, because the process is measured with uncertainty, or because certain variables used as inputs to the model may, themselves, be measured with uncertainty. The methods for dealing with these problems are important in evaluating predictive capability and in validating the model.

The validation of normative models is quite different from that of predictive models. Since normative models deal with how the energy system should develop given an objective, the issues of validation deal more with the representation of the structure of the energy system and the accuracy of its input parameters. For predictive models, validation includes evaluation of both the model's logical structure and its predictive power. Three levels of predictive capability may be identified including ability to predict, (1) the direction of a response to some perturbing factor (e.g., a decrease in GNP due to a fuel supply curtailment); (2) the relative magnitude of a response to alternative policy actions or perturbing factor; and (3) the perturbing factor itself. Validation against the requirement of the first two levels is a minimum requirement, and a model may be quite useful even if it cannot be validated at the third level. At both the second and third level, validation of a conditional form is usual, and restrictions on the perturbing factors and their range of availability must be specified. Perturbing events outside the scope of the model, such as acts of God, must, of course, be taken into consideration in evaluating predictive capability.

Energy models are formulated and implemented using the theoretical and analytical methods of several disciplines including engineering, economics, operations research, and management science. Models based primarily on economic theory tend to emphasize behavioral characteristics of decisions to produce and/or utilize energy, whereas models derived from engineering concepts tend to emphasize the technical aspects of these processes. Behavioral models are usually oriented toward forecasting uses, whereas process models tend to be normative. Recent modeling efforts (as discussed later) evidence a trend toward combining the behavioral and process approaches to energy modeling in order to provide a more comprehensive framework in which to forecast the conditions of future markets under alternative assumptions concerning the emergence of new production, conversion, and utilization technologies. In part, this trend is the result of recognizing that formulating and evaluating alternative national energy policies and strategies require an explicit recognition of technical constraints.

Methods for implementing energy models include mathematical programming (especially linear (LP) and quadratic (QP) programming), activity analysis, econometrics, and related methods of statistical analysis. Process models are usually implemented using programming techniques and/or methods of network and activity analysis, whereas the behavioral models use econometric methods.

Mathematical programming has been used in energy system modeling to capture the technical or engineering details of specific energy supply and utilization processes in a framework that is rich in economic interpretation. In mathematical programming, series of activity variables are defined representing the levels of activity in the specific processes. These are arranged in a series of simultaneous equations representing, for example, demand requirements, supply constraints, and any other special relationships that must be defined to typify technical reality or other physical constraints that must be satisfied. An objective function to be minimized or maximized must be specified (usually cost, revenue, or profit); there are many algorithms available to solve very large problems. The methodology of linear and nonlinear programming and numerous practical applications are described by Dantzig (1963) and Wagner (1969).

The linear programming (LP) technique has been used far more than other mathematical programming methods because of its efficiency in solving large piecewise linear or step function approximations. Nonlinear (e.g., quadratic) and dynamic programming techniques are also used for special purposes, but more often these methods are employed in stating conditions which would characterize economically efficient behavior.

The mathematical programming methodology possesses both interesting and useful economic interpretations. A dual problem formulated in terms of prices is associated with any LP problem formulated in quantities. The solution to the quantity optimization problem yields both the optimal activity levels in physical terms and the prices (shadow) that reflect the proper valuation of physical inputs to the real process represented by the model. Important information concerning the economic interpretation of the solution is provided. Thus the LP technique provides a natural link between process and economic analysis.

Mathematical programming models and related optimization techniques such as the calculus of variations and LaGrange multipliers are generally classified as normative techniques since they presume the existence of an overall objective such as cost minimization or profit maximization. It is possible to reflect multi-objective criteria as some weighted combination of objectives, and indeed, some objectives such as environmental control can be expressed through special constraint equations in the model. Nevertheless, the validity of this technique as a predictive tool depends on the ability to capture and represent the objectives of the players in various sectors of the energy system and in those sectors of the economy and society that affect the energy sectors. The technique is normative in that it determines optimal strategies to achieve a specific objective with a given set of constraints.

Econometrics is concerned with the empirical representation and validation of economic theories.[b] The principle method of constructing related econometric models is to specify, estimate, and simulate a system of equations. This system can reflect causality that is recursive or simultaneous in nature. Each of the equations is based on regression analysis. Such equations usually utilize economic theory to specify the relationships existing between the dependent and certain independent variables. Estimates of the related parameters then follow by combining the economic model with a statistical model of measurement and stochastic errors. Statistical distribution theory is also employed in performing hypothesis tests on the estimated parameters. Model validation is based both on such hypothesis testing, and on the predictive performance of the model when forecasts and actual observations of the dependent variables are compared. Such predictive tests are most persuasive when applied outside the sample observations used in estimating model parameters.

Econometric methods are used in modeling both behavioral and technical processes. Behavioral processes are characterized by a deci-

[b]There are many excellent econometrics textbooks including Johnson (1980), Pindyck and Rubinfeld (1981), and Theil (1971). An advanced treatment is given in Malinvaud (1978).

sion-making agent hypothesized to adjust behavior in response to changes in variables outside his direct control. For example, one could hypothesize that a household would distribute its expenditures between energy and other types of goods and services on the basis of its income and wealth and the relative prices of energy and the other products, and that the distribution would be consistent with some household objective function.

Technical processes are characterized by purely technical relations. An example would be the production of a firm in which maximum potential output is a function of the quantities of inputs available such as capital, labor, energy, and other material inputs. Given a suitable functional form for this relationship and observations on capacity output and associated inputs, econometric methods could be used to estimate the parameters of the relation. Alternatively, a technical relation might be used to derive behavioral relations concerning the firm's demand for input factors; for example, a firm could choose cost minimizing combinations of inputs to produce a given output level.

Econometric methods and engineering/process methods are sometimes alternative approaches to modeling technical processes. An example of the two approaches to modeling the supply of electricity in the United States is provided by the work of Griffin (1974) and Baughman and Joskow (1974). Griffin used an econometric approach while Baughman and Joskow employed an engineering/process approach. These two models are reviewed in the next section and are illustrative of the contrasting characteristics of each approach.

Interindustry or input-output techniques are frequently employed in energy modeling, primarily for descriptive purposes.[c] The interindustry flow table may be converted into a coefficient table measuring the quantity of input required from one sector per unit of output for another sector. The coefficient matrix represents a model of the production process. This technique provides a means of linking technical coefficients relating input requirements (e.g., energy) per unit of output with behavioral models of demand for primary factors of production (capital and labor), and demand for final goods and services. Thus, the interindustry framework provides a natural bridge between programming and econometric models of energy/economy interactions. The best known example of such an integrated model is due to Hoffman and Jorgenson (1977). The input-output approach has also been employed in energy studies by converting the inputs from the energy sector to other industry sectors from dollar flows into energy units such as the British thermal unit[d].

Although any attempt to organize the above modeling approaches into district categories is somewhat arbitrary, we have selected the following taxonomy to highlight the scope of the major methodologies:

(1) Econometric sectoral models which describe the supply or demand for specific fuels or energy forms;

(2) Econometric market or industry models which include both supply and demand aspects of market adjustments for individual or related fuels;

(3) Engineering process models which concentrate on the transformation of fuels and other primary inputs into refined products or other energy derived products;

(4) Spatial equilibrium and programming models which employ programming algorithms to describe the interaction of demand and supply in a spatial context;

(5) Resource exhaustion forms of optimization models which describe how firms or industry cartels establish prices to achieve optimal resource allocation over time;

(6) Input-output models which explain the flows of energy, goods and services among industies in establishing outputs and demands;

(7) Integrated energy models which provide hybridization of several methodologies in explaining integrated energy systems or energy-economy interaction; and,

(8) Energy balance models which yield an empirical view of future energy demand and supply equilibria.

ENERGY MODELS: METHODOLOGIES AND APPLICATIONS

Essential Steps to Modeling

Before presenting each of the methodologies outlined in the prior section, we find it helpful to review the steps essential to energy modeling.

[c]Input-output analysis originated with Leontief, for which he was awarded the Nobel prize. See Leontief (1951) for the original development, Carter and Brody (1970) for a compendium of research, and Griffin (1976) for a review of applications in energy modeling.

[d]See Herendeen (1973) and Reardon (1972) for examples and further discussion.

Developing and applying an energy model normally require the same "tool kit" that is used for constructing a typical commodity model. Such a model consists of a number of components which reflect various aspects of demand, supply, trade and price determination. Each of these components, in turn, embody basic theories of economic behavior and/or energy transformation processes. What gives an energy model its particular configuration is the kind of energy system it attempts to emulate. There is no such thing as an all-purpose energy model. Each model must describe aspects of energy behavior which are peculiar to the system of interest.

Another aspect of energy modeling is that it involves a specific modeling procedure, as illustrated in Fig. 3.8.1. The development of a model normally begins with an identification of the modeling problem to be solved. This is important since the modeling purpose must be clearly established from the outset. Modeling purpose normally leads to the selection of a modeling methodology that will effectively solve the problem at hand. In the next section, we will consider just how this selection takes place.

The most extensive modeling activities take place at the next stage. The model must be specified following the required modeling procedure. This requires a process which represents an interaction between selecting appropriate theories and establishing a data base for model implementation and application. This in turn provides the base for model estimation and calibration. In the context of energy models, parameters derived econometrically depend on statistical methodologies, while parameters employed in engineering systems normally depend on engineering experiments or on judgmental observation.

In either case, the outcome of the next step, indicated as model validation and refinement, requires two properties: first that the model's parameters are demonstrably in accord with the underlying theory, and secondly that the model's output replicates reality reasonably well. This procedure depends not only on parametric and nonparametric tests, but on other considerations as well. The overall process of validation is described in more detail later in this chapter.

The final step of modeling involves applying the model to solve the problem at hand. This process normally consists of explaining market history, of analyzing selected energy policies, or of forecasting future paths of the model's dependent variables. This step is the most important of all, and its intimate connection with previous steps should be kept in mind for, as shown in Fig. 3.8.1., feedback is involved. If the application does not prove effective, then the modeler must return to an earlier stage to identify and reconstruct the offending model components by respecification and reestimation and calibration. In the following sections, the particular character of each of these steps should become more obvious. We now turn to the major modeling methodologies outlined previously.

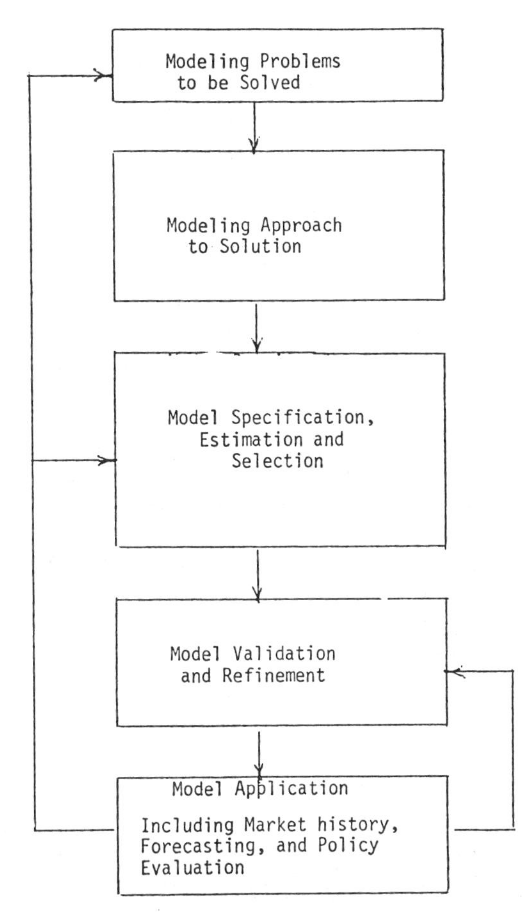

Fig. 3.8.1—Procedure for Developing and Applying Energy Models

Economic Sector Models

The most elementary forms of econometric models are not models in the complete sense. Rather they are components of models or are single sector models that relate to one particular aspect of energy market or industry as a whole. Typically, models or equations in this category focus on the price, the supply or the demand aspect of a market. The underlying methodologies used can be statistical equations which embody some complex time-series process or econometric regression equations which feature dependence on a set of economic and technological factors of an explanatory nature.

Statistical models based on time series analysis assume that the future value of a single energy variable such as demand or prices can be predicted by using the past representation of that variable. That is, by examining the underlying time series generating process of a variable, one can find a time-dependent model based on statistical theory which will permit an explanation and prediction of the time behavior of that variable, e.g. see Granger (1980) or Granger and Newbold (1977). The most elementary form of such a model is known as linear and nonlinear trend fitting. The time dependent energy variable X_t can thus be expressed as some function of time t.

$$X_t = b + a\,t \qquad (1.1)$$

$$X_t = exp\,(b + a\,t) \qquad (1.2)$$

As an example, equation (1.1) defines X_t to be a trend line and equation (1.2) represents X_t to be an exponential curve.

An alternative way of explaining X_t is to provide a structure more complex than a simple curve. This is often done by selecting an autoregressive model which generates X_t as direct function of its past values. The simplest is that of first-order autoregression

$$X_t = a\,X_{t-1} + e_t \qquad (1.3)$$

where e_t is zero-mean and randomly distributed. The more general representation is given by

$$X_t = \sum_{j=1}^{p} a_j\,X_{t-j} + e_t \qquad (1.4)$$

where a number of variable lags j up to p can be selected.

Predictions may also be made by smoothing the behavior of X_t in the form of a moving average

$$X_t = \sum_{j=0}^{q} b_j\,e_{t-j} \text{ where } b_o = 1 \qquad (1.5)$$

In this case X_t is expressed as a weighted average of past values of e_t with up to q values. This method simply finds the average of X_t taken over q periods and uses this as the forecast for the next period.

The statistical model most frequently employed combines the above two approaches to form a mixed autoregressive-moving average model (ARMA).

$$X_t = \sum_{j=1}^{p} a_j\,X_{t-j} + \sum_{j=0}^{q} b_j\,e_{t-j} \qquad (1.6)$$

When this model involves integration after X_t has been differenced a number of times, it becomes known as autoregressive integrated moving average model of order p, d and q.

$$a\,(B)\,(1-B)^d\,X_t = b\,(B)\,e_t \qquad (1.7)$$

where B is a stationary operator. Because computer software has been developed to deal easily with the related estimation and forecasting problems, this method has proven popular for explaining and predicting economic variables, e.g. see Nelson (1973). It is particularly appropriate when dealing with predictions of energy variables that are observed on a short term basis, i.e., quarterly or monthly. Thus it has been used typically to explain and predict quarterly oil price movements on the Rotterdam spot market.

Consideration of a wider set of explanatory factors for a given energy variable introduces two additional concepts to the above analysis. First, it is necessary to examine and make use of any theory that postulates relationships determining the energy variable. Second, one can expand the single variable representation to include one (simple-regression) or several (multiple-regression) explanatory variables.

Most sectoral econometric modeling efforts in the energy area have focused upon the demand for a single energy input in one particular use. Such models are used principally to provide an analysis of the basic determinants of demand as well as to forecast demand. The theory employed for this analysis derives from microeconomics. For example, demand D for a fuel can be expressed as a function of its price P, the price of competing fuels PC, and energy utilizing activity A or income (GNP).

$$D_t = B_o + B_1P_t + B_2PC_t + B_3A_t + u_t \qquad (1.8)$$

Here the variables D, P, PC and A are assumed to be independently distributed; the disturbance term u is zero-mean and normally distributed. Regression methods are used to estimate the coefficients B, e.g. see Pindyck and Rubinfeld (1981). When the variables are expressed in log form, the regression coefficients B directly measure the direct price elasticity, the cross-price elasticity, and the income elasticity of demand, respectively.

Because economic sector equations are usually developed singly rather than in more complex model form, they generally do not have broad policy applicability. Some examples of energy demand applications include that of Taylor (1975) who has surveyed and evaluated econometric equations of the short- and long-term demand for electricity in the residential and commercial sectors. And Sweeney (1975) has developed econometric equations of demand for gasoline in order to support analysis of conservation policies affecting automobiles.

Econometric Market or Industry Models

Competitive Market Models: The most basic type of model from which econometric and other energy modeling methodologies have developed is the competitive market model. Such a model initially neglects market imperfections and assumes that energy demand and supply interact to produce a price level commensurate with competitive market conditions. Such a model thus consists of a number of combined regression equations, each explaining separately, a single sector or market variable, as described above. Market models or the equivalent industry models are applicable to all energy commodity and energy use categories. Their greatest utility is in providing a consistent framework for planning industrial expansion, forecasting market price movements, and studying the effects of regulatory policy on industry.

The basic structure of such a model typically explains energy market equilibrium as an equilibrium adjustment process between demand, supply, inventory and price variables for a particular fuel, e.g. see Labys (1973).

$$D_t = d(D_{t-1}, P_t, P^c{}_t, A_t, T_t) \qquad (1.9)$$

$$Q_t = q(Q_{t-1}, P_{t-\theta}, N_t, Z_t \qquad (1.10)$$

$$P_t = p(P_{t-1}, \Delta I_t) \qquad (1.11)$$

$$I_t = I_{t-1} + Q_t - D_t \qquad (1.12)$$

Definition of variables:

D = Fuel demand
Q = Fuel supply
P = Fuel prices
P^c = Prices of substitute fuels
P_θ = Prices with lag distribution
I = Fuel inventories
A = Income or activity level
T = Technological factors
N = Geological factors
Z = Policy variables influencing supply

Fuel demand is explained as being dependent on prices, economic activity, prices of one or more substitute fuels, and possible technological influences. Other possible influencing factors and the customary disturbance term u are omitted here and elsewhere to simplify presentation function formulation. Accordingly supply would depend on prices as well as underlying production factors, for example, such as geology or resource exhaustibility, and a possible policy variable. A lagged price variable is included since the supply process is normally described using some form of the general class of distributed lag functions. Fuel prices are explained by changes in inventories, although this equation is sometimes inverted to explain inventory demand. The model is closed using an identity which equates inventories with lagged inventories plus supply minus demand.

Applications of this form of model have not been extensive because of the difficulties of dealing with regulatory policy and non-competitive influences on market behavior. One recent study by Verleger (1982), however, has shown how it can be applied to explain oil price behavior during conditions when the oil market shifts from stable supply to disruptive shortages. This model links together econometric equations for spot, consumer and crude oil prices with inventory demand, consumer demand, and shortage conditions. MacAvoy and Pindyck (1975) have built an econometric model of the natural gas industry which has been used extensively to analyze the effect on the industry of federal regulation of the wellhead price of gas and of permissable rates of return for the pipeline industry. Finally, Labys, et al. (1979)

have modeled the U.S. coal market using this approach to forecast future levels of coal demand, supply, prices and inventories.

Controlled Market Models: While the above market model can be adapted to include the influence of market regulation, it remains a competitive model. This predicament can lead to serious consequences when modeling energy markets whose structure tends to be noncompetitive. That is, their structure may vary from complete control in the form of monopoly to lesser degrees of noncompetitive behavior such as that of duopoly or oligopoly. The principal transformation that must come about in describing market behavior in controlled or noncompetitive markets is to consider price determination from the point of view of the actions of individual market participants or of government policy rather than of the workings of the market as a whole.

The econometric approaches taken to model these different market configurations are essentially similar. For example, the monopoly case involves one dominant (monopolist) producer and many (perfectly competitive) consumers. The single producer thus maximizes his own profits given the aggregate demand function for the energy commodity of interest and the supply response of the other firms in the industry.

The simplest case to envision is that of an oil cartel which sets prices to maximize profits but where the fringe S sets the quantity supplied

$$X_t = D_t - S_t \qquad (1.13)$$

where X is the quantity supplied by the cartel and D is the total market demand. Each of the right-hand variables in turn is explained by a behavioral relation.

$$D_t = b_o - b_1 P_t + b_2 A_t \qquad (1.14)$$

$$S_t = a_o + a_1 P_t \qquad (1.15)$$

Solving all three equations results in the following reduced form for the profit-maximizing price.

$$P_t = b_o^1 + \frac{D_t}{b_1 - a_1} - \frac{S_t}{b_1\ 1-a_1} \qquad (1.16)$$

Here prices can be obtained by solving the price equation along with the corresponding demand equation and the fringe supply equation.

A variant of this form of fuel model has been constructed by Blitzer et al. (1975) to analyze the behavior of the OPEC cartel in the world petroleum market. Only now prices are assumed to be given and production of the cartel and the fringe are determined. The model they postulate is as follows.

$$X_t = D_t - S_t \qquad (1.17)$$

$$D_t = d(D_{t-1}, P_t, A_t) \qquad (1.18)$$

$$S_t = s(D_{t-1}, P_t, N_t) \qquad (1.19)$$

Prices can be determined from the relation

$$P_t = \sum_{j=1}^{n} \alpha_j P_{t-j} \qquad (1.20)$$

or they can be assumed to represent an exogenous policy variable. Also assume that maximum production can be attained.

$$\overline{X}_t \geq X_t \geq O \qquad (1.21)$$

The criterion function of the cartel can be defined as

$$\pi_t = \sum_{t}^{n} g(F_t, t) + h(Y_t) \qquad (1.22)$$

where $F_t = f(P_t, X_t)$

Definition of variables:

D	=	Fuel market demand
S	=	Fuel supply from fringe
X	=	Fuel supply from cartel
P	=	Fuel prices
A	=	Income or activity level
N	=	Geological factors
K	=	Production capacity
π	=	Criterion function of cartel
F	=	Foreign reserves of cartel
Y	=	Indirect profits of cartel
θ, γ, α	=	Lag distribution parameters

Allowing for an appropriate time period of adjustment, market equilibrium is given by equating (1.17). Demand relations and fringe supply relations are formulated as in the competitive model. However, the authors interpret the supply relation (1.19) to be short run in nature since it assumes capacity to be fixed. They thus prefer to add a long run relation by first introducing capacity K into (1.19).

$$S_t = s(K_t, P_t, N_t) \qquad (1.23)$$

Capacity is shown to depend on the fuel prices P after an appropriate gestation lag and on lagged demand, thus reflecting long-run investment decision making.

$$K_t = k(P_{t-\theta}, D_{t-\gamma}) \qquad (1.24)$$

This simple model is closed assuming a long-run distributed lag of prices P given by (1.20) where the weights α_j sum to unity and investors have a "memory" of n years when forming price expectations. This is the "so-called" non-competitive market model based on the assumption of exogenous prices. The weights α_j can be said to vary according to government policy formation and their range would depend on government pricing policies. To remove the price assumption, one could compute prices as a function of a set of output policies assigned to X. The constraint equation (1.21) completes the model implying a vector of maximum production levels X in a given year.

Profit seeking or maximization is viewed through the policy criteria function (1.22). The value of π is assumed to depend on foreign reserves F gained and indirect profits Y. No actual maximization of the cartel's profits takes place in the form of a feedback control mechanism. Rather different pure-production policies are set and the criteria function evaluated over some time horizon to determine what might be an optimal policy or set of policies. Of course, the criteria function as well as the other relations would be specified more complexly in an actual model.

Demand and Substitution Models: Another form of market econometric model which has proven very effective restricts itself to energy demand and concentrates not only on interfuel substitution but also on the substitution between energy, labor and capital. The earliest forms of such models were concerned with the residential, commercial, and/or industrial sectors and were single equation, long-run equilibrium demand models focusing on a single fuel.[e] Such models were typically static and explained the demand D for a fuel as a function of its own price P, the price of competitive fuels P^c, and industrial activity A or national income.

$$D = d(P, P^c, A) \qquad (1.25)$$

While the characteristics of the fuel burning equipment are reflected in the substitution of competing fuels, the adjustment in the capital stock to changes in fuel demand is assumed to be instantaneous.

Because of the need to account for differences between short- and long-run energy demand, the above model was replaced by a dynamic, partial adjustment model. Let (1.25) now represent desired demand D^*.

$$D^* = d(P, P^c, A) \qquad (1.26)$$

Partial adjustment assumes that the changes in actual demand D from the time period t-1 to t adjust partially by λ to desired demand changes.

$$D - D_{-1} = \lambda(D^* - D_{-1}) \qquad (1.27)$$

By combining these two equations both short- and long-run demand responses can be differentiated.

Extending this model to a multiequation formulation permits a short run equation based on a fixed capital stock K to be separated from a long run equation that explicitly includes the size and characteristics of that stock, such as in equation (1.24).

$$D = D^* = d(\mathrm{P}, \mathrm{P^c}, \mathrm{A})\cdot \mathrm{U} \cdot \mathrm{K} \qquad (1.28)$$

$$K = K_{-1}(1_{-\delta}) + \Delta K \qquad (1.29)$$

$$K = k(\mathrm{P}, \mathrm{A}, \mathrm{KC}, \mathrm{P^c}) \qquad (1.30)$$

where U is the capital utilization rate, δ is the rate of retirement, and KC includes capital costs and related stock efficiencies.

More recently it has been found useful to replace the fuel substitution effect based on the cross-price elasticities P^c with an interfuel substitution model that deals with competition from other fuels explicitly and in more detail. These models generally assume that the demand for any fuel cannot be adequately assessed without quantifying the price and sometimes the capital cost and non-price competition posed by other fuels (and where appropriate their respective fuel burning devices). These models treat the demand for energy such that capital K, labor L, energy E, and all other intermediate inputs M are seen as inputs to a production activity when the output is defined as QO. The result is a production function of the following form known as *KLEM*.

$$QO = q(\mathrm{K}, \mathrm{L}, \mathrm{E}, \mathrm{M}) \qquad (1.31)$$

These models usually but not necessarily as-

[e]This explanation is based on the review paper of Hartman (1979).

sume that production is characterized by constant returns to scale and that any technical change affecting K, L, E and M is Hicks-neutral. For given input prices and an output level, energy demand can then be determined simultaneously with the other inputs, assuming cost-minimizing behavior.

The actual determination of the fuel substitution effects is based on a form of disaggregation of energy demand E. That is, one can proceed sequentially to find the interfuel substitution responses based on a homogenous energy aggregate E that depends solely on fuel inputs F_i. For example, with four fuel inputs the energy aggregate would be

$$E = e(F_1, F_2, F_3, F_4) \qquad (1.32)$$

The determination of individual fuel usage occurs in this second step. Note that the theory of separability permits individual fuel choice to occur independently of the choice of other inputs.

The econometeric modeling of this approach has been largely through the translog share model, such as that developed by Berndt and Wood (1975). By treating the price of energy as a single argument rather than including the prices of the respective fuels, the model implicitly embodies the mentioned separability. Corresponding to the energy aggregate in equation (1.32) is the unit energy cost function for the energy aggregate.

$$P_E = \phi(P_{F1}, \ldots, P_{F4}) \qquad (1.33)$$

The sufficient conditions for this particular representation are of importance. Separability occurs if the ratio of the cost shares of any two fuels is independent of the prices outside the energy aggregate, such as labor or capital prices. In effect, the ratio of the fuel cost shares depends only on fuel prices. Linear homogeneity in input prices implies that the cost shares of fuels are independent of total expenditures on energy.

The system actually estimated is based on the following translog cost function.

$$\begin{aligned} \ln P_E &= \alpha_O + \alpha_i \ln P_{Fi} \\ &\quad + \tfrac{1}{2} \sum \beta_{ij} \cdot \ln P_{Fi} \cdot \ln P_{Fj} \qquad (1.34) \end{aligned}$$

Partial differentiation of this equation yields a series of fuel cost shares S_i for the four fuels.

$$\begin{aligned} S_i = \frac{P_{F_i} F_i}{P_E E} &= \frac{a \ln P_F}{a \ln P_{Fi}} \\ &= \alpha_i + \sum_{j=1}^{4} \beta_{ij} \ln P_{F_j} \qquad (1.35) \end{aligned}$$

Given that $\Sigma S_i = 1$ and imposing the following restrictions, the final solution for the shares S_i requires estimating only three regression equations of the form (1.35).

$$\begin{aligned} \sum_i \alpha_i = 1, \sum_j \beta_{ij} &= O_1 \ \beta_{ij} = \beta_{ji} \\ &= \beta_{ji} = \beta_{ji} \ (\text{all i, j, i} \neq \text{j}) \qquad (1.36) \end{aligned}$$

Regarding the need to determine the substitution possibilities among the fuels, this can be accomplished by using the Allen elasticities of substitution *(AES)* defined for the translog as

$$\sigma_{ij} = \frac{\beta_{ij} + S_i S_j}{S_i S_j} \qquad (1.37)$$

$$\sigma_{ii} = \frac{\beta_{ii} + S_i^2 - S_i}{S_i^2} \qquad (1.38)$$

The AES are not constrained to be constant but may vary with the values of the cost shares.

Applications of the translog form of this model can be seen in the work of Berndt and Wood (1975) who analyze interfactor substitution between energy, labor and capital in industrial energy demand. This model concentrates on the derived demand for energy based on a fixed output. Emphasis is placed on determinants of outputs, substitution possibilities among inputs allowed by the production, technology, and the relative prices for all inputs. Extensions of this model dynamically can be seen in Hartman (1979) who analyzed short- and long-run price elasticities of residential energy demand. While the above research concentrates on U.S. energy demand, application at the international level can be found in the work of Griffin and Gregory (1976) and Pindyck (1979). One final application by Carson et al. (1981) combine the international demand model with a petroleum supply model to yield a demand and interfuel substitution analysis of the world oil market.

Regarding the application of the demand model in a regulated or noncompetitive context, only very limited research has taken place. One example of an attempt to include market imperfections in interfuel substitution models can be seen in the research of Fuss (1980). The model assumption normally made is that of infinitely elastic supply curves for the energy inputs. However, Fuss demonstrates that the model can still be used in a supply constraint market if the assumption is added that the producer continues to optimize subject to the constrained opportunities and the production technology. In this case market prices for the constrained inputs are replaced by shadow prices. His application deals with Canadian gas and fuel oil constraints and their impacts.

Another example can be seen in the work of Wood and Spierer (1984) who show how natural gas regulatory pricing and supply constraints can be modeled. In this approach, disequilibrium effects due to regulation or supply constraints of one or more inputs in Swiss industrial gas use are explicitly modeled by treating the regulated input as a quasi-fixed input. Systems of fuel demand equations are derived and estimated from a variable cost function conditional on the quantities of natural gas and output. An important feature of this approach is that full equilibrium values for natural gas can be evaluated and used in calculating full equilibrium price elasticities. For a regulated input, this approach eliminates biases in short-run elasticities for variable inputs due to disequilibrium caused by regulation. Another feature of this approach is that it provides a means to test for the effects of regulation by comparing actual and optimal (derived from the variable cost function) values of the regulated input.

Engineering Process Models

Energy process models concentrate on the transformation of fuels and other primary inputs into refined fuel products or other energy derived products. When used for demand determination, they explain the demand for fuels as they flow through a production or refining process. When used for supply determination, they explain the supply of various energy products or refined fuels as they undergo typical refinery processes. The particular engineering characteristics of these models have also caused them to sometimes be labeled as technico-economic models. They often attempt to couple technical aspects of energy systems with economic variables.

The methodologies necessary for constructing process models deal with modeling purposes different from that of econometric models. Using the more popular supply orientation dealing with oil refining and sometimes transportation, one wants to deliver the refined products at minimum cost subject to a fixed refinery size and a set of constraints. The latter can include product quality specifications, process capacities, and product output requirements. Linear programming (LP) is the preferred methodology for solving modeling problems of this type. Computer packages are easily available for model solution and the types and numbers of production activities and constraints required can be easily increased. The modeler can thus easily cope with a joint production problem involving multi-capital processes where "within" and "between" process changes are necessary.

In the linear programming formulation, the objective function and the constraints are assumed to be linear functions (homogeneous of the first degree). The most frequent specifications are to minimize costs, subject to production targets, or to maximize profits, subject to input resource constraints. The objective function is specified in terms of the decision variables X and the cost coefficients C that are used to evaluate them. If we have a minimization problem, the formulation in vector notation would be

$$\min CX \tag{2.1}$$

subject to

$$AX \leq B \tag{2.2}$$

$$X \geq O \tag{2.3}$$

where A is the matrix of technical coefficients and B is a vector representing the constraint values. The problem can just as easily be formulated in terms of maximization. In fact, an important property of linear programming problems mentioned earlier is that to every linear programming objective there is a dual problem.

If the primal problem is stated as above, then the dual problem is

$$\max YB \tag{2.4}$$

subject to

$$YA \leq C \tag{2.5}$$

$$Y > O \tag{2.6}$$

The most important aspect of the dual solution

values is that they represent the "shadow prices" or the "Lagrangian multipliers" of the primal.

An example of the formulation of a LP process model can be seen in the Adams and Griffin (1972) model of the U.S. petroleum refining industry. The core of that model is a typical 200,000 barrel per day refinery LP model. Major inputs to the model are the required product outputs, the prices of factor inputs, and the prices of by-products. The activity inequalities and the constraints are determined by the 12 major refining processes, the capacities of these processes, and the product quality requirements. The actual operation of the model involves simulation analysis where product demands, inventories and imports are solved from an econometric model. Medium- and long-term annual projections are prepared by solving the overall model in successive time periods to determine the impact of economic conditions and energy policies on the refining industry.

A more elaborate process model of the oil industry by Deam et al. (1974) expands this approach to include fuel transportation among some 25 worldwide geographical areas. Fifty-two types of crude oil and 22 refining centers are represented along with six types of tankers that may be selected for transport. The LP matrix for this model is quite large (about 3500 rows and 13,500 columns). The exogenous inputs to the model include future demands for products by region, refinery technology, costs of product refining, and transport of specific crudes and products. The model is solved to determine the optimal allocation and routing of crude oil and products between sources, refineries, and demand centers at some future target date. Because the model includes the transport and refining costs of crude from specific sources, it provides a basis for analyzing the relative price of these crudes in a competitive market or in a controlled market where relative prices are set to reflect the differences in transportation and refining costs among the many sources.

Among models which expand process to include electricity demands, Baughman and Joskow (1974) adopted an approach which was primarily engineering in nature, while the Griffin (1979) approach was primarily econometric. More recently De Genring and Jackson (1980) have balanced the coupling of these approaches by explaining full utilization with an econometric fuel choice model and capital equipment changes with engineering and cost relations.

Spatial Equilibrium and Programming Models

The extension of a supply process model to include fuel transportation leads us to consider how demand as well as supply interact to produce equilibrium in markets separated spatially. Solution to this type of problem involves examining spatial or interregional efficiency in energy production, transportation, distribution and utilization. The methodologies of interest here also involve linear programming and its various extensions. Because this area of modeling is very broad, only its most summary aspects are considered.

Elementary Spatial Programming: Early attempts to model spatial market equilibrium were conceived to provide an optimal allocation of international or regional trade flows. Since solving this problem normally involves transportation elements, one methodology adopted was the transportation variant of the linear programming model described in the prior section. Applied to an energy regional allocation problem, this form of spatial equilibrium model would normally involve the following components: (1) a set of energy demand points of observations and a set of energy supply points, (2) the distribution of energy activities over space, and (3) the equilibrium conditions. These components are represented in the following mathematical definition.

$$\text{Minimize } L = \sum_{i=1}^{n} \sum_{j=1}^{n} T_{ij} Q_{ij} \qquad (3.1)$$

Subject to

$$D_i \leqq \sum_{j}^{n} Q_{ij} \qquad i = 1, \ldots, n \qquad (3.2)$$

$$S_j \geqq \sum_{j}^{n} Q_{ij} \qquad i = 1, \ldots, n \qquad (3.3)$$

$$T_{ij}, Q_{ij} \geqq 0 \qquad \text{all i, j} \qquad (3.4)$$

Definition of variables:

D_i = Energy demand in region i
S_j = Energy supply in region j
T_{ij} = Transportation cost of fuel shipments or energy transmission between region i and region j
Q_{ij} = Quantity shipped between region i and region j

The model operates such that transportation costs are minimized by allowing fuels to transfer until energy demand equals energy supply in every spatially separate region. The cost minimization process is established by the objective function (3.1). The constraint relations (3.2 and 3.3) reflect the conditions that regional consumption cannot exceed the total shipment to the region and that the total shipments from a region cannot exceed the total quantity available for shipment. Relation (3.4) assures the lack of negative shipments.

Among applications of the spatial equilibrium model, Henderson (1958) was one of the first in his analysis of the competitiveness of the coal market. Coal demands and supplies were identified among some 14 regions in the United States. The objective function was solved to minimize the delivered costs (extraction plus transportation costs) of coal allocation subject to competitive market conditions. The solution of the model was then used to evaluate deviations from competitive efficiency by comparing the efficient model solution with the actual or existing coal allocation problems.

When coal again gained stature as a prominent energy source in the 1970s, a more elaborate LP model comparing the supply and demand potential of western and eastern coals was constructed by Libbin and Boehjle (1977). This model expanded the supply activities for a region embodied in (3.3) to include types of surface and underground mining as well as three quality levels based on sulfur content and heating value. Heating content and sulfur value were also embodied as additional activities in regional demand allocations reflected in (3.2). The blending of coals to meet sulfur standards was also allowed in each demand region. The solution of the model minimized the discounted total cost of meeting national coal demand subject to varying sulfur burning standards up to the year 1990. The results were then interpreted in terms of the amounts of coal that could be supplied by various western and eastern states.

Quadratic Programming Models: This more popular spatial methodology replaces the exogenously given energy demand and supply observations with observations that are endogenously determined. While other nonlinear formulations of an LP are possible, the quadratic programming (QP) model represents a specific formulation that features spatial equilibrium characteristics. A QP form of model would normally be composed of the following components: (1) a system of equations describing the aggregate demand for one or more fuels of interest in each of the included markets as well as the aggregate supply of the fuels in each of the markets, (2) the distribution of energy activities over space, and (3) the equilibrium conditions. While the fuel demand and supply equations imply a structure similar to that of an econometric market model, the equilibrium process is more adequately represented through the identification of the profits to be realized from the fuel shipments, i.e., the excess of a price differential between two points minus transportation costs. Profit maximization is assured through the use of a computational algorithm which allows fuels to transfer until demand equals supply in every spatially separated market. So that energy policy decisions can be evaluated more realistically, the equilibrium conditions and other definitional equations can be used to impose constraints on the model parameters.

An example of a system of energy or fuel demand supply equations necessary for transforming the linear to a quadratic programming model is given below. The equations are written in linear form to help specify the quadratic objective function necessary for solution. Identifying variables are assumed to be computationally embodied in the constant terms.

$$D_i = b_{oi} - b_{1i}P_i \qquad \text{for all i} \qquad (3.5)$$

$$S_j = b_{oj} - b_{1j}P_j \qquad \text{for all j} \qquad (3.6)$$

Where P_i = energy demand price in region i and P_j = energy supply price in region j. Because the formulation of this model by Takayama and Judge (1971) expresses these equations in their inverse form, the above equations can be alternatively written as

$$P_i = a_{1i} + a_{2i}D_i \qquad \text{for all i} \qquad (3.7)$$

$$P_j = a_{3j} - a_{4j}S_j \qquad \text{for all j} \qquad (3.8)$$

where $a_1, a_2, a_3, a_4 > 0$ over all observations.

The constraints imposed on the supply and demand relations are the same as in the linear programming model.

$$D_i \leqq \sum_j^n Q_{ij} \qquad \text{for all i} \qquad (3.9)$$

$$S_j \geqq \sum_{i}^{n} Q_{ij} \qquad \text{for all j} \qquad (3.10)$$

Transport costs and shipments are assumed to be nonnegative.

$$T_{ij}, Q_{ij} \geqq 0 \qquad (3.11)$$

The objective function necessary to complete the model goes beyond the cost minimization goal of linear programming. That is, the function maximizes the global sum of producers' and consumers' surplus after the deduction of transportation costs. This form of market-oriented quasiwelfare function which has been termed net social payoff (NSP) by Samuelson (1952) is defined as follows.

$$\begin{aligned} \text{NSP} = & \sum_{i}^{n} \int_{O}^{D} P_i\,(D_i)dD_i \\ & - \sum_{j}^{n} \int_{O}^{S} P_j\,(S_j)dS_j - \sum_{i}^{n} \sum_{j}^{n} Q_{ij}T_{ij} \qquad (3.12) \end{aligned}$$

The objective function for the present model is a restatement of the above after substitution of the linear demand and supply relations.

$$\begin{aligned} \text{Max (NSP)} = & \sum_{i}^{n} a_{1i}D_i - \sum_{j}^{n} a_{3j}S_j \\ & - \tfrac{1}{2} \sum_{i}^{n} a_{2i}D_i^2 \\ & - \tfrac{1}{2} \sum_{j}^{n} a_{4j}S_j^2 \qquad (3.13) \\ & - \sum_{i}^{n} \sum_{j}^{n} T_{ij}Q_{ij} \end{aligned}$$

where D_i and $S_j \geq O$.

Applications of quadratic programming possess the advantage of simultaneous interaction between prices and quantities, an adjustment not possible in linear programming. The use of a nonlinear objective function also prevents the erratic changes in spatial flows that sometimes occur in LP model adustments. Among potential applications of the QP model to energy markets, Labys and Yang (1980) have modeled the spatial allocation of Appalachian coal shipments. This model begins with a spatial allocation limited to Appalachian steam coal supplying the eastern steam coal demand regions. However, it advanced beyond the point supply and demand allocations of the LP models to include econometric demand and supply equations such that coal quantities as well as prices are solved simultaneously in the model solution. With the objective function stated to maximize net social payoff, the sensitivity of coal demand supplies is measured in response to changes in price elasticities, transportation costs and ad valorem tax rates.

An application to the world oil market employing QP has been made by Kennedy (1974). The purpose of this model is to provide simulation analysis and forecasts of the oil market based on changing exogenous factors such as tanker technology in the cost of finding and producing oil in more remote regions as well as government policies on trade, environmental restrictions and taxation. For each region and fuel or refined product, the model determines the level of oil production, consumption, price, refinery capital structure, and the pattern of world oil trade flows.

Other Programming Approaches: Thus far spatial energy models have been described in the simple context of energy allocation over space. However, the application of LP, QP or other programming algorithms to energy market problems can also be made over time. This can be accomplished by operating the LP model dynamically (Propai and Zinn, 1981) or recursively (Day, 1973; Day and Nelson, 1973). The special feature of the latter method is that it represents a sequence of constrainted optimization problems in which one or more objective functions, constraint or limitation coefficients of a given solution depend functionally on the optimal primal and/or dual solution vectors of one or more solutions earlier in the sequence. To obtain this recursive dependence of the coefficients on preceding solutions, a set of feedback functions are used. The rationale behind this approach is that it emulates a decision maker who proceeds according to a succession of behaviorially conditioned, suboptimizing decisions. The decision maker in protecting himself from errors of estimation and forecasting reviews his maximization plans each period based on current information. The sequence of "decisions" as a whole must then converge to some desired optimum over time.

One of the more well known energy applications of this method is the attempt by Day and Tabb (1972) to explain the historical adoption of mining equipment and its effect on production in the coal industry. Based on highly disaggregated mining data, the study tested the optimality of the sequence in which a series of coal mining innovations were introduced.

Another programming model which provides optimization not only over time but also over space is that of mixed integer programming. Like recursive programming it represents an application of linear programming; only the integer characteristic is introduced to accommodate combinations of 0-1 variables which can refer to the nonexistence or existence of a production facility. That is, specific allowance is made for the fact that new capital equipment suddenly comes "on-line" and is now a source of production activity. As explained by Kendrick and Stoutjesdijk (1978), the background to MIP stems from attempts to cope with a number of raw material or resource problems including shipping and transportation, industrial process, and project selection. Examples of MIP energy models include that of Langston (1983) analyzing the Gulf Coast refining complex, Kwang-Ha (1981) describing the Korean electric power industry, and Jung Suh (1982) interpreting the Korean petrochemical industry.

Regarding the application of programming algorithms to noncompetitive spatial models, these have been very few. However, Takayama and Judge (1971) have suggested that such applications are possible. Some of the modeling configurations they consider within the context of optimal spatial pricing and allocation are duopoly, oligopoly under collusion, market sharing arrangements, monopsony and bilateral monopoly. Beyond spatial allocation modeling, some extension of programming models can be seen. For example, Murphy et al. (1980) working with the PIES (1979) model have shown how gas prices can be explained in the context of a market regulated industry. Murphy (1980) also presents a noncompetitive modeling solution in a study dealing with tariff adjustment procedures and oil entitlements in the crude oil sector. Other development of programming applications in a noncompetitive context will be presented in the next section.

Resource Exhaustion Models

Resource exhaustion models represent a special class of energy models that constitute extensions of econometric models and programming models as well. Since programming models inherently feature optimization, Day and Sparling (1977) in their survey of this modeling area would include optimization models of resource exhaustion in the same class. However, they are considered separately here because of their special emphasis on modeling noncompetitive aspects of OPEC pricing and resource depletion in the crude oil market. To better explain this methodology, we begin with simple monopolist models and then advance to Stackelberg, Nash-Cournot and modified optimization models.

Monopolist Models: The optimization models of resource exhaustion that describe the crude oil market as consisting of a multiplant monopolist or cartel and a competitive fringe resemble the noncompetitive econometric market models described earlier. Net demand facing the cartel is

$$X_t = D_t - S_t \quad (4.1)$$

where D is total market demand

$$D_t = d(D_{t-1}, P_t, A_t) \quad (4.2)$$

and S is the supply of competitive fringe

$$S_t = s(S_{t-1}, P_t, N_t) \quad (4.3)$$

The principal change from the model presented earlier is that resource exhaustion comes into play for the competitive fringe as well as for the cartel

$$S_t = s(S_{t-1}, P_t, CS_t) \quad (4.4)$$

where cumulative production CS is given by

$$CS_t = CS_{t-1} + S_t \quad (4.5)$$

A similar accounting identity is needed to keep track of cartel reserves R.

$$R_t = R_{t-1} - D_t \quad (4.6)$$

The objective of the cartel is to pick a price trajectory P_j that will maximize the sum of discounted profits

$$\text{Max } W = \sum_{t-1}^{N} (1/(1 + \delta)^t) P_t - m/R_t D_t \quad (4.7)$$

where m/R is average (and marginal) production costs (so that the parameter m determines initial average costs), δ is the discount rate, and N is chosen to be large enough to approximate the infinite-horizon problem.[f] Note that average costs become infinite as the oil reserve base R approaches zero, so that the resource exhaustion constraint need not be introduced explicitly. The resulting model framework is that of a classical, unconstrained discrete-time optimal control problem, where numerical solutions can be easily obtained.[f]

The control solution to this model yields an optimal price trajectory P_t^* as well as the optimal sum of discounted profits W^* for the monopolist. One might like to compare these variables with the optimal price trajectory and sum of discounted profits that would result if the cartel dissolved (or never formed), and its member producers behaved competitively. Optimal here implies that competitive producers must manage the exhaustion of their oil reserves over time, balancing profits this year against profits in future years.

Although competitive producers cannot collectively set price, they each determine output given a price. Pindyck shows that the rate of output should be such that the competitive price satisfies the equation

$$P_t = (1 + \delta) P_{t-1} - m/R_{t-1} \quad (4.8)$$

If this were not the case, larger profits could be obtained by shifting output from one period to another. In addition, the initial price must be such that two constraints hold. First, the resulting price P and output D trajectories must both satisfy net demand at every point in time as given by equations (4.1), (4.2), and (4.3), i.e., supply and demand must be in market equilibrium. Second, as the price rises monotonically over time, the exhaustion of oil reserves must occur at the same time that net demand goes to zero. If demand becomes zero before exhaustion occurs some of the oil would be wasted and would yield no profits; profits would be greater if the oil were depleted more rapidly (at a lower price). If exhaustion occurs before demand becomes zero, depletion is occuring too rapidly and should proceed more slowly.

The computation of the optimal price trajectory for the competitive case is thus straightforward.[g] Pick an initial P_O and solve equation (4.7), over time together with equations (4.1), (4.2), (4.3), and (4.6). Repeat this for different values of P_O until D_t and R_t become zero simultaneously. Results of application of such optimizing models to OPEC behavior can be found not only in the work of Hnyilicza and Pindyck (1976) but also in studies by Cremer and Weitzman (1976).

Stackelberg Models: A first alternative to the monopolist model is to assume some interaction between the monopolist and the fringe, i.e., the case of a nonuniform cartel. This changes the model theory from that of monopoly to duopoly. The theoretical approach for explaining oil market and price behavior in this context is the Stackelberg model of the dominant firm in which the latter takes the reaction of other firms into account in its pricing policy, while the fringe or other firms take prices as given. Such an oil model is in the spirit of recent contributions to the theory of dynamic limit pricing.[h] The cartel chooses a production path or pricing policy that maximizes net revenues, and those net revenues depend on the rate of production by the competitive fringe. Unlike classical dynamic limit-pricing models, where the residual demand of the dominant firm depends only on its current rate of production, the response of the competitive fringe is a function of the entire sequence of outputs determined by the oil cartel.

Let us examine the simple example of a Stackelberg oil model presented in Aperjis (1981). OPEC is divided into two groups, one of which can be said to be dominant over the other. Using the distinction of foreign exchange absorption, the low absorber group has the largest oil reserves in OPEC and consequently, the largest potential to expand its productive capacity. The higher absorber group has less reserves and sees that the cost of an OPEC breakup will be much higher to them than to the low absorbers, because the former group has a relatively greater need for oil revenues.

Define Group 1 as the low absorbers and Group 2 as the high absorbers and assume that Group 1 dominates Group 2. According to the Stackelberg model, Group 1 will be an oil price-setter and quantity-follower, while Group 2 will be an oil price-taker and quantity-setter.[i] In other words, Group 1 sets oil prices in such a way as to

[f]This explanation is based on Hnyilicza and Pindyck (1976), pp. 6–7.

[g]To solve this problem, Hnyilicza and Pindyck (1976) have employed a general nonlinear optimal control algorithm developed by Hnyilicza. Using that algorithm, optimal pricing policies were derived in a classical control theoretical framework based on an implicit state-variable form

$$g(x_t, x_{t-1}, P_{t-1}, z_{t-1}) = 0$$

where g is a vector of nonlinear functions in the set of state variables x_t (endogenous and lagged endogenous variables, together with state variables defined for P and elements of z occurring with lags longer than one period), the price (control) variable P_t, and a set of exogenous variables z_t. The objective function can also be general in form.

[h]The description of this model appears from Gilbert (1978) p. 394.

[i]This explanation is based on Aperjis (1981), p. 32.

maximize its profits, while Group 2 takes these prices as given and sets oil production at a level which maximizes its profits. Finally, Group 1 produces enough to satisfy any residual demand for OPEC oil over that met by the production of Group 2.

A model of this behavior can be formulated in the following way. Let Group 1 determine the sequence of oil prices P^1_t which maximizes its profits according to the following maximization problem

$$\text{Maximize} \sum_{t=1}^{N} [P^l_t - C^l_t]X_t\,(1/(1 + \delta_1)^t) \quad (4.9)$$

Subject to

$$\sum_{t=1}^{N1} X_t \leq R^l_t \quad (4.10)$$

with the solution

$$MR^l_t = MC^{lt} + \lambda^1\,(1 + \delta_1)^t \quad (4.11)$$

Subsequently, Group 2 takes oil prices P^2_t and produces at that rate S_t which maximizes its profits according to

$$\text{Maximize} \sum_{t=l}^{N2} [P^2_t - C^2_t]\,S_t\,(1/(1 + \delta_2)^t) \quad (4.12)$$

Subject to

$$\sum_{t=l}^{N2} S_t \leq R^2\text{t} \quad (4.13)$$

with the solution

$$P^2_t = MC^2_t + \lambda^2\,(1 + \delta_2)^t \quad (4.14)$$

Definition of variables:

P^1, P^2 = Prices of Groups 1, 2
C^1, C^2 = Unit production costs of Groups 1, 2
δ_1, δ_2 = Discounts of Groups 1, 2
X = Production of Group 1
S_1 = Production of Group 2
R^1, R^2 = Reserves of Groups 1, 2
MR^1, MR^2 = Marginal revenue of Groups 1, 2
MC^1, MC^2 = Marginal costs of Groups 1, 2
λ^1, λ^2 = Foregone future opportunity of an additional unit of current production of Groups 1, 2.

The recognition of Group 2 by Group 1 (the cartel) will cause the latter to choose a price trajectory different from the previous monopoly model, both in magnitude and in rate of change over time. The implications of this theoretical approach can be seen, for example, in the oil modeling study of Gilbert (1978).

Nash-Cournot Models: A second alternative to the oil monopolist model is to assume duopoly market behavior in the form of a non-uniform cartel but to change the behavioral pattern from that of followers to that of bargaining or gaming. Referring to the previous example, if the low absorber group is not willing to be a quantity-follower and the higher absorber not willing to be pricetakers, then a conflict arises which can only be resolved through a process of bargaining.

Nash (1953) developed a solution to such cooperative games. As shown by Hnyilicza and Pindyck (1976), the two-part cartel can again be described in terms of the behavior of Group 1 and Group 2 with the following objectives.

$$\text{Maximize } W_1 = \sum_{t=l}^{N1} [P_t - m_1/R^l_t]X^l_t\,(1/(1 + \delta_1)^t) \quad (4.15)$$

$$\text{Maximize } W_2 = \sum_{t=l}^{N2} [P_t - m_2/R^2_t]X^2_t\,(1/(1 + \delta_2)^t) \quad (4.16)$$

Here δ_1 is assumed to be smaller than δ_2. X^1_t and X^2_t are the production levels of each group, and are determined by a division of total cartel production according to

$$X^l_t = \beta_t X_t \quad (4.17)$$

$$X^2_t = (1 - \beta_t)X_t \quad (4.18)$$

with $0 \leqq \beta_t \leqq 1$. The exhaustion of oil reserve levels for each group is accounted for by the equations

$$R^l_t = R^l_{t-l} - X^l_t, \quad (4.19)$$

$$R^2_t = R^2_{t-l} - X^2_t, \quad (4.20)$$

To these must be added the following:

$$D_t = d(D_{t-1}, P_t, A_t) \quad (4.21)$$

$$S_t = s(S_{t-1}, P_t, N_t) \quad (4.22)$$

$$CS_t = CS_{t-1} - S_t \qquad (4.23)$$

$$X_t = D_t - S_t \qquad (4.24)$$

The next step is to determine how the two groups of countries can co-operate to set oil prices, and to divide output in an optimal manner.[j] Suppose a cooperative agreement is worked out whereby the oil price and output shares are set to maximize a weighted sum of the objectives of each group.

$$\text{Maximize } W = \alpha W_1 + (1 - \alpha)W_2,\ 0 \leqq \beta_t \leqq 1,\ 0 \leqq \alpha \leqq 1 \qquad (4.25)$$

The solution to this maximization problem offered by Nash (1953) differs from the two previous ones in that the steps to optimization are more complex. Assume a bargaining game where Group 1 and Group 2 attempt to move along the set of bargaining outcomes in opposite directions; then the problem is to determine a meaningful measure of bargaining power for the two groups. Nash's approach was to introduce the notion of a "threat point," i.e., the outcome that would result if negotiations were to break down and noncooperative behavior were to ensue.

Of course the actual solution will depend on the bargaining abilities and power of the two groups. Also a bargaining solution might prevail other than the Nash solution. The mathematical optimization problem involves a control solution to the foregoing set of equations including W_1 and W_2. One obtains a solution process by repeatedly resolving the optimizations for different values of α. Actual solution values for the case of OPEC can be found in the cited work of Hnyilicza and Pindyck (1976). Another attempt to model OPEC using the Nash model can be found in the works of Salant et al. (1979) and of Salant et al. (1981). A recent attempt to apply this method to a model of the international coal market can be found in Kolstad (1982).

Modified Optimization Models: Several additional optimization approaches remain which attempt to model OPEC behavior in the oil market based on different degrees of oligopoly and the continuum of intermediate market structures between pure monopoly and perfect competition as well as on specific decision making policies within OPEC. Among these, the most well known is that of Kuenne (1980). He views resource oligopolies as a community of simultaneously competing and cooperative rivals, each of whom follows strategies that reflect its perception of the industry power structure. Those strategies are aimed at achieving a balance of multiple objectives for each rival.

Any model built to emulate this noncompetitive behavior thus should:

(1) contain an explicit formulation of the industry's power structure as perceived by each rival;
(2) retain the identities of each individual rival;
(3) embrace the set of multiple objectives for each rival; and
(4) be flexible enough to adapt to the specific personalities, goals, mores, and institutions of each specific industry analyzed.

To make this approach operational, Kuenne (1980) has developed a modified nonlinear programming format called "crippled optimization."[k] That is, each rival maximizes joint industry welfare in a manner that is "hobbled" in two dimensions. First, the firm defines an objective function consisting of its own profits plus the profits of each rival after each has been multiplied by a power structure discount factor, which has been termed a "consonance factor." Second, each firm maximizes this function subject to a set of constraints that contains its subordinate objectives and perceived restraints.

Such a model framework has been seen as having a variety of advantages. First, its flexibility permits the goals of any particular oligopolistic group of firms to be built in, whether they represent constrained profit maximization, target rates of return, increasing market shares, or whatever. Second, the model is not limited to any single goal. It can incorporate the power structure of the oil industry on a binary one-to-one basis, preserving each firm's perceived relationship to every other. Thirdly, the model can preserve the flavor of realistic oligopoly as a blend of rivalry and cooperation, avoiding the extremes of purely game-theoretical or joint profit maximization approaches. Finally, such a model has been shown to be conceptually and practically operational, permitting modifications of the more well known solution algorithms.

[j]This explanation is based on Hnyilicza and Pindyck (1976), pp. 143 – 144.

[k]This explanation is based on Kuenne (1980), p. 2.

Although the operational framework of such a model structure is too complex to review here, an example can be found in the Kuenne (1980) description of the GENESYS model dealing with OPEC behavior.

Input-Output Models

Input-output (I-O) methods have proven popular for energy modeling at the national economic level. Primary data for constructing interindustry sale and purchase accounts are collected by the U.S. Bureau of Economic Analysis on sectoral transactions. These data are expressed in terms of a common unit (the dollar) and are available for all the Bureau's census years—1958, 1962, 1967 and 1972. The data are compatible with the input-output format, which disaggregates the national economy into industries or sectors. The flow of goods and services among industries reflects systematically the relations among them. To engage in production, each industry must purchase primary commodities, semifinished goods and capital equipment from other industries. At the same time, the output produced by each industry is sold to final users or to other industries which use processed commodities or goods as inputs. Origins of the inputs and destinations of outputs are contained in the familiar input-output transactions table.

The structure of an I-O model can best be understood by beginning with a basic balance equation, which allocates the output of any industry among its various uses in the overall economy. In algebraic terms this can be expressed for a given industry, i, as

$$X_i = C_i + N_i + G_i + X_i - M_i + S_i \; (i = 1, \ldots, n) \quad (5.1)$$

where

X = Gross output
C = Consumption
N = Investment
G = Government
X = Exports
M = Imports
S = Intermediate sales
Y = Final sales

Equation 5.1 is essentially an accounting identity or a snapshot of the economy at a given point in time. It can further be generalized, thus providing us with a means of calculating key variables at different levels of scale or points of time, by explicitly defining the relationship between inputs and outputs. A linear homogeneous relationship is assumed between the amount of any given input i used in the production of good j, S_{ij} and the gross output of good j, X_j. Thus, we can define the basic "technical" coefficients of the I-O model as

$$a_{ij} = \frac{S_{ij}}{X_j}$$

The set of all technical coefficients is typically referred to as the "structural" or "A" matrix. The structural counterpart of equation 5.1 can then be expressed as

$$X - AX = Y \quad (5.2)$$

The solution of the model involves finding the vector of gross outputs X needed to yield a given level of final demands Y and is computed as follows

$$(I-A)X = Y \quad (5.3)$$

$$X = (I-A)^{-1} Y \quad (5.4)$$

where $(I-A)^{-1}$ is the "total requirements" matrix. It is often referred to as the "Leontief inverse" and embodies the direct and indirect requirements per dollar of delivery to final demand. It should also be noted that for this kind of general solution to be applied, the variables need to be expressed in terms of a common denominator—usually dollars. However, in the case of energy modeling, inputs from the energy sector to other industry sectors are converted from dollar flows into energy units such as the British thermal unit. In this format, the direct energy inputs from the oil sector to the agriculture sector, for example, are specified in the input-output matrix. Examples of this unit conversion process can be seen in Hannon et al. (1983) and Herendeen (1973).

An I-O table is like a system of double-entry bookkeeping for a regional economy. Each transaction is simultaneously recorded as both a purchase and a sale. At the level of the individual business enterprise the entries found in an I-O table correspond readily to those of an

Table 3.8.1—Income Statement and Input-Output Accounts, Southern California Gas Company

SoCal			Input-Output Table		
Account Categories		**Million 1980$**	**Account Categories**		**Million 1980$**
		Revenues			
Commercial Sales		321.2			
Industrial Sales		643.8	Interindustry Sales		1,799.9
Electric Utility Sales		835.1			
Wholesale Sales		300.2	Exports		300.2
Residential Sales		909.2	Personal Consumption		909.2
Cost & Supply Adjustment		−172.3	Inventory Adjustment		−172.3
Total Sales		2,837.2	Total Sales		2,837.2*
		Expenses			
Cost of Gas		2,196.4	Cost of Gas		2,196.4
Local Producers	55.2				
PG&E	115.8		Local Purchases	143.5	
Sell-Back	−27.5				
Pipeline Companies	2,047.8				
Federal Offshore	2.4		Imports	2,052.9	
Emergency Purchases	2.7				
Maintenance & Repairs		55.6			
Operation (non-labor)		84.5	Interindustry Purchases**		165.5
Miscellaneous Taxes		25.4			
Wages and Salaries		187.3			
Interest		65.6	Household Income		306.4
Dividends		53.5			
Retained Earnings		5.1	Other Income		69.0
Depreciation		63.9			
Income & Francise Tax		99.9	Indirect Business Taxes		99.9
Total Expenses		2,837.2	Total Gross Output		2,837.2

Source: Rose and Kolk (1983b)

ordinary income statement as shown in Table 3.8.1.[1] The left-hand portion of the table refers to the conventional accounts of an actual energy company, Southern California Gas (SoCal), while the right-hand portion refers to the corresponding accounts in an I-O table.

The upper partition of Table 3.8.1 refers to revenue or sales categories for both sets of accounts. The SoCal sales to "Commercial," "Industrial," and "Electric Utility" customers correspond to the "Intermediate" customers of the I-O table, where they are typically disaggregated on the basis of 4-digit or 2-digit Standard Industrial Classification (SIC) codes. "Wholesale" shipments are termed as "Exports" in the I-O account since the destination is outside the retail service area for which the model is built. "Residential" sales correspond directly to "Personal Consumption" of natural gas by households. The "Cost and Supply Adjustment" refers to rate changes, considered as ex-post "Inventory Changes" in the I-O table.

On the expenditure or cost side, the I-O category of "Local Purchases" refers to natural gas purchases from producers or intermediaries located within the service area. "Imports" refer primarily to purchases from pipeline companies. The dichotomy is consistent with that of "Out-of-State" vs. "Local" sources used in SoCal accounts.

Note that the difference between the purchase cost of natural gas and the sales revenue rep-

[1]The example stems from the work of Rose and Kolk (1983b).

resents SoCal's cost of maintenance and repairs, operating expenses and primary factor payments (wages, depreciation, profits, etc.). The I-O categories correspond here as well.

A 15 sector aggregation of a 66 sector retail service area I-O table for SoCal in 1980 is represented in Table 3.8.2. The entries in the table represent flows of materials and services, expressed in millions of 1980 dollars, between sectors of the regional economy. Again each cell represents both a purchase and a sale. In effect the sales distribution of a product can be read across any row. The purchase of inputs to produce the total output of the product are found in a given column. Natural gas industry sales and purchases, row 12 and column 12, respectively, have been boxed-in for easy identification.

Sectors 1 through 15 represent purchases and sales between producers, or "intermediate" uses of goods and services. For example, the entry in row 12 and column 8 indicates that the natural gas distributor, SoCal, sold $115 million of its product to the Primary Metals industry. The demand for natural gas for intermediate use by every other sector is also shown in row 12 and is consistent with SoCal accounts (compare the total intermediate input entry in row 12 with the total interindustry sales figure in Table 3.8.1). The purchase of inputs by SoCal is contained in column 12 (compare with the lower portion of Table 3.8.1).

The three columns following "total intermediate sales" constitute the final use of goods and services for consumption, investment, government and export, respectively. For example, the first of these entries in row 12 shows the $909 million in purchases of natural gas by consumers. Consumer expenditures for other goods can be read from this column as well, e.g., consumer purchases of $1,853 million of electricity from private utilities. "Total final demands" for the goods of each producing sector are shown in the third column from the right in Table 3.8.2.

The rows following the "total intermediate input" row refer to two special categories. First, are imports, including goods, such as coal, not produced at all in the region ("noncompetitive imports"), and also those quantities of goods in categories 1–15 which are produced elsewhere and shipped into the region ("competitive imports"). The value-added row represents "payments" to primary factors of production (land, labor, and capital). These payments correspond roughly to final demand components, since they are eventually spent on consumption, investment, etc.

The transactions matrix is the basis of the derivation of a set the structural coefficients, a_{ij}, defined earlier. For example, the column 8 of Table 3.8.2 shows that primary metals purchased $115 million of the output of the natural gas sector or a direct input coefficient of .0105 = 115/10,983 for each dollar of primary metals output. If we assume that the jth industry's demand for each item on the input list is similarly proportional to that industry's output, then we can solve the output equation given above simultaneously with similar equations for every other industry, in this way obtaining outputs that are in balance with all current input requirements and with final demands.

The input-output approach has a number of positive attributes for energy modeling. To begin with, it can serve as an organizing framework for an extremely large amount of data. It provides a very clear picture of the role of energy within the structure of the economy. Energy use is primarily a derived demand and, hence, a model that gives preeminence to the processing sectors of the economy is most appropriate. Secondly, the I-O model, with its emphasis on intersectoral relationships, is especially adept at determining the less obvious, or second-order, impacts of the changing energy picture.

At the same time the basic I-O model has some serious limitations. The most important is the assumption of fixed technical coefficients. Several ramifications of this assumption are a linear view of the world, a fixed technology, and zero price elasticity. A great deal of research in recent years has been devoted to relaxing this assumption. A critical evaluation of this research by Rose (1983) indicates that it has generally been successful, though often resulting in a much more complicated dynamic I-O version or a hybrid with another model form.

A second limitation is that there are the difficulties and time delays in assembling the interindustry flow data. In particular, the table currently available for the United States is based on 1972 data. However, the recent development of several accurate updating techniques offer some promise, e.g., see Almon, 1974.

Thirdly, an I-O model normally cannot be employed alone to explain energy market behavior. Rather, it can provide a disaggregated

Table 3.8.2—Input-Output Table of the Social Service Area (in millions of 1980 dollars)

	1	2	3	4	5	6	7	8	9	10	11	12
1 Agriculture	1371	*	28	2009	7	*	2	3	7	13	5	*
2 Mining	9	89	80	7	48	3211	84	122	7	18	5	144
3 Construction	69	208	7	72	29	97	37	61	106	355	116	14
4 Food, Apparel, Textiles	451	6	847	4126	358	45	213	153	696	141	10	*
5 Chemicals & Allied Prds.	407	19	481	293	536	74	444	150	273	9	6	*
6 Refined Petroleum Prds.	317	64	399	190	302	714	88	96	275	623	212	1
7 Stone, Glass, Rubber	34	6	1255	627	209	8	446	128	612	38	2	*
8 Primary Metals	17	62	2260	844	140	40	89	2015	2862	24	1	*
9 Machinery & Misc. Manufac.	40	132	981	214	52	15	74	211	7225	210	10	1
10 Transportation	142	46	484	885	248	245	265	309	762	1086	41	46
11 Electric Utilities	20	34	6	84	28	20	42	65	119	43	195	1
12 Natural Gas Utilities	17	42	4	125	84	126	97	115	45	10	835	23
13 Water & Sanitary Services	9	5	7	11	4	3	3	6	8	10	1	*
14 Services & Misc.	1056	862	3347	2540	697	187	409	961	3996	1640	136	57
15 General Government	11	9	14	25	9	7	8	14	25	30	3	23
Total Intermediate Inputs	3969	1584	10198	12063	2750	4792	2301	4509	17019	4250	1579	309
Total Imports	532	257	3317	6281	1240	2389	890	2502	4952	1170	912	2053
Total Value Added	2800	2704	10562	8443	1603	668	2277	3967	15945	9420	1298	475
Statistical Error	1	1	*	14	3	4	3	4	3	*	−294	*
Total Gross Outlays	7302	4545	24077	26802	5596	7854	5471	10983	37919	14840	3495	2838

*Less than $0.5 million.

view as to how the supply and demand patterns for fuels and energy products relate to the industrial structure and aggregate income variables of a national economy. This need to couple I-O models with other forms of models or quantitative techniques is well described by Griffin (1976). Examples cited include static I-O analysis coupled with LP optimization, dynamic I-O models utilizing process analysis, dynamic I-O models coupled with econometric models, and a combined econometric process analysis approach to a dynamic I-O model. He reviews the Hudson and Jorgenson (1974) model which embodies an energy I-O model directly within a macroeconometric model framework and the Preston (1975) description of the coupling of the Wharton energy focused interindustry model and the Wharton macroeconometric model.

Finally, when the I-O framework has been utilized independently, most of its applications have involved attempts to measure energy consumption. This usually has involved measuring the direct, indirect and income-induced energy effects of a change in final demand. Examples of such applications include the Reardon (1972) estimates of U.S. energy consumption and the Beltzer and Almon (1972) estimates of U.S. petroleum demand. More recent analyses by Park (1982) and Fritsch et al. (1977) have included the impacts of technical change and related investment on energy consumption at the international level. In addition, Rose and Kolk (1983a, 1983b) have modeled the impacts of voluntary and mandatory conservation programs on energy demand.

A significant amount of research on energy at the regional level has also made use of I-O. This includes a determination of the aggregate impact of energy development by Rose et al. (1978), the distributional impact of energy development by Rose et al. (1982), the regional impact of using energy prices by Kolk (1983) and the interregional impacts of using energy prices by Miernyk et al. (1978). Much of this research has been facilitated by nonsurvey methods of adapting national tables to the regions in question, e.g., see Morrison and Smith, 1974.

Integrated Energy Models

We have discussed the growing tendency among energy modelers to use two or more modeling methodologies in combination. In particular, as the modeling process has become more sophisticated, analysts have begun to borrow

13	14	15	Total Intermediate Sales	Personal Consumption	Investment & Inventory Change	Govern-ment & Exports	Total Final Demand	Errors & Omissions	Total Gross Output
*	390	1	3836	721	266	2479	3466	*	7302
*	27	1	3851	54	48	623	724	−30	4545
26	2961	298	4458	*	15214	4405	19619	*	24077
*	3676	7	10728	10622	1051	4390	16063	10	26802
*	448	14	3155	1315	145	980	2440	*	5596
10	1271	35	4596	2417	3	837	3257	*	7854
1	468	1	3832	677	159	843	1679	−40	5471
*	255	2	8612	237	462	1671	2371	*	10983
3	1541	8	10816	4136	5131	17835	27103	*	37919
14	2952	19	7544	4125	379	2836	7340	−44	14840
1	640	10	1308	1853	*	333	2186	*	3495
2	248	16	1300	909	−172	300	1037	*	2837
8	98	3	175	137	*	4	140	*	315
26	24236	101	40250	64532	834	23597	88964	−35	129179
9	187	4	378	685	3	45	733	−9	1101
100	39397	520	105340	92420	23524	61179	177123	−147	282316
14	6563	92	33164	14677	6683	3156	24516	*	57680
201	83212	488	144062	−6	−1393	16726	15327	*	159389
*	7	2	−251	*	*	*	*	*	−251
315	129179	1101	282315	107091	28814	81060	216965	−146	499134

skills from one another, so that the methodologies which have emerged are more difficult to classify. This phenomenon has risen not only because of the need to couple economic and engineering considerations, but also because of the interrelationships that exist between the energy sector and the growth of national economies. Although this integration has taken a variety of forms, it usually proceeds by model hybridization or model linkage.

Hybridization basically involves constructing models that combine two or more modeling methodologies. For example, it is not uncommon for the demand side of an energy model to be constructed econometrically, while the supply side might be principally engineering in character. In such cases the output of several different models might be needed to provide a comprehensive analysis of a particular energy problem. Linkage involves the coupling together of two or more models in a systematic fashion. For example, an input-output model of the energy sector can be linked with an econometric model of the macroeconomy. Analyzing feedback effects is important in this case, such as the impact of energy prices on economic growth. The growth of this modeling activity can be witnessed in the different forms that have emerged.

Most well known among these has been the linking of input-output models of the energy sector with macroeconometric models. As mentioned previously, the Hudson and Jorgenson (1974) model consists of a macroeconometric growth model of the U.S. economy integrated with an interindustry energy model. The growth model consists of submodels of the household and producing sectors, with the government and foreign sectors taken to be exogenous, and it determines the levels and distribution of output valued in constant and current dollars. The model determines the demand for consumption and investment goods, the supplies of capital and labor necessary to produce this level of output, and the relative equilibrium prices of goods and factors. The model is dynamic and has links between investment and changes in capital stock and between capital service prices and changes in prices of investment goods.

The macroeconometric growth model is linked to an interindustry energy model by estimates of demand for consumption and investment goods and the relative prices of capital and labor. The interindustry model employed is based on a nine-sector classification of U.S. industrial activity. Production submodels are developed for each sector. These submodels treat as exogenous the prices of capital and labor services determined in the growth model and the prices of competitive imports; and for each sector they determine simultaneously the sector output prices and the input-output coefficients.

The sector output prices and the demand for consumption goods from the growth model are used as inputs to a model of consumer behavior that determines the distribution of total consumer demand to the nine producing sectors. The distribution of private investment (government and foreign) is determined exogenously, and it completes the final demand portion of the model. Given final demands, the input-output coefficients may be used to determine the industry production levels required to support a given level and distribution of real demand.

The Hudson and Jorgenson model has been used to forecast long-term developments in energy markets within the framework of a consistent forecast of macroeconomic and interindustry activity. The model has also been used to analyze the impact on energy demands of alternative tax policies, including a uniform Btu tax, a uniform energy sales tax, and a sales tax on petroleum products.

Among other energy-economy models, Groncki and Marcuse (1980) report on the BESOM model which combines a programming model of energy supply with a long-run macroeconomic growth and input-output model. One example of the use of the model has been to analyze the economic impacts of fuel scarcities. Constraints can be placed on the availability of fuels and resources, and the required fuel substitutions are determined. Coefficients in the input-output model are revised to reflect the new fuel mix, and the input-output model is again solved with the revised mix. Several iterations are required between the two models in order to obtain a solution in which the energy demands and fuel mix are consistent in the two models. Impacts on the macroeconomic growth of various industrial sectors in the economy are then evaluated.

While these models analyze energy-economy interactions in a domestic context, research has also taken place in modeling international interactions. To begin with, there are the most abstract models based on trade and balance of payments theory such as that of Chichilnisky (1981). They have developed a two-sector balance of payments model which analyzes the relations between oil prices and output, employment and prices of goods in industrial economies. The industrial or developed country region is a competitive market economy that produces two goods (consumption and industrial goods) with three inputs (capital, labor and oil). It trades industrial goods for oil with a developing oil exporting region which acts as a monopolist. The general equilibrium solution of the model determines endogenously the principal variables of the industrial region: output and prices of industrial goods as well as employment of their factors and their prices. Simulations of the model are used to determine the impact of different levels of oil prices on the two regions.

Finally, a global form of model has appeared in which hybridization of energy and other submodels is a normal feature. The methodology employed is not a single or unique one. Rather it represents various approaches for combining various models in a single framework or of constructing large-scale multi-commodity models. The models to be combined have tended to be of a programming nature, often also linked to some form of econometric model. The most comprehensive effort to develop a methodology to deal with hybrid energy models is the "combined" energy model approach of Hogan and Weyant (1980). A theoretical, as well as a computational, approach is presented for reaching equilibrium solutions with a combined set of models. The most well known application of this approach is the Project/Independence Energy Evaluation System (PIES) constructed for the FEA, i.e. see EIA (1979). This model includes a macroeconomic model, an econometric demand model, and a programming model explaining fuel supplies, conversion and shipments.

The heart of PIES is the integrating model—an LP model that uses given estimates of regional demands, prices and elasticities, regional supply schedules, and resource input requirements to calculate an energy market equilibrium. The relation between the demand model and the LP submodel, which incorporates the supply schedules and conversion processes, may be summarized as follows: The demand model is used to calculate a price/quantity coordinate on the demand curve for each of the primary and derived energy products in the system. Associated with each of these coordinates are measures of the sensitivity of the quantities demanded to small changes in each of the prices in the demand model (own- and cross-price elasticities). In the first iteration of the integrating model, an LP problem is solved in which the minimum cost schedule of production, distribution, and transportation necessary to satisfy the given demand levels is calculated.

Associated with the calculated supply quantities are implicit prices. If these supply prices differ from the original demand prices, then the solution is unstable and a new problem must be structured and solved. The procedure is to calculate new demand prices that equal one half the difference between the last iteration's supply and demand price, to use the own- and cross-price elasticities to calculate the new demand quantities, and, finally, to solve a new LP problem for new production, distribution, and transportation schedules and the supply prices. The process is continued until the demand and supply prices are equal, at which point the energy market is assumed to be in equilibrium. The outputs of the integrating model are then used as inputs to certain interpretive models including a macroeconomic model, an environmental assessment model, and an international assessment model, In addition, the integrating model outputs are analyzed to determine if potential limitations exist on the availability of the necessary resource inputs.

Another area of energy-economy modeling has been the large-scale international models which examine trade between major regions of the world and attempt to include energy production and consumption within that model. Scenarios are generated with computer simulation models which explore the impact on international trade and growth of changes in energy conditions. The latter include rising oil prices as well as increased oil depletion and energy conservation. Some of these models include energy as only a minor sector of the whole model while others consider energy as the driving sector of the model. Examples of these include the Hughes and Mesarovic (1978) world integrated model (WIM), the Herrera and Scolnik (1976) version of the Bariloche Model, the Linneman (1976) MOIRA model, and the IIASA sets of models described by Basile (1979).

Energy Balance Models

Analysis and modeling of the overall energy system including supply and demand sectors as well as all fuels and energy forms has been stimulated largely by the need to develop forecasts of total energy demand. Much of the initial work in this area involved the development of overall energy balances for the United States in which forecasts for individual fuels were assembled. These forecasts highlighted many problems involving such factors as resource definition and interfuel substitution, which must be handled in a consistent manner for all fuel types and sectors and which led to increased modeling of the entire energy system. In order to produce these forecasts, a methodology had to be developed using the energy balance concept. However, the methodology never developed to the point of formal quantitative modeling. It has been included here only to complete our survey of the construction and application of energy models. Most typically these models have been separated into energy analysis models and energy gap models.

Energy Analysis Models: These models possess less of a formal structure than input-output models. According to Ulph and Folie (1977–1978), they consist of two types: (1) accounting models which attempt to give a comprehensive assessment of the energy costs involved in producing different products, and (2) net energy models which examine the ratio of energy inputs to energy outputs associated with certain processes, usually energy of food production. These methodologies are based on data which provide detailed accounts of flows of energy. As such, they could provide a basis for more formal models of the type described above.

As an accounting approach, the energy balance system focuses attention on a complete accounting of energy flows from original supply sources through conversion processes to end-use demands. The approach accounts for intermediate consumption of energy during conversion processes as well as efficiencies at various points in the energy supply system. An important characteristic of the system is that prices are not determined by the market forces of demand and supply. Rather, prices are determined by the interaction of both consumer preferences and technological considerations, the latter reflecting the sum of direct and indirect energy required to produce any commodity. Any resulting analysis assumes that energy is the only nonproduced input to any production process, which involves either ignoring other inputs such as labor, or attempting to aggregate all non-produced inputs into a singe factor called energy. Obviously such analysis ignores important differences in the nature of different inputs, the finiteness of resources, and the allocation of resources over time.

A recent review of these kinds of models can be found in Hoffman and Wood (1976). One of the first systematic attempts to account for all energy flows in a consistent manner was that of Barnett (1950). Barnett's approach involves obtaining a national energy balance of energy supplies and demands by type. The emphasis was on quantity flows expressed in physical units and a common unit, the Btu. This approach has been recently extended and refined by Morrison and Readling (1968). Some examples of an international energy application can be seen in works by Huettner (1976) and by Webb and Pearce (1975).

Energy Gap Models: These models may even have less of a technological or structural framework than energy analysis models. According to Gately (1979), they typically contain no explicit functional relationships among the variables, nor any equilibrating mechanism to ensure consistency among supplies and demands in the various markets. For any particular set of assumptions about the underlying political-economic environment, there are separate estimates of demand and supply which are disaggregated by region and fuel type. The demand estimates, for example, are developed by relating demand to aggregate economic activity and trends in energy consumption. Independent estimates of supply of major energy types are developed and compared with the demand estimates. Differences are resolved, usually in a judgmental way, by assuming that one energy type is available to fill any gap that may exist between supply and demand. This energy type is normally assumed to be imported petroleum, including crude oil and refined petroleum products. The DuPree-West (1972) study provides an excellent example of the execution of a forecast employing this methodology. Other examples include those of National Petroleum Council (1974), Exxon (1977), OCED (1977), U.S. Central Intelligence Agency (1977), and the Workshop on Alternative Energy Strategies (1977)

ENERGY MODEL EVALUATION

The intent of this chapter has been to discuss the objectives of energy market and system modeling. We now move to the step of model evaluation to determine how it can be used to improve the effectiveness of energy policy analysis. The need for evaluation and better communication is indirectly illustrated by the recent study of Greenberger and his associates (1983) who have surveyed the views of energy experts and others regarding the quality, the attention received, and the influence of 14 major energy studies of the past decade.[m] The survey indicates that while attention received and influence were positively correlated, quality tended to be negatively correlated with both attention received and influence.[n] The study offers no explanation for this disconcerting negative correlation, but throughout emphasizes the importance of timeliness and effective communication of analysis results.

The elements of energy model evaluation have been much discussed and debated. The most often cited literature includes Gass (1977) and Greenberger, et al. (1976), both of whom distinguish two fundamental aspects of model evaluation: validation and verification. Validation refers to the correspondence of the model to the underlying processes being modeled. This form of validation will include three elements: (1) the structural features of the model, (2) the inclusion of relevant variables such as policy instruments and concepts of importance for the issues to be analyzed, and (3) the predictive capability of the model. Structural evaluation is, of course, the essence of scientific analysis. It is based upon (1) the conceptual specification of the model, (2) the specification and application of the measurement process by which the model data are generated or obtained, (3) the specification and analysis of the scientific hypotheses derived from theory underlying the model and to be tested via analysis of the model data, and (4) the selection of the final model best supported by the scientific laws, principles, maintained hypotheses, and tested hypotheses which emerged from the research process. Validation of the model structure starts with the replication

[m]In addition to the survey, this study provides a comprehensive description of the objectives, context, and central results of each study, and attempts to synthesize from these experiences information and ideas for improving the credibility and contribution of policy analysis. The study is an excellent source of information, perspective, and references on the major energy studies of the past decade.

[n]See especially Greenberger et al. (1983), pp. 76–82.

of measurements and hypothesis testing, but also includes analysis and/or counter-analysis involving the variables and concepts integral to the policy issues for which the model was intended.

The second validation issue, content validity, is usually singled out from structural validity for the latter purpose. Both policy evaluation and policy analysis require that models reflect the appropriate policy concepts and instruments.[o] A policy evaluation model will be simpler than a policy analysis model in this regard in that only the policy actually implemented and being evaluated must be included. Policy analysis models are more complicated in that the policy instruments and concepts suitable for the alternative policies of potential interest and importance to the various constituencies concerned with the issue(s) of interest must be included. Further the model must be explicit concerning the resolution of "facts" and/or value judgments which are in dispute among the various constituencies. Crissey (1975) has made the evaluation of such "facts" or contention points a central feature of his approach to policy model analysis.

The third element of policy model validation is predictive capacity, determining if the scientific information and results included in the model are sufficient to discriminate among the policies being considered. If the range of scientific uncertainty spans the range of policy dispute, then the model's usefulness in policy research is very limited. Model-based studies may sometimes only report point predictions and not information on prediction confidence limits or sensitivity analysis of predictions to changes in input data and/or structural coefficients, consistent with known or conjected uncertainties in the underlying measurement processes and scientific results. The lack of this additional information may suggest an unjustified precision of analysis. Analysis of predictive power is thus an important aspect of policy model analysis quite independent of the structural validity of the components of the model.

Closely related to the various dimensions of model validity is the validity of the data associated with the model. Data validation must include not only evaluation of the measurement process by which the data component of model structure is developed, but also the processes by which the data required for model applications are obtained. While data and measurement process evaluation are closely related to model evaluation, particularly to that of model structural and predictive capability, it is probably useful to single out this aspect of validation since it typically receives too little attention in policy modeling and research.

Crissey (1975) has a similar perspective on the elements of policy model validation. He emphasizes that the credibility and utility of a policy model will depend upon its treatment of the factual, behavioral, evaluation, and structural issues in dispute. Disputed issues should be represented in the model in a manner facilitating analysis of alternative resolutions. Such issues comprise the model's contention points. According to Crissey, a *contention* point is said to be critical if changes in its resolution significantly affect the model conclusions and is a *contingency point* if changing the resolution of this contention point in combination with others results in a significant change in the model result (Crissey, 1975; pp. 83–88). This concept of model contention points provides a useful focus for structural, content, and predictive validation.

In contrast to validation, policy model verification refers to the evaluation of the actual model implemention. At issue is the correspondence of the implemented model—usually a computer program—to what the modeler intended. Verification is thus more mechanical and definitive than model and data validation. Gass (1977) has suggested that policy model verification is the responsibility of the modeler, and that evaluation should be limited to review of the verification process.

A final aspect of policy model evaluation concerns usability. This dimension of evaluation refers to both the sufficiency of documentation to support model understanding and applications, and the efficiency of the overall system. Technical documentation and materials sufficient to inform potential users of the nature of the model's structure, content, and predictive characteristics, as well as to support interpretation of model-based results, are essential for any policy model. The need for documentation to support independent application of the model,

[o]Greenberger, et al., 1976, distinguish policy evaluation and policy analysis as follows: "Policy evaluators organize a research effort around an existing program and ask how well it is achieving its intended objectives; policy analysts tend to organize their investigations around a set of policy objectives and they inquire whether there is any conceivable program or combination of programs that might achieve the desired ends more efficiently." (p. 30).

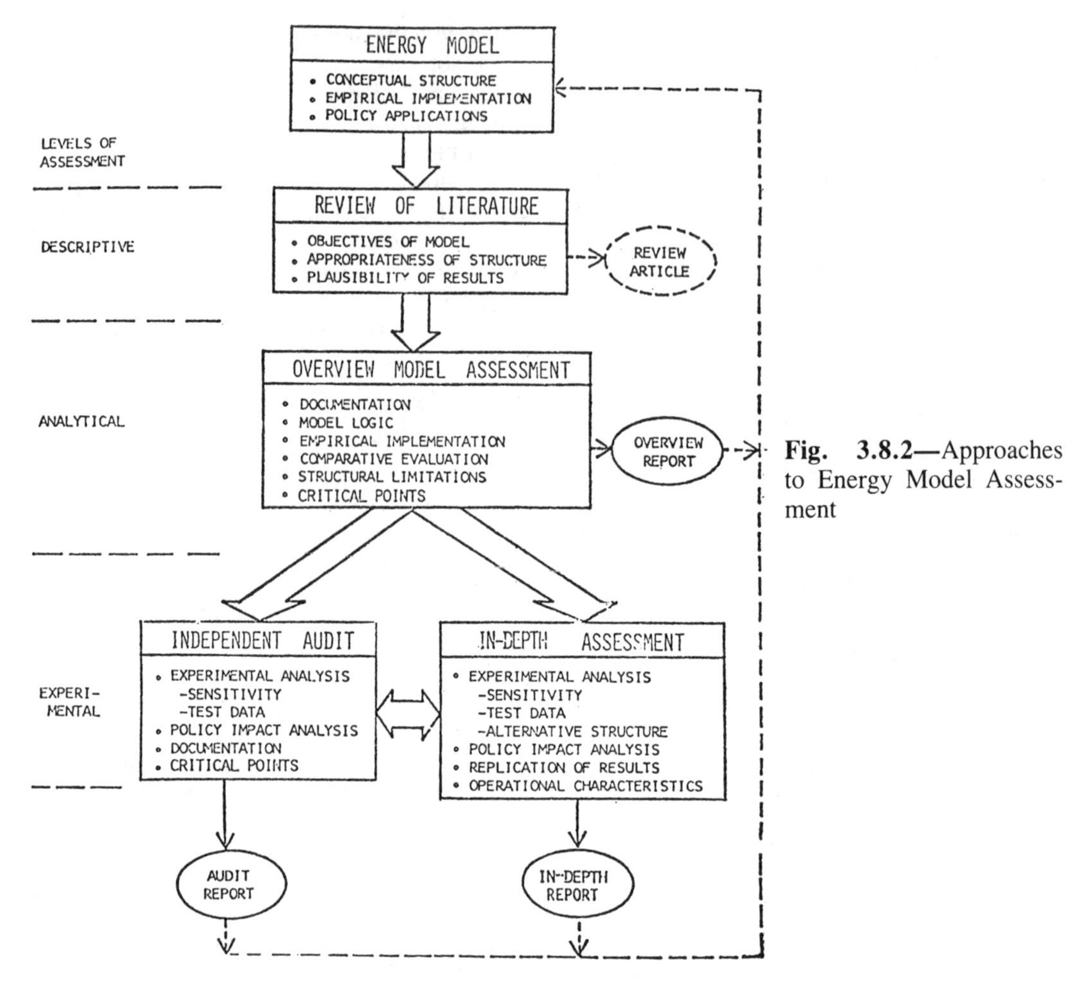

Fig. 3.8.2—Approaches to Energy Model Assessment

including user guides, system guides, and test problems will depend upon the model application environment. Of course, even if the intent is for the modeler to conduct all applications, there still should be evidence that application procedures have been developed, and that a reasonable applications practice is in effect.

Our overview of the elements of model evaluation is suggestive both as a guidelines for self-evaluation by the modeler, and for independent model evaluations. Given these guidelines, what can be said about the process of independent evaluation, an activity of increasing importance in establishing the credibility of policy models?[p]

The MIT Model Assessment Program has identified four increasingly detailed approaches to evaluation including,

Review of literature,
Overview assessment,
Independent audit, and
In-depth assessment.

The major distinction between the approaches concerns the materials used in evaluation. A summary of the relationships between these evaluation approaches is given in Fig. 3.8.2. A review of the literature for a model, or set of similar models, focuses upon model formulation, measurement and estimation issues relating to model structure, applicability for analysis of specific policy issues, and so on. Such a review may be both descriptive and evaluative. An example is the review by Taylor (1975) of electricity demand models comparing model structure with an "ideal" structure. In its various forms, literature review and analysis is the

[p]Independent evaluation is only one approach to improving model credibility and usefulness. Another approach, developed at the Stanford Modeling Forum, is to engage model builders and model users in constructing, executing, and analyzing model experiments.

traditional means of model analysis. Issues of approach, logic, measurement and interpretation are formulated and analyzed. Issues of actual implementation are less susceptible to analysis with this approach.

An overview assessment uses the underlying technical model documentation, especially the computer code, for a more precise analysis of the model's structure and implementation. An overview evaluation can identify a policy model's critical points, but it will only occasionally be able to pass judgment on the adequacy of the model's treatment of them. The overview report is a useful intermediate stage in the assessment process, but assessment of the model's validity and applicability generally requires the acquisition and analysis of experimental data.

An independent audit evaluates a model's behavior by analyzing data derived from experiments that are designed by the assessors but run by the model builders. An important element of the procedure is that the assessment group is "looking over the model builder's shoulder" while the experimental runs are being made. This is essential to the accurate interpretation of the results produced by the experiment. An audit report should use the experimental data together with the analytical material developed in previous stages of the evaluation process to determine the model's validity in as many keys areas (critical points) as possible. Audit procedures have the advantages of being relatively quick and inexpensive. With complex models, however, there will generally be some critical points that cannot be fully evaluated through an audit.

An in-depth assessment develops experimental data through direct, hands-on operation of the model. Direct operation makes it feasible to carry out more complex tests, particularly when the tests require modifications in model structure rather than single changes in model parameters and/or data. Because of the significant costs of in-depth evaluation, it is probably most efficient to conduct exploratory analysis through an independent audit before embarking on more detailed evaluation. After an in-depth evaluation has been completed, audits might subsequently be used to update the evaluation as new versions of the model are developed.

We have briefly surveyed the elements of model validation and verification, and approaches to independent evaluation. It should be emphasized, however, that the modeler has the primary responsibility for designing, conducting, and reporting the results of validation and verification efforts. When the modeling research is reported in refereed journals and monographs, the modeler will be guided by general scientific practice. However, modeling research intended to be used in policy analysis, and often not submitted for refereed publication, places an additional burden on the modeler. In addition to providing documentation consistent with "good scientific practice," the modeler must also satisfy the information needs of a (usually) less technically oriented audience interested primarily in model based results, not the modeling research itself.

The essential requirement is to provide these diverse model audiences—other modelers, policy analysts, and those affected or influenced by model based studies—with the information to judge for themselves the adequacy and legitimacy of the models, and its relevance in any particular applications.[q] No amount of independent evaluation can substitute for a well conceived and executed plan to satisfy these legitimate information requirements. Hence the prospective modeler intending to contribute to the analysis and understanding of the current and prospective energy policy issues should be aware that good models and analysis are not sufficient to ensure the credibility and utility of model based policy studies. The findings of Greenberger and his associates, mentioned at the beginning of this section, that for the major energy studies of the past decade (and many relied heavily on models), perceived quality was negatively correlated with attention received and influence, is a sobering reminder and challenge to energy modelers to consider carefully, and to satisfy, the legitimate information needs of their audiences.

CONCLUDING REMARKS

The objective of this chapter has been to survey the important steps and methods employed in formal energy market and system modeling, and to provide some suggestions for improving the credibility and utility of the resulting mod-

[q]See Greenberger et al. (1983), Chapter 10 for a general discussion of policy analysis. Although not specificially directed to modelers, this discussion contains many useful suggestions and guidelines regarding effective communication of model based studies.

els, especially those intended to support policy analysis. As is readily apparent, formal energy modeling draws on several disciplines and techniques; no one modeling approach addresses the myriad topics and issues of interest to scientists, policy analysts, and decision makers. Policy analysis especially is multidisciplinary, and successful modeling with this objective places considerable demands upon both the modeler and the relevant disciplines and techniques. In these concluding remarks, we suggest three broad areas of development that are likely to influence the next generation of energy modeling.

First, there is the need to untangle more carefully the nature of energy market structures and the role played therein by different forms of market interventions such as regulated prices, subsidies, taxes and trade controls. The potential of multidisciplinary analysis in this area is the least, unless the disciplines of quantitative political analysis and of industrial organization advance in this direction. Recent attention to market structure in energy models has been reviewed in terms of the seminal work of Pindyck (1978), Salant et al. (1981), Kolstad (1982) and others. We have thus seen the embodied structure of the petroleum market advance from simple monopoly to rather complex forms of oligopoly. A lesser amount of work has taken place regarding the role of market interventions, with the exception of the reviewed studies of Fuss (1980) and of Wood and Spierer (1984). Energy models are thus likely to improve in the direction of offering a more realistic picture of these two related aspects of market structure.

Second, we expect increasing attention to be applied to the integration of behavioral and technical/engineering information and concepts in formal energy models. There is already a suggestive literature on how this integration will take place. Brock and Nesbitt (1977) have provided a conceptual analysis of the energy system and have presented several energy models which emphasize the fundamental linkages between supply, conversion, and use in an economy context. Recently Hogan and Weyant (1982) have continued this early work, emphasizing that large-scale energy models involve integration of component models describing separate economic and technical aspects of the energy system. The relations between the component models may be characterized as a network of process models, and their joint solution as a system equilibrium solution. The network formalization is quite powerful in identifying, modeling, and analyzing information flows, especially in clarifying measurement issues (both conceptual and point in time/space/system), and in developing solution algorithms and computational facilities.

An important operational implication is that the network formalism greatly facilitates understanding of how component models (processes) may be efficiently integrated. The high costs in both resources and time of large-scale model development mean that any advances which "spread" modeling costs and reduce implementation time will increase the potential utility of resulting integrated models. Further efficiency gains will also occur to the extent that common software useful for describing and solving such network models comes into wide use. Certainly anyone interested in large-scale models for policy research and analysis will want to become familiar with this approach.

Another development relating to integration of behavioral and technical/engineering models is due to Lau (1982) and Berndt (1983), and shows considerable promise in providing a more natural description of the behavioral/process interactions in energy systems. Lau considers the conceptual problem of incorporating characteristics of a factor input in measuring that input, e.g., technical information about capital (age, boiler pressures, heat rates, etc.), education of labor, or characteristics of materials. He finds that the conditions on the underlying production function consistent with such a input "quality" adjustment are not very restrictive, most importantly that quality adjusting variables enter multiplicatively, i.e., a feature of such production functions as the Cobb-Douglas and the Translog.

Berndt and Wood (1984) have employed Lau's approach in evaluating the extent to which energy price shocks of 1973–74 and 1979–80 contributed to the revaluation of capital in U.S. manufacturing. In particular, they analyze the consequences for the value of aggregate capital when current operating costs, specifically energy costs, differ significantly from those expected at the time capital goods were purchased. They find that by 1981 the effect of the price shocks may have been to reduce the value of U.S. manufacturing capital by as much as 20%, providing an illuminating example of how to integrate behavioral/technical information in a factor demand model.

The relevance of the quality adjustment approach in integrating economic and engineering models is that it provides a means of directly incorporating technical characteristics into the model of economic behavior. This is important in its own right since often we would like to know, for example, how changes in the energy efficiency of a capital good will affect costs of production. Even more interesting is that the incorporation of such technical information directly into economic models will provide a natural linkage for the process network formulation considered by Hogan and Weyant. These techniques, then, will contribute to introducing more realistic descriptions of institutions, technologies, and markets into formal energy models.

The reader, especially the prospective energy modeler, should, therefore, not be dismayed by the extent of modeling activities of the past decade. There remains much to do in developing broader and deeper understanding of energy production, conversion, and use and the interactions of these system activities with the broader environment of economy and society. And these developments are likely to be rapid. While hopefully Vogely will not once again have to add an addendum to this edition, it is likely that by the 5th edition much progress will have been made along the lines we suggest. Certainly it remains an exciting and interesting time for energy system modelers and policy analysts.

References

1. Adams, F.G., and Griffin, J.M., 1972, "An Econometric Linear Programming Model of the U.S. Petroleum Industry," *Journal of the American Statistical Association,* Vol. 67, pp. 542–551.
2. Almon, C. et al., 1974, *1985: Interindustry Forecasts of the American Economy,* Lexington Books, Lexington, MA.
3. Anderson, D., 1972, "Models for Determining Least-Cost Investments in Electricity Supply," *Bell Journal of Economic Management Science,* Vol. 3, pp. 267–99.
4. Aperjis, D.G., 1981, *Oil Market in the 1980's, OPEC Oil Policy and Economic Development,* Ballinger, Cambridge, MA.
5. Ayres, R.U., 1969, *Technological Forecasting and Long-Range Planning,* McGraw-Hill, New York.
6. Barnett, H.J., 1950, *Energy Uses and Supplies, 1939, 1947, 1965,* Inf. Circ. No. 7582, U.S. Bureau of Mines, U.S. Dept. of Interior, Washington, DC.
7. Basile, P.S., 1979, "The IIASA Set of Energy Models," Working Paper, International Institute for Applied Systems Analysis, Laxenburg.
8. Baughman, M.L., and Joskow, P.L., 1974, "A Regionalized Electricity Model," Rep. No. MIT-EL 75-005, MIT Energy Lab, Cambridge, MA, December.
9. Beltzer, D., and Almon, C., 1972, "Forecasts of U.S. Petroleum Demand: An Interindustry Analysis," R.R. No. 4, Bureau of Business and Economic Research, University of Maryland.
10. Berndt, E.R., 1983, "Quality Adjustment, Hedonics, and Modern Empirical Demand Analysis, "*Price Level Measurement,* Proceedings from a Conference sponsored by Statistics Canada, W.E. Diewert and C. Montmarquette, eds., Minister of Supply and Services, Ottawa, Canada, pp. 817–863.
11. Berndt, E.R., and Wood, D.O., 1975, "Technology, Prices and the Derived Demand for Energy," *The Review of Economics and Statistics,* Vol. 1, pp. 259–68.
12. Berndt, E.R., and Wood, D.O., 1979, "Engineering and Econometric Interpretations of Energy-Capital Complementarity," *American Economic Review,* Vol. 69, pp. 342–354.
13. Berndt, E.R., and Wood, D.O., 1982, "The Specification and Measurement of Technical Change in U.S. Manufacturing," *Advances in the Economics of Energy and Resources,* Vol. 4, J.R. Moroney, ed., JAI Press Inc., Greenwich, CT, pp. 199–221.
14. Berndt, E.R., and Wood, D.O., 1984, "Energy Price Changes and the Induced Revaluation of Durable Capital in U.S. Manufacturing During the OPEC Decade," MIT Energy Laboratory Report (84-003), March.
15. Blitzer, C., Meeraus, A., and Stoutjesdijk, A., 1975, "A Dynamic Model of OPEC Trade and Production," *Journal of Development Economics,* Vol. 2, pp. 319–335.
16. Carson, J., Christian, W., and Ward, G., 1981, "The MIT World Oil Model," MIT Energy Laboratory Working Paper (WP81-027), MIT, Cambridge, MA.
17. Carter, A.P., and Brody,A., 1970, *Contributions to Input-Output Analysis,* North-Holland, Amsterdam.
18. Cazalet, E.J., 1975, *SRI-Gulf Energy Model: Overview of Methodology,* Standford Research Institute, Menlo Park, CA.
19. Chichilnisky, G., 1981, "Oil Prices, Industrial Prices and Outputs: A General Equilibrium Macro Analysis," Working Paper, United National Institute for Training and Research, New York.
20. Cremer, J., and Weitzman, M., 1976, "OPEC and the Monopoly Price of World Oil," *European Economic Review,* Vol. 8, pp. 155–64.
21. Crissey, B.L., 1975, "A Rational Framework for the Use of Computer Simulation Models in a Policy Context," Unpublished Ph.D. Dissertation, The Johns Hopkins University, Baltimore, MD.
22. Dantzig, G.B., 1963, *Linear Programming and Extensions,* Princeton University Press, Princeton, NJ. Dartmouth Systems Dynamics Group, 1977, "FOSSIL1: Introduction to the Model," Dartmouth College, Hanover, NH.
23. Day, R.H., 1973, "Recursive Programming Models: A Brief Introduction," *Studies in Economic Planning Over Space and Time,* G. Judge and T. Takayama, eds., North Holland, Amsterdam.
24. Day, R.H., and Nelson, J.P., 1973, "A Class of Dynamic Models for Describing and Projecting Industrial Development," *Journal of Econometrics,* Vol. 2, pp. 155–190.
25. Day, R.H., and Sparling, E., 1977, "Optimization Models in Agricultural and Resource Economics," *Quantitative Models in Agricultural Economics,* Vol. 2 in the AAEA Survey of Agricultural Economics Literature, G.G. Judge, ed., University of Minnesota

Press, Minneapolis, pp. 93–127.

26. Day, R.H., and Tabb, W.K., 1972, "A Dynamic Microeconomic Model of the U.S. Coal Mining Industry," SSRI Research Paper, University of Wisconsin.
27. Deam, R.J., et al., 1974, "World Energy Modeling," *Energy Modelling* (special issue of *Energy Policy*). IPC Sci. Technol. Press, Guildford, Surrey, UK.
28. DeGenring, P.C., and Jackson, J.R., 1980, "An Econometric-Engineering Analysis of the Impacts of the National Energy Act on U.S. Commercial Sector Energy Demand," *Energy Modeling Studies and Conservation,* ECE, ed., Pergamon Press for the United Nations, New York, pp. 303–318.
29. Deonigi, D., 1977, "A World Trade Model," *Proceedings of the Workshop on World Oil Supply-Demand Analysis,* Brookhaven National Laboratory, June.
30. DuPree, W.G. Jr., and West, J., 1972, *United States Energy Through the Year 2000,* Bureau of Mines, U.S. Dept. of Interior, Washington, DC.
31. Eckbo, P.L., Jacoby, H.D., and Smith, J.L., 1978, "Oil Supply Forecasting A Disaggregated Process Approach," *Bell Journal of Economics,* Vol. 9, pp. 218–238.
32. Energy Information Administration, 1979, *Documentation of the Project Independence Evaluation System,* Vols. I-IV, Department of Energy, Washington, DC.
33. Exxon Corporation, Public Affairs Department, 1977, "World Energy Outlook," New York, January.
34. Economic Commission for Europe (ECE), 1980, *Energy Modeling Studies and Conservation,* Pergamon Press for the United Nations, New York.
35. Federal Energy Administration, 1974, *Project Independence Report,* GPO, Washington, DC.
36. Forrester, J.W., 1961, *Industrial Dynamics,* MIT Press, Cambridge, MA.
37. Fritsch, B., Condon, R., and Saugy, B., 1977, "The Use of Input-Output Techniques in an Energy Related Model," *Input-Output Approaches in Global Modeling,* G. Bruckmann, ed., Pergamon Press, New York.
38. Fuss, M.A., 1980, "The Derived Demand for Energy in the Presence of Supply Constraints," *Energy Policy Modeling,* W.T. Ziemba, S.L. Schwartz and E. Koenigsberg, eds., Martinus Nijhoff Publishing, Boston, pp. 65–85.
39. Gass, S., 1977, "Evaluation of Complex Models," *Computer and Operations Research,* Vol. 4.
40. Gately, D., 1979, "The Prospects for OPEC Five Years After 1973/74 (A World Energy Model Survey)," *European Economic Review,* Vol. 2, pp. 369–379.
41. Gilbert, R., 1978, "Dominant Firm Pricing Policy in a Market for an Exhaustible Resource," *Bell Journal of Economics,* Vol. 9, pp. 385–95.
42. Granger, C.W.J., 1980, *Forecasting Time Series,* Holden Day, San Francisco.
43. Granger, C.W.J., and Newbold, P., 1977, *Forecasting Economic Time Series,* Academic Press, New York.
44. Greenberger, M., Brewer, G.D., Hogan, W.W., and Russell, M., 1983, *Caught Unawares: The Energy Decade in Retrospect,* Ballinger, Cambridge, MA.
45. Greenberger, M., Crenson, M.A., and Crissey, B.L., 1976, *Models in the Policy Process: Public Decision Making in the Computer Era,* Russell Sage Foundation, New York.
46. Griffin, J.M., 1974, "The Effects of Higher Prices on Electricity Consumption," *Bell Journal of Econ. Manage. Sci.,* Vol. 5, 515–639.
47. Griffin, J.M., 1979, *Energy Conservation in the OECD: 1980 to 2000,* Ballinger, Cambridge, MA.
48. Griffin, J.M., 1976, "Energy Input-Output Modeling Problems and Prospects," EPRI Report No. EA-298, Electric Power Research Institute, Palo Alto.
49. Griffin, J.M., and Gregory, P.R., 1976, "An Intercountry Translog Model of Energy Substitution Responses," *American Economic Review,* Vol. 66, pp. 845–57.
50. Groncki, P.J., and Marcuse, W., 1980, "The Brookhaven Integrated Energy/Economy Modeling System and Its Use in Conservation Policy Analysis," *Energy Modeling Studies and Conservation,* ECE, ed., Pergamon Press for the United Nations, New York, pp. 535–556.
51. Hannon, B. et al., 1983, "A Comparison of Energy Use," *Resources and Energy,* March.
52. Hartman, R.S., 1979, "Frontiers in Energy Demand Modeling," in J.M. Hollander, M.K. Simmons, and D.O. Wood (eds.), *Annual Review of Energy,* Vol. 4, pp. 433–466.
53. Henderson, J.M., 1958, *The Efficiency of the Coal Industry,* Harvard University Press, Cambridge, MA.
54. Herendeen, R.A., 1973, *The Energy Cost of Goods and Services,* Rep. No. ORNL-NSF-EP-58, Oak Ridge Nat. Lab., Oak Ridge, TN.
55. Herrara, A.O., and Scolnik, H.D., 1976, "Catastrophe or New Society: A Latin American World Model," International Development Research Center, Ottawa.
56. Hnyilicza, E., and Pindyck, R., 1976, "Pricing Policies for a Two-Part Exhaustible Resource Cartel: The Case of OPEC," *European Economic Review,* Vol. 8, pp. 136–154.
57. Hoffman, K.C., and Jorgenson, D.W., 1977, "Economic and Technological Models for Evaluation of Energy Policy," *Bell Journal of Economics,* Vol. 8.
58. Hoffman, K.C., and Wood, D.O., 1976, "Energy Systems Modeling and Forecasting," *Annual Review of Energy,* J.M. Hollander, ed., Annual Reviews, Inc., Palo Alto, pp. 423–453.
59. Hoffman, K., Groncki, P.J., and Graves, W.L., 1979, "Energy Policy Analysis Forum: World Oil Analysis," Report from a February 7 meeting, Brookhaven National Laboratory.
60. Hogan, W.H., and Weyant, J.P., 1980, "Combined Energy Models," *Advances in the Economics of Energy and Resources,* J.R. Moroney, ed., JAI Press, Greenwich, CT, pp. 117–150.
61. Holloway, M.L., 1980, *Texas National Energy Project: An Experiment in Large-Scale Model Transfer and Evaluation,* Vols. 1–3, Academic Press, New York.
62. Hudson, E.A., and Jorgenson, D.W., 1974, "U.S. Energy Policy and Economic Growth, 1975–2000," *Bell J. Econ. Manage. Sci.,* Vol. 5, pp. 461–514.
63. Huettner, D.A., 1976, "Net Energy Analysis: An Economic Assessment," *Science,* Vol. 192, pp. 101–104.
64. Hughes, B.B., and Mesarovic, M.D., 1978, "Analysis of the WAES Scenarios Using the World Integrated Model," *Energy Policy,* pp. 129–139.
65. Johnson, J., 1980, *Econometric Methods,* McGraw-Hill, New York.
66. Johnson, S.R., and Rausser, G.C., 1977, "Systems

Analysis and Simulation: A Survey of Applications in Agricultural and Resource Economics," *A Survey of Agricultural Economics Literature, Vol. 2,* Quantitative Methods in Agricultural Economics, G.G. Judge et al., eds., University of Minnesota Press, Minneapolis, pp. 157–304.

67. Jung Suh Suh, 1982, "An Investment Planning Model for the Refining and Petrochemical Industry in Korea," Ph.D. Thesis, University of Texas at Austin.
68. Kendrick, D., and Stoutjesdijk, A., 1978, *The Planning of Industrial Investment Programs: A Methodology,* Vol. I in the Series, The Planning of Investment Programs, A. Meeraus and A. Stoutjesdijk, eds., The World Bank, Washington, DC.
69. Kennedy, M., 1974, "An Economic Model of the World Oil Market," *Bell Journal of Economics and Management Science,* Vol. 5, pp. 540–577.
70. Kolk, D., 1983, "The Regional Employment Impact of Rapidly Escalating Energy Costs," *Energy Economics,* January.
71. Kolstad, C., 1982, "Noncompetitive Analysis of the World Coal Market," Working paper, Energy Laboratory, Los Alamos.
72. Kuenne, R.E., 1980, "Modeling the OPEC Cartel with Crippled Optimization Techniques," paper presented at the Conference on World and World Region Energy Studies, International Congress of Arts and Sciences, Harvard University, June 16.
73. Kuh, E., and Wood, D.O., 1979, "Independent Assessment of Energy Policy Models" (EA-1071), Electric Power Research Institute, Palo Alto, CA, May.
74. Kwang Ha Kang, 1981, "An Investment Programming Model of the Korean Electric Power Industry," Ph.D. Thesis, University of Texas at Austin.
75. Labys, W.C., 1973, "A Lauric Oil Exports Model Based on Capital Stock Supply Adjustment," *Malayan Economic Review,* 18, pp. 1–10.
76. Labys, W.C., 1978, "Commodity Markets and Models: The Range of Experience," *Stabilizing World Commodity Markets: Analysis, Practice and Policy,* F.G. Adams, ed., Heath Lexington Books, Lexington.
77. Labys, W.C., 1982, "Measuring the Validity and Performance of Energy Models," *Energy Economics,* Vol. 4, No. 3, July, pp. 159–168.
78. Labys, W.C., ed., 1975, *Quantitative Models of Commodity Markets,* Ballinger, Cambridge.
79. Labys, W.C., Paik, S., and Liebenthal, A.M., 1979, "An Econometric Simulation Model of the U.S. Steam Coal Market," *Energy Economics,* Vol. 1, pp. 19–26.
80. Labys, W.C., and Pollak, P.K., 1984, *Commodity Models for Forecasting and Policy Analysis,* Croom-Helm, London.
81. Labys, W.C., and Yang, C.W., 1980, "A Quadratic Programming Model of the Appalachian Steam Coal Market," *Energy Economics,* Vol. 2, pp. 86–95.
82. Langston, V.C., 1983, "An Investment Model for the U.S. Gulf Coast Petroleum Refining Complex," Ph.D. Thesis, University of Texas at Austin.
83. Lau, L.J., 1982, "The Measurement of Raw Material Inputs," ch. 6, *Explorations in Natural Resource Economics,* V.K. Smith and J.V. Krutilla, eds., John Hopkins Press for Resources for the Future, Inc., pp. 167–200.
84. Leontief, W.W., 1951, *The Structure of the American Economy 1919–1939,* Oxford University Press, New York.
85. Libbin, J.J., and Boehjle, X.X., 1977, "Programming Model of East-West Coal Shipments," *American Journal of Agricultural Economics,* Vol. 27.
86. Linneman, H., 1976, "MOIRA: A Model of International Relations in Agriculture—The Energy Sector," Working Paper, Institute for Economical Social Research, Free University, Amsterdam.
87. MacAvoy, P.W., and Pindyck, R.S., 1975, *The Economics of the Natural Gas Shortage 1960–1980,* North-Holland, Amsterdam.
88. Malinvaud, E., 1978, *Statistical Methods of Econometrics,* Revised Edition, Rand McNally, Chicago.
89. Miernyk, W. et al., 1978, *Regional Impacts of Rising Energy Prices,* Ballinger, Cambridge.
90. Morrison, W., and Smith, P., 1974, "Nonsurvey Input-Output Techniques at the Small Area Level," *Journal of Regional Science,* Vol. 14, No. 1.
91. Morrison, W.E., and Readling, C.L., 1968, *An Energy Model for the United States: Featuring Energy Balances for the Years 1947 to 1965 and Projections and Forecasts to the Years 1980 and 2000,* Inf. Cir. No. 8384, U.S. Bureau of Mines, U.S. Department of Interior, Wasington, DC.
92. Murphy, F.H., 1980, "The Structure and Solution of the Project Independence Evaluation System," Energy Information Administration, Department of Energy, Washington, DC.
93. Murphy, F.H., Sanders, R.C., Shaw, S.H., and Thrasher, R.L., 1980, "Modeling Natural Gas Regulatory Proposals Using the Project Independence Evaluation System," *Operations Research,* Vol. 29, pp. 876–902.
94. Nash, T.F., 1953, "Two-Person Cooperative Games," *Econometrica,* Vol. 21.
95. National Academy of Sciences (NAS), 1982, *Mineral Demand Modeling,* Committee on Nonfuel Mineral Demand Relationships, National Research Council, National Academy Press, Washington, DC.
96. National Petroleum Council, 1974, *Emergency Preparedness for Interruption of Petroleum Imports into the United States,* National Petroleum Council, Washington, DC.
97. Nelson, C.R., 1973, *Applied Time Series Analysis for Managerial Forecasting,* Holden Day, San Francisco.
98. Organization for Economic Cooperation and Development, 1977, *World Energy Outlook,* OECD Secretariat, Paris.
99. Park, Se-Hark, 1982, "An Input-Output Framework for Analyzing Energy Consumption," *Energy Economics,* Vol. 4, pp. 105–110.
100. Pindyck, R.S., 1978, "Gains to Producers from the Cartelization of Exhaustible Resources," *Review of Economics and Statistics,* Vol. 60, pp. 238–51.
101. Pindyck, R.S., 1979, *The Structure of World Energy Demand,* The MIT Press, Cambridge and London.
102. Pindyck, Robert, and D. Rubinfeld, 1981, *Econometric Models and Economic Forecasts,* Second Edition, McGraw-Hill Publishing Company, New York.
103. Preston, R., 1975, "The Wharton Long-Term Model: Input-Output Within the Context of a Macro Forecasting Model," *International Economic Review,* pp. 3–19.
104. Propoi, A., and Zimin, I., 1981, "Dynamic Linear Programming Models of Energy, Resource and Economic Development Systems," RR-81-14, Interna-

tional Institute for Applied Systems Analysis, Laxenburg.

105. Reardon, W.A., 1972, *An Input/Output Analysis of Energy Use Changes from 1947 to 1958 and 1958 to 1963*, Rep. submitted to Office of Science and Technology, Executive Office of the President, by Battelle Mem. Inst.

106. Rose, Adam, 1983, "Technological Change and Input-Output Analysis: An Appraisal," West Virginia University College of Mineral and Energy Resources (mimeo), p. 41.

107. Rose, A., et al., 1978, "The Economics of Geothermal Energy Development at the Regional Level," *Journal of Energy and Development*, Vol. 4, No. 1, 1978.

108. Rose, A., and Kolk, D., 1983a, "A Policy Simulation Model for Natural Gas Utilities," *Modelling and Simulation*, Vol. 14, part 1.

109. Rose, A., and Kolk, D., 1983b, *Forecasting Natural Gas Demand in a Changing World*, Mineral Resource Economics Monograph, No. 15, West Virginia University, Morgantown.

110. Rose, A., et al., 1982, "Modern Energy Region Development and Income Distribution," *Journal of Environmental Economics and Management*, Vol. 9, No. 2.

111. Salant, S., Sanghi, A., and Wagner, M., 1979, "Imperfect Competition in the International Energy Market: A Computerized Nash-Cournot Model," Report submitted by I.C.F., Inc. to the U.S. Department of Energy, May.

112. Salant, S., Miercort, F., Sanghvi, A., and Wagner, M., 1981, *Imperfect Competition in the International Energy Market*, D.C. Heath, Lexington, MA.

113. Schinzinger, R., 1974, "Integer Programming Solutions to Problems in Electric Energy Systems," *Energy Policy Evaluation*, D.R. Limaye, ed., Heath Lexington Books, Lexington, MA.

114. Sweeney, J., 1975, *Passenger Car Use of Gasoline: An Analysis of Policy Options*, FEA, Washington, DC.

115. Takayama, T., and Judge, G., 1971, *Spatial and Temporal Price and Allocation Models*, North-Holland, Amsterdam.

116. Taylor, L.D., 1975, "The Demand for Electricity: A Survey," *Bell J. Econ. Mange. Sci.* 6(1), pp. 74–110.

117. Theil, H., 1971, *Principles of Econometrics*, Wiley, New York.

118. Ulph, A., and Folie, M., 1978, "Gains and Losses to Producers for Cartelization of an Exhaustible Resource," C.R.E.S. Working paper R/WP26, Australian National University, Canberra.

119. Ulph, A., and Folie, M., 1977–78, "Role of Energy Modeling in Policy Formulation," *Energy Systems and Policy*, Vol. 2, pp. 311–340.

120. United States Central Intelligence Agency, 1977, "The International Energy Situation Outlook to 1985." McLean, VA, April.

121. US Department of Commerce, 1974, *Input-Output Structure of the U.S. Economy, 1967, Volume 3: Total Requirements for Detailed Industries*, GPO, Washington DC.

122. Verleger, P.K., 1982, *Oil Markets in Turmoil*, Ballinger, Cambridge.

123. Volgely, W.A., 1976, *Economics of the Mineral Industries*, American Institute of Mining, Metallurgical, and Petroleum Engineers, Inc., New York.

124. von Neumann, J., and Morgenstern, O., 1944, *Theory of Games and Economic Behavior*, Princeton University Press, Princeton, NJ.

125. Waelbroeck, J.L., ed., 1976, *The Models of Project LINK*, North-Holland, Amsterdam.

126. Wagner, H.M., 1969, *Principles of Operations Research*, Prentice-Hall, Englewood Cliffs, NJ.

127. Webb, G., and Pearce, D.C., 1975, "The Economics of Energy Analysis," *Energy Policy*, Vol. 3, pp. 318–331.

128. Wood, D.O., and Spierer, C., 1984, "Modeling Swiss Industry Interfuel Substitution in the Presence of Natural Gas Supply Constraints," M.I.T. Energy Laboratory Working Paper (WP84-011), Cambridge, MA.

129. Workshop on Alternative Energy Strategies, 1977, *Energy: Global Prospects 1985–2000*, McGraw-Hill, New York.

3.9

Mineral Models

Walter C. Labys,* Frank R. Field,† and Joel Clark‡

MINERAL MODELING

Mineral models provide a systematic and comprehensive approach for analyzing and forecasting the behavior of mineral markets and industries.[1] They also permit the analysis of a wide range of policy decisions. Obviously such models cannot capture all of the intricate relationships that shape the behavior of a mineral market. However, such models have become extremely useful and in many cases support the decision making approach of judgmental analysts. They tend to narrow the range of likely outcomes and thus reduce the risks and uncertainties of investing, trading, and speculating in minerals. Mineral models provide not only answers to "what-if" questions, they also inform analysts of the benefits and costs associated with mineral investment decisions. More recently they have been applied to specific mineral issues of the possible impact of buffer stock stabilization schemes and of import supply restrictions.

Mineral Models Defined

Because mineral models are so diverse, we must begin with a fairly broad definition. A mineral model is a formal representation of a mineral market, industry or system where the behavioral relationships included reflect the underlying economic and engineering factors as well as the political and social institutions. Often, but not always, a mineral model will consist of some combination of the following components: demand, supply, inventories and prices.

Modeling of demand behavior normally begins with an interpretation of basic demand responses to prices and income (i.e., the associated price and income elasticities). Price elasticities vary among different minerals and among their diverse uses. They tend to be lower for minerals without close substitutes than for those where such substitutes exist. In addition, these elasticities vary with time as well as with the actual excess demand or excess supply. Variations like these, coupled with the relative instability of prices for a number of minerals, make explanations of demand behavior difficult.

The degree to which demand responds to changes in income depends on the income elasticity of the end products to which the minerals serve as inputs. Those products with a relatively low income elasticity tend to experience proportionally smaller swings in sales than those with higher elasticities. The transmission of demand cycles to minerals producers comes through a series of stage-of-process effects.

Demand measurement also is influenced considerably by mineral substitution patterns. The cross-price elasticity of a competitive or complementary commodity also varies with minerals, markets, and time. Where minerals easily substitute for one another, as in the case of copper and aluminum in electrical conductor applications, cross-price elasticities are relatively high.

Modeling of mineral supply behavior normally involves consideration of resource development, economic conditions, geologic

*Professor of Mineral and Energy Resource Economics, West Virginia University.
†Research Assistant, Materials System Laboratory, Massachusetts Institute of Technology.
‡Associate Professor of Materials Systems Massachusetts Institute of Technology.

conditions, technology, and investment decisions. Concerning resource development, the production of a mineral commodity embodies the activities involved in finding it, in determining the size and physical characteristics of deposits, and in developing the deposits to the point where mining and processing can begin. The first two of these activities are generally-referred to as exploration. Exploration in an elementary sense can be considered as an input similar to other inputs essential to mining and processing. A modeler may attempt to optimize allocations among these inputs, making appropriate adjustments at the margin, where changes among inputs are possible. A major factor in determining the profitability of exploration is the fact that the location, size, and grade of mineral deposits are rarely known accurately and forms of uncertainty are involved.

Economic conditions relate to the process of producing a single or joint output through acquiring and transforming several inputs. Most obvious of the identifiable variables in this relationship are the market price of the product and the prices of the inputs utilized. The latter prices are normally associated with the variable costs of production, i.e., those that vary with the rate of production. In addition, it may be necessary to include the fixed costs, which are independent of the rate of output in the short run. Here we have primarily an investment in fixed or quasi-fixed factors such as rental of leased property or development of a new mine facility. Depreciation and depletion costs associated with owning and maintaining a fixed productive facility such as a mine also enter the picture.

The incorporation of these factors into a mineral-supply model involves consideration of important lags: (1) an implementation lag, which is the time lag between a change in price and the reaction by decision makers; (2) a technological or developmental lag, which is the time required to place new techniques or new mining capacity into full production, and (3) an exploration lag, which is the time between the decision to explore for new deposits and the utilization of the deposits in production. Geological conditions refer to modes of mineral occurrence or variations in quality. As a reserve becomes depleted, costs of production normally rise. When current and outdated prices exceed current expected extraction costs, this signals future capital gains. This normally leads to increased exploration, although it also can instigate a slowdown in the extraction rate or the development of substitutes for the scarce resource.

Modeling inventory behavior is more important in the short run, as well as when inventories are greater or smaller than their average levels. The modeling of inventory behavior, particularly in international markets, is made difficult by both the outright lack of data and the disaggregation of data as categorized by the types of stockholding. Inventories can exist in the form of working stocks, buffer stocks, or strategic stockpiles. They can be held by producers, consumers, or intermediaries, each of whom may hold them for transactions, precautionary or speculative purposes.

Modeling price behavior is difficult because of the problem of sometimes ascertaining which of several prices actually represent the market equilibrating process. Although the spot price on the market with the largest volume of trading often serves this function, other prices also need to be taken into account, i.e., contract prices, forward prices, futures prices, administered prices, arm's-length prices, cost-plus prices, and occasionally government-regulated prices. Prices are indicators not only of market development but also of complex adjustments taking place in the international system, such as inflation, exchange rate alignments, speculation, and military, natural or political crises.

An important aspect of modeling price behavior is determining the nature of market structure and the related adjustment mechanisms that move a market toward equilibrium. Price may not be the result of purely competitive adjustments. Consequently, market structures must be investigated to see if they reflect oligopolistic conditions. These appear to be more frequent when only one or a few large suppliers effectively control the market, as in the case of cobalt or diamonds. Labys (1980a) reports that many mineral markets differ widely in their competitive character, and only by assessing this aspect of market behavior can we move toward constructing more realistic models.

Modeling Purposes

There is no such thing as an "all-purpose" mineral model. Modeling has been shown to involve exploring industry structure, analyzing different forms of policy, and forecasting market behavior. Each modeling application presents the mineral analyst with different problems, and the mineral analyst must pursue different objectives in the attempt to solve them. While there is a risk of oversimplification in indicating that modeling has several specific purposes, such an enumeration does offer a perspective as to why mineral models are constructed. The following purposes have been derived from recent mineral modeling efforts:

(1) market stabilization analysis involves finding those control mechanisms or forms of market organization which lead to more stable demand, supply, price, or equilibrium situations.

(2) Market planning analysis involves the forecasting of long-run mineral demand or supply, depending on the policy problems, product technologies, or end-use strategies foreseen. It also includes analysis of mineral material flows between intermediate sectors of the industry.

(3) Supply restriction analysis is enhanced by an accompanying probabilistic theory describing the possibilities of sudden disruptions used to assess future demand constraints.

(4) Industrial process, or engineering, analysis describes the relationships between GNP, end products, technical transformation processes within industries, and minerals input demand requirements.

(5) Spatial flow analysis requires the application of spatial economic theory to a mineral market so that the relationships between demand, supply, transportation costs, and industry location can be determined.

(6) Interindustry analysis requires the utilization of the input-output framework to determine the relationship between growth in final demands and the consumption of minerals that will follow from intermediate mineral supply, utilization and conversion.

Modeling Procedure

The procedure normally followed in mineral modeling is to construct a model whose applications will meet the intended purposes. This procedure can be expected to vary slightly from model to model, as will be seen later in this chapter. However, the procedure generally does involve a number of specific steps: modeling purpose, adopting the correct methodology, model specification, model estimation, and model validation. While each of these steps is pertinent to most of the modeling procedure, it should be remembered that model application is by far the most important.

After the model builder has determined the model's purpose, he must decide which modeling methodology is most suitable for achieving that purpose. Given a particular methodology, the model builder must then decide how to specify the model. This normally involves selecting a number of behavioral relationships within the industry of interest, according to some given theory. A number of important considerations must be taken into account in constructing the actual model, such as:

—the primary, intermediate, or final levels of supply or demand
—temporal or spatial configuration
—simultaneous or recursive casuality
—short, medium, or long time span
—single commodities or multi-commodity orientation
—competitiveness of the market structure.

Most often these considerations dictate the nature of the modeling methodology to be adapted.

Choice of Methodology

The methodologies applied to modeling mineral markets and industries are quite diverse. Alternative methodologies concentrate on explaining different aspects of supply and demand patterns, policy, and forecasting. A model depends on the economic and engineering behavior of interest as well as on the underlying purpose of the model.

In this chapter we have organized the methodologies employed in mineral modeling in three groups: econometric models, engineering models, and input-output models. Although there is overlap and commonalities among these groups, there are also distinctions among them which justify this classification. For example, econometric models relate more to classical mineral demand, supply, and price adjustment, while input-output models refer more to the primary, intermediate, and final utilization of minerals throughout the economy. Engineering models, on the other hand, are more concerned with the transformation of specific mineral inputs into manufactured goods. We now turn to a detailed discussion of each of these methodologies.

ECONOMETRIC MODELS

The most basic type of microeconometric structure from which mineral models have developed is the market model. The earliest forms of commodity models were developed by agricultural economists who recognized the relationship between the theory of microeconomic equilibrium involving market demand, supply and price adjustments and the then new methodology of econometrics which concentrated on the identification and estimation of demand and supply equations as well as their simultaneous interactions. The technology of constructing econometric models of agricultural markets as it has evolved over time has thus provided the methodological background from which econometric mineral market models have evolved.

From the start, however, it was obvious that special requirements exist in modeling mineral markets which required additional attention, e.g., see Labys (1977) and Vogely (1975). Consider the basic or endogenous variables which determine the nature of mineral market behavior: demand, supply reserves, inventories, and prices. Mineral demand tends to be derived and thus requires focusing on the technology of the products using minerals. This requires using derived demand rather than primary demand theory. Production also requires differentiation between primary production based on ore conversion and secondary production contingent on scrap conversion adjustments. Reserves must also be accounted for, since a feedback relationship exists between reserve formation and reserve depletion (demand). Inventories are also more complex to explain since they represent disequilibrium market adjustments rather than seasonal carryover. And prices are often determined by long-term contracts rather than by spot markets.

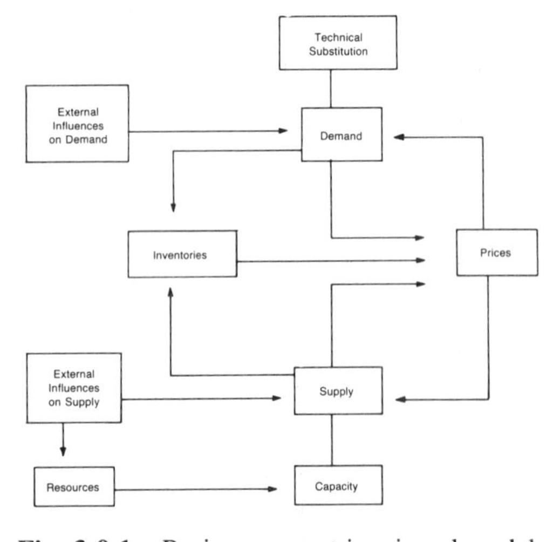

Fig. 3.9.1—Basic econometric mineral model.

The construction of econometric mineral models requires that specific theories be employed to explain the complex behavioral relationships stated above. Each of the above endogenous variables must be related to a set of explanatory variables, some of which also are endogenous while others are exogenous or independent. This form of equation formulation involves regression analysis in which parameters for each of the variables in the equation are estimated and the errors resulting from the inability to explain the endogenous variables completely are assessed. This stage of construction also requires that these equations be interrelated in a market or model structure that can be considered simultaneous or recursive.

The fact that mineral models are likely to be recursive or dynamic can be understood from Fig. 3.9.1 which represents the structure of a typical mineral market model. The technical and geological factors underlying mineral project development can be considered an external influence on mineral supply. However, the time period between the initial investment and exploration, discovery and actual mining requires that external influences be represented by a time lag. Causality in the model is also influenced by feedback, not only by the impact of depletion or demand on reserve levels but also by the simultaneous relations between prices and quantities. As Fig. 3.9.1 shows, demand and supply clearly influence price, but the effect of prices, in turn, on both supply and demand must also be included.

Other considerations also enter the construction of a mineral market model. Model specification requires that the market structure be carefully identified as being competitive or embodying varying degrees of noncompetitive behavior. The degree of aggregation of end uses and mineral product categories is important. Substitution and technical change are also important regarding demand. Reserve dynamics, rates of exhaustion and degrees of processing and refining also are significant concerning supply. Model estimation demands attention to data problems as well as to lag structures and feedback. Model simulation also requires using appropriate computer algorithms as well as attention to assumptions and exogenous variable projections in generating forecasts or mineral policy scenarios.

The general uses of econometric minerals models vary. Governments have tended to use them for assistance in mineral policy analysis and planning. For example, price forecasts can be used as an aid to investment planning. Models have also been employed to analyze market stabilization mostly in the form of buffer stocks or national stockpiles. This became most evident during the recent attempts of UNCTAD to establish an Integrated Program for Commodities with special stabilization directives for copper and tin. Models have also been constructed to evaluate alternative federal policies on possible restrictions initiated by producer countries on supplies of aluminum, bauxite, chromium, cobalt, copper, manganese, and platinum/palladium. Econometric mineral models have also been linked with econometric macro models to determine the impact of mineral investment, production and trade on economic growth.

Private industry has focused on two used in particular. First, market models have been used to forecast short-term demand and price conditions for such purposes an annual revenue forecasting, budgeting and financial planning, inventory management, short-run production decisions, and annual contract negotiations. Second, models have been employed in evaluating new mineral products to determine the timing and magnitude of an investment and to provide a price scenario. The growing importance of project financing, as distinct from recourse financing, in new mining ventures has been a particular stimulant to such efforts. As Crowson (1981) points out, however, mining companies rarely use pure econometric models, tending instead to favor hybrid forecasting models. Nevertheless, econometric equations typically play a very important role in such models.

Theoretical Framework

Competitive Markets: Focusing on the price adjustment between demand and supply, which serves to clear the market, a competitive market model comprises of four relations. Typically more complex structures are employed, as described in Labys (1973) and Labys and Pollak (1984).

$$D_t = d(P_{t-s}, P^c_{t-\theta}, A_t, T_t) \quad (1)$$

$$S_t = q(P_{t-s}, G_t, Z_t) \quad (2)$$

$$P_t = p(I_t/D_t) \quad (3)$$

$$I_t = I_{t-1} + S_t - D_t \quad (4)$$

Definitions of variables

D = Mineral demand
S = Mineral supply
P = Mineral prices
P_{t-s} = Prices with lag distribution
$P^c_{t-\theta}$ = Prices of substitute minerals with lag distribution
I = Mineral inventories
A = Income or industrial activity
G = Geological influences
Z = Policy variable influencing supply
T = Time or secular

Mineral demand is explained as being dependent on current and lagged prices as well as on external factors, such as the prices of one or more substitute commodities, and general economic activity. The subscript "t" implies that the behavior of these variable is normally observed over time, e.g., years, quarters, months. Mineral supply depends on lagged prices as well as on external factors, including geological factors such as reserve seam thickness and ore quality, and policy variables. Lagged prices are used since the supply process normally is described based on the past behavior of prices, reflecting the long lead times typically required for exploration and mine development.

Mineral prices are explained by demand and inventories, although this equation is sometimes inverted to explain demand. The model is closed using an identity which equates inventories with lagged inventories plus supply minus demand. The simultaneous solution of such a model leads to what can be considered "market equilibrium" values for the dependent variables. While such a model typically describes a world mineral market or a market without trade, variables and equations describing mineral imports or exports can easily be added.

Several attempts have been made to add considerably to the sophistication of the competitive mineral market model. Labys (1980) considered dynamic reserve and inventory adjustments in a disequilibrium adjustment framework; CRA (1980) extended the supply sector to include long run geostatistical factors in reserve formation and depletion; Ogawa (1982) employed rational expectations to better explain short run price and inventory adjustments; and Ghosh (1981) et al. introduced optimal control to improve the analysis of market stabilization policies.

Monopolist Markets: Not all mineral markets possess a competitive structure leading to equilibrium price adjustments. Mineral markets can be dominated by one or a few buyers and sellers, and market models will then embody some degree of monopolistic or oligopolistic behavior. At one extreme is the pure monopoly model, which may be appropriate for markets dominated by a single seller. In these markets the sellers are not price takers; they set prices to maximize their net revenues, given their costs and the nature of demand facing them. The principal transformation that must come about in modeling such markets is among other things to introduce a price equation which in effect describes the monopolist's price-quantity locus.

The change in specification required in the competitive model to reflect this behavior involves principally the replacing of the supply relationship with an arbitrary cost function.

$$D_t = d(P_t, A_t) \qquad (5)$$

$$C_t = c(S_t), MC = c'(S_t) > 0 \qquad (6)$$

$$D_t = S_t \qquad (7)$$

There is an assumed equivalence between demand and supply, with total costs C increasing as supply S increases. Since the monopolist sets prices to maximize profits, the model framework must also contain a profit relationship:

$$\pi_t = (P_t \cdot S_t) - c(S)_t \qquad (8)$$

Since the cost relation cannot always be observed, the model is often solved so that the maximum value of price is expressed as a function of quantity

$$P_t = MC - S_t/\alpha \qquad (9)$$

where α is the price elasticity of demand. This equation can then replace the market clearing equation in the competitive model.

Intermediate Markets: This view of mineral markets as behaving somewhere between pure monopoly and pure competition has been developed further. In particular Pindyck (1978) has considered the likely impact of producer cartels on mineral market performance. Such a model begins with a single dominant producer or a small group of like-minded dominant producers, and another, and more diverse, group of small "fringe" or "second-tier" producers. Assume the producer to be a cartel or a less than perfect monopolist. Net demand facing the cartel is

$$X_t = D_t - S_t \qquad (10)$$

where D is total market demand.

$$D_t = d(P_{t-s}, P^c_{t-\theta}, A_t) \qquad (11)$$

and S is the supply from the competitive fringe.

$$S_t = s(P_{t-s}, G_t) \qquad (12)$$

An important consideration in mineral markets is resource depletion which would affect the fringe as well as the cartel. Define cumulative production CS for the fringe to be

$$CS_t = CS_{t-1} + S_t \qquad (13)$$

Cartel reserves R are given by

$$R_t = R_{t-1} + D_t \qquad (14)$$

The objective of the cartel is to pick a price trajectory P_t that will maximize the same of discounted profits

$$\text{Max } W = \sum_{t=1}^{N} (1/(1 + \delta)^t)\, [P_t - m/R_t]D_t \qquad (15)$$

where m/R is average (and marginal) production costs (so that the parameter m determines initial average costs), δ is the discount rate, and N is chosen to be large enough to approximate the infinite-horizon problem. Note that average costs become infinite as the reserve base R approaches zero, so that the resource exhaustion constraint need not be introduced explicitly. The resulting model framework is that of a classical, unconstrained discrete-time optimal control problem, where numerical solutions can be easily obtained.

Mineral Applications

The development of econometric models suitable for analyzing mineral markets possessing competitive behavior has been numerous. As described by Gupta (1982, pp. 91–98), one of the first was the Desai (1966) tin model which explained tin price fluctuations on a world basis. The model was disaggregated on the demand side into three regions: the United States, OEEC and Canada, and the rest of the world. The total demand for tin in the former two regions was further disaggregated according to two end-use categories—tinplate and non-tinplate—to capture more accurately the influence of the relevant activity variables and technological changes in the end-uses. The immediately relevant variables relating to the use of tin for tinplate and non-tinplate were linked with larger macro-variables, such as GNP and industrial production. Price variables did not contribute to the explanatory power of either the supply or demand functions, and hence were excluded.

The tin model was used to study how various conditions within the tin market could be improved to stabilize the revenues of developing tin exporting countries. Policy analysis employing stochastic simulations were utilized in this respect. Other simulations carried out were aimed at investigating the possibility of reducing fluctuations in price and revenue received by tin producers, through the operation of an international buffer stock and the restriction of output by the International Tin Council.

The basic model structure as well as the model purpose was later expanded by Smith and Shenk (1976) who compared buffer stock operations by the International Tin Council with strategic stockpiles releases by the U. S. General Services Administration. The U. S. government also sought to determine the likely impact of a possible embargo of shipments of minerals to this country. Under its auspices, Charles River Associates (1976) constructed market models for aluminum/bauxite, chromium, cobalt, copper, manganese, and platinum/palladium to evaluate alternative federal policies on possible restrictions initiated by producer countries on supplies of these minerals.

The copper market has also been subject to several modeling efforts. Most noteably, Fisher, Cootner and Bailey (1972) built a world copper model which was recognized as one of the first major econometric mineral modeling efforts. Their model divided the world copper market into the United States, where prices are administered by U. S. producers, and the rest of the world, where prices are determined by free market forces of demand and supply at the London Metal Exchange (LME). Since the LME price is a free-market price, it also plays a role in determining the U. S. producer price in the long run, as well as providing a link between the two markets. Interregional trade between the United States and the rest of the free market world, that depends on the differential between the two market prices, provides a further link between the two markets. The model was relatively disaggregated by incorporating different supply equations for the major copper-producing areas (United States, Chile, Canada, Zambia, and the rest of the world). The demand equations were, however, not disaggregated to end-use categories. Neither were resources, capacity, technological variables, and prices of coproducts included in the model.

The principle application of the model was in answering several then important policy questions: (1) How would possible cartel behavior on the part of major copper exporting countries affect the market; (2) What would be the effect on Chilean revenues and world prices of increases in Chilean exports; and (3) How would major new copper supplies affect the LME price?

A major concern is utilizing econometric models to explain the wide and frequent price and quantity fluctuations in mineral markets has been the disequilibrium compared to the equilibrium characteristic of these markets. This can be most strongly seen in the role that stocks either in the form of capital assets (including capacity and reserves) or inventories play in mineral markets. CRA (1978) thus extended the supply sector of the Fisher, Cootner and Bailey (1972) copper model to include long run adjustments in exploration and discovery as well as subsequent mining capacity formation. This long run adjustment process was combined with a short-run inventory adjustment process in a distinctly disequilibrium form of copper model by Labys (1982b). Such an approach to modeling the copper market was suggested by Richard (1978) with his continuous time-differential equation approach to econometric copper modeling.

An example of an econometric model that is based on the copper market and considers disequilibrium adjustments can be drawn from a study by Labys and Kaboudan (1980). This model based on quarterly data recognizes the role that copper inventory behavior plays in stock adjustment as a particular vehicle that explains short run disequilibrium adjustments around the flow adjustment or equilibrium time path. In addition disequilibrium is considered as originating from expectations which can't be realized in a single period, time gaps which establish this delayed response, spillover effects extending from the demand for final goods in which primary and fabricated copper products are embodied, and the inability of copper capacity to be adjusted in periods as short as quarters or months.

The model consists of a domestic sector for the United States and the rest of the world sector. The sample period selected begins the first quarter of 1966 and ends the last quarter of 1977. Because quarterly data are not available for every producing and consuming country, and because copper is traded on the U. S. market at the domestic producer price and on world markets at the London Metal Exchange (LME) price, the choice of two sectors is a logical one. The demand sector for the United States is divided into primary and secondary copper demand as well as copper exports. Rest of world demand is that for combined or refined demand. The supply sector of the United States includes both primary and secondary supply, while rest of the world supply includes only the total of both of these supplies. Disequilibrium refined price adjustments center about the U. S. producers price and corresponding refined inventory adjustments. Copper scrap prices are also determined within the model, based on LME refined price adjustments. Inventories, a crucial variable in market adjustments, include stocks held by producers, by consumers, by the GSA stockpile, and by the LME in European warehouses. The model is closed with appropriate trade balance and inventory equations and identities. All equations are estimated by required distributed lag methods and two-stage least-squares where appropriate.

The quality of a model of this type has been confirmed not only by validation in the form of conventional statistical tests but also by means of the mean average percent error (MAPE) measured over the given sample period. For example, simulation results for U. S. primary demand, primary supply and prices have yielded a MAPE of 7.2%, 12.1% and 7.5% respectively. The quite reasonable accuracy of these simulations is reflected in Fig. 3.9.2, 3.9.3, and 3.9.4 which compare the actual and estimated values of the variables over the sample and the forecast periods. Regarding the latter, the model

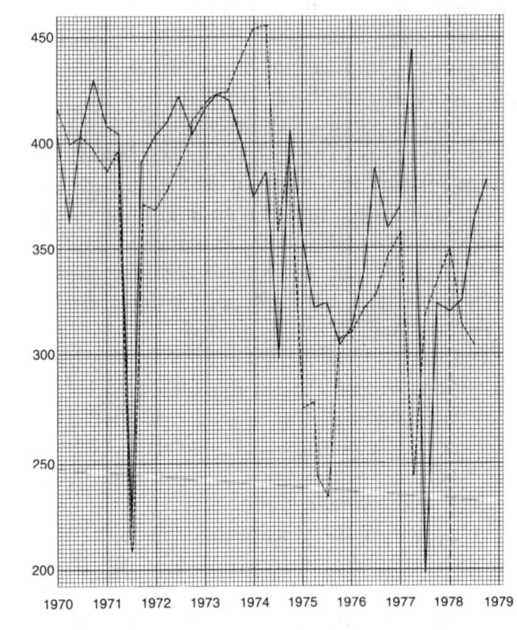

Fig. 3.9.3—Estimated and forecast U.S. primary copper supply.

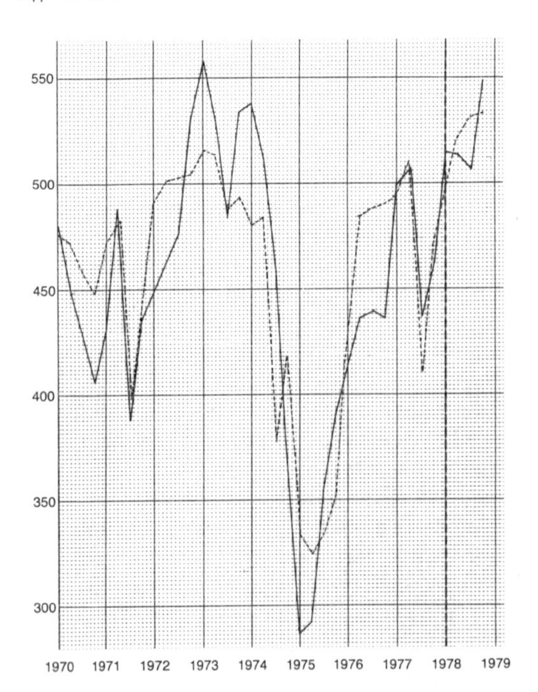

Fig. 3.9.2—Estimated and forecast U.S. refined copper demand.

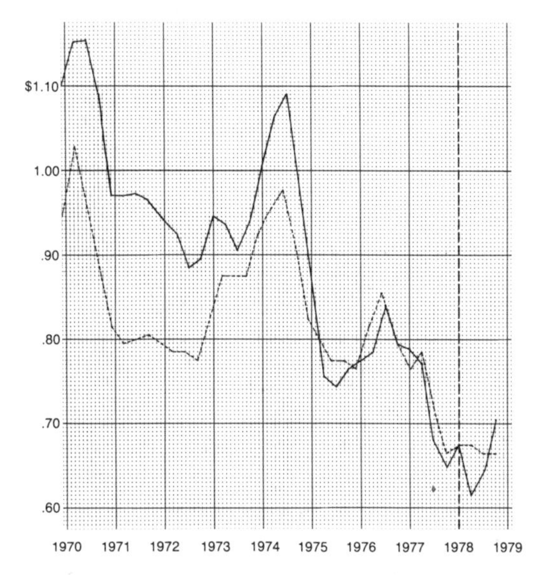

Fig. 3.9.4—Estimated and forecast U.S. producer copper prices.

has been employed to produce forecasts four quarters into the future for 1978. Availability of actual data for comparison purposes provides a basis for confirming the forecast accuracy of the model. As shown in the cited figures, the computer MAPE is 11.5%, 10.4% and 8.2% for demand, supply and prices, respectively.

The development of mineral econometric models also has been subject to the influence of the renewed interest in rational expectations. This form of expectations equivalent as compared to adaptive or extrapolative expectations has been seen as useful for lessening the costs of forecasting and of market stabilization. This prospect has been investigated theoretically by Ogawa (1983) and by Newberry and Stiglitz (1981). Ogawa (1982) also advanced a basic model of the international copper market to include rational expectations in price formation influencing the quantity variables. This same adaption was used in a series of mineral models constructed by Smithson et al. (1979) at the Canadian Ministry of Natural Resources. Econometric models were constructed for aluminum, copper, nickel, and zinc. These models were used for mineral policy simulation analysis as well as for providing demand, supply, and price projections to 1988.

Mineral econometric models have also assisted in the operation of macroeconometric models which have been used to measure economic growth and to plan in developing countries. This was a rational consequence of the goals of recent years to utilize mineral exports as a means of accelerating economic growth in resource-rich developing countries. Again the copper market was of interest in the linking of copper model to macroeconometric models for Chile (Lasaga, 1981) and for Zambia (Obidegwu and Nziramasanga, 1981).

Among the development of econometric models relating to monopolist market configurations, one of the few is that of Burrows (1971) who modeled cobalt producers as acting as an effective cartel. Unlike copper, tin, and many other mineral commodities, the production of cobalt is highly concentrated. One company, Union Miniere Haut Katanga (UMHK) produces more than 60% of the world output, the rest being produced by various companies in Canada (8%) and many other countries. Such a concentration on the supply side rightly warrants allowance for market imperfections in model specification. The general structure of the model is derived by treating UMHK as a price setter following profit-maximization principles (given the supply response of all other producers).

Profit maximization, given the world demand for cobalt and the supply response of the other producers at the prices set by UMHK, yields the price-determining equation for cobalt. Although the consumption structure of cobalt is fairly detailed according to end-uses in the United States, the model lacks determination of the rest of the world's (ROW) cobalt consumption and UMHK production behavior, which were later included in Adams (1972). The U. S. government's General Services Administration stockpiles (GSA) are explicitly introduced in the price equation, these being looked upon as potential sources of supply by the producers of cobalt.

Regarding intermediate market configurations, Pindyck (1978) has applied his model to determine optimal price and quantity paths that would result from cartel behavior on the part of producers organizations in the copper and bauxite markets. Other attempts to model intermediate market structures have not emphasized optimization but rather various forms of producer pricing systems. In an intermediate mineral market, producers price is often modeled simply as a constant markup over costs. A certain amount of material in such markets will be sold at open market transactions prices. In times of very "weak" markets (low demand at the list price relative to capacity), discounts will be large, and much of the material will move at discount prices. In contrast, when shortages occur (high demand at the list price relative to capacity), producers are apt to allocate sales at the list price. The prices of material sold by consumers and fringe producers on the open market will then include a substantial premium.

In modeling intermediate markets closer to monopoly, we have shown how the market equilibrium equations can be replaced with a list price or administered price equation in which price is modeled as a function of cost. (This procedure, of course, requires that accurate data be collected on the production costs of the dominant producer or producers.) Depending on the modeler's perception of the relative strengths of the monopoly group of producers and the fringe producers, prices may be modeled in one of two ways. If the monopoly group is relatively strong, the list price can be a function of cost and transactions price; this depends on a measure of excess capacity. If the monopoly group is relatively weak, the transactions price can be a function of demand and inventory, and the list price a function of the transactions price and a measure of cost pressures. Examples of such models include those constructed by CRA for nickel (1974), molybdenum (Burrows, 1974) and aluminum (Burrows, 1972) and by Gupa (1982) for zinc.

It is important to realize that the character of intermediate markets can change over time for both cyclical reasons and for longer-term structural reasons, such as the entry of new producers. Thus, mineral markets can be portrayed as a spectrum ranging from perfect competition to pure monopoly, with the markets for individual metals constantly shifting back and forth along the spectrum. There are thus a wide variety of specifications possible for econometric mineral models. A further discussion of these and other aspects of econometric models can be found in the chapter on energy modeling.

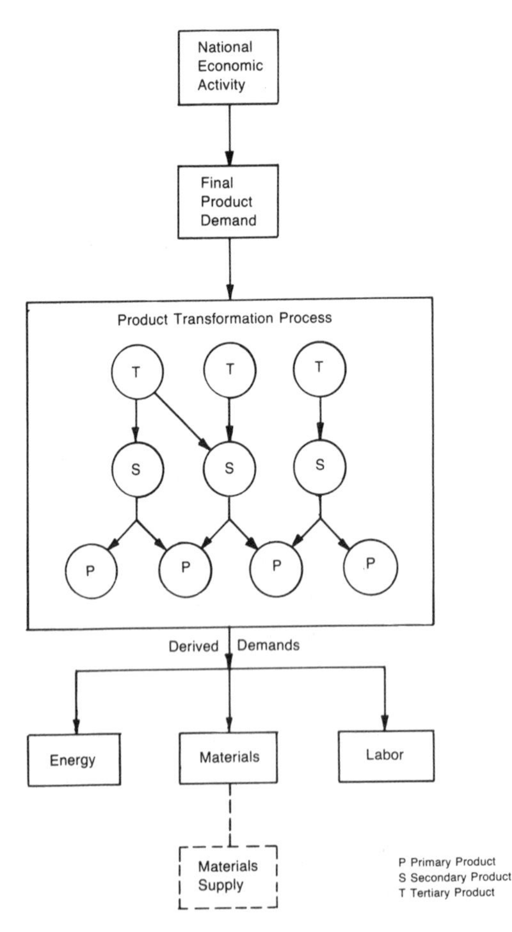

Fig. 3.9.5—Flow diagram of an engineering process model.

ENGINEERING MODELS

The study of mineral engineering models should begin with the realization that such models do not possess a single, unique specification or form. They can take many different forms, depending on the modeling purpose intended. They also tend to incorporate economic information and variables to a lesser or greater degree. We begin with the most typical model forms and then expand the explanation to include more complex forms and methodologies.

Range of Models

The most recent basic form of engineering model is one that deals explicitly with engineering parameters and that used engineering analysis to deduce the structure and/or data that are used to formulate and test the model. An engineering model depicts the technical structure of the production process and permits a linkage to be made between final product demands and derived material demands. The characterization can better be explained in terms of Fig. 3.9.5.

In that figure national economic activity is used to explain final product demand, assuming the higher the activity, the greater the demand. The engineering model explains the product transformation process, i.e., how primary products derived from materials can be transformed into secondary and then tertiary products (or greater) until final product demand is met. The supply of materials and analyses of their production are also available through the use of engineering models.

To help distinguish this class of models from econometric models, we focus on the way in which the production transformation is presented. In econometrics, it is normally embodied in the economist's statistically derived production function. The engineering approach to production function formulation requires that we construct mathematical equations that represent the various technological production possibilities. Usually, the overall production flow is disaggregated into elementary process routes, and input-output coefficients are derived for each stage of each route, based upon engineering data. For each process, mathematical programming techniques can be used to select from among the various production possibilities the ones that achieve an optimal result, according to preset criteria. Alternatively, system simulation techniques can be used to "run" (rather than solve) the model, as dictated by a set of decision rules.

The purposes for constructing an engineering model may be as varied as those for constructing any other kind of mineral model. Once the model is constructed and validated, it can be manipulated to analyze various aspects of mineral market behavior. One approach is to analyze the effects of alterations in policy, technology, or economic usefulness on mineral supply and demand. Alternatively, one can begin with mineral demands and analyze their impacts on inputs or related variables, such as particular industrial sectors, geographic sectors, or national economies.

Another purpose is to select mineral inputs by beginning with a product configuration and examining (or optimizing) the input or derived demand it generates. Finally, the models can be operated on a data base extending into the future. One can then forecast the behavior of individual mineral markets, including details of application or production, such as product shape or form, end use, and spatial or geographic region.

The structure of engineering models can be organized in any one of *four* ways, although in many cases the alternative structural frameworks overlap. The most common approach is to focus on a single mineral or material system, such as copper or phosphate. Such a structure allows one to analyze directly the demand for the mineral or material in its end use markets or to examine the demand for materials used in some part of the production process. Alternatively, the effects of economic variables (such as raw materials costs or taxes) on the production of materials can be explored with such models.

When studying materials demand, an alternative way to structure the model can be to focus on the industrial sectors of the economy and work backwards to the demand for competing or complementary materials. Given the end use sector of the industrial product, one analyzes the entire materials selection process, including product design and the tradeoffs between costs and properties. For example, it is possible to analyze the demand for superalloys used in the production of jet engines for aircraft in such a manner, and eventually arrive at an estimation of the demand for the constituent materials, such as cobalt, chromium, tantalum, and tungsten.

Thirdly, it is possible to structure the model in terms of the intermediate processes that are used to produce the final materials or products used by industrial sectors. For example, one could focus on rapid solidification processes, which, although currently still in the development stage in most applications, are expected to have a substantial impact on material use patterns in the next 5 to 10 years. Such a focus could become increasingly important in the future because of the number of material-processing innovations that are currently under development and that are expected to be used to produce a variety of diverse engineering materials across a wide spectrum of end use applications.

Finally, the distribution of mineral commodities involves selecting optimal spatial or regional allocations with respect to these intermediate processes. The location of refineries of fabricating plants should be decided on the basis of total cost minimization or profit maximization. While this goal relates to domestic material shipments, there also is a need to rationalize materials shipment patterns on an international scale. Thus spatial equilibrium modeling approaches have received interest in mineral modeling exercises.

Whereas the econometric and input-output forms of market models generally depend on a single methodology, minerals engineering models can be classified or grouped in several distinct ways. However, such groups tend to obscure the difficulties of clearly characterizing the nature of such models. One problem is that the concept of an engineering model is not well defined, and its formulation has been construed in various, often conflicting, ways. Another problem is that there often is overlap between the various forms of engineering models themselves. With these limitations in mind, we have arbitrarily selected the following classification: (1) optimization models including linear programming, process optimization and spatial optimization; and (2) system models including system simulation and system dynamics.

OPTIMIZATION MODELS

Optimization involves "choosing the best solution" from among a set of feasible alternatives. Choice implies selection criteria, and in quantitative models these criteria must be specified explicitly in mathematical terms. This is accomplished through what is known as the objective function. We usually seek to maximize or minimize this function, subject to a set of mathematical constraints that describe the conditions under which the decisions must be made. The model is solved, usually with the use of a solution algorithm on a computer, to find the solution that satisfies all the constraints and gives the best possible value for the objective function. Unlike some other modeling techniques, this formulation leads to deterministic and prescriptive results.

The major optimization techniques that have been employed in materials and mineral market models are variants of the general approach known as mathematical programming. While some of the techniques are based on linear and nonlinear programming, there also exists a special class of techniques related to process modeling itself that is known as process optimization. Another important class of optimization models comprises spatial equilibrium models. Because of its popularity and ease of exposition, we begin with linear programming and then discuss process optimization and spatial optimization models.

Linear Programming

Linear programming is most often the preferred technique because it is less expensive to use than the others, and because considerable technical detail about the system can be included by means of a few simplifying assumptions. Also, a number readily available computer packages have been designed to solve linear programming problems.

Theoretical Framework: In the linear programming formulation, the objective function and the constraints are assumed to be linear functions (homogeneous of degree one). The most frequent specifications are to minimize cost, subject to production targets, or to maximize profits, subject to input resource constraints. The objective function is specified in terms of the decision variables (X_j)and the cost coefficients (C_j) that are used to evaluate them. If we have a maximization problem, the formulation would be

$$\max \overline{C}\,\overline{X} \tag{16}$$
$$\text{subject to } \overline{A}\,\overline{X} < \overline{b},\ \overline{X} > 0 \tag{17}$$

where A is the matrix of technical coefficients and b is a vector representing the constraint constants. The problem could just as easily be formulated in terms of a minimization problem. In fact, an important property of linear programming is that to every linear programming objective, there is a dual problem. If the primal problem is

$$\max \overline{C}\,\overline{X} \tag{18}$$
$$\text{subject to } \overline{A}\,\overline{X} < \overline{b},\ \overline{X} > 0 \tag{19}$$

then the dual problem is

$$\min \overline{Y}\,\overline{b} \tag{20}$$
$$\text{subject to } \overline{Y}\,\overline{A} > \overline{C},\ \overline{Y} > 0 \tag{21}$$

The relationship between the primal and the dual is spelled out in the fundamental theorems of linear programming, which are not developed here. However, one way to illustrate these theorems and the primal-dual relationship is to express the linear programming optimization in terms of the classical constrained optimization problem solved by means of Lagrangian functions; the variables of the dual problem (Y) are the Lagrangian multipliers of the primal problem.

Linear programming has shown itself to be amenable to a wide range of applications and a variety of extensions that serve to diminish its inherent limitations. These will be discussed briefly. Certain problems involve mixed equality and inequality constraints in both directions. When the problem has been clearly specified, a wide array of mixed constraints can be incorporated. For particular problems, it is possible to specify activity variables in integer form and provide corresponding algorithms for certain problems if linear programs exist. The problem of linear specificity of the constraints and the objective function can be overcome to a degree, but at the cost of computational complexity. nonlinear constraints can be expressed as linearized segments, where a linear function is specified for each segment and the segment endpoints are included as constraints. It is difficult to deal with a nonlinear objective function beyond a quadratic specification. Its solution is handled through the fundamental properties of the primal-dual relationship. Multiple objective functions can only be accommodated by means of a weighted average where a single value is optimized.

Mineral Applications: As an illustration of an elementary linear programming problem in which costs are to be minimized, consider the refining of steel in an open hearth furnace. The product (raw steel) is made from steel scrap and hot metal (molten pig iron). The cost per ton of hot metal is $100. The technological relationships and constraints for raw steel are:

(1) a minimum of one unit of hot metal for every three tons of scrap,

(2) the scrap and hot metal combine approximately linearly to make raw steel,

(3) the process loss is about 5% from scrap and 1% from hot metal, and

(4) a minimum of 250 tons of raw steel must be made during a production run.

The objective function that we must minimize is specified in terms of the decision variables (X_1 = number of tons of steel scrap, at $90 per ton; and X_2 = number of tons of hot metal, at $100 per ton):

$$\text{Minimize cost} = 90X_1 + 100X_2 \quad (22)$$

The constraints are

$$-1X_1 + 3X_2 > 0 \quad \text{(compositional requirement)} \quad (23)$$

$$0.95X_1 + 0.99X_2 > 250 \quad \text{(production requirement)} \quad (24)$$

$$X_1, X_2 > 0 \quad \text{(nonnegativity constraint)} \quad (25)$$

The problem is to minimize the cost of producing raw steel. The cost function is found by summing the quantities of scrap and hot metal used, multiplied by their respective costs. The first constraint is that a minimum of one ton of hot metal must be used for every three tons of scrap (i.e., $X_2 > X_1/3$). The second constraint accounts for process losses and the minimum production run. The solution to the problem could be found by inspection, graphically, or—if the situation were more complicated—with a computer solution algorithm.

Mathematical programming models of the production process(es) of firms or an industry have been in use for over 30 years. The modern linear programing approach was formulated by Dantzig around 1947 and can be reviewed in Dantzig (1949) and by Dorfman et al. (1950). One of the first applications of a process optimization approach to a nonfuel mineral industry was the linear programming model of the iron and steel industry published by Fabian (1958). The most recent detailed reviews of optimization models in a commodity market context has been provided by Labys (1978 and 1982). Specific descriptions of process models can be found in Sparrow and Soyster (1980).

The previous example illustrated a simple linear programming formulation designed to find the least cost combination of raw material inputs for a single production operation. Any example of a model in use would be much more complicated and difficult to summarize in detail. One that we can summarize, however, aims at finding the least cost combination of production technologies in an integrated steel plant, subject to demand, technological, and raw material constraints, as developed by Ray and Szekely (1973). Their formulation allows the analyst to find the optimal mix of three process routes (blast furnace/basic oxygen furnace, blast furnace/open hearth furnace, electric furnace), depending on the prices of the factor inputs, such as scrap and hot metal. Such a formulation is of interest for studying the demand for materials that are used in the production sequence. It is possible, with the use of sensitivity analysis, to analyze the effects of changes in technological constraints (such as coke rate in the blast furnace) and in the prices of factor inputs on the optimal use of materials by the steel plant.

Another example of a linear programming model is the copper industry model developed by Soyster and Hibbard (1980). The model provides an engineering representation of the flow of material from mining through benefication, smelting, and refining to intermediate fabrication and end use, in terms of real physical units. The mining sector is disaggregated by individual mines, and the production function is optimized to minimize the total cost of production, given a set of exogenously defined assumptions about the demand function, the required rate of return, import prices, scrap recycling, and investment constraints. The model is used to determine the least cost combination of primary domestic and foreign sources and of secondary sources of copper over 10 two-year time periods.

Process Optimization

Process optimization is a subsystem of mathematical programming in which an intermediate step is included in the formulation to improve the approximation of production function. Process optimization is the most widely used application of mathematical programming found in the minerals and materials industries and normally depends on linear programming as the underlying solution algorithm.

Theoretical Framework: The actual specification of the model also resembles that of linear programming; it features an objective function, activities, and constraints. Model solution usually involves minimizing an objective function specified in terms of production costs, subject to such constraints as the time sequence of production, regional capacities, the demand for the product, and technical relationships among the production variables.

The activity component is the distinguishing feature of process optimization. It can be considered as an intermediate step that describes how decision variables (resources in the production problem) are combined in fixed proportions by the production technologies to produce an output. The process optimization approach, emphasizing the activity component, is in some ways similar to the input-output approach. But, instead of a single technology for every process, linear programming allows for multiple processes or activities.

RESOURCES → ACTIVITIES ⟶ OUTPUT(S)

These activities define technologically possible alternatives in physical terms only (e.g., different energy, materials, and labor requirements) and do not necessarily yield economically (or socially) efficient solutions. Economic considerations are introduced by the cost function, which shows the minimum cost of producing various levels of output, given factor prices and technologies.

The constraints are formulated to capture the various technological and institutional relationships that exist in the production sequence of the minerals or materials industry being modeled. They can be classified into six general types (Sparrow and Soyster, 1980):

(1) *Accounting constraints,* which insure that the total use of a purchased input is equivalent to the total purchase of the input;

(2) *Capacity constraints,* which insure that the level of an activity does not exceed the industry's capacity;

(3) *Material balance constraints,* which insure that the input requirements for intermediate products equal production from activities that occur at previous stages;

(4) *Demand constraints,* which insure that activities will produce enough final goods to satisfy final demands;

(5) *Constraints* on the availability of resources; and

(6) *Constraints* that incur nonnegative or integer values for parameters that require such values to conform with physical reality.

It should be noted that once a cost function is formulated to help decide the minimum cost of producing various levels of output, one can also find the associated average and marginal cost curves. The cost function can be used to analyze the effects of changes in the prices of inputs on the demand for that conditional input for a given mix of technologies. From the cost function, we can also derive conditional input demand functions. These show how much of a particular input will be used as a function of input prices and output. Note that the input demand function depends upon output. Once an input price changes, output is also subject to change. If we have some way of predicting changes in output, we can get a good idea of how input demand will change as input prices change, and vice versa.

Mineral Applications: A number of optimization models of the iron and steel industry, mostly using linear programming or a variant thereof, have been published in the past 25 years. In this section we present for illustrative purposes a brief summary of one of these models, a linear programming formulation by Tsao and Day (1971) that treats the U.S. iron and steel industry as a whole. (This illustration, which involves certain processes that are out of date, was selected because of its simplicity in addressing optimization modeling.)

Tsao and Day built a short run (fixed capacity) model of the U.S. carbon steel industry, which was used to compare the optimal and actual behavior of the industry over the period 1955–1968. Alternative arrangements of processes are included in the technology matrix, with optimization indicating the desirable proportions of each. The technology matrix contains five segments, which correspond to the sequence of production. It starts with the primary inputs (i.e., iron ore, coal) and advances through five stages of intermediate and final materials production: coking, iron making, ingot steel production, preliminary shaping, and final finishing. In each segment, average industry practice is represented by standardized activities.

In addition to the technology matrix, the model also contains a set of auxiliary activities, which represent the purchase of inputs and the sale of final outputs. The constraint on input purchases prevents the total use of a given input (such as oxygen) from exceeding the amount purchased. Sales constraints are included to insure that production of finished products meets sales requirements. In the subsequent discussion we briefly outline the activities and constraints in the first two stages of the model, coking and iron melting.

In the coking stage, two processes are modeled: the beehive process and the byproduct oven process. The former is included, even though technologically obsolete, to account for the historical evolution of the industry over the time period modeled. Coal inputs are of three types: bituminous, subbituminous, and lignite. The first two are in coke ovens, and the third is used in making direct reduced iron (DRI). The constraint equations corresponding to this configuration are

$$a_{1,1} X_1 + a_{1,2} X_2 < X_{49} \quad \text{(bituminous coal supply)} \qquad (26)$$

$$a_{2,3} X_2 < X_{50} \quad \text{(subbituminous coal supply)} \qquad (27)$$

$$a_{3,12} X_{12} < X_{51} \quad \text{(lignite coal supply)} \qquad (28)$$

where $a_{1,1}$ is the net amount of bituminous coal input per ton of beehive coke output, X_1 is the beehive coke output, X_2 is the byproduct coke output, and X_{12} is the direct reduced iron output. The activities X_{49}, X_{50}, and X_{51} represent the amounts of the respective types of coal that are purchased.

The constraints of the two coke processes are due to the oven capacities, b_4 and b_5, given by

$$X_1 < b_4 \quad \text{(beehive oven capacity)} \qquad (29)$$

$$X_2 + X_3 < b_5 \quad \text{(byproduct oven capacity)} \qquad (30)$$

where b_4 and b_5 are the total capacities of the two types of ovens.

Finally, the total amount of coke produced must be at least as great as that required by the next stage of production (iron making). The total industry demand for coke is represented by b_6 and the constraint is:

$$\sum_j X_j > b_6 \quad \text{(coke demand)} \qquad (31)$$

This last constraint defines the input-output balance between the first two stages.

Iron making requires three basic input materials: iron bearing materials, limestone, and coke. The model accounts for five sources of iron bearing materials enumerated in the constraints below.

$$a_{7,4} X_4 + a_{7,8} X_8 < X_{52} \quad \text{(domestic iron supply)} \qquad (32)$$

$$a_{8,5} X_5 + a_{8,9} X_9 < X_{53} \quad \text{(imported ore supply)} \qquad (33)$$

$$a_{9,6} X_6 + a_{9,10} X_{10} < X_{54} \quad \text{(sintered ore supply)} \qquad (34)$$

$$a_{10,7} X_7 + a_{10,11} X_{11} < X_{55} \quad \text{(pellet supply)} \qquad (35)$$

$$a_{11,12} X_{12} < X_{56} \quad \text{(prereduced ore supply)} \qquad (36)$$

The variable $a_{i,j}$ represents the amount of the ith iron bearing material required per ton of output from the jth iron producing process (activity). For instance, $a_{7,4}$ is the amount of domestic ore required to produce a ton of hot metal in the blast furnace. The variables $X_4 - X_7$ represent hot metal production from the blast furnace; $X_8 - X_{11}$ represent cold iron production from the blast furnace; and X_{12} represents direct reduced iron production.

All of the iron producing activities use coke, leading to the following constraint

$$\sum_j (A_{6,j}) X_j < \Sigma_j X_j \qquad (37)$$

where $a_{6,j}$ represents the amount of coke required per ton of iron produced by the jth iron making activity. The left hand side of the equation defines the total intermediate demand, and the right hand side of the total amount of coke produced within the industry. This constraint replaces the previous constraint on coke demand, which is redundant because of the endogenous nature of the coke supply in the model.

There also are constraints and activities associated with (1) limestone supply, (2) blast furnace capacity, (3) direct reduction capacity, (4) hot metal demand, and (5) cold metal demand (these will not be described here). The complete technology matrix is formed by linking these five submatrices in a sequential manner by the appropriate input-output relationship. For instance, equation 31 connects the first two stages. The output of one stage is the input for the next stage, until the final stage is reached. Therefore, changes in an input variable reverberate throughout the model.

The objective function is formulated by assuming that the industry as a whole allocates resources so as to minimize the short run variable costs of production. If Q is a vector that represents the unit costs corresponding to the elements in X, and C is the short run variable cost, the objective function is:

$$\text{Minimize } C = \sum_j Q_j X_j \qquad (38)$$

and the constraints are of the general form previously referred to. The solution procedure starts by solving the system of simultaneous equations given by $AX < b$ and $X > 0$. Among the set of solutions satisfying these constraints, the one that minimizes costs in equation 38 will be chosen.

Spatial Optimization

Process optimization models can be expanded to include spatial optimization, or spatial optimization can take place at a single stage of process. Spatial optimization involves the examination of spatial equilibrium or interregional efficiency in mineral production distribution and utilization. Attempts to analyze equilibrium between markets over space and time have been made to analyze mineral trade between regions in a domestic as well as in an international context. The principal modeling methods employed for this purpose have been linear and nonlinear programming forms of transportation models which have the specific advantage of recognizing transport costs and political constraints in the formulation of an equilibrium approximating competitive trade. A history of the evolution of this form of modeling can be found in Takayama and Labys (1984).

It is intuitively obvious that a commodity produced in several regions will probably have different factor requirements in each location. These disparities will lead to measurable cost differences, which are likely to influence the comparative advantage of each producer. Both the classical and Hecksher-Olin theories of international trade rely on production costs and demands to explain trade flows. Consideration of location theory requires an explicit recognition of the importance of transportation costs, particularly in the analysis of movement of relatively low-value goods where transportation costs are a significant cost component.

It was Samuelson (1952) who recognized that a programming solution could be employed to determine the competitive equilibrium with respect to the quantities transported to each location. This fundamental algorithm was subsequently developed by a number of researchers who constructed programming models of international trade in a number of primary commodities. Most notable is the formal spatial equilibrium approach advanced by Takayama and Judge (1971). Mineral applications have followed from recognition of the importance of trade flows in determining the supply of a mineral commodity to consumers. This conceptualization is presently relevant for the United States because of its high mineral import dependence.

Theoretical Framework: The specifications of the structure of a mineral trade model depends

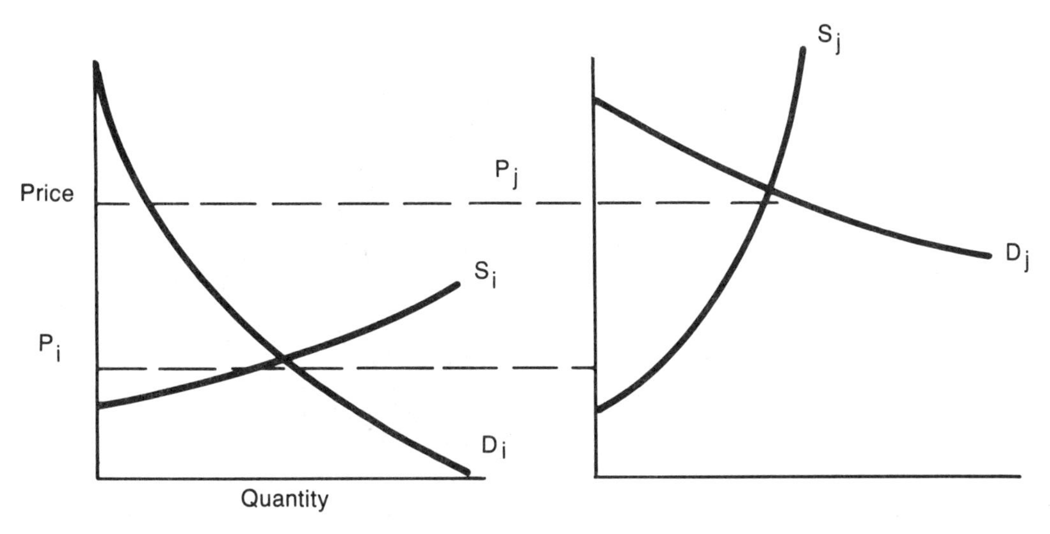

Fig. 3.9.6—Supply-demand balance *not* allowing transportation.

on the maximization of net social surplus. Consider the case of one-commodity and two-region system shown in Fig. 3.9.6. Local equilibrium in the absence of the interregional transportation results in local equilibrium prices. By allowing transportation between regions, Fig. 3.9.7 shows that prices adjust to a common level that is equivalent to the transportation costs between the regions. Consumers' surplus is the area under the demand curve and above the equilibrium price. Similarly, producers' surplus is the area bounded by the supply curve and the eqilibrium price. The total social surplus is a subset of the total surplus, and is the result of the increased surplus related to trade, graphically shown as the areas Q_{ij} in Fig. 3.9.8. Because local social surplus with transportation is unchanged from the no-trade case, it can be seen that maximizing total social surplus is equivalent to maximizing the net social surplus accruing from trade.

Of course mineral trade flows typically need to be considered over more than two regions. Multiple-region models of mineral trade emphasize interrelations or simultaneities among

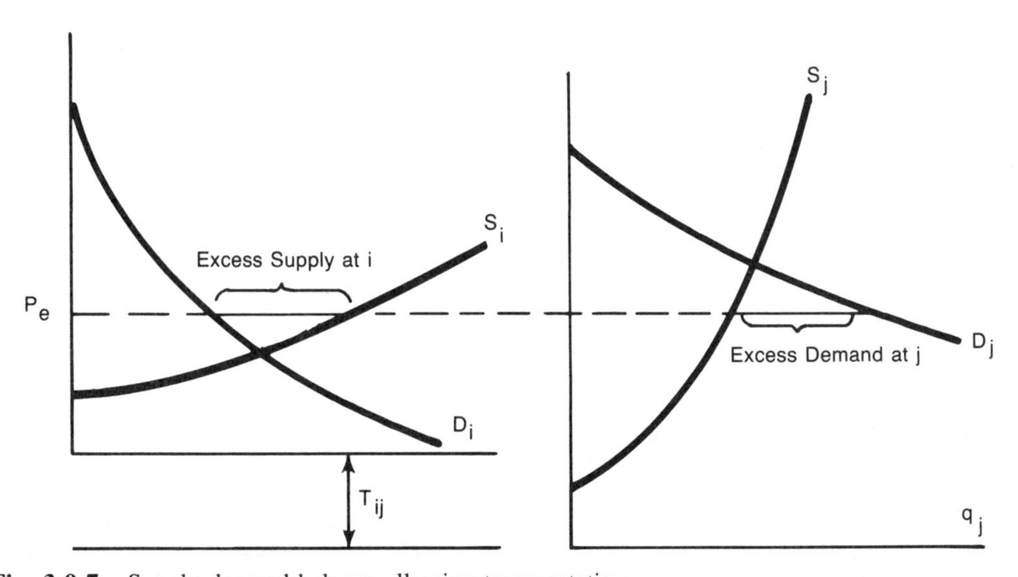

Fig. 3.9.7—Supply-demand balance allowing transportation.

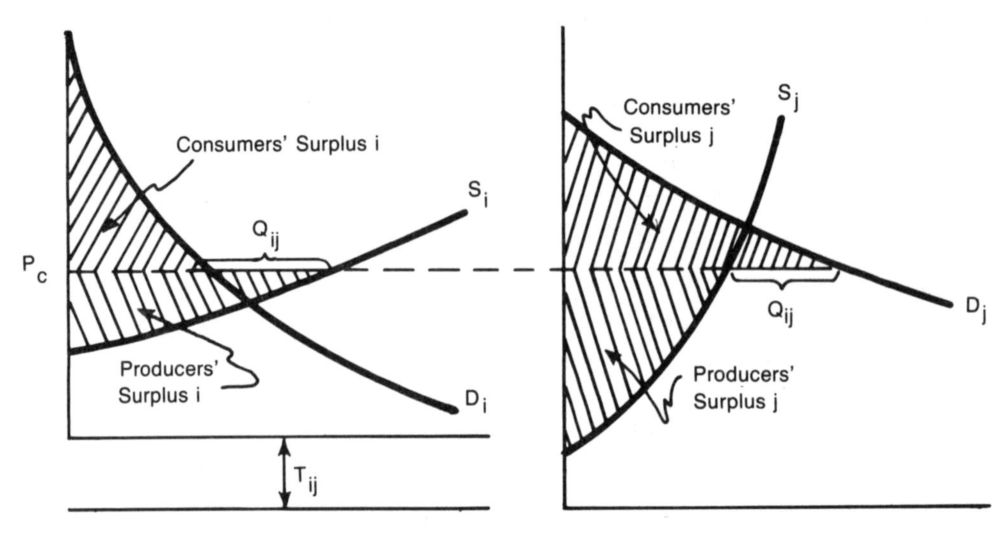

Fig. 3.9.8—Consumers' and producers' surplus.

countries through world trade. Here the aggregate rest of the world region of the two-region models is divided into two or more trading regions. Each region may be an individual country or a group of countries, sometimes contiguous, with relatively homogeneous geological conditions, level of economic development, and policy interventions. All regions are assumed to be large trading regions, that is, the actions of each region acting independently can affect the world market prices for its imports or for its exports. Nevertheless, most models assume that mineral production and marketing are characterized by a large number of small firms, each of which is a price taker or perfect competitor.

The construction of an n-region mineral trade model derives from the basic Koopmans-Hitchcock transportation cost minimization LP model. However, the applicability of the model can be improved by replacing the typically, exogenously given demand and supply estimates with estimates that are endogenously determined. The mathematical solution to such a model typically follows from the Takayama and Judge (1971) market spatial equilibrium model based on quadratic programming. This model normally is specified to include the following components: (1) a system of equations describing the aggregate demand for one or more commodities or interest in each of the included markets as well as the aggregate supply of the commodities in each of the markets, (2) the distribution activities over space, and (3) the equilibrium conditions. While the demand and supply equations imply a structure similar to that of an econometric market model, the equilibrium process is more adequately represented through the identification of the profits to be realized from the flow of commodities, i.e., the excess of a price differential between two points minus transportation costs. Profit maximization is assured through the use of a computational algorithm which allows commodities to transfer until demand equals supply in every spatially separated market. So that policy decisions can be evaluated more realistically, the equilibrium conditions and other definitional equations can be used to impose constraints on the model parameters. Quadratic programming also possesses the advantage of simultaneous interaction between prices and quantities, an adjustment not normally possible in linear programming. The use of a nonlinear objective function also prevents the erratic changes in spatial flows that sometimes occur in linear model solutions.

An example of a spatial optimization model of this type begins with the system of equations describing demand D_i in region i and supply S_j in region j.

$$D_i = b_{oi} - b_{1i}P_i \quad (1)$$

$$S_j = b_{oj} - b_{1j}P_j \quad (2)$$

where P_i = commodity demand price in region i and P_j = commodity supply price in region j. Because the formulation of this model by Takayama and Judge (1971) expresses these equations in their inverse form, the above equations can be alternatively written as

$$P_i = a_{1i} + a_{2i}D_i \qquad \text{for all } i \qquad (3)$$
$$P_j = a_{3j} - a_{4j}S_j \qquad \text{for all } j \qquad (4)$$

where a_1, a_2, a_3, $a_4 > 0$ over all observations.

The constraints imposed on the supply and demand relations reflect the conditions that regional consumption cannot exceed the total shipment to a region and that the total shipments for a region cannot exceed the total quantity available for shipment.

$$D_i \leqq \sum_j^n Q_{ij} \qquad \text{for all} i \qquad (5)$$
$$S_j \geqq \sum_i^n Q_{ij} \qquad \text{for all } j \qquad (6)$$

Transport costs and shipments between regions i and j are assumed to be non-negative.

$$T_{ij}, Q_{ij} \geqq 0 \qquad (7)$$

The objective function necessary to complete the model would maximize the global sum of producers' and consumers' surplus after the deduction of transportation costs. This form of market-oriented quasi-welfare function has been termed net social payoff (NSP) by Samuelson (1952) and is defined as follows.

$$\text{NSP} = \sum_i^n \int_0^D P_i(D_i)dD_i - \sum_i^n \int_0^S P_j(S_j)dS_j - \sum_i^n \sum_j^n Q_{ij}T_{ij} \qquad (8)$$

The corresponding objective function is

$$\text{Max (NSP)} = \sum_i^n a_{1i}D_i - \sum_j^n a_{3j}S_j - ½ \sum_i^n a_{2i}D_i^2 - ½ \sum_j^n a_jS_j^2 - \sum_i^n \sum_j^n T_{ij}Q_{ij} \qquad (9)$$

where D_i and $S_j \geqq 0$.

Mineral Applications: The spatial optimization model has been applied less frequently in mineral market analysis than in agricultural or energy market analysis. Some examples limited to a linear programming formulation include the model of world nickel industry by Copithorne (1973) and model of the world copper industry by Kovisars (1975). In constructing a model of the world zinc market, Kovisars (1976) introduced a temporal dimension by the successive adjustment of demand, cost, and capacity variables using the results of the previous simulation. Being production-based, the model employs static linear programming to solve for the regional supplies and transfers each period. Of some interest is the use of an integrating model similar to that proposed by Hogan and Weyant (1980) for energy modelling to integrate demand estimates over time together with the programming stipulated supply schedules and resource requirements. Concerning the application of quadratic programming, these models have been more widely applied to the fuel minerals, principally the world oil refining model of Kennedy (1974), the U.S. regional coal trade model of Labys and Yang (1980), and the electric utility model of Uri (1976).

A domestic spatial optimization model can be seen in Trozzo (1966) who employed linear programming to determine the optimal location of coking and pig iron facilities in the United States (disaggregated into 32 producing regions). Steelmaking capacity and location are taken as given, but it is recognized that the demand for steel and factor supply are local characteristics that must be considered as an investment choice. Nelson (1970) extended the framework developed by Trozzo to include the steelmaking and finishing segments of the industry. He also combined the production sector with estimates of industry demand forecasts to simulate the optimal path of production and investment on a regional basis in the United States over 20 years. The regional models of Trozzo (1956) and of Nelson (1970) are useful for analyzing the temporal and spatial aspects of the demand for input materials.

More recent applications have attempted a stronger integration with econometric methods, particularly on the demand side, and have emphasized dynamic model solutions. For example, such a model has been constructed by Hibbard, Soyster and Kelley (1979) for long range aluminum forecasts including the stage of process from mining through refining, smelting and fabrication with special emphasis on scrap recycling. The aluminum supply model is driven by a set of econometric demands derived from historical data. The model results projecting industry investments and operations from 1977 until 1994 are reported and discussed, particularly as they relate to mineral policy. In addition, Hibbard et al. (1980) have constructed a similar model for the copper market known as MIDAS-II. This model features disaggregation to include individual production characteristics about each mine, smelter, refinery and electromining facility of U.S. companies and some international companies. It also gives recognition to modeling noncompetitive aspects of international mineral market behavior in alternatively considering copper market structure in the form of an oliogopoly dominated by a few large firms.

Other applications of spatial and process optimization have concentrated on the modeling of investment analysis. Dammert (1980) employed the mixed integer programming method to study copper investment, allocation, and demands in Latin America. Dammert also showed how reserve levels and reserve limits could be coupled with depletion in the mining and processing of different ore goods. A similar approach has been taken by Brown et al. (1981) in their modeling of worldwide investment analysis in the aluminum industry. Linear complementarity programming methods have also been applied to spatial investment analysis. An important application of this approach to the minerals industry can be witnessed in the world iron and steel industry model (WISE) constructed by Hashimoto (1979).

SYSTEM MODELS

System Simulation

A system simulation model is not always distinguishable from other modeling methodologies. To distinguish it from the previous modeling approaches, it is useful to contrast the simulation approach with other, the causal, modeling methodologies considered, i.e., econometric and optimization. We can categorize an econometric model as one in which (1) relationships between economic variables are hypothesized, and (2) statistical procedures are applied to determine whether or not the hypothesized interdependence can be accepted. The relationships specified are sometimes dynamic and sometimes not, and econometric models may comprise a large number of equations. A process optimization model, on the other hand, is one which determines the optimum mix of alternative substitutible inputs to produce a well-characterized output. Optimization models generally ascribe or implicitly attribute rational economic decision making to the decision maker being modeled, and can be either statistically or dynamically specified.

Theoretical Framework: In its simplest form, a system simulation model is nothing more than a set of hypothesized relationships between variables, typically dynamically specified, in which the time behavior of the interrelationships is revealed via simulation. (By ''simulation,'' we mean the process of integrating the system of differential equations that are constructed from these hypothesized relationships.) Thus, a system simulation model can be distinguished most clearly from the other causal modeling approaches as one not requiring parametric validation in the statistical sense, nor attribution of rational economic decision making to the actors being modeled.

Typically, a system simulation model will interconnect submodels that capture generally accepted behavior on a small scale in an attempt to simulate overall performance on a large scale. As practiced in some forms, the model will often articulate global constraints, generally accepted to exist, but quantified only approximately (usually due to knowledge limitations) that have important implications for the time behavior of the model variables. For example, constraints that trigger turning points, changes from growth to decline or vice versa, or changes from growth to steady states, are favorite targets of analysis.

Probably the greatest strength of the system simulation methodology is its flexibility. One is not constrained by data or economic rationality in constructing the model. Because of this flexibility, however, the human resource requirements for system simulation modeling are uniformly stringent. Although all of the methodologies that are reviewed in this chapter place great demands on the modeler, optimization and system simulation models are more demanding than the other techniques, partly because they are potentially open to greater abuse and partly because of the detailed working knowledge that is required of such matters as the technology of materials production and utilization, of microeconomics, and of the particular systems analysis technique being used. An understanding of only the techniques of optimization or systems simulation is not sufficient.

Although these techniques are the most demanding in terms of their human resource requirements because of their interdisciplinary nature, they are the easiest to use. It is relatively easy to learn to manipulate one of the many "canned" optimization or system simulation computer packages that are now available commercially. It is also easy to construct a bad model with these tools, and the literature abounds with poorly constructed optimization and systems simulation models of the materials industries. Usually, the deficiencies in the model can be traced to a lack of understanding of the technology, the economics of the industry, or the systems analysis technique, in that order.

As far as the capability of the approach to deal with anything other than economic rationality is concerned, it is useful to observe that political processes often produce solutions to economic problems that are politically acceptable, but not economically rational. To the extent that the systems simulation approach to modeling can reflect political realities, it might be preferred to the econometric or optimization approach to a description of reality. When government policy makers are likely to influence model variables, system simulation may be the only fruitful modeling approach. The limitation is that quantitative political analysis is just beginning to emerge as a useful source for policy modeling.

The system simulation methodology does not preclude the use of other modeling techniques to derive certain parameters or subsectors of the model. A simulation model may be composed of components econometrically derived, parts based upon optimization or process models, or parts derived from other techniques. In such a model it is not advisable to distinguish the three exclusive approaches to causal modeling as separate and mutually exclusive.

In summary, the system simulation model is essentially composed of a set of mathematical equations expressing causal relationships, which are usually converted into a computer model. The results of simulations (the output of the model) are simply the implications of the assumptions upon which the equations are based. Therefore the efficacy of the approach is determined completely by the validity of the specification of the causal relationships and the accuracy of the data (usually empirical) used to estimate the parameters of the model.

Mineral Applications: A wide range of engineering/economic materials systems have been reduced to system simulation models, reflecting the ease with which nonoptimizing decision criteria can be incorporated in the model. In this section, a model of the magnesium industry (Busch, et al., 1980) will be presented.

The world magnesium market is dominated by the United States, which produces over 67% of the total global demand. This productive capacity is held by three companies; Dow Chemical, AMAX Specialty Metals, and Northwest Alloys. Of the three, Dow's productive capacity (130 ktons/year) is more than two thirds of total U.S. production. Within this market, it is unlikely that competitive pricing occurs, considering Dow's considerable market power. In fact, observers claim that Dow manages prices in such a way as to maintain the marginal producers without giving them incentives to expand capacity.

Given this market structure, engineering simulation modeling was a logical choice for analysis. Using system simulation, the engineering parameters of magnesium production and the peculiarities of magnesium pricing can be consistently presented and analysed. The corresponding magnesium model is broken down into five major sectors; traditional U.S. demand, potential U.S. demand, world residual demand, U.S. and world production costs, and world prices. Each is briefly described.

Traditional U.S. demand consists of military uses, aluminum alloying, etc. In each case, a regression model, employing economic indicators and lagged price was developed to describe the magnesium demands of each sector. Regression models were used because the industries/sectors being modeled were well developed and likely to perform in the present as they had performed in the past.

Potential U.S. demand was modeled for two major sectors—automobiles and steel desulfurization. Automotive demand was modeled by

(1) Determining those automotive applications (on an engineering basis) which had seen or were candidates for magnesium use,

(2) Estimating the mass of magnesium required for part, and

(3) Determining for each part, through interviews, automakers attitudes toward increased use of magnesium on the basis of price fluctua-

tions in the magnesium and aluminum markets.

With this information, a potential demand curve for magnesium in automotive applications could be defined. Application of magnesium in a particular part was modeled as dependent upon the ratio of the price of magnesium and aluminum, averaged over time. The time averaging was necessary to capture producers preferences for a stable price advantage over time. Changes in automotive demand were smoothed over time to resemble typical "S-curve growth."

Desulfurization of steel by magnesium is another area of potential demand that was included in this model. Market projections by magnesium producers were used to estimate the total tonnage of steel that might be processed with magnesium desulfurization by the year 2000. Engineering techniques were used to determine the amount of magnesium required to desulfurize a ton of steel. Finally, the growth of this application was tied to the rate of capacity utilization in the steel industry, as estimated by econometric and judgmental techniques.

World residual demand was modeled by estimating the total world demand for traditional and potential applications of magnesium. As for the U.S. demand models, the traditional demands were estimated from regression equations and the potential applications from judgmental engineering estimates. The world potential demand was far less disaggregated than the U.S. models, reflecting the difficulty associated with interviewing producers in many countries as well as the emphasis of the model on a perspective. World supply was estimated to grow at a fixed rate over the model period, although this parameter was readily modifiable to explore varying scenarios. Nevertheless, production shortfalls for foreign producers have been a feature of the magnesium market over the past 10 years and are expected to persist. Thus, the model assumed that U.S. producers would continue to make up this shortfall, selling magnesium at world rather than U.S. prices, which are considerably higher.

The fourth sector of the model was the production cost sector. Based on engineering flow streams, production costs for each of the current producers were computed. Furthermore, a potential new plant, employing the latest technologies, was also included to estimate the effects of new entrants on the market. Finally, production costs for non-U.S. magnesium producers were estimated based upon the production techniques employed by Norsk-Hydro, the largest nondomestic producer.

Each of these cost models estimate the cost per ton of magnesium produced, given the plant production target. These costs are disaggregated by feedstock, labor, capital, and energy costs, thus simplifying cost projections by enabling the use of specific cost inflators for each relevant cost.

The final sector of the model is the pricing sector. Within this sector, the interaction between the other model sectors is accomplished. Prices are supplied to demand sectors which estimate the volume of magnesium demanded. Production requirements are fed into production cost models to yield the costs per ton of magnesium produced. It is the producing sector that closes the circle by estimating market prices based upon the costs of production.

As described above, the nature of the magnesium market precludes the simple use of the competitive model of pricing. Other models must necessarily be considered. The interesting feature of this model is that three different pricing schemes may be employed, depending upon the scenario being studied. These schemes are (1) cost-plus pricing, (2) historical pricing, and (3) marginal cost pricing. The first two pricing schemes are based upon Dow's production costs, reflecting their considerable market power. Cost plus merely applies a fixed markup to Dow's production costs while historical pricing employs a regression model based on Dow's costs and the degree of capacity utilization in the industry. Marginal cost pricing traces a curve

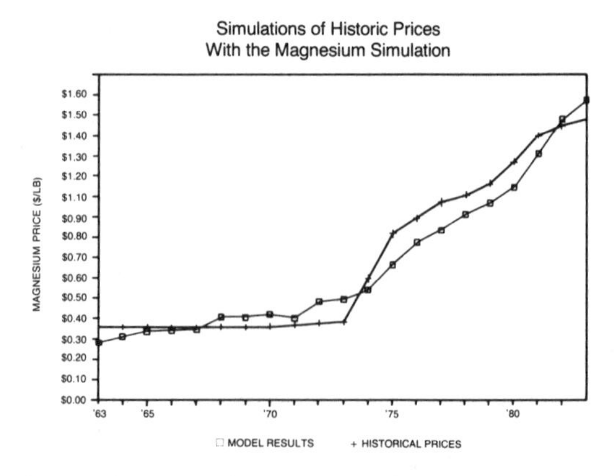

Fig. 3.9.9—Simulations of historical prices with magnesium simulation model.

through the production costs of each of the producers, indexed by the production capacity of each increment in production. The marginal cost pricing scheme includes the potential new producer, thus enabling an analysis of the effects of the new producer's costs on prices. Fig. 3.9.9 illustrates the correlation between the cost-plus pricing mechanism and published magnesium prices over the period from 1963–1983.

This model has been used to analyse the profitability of alternative investments in the magnesium industry. The long time horizon of the model (1983–2001) enables long term planning while the flexibility of the model structure enables examination of a wide range of investment and production scenarios, considering not only present magnesium applications and technologies, but also potential developments in the industry.

System Dynamics

System dynamics is a special class of system simulation models that has its basis in control theory and systems engineering. It represents a theory of system structure and is a framework for identifying, depicting, and analyzing nonlinear feedback relationships. System dynamics is included here as a separate category because of its special focus (control theory) and because the methodology (Forrester, 1961) has been the subject of considerable attention and controversy in the past decade.

System dynamics was originally developed more than 20 years ago as a means of analyzing the management problems associated with production scheduling and product distribution in industrial firms. The method was later applied to the problems of world growth in *World Dynamics* (Forrester 1971) and problems of commodity markets in *Dynamic Commodity Cycle Models* (Meadows, 1970). It is these latter two approaches which have received most attention in mineral-related applications.

Theoretical Framework: In general, the systems dynamics approach provides a framework within which the internal operations of systems can be examined in a coherent and consistent manner. All systems feature variables that are defined in terms of "levels" and "rates." Levels are the state variables in the system and are the major structural determinants. A level may be thought of as a number that represents part of a system at a instant of time. For example, in a material demand model inventories of materials or production capacities might be represented as levels.

Systems dynamics models move forward in time in fixed, finite intervals. The rates of movement control the fluctuations of the state variables and define the amounts by which the levels will change during a particular interval of time. The levels and rates form the basis for a set of time dependent differential equations. Once the initial conditions of the system (the boundary conditions) are defined, the model equations are solved as the model is "run" over time.

Other variables and equations in system dynamic models are used to define and modify levels and rates. These variables are called auxiliaries, delays, table functions, and constants. The auxiliaries are used to define the rates. Delays represent time lags in the system and are of two kinds: material delays (representing lags in the material flows) and information delays. Table functions represent functional relationships between two variables and are the primary mechanism by which nonlinear relationships are incorporated into the model. They are usually graphical representations of functional forms; the shape of the graph provides the information necessary to specify the value of a system variable or parameter at a point of time. Constants are coefficients or multipliers, which remain constant over time.

Mineral Applications: System simulation techniques designed to construct models of materials industries and consuming sectors have included the system dynamics approach. The following example based on the copper industry illustrates the use of this approach to construct a model of the demand substitution between copper and aluminum within the dynamic interactions of specific markets and product applications.

The approach is based on an analysis of the fraction of total consumption that can be substituted on a technically feasible basis, the amount of each competing material required for each functional use, and the costs of using each material. Parameters in the model are estimated by a combination of engineering, judgmental, and statistical analyses. Fig. 3.9.10 shows the important relationships postulated to exist among the variables in the model. Consumption of each material is divided into substitutible and nonsubstitutible components.

The nonsubstitutible fraction is derived from an engineering and judgmental analysis of the

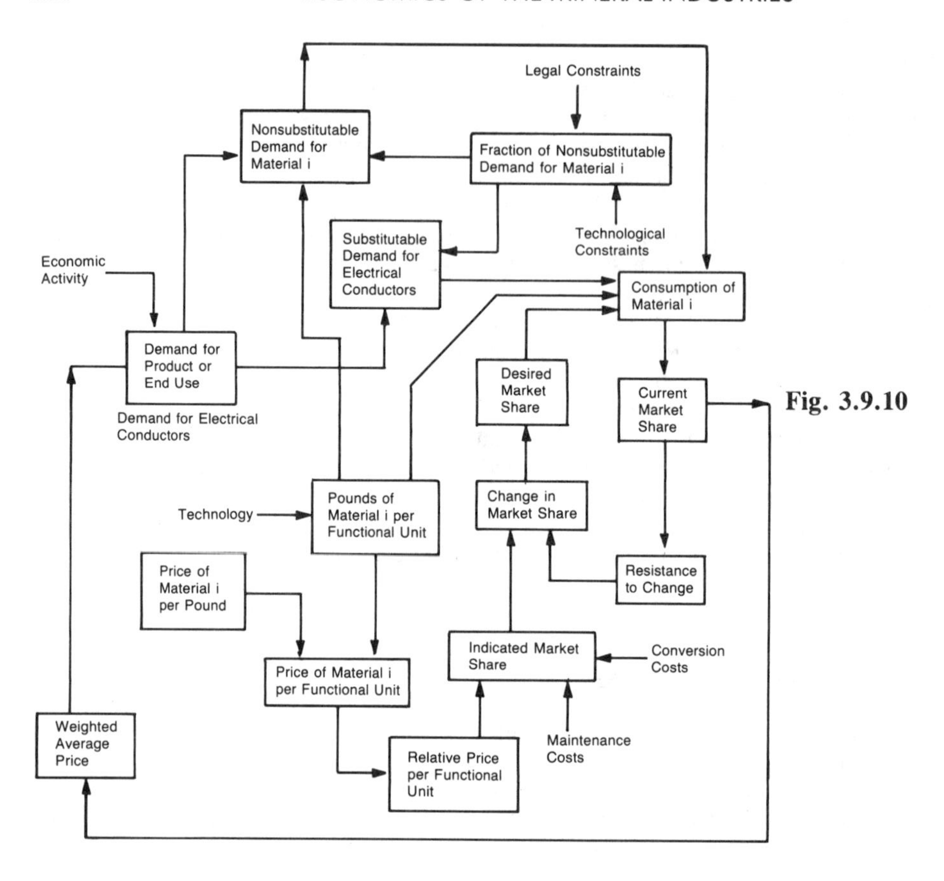

Fig. 3.9.10

particular material within the price limits expected to exist within the time frame of the analysis. Legal and technological constraints are taken into account. For instance, the fraction of the building wire market that can only be satisfied by copper because of prohibitions against the use of aluminum conductors and terminations is estimated from information such as that provided by the 1972 study of the National Materials Advisory Board (National Research Council, 1972). Technological constraints, such as NEMA performance standards and size restrictions that prohibit the use of aluminum stator windings for many designs of medium size (1–125 hp) ac induction motor, are also estimated on a judgmental basis.

The demand for the product (or end use) is usually based on statistical relationships. For instance, the copper-aluminum electrical conductor substitution model uses an equation for the demand for electrical conductors that is based on the total demand for electrical conductors, economic activity of industrial sectors that consume electrical conductors, and the average price of electrical conductors.

The substitutible demand is determined by subtracting the nonsubstitutible component and multiplying the remainder by the user's desired market share. Resistance to change affects how quickly the desired market share increases or decreases towards indicated market share. Resistance to change might be expressed in years required to make the changes. It is a constant that may be different for different end uses. The effect of current market share acts to reduce the constant as market share rises, or as adoption spreads. When the substitution is just beginning and market share is low, consumers are wary of substituting. They generally prefer to wait until the market share has grown and the application has been tested. This hesitation leads to the "bandwagon" phenomenon, which begins with slow growth as a few innovative users perceive a cost or technological advantage, then goes through a phase of fairly rapid growth as more customers perceive the successful application, and finally a leveling off phase as maximum market share is achieved. (This is also referred to as "S-shaped" growth.)

Indicated market share depends on relative

price per functional unit and on conversion costs. When relative prices (determined by the difference between material price per functional unit and competing material price per functional unit) are equal, the indicated market share equals the current market share. Customers have no incentive to change, so they continue to use present materials. As prices fall in relation to those competing materials, the indicated share increases until it reaches 1.0, and all customers of the competing material have a rational reason to switch to the material in question. Conversely, as prices rise relative to competing materials, indicated share falls until it reaches minus 1, where all consumers of the material should rationally switch to the competing material.

The price differential that drives substitution in the model is based not on current prices but on an average of prices. Such an averaging reflects consumer desire to be sure that price changes are permanent before making a switch. The length of the averaging time should depend on the cost of conversion. If conversion costs are high, price changes must last longer in order to compensate for those costs.

Several complicating factors can be introduced into the model. First, conversion costs may differ, depending upon the direction of the change. Second, more than one competing material can be considered, although this may not be important if end use disaggregation is detailed enough. And finally, availability or security can be introduced as a reason to substitute materials, although decision rules governing such a substitution must be formulated on a judgmental basis.

In summary, this model of the substitution process captures a number of relevant considerations: legal and technical constraints that make some demand unsubstitutible, resistance to change and bandwagon effects, break-even costs based on a functional unit basis, conversion costs, and the need for a price change to persist in order to elicit substitution.

INPUT-OUTPUT MODELS

Input-output is a conceptualization of the interdependence of economic production units that has come to be used over the years both as a modeling device and as an element in national economic accounting systems. The input-output (I-O) framework as such cannot be employed directly to model mineral market behavior. However, it does provide a disaggregated view as to how the demand and supply patterns for different minerals relate to the interindustry structure and aggregate or macroeconomic variables of a national economy. The construction of input-output minerals models thus requires that some mathematical or econometric model be used in conjunction with an appropriately organized I-O table.

Theoretical Framework

The traditional input-output model has influenced the field of mineral modeling in two ways. First, it has made clear the usefulness of depicting the economic system by means of detailed categories. For such a purpose each producing sector is defined as a component of the system having a homogeneous output for a given technology. Second, it requires that production must satisfy not only final demand but also intermediate demand needed directly and indirectly to yield final demand. The main contribution of the traditional input-output structure is that it allows the list of final demands to be transformed into a vector of sectoral outputs.

The basic input-output structure is developed by dividing the intermediate and final demand activities of an economy into a number of sectors, which are arrayed in matrix form. The distribution of the sales and purchases of each industry is then estimated for each sector during a one-year period. Final demand can be further disaggregated into the components used in the national income accounts. Thus, the total final demand for the output of an industry is the sum of those components:

$$Y = C + I + G + T \qquad (39)$$

where*

$Y = [y_i]$ Final demand for the output of industry i ($i = 1, \ldots, n$),
$C = [c_i]$ Personal consumption expenditure for industry i output,
$I = [i_i]$ Private investment expenditure for industry i output,
$G = [g_i]$ Government expenditure component for industry i output,
$T = [t_i]$ Net export (exports minus imports) of industry i output.

*Square-bracketed, lower- or upper-case subscripted variables denote vectors or matrices.

The gross output of an industry is the sum of its sales to other industries and to final demand:

$$X = SL + Y \tag{40}$$

where

$X = [X_i]$ Gross output of industry i,
$S = [s_{ij}]$ Sales of industry i to industry j $(j = 1, \ldots, n)$,
$L =$ n dimensional unit vector.

Analogously, the gross output of an industry is the sum of its purchases from other industries and of value added:

$$X = S'L + V \tag{41}$$

where

$V = [v_i]$ value added by industry i.

Gross national product is measured as the sum of final demand (expenditure approach) or the sum of value added (income approach).

Up to this point, the input-output table is essentially a system of accounting identities. However, in situations where producers are regarded as having only a limited choice regarding factor input intensities and where adjustments to shifts in demand take the form of output quantity rather than price adjustments, the transactions table can be utilized to develop a general set of production or "technical" coefficients. Specifically, the set of technical coefficients can be derived from the S matrix. A technical coefficient is defined as the dollar input purchases from industry i per dollar output from industry j, or

$$A = [a_{ij}] \tag{42}$$

where

$$a_{ij} = s_{ij}/x_j$$

Substituting the value of x_{ij} from equation (42) into equation (40) yields the results:

$$X = AX + Y \tag{43}$$

This is equivalent to

$$(I - A)X = Y \tag{44}$$

where $I =$ an identity matrix.
From equation (44), one can find the "total requirements matrix," X

$$X = BY \tag{45}$$

where

$$B = [b_{ij}] = [I - A]^{-1}$$

Each b_{ij} represents the dollar output of industry i required, both directly and indirectly per dollar of final demand from industry j.

There are several important aspects of applying the completed tables. First of all, use of the I-O table for the economic analysis of minerals is limited largely by the national accounting purposes for which the tables are intended. Second, the I-O table is organized on the basis of dollar transactions, including the technical coefficients of the A matrix. While the dollar (or other currency) amounts can be used as surrogates for the underlying physical reality, there is no system of price deflation, for example, that will both preserve the physical constancy of dollars within input-output cells over time and yet reproduce an independently deflated constant currency GNP. There is also the problem that any given physical quantity of material is in reality marked up (or down) in value as it proceeds through successive stages of processing and manufacture. This produces anomalies, particularly in the case of circulating industrial scrap. Because it is marked down in value before being returned to earlier stages of processing, the recirculated scrap shows up as an input at only a fraction of its true value.

Thirdly, I-O tables as currently constructed make no provision for differences in total input requirements according to the particular kind of final demand. Whether a final product is destined for personal consumption, government, inventory, investment, or export, it is assumed to be exactly the same product and to generate the same direct and indirect unit input (including mineral) requirements. To the extent, therefore, that an I-O table seems to reveal differences in demand for mineral commodities that relate to differences in the kind of final demand, it will only be because of differences among the various kinds of demand in their respective product mixes. It follows that the amount of the calculated difference in inputs for the different kinds of final demand will depend heavily on the amount and kind of sector detail the table offers.

What makes input-output useful for mineral modeling purposes is the fact that the matrices can be treated as a series of producing sector requirements equations. These requirements (the independent variable) are in each case dependent, according to parameters stated in the matrix, upon exogenous final demand variables as well as upon other intermediate-industry variables further up the production chain. In addition some of the producing sectors are specifically mineral production and/or processing sectors. Thus some other configuration or levels may be substituted to determine the implications of the alternative exogenous conditions on the output requirements of the mineral sectors.

The effect of assuming different values for the parameters can also be investigated. However, given the requirement for complete consistency between inputs, outputs, and final demand, only the final demand "bill of goods" may in fact be treated as a fully independent set of variables. Other exogenous alterations (Rose, 1984), though useful for many practical purposes, may be tainted by the inconsistencies they introduce into the overall matrix. A rebalancing of the matrix may be employed to eliminate formal (accounting) inconsistencies, but will not necessarily preserve the kind of consistency among input-output parameters that was automatically provided in a historical recording.

The coefficients of such a matrix are linear and static, and imply constant returns to scale. Because of this static character, dynamic phenomena such as capital requirements must be handled in auxiliary fashion. Moreover, for the table to be balanced, the variables need to be expressed in terms of money, and the coefficients therefore represent value relationships. However, they may sometimes be treated as if also representing physical quantities, and subsidiary physical relationships may readily be added.

Mineral Applications

Input-output models are increasingly being incorporated into larger modeling frameworks and being made "dynamic" by being hybridized with other modeling techniques. The most frequent application to date has been to automatically link some econometrically determined variables with an I-O matrix. This linkage has taken mostly in the modeling of mineral demands and in the preparation of mineral demand projections. For a complete system one has to have a bill of final demand, derived externally from assumptions (or projections) of GNP. The external projections may already provide a good deal of the final consumption detail. It remains to convert from the original classification to the input-output classification and, as a rule, to constant currency of the same vintage as the input-output table. The rest is a matter of applying the inverse coefficients and reaggregating by mineral-supplying industries. The procedure can also be applied to determining the differential impact of changes in the level of final demand for particular products or changes in the level of output of particular end uses. There are technical flaws in the approach if the industries chosen for the simulation are not final enough, but the errors involved are generally not important.

One of the simplest of these mineral demand models is that designed by the Federal Emergency Management Agency (FEMA) to explain how future consumption of ferrous and nonferrous metals can be forecast by combining projections of material consumption ratios with input-output forecast of output of individual industries. Components of the modeling structure are given in the flow diagram of Fig. 3.9.11. They include an estimate of GNP derived from a macro forecasting model, an input-output table that

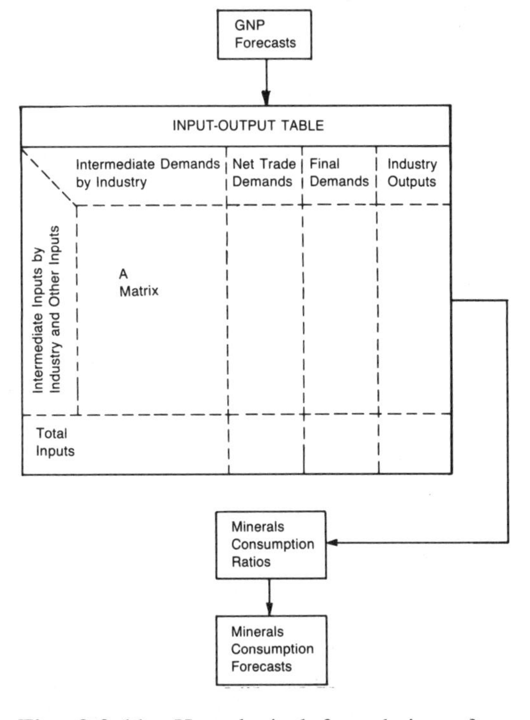

Fig. 3.9.11—Hypothetical formulation of an input-output mineral demand model.

converts the economic estimates to industry production estimates, and projected ratios reflecting materials consumption per unit of industrial output for individual industries.

The application of this model as reported by Kruegar (1976) employed an input-output table derived from one prepared by the Bureau of Economic Analysis of the U.S. Department of Commerce. Interindustry transactions are represented in dollars, with some 80 industries appearing in the actual table. The sum of final demands equals the sum of value added and total GNP expenditures, and the gross inputs for any industry are equal to the gross output of that industry.

GNP estimates are first disaggregated by industry to become a final demand vector, D, by utilizing a specially designed table embodying "demand impact transformation." Total direct plus indirect industry output requirements derived according to the relation

$$X = (I - A)^{-1} Y \qquad (46)$$

are then converted to mineral consumption by employing mineral consumption ratios (*MCR*). Historical *MCR*s have been calculated from past levels of mineral consumption and industry output for individual industries. Forecasts of these ratios MCR_i^*, together with the forecasts of the corresponding industry outputs X^*i, provide the mineral consumption forecasts, C_i^*.

$$C_i^* = MCR_I^* \times X_i^* \qquad (47)$$

Summing over all industries yields total consumption for each mineral. Thus far, the methodology has been applied to the planning of national stockpiles for a number of commodities, e.g., aluminum, chromite, lead, manganese, palladium, silver, tin, tungsten, vanadium, and zinc.

A much larger input-output model for the United States was similarly applied to determine mineral requirements and pollutant emissions in a project at Resources for the Future, reported on in *To Choose a Future* (Ridker and Watson, 1980). EPA's SEAS model which has as its core one of the earlier versions of the INFORUM (dynamic I-O) model developed by Almon et al. (1974) was modified for this purpose. Both SEAS and INFORUM incorporate procedures (mainly consumption functions) for moving the input-output matrix automatically forward in time, but this refers mainly to the mode of arriving at the details of final demand (including as one step the derivation of various price implications from the input-output solution). In this case assumptions have been made independent of the model for the rate of growth of GNP.

Based on an 185-sector input-output matrix, the bill of final demands is generated automatically, subject to a limited number of exogenous specifications. Most of the final demands are derived through regression equations for categories of personal consumption (consumption functions), linked to the input-output categories by a "bridge" matrix. There is provision, however, for judgmental overriding of the automatic determinations. There is a special matrix to convert residential and public construction into the input-output categories, and another one to convert the initially limited number of categories of government expenditures into the require array of input-output sectors. There is also a capital coefficient matrix to convert levels of both normal and pollution abatement investment into corresponding current flow purchases from each of the input-output sectors. The actual computation of the demand for minerals can be expressed in dollars or physical quantities. Mineral detail in this respect includes a number of "side-equations": 14 for mining and drilling, 7 for cement and stone, 26 for ferrous metal processing, and 30 for nonferrous metal processing.

An extension of this approach to the international level can be seen in the Leontief et al. (1982) model effort (IEA/USMIN). The minerals projections feature production as well as consumptions variables and are based on a systematic integration of the factors which determine domestic production, such as the level of final demand, import dependence, recycling rates and materials substitution. The consumption/production projections are generated for some 26 nonfuel minerals. Consumption includes final demand categories exports, imports, and changes in inventories. Production includes mine output, by-product output, imports, and releases from government stockpiles. Using technological updating for the 1972 I-O coefficients, mineral projections were prepared to the year 2000 for the United States and to the year 2030 on a global basis.

Projections and simulations of the foregoing types are the most common applications of input-output tables for mineral analysis, but they are not the only kinds. One type of application, which uses input-output more as an analytical tool than as a model, is to determine the relationship of mineral consumption in aggregative form to its various macroeconomic and demographic determinants. Input-output tables are particularly valuable for this purpose, since they necessarily account for the whole array of inputs into each consuming sector and thus produce aggregative data for minerals in general, as well as for major classes of minerals, at various stages of extraction and processing. By also permitting the determination of the direct and indirect mineral requirements related to any general category of demand (e.g., investment, durable goods) or of economic activity (e.g., construction, transportation), they make it possible to expose relationships that are not easily ascertained through any other statistical system. They are particularly useful in comparing differences in the intensities of mineral consumption for various ultimate purposes at different times and among different countries. A project involving these sorts of analyses, partly sponsored by the Bureau of Mines, was carried out at Resources for the Future.

An application for which input-output analysis is *not* suitable is determining the interrelationships among the demands for different minerals. When minerals are joint inputs into any given product, there is no way of determining from the input-output table the extent to which they are complementary or competitive, and therefore the direction or size of change in any one such input that is implied by any given change in another. The tables do serve as a very useful display, on the other hand, of these various coordinate relationships, thus facilitating external analysis. Such a display of interrelationships is not the least of any model's utility.

MODEL IMPLEMENTATION

The specification, estimation and simulation of a mineral model is a straight-forward quantitative exercise compared to the difficulties of formulating and implementing the model to emulate mineral reality very accurately. In this section we thus focus on the most important aspects of these difficulties: model validation, data requirements, substitution, technology, aggregation, imperfect market structures, and selection of methodology.

Validation

We cannot overemphasize the importance of validating mineral models. Without some idea of the performance of a particular mineral model, the model user cannot establish a level of confidence in the model's ability to analyze policy or to forecast the future. Econometric models are perhaps the easiest to validate. A battery of parametric and nonparametric tests, many of a statistical nature, can provide a good idea of the quality of a model. These range from tests of parameter significance and goodness fit to measures of deviations of actual from observed values and of capturing turning points. These and other tests of mineral model validation can be reviewed in a recent study by Labys and Pollak (1984).

In contrast to econometric models for which tests are available, the dominant criteria used to validate optimization, simulation, and input-output models usually are (1) how well the model replicates historical data, and (2) the sensitivity of the output to changing parameters and relationships.

Unfortunately, replication of historical data is an inadequate measure of the validity of any model, as it is usually possible to manipulate the variables and relationships so that a reasonable fit with historical data can be obtained. Moreover, even if the model passes such a test, one still cannot be confident that its projections will be accurate. This is particularly true when historical data are used to estimate coefficients in the model. In such cases, as Greenberger et al. (1976) point out, "the fact that the model traces past behavior reflects more on its ability to interpolate than extrapolate." There is some hope, however, in the fact that some rigidities in market structure will permit a certain replication of past behavior.

Several other validation approaches, nonetheless can be applied to optimization and simulation models. Labys (1982a) offers the suggestion of conducting sensitivity analysis based on some form of experimental design. Two advantages can be seen for this technique. First, the nature of the response of a model to its coefficients or variables can be stated in terms of statistical confidence. This, together with the insights a modeler can gain concerning the impacts of different coefficients or variables on model output, can result in a narrowing of confidence bands for that output. Second, the nature of this response can be compared to the kind of demand responses one would expect in the real world. In that same study, Labys also specifies a number of coefficient and other tests suitable for testing the validity of mineral-related input-output models.

Given the weaknesses of validation techniques for anything but econometric mineral models, the modeler at last resort can subject his structural relationships, data, assumptions, and output to a critical review by experts in the relevant technology, economics, and mineral fields. If such experts judge that the model, the data, and its output are "reasonable," it may then be assumed that it is an adequate basis for making projections. However, there can be no guarantee of this. There is no absolute measure of the "validity" of any model; there can only be a subjective level of confidence in the use of a model for specific purposes. Unfortunately, using the test of expert judgment is usually difficult or impossible because documentation of mineral models is often inadequate. Although professionally competent and honest modelers will always specify the assumptions that have been made, the sources and limitations of the data, and the uncertainties and caveats that are known or suspected, it is very difficult to document such limitations and uncertainties. The problem of inadequate documentation is a shortcoming that will have to be rectified before optimization and system simulation materials models reach their full potential.

Data Requirements

Compared to the other modeling methods discussed in this report, econometric models tend to require fewer data because they usually deal with aggregate relationships rather than composites of individual relationships. However, an important characteristic of econometric models is that aggregate relationships are estimated on the basis of past behavior. This requires time series data to be available for a considerable period in the past, but exactly how many is not a clear-cut issue. Too brief a history will clearly impede the analyst's ability to develop statistically valid relationships. An analyst usually wants to have information on at least two or three business cycles so that he can understand how the relationship under consideration may have been influenced by events of a cyclical nature. On the other hand, most forms of econometric analysis attach similar weights to the most recent and the most distant data. To the extent that structural change may be occurring, errors may develop if the estimation period is too long.

Mineral optimization and simulation models require data concerning a large number of properties and various categories of costs, particularly regarding supply or production. For process models, the derivation of these data requires engineering analysis of the production processes. The extent of the analysis is determined by the level of disaggregation of the technology itself as well as the market structure. At the highest level of aggregation (i.e., aggregate process inputs at the industry level), the data are relatively easy to obtain from government and industry publications, although the reliability is often suspect because of data collection difficulties. The detailed data (e.g., plant specific data) are very difficult to obtain because the analyst is required to sift through a variety of potential sources and then must visit production sites to get such data. This task is often complicated by the proprietary nature of the data.

In general, the data requirements for optimization and system simulation models tend to be more extensive than those for econometric models in that relatively more detail is included in these models. On the other hand, engineering models have the advantage of not necessarily requiring time series data, although components of system simulation models may utilize such data for purposes of statistical estimation.

Most input-output mineral models require some supplementation of the official or published country input-output models. Mineral demand projection or modeling rarely uses more than a handful of end use product or industry demands as the starting point, and frequently rests on nothing more than an aggregate for the GNP (or GDP). For input-output manipulation, this limited number of variables has to be restated as a detailed bill of final demands. If some amount of end use detail is exogenously assumed, that much can be converted from, say, standard industrial classification (SIC) to input-output classification by means of "bridge" or "crosswalk" tables that link the two classification schemes.

Regarding model output, there is a similar need to transform the total dollar (or other currency) mineral requirements generated by any given final bill of demand into quantities of crude or processed minerals. Even the most detailed recent U.S. table, for example, distinguishes only one specific kind of ore (copper) and four kinds of refined metals (primary copper, lead, zinc, and aluminum). Iron ore is lumped with ferroalloy ores, pig iron with all the other products of steel mills, scrap copper with all the other kinds of nonferrous scrap, and so on. In other words, for many mineral modeling purposes there has to be a supplementary data set (in effect, sets of parameters or equations) by which aggregates such as "chemical and fertilizer mineral mining," "electrometallurgical products," and "primary nonferrous metals," can be disaggregated. An alternative approach is to expand the official input-output table by disaggregating the mineral cells into smaller cells. This permits varying the relationship between the subcomponents and all of the cells outside the particular mineral industry, but it imposes additional data collection and estimating requirements (distribution of purchases and sales) and "cements in" the base table relationships.

There is an implicit requirement, of course, that the physical amounts assumed to correspond with the dollar amounts in the reference table actually do correspond. One cannot simply divide the dollar output of copper, say, by the quoted price of copper in the reference year. The actual amounts may be significantly different. Reports to the Census Bureau of the dollar value of shipments will not necessarily match those of the Bureau of Mines for the same year as to physical volume of production or shipments. At the same time, the physical production or shipment data reported to the Census Bureau, which should be consistent with the input-output table, may not be the most useful data to have as model output. The reference table's data specifications simply have to be investigated sufficiently to determine whether or not a model based on that table generates a valid index for the output series that is really desired.

A data requirement that is not essential to the use of existing tables but that is perceived by practitioners as critical to the utilization of input-output tables for projection purposes relates to the projection of changes in the "technical" (A-matrix) coefficients. Given the fact of continuing technological change, not to mention changes in product mix within input-output sectors, it is inevitable that these coefficients will in fact change over time. A simplistic way of making the adaptation is merely to extrapolate the trend in the coefficients over successive input-output tables. This has severe drawbacks such as the lack of coefficient data over time, their dependence on stages of the business cycle, and the lack of suitable price deflators. The alternative way of projecting the technical coefficients is to bring to bear independent information on the relevant technological trends. Not only does this impose a massive information requirement, but it also necessitates overcoming the difficulty of translating a limited number of specific physical trends into aggregate monetary values that usually encompass more than those few interindustry trading relationships.

Substitution

Mineral substitution, in practice, can take place at various levels of process. Conventional substitution occurs when one material replaces another because of cost. The basis for modeling this type of substitution is the assumption that producers of a given output are free to vary inputs in some systematic manner and do so in a way that minimizes the total cost of the output. The incorporation of such substitution effects in mineral models normally takes place by (1) the estimation and use of parameters, such as price elasticities and time responsiveness of demand, or (2) the analysis of the relationships between costs, properties, and demand for materials.

Concerning parameter estimation, econometric mineral models have been relatively powerful at estimating the direct-price and income elasticities of demand. For example, Clark and Mathur (1981) have illustrated how econometric procedures can be used to measure the elasticity of substitution between copper and aluminum in electrical conductor applications. In contrast, optimization and system simulation models can not give us such ostensibly precise information, since they are not based on large, historical data sets and are not statistically estimated.

However, there are two drawbacks to statistical techniques. One is that actual demand functions are usually nonlinear. The second is that statistical techniques can validly measure elasticities only within and around the limited range defined by historical data. They do not apply very well in instances where there is a large change in the price of material that takes it outside the realm of historical data. Engineering estimates, in contrast, can be designed to provide a good idea of the approximate range of the price responsiveness of demand, even when outside the range of historical experience. For example, King and Reddy (1981) illustrate a procedure for estimating the price elasticity and time-responsiveness of demand for manganese by engineering methods, and for incorporating such estimates directly into system simulation models. The method involves engineering analysis of specific measures that might be implemented to conserve or substitute for manganese at progressively higher prices than the 1980 level. Estimates are also made of the time required to carry conservation and substitution efforts to various stages of completion.

A second form of substitution simply is to use less of a given material in response to rising price—in other words, conservation. This is sometimes called technological substitution, since it is hypothesized that metal consumers are always trying to use less material, irrespective of price, out of a fundamental desire to minimize their production costs. The intensity of their efforts to do so, however, may be related to the extent of change in the market price. In the Chase (1982) econometric zinc model, for example, the reduced consumption of zinc for automotive die castings was attributed to the very high level of zinc prices in the 1973–74 period as well as to government actions that encouraged the production of lighter vehicles. During the late 1970s a technique known as thin-walled die casting became widespread and permitted the manufacture of metal parts with a reduced zinc content. While it cannot be proved that the advent of thin-walled die casting technology was caused by the very high price of zinc and the acute shortage of 1973–74, the fact is that these events followed one another, and the model thus hypothesizes a logical connection between the two. Engineering models also can capture the effect of technological change on materials substitution in the production process. As production technologies compete with one another for market shares, the consumption of input materials shifts and materials substitution can be accounted for indirectly.

Technology

It is often desirable that a model have the capability to address three general types of issues involving technology and the demand for materials. These are (1) technological change: the effects of a change in the technology of production; (2) technological progress: the efficiency of use of the effects of a change in materials by consuming industries; and (3) quality substitution: the effect of a change in the patterns of end use on the technological constraints.

An example of the first category is the substitution of the AOD process for the electric arc/double slag practice for making stainless steel. The advent of the AOD process, which makes efficient use of high carbon ferrochromium as a substitute for a large percentage of the more expensive low-carbon ferrochromium in the furnace charge, has a profound effect on the demand for raw material (including gases) by the stainless steel industry. Similar events, such as the replacement of the open hearth furnace by the oxygen and electric furnaces, and the substitution of continuous casting of ingot casting, have periodically shifted the requirements for various forms of raw materials in the carbon steel production process.

Two notable examples of change in technological progress regarding the use of materials have occurred in the automotive industry. First the use of copper in automobile radiators (on a per unit basis) has decreased over time as more efficient designs have permitted the use of thinner gauge sheet. Second, the advent of new foundry technology in the mid-1970s allowed automotive engine blocks to be manufactured from thinner-walled cast irons, resulting in significant weight savings.

The third category may be illustrated by observing the effects of the changes in consumer preferences on the demand side for materials in the automotive industry. The shift toward smaller, lighter, and thus more fuel efficient automobiles in the United States resulted in an increase in the demand for lightweight materials, such as fiber reinforced plastics, aluminum castings, and lightweight microalloyed steels, and a concurrent reduction in the demand for low carbon steels on a per vehicle basis.

Engineering process models are, in principle, ideally suited to addressing the first category of technological change. Both the optimization and the system simulation approach, when used to formulate models of the production sequence, inherently account for direct and indirect use of materials. For example, the system simulation models of the U.S. steel industry developed by Elliot and Clark (1977) for the Bureau of Mines have been used to analyze the effects of changes on technology variables (e.g., continuous casting versus ingot casting, open hearth versus oxygen and electric furnaces) on the requirements of industry for raw materials and fuels. If these models were to be used to forecast the demand for material inputs, it would also be necessary to forecast technological change.

In the situation where the change is occurring continuously over time, as in the substitution of continuous casting for ingot casting, predicting the future course of events is relatively straightforward. However, when the change occurs in a discontinuous fashion, it is quite difficult to use these models or any other models for forecasting purposes. For example, we could construct engineering process models for any of several technologies for direct steel making that are currently under development. However, since we do not have reliable information concerning the date when large scale applications of these processes will become technologically or economically feasible, it would be foolish to attempt to use these models for short or medium term forecasting. The best we can do in such a case is to analyze the implications of assumptions about future production via the new technology on the consumption of materials.

Econometric models also have a capability for incorporating technological change. This can be accomplished by employing a time proxy or some other explicit technological variable to take account of technological change. Such a proxy assumes that this change continues to occur on the same basis and to the same degree as it has during the past. The passage of time thus serves as a proxy for technological change. Where this change does not occur continuously but rather in the form of abrupt shifts, dummy variables can sometimes be employed to capture the impact of the particular technological change. Or the familiar S-shaped time-function can be introduced to embody the rate of adoption of particular innovations or technologies.

Imperfect Market Structures

Market structure—the number of firms in the industry and the degree of cooperation among them—is an important determinant of mineral market behavior. One frequently encountered modeling complication results from the fact that the structure of the markets in which mineral commodities are bought and sold are rarely perfectly competitive, e.g., see Labys (1980a). These markets can be characterized by few sellers—a single firm may be dominant in one market (for example, a single producer dominates the molybdenum market) and a single producer country may dominate another (for example, Zaire dominates the cobalt market). In addition, there may be only few uses for certain mineral commodities; for example, a large proportion of metallic sodium output goes into the production of lead-based antiknocks, and there are few firms in the antiknock market.

Just as mineral markets are seldom perfectly competitive, they also are rarely controlled by a single monopolist or monopsonist. Most mineral markets are intermediate in structure. Intermediate market structures are the most difficult to model because neither simple monopoly nor perfect competition theories (market structures whose behavior is best understood) are applicable. The result is that each mineral market must be studied individually to determine how prices are set and to understand the institutions that govern the way in which the market operates. Any model that ignores institutional reality in a particular mineral market will be a poor tool for forecasting future conditions in that market or for assessing the impact of proposed government policies and regulations. The latter can take the form of taxes and subsidies, quotas, price controls, or purchases for and releases from a stockpile.

Mineral markets are also characterized by various degrees of vertical integration. For example, some copper mining firms are integrated forward into smelting, refining, and fabrication, whereas others sell concentrates to custom smelters. The degree of vertical integration affects pricing and cost and therefore the quantity of a mineral demanded. Vertical integration makes it difficult to model mining without modeling metal manufacturing (smelting and refining), since the demand for the ore is essentially set at the refinery level. Finally, market structure can be noncompetitive in a spatial context for example, where one country or one region represents a sole world mineral source.

Econometric models have been shown as capable of explaining mineral market behavior of a monopolistic or oligopolistic nature. For example, Burrows (1971) developed a weak-monopolistic model of the cobalt industry capable of explaining industry demand and supply as well as price formulation. More elaborate explanations of oligopoly behavior have been based on econometric models but also have employed various optimization methodologies, e.g., see Labys and Pollak (1984). These methodologies have usually emulated more complex oligopolistic market structures, such as that of Stackelberger and Nash-Cournot bargaining theories. Most applications in this respect have involved world oil markets.

Several spatial as well as process programming models also have attempted to address mineral market behavior under noncompetitive market conditions reflecting possible cartel behavior. This work has proven significant not only because of the threat of cartel formation but also because of the increased recognition of the noncompetitive nature of price and quantity adjustments in a number of mineral markets, e.g., see Labys (1980a). For example, Hibbard and Soyster of the Virginia Polytechnic Institute constructed highly detailed and disaggregated dynamic linear programming models of the copper (Hibbard, et al., 1980) and aluminum (Hibbard, et al., 1979) industries. It was Soyster and Sherali (1981), however, who investigated the structure of the copper industry and its corresponding model (MIDAS-II) under the conditions of oligopolistic market structure. Although the authors employed a linear programming framework, Takayama and Judge (1971) already had stipulated that the quadratic programming class of models might be redefined to deal with the problem of optimal spatial pricing and allocation for duopolists, oligopolists under collusion, market sharing arrangements, monopoly or monopsony, and bilateral monopoly.

Soyster and Sherali based on the work of Murphy, modified their programming algorithm to achieve Nash-Cournot (1951) equilibrium conditions. While traditional approaches to reach such an equilibrium have usually relied on the complementary pivot theory of Lemke and Howson (1964) and of Scarf (1973), the modelers instead relied on solving a sequence of convex programming problems based on the underlying production conditions, i.e., Sherali et al. (1980). The results of their analysis showed that the incorporation of market structure significantly changed the spatial temporal and process solution of the model.

Selection

One of the most difficult tasks in model implementation is deciding which single methodology or which combination or hybridization of methodologies will best solve the modeling problem at hand. It is important to realize that there is no such thing as an "all-purpose" model. Different model structures lead to the solution of different kinds of mineral modeling problems. At the same time, however, some relation does exist between the requirements of model specification and the methodologies employed. Specifications that require temporal, process or spatial configurations might employ econometric, engineering process, or spatial equilibrium models respectively.

To help make these and more difficult decisions of model selection, we have compared the major methodologies of interest on the basis of their relative strengths and weaknesses. Obviously there are many factors that could be included in making such a comparison. The factors we have selected are incorporated in Table 3.9.1 and represent some of the more crucial decision-making points in constructing a mineral model. Among these, one important group helps to identify shifts in mineral consumption and production. Here we refer to the embodiment in models of technical compared to technological change as well as structural change. These factors are also inexorably related to materials substitution. One should also be concerned with the potential of a model to deal with short-run as compared to medium- or long-run time spans and forecasts. Model linkages either upstream or downstream as well as with other models reflects the potential of a methodology for hybridization. There also should be some potential for including macroeconomic effects. In this content, a methodology should be able to embody the impacts of cyclical change not only in the form of business cycles but also in the form of market instability or supply disruptions.

Econometric Models: We should first emphasize that, although econometric models are relatively strong in specific areas, this does not mean that econometric models will always be successful in these applications. For example, the nonavailability of data at an acceptable cost may preclude the use of econometric techniques in certain circumstances. Likewise, because econometric techniques are relatively weak in certain applications, it does not mean that they cannot be used for these applications. It merely indicates that great care must be taken, and the model builder must be prepared to modify the results of his forecast to take account of the technique's limitations.

As shown in Table 3.9.1, econometric mod-

Table 3.9.1—Strengths and Weaknesses of Mineral Modeling Methodologies

Strengths and Weaknesses of Econometric Models

Strengths	Weaknesses
Linking mineral demand with macroeconomic trends	Linking mineral demand with specific engineering decisions by minerals consumers
Short- and medium-term forecasting	Long-term forecasting
Analyzing cyclical instability	Analyzing structural change
Assessing impact of supply disruptions (strikes, cartels)	Assessing impact of demand disruptions (price control, rationing)
Projecting generalized technological and institutional change	Projecting specific technical change
Analyzing marginal reversible substitution	Analyzing radical irreversible substitution
Management of stockpiles (acquisition and disposal)	Managment of conservation and substitution programs

Strengths and Weaknesses of Engineering Models

Strengths	Weaknesses
Analyzing mineral demand in a well-defined range of specific industry applications	Analyzing diversified mineral demand in loosely defined applications
Analyzing nonmarginal substitution	Measuring price elasticity of demand
Analyzing technological change	Measuring income elasticity of demand
Long-term forecasting Analyzing structural change Analyzing specific consumer decisions of process technology changes	Analyzing business cycle and other macroeconomic impacts on demand

Strengths and Weaknesses of Input-Output Models

Strengths	Weaknesses
Accounting comprehensively for all materials	Ascertaining specific mineral requirements under changed assumptions or circumstances
Analyzing the effect of structural changes	Analyzing the effect of business cycles
Analyzing the impact of assumed technical or technological change	Accounting for the effects of unspecified ongoing technical change
Analyzing the effects of major macroeconomic change on mineral requirements	Analyzing the effects of minor macroeconomic change on mineral requirements
Requirements forecasting	Price forecasting.
Ease of use and understanding (nondynamic models)	Difficulty of construction

els are perceived as being strong at linking mineral demand with macroeconomic trends. Econometrics deals with aggregate and generalized relationships of this kind rather well. On the other hand, linking mineral demand with specific engineering decisions made by narrowly defined classes of mineral consumers is much less amendable to econometric treatment.

Thus the econometric model builder must supplement his generalized statistical analysis with specific investigations of the views of mineral consumers, especially where there are relatively few consumers or few applications for a mineral.

Econometric models are strong for short- and medium-term forecasting. The basic institutional structures that they embody, both in the national economy and in the industry, tend to remain relatively stable or move in a predictable manner. For very long term forecasting, such as a 15- to 25-year time frame, serious forecasting errors can result from major structural changes in the economy as well as in particular mineral markets. When econometric models are employed for such purposes, they usually are supplemented by the skills of "futurists" and other specialists in technological forecasting.

Econometric models are also perceived as being relatively strong in analyzing cyclical instability where the behavior of a specific mineral market in response to cyclical changes can be established from an analysis of past cycles. On the other hand, the econometric model is not very useful for handling structural change. A good example of this is when a market changes from being one characterized by monopoly conditions and administered prices to one that is competitive. Examples of such changes would be the zinc market in the mid-1970s, the aluminum and nickel markets in the late 1970s, and, potentially, the molybdenum market in the early 1980s. The willingness on the part of the econometric model builders to modify the price adjustment mechanism in the light of any perceptions of structural change in the market may be critical to the success of the model.

Econometric models are also relatively strong in assessing the impact of supply disruptions because of the central role of a market-clearing price mechanism. On the other hand, econometric models are less satisfactory for assessing the impact of demand disruptions, particularly rationing schemes. This is because supply disruptions tend to change the aggregate quantity of material supplied, whereas demand disruptions tend to change the rules under which the marketplace operates.

Econometric models tend to be strong in predicting generalized technical progress because such trends are typically well established over the historical period to which the model refers. By contrast, specific technological changes represent a break with past trends, and there is no way that an econometric technique can easily accommodate these, other than through the use of an engineering model adjunct to the econometric formulation. Clearly, considerable skill on the part of the econometric model builder is necessary to distinguish between developments that represent the trend of technical progress and those that represent a fundamental break with the past.

Because most econometric models are based on the assumption of perfect competition, they tend to be strong in analyzing marginal and reversible substitution of the classic economic kind. By contrast, radical substitution triggered by changes in price outside the realm of historical experience, and substitution involving asymmetric or irreversible changes, is much harder for the model to handle.

Finally, econometric models are relatively strong for developing policies for managing economic stockpiles, including, for example, international buffer stocks, since the objectives of the stockpile managers can be incorporated in a set of economically rational statistical relationships. On the other hand, good management of political (strategic) or military stockpiles is much more difficult to achieve with an econometric model. While such stockpiles are governed by clear behavioral rules, these rules are not a function of economic pressures but operate irrespective of market conditions.

Engineering Models

System stimulation models are strongest for exploring the effects of changes in technology, policy, or economic variables on the markets for materials. Because of their capability for incorporating considerable engineering detail related to the production and use of materials, these models are particularly well suited to analyzing the effects of changes in variables that are directly related to production costs and/or design decisions. Examples are listed below.

Technology: (1) Analyzing the effects of a change in the design of a product on the demand for specific materials; e.g., the effects of redesigning automobiles produced in the United States on the demand for materials. (2) Analyzing the effects of a new production process, such as direct casting of carbon steel sheet, on the cost of production.

Policy: Analyzing the effects of changes in legal restrictions or design codes on the potential demand for a material, e.g., the effects of removing restrictions on aluminum wire in buildings on the relative demand for copper and aluminum electrical conductors.

Economic: Analyzing the effects of changes in output of a specific product on material requirements, e.g., the effects of change in the production of specific jet engines on the consumption of constituent superalloys.

The ability to incorporate engineering detail also allows us to investigate the effects of changes in the variables listed on substitution. These models have been used to analyze the effects of changes in factors that influence the costs or properties of finished materials on their competitive position vis-a-vis substitutes. For instance, the system dynamics copper-aluminum substitution model provides a framework for analyzing the effects of changes in the prices of competitive materials, and the costs of fabricating, maintaining, and using them, on the total costs of producing a functional unit. The flexibility and the ability of the system simulation structure to incorporate detail also allow the analyst to include information about time lags, asymmetries, and irreversibilities associated with the substitution process within the model.

While these models are strong in incorporating disaggregate detail, they are relatively weak in measuring parameters related to the supply of and demand for materials. They cannot be used to measure directly (1) elasticities of supply or demand, elasticities of substitution, direct or cross-price elasticities, income elasticities, or (2) time lags associated with the independent variables that determine supply and demand. Thus the system simulation models are dependent upon other techniques, such as econometrics or engineering analysis, for estimates of these parameters.

These models are also relatively weak in short- and medium-term forecasting, with one exception. The most straightforward specification of a demand function (which uses an income variable estimated by engineering analysis and which usually neglects price effects) can provide accurate short-term forecasts when the material requirements per functional unit are known, there are few applications within an end use sector, and the short-run price elasticity of demand is negligible. Other system simulation models are not usually used for forecasting, except in the long term. These models are relatively strong for long-term forecasting only because the task is impossible for any other type of model.

The strength of the simulation approach is that it allows us to incorporate future technological change that we have good information about. Therefore, if we are forecasting the demand for copper, we can specify that at a certain point in the future the use of fiberoptics technology will reduce the demand for copper in the telecommunications industry by a specified proportion. Alternatively, we could specify an engineering production function for a new production technology, such as the liquid dynamic compaction process for producing rapidly solidified metal powders, to simulate future costs of production. Of course, if the past is a guide to the future, we can be assured that there will be numerous changes in technology, as well as in other variables, that we are not able to anticipate. Thus we cannot be confident that any of the methodologies currently at our disposal can provide useful long-term forecasts.

Another factor to be considered is model linkage. System simulation models are relatively strong in linking either the demand for materials with specific engineering decisions made by consumers of materials, or the costs of production with specific production technologies. This is because the models can be made as detailed or as disaggregated as one desires. On the other hand, they are less useful for linking supply/demand relationships with macroeconomic trends. This is also a result of the disaggregate nature of the models. Thus a system simulation model of the aluminum industry might be capable of representing the effects of decisions by aircraft manufacturers to use significant quantities of aluminum-lithium alloys in the future, but it would not be useful for relating the demand for aluminum to changes in the gross national product or the aggregate industrial production index.

There are two major weaknesses in system simulation models, both of which are correctable but seldom are corrected. First, the demand for input materials is usually calculated simply as the requirement for the input material per unit of output. The price of the input material is implicitly assumed to be unimportant in the range of interest. This may be a good assumption in some cases, but it generally is not. There are two ways to handle this problem. One is to use engineering analysis to estimate the response of the production process to changes in the price of the input material, as demonstrated by King and Reddy (1981) for manganese used in steelmaking. The second is to include an optimizing routine that allows the model to alter the mix of input materials, depending on relative prices. This has been use by Clark et al. (1981) to model the demand for scrap and hot metal in steelmaking.

A second weakness is related to the fact that the demand function for finished materials must either be supplied exogenously or estimated simultaneously with the supply function. The latter option is seldom taken because simultaneous estimation is time-consuming and often expensive. The problems with using an exogenously specified demand function are twofold. First, there is the likelihood of an identification problem if both the supply and the demand function have some price sensitivity. Second, an incorrect demand function is often used because there is nothing else available. If process analysis has been used to estimate a supply function for primary copper in the United States, for example, it is appropriate to combine this function with the residual demand curve facing U.S. primary producers (netting out supply and demand in the rest of the world) to simulate market clearing prices. However, since it is difficult to derive such a residual demand function, it has been the practice in the past to estimate the demand for copper in the United States with an econometric model. This is not appropriate when demand includes the demand for imports and scrap as well as the demand for domestically produced primary copper.

The point is that unless these models are combined with proper and compatible demand functions, they are not suitable for simulating market behavior. If they are combined with appropriate demand functions, they are as capable of providing simulations of the dynamic paths of prices, supply and demand as econometric models are. Thus a properly structured system simulation model is useful for analyzing issues similar to those that have been the province of econometric market models in the past. These issues include (1) the impact of supply disruptions; (2) analyzing cyclical instability; (3) analyzing policies on the management of stockpiles; and (4) analyzing trade policies.

Optimization models have many of the same strengths and weaknesses that engineering system simulation models do. They also have some characteristics that are quite different because of the nature of the solution procedure. In contrast to system simulation models, which usually use judgmental methods to project trends in the mix and location of production processes, process optimization models are concerned with the choice of the economic mix of resources and technologies. In order to model such a choice it is usually necessary to assume that the firms in the industry act collectively (as if they were a single firm) to minimize cost. Thus optimization is the normative process, indicating the preferred decisions or directions the industry should take, given the assumptions made in the analysis about the form and nature of the constraints and the demand target (objective function). Since this procedure provides estimates of the lowest feasible bound to the cost of supplying the final products, its major strength is that it can provide a benchmark for evaluating the actual functioning of firms or industries.

However, the process optimization approach has limited utility for forecasting or even predicting the state of the system being modeled. In formulating an optimization model, it is usually necessary to assume that firms and other resource owners either minimize costs or maximize profits over a specific period of time. In reality, we know that firms usually have other objectives—such as greater market share, employment, growth, or even short-term survival—in addition to profits. Moreover, the relative importance of the various factors, and the time frame in which decisions are made, varies from firm to firm.

Further complicating the problem is the fact that it has proven difficult to build robust optimization models of imperfect competitive markets and yet a large majority of the materials markets is not competitive in the economic sense. It is the nature of most of the minerals and materials markets that a small group of firms or countries have a significant concentration of market power. Thus, given the nature of the problems (a large array of possible motivations among firms or countries, different time horizons, imperfect competition), it is not surprising that process optimization models have not proven to be useful for forecasting or simulating the state of the system.

Regarding the spatial equilibrium form of optimization model, this methodology to some extent represents a hybridization of the econometric and the engineering process approaches. It thus possesses some combination of the strengths and weaknesses of the separate methodologies. Among its strengths, demand as well as supply elasticities are embodied in the included econometric equations. This provides a linkage with macroeconomic activities and helps explain mineral substitution. Where a temporal structure is employed, it can analyze market instability and market disruptions. Substitution analysis can take place not only in the econometric sense but also by including technological factors in the stage of process analysis that is convenient to mathematical programming formulations. Its extensions from QP to LCP and MIP have made the approach particularly adept at including mineral investment decisions of a long-run nature. Concerning its weaknesses, the approach tends to be aggregative, thus omitting detail on the many derived demands for a mineral. Structural change also cannot easily be handled, since the econometric equations are restricted to the particular time period of estimation. This also limits its potential to include technological change in forecasting future demands and supplies.

Input-Output Models: Before referring to the tabulation of Table 3.9.1, it is important to realize that a major advantage of I-O for modeling is its comprehensiveness. This is the only model in which the aggregate of all mineral materials (among other materials and inputs into productive activity) is accounted for. It is thus a particularly valuable source for making analyses of total mineral utilization and observing the mix of such utilization (at least on a monetary value basis) according to different categories of minerals. Input-output tables also provide a ready way of observing the separate implications for mineral consumption of different kinds of final demand, different end use industries, and, in fact, any other specific sector. The susceptibility of the I-O matrix formulation to manipulation by matrix algebra also makes possible the calculation of all indirect and direct mineral requirements.

The very comprehensiveness of the system is, at the same time, a severe drawback if the mineral modeler cannot work exclusively with a published input-output table. If it is necessary to adjust available I-O matrices for desired evaluations of different assumptions, a great deal of work has to be done in order to end up with full consistent alterations, either of the transactions matrix or of the technical coefficients. Worse still is the situation where (for example, in many foreign countries) an input-output table would have to be compiled from scratch. Even where reliable tables exist, considerable work may have to be done to expand such tables with newly acquired data or estimates in order to provide the detail necessary for adequate mineral market analysis. The shortcut of estimating mixes within more aggregative input-output sectors sacrifices the opportunity for precise calculation of indirect flows and requirements that the input-output system potentially offers.

Even the most detailed I-O tables, if used mechanically to determine the implications for mineral inputs of different levels of output, suffer from problems of aggregation. An often cited example relates to mineral requirements for containers. The most recent detailed U.S. table (496-order) contains sufficient industry breakdown to distinguish among metal barrels and drums, metal cans, and glass containers, but it does not distinguish among aluminum cans, tinplate cans, and hybrids. Thus use of the table to evaluate the mineral requirements implied by growth in the consumption, say, of malt liquors would generate some amount of derived demand for steel and tin. But the actual metal input into the portion of beer consumption that is purveyed in cans is aluminum.

The coefficients in any given I-O table may suffer from the accidental effects of stage of the business cycle. For cyclical reasons or otherwise, input-output ratios may be affected by the extent of capacity utilization, but even when this is not the case it may be invalid to assume that they do not change with scale. Unless the tables are used to model only minimal changes, it is almost inevitable that scale differences will imply significant changes in the production process and efficiency mix. This is in addition to the technological change that takes place over time. Thus for an input-output table to be valid as a model for another time and place, various adjustments have to be made.

Since even the most homogeneous of input-output sectors has some element of heterogeneity, the problem of external adjustment is inevitably complicated by the need to move back and forth between physical analysis and value composites. This affects the handling both of substitution and of technological change. Normally physical terms provide some basis for introducing such assumptions into a model, but the unit values of these physical units, as they combine to make up the sectoral value composites, are rarely ascertainable. It is necessary to fall back upon indexing assumptions, with unknown and sometimes wide degrees of hazard.

It follows from the randomness of given I-O tables in relation to the business cycle (and from the randomness in the coefficients that is to be expected from other causes as well) that the tables have serious drawbacks as models for predicting even marginal change. However, the more nearly homogeneous the inputs and outputs linked by an coefficient, the less the particular coefficient is going to be affected by purely economic fluctuations—i.e., the closer it will come to being a purely technical coefficient, valid for limited amounts of change within a limited time period. So far as radical change is concerned, no model that is fitted to a particular period in history can be expected to be valid for other periods without appropriate adjustment.

Once an I-O matrix has been established and "inverted," the cost of utilizing it for answering many kinds of mineral requirements questions is quite small, far less than for most other kinds of detailed models. To determine the effect of different arrays of final demand on the consumption of any given mineral (assuming the mineral is represented by a single input-output sector or that connecting ratios have been established), it is unnecessary to run the whole model (matrix). Since the predetermined inverse coefficients can be utilized, evaluation of the implications for a particular mineral or minerals of changes in just one or two final demand (or intermediate production) variables takes hardly any time at all.

The comprehensiveness of input-output formulations should not be taken to signify that they are full-equilibrium models. The base tables are in equilibrium because they represent a recording of history. At least they represent as much of an equilibrium as ever exists in fact rather than in theory. Basically, they are exclusively requirements models. There is nothing of the supply or price interaction that can be obtained only by linking the model with a feedback system. I-O models, therefore, are most often operated only in conjunction with some other modeling methodologies.

It is often cited as a drawback of I-O analysis as a modeling device that it is static rather than dynamic. This can be overstated. Even if there is no attempt to provide automaticity of movement from one period to the next, the need to use successive static models to represent successive cross-sections of time is far from being a fatal defect. In fact, it may be an advantage to be able to enter deliberate assumptions and parameters arrived at through independent analysis and investigation rather than to rely upon the self-perpetuating system that may propagate growing errors.

One of the advantages of I-O analysis, in fact, is its transparency to the user. The direct relationship of each dependent variable to each independent variable is plainly laid out in the basic transactions table for anyone to see. The unit of measurement is the same for all variables, and the parameters are always in terms of dollars or similar equivalents in other currencies. Once explained, the meaning of the inverse (or "total input") coefficients is plain. Because such coefficients are derived from the empirically established direct coefficients, they are above suspicion.

As an analytical tool for evaluating the determinants of mineral consumption at any given time in any given country, input-output matrices, if sufficiently disaggregated, can be of great value. If price deflation problems can be overcome, these matrices can also be very valuable for comparing differences in the determinants of mineral consumption between one period and another in any given country, and, if, in addition, exchange rate problems can be overcome, they can be useful in comparing differences among countries.

But input-output matrices suffer from being point observations, not even averages of several years' running. For dynamic simulation or projection purposes they need laborious and very costly adaptation, unless confined to simulation of the effects of relatively marginal changes, without significant cyclic, scale, or technological implications. Despite their great detail these matrices often suffer severely from problems of aggregation, and where they do not already exist, they are rarely worth developing for purposes of mineral requirements analysis alone.

POSTSCRIPT

We trust that the reader has followed the formulations and the comparisons of the various mineral modeling methods. Mineral modeling represents an active and growing interest of mineral economists. The hope is for modeling of a more theoretically sound and more empirically valid nature. This chapter will fulfill its purpose if that goal is realized.

NOTES

[1]Thanks are due to the following authors whose materials have been employed to enhance the mineral application sections of this text: W.C. Labys and Peter K. Pollak, *Commodity Models for Forecasting and Policy Analysis* (London: Croom-Helm, 1984); S. Gupta, *World Zinc Industry* (Lexington: D.C. Heath and Co., 1981); National Academy of Sciences, *Mineral Demand Modeling* (Washington, DC: National Academy Press, 1982); C.S. Tsao and R.H. Day, "A Process Analysis Model of the U.S. Steel Industry," Management Science *17*, pp. 588–608, 1971; F.R. Field, J.V. Busch, R.E.O. Piret and J.P. Clark, "A System Simulation Model of the World Magnesium Industry," Working Paper No. 21, Materials Systems Laboratory, M.I.T.

References

1. Adams, F.F., 1972, "The Impact of the Cobalt Production from the Ocean Floor: A Review of Present Empirical Knowledge and Preliminary Appraisal," A study prepared for the United Nations Conference on Trade and Development, Wharton School of Finance, Philadelphia, PA.
2. Almon, C., Jr., M.B. Buckler, L.M. Horwitz and T.C. Reimbold, 1974, *Interindustry Forecasts of the American Economy*. Lexington Books, Lexington, Mass.: D.C. Heath and Company.
3. Brown, M., A. Dammert, A. Meeraus and A. Stoutjeskijk, 1982a, *Worldwide Investment Analysis: The Case of Aluminum*, Preliminary Report, The World Bank.
4. Burrow, J., 1972, "Analysis and Model Simulations of the Non-Ferrous Metal Markets: Aluminum." Special Report, Cambridge, Mass.: Charles River Associates.
5. Burrows, J., 1971, *Cobalt: An Industry Analysis*. Lexington, Mass.: Heath Lexington Books.
6. Charles River Associates, 1978, "The Economics and Geology of Mineral Supply: An Integrated Framework for Long-Run Policy Analysis." Report N. 327, CRA, Boston, MA.
7. Charles River Associates, 1976, *Modeling Analysis of Supply Restrictions in the Minerals Industry*. Cambridge, Mass.: Experimental Technology Improvement Program.
8. Charles River Associates, 1976, *Econometric Model of the World Nickel Industry*. Boston, Mass.: Charles River Associates.
9. Clark, J.P. and A. Church, 1981, "Process Analysis Modeling of the Stainless Steel Industry." Paper presented at the Workshop on Nonfuel Minerals Demand Modeling held at Airlie House, Warrenton, VA, June 1–2, 1981 (unpublished). Board on Mineral and Energy Resources, Committee on Nonfuel Mineral Demand Relationships. Washington, DC: National Academy of Sciences.
10. Clark, J.P. and S.C. Mathur, 1981, "An Econometric Analysis of the Substitution Between Copper and Aluminum in the Electric Conductor Industry." Paper presented at the Workshop on Nonfuel Minerals Demand Modeling held at Airlie House, Warrenton, VA, June 1–2, 1981 (unpublished). Board on Mineral and Energy Resources, Committee on Nonfuel Mineral Demand Relationships. Washington, DC: National Academy of Science.
11. Clark, J.P., J. Tribendis and J. Elliott, 1981, "An Analysis of the Effects of Technology, Policy and Economic Variables on the Ferrous Scrap Market in the United States." Working Paper, Department of Material Science and Engineering, Massachusetts Institute of Technology.
12. Copithorne, L.W., 1973, *The Use of Linear Programming in the Economic Analysis of a Metal Industry: The Case of Nickel*, monograph, Department of Economics, University of Manitoba.
13. Crowson, P.C.F., 1981, "Demand Forecasting: A Mining Company's Approach," Paper presented at the Workshop on Nonfuel Minerals Demand Modeling held at Airlie House, Warrenton, VA, June 1–2, 1981 (unpublished). Board on Mineral and Energy Resources, Committee on Nonfuel Minerals Demand Relationships. Washington, DC: National Academy of Sciences.
14. Dammert, A., 1980, "Planning Investments in the Copper Sector in Latin America," in W. Labys, M. Nadiri and J. Nunez del Arco (eds.), *Commodity Markets and Latin American Development: A Modeling Approach*, New York: National Bureau of Economic Research.
15. Dantzig, G.B., 1949, "Programming in Linear Structure," *Econometrica* 17:73–74.
16. Desai, M., 1966, "An Econometric Model of the World Tin Economy." *Econometrica* 34, 105–134.
17. Dorfman, R., P. Samuelson and R. Solow, 1950, *Linear Programming and Economic Analysis*, New York, NY: McGraw-Hill Book Company.
18. Elliott, J.F. and J.P. Clark, 1977, "Mathematical Modeling of Raw Materials and Energy Needs of the Iron and Steel Industry in the USA." Open-file Report 32–78. Washington, DC: U.S. Bureau of Mines.
19. Fabian, R., 1963, "Process Analysis of the U.S. Iron and Steel Industry." In A.S. Manne and H.M. Markowitz (eds.) *Studies in Process Analysis: Economy-Wide Production Capabilities*, Cowles Foundation for Research in Economic Monograph No. 18, John Wiley & Sons, New York.
20. Fisher, F.M., P.H. Cootner and M. Bailey, 1972, "An Econometric Model of the World Copper Industry." *Bell Journal of Economics and Management Science* 3, 568–609.
21. Forrester, J., 1961, *Industrial Dynamics*. Cambridge, Mass.: MIT Press.
22. Forrester, J., 1971, *World Dynamics*. Cambridge, Mass.: Wright-Allen Press.
23. Ghosh, S., C.L. Gilbert and A.J. Hughes, 1981, "Optimal Control and Choice of Functional Form: An Application to a Model of the World Copper Industry." *Discussion Paper Series 8109/E*. Erasmus University, Institute for Economic Research, Rotterdam.
24. Greenberger, M., M. Crenson and B. Crissey (1976) *Models in the Policy Process*. New York, NY: Russell Sage Foundation.
25. Gupta, S., 1981, *World Zinc Industry*. Lexington, Mass.: Heath Lexington Books.
26. Hashimoto, H., 1981, "A World Iron and Steel Economy Model: The Wise Model." Staff Commodity Working Paper No. 6, World Bank, Washington, DC.
27. Hibbard, W.R. et al., 1979, "An Engineering Econo-

metric Model of the US Aluminum Industry," *Proceedings of the AIME*, New York.

28. Hibbard, W.R., A.L. Soyster and R.S. Gates, 1980, "An Disaggregated Supply Model of the US Copper Industry Operating in an Aggregated World Supply/Demand System," *Materials and Society*, 4(3), 261–284.
29. Hibbard, W.R. et al., 1982, "Supply Prospects for the U.S. Copper Industry: Alternative Scenarios: Midas II Computer Model," *Materials and Society*, Vol. 6, No. 2, pp. 201–210.
30. Hogan, W.H. and J.P. Weyant, 1980, "Combined Energy Models," Discussion Paper E80-02, Kennedy School of Government, Harvard University.
31. Kennedy, M., 1974, "An Economic Model of the World Oil Market," *Bell Journal of Economics and Management Science*, Vol. 5, No. 2, pp. 540–577.
32. King, T.B. and B.J. Reddy, 1981, "Analysis of the Effect of Price Increases on the Demand for Manganese." Paper presented at the Workshop on Nonfuel Minerals Demand Modeling held at Airlie House, Warrenton, VA, June 1–2, 1981. Board on Mineral and Energy Resources, Committee on Nonfuel Mineral Demand Relationships. Washington, DC: National Academy of Sciences.
33. Kovisars, L., 1976, "World Production Consumption and Trade in Zinc—An LP Model," U.S. Bureau of Mines Contract Report J-0166003, Stanford Research Institute, Stanford.
34. Kovisars, L., 1975, "Copper Trade Flow Model," *World Minerals Availability*, SRI Project MED 3742-74.
35. Krueger, P.K., 1976, "Modeling Futures Requirements for Metals and Minerals." *Proceedings of the XIV Symposium of the Council for the Application of Computers and Mathematics in the Minerals Industry*. University Park, PA: The Pennsylvania State University.
36. Labys, W.C., 1981, "A General Disequilibrium Model of Commodity Market Adjustments." National Science Foundation Report, Department of Mineral and Energy Resource Economics, West Virginia University, Morgantown.
37. Labys, W.C., 1980b, "A Model of Disequilibrium Adjustments in the Copper Market." *Materials and Society* 4, 153–164.
38. Labys, W.C., 1978, "Commodity Markets and Models: The Range of Experience," in F.G. Adams and S. Klein (eds.) *Stabilizing World Commodity Markets: Analysis, Practice and Policy*. Lexington, Mass.: Heath Lexington Books.
39. Labys, W.C., 1982a, "Measuring the Validity and Performance of Energy Models," *Energy Economics*, Vol. 4, pp. 159–168.
40. Labys, W.C., 1973, *Dynamic Commodity Models: Specification, Estimation and Simulation*. Lexington, Mass.: Heath Lexington Books.
41. Labys, W.C., 1977, "Minerals Commodity Modeling: The State of the Art." *Proceedings of the Mineral Economics Symposium on Minerals Policies in Transition*. Washington, DC: Council of Economics of the AIME, 80–106.
42. Labys, W.C., 1980a, *Market Structure, Bargaining Power and Resource Price Formation*, Heath Lexington Books.
43. Labys, W.C., 1982b, "A Critical Review of International Energy Modeling Methodologies," MIT Energy Lab Working Paper, No. 82-034WP, Massachusetts Institute of Technology.
44. Labys, W.C., and M.A. Kaboudan, 1980, "A Short Run Disequilibrium Model of the Copper Market," NSF Project Report DAR 78-08810, Working Paper No. 16, Department of Mineral and Energy Resource Economics, West Virginia University, Morgantown.
45. Labys, W.C., and C.W. Yang, 1980, "A Quadratic Programming Model of the Appalachian Steam Coal Market," *Energy Economics*, Vol. 2, pp. 86–95.
46. Labys, W.C. and P.K. Pollak, 1984, *Commodity Models for Forecasting and Policy Analysis*, Croom-Helm.
47. Lemke, E.C. and J.T. Howson, Jr., 1964, "Equilibrium Points of Bimatrix Games," *Journal of Society of Industrial Application of Mathematics*, Vol. 12. pp. 413–423.
48. Leontief, W., J. Koo, S. Nasar and I. Sohn, 1982, "The Production and Consumption of Non-Fuel Minerals to the Year 2030 Analyzed Within an Input-Output Framework of the U.S. and World Economy." Technical Report, Institute for Economic Analysis, New York University.
49. Lofting, E., 1979, *The Input-Output Structure of the U.S. Mineral Industries*.
50. Meadows, D.L., 1970, *Dynamics of Commodity Production Cycles*, Cambridge: Wright-Allen Press.
51. Nash, J., 1951, "Non-Cooperative Games," *Ann. Math.* 54(2), 286–295.
52. National Academy of Sciences (NAS), 1982, *Mineral Demand Modeling*, Committee on Nonfuel Mineral Demand Relationships, National Research Council, Washington, DC, National Academy Press.
53. National Research Council, 1972, "Mutual Substitutability of Aluminum and Copper." NMAB 286, National Materials Advisory Board, Washington, DC: National Academy of Sciences.
54. Nelson, J.P., 1970, "An Interregional Recursive Programming Model of the U.S. Iron and Steel Industry 1947–67," Ph.D. Thesis, University of Wisconsin.
55. Newberry, D.M. and J.E. Stiglitz, 1981, *The Theory of Commodity Price Stabilization*, Clarendon, Oxford University Press.
56. Nziramasanga, M. and C. Obideguri, 1981, "Primary Commodity Prices Fluctuations and Developing Countries: An Econometric Model of Copper and Zambia," *Journal of Developing Countries* 9.
57. Ogawa, K., 1982, "A New Approach to Econometric Modeling: A World Copper Model." NSF Report, Department of Economics, University of Pennsylvania.
58. Ogawa, K., 1983, "A Theoretical Appraisal of Price Stabilization Policy Under Alternative Expectational Schemes." Discussion Paper No. 9, Faculty of Economics, Kyobe University, Japan.
59. Pindyck, R.S., 1978, "Gains to Producers from the Cartelization of Exhaustible Resources," *Review of Economics and Statistics* 60, 238–51.
60. Ray, W.H. and J. Szekely, 1973, *Process Optimization: With Application in Metallurgy and Chemical Engineering*, New York, NY, John Wiley & Sons.
61. Richard, D., 1977, "A Dynamic Model of the World Copper Industry," Working Paper, International Monetary Fund, Washington, DC.
62. Ridker, R.G. and W.D. Watson, 1980, *To Choose a Future*, Baltimore, MD, The Johns Hopkins University Press.
63. Rose, A., 1984, "Technological Change and Input-

Output Analysis: An Appraisal,'' *Socio-Economic Planning Sciences*, forthcoming.
64. Rose, A. et al., 1983, ''Non-Renewable Resources and the Development of Arid Lands,'' in UNITAR, *Alternative Strategies for Desert Development and Management*.
65. Samuelson, P.A., 1952, ''Spatial Price Equilibrium and Linear Programming,'' *American Economic Review*, Vol. 42, pp. 283–303.
66. Scarf, H., 1973, *The Computation of Economic Equilibria*, Yale University Press, New Haven, Conn.
67. Smith, G.W. and G.R. Shink, 1976, ''The International Tin Agreement: A Reassessment,'' *Economic Journal*.
68. Smithson, C.W. et al., 1979, *Mineral Markets: An Econometric and Simulation Analysis*, Ontario, Canadian Ministry of Natural Resources.
69. Soyster, A. and H.D. Sherali, 1981, ''On the Influence of Market Structure in Modeling the U.S. Copper Industry,'' *International Journal of Management Science*, pp. 381–388.
70. Sparrow, F. and A. Soyster, 1980, ''Process Models of Minerals Industries.'' *Proceedings of the Council of Economics of the AIME*, New York, NY: American Institute of Mining, Metallurgical, and Petroleum Engineers, 93–101.
71. Takayama, T. and G.G. Judge, 1971, *Spatial and Temporal Price and Allocation Models*, Amsterdam, North-Holland Publishing Company.
72. Takayama, T. and W.C. Labys, 1984, ''Spatial Equilibrium Analysis: Mathematical and Programming Model Formulation of Agricultural, Energy and Mineral Models,'' in P. Nijkamp (ed.) *Handbook of Regional Economics*, Amsterdam, North-Holland Publishing Company.
73. Trozzo, C.L. 1966, ''Technical Efficiency of the Location of Integrated Blast Furnace Capacity,'' Ph.D. Thesis, Harvard University.
74. Tsao, C.S. and R.H. Day, 1971, ''A Process Analysis Model of the U.S. Steel Industry,'' *Management Science* 17, 588–608.
75. Uri, N., 1976, *Toward an Efficient Allocation of Electric Energy*, Lexington, MA, Heath Lexington Books.
76. Vogely, W.A. (ed.), 1975, *Mineral Materials Modeling*, Washington, DC., Resources for the Future and Johns Hopkins University Press.
77. Yaksick, R. (ed.), 1981, ''Collected Papers from the Workshop on Nonfuel Minerals Demand Modeling,'' Airlie House, Warrenton, VA, June 1–2, 1981 (unpublished). Board on Mineral and Energy Resources, Committee on Nonfuel Mineral Demand Relationships, Washington, DC, National Academy of Sciences.

Part 4

STRUCTURE AND PERFORMANCE OF THE MAJOR MINERAL SECTORS

4.10

The Metals

John E. Tilton*

INTRODUCTION

Mineral commodities are normally separated into three generic classes—metals, nonmetals, and energy minerals including oil and gas as well as the solid fuels. Metals, the focus of this chapter, encompass a large number of different substances. The U.S. Bureau of Mines, for instance, has commodity specialists following trends in over 40 metal products of importance to the country's economy and well being.

Ranging from aluminum to zirconium, the metals display an incredible degree of diversity. Some such as lead are heavy, others such as magnesium are light. Some such as copper are good conductors of electricity. Others such as silicon are semiconductors. Mercury is found in liquid form, while some metals melt only when heated to extremely high temperatures.

Iron and steel, aluminum, and copper are consumed in particularly large tonnages in a multitude of end uses, while many minor metals are needed in only small amounts in a few highly specialized applications. The use of some metals can be traced back into history for millennia, indeed back to the bronze and iron ages, while the commercial consumption of aluminum and other newer metals is less than a hundred years old.

Some metals are extracted from large open pits, others are dug out of deep underground mines, and still others are processed from the sea. Mining and processing can be relatively uncomplicated and inexpensive, though in most instances highly sophisticated technology is necessary and the costs are high. Some metals are produced mainly as byproducts of other metals. Some are recovered in large quantities from the scrap of obsolete equipment and demolished buildings. Some are mined in only a few locations and traded worldwide, others are produced in many different countries. Some are sold by numerous firms at fluctuating prices determined on competitive commodity exchanges, others are produced by only a handful of firms and sold at stable producer prices.

This diversity makes the metals interesting, indeed fascinating, to study. Yet, it also poses problems, for each metal in its own way is unique. There is no general model or economic analysis applicable to all metals. Rather each must be considered individually, so that the analysis or model takes explicit account of its particular features.

This means that a single chapter cannot begin to cover comprehensively the economics of all metals, and no attempt to do so is made here. Instead, we will concentrate on illustrating the usefulness of relatively simple economic principles, particularly those associated with supply and demand analysis, in understanding the behavior of metal markets. The next section begins by exploring the nature of metal demand. It is followed by an investigation of metal supply—from individual product production, from byproduct and coproduct production, and from secondary production. The final section then illustrates the usefulness of the concepts introduced in earlier sections by using them to analyze

*The author is Research Leader, Mineral Trade and Markets Project, International Institute for Applied Systems Analysis, Laxenburg, Austria, and Professor of Mineral Economics, The Pennsylvania State University. He is grateful to Walter Cruickshank, Stephen Dresch, Roderick Eggert, Ruthann Moomy, Marian Radetzki, and George Yocher for their comments on an earlier version of this chapter.

the causes and consequences of metal market instability, to appraise the market impacts of public stockpiling, and to assess the "incentive price" technique for forecasting long run metal prices.

There is one particularly important conclusion that the following pages should make clear. The simple tools of economics can provide powerful insights into the operation and behavior of the metal markets, but only if the analyst applying these tools has a firm understanding of the important technological and institutional relationships governing the metal market he is examining, and can tailor his analysis so as to take these relationships explicitly into account. Studies by good economists who apply their theoretical concepts in ignorance of important technological and institutional constraints are almost inevitably sterile and misleading. The same can also be said for commodity specialists, who may know well the relevant institutions and technologies but who lack a basic understanding of economic principles. Good analysis requires knowledge of both economics and the particular metal of interest.

METAL DEMAND

Metals, at least in their unwrought form, rarely are final goods. The only exception that comes readily to mind is the hoarding of gold and other precious metals as a store of value, and even here one might argue that it is the goods and the services these metals will eventually buy that are of interest to the hoarder, not the metals themselves.

Rather metals are in demand because they possess certain qualities or attributes, such as strength, ductility, heat conductivity, resistance to corrosion, that are needed in the manufacturing of final consumer and producer goods. This means that the demand for metals depends on the demand for final goods, and for this reason is often characterized as a derived demand. Since demand is really for a set of attributes, rather than for a metal per se, in many end uses one metal can replace another, or even a nonmetallic material, such as a plastic or ceramic.

The importance of material substitution is highlighted in the discussion that follows on the major determinants of metal demand. We then review three economic concepts—the demand function, the demand curve, and demand elasticities—and their uses in metal demand analyses.

Major Determinants

Literally thousands of factors affect the demand for metals—poor weather in the mid-West of the United States and the resulting consequences for agricultural income and farm equipment sales, the rising price of petroleum and the stimulus it provides for oil exploration, a decision by the French government to modernize part of its naval forces, a World Bank loan to Brazil to build a dam and hydroelectric power station.

Clearly, however, some factors are more important than others. We would expect, for example, the price of aluminum to have a greater impact than the price of oil on aluminum demand, even though the latter presumably does have some influence. Higher oil prices, for instance, encourage automobile manufacturers to substitute aluminum for heavier materials to increase gasoline mileage.

In analyzing metal demand, it is not possible to take account of all possible determinants. There simply are too many. Moreover, the effects of most are so trivial they can be safely ignored, and indeed should be ignored so as to avoid needlessly complicating the analysis. The problem is deciding which factors are of such importance they need to be considered. The answer depends not only on the metals of interest, but also on the purpose and time horizon of the analysis. Technological change, for example, is not likely to alter greatly the demand for zinc over the next three months, as new innovations normally take a number of years to introduce and diffuse. So in assessing demand for the next quarter, we can usually safely ignore technological change. On the other hand, an analysis of zinc demand in the year 2020 would need to consider carefully the effects of new technology.

The choice of which factors to consider and which to ignore is important, and will to a large extent determine the quality of the analysis. In this regard, it is useful to review those determinants often considered in metal demand studies.

Income: Metals are used in the production of consumer and producer goods. So changes in the output of either have a direct and immediate impact on metal demand. In this connection, two types of changes in aggregate production or income are often distinguished: the first encompasses relatively short run changes that come about largely as a result of fluctuations in the business cycle; the second covers longer run changes caused by secular growth and structural change in the economy.

As income is one of the most important variables affecting metal demand, its influence is almost always taken into consideration. In many studies, gross domestic product (GDP) or industrial output is employed for this purpose. More disaggregated measures of income are also used. For example, in assessing the demand for copper wire, we might use the production of electrical and electronic equipment to capture the effect of income fluctuations.

Own Price: A metal's own price is also normally an important determinant of demand. Demand tends to fall with an increase in price, and rise with a decline in price. There are two reasons for this inverse relationship. First, a higher metal price increases the production costs of the final goods in which it is used. If these costs are passed on to the consumer in the form of higher prices, then demand for the final goods, on which the demand for the metal is based, will fall. This is because consumers with given incomes will now be able to buy less (the income effect), and because they may now shift their consumption in favor of other goods whose prices have not risen (the substitution effect). Second, manufacturers are motivated to substitute other materials for the higher price metal in the production of their final goods. Indeed, this latter response by producers usually has a much greater impact on metal demand than that induced by consumers through a reduction in the demand for the final goods. This is because in most of their end uses metals account for only a very small percentage of total production costs. The cost of the steel in an automobile, for instance, represents less than one tenth of the latter's price, so an increase of 10% in the price of steel raises the price a consumer must pay for a new automobile by less than one percent.

It is important, however, to note that material substitution by producers takes time. New equipment is often necessary, personnel may have to be recruited or retrained, and production techniques must frequently be altered. Consequently, the initial effect on demand of a change in a metal's price may be quite modest, and a number of years may be needed before the full effect is realized.

Prices of Substitutes and Complements: The demand for a metal may be affected by prices other than its own. Most metals compete with other materials for their end use markets, and so a fall in the price of any such substitute can adversely affect the demand for the others. Wood, brick, aluminum, and plastic, for example, have all been widely used in home construction as external siding. In recent years the decline in the price of plastic siding has encouraged its widespread use, and aluminum has been all but eliminated from this once important market.

In some instances, a fall in the price of one material may actually increase demand for another. In such cases, we say the two materials are complements. For example, a fall in the price of steel tends to increase the use of tinplate, since tinplate is composed primarily of steel. This, in turn, stimulates the demand for tin. Consequently, in end use markets for tinplate, steel and tin are complementary materials.

As with changes in own price, changes in the prices of substitutes and complements affect metal demand primarily by inducing producers to alter the nature of their manufacturing processes. Consequently, some time is required for the full impact of price changes on demand to be realized.

Technological Change: New technology can alter demand in several ways. First, it can reduce the amount of metal required in the production of specific items. For example, the amount of tin consumed in manufacturing a thousand beer cans fell from over two and a half pounds per thousand in the 1930s to under half a pound by 1957, largely as a result of the development of electrolytic tinning and its widespread use in place of the older, less efficient, hot-dip process for making tinplate (Demler, 1983).

Second, new technology can affect the ability of a metal to compete in particular end use markets. This is nicely illustrated by the waterpipe market for home construction, where innovations in the production of polyvinyl chloride (PVC) plastic pipe have allowed this material to capture a sizable market share over the last 20 years. In the process, the demand for copper and other traditional pipe materials has suffered. In contrast, the demand for tin, needed to manufacture the organotin chemicals required in the production of PVC plastic, has been stimulated (Gill, 1983).

Finally, new technology can change the number and size of end use markets. The advent of the automobile, for instance, gave rise to a major new market for steel. The same development, however, led to a contraction in steel's use for the production of carriages and horse shoes. Germanium, whose widespread use in the production of transistors and diodes in the 1950s was the direct result of new technology, suffered during the 1960s as new planar technology and other developments made silicon chips the preferred material for transistors, integrated circuits, and other semiconductor devices.

Since measuring technological change is difficult, some studies ignore this particular determinant. This may not be serious when assessing demand over a very short period, for as pointed out earlier, the introduction and dissemination of new technology takes time. Over the longer term, however, it is much harder to rationalize the exclusion of this variable. Other studies simply assume that technological change is closely correlated with time. This allows the use of a time trend to capture the effects of technological change. While such a procedure may be acceptable in some situations, in most the influence of technological change is too random and discrete. The tremendous impact that electrolytic tinning had on the demand for tin was basically a once-and-for-all event. The effects of such major innovations are not likely to be closely correlated with time, and should be explicitly and individually taken into consideration.

Consumer Preferences: Changes in consumer preferences alter the number and magnitude of end use markets in which metals are consumed. Over the last decade the American public has experienced an on-again-off-again love affair with the small car. Small cars are imported in large numbers, and in any case use less steel, copper, aluminum, and other materials. So when preferences shift towards small cars, the domestic demand for metals by the motor vehicle industry tends to fall.

Consumer preferences may vary over time and among countries for a number of reasons. The age distribution of the population, for example, can be important. Between the ages of 18 and 35, many individuals are engaged in setting up new family units, and spend a relatively large proportion of their income on housing, automobiles, refrigerators, and other consumer durables needed in establishing new homes. Over the last several decades, the population in the United States and other developed countries has grown older. As a larger proportion of the total falls into the over 35 age bracket, preferences are likely to shift to less material intensive goods, reducing metal demand.

Per capita income and the overall level of economic development also influence consumer preferences. The poor have to spend their limited incomes almost entirely on basic necessities, while the rich can indulge in more luxuries. The rich also tend to save a large portion of their income. So a shift of income in favor of the poor is likely to reduce the amount of total income invested. Since investment stimulates the construction, capital equipment, and other material intensive sectors of the economy, such a redistribution may reduce the demand for metals and other materials.

New technology by making new and better products available also causes shifts in consumer preferences. The rapid growth of the airline industry over the last 50 years has substantially increased the use of aluminum and titanium in this market, while reducing the consumption of steel in railroad passenger cars and ocean liners.

Finally, even if the age distribution of the population, per capita income, income distribution, and the quality and choice of product dictated by existing technology remain constant, consumer preferences can change simply in response to shifts in personal tastes. In some instances, these shifts are influenced by advertising and psychological considerations that are not fully understood.

Normally, consumer preferences evolve slowly over time, as the demographic, income, and other important factors just discussed seldom change quickly. There are, of course, exceptions, as again the surge in consumer preference for small cars during the 1970s illustrate. Still, changes in consumer preferences usually have a much greater impact on metal demand over the longer term.

Government Activities: Government policies, regulations, and actions constitute another major determinant of metal demand. This is perhaps most dramatically and starkly illustrated when government policies lead to war. At such times, a substantial portion of a country's resources are redirected towards the production of arms and defense related activities. The demand for aluminum, nickel, cobalt, molybdenum, titanium, and other metals surges with the output of ships, tanks, aircraft, ammunition, trucks, and other military vehicles.

In peace time, government activities also influence metal demand in a number of ways. Changes over time in government expenditures on education, defense, research and development, highways, and other public goods alter the output mix of the economy. Fiscal, monetary, and social welfare policies affect income distribution, and the overall level of investment and economic growth. Worker health and safety legislation, environmental standards, and other governmental regulations may proscribe certain materials in particular end uses. Local building codes, for example, for many years, retarded the use of plastic water pipe in parts of the United States, helping to maintain the demand for copper in this particular market (Gill, 1983).

Because government actions and their effects on metal demand are not always easy to identify and quantify, they are often ignored. Unfortunately, they are also often important, and on occasions produce substantial shifts in metal demand even in the short term.

The Demand Function

The relation between the demand for a metal and its major determinants, such as those we have just discussed, is given by the demand function. This economic relationship is often expressed mathematically. In some analyses, for example, demand during year t (Q_t^d) is assumed to depend on only three variables—income during year t (Y_t), own price during year t (P_t^o), and the price of its principal substitute during year t (P_t^s),

$$Q_t^d = f(Y_t, P_t^o, P_t^s) \qquad (1)$$

Several things about Eq. 1 are worth noting. First, it is a rather simple demand function, indeed too simple to be useful in most instances. This is in part because it recognizes only three variables affecting demand. In most situations, as we have seen, there are other important determinants that belong in the demand function.

Second, Eq. 1 considers only the immediate or short run effects on demand of changes in its determinants. For the income variable, this is not a serious shortcoming, since metal demand tends to respond rather quickly to changes in the overall level of economic activity. This is not the case for prices. Producers take time to substitute one material for another, and in other ways to respond fully to a change in a metal's own price or that of its principal substitutes. This means that demand this year depends not only on prices this year, but also on prices a year ago, two years ago, and so on for as far back as past prices affect current consumer demand. Thus, at best, Eq. 1 provides an indication of the short run response of demand to changes in price.

Third, while Eq. 1 identifies important variables presumed to influence demand, it does not specify the nature of the relationship. Normally, a rather simple specification between demand and its determinants is assumed. A linear or log linear relationship, similar to those shown in Eq. 2 and 3, are particularly popular, primarily because they are relatively simple and easy to estimate. Unfortunately, such specifications entail rather strong assumptions about the nature of the demand function, whose validity is often difficult to assess.

$$Q_t^d = a_o + a_1 Y_t + a_2 P_t^o + a_3 P_t^s \qquad (2)$$

$$\log Q_t^d = b_o + b_1 \log Y_t + b_2 \log P_t^o + b_3 \log P_t^s \qquad (3)$$

The Demand Curve

In analyzing metal markets, we at times focus on one particular variable and try to assess how it alone affects demand. For example, if the U.S. economy is expected to grow by five percent over the coming year, aluminum firms need to know how this will alter their demand.

Another variable whose influence is often of special interest is price, particularly a commodity's own price. The demand curve, which is frequently encountered in mineral analyses, economic textbooks, and elsewhere, portrays the relationship between price and demand. More specifically, it shows how much of a commodity can be sold at various prices over a year or some other time interval, on the assumption that income, the prices of substitutes, and other determinants of demand remain fixed at certain designated levels.

Normally, demand curves are drawn with a downward slope, like those shown in Fig. 4.10.1 and 2. Intuitively, one would expect demand to fall as price rises, and in standard economic

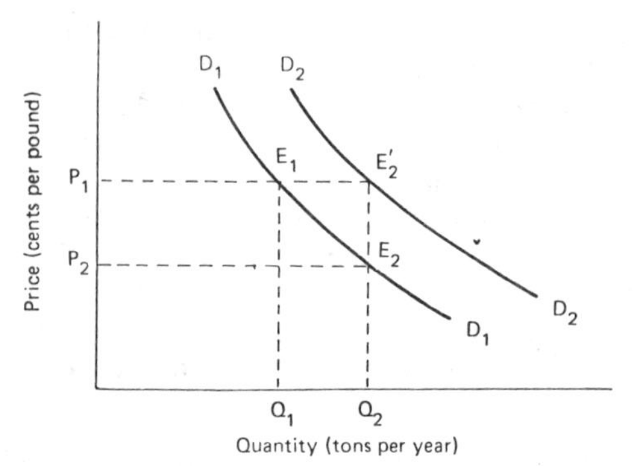

Fig. 4.10.1—Movements along and shifts in the demand curve.

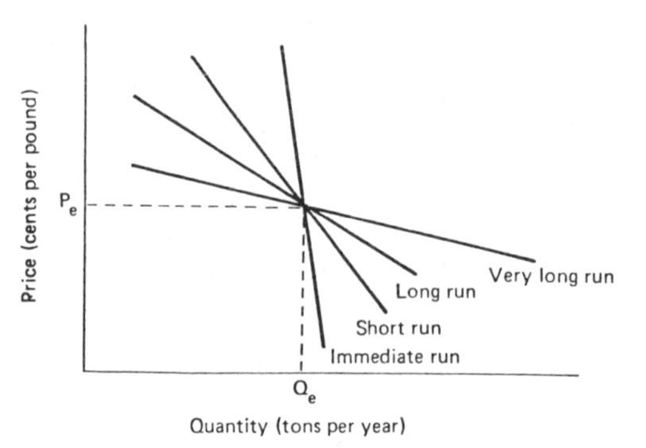

Fig. 4.10.2—The demand curve in the immediate, short, long, and very long run.

textbooks the downward slope is derived from the theory of consumer behavior and the theory of the firm. Still, there can be exceptions. In special circumstances, the demand curve can, at least over a significant segment, be vertical (implying that consumers want a particular amount of the commodity, no more and no less, regardless of its price), horizontal (implying that above a particular price consumers demand none of the commodity while below that price their demand is insatiable), and upward sloping (implying that consumers actually increase their demand as price goes up). Such situations are rare, but when they do occur are likely to be of considerable interest and importance.

Several other characteristics of the demand curve are also important to remember:

(1). A movement along the curve reflects the effect of a change in a commodity's own price. A change in any of the other variables influencing demand causes a shift in the curve itself. In Fig. 4.10.1, for example, demand can increase from Q_1 to Q_2, because price falls from P_1 to P_2, causing a movement along the curve DD_1 from point E_1 to E_2. Or, the same increase can occur because the demand curve shifts from DD_1 to DD_2, causing the equilibrium point to move from E_1 to $\acute{E}_2$. Such a shift in the demand curve can occur in response to a rise in income, a new technological development, increase in the price of a substitute commodity, or a change in one or more of the other demand variables. In using demand curves, it is important to keep the distinction between a movement along the curve and a shift in the curve clear.

At the same time, it should be recognized that a commodity's own price is not always independent of the other demand variables. This makes it difficult at times to isolate price effects, and complicates the use of demand curves. For example, a reduction in the price of aluminum sheet, which is widely used in beverage and food containers, may cause the producers of tinplate to lower their prices to remain competitive. While the fall in the price of aluminum sheet increases demand, the increase is less than would have been the case had the producers of tinplate not also reduced their price. This development can be portrayed as simply a movement down the demand curve for aluminum sheet, or as both a movement down the curve and a leftward shift in the curve. This is because the change in aluminum sheet price has both a direct and indirect effect on demand. The indirect effect results from the fact that the change in aluminum sheet price causes a change in one of the other demand variables, in this case the price of the substitute material tinplate. The first approach attributes both the direct and indirect effects to the change in the aluminum sheet price; the second attributes only the direct effect to the change in this variable. Which approach is better depends largely on the purpose of the analysis.

(2). The same commodity may have many different demand curves. At the most aggregate level is the total demand curve, which indicates how much all buyers are willing to buy, or alternatively how much all sellers can sell, at various price levels. On the buyers' side of the market, we can define demand curves for individual buyers, for regional or national markets, for a country's imports, and for particular consuming sectors or industries. We can distinguish between the demand curve for consumers and the demand curve for speculators and hoarders. On the sellers' side of the market, a similar breakdown is possible. So for refined copper, we can identify a demand curve for the world as a whole, for the United States, for the telecommunication sector, for the American Telephone and Telegraph Company, for speculative stocks, for U.S. imports, for U.S. exports, for U.S. producers, for the Newmont Mining Corporation, and so on.

(3). The demand curve—and the demand function as well—indicate the demand for a commodity, and not its consumption or production (even though the horizontal axis on the demand curve is sometimes identified as output). Demand is the quantity of a commodity that can be sold at a particular price in a given market over a year or some other time period. If the U.S. government is selling tin from its strategic stockpile, or if speculators or other private stockholders are drawing down their inventories, production may be considerably below demand. Even in the case of a demand curve specifically for producers, production will be less than demand when producers are liquidating their inventories, and more than demand when they are building up inventories. Similarly, consumption will be above demand when consumers and other buyers are decreasing their stocks, and below demand when they are increasing their stocks.

Over a number of years, the differences between consumption, production, and demand are small, and can safely be ignored. This is because inventory changes over say 10 years will largely cancel out, and any remaining differences will be small compared to cumulative demand over such a period. Usually, however, demand curves indicate how much of a commodity is needed over a year or shorter period, and so changes in stocks can cause sizable discrepancies among production, consumption, and demand.

(4). The demand curve does not indicate how the effect of a price change varies with respect to time. Rather it assumes, explicitly or implicitly, one specific adjustment period. In this connection, economists typically distinguish between the *short run*, a period sufficient for firms to adjust output by altering their labor, raw material, and other variable inputs, and the *long run*, a period long enough for firms to vary their fixed inputs, such as plant and equipment, as well as variable inputs.

In examining metal markets, it is useful at times to consider what we will call the *very long run*, which provides time not only for all inputs to change, but also for the development and introduction of any new technology induced by price changes.[a] At the other end of the spectrum, the *immediate run* is needed in addressing certain mineral issues. It provides so little time for adjustment that firms find it infeasible to alter their output. Only changes in inventories are possible.

As illustrated in Fig. 4.10.2, the responsiveness of demand to price increases with the adjustment period. In the immediate run price has very little effect, and the slope of the demand curve is very steep. What response there is comes about because price has some influence on the level of stocks that consumers and others desire to hold. In the short run, some material substitution can occur. On occasions, for example, when strikes in the nickel industry have made it difficult and expensive to obtain this metal, specialty steel manufacturers have used cobalt instead.

Material substitution, however, often requires altering the production process, retraining personnel, and acquiring new equipment. So the opportunities for consumers to resort to substitution in response to changes in material prices are appreciably greater in the long run. Their range of options is further enhanced in the very long run by the new technology induced by material price changes. This technology may also help consumers stretch their material use, allowing them to produce more from a given quantity of steel, aluminum, or chromium. Innovations stimulated by high cobalt and silver prices during the late 1970s and early 1980s have reduced the need for these metals in recent years. Because the full response of demand to price changes is realized only over the very long run, this demand curve exhibits the gentlest slope in Fig. 4.10.2.

Just how long the intermediate, short, long, and very long runs are in practice is complicated by the fact that no one answer is valid for all situations. The lag between an infusion of variable inputs and increased output depends on the manufacturing process, and may even vary over time for the same process. Similarly, new capacity can be built more quickly in some industries than others. Normally, we would not

[a]As pointed out earlier, the indirect effect on demand that a price change produces by inducing a change in technology could be treated as a shift in the demand curve rather than a movement along it. In this case, there would be no difference between the long and very long run demand curves. However, neither would reflect the full impact on demand of a change in price over the very long run.

expect the immediate run to last for more than several months, and the short run for more than about several years. The shift from the long to the very long run is more difficult to pin down. Some of the new technology induced by a price change may occur quickly, indeed within a year or two, but other developments may take decades before coming to fruition.

The time dimension introduced by the immediate, short, long, and very long runs should not be confused with the time interval over which demand is measured. All of the curves drawn in Fig. 4.10.2 presume that demand is in tons per year. The very long run demand curve does not indicate how much of a commodity will be demanded over a very long period, for example, over the next 20 years. Rather, it indicates how much will be demanded per year in 20 years time as a consequence of a price change today, assuming price stays at the new level and all other determinants of demand also remain unchanged.

Since neither of these conditions will hold for 20 years, it is best not to think of the very long run demand curve as showing annual demand 20 years from now. What it shows is the new equilibrium towards which demand is moving over the very long run in response to the price change. Long before this equilibrium is actually reached, price and other determinants will change again, causing the trend in demand to shift course and follow a new path towards a different equilibrium.

(5). The downward sloping demand curve, as commonly drawn, implies that the relationship between price and demand is continuous and reversible. Continuity means that the demand curve is smooth, like those drawn in Fig. 4.10.1 and 4.10.2, without any kinks or breaks. Reversibility means that if price, after an upward or downward movement, returns to its original level, demand will also return to its original level. In other words, one can move up and back down the same curve in response to price changes without causing the curve itself to shift.

For the immediate and short run demand curves, reversibility seems reasonable. If the desired level of stocks that consumers and others wish to hold declines by a certain amount as price rises, the reverse is likely when price eventually falls. Material substitution that can occur in the short run by its nature involves changes that can be made quickly with minimal costs and disruption. After such a switch, it should be relatively easy to switch back to the original material.

Reversibility is less likely with the long run demand curve. Here, material substitution will entail new equipment, lost production, and other conversion expenses. As a result, a firm will not switch back to a material until its price falls considerably below the level at which its replacement became attractive. In the very long run, the assumption of reversibility is even more doubtful, for now price induced innovations may substantially change the underlying technical and economic conditions governing the demand for a material.

Metal analysts and others often claim that if a material loses a particular market that market will be lost forever. Such statements suggest that a material can not recapture a market lost as its price rises, even if subsequently price returns to its previous level. In other words, after moving up the downward sloping demand curve, an industry may not be able to reverse itself and move back down the same curve, as the conventional demand curve implies.

The assumption of continuity may also not hold, particularly for those metals and materials whose consumption is concentrated in a few major end uses. Over a wide range price may rise with little or no effect on demand. Then, at a particular threshold an alternative material becomes more cost effective in a major application, causing demand to drop sharply. Such discrete jumps or breaks may be found in both short and long run demand curves. They are particularly likely to characterize the very long run demand curve, as price induced innovations by their nature are discrete events. They either do or do not occur. When they do occur, they can have a substantial impact on demand.

Demand Elasticities

In addressing many mineral issues, we need a measure of how sensitive demand is to a change in price. In the mid-1970s, for example, there was widespread concern that the International Bauxite Association might become a cartel and sharply increase world bauxite prices, just as the Organization of Petroleum Exporting Countries had raised the price of oil. To operate successfully a cartel must be able to raise price without a large loss in market demand. So there was at that time much interest in the possibility that aluminum might be economically produced from alunite and other nonbauxite ores, and more generally in the overall responsiveness of bauxite demand to higher prices.

The measure economists used for this purpose is the elasticity of demand with respect to own price, or simply the price elasticity of demand. As Eq. 4 indicates, it is defined as the negative of the partial derivative of demand with respect to own price ($\partial Q_t^d/\partial P_t^o$) times the ratio of own price to demand (P_t^o/Q_t^d). Since an increase in price normally produces a decrease in demand, this derivative is itself negative, making the price elasticity a positive number.

$$E_{Q_t^d,\, P_t^o} = -\frac{\partial Q_t^d}{\partial P_t^o} \cdot \frac{P_t^o}{Q_t^d} \tag{4}$$

$$= -\frac{\text{percent change in } Q_t^d}{\text{percent change in } P_t^o} \tag{4a}$$

For those who have forgotten their calculus (or would prefer to), the price elasticity of demand can be easily remembered as the percentage increase in demand resulting from a one percent reduction in price. If the increase in demand is greater than one percent, the elasticity is also greater than one, and we say that demand is elastic. When the elasticity is less than one, demand is inelastic.

Since the derivative of demand with respect to price is equal to the inverse of the slope of the demand curve, where two curves cross, the elasticity of demand will be lower for the curve with the steeper slope. This means that demand at the point where the curves intersect in Fig. 4.10.2 is most elastic in the very long run, and becomes increasingly less elastic in the long, short, and immediate runs. This, of course, is exactly what we would expect, for consumers have more opportunities to increase or decrease the usage of a material in response to a price change the longer the period they have to adjust.

If the relationship between demand and price is linear, as is assumed in Fig. 4.10.2 (and earlier in Eq. 2), the slope of the demand curve is the same at all points. This means that the price elasticity of demand decreases as one moves down the demand curve, and the ratio price to demand falls. Consequently, other than at their intersection point, we must be careful in comparing the demand elasticities of two curves. The steeper curve will not necessarily have the lower elasticity everywhere.

At times, the relationship between demand and price is assumed to be linear in the logarithms, as in Eq. 3. In this case, the elasticity does not vary with the level of price and demand; a one percent decrease in price produces the same percentage change in demand over the entire demand curve. The latter, if drawn using logarithmic scales, is a straight line whose slope alone determines the price elasticity of demand. So one can easily compare the elasticities of two curves, even at points where they do not intersect. These properties make the logarithmic relationship popular in analyzing material demand. However, as stressed earlier, its use is appropriate only if there are good reasons to believe it reflects the true relationship between demand and its determinants.

Up to this point, we have considered only the price elasticity of demand. It is possible to define a separate elasticity for every variable affecting demand, though in practice we normally encounter only two others—the elasticity of demand with respect to the price of substitutes, and the elasticity of demand with respect to income.

The elasticity of demand with respect to the price of a substitute, often called the cross (price) elasticity of demand, measures the percentage increase in demand for a material caused by a one percent increase in the price of a substitute. It too will be larger in the long and very long run than in the immediate and short run, since the opportunities to respond to a change in a substitute's price grow with the adjustment period.

The income elasticity of demand similarly measures the percentage increase in demand caused by a one percent rise in the GDP or some other measure of income. Since the demand for final goods, and in turn the demand for the raw materials needed to produce these goods, responds fully to a change in income rather quickly, the income elasticity does not increase with the adjustment period. There is consequently no need to distinguish among the immediate, short, long, and very long runs, as is the case for own and cross price elasticities.

Another distinction, however, is significant. Earlier we noted that a change in income can be separated into two parts: a cyclical component caused by short term fluctuations in the business cycle, and a secular component caused by long term growth trends. Which of these is primarily responsible for an income change will affect the magnitude of the demand response and the size of the income elasticity.

The demand for materials is particularly responsive to income changes caused by business cycle fluctuations. Metals and other materials are consumed primarily in the capital equipment, construction, transportation, and consumer durable sectors of the economy, which use them to produce automobiles, refrigerators, homes and office buildings, new machinery, and other such items. These sectors boom when the economy is doing well, and they suffer severely when it falters. Since small fluctuations in the business cycle cause major change in their output and in turn the demand for materials, the income elasticity is normally greater than one when the business cycle is responsible for changes in income.

When income changes are the result of secular growth trends, the traditional and still very common presumption is that metal demand grows or declines in direct proportion with income. The income elasticity of demand in such situations is thus one.

In recent years, this assumption has come under attack, in part because the consumption of steel, copper, and other metals has not kept pace with income growth in many countries. This, though, could have nothing to do with rising income. It could simply be the result of technological advances that on one hand permit firms to produce more with the same or less material, and on the other reduce the need for the older and more traditional materials by increasing the variety of new composites, plastics, and other materials available.

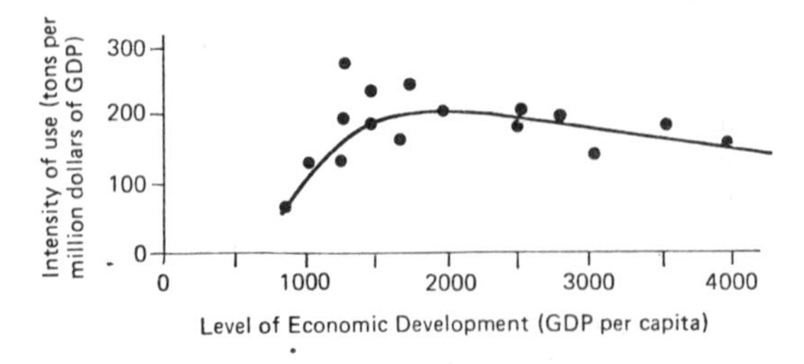

Notes: GDP is measured in constant (1963) dollars. Points shown in the figure are five year averages, through which a free hand curve has been drawn.
Source: OECD (1974), p. 58

Fig. 4.10.3—Relationship between intensity of steel use and per capita income in the United States 1888–1967.

Still, as income grows, the desired mix of final goods may also change, affecting material usage and causing the income elasticity of demand to deviate from unity. Indeed, Malenbaum (1975, 1978) and others writing over the last two decades on the intensity of material use provide a rationale for expecting just such a shift.

They contend that countries in early stages of economic development with low per capita incomes are largely agrarian. Their intensity of material use, defined as the amount of material consumed per unit of GDP, is quite low. As such countries begin to industrialize, they invest in basic industry, infrastructure, and other material intensive projects, which cause their intensity of use to rise. As development proceeds, the demand for factories, water and sewer systems, roads, housing, schools, and automobiles is gradually satisfied, and the composition of final production shifts away from manufacturing and construction and toward services. For this and other reasons, they believe the relationship between the intensity of material use and per capita income follows an inverted U-shaped curve similar to that shown in Fig. 4.10.3 for steel use in the United States.

This implies that the income elasticity of demand is greater than one for developing countries operating on the rising portion of the intensity of use curve, and less than one for developed countries on the declining portion, if the rise in income is also accompanied by economic development and higher per capita income. It is possible for income to increase solely as a result of population growth with per capita income remaining stagnant. In this case, there is no movement along the intensity of use curve and no change in the ratio of material usage to GDP. A one percent increase in income causes a one percent increase in material consumption and the income elasticity of demand is one.

Canavan (1983), Landsberg (1976), Radcliffe (1981), Vogely (1976), and others have raised some serious questions regarding the intensity of use hypothesis, though most of the criticisms concern its use for forecasting. The evolution it anticipates as a country develops and per capita income rises in the importance of material intensive goods in overall GDP, while far from proven, certainly seems plausible.

In summary, the income elasticity of demand for metals and other materials depends on several considerations. When the business cycle produces a change in income, the elasticity will normally be greater than unity. When secular growth causes the change in income, the elasticity is likely to be greater than unity only if growth is concentrated in developing countries on the upward sloping portion of their intensity of use curve and if their per capita income is also growing. It will be at or near unity if growth is concentrated in countries at the top of their intensity of use curves or if per capita income is stagnant. Finally, it will be less than unity if growth is concentrated in developed countries on the downward sloping portion of their intensity of use curves and if their per capita income is also growing.

Since the full impact of a change in income is quickly transmitted to the demand for metals and other materials, the income elasticity of demand is the same in the immediate, short, long, and very long run. This is not the case, however, for the elasticity of demand with respect to own price or the price of substitutes. With price changes, the longer the adjustment period, the greater the demand response. Consequently, these elasticities are often less than one in the immediate and short run, and greater than one in the long and very long run.

METAL SUPPLY

Metals come initially from ores extracted from mineral deposits. Some, such as bauxite and most iron ores, contain only one metal worth recovering, and their exploitation results in a single, individual product. Others contain several valuable metals. For instance, molybdenum and gold are often found in porphyry copper deposits, and sulfide nickel mines may produce copper as well.

Where joint production occurs, the resulting metals may be main products, coproducts, or byproducts. A main product is so important to the economic viability of a mine that its price alone determines the mine's output. A byproduct on the other hand is so unimportant, its price has no influence on mine output. When prices of two or more metals affect output, the metals are coproducts.

Once processed and consumed in final goods, metals are often recovered and reused after the final goods come to the end of their useful life and are scrapped. Most of the gold ever mined, for example, is still in use today. The recycling of metals is called secondary production, not because recycled or secondary metals are in some way inferior, but because the scrap from which they are made is not the original or primary source of the metal.

This section examines metal supply. It begins by assuming that all metal supply comes from the primary production of individual products. It then relaxes this assumption, and considers how the recovery of byproducts and coproducts and the production of secondary metal supplement supply.

Individual Products

In examining metal supply, we again want to ignore the multitude of factors whose influence is minor, and concentrate on the few most important variables. Just which variables are worthy of consideration and which can be safely neglected varies with the metal, the source of supply, the time of adjustment, and other factors, and calls for considerable judgment on the part of the analyst. As with demand, the choice is important, and greatly influences the quality of analysis.

While no single list is appropriate for all situations, the following variables are often important determinants of metal supply:

Own Price: Firms have an incentive to increase their output up to the point where the costs of producing an additional unit just equals the extra revenue they receive from selling that unit. Consequently, a rise in a metal's price normally increases its supply, while a fall reduces its supply.

In the short run, however, the response of supply to a change in price may be constrained by existing capacity. It takes time to develop new mines and build processing capacity, and so producers may need five to seven years to respond fully to a price increase. An even longer time may be required to adjust fully to a price decrease. Mining and metal processing are capital intensive activities, requiring equipment and facilities with long productive lives. Firms will remain in production, despite a fall in price below average costs, as long as they are recovering their variable or out-of-pocket costs. Only when existing plants and equipment need to be replaced, will they cease production.

Input Costs: The costs of labor and other inputs used in metal mining and processing also affect profitability, and in turn metal supply.

For example, the rise in world oil prices during the 1970s sharply increased the costs of producing aluminum in Japan. Aluminum smelting consumes large quantities of electric power, and in Japan the needed electricity is generated from oil fired plants. Again, the long lags in adjusting capacity to new conditions mean that the full effect of a change in costs on supply may take a number of years.

Technological Change: Advances in technology that reduce the costs of mining or processing also affect metal supply. For example, the mining of copper from large open pit porphyry deposits became feasible in the early years of the 20th century as a result of the introduction of the flotation process for concentrating such low-grade ores. Advances in earth moving capability—more powerful blasting techniques, bigger trucks, and stronger shovels—have since helped keep low-grade deposits economic at constant or even lower real prices. In the future, new technology may augment the supply of cobalt, copper, nickel, and perhaps manganese by permitting the commercial production of these metals from potato shaped nodules lying on the deep ocean floors.

Strikes and Other Disruptions: Industry-wide strikes have closed down the U.S. copper industry and the Canadian nickel industry for months. Inadequate precipitation in the Pacific Northwest has curtailed hydroelectric power generation, and in turn aluminum production, in that region. Rebel invasions into the Shaba Province of Zaire, the world's largest producer of cobalt, have on several occasions disrupted world supplies of this important metal. Turmoil in neighboring Angola has similarly at times prevented Zairian and Zambian copper from moving over the Benguela railroad to ocean ports and world markets. Strikes, mine accidents, natural disasters, civil disturbances, and other such disruptions can affect the supply of a metal by interrupting either its production or transportation.

Government Activities: Government actions influence metal supply in a variety of ways. Environmental regulations and state imposed severance taxes tend to increase costs and reduce supply. Abroad some countries require that mining companies purchase certain supplies from domestic producers, process ores and concentrates domestically, and employ nationals for managerial and technical positions, even though these restrictions may reduce efficiency and increase costs.

Alternatively, governments stimulate metal supply by subsidizing new mines and processing facilities. The United States, for example, provides low interest loans for the purchase of U.S. mining equipment, and offers firms operating abroad insurance against expropriation and other political risk. Almost all governments aid ailing industries, including steel, copper, and other metal firms, to keep them from shutting down.

Market Structure: Where a few firms account for most of a metal's production, they may maintain a producer price. As discussed later, this alters the nature of supply.

In addition, over the last 30 years the number and importance of state owned mining companies have grown. State enterprises now control 20 to 40% of the bauxite, copper, and iron ore mine output outside the socialist countries (Radetzki, 1983b). In their production and marketing decisions, such firms may be less concerned about the profits and more concerned about maintaining employment, foreign exchange earnings, and other public goals. If so, their market supply is likely to respond less to price signals, particularly low price during market recessions.

The relation between the supply of a metal and its principal determinants, such as those just discussed, is given by the supply function. Normally, it is expressed mathematically. Equation 5, for example, is the function for a metal whose supply (Q_t^s) depends on its price (P_t^o), the wage rate paid by producers (W_t), the cost of energy (E_t), and strikes (S_t). This is a rather simple supply function. It does not consider

$$Q_t^s = g\,(P_t^o, W_t, E_t, S_t) \qquad (5)$$

technological change and certain other variables that often affect supply. It contains no lagged values of the price or cost variables, and hence takes account only of their short run influence. Finally, its exact specification is not indicated.

The relationship between a metal's price and its supply is often of special interest, and is portrayed by the supply curve. The latter shows how much producers will offer to the market place at various prices over a year or some other time period, on the assumption that all other variables affecting supply remain at some specified level.

The supply curve is normally drawn sloping upwards, indicating that supply increases with price. This positive relationship seems plausible for reasons already mentioned, though it can be derived in microeconomics from the theory of the firm.

In special circumstances, however, the curve can over relevant portions be horizontal (implying that sellers are willing to provide the market with as much as they have to offer at a particular price and with nothing below that price), vertical (implying that sellers will provide the market with a given quantity of metal, no more and no less, regardless of the price), or downwards sloping (implying that sellers will offer more to the market the lower the price).

Such behavior can occur for various reasons. Firms may maintain a producer price at which they are prepared to sell all of their available supplies. In other instances, a change in price may not alter the output that maximizes profits for producers. For example, if firms are already operating at full capacity, increasing output in the short run may be extremely (or infinitely) expensive. So even though price rises, firms cannot increase profits by expanding production. Some firms, particularly state enterprises, may also weigh heavily factors other than profits in making their output decisions. These firms may continue to produce at or near capacity, even though it would be more profitable to reduce output, in order to avoid laying off employees. Some may even attempt to increase production at such times if they feel responsible for maintaining their country's foreign exchange earnings.

While such situations do occur, they are unusual. Normally the supply curve is upward sloping, and in this important respect differs from the demand curve. Other characteristics of the supply curve, however, are the same or similar to those discussed for the demand curve.

For example, a movement along the supply curve reflects a change in price, while a shift in the curve itself reflects a change in one of the other determinants of supply. As with demand, a change in price may affect other determinants, and have both a direct and indirect effect on supply. For example, when prices and profits are up, firms are likely to find it more difficult to resist demands from labor for higher wages.

There are also many different supply curves. For refined copper, separate curves are possible for the supply of all producing firms, for the supply of U.S. producing firms, for the supply of Newmont Mining Corporation, for the supply of U.S. exports, for the supply of U.S. imports, for the supply from U.S. government stockpiles, and so on.

Supply reflects how much sellers are willing to offer in the marketplace, and so like demand should not be confused with consumption or production. Where the supply and demand curves intersect, the quantity desired by buyers and the quantity offered by sellers are equal, and at that price the market clears.

As normally drawn, the supply curve assumes that the relationship between price and supply is continuous and reversible. In practice, neither condition always holds. Some mines and smelters operate on a very large scale. When they begin or stop production, supply experiences a discrete jump. Similarly, when price goes up, it may induce higher wages, shifting the supply curve to the left, which makes it impossible to move back down the original curve. Alternatively, higher prices may stimulate new technology and lower costs, causing the supply curve to shift to the right. In this situation, should price return to its initial level, supply would be greater not less than originally.

The supply curve also assumes that metal producers have a certain amount of time to adjust to changes in price, input costs, and other determinants. Here, as with demand, it is useful to distinguish four adjustment periods and in turn four types of supply curves—the immediate, short, long, and very long run.

In the *immediate run*, firms do not have time to alter their rate of production. Consequently, supply cannot exceed current output plus available producer inventories or stocks. This does not mean, however, that producers must provide to the marketplace all of their output. If demand is weak, they can build up inventories for sale at a later time when market conditions have improved. So the immediate run supply curve is not everywhere vertical or nearly vertical, as we might first think.

Before assessing the general shape of the immediate run supply curve, we need to distinguish two types of metal markets, producer markets and competitive markets, for the supply curve is different for each. Firms in producer markets quote the price at which they are prepared to sell their product. These markets, normally characterized by a few major sellers, have relatively stable prices, though when demand is weak, actual prices may fall below quoted pro-

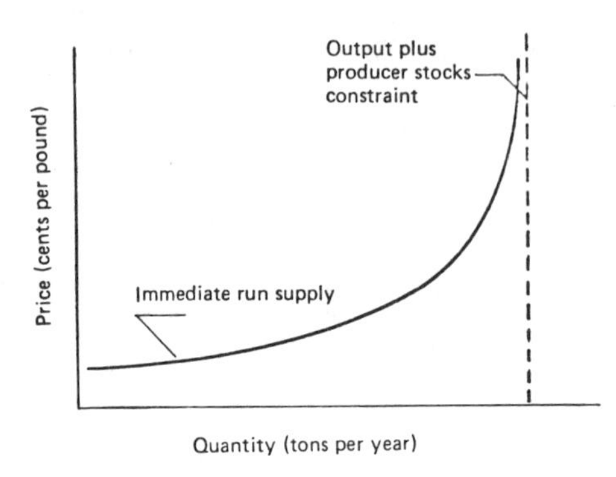

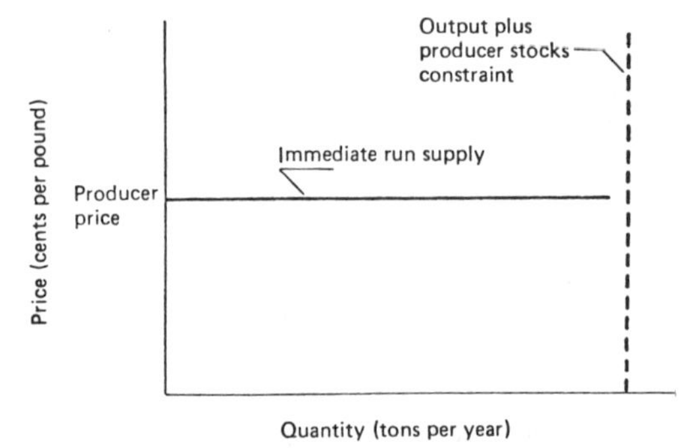

Fig. 4.10.4—Supply curves in the immediate run.

ducer prices as a result of discounting and other concessions. Steel, aluminum, nickel, and magnesium are a few of the metals sold in producer markets.

In competitive markets, price is determined by the interplay of supply and demand, and is free to fluctuate as much as necessary to clear the marketplace. Many buyers and sellers are typically active in competitive markets, and price is often set on a commodity exchange, such as the London Metal Exchange (LME) or the New York Commodity Exchange (Comex). Tungsten, manganese, and silver are metals sold in competitive markets.[b]

Producers are price takers in competitive markets, and have no influence over the going market price. Nevertheless, they still control their own supply. While reducing output is not feasible in the immediate run,[c] they need not, as pointed out before, supply what they produce to the marketplace. Alternatively, when prices are high and likely to fall in the future, firms may supply some of their available inventories in addition to current production.

An immediate run supply curve for producers selling in a competitive market is shown in Fig. 4.10.4a. At very low prices, it indicates that no supply is forthcoming as production is withheld from the market in anticipation of higher prices in the future. At some point as price rises, supply begins to come onto the market. Once this threshold is reached, supply at first expands greatly in response to higher prices as more and more current production is offered for sale. Eventually, however, further increases in supply must come from inventories. Since producers will deplete their stocks only at high prices, and since stocks can normally augment supply by only modest amounts compared to current output, the supply curve shown in Fig. 4.10.4a becomes quite steep at high prices, and finally vertical. As the curve approaches the constraint imposed by current production plus producer stocks, firms are unwilling or unable to add further to supply by drawing down inventories.

An immediate run supply curve for a producer market is illustrated in Fig. 4.10.4b. It is simply a horizontal line at the producer price that extends from zero to an amount equal to current production plus the stocks producers are willing to sell. Since firms can not increase their output in the immediate run, current production plus available producer stocks impose a constraint on supply. Fig. 4.10.4b shows the curve stopping slightly before this barrier, as producers usually are not willing to sell all their available stocks.

The supply curve in Fig. 4.10.4b assumes that firms faithfully adhere to the producer price. If this is not the case, if some or all firms discount

[b]For many years copper was unusual in that the major firms in North America maintained a producer price, while elsewhere the producers sold at prices closely tied to the competitive LME copper market. At times, the U.S. producer price would deviate considerably from the LME price, causing economists and others (for example, see McNichol, 1975) to ponder how this was possible given the ease of shipping copper to and from North America. In the late 1970s, however, many North American producers began selling their copper on the basis of the Comex price, which closely parallels the LME price. Others changed their quoted prices so frequently in response to fluctuations in the Comex price that a producer price in fact ceased to exist.

[c]It is, of course, always possible for firms even in the immediate run to reduce or stop production. Management can simply close down operations. This, however, is not a feasible or reasonable option in the immediate run, because labor and possibly other variable inputs require some notice before they can be laid off. So in the immediate run even variable inputs are to some extent fixed. In addition, firms will often want to finish goods in the process of being produced, since they already have an investment in these goods.

the producer price at times of weak demand, the curve is not perfectly horizontal, but instead drops somewhat at lower quantities reflecting these price concessions.

The curves shown in Fig. 4.10.4 indicate that in both competitive and producer markets metal supply is quite responsive to price until supply approaches current output plus producer stocks. At this point higher prices attract little or no additional supply into the market.

To assess the responsiveness of supply to price, economists use the elasticity of supply, defined as the partial derivative of supply with respect to price ($\partial Q_t^s/\partial P_t^o$) times the ratio of price to supply (P_t^o/Q_t^s). This measure reflects the percentage increase in supply produced by a one percent rise in price.

$$E_{Q_t^s} \times {}_{P_t^o} = \frac{\partial Q_t^s}{\partial P_t^o} \cdot \frac{P_t^o}{Q_t^s} \quad (6)$$

$$= \frac{\text{percent change in } Q_t^s}{\text{percent change in } P_t^o} \quad (6a)$$

We say that supply is elastic when the elasticity is greater than one, and inelastic when it is less than one. Where the supply curve is vertical, (as on the right side of Fig. 4.10.4a) or where it simply ends (as on the right side of Fig. 4.10.4b), supply is completely unresponsive to price and the supply elasticity is zero. Where the supply curve is flat or horizontal, supply is highly or infinitely responsive to price and the elasticity is very large.

In the *short run*, producers have time to change their output but not capacity. So the supply curve is constrained by existing capacity rather than current production. If the industry is not operating at full capacity, this means the supply curves shown in Fig. 4.10.4 for both producer and competitive markets are extended in the short run by the amount of available idle capacity.

In the *long run*, new mines can be developed, and processing facilities built. Firms can also expand the capacity of existing operations. Consequently, the relatively flat or elastic portions of the supply curve encompass far more output than in the immediate or short runs. Only after all known deposits of a metal are in operation does the supply curve stop or become vertical.

In the *very long run*, even the constraint imposed by existing known metal deposits no longer holds, as firms have the time to conduct exploration and find new deposits. New technology induced by the exhaustion of known deposits and higher metal prices may also permit the exploitation of new types of deposits. Concern over the depletion of the high grade iron ore in northern Minnesota after World War II, for example, led to new techniques permitting the mining of taconite, a lower grade but abundant iron bearing rock. This greatly increased the number of iron ore deposits in Minnesota and elsewhere.

So in the very long run no barrier or constraint forces the supply curve to terminate or become vertical.[d] Consequently, the very long run supply curve, in contrast to the other supply curves we have considered, may be relatively flat or elastic over its entirety with no terminal point. In fact, for a number of metals sold in competitive markets, it may become more elastic at higher prices and quantities, because more costly sources of supply or deposit types are found in greater numbers and contain on average more metal. Large porphyry copper deposits containing about four tenths of a percent copper, for example, are fairly abundant and many have been found over the last 30 years. Should the price of copper reach the level needed to make such deposits attractive, supply would expand greatly. A similar situation exists with iron ore and taconite deposits, with nickel and laterite deposits, and with aluminum and nonbauxite ores.

The major differences just discussed among the four time-related supply curves are highlighted in Fig. 4.10.5a for competitive markets and in Fig. 4.10.5b for producer markets. It is particularly important to notice how the supply curves in both types of markets expand over increasingly greater quantities as the time permitted for adjustment increases, allowing first the output (plus producer stocks) constraint, then the capacity (plus producer stocks) constraint, and finally the known deposits (plus producer stocks) constraint to be overcome. So in the very long run, no constraint on supply may exist, and supply may be elastic over the entire range of plausible outputs.

The relative vertical positions of the curves

[d]It is true that the earth is finite, and so the amount of copper and other metals it contains is fixed. But the quantity of every metal found in the earth's crust compared to the amount supplied *annually* (which is what the supply curve relates to price) is so enormous (Tilton, 1977) that this ultimate constraint simply is not relevant.

shown in Fig. 4.10.5 are based on certain presumptions about the influence of production costs on supply. In the case of producer markets, for example, we assume the producer price (which determines the height of the supply curve) is set on the basis of average total production costs at a standard or representative level of capacity utilization. These costs are likely to be similar in the immediate and short run, since capacity is fixed. In the long run, as new and presumably higher cost deposits are brought into production, average total costs will probably rise. In the very long run, though, new discoveries and the exploitation of new types of deposits should keep average total costs below those of the long run. Consequently, in Fig. 4.10.5b the immediate and short run supply curves are drawn at about the same height, the very long curve somewhat higher, and the long run curve even higher.

Unfortunately, we still have much to learn about how producer prices are set, and the extent to which they are actually tied to production costs. While average total costs may often be

a. Competitive Market

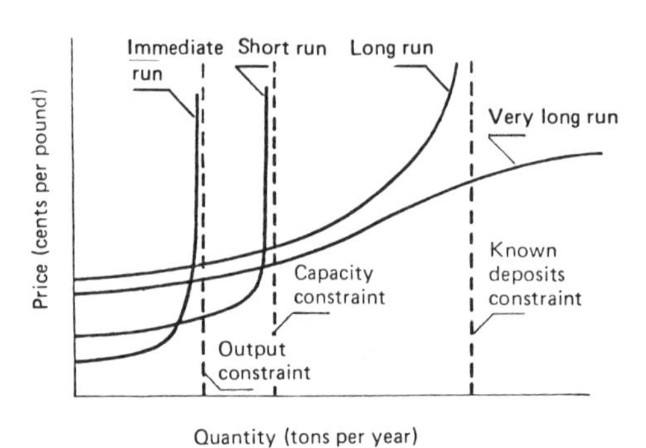

b. Producer Market

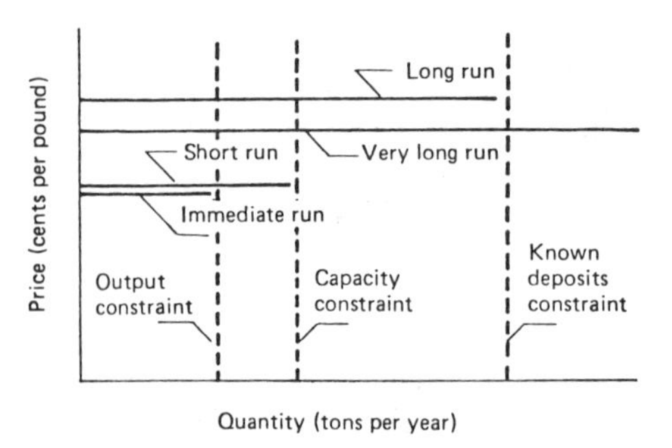

Fig. 4.10.5—Supply curves in the immediate, short, long, and very long run.

the major determinant, there are presumably instances when this is not so, and when other factors are important. So the relative heights of the curves in Fig. 4.10.5b should be considered simply as plausible and illustrative.

For competitive markets, the immediate run supply curve is drawn below the short run curve in Fig. 4.10.5a until output approaches the output constraint. Firms have two options available in the short run for reducing their supply—they can cut production in addition to building up inventories—and so are more likely to supply less at any particular price when demand is low.

The short run curve, in turn, lies below the long run curve until output approaches the capacity constraint. This is because firms in the short run, in contrast to the long run, have an incentive to continue production when price is below their average total costs so long as they are recovering out-of-pocket or average variable costs.

The long run curve in Fig. 4.10.5a is also shown above the very long run curve, for in the very long run the discovery of low-cost deposits and the development of new deposit types should reduce costs and in turn the price required to elicit any level of supply.

Economists have much more to say about the relationship between prices and costs in competitive markets, than in producer markets. In the short run, according to microeconomic theory, competitive firms will have, as pointed out earlier, an incentive to produce and supply a commodity as long as price covers average variable costs. Since variable costs do not include the costs of capital and other fixed inputs, in the short run firms may remain in production even though they are losing money. This is because fixed costs must be paid whether a firm produces or not, and so losses are minimized by staying in business as long as price is above variable costs.

This suggests that one might estimate the short run supply curve for a metal sold in a competitive market by determining the average variable costs and capacity associated with each operating mine. This information can then be arranged as in Fig. 4.10.6, so that mine A with the lowest average variable costs (OC_a) and capacity (OQ_a) is shown first, mine B with the second lowest variable costs (OC_b) and capacity ($Q_a\ Q_b$) next, and so on. The resulting curve traces out a short run supply curve. It approximates the short run marginal cost curve for the industry, and shows how costs increase as the industry expands output by bringing back into operation mines with increasingly higher production costs.

While this procedure is used by mineral firms and others, and can provide useful insights into the nature of the short run supply curve, it is predicated on several assumptions. All producers must have similar shutdown and start up costs, and share similar views regarding future price movements. Otherwise, some mines may remain operating while lower cost competitors shut down because they have higher shut down and start up costs, or because their managers anticipate a rapid recovery of prices. In addition, governments must not provide subsidies or other public assistance when prices decline to keep marginal mines from closing. And, all producers must be primarily interested in maximizing profits. In practice, of course, these conditions may not be met. State owned enterprises in foreign countries, for example, may remain in production even when the market price drops below their variable costs because employment and foreign exchange earnings are more important than profits. When this is the case, the actual short run supply curve lies below that shown in Fig. 4.10.6.

Long run supply curves for metals sold in competitive markets can also be derived using a similar procedure. The U.S. Bureau of Mines (Bennett, 1973; Babitzke, 1982) and others have estimated the average total costs, including a competitive rate of return on invested capital, associated with all known deposits and their annual production capacity for copper and other metals. With this information, the industry's long run supply is approximated in a manner similar to that illustrated in Fig. 4.10.6. Though, of course, the long run supply curve is based on average total costs, not average variable costs, and must take account of all known deposits, including those that have not yet been placed into production.

Again, this approach can provide useful insights into the nature of supply, but it does presume that undeveloped deposits will be brought into operation in order of their average total costs, and only after price rises to a level that covers these costs. In practice, this is not always so. Some deposits come on stream sooner, because host governments are willing to provide expensive infrastructure and in other ways subsidize their development. On the other hand,

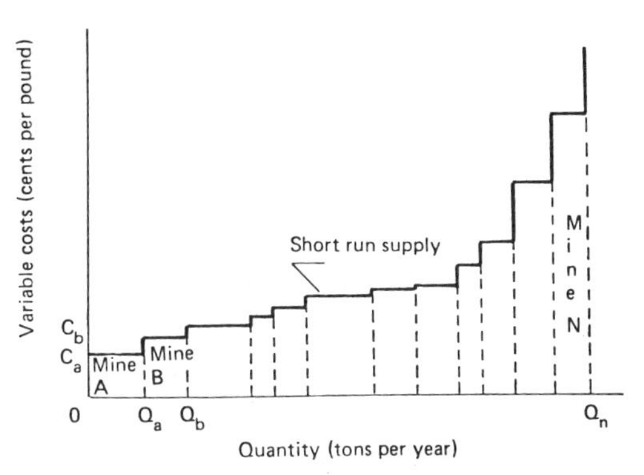

Fig. 4.10.6—Short run supply curve derived from capacity and variable cost data.

political risk, heavy taxation, and other adverse public policies may delay the development of other deposits.

Byproducts and Coproducts

So far we have taken account of only the primary supply of individual products. Often, however, two or more metals are recovered from the same mine and ore body. For example, gold, silver, platinum, molybdenum, selenium, and a number of other metals are at times contained in copper ores. On the other hand, copper is recovered in substantial quantities from the major sulfide nickel deposits in Canada and elsewhere. Similarly, lead and zinc are often found and mined together.

In such instances, as pointed out earlier, we can distinguish metals mined as main products, coproducts, and byproducts. A main product is by definition so important that it alone determines the economic viability of a mine. Its price is the only metal price affecting the mine's output of ore and main product. When two metals must be produced to make a mine economic, both influence output, and they are coproducts. A byproduct is produced in association with a main product or with coproducts. Its price has no influence over the mine's ore output, though normally as we will see, it does affect byproduct production.

Main product supply is quite similar to that for individual products discussed in the previous section, and so is not considered further here. Instead, this section, drawing on Brooks (1965), highlights the differences between the supply of byproducts and coproducts on the one hand and the supply of individual products on the other.

Before proceeding, however, we should point out that some metals, such as gold, silver, and molybdenum, are main products at some mines, coproducts at other mines, and byproducts at still other mines. To determine the total primary supply for such metals, one needs to assess main product, byproduct, and coproduct supply and then add them together. Moreover, gold and other metals may at some mines be a byproduct at times, and a coproduct or even main product at other times, if the price of gold and associated metals varies greatly.

Byproduct Supply: There are two important differences between the supply of byproducts and the supply of individual or main products. The first is that byproduct supply is limited by the output of the main product. The amount of molybdenum recoverable as a byproduct of copper production, for instance, can not exceed the physical quantity of molybdenum actually in copper ore. As production approaches this constraint, the byproduct supply curve turns upward and becomes vertical. This is because a higher byproduct price does not increase the output of ore or of main product. Otherwise, it would not be a byproduct. So at some output, supply even in the very long run becomes unresponsive or inelastic to further increases in byproduct price.

This characteristic of byproduct supply is illustrated in Fig. 4.10.7. Both the long and very long run supply curves, which are shown as the same curve in Fig. 4.10.7, are quite elastic with respect to price until output approaches the byproduct constraint, which is imposed by main product output. Thereafter, little or no increase in supply is possible, as most of the byproduct contained in the available ore has been recovered. Since normally byproduct producers

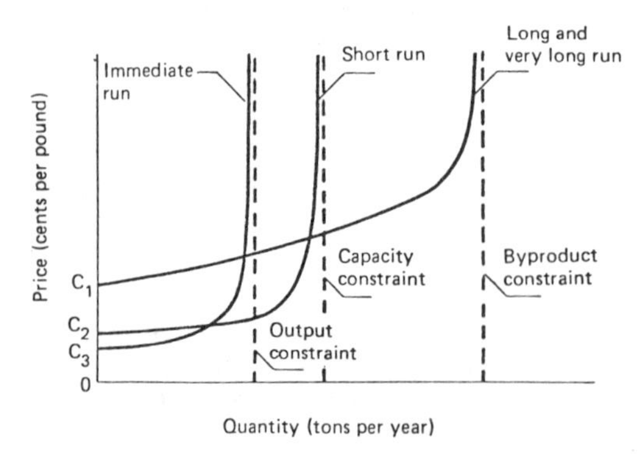

Fig. 4.10.7—Supply curves for byproduct metal in a competitive market.

are competitive firms, in the sense that they have no control over market prices, Fig. 4.10.7 shows the byproduct supply curve for only the competitive market. The same constraint, however, would apply in a producer market, causing the supply curve to end as output approached the amount of byproduct contained in the available ore.

Since it is the price of the main product, rather than the byproduct, that stimulates supply in the very long run through exploration and development of new technologies for exploiting alternative ores, the long and very long run byproduct supply curves may actually be the same, as shown in Fig. 4.10.7. This, however, need not be the case. A high byproduct price, for example, may encourage new technology that allows a greater recovery of the byproduct contained in the main product ore. In this case, the byproduct constraint becomes binding at a greater output in the very long run, causing the long and very long run byproduct supply curves to deviate from each other.

The constraint imposed by main product output shifts in response to changes in main product demand and supply. If the demand for copper goes up, for example, this causes the price and output of main product copper to rise, increasing the amount of ore from which byproduct molybdenum can be recovered. For this reason, the supply function of a byproduct should normally contain the output (or price) of the main product, in addition to its own price and the other important supply variables discussed earlier. Since changes in main product output are independent of the byproduct price, they reflect a shift in the byproduct curve itself, rather than a movement along it.

While the output of the main product places an upper limit on byproduct supply, this limit may not be binding in the immediate and short run. Main product ores may contain more of the byproduct than is needed or demanded. In this situation, the equipment and other facilities required for byproduct processing may be insufficient to treat all the available ore, so that capacity rather than main product output limits byproduct supply. This constraint, however, applies only in the immediate and short runs, as new capacity can be installed in the long run. Such a situation, where capacity rather than main product production limits byproduct supply in the immediate and short runs is illustrated in Fig. 4.10.7. The immediate run supply curve in Fig. 4.10.7

is further constrained by current output (plus available producer stocks), implying that the industry is not operating at full capacity. Otherwise, the output and capacity constraints would be the same, and the immediate and short run curves would both become inelastic at about the same output.

There is, as noted earlier, a second important difference between byproduct and individual (or main) product supply, namely, that only costs specific to byproduct production affect byproduct supply. Joint costs, those necessary for the production of both the main product and the byproduct, are borne entirely by the main product, and do not influence byproduct supply.

This means that byproduct supply curves for competitive markets, such as those drawn in Fig. 4.10.7, reflect the marginal costs of byproduct production exclusive of all joint costs. As a result, byproduct supply until production approaches the constraint imposed by main product output is often, though not always, available at lower costs than the same metal from main or individual product supply.

Since joint costs are borne by the main product, it is sometimes assumed that byproducts are basically free goods, and that the byproduct supply curve is simply a vertical line at that output reflecting the amount of byproduct in the main product ore. For this to be true, however, two conditions must be satisfied: the production of the main product must require the separation of the byproduct, and no further processing of the byproduct must be necessary after separation. Only if both of these conditions are met are all byproduct costs joint, and so attributable to the main product. In practice, the first condition often is satisfied. Though there are exceptions; for example, in some end uses the separation of antimony from lead is not necessary. Still, it is the second condition that normally gives rise to specific byproduct costs. Antimony, molybdenum, gold, silver, and other byproducts require further upgrading and processing after separation from the main product before they can be sold on commodity markets or used in the production of other goods.

The existence of specific costs means that byproducts are not free, and that they will not be recovered and supplied to the market unless their price covers these costs. In Fig. 4.10.7, the lowest cost byproduct producer has specific costs equal to OC_1 cents per pound. Once byproduct capacity is in place, the market price can fall below specific costs and production will continue in the short run, as long as price covers the minimum variable or out of pocket costs specific to byproduct production (OC_2). In the immediate run, the price can even decline further (OC_3). Over the long run, however, capacity will not be replaced and byproduct production will cease if price remains below the minimum specific cost (OC_1).

Since byproduct production tends to occur first where main product ores are particularly rich in the byproduct metal or are for other reasons less costly to process, the marginal costs specific to byproduct production usually rise with output. For this reason, the long (and very long run) supply curve shown in Fig. 4.10.7 has an upward slope over its relatively flat or elastic segment.

Coproduct Supply: Coproducts are in many respects between byproducts and main products. Their price influences mine output, but so do the prices of associated coproducts. Joint production costs must be shared, as no single coproduct can support them alone. This means that a coproduct's price must cover its specific production costs plus some but not all of joint costs.

Consequently, a coproduct's supply function includes its own price, the price of other coproducts, its specific costs, and joint costs, as well as possibly other factors. A change in any of these supply determinants, other than own price, causes a shift in the supply curve. An increase in specific or joint costs, for example, shifts the curve upward, while an increase in the price of an associated coproduct shifts it downwards.

Coproduct supply curves have the same general shape as those illustrated for individual products in Fig. 4.10.5. In the immediate, short, and long runs, coproduct supply is similarly constrained by current output, capacity, and known deposits. Since a coproduct must bear only a part of joint production costs, supply may be available at lower costs than from main or individual product output. This reduces the height of the supply curve in competitive markets, and possibly in producer markets as well. In addition, the long run supply curve until the known deposits constraint is approached and the very long run supply curve have a tendency to become more elastic or flatter as the coproduct's price increases. This is because a one percent increase in the coproduct's price produces a greater increase in overall mine revenues, and hence the incentive to develop additional supply, the higher the coproduct's price is initially.

Secondary Production

Secondary production adds to the supply of metals by recycling new and old scrap. New scrap is generated in the manufacturing of new goods. When the telephone company installs the phone lines for a new apartment building, a certain amount of copper wire scrap results. The skeleton that remains after the round tops for soft drink cans are stamped from a rectangular sheet of aluminum is also part of the supply of new metal scrap.

Old scrap comprises those consumer and producer goods that have come to the end of their useful lives. They are obsolete, worn out, or for some other reason no longer of use. When an apartment building is torn down, its phone lines are recovered and added to the supply of old copper scrap. An empty soft drink can is also part of old scrap supply.

Secondary production is an important source of total supply for many but not all metals. In recent years, the recycling of old scrap alone has accounted for about 20% of U.S. aluminum consumption, 30% of U.S. copper consumption, and 50% of U.S. lead consumption. On the other hand, little or no beryllium, columbium, or germanium is recovered from old scrap. These metals are dissipated when used or for other reasons are too costly to recycle.

Secondary metal supply differs in several respects from primary metal supply, particularly that for individual or main products. For example, secondary producers are for the most part highly competitive, and rarely support a producer price. So in considering secondary supply, we need not be concerned with supply curves for a producer market.

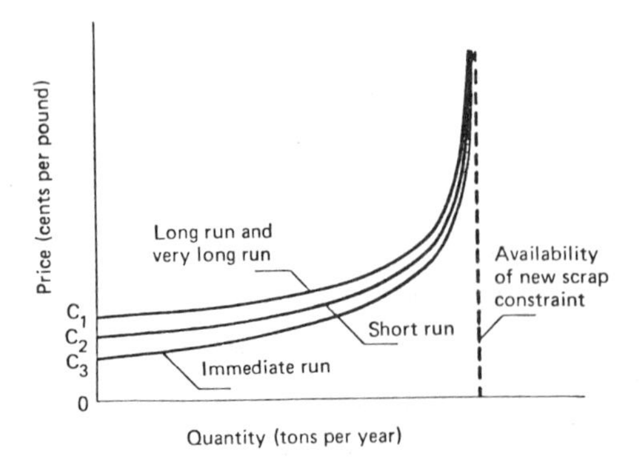

Fig. 4.10.8—Supply curves for secondary metal from new scrap in a competitive market.

In addition, it is the availability of scrap, rather than the availability of known deposits, that limits secondary supply in the long run. Since the important factors determining the availability of new and old scrap differ, secondary supply from these two sources are best considered separately.

Secondary Supply from New Scrap: The amount of new scrap available for recycling depends on three factors—current overall metal consumption, the distribution of this consumption by end uses, and the percentage of consumption resulting in new scrap for each end use. A rise in overall copper consumption, due to an increase in GNP or a change in consumer preferences, increases the availability of new copper scrap, causing both the constraint and the long run supply curve for secondary copper metal produced from new scrap to shift to the right. On the other hand, improved manufacturing techniques that reduce the percentage of copper scrap generated in the manufacturing of electric wire or other fabricated products shifts the constraint and the curve to the left. A change in the allocation of copper so that more goes into goods whose production results in little scrap will also tend to shift the constraint and the curve to the left.

This means that the supply function for secondary metal from scrap, in addition to technological change and other determinants discussed earlier, should include a variable for the availability of new scrap. Often metal consumption is used as a proxy for the availability of new scrap. This is appropriate, however, only if its distribution by end uses and the proportion of consumption resulting in new scrap for each end use remain unchanged. Otherwise, these variables as well belong in the supply function.

The shape of the long run (and very long run) supply curve for secondary metal from new scrap is shown in Fig. 4.10.8, and reflects the cost of collecting, identifying, and processing new scrap. The scrap which is the least costly to recycle will be processed first. These costs, given in Fig. 4.10.8 as OC_1 cents per pound, determine the point where the curve intersects the vertical axis. As most new scrap is relatively inexpensive to recycle, the slope of the curve rises from this point very gently over the range of possible outputs. Only as the constraint imposed by the availability of new scrap is approached does the supply curve turn upward.

The low cost of recycling new scrap compared to alternative sources of supply means all or almost all new scrap is recycled. So over the range of normal prices little additional supply from new secondary is possible, making supply price inelastic. However, at very low prices, those approaching the cost of recycling new scrap, supply is, as Fig. 4.10.8 illustrates, quite elastic with respect to price.

The fact that almost all new scrap is recycled means that the constraints limiting supply in the immediate and short run, namely output and capacity, are not likely to differ significantly from the constraint in the long run imposed by the availability of new scrap. So although the immediate and short run curves lie below the long run curve, they turn upward and become vertical at about the same output.

If current technology allows the full recovery of the metal content of new scrap, the very long run supply curve is also constrained by the availability of new scrap. In this case, it and the long run curve coincide, as illustrated in Fig. 4.10.8. If this is not the case, if a high metal price induces over the very long run new technology that allows more metal to be recovered from the available scrap, the constraint on supply would be further to the right in the very long run. It would, however, not be eliminated. At some output the supply of secondary metal from new scrap, like that for byproduct supply, becomes inelastic to price even in the very long run, and in this regard differs from the supply of independent and main metal products.

Secondary Supply from Old Scrap: The availability of old scrap during any particular year depends on (a) the *flow* of metal containing products reaching the end of their service life during the year, and (b) the *stock* of metal containing products no longer in use or service at the beginning of the year, but which have not yet been recycled. The number of old automobiles available for recycling, for example, includes those scrapped during the year as well as those scrapped in earlier years but which for one reason or another have yet to be recycled. Some are rusting away in their owners' backyards, others have been abandoned in remote areas.

The flow of old metal scrap depends on the number and types of goods in use throughout the economy at the beginning of the year, their metal composition, their age distribution, the mean age at which they come to the end of their service life, and the frequency distribution around this mean. Since these factors together determine the flow of old scrap, they can be included in the supply function for secondary metal produced from old scrap in place of the latter.

The stock of old metal scrap depends on the accumulated past flows of products coming to the end of their service life. From this total, the quantities already recovered through recycling must be subtracted. This means that over time the stock of old scrap will grow if the amount recycled is less than the incoming flow.

Even though it depends on two flow variables—the accumulated flow of old scrap and the accumulated recycling of this scrap in the past—the stock of old scrap is as stated a stock and not a flow. Consequently, what is recycled this year is not available for recycling in the future. This has several important implications, and makes it necessary to distinguish between the supply of secondary metal produced from the flow and from the stock of old scrap.

A long run supply curve for the former is shown in Fig. 4.10.9a. At the price P_1, this curve indicates that the quantity q_1 of the secondary metal is recovered from the incoming flow of old scrap. The remainder of the incoming flow of old scrap is not recycled, but rather added to the stock of old scrap available for recycling in the future.

The curve begins at the vertical axis at a fairly low price, reflecting the fact that some old scrap is of very high quality. It can be recycled at a relatively low cost (OC_1) per pound of contained metal. However, in contrast with new scrap, costs rise notably as more and more of the old scrap flow is recycled. This is because some scrap is scattered geographically, and so collection costs are high. Some of it is a mixture of various scrap types, requiring expensive identification and sorting techniques. Some of it is highly contaminated with rubber, glass, wood, and other waste material, making treatment costs high. Indeed, in some uses, such as lead in gasoline, the metal is so dissipated after use that recycling is simply too costly to contemplate. It is for such reasons that much of the old scrap flow is not recycled for many metals.

Fig. 4.10.9a also illustrates a short run supply curve for secondary metal produced from the flow of old scrap. This curve is drawn beneath the long run curve at prices below P_1, on the

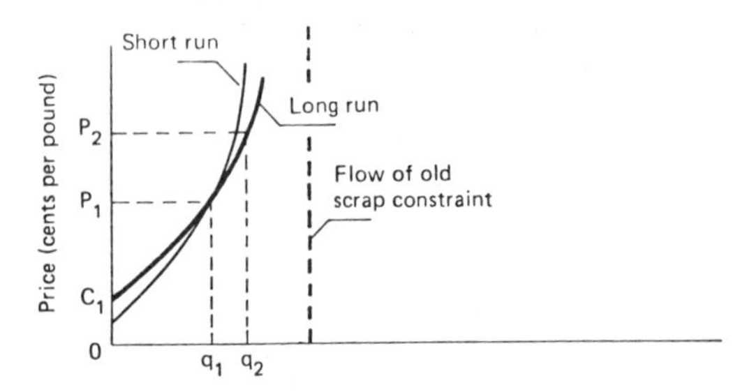

b. Secondary Metal From the Stock of Old Scrap

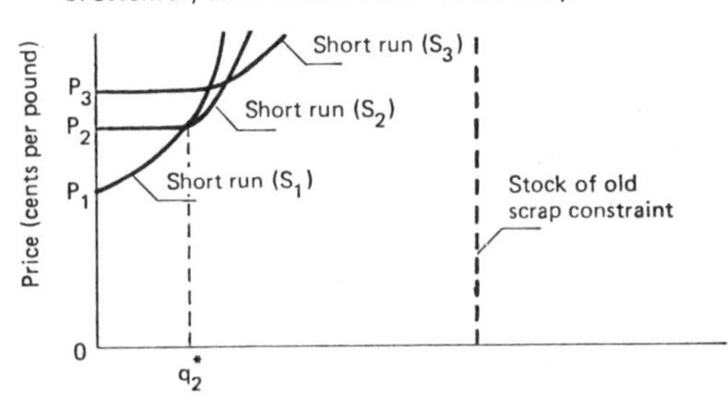

c. Secondary Metal From All Old Scrap

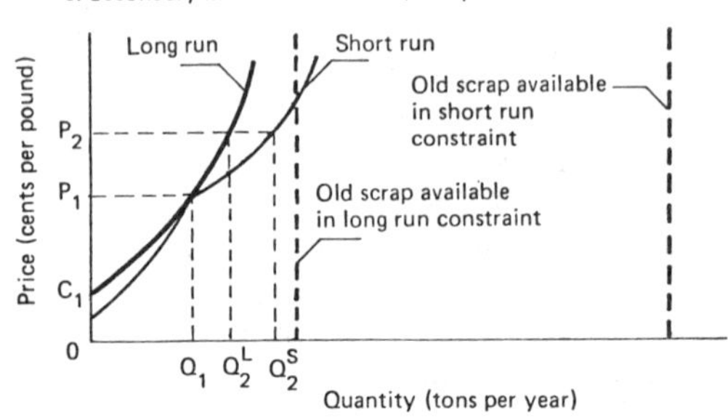

Fig. 4.10.9—Supply curves for secondary metal from old scrap in a competitive market.

assumption that secondary producers will continue to operate and supply the market in the short run as long as they can cover variable costs. Since fixed costs tend to account for a relatively small share of total production costs, particularly in comparison with primary production, the short run curve lies relatively close to the long run curve.

At prices higher than P_1, Fig. 4.10.9a shows the short run curve above the long run curve, implying that producers take advantage of capacity constraints in the short run to realize higher prices and profits. However, capacity tends to be more flexible in secondary compared to primary production. It is easier to increase output by increasing the number of shifts and by augmenting labor in other ways. For this reason, the short run supply curve above the price P_1 is drawn quite close to the long run curve.

This, though, need not necessarily be the case. Under certain circumstances, the short and long run curves may lie quite far apart. For example, at particularly high metal prices, certain products that account for a significant part of the total old scrap flow may become economic to recycle. The capacity to handle these products, however, may not exist, causing the short run supply curve to become vertical considerably before the long run curve.

Conversely, short run supply can exceed long run supply at relatively high prices, with the reverse being true at low prices, as a result of the "accelerated scrapping" phenomenon. The latter occurs when products close to the end of their service life are scrapped earlier than they otherwise would be as a result of high metal prices. For example, when machines become old and obsolete, they are at times held in reserve, and used only occasionally during peak production periods. Eventually, they may be kept for emergencies, for example when newer equipment fails. In some cases, they are stored away, and cannabilized for their parts. High metal scrap prices encourage the premature recycling of such equipment. Conversely, when metal prices are particularly low, the costs of keeping such equipment in terms of the scrap revenues foregone are quite modest, which may prolong the period before they are eventually scrapped. This implies that in some circumstances the constraint imposed by the flow of old scrap may not be invariant in the short run to price, but rather may increase with price at least over a range.

So far we have focussed our attention entirely on secondary supply from the flow of old scrap. Fig. 4.10.9b illustrates three short run secondary supply curves from the stock of old scrap. The first curve S_1 indicates that at the price P_1 no metal is recovered from the stock of old scrap. This price simply does not cover recycling costs. At higher prices, however, some of the available old scrap stock can be economically processed. At P_2, for example, the old scrap stock will produce an amount of metal equal to q_2^*.

Over the long run, however, this output is not sustainable, because the old scrap stock recoverable at costs under P_2 is depleted. So if price remains at P_2, the short run curve soon becomes truncated in a manner similar to that illustrated by the curve S_2. Above price P_2 this curve lies somewhat to the right of the original curve, reflecting the fact that the flow of old metal scrap with recycling costs above P_2 has not over the intervening years been recycled but rather added to the stock of old scrap. Below price P_2, however, the new supply curve S_2 indicates that no supply from the stock of old scrap is now forthcoming, as the material that can be recycled economically at this price has been exhausted. Similarly, if the price rises to P_3, the short run supply curve will soon approach the shape indicated by the S_3. This means that normally there exists no long run supply curve for secondary metal from the stock of old scrap.[e]

Although the size of the stock of old scrap does impose a constraint on the ultimate supply of secondary metal from this source of scrap, this constraint is not nearly as binding as is the availability of scrap in the case of supply from the flow of old scrap or from new scrap. Long before the stock of old scrap is exhausted, processing is constrained by high costs. Most of the available old scrap stock has not been recycled precisely because at prevailing prices it has been uneconomical to do so. Other sources of supply—individual product, main product, coproduct, byproduct, secondary from new scrap, and secondary from the flow of old scrap—for the most part have been cheaper.

It is also worth noting that the short run secondary supply curve from the stock of old scrap depends over time on the market price and the short run supply curve for old scrap flows. If price falls below P_1, some of the incoming flow of scrap with recovery costs below P_1 will not be recycled. Rather it will be added to the stock of old scrap, shifting the short run secondary supply curve from the stock of old scrap (S_1) downward as well as to the right over time. Alternatively, if the market price remains at P_1, the curve S_1 shifts to the right, but not downward. If price rises above P_1, as explained earlier, the curve S_1 moves to the right at prices above the prevailing market price. While below this price, the curve moves to the left as the scrap with processing costs below the market price is recycled. This continues until the supply curve intersects the vertical axis at the prevailing market price.

As Fig. 4.10.9c illustrates, the secondary supply curve from all old scrap can be derived by combining, or more precisely by adding horizontally, the secondary supply curve for the old scrap flow and for old scrap stocks. The long run curve is simply that for secondary from the old scrap flow shown in Fig. 4.10.9a, as there is no long run secondary supply curve from the stock of old scrap.

The short run curve is derived by adding the appropriate short run curve for the stock of old scrap, assumed to be the curve S_1 in Fig. 4.10.9b, to the short run curve for the flow of old scrap. Since the short run curve for the stock of old scrap intersects the vertical axis at P_1, implying that below this price no secondary metal from old scrap stocks is forthcoming, the short run secondary curve for all old scrap below this price is simply the short run curve for the flow of old scrap.

Fig. 4.10.9c highlights two interesting facets of secondary supply from old scrap. First, the constraint posed by the availability of scrap is actually less binding in the short run than in the long run. This, of course, is because the stock of old scrap, if exploited in the short run, is not available in the long run.

Second, an increase in price, for example from P_1 to P_2, may actually produce an increase in supply that is greater in the short run than the long run. In the short run, some of the stock of old scrap may be recycled, adding to supply. It is presumably for this reason that efforts to measure the price elasticity of secondary metal supply from old scrap (Bonczar and Tilton, 1975; Fisher, 1972) have found the elasticity to be greater in the short run than in the long run, which is just the opposite from what one normally finds with other sources of metal supply. However, as Fig. 4.10.9c indicates, this unusual result should be expected only if the market price is above the price at which secondary supply from old scrap stocks is forthcoming, that is, above the price P_1 in Fig. 4.10.9. When this is not the case, the figure suggests a change in price will produce a greater increase in supply in the long run than in the short run. Though as pointed out earlier, even here the phenomenon of accelerated scrapping can cause the response of supply to an increase in price to be greater

[e]It is possible that at very high prices large reservoirs of old scrap may become economical to process. For example, at some price certain metals found in dumps and landfills could presumably be extracted profitably. In such instances, the depletion of the available stock of old scrap could take a number of years, and make a long run supply curve possible.

in the short than the long run.

Nothing has been said so far about the immediate run and very long run supply curves for secondary metal from old scrap, in large part because the curves are not particularly unusual. The immediate run curve has the general shape illustrated for the competitive market in Fig. 4.10.5. At very low prices, producers will save much or all of their current production, in hope of higher prices in the future. As price increases, however, and supply approaches the constraint imposed by current production, the supply curve turns upward and becomes quite inelastic.

The very long run supply curve has the general shape of the long run curve, and is similarly constrained by the metal content of the flow of old metal scrap. Indeed, in some circumstances, the two curves may coincide. Where high metal prices, however, stimulate old scrap recovery technology and thereby permit a greater proportion of the flow of old metal scrap to be economically recycled at any particular price, the very long run curve will lie somewhat to the right, and perhaps below, the long run curve.

a. Byproduct and Main Product Supply Curves

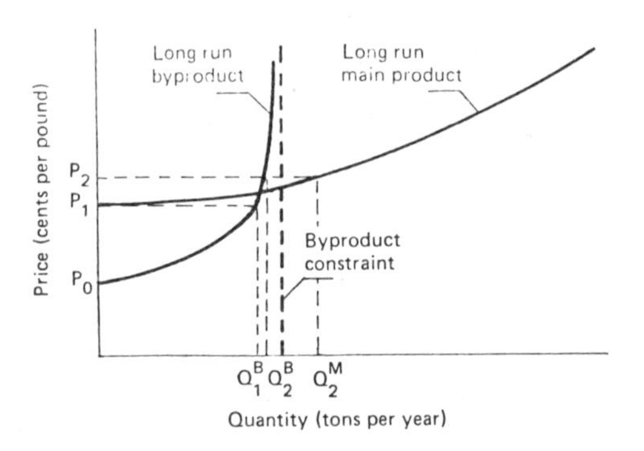

b. Total Supply Curve

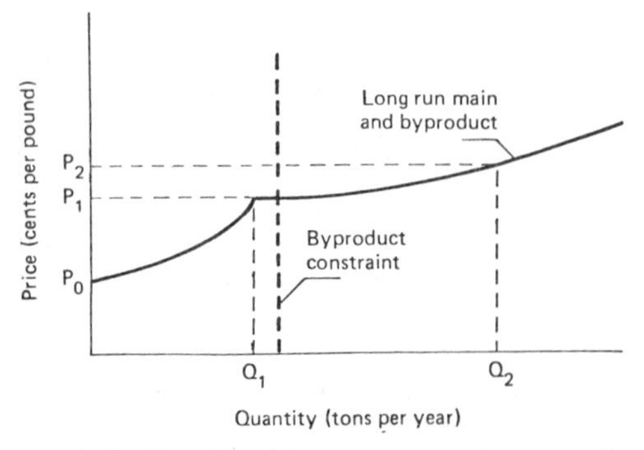

Fig. 4.10.10—Total long run supply curve for a metal produced as a byproduct and a main product.

Total Supply

We have now examined the nature of supply for individual products, main products, coproducts, byproducts, secondary from new scrap, and secondary from old scrap. These various sources are all potential contributors to a metal's total supply.

To derive the total supply curve, we must add horizontally the individual curves for all significant sources (in a manner similar to that just used to derive the secondary supply curve for all old scrap from those for the flow of old scrap and for the stock of old scrap). For example, if the byproduct and main product supply curves for a metal are as shown in Fig. 4.10.10a, and if other sources of supply are unimportant and can be ignored, the horizontal summation of these two curves gives the total supply curve shown in Fig. 4.10.10b.

Among its many uses, the total supply curve provides an indication of the relative competitiveness of the different sources of supply, and in turn their relative importance at different market prices. For example, byproduct production is initially the most competitive and cheapest source of supply for the metal whose total supply curve is shown in Fig. 4.10.10b. When market price is below P_1, it is the only source of supply. As the market price rises above P_1, however, main product production begins and becomes increasingly important. At the price P_2, main product supply exceeds byproduct supply.

A market price sufficiently high to call forth main product production permits most byproduct producers to realize a price considerably above their reservation price, that is the price at which they are willing to sell their byproduct output and which normally reflects their production costs. This difference between costs and market price is economic rent. It accrues to all producers whose costs, including a normal rate of return on invested capital, are below the prevailing market price.

In deriving the total supply curve, a question sometimes arises as to whether the secondary supply curve for new scrap should be included. New scrap, after all, is generated in the manufacturing process, and depends on other sources of supply. If metal fabrication becomes more efficient and generates less new scrap, this does not mean total metal supply has declined.

Whether secondary supply from new scrap should be counted as part of total supply depends on the purpose for which total supply is being considered. In assessing the extent to which the United States is vulnerable to interruptions in supply from certain foreign producers, for example, we would not want to include secondary supply from new scrap. To do so, would ignore the fact that this metal is generated from other sources of supply, and so would not be available in their absence. Including secondary from new scrap in total supply would underestimate U.S. dependence on foreign producers. On the other hand, in assessing the competitiveness of secondary metal markets, we would normally want to consider secondary supply from new scrap.

Regardless of how secondary supply from new scrap is treated, it is important that total supply and demand be consistent in this regard. If the demand curve takes account only of metal actually contained or embodied in final products, and excludes metal that ends up as new scrap and is recycled, then total supply should also exclude new scrap. Conversely, if demand includes the demand for all metal, including that which ends up as new scrap, then the total supply curve should include new scrap as well.

APPLICATIONS

This section illustrates the usefulness of the supply and demand principles we have examined in the preceding sections. It uses these concepts to analyze metal market instability, the market impact of government stockpiling, and the "incentive price" technique for forecasting long run metal prices.

Market Instability[f]

Metal markets are well known for their instability, for their feast or famine nature. In an effort to stabilize commodity markets, particularly for the benefit of producers in the developing countries, the United Nations Conference on Trade and Development (UNCTAD) has over the last 10 years pushed for the creation of an Integrated Program for Commodities. Among other measures, this program proposes to establish a common fund on which international commodity agreements can draw to support market stabilization measures. While the proposed program has encountered a number of difficulties, it does reflect the concern on the part of both producing and consuming countries over the instability that plagues mineral markets.

Nor is this instability new. One of the major driving forces behind the multiple mergers in the American steel industry at the turn of the last century, which culminated in 1901 with the creation of the U.S. Steel Corporation possessing at that time some two thirds of the country's steelmaking capacity, was a desire to control the volatile steel market (Temin, 1964). Gyrations in the steel industry during the 1880s and 1890s had created severe problems for all producers.

A highly concentrated market structure where one or a few major producers dominate the market and set a producer price does not, however, eliminate market instability (though as we shall see, it does alter the ways in which market instability manifests itself). This is because the following three characteristics of short run metal supply and demand, which are responsible for market instability, are present no matter how concentrated the market.

First, as output approaches the capacity constraint total supply becomes increasingly price inelastic. In a competitive market, as shown earlier, the short run supply curve turns upward and at some point becomes vertical. In a producer market, the curve simply ends, when major producers no longer have sufficient supply to satisfy demand at the producer price.

Second, demand also tends to be price inelastic. So the slope of the demand curve is quite steep.

Third, demand is highly elastic to changes in national income over the business cycle. The consumption of most metals is concentrated in four sectors—construction, capital equipment, transportation, and consumer durables—whose output is particularly sensitive to fluctuations in the business cycle. During a recession, these sectors suffer far more than the economy as a whole. During a boom, their sales soar. As a consequence, the demand curve for most metals shifts considerably over the business cycle.

These characteristics of short run supply and demand are illustrated in Fig. 4.10.11a for a

[f]This section draws upon Tilton and Vogely (1981) and Tilton (1977, Chapter 5).

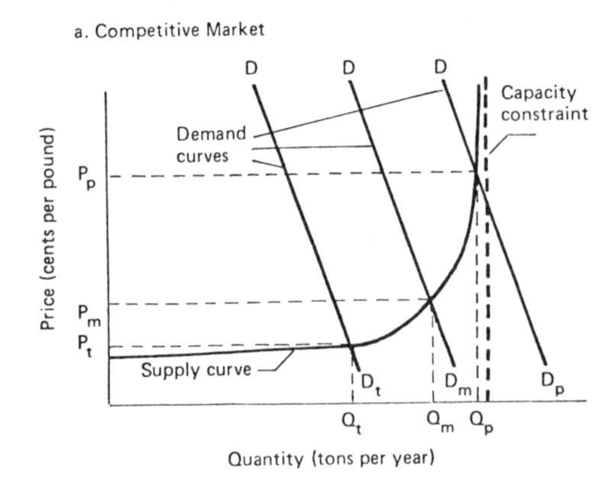

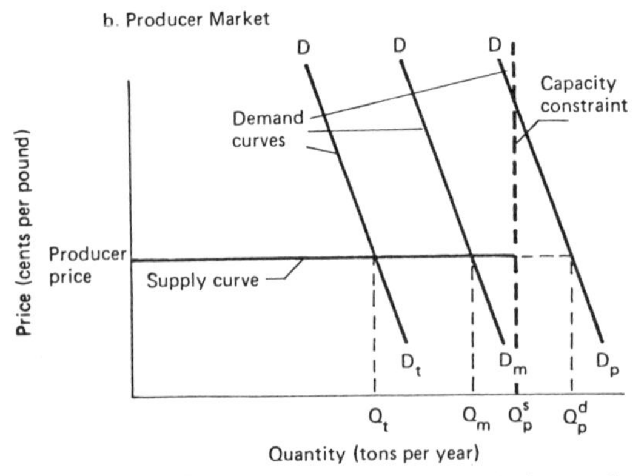

Fig. 4.10.11—Short run supply and demand curves.

metal sold in a competitive market and in Fig. 4.10.11b for a metal sold in a producer market. In both instances, it is assumed that supply comes from individual or main product production. This simplifies the analysis, but does not alter the conclusions, since total supply regardless of the combination of sources from which it is derived is at some output constrained in the short run by the available production capacity. As supply approaches this constraint, it becomes inelastic to price.

The two characteristics of metal demand—its low elasticity with respect to price and its high elasticity with respect to income—are portrayed in Fig. 4.10.11a and 11b by the steep slope of the demand curves and by the shifts in the demand curves over the business cycle. The curve DD_t reflects demand at the trough of the cycle, the curve DD_m at a mid point of the cycle, and the curve DD_p at the peak of the cycle.

One of the important consequences of market instability for copper, tungsten, and other metals sold on competitive markets is the severe fluctuation in market price. It varies in Fig. 4.10.11a from a high of P_p at the peak of the business cycle to a low of P_t at the trough. The quantity of metal that producers supply to the market also varies greatly, from a high of Q_p (approximately the maximum possible given the industry's production capacity) to a low of Q_t. At the latter level, producers are burdened with either shutting down much of their capacity or adding a large part of their output to their inventories.

Fig. 4.10.11a also indicates that when market instability is caused by shifts in the demand curve (rather than by shifts in the supply curve, which is more typical for agricultural commodities), the quantity sold and price move together. When one is down, so is the other. Consequently, total revenues and in turn profits tend to be highly volatile.

For metals sold in producer markets the situation is somewhat different. If all firms faithfully adhere to one producer price, there is no price instability. Even if this is not the case, if some open or secret discounting occurs, or if the producer price is reduced when the market is weak, price instability will generally be less than in competitive markets.

However, physical shortages, where the available supply is insufficient to satisfy demand at the prevailing price, can occur in a producer market. In Fig. 4.10.11b, for instance, the quantity demanded during the peak of the business cycle is Q_p^d while the maximum amount the industry can supply is only Q_p^s. The short fall requires that producers allocate or ration their limited supply to customers on the basis of past purchases or some other criterion.

Despite the greater price stability, firms in producer markets still suffer from sizeable fluctuations in total revenue and profits over the business cycle, as a result of the instability in metal demand and the impact that this has on sales. In these respects, the adverse effects of market instability are similar for producer and competitive markets.

Public Stockpiling

The U.S. government has for several decades maintained stockpiles of copper, lead, cobalt, tungsten, and a number of other metals for strategic purposes. More recently, France, Japan, and other industrialized countries have also contemplated such stockpiles, and in a few cases accumulated certain commodities.

This section examines how public stockpiling can affect metal markets. It begins by looking at the effects in the immediate run.

Earlier we distinguished between the immediate run supply curve in a producer market and in a competitive market. As shown in Fig. 4.10.4, the curve for the producer market tends to be a horizontal line at the producer price that simply terminates once the constraint imposed by current output is reached. The curve for the competitive market, in contrast, starts at a lower price and rises gradually until the constraint imposed by current output is approached. It then turns upward and becomes vertical. These curves are reproduced in Fig. 4.10.12.

In analyzing the effects of stockpiling, it is useful to separate total demand into three subcomponents or curves. The first is the consumption demand curve, which indicates the amount of metal demanded at various prices over the year for actual consumption. In Fig. 4.10.12a and 12b this curve is shown with a relatively steep slope, since for reasons discussed earlier the demand for consumption tends to be quite unresponsive to price changes in the immediate run.

The second is the inventory demand curve, which shows the amount of metal demanded at various prices by fabricators, speculators, and others for inventory adjustments.[g] As the price of a mineral commodity rises, inventory demand is likely to fall, in part because carrying costs increase. In addition, the higher the current price, the greater and more widespread the expectation that price will fall in the future. For this reason, Fig. 4.10.12a and 12b portray the inventory demand as moderately responsive to price. They also indicate that at high prices fabricators and others may on balance reduce their inventories, causing inventory demand to be negative.

The third is the stockpile demand curve, which shows the amount of the metal demanded by the government during the year in question for net additions to the public stockpile. The shape of this curve, which is not shown in Fig. 4.10.12a

[g]The inventory demand of producers is taken into account by the supply curve, as the amount producers supply to the market at various prices depends on their production and their demand for inventory. Consequently, it is not included in the inventory demand curve.

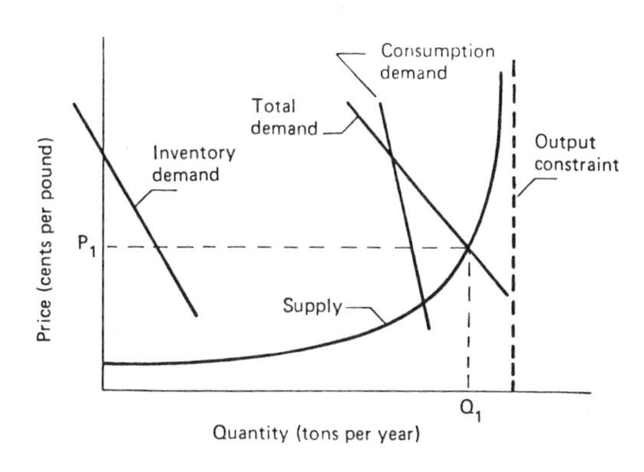

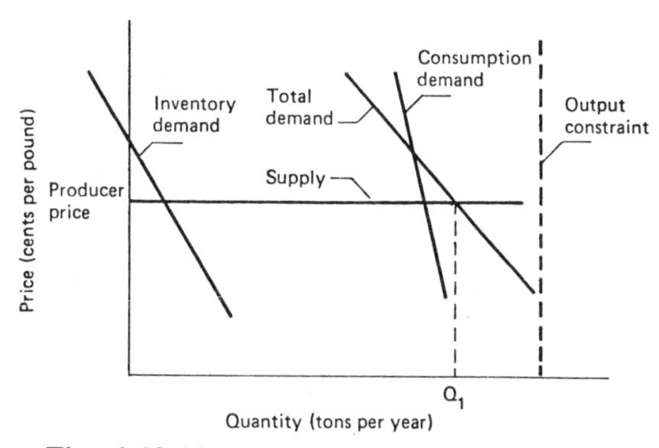

Fig. 4.10.12—Immediate run supply and demand curves.

and 12b, can be determined only with specific information about the purpose and operation of the public stockpile. It too can be negative, implying that the government wishes to dispose of some of its stockpile.

By adding these three individual demand curves horizontally, we can derive the total demand curve. If no public stockpile exists, only the demand curve for consumption and the demand curve for inventory need be added. Such a total demand curve is shown in Fig. 4.10.12a and 12b.

To assess the immediate effects of stockpiling, we need to know how the total demand curve, illustrated in Fig. 4.10.12a and 12b, is altered when stockpiling occurs. This requires that we consider the nature and shape of the demand curve for net additions to the public stockpile in more detail.

Public stockpiles are created for a variety of

reasons. As noted earlier, the United States maintains strategic stockpiles, which in principle are to be used only for military emergencies. The government has, however, on occasions used them to provide relief to distressed domestic mineral producers, to assist fabricators during shortages, and to discourage domestic producers from raising prices. It has also accumulated stocks to encourage the expansion of domestic mineral production. Other objectives are also possible. The International Tin Council, for example, operates a buffer stock to reduce the volatility of tin prices.

What is of particular importance in assessing the nature of the stockpiling demand curve is the rate of accumulation, not the current or desired size of the stockpile, and how this rate varies with the price of the metal. In this respect, a public stockpile will at any particular time be in one of three phases: the acquisition phase during which the government is purchasing the commodity and building up its stockpile; the disposal stage during which the government is selling the commodity and drawing down its stockpile; and the holding phase during which the government is neither buying nor selling the commodity but merely maintaining its stockpile at a given level.

If the rate of acquisition or disposal is fixed and does not depend on the price of the metal, the stockpile demand curve is simply a vertical line located to the right or left of the vertical axis shown in Fig. 4.10.12a and 12b by the amount equal to the rate of acquisition or disposal. Under these conditions, and if government stockpiling alters none of the other demand or supply curves affecting the market (an assumption that is relaxed below), the immediate effects of public stockpiling can be readily appraised by recalculating the total demand curves shown in Fig. 4.10.12a and 12b so that they include the demand for net additions to the public stockpile.

When the stockpile is in the acquisition phase, the total demand curve is shifted to the right by an amount equal to the rate of stockpile accumulation. For metals sold in competitive markets, this tends to increase the equilibrium price. Though the supply curve does not shift, the higher price will elicit an increase in supply. This increase, however, will not equal the rate of stockpile accumulation, since the higher price will reduce the demand for consumption and for additions to inventories. Fig. 4.10.12a also indicates that the magnitude of these effects depends on market conditions at the time, and in particular where on the supply curve the industry is operating. When supply is substantially below capacity, acquisitions for a public stockpile have little effect on price, and hence on the demand for consumption and net additions to inventories. The major effect is simply an increase in supply and presumably production. In contrast, when the industry is operating at or near capacity, price increases significantly, and the demand for the public stockpile is largely accommodated by a reduction in the demand for consumption and additions to inventories.

For metals sold in producer markets, purchases for a public stockpile similarly shift the total demand curve rightward by an amount equal to the rate of acquisition. According to Fig. 4.10.12b, this does not affect the price, nor the demand for consumption or inventory additions. The quantity supplied by producers simply expands to provide for the additional demand, assuming that existing capacity is sufficient to accommodate this increase. If this is not the case, government stockpiling produces a physical shortage. This presumably causes actual consumption and additions to inventories to fall below their demand at the producer price.

During the disposal phase, when the government is selling from the public stockpile, the immediate run effects are just the reverse of those for the acquisition phase. The total demand is shifted leftward by the rate of disposal. This tends to put downward pressure on price, to reduce production, and to stimulate demand for consumption and inventory additions. The relative magnitude of these effects, again, depends on how closely to full capacity the industry is already operating and on the manner in which prices are set. During the holding phase, when the government is neither buying nor selling for the stockpile, the total demand curve remains unchanged and the market is not disturbed.

So far our analysis rests on two rather restrictive assumptions: The first is that the rate of government acquisition or disposal for the public stockpile is invariant to the market price. The second is that the immediate run supply, inventory demand, and consumption demand curves shown in Fig. 4.10.12a and 12b do not shift as a result of public stockpiling.

In practice, the rate of acquisition or disposal of public stockpiles is often influenced by market conditions including price. This is clearly the case for the buffer stockpile operated by the International Tin Council. It is also frequently true, as noted earlier, for the strategic stockpiles maintained by the U.S. government.

This means the stockpile demand curve is seldom a vertical line, but instead tends to slope downward. The lower the metal's price, the more likely stockpile accumulations will occur and the larger they are likely to be. Unfortunately, it is not possible to assess the nature of this relationship in a general manner, for it varies greatly depending on the purpose of the stockpile, the metal in question, and numerous other considerations. Moreover, since stockpile demand depends on only one or a few public bodies it is much less likely that the relationship between this demand and price can be represented by a smooth continuous curve similar to that posited for the demand for consumption or for private inventory additions. This means that the stockpile demand curve must be assessed on a case-by-case basis, and even this may prove difficult.

In tin, for example, the U.S. government has in recent years indicated a willingness to dispose of excess stocks from the strategic stockpile. At no price, no matter how low, does the government appear willing at the present time to purchase tin. Nor does it appear willing to sell tin at depressed market prices, and for this reason has on occasions rejected all bids. In addition, for political reasons the government has recently reached an agreement with Malaysia, Thailand, and Indonesia that limits tin stockpile releases to 3000 tons per year. This suggests that the demand curve for net additions to the U.S. tin stockpile coincides with the vertical axis in Fig. 4.10.12a and 12b until price reaches that level at which the government is willing to sell. At this point, the demand curve moves leftward following a horizontal path until the 3000 ton limit is reached, and there turns upward and again becomes a vertical line at higher prices.

Whether the preceding is a completely accurate picture of the immediate run demand curve for net additions to the U.S. tin stockpile can, of course, be debated. It does, though, illustrate the need for a case-by-case approach in assessing such curves. It also demonstrates one possible way in which stockpile demand may vary with price.

The second assumption holds that government stockpiling does not shift any of the supply and demand curves shown in Fig. 4.10.12a and 12b other than the total demand curve. This seems reasonable for the consumption demand curve. In the immediate run, the demand for consumption is largely the result of the overall level of economic activity and consumer tastes, which determine the mix of final goods and services. Neither of these factors is likely to be affected by government stockpiling. The latter may alter the price of a mineral commodity, and thereby either encourage or discourage its consumption, but this reflects a movement along the consumption demand curve, not a shift in the curve.

The supply curve, it will be recalled, shows the amount producers will offer to the market at various prices. In competitive markets there is no reason to believe that government purchases or sales for stockpiling purposes affect the amount firms are willing to supply other than by altering the market price. So here, too, the assumption appears reasonable.

In producer markets, the entry of the government into the market to acquire stocks may prompt firms to raise the producer price, which would shift the immediate run supply curve upward by the amount of the price increase. Conversely, the disposal of government stocks would cause a drop in the producer price and the supply curve. (So the list of possible effects of government stockpiling on commodities with producer prices should be expanded. Earlier this list included a change in the quantity supplied and the possibility of a physical shortage. With a shift in the supply curve, price also changes, which in turn affects the demand for consumption and inventory additions.[h])

Government stockpiling activity may also shift the inventory demand curve. During the acquisition phase, when the government is accumulating stocks, expectations of higher future prices and of physical shortages may encourage the private sector to build up its inventories. Such a rightward shift in the inventory demand curve augments the shift in total demand caused by public stockpile acquisitions, causing an accentuation in the immediate run effects of the latter. Similarly, the effects of stockpile disposals may be accentuated by a leftward shift in the inven-

[h]Even without a shift in the supply curve, price and in turn the demand for consumption and inventory additions could change, if some firms discounted the producer price when demand was weakened, so that the supply curve was not completely horizontal as shown in Fig. 4.10.12b.

tory demand curve.

The influence of government stockpiling on the inventory demand curve depends to some extent on the objective of the stockpile. For example, stockpiles designed to reduce short term market instability or to augment available supplies in the event of import disruptions are likely to reduce the inventories held by the private sector. In this case, any shift in the inventory demand curve accentuating the immediate effects of government stockpiling would be reduced and perhaps eliminated.

As a result of the influence of public stocks on inventory demand, a government with large stockpiles can affect conditions in mineral commodity markets without actually buying or selling. The announcement in the early 1980s that the U.S. government planned to dispose of its excess tin stocks, for instance, had a dampening effect on that market, even though actual government sales for some time following this announcement were trivial. This occurred because the expectation of stockpile disposals shifted leftward the inventory demand curve, and in turn the total demand curve.

Metal traders and others often maintain that the mere existence of a larger government stockpile may overhang the market and depress prices, presumably for this reason, even though the stockpile is in a holding phase. Such an overhang effect, however, is not likely unless the private sector anticipates a shift out of the holding phase and into either a disposal or acquisition phase in the near future. Nor is it likely to persist over an extended period unless the government does move out of the holding phase. Still, in the immediate run the overhang effect can have a substantial impact on metal markets.

During the acquisition stage, we have seen that public stockpiling tends to increase commodity prices and expand production. In producer markets, it may also promote physical shortages. These effects, which are realized immediately, influence investment behavior in two ways.

First, producers are more likely to build new capacity and less likely to shut down existing facilities. Over time this shifts the immediate run supply curve rightward (or keeps it from moving leftward as much as it otherwise would). The extent of this shift depends on both the rate and duration of government accumulation. Little new capacity is likely to be added as a result of stockpiling that is over within a month or two. However, stockpiling that continues for several years may have an appreciable effect.

Second, metal consumers experiencing higher prices and possible shortages as a result of public stockpiling will look for alternative materials. If public stockpiling continues over an extended period, some fabricators are likely to turn to other materials, causing the consumption demand curve to shift leftward.

Thus, while the acquisition phase of public stockpiling initially stimulates the market—pushing prices up, expanding supply, and raising the possibility of physical shortages—if it persists for several years it produces shifts in the supply and consumption demand curves that tend to offset these effects. Conversely, during the disposal phase, the initial market impact is depressing. However, over time the induced shifts in the supply and consumption demand curves will moderate this impact.

If the government, after building up a stockpile for several years, decides that its size exceeds its requirements and shifts from an acquisition to a disposal phase, the depressing effect on the market will be accentuated by the shifts in the supply and consumption demand curves induced by the earlier acquisition phase. In examining the market impact of U.S. strategic stockpiles, Gauntt (1980) identifies a number of such instances where the induced effects of earlier behavior reinforce the immediate effects and thereby compound the market disruption of public stockpiling.

The preceding indicates that assessing the market consequences of public stockpiling is a complicated task. Even if the anticipated size of a stockpile is small, it can for a time have a significant impact depending on the speed with which the government acquires or disposes of stocks. Moreover, the direct effect of government purchases or sales may be accentuated, or moderated, by changes in the inventory behavior of the private sector induced by the current stockpiling activity and by the shifts in metal supply and demand resulting from earlier stockpiling efforts. In addition, both the magnitude and the nature of the consequences of public stockpiling may be influenced by the purpose or objectives of the stockpile, the manner in which it is operated, and the prevailing market conditions at the time purchases or sales are undertaken.

Forecasting Long Run Metal Prices

Firms studying the feasibility of developing new mines or of expanding existing facilities make forecasts of long run metal prices. Similarly, manufacturers contemplating the substitution of one material for another rely on price forecasts. Governments whose tax receipts and foreign exchange earnings follow trends in metal export markets, and local communities whose economy depends on mineral production, also have an interest in future price forecasts.

The "incentive price" technique is one method of making such forecasts. It assumes that over the long run a metal's price must approach that level which provides the necessary incentives to ensure production capacity just sufficient to satisfy demand.

As long as entry into the industry is not severely restricted, this condition is likely to be met. When market price is above the incentive price, new firms will enter the industry and existing forms will expand their capacity. The resulting increase in supply will eventually reduce price, and this downward trend will continue until the incentive price is reached and the expansion of existing capacity ceases.

Conversely, when the market price is below the incentive price, neither new firms nor established producers will be motivated to invest in new capacity. As existing facilities become obsolete and are retired from service, and as demand grows over time, the market price will rise. This upward movement will continue until the incentive price is reached and once again firms are willing to build new capacity.

Success in applying the incentive price technique lies in identifying the incentive price itself. While this can be extremely difficult, for some metals due to the nature of their long run supply it may be relatively simple.

Fig. 4.10.13, for example, portrays the long run supply curve for a metal produced largely as an individual or main product, and sold in a competitive market. This curve is derived from the average total costs and annual production capacities associated with available deposits, both those actually in operation and those that are undeveloped but known to exist, along the lines discussed earlier. The undeveloped deposits, which are denoted by the shaded areas in Fig. 4.10.13, are among the higher cost sources of supply, since lower cost deposits tend to be exploited first.

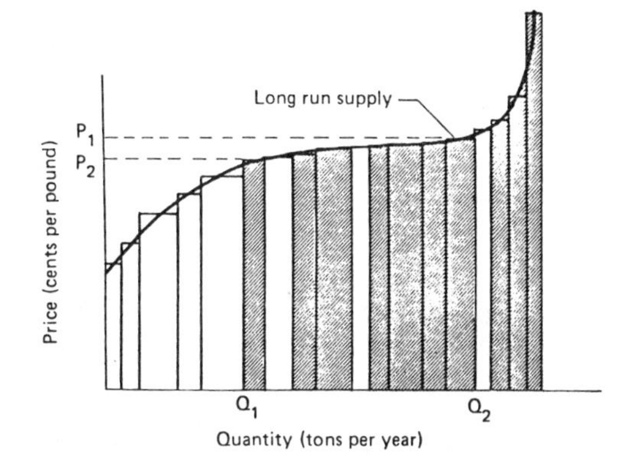

Note: Shaded areas represent known but undeveloped deposits.

Fig. 4.10.13—Long run supply curve for a metal with many undeveloped deposits of similar quality.

Because a very wide range of supply (from Q_1 to Q_2) is available in the long run over a narrow band of prices (from P_1 to P_2), the incentive price for the metal portrayed in Fig. 4.10.13 may be relatively easy to approximate. The long run supply curve has a relatively flat segment, such as that shown, when a number of large deposits of comparable quality and costs are known to exist and so are available for development. For example, many large porphyry deposits containing about 0.4% copper become attractive to develop at a copper price in the vicinity of 150 cents (in 1980 dollars) per pound.

If this flat segment of the long run supply curve is broad enough, it may cover all likely points of intersection with the long run demand curve. The incentive price can then be approximated from the estimated average production costs, including an adequate rate of return on invested capital, associated with those marginal deposits determining the relatively flat and wide portion of the long run supply curve.

Using this procedure, Radetzki (1983a) has estimated that the incentive price for aluminum is between 80 and 100 cents per pound, and that for copper between 120 and 150 cents per pound (both in 1980 dollars). At the time of his study, prices for both of these metals were substantially below these figures, allowing him to predict that aluminum and copper prices would rise over the long run.

The incentive price technique is a useful forecasting method, when the incentive price itself is easily and accurately determined. In the following circumstances, however, this is not the case.

First, when entry into the industry is restricted and one or a few firms set a producer price, the long run price may not approach the average production costs of the marginal deposits. In such situations, the determinants of long run price are more difficult to identify and assess.

Second, the long run supply curve may not have a broad flat segment. This depends on the quality differences among known deposits. If there is no tendency for a sizeable number of large deposits to share the same quality—the same ore grade, ease of access, and processing costs—there will be no narrow price band over which large quantities of supply are forthcoming in the long run.

Third, even if the long run supply curve possesses a broad flat segment, the range of possible intersection points with the long run demand curve may fall outside this segment. Some forecasters, for example, are now suggesting that large porphyry copper deposits may not be needed in the foreseeable future, in part because the growth in copper demand has slowed in recent years, and in part because more copper supply will be coming from secondary and coproduct production. As a consequence, the long run demand curve may intersect the long run supply curve before the latter flattens out in response to the availability of porphyry copper deposits of comparable quality.

Fourth, the discovery of new deposits and especially the development of new technology can cause the very long run supply curve to lie appreciably below the long run supply curve. Forecasting on the basis of the incentive price implied by the long run supply curve will as a result substantially over estimate the future price in some instances.

These considerations reflect violations of the four necessary conditions for obtaining reliable forecasts with the incentive price technique. It is important to consider whether these conditions are satisfied, before using the technique to predict future metal prices.

Market instability, government stockpiling, and long run price forecasting are only three examples of the usefulness of economic principles in analyzing metal markets. Many other possibilities exist. We could, for instance, have assessed the likelihood of a successful cartel in the world bauxite industry, the impact of another invasion of the Shaba province in Zaire on the cobalt market, the consequences for land based nickel producers of seabed mining, or the costs to domestic consumers of protecting the U.S. steel industry.

The three examples, however, suffice to demonstrate the usefulness of relatively simple economic principles of supply and demand for the analyst investigating metal markets and industries. The examples also indicate that these conceptual tools must be combined with specific knowledge about the nature of metal supply and demand. It is not enough to know that the demand curve in most markets is downward sloping. Some idea of just how responsive or unresponsive demand is to price, and how this responsiveness varies from the immediate to the very long run, is essential. Some applications even require that the total demand be broken down into its various components, such as the demand for consumption, for inventories, and for government stockpiles.

Similar information is also needed on the nature of supply. To what extent, for example, does supply come from individual and main product output, from secondary production, from byproduct and coproduct production? How does market structure and the manner in which prices are determined affect supply? How responsive is supply to a change in price? How does this responsiveness change as supply approaches the constraint imposed by current production in the immediate run, by existing capacity in the short run, and by the availability of mineral deposits in the long run? How does the output or price of a main product affect the supply of its associated byproducts? How does secondary supply vary with metal consumption, prices, and the flow of old scrap?

No single set of answers to such questions is valid for all metals. The answers may even change for the same metal, as technology and other conditions evolve over time. So the good analyst of metal markets must know his economic principles, but he must also know how technology, institutions, market structure, and government policies shape metal supply and demand.

References

1. Babitzke, H.R., Barsotti, A.F., Coffman, J.S., Thompson, J.G., and Bennett, H.J., 1982, *The Bureau of Mines Minerals Availability System*, U.S. Bureau of Mines, IC 8887.
2. Bennett, H.J., Moore, L., Welborn, L.E., and Toland,

J.E., 1973, *An Economic Appraisal of the Supply of Copper From Primary Sources*, U.S. Bureau of Mines, IC 8598.

3. Bonczar, E.S., and Tilton, J.E., 1975, *An Economic Analysis of the Determinants of Metal Recycling in the United States: A Case Study of Secondary Copper*, Final Report to the U.S. Bureau of Mines, Department of Mineral Economics, Pennsylvania State University, University Park.
4. Brooks, D.B., 1965, *Supply and Competition in Minor Metals*, Resources for the Future, Washington, D.C.
5. Canavan, P.D., 1983, *The Determinants of Intensity-of-Use: A Case Study of Tin Solder End Uses*, Ph.D. Thesis, The Pennsylvania State University, University Park.
6. Demler, F.R., 1983, "Beverage Containers," in *Material Substitution: Lessons from Tin-Using Industries*. J.E. Tilton, ed., Resources for the Future, Washington, D.C.
7. Fisher, F.M., Cootner, P.H., and Baily, M.N., 1972, "An Econometric Model of the World Copper Industry," *The Bell Journal of Economics and Management Science*, Vol. 3, No. 2, Autumn, pp. 568–609.
8. Gauntt, G.E., 1980, "Market Stabilization and the Strategic Stockpile," *Materials and Society*, Vol. 4, No. 2, pp. 203–209.
9. Gill, D.G., 1983, "Tin Chemical Stabilizers and the Pipe Industry," in *Material Substitution: Lessons from Tin-Using Industries*, J.E. Tilton, ed., Resources for the Future, Washington, D.C., pp. 76–115.
10. Landsberg, H.H., 1976, "Materials: Some Recent Trends and Issues," *Science*, Vol. 191, No. 4228, pp. 637–641.
11. Malenbaum, W., 1975, "Law of Demand for Minerals," *Proceedings of the Council of Economics*, 104th Annual Meeting of AIME. New York, pp. 147–155.
12. Malenbaum, W., 1978, *World Demand for Raw Materials in 1985 and 2000*, McGraw-Hill, New York.
13. McNicol, D.L., 1975, "The Two Price Systems in the Copper Industry," *The Bell Journal of Economics*, Vol. 6, No. 1, Spring, pp. 50–73.
14. OECD, 1974, *Forecasting Steel Consumption*, Organization for Economic Cooperation and Development, Paris.
15. Radcliffe, S.V., Fischman, L.L., and Schantz, Jr., R., 1981, *Materials Requirements and Economic Growth: A Comparison of Consumption Patterns in Industrialized Countries*, A Report Prepared by Resources for the Future for U.S. Bureau of Mines, Washington, D.C.
16. Radetzki, M., 1983a, "Long-Run Price Prospects for Aluminum and Copper," *Natural Resources Forum*, Vol. 7, No. 1, pp. 23–36.
17. Radetzki, M., 1983b, *State Enterprise in International Mineral Markets*, International Institute for Applied Systems Analysis, CP-83-35, Laxenburg.
18. Temin, P., 1964, *Iron and Steel in Nineteenth Century America*. Cambridge University Press, Cambridge.
19. Tilton, J.E., 1977, *The Future of Nonfuel Minerals*. Brookings Institution, Washington, D.C.
20. Tilton, J.E., and Vogely, W.A., eds., 1981, "Market Instability in the Metal Industries," Special Issue: *Materials and Society*, Vol. 5, No. 3, pp. 243–346.
21. Vogely, W.A., 1976, "Is There a Law of Demand for Minerals?" *Earth and Mineral Sciences*, Vol. 45, No. 7, pp. 49, 52–53.

4.11

The Economics of Coal and Nuclear Energy

Richard Newcomb and Michael Rieber*

INTRODUCTION

The solid fossil and nuclear fuels, but especially uranium, are given prominent, indeed leading, roles by energy experts in most long range estimates of world energy futures. Optimistic forecasts are inspired partly by the disproportionately large solid fuels resource base compared to present rates of consumption or to oil and gas resources, and partly by the vision of inexpensive high technology applications. Such a view overlooks the fact that while resource-use disparities and new technologies give allure to the promise of the solid fuels, they do not guarantee cheap solutions to the basic problems of solid fuel uses: viz., their relatively difficult conditions for efficient ignition and control, including the management of environmentally acceptable residuals disposal.

Contrary to the varied versions of this optimistic hypothesis, this paper shows that a great many persistent economic and technical problems interfere with such a simple inference. These act as constraints which qualify or dispute much of the national energy policy based on existing modeling of solid fuel requirements and optimistic engineering evaluations. The assumptions and techniques of such exercises are sometimes at fault. As often as not, however, the distortions created in open market fuel competition due to misdirected national energy policies themselves are to blame for misdirected efforts. A number of interesting implications result. Among them are the distinct possibility that the low technology applications will dominate, so that more direct coal and less synfuel or nuclear use will occur; that, accordingly, exhaustion rates for uranium will be low, that the most extensive growth of solid fuel through 2000 will be in conventional pulverized coal cumbustors, and that the traded share of the world coal burn will be much larger than predicted. These trends will place much more pressure on oil and gas prices than forecast by most experts.

On the way to these findings the paper demonstrates that solid fuel resource adequacy *in situ*, balanced against the projections of demands as capacity requirements, mean little as conventionally measured by years to exhaustion or potential supply schedules. Indeed, all stock concepts of demand requirements or of fuel resources must be tied to the paths of the unit costs and the deliverable prices of all substitute fuels and systems over time. This requires a dynamic analysis of supply and demand, integrated with geological and engineering constraints, to provide insights into the multiple modes of market and product substitution. Market forces rather than technical promise will determine the allocation of the solid fuels. Moreover, the equilibrium paths of individual fuel prices and market shares are not easily simulated by general or network models, and decision makers will continue to rely on the partial, local comparison of alternatives in their choices of fuel systems at any time.

The economics of the solid fuels differs from that of other energy sources in important ways relating to their form in use, conversion efficiency, (dis)economies and externalities. Petroleum and gas are in a more convenient form for

*University of Arizona, College of Mines, Tucson.

the wide variety of uses which, prior to their discovery in large natural deposits, had to be served by solid fuels directly or solid fuel conversions. In consequence, coal gasification and oil shale retorting technologies appeared early in the 19th century wherever high values permitted their conversion expense. However, with the discovery of petroleum and natural gas and throughout the subsequent one hundred years of declining oil and gas prices, the demand for synthetic fuel conversions withered. Indeed after World War I, the direct demands for anthracite and bituminous coals, except for metallurgical uses, also atrophied. By the end of the secular decline in fuel prices in 1973, the only significant markets remaining to coals in developed countries were metallurgical and very large boiler utilizations limited largely to the generation of electric power. Wood remained a dominant solid fuel in developing countries wherever transportation and alternatives were relatively costly. From 1930 to 1970 the solid fuel of promise was uranium, but doubts about its ability to compete with cheap oil were continually raised. In 1973, nuclear prospects improved with the rise in oil and gas prices.

Technical Efficiencies and Roundaboutness

The nature of nuclear reactions restricts the peaceful uses for uranium principally to power generation. Until commercial small scale modular reactors are successfully marketed, nuclear energy for process heating is not plausible. As in the case of coal, there are significant dissipations in British thermal units (Btu) in the transformation of nuclear fuels to electric power and its delivery, limiting further the solid fuel advantages over more direct utilizations of oil and gas by users. In the case of conventional reactors, without fuel reprocessing some 98% of the potential Btu is unused or dissipated in conversion. Further economic losses occur in transmission. The resulting overall inefficiency is so severe it could only be offset by the highest value uses of electrical energy or extremely advantageous locational factors were it not for the small amounts of uranium mined and procured relative to the electricity produced. A dramatic example of such a use is the success of the nuclear submarine. Nevertheless, because of the larger known resource and reserves of the solid fuels compared to crude oil and gas, developments in the nuclear power and coal synfuel technologies continue to hold promise.

Greater inefficiencies and roundaboutness in conversion imply higher capital investments in solid fuel delivery systems than in competitive systems using oil and gas. Clearly, great technical sophistication is also required in all refining and conversion stages. In consequence, from the outset it has been known that high technology solutions offsetting these disadvantages would be required to keep the solid fuels competitive with their hydrocarbon fuel rivals. In the early stages of development, without subsidized exploration, fuel enrichment, reprocessing, waste disposal, and the subsidized development of a second generation of breeder reactors, the conventional nuclear power plants constructed in the United States could not become a truly cost effective alternative to oil or gas fired power plants or to direct utilizations of oil and gas by users. Similarly the promise of renewed coal uses had to be linked to a substantial improvement in techniques for coal gasification or liquefaction, and clearly this required considerable subsidy. In the case of strategic necessity, such as in Germany during World War II or South Africa during post-war embargoes, coal synfuel subsidies for development were "justified." It is not clear that peaceful commercialization, such as the U.S. Project Independence or similar proposals advanced, can be so justified.

Another view of conversion inefficiencies can be gained by considering the greater waste disposal problems generated by solid fuel systems in contrast with alternative direct uses of oil and gas. Bituminous coals have ash disposal and emissions control problems which are significantly more expensive to correct than those of oil or gas. These problems increase with the employment of bituminous coals of lower rank, sub-bituminous coals and lignites. The residuals management and disposal problems for the uranium cycle are complicated by high-level, long-lived radioactivity. In all solid fuel technologies, and especially in nuclear power generation, the problems of safe conversion appear greater than for oil or gas systems.

Delivery Systems and Spatial Considerations

Solid fuel transportation problems and costs differ from those of oil and gas transportation. Of course, coal in slurry takes advantage of some pipeline transmission economies, and both coal and uranium have some cost advantages in certain forms of storage. High transport costs in general and high rail costs in particular render coal resources more sensitive to locational factors than oil or gas resources. While nuclear power is tied least of all to the location of uranium resources, conventional light water reactors are tied to water sources, and siting problems are complicated further by considerations of reactor safety. So facility location remains an important factor in the ultimate cost of nuclear power. Economies of scale in transportation and transmission add to those of power generation. In sum, scale economies become a significant factor in the investment decision and, together with capital intensity, they imply that the prospects for solid fuel investments depend importantly on the rate of growth in demand for electric power. The faster power systems expand, the better scale economies can be exploited to lower the average cost of power in the long run. This rate and the rate of technological change are the major determinants of solid fuel futures, given the level of rival fuel prices. As long as the real prices of oil and gas were falling over time, only the prospects of nuclear energy too cheap to meter could keep the nuclear power alternative alive, while little coal could be mined and moved cheaply enough.[1] The cost of environmental controls added further to coal's problems. Clearly, low prices for rival fuels also ruled out synfuel conversions. When, on the other hand, the prices of oil and natural gas rose significantly and remained high, the prospects for these solid fuels and their conversion obviously improved. In the analysis that follows, these prospects are reviewed, assuming no drastic fall in oil and gas prices.

Outline

Part One describes briefly those characteristics of the solid fuel technologies and resources which lend credibility to conventional beliefs about their future prospects. Closer analysis leads to questions about this credibility. Part Two analyzes the structure of supplies and demands which will determine the outcome of future interfuel competition, the efficiency of the coal industry and the reasons for the poor performance of existing modeling efforts. Part Three analyzes solid fuel market organization, evaluates the performance of the nuclear industry, and compares coal and nuclear power costs. These are reviewed in the conventional manner. Part Four concludes by summarizing the policy implications of these analyses.

THE CHARACTERISTICS OF SOLID FUEL RESOURCES, TECHNOLOGIES, AND DELIVERY SYSTEMS

Coal Resources, Reserves and Potential Supply

Geologists and economists have in recent years developed measures of potential mineral supply.[2] Known resources of major coal producing regions in the world approximate one trillion metric tons, one third occurring in North America and much of the remainder in China, Russia, and Australia. Lesser but significant resources exist in South American, European and African coal basins. These deposit inventories are not homogeneous as the penchant of economists for dealing with aggregates might incorrectly imply. Petrologists distinguish coals as heterogeneous rocks whose carbon content is qualified by a vector of characteristics, varying with *situs*, the conditions of seam formation and situation, giving coals their rank and grade.

As an illustration, in the United States coals are classified according to supply region by five categories of carbon content. These range from 26 million British thermal units (MBtu) per ton or greater to 15 MBtu or lower as the coal ranks progress from anthracite to sub-bituminous. Similarly, coals are divided into eight categories of sulfur per MBtu raw, from the highest sulfur content of 7.0% and above to the lowest sulfur coals of 1.0% or below. Steel companies categorize coal reserves further by their coking, volatility, and agglomeration properties, and utilities categorize coals locally by other characteristics affecting their beneficiation or use (e.g. washability, and chloride content). Because of sulfur and the large number of accessory trace elements in coals, including cadmium, nickel, and arsenic, a variety of environmental concerns have been raised over coal combustion. Despite the notable removal of visible ash in modern utilizations, coals are generally classed as "dirty" fuels compared to most gas and oil products. Nonetheless, extant studies on coals imply that most accessory trace elements in coals (i.e., those measured in terms of parts per million) are clay related and therefore inexpensive to remove by ordinary coal preparation prior to burning and by electrostatic precipitation after combustion. Thus, save for sulfur and nitrogen oxides, the carbon content of coals prepared for combustion is relatively clean, (cf. below).

Important characteristics which affect mining ease are seam thickness, depth, slope, and friability; those which further affect the qualities of beneficiated product are sulfur and the accessory elements. The latter can greatly affect the value of a coal to certain users. Except in a few major seams and regions of the developed countries, coal resource endowments regionally are not sufficiently distinguished on the basis of these characteristics to permit firm or statistically robust estimates of potential supply. Nevertheless, the subset of known deposits in major (currently mined) seams has been evaluated by engineers well enough in the principal producing regions of the world to compare the approximate mouth-of-mine (MOM) prices of representative coals which compete with one another or rival fuels for consumption at any place or time.

Mine productivities within a likely range are dominated by the feasible mix of capital equipment, corresponding labor inputs and the conditions of seam which dictate the size and employment of standard surface or underground mining components. Given regional factor prices, this permits the process engineering evaluation of major coal seams to determine the present value for the typical deposits composing the resource base. Reduced to costs MOM for the standard ton, deposits can be divided into two subsets. Those economically producible over the range of feasible technologies at a designated price are designated "reserves." Invariably, the seam thickness, size of mine, and accessibility to surface mining techniques are the most critical factors establishing the range of these costs. Next an exploitation submodel assigns reserves in the demonstrated resources, after the elimination of inaccessible and depleted reserves, to known and inferred deposits. For the former, an estimate of existing mine production capacities and reserves committed to them designates the short run supply for coals. The last step, potential supply, is reached by comparing estimated exploration and development costs, which permit the ordering of deposits by estimated future costs of mining. Mineral processing costs are added by region as a constant. Long run potential supply for coals thus involves three aggregation problems, each of which may also require an exercise in statistical or econometric estimation.

The simplifying assumptions required for the construction of regional potential supplies are heroic, given the paucity of data in most regions, even for major seams. In minor seams, or regions without experience, the elaboration of supply can stretch credibility further. Even when accomplishable, estimates of MOM costs cannot define regional comparative advantages unless transportation costs to the major consuming centers and the costs of preparation and adoption can be integrated into information on demands.

Coal Processes and Costs

In countries which early developed coal-based industrial sectors, the demands for coals were finely classified among metallurgical, industrial, transportation, residential, and electric power utilizations. Intricate distribution systems evolved along with a variety of boilers, stokers, and other user technology to make anthracite and bituminous coals the principal energy source of modern economies. The substitution of readily available oil and gas after 1920, however, in these countries dramatically switched industry away from solid fuels. User technology and delivery infrastructure withered, so that the end of the era of cheap alternatives 50 years later has marked a period of relatively expensive transition for the future. Except for electric power generation and metallurgical applications, the return to coal-based energy sources is further complicated by environmental control technology and the expense of emission controls. Nonetheless, the rising price of rival fuels has revived coal's prospects as a major, or even dominant, source of energy. In view of many experts, coal's role will be to provide growth until scientists can fulfill the promise of cheap nuclear power expansion. In the view of some, the coal revival is a technical retrogression and an admission of the currently prohibitive costs and inadequacies of nuclear power. To this extent, a return can only be temporarily and reluctantly conceded. In contrast, others can see coal economies not only as dominating the nuclear choice with direct-coal-based power, but as presaging the return of coal-based technologies in myriad synthetic fuel applications for industry and transportation as well as in residential and petrochemical uses. Neither extreme is commercially defensible. Instead, as this paper shows, the economic and technical aspects of solid fuel substitution and competition remain complex, precluding any simple choice by modeling of the field prices per MBtu of rival fossil or solid fuels.

In the case of coal conversion, a host of processes are under investigation. Low-Btu processes are being researched by Westinghouse, the U.S. Bureau of Mines, Union Carbide/Batelle, Combustion Engineering, Pittsburgh and Midway and Lurgi, for both agglomerating and non-agglomerating coals. High-Btu commercial Lurgi plants are in operation abroad. Other high-Btu gas processes include IGT's HYGAS, the CSG-CO_2 Acceptor Process, BCR's BI-GAS, the Kellogg Molten Carbonate Process, and the U.S. Bureau of Mines' Synthane and Hydrane processes. There is little doubt that the forecast differentials between the price of delivered coals and natural oil or gas will be high enough to find in many places forms of gasification or liquefaction economic in moderate-technology industrial and residential applications. Examples include fluidized bed combustion for district heating or apartment complexes and small scale Wellman-Galusha or Koppers-Totzek gasifiers. Among liquefaction processes are the Gulf and Given, SRC, H-COAL and COED processes, ARCO-SEACOKE, CONSOL and the U.S. Bureau of Mines Process, Fischer-Tropsch and the catalytic processes. Few coal-oil slurry and low grade liquefaction processes appear promising. In all such cases, applications will be remote from oil and gas sources, ports or pipelines and close to coal fields. Such locations have been reduced by the development of alternative infrastructure in many countries during the period of cheap oil and gas. Some experts envision coal-oil price differentials rising sufficiently to encourage the use of synfuels in higher value applications, such as transportation. The use of coal for petrochemicals, however, appears unlikely. For most low value utility use, synfuel steps are too costly compared to modern direct-coal alternatives. The major drawbacks for synfuels at all levels are the loss of substantial Btu and the very extensive environmental control costs incurred in the conversion processes.

Possible exceptions of interest to utilities are entrained gasification-combined cycle with fuel cells, moving-bed gasification-combined cycle with fuel cells or 1427°C turbine, atmospheric, and pressurized fluidized-bed combustion and entrained gasification combined with 1427°C or 1093°C turbine. There is little indication in most applications that synthetic fuels from coal or oil shales can become commercially viable at present differentials, or that they would be viable even at much higher levels. On the other hand, at present price differentials, many environmentally admissible coals can be secured for direct-coal-based electric power generation at half the cost of natural gas or oil. In metallurgical uses, substantial world growth is not expected. For these reasons, a critical issue is whether present price differentials are maintained in the oil and gas markets. It may be argued that most non-OPEC nations with indigenous fuel resources are at least as interested in maintaining the differentials as are the OPEC and other oil exporting nations. In this view, oil prices are unlikely to be permitted to fall in real terms for long. In the absence of a dramatic return to cheap oil, attention focuses on the trade-off between the direct use of coal versus nuclear energy in the largest steam boiler applications.

Coal Delivery Systems

Sponsored research and high optimism over the technical prospects of massive synfuel output and its movement, or the movement of coal in slurry form, particularly from resources in the Western United States remote from major markets, led to renewed interest in pipelines and unit train transportation. Such movements for the competitive form of coal were compared with electric transmission by extra-high voltage lines with alternating or even direct current. Historically, competition between pipeline and rail involved regions where rail lines existed. It is now clear that the building of rail facilities from scratch can seldom be warranted on the basis of coal traffic alone. Thus, where no rail exists new coal slurry pipelines appear to be the efficient choice. The other interesting case is that of multiple rail line uses or the improvement of existing rail versus new slurry pipeline over distances long enough to garner the economies of bulk rail movements and in volumes large enough to garner the economies of pipelines.

None of the considerable study given to the question of coal transportation modes by either the industry or the government has proved operationally useful to individual utilities. The basic problem is that the "hands on" comparisons compatible with specific utilities' decision making and accounting procedures are quite different from the generalized optimization models which are applied to the study of the industry's coal supply problems. Coal slurry pipelines require multi-plant demands often beyond the requirements of single utilities. Such large scales bring accompanying large risks. To be useful to utility decisions, the analysis of these risks should reflect the utility's capital endowments and its accounting statements. For these reasons, route-specific comparisons which take into consideration levels of coal comminution and washabilities, in addition to sulfur and MBtu content, are most relevant to the discussion of ultimate resource adequacy and use. In general, precise analyses of costs in such cases have resulted in rail being favored over pipeline.[3]

Transportation often adds as much to the price of delivered coal as the cost of mining. This encourages models of coal markets in which the efficient allocation of multiple sources to sinks takes place constrained by transportation costs, the long run supply functions and the elasticities of projected coal demands.

Coal Prices, User Costs and Deliveries

To complete the representation of coal markets as a matrix of equilibrating shipments, information is necessary on the potential costs of producers and users and on the significance of economic rents. The latter occur as a result of market growth, decline and cycles. In the United States, the modeling of mine costs and of conventional power generation with different design options is almost a continuing process for mining companies, utilities and the public agencies responsible for energy policy. One typical computation program is employed to evaluate mine costs and power systems. Facility costs are simulated by individual components for equipment, labor and materials, adjusted by appropriate cost indexes and summed to total capital costs. The cost indexes are calculated from functional relations in the case of capital, and from wage rates, labor productivity and overtime considerations in the case of labor. Data on literally thousands of variables and parameters are kept current for regions of the country. Such programs generate site-specific fixed and operating costs per ton and per kilowatt-hour (kwh). The information is input into a second kind of model used widely to conceptualize the spatial and temporal fuel allocations given different coal-nuclear-electric utility choices. Different configurations of demand, environmental regulations and costs of the technical options themselves are then assumed. Such long run evaluations conceive of the potential fuel sellers and utilities determining both future coal suppliers and shipments $(x_j, x_i) = x$, and market-clearing shadow prices $(k_j, k_i) = k$. These force patterns of shipments and prices as solutions, $z' = (x', k')$ with their associated slack variables, $w' = (u', v')$, which satisfy a joint primal-dual problem:

$$\begin{aligned} w - Mz + q &= 0 \\ w'z &= 0 \end{aligned} \qquad (1)$$

where $w > 0$, $z > 0$. $(c', b') = q'$ then gives the vector of primal and dual constants representing fixed costs and costs of transportation given in c, while fixed capacity levels are given in b. M = the copositive matrix defining the parameters of the solution variables. These include own-price slopes representing variable costs.

Decomposed, the system can be written as a linear complementarity program (LCP):

$$\text{Find } z > 0,\ w > 0.\ \begin{bmatrix} x \\ k \end{bmatrix} \begin{bmatrix} u \\ v \end{bmatrix} \text{ to satisfy}$$

$$\begin{bmatrix} u \\ v \end{bmatrix} = \begin{bmatrix} -c \\ b \end{bmatrix} + \begin{bmatrix} Q & B \\ -A & 0 \end{bmatrix} \begin{bmatrix} x \\ k \end{bmatrix}$$

$$\begin{bmatrix} u \\ v \end{bmatrix}' \begin{bmatrix} x \\ k \end{bmatrix} = 0 \qquad (2)$$

Following Henderson's pioneering analysis in *The Efficiency of the Coal Industry*,[4] such a formulation of hypotheses capable of verification empirically in the coal markets over time can be easily reduced to a linear program (LP) by (1) setting $Q = 0$, eliminating the slopes of all structural supply and demand equations formulating market conditions, and (2) setting $B = A$, resulting in the simplification of the other general problem constraints.

Alternatively, with $Q = 1/2(Q + Q')$, the model becomes a quadratic program (QP). This offers a more elaborate structure for supplies and demands which collapses a multitude of linear programming constraints into simple pairs of parameters. This replaces literally hundreds of process analyses runs with high and low estimates for given scale plants in a simulation and accounts for the considerably lower expense of QP allocation models. So employed, the framework makes possible the simulation of shifts in supply patterns of a wide variety of technological changes over the long run.

Fuel Substitution

Substitution among fuels conceptually takes place in existing plant technologies by fixing the relation of the vector of the inputs, x, to busbar output, y, kwh. Given input prices, r, the utilities are presumed to minimize the cost per kwh, $C = xr$, subject to $y = f(x)$ as designated by process analyses. For economists this is equivalent to differentiating with respect both to x and the Lagrangian k in the objective function:

$$V = [c - k(y - f(x))] \qquad (3)$$

which yields $n + 1$ equations of the form: $i = 1, 2, \ldots n$

$$\begin{aligned} \partial f/\partial x_i - (1/m)r &= 0 \\ y - f(x) &= 0 \end{aligned} \qquad (4)$$

The n proportionalities plus the production function determine the selection of fuel and marginal cost as functions of the fuel prices and output. Thus, economists obtain the derived demand relations for coal, oil or gas. Similarly, by differentiating with respect to r, they obtain all the own-price variations.

Substitution of energy forms in the case of multiple industry analysis with multiple modes of production, a matrix of productivities or input/output coefficients, A, gives the intensities of use for each of i inputs in the production of a unit of every jth output. Given product prices, p, this yields a Leontief system of equations:

$$(I - A)xr = yp \qquad (5)$$

To represent technical change, new activities can be defined and "costed" to expand sectors in the system so that several activities, old and new, can be used to produce the same output. In this case, nuclear or synfuel industries can be simulated as well as coal-fired technologies and switching can occur.

Because the models of equations (1)–(5) incorporate the theories economists require in their analyses of likely future market shares for coal and nuclear power, they are employed in Parts Two and Three as simple frameworks for defending or disputing the conclusions of energy economists or policy analysts about fuel choices, and for relating these opinions to the industrial or engineering analyses and decisions determining future fuel facilities.

U.S. Uranium Resources and Reserves

Uranium resources and reserves are categorized by the forward cost of their estimated or potential availability. They are subgrouped by the confidence in the estimate and by the type of extraction. Given mining methods, the higher the uranium price the larger the published reserve, and conversely. Similarly, the higher the uranium price the lower the cutoff grade, other things remaining constant. Unlike endowments, reserve and resource estimates are functions of costs, prices and technology as well as geology. At best, reserve-resource estimates are a time tagged photograph of a dynamic situation. It is essentially incorrect to compare reserves and resources at a given date with requirements, needs or demands over some future period to arrive at either an excess supply or demand crisis. A reduction of reserves in all cost classes is due to depletion of higher ore grades, estimating methodology changes, inflation, and environmental and other mining cost increases. These need not be offset by successful exploration and development, particularly in the low cost categories. As uranium prices fall, exploration as measured by drilling also falls. If the price rises, or is anticipated to do so, exploration activity will increase. Some new reserves will be found. It is the dynamic reserve—demand relationship, linked by expected prices and costs, that is important; not the drawdown from an arbitrary stock.

Table 4.11.1 provides the mean and a conservative estimate of U.S. uranium reserves and resources by forward cost category effective October 1980. Reserves are defined by direct sampling. Probable potential resources (estimated additional resources category I) occur in known productive uranium areas as extensions of known deposits or in undiscovered deposits in known geologic trends or areas of mineralization. A subtotal of the two categories provides a conservative estimate of availability. The possible potential category (estimated additional resources category II) occur in undiscovered or partly defined deposits within similar or the same productive geologic provinces or subprovinces. Speculative potential resources are estimated in undiscovered or partly defined deposits in heretofore unproductive formations or geologic settings in productive and unproductive provinces or subprovinces.

The cut-off grade is 0.01% U_3O_8. Forward costs include operating and capital costs but exclude sunk costs, interest, profits, and income taxes. Exclusion of interest and profits in the forward costs suggests a shift up in actual cost and down in the estimate in at least the highest cost category. For example, a 15% rate of return on investment increases the actual cost of the $50 per lb forward cost category by 1.4 to 1.6 times.[5] Finally, the exhibit excludes an estimated 140 thousand tons of U_3O_8 recoverable from phosphate and copper mining through year 2000. The level of phosphate production is based on 1980 production rates, that for copper is not specified. Both have suffered production declines since 1980.

Tables 4.11.2 and 4.11.3 provide alternative views of U.S. uranium resources and reserves. Clearly, the type of mining affects the cost category. The quantities tabled are the mean of the probability distributions representing each cost class. The decline in average ore grade with cost reflects the price required to mine successively lower ore grades. Finally, it may be noted that mean estimates of reasonably assured resources (reserves) fell steadily in each forward cost category from 1979 through 1983. In 1983, U.S. uranium reserves, by cost category, were 180,000 short tons U_3O_8 at $30 per lb, 570,000 tons at $50 per lb, and 885,000 tons at $100 per lb.[6] The tonnages are cumulated.

Immediate additional sources of uranium are enrichment plant tails and mill tailings. By 1980, the former amounted to 258,000 tonnes of yellowcake equivalent in government stockpiles at the plants.[7] Future output depends not only on enrichment plant activity, but on the tails assay. Mill tailings are recoverable by leaching.

The tails from uranium enrichment contain low grade U_{235}. Assuming this was re-enriched, at a 0.2% tails assay the equivalent of 18,900 tons of U_3O_8 would be recovered based on 1980

Table 4.11.1—Uranium Resources of the United States[1]—1980
(000 tons U_3O_8)

Forward Cost Category	Mean	95th Percentile[2]
$30/lb U_3O_8		
Reserves	645	567
Probable	885	659
(Subtotal)	(1530)	(1226)
Possible	346	194
Speculative	311	155
Totals	2187	1731
$50/lb U_3O_8		
Reserves	936	821
Probable	1426	1102
(Subtotal)	(2362)	(1923)
Possible	641	346
Speculative	482	251
Total	3485	2771
$100/lb U_3O_8		
Reserves	1122	971
Probable	2080	1646
(Subtotal)	(3202)	(2617)
Possible	1005	521
Speculative	696	378
Total	4903	3875

Source: U.S. Department of Energy, *An Assessment Report on Uranium in the United States of America*, Grand Junction Office, GJO-111 (80), October 1980, Table 1, p. 1.
[1]In place quantities. Processing losses may range from 5–15%.
[2]A 95% confidence in the existence of at least the amounts shown.

Table 4.11.2—Estimated U.S. Recoverable Uranium for Reasonably Assured Resources—1981 (000 tons U_3O_8)

Forward Cost Category ($/lb U_3O_8)	Underground	Open Pit	Non-Ore	Total
20.01–25.00	3.402	22.540	18.840	44.791
25.01–30.00	34.874	23.425	51.445	109.744
30.01–35.00	72.298	28.894	14.620	115.811
35.01–40.00	47.126	23.862	—	70.988
40.01–50.00	17.671	16.567	—	34.238
Total	175.370	115.287	84.905	375.562

Source: Energy Information Administration, *World Uranium Supply and Demand: Impact of Federal Policies*, DOE/EIA–0387, March 1983, Table 4, p. 26.

U.S. cumulative tailings. If a sufficiently high price for U_3O_8 offset the additional enrichment charges, at 0.1% tails assay 70,170 tons would be recovered. Based on 1980 DOE estimates of enrichment from 1980 through 2009, 716,000 tons of uranium will be in the tails at a 0.20% tails assay. Further enriched to a 0.10% tails assay, 138,000 tons U_3O_8 equivalent is obtainable.

Uranium mill tailings amounted to 2600 tons of recoverable U_3O_8 estimated at $30 per lb in 1980, 6,600 tons at $50 per lb and 9,500 tons at $100 per lb. The projected 1980–2009 accumulation equals 23,200 tons of U_3O_8.[8]

The 1980–2009 estimates for both enrichment tails and mill tailings are based on U.S. nuclear power program projections unlikely to be achieved. They and the $100 per lb category have limited application unless imports are foreclosed.

Low grade uranium resources in the United States include the Chattanooga shales, phosphate by-product, seawater and nonferrous metal by-products.

Phosphate output could yield 4,400 tons U_3O_8 in 1980, 6000 tons by 1985 and 5,000 tons by 2000. Cumulative production could be 113,000 tons through year 2000. However, as a by-product, the uranium production depends on the demand for phosphatic fertilizers, currently in worldwide recession. It is estimated that a uranium price of $40 per lb U_3O_8 (1980) would make recovery profitable,[9] but as mining and milling costs need not be included and as the uranium recovery beneficiates the phosphate, the estimate is too high.

Possible U.S. output of uranium from nonferrous metals include copper (500 to 1000 tons per yr), beryllium (17 tons per yr) and aluminum (a few hundred tons annually).[10] As all are by-products, production depends on the reworking of prior waste dumps and future output of the primary product. Production cost estimates are not available.

Seawater contains perhaps five billion tons of U_3O_8. Estimates of recovery costs range from $1400 per lb through $300 per lb to $100 per lb or less for optimized plants and new technology. Japan is already working on conceptual designs for a 1000 ton per year plant.[11]

Resources exist below the 0.01% cut-off. For example, the Chattanooga shales contain almost five million tons of U_3O_8 in Gassaway. To move these into the the reserve category efficient underground mining would be required as well as high uranium prices and adequate prices for the by-products. At an estimated 100,000 tons of shale containing 65 ppm U_3O_8 mined daily, annual output would be 1360 tons of U_3O_8, 10,650 tons of vanadium, 19.3 million barrels of oil, 171,500 tons of ammonia, and 790,000 tons of sulfur.[12] While by-product values contribute to profitability and to the possibility of uranium output, neither sulfur nor vanadium, produced in those quantities are likely to command high credits, the oil will be low valued as the grade is not high, and the ammonia will command the hydrogen value unless the fertilizer market is growing.

Uranium Potential Supply

From 1978 through 1982, western world uranium production exceeded requirements. In 1982,

Table 4.11.3—U.S. Recoverable Uranium from Estimated Additional Resources—1981 (000 tons U_3O_8)

Forward Cost Category ($/lb U_3O_8)	UNDERGROUND			OPEN PIT			SOLUTION MINING			Total Probable Potential Resource	
	Avg. Grade (%)	U_3O_8 Contained	U_3O_8 Recoverable	Avg. Grade (%)	U_3O_8 Contained	U_3O_8 Recoverable	Avg. Grade (%)	U_3O_8 Contained	U_3O_8 Recoverable	Contained	Recoverable
20.01–25.00	0.287	141.5	141.5	0.205	50.0	46.5	0.123	32.5	30.5	224.0	208.5
25.01–30.00	0.240	145.0	135.0	0.153	35.5	33.5	0.096	46.0	43.0	226.5	211.5[1]
30.01–35.00	0.218	64.5	60.0	0.135	25.0	23.0	0.107	15.5	14.5	105.0	97.5[2]
35.01–40.00	0.179	55.5	51.5	0.142	12.5	12.0	0.094	15.5	14.5	83.5	78.0[3]
40.01–50.00	0.163	27.5	26.0	0.124	3.0	2.5	0.098	15.0	13.5	45.5	42.0
Total	0.23	434.0	404.0	0.164	126.0	117.5	0.103	124.5	116.0	684.5	637.5

Source: Energy Information Administration, *World Uranium Supply and Demand: Impact of Federal Policies*, DOE/EIA–0387, March 1983, Table 5, pp. 27–28.
[1]Estimated additional 20 thousand tons available as byproduct through 2010.
[2]Estimated additional 35 thousand tons available as byproduct through 2010.
[3]Estimated additional 20 thousand tons available as byproduct through 2010.

production exceeded requirements by 10,000 tons of metal. Worldwide, four to five years of supplies are already held by the world's power stations. Spot prices, about five percent of the world market, fell from $40 per lb in 1980 to $17 per lb of yellowcake by March 1984.[13] In the United States, most utilities are making new purchases on the spot market. In the rest of the world, most utilities purchase on 10 year contracts at fixed prices. The United States contract price was about $30 per lb in 1982 and $25 per lb in 1984.

With the decline in prices, world capacity utilization in uranium mining is about 76 percent. Production has shifted from underground to open cast mines due to the lower cost of the latter. The beneficiaries are Canada and Australia. In the United States, where most of the mines are small and high cost, output fell from 14,000 tonnes in 1978 to 5,000 tonnes in 1983. The shifts are represented by the change in production shares among western world nations (Table 4.11.4).

Table 4.11.5 provides the most recent estimates of noncommunist reserves and additional resources. In the reasonably assured resource category, with the exception of Namibia, India, and Brazil in the $50 category and Brazil and Other Europe in the $30 per lb category, all areas and countries have reduced reserve estimates since year-end 1981; some quite sharply. Given the decrease in yellowcake prices and the more conservative estimates of nuclear power

Table 4.11.4—Uranium Production Shares—Western World (%)

	1978	1982	Difference
United States	42	25	(17)
Canada	20	19	(1)
South Africa	12	14	2
Namibia	8	9	1
Niger	6	10	4
France	6	7	1
Gabon	3	2	(1)
Austrailia	2	11	9
Other	1	3	2
Total	100	100	
Total Output (000 tonnes)	34	42	8

Source: *The Economist* (London), 19 May 1984, p. 87.

Table 4.11.5—Noncommunist World Reasonably Assured Resources and Estimated Additional Resources (EAR-I)—31 December 1982 (000 stort tons U_3O_8)

	RAR		EAR-1	
Forward Cost ($/lb)	30	50	30	50
United States	180	570	42	114
Canada	229	240	235	298
Other North America	4	39	5	29
South Africa	248	407	129	191
Niger	208	208	69	69
Namibia	155	176	39	69
Other Africa	84	98	4	27
Australia	408	437	480	512
France	73	88	35	43
Other Europe	37	103	19	86
India	41	55	6	25
Other Asia	13	29	0	0
Brazil	212	212	120	120
Argentina	24	30	9	9
Total (rounded)	1920	2700	1190	1600

Source: U.S. Department of Energy, *United States Uranium Mining and Milling Industry*, DOE/S-0028, May 1984, Table 9, p. 27.

plant construction to year 2000, this is hardly surprising. The anomoly of Brazilian reserve increases is due to a major governmental exploration and development effort aimed, as with ethanol, at fuel security if not self-sufficiency.

In Australia, uranium is currently mined at the Ranger and Nabarlek mines. These produced about 4.5 million tons of yellowcake in 1982 and 1983. The Jabiluka mine has been held up by government action while the Roxbury Down mine is still under development. The reluctance to mine and export more uranium is partly political, partly moralistic, and may be the reluctance of a former cartel member to expand into a glutted market. The new Australian mines, if opened, would account for about 10% of the 50,000 tonnes of capacity which could be added to production over the next five years.

South Africa usually mines gold and uranium jointly, spreading the costs, and making it impossible to state an unambiguous uranium mining cost. As a joint product, however, the production decision is tied to the gold as well as to the uranium market. South Africa may produce despite adverse market conditions. The only simple uranium mine in South Africa, Beisa, failed. Total production in 1983 was 7000 tonnes with 14% less expected in 1984.

In Namibia, Rio Tinto is producing (1984) at 90% of its 5000 tonne capacity indicating a low cost mine.

Despite current prices and the related reduction in reserves Canada plans to increase uranium output by about 25% over 1983 levels. The new mines are open cast, with high-grade ore, and low costs. Most of the output was sold on long-term contract before prices fell, thus there may be some renegotiations. While most underground, high-cost, mines with long-term supply contracts have continued operation, at least three have closed. Exploraiton activity has been curtailed.

Uramium Processes and Costs

Uranium Mining: Table 4.11.6 provides an estimate of anticipated uranium costs in 1990 (1981 dollars). The real cost of money, taxes, and a marginal return for risk are included. As 1984 prices are below estimated 1981 operating, full, and most producer's marginal forward costs, some explanation is required.

Three cost measures are common in the uranium industry. Average production costs equal the sum of forward and sunk capital costs plus operating costs. Average forward marginal costs equal forward capital costs and the return on those costs plus operating costs. Average full production costs are the sum of sunk and forward capital plus the return on each and operating costs.

The costs shown are averages. High cost countries may have some low cost mines. Even with imports or low prices some portion of the producing industry in any country is likely to survive. At best, such a cost table provides an indication of cost trends for individual producers. Given adequate supplies at the costs quoted and free commodity trade, the estimates provide an indication of the direction of trade (imports and exports) in yellowcake.

In the short-run, rather than close immediately because of inadquate prices, a firm need cover only operating costs. Where these cannot be covered even in the short-run firms will close. In the United States this has already occurred. Those firms with better deposits, in the United States and elsewhere, will cover operating and at least some sunk capital costs. Where all of the latter are covered, some return on that capital may be expected.

If long-term prices are expected to be inadequate, no forward costs will be incurred. Costs remain low, but new mine output will not be available. It should be noted, however, that costs incurred for mine expansion are apt to be considerably less than for a new mine opening. If forward costs are based solely on the latter they are overstated.

Forward costs exclude consideration of costs already incurred. Producers would prefer that sales prices be greater than full production costs, but if they are to expand they must at least cover forward costs. Here, however, as in full production costs, the return on capital is a problem. Taking the ratio of the respective returns to the associated year total costs indicates that ratios of returns vary widely. As user costs or Ricardian rent enters into such returns, they should be highest in otherwise low cost producing areas (e.g., Saskatchewan and Australia). For 1990, they should be highest in those areas which are anticipated to be low cost or which can expand output at low cost. Though generally indicated, these results are not found to be uniform.

Return on investment is a target which may be exceeded or missed. If the market for yellowcake remains poor, return on investment wil

Table 4.11.6—Uranium: Average Production Costs
($/lb U_3O_8: 1981 Dollars)

Producer	Capital Costs				Operating Costs		Return on Capital				Total Costs	
	Sunk		Forward				Forward		Sunk			
	1981	1990	1981	1990	1981	1990	1981	1990	1981	1990	1981	1990
South Africa and Namibia	3	3	2	2	11	11	3	3	12	12	31	31
Niger and Gabon	6	3	2	6	15	22	2	14	7	10	32	55
Australia	4	1	2	4	20	8	1	15	3	4	30	32
United States	6	3	2	4	20	18	1	9	6	12	35	46
E. Canada	6	4	2	4	19	19	3	9	10	15	40	51
Saskatchewan	5	2	1	4	8	4	1	9	4	6	19	25
Other	5	5	3	4	15	7	4	8	7	14	34	48

Source: D.W. Weaver, "The Cost of Uranium," in Nuclear Assurance Corporation, *The Future of the U.S. Uranium Industry*, Uranium Colloquim 1981, Grand Junction, Colorado, October 1981, figures 9–10, pp. 13–14.

fall and yellowcake costs will fall. The drop will occur as producers forego a return on such capital and part of the return on forward capital. Some firms will close, future supply will diminish and market equilibrium will be restored. Absent subsidy or protection, profits can be negative, properly leading to supply reduction.

Uranium Processing Alternatives: If uranium resources are inadequate, if the price of uranium precludes nuclear development, if fuel substitutes are unavailable at lower prices, technologically advanced nuclear alternatives may be considered.

Spent fuel reprocessing reduces the drawdown of uranium reserves and resources, decreases the amount of material that must be placed in final storage as spent fuel, decreases enrichment requirements, and increases the risk of plutonium spillage, proliferation, and diversion.

After discharge from a power plant, spent nuclear fuel contains about 0.8% U_{235}. After reenrichment to 2.5–3.0% U_{235} this uranium can be recycled to a power plant for a second cycle. About 95% of the original uranium charge can be so recycled. Additionally, the power plant reactor yields about one percent plutonium. The remaining four percent is high level radioactive waste.

The reprocessing steps include chopping up the spent fuel assemblies, leaching with nitric acid to produce liquid uranium and plutonium nitrates plus liquid high level radioactive wastes, organic solvent extraction of uranium and plutonium from solution, conversion of the uranium to HF_6, and conversion of the plutonium to a solid form. The radioactive waste would be solidified for permanent burial.

The plutonium recovered by reprocessing (10 to 15 tonnes per yr for a 1500 tonne per yr capacity plant) might be used with uranium in a reactor as a mixed oxide fuel. Alternatively, it can provide the initial charge for a breeder reactor. Coupled with reprocessing, breeders could increase the energy utilization of uranium by a factor of about 70.[14]

The free world reprocessing of spent uranium takes place in five West European countries, in Japan and in India. Present capacity is 1215 tonnes per year. To 1990, planned capacity totals 5505 tonnes per year.[15] This amounts to the annual spent fuel output of about 183 typical 1000 MWe reactors.

By late 1972 reprocessing costs (excluding high-level radioactive waste disposal) was estimated at about $40,000 per ton. Given AEC estimates of nuclear power growth, one could estimate the need for a 1500 tonne per year reprocessing plant every 15 months between 1982 and 1990.

In the United States, four commercial reprocessing plants have been suggested. The only one that operated was closed in 1976 as uneconomic. Of the rest, one was never constructed; another never operated due to technical problems; the last was partially completed, came under the reprocessing ban in 1977, and closed in 1983. The United States, like any other military nuclear power, operates noncommercial reprocessing facilities for defense purposes.

The minimum estimated capital costs (1982 dollars) for a plant capable of processing 1500 tonnes per yer of spent fuel is $1.2 billion.[16] Completion of the Barnwell facility would take a minimum of 8 to 10 years. It is not clear that the capital cost cited correctly anticipates commercial finance charges for the project. Accepting the estimate suggests that for a 25 year life, undiscounted capital costs are $32,000 per tonne processed.

For commerical reprocessing to be economic, reprocessing costs must be exceeded by the value of the products plus any saving in short and long-term spent fuel storage. The value of both the uranium and plutonium is approximately the amount saved by foregoing the use of natural uranium and its attendant costs from mining through enrichment. However, the plutonium must have a commercial market, either as a mixed oxide fuel for LWR's or as a breeder reactor charge. The amount that would be produced far exceeds any commercial breeder requirements. In the United States, both options are currently foreclosed.

Short-term spent fuel storage capacity is likely to be generally short until the 1990s. Some utilities will have too little space earlier. As they should be willing to pay others to store their assemblies, they should be willing to pay about the same effective amount for reprocessing. Long-term high-level waste storage is a charge on the utilities. A reprocessing facility would eliminate most of this for the utility and would in turn be obligated to pay for long-term storage. However, the amount to be disposed of after reprocessing is considerably less than without reprocessing. The cost passed on to the reprocessing client should be much less than the client's own disposal costs.

In 1979, Bechtel National Inc. estimated the value of reprocessed uranium at $130 per kg and plutonium at $210 per kg. At the time, natural uranium was $43 per lb. In 1983, they estimated that with natural uranium selling for less than $25 per lb, the combined product value was less than $240 per kg.[17] As the plutonium cannot be used and must be stored, it need have no value and may bear a cost. If used in some distant future, its sale price then must be discounted now. The uranium value alone could approximate $75 per kg. It is estimated that commercial costs at Barnwell (1983 dollars) would be $300 to $350 per kg of spent fuel. Savings in storage costs are unlikely to make up the difference. For reprocessing to be economic, natural uranium prices must rise and plutonium must be acceptable as a reactor fuel.

On a uranium basis alone, if Barnwell is to be commercially feasible, natural uranium prices must be about $100 per lb.[18] Although price range data touch this level, it is unlikely that such prices will be reached in this century unless a major closure of Middle East oil develops. Based on an estimate of 132 gigawatts of nuclear power produced by year 2000, a DOE estimate of greater than three million tons of domestic uranium available at less than $100 per ton, and a requirement of only one million tons,[19] the excess supply precludes a $100 per pound yellowcake price until sometime in the next century. In brief, reprocessing costs set a limit on both uranium prices and reserve estimates.

Reportedly, overseas reprocessing service charges range from $600 to $1100 per kg of spent fuel.[20] Assuming a free market (i.e., governments are not paying such prices simply to disguise a subsidy), if the price is cost determined, the plants are inefficient compared to Barnwell or the Barnwell costs are grossly underestimated. Given current market prices for natural uranium, payment of such reprocessing charges suggests ties by long-term contract, use of the recovered plutonium for LWR's or bombs, the desire to preserve a scarce domestic uranium resource (and the assumption that in the future uranium will no longer be freely traded), and the expectation of a significant increase in natural uranium prices in the near future. The last implies a surge in nuclear power generation and no significant additions to uranium resources.

In a breeder reactor, natural uranium (fertile) is converted to plutonium which, after discharge, is reprocessed to extract the converted plutonium, the now irradiated natural uranium (fissile), and the residue of the original plutonium charge. This recycle continues until most of the natural uranium is so consumed. In total, the breeder produces more fuel than it consumes. In the process, the facility also generates electricity. A breeder reactor is desirable if its capital and operating costs are found to be competitive. There are only a few breeders. In the USSR, the BN350 (350 MWe) achieved criticality in 1972. In France, the first Phenix reactor (250 MWe) went critical in late 1973. These were followed by England's PFR reactor, the SN300 supported by Germany-Benelux and a second Phenix.

A breeder reactor is symbiotic with several light water reactors. The combined power system should be almost self-sustaining in terms of fuel. The breeder generates fuel for the fissile fuel reactors and in turn is charged with the fertile fuel from those reactors (U_{238}). Almost no additional fissile fuel is required. The breeder charge can be made up of depleted uranium from LWR's (or thorium) or from the tails of the enrichment plants. In either case, a waste product which gives rise to either disposal problems or inventory costs now has a positive value. Success, however, also requires the handling and use of plutonium as a mixed oxide fuel.

In the United States, the breeder was plagued by early cost overruns which constantly increased the target price for uranium if the breeder was to be commercially successful. Furthermore, the breeder gain, the amount of fuel generated for use in fissile (LWR) facilities, was less than anticipated. Alternatively, the fuel doubling time was lengthened. Again, this required higher uranium prices if the breeder was to be commercial.

The U.S. breeder reactor has cost a minimum of $1.6 billion since 1972. When finally cancelled in October 1983, additional costs of $4 to 8 billion were estimated. The reactor was many years from completion; parts had been produced and assembled, the site had been cleared, but nothing had been built. Industry support was not great and might have increased both costs and the government subsidy. The projected overrun was not significantly greater than that of the worst of the current reactors.

If uranium is depleted, thorium can extend nuclear power by augmenting uranium resources. Current reactor systems that can use thorium are heavy water reactors (Candu), high temperature gas cooled reactors, and fast breeder reactors. In the thorium cycle, U_{233} provides neutrons which are captured by Th_{232} yielding more U_{233} and heat. If the amount of U_{233} is increased between fuel loadings, the reactor has acted as a breeder reactor. Exploration for thorium has been less than that for uranium. Table 4.11.7 provides an early estimate. The estimates are undoubtedly conservative.

In the future, nuclear fusion may replace both fossil and fissile fuels in the production of electricity. As a by-product of the process, hydrogen may be produced leading to a partial approach to a hydrogen economy, at least for some transport modes. Plant size could range from small (100 to 250 MWe) to utility standard (500 to 1000 MWe).

Fusion reactions can occur among many light atomic weight elements: hydrogen, deuterium, tritium, lithium, and boron. Deuterium and tritium are isotopes of hydrogen: the former occurs naturally (deuterium oxide, heavy water, is used as the moderator in the Candu fission nuclear reactors), the latter is radioactive and would be made from lithium in the reactor. In its simplest form, fusion is the joining of the nuclei of hydrogen atoms to form helium. As the mass is reduced, energy is produced. The helium is nonradioactive, nontoxic, and currently valuable. It would be less valuable subsequent to large scale fusion commercialization.

Two basic approaches to controlled nuclear fusion are plasma (Tokamaks and mirrors) and inertial (laser and particle beam). The latter may also have military uses. The former appears to be more technically advanced. To provide the reaction using the plasma approach sufficient energy must be provided to meet a threshold temperature. Ideally, the fusion materials would yield the highest energy gain per reaction. The conditions are met by a combination of deuterium and tritium and with a threshold plasma temperature of 4keV (45 million degrees Kelvin) and a maximum energy gain per fusion energy gain per fusion of 1800.[21] To date, temperature levels of 80,000,000°K have been attained.[22] The engineering problem is to sustain and confine the temperature long enough so that sufficient fusion occurs to both offset the energy required to maintain the plasma and generate the temperature as well as yielding a significant energy output. As the more dense the plasma the shorter the required sustaining time, an input energy trade-off is involved.

The advantages claimed for fusion include: the availability of fuel sources (water and lithium); fewer coolant problems; no possibility of criticality (accidents); the elimination of fuel mining, milling, enrichment, encapsulation, and reprocessing; and no chemical or particle emission. However, at least initially, large quantities of radioactive tritium must be produced and handled; deuterium oxide must be separated from water. Aside from the tritium, neutron irradiation of equipment on site lead to activated structural materials. Both, however, contain much less radioactivity and for a shorter time span than do the fission products and actinides of current reactors. The waste disposal problem, while not eliminated, would be substantially reduced.

While not in short supply at present, it has been estimated that about three-fourths of the current U.S. lithium reserve base will be consumed under present usage patterns by year 2000. The commercialization of rechargeable lithium batteries would, in the absence of recycling, speed the process. Fusion power in the United States would require about 0.5 million tons of natural lithium by year 2050.[23] The current resources would be strained. While other sources might be developed along with processes to obtain the material, these are more costly than current sources. Table 4.11.8 provides an estimate of world lithium resources.

A shift from lithium (as a source of tritium)

Table 4.11.7—Thorium Resources
(000 metric tons)

	Reasonably Assured	**Estimated Additional**
Australia	18,500	—
Brazil	58,200	3,000
Canada	—	250,000
Denmark	15,000	—
Egypt	14,700	280,000
India	320,000	—
Iran	—	30,000
Liberia	500	—
South Africa	11,000	—
Turkey	500	—
United States	52,000	270,000
Total	490,400	833,000

Source: OECD, *Uranium: Resources Production and Demand*, OECD Nuclear Energy Agency and International Atomic Energy Agency, December 1977, Table 12, p. 125.

Table 4.11.8—Identified World Lithium Resources (000 short tons contained Lithium)

	Reserve Base	Other	Total
Chile	1000	3000	4000
Zaire	200	1800	2000
U.S.	400	400	800
Canada	200	300	500
Eastern Block (USSR)	200	200	400
Rest of World	100	200	300
Total	2100	5900	8000

Source: Singleton, R.H., *Lithium*, U.S. Bureau of Mines, Mineral Commodity Profiles, September 1979, Table 4, p. 9.

to a reactor fueled entirely by naturally occurring deuterium involves an increase in the threshold plasma temperature of 50keV and a reduction in the maximum energy gain per fusion to only 70 to 80.[24] The higher number is obtained if tritium is an output leading to some possibility of a joint program consisting of deuterium and deuterium/tritium reactors. Given, however, the increased temperature and decreased gain, produced electricity costs would necessarily rise.

Costing fusion power is impossible at this time. The technology to start and sustain the reaction process is very complex. Fusion power cannot be expected until early in the next century. The Magnetic Fusion Energy Engineering Act of 1982 (P.L. 96-386) establishes the goal of operating a fusion demonstration plant by year 2000. The Secretary of the Department of Energy is to establish a magnetic fusion engineering center. A government expenditure for R and D (subsidy) of about $20 billion to year 2000 has been authorized but not appropriated.

THE STRUCTURE OF SOLID FUEL SUPPLIES AND DEMANDS

World Coal Markets

The extent of the present (1977) annual coal market burn in millions of metric tons of coal equivalents (mtce) is approximately 2,500. At 25 million Btu per metric ton, world coal consumption is 62 Quads. The fact that this is only a trivial portion of known world coal reserves of 3,000 trillion tons gives little indication of coal's actual short run or long run supply capabilities or its competitive position vis-a-vis rival conventional and nuclear fuels. To simulate the shares of fuel in planned consumption, economists have employed the static frameworks described earlier and the assumptions of competitive market equilibrium to model demands and supplies regionally as a function of prices and transportation costs.

Unfortunately, the dynamics of investment determining the exploration, development of mines and infrastructure are very complex. Other countries of the world collect little data beyond those on the volume of current production, consumption and trade. Furthermore, most countries have long range national energy policies based on explicit assumptions permitting one to place some bounds on the future demands of coal, the data to test the consistency of such projections or relate them to price are deficient. In consequence, most national agencies do not model markets, but instead project total energy use on the basis of established rates of growth in GNP and then attempt to balance this crudely against hypothetical supply sources by fuel. When such forecasts base-load the relatively inelastic sources of crude oil and gas, the difference between total demands and the latter supplies, or "solid fuel gap," ignoring conversions, is determined as a residual solid fuel share. Even when all shares are made more or less consistent with planned electric power expansion, such projections neglect most of the important interactions of real markets.

The actual growth of solid fuels is a function of rival fuel prices and trends in the substitution (1) of coal for conventional fuels and (2) for nuclear fuels in the generation of electric power. In the discussion that follows, U.S. estimates and those of its free-trade partners are discussed for the 20 year period extending from the first long term impacts of higher fuel prices in 1973 to the turn of the century, circa 2000. The framework is the theory sketched in the previous paragraph.

The most optimistic scenarios, are those of MIT's World Coal Study (WOLCO).[25] These foresee the increase of world coal demands through year 2000 from 2,500 mtce to 5,000 or 7,000 mtce. They do this on the basis of cost engineering comparisons, including the costs of transportation, but without consideration of market allocation. The resulting flows can be distributed in a matrix representing the 22 con-

Table 4.11.9—Principal World Trade for Coal
by Major Producing and Consuming Regions, 1982 (est.)
(in million of net tons)

Supply Region	(NA)	(LA)	(PB)	Demand (Af)	(So)	(Eu)	(Ot)	Total Supply
(NA) North America	18.4	5.1	38.1			52.3		113.9
(LA) Latin America								
(PB) Pacific Basin			43.8				13.7	57.5
(Af) Africa		2.0				27.4		29.4
(So) Soviet Block					25.6	20.4		46.0
(Eu) Europe						12.6		12.6
(Ot) Other					10.0		14.7	24.7
Total Demand	18.4	7.1	81.9		35.6	112.7	28.4	284.1

Source: Chase Manhattan Bank, New York, "The Coal Situation," 1982.

suming and producing regions the studies identify. The current (1982) situation is shown in Table 4.11.9. The U.S. share of world coal production is 700 mtce or 28%. The exports and imports of coal worldwide currently are 14% or 284 mtce. Approximately 40% of all trade originates in the United States. Trade among Soviet bloc, African-European and Australian-Pacific countries divides the remaining 60% of tonnage roughly in equal parts.

Table 4.11.10 summarizes the "high coal" trade estimates of the WOLCO study for the year 2000. The diagonal entries reflect exchanges between countries within an aggregate region.

Only 994 mtce of the 7,000 mtce consumed enters world trade. Because WOLCO does not concern itself with coal prices, it misses the fact that traded coals can be delivered from the United States to Europe for one half to one third the cost of domestic production. Similar appraisals show very strong comparative advantages from coals in China, Australia, and South Africa over the domestically produced coals in other regions. It is incredible, therefore, that of the 7,000 mtce estimated burn in consuming centers, WOLCO predicts 6,000 mtce or 86% are expected to be high cost coal produced within consuming countries, leaving only 14% in the countries of comparative advantages to enter world trade! The broad mineral appraisals of world coal basins confirm that the export forecasts shown in Table 4.11.10 are far too low if open market competition determines trade.

Only in the United States and Australia are the energy and utility coal sectors dominated by private rather than public enterprise or planning. In most countries, the nuclear, coal, oil and gas industries are essentially nationalized, and operate by public plan. Reluctance to change plans,

Table 4.11.10—Principal World Trade for Coal
by Major Producing and Consuming Regions, 2000 (est.)
(in millions of net tons)

Supply Region	(NA)	(LA)	(PB)	Demand (Af)	(So)	(Eu)	(Ot)	Total Supply
(NA) North America			103			65	47	215
(LA) Latin America					55	7	50	112
(PB) Pacific Basin			75			43	161	279
(Af) Africa			8			41	50	99
(So) Soviet Block			6			35	30	71
(Eu) Europe						102	16	118
(Ot) Other[1]							100	100
Total Demand			192		55	293	454	994

Source: Adapted from World Coal Study, M.I.T., 1980.
[1]Unspecified Exports.

once confirmed, is evidenced chiefly by four tendencies in national planning projections: (1) to overstate growth rates for both total energy and electricity's share in national planning projections, (2) to understate the rate of switching from oil and gas to coal in existing capacity shares, (3) to understate nuclear costs and, (4) to overstate nuclear plant projections.

In international markets, the current forecasts no doubt overstate the demands for energy. Even so, they most certainly understate the share that coal will have in total energy flows and in trade.

Inter-regional Coal Competition

Contemporary studies of bituminous coal markets follow Henderson's seminar analysis by stressing the regional character of competition among the boiler fuels.[26] As shown earlier within this framework, coal petrology, technology and demands are treated as spatial constraints, subject to which total delivered resource costs to consuming centers determine the efficient patterns of shipments and prices. The best of these models share two common predilections whenever high coal demands are assumed: (1) an impressive expansion of western surface mined coals occurs, shifting production significantly away from the traditional mining centers of Appalachian and Interior coal basins; (2) MOM prices rise more rapidly in eastern seams in accord with their higher rankings, coking qualities, and rates of depletion. The last implies higher extraction costs per ton and per Mbtu.

The suggestion that the speed with which shifts of capital in transportation and production over time are a function of the richness of remote resources and the rates of change in depletion and technology is as old as Jevons.[27] The emphases on the expansions of coal transportation networks and the role in this played by tariffs (economic rents) also have historical precedents.[28] All this suggests that the static treatment of producers' surplus in the present multi-market models is inadequate to the long run analysis of dynamic coal markets. Notable examples of multi-market modeling are the efforts of the Department of Energy in the National Coal Model, NCM;[29] its commercial analogue, ICF's Coal Electric Utility Model, CEUM;[30] the Charles River Associates modeling for the electric Power Research Institute, CRA-EPRI;[31] Zimmerman's work with Data Resources, Inc., Z and DRI-Z;[32] Newcomb's Quadratic Program (QP) for EPRI;[33] and the Argonne Laboratory QP.[34] Notable among analyses of railroad costs and capital expansions are the studies of coal supply alternatives by Rieber and Soo.[35] The Department of Energy predicts a substantial penetration of eastern markets by western coals in its National Coal Model as the industry's capacity approaches targets of 1.0 to 1.6 billion tons, anticipated from 1985 to 1995. Such high coal scenario shifts (Fig. 4.11.1) are repeatedly confirmed by others including CEUM, Z, DRI-Z, and CRA-EPRI among LP versions of the model, and by the Argonne National Laboratory among QP versions. All apply Dantzig's fundamental programming theorem[36] to geostatistical long run potential supply schedules relating geologic parameters and differential depletion factors to the regional elasticities of supply. Because of the importance of transportation, the problem of industry efficiency is spatial, and all study must be regionalized. The figure shows i supply and j demand regions, making use of the notation of Table 4.11.11.

The structures of demands in the j consuming regions focuses attention on consumption in the ($j = 3$) region of the northeastern U.S. from Chicago through the Ohio River Basin to the north and central Atlantic, where two thirds of the

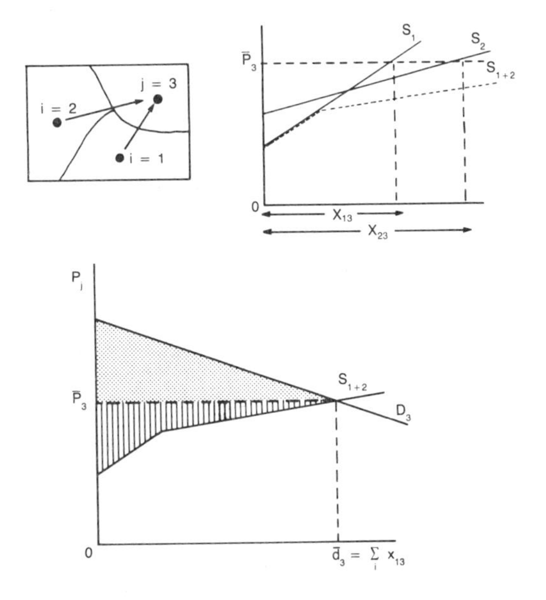

Fig. 4.11.1—Single Final Market Equilibrium. Two producing plus one consuming region.

Table 4.11.11—Variables of Coal Models

Matrix	Vector	
x	d, y	quantities demanded
	s, x	quantities supplied
	x_{ij}	shipments from i to j
p	p_j	own price of demand
	p_i	own price of supply
Q, c	q, c	coefficients and constants of supply and demand
A	a	coefficients and constants of balance equations
	t_{ij}	transportation from i to j
B, g	b, g	parameters and constants of dual relations
w	u	slack variables of primal
	v	slack variables of dual
z	x	solutions of primal
	k	shadow prices, solutions of dual

U.S. electric power is consumed and planned.

The geostatistical structure of supplies from the i producing regions, separated here in Appalachian ($i=1$) and western ($i=2$) regions for simplicity, is the major factor determining shifts. Supply functions are estimated by engineering simulations, statistically, and by indexing costs to geological parameters such as seam thickness and coal qualities for a variety of surface and underground techniques of mining and cleaning. In the eastern region, supply is characterized by lower intercepts to reflect low surface mine entry costs and low transport costs, t_{ij}, to the major consuming centers. Slope terms are high to reflect the limitations of Appalachian mining in hilly terrain and thin seams, which force high variable costs on the industry.

The ability of the western region to penetrate the northeastern markets is predicated on the entry and variable costs of its large flat areal surface mines in thick seams being low enough to offset higher transportation costs to eastern markets, i.e., $p_j = p_i + t_{ij} + r_i$. The producer with cost advantages gains a surplus or economic rent, r, but because of its static nature, the models do not relate rents directly to mine or transportation investment, or to changes in the supply capacities. Instead, all future capabilities for production are built into the long run supply functions, assumed to be always in excess demand.

Coal Industry Efficiency

In the joint primal-dual LCP notation (Table 4.11.11) the efficient allocation of shipments and determination of shadow prices come as the solution to the maximization of the sum of the areas under the demand functions less the areas under aggregate supply including transportation over all the j markets:

$$c'x - 1/2\, x'Qx - cx - 1/2Qx - T'x$$

This is simply Samuelson's net social product,[37] NSP, for the Lagrangean $L(x_i, X_j, k_i, k_j) > 0$ subject to:

$$\begin{bmatrix} -c \\ b \end{bmatrix} = \begin{bmatrix} Q & B \\ -A & O \end{bmatrix} \begin{bmatrix} x \\ k \end{bmatrix}$$

In this QP formulation, Q is the expansion $1/2(Q + Q')$, and symmetry is imposed such that $B = A$. Notationally, c and Q give the x-constraints for demands and supplies represented above by their respective equations, while B, b and A give the k-constraints for the dual shadow prices and equilibrium equations. In the LP formulations of the coal models frequently employed actual levels of capacity in the past 10 years are compared with those predicted by the primal of the models for 1985 in moderate (1 billion tons) and high (1.7 billion tons) coal scenarios. When this is done, as in Table 4.11.12, the models predict western capacity optimally grows three-to-six-fold. This forecasts capacity growth in the West from the level indicated by 1976 production of 295 million tons to as much as 1,414 million. In contrast, eastern capacity is expected to grow only by one third, from 355 million to at most 461 million tons. Actually, capacity in 1981 reached 1,050 million tons, and so has exceeded the lower 1,000 million ton target level for 1985. In 1982 production reached only 820 million tons. As a result, there is some 230 million tons of excess capacity in the industry today, 150 (65%) of it in the Appalachian and 50 (22%) in the Interior basins. The East is the only area which reached and exceeded its target capacities. The shortfall is entirely in the West, and the predicted penetration of eastern markets is not taking place.

Table 4.11.12—Actual and Simulated Coal Capacities and CIF Prices by Region in Base and Target Years in annual short tons (000,000) and constant $/MBtu and $/ton

	Actual			Simulated		
Capacities[1]	**1975 (1) DOI**	**1982 (2) DOE**	**1982 (3) Est**	**1985 (4) DOI**	**1985 (5) ZIM**	**1995 (6) ZIM**
NATIONAL (tons)	na	na	1050	1100	1000	1750
Production:						
East						
(Appalachia)[3]	355	425	575	459	461	327
West						
(Interior)[4]	182	199	249	184	84	300
(Far West)[5]	110	196	226	451	439	1114
	292	395	475	635	523	1414
TOTAL	647	820	1050	1094	984	1741
CIF PRICES[2]	**1976 (7) DOE**	**1981 (8) DOE**	**1981 (9) Est**	**1985 (10) DOI**	**1985 (11) ZIM**	**1985 (12) ZIM**
NATIONAL ($/MBtu)	.83	1.44	1.52	1.82	na	na
EAST						
(Atlantic)[6]	.99	1.56	1.65	2.01	2.11	2.93
WEST						
(Central)[7]	.81	1.55	1.65	1.99	1.99	2.07
(Mountain)[8]	.53	.97	1.03	.97	1.14	1.23
NATIONAL ($/ton)	18.30	32.31	34.25[9]	40.95	na	na
EAST						
(Atlantic)	24.67	38.95	41.29	50.25	52.72	73.30
WEST						
(Central)	19.09	36.48	38.67	46.77	46.64	48.78
(Mountain)	9.58	17.38	18.42	17.46	20.52	22.19

SOURCES:
(1) U.S. Department of Interior, actual 1975 regional production.
(2) National Coal Association, actual 1982 regional production.
(3) Author, based on estimates of actual capacity by two major U.S. coal firms, 1982.
(4) CEUM, Model version no. 1, 1979; OSM *Regulatory Analysis*, Table 45, ICF., Inc., by Klein and Van Allen.
(5) Zimmerman, M.B., *The U.S. Coal Industry*, Table 3.17, assuming increased transport costs and 1.2 lbs/MBtu SO_2 constraints, 1981.
(6) Zimmerman, Table 3.15, op. cit. assuming changing relative prices and NSPS, 1981.
(7) DOE, Form #423, Cost and Quality of Fuels from Electric Utility Plants in 1976; Sept., 1982.
(8) Idem, op. cit., in 1981; Sept., 1982.
(9) Constant 1982 dollar value of (8) using GNP inflator; Council of Economic Advisors' Annual Report, 1983.
(10) CEUM, Model version no. 3, 1979; OSM, op.cit., Table 46, by Klein and Van Allen, projections in 1978 prices inflated to 1982 as in (9).
(11) Zimmerman, op.cit., Table 3.22, in 1977 constant prices inflated to 1982 constant dollars for high sulfur coals.
(12) Zimmerman, op. cit. idem, for low sulfur coals.

NOTES:
[1] Capacities are based on mine production in base year (1975) and on industry estimates in target year (1982) when they reached the 1,000 level of 1985 targets simulated by DOI and ZIM. 1995 level of 1,700 is given by ZIM simulation only.
[2] Prices are Form 423 actual and DOE 1976 data for base year and 1981 data for target year, inflated to 1982 dollars for comparison with simulated prices, which are also in 1982 dollars.
[3] Al, EKy, Md, Oh, Pa, Va, WV.
[4] In, Il, Io, Ky, Ark, Ka, Mo, Ok.
[5] Az, Co, Mon, MN, ND, Tx, Ut, Wash, Wy.
[6] For DOE data, weighted average of New England, Mid-Atlantic, South Atlantic; for ZIM and CEUM, Mid-Atlantic; average MBtu/ton is 25.0.
[7] For DOE data, weighted average of East Central; ZIM and CEUM, West Central; average MBtu/ton is 23.5.
[8] For DOE data, weighted average of West Central, Mountain; for ZIM and CEUM, Mountain, average MBtu/ton is 18.0.
[9] Actual 1982 national average, DOE, op. cit., is $38,87; average MBtu/ton is 22.5.

This deviation of the actual industry course from that predicted has been attributed variously to factors discriminating against the eastward movement of western coals, such as higher taxes and railroad tariffs absorbing economic rents in the West, or the adoption of best available cleaning technology (BACT) which favors the higher sulfur eastern coals. SO_2 removal on a Btu basis (not ton) also favors the East. If the models were accurate, of course, the extent to which price rises in the East are higher than predicted due to these increased costs would be one test of industry inefficiency. Given the geostatistical nature of regional supply inelasticies, the upward adjustments in tariffs forcing the inefficient expansion of eastern basin suppliers should have raised coal prices in the models to levels even higher than those forecast. On the contrary, prices have fallen in the East in real terms to levels significantly below the DOE projections. The fact that minimal investments in transportation were required to achieve the eastern expansions at lower prices implies excess railroad capacity existed along most routes.

These errors in the primal forecasts clearly signify a need to reevaluate the long run supply models. Price[38] repeats the call of the model builders for correction of the data base relating to supply and the further refinement and disaggregation of the supply constraints. While much work can be recommended along these lines, it is likely to be both expensive and ineffective. This can be seen from the nature of the simultaneous errors in the dual, which clearly signify the need to return to the short run market analyses of mine, transport, cleaning, and related aggregate capacities of the industry regionally and to changes in these. As Henderson showed, this is not an expensive exercise in comparative static studies either from the data collection or modeling viewpoint. Indeed, it is perfectly consistent and compatible with the contemporary static model structure.

The imposition of short run constraints on the long run model provides an analytical approach to the distinction between secular and cylical trends. In secular growth, as DOE targeted in its forecasts, the spot and new contract prices will be driven above long run regional average costs. This happened from 1972 to 1978 as the price of coal rose from \$11 to \$26 in Appalachia, providing substantial profits and rents. Henderson observed the industry from 1947–51[39] in secular decline rather than growth. Thus, the direction of secular trends was reversed. The negative profits in the industry should have signaled disinvestment in the East, shifting the regional supply functions in the direction of more efficient primal program solutions. The extent to which the industry failed to do this can be measured by the divergence from optimal, $L(\bar{x}_i,\bar{x}_j,\bar{k}_i,\bar{k}_j)$, in actual regional outputs, shipments, MOM and delivered prices. These were found by Henderson to be:

$$\textstyle\sum_i \sum_j (1/2)|\bar{x}_{ij} - x_{ij}|/x_{ij} = 0.203 \qquad (6)$$

$$\textstyle\sum_i |\bar{\mathrm{k}}_i - \mathrm{p}_i|/\mathrm{p}_i = 0.069 \qquad (7)$$

$$\textstyle\sum_i |\bar{\mathrm{k}}_j - \mathrm{p}_j|/\mathrm{p}_j = 0.015 \qquad (8)$$

To avoid double counting, half the misallocated shipments (6) are used. The discrepancy between short and long run equilibrium prices is due largely to lags in intermediate run adjustments on the supply side. The misallocation is severe, as evidenced by the fact that with only a 1.5% rise in delivered prices, as shown in equation (8), the observed mouth-of-mine prices in the East fell 6.9% *below* the competitive levels, as evidenced by equation (7).

Because the model remains static, it incorporates no measures of capital investment. However, the role of economic rents in the shift of mine capacities and infrastructure can be analyzed descriptively. In the 1950s the industry needed to retire excess capacity. The key to the (dis)investment interpretation is the change in profits. In the East, producers forego rents and lower MOM prices to permit the absorption of a larger than optimal transportation bill. This has the effect of maintaining production in Appalachia above optimal levels and, conversely, lowering output below optimal levels in the West. The current (1980s) fall in the real price of coal similarly signals rents foregone and provides both demonstration and measure of the extent to which chronic (long run) excess capacity in the industry leads to short run distortions. Radical modifications are required of contemporary models integrating short run constraints into the framework to correct and revalidate the models.

In the absence of such modification, simple approximations can be made of the extent to which existing models require correction. The industry attained its 1985 capacity projections in 1981 (Table 4.11.12). If longer term targets are realistic, little secular inefficiency exists. The overstatement of shifts westward can be measured by applying the form of equations (6) and (8) to the divergence of actual and simulated model capacities. For example in CEUM, this is 19% = 0.5(406)/1094. Its overstatement in prices is 20% = (1.52 − 1.82)/1.52. Because CEUM does not define short run capacities, we cannot attribute its price overstatements to errors in structure or information. If CEUM were validated during a base period when both long and short run markets were in equilibrium, then the divergence might have represented current excess capacity. Most of the inefficiencies would then be attributable to the eastern basins. We do not have MOM price comparisons, but transportation rates rise significantly after 1977. Currently coal spot prices have fallen below $30.00 per ton in the Appalachian basins. This is $26 below the marginal cost of new underground capacity per ton, and $10 below average long run contract prices. This implies that, as in Henderson's study, economic rents are being foregone, absorbing the excess in the total transportation bill during cyclical short run excess capacity.[40]

Coal Forecasts and Performance

Examining the structures of long run and short run shifts clarifies the weaknesses of existing coal models and suggests the need for intermediate analysis. The most important difference between the early use of Dantzig's fundamental program by economic theorists such as Henderson and the applications by contemporary coal supply modelers is that the latter use divergences to validate their models for long run forecasts. This precludes the use of supply and demand equations as indicators of the divergence of short run from longer run equilibria. Thus, Henderson's analysis of short and long run impacts cannot be defined in the contemporary models. In the period 1972–78 the coal industry reversed its secular trend from that of a transition fuel with a declining share to that of a major substitute for oil with a growing share of energy demands. The short run marginal costs and equilibrium prices (often indicated by the spot market rather than the average long term contract prices) were thus above the long run equilibrium during the initial stages of secular growth. The contemporary coal models, by using for validation base years when excess demands obtained, over-estimated significantly the expected rise in eastern coal prices, confusing short run price trends with depletion. In other words, actual prices from 1973 to 1978 contained considerable economic rents, especially in the East. Presumably these were collected by the resource owners. In the West, because of the elastic characteristics of supply, producers' rents were smaller. Furthermore, due to buyer concentration and singular transport modes, it is much more difficult for resource owners to collect these rents.

Most of the contemporary models employ algorithms and constraints beyond those called for by the Dantzig and Cottle[41] fundamental program. This is done to extend the number of shipments possible in optimal patterns of trade, and thereby to validate more closely the larger number of actual exchanges that occur in the real industry beyond those obtainable through the use of Simplex or Kuhn-Tucker conditions. So modified, the models over-determine regional shares by imposing these additional constraints. When these are fixed in the forecast simulations they can easily lead to continued overstatement of the shifts between regions when the latter are viewed as adjusted long run equilibria. In addition to ease of validation, the added constraints have the merit in the LP versions of the model of increasing the stability of comparative static exercises. They add greatly to the robustness of the models under policy simulations and so to their apparent credibility. However, the cost is high to the users because they over-determine the result spatially, and inadvertently contribute to the repeated confirmation of any widely held hypothesis.

Finally, lacking an integrated treatment of rents, the models fail to associate retained earnings and cash flows with the availability of funds to finance expansion. In the East, this ability is notable, especially in the small surface mine sector, where equipment loan chattles permit producers to double or triple their capacity in a short period if they have demands. Data on West Virginia and Kentucky mines show how contractors literally "swarm" as spot coal prices rise. When prices fall, much of this capacity can shift to road construction. This accounts for the accelerated demands for secondary road improvement in those states whenever coal demand falters. Of course, such expenditure is part of the areal extension of the small surface mine sectors to the extent it provides haulage roads. Underground, changes in productivity also affect capacity.

Estimates of the industry's change in balance

sheet investment from the Census of Manufacturers show this had declined to $1.8 billion in 1953. Wall Street estimates this rose to $27 billion using 1982 financial statement information. Income in the period increased from $2.2 billion in 1953 to $22.1 billion. While considerable inflation occurred, cash flows in 1981 were in excess of $4 billion. Most of this cash was generated in the East, and helped provide the means for very rapid expansion. In the West, the major producers exhibited negative cash flows on most new properties, while the older surface mines remained locked into long term contracts with larger buyers and collected smaller rents. Because existing contracts and capacities were ignored, the producers' surplus available to the West was overstated in the coal models. Profits which were passed forward to the consumer by the utilities in older contracts were not available for expansion of mines or infrastructure, even when leases on government reserves were accelerated. This is one reason for the low bids complained of in the Linowes Commission testimony.

The shortcomings of contemporary attempts have here been exemplified by their overstatements of western shifts in production and eastern MOM prices over the past 10 years. The important criticism remains that dynamic changes in infrastructure are not explicitly incorporated into the analysis. This is a complex problem. When comparative statics is inadequate the use of optimal networks has to be continually checked by resorting to modest, reduced scale spatial and temporal analysis.

Coal Substitution for Conventional Fuels

The current world capacity of large boilers to consume oil is 10 million barrels per day. While older boilers designed for coal were switched to oil during the period of very favorable oil prices, their advanced age and small size, as well as the elimination of space for storage and other requirements of direct coal utilization, have made it costly to refit them for clean coal utilization. Nonetheless, at current oil price levels of $25 to $30 per barrel, it is economic to convert some of the old and most new larger oil-fired utility boilers to coal, even at the loss of significant rated capacity. Such conversion or expansion of coal in boilers in the United States has reduced the share of oil in utilities by more than half since 1973. As it proceeds, it places a significant pressure on the Organization of Petroleum Exporting Countries (OPEC) to cut prices or production. Some energy economists foresee the future collapse of oil prices. However, assuming that reduced OPEC production offsets the continued growth of shipments from non-OPEC producers, stable oil prices will insure a shrinking market for residual oil. Only about three dollars separate the higher refinery product distillates from residual oil costs per barrel at a modern refinery. Prices of various oil product grades would tend to move together. This adds

Table 4.11.13—Coal and Oil, Base Case New Plant Comparisons
(in constant 1981 dollars)

Region	Europe		Japan	
	Coal[1]	Oil[2]	Coal[1]	Oil[2]
Total Busbar Costs (Mills/kwhr)	46.7	68.9	50.3	74.1
Capital[3] ($/kw)	1100	780	1175	800
O&M ($/kwyr)				
Fixed	28	5	35	6
Variable	0.0014	0.0005	0.0014	0.005
Fuel ($/MBtu)	2.56	6.00	2.69	6.50
Breakeven Oil Price[4] ($/bbl)		23.10		25.16

Source: Chase Manhattan Bank, New York, "The Coal Situation," 1982.
[1]Coal Plant Size: 1200 MW (2 600 MW units); Net Heat Rate, 10,000 Btu/hr.
[2]Oil Plant Size: 1200 MW; Net Heat Rate, 9,500 Btu/hr.
[3]Capital Cost: 6.5% real; 30 yr. life.
[4]Delivered price of oil required to make coal a break-even proposition for new electric power plant at 65% capacity.

Table 4.11.14—Developed Country Real Levelized Cost of Generation: Coal/Oil (in mills per kwhr, 1981 dollars)

	Europe		Japan	
Capacity	Coal	Oil	Coal	Oil
65%	46.7	68.9	50.3	74.1
55%	50.3	70.9	54.2	76.2
45%	55.5	73.9	60.0	79.3

Source: The Chase Manhattan Bank, New York, "The Coal Situation." 1982.

to the forces creating excess supply in the upper distillates associated with the significant long run elasticities of transportation and residential demands. Process modeling of oil and coal plants evaluates the switching point in terms of delivered oil costs at which existing or newly planned facilities would cease to adopt coal and return to oil-fired boilers.

A number of well regarded energy economists hold the view that oil prices will continue to decline and be maintained, slowing or even reversing the return to coal. Process evaluation models "predict" the switching field price lies between $23 and $25 per barrel, assuming that the developed country power plant is accessible to coastal deliveries (Table 4.11.13).

Economists who hold the view that OPEC will be ineffective over the long term logically could question the expanded coal or nuclear power forecasts. Nonetheless, developed world policy economists generally cannot explain their predictions of low direct coal use and world trade on the basis of oil substitution, because invariably the energy policies of their countries are based on the presumption that oil prices will not fall, but rather rise in real terms over the next 30 years. This changes the focus of the fuel selection problem to process analyses comparing clean coal systems to conventional nuclear reactors or further technological developments. In both this and the previous case, the busbar costs of power at given sites are the relevant decision variables.

There is very little controversy over the use of process analyses models to rationalize plant decisions or to forecast these in developed economies. Table 4.11.14 compares the present (1983) annualized cost of power generation in 1981 constant mills per kwhr for oil and coal in Europe and Japan. The results are from recent Chase Manhattan Bank comparisons similar to those already described. By far the most important assumptions made are the scale of plant and the base loading of plant capacity as a percentage of total capacity, so these must be fixed in the comparisons. Table 4.11.13 demonstrates that for the unit costs of oil plants to equal the cost of generation at coal plants, the c.i.f. price of oil delivered has to fall below $23.10 per barrel in Europe and $25.16 per barrel in Japan. Average prices c.i.f. currently have been approximately $30 to $31 per barrel. Thus, in 1981 dollars, the real delivered cost of residual oil would have to fall by 41% in Europe and 39% in Japan. Of course, oil prices would have to remain at this low level over the 30 year life of the facility to maintain costs comparable to coal-fired power. Since in 1984 the dollar has appreciated 35% against European and Japanese currencies, delivered oil costs there are higher now in real terms.

World Uranium Markets

Uranium supply was presented in Part One. The difficulty attendant on its estimation has been described elsewhere.[42] For purposes of this study, reported potential supply, but not production, is accepted as given while the analysis of demand provides the means for the estimation of the adequacy of that supply.

Uranium is not a unique mineral commodity. Its pricing reflects the same long-run reactions to excess supply and demand as do other fuels. Long-run supply and demand react consistently to changes in price.

The U.S. uranium industry has witnessed two booms. The first, which peaked in 1960, was based on U.S. military needs. Both imports and, under an incentive program, domestic production rose rapidly. Imports exceeded domestic production from 1952 through 1959. In 1960, domestic production was about 35 million and imports were about 37 million pounds of U_3O_8.

The government's price and purchase guarantee program was terminated in 1970, but stockpiles were so large that military requirements were covered well beyond the 1980s. U.S. production fell to about 21 million pounds in 1965, rising to 25 million pounds in 1968 when imports ceased. Imports were not resumed until 1974.

Exports of U.S. yellowcake began in 1965. They were insignificant until 1978 when they reached six million pounds. Domestic production was relatively flat or slightly falling from 1967 through 1974. The oil embargo, the ordering of new reactors, the government and industry forecasts of future demand, and especially the OPEC induced price increases led to more long term contracting for supplies. It may be noted that the resultant market price increases were unlikely to be cost based if, as suggested by DOE,[43] the industry was in recession. Excess mining and milling capacity would be the rule. But, as oil prices rose, competing fuels (coal and uranium) became more valuable, uranium reserve values were recalculated, costs and prices rose.

U.S. production rose to a peak of 44 million pounds in 1980 and fell rapidly to 10.6 million pounds in 1983. Imports, which were only 3 million pounds at the peak, rose to 17 million pounds in 1982. U.S. exports in 1982 were 4 million pounds, principally from utilities eliminating unwanted stocks.

The production break after 1980 indicates industry awareness that the forecasts for nuclear power were wrong. Plants were cancelled or delayed and new U.S. orders ceased in 1978. The growth of imports reflected the availability of cheaper foreign sources. Although most U.S. utilities obtained most of their material from domestic sources in 1982, most new uranium supply contracts were with foreign suppliers.

As the cancelled or delayed nuclear facilities no longer needed their contracted uranium, a secondary market rapidly developed, complementing U.S. exports, and competing with primary production. In 1982, secondary market sales of U.S. produced uranium for delivery in 1982 totalled 7.7 million pounds, or one fourth of the 30.8 million pounds of primary plus secondary market deliveries.

Though reactor orders grew rapidly in 1964–1970, construction times suggest that actual uranium delivery did not take place before 1970. The remainder was in the form of committed reserves. After 1970, reactor orders slipped and electric power generation forecasts were regularly reduced. The rate of increase in demand represented by forward committments slackened. This was aided by releases from the government stockpile.

The increase in planned nuclear power facilities generated an increase in uranium exploration. From 1970 to 1974, world non-Communist resources estimated to be available at costs up to $10 per lb U_3O_8 increased from 1.72 to 2.30 million tons. Of this, 48.8 and 51.3%, respectively, were reserves.

By 1974, Canada had a growing stockpile of U_3O_8 and a soft market with excess capacity at both mines and mills. Australia also had a surplus of low cost U_3O_8 with resources far exceeding reserves. Niger and (then) South-West Africa were rapidly expanding low-cost capacity. The 1964 limitation on U.S. enrichment of foreign uranium which, given the shortfall in foreign capacity, amounted to an embargo preserved the U.S. market for domestic producers. The embargo was not even partially lifted until 1977.

In the U.S. exploratory drilling, which hit its peak in 1968-1970, but fell precipitously in 1971, rose slowly through 1974. Development drilling, needed for mine planning followed much the same pattern. The drop marks the first U.S. crisis of confidence in nuclear power with only a partial recovery.

In 1971, the yellowcake price reached a low of $5.45 per lb. Spot prices for immediate delivery from 1969-1972 were in the $6 to 6.25 per lb range and expected to go to $8 to 8.25 per lb by mid-1977. In response to a tender for the Washington Public Power Supply System the following bids were received for U_3O_8 delivery in[44]

1972 – $6.10 – 6.50/lb
1975 – $7.57 – 8.25/lb
1976 – $7.73 – 8.50/lb
1977 – $7.95 – 9.00/lb.

This may have been the basis on which Westinghouse coupled its reactor sales proposals with fuel supply contracts. The increase in prices hardly covers an anticipated consumer price index inflation. The U.S. price was generally above world prices at that time. Yellowcake inventories in 1972 were about 20 million pounds with potential production through 1975 exceeding unfilled requirements by over eight million pounds. As nuclear power construction schedules slipped, the excess grew.

By the end of 1973 the situation had changed. Bid invitations received few offers from U.S. suppliers, quoted prices for 1979 delivery were over $12 per lb with little of the offerings covered. Suppliers were refusing to bid preferring to negotiate price. Where past prices for 1980 delivery had been $7.08 per lb in 1973 dollars, new prices were in the $11 to 12 per lb range. By early 1974, price expectations for the mid-1980s were in the $15 to 20 per lb range with some expectations of $30 per lb. U.S. mines had produced some 13,000 short tons of concentrate in 1973 but just 7,400 tons had been enriched. Uranium was in substantial surplus, but producers were in no rush to enter into contracts. Mid-year 1974 prices for early 1980s delivery were about $20 per lb.[45] Table 4.11.15 provides an indication of the changing price expectation.

A number of reasons may be adduced to explain the 1973-74 price surge. All may have played some role. They include a sudden excess demand at then current prices, a lack of competition in the U.S. market, cartelization of world uranium supplies, and the actions and goals of the AEC. It is more likely that the major variable explaining both the level and timing of the price increase was the OPEC oil price increase. Uranium is not a unique fuel; when the price of a rival increased, its own price could increase.

The soft U.S. market of 1970-1 led to a curtailment of uranium exploration. There was some concern that supplies would be inadequate by the late 1970s and early 1980s so that prices would rise. Nevertheless, utilities were contracting forward for only 76% of their initial cores and only 39% of their fourth reload. In part this was due to known construction schedule slippages and a reduced rate of ordering. In part it was due to the practice of some reactor suppliers tying initial core loadings, and often the first three reloads, to the nuclear steam system bid. Utility inventory levels of natural and enriched uranium were high.

Assuming a 0.3% enrichment tails assay, cumulative requirements from 1973 through 1985 of U_3O_8 were expected to be 474,000 tons, to 1990 the expectation was 902,000 tons. If an eight year reserve is added, the 1990 total becomes 1.988 millions tons. AEC argued for a 1990 potential resource base of 2.664 million tons which, when added to requirements and reserves totalled 7.652 millions tons of yellowcake by 1990 or a 1972 based supply price of $15 to 20 per lb yellowcake.[46] Even if such addition was not fortuitous, the calculations could have been made in 1971 or 1972. They do not explain timing.

The AEC forecast can be translated into a requirement for about 200 to 240 million tons of ore delivered to the mills. This depends on the ore grade and the enrichment tails policy. A lag of eight years was anticipated between successful exploration and the operation of new milling capacity. However many independent mining operations existed, ore concentration was in relatively few hands, most of which maintained mining operations. Of the 16 milling firms in 1972, the largest eight accounted for over 77% of the nominal U.S. capacity, the top four for almost 52%.[47] One might expect an absence of individual price competition.

Evidence of a supply cartel for yellowcake with South African, French, Canadian, and Australian membership existed by late 1973.[48] The cartel must have existed earlier, but like the Arab Congresses of the late 1960s was impotent. It has obviously lost power once again.

Table 4.11.15—Representative Uranium Cost and Price Expectations

Year of Forecast	Current Spot Price	Expected Price		Cost/lb U_3O_8[1] (1967 $)	Expected Cost/lb Yr T+10 (1967 $)
		Year of Forecast	Yr T+10 (1967 $)		
1969	N.A.	6.20	8.45	7.97	5.81
1973	2.68	6.40	9.87	7.84	4.53
1976	4.94	39.70	33.75	18.50	5.04
1977	5.15	42.40	53.75	27.55	4.80
1978	7.13	43.20	37.64	17.98	5.26

Source: U.S. Department of Energy, *Investment in Exploration by the U.S. Uranium Industry*, DOE/EIA-0362, September 1982, Tables 5–6, pp. 22–23.
[1]Estimated.

Again, like reduced competition in U.S. milling, the issue is not one of existence or ability to raise price, but rather why 1973 was the reasonable year.

The impact of the AEC on prices was twofold: (1) it slowly ran down its stock of 50,000 tons of U_3O_8, rather than auction it off to limit initial price increases. (2) Given the shortfall of foreign enrichment facilities, the embargo on enrichment of foreign uranium for domestic use constituted an embargo on foreign uranium imports in general. The oversupply in foreign markets could not aid the domestic market even without a cartel. Domestic producers could raise prices without spurring imports.

As a fuel, uranium is limited to electric power generation. It competes with coal, residual fuel oil, and natural gas. In the early 1970s, natural gas prices were controlled. Those with existing supply contracts found the price low. For new contracts, if supplies were unavailable, the price might be considered infinite.

The cost of coal, particularly eastern coal, increased following the 1969 imposition of mine health and safety regulations. Productivity fell, mines closed and costs rose. By 1973, the Act had been court tested. The Clean Air Act and amendments of 1970, also court tested by 1973, curtailed the availability of compliance coal, particularly in the East.

In the United States, the percent of residual fuel oil produced from a barrel of crude oil had been falling steadily since before 1959. This reflected the greater returns obtained by maximizing the lighter petroleum product fractions. Refineries were no long geared to heavy oil output.

The reduction of competing coal in 1973 in the world's largest single market permitted OPEC oil price increases in 1973, increases which had been only rhetorically argued for almost ten years. Residual oil could not fill the gap nor could price controlled natural gas. U.S. government policy was hardly antagonistic to the oil price increase which created a price umbrella over high cost domestic oil (the object of a prior embargo to protect it from cheap foreign oil), over U.S. coal and over nuclear power. Given a lack of coal and domestic oil and gas competition, as OPEC oil prices rose the few U.S. uranium producers recalculated the value of their reserves and set prices accordingly. At the new higher values, it is logical that ore cut-off grades fell.

Nuclear Capacity Projections

The demand for uranium is derived from the demand for nuclear power viz, the generation of electricity. By year-end 1983, the U.S. had 80 operating reactors with 50 more still under construction. Of the latter, four were deferred indefinitely and 11 more may join the 102 previous cancellations. All U.S. reactors ordered in 1974 or later have since been cancelled or shelved. There have been no new reactor orders since 1978. Note that this predates Three Mile Island and the costly surge in concern for reactor safety.

Uranium supply-demand balance estimates are heavily dependent on the scenario developed. The following may be considered exemplary. Starting with a forecast of 205 gigawatts of nuclear power in 1985 and 433 gigawatts in 2000, the scenario suggests 70% nuclear power plant capacity factors, flexible enrichment contracts, a U.S. enrichment tails assay of 0.2% and a Eurodif tails assay of 0.18%.[49] It may be noted that the gigawatt estimate is very high, capacity factors in the U.S. are much lower than assumed here, flexible contracts are not the rule and a 0.2% tails assay would increase enrichment costs per unit while reducing feed.

However, based on the above, from a U.S. demand for uranium of about 15,000 short tons in 1981, demand rises to about 25,000 tons in 1990 and 29,000 tons in 2000. Non-U.S. demand is 30,000 short tons, 32,000 tons and 46,000 short tons for the same years.[50] As U.S. inventories are expected to peak by 1985 at about 3.5 times demand for that year and Eurodif inventories are expected to peak by 1990 at about seven times demand for that year, the U.S. is expected to have an excess demand for U_3O_8 by 1989 while non-U.S. supplies are expected to be in excess through 1990.[51]

It is notable that not only is trade not considered, but prices are absent in this supply/demand forecast. Implicitly, uranium demand is treated as a final consumption item rather than a commodity the demand for which is derived from the demand for electricity and subject to the competitive constraints of alternative fuels. This is particularly true of year 2000 estimates which must be based on power plants not yet started.

Tables 4.11.17 and 4.11.18 represent the most recent DOE/EIA projections of nuclear power through year 2000. The U.S. estimate is contained in the OECD nations in Table 4.11.18.

Table 4.11.16—Low, Mid, and High-Case Projections of U.S. Nuclear Capacity in Commercial Operations, 1985–2000 (Net GWe)

Year of Commercial Operation	Cumulative Design Capacity		
	Low	Mid	High
1982	57.3	57.3	57.3
1985	72.1	80.1	89.9
1990	117.1	114.0	121.0
1995	113.0	122.4[1]	127.3[1]
2000	109.0	130.0	139.6

Source: Energy Information Administraiton, *Commercial Nuclear Power: Prospects for the United States and the World*, DOE/EIA–0438, November 1983, Tables 5–6, p. 14.
[1]The Clinch River Breeder Reactor is included (350 MWe) in 1995.

Table 4.11.17—Low, Mid and High-Case Projections for Non-Communist World Nuclear Capacities 1985–2000 (Net GWe)

Year of Commercial Operation	Design Capacity		
	Low	Mid	High
1982 (December 31)			
OECD	135.0	135.0	135.0
Non-OECD	6.2	6.2	6.2
Total	**141.2**	**141.2**	**141.2**
1985			
OECD	179.0	193.0	205.0
Non-OECD	10.0	12.0	13.0
Total	**189.0**	**205.0**	**218.0**
1990			
OECD	257.0	271.0	291.0
Non-OECD	21.0	23.0	29.0
Total	**278.0**	**294.0**	**320.0**
1995			
OECD	274.0	309.0	341.0
Non-OECD	28.0	36.0	45.0
Total	**302.0**	**345.0**	**386.0**
2000			
OECD	286.0	349.0	407.0
Non-OECD	41.0	57.0	75.0
Total	**327.0**	**406.0**	**482.0**

Source: Energy Information Administration, *Commercial Nuclear Power: Prospects for the United States and the World*, DOE/EIA–0438, November 1983, Table 14, p. 37.

For the United States, the low estimate includes some decommissioning. But as the number of reactors increase from 73 in 1982 to 122 in 1995, it must be assumed that all of those under construction in 1982 were expected to be completed. This is not longer supportable. Furthermore, reactors due for commercial operation by year 2000 must be announced at least by 1989, and probably sooner. It may be concluded that the U.S. low forecast is too high, the mid- and high-case forecasts are irrelevant. However, most policy assumptions concerning U.S. resource adequacy, imports, prices, and tariff options are based on the mid-level case.

Like U.S. projections, projections of nuclear capacity for the rest of the world tend to be lower the more recently they are made. If this continues, while plant cancellations are unlikely, rapid nuclear expansion is also unlikely. The low end of the capacity range is likely to be the most accurate especially for non-OECD nations which must borrow internationally to finance the reactors.

DOE estimates of the adequacy of uranium reserves and resources are based on the tonnage requirements for the material by utilities. The latter depends on the number of utilities and the weighted average of their operating characteristics. In no sense does DOE make an actual forecast of uranium demand. Thus, successive projections of excess requirements are endemic to the system. Furthermore, successive projections of future requirements must continually be reduced; need or requirements are not demand.

Short and intermediate term projections (through year 2000) are based on reactors operating, under construction or on order. DOE uses the same general methodology to estimate both foreign and domestic reactors.[52] The following refers to U.S. reactors. In the near-term estimate (through 1990), all reactors built or at least 45% completed are included with estimates of the completion dates. The latter helps determine the low to high scenarios by shifting online dates. Capacity factors are assumed to increase from the current 55% average to 60% by 1990. Nuclear is expected to replace much oil and gas fired generating capacity. As pointed out previously, at present oil prices coal rather than nuclear may replace oil and gas. Second, there is hope, but no precedent for assuming that U.S. plants will achieve the required capacity factor. If they fail to do so, uranium requirements are reduced irrespective of the accuracy of the projection concerning plant numbers. Finally, there is significant evidence that plants are cancelled or indefinitely delayed far beyond the 45% completion point.

For the intermediate term projection the near term projection is taken as given. Only reactors under license and/or construction are considered except for the high-level case which contemplates eight new orders and completions. The low case assumes that all reactors currently less than 30% complete and all indefinitely delayed reactors are cancelled. The mid-case assumes that all reactors currently considered are completed and further cancellations are matched by later new orders. Average time from reactor order to commercial operation is 126 months, capacity factors rise to 65% by year 2000, and as a 30 year life is anticipated, in all cases retirements take place. Again, there is little warrant for such an arbitrary determination of cancellations without an analysis of inter-fuel competition. The further increase in the capacity factor is not warranted. And, if either useful life is decreased or construction lead times extended, the resultant uranium requirement projections are overstated.

The long-term projection (2000-2020) is simply computer arithmetic. The initial condition is the intermediate term projection. First GNP growth rates are determined as a function of the growth rates of labor-age population, labor force participation and labor productivity. Next, delivered energy growth is determined as a function of GNP growth, income elasticity, price elasticity, and real aggregate energy prices. No mention is made of substitution among factors of production or shifting shares in the form of economic activity. Third, the share of electrical power is based on the base year share and the use of a logistic curve to asymptotically approach a predetermined maximum electricity saturation (25 to 30%) of the energy market at a predetermined rate (15 years half-time). Finally, the nuclear share of electrical energy is handled as above, but with saturation asymptotes ranging from 20 to 30 percent. The capacity factor is flat 65 percent. Given the heroic assumptions required for the last two steps, the first two are superfluous.

The Demand for Concentrate Capacity, Imports and Inventory

U.S. utility requirements are measured by contracted or projected deliveries of uranium to enrichment facilities, not by a formal demand analysis for electric power and/or nuclear power shares. In 1983, requirements were 14,000 tons. By year 2000, DOE projects annual requirements at about 30,000 tons. Domestic production by year 2000 is expected to be only 14 to 21,000 tons necessitating imports and increasing yellowcake prices.[53] Imports are expected to account for 32 to 46% of the U.S. market by year 2000. Nuclear fuel requirements and price are likely to be lower than noted above as 10 gigawatts of nuclear power were cancelled and 2.2 GWe more delayed beyond year 2000.[54] Since 31 May 1984, additional units have reached the stage of probable cancellation or abandonment. The result will be to further increase utility inventories, delaying possible yellowcake price increases and the import time horizon.

Interestingly, the 10% reduction in year 2000 capacity estimations due to cancellation is to reduce requirements by only 4 to 5 percent. As the DOE anticipates 109 to 140 GWe of U.S. nuclear capacity by year 2000, the stated reduction is low.

Assuming an excess U.S. demand for domestic uranium production (at 1984 prices) either new U.S. mining capacity will be needed in the 1990s or a considerable amount of uranium will be imported.

If a high nuclear power scenario is assumed (205 gigawatts in 1985 and 433 in 2000) it can be argued that both Canadian and Australian sources will be cheaper than those in the U.S. Production costs in Australia are expected to remain in about $30 per lb through 1995 while forward costs are expected to rise to about $55 per lb. In Saskatchewan, production costs are expected to move from $19 per lb in 1980 to $25 per lb in 1995 with forward costs increasing from $31 to $42 per lb. By contrast, even if imports are assumed to cover 25% of demand, U.S. production costs are expected to rise from $35 to $58 per lb from 1980 to 1995 and forward costs from $45 to $98 per lb over the same period.[55]

If these data are reasonable, unless Canadian and Australian production is less than their domestic consumption plus U.S. excess demand, world prices cannot rise to projected U.S. levels in the absence of a cartel or other market restriction. Furthermore, imports will take an increasing share of the U.S. market in the absence of protectionist trade restrictions. Finally, if yellowcake prices rise sufficiently to allow full forward cost recovery and if those costs are passed to the electric consumer, either consumption levels may fall or alternate fuels will be used.

As of January 1983, unfilled U.S. uranium requirements for reactors in operation, under construction or on order cumulated to 76.7 million pounds of U_3O_8 equivalent by 1991 and 96.5 million pounds by 1993.[56] The figures include future requirements less inventories and deliveries under existing commitments and includes the desired level of inventory coverage. As reactors are rescheduled or cancelled, the figure will decrease.

Based on U.S. producer estimates of uncommitted uranium available at various prices, EIA estimates that unfilled requirements can be satisfied domestically from uncommitted supplies at prices below $40 per lb through 1987. By 1989, domestic supplies will be insufficient even at $60 per lb.[57] While 1983 weighted average floor prices for 1989 delivery exceed this price, contract prices do not. It may be noted that the market does not expect the type of price surge implied by the excess demand. This suggests that imports will increase and/or that the number of reactors is overstated. It also suggests that mines will reopen and/or expand in anticipation of price increases. The mines may be foreign or domestic.

Table 4.11.18 suggests the importance of uranium imports to the United States. It may also be noted that of the 44 million pounds supplied in 1982, 59.5% went to stocks. Indeed, stock additions virtually equaled domestic production. In the absence of a surge in demand, continued stock additions can only result in foreclosure of domestic production. As import prices are lower than domestic prices, without a federal restraining policy, imports will dominate the U.S. market. However, a call for protection of the mining industry cannot be based on either military or civilian need if stockpiles are already excessive.

The size, or activity level, of the secondary market is also reflected in the table. Cycling material equals 22.5% of total material though it is not counted in the total. If buyer exports are added, the secondary market is about 26.8% of the total market.

The quantity weighted average price per pound of foreign origin uranium oxide was well below U.S. weighted average contract and market prices in both 1982 and 1983. As this is expected to continue, in the absence of either a cartel or U.S. import restrictions, exports of uranium by U.S. producers should fall steadily in the future. Certainly, the share of delivered uranium from

Table 4.11.18—U.S. Supply-Demand Balance, 1982
(millions pounds of equivalent U_3O_8)

Supply		
Production	25.7	
Captive Production	1.2	26.9
Supplier Imports	11.1	
Buyer Imports	6.0	17.1
Total		44.0

Demand		
Estimated Enriched Uranium		11.5
Supplier Stock Additions	12.0	
Buyer Stock Additions	14.2	26.2
Supplier Exports	4.4	
Buyer Exports	1.9	6.3
Total		44.0

Cycling Material	
Interior-Supplier Sales	3.6
Inter-Buyer Sales	3.2
Buyer Sales and Loans to Suppliers	3.1
Total	9.9

Source: EIA, *1982 Survey of U.S. Uranium Marketing Activity*, DOE/EIA-0403, September 1983, Fig. 2, p. 5.

foreign sources has increased rapidly since 1977.

In 1982, foreign feed delivered to U.S. enrichment facilities for domestic reactors came from Canada (33%), Australia (30%), South Africa (29%) and Africa via France (5%). The remaining sources (3%) are unknown.[58]

The fearful impact of low cost foreign uranium supplies on the domestic mining and milling industry has produced two recent import chilling studies.[59,60] The impact of low-cost uranium on nuclear power costs and competitive advantage is barely discussed.

The first study investigates deferring the phaseout of import restrictions, provision of import ceilings of 20, 30, or 40% of domestic requirements, import tariffs to offset cost differentials, changing enrichment services transactions, establishment of an international commodity agreement, and a federal stockpile to support the industry. All but the deferral of import deregulation and the commodity agreement (which would be too costly) are expected to raise domestic yellowcake prices and output.

The second study furthers the analysis of import ceilings and tariffs. Table 4.11.19 indicates the anticipated impact on expected prices (midcase requirements) for each scenario.

Interestingly, DOE assumes that even the most extreme import controls will have little impact on nuclear busbar costs. For example, if imports are limited to 20% of requirements and the nuclear share of a utility's total electric generation is 75%, the increase in the electricity price in only 0.15 cents per kwh or 2.0% of the bill for an average residential customer. A 25% tariff with a nuclear share of 75% increases electric prices by 0.18 cents per kwh and the monthly bill by only 2.4%.[61] Assuming that this were true, substantial increases in yellowcake prices have negligible effects on busbar costs or on the competitive position of nuclear power. Yet, given the excess of nuclear power capital costs over those of coal, claims of nuclear power superiority were based on low fuel costs. It is shown below that busbar costs are sensitive even to the way they are calculated; they are quite sensitive to yellowcake costs.

Based on 1983 requirements, total inventories of uranium in all forms and including stocks held by DOE are equivalent to over nine years of forward requirements.[62] The stocks are held as natural U_3O_8, natural UF_6, natural UF_6 under usage agreements, enriched UF_6, and fabricated fuel. DOE stocks are used for government purposes, the remainder are for utility usage. Utilities hold about the same levels of U_3O_8 and natural UF_6 which, together, equaled 65.3% of all utility inventories in 1983. The form of holding is not surprising; these are the forms that are easily traded.

Based on a January 1983 utility survey, most reporting companies with an inventory policy wanted an inventory of only U_3O_8. A few more companies wanted less than one year of inventory coverage. Desired U_3O_8 inventory coverage ranged from one to two years (70% of the firms).[63] Given stock levels versus desired levels, decreased purchases may be expected. Given the cost of holding inventory, sales competing with new production may also be expected.

SOLID FUEL MARKET ORGANIZATION

Coal Market Structure and Performance

The coal industry, despite the favorable effects of growth in demand greater than forecast, faces the traditional problems of the past: (1) chronic excess capacity, (2) sensitivity to interregional competition and (3) continued inefficiencies due to the restrictions on supply patterns placed by public policy and market imperfections. Historically, excess capacity in the coal industry has been acerbated by intense competition on the sellers' side putting pressure on prices and profits. Except in rare instances, coal MOM prices rise by less than the rate of inflation secularly even during periods of substantial growth. Cyclically, in the short run, however, spot prices tend to fall and rise more sharply than the fluctuations in demand. In the early 1970s, as explained previously, this gave rise to the impression that the industry was becoming too concentrated[64] or showing signs of severe depletion. Under conditions of excess capacity, the eastern surface mine sector contracts rapidly, but the underground mine sector shows less flexibility. Prices, therefore, tend to fall faster than output and rents are foregone as operators attempt to absorb freight rates and other costs to maintain or extend their shipments beyond their customary selling areas. In the West, where longer term contracts dictate most mine investment and production, prices tend to rise and fall less dramatically, little spot market coal is sold, and the appearance of a concentrated industry is more pronounced. This appearance

Table 4.11.19—Projected U.S. Market Prices for Uranium (Constant 1983 dollars/lb of U_3O_8)

Year	Current Policy	Policy Alternatives[1] Import Ceiling 20%	30%	40%	Import Tariff (25%)
1982	20.80				
1983	23.00				
1984	20.20	20.70	20.50	20.20	22.30
1985	23.10	21.50	21.00	20.50	25.30
1990	43.60	70.70	52.50	44.30	72.70
1995	48.20	94.60	62.50	50.00	76.90
2000	73.10	106.70	81.30	76.20	83.40

Source: U.S. Department of Energy, *United States Uranium Mining and Milling Industry*, DOE/S–0028, May 1984, Table D-1, p. 93; Table D-9, p. 101.
[1]Mean projected values.

is deceptive, however, because concentration is much higher among buyers of coal than among the sellers. Characteristically, only one or two large utilities in a region are negotiating long term contracts at a given time, and there are many alternative coal sources. Prices tend to be negotiated down to market or cost, whichever is lower. Without regulation, oligopsony pressure often drives delivered prices for steam coals below equilibrium levels at which mine reclamation and rail maintenance costs cannot be fully covered. At the same time, oligopsony combined with the high cash flows of mine operations, prevents excess capacity in a region from putting direct pressure on new investment. New mine equipment and even rail cars can be financed by the utility and capitalized in its rate structure. Thus, the element foregone in delivered price is a combination of user cost and transportation charges. In eastern surface mining, characterized by sharply rising cost functions, rents remain adequate when prices fall, but reserves shrink correspondingly as output falls: a traditional depletion effect. In the rare cases when short run excess demand obtains, rapidly rising spot prices reflect the recovery of rents and rising costs. This lends credence to concerns over seller market power and performance even while competition grows and the ranks of active firms swarms. These anomalies in coal markets are often missed by analysts.

Outside the United States and Australia, the coal industries of the world are almost all nationalized. Prices are maintained well above the international import price. In Europe, bituminous coal imports are strictly regulated and taxed. In most OECD countries including Japan, domestic mines are subsidized. As a result, however, safety and environmental costs are seldom neglected and standards are higher.

Uranium Market Structure and Performance

U.S. DOE estimates (1980) of U.S. nuclear power growth and the associated domestic uranium requirements are combined in Table 4.11.20.

Table 4.11.20—Projected Nuclear Power Growth and Domestic Uranium Requirements[1]

	Low Case Power (GW_e)	U_3O_8 (10^3 T)	U_3O_8 Usage (T/GW_e)	Mid-Case Power (GW_e)	U_3O_8 (10^3 T)	U_3O_8 Usage (T/GW_e)
1985	85	23.2	272.9	96	23.6	245.8
1990	125	26.9	215.2	129	28.9	224.0
1995	142	31.4	221.1	155	35.0	225.8
2000	160	37.3	233.1	180	42.0	233.3

Source: U.S. Department of Energy, *An Assessment Report on Uranium in the United States of America*, Grand Junction Office, GJO-111(80), October 1980, Tables 53–54, pp. 122–123.
[1]Based on 1980 enrichment plant studies and enrichment tails assay of 0.2%.

Based on the DOE methodology described earlier, U.S. uranium production and uranium prices are expected to increase steadily from 1985 through 2000, especially if imports are limited by quotas or tariffs.[65] Nonetheless, DOE expects no impact on consumer electric bills and, by implication, no impact on demand for electricity. The price impacts on the nuclear share are expected to be trivial.

The published low point for U.S. uranium exploration, mining and milling is 1983. Exploration included only 63 companies, mining counted 15 companies and milling 13 companies. Expenditures on each activity fell faster than the number of companies.[66] The numbers alone reflect a rather loose oligopoly situation, but an analysis of competitive strength would require data on both reserves and mining cost by company and location. Mine or mill capacity figures merely indicate near term capability. The competitive lag due to construction is about two years for mine expansion or new mills and five for new mines.

Of 20 U.S. uranium exploration companies listed for the period 1970-1979,[67] only five derived more than 50% of their revenues from the sale of concentrate or uranium related products. Eleven more were firms principally engaged in the fuel sector, of which 10 were oil companies. Seven companies were mineral diversifiers principally engaged in the mining sector. The remainder ranged from a railroad, through a reactor supplier to an electric utility. From 1980 through 1983, rated U.S. mill capacity fell from 51,050 to 29,250 short tons per day. Only one mill was dismantled, 11 more (of 24) have been inactivated. Those remaining open are likely to be at the better properties.[68]

The U.S. mining and milling sector cannot be considered highly concentrated though it is far less competitive than coal. In the rest of the world, uranium mining and milling involves far fewer companies; production and export licenses are usually firmly under government control. The result is a reduction in the number of decision making entities. However, the number of companies with significant tonnages in the $30 per lb and $50 per lb reasonably assured resource class has increased, as have those in the estimated additional resources category.[69] In the absence of overt national restrictions on uranium exports, covert acquiescence by major uranium consuming countries, and greatly increased prices of oil and coal, a reemergence of a potent uranium cartel seems unlikely.

The first competitive bottleneck in the uranium fuel cycle is enrichment. This is firmly under government control and the number of facilities worldwide is strictly limited. Currently, enrichment capacity is in excess supply, leading to more generous contract terms and possible subsidization of enrichment in non-U.S. facilities. The market is clearly oligopolistic. If, however, laser enrichment technology proves feasible, costs should be severely reduced, many countries will be able to adopt the technology, nuclear proliferation will increase, and both gaseous diffusion and cascade centrifuge enrichment plants will close.[70]

Conversion of yellowcake to UF_6 is by decomposition of uranium nitrate to UO_3, reduction to UO_2 and two step flourination to UF_6. In 1976, five plants accounted for the noncommunist world capacity of 32,600 tonnes per yr. Of these, two U.S. companies accounted for 17,200 tonnes. As new plant capacity can be added in two years[71] including design and construction, this concentration should constitute neither a supply bottleneck nor a basis for monopoly pricing.

Waste disposal is the second competitive constriction point. Safety and proliferation arguments both suggest that reprocessing and/or ultimate disposal be closely regulated. The number of firms will be few.

In the United States, there have been six suppliers of nuclear reactors, of which four have been important. In the noncommunist world, including U.S. and Russian exports, there are 31 reactor suppliers. The effective total is somewhat less as a few companies act jointly in specific countries and a few do not produce light water reactors.[72]

Primary contractors or architect-engineers numbered 15 in the United States in 1983. Subcontractors number in the hundreds. Of the AE firms, the Bechtel Corporation has been by far the major contractor for operating plants. In 1981, of 107 plants under construction or on order, Bechtel accounted for 25%, Stone and Webster, and Sargent and Lundy the next largest contractor, accounted for 14% each. Utility companies and TVA, acting as prime contractors accounted for 24% indicating that the number of primary contractors is not necessarily a competitive limitation.[73]

Price Formation: In the U.S., uranium is sold on contract or at the market. The former is an agreement which may include an escalation clause. The latter is a spot market price, taken at or before delivery, and often contains a minimum or floor price clause. In the United States, there is neither a formal spot nor a futures market for uranium either in the natural or at any enriched state. Contracts are either bilateral agreements between a producer and a utility (or an agent buying for his own account) or they are arranged through power plant manufacturers or brokers (agents). The major broker is the Nuclear Exchange Corporation (NUEXCO). It is their published exchange value that is usually reported as the spot price. Actually the price is an estimate of a market price for immediate delivery as of the last day of the previous month. In the absence of a formal futures market, contracts for future delivery can vary widely for equivalent deliveries among different customers.

The absence of formal spot and futures markets is consistent with the fewness of both buyers and sellers. The uranium market is simply not as competitive as aluminum, copper, coal, or some petroleum products.

Unlike European practice, contract prices comprise a declining portion of U.S. uranium sales. In 1982, they were 42% of total quantities delivered. By 1991 they will be only 20%. Contrastingly, market price arrangements will rise from 53 to 60% over the same period while other (mostly captive production) will rise from 5 to 20% by 1991.[74]

Table 4.11.21 shows anticipated contract and market prices to 1991. Contract prices in year of delivery dollars, which includes escalation factors and the market conditions for the year in which the contract was made, are expected to rise on average from \$38.37 per lb U_3O_8 in 1982 to \$58.91 per lb in 1991.[75] The increase is not steady and is subject to a wide variation. While the entire price range rises, the proportion of settlements (not pounds of yellowcake) below the weighted average price tends to rise in later years. Expected market prices, as evidenced by weighted average floor prices also rise, but the low end of the price range is steadily below the low of the contract price range from 1988 on. As with contract price ranges, these price ranges reflect varying anticipations of inflation.

Secondary markets usually provide an unrestricted, free market, test of existing price quotations. In 1982, total interbuyer sales averaged \$18.90 per lb. Of this, spot sales averaged \$19.56 per lb, buyer to producer contracts averaged \$21.23 per lb, and producer to producer sales averaged \$21.55 per lb. These should be compared with weighted average floor prices for the market price option.

Increasing uranium prices do not help make the domestic nuclear industry more competitive either at the level of mine and mill or at the reactor. Absent restrictions, increased imports may be expected. The width of the price ranges indicates that not all reactors or utilities will be equally affected by price increases. Some may

Table 4.11.21—U.S. Uranium Weighted Average Contract and Market Prices 1982–1991 (\$/lb U_3O_8—Year of Delivery)

	Contract Prices			Market Prices	
	Reported Price	Range	% Below Range	Floor Price	Price Range
1982	38.37	10–70	45.5	51.27	16–63
1983	35.62	10–70	50.0	53.49	17–70
1984	44.84	25–75	45.0	55.93	20–70
1985	50.00	30–110	49.5	61.05	25–76
1986	48.98	30–110	49.0	62.84	27–82.5
1987	52.20	30–70	42.0	65.50	29–92
1988	46.65	35–75	69.0	70.74	31–97.5
1989	49.59	40–80	69.0	75.05	33–108.5
1990	57.16	45–85	60.0	72.39	35–112.5
1991	58.91	50–90	67.0	76.85	38–121

Source: Energy Information Administration, *1982 Survey of United States Marketing Activity*, DOE/EIA-0403, September 1983, Tables 6, 7, pages 13, 15; Figures 4, 5, pages 15, 17.

remain low cost, some may be very high cost, and some may have widely varying fuel costs over a series of years.

Calculating the Full Costs of Nuclear Power Versus Coal

An alternative to the DOE method of assuming a nuclear power penetration rate into electric production (and electric production into energy consumption) as described earlier is to estimate the competitiveness of nuclear power in particular and its impact on electric power costs in general. The analysis must be based on current and expected cost and operating characteristics rather than a weighted average of prior plants.

Ultimately, the decision between coal and nuclear fuels rests on the competitive busbar costs expressed as mills per kwh. For the comparisons to be viable, both the fuel cycle and the capital costs must be reasonable, defensible, and current. It is the recent, rather than the average, experience that counts. Furthermore, it is necessary that the comparisons be unbiased. The following provides a partial checklist.

Plant size for current nuclear projects is somewhat over 1000 MW_e. Yet, at the end of 1980, of the 11 largest all coal fired power plants, the maximum size (4 units) was 1300 MW_e and the average (36 units) was only 839 MW_e.[76] Industry average size for base load plants is somewhat less. It is not more reasonable to compare coal and nuclear plants at 1000 MW_e than at 839 MW_e or less. Reasonable comparisons compare the normal unit for each. If there are relative economies of scales they will be reflected in the sizes chosen. If there are advantages to constructing two plants at a single station this will be noted. If there are advantages to modular sizes tracking the growth in electric demand, these too will be reflected.

Plant capital cost should be complete. For coal fired facilities this should include the latest requirements in air pollution control, however solved. For nuclear plants the latest safety and emission requirements should be included. For both, the present value of decommissioning costs must be estimated and offset by salvage values.

The longer a plant lasts the more power is produced and the greater the revenues generated to offset capital costs. Alternatively, levelized capital cost recovery is lower for longer lived plants. If nuclear or coal plants last less than 30 years as base load facilities, this should be anticipated in the comparison structure. Furthermore, if after aging a plant is used for cycling power, the cost advantages should be considered only after measurement of its base load generating life.

A nuclear power plant is composed of two major circuits, the primary or nuclear circuit and the secondary or steam circuit. The former is new, complex and has been subjected to scale up at a rate that exceeded experience at smaller sizes. The latter operates at about half the temperature of a coal fired plant and at one fourth to one third the pressures. In consequence, pump and other component sizes are much larger than those in a fossil fuel plant. The lack of experience has led to problems. Cooling problems arose because the U.S. reactors were originally designed for a submarine operating within an unlimited coolant. Despite the problems, U.S. reactors were assumed to have an economic life of 30 years. In France, the United Kingdom, and perhaps West Germany, similar reactors are assumed to have an economic life of 20 to 21 years. As yet, however, there is no actual longevity experience. None of the original U.S. commercial reactors have attained 30 years. Coal fired plants last at least 30 years, several built in the 1920s are still operating, but not as base load plants.

Siting is an important consideration. Nuclear power plants have limited locational possibilities. A coal fired plant can be located anywhere a nuclear plant can be sited; the converse is not true, but the location need not be optimal for the coal facility. Unbiased comparisons are based on the optimal location for each type of plant given the market to be served. For coal, this reflects not only transmission distances, but coal sources and coal transport modes as well.

There are some additional cost considerations. These include (1) relative maintenance costs as the plants age, (2) the present value of the cost of nuclear decommissioning versus fossil fuel plant scrapping, (3) the probable useful life of the plants, and (4) waste disposal costs.

Given size and longevity, the amount of electricity actually generated depends on the capacity factor of the plant. The amount of fuel used and waste generated depends on the availability factor. Both factors vary over the life of the plant. It is unlikely that the factors reach a peak at some point after start up and remain there until decommissioning or shift to cycling or standby status. The practice of estimating generation and fuel consumption by extrapolating to the half life of the plant is essentially incorrect if the aging of coal and nuclear plants differs.

The amount of nuclear power available depends not only on plant capacity but also upon plant availability. If capacity and availability factors are high, fewer plants need be built to obtain a given amount of electricity than if they were low, less uranium need be supplied. High factors mean high output spreading capital costs in any accounting period over more units sold, reducing the required price per unit sold and, perhaps, fostering demand.

A plant availability factor is the percent of the total time in a given period that a plant or unit is producing electricity. It is equal to the time the generator was on line divided by the total time during the period. Outage time, both forced and scheduled, and time spent on standby status are subtracted from total available plant operating time to obtain generating time. The capacity factor is the percent of the total electrical energy actually produced by a plant or unit during a specified period compared to that which might have produced had it operated at the licensed design power level for the entire period. It is equal to the MW_e actually produced during the period divided by the product of the licensed design power level (MW_e) times the number of hours in the reporting period.

Clearly, the capacity factor will be less than the availability factor. Logically, the former is used for estimation of revenue generation to offset capital costs. The latter is misquoted as a measure of reliability and properly in the estimation of fuel costs. In the early years of nuclear power AEC and others claimed an availability factor of 80%. This was later reduced to 75% and estimated costs were based upon this. Yet from start-up through 31 March 1974, the quarterly weighted average of availability factors in the United States never exceeded 73%. It has been as low as 61 percent. Capacity factors were reportedly as low as 54% from start-up through mid-1974. The high was about 57% in 1973. The low level of these factors was in part responsible for unanticipated high nuclear power costs.

For nuclear plants, load factors probably rise through the fifteenth year to a peak of 70% and fall at a rate of about 2% per annum to a minimum of 30 to 40 percent. Given U.S. experience, the high is not likely to be attained, though Europeans can apparently achieve this level. There is insufficient experience to gage the decline but it is common in electric generation and there is no apparent reason why nuclear power is exempt. Common practice, however, is to assume for costing purposes that nuclear power plants maintain a 65% load factor throughout their economic lives.

Availability factors and capacity factors for U.S. fossil (mainly coal) and nuclear power generation in plants over 400 MW_e were reported to be similar in 1980. Both fuels yielded 59.5% capacity factors. The equivalent availability factor was 69.5% for fossil and 64.5% for fissile fuels.[77] Nuclear power plants are base loaded. The fossil plants, particularly older coal plants and oil and gas fired plants, are likely to be on standby or intermediate load. Furthermore, given their lower fixed costs, coal fired plants are apt to be dispatched after all nuclear plants in a given utility system. Thus, if the fissile estimates of availability and capacity are accepted, the fossil estimates may be too low as a measure of reliability of recently constructed, base loaded, coal fired plants.

Though other nations have not been free of nuclear malfunctions, European reactors appear to function better than those in the United States even though they may be of the same design. In the United States, pressurized water reactors (PWR) averaged 57% of capacity in 1981. In France the capacity factor was 66.8% and in Germany 78.9%. Similarly, it has been found that the U.S. PWRs are down for repairs and checks two to three times more often than are German PWRs and 1.5 times more often than the French PWRs. As these are base load generators, offsetting power must be purchased or more expensive standby plants must be operated. If the designs of the reactors are similar, the differences may be due to quality control in construction and management during operation.

For 106 pressurized water reactors, listed in 14 countries excluding the east bloc, the average reactor cumulative load factor to the end of 1983 was 56.7 percent. In the United States, the average was 54.8 percent. Those with ratings were Brazil, Italy, Spain, and Sweden with a total of six reactors. For boiling water reactors, the world average for 57 units in 10 countries was 56.6% with the United States at 56.1%. The countries with lower ratings were West Germany, India, Italy, and Taiwan with a total of 11 reactors.[78] As the United States accounted for 45% of the PWRs and 40% of the BWRs, closeness of the United States to the world average experience is to be expected.

In the United States the 10 plants exhibiting the highest lifetime capacity factors approached the European experience with factors of 70 to 79% to 1983. The best European experience is 85% for a PWR of 364 MW_e and 80% for a BWR of 336 MW_e. Of the 10, eight are PWRs, the average size was 541 MW_e and the size range was 497 to 845 MW_e. Of the nine worst reactors, six were PWRs, the average size was 857 MW_e, and the size range was 538 to 1090 MW_e. The best reactors averaged 9.4 years of operation with an average of 2.1 reactors operated by the same company. The worst reactors were somewhat newer and averaged 7.1 years of operation with an average of only 1.78 reactors operated by the same utility.[79]

Capital Costs: In earlier years, it was the cost of uranium that limited nuclear competition; currently it is capital costs. A major factor in the nuclear decision is reactor construction costs. As estimated by the OECD Nuclear Energy Agency in early 1983, these were (in terms of installed capacity):[80]

France	\$680/kw
Italy	\$812/kw
Belgium	\$876/kw
Netherlands	\$1,148/kw
W. Germany	\$1,213/kw
Britain	\$1,298/kw
U.S.	\$1,434/kw
Japan	\$1,438/kw

The figures are difficult to interpret. Given the costs of U.S. plants currently under construction, the U.S. estimate is too low. Furthermore, an element of subsidy, at least in the cost of capital, is present in all of the European plant costs; they too are too low; by how much is not known. Finally, as these plant costs are weighted averages, vintaging of plants and among nations is an important consideration. Earlier plant costs were less than later ones.

Nuclear plant capital costs at the inception of the U.S. program, when turnkey procurement at a fixed price was the rule, were only slightly higher than those for coal. In the early 1970s, excluding interest, a 1000 MW_e plant could be completed for less than \$500 million. With the assumed fuel cost savings such a facility was considered cheaper than coal. If a more realistic fuel cycle costing was used they may have been equally competitive. By the end of the 1970s, with completions costing in excess of one billion dollars and much increased fuel cycle costs, nuclear power was unlikely to be competitive with coal. Those presently under construction cost up to \$4 to 5 billion or twice the average cost of five years ago. Current capital costs have risen despite falling inflation and interest rates.

Unless costs are prudently managed they cannot be passed on to the consumer. As virtually all of the costs of the completed and cancelled plants have been passed on it must be assumed that the final one or more billion dollars per unit cost completion estimate represents the nature of U.S. nuclear power construction costs. That these are four to ten times the amount stated at the hearing before the state public service commissions for certificates of necessity, and before the Atomic Safety and Licensing Board, may be regarded as irrelevant. In the nuclear/coal comparisons almost invariably made at those times no requirement was placed on the utility, its prime contractor, or its nuclear steam source supplier to abide by those costs or the assumed load factor. It is only within the first months of 1984 that public service commissions began placing caps on plant costs that can be passed on to the consumer. Invariably, these cap costs are high.

Current experienced nuclear power construction costs bear no resemblance to their original estimates or to those conjured by studies such as ORBES (cf prior paragraph). For plants completed in late 1983 and early 1984 or soon to be completed, cost estimates range upwards from \$3000 per kwh installed capacity to as high as \$7800 per kwh. At these costs, to be competitive with coal, nuclear fuel costs must be very low.

Unless it can be shown that U.S. nuclear plant costs will decline in the near to mid-term, the estimated costs of current construction provides the appropriate view of future nuclear-coal competition. A detailed estimate of the nuclear costs is presented in Table 4.11.22. Recent experience has shown that even in the 95% completion category the high end of the stated cost range is too low. As one moves to the lower stages of completion uncertainty grows, but it is unidirectional towards increased costs. Part of the reason for increased capital costs is construction delay; labor, finance, and materials costs are all increased. Nuclear plant construction delay is not a new phenomenon. Of the 30 plants originally scheduled for operation prior to the summer of 1973, 29 units were late. The number of reasons cited for delay totalled 89 with most plants citing more than one. Of the total cita-

Table 4.11.22—Estimated Costs of Nuclear Plants Under Active Construction in the United States[1]

Percent[2] Cost Complete	Number of Plants	Total Rating (MWe)	Average[3] Expected Cost ($/kW)	Expected Range ($/kW)
95% or more	6	6413	3784.1	830–4500+
90–95%	9	9988	2120.7	1100–2900
80–90%	9	9802	2212.4	1500–3000
50–80%	15	16398	2299.2	1050–3700
45–49%	6	7070	2489.4	1700–3100+
about 20%	3	3530	2969.6	2700–3100+

Source: Office of Technology Assessment, *Nuclear Power in an Age of Uncertainty*, OTA-E-216, February 1984, Table 3A, p. 76.
[1] Cost data December 1983, in mixed dollars
[2] Construction completion October 1982
[3] Weighted average
+Costs are likely to go higher

tions, 25.8% were equipment related, 37.1% were labor related, 27.0% were regulatory, 5.6% represented legal challenges; weather and local authorities accounted for the remaining 4.5%. A Federal Power Commission study of construction delays in 28 plants due to become operational in 1974 found that 229 plant-months of work were lost due to low labor productivity and shortages, late deliveries, component breakdown and similar economic or technical failures. Only 32 plant-months of delay were caused by regulatory changes or public law suits.[81]

Nuclear Fuel Cycle Costs: It is true but trivial that nuclear fuel costs are less than coal costs. At issue for competitive argument is, are they sufficiently lower? Because nuclear power plants cost more per kilowatt to build than do fossil plants, to be economic the fuel cycle costs must be low enough to offset the difference. The amount that must be offset is inversely proportional to the relative load factors and directly proportional to the relative capital costs. Furthermore, because the ratio of fuel cycle to total costs is higher in fossil than in fissile plants, the cost of nuclear over fossil fuel costs is directly proportional to the assumed load factor.

Between exploration and burnup in a light water reactor, uranium (U_3O_8) must be mined, milled (yellowcake), converted to UF_6, enriched from 0.711% U_{235} to 3% or less U_{235}, processed as UO_2, and fabricated into rods. Mining/milling and enrichment are the most costly steps. Waste management or recycling the spent fuel, inventory charges, safeguarding and insurance are additional costs.

Table 4.11.23 shows the calculations needed to estimate the amount of uranium required annually for the operation of an LWR subsequent to the initial loadings. As the initial step is equal to the mill output, this corresponds to the station's demand for yellowcake. As one kg U equals 1.179 kg U_3O_8, one kg U equals about 2.6 lbs. of U_3O_8 (yellowcake).

Table 4.11.24 provides an estimate for weighted average LWR consumption of yellowcake based on the equations provided. The amounts are sensitive to the assumptions listed. If reprocessing is excluded, the recycled material in enrichment must be substituted for by slightly more than the 22,636 kg U listed (i.e., by 27,061 kg U). This must be worked through the conversion to the amount of yellowcake required. Similarly, separative work units must be increased as the new material is 0.711% U_{235} rather than the 0.85% in the recycled material.

Given the material moved and the cost component estimates, a projection of fuel cycle costs of 4.97 mills per kwh for 1980 was made in 1974.[82] If the AEC load factor, burnup rate, and efficiency assumptions, but not their cost component assumptions, had been used, a fuel cycle cost of 6.87 mills per kwh resulted for 1980.

Both estimates would have been higher without the embedded reprocessing assumption. For comparison, in 1973 the AEC projected that 1981 nuclear power generation costs would be 15.20 mills per kwh, the sum of 11.70 mills for capital, 2.50 mills for fuel and 1.00 mills for operation and maintenance. AEC capital costs in 1980 dollars for a 1000 MW_e plant were $608 million.[83] Given the higher fuel cycle cost estimates, even if capital costs had not surged, nuclear power after the first set of turnkey plants was unlikely to be more than marginally competitive with coal.

Actual calculations begin with the reactor, or

with the sum of existing and anticipated reactors. Both costs and uranium consumption are sensitive to the specific operating assumptions made. Increasing the load factor (availability or capacity) increases the amount of uranium consumed and reduces unit generating costs. In-

Table 4.11.23—Equations for Deriving the Annual Uranium Flow for a Typical[1] 1000 MWe Light Water Reactor

		Equations for Deriving Kilograms U/yr.	U_{235} Percent Weight[3]	Estimated SWU's/yr
Conversion	(in)	U = N/.995	0.711	
(0.5% loss)	(out)	N	0.711	
Enrichment	(regular in)	N = (5.479)B	0.711	
	(regular out)	B = Enr − S	3.0	B(4.306)
	(recycl—ed in)	E	0.85	
	(recycled out)	S	3.0	S(4.306)
	(out)	Enr + P − r	3.0	Enr(4.306)
Recycled U		r = (0.02P) + (0.05F)	3.0	
Fuel Preparation	(in)	P = F/.975	3.0	
(2% recycled)	(recycled)	(0.02)P	3.0	
(0.5% loss)	(out)	F	3.0	
Fabrication	(in)	F = R/.945	3.0	
(5% recycled)	(recycled)	(0.05) F	3.0	
(0.5% loss)	(out)	R	3.0	
Reactor	(in)	$R = \frac{(e)\ (k)\ (8760 \text{ hrs.})}{(b)\ (eff)\ (24\text{hrs/day})}$	3.0	
	(out)	D = (0.97)R	0.85	
Reprocessing[2]	(in)	D	0.85	
(1% loss)	(out)	C = (0.99)D	0.85	
Conversion[2]	(in)	C	0.85	
(0.3% loss)	(out)	E = (0.997)C	0.85	
Enrichment[2]	(in)	E	0.85	
	(tails)	E − S	0.2	
	(out)	S = E/4.583	3.0	S(4.306)

Source: Rieber, M. and R. Halcrow, *Nuclear Power to 1985: Possible versus Optimistic Estimates*, Washington, D.C., National Technical Information Service, NTIS PB 248-061/AS, November 1974, Table III-9, pp. 70–71.

Notation
U = kilograms of natural uranium entering conversion (leaving the mill)
N = kilograms of natural uranium entering enrichment (leaving conversion)
B = kilograms of enriched uranium leaving regular enrichment
E = kilograms of spent uranium entering (re)enrichment (leaving (re)conversion)
S = kilograms of enriched uranium leaving (re)enrichment
Enr = kilograms of enriched uranium leaving total enrichment
r = kilograms of recycled uranium from fuel preparation and fabrication
P = kilograms of enriched uranium entering fuel preparation
F = kilograms of enriched uranium entering fabrication (leaving fuel preparation)
R = kilograms of enriched uranium entering the reactor (leaving fabrication)
D = kilograms of spent uranium entering reprocessing (leaving the reactor)
C = kilograms of spent uranium entering (re)conversion (leaving reprocessing)
e = nuclear plant size (megawatts of electricity)
k = nuclear plant availability factor
b = levelized nuclear core burnup rate (MW(t)days/MTU)
eff = the nuclear reactor's thermal to electrical conversion efficiency (MW(e)/MW(t))

Notes
[1]Combined characteristics one-third BWR and two-thirds PWR.
[2]If no reprocessing is permitted, these steps are omitted.
[3]Enrichment and tails levels are variable.

Table 4.11.24—Uranium Flow for a Typical 1000 MWe Light Water Reactor

		Uranium Kg/Year	Percentage Weight of U_{235}	SMU/Year
Conversion	(in)	104,192	.711	
(0.5% loss)	(out)	103,671	.711	
Enrichment	(regular in)	103,671	.711	
	(regular out)	18,922	3.0	81,478
	(recycled in)	22,636	.85	
	(recycled out)	4,939	3.0	21,267
	(out)	23,861	3.0	102,745
Recycled U		1,761	3.0	
Fuel Preparation	(in)	25,622	3.0	
(2% recycled)	(recycled)	512	3.0	
(0.5% loss)	(out)	24,981	3.0	
Fabrication	(in)	24,981	3.0	
(5% recycled)	(recycled)	1,249	3.0	
(0.5% loss)	(out)	23,607	3.0	
Reactor	(in)	23,607	3.0	
	(out)	22,934	.85	
Reprocessing	(in)	22,934	.85	
(1% loss)	(out)	22,705	.85	
Conversion	(in)	22,705	.85	
(0.3% loss)	(out)	22,636	.85	
Enrichment	(in)	22,636	.85	
	(tails)	17,697	.20	
	(out)	4,939	3.0	21,267

Assumptions:
Load Factor = .65
Burnup = 30,000 MW(t)D/MTU
Efficiency = 33.5 percent
104,192 KgU = 122,871 KgU_3O_8 = 270,930 lbs U_3O_8

Enrichment tails assay = 0.20 percent U_{235}

Source: Rieber, M. and R. Halcrow, *Nuclear Power to 1985: Possible versus Optimistic Estimates*, Washington, D.C., National Technical Information Service, NTIS PB248-061/AS, November 1974, Table III-7, p. 70.

creasing thermal to electrical efficiency reduces uranium consumption.

It should be noted that the relationships cited for the reactor requirement for enriched uranium hold best as the reactor reaches steady state irradiation (burnup rates) over time. Boiling water reactors have a lower burnup rate than do pressurized water reactors. Efficiency, however, tends to be higher for BWR than for PWR.

Enrichment requires work (separative work units—SWU) to separate uranium into an enriched product with a higher concentration U_{235} and a low-level waste (tails). An increase in the amount of U_{235} allowed in the waste increases the amount of natural uranium feed required to produce a given amount of enriched uranium (to each specified enrichment level), but the amount of separative work and its cost is reduced. Futhermore, the higher the enrichment level desired, the greater the feed and the higher the residual U_{235} level in the reactor waste (given reprocessing), the less is the enrichment feed required.

In general, demand for enrichment services decreases as the number of projected plants decrease, as availability factors remain low, as core burnup rates remain low, and as tails assays are increased. Enrichment service requirements increase if reprocessing spent fuel and the use of plutonium mixed oxide fuels is prohibited, if the breeder is not developed, and if enrichment services in the rest of the world prove inadequate or high cost.

Existing enrichment plants in the United States, USSR, and Europe have a capacity of about 45 million SWU per yr or about twice world demand. In the United States, enrichment facilities are currently running at under 50% of the 27 million SWU per yr available. U.S. enrichment facilities are located at Oak Ridge, Tenn.; Paducah, Ky.; and Portsmouth, Ohio. The last is a gas centrifuge operation; the others are gaseous diffusion plants.

Unfortunately, gaseous diffusion is a consumer of large amounts of electricity. As electric rates rise so do diffusion costs. As these costs must be covered by revenues, U.S. SWU charges have risen to about $139 per SWU. With a strong dollar and cheaper foreign enrichment prices, (Eurodif charges $100 per SWU), U.S. utilities have been importing enriched uranium, enrichment services, and purchasing enriched uranium inventories from other domestic utilities.

DOE, which runs the financially troubled enrichment program, has taken a number of short-term steps to alleviate its problems. These include price cuts of 2 to 8% to encourage sales and the establishment of long-term contracts at fixed prices. In the past, year to year inflation raised enrichment prices by 10 to 15% per yr. Finally, a 70% take without penalty has replaced the 100% take or pay requirement. Enrichment prices are not likely to rise in the near future so that one part of nuclear utility operating costs is likely to be stabilized.

Long-run enrichment cost reductions may come from completion of the Portsmouth centrifuge facility, now only one fourth complete. Electric usage is about 0.10 that of gasous diffusion but the latter are already built, the costs are sunk. The Portsmouth plant, if completed, would have to recover the additional capital costs.

Perhaps a step beyond the centrifuge is lasar separation. This process attaches a positive charge to vaporized U_{235}, but not to the U_{238} so that separation can be accomplished electrically.

Despite capacity, present U.S. enrichment plans include $1.5 billion to upgrade three U.S. gaseous diffusion plants, $2 billion already spent for the Portsmouth centrifuge facility, and about $5.5 billion more slated to be spent at Portsmouth. U.S. enrichment capacity will be in excess supply for a considerable period of time. The stockpile of enriched uranium already in utility hands has led to the creation of a market for its exchange. Until this inventory is worked off, demand for enrichment will not increase substantially.

Coal and Nuclear Externalities

Safety and environmental proponents point to mining health and safety through toxic exhausts to waste disposal for both coal and nuclear power production. Without taking sides on the safety and environmental issues, it is possible to suggest that all costing of the fuel cycle starts at the mine and continues through ultimate disposal. If any portion of this is subsidized, costs are reduced at the plant level by distribution to the taxpayer.

Government support of the U.S. nuclear industry includes research and development of fission and fusion technologies, waste disposal, enrichment, and until recently the breeder reactor. Additionally, enrichment services are provided at cost. Finally, the Price-Anderson Act limits the liability of utilities in the event of an accident, thereby minimizing insurance premiums. It is possible that for large plants, such premiums would be unavailable without the liability limitations.

Economists and engineers have had ample research support and time to study the problems associated with coal and nuclear energy utilization in the light of environmental and safety standards. These problems are dealt with elsewhere in this volume. For nuclear power, safety regulations were a significant source of higher construction and operation costs in the 1970s. Additional consideration in the 1980s must be given to the problems of uranium tailings, plant decommissionings, spent fuel reprocessing and spent fuel disposal. For coal production the initial concerns of safety in mining were met in the 1970s with the implementaltion of the Coal Mine Health and Safety Act, and for surface mine control of environmental hazards by the Surface Mining Control and Reclamation Act. At the time, considerable fear was expressed by the industry that these acts would so curtail the production of coal, disqualify mine reserves and raise the cost of mining that most coals could not compete with oil or natural gas, even at highly taxed post-OPEC price levels. These fears have proved unfounded. Both underground and on the surface labor productivities have largely recovered from the initial impacts and confusion created by this legislation. The cost per ton has been significant, but hardly sufficient to change coal's status as the cheapest fuel for electric power generation. Environmental concerns over coal combustion have not been so successfully or inexpensively treated.

The efficiency of modern large boiler coal applications owes much to the development of pulverized coal-firing technologies. These permit boilers to maximize flexibility in furnace design and scale up remarkably. Pulverized coal particles are small, and high levels of heating value in combustion result, along with high levels of solid and gaseous pollutants. Large particle release, the focus of early concern, is neccessarily controlled in emissions by electrostatic precipitation and fabric filtration. Trace elements of toxic material other than sulfur oxides (SO_x) and nitrogen oxides (NO_x) are commonly clay associated, and can generally be removed by inexpensive mineral preparation techniques prior to burning. However, somewhat less than half of the sulfur can generally be so removed, leaving that which is bound organically in the maceral with the potential to damage vegetation and create adverse health effects for humans in resulting smog or acid rain.

Emissions Control: Gaseous SO_x emissions, have been historically "controlled" by dispersion through high stacks. This is no longer an admissable solution. Emissions are today reduced by flue-gas desulfurization (FGD), or "scrubbing." This "best available control technology" (BACT) with its associated sludge ponds or waste removal equipment accounts for almost half the present high capital cost of new large coal-fired plants (cf. previous). The BACT reliability applied to high sulfur coals is such that operating costs are also increased. New facility applications have proved cost-effective, but retrofitting costs have been far higher and more controversial in the industry. This has impelled utilities toward the very low sulfur, compliance, coals wherever possible, and toward variances for older facilities, especially in the East. Acid rain, as a result of the industry's foot-dragging, has emerged as one of the highest priority environmental issues in the 1980s.

Atmospheric conversion of gaseous SO_x and NO_x into sulfuric and nitric acids are thought to be largely traceable to coal boilers in the eastern United States, although auto emissions are also a prime offender for NO_x, as are the other fossil fuel emissions. During combustion, NO_x forms in high temperature regions around the flame, from both fuel-bound and atmospheric nitrogen sources, as a function of temperature and the amounts of oxygen fed in the furnace. The higher the temperature the greater the atmospheric NO_x formation. Low temperature combustion techniques, such as fluidized-bed, greatly reduce the NO_x conversion compared to pulverized coal boilers. They therefore have the potential of meeting standards of emissions at lower control costs.

It is generally known that on new small boiler installations, whether fluidized-bed combustion (FBC) or FGD systems are employed, coal-based power that meets standards in these ways can be economically produced, i.e., at costs below that of natural gas or oil used as alternative boiler fuel. For large installations, FGD also meets SO_x standards below the costs of conventional oil or gas fired utility plants, but FBC has yet to demonstrate a similar low cost. The acid rain contribution of existing plants, even if accepted at the high levels charged by the critics of coal, can readily be halved if conventional mineral preparation techniques are managed prior to all combustion. Since much coal beneficiation is achievable at relatively low cost, it is clear that the principal concerns of critics of environmental control can be met economically in the case of coal.

Uranium Tailings: New EPA regulations pertaining to uranium and thorium tailings will raise mining costs somewhat. Under the Uranium Mill Trailings Radiation Control Act of 1978, required disposal of mill tailings must be effective for 1000 years in most cases. Thick earthern coverings will be required. Incorporated also under the Solid Waste Disposal Act are rules that require background or drinking water levels to be achieved. Thus, clay or synthetic liners for ground water will be required. Twenty-seven existing and 24 inactive milling sites are affected. For the active sites, the cost will be about \$260 million; by 2000 the control cost is estimated to range from \$310 to \$540 million. The new regulations have been challenged by the American Mining Congress in the courts.[84]

Decommissionings: Assuming a plant life of 30 years, 20 nuclear reactors will be decommissioned by 1998; by 2010, about 70 reactors will be decommissioned. The small Shippingport reactor is the first. Demolition should begin in 1985 with removal and site restoration completed by 1988. The predicted cost is $80 million. Safe storage, while much of the radioactivity decays, is another option. Disposal of on-site spent fuel remains a problem until a national repository is in operation. Decommissioning of multiple plant units can be delayed until the last of the set is retired. If the steam generators, power turbines and other equipment are salvagable, only the core (about 10% of the initial cost) and the fuel need be removed. The remainder could be used for a new plant.

Decommissioning of a small, government owned, plant may not be a complete guide to future costs. Utilities must set aside reserve funds for such future action. While most states allow the utility to use these funds for other purposes, California, Maine, and Pennsylvania require escrow accounts. At issue is whether without such accounts utilities will have sufficient funds for the activity when a decision must be made.

The present value of decommissioning costs does not adequately appear in capital cost estimates for nuclear power plants. Yet, unlike fossil fuel facilities, radioactive plant structures and materials do not have a salvage value. Rather they bear a dismantling and storage cost. No estimate of such costs is possible. There has been no experience with large scale plants. It has been contended that a refurbishing and continuation of plant life is an alternative to dismantling. The following represent increasing levels of decommissioning and of costs. At any level, economically salvagable equipment and all reactor fuel elements are removed. Some equipment may be decontaminated. The remainder and the waste normally shipped during operation must be sent to depositories. There has been little U.S. action on the permanent storage of high level fuel waste, there has been none on the deposition of contaminated structures which would require far more room.

However the dismantling is to be handled, the following restoration measures must be taken:

(1) At the lowest level there would be minimal dismantling and relocation of radioactive equipment. All radioactive material would be sealed in containment structures, primarily existing ones, which would require perpetual, continual surveilance for security and effectiveness.

(2) At the next level some radioactive equipment and material would be moved into existing containment structures to reduce the extent of long term containment. Surveilance as in the lowest level would be required.

(3) At the third level radioactive equipment and materials would be placed in a containment facility approaching a practical minimum volume. All unbound contamination would have been removed. The containment structure would be designed to meet minimal perpetual maintenance surveillance and security.

(4) At the highest level all radioactive equipment and materials would be removed from the site. Structures would be dismantled and disposed of on-site by burial or offsite to the extent desired by the tenant.

The cost estimation must include the amounts associated with the desired level of decommissioning and the cost of surveilance, if required by the method chosen. The present value of such estimates must be included in the rates charged for electricity if final costs are to be met. The longer lived and the higher the plant capacity factor the lower will be the necessary rate increase. Finally, utility cash flow is adversely affected if the portion of the rate charged for decommissioning must be kept in escrow to guarantee the timely availability of the funds.

Spent Fuel Disposal: Both coal and nuclear power plants produce fuel cycle waste. Those from nuclear plants are smaller in volume, higher in toxicity and longer lived. Waste disposal must, therefore, be added to the fuel cycle cost. If nuclear power plants, like coal-fired plants, were required to establish their own disposal facilities, rather than rely on the government, waste disposal (like decommissioning) would be an addition to capital costs. Alternatives to high level radioactive waste disposal include reprocessing and the breeder reactor; both of which reduce the demand for virgin uranium.

As utilities burn about five percent of the fuel charge, without reprocessing a large amount of waste product is produced. Currently this is a liability to the utility. In earlier years, it was commonplace to ascribe a plutonium credit to nuclear costing exercises. Without reprocessing the credit is no longer taken. The spent fuel is also a physical liability. It must be protected and stored in cooling ponds. The waste is increasing while the ponded storage is finite. Excess space can be rented, but that is a diminishing asset.

The difference between storage and disposal is essentially that of retrievability. In general long-term disposal of high level waste involves solidification in a noncrystalline material (borosilicate glass or a ceramic synthetic rock), packaging in noncorrosive containers (mental or concrete), backfilling with a buffer material (bentonite clay), and residence in a geologically stable, water free environment (granite and basalt, salt domes, volcanic tuffs, and some strata of clays).

A large nuclear reactor produces over 30 tons of high level spent fuel waste and a smaller amount of low level wastes. Including the military, the high level radioactive wastes may be in the form of rods, caustic liquids, powder, and semi-sludges.

Above ground temporary storage does provide the opportunity to reprocess the uranium in times of excess demand. However, if held by the government, this offset to reserve depletion might be sacrificed to the establishment of satisfactory market prices for producers.

If waste disposal is not to be subsidized, the price paid by utilities must equal their share of the levelized cost of providing the repository plus its operation until permanent closure. Additional costs include waste preparation and concentration, nitrification, and transportation. Compared to reprocessing, utilities should be willing to pay the sum of reprocessing costs (current European levels are about $700 per kg) plus waste storage costs less the value of the returned reprocessed fuel.

In the United States, one waste disposal site is to be in operation by 1998. By January 1985, DOE is to select three possible sites for complete geologic examination. The 1998 operational deadline coincides with the date the government is obligated to begin taking responsibility for the almost 950,000 cubic feet of spent fuel rods expected to be generated by them. There is, however, some delay. Rather than receiving 3000 tons of spent fuel, the repository will receive only 400 tons in 1998. The initial tonnage level has been delayed to 2003.

Presently, utilities are subject to a special tax, which yields somewhat less than $400 million per year, to pay for storage costs. It is in exchange for these funds that the government promised waste transfer by 1998, even if it must be moved to above ground monitored retrievable storage.

The result of the delay is two-fold. First, the cost of storage to a utility can only be partially estimated, it cannot be ascertained. Any assessment of this factor adduced to nuclear power generating costs is, on the basis of past estimating, likely to be understated. Second, the existence of interim storage favors eventual recycling (reprocessing) of spent fuel, reducing the demand for uranium.

Recently, the government of China has bid for the right to store high level wastes produced by European reactors. Aside from the issues of proliferation should China develop a reprocessing facility and the ethics of storing a developed country's wastes in a less developed nation, the offer provides a partial benchmark for storage costs. The Chinese appear to be willing to store about 4000 tons of waste for $1510 per kg.[85] Reduced to reactor scale of about 30 tons of high level waste per year, storage costs are about $4.1 million per year.

SOLID FUEL POLICY IMPLICATIONS

A central issue underlying much energy policy is the adequacy of individual fuel stocks for meeting far future requirements. As described in Part One, resource appraisal is the beginning of fuel policy. Reserves are constrained by environmental, national, and locational considerations limiting their size and availability. The conventional measure of reserve adequacy is the time to exhaustion, given offsetting stock requirements. Sometimes the more sophisticated stock concept of potential supply is employed to order reserves. Neither measure is adequate to the task of analyzing economically the future flows of quantities supplied and demanded as a function of prices to assure balance.[86]

On the demand side, it is insufficient to project requirements merely on the basis of population and growth, relating requirements by a functional form, and perhaps correcting by the stock of material on hand or the consumption saturation level already achieved. Such a procedure suggests that the future mirrors the past and leads to incorrect measures of exhaustion rates. What is provided is a measure of historical needs, not future demands. Without a concept of the path of equilibrium prices over time, cumulative demand cannot be discussed.

The problems with the conventional analysis are amply defined in other sections of this book, but they may be summarized briefly here:

(1) the stock ratios estimated by geoscientists neglect the market forces interrelating the supplies of all competing materials as annual flow variables to their relative prices.

(2) They neglect many of the dynamic elements in the behavior of agents and the formation of supply and demand balances over time. These create lags and divergencies between short run and long run equilibrium price paths.

(3) They trivialize energy demands and their inter-relations by the neglect of substitution and technical change.

(4) Finally, they neglect delivery system costs and the efficiencies of international trade.

Oil and natural gas are premium fuels economically precisely because they are rather immediately usable in their convenient forms, while the solid fuels are not so easily used. Moreover, the latter do not lend themselves to controlled combustion, emissions or waste disposal as easily as oil and gas. The solid fuel delivery systems are more complex, capital-intensive and round-about. For all these reasons, it is an essential task of economic analysis, as developed for the solid fuels in Part Two, to relate stock concepts of potential supply to the unit cost and delivered price functions describing the flows of supply and demand in the actual markets. The adequacy of reserves and resources can then be discussed in terms of the rate of draw-down or flow over a specific period, i.e., a consistent estimated or experienced rate of exhaustion.

The demands for energy are derived demands. A derived demand can often be satisfied by several means, and by direct or indirect processes. The demand for electricity can be satisfied, in whole or in part, by a wide variety of fuels and methodologies. No fuel has an inherent superiority. Which fuels will be used and in what shares cannot be arbitrarily assumed. Reasonable estimation of future fuel use requires an intensive examination of the engineering, costing, and competitive aspects of fuel production, trade, conversion, and consumption in order to predict the outcome of interfuel competition.

Minerals have two additional characteristics, quality differentials as well as exhaustion and depletion. The first, which is shared with a number of manufactured items does not refer to ore grade, depth, or overburden. Those features are imbedded in both the production function and costs. Rather, quality refers to by-products or residuals, which must be removed, supplying either a credit or debit on disposal. Exhaustion and depletion are unique to minerals. They may occur with renewable resources by poor management or erroneous policies, but they are not a necessary feature of renewable resources. The issue of exhaustion and depletion leads to the policy question of the adequacy of reserves and resources. Because these are not fixed stocks, analyses which measure the drawdown of the material based on some set of requirements leading to a shortage or crisis are essentially wrong.

World coal reserves and resources are in excess supply irrespective of any projected level of prices and consumption rates for the foreseeable future. World coal resources appraisal demonstrates it to be a reasonable assumption that coal prices will remain low in real terms for many decades. The situation can be altered radically only if coal transport costs dramatically increase, if coal trade remains severely restricted, if coal properties are not permitted access to markets, or if coal combustion technologies are severely restricted. All of the above are policy issues in coal producing and consuming countries, and the manipulation of dynamic spatial models gives insight into their economic and resource effects.

Nuclear power is similarly subject to policy determinations. Here, however, the reserve/resource base is not as extensive or as elastic as coal's; therefore, the extent of increases in the price and cost levels of uranium and nuclear power are central to the question of anticipated shortages. Analysis suggests that demand (to be distinguished from requirements) is unlikely to reach the levels claimed, because nuclear power is being priced out of the market. If it remains expensive, national reserves and resources will prove adequate. Even if nuclear costs decline, reserves will prove inadequate only if policies continue to exclude world trade. Even with import restrictions, reprocessing costs provide a ceiling on both uranium prices and the rate of drawdown. Such a ceiling precludes both a major expansion in reserves/resources and a critical uranium shortage.

The analysis that follows deals more directly with processing, cost engineering, and trade issues than with resources and reserves. Nevertheless, in comparative fuel economics, such induction is the only way to proceed in order to obtain an estimate of supply-demand balances.

Interfuel Competition: Coal and Uranium

Although several fuels may be used to generate electricity, coal clearly dominates natural gas and residual oil and is likely to grow even more competitive in the future. Except for very limited applications, synthetic fuels or energy from renewable resources are not competitive with the direct use of the fossil fuels in electric generation. The major area of inter-fuel competition is direct coal use and uranium.

In February 1964, the New Jersey Central Power and Light Company announced that anticipated costs for electric generation at its planned Oyster Creek nuclear plant would be 4 to 4.5 mills per kwh. This was less than the cost obtainable from any other fuel. Given the expectation of increases in the demand for electricity of 6 to 7 per annum, the demand for uranium, enrichment services and reactors surged. Both the industry and the nuclear share in the industry were expected to grow rapidly. What was not often discussed was that the first set of nuclear plants, including Oyster Creek, were turnkey operations. Both the reactor suppliers and the architect/engineers had a vested interest in cost saving and, perhaps, loss leader pricing.

As the new cost-plus reactors came on line, reactor orders were delayed, construction was delayed, and demand for uranium was reduced. Estimates of 1980 U.S. nuclear power capacity were reduced from 1970 onwards.

U.S. coal-nuclear cost comparisons, to the extent to which they are based on a trended increase in coal prices, are essentially wrong. These prices have been constant or declining on an inflation adjusted basis. In the West, in Wyoming and Montana alone, excess annual coal mining capacity is about 65 million tons. The Linowes Panel has suggested that increased coal leasing in the face of excess supply and low prices may have cost the government millions of dollars in potential revenues.[87] The emphasis is on current low prices, excess supply, and additionally available coal supplies from the new leases. Excess leasing also reduces the value of existing leases.

In the East, excess capacity is 150 million tons. History has shown that as coal prices rise new supplies rapidly become available by mine expansion, new mine openings and the reopening of inactive mines. To a significant extent, the increases in coal costs as delivered are nominal and due to inflation, rail transportation increases and the costs of pollution control. The first affects nuclear power as well, the second may be mitigated by developing alternative transport. Even the threat of coal slurry transport impinges on rail transport contracts. For new coal fired plants, siting and coal selection also serve to reduce prospective transport costs. Pollution control will remain a problem. The associated costs are likely to rise whether the solution is western low sulfur coal (transport), stack gas cleaning (regulation), or coal preparation. In the East, the last may be partially offset by reduced transport costs, i.e., more carbon per ton moved, alternative transport modes, and the firing of coal-oil or coal-water mixtures.

By 1974, it was possible to show that based on 1973 data nuclear fuel cycle cost projections for 1980 should have been about double those claimed by the AEC.[88] While the escalation of capital costs was hardly foretold, the very wide divergence among companies and reports in claimed costs indicated that nuclear costing was an uncertain art rather than an engineering science.

Assuming that long term oil prices do not fall, the essential comparisons between nuclear and coal-fired plants are neither as simple nor as uncontroversial as the coal vs. oil case for a number of reasons.

Environmental and safety comparisons are more difficult between coal and nuclear use because of the different nature of the hazards involved. However, both may be measured by their contribution to costs. The much higher capital and fixed costs for nuclear make ultimate costs per kwhr far more sensitive to some changes, e.g., in the real rates of interest. The most recent extensive modeling of plant costs in an independent study, employing the DOE CONCEPT and OMCOST engineering models, is that conducted by the Environmental Protection Agency for the Ohio River Basin (ORBES) in 1980–83. Results are given in dollars roughly comparable to those of the Chase Manhattan coal-oil tradeoffs. The ORBES assumptions reflected historic 10 year construction times for nuclear plants of 1,000 megawatts scale and five years for coal plants of the same size. These comparisons assume an inflation rate of 1.9% and average real cost of capital of about 8 percent. If either rate increases the coal advantage increases. The CONCEPT variations in site prove significant among regions, but not among states in adjacent regions.

The simulations demonstrate coal to be cheaper by half than nuclear. In constant 1980 dollars the busbar price for electricity is 5.1 cents per kwhr for coal and 10.2 cents for nuclear. As one might expect, the component of cost that dominates the price at the busbar is the capital plus fixed cost charge, which, under public utility accounting, is treated as capital. Thus, changes in the scale of plant and the utilization rate swamp other changes in operating and maintenance costs. Secondly, because the interest charges on construction are a major capital cost that compounds, higher costs of capital or longer times to completion have similar busbar effects. The elasticity of capital costs with respect to plant size in the ORBES nuclear applications is -1.6 and their elasticity with respect to construction time is approximately $+1.2$. Translated to busbar cost elasticities, the impacts of falling operating and rising construction delays are higher for nuclear than for coal because the capital component for nuclear is so much higher for equivalent scale plants equivalently loaded. This turns out to be true irrespective of the experienced range in almost every other source of variation, including the costs of delivered coal (which offset somewhat the disadvantage of nuclear's higher capital intensity) and the capital costs of safety regulations. High transportation costs of coal are assured when only sites farthest from coal fields are simulated in such studies.

Engineers and accountants employ various devices to equalize coal and nuclear costs. Suppose, in this case, the following constraints be accepted:

(1) Unit costs are computed for nuclear reactors scaled at 1,000 megawatts, approximately twice the scale employed for coal boilers of 650 megawatts.

(2) Lifetime capacity factors are increased for the larger nuclear reactor size to make them equal to those experienced in the smaller coal boilers.

(3) Nuclear construction times experienced are decreased to those experienced by coal (5 years).

(4) Most disposal and decommissioning costs are not considered.

By these assumptions the busbar costs of nuclear can be lowered 25% and coal's raised 33% to equalize the two costs. Even so, advantages will reappear for coal as plant sites closer to mouth-of-mine coal are employed in the comparisons (Table 4.11.25).

In a final ORBES study by Newcomb and Bancroft.[89] instead of assumption (3) the longer experienced construction times for nuclear (10 years) are used. Thus, the advantage returns to coal despite the smaller assumed coal plant scales, and nuclear busbar costs rise to 10.2 cents per kwhr in 2000. This advantage for coal more than doubles when existing tax subsidies are added to the costs utilities pay. The tax subsidy for nuclear is about 3.7 cents per kwhr compared to 1.2 cents for coal.

These comparisons, unfavorable as they are for nuclear, still understate considerably the advantages of coal (cf. following). As updated CONCEPT projections show[90] the ORBES simulations neglect many of the high actual costs of construction experienced by utilities during the period of regulatory review following the Three Mile Island (TMI) accident. DOE indicates the "mean" lifetime nuclear capacity factor achievable is 52% over 30 years, and this

Table 4.11.25—Coal and Nuclear, Base Case New Plant Comparisons, 1985 (in constant 1975 dollars)

Region	Illinois		Ohio	
	Coal[1]	Nuclear[2]	Coal[1]	Nuclear[2]
Total Busbar Costs (Mills/kwhr)	50.0	49.6	50.4	50.3
Capital[3] (Mills/kwhr)	34.4	38.3	34.8	39.0
O & M (Mills/kwhr)				
Fixed[4]	0.6	2.5	0.6	2.5
Variable	3.7	2.5	3.7	2.5
Fuel (Mills/kwhr)[4]	11.3	6.3	11.3	6.3

Source: ORBES Final Report, Newcomb and Bancroft, 1980.
[1]Coal Plant Size: 650 MW; 65% Capacity; 5 yrs construction period.
[2]Nuclear Plant Size: 1000 MW; 65% Capacity; 5 yrs construction period.
[3]Capital Cost 8% real; 30 yr. life; real escalation rate 1.9%; annual inflation rate, 6.5%.
[4]Inventory carrying charge for fuel. Fuel cost for coal assumed to average $27/ton for IL and OH by Teknekron USM model in 1985.

mean achievable rate must be reduced further for reactor sizes over 600 megawatts. A host of major safety issues difficult to quantify must also be added to the expected costs of reactors in operation. These issues include the predictability of reactor control during disruptive events, equipment malfunctions, pipe degradation (especially embrittlement), containment behavior, and technical problems related to the immediate safety and integrity of the plants. DOE underestimates the cost of remedies, which it expects will add only $90 million in capitalized cost to new designs on average. This is trivial compared to the experienced losses in actual operating costs and repairs to existing reactors or those actually under construction. In contrast, the industry estimates these non-included items can add as much as $1 billion to the cost of a plant. The Nuclear Regulatory Commission listed 16 unresolved safety issues in 1981. Controversy continues concerning the remedial measures to be required. All such issues have greatly added to the uncertainties and the variances in costs recorded for units under actual construction. Finally, ORBES costs do not include major unresolved issues of waste management or plant decommission, which are also treated in more detail later.

As a result of overruns and delays, financial institutions have shortened the mortgagable life of nuclear plants from 30 to 15 years. There is also serious concern over whether the limits to nuclear accident liability of the Price-Anderson Act of 1957 are adequate to cover the potential compensation liabilities for off-site victims. This is now set very low at $560 million. Even the higher on-site insurance typically carried by utilities covers only 1% to 3% of the losses actually sustained in the TMI case. Thus uncertainties have added much more to the costs of delays and other problems of the industry than CONCEPT adjustments accommodate. These high costs and uncertainties have occasioned the cancellation of 82 plants contracted for between 1972 and the TMI accident.

Despite these dim prospects, the revised CONCEPT estimates continue to assume higher capacity rates for nuclear than for coal, and continue to neglect the higher variances in nuclear operating costs associated with safety requirements. Higher nuclear variances (risks) are suppressed by the capitalization of hypothetical "resolutions" to safety issues, while scale differences are increased to equalize the estimated

Table 4.11.26—Coal and Nuclear, Median Case New Plant Comparisons, 1995 (in constant 1980 dollars)

Region	Illinois, Ohio, Wisconsin, Minnesota, Michigan, Indiana	
	Coal[1]	Nuclear[2]
Total Busbar Costs (Mills/kwhr)	42.4	43.5
Capital[3] (Mills/kwhr)	17.7	29.8
O & M (Mills/kwhr)	5.3	5.2
Fuel (Mills/kwhr)[4]	19.4	8.5

Source: DOE/EIA, 035612 Vol. 2, Nov. 1982.
[1]Coal Plant Size: 1200 MW (2–600 MW units); 65% Capacity, 5 yr. construction.
[2]Nuclear Plant Size: 1200 MW; 65% Capacity, 7 yr. construction.
[3]Capital Cost 4.6% real; 30 yr. life; 1.9% real escalation rate; 6.5 annual inflation rate.
[4]Fuel cost for coal assumed $51/ton in 1995 by other DOE studies.

EIA coal and nuclear costs in any selected region (Table 4.11.26). These assumptions are made to support EIA predictions of expanded nuclear orders. Yet utility managements continue to reject nuclear options in practice because of the higher costs and risks experienced for nuclear investments, which continue to be neglected in the EIA evaluations. Thus, Arthur D. Little's Heuchling declares such studies to be "empty exercises," and many private engineering consulting firms continue to estimate life-cycle capital costs to be realistically twice as cheap for coal.

Using data provided by EIA, rather than their analyses, yields a result which supports the ORBES study. Table 4.11.27 provides comparative capital and electric production costs in the United States for coal and nuclear power. Given the data limitations only 1975–1979 are useful. The nuclear power data officially presented for the years 1978–1980 are seriously deficient and are subsequently unavailable.

The year 1975 was chosen as a starting point simply to eliminate the earliest smaller reactors. Those initially operating in 1975 were begun at least by 1969. It may be noted that for the first three years tabled, nuclear power plant capital costs were 40% to 160% greater than those of coal.

For those three years production costs (operation, maintenance and fuel) averaged 21.68 mills per kwh and 12.56 mills per kwh (adjusted) for nuclear power, but 17.74 mills per kwh and 16.19 mills per kwh (adjusted) for coal.

Table 4.11.27—U.S. Comparative Capital and Operating Costs—1980

Year of Initial Operation	Nuclear Power					Coal Fired[1]				
	Plants (No)	Production Expenses (Mills/KWH) Avg.	Production Expenses (Mills/KWH) Adj. Avg[2]	Capital Cost ($/KW Cap.) Avg.	Capital Cost ($/KW Cap.) Adj. Avg[2]	Plants (No)	Production Expenses (Mills/KWH) Avg.	Production Expenses (Mills/KWH) Adj. Avg[2]	Capital Cost ($/KW Cap.) Avg.	Capital Cost ($/KW Cap.) Adj. Avg[2]
1975	5	10.00	8.80	470.2[3]	462.5	7	18.27	14.38	284.7	274.2
1976	4	41.78	15.22	569.3[3]	594.0	8	17.55	17.72	405.8	386.0
1977	4	16.21	14.61	742.5	784.5	6	17.37	16.25	341.0[3]	298.0
1978	1	12.44	12.44	671.0	671.0	7	14.37	14.42	424.3[3]	427.5
1979	1	9.50	9.50	557.0	557.0	4	17.38	16.86	512.0	545.0
1980	0	([4])	([4])	([4])	([4])	8	19.35	19.41	617.8	599.0
Average		2.026	12.00				17.43	16.56		

Source: Energy Information Administration, *Thermal Electric Plant Construction Cost and Annual Production Expenses 1980*, DOE/EIA-0323 (80), June 1983.
[1]Principally coal but may have dual firing capacity.
[2]Average adjusted by eliminating extreme case at each end of range. Note: The averages are not weighted by plant size.
[3]One observation less than the number of plants indicated.
[4]Plants were not listed with sufficient data to be of use.

On balance, with the addition of capital costs, coal fired plants of the 1975–1977 initial operation dates must have produced cheaper electricity in 1980 than did nuclear power plants of the same dates.

In this review, detailed inquiry is made into the curiosity of continued predictions of increased nuclear adoptions. In the United States, they must be attributed to the understating of the costs of nuclear technologies and overstating those of direct coal systems. This explains the expected fault of policy analyses by national agencies, and their consultants, which have failed chronically to account for the high costs of nuclear or of synthetic fuels in power generation. The experienced failure of existing nuclear power plants to meet operating and safety specifications have compounded both the high capital costs and errors in operating cost estimation.

In order to meet EPA standards, coal has been forced to add advanced coal cleaning systems. However, even with emissions controls, coal-based capital costs to utilities remain less than half that of nuclear reactors ($1,000 per kwhr capacity). Moreover coal plants operate with lower than estimated fuel costs (under $2.50 per MBtu), and at higher rates of capacity (65%) and lower variable operating and maintenance costs (below 1.5 mills per kwhr). Even with the substantially greater subsidies, therefore, nuclear plants must be larger and run at higher capacities than direct coal systems. In the United States, these objectives have not been sustainable in practice.

Both regionally and nationally neither electric power generation in general nor nuclear power in particular are likely to increase as rapidly as previously forecast.

In the United States, successive DOE predictions of 10 year annual growth rates have declined yearly, while average actual growth rates for 1974–1983 were less than the 10 year average growth rate prediction made in 1984. The newly projected rate is unlikely to be achieved.

In the northeastern United States and the northern tier of states in general, purchases of Canadian surplus power can competitively substitute for new U.S. plants, both coal and nuclear. For financially burdened utilities, the avoidance of the front-end capital costs is desirable. Hydro-Quebec already exports almost 24 billion kwh annually of interruptible power to New York State. Starting in 1986 it will sell three billion kwh per yr to the New England Power Pool for an 11 year period and an additional seven billion kwh per yr of firm power starting in 1990 and running to 2000. Vermont will get a smaller amount of power, also on a firm basis starting in 1985. The rates are 3.3 to 4.0 cents per kwh initially and are pegged to oil and coal prices in the last five years of the agreement. Northern States Power has a trade or swap agreement for summer sales and winter purchases from Manitoba Hydro. Western Canadian power has been available for years.

At least in the United States, utility peak load capacities far in excess of reserve margins are likely to keep the demand for both coal and nuclear power stations low for the next decade. The accepted reserve margin for electric power generation in the United States is 20 percent. This is likely to be exceeded if each utility attempts to maintain the margin rather than averaging margins for the pool or electric reliability council. In the early 1960s, the margin was over 30%, but by 1969–70 it had fallen to less than 20%. In part this led to the construction and ordering boom of the early 1970s. However, due to the size of the plant increments, to the effort by each utility to individually meet the margins and to the reduction in consumption rates as a consequence of price increases, utility margins rose to 40% in 1982. They were above 30% in 1983. Until they are reduced to 20% or less, utility construction is unlikely.

Price induced "conservation," the movement down the demand schedule, provides a limit to the ability of electric utilities to pass the impact of high cost power generation through to the consumer or ratepayer. Cogeneration provides another offset or limit. The joint production and sale of electricity and heat by energy intensive industries is not new. It was pervasive in the United States until the 1920s. In Europe it is relatively common. The major industrial groups concerned included pulp and paper, oil refining, chemicals, primary metals, and food processing. Additionally, small cities, large towns, and public institutions (universities, hospitals, apartment complexes) can all produce and utilize both self-produced electricity and self-produced heat.

Two problems arise:

(1) The balance of heat and electrical production is rarely that of consumption at a given time; neither heat nor electricity is easily or cheaply stored.

(2) On a joint basis, cogeneration must be

cheaper to the producer/user than the separate purchase of each item by the user.

Ideally, groups of firms would arrange a heat/electricity—production/consumption balance among themselves. This would minimize transmission and, probably, production costs. It would tend to destroy the electric utility industry and its monopoly on power sales. In the United States it would be regulated.

In the United States cogeneration is assisted by the Public Utility Regulatory Power Act which requires utilities to purchase excess privately generated electricity at the avoided cost of new capacity (a mix or the latest addition). An expected result is that producer/consumers maximize electric output subject to the constraint that the required heat or steam demand level is satisfied. A significant amount of electricity is then not purchased from utilities by such producers and a large amount of electricity may be sold to utilities at prices above their weighted average cost.

Cogenerators can use a wide range of fuels, but like utilities concentrate on oil, gas, and coal. The required equipment is standard, the engineering is not restrictive. In future, sizes may range to those applicable to private homes. Capital costs are about $800 to $1200 per kw of installed capacity. Some operating systems currently cost from $275 per kw for a 105,000 kw installation to $440 per kw for a 22,400 kw unit. The capital costs depend partly on the fuel proposed; gas is cheapest followed by oil.

Given its capital costs, unless future nuclear fuel costs and the resultant busbar costs are expected to be much lower than combined net electric stream costs for cogeneration, the latter is an effective substitute for nuclear power. Currently, about 7% of U.S. electricity is produced via cogeneration. This figure may rise to 15% by year 2000, about the level of nuclear power.

Given the cost problems in the U.S. nuclear power sector, the decline in the demand for uranium is unsurprising. In 1972, the AEC predicted that by year 2000 U.S. nuclear power generation would be 885–1500 gigawatts. In 1982, DOE estimated that by year 2000, generation would be 145 to 185 gigawatts. In 1983, DOE revised the estimate to 132 gigawatts.[91] By 1984, it is clear that the figure must be further reduced.

By 1974, based on 1973 data, it was possible to conclude that AEC projections of nuclear power for 1983 and after were simply not possible. Furthermore, given the construction delays already endemic by 1973, the increasing rate of construction would lead at least to longer construction delays.[92] Two results follow:

(1) increased construction delays would increase capital costs and
(2) the future demand for uranium would be less than that forecast by AEC and others.

Due to revisions in reactor and enrichment services schedules, mutual agreements to cancel or reduce some delivery commitments, buyers' adjustments to their planned captive production, and litigation settlements, domestic uranium deliveries and future commitments have been falling. The first two reasons are the most important. By the end of 1983, future commitments for uranium deliveries by domestic suppliers to domestic buyers through 2001 were only 255.5 million pounds of oxide (U_3O_8).[93] Of this, 20% represents optional commitments which need not be delivered. Commitments for 1983 through 1991 were only 169.4 million pounds. If 1982 production levels were continued, this last would be worked off in just over six years.

Optional uranium delivery commitments between U.S. producers and U.S. buyers have increased each year (1981–1983) and for each forward year.[94] For the United States, total domestic forward delivery commitments of uranium oxide are declining. Of this, firm commitments are declining while optional deliveries and captive production are increasing. Imports are also increasing.[95]

Unlike the United States, a number of European and other countries are increasing the share of nuclear power in electric generation. In part this is due to protectionism and to their relatively higher cost of domestic competing fuels, coal and oil. In part it is due to a desire, less evident in the United States, to limit dependence on imported fuels. However, limiting dependence is only partially achieved if either or both yellowcake and enriched uranium must be imported. Competing fuel costs may be high if measured in terms of domestic supply, they are likely to be significantly lower if measured against imported coal.

In the United States, fuel cycle costs, which were never as low as those claimed by AEC, rose far more rapidly in the 1970s than did coal, so that nuclear power was simply becoming noncompetitive. In Europe with high cost coal, in Canada with little coal, and in Japan with almost no coal, nuclear power remained more competitive. In Australia with both low cost coal and low cost uranium, nuclear power is not a factor for the foreseeable future.

Throughout the developed world utilities planning or implementing investment decisions for new or replacement facilities are selecting conventional or advanced direct coal fired boiler systems over oil, gas or indirect coal (synfuel) options. Moreover, with the exception of France and Japan, direct coal systems dominate nuclear alternatives.

Table 4.11.28 provides an estimate of nuclear power shares of electric generation in (primarily) West European countries. The shares appear to be independent of both domestic coal availability and overall power consumption.

In 1982, Taiwan, Spain, Japan, both East and West Germany, and Canada all planned for major nuclear expansion. Poland, Portugal, Rumania, and Egypt, which had no nuclear facilities, also had ambitious nuclear development plans. Austria will remain non-nuclear. Nuclear development in Sweden and the Netherlands is on hold. China is ready to sign contracts with French and British firms for its second nuclear power plant (1.8 million kw).

As nuclear costs have risen, Spain, France, and Japan have all reduced the scale of their nuclear programs. Unlike the United States, however, there have been no abandonments. In

Table 4.11.28—Western World Electric Power Generation—
Selected Countries, 1983

	Total Electricity Generated (Billion KWH)	Percent Nuclear Power
United States	2365.1	12
Japan	583.3	15
West Germany	368.8	14
France	282.5	40
United Kingdom	277.7	14
Italy	181.8	2
Spain	110.7	9
Sweden	103.4	37
Norway	92.8	0
Netherlands	64.9	6
Belgium	50.8	25
Finland	39.1	36
Greece	23.4	0
Portugal	13.9	0

Source: *New York Times*, 23 January 1984.

the United States some of the prominent engineering consulting firms are supporting the utility decision to switch to coal from previous designations of oil, natural gas or nuclear facilities. Currently 276 large power generating stations are at the margin of investment, ranging from plants under contract to those at the planning stage. Indeed, some dramatic switches from nuclear to coal-fired boilers are in the making at plants where the construction of nuclear units is well underway. Thus, in the area of major new installations and additions, the once dominant fuels, oil and natural gas, have been largely delegated to peak load facilities, and the nuclear option, once the preferred base-load alternative, has all but disappeared. This switch by utilities back to coal has been in evidence for an extended period of time.

Despite this turn of events, energy economists and government forecasters in developed countries continue to predict a limited role for coal in future power generation. Even the more sanguine forecasts of increased coal use predict a parallel growth of nuclear and synfuel applications in their high coal scenarios. The U.S. Department of Energy's Information Administration (EIA) predicted an increase in the number of U.S. reactors through year 2000 from 73 to 125, and in the share of electricity generated by nuclear power at the time from 12.6% to 19.0 percent. Industry organizations such as the Electric Power Research Institute (EPRI), charged with monitoring new technologies and speculating on the eventual cost of new systems for power generation, continue to predict the relative decline of direct coal applications in favor of nuclear reactors through year 2000, and an absolute decline thereafter. In comparing options, direct coal combustion is a last choice among a large number of future synfuel or combined cycle options.

A number of implications can be drawn from the economic analysis of the competition between the two principal solid fuels; nuclear is considered first followed by coal.

Nuclear Energy Policies

The commercial use of uranium is almost entirely limited to the production of electricity. Reserves and resources are traditionally measured with a cost component. The demand estimates needed to determine the adequacy of the reserve/resource are poorly done. They are in fact estimates of hypothetical need.

Neither electricity nor uranium are unique commodities. Both have substitutes. Both apparently are subject to the same laws of demand as are other commodities. If electric costs rise, at constant income levels quantities demanded will fall. The demand curve may shift out over time, but each period demand curve remains downward sloping. The long term locus of equilibrium prices and quantities can be up or downward sloping. The nuclear portion of the demand for electricity depends on production costs relative to substitutes. In the United States the substitute is principally coal. As uranium costs rise or as power plant construction costs rise, unless it can be shown that corresponding coal costs have increased faster, the share of nuclear power will diminish and a given resource/reserve of uranium will be drawn more slowly. The adequacy of the reserve will increase. It may be noted, however, that as uranium demand shifts downward, cut-off grades rise, and marginal mines close. Prices will drift downward, reserves shift back into the resource category, and the claim will be made that reserves are inadequate. Conversely, if uranium prices rise, newer high cost mines are opened, existing mines expand, and it is commercially desirable to mine lower grade ores where the tonnages are sufficient. The claim is then made that the industry is troubled by high costs and low ore grades.

What is important to note is that high fuel costs at any production stage do not justify high energy prices unless it can be proven that the necessary quantities demanded equal supply quantities at those prices. It is entirely possible that with higher prices, excess supply will be evidenced.

In this paper, the cost of nuclear power is developed and found high relative to coal. The resulting demand based analysis greatly increases the adequacy of uranium reserves. Coal-based electric power in contrast, is found to be low cost relative to oil and natural gas. Thus, demand-based analysis predicts faster depletion rates for coal. Yet coal prices rise very slowly in view of the highly elastic supply of coal from the three major producing basins. Of course, free international trade has to be re-established for the role of traded coal to increase substantially.

Nuclear Prospects: At least in the United States, the growth in demand for electricity is likely to be low through year 2000. The role of nuclear power is unlikely to achieve its DOE logistic curve assigned share of electric ouptut. In the absence of major changes in both regulation and construction practices, capital costs are unlikely to be sufficiently low for the low cost of nuclear fuel to be the decisive competitive factor. Abroad, nuclear plants are not being cancelled, but the growth rate of new orders has slowed significantly in OECD countries. Third world countries, beset by debt, are unlikely to undertake, or be able to finance billion dollar nuclear construction projects even if these are cost competitive in the long run.

As a result, unsubsidized demand for uranium will not increase rapidly from existing levels. Given the current potential supply schedule, in the absence of trade restrictions, cartels and other market impediments, uranium prices cannot increase significantly. A further result is that apparent reserves will decrease and ore grades will increase. Mines, mills, and exploration firms will close or reduce operations.

An extreme U.S. scenario is one in which nuclear capital costs and nuclear reliability are assumed to equal the best European practice. Either on a commercial or a government regulated basis the plants, here and abroad, are assumed competitive with alternative fuels at the busbar. The issue then becomes one of the adequacy of uranium reserves and resources. As the opening of a mine and the construction of mills, conversion, and fuel fabrication facilities require the same time as that needed for European nuclear plant construction, the issue of time lags is not important. As in western U.S. coal, mines and facilities are contracted for at the time of power plant construction.

In such an extreme scenario, two cases can be distinguished: free trade and trade restrictions. The former reduces the rate at which existing mines are depleted, but does little for the U.S. mining industry in the near future. Some U.S. mines remain profitable.

Over time, the reserves at existing mines are drawn down, but as prices rise mined ore grades decrease, additional properties become reserves, mining costs increase and exploration becomes valuable. Reserves increase in each forward cost category, but most likely in the highest categories. Given reserve sizes, price related additions to reserves and world trade, the rate of price increases need not be rapid.

The cap on uranium prices is set by reprocessing and the use of mixed oxide fuels. As the cap is reached, the share of virgin uranium decreases. As the cap is equal to the projected high price for yellowcake in year 2000, it is unlikely that a uranium shortage will exist in the United States or elsewhere, with or without free trade.

At various times, government concerns about the over-estimated dependence on foreign uranium sources have led to subsidies to domestic uranium producers in the form of stock-pile acquisition, guarantees and manipulations. With the present over-estimation of nuclear power demands recognized, most experts acknowledge that world prices will favor the growing importation of whatever fuel is necessary.

DOE projections in early 1984 suggest that under current policy U.S. uranium production will continue to decline through 1985 falling to a range of 6.0 to 9.5 million short tons of U_3O_8. By year 2000, production is expected to increase to a range of about 13.5 to 22.0 million short tons. Imports may grow as high as 41 to 54% of the domestic market by 1987, levelling to 32 to 46% of domestic requirements in the 1990s.

A wide variation in price has been forecast. Price in year 2000 ranges from about $25 per lb to $118 per lb of yellowcake. U.S. yellowcake costs will rise as lower grades of ore are mined, as underground mines become deeper, and as mine health and safety as well as environmental regulations become more stringent.[96] The highest U.S. costs lead to both increased import pressures and expections of higher prices.

Driving both the production and price estimates are the reactor capacity estimates. For year 2000 the range in the United States is 110 to 140 gigawatts. For the rest of the world, it is 327 to 428 gigawatts. As shown before, the low end of the U.S. range is most likely, implying that 30 of the 50 reactors currently under construction in the United States will be completed. The high end implies that of those reactors now under construction, all will be completed, and 10 more will be ordered. With a 12 year construction horizon, 1988 is the last year these can be ordered if the high projection is to be reached.

That scenario is unlikely. Unless uranium imports are precluded or otherwise limited, U.S. production will not increase as suggested and uranium prices will not reach the upper end of the price range. At least in the United States and possibly in the rest of the world reactor capital costs will, in the absence of greatly reduced yellowcake and enriched uranium prices or a rapid increase in coal and hydrocarbon prices, preclude the upper end of reactor installation capacities. In the United States and the rest of the noncommunist world, even the lower end may not be achieved.

European and Japanese nuclear cost and operating characteristics provide some indication that the U.S. experience need not preclude a nuclear future. It was noted previously that European capacity and availability factors were higher than those in the United States. Their capital costs are lower than those here. In part this is due to standardization. Most European countries appear to have limited design configurations and sizes. In the USSR, where the program is limited by inadequate manufacturing capacity, a 1000 and a 1500 MW_e reactor is produced. These are grouped in units of 6000 MW_e. In France nuclear power plants are standardized at 900 and 1300 MW_e. In both countries the regulatory practices are less public than in the United States and more under the control of the manufacturers and the centralized utilities. The result is capital saving due to reduced construction time and a learning curve resulting from the production of many similar units.

U.S. nuclear power plants are built on an essentially cost plus basis at all levels of contracting with overruns presumably to be paid for by the electricity consumer. Each utility is in charge of its own projects or in a consortium. These hire an architect/engineering firm as prime contractor which, in turn, deals with a host of subcontractors. Any managerial or engineering talent is spread very thin; cost control is difficult. Furthermore, it is not clear that the architect/engineering firms have any liability for excess cost or reliability problems. If such risks were to be assumed by the AE firms, contract costs would be accordingly adjusted upwards.[97] Nuclear power capital costs would rise. There is some evidence that currently, in the face of reduced orders, AE firms are not only willing to accept some plant performance responsibility, but are willing to assist in financing.

In contrast, Japanese reactor manufacturers provide the entire plant. There are only eight utilities of which two are quite large. Both the manufacturer and the utilities have sufficient orders to develop a learning curve approach to lower costs and reliability. The situation is even simpler in France where the sole utility, Electricite de France, purchases reactors from and cooperates with the producer, Framatome.

Minimizing the number of producers and consumers can easily lead to bilateral monopoly with attendant super-normal profits and non-competitive costs both to be borne by the electricity consumer. In both Japan and France, this possibility seems to be reduced by the efforts of the governments to keep electricity costs low.

As European nuclear plants must compete with coal, it is useful to note that European domestic coal costs far exceed those in the United States. Except for Poland, European countries are not major coal exporters. In France, nuclear power currently costs about half that of coal fired electric power.

Finally, capital costs are sensitive to the cost of capital. Unlike many U.S. utilities, utilities abroad may include capital costs in the rate base as they are incurred. The cash flow is greater and interest changes are less. There is an intergenerational consumer equity question. It is also possible that governments aid in financing the capital costs. If true, U.S. nuclear power plant cost comparisons with the rest of the world are biased.

European and Japanese nuclear construction has shown that plant costs need not reach the recent U.S. scale. Though legal restrictions are likely to preclude the supplier-utility cooperation available elsewhere, a return to the fixed price turnkey plant with acceptable performance guarantees might go far towards maximizing quality assurance and minimizing cost. What is likely to disappear are understated capital costs, stretched construction schedules, and excessive interest charges.

Although commonality of design and engineering can lead to both cost economies and higher standards when spread over many reactors, they can also lead to generic problems and widespread outages. The latter affects system reliability and utility inter-ties.

In the United States, almost 60 utilities have nuclear projects. As a result many U.S. firms with only one or two plants have a very short learning curve, can have only a small cadre devoted to nuclear construction and operation and can attract few well qualified staff at the higher engineering and management levels. Capital cost overruns, operating problems reflected in capacity factors, and safety problems result. Recent fines levied by the NRC following Three Mile Island only point up the recent safety issues. Prior to 1979, public pressure was less and the NRC perhaps more complacent. Few would regard the Commission as anti-nuclear. Its requirements for safety and the changes mandated in construction are apt to be minima rather than maxima. As design often is only slightly ahead of construction, changes are endemic.

On the supply side, the nuclear power steam source supplier provides only the reactor, perhaps 20% of the total construction cost. The remainder of the design is unique and particularized to the local situation. There will be many subcontractors. Despite the use of a very limited number of prime contractors, learning curves are not well developed in such a situation, both cost control and quality assurance are difficult to maintain.

A European solution would be to monopolize nuclear construction and/or operation to reduce cost and increase safety. Alternatively, the nuclear portion of the industry could be at least regionalized. An industry in this form would offer career incentives to technical and managerial personnel, could standardize designs and sizes to provide learning experience and economies of scale, could experiment with smaller, demand following, modular sizes, and could try new designs in an industry with a rigid technology.

Given the power of each state's public utility commission, a European solution appears unlikely. It may be noted, however, that regional power pools and electric relaibility councils exist and might be strengthened. Unless plants fired by other fuels were put on the same regulatory control basis, it is difficult to see how balance within a region or a state could be maintained.

Given high capital costs, maintenance of a viable, high cost, uranium industry in the United States may not be possible in the absence of trade restrictions and subsidy. Both actions are limited by consumer subsitution.

High uranium prices are tenable only if they can be successfully passed on to the consumer; demand must be ineleastic. Furthermore, the electric utility industry must be growing and the nuclear share must be growing within that. If not, demand for uranium does not rise sufficiently to wipe out excess supply.

An increasing nuclear share is unlikely. Indeed, cancellations of existing plants continue to grow, even when construction is well advanced. Furthermore, of the U.S. power plant cancellations since the early 1970s, the brunt has been borne by nuclear power.

In the context of maintaining a viable uranium industry, government actions with respect to an earlier uranium embargo, proposed limitations on uranium trade, and the drawdown of its stocks are understandable. Similarly, limited domestic competition in milling and fuel fabrication with a monopoly in enrichment are a means to price stability at levels acceptable to producers. An international cartel or commodity agreement would only be a welcome addition. Unfortunately, given the amount of low cost uranium currently available, the latter would become merely a buying agency unless production controls were implemented, in which case it would approach cartel status.

Nuclear Power in the Near Term: European and Asian nuclear power plants currently under construction are likely to be completed. Those on order may be cancelled or delayed. It is in the United States, however, that the situation is critical. It may be argued, however, that if technology and cost control provide a firm nuclear future, near-term nuclear power competitiveness may depend upon selected bankruptcies.

Bankruptcy as Optimal Public Policy: In the United States, 20 or more electric utilities may be unable to raise sufficient funds to complete current plant construction. Perhaps the same number may soon be in the same position. If a power plant is cancelled, the issue arises as to who pays and how much. If a utility is in financial difficulty, credit becomes more expensive and less available; skipped or reduced dividends also increase finance costs. Raising electric rates to meet rising capital costs diminishes electric consumption making the plant in question redundant. Currently, the issue of whether customers should pay for abandonments, cancellations or cost overruns at nuclear power plants is being seriously debated.

Excessive construction costs, if capital recovery is to be made, require an increase in electric rates charged the consumer. As this is not done on a marginal basis (most recent plant supplying the most recent demands) averaging suggests that rate increases will be somewhat moderated, but even less expensive electric sources will be affected.

As rates increase, theory suggests that electricity users demand less power. Those industries that are energy intensive may relocate or cogenerate their own power. In the United States, PURPA favors the retention of cogenerators in a high cost electric region as their power sales to the utilities is on the basis of the cost of avoided power production. As consumers reduce demand, relocate, or cogenerate, the rate necessary to cover costs can only rise as the number and size of consuming units falls.

For a number of utilities bankruptcy is a distinct possibility. Constructors, contractors, and suppliers have already been paid and made their profits. The funds came from the ratepayer and the financial community. Interest charges due the latter as well as principal on obligations falling due are a basis for a Chapter 11 filing. Of interest here is the impact on the demand for uranium.

The last major utility bankruptcy was 40 years ago. There is little precedent upon which to base suggestions of the outcome. It is very unlikely, however, that the courts would order a utility closed down or that they would order completion of a cancelled or abandoned plant. They might order the operation of a completed, all but completed, or all but licensed plant. It is not clear whether the public service commission or the court would oversee the utility or who would have final control over the rates required to repay creditors.

The following is merely a possible scenario. Secured creditors must be compensated; unsecured creditors discounted. Share holders, both common and preferred, see most if not all of their equity eliminated or seriously diluted. Creditors may obtain equity for debt or may face a delay or discount in collection of principal. Interest may be foregone. As this reduces the value of the obligation, a market in instruments is likely. Ratepayers will pay higher electric bills, but may avoid this over time by relocation, cogeneration, and reduced demand. There is a limit to what they can and will be forced to pay.

Rates must rise somewhat if only because the utility cannot purchase fuel, regional power, or services on credit; it is likely to pay in advance if not at sale. If the state operates the utility this may not occur as the state's credit rating is substituted for that of the utility. Alternatively, the creation of new categories of debt following bankruptcy may actually decrease borrowing costs and rates.

The write down or elimination of original stockholder equity and the substitution of equity for debt reduces the capital load to the new owners. It is possible that with a moderate rate increase the utility will become profitable. It is also likely that:

(1) future public service commissions will be reluctant to grant certificates of necessity without extensive demonstration of need at the utility level if any excess reserve capacity exists within the relevant reliability council,
(2) financial institutions will be reluctant to lend to the utility market,
(3) construction costs will rise.

The results are likely to be fewer plants, scrupulous attention to commercial fuel cost comparisons, and smaller plants which track demand growth despite possible scale diseconomies.

Bankruptcy might lead to brownouts, blackouts, and loss of regional jobs and industry. Alternatively, after a interregnum managed by a referee or the public service commission, a new buyer might be found. As the capital charge is likely to be significantly lower, the operation might well be profitable to the new owner. Ratepayers and bondholders may suffer, stockholders will suffer. Profits, however, are always measured as positive or negative. There are no guarantees. Stockholder protection is not insured in an industrial society.

A possible alternative to bankruptcy is governmental purchase of utility stock at the going rate and the creation of a public utility district. Government ownership of utilities, particularly in the northeast, has a long history. In the present situation the government credit rating would substitute for the utility rating, share holders and bond holders would be saved from further losses, and revenue bonds or tax assessments could be used to substitute for rate increases. It may be noted that consumers in utility service areas unaffected by bankruptcy or government salvage may object to the spreading of tax assessments and the tax implications of the bonds.

Despite nuclear power capital costs in excess of 5 to 10 times the original estimate, utilities are under considerable pressure to complete the plants.

In some states, a utility can recoup its investment only upon project completion. In others there is no clear law on recoupment if the project is abandoned. Most states allow a partial recoupment, but only over time (5 to 10 years) and with no interest costs added to the rate base. In New York State alone, full recovery, including interest, is possible upon abandoment. It should be noted that even if all or some invested capital is recouped by the company upon cancellation, if there is a delay, interest on the borrowed funds must still be paid; there is a cost to the utility.

Abandonment or cancellation permits a tax write-off of the value spent. This can be subtracted from taxable income. The write-off can also be deducted from retained earnings. Thus the utility can pay tax free dividends. The U.S. Treasury is likely to be the largest loser from an abandonment. Alternatively stated, utility abandonments are subsidized by the taxpayer. The unwillingness to abandon high cost construction lends some credence to the use of uranium requirements. It does not, however, lead to successful inter-fuel competition.

In general, a utility prefers to obtain external funds by obtaining either equity capital or 30 year mortgage financing. Both avoid the uncertainty of repeated refinancing. For nuclear exposed utilities the costs of equity financing have risen. A review of 30 such utilities indicates that the average return to the investor was about 3.2% higher than the average yield on similar companies without such exposure. Even preferred stock yields are about 2% more than utilities are willing to pay.[98] Mortgage terms have also become more stringent; 15 year mortgages are becoming more common. Finally, debt generally costs nuclear related utilities two percentage points more than they have paid, even in the recent past. The result has been more application to banks for project loans, credit lines, and short-term credits. This is limited, however, by the willingness of banks to accept exposure.

Although it presents problems in the areas of cost control and inter-generational equity, the cost of capital is lower if a utility can recover its cost of work in progress. A second method which partially protects stock and bondholders is AFUDC.

Where a utility is not permitted to cover interest and some other expenses by charging ratepayers for "allowance for funds used during construction" (AFUDC), application must be made to the money markets or, in the last resort, by diluting stock and reducing dividends. The smaller the utility the more often it is likely to seek the markets as construction costs rise. The more often the application is made and the more exposed the position, the greater will be the cost of debt and the more costly the plant. In the event the plant is not complete, offsetting revenues will not be obtained. Either the ratepayer must make up the difference, the stockholder must endure the loss, or bankruptcy may ensue.

If AFUDC is permitted, those funds received from the ratepayers appear as income to the utility. When the plant is completed, the interest generated is included in the rate base. Until then it is actually a form of debt. Use of the money market is avoided, but the utility's income statement looks better as the debt increases. In some cases, AFUDC may equal half a company's net income. If the plant is not completed, it is difficult to repay the debt. If the public service commissions do not permit rate increases, bankruptcy may ensue. At the least, afflicted utilities would lack the cash to repair and maintain existing plants.

For nuclear power it may be concluded that it will grow very slowly to year 2000. The draw down of uranium stocks will be correspondingly low. With yellowcake prices remaining low little exploration will be undertaken while reserves will be recalculated and reduced.

Efforts to help the domestic mining industry by protectionist means will not increase the competitive demand for reactors. In the long run, such efforts must fail as consumers reduce demand, cogenerate electricity, or move to non-nuclear areas. At best protection is a short-run aid. In the absence of sharply reduced capital costs and substantially increased load factors subsidization provides the only long-term relief (e.g. as in farm subsidies).

The development of the breeder reactor and/or fuel reprocessing (with or without the use of mixed oxide fuels) places a cap on yellowcake costs, but assists the reactor industry at the expense of the mining industry. The development of a thorium fueled reactor would greatly increase nuclear fuel availability under any scenario. Both reprocessing and a thorium option reduce waste disposal problems. Reprocessing, however, eliminates much of the bulk going to permanent disposal.

Perhaps the best means of maintaining a nuclear future is the development of standardized, small sized (300–600 MWe), incremental demand following reactors. The evidence suggests that these have the highest load factors with which to spread costs.

Coal Energy Policies

Unlike the nuclear markets, coal markets in the United States have been less affected economically by direct public policy, if we define this as subsidies or regulation impacting on markets, trade and the conditions of production, including environmental controls, safety and related worker or user conditions. Such a definition includes also the impacts of coal leasing agreements, sales, and transfers. The government remains the largest holder of coal reserves in an actively competitive industry and has a large stake in coal rents. Nonetheless, on the coal supply side, the government has released reserves in a fashion which has lowered the price of coal below equilibrium and at times below the full costs of production in some regions. Chronic excess supply hastens the exit of weaker firms, but alarms those who fear the concentration of producer power in the coal industry. On the other hand, the industry contends that the constraining impacts of government restrictions and regulations on coal production has significantly raised costs. Thus U.S. coal policy is criticized for lowering coal prices dramatically while restricting coal's growth. Clearly such a combined result is unlikely. Restrictions have not prevented a return of the industry to chronic excess supply. What appears to be happening is an increase in competition. In the face of higher and rising costs for alternatives, a vigorous resurgence of steam coal utilization is likely to continue. At the same time, the price of coal-based power has not soared as predicted, and environmental controls have not proved debilitating.

This is unexpected in view of the initial objections of the industry to the regulations and controls imposed. The reason for the pradoxical results remain the same as observed by Henderson in the 1950s before predictions were common of either coal's revival or demise: North America's coal supply is so elastic that virtually no surge in demand is great enough to long outpace the ability of a producing region to oversupply its market. This is partly due to a mature transportation infrastructure in the United States. Traditionally, the control of excess supply has been the domain of the labor unions, who are largely concentrated in underground mining. In the 1970s and 1980s, with the growth of non-union surface mining, constraining supply has become the role in part of environmental regulators.

It is becoming clear that in the United States the industry remains so competitive that producers' surplus is often passed to consumers.

Thus, the environmental concerns of the major producing regions must receive even more careful regulation in the future, if the external costs of production are to be covered fully. Similar policy concerns appear to be prominent in the two other major producing regions with export potential: China and Australia. However, the need to provide a return to infrastructure investment requisite to the development of the interior coal provinces of those countries may sustain prices at present competitive levels for many years.

In the rest of the world, rather different concerns have appeared. In developed countries such as England or Europe, the structure of high tariffs, subsidies, and high energy prices is encouraged by fossil fuel policies. In underdeveloped countries, pricc policies are mixed, but subsidies often favor nuclear and oil.

However, some countries are reexamining carefully their planned reliance on the newest solid fuel technologies, and are returning to pulverized coal combustion for the large scale generation of electric power. Japan alone appears to be hoping that greater reliability and faster construction times may come with the return of reactor designs to smaller scales. However, many cost advantages are lost at smaller scales. This may leave direct coal combustion cheaper and lower risk internationally for many years. A similar conclusion holds for synfuels where fluidized bed combustion and coal gasifiers may be competitive in small scale applications, but prove to be poorly suited to large base-load power generation.

Clearly developing countries with large coal reserves will reexamine their potential for immediate development in large boiler applications. If, like Indonesia or Mexico, they also export oil and natural gas, they will be very keen to substitute these fossil fuels out of domestic large boiler uses and switch to coal. Coal imports at certain locations can serve them as well or better than the development of domestic coals. Oil and gas importers, such as India, will not use these fuels in the generation of large scale power requirements, but will rely on coal. *A fortiori*, one can argue that unless infrastructure costs for moving coal to markets are very high relative to the scale of demands, oil and gas will disappear from large boiler use except for peaking requirements.

This implies major revisions in OECD and developing country energy policies, which now constrain international coal shipments to minimal levels (Table 4.11.9). To date, most developed countries have favored domestic coal expansions over imports, consistent with their high internal energy price policies and visions of the eventual dominance of nuclear power. If these countries return energy allocation to free market forces, the presumption is that coal imports will become increasingly attractive as nuclear power alternatives become less attractive. This holds for most of the OECD countries including Japan, because their industrial regions are quite accessible to coal imports and they are deficient in low-cost domestic coal resources. Developing countries likewise can save enormous sums of hard currency by switching to coal imports in the short run, even if, eventually, they complete infrastructure plans to develop less accessible domestic coal resources in the long run.

Process evaluation comparisons and spatial modeling can aid in the evaluation of alternative coal sources and power plant location following the current practice of developed country utilities. Such reassessments begin whenever oil prices show signs of instability. New plants can often be designed to accommodate coal instead of oil and it is easy to obtain coal without hard currency. The substitution of oil out of large utility markets may become a major element in the stabilizing of world oil prices. Indeed, it may force an oil price decline. In short, given the state of the solid markets and technology, few countries will ignore direct coal alternatives in their large boiler applications for the next 20 years.

Utility Costs and Rates: The assumed ability to pass costs endlessly through to the consumer is likely to prove limited, but it has to date contributed to cost overruns and excessive capital costs. Two alternatives may be suggested, neither will prevent cost increases, but both will contribute to more realistic initial cost assessments and each may lead to higher capital cost control during power plant construction.

Public service commissions could set a cap on costs that may be recovered through rates. At any higher cost, stockholders absorb all or a major portion of the excess. Alternatively, the commissions can refuse a certificate for any construction that is not a fixed cost turnkey operation. Primary contractors would have an interest in cost control and realistic bidding. The latter would provide both realistic inter-fuel comparisons and a reasonable basis on which to decide whether the facility is needed.

Coal Prices and Finance: The financial implications for the solid fuel industries are complex. The nuclear industry, like the electric industry in general, has been adversely affected by demand levels far below those projected. In the United States, the excess capacity does not bring a price reduction to restore equilibrium. Rather, regulations encourage electricity prices which are related to production cost. To the extent possible investors returns, while relatively low, are guaranteed. As a result, new construction costs in times of excess supply are driving some utility rates up by far more than the general inflation rate. Theory suggests a resultant shift of consumption will occur away from these regions, away from electric power, and towards substitute energy forms.

An alternative policy would be the deregulation of utility pricing, returning power to the status of a normal market commodity. Rates would then rise with excess demand and fall with excess supply. Administration of service and price behavior would remain the provenance of the commissions, but there would be more flexibility in their rulings. The rigid links protecting investors would be severed. Capital costs might rise, but construction of plant in excess of profitable expectations would be unlikely.

SUMMARY

Few areas of prediction have produced larger errors relative to the size of forecasting efforts than energy economics. This has led some analysts to speculate that the long term forecasting of fuel demand and shares by sources may be a practical impossibility. Accordingly, our approach in this review of the solid fuels and their roles has placed emphasis on the past errors in solid fuels analysis. This favors coal prospects as it weights the future role of nuclear power less favorably. However, other assumptions reversing the roles and the record for nuclear power and the synfuels are clearly conceivable. Unfortunately for nuclear, many of the optimistic scenarios imply costs and price patterns for conventional coal, gas and oil uses which are inconsistent with current resource bases or with the goals of efficient domestic markets and free trade. For example, it is less likely in this view that nuclear will grow as predicted to coal's disadvantage than it is that coal uses will slow in the face of oil price declines. The implications for the adequacy of solid fuel reserves and resources spelled out here are not the only plausible ones; they appear, however, to be the most likely.

Critics of DOE energy policies generally, or of one or another solid fuel scenario for electric power expansion in particular, should take heart in the fact that, if deregulation is pursued as vigorously as proposed in this review, the effects of DOE or economists erroneously forecasting the share for a favored fuel will be minimized and greatly reduced. The reason is simply that utilities will have very little to gain in practice by considering the national modeling exercise which purports to find "optimal shares" for fuel sources, "optimal transportation networks," or the like. With fewer subsidies, and far less chance that inefficient or noneconomic solutions can be passed on to consumers, the investment decisions of utilities will be reduced to the much more limited problems of selecting a future type of capacity among a limited set of alternatives.

In the case of coal alternatives, given different characteristics with respect to sulfur, ash and moisture, decisions will depend in large measure on the available discounted funds differential between some feasible set of local coals and the interregional transport costs of other desirable distant coals.

Expected increases in coal production and consumption as well as the shift among coal mining regions will define quite well all the important remaining questions concerning coal carrying capacity, both the overall coal hauling ability and capacity along specific routes, in the comparing of relative advantages of particular fuels or modes. Except for bottlenecks, particularly in the east but projected for some western lines, overall transport capacity will not be seriously questioned, particularly if coal consumption increases smoothly over the next two decades. There is, however, little question that some existing rail routes are seriously inadequate. Therefore, change characterizes the actual and potential transport network. For instance, as western coal development proceeds, new routes are being developed or augmented, e.g., in the Great Lakes for northeastern and north central U.S. shipments, as well as shipments to the predominantly oil and gas burning south. Again, as sulfur regulations change, the specific form of their implementation changes the competitive position of western coal moving in interregional markets.

This review has found no lack of previous or current studies relating to coal choices or to coal transportation models for energy supply analysis. Most of the recent EPA funded regional assessments have major sections devoted to coal transportation. The coal slurry pipeline issue developed a plethora of cost and route specific studies. Finally, there is a considerable body of technical data and analyses by both independent researchers and corporations devoted to particular coal decision factors.

There is no need to survey these results, nor is an undiscriminating repackaging of existing lore warranted. What is needed by the decision-making utility is a partitioning, synthesis and extension of existing work. This should include analyses of new and foreseen developments in technology. The final result need be related only to a set of reasonable and defensible forecasts or scenarios articulated for the individual user. To be useful, the analysis must be soundly based on engineering design and costing coupled with the economic and financial analysis of existing utility markets. While no single methodology or systemization may be expected to provide all the answers without undue simplification, a series of connected, interrelated methodologies can be displayed which are sufficiently detailed and transparent to provide the user with a hands-on capability for comparison among facility alternatives. Such analyses can easily provide the answers concerning total carrying capabilities and route/mode system specification. In this ground-up approach, the primary concerns relate to industry structure, the availability of the firm's funds, the levels of investment it requires over time, route capacity and capability, and long-range forecasts, including construction times of future expansion as a transportation company or a specific utility would see them.

While disagreement concerning assumptions and data might easily be expected in economic and financial considerations, one might believe that at least in two areas, the choice of engineering designs and engineering costing, a large measure of commonality among users will prevail. As pointed out here, public studies in both areas have been ill defined and subject to widely conflicting claims. In an era of over regulation, this has led to confusion and a component of argument at cross purposes. With deregulation, all of the available data and assumptions on a set of comparable bases are provided an individual firm, and it will determine which arguments are real, provide a common basis for comparisons among choices, set a rational basis for estimating capacities, and provide the initial conditions from which to appraise its future.

Specific difficulties of analysis will remain, and a summary of major problems in selecting fuels may be useful. In the cost engineering tasks, as the history of large errors and cost overruns attest, serious problems of estimation exist. Some of these relate to site specific differences. In the financial analysis, when the experience of utilities, rather than engineering costs, is the basis for estimates, the problem of distinguishing user costs from the value added from inputs will remain. Both of these types of problems are worrisome in the negotiation of contracts and in planning for particular utilities.

There are, of course, many important sources of variance in estimates, but frequently the difficulties can be reduced to accounting problems associated with joint costing, joint production, diseconomies of a technical nature (e.g., scale) or of market organization (e.g., restrictions in the factor markets). There are always problems in the delineation of rents or user costs.

In the global optimization programs sponsored by DOE, it is simply not possible to deal with these problems in the technical constraints. This is because the mathematical means of describing the technical constraints must meet the requirements of the DOE programs. In contrast, the individual firm will easily deal with such difficulties by the careful selection of accounting conventions and in its costing formulation.

In the measurement of risks, attempts to identify fully the general nature of variances in line performance or financial flows will remain extremely difficult. Special care in the formulation of important risk factors in decision models must be developed, therefore, to support the individual utility's decision making. On the other hand, precise real-life decision making by economic agents under conditions of uncertainity is itself highly varied. Consequently, a system that is to be of use to many utilities, yet efficient for each utility's use, would seem desirable and might provide considerable flexibility in the ways that risk can be represented. Examples of the variety of formulation range from analysis of sensitivities by differing assumptions (scenarios) to the use of conditional probabilities or conventional stochastic simulation. Sometimes it will aid analysis if the expected utilities of the individual agent can be specified. In other cases, the computation of risk adjusted values can be employed. The use of Markowitz capital cost equations and risk evaluation offer yet other approaches. How these various techniques might be judiciously applied to the highly specified risks and uncertainties are important questions. Different approaches will be required for financial options and the specific business environments faced by regional utilities.

In public utility pricing and costing areas, many variations in estimates of project costs come from the use of different accounting conventions. Often this occurs without the realization on the part of engineers or economists doing the work that the cash flows and rates of returns are particularly sensitive to the method of accounting selected. This problem is further complicated whenever the choice of mode must be integrable with the eventual calculation of system busbar costs.

Finally, for the problem of distinguishing user costs, all analysis should be cost based. Transportation charges are tariffs (rates or prices) whether the system is owned by the utility or the service merely contracted. Costs merely set a level below which long-term prices cannot fall. Given competition among only a few alternatives, the least cost competitor should win the contract, but need charge only a small sensible discount below that lowest price which can be charged by the next to the least cost competitor. Depending on its bargaining position, the utility may achieve large cost savings by choosing a particular mode, or it may achieve virtually none. Nevertheless, deregulation in most cases encourages the utility to develop a cost analysis as the basis both for its bargaining position and for a reasonably close approximation to its final fuel choices and power prices.

All this is not to say that such an emphasis on decision making by individual utilities will eliminate the need for national or regional forecasting. Forecasts will continue to be made, and these will contain a large element of risk. Nevertheless, commercial decisions on a current basis will best reflect the experts' future expectations, and this is the strength of the solid fuel policy analysis presented in this review.

Notes

1. Among the early and better expositions of coal's problems are Richard L. Gordon's, *U.S. Coal and the Electric Power Industry*, RFF and the John Hopkins University Press, Baltimore, Md., 1975, and *Coal in The U.S. Energy Market*, Lexington Books, MA, 1978. Cf. also his *The Economic Analysis of Coal Supply: An Assessment of Existing Studies*, Electric Power Research Institute, May 1975 to June 1979.
2. The best source describing these techniques is DeVerle Harris' *Mineral Resources Appraisal*, Vol. 1, Oxford Geological Sciences Series, Oxford University Press, Oxford, 1984, covering the concepts and methods of measuring mineral endowment, resources and potential supply. The U.S. uranium potential is covered in case studies. The key questions of statistical inference and optimizing techniques in all potential supply appraisals, including the case of coal, are discussed in *Future Resources: Their Geostatistical Appraisal*, Richard Newcomb (editor). Papers and Proceedings of the Conference on Geostatistical Appraisal of Future Mineral Resources, West Virginia University Press, Morgantown, WV, 1982.
3. Cf. the analysis of coal slurry and rail costs by Michael Rieber and S.L. Soo in *Comparative Coal Transportation Costs: An Economic and Engineering Analysis*, Volumes 1–3, National Technical Information Service, (1977), NTIS PB 274-379/AS, PB 274-380/AS, PB 274-381/AS, and *Coal Slurry Pipelines, A Review and Analysis of Proposals, Projects, and Literature*, Electric Power Research Institute, EA 2546 (1982).
4. Henderson, J.M. *The Efficiency of the Coal Industry*, Harvard University Press, Cambridge, MA, 1958.
5. U.S. Department of Energy, *An Assessment Report on Uranium in the United States of America*, Grand Junction Office, Colo., October 1980, GJO-111(80), p. 3.
6. U.S. Department of Energy, *United States Uranium Mining and Milling Industry*, DOE/S-0028, May 1984, Table 3, p. 20.
7. U.S. Department of Energy, *An Assessment Report . . .*, Table 51, p. 119.
8. *Ibid.*, Table 52, p. 119; p. 118.
9. *Ibid.*, p. 117.
10. *Idem.*
11. *Idem.*
12. *Ibid.*, p. 116.
13. *Wall Street Journal*, 21 March 1984.
14. Comptroller General, *Status and Commercial Potential of the Barnwell Nuclear Fuel Plant*, U.S. General Accounting Office, GAO/RCED-84-21, March 1984, p. 1.
15. *Ibid.*, p. 34.
16. *Ibid.*, p. 11.
17. *Ibid.*, pp. 18–19.
18. *Ibid.*, p. 20.
19. *Idem.*
20. *Ibid.*, p. 20.
21. Dean, S.O., "Overview of Magnetic Fusion," in Dean, S.O. (ed.), *Prospects for Fusion Power*, New York: Pergamon Press, Inc., 1981, p. 3, Table 1.1.
22. *Ibid.*, p. 1.
23. Singleton, R.H., *Lithium*, U.S. Bureau of Mines, Mineral Commodity Profiles, September 1979, p. 24.
24. Dean, S.O., *Op. Cit.*, p. 3, Table 1.1.
25. Wilson, C.L., *Coal-Bridge to the Future: Report of the World Coal Study*, Vol. 1, Harper and Row, Ballinger, MA, 1980.
26. Henderson, James M., *The Efficiency of the Coal Industry*, Harvard University Press, Cambridge, MA, 1958.
27. Jevons, S., *The Coal Question*, London, 1865.

28. Taussig, F.W., "A Contribution to the Theory of Rail Rates," *Quarterly Journal of Economics*, 1891.
29. U.S. Department of Energy, *The National Coal Model: Description and Documentation*, Federal Energy Agency, by ICF, Inc., Washington, D.C., 1978.
30. ICF, Inc., "Coal Electric Utilities Model Documentation," Federal Energy Agency, Washington, D.C., 1980.
31. Charles River Associates, Inc., *Coal Price Formation*, Electric Power Research Institute, EPRI-EA 497, 1977.
32. Zimmerman, M. G., The U.S. Coal Industry: *The Economics of Public Choice*, MIT Press, Cambridge, MA, 1981.
33. Newcomb, R.T., and J. Fan, *Coal Market Analysis Issues*, Electric Power Research Institute, EPRI-EA 1575, 1980.
34. Argonne National Laboratory, *An Integrated Assessment of Increased Coal Use in the Midwest: Impacts and Constraints*, ANL/AA-11, 1977.
35. Rieber, M., and S.L. Soo, *Comparative Coal Transportation Costs*, U.S. Bureau of Mines and Federal Energy Administration, Vols. 2 & 3, Contract BM/JO1-66163, 1982.
36. Dantzig, G., "Maximization of a Linear Function of Variables Subject to Linear Inequalities," in T. Koopmans (editor), *Activity Analysis of Production and Allocation*, John Wiley, New York, 1951.
37. Samuelson, Paul, "Spatial Price Equilibrium and Linear Programming." *American Economic Review*, 1952.
38. Price, J. et al., *A Review of Coal Supply Models*, for the U.S. Department of Energy, Contract De-AC01-81FE-16115, Resource Dynamics Corporation, McLean, VA 1982.
39. It was not so apparent over the immediate post-war period that the coal industry was in dramatic secular decline. In the light of historical confirmation of this corresponding dynamic trend, the foregoing of rents in Appalachia was not so much a sign of inefficiency as an efficient way to close out Eastern mine reserves at variable costs per ton with the object of shutting down the industry in the long run.
40. In contrast with the 1950's, The secular trend in the coal industry today is expansionary. This defines the current slump as an indication of new capacity addition outpacing demand advances. In view of these dynamic prospects, it is inefficient for the industry's producers' surplus to be passed on to users to the extent profits provide capital for future growth and signal efficient allocation. Cf., Richard Newcomb, "Modeling Growth and Change in the American Coal Industry," *Growth and Change*, Vol. 10, Lexington, KY, 1979.
41. Cottle, R. and G. Dantzig, "Complimentary Pivot Theory of Mathematical Programming," in *Linear Algebra and Its Applications*, American Elsevier, 1968.
42. See Harris, D.P., *Mineral Resource Appraisal*, New York, Oxford University Press, 1984; Harris, D.P., "Mineral Resources Appraisal and Policy—Controversies, Issues and the Future, *Resources Policy*, July 1984; Harris, D.P. and L. Chavez, "Modeling Dynamic Supply of Uranium—An Experiment in the Integration of Economics, Geology, and Engineering," 18th International Symposium on Applications of Computers and Mathematics in the Mineral Industries, London, Institute of Mining and Metallurgy, March 26–30, 1984; Harris, D.P. and L. Chavez, "Crustal Abundance and a Potential Supply System," *Systems and Economics for the Estimation of Uranium Potential Supply*, Part III, U.S. Department of Energy, 1981; Harris, D.P., "Crustal Abundance Modeling of Mineral Resources: Some Recent Investigations," 27th International Geological Congress, C. 20.1.3 Geostatistics, Moscow, Russia, August 10, 1984.
43. Energy Information Administration, *1982 Survey of United States Uranium Marketing Activity*, DOE/EIA-043, September 1983, p. 2.
44. Atomic Industrial Forum, *Nuclear Industry*, May 1972, p. 31.19.
45. Rieber, M. and R. Halcrow, *Nuclear Power to 1985: Possible Versus Optimistic Estimates*, National Science Foundation, November 1984 National Technical Information Service, NTIS PB-248-061/AS, Section IVB.
46. *Ibid.*, p. 122.
47. *Ibid.*, Table IV-7, p. 126.
48. *Ibid.*, p. 127.
49. Bleistine, P.A., "Current Demand Estimates, 1981–2000., *The Future of the U.S. Uranium Industry*, Nuclear Assurance Corporation, Uranium Colloquim 1981, Grand Junction, Colo., October 1981, pp. 1–2.
50. *Ibid.*, figures 3–4, pp. 7–8.
51. *Ibid.*, pp. 3–5.
52. Energy Information Administration, *Commercial Nuclear Power: Prospects for the United States and the World*, DOE/EIA-0438, November 1983.
53. U.S. Department of Energy, *United States Uranium Mining and Milling Industry*, DOE/S-0028, May 1984, p. XV.
54. *Ibid.*, p. xx.
55. Hahne, F.J., "Current Supply Estimates—1981–2000," Nuclear Assurance Corporation, *Op. Cit.*, figures 3–5, pp. 7–9; p. 3.
56. Energy Information Administration, *1982 Survey . . .*, *Op. Cit.*, p. 37.
57. *Ibid.*, p. 41; figure 7, p. 43.
58. *Ibid.*, p. 26.
59. Energy Information Administration, *World Uranium Supply and Demand: Impact of Federal Policies*, DOE/EIA-0387, March, 1983.
60. U.S. Department of Energy, *United States Uranium Mining . . .*, *Op. Cit.*
61. *Ibid.*, Table 20, p. 64.
62. Energy Information Administration, *1982 Survey . . .*, *Op. Cit.*, p. 32.
63. *Ibid.*, pp. 34–35.
64. Duschesneau, Thomas D., *Competition in the U.S. Energy Industry*, Ford Foundation Energy Policy Project, Lippincott, Ballinger, Cambridge, MA, 1974.
65. U.S. Department of Energy, *United States Uranium Mining and Milling Industry*, DOE/S-0028, May 1984, figures 10, 16, 17; pp. 51, 61, 62.
66. *Ibid.*, Table 16, p. 37.
67. U.S. Department of Energy, *Investment in Exploration by the U.S. Uranium Industry*, DOE/EIA-0362, September 1982, Tables 3–4, pp. 20–21.
68. U.S. Department of Energy, *United States Uranium Mining . . .*, *Op. Cit.*, Table B-2, p. 80.
69. *Ibid.*, Table 9, p. 27.
70. *The Economist* (London), 30 June 1984, p. 71–2.
71. Duret, M.F., *et al.*, *The Contribution of Nuclear Power to World Energy Supply, 1975 to 2000*, Conservation Commission, Report on Nuclear Resources, World Energy Conference, Study on World Energy Supply, Ottawa, Canada, July 1977, p. IV–5.

72. Energy Information Administration, *Commercial Nuclear Power . . ., Op. Cit.*, Tables C–G, pp. 69–90.
73. Office of Technology Assessment, *Nuclear Power in an Age of Uncertainty*, OTA-E-216, February 1984, Table 17, p. 114; Table 23, p. 180.
74. Energy Information Administration, *1982 Survey of United States Uranium Marketing Activity*, DOE/EIA-0403, September 1983, Table 10, p. 19.
75. *Ibid.*, Table 6, p. 13.
76. Energy Information Administration, *Thermal Plant Construction Cost and Annual Production Expenses—1980*, DOE/EIA-0323 (80), June 1983, Table 4, p. 12.
77. Office of Technology Assessment, *Op. Cit.*, Figure 21, p. 89.
78. *Ibid.*, Table 26, p. 195.
79. *Ibid.*, Table 19, p. 117.
80. *New York Times*, 4 December 1983.
81. Rieber, M. and R. Halcrow, *Nuclear Power to 1985: Possible Versus Optimistic Estimates*, National Science Foundation, November 1984, National Technical Information Service, NTIS PB-248-061/AS, p. 4.
82. *Ibid.*, Section III.
83. *Ibid.*, Table I-3, p. 9.
84. *American Mining Congress Journal*, 15 October 1983, p. 3.
85. *New York Times*, 8 February 1984.
86. Both reserves and resources are at least implictly defined in terms of cost, though a geologist should be able to make a non-cost estimate of endowment. The shift of material from the resource to reserve category is clearly price-related, as is the search for new reserves. The potential supply function represents an hierarchical ordering of reserves and resources on the basis of retrieval cost by the opening of new deposits and the expansion of existing deposits. As known deposits are depleted (become more expensive to mine) they are shifted up the hierarchy of potential supply. As newly discovered deposits prove to be low cost, they are inserted in their proper order. As price increases, higher cost deposits can be commercially exploited and the chance of profit from the finding on new reserves encourages more exploration.
87. "Linowes Panel Faults Coal Leasing Policy," *American Mining Congress Journal*, 25 January 1984, p. 2.
88. Rieber, M. and R. Halcrow, *Nuclear Power to 1985: Possible versus Optimistic Estimates*, National Science Foundation, November 1984, National Technical Information Service, NTIS PB-248-061/AS, Section IB.
89. Newcomb, R. and Bruce Bancroft, "Capital Requirements and Busbar Costs for Electric Power in the Ohio River Basin: 1985 and 2000," *Ohio River Basin Energy Study*, for the Environment Protection Agency, University of Illinois, Urbana, IL, 1980.
90. DOE/EIA 035612, "Projected Costs of Electricity from Nuclear and Coal-Fired Power Plants," Vol. 2, November, 1982.
91. Comptroller General, *Status and Commercial Potential of the Barnwell Nuclear Fuel Plant*, U.S. General Accounting Office, GAO/RCED-84-21, March 1984, pp. 21–22.
92. Rieber, R. and R. Halcrow, *Op. Cit.*, Section IIA.
93. Energy Information Administration, *1982 Survey . . ., Op. Cit.*, p. 7.
94. *Ibid.*, Table 4, p. 9.
95. *Ibid.*, Figure 3, p. 10.
96. "DOE Predicts Downturn, Then Upward Trend for Uranium Industry," *American Mining Congress Journal*, 21 June 1984, p. 11.
97. Hellman, R. and C.J.C. Hellman, *The Competitive Economics of Nuclear and Coal Power*, Lexington Books (1983), pp. 4–6.
98. *Wall Street Journal*, 12 February 1984.

4.12

The Oil and Gas Industry: Regulation and Public Policy

Robert T. Deacon and Walter J. Mead*

Oil and gas together are the most important energy sources consumed in the United States. In 1983, 67% of total energy consumption was represented by these two resources (43% was oil and 25% natural gas), while coal accounted for only 23% of U.S. energy use. All other sources (hydroelectric, nuclear, geothermal, wood and other waste products) accounted for only 10% (U.S. Department of Energy, 1983, p. 1).

In this chapter we address the issue of regulation in the oil and natural gas sectors of the U.S. energy industry. In addition, we analyze the cartel characteristics of the Organization of Petroleum Exporting Countries (OPEC). If OPEC is an effective cartel, then U.S. oil as well as substitute energy prices, production, and consumption will be affected by cartel action. Further, as the nation learned in the 1970s, major changes in oil prices produced major disturbances in other sectors of the U.S. and world economies.

Consequently, it is important to determine whether the roughly ten-fold increase in nominal prices that occurred in the 1970s was the result of cartel action, the result of normal economic forces adjusting to the fact that the world was running out of cheap oil, or some combination of the two. Furthermore, the presence of OPEC-associated oil price increases is related to regulations that were imposed on the energy sector during the decade of the 70s.

THE RATIONALE FOR GOVERNMENT INTERVENTION

Economists have traditionally judged the state of an economy to be efficient if it is impossible to improve the welfare of any one member of society without harming someone else. Any economy that satisfies this condition is said to be Pareto efficient, after the Italian economist Vilfredo Pareto who first gave rigorous attention to such questions (see Chapter 4.10). It is easy to see why an economy that fails to satisfy this condition could not be efficient; by definition, it would be possible to alter the allocation of resources in a way that improves everyone's welfare.[1] Much of the appeal of this weak efficiency criterion is that it avoids comparisons of levels of well being among different individuals. The cost of such generality, however, is that in any given economic system a wide array of different Pareto efficient states will typically be attainable. In the simple Robinson Cursoe economy of the undergraduate textbook, for example, *efficiency* could just as easily be attained in the case where most of the island's resources are owned by Friday as in the case where they are under Crusoe's control. However, the distribution of welfare among the two inhabitants would obviously be quite different in the two situations. Such *equity* issues cannot be settled on objective or scientific grounds. To decide which distribution of welfare is socially "best" an explicit value judgment is needed.

It is useful to keep in mind the preceding distinction between efficiency and equity issues when analyzing the role of government in a market economy. Decentralized market processes will not necessarily result in a distribution of welfare that society considers just. For this reason a potential role for government lies in the redistribution of income among individuals. In some circumstances, however, market pro-

*Professors of Economics, University of California, Santa Barbara.

cesses can solve the efficiency problem. A fundamental result from modern welfare economics is that, in an economy where all *markets are competitive*, and where *markets exist* to allocate all goods and services valued by society, an equilibrium allocation of resources is Pareto efficient.[2] In such a world the competitive behavior of various market participants will solve the efficiency problem without any centralized direction or coordination. The efforts of government in this case could be confined to the ascientific problem of achieving a socially just distribution of welfare.

In actual economies, however, markets for some valuable goods and services simply may not exist. Such phenomena are associated with the concept of externalities, as follows. In other instances, markets may be present but competition absent. In either case, the resulting economic equilibrium will not be Pareto efficient. Thus, these two additional dimensions, correcting for imperfect competition and for market failure, must be included among the potential functions of government in a market economy. It must be point out, however, that government intervention to correct such problems is by no means costless. Among the most obvious costs are those reflected in the already large budgets and staffs of federal, state and local government regulatory agencies. Less obvious, but perhaps more important, are the costs incurred in the private sector by those forced to comply, and the inefficiencies that can arise from the politicization of economic activity that often accompanies regulation. The economic losses become particularly severe when political prowess and legal expertise supplant productive efficiency as criteria for economic survival. As the analysis in this chapter points out, such instances are common in the area of petroleum industry regulation. Recognizing the costs of such intervention, then, leads to the prescription that intervention is warranted only for those market failures and distributional imperfections that are large relative to the full costs of correcting them.

Each of the preceding three governmental objectives (redistribution of income, correction for market failure, and promotion of competition) has inspired public policy toward the oil and gas industry. The redistribution motive, away from producers and toward consumers, is apparent in the natural gas price regulations that have been in place since the 1950s, and in the crude oil price controls and the Windfall Profit Tax imposed since 1973. Paradoxically, the market demand prorationing regulations and crude oil import controls that characterized federal energy policy prior to 1973 effectively redistributed income in the opposite direction—away from consumers and toward producers and resource owners.

Historically, the promotion of competition has also been an important component of public policy toward the oil and gas industry. The most famous episode was Theodore Roosevelt's successful campaign to break up the old Standard Oil trust in 1911. In more recent times the pursuit, or even the threat, of antitrust activity and vertical dissolution has, no doubt, had an effect on the way that oil and gas producers do business. As of 1984, the tradition may be in a reversal phase, as evident by government approval of the Texaco-Getty acquisition. Downstream from the wellhead the issue of competition arises in a different form. It is often claimed that pipeline transportation and natural gas distribution are "natural monopolies," activities in which the structure of cost precludes effective competition. Here the policy approach has been to regulate prices and operating levels with the intent of preventing the exercise of monopoly power.

Finally, the production and consumption of fossil fuels often has adverse effects on the environment. Production, transportation, and refining of crude oil can result in significant air pollution. Marine transport and offshore operations sometimes result in oil spills. The eventual combustion of refined petroleum products in homes, automobiles, and power plants is the most important source of air pollution in most regions. The economic issue is not that such environmental degradation occurs, but that, due to market failure, the firms and individuals responsible for it have little incentive to incorporate such damages in their decision-making. Without government action, the cost that pollution imposes on society will not be reflected in their profits and losses. The notion that these factors are "external" to the relevant decision maker, i.e., borne by some other party, has resulted in the term "externality" to describe such phenomena.

At a more fundamental level, however, this problem can be traced to the fact that alternative uses of the environment, whether as a dump for waste products or as a medium for human enjoyment, are not priced and allocated by the market system. If they were, then those wishing to use an air shed for disposal of the waste products of combustion, for example, would be forced to bid against all other potential users. In this fashion, the value of these alternative uses would be reflected in the polluter's decisions. Historically, the government response to such problems has been to regulate, in detail, the ways in which energy products are produced and used in order to control these external effects. However, there are alternative methods of control.

In judging the appropriate role of government in any sphere of economic activity, a necessary first step is to accurately assess the magnitude of the problem in question. This is clearly necessary to determine whether any intervention at all is warranted. Thus, for example, an analysis of the degree of competition present in the energy industry, or the severity of a particular environmental problem associated with the use of energy, are logical prerequisites to the formulation of actual policies. An efficient government policy will avoid regulation in cases where the costs of dealing with a problem exceed the damage to be controlled. Second, governments can typically select among a range of policy instruments to achieve a given goal. Environmental protection can be pursued through detailed regulation of behavior, reliance on the judicial system, a tax-subsidy approach, or the implementation of market-like institutions. Similarly, redistributive objectives can be approached by regulating prices, by imposing taxes, or by altering the level of competition. Because the costs of achieving a particular goal can vary widely with different instruments, it is important to formulate efficient policy responses.

Attempts to implement the preceding concepts, either in the pursuit of efficiency or a more socially desirable distribution of income, always encounter difficulties. Many of these arise from inherent problems in the identification of effects and the estimation of relevant magnitudes. The problem of formulating policy to promote *competition* is a case in point. There is usually little qualitative difference between the behavior of monopolies and competitive industries; both respond to shifts in market conditions, cost changes, and technical advances in much the same fashion. Moreover, disclosure rules and differences between accounting conventions and relevant economic concepts often make comparisons of profitability difficult or inconclusive. As a result it is generally difficult to establish the presence of monopoly power or to determine its strength.

The identification and estimation problems that arise with *externalities* are, if anything, even more severe. By definition, such analysis requires an assessment of the value of goods and services that are allocated outside of markets. The kinds of externality problems associated with energy can be enormously varied, ranging from air pollution and human health, to recreation, scenic amenities, and national security.

In the area of *redistributive policy*, answers to seemingly simple questions can also prove elusive. As pointed out elsewhere in this chapter, the legal liability for paying taxes has little to do with the distribution of true tax burdens. Taxes imposed on one set of agents will often be shifted, through price changes, to other segments of the economy. Even regulation of prices in order to benefit consumers may not have clear-cut distributional effects. Price controls may reduce available supplies to the point where consumers are actually made worse off.

It should also be pointed out that the neat conceptual distinction, between questions of efficiency in the allocation of resources and equity in the distribution of welfare, often becomes blurred in practice. Policies intended purely for redistribution typically have important efficiency effects as well. This is clearly true of price regulation, and it appears to be true of all important taxes as well. Similarly, policies designed to correct for inefficiencies, whether due to market failure or imperfect competition, will have important distributive effects. For example, an efficient pollution control policy could theoretically be implemented either by regulation or by a tax-subsidy scheme. However, the distribution of income under these two alternatives would be quite different. Likewise, the oil import quotas adopted by the United States in the late 1950s, though ostensibly intended to promote national security, had the added effect of enriching domestic energy producers. These examples are intended to point out the danger that those who seek income redistribution for their own benefit may do so under the guise of policies apparently addressed to security or efficiency issues.

The following three sections survey government policy toward the oil and gas industry in each of the three areas of government intervention, redistribution, correction for market failure, and promotion of competition. Because the redistributive and market failure aspects of government policy are treated extensively elsewhere in this chapter, they are only briefly discussed here. The discussion of policy toward competition is, however, more extensive.

GOVERNMENT POLICY TO REDISTRIBUTE INCOME: A SURVEY OF ISSUES AND POLICIES

Taxes and subsidies are the traditional means by which income is transferred between segments of society, although price ceilings and price floors are also frequently important in altering income distributions. Both of these classes of policy instruments are important in oil and natural gas markets. In general, taxcs transfer income away from those in the private sector who bear the tax burden, toward the recipients of government expenditure. Viewed in this light all tax levies on oil and gas production, whether in the form of severance taxes, property taxes, windfall profits taxes, or corporate income taxes, are redistributive measures.

Of course, all sectors of economic activity are taxed and several important taxes are levied broadly across virtually all individuals and businesses. When discussing redistributive policies toward a particular industry, therefore, it is appropriate to focus on those aspects of tax policy that are special or unique to that industry. Oil and natural gas taxation is the object of a variety of special considerations. The production or severance taxes that are levied on oil and gas produced in most states are not found outside the natural resource sectors of the economy. Likewise, the federal crude oil windfall profit tax appears to have no counterpart in other markets. Furthermore, application of the corporate income tax to oil and gas production has historically been characterized by very favorable treatment in the form of generous deductions and exemptions.

The actual pattern of distribution that results from energy taxation is often difficult to determine. This is a theme stressed repeatedly in subsequent sections of this chapter. The specification of legal liability for tax payment is typically irrelevant in assessing the true distribution of the burden. Determination of who bears the tax burden depends on the definition of the tax base (e.g. whether on production, income, or assets in a given industry), and on supply and demand conditions in the market for the taxed good. Thus, for example, the distribution of the tax burden will change if market conditions change, even if there is no alteration in the tax law. Similarly, the distribution of burdens, e.g. between producers and consumers, could be quite different for taxes levied on crude oil versus natural gas. In subsequent sections, the general question of ''tax incidence'' in oil and gas industries is examined in much greater detail.

Redistribution of income, away from crude oil producers and toward consumers, was the apparent motive from the crude oil and refined product price controls that dominated U.S. energy policy in the 1970s. The original controls adopted in 1971 were economy-wide and reflected a national concern over inflation. However, the rapid crude oil price increases of 1973–1974, repeated in 1979–1981, and the ensuing wealth transfers from consumers of energy products to the owners of crude oil reserves, created political pressure for specific controls on crude oil prices. When these controls were removed in 1981 the redistributive policy was continued in the form of a Windfall Profit Tax, a levy designed to capture a portion of the gains to producers that accompanied decontrol. The history of natural gas price controls is much longer, beginning in the 1950s and continuing to the present. In both cases, the redistribution obtained via price controls came at the expense of substantial resource misallocation.

Aside from the central redistributive instruments of taxation and price controls, the entire spectrum of government policy toward the petroleum and natural gas industries is replete with special treatment for favored classes of producers. Examples are the special considerations given to independent producers and small refiners in taxation, environmental regulation, price controls, import restrictions, and other policies. Again the apparent intent is to redistribute income, in this case by granting a competitive edge to small operators. It is important to emphasize, however, that such transfers, indeed all forms of redistribution, are non-neutral. Such policies always affect incentives to invest and produce, and accordingly alter the activity and even structure of affected industries. In such cases the apparent quest for fairness can result in a proliferation of small firms and a degree of inefficiency in the operation of the industry.

The importance of distributional issues as a force in national energy policy can hardly be overemphasized. In summarizing the recent history and current status of government policy, Kalt (1981) concluded

> "The domestic energy 'crisis' is, far more than anything else, a quarrel over income distribution—and U.S. energy policy is the outcome of this quarrel." (Kalt, 1981, p. 243)

In subsequent sections, the specific analysis of U.S. energy policy regarding taxation, price controls, and other measures serves to reinforce this conclusion.

INTERVENTION TO CORRECT FOR MARKET FAILURE: AN OVERVIEW OF POLICY APPROACHES

The production and use of oil and gas are characterized by a variety of externality and common property problems, phenomena that arise because markets fail to exist for certain resources or activities. A variety of air and water pollution problems can be attributed to the use of fossil fuels. In the absence of policy to the contrary, or action wherein involved parties either contract to internalize or bring legal action to force internalization, these social costs will not be reflected in the profits and losses of firms responsible. It has also been argued that reliance upon foreign supplies of crude oil imposes a national security cost on the United States, a cost that is not fully recognized or borne by those who import petroleum from abroad. A third area of possible market failure arises from the fact that property rights to subsurface oil and gas reservoirs are often difficult to specify and enforce. In the absence of public policy or a redefinition of property rights, the consequence can be over-rapid and wasteful development of reserves.

Each of these externality or market failure phenomena has given impetus to important government involvement in the industry. At a conceptual level such problems can be approached in a variety of ways.[3] Traditionally, however, governments in the United States have attempted to deal with these issues by adopting detailed regulations that specify the types of operations and activities that are or are not permissible. This *regulatory approach* is nowhere more evident than in environmental policy, where it is exemplified by standards relating to requirements for air pollution control equipment, the sulfur content of fuels, auto-emissions and emission controls, and levels of treatment required waterborne wastes.[4] A strict reliance on regulatory solutions to problems of market failure also characterizes policies toward national security and common property.

A distinct alternative to the regulation of externality problems is the use of *corrective taxes* to modify behavior. Intuitively, such taxes perform exactly the same function that equilibrium prices would accomplish if markets for the goods and services at issue existed. Thus, in the field of environmental quality, taxes on pollution would encourage firms to "economize" on the use of air and water environments for waste disposal; enhance the profitability of investments in pollution control or treatment; and raise production costs and final product prices, thereby discouraging consumption of the products ultimately responsible for these environmental effects.

It is sometimes possible to approach these market failure problems in a third way, by creating *market institutions* to promote efficiency. Again, relying on an environmental quality example, application of this method might involve specifying an overall level of emissions permitted in a given watercourse or airshed, and then defining transferrable rights to emit such wastes up the specified limit. Ordinary market processes, i.e. purchases and sales of permits among users, would allocate emission rights to those firms that can use them most efficiently. In this case the price of an emmission permit would perform precisely the same function, in altering incentives, that the pollution tax accomplishes.

Although each of these three policy alternatives has been utilized in various contexts, the regulatory approach has been by far the most important. The legislation of standards and imposition of regulations is the basis of all major federal air and water pollution control policy. It has also been at the center of government attempts to deal with the national security and common property issues described previously. In situations where external effects are predictable and costs can be measured, many economists have argued for the use of corrective taxes instead of direct regulation. One of the major disadvantages of regulation is that it is inflexible and it generally fails to efficiently allocate the activities of those who are regulated. Corrective taxes, on the other hand, incorporate the efficiency of the price system into the responses of those affected.[5]

At present, pollution taxes are most common in the field of local water treatment where agencies sometimes charge firms according to the volumes and characteristics of wasteloads they impose on treatment systems.[6] Although such examples are relatively rare, they have demonstrated the feasibility of the corrective taxation approach for at least some externality problems. Moreover, as noted elsewhere in this chapter, recent experimentation with market-like processes has sparked interest in the use of this third approach to environmental problems.

The regulatory approach has also been dominant in policies adopted to cope with the national security issue. The mandatory import quotas adopted in 1959 placed volumetric limits on crude oil imports to the United States. Because import permits were transferrable, however, this policy allowed the use of market processes to accomplish the allocation task. These quotas naturally reduced imports and raised domestic petroleum prices above worldwide levels. Exactly the same effect on prices and patterns of trade could have been accomplished by placing a tariff, or "corrective tax," on foreign imports. The resulting distributions of income that would result from these two alternative policies are obviously very different. Revenue that would accrue to government with a tariff policy is, in the case of quotas, received in the form of low priced foreign supplies by those allocated rights to import.

The tendency to address allocation problems through regulation is also apparent in policies adopted to deal with the common property nature of some oil and gas reservoirs. State imposed production prorationing schemes, with detailed specifications regarding well spacing and production rates, have dominated government conservation policy. An alternative regime is unitization, wherein a single operator is designated to manage production for an entire reservoir. The unitization alternative is more in the spirit of market-based approaches to the externality problem; historically, however, it has been given less emphasis than regulation.

While in theory, externalities may be corrected by governmental policy action as already suggested, severe limitations constrain this approach in practice. Governmental intervention has its own market failures. Congress is a political body that is concerned primarily with votes (re-election of its members) and only distantly with the economist's goal of efficient resource allocation. Any legislative action by Congress must pass thorough hearings in which various interest groups are afforded an opportunity to voice their objections. Action emerging from Congress in the form of corrective legislation may bear no resemblance to the optimum action needed to internalize externalities. Further, legislation must be administered. Political scientists point out that a bureaucracy tends, in time, to serve the interest of the group to be regulated. As a consequence, further distortion of intended internalization can occur.

INTERVENTION TO CONTROL MONOPOLY POWER

In addition to externalities giving rise to resource misallocation, the presence of monopoly power may produce the same result. Theoretically, a profit maximizing monopolist will restrict his production below competitive levels in order to raise prices above competitive levels and earn monopoly profits. If an industry is competitive, such private monopoly behavior is impossible. To assess the role for government intervention in this arena, it is necessary to determine whether the oil industry is effectively competitive.

In 1970, the U.S. average service station price of regular grade gasoline (including taxes) was 35.7¢ per gallon. By 1981, the nominal pump price had risen 3.7 times to $1.31 per gallon. During the same period, crude oil prices (U.S. wellhead) rose ten-fold from $3.18 to $31.77 per barrel and this increase was largely responsible for the rise in gasoline prices.

Understandably irate consumers were quick to blame monopoly in "big oil," both on the part of OPEC and the large international oil companies. Oil company profits rose, price controls on crude oil and its products plus natural gas price controls notwithstanding. Periodically, during the decade of the 1970s, gasoline shortages appeared and consumers found themselves in long time-consuming lines at service stations. Natural gas shortages persisted throughout this period as well. Economists' explanations that shortages were due to price controls appeared unconvincing.

The late Senator Henry Jackson gained political points by repeatedly denouncing the "obscene profits" being earned by "big oil companies." The Federal Trade Commission responded by filing an antitrust action against the eight largest U.S. oil companies. The man in the street found confirmation in these events that monopoly was the cause of high energy prices and shortages. Yet almost every systematic search for evidence of monopoly conducted in recent years by professional economists has concluded that the domestic petroleum industry is effectively competitive. Similar conclusions are reached for the natural gas and coal industries, as well as for the energy industry in total.[7] Only in the relatively new uranium industry is there much evidence of a level of economic concentration that might lead to monopoly power.

The International Oil Market

The "Seven sisters" as a Cartel: Historically, the presence of some monopoly power in international petroleum is reasonably well established. In a 1952 study, the Federal Trade Commission (1952) concluded that

> the outstanding characteristic of the world petroleum industry is the dominant position of seven international oil companies. The seven companies that conducted most of the international oil business include five American companies [using current names they are Exxon, Standard Oil of California, Mobil Oil Company, Gulf Oil Corporation, and Texaco] and two British-Dutch companies [British Petroleum and Royal Dutch Shell] . . . Control is held not only through direct corporate holding by parents, subsidiaries, and affiliates of the seven, but also through such indirect means as interlocking directorates, joint ownership of affiliates, intercompany crude purchase contracts, and marketing agreements . . . In 1949 the seven international petroleum companies owned 65% of the world's estimated crude oil reserves.

Adelman (1973) showed that in 1950 these same seven firms, plus the French government oil company CFP controlled 100% of world crude oil production, excluding communist countries and North America. Despite these high concentration ratios in production, Adelman (1972) testified that "the cartel ended with the outbreak of war in 1939 and has never been revived."

Beginning around 1950, the entry process brought about a substantial erosion of any market power which the "seven sisters" might have exercised. A recent U.S. House of Representatives committee report (1974) detailed the entry process as follows:

> In 1940, the seven sisters and the CFP were the only significant presence in the Middle East. By 1950, 10 U.S. independents were present though their production was minuscuel. By 1955, seven more U.S. firms were active, the majority of them brought in through the Iranian Consortium. Two more U.S. firms, and two Japanese companies had entered by 1960, to be joined in the next five years by eight more U.S. firms and seven other foreign companies. Between 1965 and 1970, 31 companies entered, leaving (after mergers) a total of the big seven, 24 U.S. independents, 31 foreign companies, and 13 government entities. The 75 companies had grown to almost 100 by 1974.

While private monopoly power in the international oil industry was declining, foreign intergovernmental monopoly power was taking its place. The power of OPEC had replaced any private monopoly by around 1970. Today oil production is totally controlled by host governments in all major producing countries except Canada, Australia, and the United States.

OPEC as a Cartel: Internationally, the competition issue is no longer a question of private monopoly power. Rather, it is whether OPEC is an effective intergovernmental cartel, able to force its member states to reduce output in order to attain higher crude oil prices. There is no doubt that OPEC members hold periodic meetings, debate price policy, and even announce one or more prices to be followed. These events alone are without economic meaning, however, unless accompanied by enforceable output quotas that are consistent with the desired prices.

In the United States, "market demand prorationing" was introduced in the mid-1930s and enforced by the power of the federal government with its courts and enforcement agencies. Under this system, states were empowered to determine allowed levels of production for each producer. Any producer not in compliance with these "allowables" was denied access to interstate oil markets.

Within OPEC, there is no system of enforceable prorationing. A former advisor (Dr. Fereidon Fesharaki, 1979) to the Prime Minister of Iran reported that the U.S. style market-demand prorationing had indeed been tried in OPEC and failed.

The strong conflicts and different interest among OPEC members make effective cartel control virtually impossible. Two of the member countries with the largest oil reserves have been at war since 1980. International political allegiances divide member countries, with Libya strongly allied with the USSR and Saudi Arabia allied with the United States. Some are Arab countries that could unite in the 1972–1973 Arab-Israeli war to reduce output, while non-Arab members increased output. Some countries have abundant cash reserves and oil income, while others are heavily in debt. Pindyck (1978) has indentified "saver countries" and "spender countries" and Moran (1978) wrote of a "core of balancer" countries and "competitive fringe" countries. The first group of countries is able to reduce output if national policy dictates, while the latter can do so only with great pain. Finally, widely differing oil reserves creates differing interests among OPEC members. Saudi Arabia, with reserves estimated at 165 billion barrels, must consider its long-run interests. In contrast, Algeria with reserves estimated at only 8 billion barrels, has very little interest in the fact that monopoly prices today bring forth future energy supplies that are close substitutes for oil, thereby depressing future oil prices. There is no single price policy scenario that is optimal from both the Saudi Arabian and the Algerian perspectives. Given these different interests, it is not surprising that OPEC meetings have failed to produce the essential, agreement and enforcement of output quotas.[8]

There have been two major oil price shocks since 1970. As of mid-1971, prior to the imposition of price controls by President Nixon on August 15, 1971, the price of crude oil in the United States was approximately $3.25 per barrel. As a result of (1) the Arab-Israeli war of 1972–1973, (2) the shift of control of oil production in OPEC countries from producing companies to host governments, culminating during the Arab-Israeli war and (3) a general reappraisal of remaining world oil reserves relative to increasing demands on those reserves, the price of oil moved up nearly four-fold to a peak of $15 per barrel by November, 1975.

Between the first oil price shock and the second shock corresponding with the Iranian Revolution late in 1978, nominal crude oil prices declined. In mid-1978, imported crude oil was selling in the United States for about $12.50 per barrel. Prior to this second shock, Iran and Iraq together were producing 9.8 million barrels per day. During the height of their hostilities which started in late 1980, their total production declined involuntarily to approximately one million barrels per day, but in 1983 has stabilized at about 3.3 million barrels per day. This major supply reduction, independent of OPEC policy, is probably sufficient to account for the threefold price increase occurring between mid-1978 and peak prices of $39 per barrel in February 1981.

The Arab-Israeli war of 1972–1973 with its attendant rather small output curtailments may be sufficient explanation for the sharp crude oil price increases occurring during that war. However, another rationale must be provided to explain why prices increased further through 1975, and did not decline to pre-war levels.

Johany (1978 and 1980) observed that during the 1950s and 1960s there was a progressive awareness on the part of international oil companies holding oil concessions in the Middle East that their property rights were in jeopardy. Nationalization was the apparent wave of the future. As the concessionaire private companies producing oil in the Middle East became increasingly fearful of losing their property rights, they rationally raised their discount rates. This means that production was shifted from the future to the present. They increased oil production from Middle East reserves as fast as economically feasible. From 1960 through 1970 the compound annual growth rate in oil production from the Middle East was 10.9 percent. From 1970 through 1973 the compound annual growth rate was 15.0 percent. This is rational economic behavior for firms convinced that they faced imminent loss of property rights. The threat of revolution in the late 1960s and early 1970s may also have stimulated more rapid production. Reflecting these output increases, crude oil prices in real terms declined 6% in 20 years from 1950 through 1970. By the end of 1973 a total shift in property rights had occurred. Host countries were in complete control of output. By the year end 1973 the relevant discount rate became that of the host country rather than the concessionaire company. Discount rates declined not only because property rights became secure in the hands of the host country, but also because the opportunity cost of money for the host country was relatively low. With a decline in the relevant discount rate, production from these nonrenewable resources shifted from the present toward the future. The record shows that from 1973 to 1977, Middle Eastern oil output increased at a compound annual rate of only 0.7% in contrast to the 15% compound annual rate from 1970 through 1972 when privately owned oil companies controlled most of the output. As a consequence, crude oil prices increased approximately four-fold from 1973 to 1977.

During the 1970s, the popular press appeared to uncritically accept the OPEC cartel concept. Politicians in the United States advanced the cartel argument, probably because it was a convenient non-voting party to blame for such persistent economic problems as inflation and balance of payments deficits. However some economists also accepted the cartel or monopoly interpretation. (Adelman 1978, Lichtblau, 1981 and Salant, 1979)

Differences in opinion among economists are probably due mainly to definitional problems. Here, a cartel is defined as a collusive combination of a small number of producers or sellers of a homogeneous product acting in harmony in order to achieve some price or other economic objective. Output control is an essential ingredient for effective cartel (monopoly) power. In the absence of collusive control over output involving the major producers, the appropriate market classification would appear to be a dominant firm price leadership model in which other sellers take price as given by the market and independently determine their output level. Saudi Arabia is clearly the dominant producer and on occasion is joined by Kuwait and other smaller Gulf states in managing output to attain agreed-upon price objectives.

Prior to the March 1983 OPEC meeting in London, all attempts to establish production quotas for OPEC members failed even at the meeting stage. In addition to the diversity of interests preventing agreement, the Saudi Arabian Oil Minister had repeatedly stated that for his country, control over oil production is a matter of "sovereign right" and had refused to discuss Saudi Arabian production levels, insisting that "output policy was Saudi Arabia's business only." (See for example, press reports in the *Wall Street Journal*, May 28, 1981, p. 3, and *Oil and Gas Journal*, Newsletter, Vol. 79, September 28, 1981, p. 1)

By the March 1983 meeting, the U.S. refinery average acquisition cost for imported crude had declined from a peak of $39 per barrel as of February, 1981, to about $30.75 per barrel, in nominal terms. With OPEC in considerable disarray and intense pressure on Saudi Arabia, the London meeting produced output quotas on paper but no credible enforcement arrangements. Quotas were specified for all member countries except Saudi Arabia, the latter agreeing to a "swing producer" role in which it would set its own production to make up the difference between the sum of all other quotas and the amount of 17.5 million barrels per day. In fact, through the balance of 1983, OPEC output averaged 18.5 million barrels per day with some countries violating their quotas and Saudi Arabia being unwilling to provide a compensating curtailment. This actual level of output was well below the 24.8 million barrels per day of OPEC production as of February 1981.

The London meeting set a $29 price for the marker crude (Saudi Arabian light, 34 degrees API gravity). Quota violations notwithstanding, the spot price corresponding to the marker crude has held within a range of about $2 per barrel below the agreed price. This price evidence provides the strongest support for the point that OPEC has exercised considerable cartel power during the 1981–1983 weak market.

Correspondingly, output evidence in this period supports the cartel thesis. Table 4.12.1 shows that OPEC members lost market share relative to non-OPEC producers, with OPEC output declining from 42.8% to 33.8% of world production. This pattern is typical cartel behavior and supports the well known rule that it is more profitable to be outside of a cartel than to be a member. The market share patterns also support the "dominant firm" hypothesis. The net cutback within OPEC amounted to 6,523 million barrels per day and 70% of this reduction was borne by Saudi Arabia. The primary burden of enforcing a price policy within a cartel falls on the dominant producer. Iran and Iraq expanded production in absolute terms while eight OPEC members gained market share within the Organization.

The record of price movements since the Iranian Revolution suggests that the crude oil price increase from about $13.50 per barrel in January, 1979, to $35.50 in January, 1981 was due

Table 4.12.1—Changes in Output and Market Shares Within OPEC, 1981 to 1983

	February 1981		August 1983	
OPEC	Output Millions of Barrels/Day	Market Share (Percent) OPEC/Total	Output Millions of Barrels/Day	Market Share (Percent) OPEC/Total
Iran	1,500	6.0/2.6	2,500	13.7/4.6
Iraq	700	2.8/1.2	900	4.9/1.7
Indonesia	1,600	6.4/2.8	1,300	7.1/2.4
Venezuela	2,195	8.8/3.8	1,730	9.5/3.2
Ecuador	235	0.9/0.4	240	1.3/0.4
Algeria	900	3.6/1.6	700	3.8/1.3
Kuwait	1,565	6.3/2.7	1,182	6.5/2.2
Gabon	175	0.7/0.3	150	0.8/0.3
Qatar	482	1.9/0.8	300	1.6/0.6
United Arab Emirates	1,605	6.5/2.8	1,150	6.3/2.1
Nigeria	1,943	7.8/3.4	1,308	7.2/2.4
Libya	1,650	6.6/2.8	1,100	6.0/2.0
Saudia Arabia	10,265	41.4/17.7	5,732	31.3/10.6
Total	24,815	100.0/42.8	18,292	100.0/33.8
Non-OPEC	33,116	57.2	35,834	66.2
Total World	57,931	100.0	54,126	100.0

Source: *Oil and Gas Journal*, June 1, 1981, p. 192, and November 7, 1983, p. 160.

to the market, not OPEC. However, the weak market of 1981–1983 indicates that while OPEC was unable to prevent a price decline to about $28.75 per barrel by mid-1983, its major output reductions were able to stabilize crude oil prices at levels that may be in excess of what competitive markets would have determined.

The record of average international and official OPEC prices during the strong market of 1979–1980 is shown in Fig. 4.12.1. If OPEC had been an effective cartel, the record would show either identical official OPEC and actual prices, or official OPEC prices leading the average international price upward. Neither case is present. To the contrary, average market prices clearly lead official OPEC prices upward throughout 1979–1980. The evidence does not support the cartel hypothesis. This may be the first case in which an alleged cartel held prices below competitive levels.

The weak market beginning in March, 1981 is shown in Fig. 4.12.2. Throughout the period ending in mid-1983, the average market price was declining, even as the official OPEC price increased in September, 1981. Then beginning in March, 1982, the average price fell below the official price and the latter became irrelevant for the remainder of 1982, except for Saudi Arabian production forced on the ARAMCO companies. The weak market further forced two sharp cuts in the official price in January and February, 1983. OPEC was powerless to prevent these cuts. However, the production cutbacks noted earlier clearly prevented even greater price reductions.

Attempts to identify by econometric techniques the price determining roles of OPEC relative to market forces have confirmed findings reported in the preceeding paragraph. Verleger used regression analysis to establish that oil product prices in spot markets are closely correlated with crude oil producer prices. He concluded that "the official price for crude oil set by OPEC countries is determined by the prevailing prices of the products derived from the crude on the major world petroleum product markets . . . all producers are following the market." (Verleger, 1982, p. 181.) His analysis was based on data in the initially weak market, and later very strong market from the first quarter of 1975, through the third quarter of 1980, then expanded to include the weak market through 1982.

In sum, the first major price shock of the 1970s can be explained by the Arab-Israeli war, plus the shift in control over foreign oil pro-

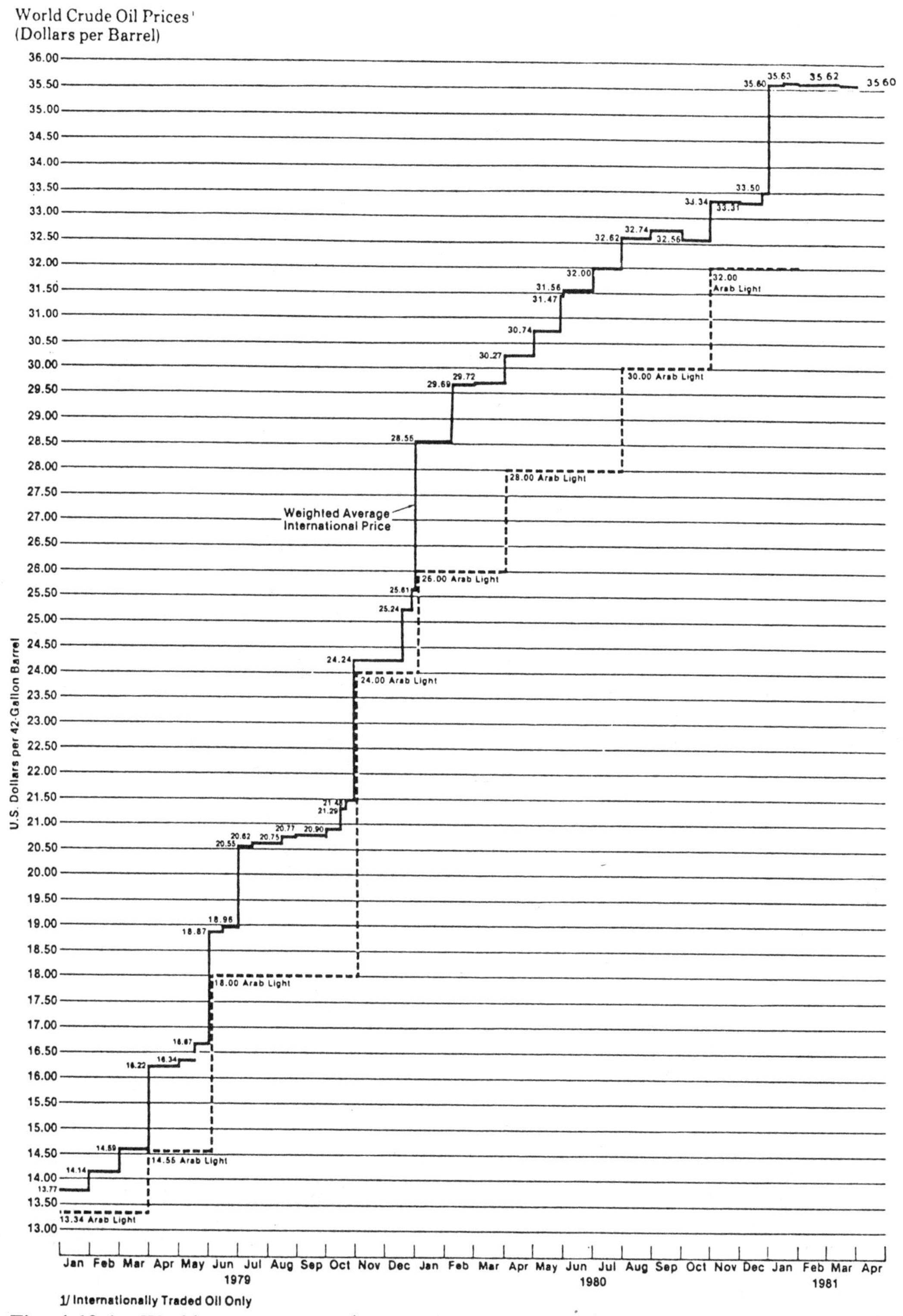

Fig. 4.12.1—World crude oil prices[1] (dollars per barrel). Source: U.S. Dept. of Energy, EIA, 1981, Weekly Petroleum Status Report, March 27, p. 20.

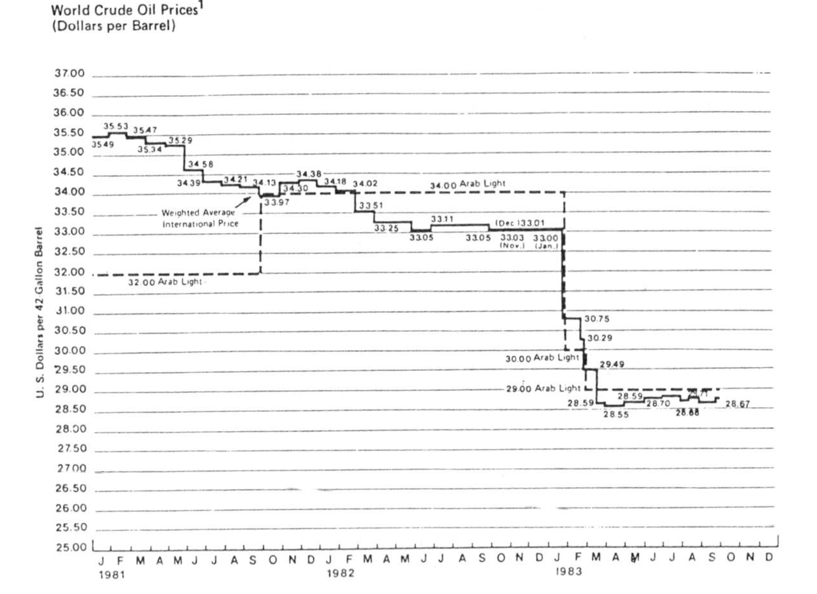

1 Internationally traded oil only. Average price (FOB) weighted by estimated export volume.
Note: Beginning with the May 1, 1981 issue of the Weekly Petroleum Status Report, the world crude oil price is based on a revised crude list.
Additions: Saudi Arabia's Arabian Heavy, Dubai's Fateh, Egypt's Suez Blend, and Mexico's Maya. Omissions: Canadian Heavy. Replacements: Iraq's Kirkuk Blend for Iraq's Basrah Light.
The above graph shows an estimated world crude oil price based on this revised list beginning January 1, 1981.

Fig. 4.12.2—World crude oil prices[1] (Dollars per barrel). Source: U.S. Dept. of Energy, EIA, 1983, Weekly Petroleum Status Report, October 14, p. 18.

duction from the international oil companies, who were correctly forced by anticipated events to heavily discount future production (to increase present production), to host governments where property rights were relatively secure and discount rates consequently lower. The second price shock was due to major involuntary output reductions first by Iran and then Iraq. No cartel theory is needed to explain the 1979–1980 price increases. Further, the price record in this period is inconsistent with a cartel theory. However, the declining prices observed in 1981–83 would probably have been even more severe without OPEC efforts to curtail output. There is still no enforceable system of "market demand prorationing" within OPEC. In its absence, one cannot be sure that the observed output reductions are due to cartel action rather than to individual producer action as each country pursues its individual and diverse interests.

Unfortunately, important regulations have been legislated based in part on the assumption that OPEC is, and from the beginning has been, an effective cartel.[9] The following are examples: (1) On August 15, 1971, President Nixon instituted a wage-price freeze intended to control rising inflation. After four phases, these controls were terminated on April 30, 1974. However, controls were continued on crude oil and oil products until decontrolled by President Reagan in January, 1981. Part of the rationale for continued oil price controls was the rhetorical question–why should the United States allow OPEC to set our oil prices? Continued price controls allowed price increases, but maintained oil produce prices below those consistent with OPEC prices. (2) The "Windfall Profits Tax" (it is actually an excise tax) introduced during the Carter Administration was rationalized in part on the argument that OPEC prices created unexpected and unearned gains for U.S. producers and equity considerations justified their transfer to the government. (3) Natural gas is a substitute for oil and its price is related to oil prices. Thus, part of the justification for bringing intrastate natural gas under price control in the Natural Gas Policy Act of 1978 was to avoid a further "spike" in gas prices. Additionally, this OPEC argument is used in the current Congress to defeat a natural gas decontrol bill. (4) The multibillion dollar synfuels subsidy program was supported largely as a means of gaining independence from the OPEC cartel and its ability to cut off U.S. supplies. (5) A policy considered by both the Carter and Reagan administrations calls for a tariff on crude oil, not only to reduce OPEC imports, but also to limit the ability of OPEC to raise crude oil prices. While this policy has not been enacted, it is periodically under Congressional consideration.

Competition in the Domestic Petroleum Industry

As in the international case, if competition is effective in the domestic oil industry, then the need for government regulation to protect consumer interests is reduced. However, in addition to regulation as a means of consumer protection, the government has a second and perhaps more attractive alternative, that of antitrust policy. As a continuing prelude to the discussion of regulation in the domestic petroleum industry, we review the evidence indicating the presence or absence of monopoly.

The structure of the U.S. petroleum industry has changed radically over the last century. In 1880, the Rockefeller Company controlled between 90% and 95% of oil production in the United States. By 1911, when a U.S. Supreme Court decision forced the breakup of the Rockefeller Standard Oil Company, the market share of this firm had fallen to 64 percent. This decision created several Standard Oil companies but the dominant survivor became the Standard Oil Company of New Jersey, currently known as Exxon.

The trend toward lower levels of concentration (increasing evidence of a competitive structure) has continued. Table 4.12.2 shows concentration ratios for various segments of the U.S. petroleum industry as well as other segments of the energy industry. As a benchmark, Table 4.12.2 shows that for 292 U.S. manufacturing industries in the aggregate, the four largest firms accounted for 41.5% of industry output in the year 1970 and the eight largest firms account for 54.3 percent. Most industrial organization economists conclude that manufacturing in the United States is effectively competitive. Thus if concentration ratios in the domestic oil industry do not significantly exceed these average concentration ratios, then we may conclude that the petroleum industry is effectively competitive.

Table 4.12.2 shows that as of 1982, concentration ratios for the four largest and the eight largest firms in various segments of the petroleum industry were substantially below the average concentration ratios in U.S. manufacturing. Furthermore, concentration ratios clearly are in a downward trend. Only uranium oxide concentrate production ratios for the big eight firms approximate the average of all U.S. manufacturing. If we combine concentration ratios for oil, gas, coal and uranium oxide concentrate production and compute concentration ratios for energy production in total we find that the concentration ratios again are very low. This evidence supports the conclusion of a report by the Federal Trade Commission (1974), p. 148 as follows:

> With regard to the three fossil fuels—crude oil, natural gas, and coal—concentration in production is moderate and lower than concentration in many other industries.

However, a red flag of caution may be raised at this point. Concentration ratios as discussed above presume that the big four firms and the big eight firms are independent competitors. In fact, the major firms in the U.S. oil industry are partners in a series of joint ventures all over the world. For example, Caltex is a joint venture owned by the Standard Oil Company of California and Texaco to produce and distribute petroleum products in the Far East. Similarly, Standard Oil Company of California, Texaco, Exxon, and Mobile are partners in Aramco, the major producer of crude oil in Saudi Arabia. An analysis of joint ventures (Mead, 1968) found that for the 32 largest oil companies in the United States, Exxon had joint ventures with 26 of the other 31 possibilities. Mobil was found to have 300 joint ventures with 28 of the other 31 firms, and Texaco had 286 joint ventures with 26 out of 31 possibilities. Among the big 11 firms, all had joint ventures with all other of these big eleven.

Superficially, this joint venture evidence suggests that such partnership arrangements would facilitate collusion and reduce competition. However various scholars have searched for evidence of either collusive behavior or any reduction in competitive vigor arising from joint ventures and found that such evidence is lacking.

It has also been noted that oil companies jointly bid for government oil and gas leases. Gaskins and Vann (1979) have suggested that such practices might lead to failure on the part of the government to collect its full economic rent from outer continental shelf oil and gas leases. This suggestion led the Department of the Interior to issue a regulation on October 1, 1975 banning joint bidding among firms that individually produce worldwide more than 1.6 million barrels of oil and natural and gas equivalent per day. This ban was formalized in legislation as the

Energy Policy and Conservation Act (PL. 94–163, December 1975). The ban in effect prohibits joint bidding among the six largest U.S. oil companies.

A recent analysis (Mead, Moseidjord and Sorensen, 1983) established that for leases sold in the Gulf of Mexico over the years 1954 through 1969, firms bidding jointly for such leases earned a higher after-tax rate of return on their investments (11.74%) than was earned by firms bidding solo (10.10%). However, the analysis also showed that firms bidding jointly paid 70% more for their leases than did firms bidding solo. Further, the leases obtained by joint bidders are of much higher quality. The average gross value of oil and gas production from the joint bid leases through the year 1979 was 40% higher than for solo bid leases. Also, leases won by joint bidders showed a much lower percent dry relative to solo bid leases. The data suggest that the joint bidding device is used by firms to spread the risk associated with higher bonus bids for

Table 4.12.2—Concentration Ratios for Energy Industry Firms
Percent

Industry Group	1970	1975	1978	1980	1982
292 U.S. Manufacturing Industries†					
Big 4	41.5				
Big 8	54.3				
Proven Crude Oil (Liquids) Reserves‡					
Big 4	N/A	36.3%	36.4%	31.1%	29.8%
Big 8	N/A	55.6%	55.1%	46.3%	44.5%
Proven Natural Gas Reserves‡					
Big 4	N/A	26.9%	25.6%	23.2%	23.4%
Big 8	N/A	39.7%	38.6%	36.3%	35.7%
Crude Oil Production‡					
Big 4	26.3%	26.0%	23.5%	25.3%	25.1%
Big 8	41.7%	41.2%	40.6%	40.8%	39.1%
Natural Gas Production‡					
Big 4	25.2%	24.2%	22.6%	19.0%	16.7%
Big 8	39.1%	36.4%	34.0%	30.2%	27.7%
Oil Refinery Runs‡					
Big 4	34.2%	32.9%	32.3%	31.2%	29.4%
Big 8	61.0%	57.7%	55.8%	53.9%	51.8%
Gasoline Sales‡					
Big 4	30.7%	29.5%	29.7%	28.0%	26.6%
Big 8	54.6%	50.3%	49.4%	49.0%	47.3%
Interstate Oil Pipeline Ownership‡					
Big 4	33.7%*	32.4%	30.6%	30.8%	31.7%
Big 8	55.6%*	53.9%	54.8%	54.8%	53.2%
Bituminous Coal Production‡					
Big 4	30.7%	26.0%	21.4%	21.4%	20.6%
Big 8	41.2%	35.7%	28.7%	30.4%	30.2%
Uranium Oxide Concentrate Production					
Big 4	54.7%	58.2%	48.3%	37.0%	37.4%
Big 8	79.8%	81.5%	71.6%	54.8%	54.6%
Energy Production (Btu basis)					
Big 4	19.3%	18.0%	17.4%	14.4%	14.1%
Big 8	31.6%	29.2%	27.9%	24.3%	23.8%

†These benchmark ratios are from Mueller, W.F., and Hamm, L.G., 1974, "Trends in Industrial Market Concentration, 1947 to 1970," *Review of Economics and Statistics*, Vol. 56, No. 4, November, pp. 511–520.

‡Energy industry ratios are from American Petroleum Institute, 1982, *Market Shares and Individual Company Data for U.S. Energy Markets: 1950–1981*, API, Washington, D.C. pp. 5-14 and 128.

*Data for 1972.

better quality tracts and to facilitate entry, especially by smaller firms.

Joint bidding appears to advance the interests of small relative to large firms. The smallest firms winning joint bid leases realized a 13.3% rate of return on their leases compared to a 9.5% on their solo leases. This 40% increase in return is unmatched by either the big eight (10%) or the big nine through 20 firms (11%). These findings support the earlier conclusion of Wilcox (1975, p. 123) that "joint bidding is procompetitive up to the point where one of the largest eight firms bids jointly with one of the largest 16 firms."

Effective competition in any industry requires reasonably free entry by new firms. Barriers to entry into the petroleum industry are not significantly high. There are approximately 9,000 producers of crude oil in the United States. Most of the undeveloped oil resources of the nation are owned by the federal government, either on the outer continental shelf or in the public lands, largely in the western part of the nation. These prospective oil producing resources are periodically offered for lease at competitive auctions, in the case of the outer continental shelf, or either competitive auctions or first-come first-served allocations from the public domain lands. Thus, access to crude oil production is available without significant barriers.

In the case of oil refining, Boyce (1983) has shown that a 100,000 barrel per day refinery can be constructed at a capital cost of $130 million and that a refinery of this size captures most of the required economies of scale. While this may appear to be a large sum of money, it is not an effective entry barrier.

Performance data tend to support the conclusion that the U.S. petroleum industry is effectively competitive. Fig. 4.12.3 shows the after-tax rates of return earned in the U.S. petroleum industry compared to all U.S. manufacturing industries over the years 1960 through 1982. Except for the two periods in the 1970s when international developments brought about sharp crude oil price increases, rates of return on equity capital investments in the oil industry have been approximately normal. Over this entire 23-year period the average rate of return on equity investments in petroleum has been 12.1% compared to 11.6% in all U.S. manufacturing industries.[10]

In sum, the evidence reviewed suggests that the domestic petroleum industry is effectively

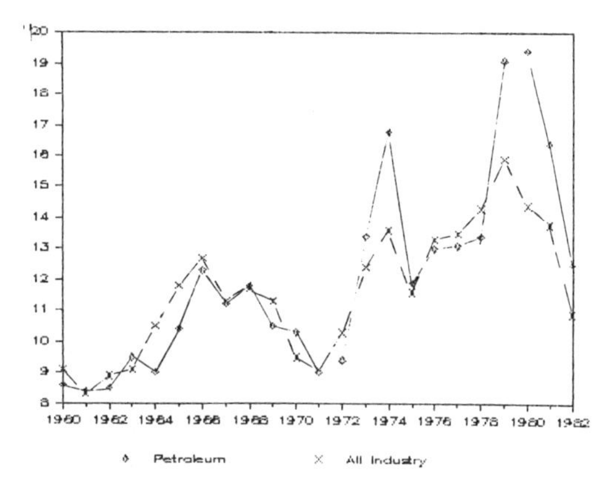

Fig. 4.12.3—Rates of return on equity for petroleum industry and all U.S. manufacturing industries (percent).

competitive. This conclusion is consistent with a staff study prepared by the Federal Trade Commission and a wide variety of academic studies conducted over the past decade.

It has been noted frequently that the domestic petroleum industry is vertically integrated with many large firms individually producing crude oil and natural gas, owning pipelines to transport these resources to refineries and markets, owning refineries to produce a variety of petroleum products, and owning retail gasoline and petroleum product service stations. Studies by Teece (1976), Liebeler (1976), Mancke (1976), and Mitchell (1976) all conclude that vertical integration of the kind found in the domestic petroleum industry is both socially desirable and free of anti-competitive results. The classical study of vertical integration and its competitive effects in the petroleum industry prior to 1959 was by DeChazeau and Kahn (1959). One of the authors, A.E. Kahn (1974), recently reviewed the earlier findings and testified as follows:

> I have no doubt at all that perverse and anti-competitive government policies bear a heavy responsibility for our current problems, and that the oil industry itself has exerted a powerful political influence in promoting these policies . . . In my judgment, the most serious anti-competitive element in our domestic petroleum industry has been the complicated network of government controls on production in the crude oil market . . . I have argued in my book that if the crude oil market were derigged (government anti-

> competitive policies eliminated) and made effectively competitive, I would see no objection to continuation of the vertical integration. It is the combination of the two that is objectionable.

The Kahn testimony suggests the true nature of the problem. The evidence reviewed in the foregoing indicates that the U.S. petroleum industry is effectively competitive. The problem is that monopoly powers have been introduced by the U.S. government over many years of its energy policies. In effect, the government has enforced monopoly actions that the petroleum industry, with its effectively competitive structure, was unable to introduce or enforce. Monopoly has one purpose—to reduce output in order to obtain a price above competitive levels. Two government regulatory policies have accomplished this objective. First, in the 1930s the federal government passed legislation enabling the states to control oil production within their borders and additional federal legislation was enacted to enforce these output restriction rules. This policy is known as market demand prorationing and is reviewed hereafter. Second, competition from foreign crude oil imports was restricted by action of the federal government in 1959 through its mandatory oil import quota program. These two policies together restricted crude oil supplies below their competitive levels and caused prices to rise above competitive levels. These are precisely the objectives of monopoly. They came about through government regulation, however, not by inadequate competition in the domestic petroleum industry.

REGULATIONS RATIONALIZED BY RESOURCE CONSERVATION OBJECTIVES: THE CASE OF MARKET DEMAND PRORATIONING

The term "market demand prorationing" (MDP) refers to government control over private oil production. In essence, MDP uses the power of government to control (reduce) output per well. Market demand prorationing was introduced in the United States in the 1930s and rationalized on the basis of resource conservation to be explained subsequently. An obvious byproduct of prorationing is that where output is reduced through government controls, price will rise above competitive levels.

The root of the resource conservation problem in crude oil is the common property problem, a pure case of an external cost. In the United States where land is for the most part privately owned, an oil reservoir is likely to be owned by several parties. But crude oil is a migratory resource and disregards property lines. The larger the reservoir, the more ownership claims are likely to exist for the same oil. The ownership claims are ordinarily defined on the surface of the earth and do not extend in straight lines to the center of the earth. The "law of capture" prevails. This law states that the owner of surface land under which oil is found has legal title to the oil, even though it migrates from a larger oil reservoir underlying other surface ownerships. The common property problem provides an economic incentive for each landowner to develop a new discovery as rapidly as possible and to produce oil from the reservoir at a physical maximum, independent of reservoir engineering rules indicating an efficient level of production. Following this incentive, the first and fastest producer captures not ony his own oil but some of his neighbor's as well. Failure of one owner alone to follow this private incentive would allow one's neighbors to drain one's own oil.

Ultimate production from any reservoir is sensitive to the daily rate of actual production from that reservoir. For levels of production up to 100% of the maximum physical efficiency rate (MER), ultimate production is unaffected. However, as individual landowners attempt to both protect their own interests and capture oil from others, production is quickly pushed well above the MER. Reservoir pressure is rapidly depleted and some otherwise producible oil is forever lost, unrecoverable under ordinary economic conditions. Thus, ultimate recovery declines, leading to an unnecessary sacrifice of valuable natural resources.

This common property problem has always existed where property rights are undefined. Beginning in the 1920s in the United States, and coming to a head in the 1930s, events pushed this problem to a critical level and led to intervention by the federal and state governments. In the 1920s enormous new oil fields were found, mainly in the southwestern part of the nation. These huge discoveries culminated in 1930 with discovery of the vast East Texas Field. The common property problem led to rapid drilling and production involving economically wasteful excess investment in well development and production.

In addition to the sacrifice in ultimate recovery as outlined, from 1919 to 1929 U.S. oil production increased from 378 million barrels to 1,007 million, an increase of 166 percent. Production in the state of Texas, where large fields were found, increased from 79 million barrels in 1919 to 297 million barrels 10 years later, an increase of 274 percent. This increase in output led to significant price decreases, even during the "roaring twenties." With the beginning of the worldwide depression in 1929, crude oil prices tumbled. By June of 1933, the average price of mid-continent crude oil (36 degree API gravity) had declined to $0.25 per barrel from $2.29 exactly seven years earlier.

These developments led to widespread demands from oil-producing states for Congressional regulatory action. Political activity culminated in August 1935 when Congress approved the "Interstate Compact to Conserve Oil and Gas." This act authorized individual states to restrict output of oil and gas within their respective states. Congress later passed the Connally "Hot Oil Act" authorizing the federal government to prohibit the movement of crude oil in interstate commerce when such oil is produced in violation of state prorationing laws.

Article II of the Interstate Compact states that "the purpose of this compact is to conserve oil and gas by the prevention of physical waste thereof from any cause," and Article V specifically denies any price-fixing intent. This Article states that "it is not the purpose of this Compact to authorize the state joining herein to limit the production of oil or gas for the purpose of stabilizing or fixing the price thereof, or create or perpetuate monopoly. . . ." However, it is an economic fact of life that any governmentally mandated reduction in output, whether or not rationalized in terms of resource conservation, also has a price effect. Prices will be higher as supply is reduced.

Nowhere in the legislation is conservation defined. Instead, the Act speaks of various types of physical waste that member states are bound to avoid. From an economic perspective, resource conservation may be defined as "action designed to achieve or to maintain, from the point of view of society as a whole, the maximum present value of natural resources." (McDonald, p. 71) This definition requires consideration of costs and product values both in the present and the future. The Interstate Compact legislation clearly arose out of a real need to solve the common property problem and its attendant waste of natural resources. However, the political forces supporting the Compact could not have been innocent of the price-supporting results. If the domestic industry had been an effective monopoly, it would undoubtedly have solved the common property problem and, in addition would have reduced output in order to attain monopoly price objectives. However, as we have seen above, its competitive structure precluded private monopolization.

McDonald (1971, p. 41) has pointed out that regulations arising out of the Interstate Compact Act are of two types. First MER (maximum efficient rate) restrictions are applied to individual oil reservoirs (not wells) and are intended to prevent rates of production that are too high and significantly reduce ultimate oil recovery. These MER restrictions are ordinarily accompanied by allocations of total allowed production among *wells* in each reservoir. The allocations are commonly based on acreage and estimated recoverable reserves per well. Second, market demand restrictions are applied to individual reservoirs or areas within states or even to entire states, with the result that the state limits output to some estimate of demand. Since demand in turn is a function of price, a desired price is obviously implied. McDonald (1971, p. 48) has identified the "typical administrative procedure" followed by market demand prorationing states as follows:

> Once a month in a public hearing, with Bureau of Mines demand forecasts and data on inventories before it, the regulatory commission receives 'nominations' (statements of intention to buy) from the principal oil purchasers in the state. Using these data, and any other information considered relevant, including its own demand forecast, the commission then determines the probable market demand for the calendar month shortly to begin. The quantity determined is next related to the 'basic' state allowable, which is the sum of: (a) maximum production of fields or wells assigned special allowables or permitted to produce at capacity, such as water-flood fields and stripper wells, (b) the MERs in fields subject to MER restriction, and (c) the maximum production permitted in other fields under a statewide

depth-acreage schedule of allowable production per well. If the quantity determined to be the market demand exceeds the basic allowable, then the latter—in total, and in each of its separate components—limits production in the state. If the quantity determined to be the market demand is less than the basic allowable, then the maximum production of fields and wells exempt for one reason or another from market-demand restriction is subtracted from the estimated market demand, and the remainder is allocated to the non-exempt fields and wells in proportion to their respective basic allowables. The proportionate allocation to non-exempt fields and wells in some states (e.g., Texas) is accomplished by means of a "market-demand factor" a decimal fraction representing the ratio of the total non-exempt share of market demand to the total non-exempt basic allowable, which is multiplied by each well's basic allowable to compute its effective allowable for the month.

The nominations from principal oil purchasers referred to by McDonald turn out to be, for the most part, submitted by integrated oil companies plus a few independent refineries and pipeline companies. By and large, those firms and individuals used as a source to estimate market demand are the same as the producing firms. Hence the producers are in a commanding position to influence, if not determine, the level of production selected and enforced by the state regulatory commission. If such firms control both supply and demand, then they effectively control price *through the power of government*. Enforcement is by government, and there is no question of antitrust vulnerability. Nor is there any chance that ordinary market forces will cause this monopoly power to disintegrate. Problems for producers arise out of the fact, made clear by McDonald above, that some producers within a state are favored by the regulation and are exempt from output cutbacks. Furthermore, not all states have adopted market demand prorationing systems. The major exception is the state of California, which in 1933 was the third largest crude oil producing state behind Texas and Oklahoma, and accounted for 19% of total U.S. production.

During the 1960s when crude oil markets were relatively weak, state regulatory agencies reduced market demand factors severely with the result that prices in the United States were held up well above what competition would have determined. Table 4.12.3 shows both market demand factors for the major producing states in the year 1962 and a ratio of actual output to productive capacity for the same states.

Table 4.12.3—Annual Average Market Demand Factors and Average Ratio of Crude Oil Output to Productive Capacity for Select States, 1962 (percent)

State	Average 1962 Market Demand Factor	Ratio of Crude Oil Output to Productive Capacity
Texas	27	59
Louisiana	32	69
Oklahoma	35	82
New Mexico	50	83

Source: McDonald, 1971, pp. 164–165.

In effect, the Texas regulatory agency (the Texas Railroad Commission) mandated that non-exempt wells could produce only 27% of their state determined allowables. While numbers for other states indicate similar cutbacks, the percentages are not strictly comparable due to the fact that different rules are used to compute allowable production levels. Closer comparability is attained in the ratio of actual crude output to productive capacity, also shown in Table 3. Here again we find Texas requiring the largest cutback and New Mexico the least. These severe cutbacks have very little to do with any resource conservation objective. Instead, from an economic point of view, their primary effect is to attain a price level that is satisfactory to the producers.

While the system may be preferable to unlimited rights to produce given the community property problem, market demand prorationing has been criticized by economists Lovejoy and Homan (1967) and McDonald (1971) from the point of view of negative effects for resource conservation and monopolistic consequences. First allowable production schedules have been selected primarily on the basis of administrative feasibility and have tended to be based on such factors as well depth and well spacing—the so-called depth-acreage allowable schedule. The system allows higher production quotas for deeper wells, but depth has very little to do with conservation. The rule apparently is followed because the deeper wells are thought to be higher cost and such wells are favored relative to be more shallow wells. This is more a political judgment than it is relevant to conservation. Also, wider well spacing yields higher production allowables per well. While this administrative procedure is more closely related to conservation objectives, it is commonly independent of specific reservoir characteristics that should be used to determine the maximum efficient level of production and the allowable production rate.

Second, the depth-acreage administrative procedure is largely arbitrary. Third, although specifically denied in the legislation, market demand prorationing is obviously a price stabilization system enforced by the government for the benefit of producers. It allows the government to accomplish on behalf of producers what the antitrust laws of the same government make illegal if undertaken by the producers. Hence, monopoly power. which we find lacking in the U.S. petroleum industry, is in fact enforced by the U.S. government.

Fourth, the exemption from market demand prorationing allowed for stripper wells and certain other categories of production bears no relationship to conservation. The policy is justified instead by other factors. They include political power, ideas of equity held by state legislators and administrators, incentives for producers to behave in ways that are believed to be optimal by legislators and administrators, and finally political power of differing groups within the industry.

Fifth, the MER is computed on the basis of physical factors and is independent of costs or market prices prevailing either in the present or expected for the future. Thus, in no sense does the system maximize the present value of the nation's oil resources as demanded by an economic definition of resource conservation.

McDonald (1971, p. 188) concisely summarizes his criticism of the MER and market demand prorationing system as follows:

> The prevailing system of conservation regulation, particularly in the market-demand states, induces or permits certain inefficiencies in the exploitation of petroleum resources. The margins of both development and exploration are needlessly contracted, so that the long-run supply of oil is reduced below its optimal level. Consequently, price is higher than the optimum at every point in time, and the resource-life of oil and gas—the time remaining before substitute energy sources absorb their markets—is unduly shortened.

From an economic perspective, the MER and market demand prorationing system has an attractive alternative requiring far less regulation. "Unitization" has been proposed as a means of correcting the common property problem. Under unitization, a reservoir that might be subject to multiple ownership would be developed, managed, and controlled by one operator on behalf of all of the reservoir owners. Costs, revenues, and profits would be apportioned among the owners on the basis of their estimated ownership interests. The incentive of the operator would be identical to that of all other nonoperating owners and should also correspond with the welfare of society at large. The only regulation required under unitization would be legislation requiring unitization of a reservoir subject to more than one ownership interest.

As indicated, the MER system, and especially its market demand prorationing aspect, raised crude oil prices in the United States above competitive levels and above world market levels. As a consequence, the U.S. market in the 1950s and 1960s became very attractive for foreign production, and imports into the U.S. steadily increased. In effect, U.S. legislation was propping up the world price of oil. The same industry interests that benefited from market demand prorationing later complained about increasing imports and competition from abroad. Complaints, largely from independent oil producers joined by coal producers, led to the imposition of first voluntary, and later mandatory, oil import quotas by President Eisenhower in 1959. Thus import quotas became a politically necessary extension of the market demand prorationing system.

Biologist Garrett Hardin has coined the phrase, "You can't do just one thing." The meaning of this phrase is that legislation enacted to accomplish some specific objective (often desired by a particular beneficiary group) produces unforeseen and unwanted consequences requiring additional legislation to correct. Thus, market demand prorationing created conditions that, from a political perspective, demanded and led to import restrictions.

REGULATION FOR ENVIRONMENTAL PROTECTION

Because U.S. environmental policy toward the mineral industries is treated extensively by other contributors to this volume, the present discussion is brief. The production, processing and use of fossil fuels often alters the landscape in undesirable ways, contributes to significant water pollution, and represents the major source of most air pollutants in the United States and other industrialized nations. To circumscribe the discussion in the present section however, only those environmental concerns associated with production and transportation of crude oil and natural gas are addressed. The environmental concerns that result from refining, use of oil and gas to generate electricity, and the end use of refined products, are outside the scope of this chapter. In the balance of this section, the current federal approach toward environmental protection is discussed. Recent experience with the use of market-based approaches to pollution control, and the role that these policies have played in oil and gas development, are surveyed next. Finally, the special environmental problems and the social costs that result from marine oil spills are outlined.

Current federal policy toward air and water pollution is, for the most part, embodied in the Clean Air Act Amendments of 1970, the Federal Water Pollution Control Act of 1972, and subsequent revisions of these laws. Both laws reflect the heavy federal reliance on technology-based standards and regulations to achieve environmental quality goals. The 1970 amendments to the Clean Air Act required the Environmental Protection Agency (EPA) to identify critical pollutants and to establish health-oriented ambient air quality standards. The act then required individual states to submit, to EPA, implementation plans that would achieve those air quality goals. In addition, EPA issued technological pollution control standards that were to be met by all *new sources* of air pollution. When, in 1977, most major metropolitan areas of the United States were found to be not in compliance with the national standards, new amendments to the Clean Air Act were adopted. These provisions identified *nonattainment regions*, where national ambient air quality goals were not being met, and further specified that in all other areas, the *prevention of significant deterioration* in air quality was to be sought. In the latter areas, any introduction of new emission sources or expansion of old sources was required to undergo a review to determine the status quo level of air quality, to demonstrate that the increment in emissions would not violate ambient standards, and to ensure installation of the "best available control technology" (BACT) to curtail airborne emissions. In nonattainment regions, additional requirements specified that any changes result in no increases in emissions, and that any potential new sources use control equipment that yields the "lowest achievable emission rates."

The most important item of water pollution control policy in the United States is the Federal Water Pollution Control Act of 1972, as amended in the 1977 Clean Water Act. The 1972 act required EPA to determine, for each major water pollution source in the United States, the "best practicable technology" (BPT) currently available, and the "best available technology" (BAT) economically achievable for controlling emissions. These BPT and BAT waste control standards were then to be imposed on industrial dischargers by 1977 and 1983 respectively. The 1977 amendments made modifications to the BAT designation. It also postponed compliance deadlines for the control of certain toxic pollutants.

Both of these federal statutes impose significant regulatory constraints on the crude oil and natural industries. Production and transportation of oil and gas are also affected by a variety of other federal environmental regulations, including: the *Endangered Species Act* (1973); *Marine Mammal Protection Act; National Environmental Policy Act* (1970); *Coastal Zone Management Act* (1972); *TransAlaska Pipeline Act* (1973); *Deepwater Port Act* (1974); *Comprehensive Environmental Rehabilitation, Compensation, and Liability Act of 1980* (CERCLA); and others. Considerations of space do not permit a detailed examination of these individual laws. It should be noted, however, that they represent a continuation of the federal reliance on regulation, as an approach to environmental protection.

Economists have presented strong reasons for switching the emphasis of environmental policy to approaches that rely more on economic incentives than centrally dictated regulation of technology and production methods. A growing public interest in alternatives to the technological standards approach appears to stem from a recognition that the current policy directive is often ineffective and generally has undesirable side effects. Air pollution regulations in nonattainment regions have, for example, severely curtailed economic growth and change. Moreover, the administrative difficulty and inevitable arbitrariness required to interpret such vague prescriptions as "best available control technology" and other designations have hindered progress toward national environmental goals.

Market-Based Approaches to Pollution Control

It has long been evident to economists that technology based regulations do not allocate pollution control effort efficiently. In particular, such policies often result in wide variations in marginal abatement costs among firms.[11] In such situations, the total cost of achieving any given level of pollution control can be reduced by reallocating pollution control effort among emitters. Several recent modifications to the traditional technological standards approach have gone part way toward introducing market-like incentives into the system. Among these innovations are the "bubble" concept, the use of air pollution "offsets," and allowances that permit firms to "bank" emission rights.[12]

In 1979 EPA announced a so-called "bubble" policy, whereby all of the airborne emissions from a given plant are considered to be a single source of pollution. This interpretation allows firms the option of relaxing abatement at sources where control costs are high and substituting tighter controls where costs are lower. In effect, the bubble concept gives the firm some latitude in adopting a cost-minimizing abatement strategy in a given plant. Recent extensions of this approach have allowed such tradeoffs to take place between plants owned by different firms. An additional source of efficiency in pollution abatement results from the use of air pollution "offsets" in nonattainment regions. Firms that achieve certified reductions in emissions receive "Emission Reduction Credits." These credits may be sold to other firms in the region who expect increased emissions due to expansion. Alternatively, the firm earning the credits may "bank" them for use in a future period. The interfirm trading of credits allows the emissions of firms with high abatement costs to be offset by those who can control pollution at lower cost. Currently, use of this offset procedure requires a greater than one-for-one reduction in discharges, so that decreases in overall emission levels result.

The offset policy for airborne discharges in nonattainment regions has been important to the oil and gas industry. One of the first applications of the offset procedure arose from a proposal by Standard Oil of Ohio (SOHIO) to construct a marine tanker terminal in Long Beach, California.[13] Because ambient pollution concentrations exceeded federal standards, operation of such a facility would not have been possible under strict application of the Clean Air Act. To enable the project to go forth, SOHIO entered into an agreement with Southern California Edison, an electric utility. The terms of the agreement called for the utility to install a sulfur dioxide scrubber to offset expected emissions from the planned terminal. Application of the plan required approval from EPA, the local air quality management district, the state air resources board, and a variety of agencies involved in land use planning. The project was eventually abandoned in 1979, and a principal reason cited was the time required to obtain permits. The case is noteworthy, however, in that it demonstrates the potential use of air pollution offsets to enable petroleum development in nonattainment regions.

A more recent and much more novel use of the offset concept was recently granted to the Atlantic Richfield Company (ARCO) in Santa Barbara County, California. In this case emissions from natural oil seeps, rather than a man-made source, were controlled to permit expansion elsewhere.

Although the offset, bubble, and banking provisions represent positive steps, a variety of obstacles to efficiency in pollution control remain. In the case of offsets, the scope for trading emission rights is limited by the requirement that all sources continue to apply EPA mandated control technologies. Another drawback is the "thinness" of markets for emission offsets. Often only one or a few major sources of potential emission reductions exist in a given region, and this can make negotiations lengthy and transactions costs high. Moreover, owners of firms may be unwilling to make long run decisions regarding bubbles, banks, or offsets, if they fear that these policies may be changed by the regulators or overturned in the courts. Despite these problems, however, the fact that these options are now being widely applied is evidence of a trend toward efficiency in the implementation of environmental policy.

Pollution from Marine Oil Spills

Marine oil spills are among the most publicly visible forms of pollution that result from oil and gas production. Moreover, their source is typically obvious, and responsibility is often easily assigned. As a result they can assume a political importance that is disproportionate to the environmental damage caused. The Santa Barbara oil spill in 1969 helped crystallize public sentiment for the cause of environmental protection and has been viewed by many as a turning point for the environmental movement in the United States. Several other notorious oil spills have occurred during the last two decades, including the wreck of the Torrey Canyon in the English Channel in 1967, the grounding of the Amoco Cadiz off the northern coast of France in 1979, and the spill that resulted from a blowout on a Pemex production platform in the Bay of Campeche, Mexico.[14] In the case of two of these incidents, the Santa Barbara and Amoco Cadiz spills, economic analysis has been conducted to assess the nature and magnitude of damages that resulted.

Before proceeding with a discussion of economic damage estimates or policy analysis, however, it is appropriate to examine the size of the marine oil pollution problem, and to assess the importance of production and transportation activities in contributing to this form of pollution. According to estimates made by the National Academy of Sciences in 1975, total direct emissions of oil into the marine environment amounted to about 4.91 million tons per year.[15] Estimated volumes for individual sources of oil pollution are shown in Table 4.12.4. To provide a context for the overall volume of oil released, 4.91 million barrels represents about 0.2% of annual worldwide production. Offshore production operations are credited with only about .08 million tons per year, or about 1.6% of all direct pollution. Though spills associated with blowouts can be spectacular, collectively they are a relatively minor source of marine pollution. Marine transport accidents are estimated to account for an added 0.30 million tons per year, or about 6.1% of direct emissions. Another way to place pollution from spills in perspective is to note that estimated annual oil pollution is 40 times as great as the volume of oil lost from Torrey Canyon, and 23 times as large as the Amoco Cadiz spill. It should be recalled, however, that the economic damages associated with tanker or platform spills can be disproportionately large because the resulting pollution is highly concentrated.

A major source of avoidable oil pollution is associated with the handling of ballast in crude

Table 4.12.4—Estimated Annual Volume of Marine Oil Pollution

	Millions of Metric Tons	Percent of Direct Pollution
Marine Operations:	2.13	43.4
Tanker Operations	1.08	22.0
Bilges, bunkering	0.50	10.2
Tanker and non-tanker accidents	0.30	6.1
Other	0.25	5.1
Non-Marine Operations	2.78	56.6
River runoff	1.60	32.5
Municipal wastes, urban runoff	0.60	12.2
Coastal refineries, industrial wastes	0.50	10.2
Offshore oil and gas production (includes platform spills)	0.08	1.6
Total Direct Pollution	4.91	100.0
Indirect Pollution[a]	1.20	
Total	6.11	

[a]See footnote for further explanation.
Source: *U.S. Council on Environmental Quality* (1977), p. 151.

oil tankers. One method of separating oil from ballast or tank washings is the so-called "load on top" method. With this procedure ballast is allowed to settle while the ship is underway. This concentrates any remaining oil and allows it to drawn off into a separate tank where new cargo may be loaded on top of it. In 1975 it was estimated that over 80% of all crude oil transported was carried in vessels that used the load on top method.[16] Yet these tankers accounted for only about 310,000 tons of emissions, or less than one-third of oil pollution resulting from bilge or ballast discharges. The remainder of such discharges,, estimated at 770,000 tons was attributed to the minority of vessels that do not use the load on top technique.

In 1973, the United Nations Intergovernmental Maritime Consultative Organization (IMCO) drafted a treaty on marine oil spills. This treaty, which was not ratified by member countries until 1983, imposed strict rules regarding nearshore oil discharges, and initially required all *new vessels* over 70,000 deadweight tons (DWT) to be fitted with separate ballast tanks, smaller segregated cargo chambers, and automatic monitoring devices. By January 1986, all new crude oil tankers of 20,000 DWT or larger are to be fitted with segregated ballast tanks and crude oil washing equipment, and regulations regarding existing tankers over 40,000 DWT are to be imposed. Prior to its ratification, the U.S. Coast Guard had enforced its provisions for ships traveling in U.S. waters.

The first major spill that received detailed attention from economists began on January 28, 1969 with the blowout of a well being drilled by Union Oil in the Santa Barbara Channel.[17] It took approximately two weeks for the flow of oil to be substantially reduced, by cementing the bore hole and the ocean floor immediately around the well. Following abatement of the spill, an extensive clean-up effort was undertaken both by Union Oil and by local government agencies. In comparison with other well-known spills, the Santa Barbara incident was relatively small. An estimated 77,000 barrels of oil was released,[18] or about one-twentieth the volume lost in the Amoco Cadiz spill in 1978.

The full social cost of the Santa Barbara spill has been placed at $16.5 million (1970 dollars). Of this total, about $10.6 million was borne by Union Oil and its partners in the operation, in oil well control efforts, clean up costs, and in the value of the crude oil lost. The second largest category of damages was lost recreational opportunities at local beaches; these costs were placed at $3.1 million. The local businesses that depend on the tourist trade also suffered losses in income, but available evidence suggests that this business was largely diverted to other nearby areas that were unaffected by the pollution. Losses were also recorded in the commercial fishing industry, due in part to an oil barrier that was placed across the mouth of the harbor and that effectively kept the fishing fleet in port. Owners of private boats and beachfront property also suffered losses in property values. Together, damages to fisherman and property owners were estimated at $2.0 million. In summary, over half of the estimated costs were directly borne by the party responsible, Union Oil, in the form of clean-up and control efforts; lost recreational opportunities accounted for most of the remaining damage.

Prior to the Santa Barbara spill government relied on the industry's profit motive, rather than regulation, to minimize spills and operating accidents. Following the Santa Barbara incident, the U.S. Geological Survey tightened offshore inspections and issued new operating regulations regarding the design and operation of "down hole" shut off valves. In addition, EPA imposed several "housekeeping" regulations intended to minimize the chronic small discharges that typically accompany routine operations.

The second oil pollution case for which economic results have been compiled began on March 16, 1978 when the Amoco Cadiz, carrying 216,000 tons of crude oil and 4,000 tons of bunker fuel (1.6 million barrels of oil), ran aground on the northwest coast of France.[19] Stormy seas, unusually high tides, and strong onshore winds combined to prevent transfer of the cargo to other ships, and tended to drive the spilled oil into sensitive intertidal areas and estuaries. The costs of the spill were heightened by the fact that it occurred in an area prized for its scenic and recreational resources, and one that relies on fishing and oyster culturing for much of its income.

It is interesting that many of the findings from the Santa Barbara spill, regarding the sources of damages, reappeared in the economic assessment of the Amoco Cadiz mishap. The total social costs of the spill were placed at $195 to $284 million (1978 dollars).[20] Of this total, the major portion ($145 to $165 million) was attributed to clean-up efforts and to the value of the lost ship and cargo. As was the case in Santa Barbara, recreational losses represented the second most important category of costs, accounting for $13 to $82 million in damages. Also, losses of business to the local tourism industry were substantially offset by increases in tourism in substitute areas. The Amoco Cadiz study team also attempted to identify the temporal distribution of costs. According to their estimates, the vast majority of all costs, especially those associated with clean-up and lost recreation, were confined to the one year period following the spill. Damages suffered by the local oyster culturing industry were, however, expected to be long-lived.

Conclusions

The preceding discussion of selected environmental topics points to the need for a more unified and better informed approach to the problem of environmental regulation. Policies are often adopted and regulations imposed on the basis of little quantitative knowledge of the benefits and costs involved. Moreover, there is growing evidence that the environmental protection provided under the traditional standards—regulatory approach could be achieved at lower cost by the use of market-based policies. Finally, although it has not been emphasized in any detail, the current apparatus for reaching local decisions on petroleum development projects needs to be reevaluated. The environmental impact reporting process, as mandated by the National Environmental Policy Act and similar state measures, is cumbersome and time-consuming at best. As currently implemented, this process seems to serve primarily as an adversary proceeding, and an arena for legal maneuvering, rather than an efficient approach to the important problem of environmental protection.

REGULATIONS RATIONALIZED BY NATIONAL SECURITY CONSIDERATIONS.

In the name of promoting national security, voluntary oil import quotas were introduced in 1956 by President Eisenhower. This voluntary approach to import restrictions failed and was replaced in 1959, again by Eisenhower, by mandatory import quotas. This system lasted until 1972 when it became unworkable. National security was also the justification for government entry into a strategic petroleum reserve storage system in 1976.

Mandatory Oil Import Quotas.

Under the Mandatory Oil Import Quota system, crude oil imports into PAD Districts I-IV (the United States east of Nevada and New Mexico) were limited to 12.2% of estimated domestic production. The problem with this limitation was that as production in the United States declined due to declining discoveries and reserves, the system required reduced imports. For District V (the West Coast plus Nevada and New Mexico), imports were limited to the difference between estimated demand for a calendar year and estimated U.S. and Canadian supplies produced in or shipped into this district (*Cabinet Task Force on Oil Import Control*, 1970, p. 10). Thus for District V, the level of imports was flexible. The system allowed domestic producers first access to the domestic market at prevailing prices and provided 100% protection against imports.

Imports from Canada and Mexico were exempt from control as "overland" imports. The Mexican connection led to a strange system that became known as the " Brownsville turnaround" wherein imports from South America primarily were landed in Brownsville, Texas, trucked across the Mexican border, then immediately turned around and re-entered on the same truck as overland oil and therefore exempt imports from Mexico.

The import quota system was a political response to pleas for protection by independent oil producers and coal producers. The administration of the system similarly became political. Various exemptions were quickly granted to a variety of special interest groups. Small refiners argued successfully for preferred access to quota tickets on the basis that there was unfair competition between them and the "big oil companies" with their larger, more efficient refineries. This led to a "sliding scale" wherein small refiners were given a disproportionate share of valuable import tickets. A ticket to import lower cost foreign oil rose in value to about $1.25 per barrel imported. This sliding scale became known as the "small refiner bias." When the quota system was removed by President Nixon in 1972, the small refiner bias was continued in the price control system until its demise under

President Reagan when oil prices were decontrolled in January 1981.

The effects of the import quota system included the following. First, small refiners were subsidized relative to their larger competitors. This subsidy, lasting from 1959 through 1981, led to the construction of inefficient small refineries and to consequent waste of scarce economic resources.

Second, the price of oil, artificially increased under the market demand prorationing system, was supported by this government restriction of competition at about $1.25 per barrel above the world price, and the price that would have prevailed in the United States in the absence of the quota system. Third, exemptions to the quota system were immediately introduced, leading to widespread criticism among users of crude oil and some petroleum products.

Fourth, because imports were artifically restricted, domestic production of this non-renewable resource was artifically stimulated, contributing to the energy crisis which developed in the 1970s. This result also had national security implications. The United States became dependent on foreign crude oil sources earlier than would have been the case in the absence of the quota system. The quota system led to what S. David Freeman called a policy of "drain America first." Fifth, resource misallocation occurred as a result of oil being overpriced under the quota system. This led to substitution of other energy sources for the higher priced oil in the 1960s.

Sixth, the consumer cost of the import quota system was very high. The Cabinet Task Force estimated that the various consumers of petroleum products would have saved $4.85 billion in the year 1969 without the quota system, (1970, p. 23). Bohi and Russell (1978, p. 285) conducted a thorough review of the quota system and concluded that in 1969 the total direct cost to all U.S. consumers, including their loss of consumer surplus, was $5.9 billion. This study further estimated that the annual cost increased steadily from $3.2 billion in 1960 to $6.6 billion in 1970.

The import quota system became unworkable in the late 1960s and early 1970s under the burden of a multitude of exceptions, declining U.S. production paired with an increasing need for imports, and sharply rising world oil prices. The quota system was effectively phased out by President Nixon over the years 1970 through 1972. However, it was not until December 11, 1983 that President Reagan removed the import licensing system as the final vestige of the import quota system.

The rationale for introducing the quota system in 1959 was national security. Its proponents argued that with imports increasing, the United States would become severly dependent upon unreliable foreign sources to the detriment of our national security. In fact, the quota system produced the opposite result by accelerating domestic production in lieu of imports. The nation became severely dependent upon imports beginning in the 1970s with the feared national security consequences. In 1977, crude oil imports accounted for 44% of total U.S. crude oil requirements.

Apart from the rationale offered by its proponents, the major argument for the quota system remained unstated. Independent oil producers and coal producers in the United States, like sellers of any other product, preferred higher prices and profits to lower prices and profits. By restricting competition, both prices and profits would obviously be higher.

The major international oil companies that were importing crude oil uniformly opposed mandatory oil import quotas when they were proposed in 1956 and 1957. For example,, T. S. Peterson, President of Standard Oil Company of California, testified at the Office of Defense Mobilization Hearings concerned with the oil import restrictions in 1956 as follows (Office of Defense Mobilization, pp. 542–550):

> It is our carefully considered opinion that the contention of injury to the domestic industry is without substance. (p. 547) (1) Standard Oil of California, together with other companies, is continuing to invest large sums of money in domestic oil exploration. (2) Petroleum is not an inexhaustible natural resource and this country, with only 20% of the free world reserves of petroleum, currently accounts for 51% of production. There is mounting evidence that, despite the strides made in geological and petroleum engineering technology, the discovery and development of additional reserves in the United States is becoming more difficult and certainly more costly. (p. 546–547) (3) There is a serious question in my mind that the economic well-being and strength of this

> country—the bases of our national security—would be promoted by measures of control set up to promote a greater dependence on increasingly costly domestic reserves. To the contrary, consideration seems to be warranted for a gradual increase in the proportion of imports to domestic requirements. (4) Those who contend that imports pose a threat to our national security base their concern on the narrow point that our country's military resources, in the event of war, would be restricted to those of the United States and Canada. It is a view that can have merit only if we assume that, in a World War III, our defense lines would be our own shores. I want to reiterate the conviction that the facts demonstrate that imports are not injuring the domestic industry in any reasonable sense of the term. Consequently, I submit that on that score the national security has not been impaired. (p. 510)

In spite of this clear statement on behalf of free trade, and similar statements by heads of other major companies, Standard Oil of California and other major international oil companies became supporters of continued import quotas during the 1960s. This shift of position may have been due to the fact that companies made large-scale location-specific investments in refining and other facilities under the rules of the quota system. Such investments would be less than optimal if the quota system was removed.

In sum, the oil import system was a major regulatory effort. It lasted as a mandatory program from 1959 through 1972 and imposed major costs on consumers leading to more rapid depletion of the nation's oil resources, thereby contributing to the energy crisis of the 1970s.

Strategic Petroleum Reserves

The national security issue in crude oil supplies, and a national interest in establishing a system of stored petroleum reserves for emergencies is an old one. During the 1956–1957 Arab-Israeli war and again during the similar outbreak of hostilities in 1967, the Suez Canal was closed. While these events did not cut off oil supplies, they did close this important waterway for oil tankers bringing crude oil from the Persian Gulf into the Mediterranean and then on to both Western Europe and the United States. Closure of the canal required a much longer and more expensive delivery system around South Africa into Western Europe and the United States, or eastward from the Persian Gulf halfway around the world to the west coast of the United States. Thus the canal closure delayed crude oil deliveries and led to a crisis atmosphere.

Fortunately, the United States had idle crude oil productive capacity mandated under the prevailing system of market demand prorationing discussed earlier. In 1956, Texas market demand factors were at 52%, and in 1967 they were at 41 percent. Thus on both occasions U.S. production of crude oil could be, and was, increased sharply, and some exports from the United States to Western Europe occurred. However, in 1972 market demand factors in Texas as well as Louisiana and other states were raised to 100% and have remained at that level through the present time. Thus idle productive capacity in the United States has been non-existent since 1972, and this element of national security for crude oil supplies is no longer available.

As previously mentioned, the Arab-Israeli war starting in the fall of 1972 was accompanied by an OPEC embargo of the United States and several Western European countries. After the embargo ended, OPEC threatened the United States, on many occasions, with the use of its "oil weapon" for political purposes. The possibility of curtailed foreign supplies of crude oil, whether due to an intended embargo or to wartime conditions, led the Congress to move toward establishment of a strategic petroleum reserve.

Given the fact that by 1975 crude oil imports accounted for 33% of U.S. crude oil supplies, the possibility of import supply disruption, and the enormous costs that a serious supply disruption would impose upon the nation, it became apparent that import dependence imposed external costs on the nation. That is, the price which U.S. consumers paid for crude oil reflected its current supply and demand, and did not reflect the potential external costs of supply disruption. This condition led Brannon (1974, p. 21) to recommend a tariff on imported oil in an amount that would approximate the external cost of a potential import supply disruption. He further proposed that the proceeds be used to establish a "stockpile" of crude oil.

One may ask, would private enterprisers in the oil industry maintain a reserve supply of crude oil, not only for their normal refinery needs, but in addition, for national security needs? In the event of a cutoff of crude oil supplies, oil prices would immediately rise and firms holding large reserves would be able to sell such reserves at high prices, thereby recouping not only their original cost but a handsome reward for holding large inventories in anticipation of these events.

Indeed, this reserve stock might well be maintained if investors believed that government would in fact allow them to sell at market clearing prices. However, experience has been clear that the U.S. government in particular does not allow sharp oil price increases to occur. Throughout the 1970s, when supply disruptions led to price increases, price controls were maintained in order to prevent what was commonly called "windfall profits" from accruing to oil companies. One element of price controls was specifically designed to preclude any firm profiting from price increases on inventories of crude oil which they might have held. The "entitlements" program was designed to equalize the cost of crude oil to each and every refiner, regardless of any advance plans which might have been made to provide large crude oil reserves or inventories to meet their refinery needs during a national emergency. Under the "entitlements program," firms with low cost crude oil supplies were forced to make payments to firms without such supplies. Furthermore, under the "allocations program," firms with large supplies of owned crude oil were required to sell some of their supplies to other refiners lacking similar foresight. Thus, no rational private entrepreneur would hold emergency reserves in anticipation of subsequent compensating gains. As a consequence, if reserves are to be maintained for national emergencies, the cost of such reserves must be borne by government.

The idea of a strategic petroleum reserve was embodied in the Naval Petroleum Reserves (NPR) established around the turn of the century. The Elk Hills and Buena Vista fields in California (NPR 1 and NPR 2), Teapot Dome in Wyoming (NPR 3) and the Alaskan North Slope reserves (NPR 4) contain both discovered and potential reserves of a large scale. All except NPR 4 were developed and producing oil to some degree. The largest known reserve (Elk Hills, NPR 1) contains an estimated 1.3 billion barrels of crude oil and was partially developed at the time of the 1972–1973 Arab oil embargo. During this embargo, jurisdictional disputes over ultimate control of the Elk Hills field, followed by Congressional indecision, prevented any production from Elk Hills to meet U.S. crude oil needs at that time. Subsequently, when a so-called "West Coast oil glut" developed after 1977, when Prudhoe Bay Alaskan oil started to flow into the West Coast in large quantities, Congress authorized production from Elk Hills, thereby adding to the West Coast glut.

The Elk Hills field was 81% owned by the federal government (19% by Standard Oil Company of California), and controlled by the Department of the Navy. If the field were fully developed and placed in a shut-in condition, it would serve as an important element in a U.S. strategic petroleum reserve capacity.

The advantage of an *in-situ* reserve like Elk Hills is that the oil is already in a secure position where it has remained for thousands of years. The cost of maintaining the system would consist of the "user cost," i.e., any sacrifice in value due to delayed rather than present use, plus interest and maintenance cost on standby production and transportation facilities. The principal disadvantage of Elk Hills as a strategic petroleum reserve is that annual production from the reserves would be constrained by reservoir production efficiency considerations, in contrast to production from steel tank storage or salt domes, which would not be similarly constrained. (Mead, 1971).

Congress established the present Strategic Petroleum Reserve (SPR) through passage of the National Emergency Petroleum Storage Act of 1976. The primary method of storage under this system is to take oil previously recovered from such fields as Elk Hills or purchased in the open market and pump the oil back into the earth in salt domes for long-term storage. The primary cost of salt dome storage is interest on the stored oil plus a small charge for interest and maintenance of standby production and transportation facilities, minus any increase in value of the oil during storage.

The goal established for the SPR was one billion barrels of stored crude oil by 1980. The record shown in Table 4.12.5 falls far short of this goal. We find that in 1980, only 108 million barrels were in inventory by year end. By September 1983 only about one-third of the 1980 goal had been attained. Energy Secretary Donald Hodel testified before a Congressional Committee in January of 1984 that "the Administration since its beginning has maintained a strong commitment to the continued construction and de-

Table 4.12.5—Strategic Petroleum Reserve (Year End Stocks—Mill Bbls.)

1977	7
1978	67
1979	91
1980	108
1980 Goal—1,000	
1981	230
1982	294
1983 (September)	352

velopment of 750 million barrel SPR by January 1985 (*Oil and Gas Journal*, January 30, 1984, p. 73).

In its first three years of operation from 1977 through 1980, the SPR was plagued by management problems which prevented attainment of the stated goal. With 351 million barrels in storage, the SPR would currently replace 103 days of total U.S. imports or 133 days of OPEC imports, or 405 days of Arab OPEC imports. Thus the SPR provides a valuable shield against any planned embargo of the United States, or against a cutoff which might result from a Middle Eastern war.

From an international perspective, there is an externality that leads to sub-optimal investments in reserve stocks of petroleum. All nations benefit as a result of any one nation building up significant reserves. This is true because, given a cut-off, the larger the reserve stocks existing at the time of the cut-off, the less the impact of the cut-off. The reduced flow of crude oil will be reallocated toward nations that have not provided reserve supplies. Ordinary market forces will produce this result. This is the free-rider problem. Hogan (1983, p. 50) has pointed out that "the actions of governments reflect an awareness that an oil stockpile is an international public good. Each country recognizes that in maintaining a large stockpile it will absorb most of the direct costs while other countries will reap many of the benefits." From the point of view of any one major importing country, its private benefit is served when it urges all others to invest in stockpiles, while making only minor investments itself. The solution to this externality problem is cooperative international action among importing nations which begins with an acknowledgment of the problem, then negotiates a common agreement to provide stockpiles, with penalties for non-compliance.

TAXATION OF OIL AND GAS

The obvious reason to tax crude petroleum and natural gas is to transfer revenue to the public sector to support the operations of government. However, because taxes alter the incentives of market participants, tax policy can affect rates of energy production and consumption, levels of imports, patterns of interregional trade, prices paid by consumers, and incomes of factors employed in energy production. Current taxes, by altering production rates and investments in exploration and development, can also affect levels of energy reserves and consumption possibilities in the future. As a consequence, energy taxation can and has been used by governments to pursue energy policy goals. The nature of effects that oil and gas taxes have on market outcomes depend on several factors, including: the tax base, e.g. whether the tax is levied on the value of production, income, or property; the jurisdictional level, whether federal, state, or local; and the market setting in which the tax is levied, whether regulated or unregulated, isolated or closely integrated with other markets.

Major Taxes on Oil and Gas

The major taxes imposed on the U.S. energy industry may be segregated into three broad categories: taxes on crude oil and natural gas *production*, taxes on *income*, and taxes on *sales* of final products. Production taxes are levied under several different names, and by all three levels of government. The most common production tax is the state *severance tax* on crude oil and natural gas. Such taxes are typically expressed on a per unit basis (e.g. in cents per barrel of crude oil produced) or as a fraction of the wellhead value of output. Severance taxes for oil and gas production are levied by all major oil producing states.[21] Rates vary widely, from nominal levels in California and Kansas, to 15% in Alaska.[22] Several states allow exemptions or reduced tax rates under certain circumstances, e.g., for small producers, for production from wells with low production rates or high operating costs, or for lease fuel used in pressure maintenance or enhanced recovery projects. In part, the purpose of such exemptions is to prevent the tax from inducing premature abandonment. Exemptions do little, however, to offset the tax disincentive for exploration and development.

It is common for state and local governments to impose *property taxes* on crude oil and natural gas resources in their jurisdictions. In most cases, however, the base of such taxes is actually the wellhead value of production. This is true in Colorado, Texas, Utah, and Wyoming. Among the major producing states, only California levies such taxes on the assessed value of reserves. Property taxes that use the value of production as a base have the same economic effects as severance taxes. Property taxes levied on the value of reserves, as in California, however, have much different economic impacts. Because the former are more important empirically, the discussion in the remainder of this section treats all property taxes as taxes on production.

The federal Windfall Profits Tax (WPT) that was implemented on April 2, 1980 is, despite its name, effectively a tax on production. It is much more complicated than most state production taxes, however. The stated intent of the tax was to capture a portion of the "windfall" that resulted from the decontrol of U.S. crude oil prices. For a given barrel of oil, this windfall is computed as the difference between the removal price (roughly equivalent to the wellhead price) and an adjusted base price that approximates the ceiling price under federal controls. Base prices vary for different categories of crude oil. In 1982 the range of base prices extended from $14.80 per barrel, for "tier one" oil from the Prudhoe Bay region of Alaska and other so-called "old oil" to $21.04 for "tier three" newly discovered oil.[23] The tax rates applied to this windfall are also non-uniform; special reduced rates apply to the first 1,000 barrels per day produced by independent producers, to oil produced from secondary and enhanced recovery processes, to stripper oil, newly discovered oil, and heavy oil (below 16° A.P.I.). As a consequence of differences in base prices and tax rates, the levies on various crude oils vary dramatically. The figures on removal prices and tax rates presented in Table 4.12.6 indicate this range of variation. When expressed as a fraction of the wellhead value, effective tax rates varied from 10% for heavy oil, to 38% for tier one oil produced by integrated firms.[24] In addition a variety of exemptions, for production from federal lands and from properties north of the Arctic Circle, are allowed. Finally, the WPT incorporates a provision that limits the taxable wind-

Table 4.12.6—Federal WPT Liability, by Oil Tier and Tax Rate, 1981 (dollars per barrel)

	Removal Price (1981)	Base Price (1981)	WPT[a] Liability (1981)	Percent of Price
Total, all returns	31.77	15.88	9.07	28.5
Tier one:[b]				
Taxed at 70%	33.75	14.61	12.78	37.9
Taxed at 50%	34.64	14.92	9.28	26.8
Tier one, Sadlerochit:				
Taxed at 70%	23.16	14.70	5.13	22.1
Taxed at 50%	34.45	14.76	8.86	25.7
Tier two:				
Taxed at 60%	34.18	17.67	9.43	27.6
Taxed at 30%	35.07	18.14	4.87	13.9
Tier three (taxed at 30%)				
Newly discovered	35.60	20.76	4.18	11.7
Incremental tertiary	33.72	20.58	3.69	10.9
Heavy oil	25.13	16.41	2.59	10.3

[a]Tax liability before adjustments.
[b]Other than oil produced from the Sadlerochit (Prudhoe Bay) reservoir in Alaska; see U.S. Internal Revenue Service (1982) for further detail.
Note: The lower tax rates apply to independent producers; non-independent (integrated) firms are subject to the higher rates.
Source: *U.S. Internal Revenue Service* (1982), p. 41 ff.

fall to 90% of the net income attributable to a barrel of oil. Despite these complications, however, for a given base price, removal price, and tax rate, the producer's tax liability is proportional to the rate of production. Hence, the economic impacts of the tax are essentially the same as the impacts of a production tax.[25]

Producers of crude oil and natural gas are subject to the federal corporate income tax and to corporate income and franchise taxes in most states. The marginal federal tax rate is 46%, but a variety of special provisions for the petroleum industry, including depletion allowances for small producers, immediate deduction of intangible drilling expenses, and credits for taxes paid abroad, complicate any attempt to assess the true impact of the tax. These special provisions are of sufficient importance to be afforded special treatment in a subsequent section. All major producing states except Wyoming levy significant corporate income or franchise taxes. Rates vary markedly among states, and California and Alaska top the list with maximum marginal levies of 9.6% and 9.4% respectively.[26] States also vary significantly in provisions for exemptions and deductions. For example, Oklahoma and Louisiana allow percentage depletion for all producers at rates of 22% and 38% of gross revenues respectively. California, on the other hand, like the federal government since 1975, permits only small producers to take percentage depletion deductions from taxable income.

Corporate income, as defined for tax purposes does not allow deductions for a normal rate of return on equity capital invested in the firm. (Interest deductions for capital acquired through debt are allowed, however.) As a consequence, the tax on corporate income is largely a tax on the return to corporate capital. To the extent that the amount of corporate capital invested in energy production is sensitive to the after-tax rate of return in such investments, the various state and federal corporate taxes now in existence tend to suppress such investments. In the case of oil and gas production, most capital takes the form of hydrocarbon reserves and the equipment required to produce them. By lowering the after tax return to these investments, the corporate tax discourages the formation of such capital.

In addition to taxes on income and production, there are also taxes on sales of final products. The most important such taxes are the federal levy on motor fuels, and state taxes on sales of gasoline and diesel. In recent years, state motor fuel excise taxes have ranged from a low of 5¢ per gallon in Texas to 13.5¢ per gallon in Hawaii.

Resource Allocation and Tax Incidence

The distribution of the economic burden of taxation seldom coincides exactly with the legal liability for tax payment.[27] Taxes levied on energy products will generally alter the behavior of both producers and consumers, and may lead to changes in levels of investment, rates of production, and prices of inputs and outputs. For example, a severance tax levied on natural gas will reduce producers' profits, and lower incentives for exploration, development, and production of gas resources. As a result of reduced profitability, part of the tax will be shifted backwards from producers to landowners in the form of reduced royalties and leasehold bids. Similarly, decreases in production will tend to reduce employment and incomes among inputs in the production process. Output reductions may also cause higher prices to final consumers, a phenomenon known as forward shifting. In summary, the ultimate impact of the tax on the distribution of income among various market participants may be far different than the pattern of legal liability specified in the enabling statutes.

The pattern of shifting that results from a tax levied on producers, whether backwards in the form of lower input prices, or forward in the form of higher prices to consumers, depends on the structure of the market in which the tax is levied. Of particular importance are the sensitivities of production and consumption decisions to changes in price. These sensitivities depend in part on the physical attributes of the taxed resource, on the jurisdictional setting (e.g. state versus federal) in which the tax is imposed, and upon the amount of time allowed for production and consumption responses to occur. An additional complication arises when the economic forces that would otherwise govern price responses are suppressed or controlled by government regulations. Finally, it should be recognized that the myriad of federal, state, and local taxes now imposed on oil and gas production can interact with one another in complex ways.

Before examining the incidence of individual taxes, it is useful to present some basic economic principles that can guide this evaluation.

To provide a specific context in which to develop these concepts, consider a severance tax levied on the production of natural gas in a particular state. The tax may be viewed either as a reduction in the net-of-tax revenue of the firm, or as an increase in operating costs. With either view, if the tax does not result in higher prices to consumers, then some exploration and development prospects that would have been undertaken in the absence of the tax will now be uneconomic. A reduction in such investments will lead to reduced outputs, particularly in the long run. The sensitivity of output to changes in the net-of-tax price, a concept commonly referred to as the price elasticity of supply, will generally depend on the geologic and cost characteristics of the resources subject to taxation. The more important are economically marginal resources in the taxing jurisdiction the greater will be the responsiveness of supply to changes in the after-tax price. The magnitude of the effect on output will also tend to be greater the more time is allowed for the response to occur. In the short run, the producer's options for adjusting production are largely limited to decisions of whether or not to shut in and abandon producing wells. In the long run, investments in exploration and development can be reallocated away from the jurisdiction levying the tax and toward lower tax areas.

The question of whether or not prices rise when the tax causes output to decline largely depends on the response of consumers to reduced supplies. If consumers can readily obtain untaxed substitute supply sources, then the ability of producers to pass tax increases forward to consumers will be limited. If there are few such possibilities, however, then competition for available supplies will be intense, and prices will rise. The sensitivity of consumption to changes in price is termed the price elasticity of demand, and its magnitude depends primarily on possibilities for substitution.

As the preceding discussion indicates, the question of whether production taxes will be borne primarily by producers or by consumers depends on which side of the market is more price sensitive, i.e. on whether supply or demand exhibits the greater price elasticity. In general, the side of the market that is the more price sensitive will be able to shift the major portion of any tax to the other side, and thus avoid the burden. This conclusion must be modified, however, in situations where prices are set by government regulation rather than by the forces of supply and demand.

This incidence framework can be directly applied to the analysis of production taxes on crude oil. The market for crude oil is worldwide in scope; petroleum resources are geographically dispersed both within the United States and abroad, and transport costs are low relative to delivered value. Because virtually all consumers have access to a wide array of alternative supply sources, crude oil prices are determined by the forces of worldwide supply and demand. If, for example, a severance tax were levied in California, producers in that state would be unable to increase prices to the refiners who purchase from them. Any attempt to do so would cause refiners to purchase untaxed supplies from outside the state. In the terminology previously developed, the price elasticity of demand for the petroleum output of a particular state is very high. Since prices cannot rise to offset production taxes, the burden of crude oil production taxes lodges in the supply side of the market, with the owners of capital, labor, and mineral rights employed in oil production. Regarding effects on production and trade patterns, such taxes lead to lower production and increased imports into the states that impose them. Furthermore, the same conclusion applies to those state and local property taxes that are levied as a fraction of the value of production.

The preceding conclusions also largely apply to the federal Windfall Profit Tax. The major differences that occur in the case of the WPT arise because it is applied nationwide, it has a multitude of special provisions and separate rates, and it contains stated provisions that the tax will be removed after total revenues have reached \$227 billion. Despite the fact that the WPT applies nationwide, rather than on a state by state basis as is the case with severance taxes, the conclusion that no significant crude oil price increases will result from it remains true. Foreign oil is already imported into most regions of the United States, and the ability of domestic consumers to substitute imports for U.S. production prevents domestic producers from passing the tax through. For this reason, the effect of the tax is to reduce the after-tax price received by producers, and to lower the incentive to develop and produce domestic crude oil. In these respects it is much the same as other production taxes. The WPT does contain a provision that prevents the tax from exceeding 90% of the net income received by the producer, and this feature may mitigate the tendency for premature shut-in and abandonment.[28] The "net income" limit, however, excludes depletion, intangible drilling expenses, tax payments, and some production costs. Hence, it does little to soften the impact of the WPT on incentives to explore and develop. As noted earlier, there is a wide range of WPT rates for various categories of crude oil and classes of producers, and these differences may redirect economic incentives regarding types of crude oil resources or production methods. There is, however, no systematic evidence on the degree to which these differences have affected operations in the industry.

Provisions of the legislation that implemented the WPT call for the tax to be removed once total receipts have risen to $227 billion. Although this removal is by no means certain, speculation regarding its expiration may encourage producers to postpone some development, or to reschedule production from producing properties, in order to avoid the tax. The possibility for such reallocations would, of course, rise in importance as the tax revenue goal is approached. The impact of such reallocations would be to reduce domestic production and to increase imports in the short run; after expiration of the tax, these short run effects would then be reversed.

McDonald (1981) concurs with the conclusion that the burden of the WPT is confined to the supply side of the market, and goes on to point out that, in the short run, the burden will be borne mainly by operators. In the long run, however, producers will be less aggressive in bidding for private leases and lease payments and royalty shares will fall accordingly. The federal government is the largest owner of potential new oil and gas reserves. Consequently, for new leases, the government will lose in lease bonus and royalty income approximately what it collects in the form of WPT from its leases sold since the WPT was imposed in 1980. In the case of new leases sold by states, their lease income will decline as well, and flow to the federal government in the form of WPT revenue. Thus, at least a portion of the long run burden of the tax will fall upon owners of mineral rights.

In addition to affecting the general level of reserve development, the WPT and other production taxes tend to alter the time pattern of production. If expectations of future prices (net of production costs) lead producers to anticipate that the present value of tax liabilities for future production will be lower than they are currently, then the time pattern of production will be tilted away from the present and toward the future. This would, of course, reinforce the tendency of the tax to reduce current production. Alternatively, if future price increases (again net of production costs) are expected to be lower than the rate of time discount, then production would be raised in the short run and reduced in the future.[29]

Conclusions regarding the effects of production taxes on crude oil cannot be applied directly to the case of natural gas.[30] Market responses to natural gas production taxes may differ for two reasons. First, natural gas reserves are geographically more concentrated than are crude oil reserves, and natural gas transportation costs are typically much larger in relation to delivered prices than is the case with crude oil. These features tend to insulate regional natural gas markets from one another, and thus limit opportunities for natural gas buyers to shop for alternative supply sources. In particular, natural gas buyers often cannot easily substitute imported supplies for domestic natural gas if domestic prices rise. As a consequence, the demand for domestic natural gas is less price sensitive than is the demand for crude oil.

Second, as explained earlier in this chapter, the prices that natural gas producers may charge, and the prices that their buyers will pay, are heavily regulated. In addition to these regulations, the degree to which buyers are willing and able to absorb the price increases that may accompany production taxes is influenced by the controls applied to common carrier pipelines and to public utility gas distributors. Federal regulations regarding rates charged by interstate pipelines allow production taxes to be passed through in the prices that pipelines charge their customers. Downstream from the pipeline, the public utilities that distribute natural gas to final consumers are regulated by state public utility commissions. These utilities are typically allowed to include such taxes in the fuel costs used to determine rates charged to the public. Further, the regional monopoly position of most utilities prevents final consumers from shopping elsewhere when local natural gas prices rise.

All of these points indicate that the price elasticity of demand for natural gas is likely to be lower than in the case of crude oil. As a result, taxes imposed on gas producers are more likely to be passed forward to consumers. The incidence of such taxes is, therefore, expected to be shared by consumers, and by the owners of capital, labor, and mineral rights used in natural gas production. Accordingly, the effects of such taxes on natural gas production will be smaller than would be the case if the entire burden of the tax were borne by suppliers.

The corporate income tax, despite its name, is considered by economists to be a tax on equity capital invested in corporations. The base of the tax is corporate income net of variable production costs, allowed deductions for depreciation and depletion, and interest on capital acquired through debt. Firms are not, however, permitted to deduct a normal or competitive return for equity capital invested in their operations. Thus the tax is actually levied on the normal return to equity capital, plus any true economic profits the firm may earn. In the long run, economic profits (returns above normal competitive levels) do not persist. Thus the corporate levy becomes, in the long run, a tax on the return to equity capital invested in corporations.

Viewed as a levy on capital, the corporate income tax reduces the after-tax return to investors. The impact on capital formation depends on the sensitivity of such investments to levels of after-tax returns. If investment is significantly reduced as a result of the tax, then output in the corporate sector will fall and prices rise. In this case a part of the burden of the tax would be shifted forward to consumers of corporate sector output. If such shifting is not possible then the burden of the tax will fall upon the owners of corporate capital and the inputs used in investment.

Proven reserves of crude oil and natural gas are among the most important capital assets owned by the petroleum industry. The impact of the corporate tax on the exploration and development of new crude oil and natural gas reserves depends on the sensitivity of such activities to the after-tax rate of return. Reductions in such investments will slow the rate at which reserves are formed, and thus constrain production. If reduced production were to result in price increases, then all or part of the tax burden would be shifted from owners of capital to consumers of petroleum products. Accordingly, the net reduction in after-tax returns and levels of investment would be mitigated. For reasons explained earlier, crude oil prices are determined on international markets, and are largely independent of the U.S. tax system. Hence, the burden of the corporate tax on crude oil is borne by suppliers and royalty owners. In the case of natural gas, however, some possibility for price increases and shifting of the tax burden is present, particularly in the long run. Most of the effects of the income tax on production are confined to decisions regarding inputs that can be varied only in the long run, e.g., exploration, development, initiation of enhanced recovery, etc. Because production costs are deductible from taxable income, the question of whether or when to shut in a producing well will be largely unaffected by income taxes.[31]

The effects of state corporation income taxes are essentially the same as those that result from the federal tax. Among major producing states, California, Alaska, and Louisiana levy substantial corporate taxes. In judging the effects of these levies on investment decisions, however, it must be remembered that state taxes are deductible from the corporate income taxed at the federal level. For example, the eight percent corporate levy in Louisiana reduces a corporation's income *net of federal tax* by eight percent. However, because the federal tax rate is 46%, this only represents about four percent of the firm's before-tax income.

From the preceding discussions, it is apparent that production taxes and income taxes have somewhat different effects on resource allocation in the petroleum industry. Accordingly, the impacts of such taxes on production and investment might differ even if revenue yields were equal.[32] Because production taxes do not allow development outlays and operating costs to be deducted, they can curtail or eliminate production from marginal resources. Income taxes, on the other hand, result in tax liabilities only when an investment prospect yields positive net revenues. For these reasons, it has been argued that a production tax will have greater effects on investment and production than would an income tax designed to yield the same amount of tax revenue.[33]

Excise taxes on refined petroleum products represent the third category of tax levies examined. Here, application of the incidence framework indicates that the burden of these taxes will be shared by consumers and refining firms. Such taxes cannot be shifted back to crude oil producers, because crude prices are determined in international markets. In the short run, a share of the burden of such taxes may fall on refiners in the form of lower returns to investments in refining. In the long run, any tax induced reduction in refining margins that lowers returns below competitive levels will lead to reductions in refining capacity, and to increases in refined product prices. Only when prices have risen enough to restore competitive returns to refinery investments will this process cease. If, as seems likely, the long run supply of refining capacity is highly price elastic, then consumers will ultimately bear the entire burden of such taxes.

Special Income Tax Treatment of Oil and Gas

Three aspects of corporate income taxation are of particular importance to the oil and gas industries. These are the rules for deducting exploration and development costs from taxable income, allowances for reservoir depletion, and the treatment of taxes paid to foreign governments. Of these three, the last is not discussed here. An overview of U.S. policy and the economic issues involved in foreign tax treatment is available in National Academy of Sciences (1980).

In 1917, the U.S. Treasury ruled that the incidental expenses of drilling wells, outlays that do not become a part of the capital invested in the property, might be deducted from taxable income as ordinary operating expenses. In the 1950s, the definition of incidental expenses was expanded to include all expenditures of an "intangible" nature, that is, not directly represented by an item of capital installed on the property. Among these intangible outlays are expenses for fuel, labor, supplies, rentals, and repairs. Such expenses often represent most of the total cost of drilling and equipping producing wells, especially when firms elect to hire contract drillers rather than to undertake their own exploratory drilling. Geological and geophysical costs cannot be deducted as incurred. They can, however, be expensed for dry holes upon abandonment. Among the costs that cannot be immediately deducted from taxable income are the "tangible" outlays and the geological and geophysical expenditures attributable to producing wells.

These provisions allow immediate tax deductions for outlays that represent capital investments, expenditures for petroleum reserves that will yield a flow of income in the future. If treated in a manner commensurate to investments in other sectors, these expenditures would be subject to amortization, i.e., recovery over the period of time during which they yield taxable income for the owner. By allowing such outlays to be immediately deducted, corporate income tax liabilities are delayed; the effect on the firm is equivalent to an interest-free loan from the government. Accordingly, these expensing provisions encourage exploration and development activities.

For those expenses that must be amortized, there are also special provisions for oil and gas producers. The reduction in value that occurs as reserves are withdrawn from a reservoir is termed depletion. It is the mineral resource counterpart to the depreciation that applies to other forms of capital, and for tax purposes depletion may be deducted from reported income. In practice, depletion allowances for oil and gas producers may be computed in two very different ways. With the first method, "cost depletion," any exploration and development outlays not immediately expensed are deducted as production proceeds. The fraction of this cost recovered in a given year is set equal to the fraction of estimated reserves withdrawn in that year. This method of computing depletion is similar in spirit to depreciation allowances for other forms of capital. The second depletion method is termed "percentage depletion." This practice applies a simple rule of thumb that sets depletion in a given year equal to a specified fraction of the *gross revenue* the property produces in that year. As applied historically, percentage depletion allowances could easily exceed the unexpensed cost of exploring and developing a reserve.

The Revenue Act of 1918 allowed producers the option of basing depletion for a productive property on its "discovery value," if that value exceeded exploration and development costs. (Recall that, for non-productive properties, exploration and development costs could be immediately expensed.) During the next eight years this generous allowance was reduced until, in 1926, Congress adopted a standard rule of thumb. This rule allowed firms to set depletion equal to 27.5% of the gross income from producing properties, up to a limit of 50% of net income. Intangible expenses were still immediately deductible, and producers retained the option of switching to cost depletion when it was to their advantage.

Percentage depletion was allowed for all producers until passage of the Tax Reduction Act of 1975. This act limited the use of percentage depletion to "independent" producers (firms that were engaged in no substantial retailing activity), and severely constrained even the independents' use of it. In 1970, the depletion allowance had been reduced from 27½ to 22% of gross revenue. The Tax Reduction Act of 1975 provided for the phased reduction of that figure to 15 percent. A limit was also placed on the volume of crude oil and natural gas eligible for percentage depletion. After a series of phased reductions, that limit has been placed at 1000 barrels of crude oil and six thousand cubic feet of natural gas per day.[34]

As a consequence of the Tax Reduction Act of 1975, the *federal* percentage depletion allowance is now largely a matter of historical interest. In some large producing states, however, percentage depletion is still an important item in the computation of state income tax liabilities. Louisiana, for example, levies a state corporate tax with a maximum marginal rate of eight percent and allows percentage depletion, at 38%, for all producers. Similarly, Oklahoma permits percentage depletion, at 22%, for all petroleum producing corporations operating in the state.

To the extent that percentage depletion reduced a firm's corporate tax liability, it made investments in petroleum development more attractive, on an after-tax basis, than they would otherwise have been. Qualitatively, then, the elimination of percentage depletion tended to reduce investments in the oil and gas sector of the economy. It is of interest to note, however, that the tax advantage of percentage depletion was largest for properties that had the greatest excess of gross revenue over expensible costs. It is just such investments that would most likey have been undertaken in the absence of percentage depletion. The small producer exemption in the 1975 Tax Reduction Act may also affect resource allocation by encouraging the formation of small firms, and by generally providing the industry with incentives to arrange its legal organization to qualify for small producer preferences. To date, however, no systematic evidence on the importance of this effect has been presented.

Summary

In summary, it is important to emphasize that all taxes affect resource allocation, including decisions to explore and produce, and the intertemporal pattern of resource development. The burden of taxation is often hard to identify, and seldom coincides with the legal definition of tax liability. In the case of crude oil, an elastic foreign supply prevents taxes from being passed on to final consumers. Thus taxes on petroleum tend to lodge in the supply side of the market. Natural gas markets, on the other hand, are more regionalized so that some scope for passing tax payments through to final consumers is present. Since the inception of the corporate income tax, the petroleum industry has received favorable treatment in its application. This has come in the form of a generous percentage depletion allowance, which was largely phased out in 1975, and ongoing provisions for expensing intangible drilling and exploration expenses. By reducing development costs below what they otherwise would have been, these policies have tended to accelerate production and encourage more rapid depletion of the nation's reserves.

USE OF TAXES AND SUBSIDIES TO STIMULATE SUPPLY OF NEW ENERGY SOURCES

It is abundantly clear that the United States cannot long continue to produce 3.2 billion barrels of oil per year and 18.7 trillion cubic feet of natural gas. In a physical sense, these resources are absolutely limited. As a stock resource they can be produced now or in the future, but unlike the renewable resources, including wood, they cannot be produced indefinitely. As an economic resource, at higher prices (or lower costs) more oil and gas resources will pass into the *reserve* category becoming producible under current price and cost conditions. This raises the question of substitute energy sources in the form of coal from vast U.S. reserves, either oil or gas produced from coal, oil from oil shale, or alternative energy forms directly from solar, or indirectly from wind, biomass, tidal action, and the like.

The reason that energy resources alternative to those in present use have not become available is that their costs of production exceed present prices for conventional energy resources. Studies currently available indicate that even at the high crude oil prices reached in early 1981 (about $39 per barrel) alternative and exotic energy sources were uneconomic. Cost studies prepared by the Bechtel Company for alternative energy sources are shown in Table 4.12.7. Oil is currently being produced from Canadian tar sands and oil sands. While some oil has been produced commercially from oil shale in China, Russia, and Brazil, there is currently no commercial shale oil production in the United States. Two plants are currently being built. One "semi-works" plant by Chevron will begin small-scale production within a year. Union Oil Company of California is building a production plant in Colorado which may begin operating within three years. The latter is possible only with a generous oil purchase subsidy agreement with the U.S. government wherein the government will purchase all shale oil produced by Union at prices well above current crude oil prices. All other oil and gas substitutes shown in Table 4.12.7 clearly involve private costs substantially above current prices for either oil or natural gas. Given these cost estimates, one must conclude that alternatives to oil and gas are not economically feasible under current price-cost conditions.

Under what conditions might it be in the public interest to provide a government subsidy such that private enterprise would find it profitable to produce coal and gas substitutes?[35] From an economic perspective, a subsidy would be in the public interest if any one of the following three conditions prevailed:

Table 4.12.7—Estimated Private Costs for Alternative Synthetic Oil and Natural Gas (Constant 1981 Dollars)

	Cost Including 15% ROI After Taxes	
Energy Source	**$/bbl. of oil equivalent**	**$/million Btu of gas equivalent**
Tar sands	27	4.60
Oil sands	32	5.60
Oil shale	36	6.20
Coal liquifaction	53	9.10
Coal gasification High Btu gas	60	10.40
Beomass (Wood to high Btu gas)	85	14.70
For comparison		
Crude oil price, U.S. Wellhead average (August, 1983)	26.02[A]	
Natural gas, average wellhead marked production (July 1983)		2.52[B]
Delivered to electric utility plants		3.70[B]

Source: Bechtel Co., unpublished data.
[A]USDOE, *Monthly Energy Review*, November 1983, p. 82.
[B]USDOE, *Monthly Energy Review*, November 1983, p. 90.

(1) There was a large net external benefit from producing an energy substitute. This net benefit must exceed the difference between private costs and revenues.

(2) Due to government interference, market prices are giving incorrect price-cost signals to private firms. Rather than introducing a subsidiary in this case, consideration might be given to removing the existing government interference.

(3) Private monopoly exists in one or more of the alternative energy sources, causing cost estimates to be artificially high.

The externalities that might exist could be in the form of national security benefits or technological spillovers including failure of private enterprise to collect all of the benefits from its research and development expenditures.

Several other factors are occasionally mentioned that might justify government intervention with subsidies to new energy sources. It is often claimed that the capital costs for entry into a new energy technology are so high that private capital is reluctant or unable to flow into the new enterprise. This argument assumes an imperfect capital market. Most economists believe that the highly developed U.S. capital market is fully capable of supplying capital funds providing that the new energy source is economically feasible. One also frequently hears that uncertainty of a payout to private enterprise is so great that private decision-makers will refuse to accept anticipated high risks. If this is true, the same factors should preclude government from committing society's scarce capital to the same enterprise. Uncertainty and high risk are not sufficient arguments for a subsidy in the absence of a net external benefit.

In the case of a technological spillover mentioned above, the argument is that benefits from private investments in research and development of a new technology become quickly known, and the firm making such investments cannot capture all of the benefits that flow to society. This in fact is one reason supporting the nation's historic commitment to 17-year patents for new inventions. The technological spillover argument is more relevant to basic research that produces non-patentable processes than it is to applied and technological research that can be patented and commercially exploited.

Market price signals have been and still are incorrect in both crude oil and natural gas prices. Since the 1920s, crude oil and natural gas have received tax subsidies in the form of percentage depletion allowance and the right to expense intangible drilling costs (in lieu of capitalizing all investments in oil and gas development). These two subsidies together have artificially increased the after-tax rate of return for investments in and production of crude oil and natural gas, in turn causing excess capital flows into oil and gas exploration, and supplies to be higher than they would be in the absence of such subsidies. Consequently, market prices have been artificially low. As indicated earlier, market demand prorationing had the opposite effect. By restricting production of crude oil, through 1972 market prices were held *above* competitive levels. Similarly, oil import quotas in effect from 1959 through about 1972, raised domestic crude oil prices well *above* competitive levels. Then crude oil price controls introduced, in 1971 and maintained through January, 1981, held crude oil prices in the United States well *below* world oil price levels. Natural gas price controls, introduced in 1954 and still in effect under the Natural Gas Policy Act of 1978, artificially depress natural gas prices. Thus market prices have been incorrect as a result of a series of federal government policies. The solution to this problem is not to introduce new subsidies attempting to offset past intervention. Rather, it is to eliminate present government intervention that distorts market prices. In fact, the percentage depletion allowance subsidy has been completely removed for integrated oil companies as of 1975, but remains in effect, although somewhat reduced for independent oil producers. The right to expense intangible drilling costs remains in full effect today. Market demand prorationing ended in 1972. Oil import quotas ended in 1972. Price controls on crude oil ended in 1981, and natural gas price controls continued to exist although the issue is currently being debated in Congress.

Regarding monopoly in the energy industry, we have explored this problem at an earlier point and concluded that there is no evidence that monopoly power exists in any sector of the energy industry. In a recent review of government subsidies to non-conventional energy sources, Joskow and Pindyck (1979) concluded that

> government subsidies are a costly and unnecessary alternative to dealing with the problem directly. The removal of price controls—and the guarantee that controls will not be imposed on the prices of nonconventional energy supplies produced by the private sector in the future—would eliminate the one form of market imperfection that is indeed significant and serious. The removal of controls . . . together with a revision of those environmental regulations that are unnecessary and unreasonable, and the clarification of environmental standards and regulations that would apply in the future, would permit private firms to develop new energy technologies at a socially optimal rate. There would be little or no need for the government to subsidize the commercialization of these technologies. (Pindyck and Joskow, p. 17)

Apart from the economics of subsidies to synthetic fuels, political reality apparently required government intervention in this area. President Carter recommended and the Congress legislated that some of the proceeds of the Windfall Profit Tax imposed on crude oil production be earmarked for a newly created Synthetic Fuels Corporation (SFC). The SFC was slow in getting started and during the Reagan Administration has in fact allocated funds to relatively few projects, including oil shale, coal liquifaction, coal gasification, and miscellaneous other projects. President Reagan in 1982 announced his intention to close out the SFC on the grounds that continued subsidies would not advance the general welfare.

In addition, federal energy legislation enacted in 1978 introduced a series of new subsidies to stimulate development of new energy sources, and to subsidize energy conservation which has been advocated as a substitute for additional energy production. The Energy Tax Act of 1978 provides new credits against income tax liabilities for expenditures or investments in solar energy, geothermal energy, heat exchangers, recycling equipment used to recycle solid waste, shale oil equipment for use in retorting shale, equipment used for producing natural gas from geopressured brine, and wind energy production. Tax subsidies were enacted to encourage production of gasolhol as a substitute for gasoline. In addition, tax credits were added to subsidize home insulation or including installation of thermal windows or thermal blankets for swimming pools.

Passage of these new subsidies and tax stimulants may have had wide political support. However, there is no evidence that they were analyzed and evaluated in terms of the three economic justifications listed earlier.[36]

PRICE CONTROLS TO REDISTRIBUTE INCOME

Every economic decision has two *direct* major areas of consequences. One is resource allocation among competing uses of such resources. Government regulations normally alter resource uses and affect resource conservation and the living standards of the people. The other effect is on income distribution. Further, there are commonly *indirect* interaction effects between these two areas. As political scientists have pointed out, there will be gainers and losers from government regulations. However, it is not necessarily a zero-sum game. Often, the losses exceed the gains, with the difference being accounted for by administrative costs or differences in values between winners and losers. In any event, identifying winners and losers is a complex process and correctly evaluating the dollar amounts gained or lost by various interest groups is typically impossible.

The Case of Natural Gas

In 1938 Congress passed the Natural Gas Act authorizing the federal government (the Federal Power Commission) to regulate natural gas flowing in interstate commerce. In 1954, the Supreme Court, through its famous *Phillips vs. Wisconsin* decision, decided that the 1938 legislation required the government to control the wellhead price of natural gas moving in interstate commerce, in addition to prices charged by interstate pipelines for transportation of natural gas.

Neither the 1938 law nor the 1954 Court decision applied to gas limited to intrastate commerce. President Carter recommended and Congress in 1978 passed the Natural Gas Policy Act, which extended natural gas price controls to intrastate gas. This same 1978 legislation also provided for phasing out certain classes of natural gas price controls. However, so-called "old gas" was not included, and under this legislation, price controls are authorized in perpetuity.

In general, consumers that receive and use natural gas at prices below market clearing levels would gain by price controls. Beginning in the late 1950s, prices were set below market clearing levels, with the effect that shortages were created. However, administration of the Act favored residential consumers over industrial gas users. As a consequence, those who had natural gas hook-ups at less than market prices were gainers. In the 1970s, the shortage of new gas became sufficiently severe so that in some areas of the nation, new hook-ups of natural gas were denied. These potential consumers were losers because they received no gas at low prices.

Industrial customers were not favored in the regulatory system. As a consequence, some industrial users were cut off, and many others were denied access to new supplies. These potential industrial users were forced to either do without or, more commonly, to shift to less desirable and higher cost energy sources. The ultimate losers were the consumers of these products, who were forced to pay higher prices.

The major losers under natural gas price control were the owners of natural gas reserves, both lease-holders and royalty owners. However, prior to the 1978 legislation, these losers were limited to suppliers of gas in interstate commerce. With the passage of the 1978 Act, approximately 33 tiers of price controls were mandated, effective in January, 1979. Prices were established ranging from a minimum of $0.204 to $2.243 per thousand cubic feet (Mead, 1979). Congress made economic distinctions between favored and unfavored groups. For example, for the same type of "rollover gas," gas produced by large firms must be sold in 1979 for $0.607 per thousand cubic feet (Mcf). In contrast, smaller firms were allowed to sell the same class of gas for $0.715 per Mcf. And even more favored, Indian tribes or state governments were permitted to sell this gas for $2.096 per Mcf. However, in 1979, the market clearing price of natural gas was estimated to be about $6.00 per Mcf. Thus, all three classes identified above were losers in varying degrees relative to more favored consumers who were able to buy price controlled gas. Thus income was redistributed from natural gas producers to those consumers who received natural gas supplies.

When "new gas" was decontrolled, as prescribed in the 1978 legislation, resource owners for the first time since the 1950s received at least full value for their qualified new gas. The system of controls allowed pipelines transporting natural gas, and public utilities distributing the gas, to "roll in" high cost and low cost supplies and sell gas at a computed average price. This "roll in" procedure led buyers of natural gas at wellhead to pay prices well above what pure competition would have determined, knowing that they could roll in high cost gas mixed with low cost gas and sell it all at an average price. Thus, with decontrol of new gas the resource owner actually received more than a competitively determined market value, and he moved into the winner's column. Prior to the 1978 legislation, when all interstate gas prices were controlled at below competitive levels, new leases purchased reflected artificially depressed market prices. The largest seller of new leases was the federal government for its new OCS leases. Bids for these leases were calculated on the basis of controlled market prices minus all costs discounted to present values at the time of the lease auction. Thus the federal government, which controlled gas prices, actually suffered the loss attributable to the resource owner.

The Case of Crude Oil and Refined Products

Directly and indirectly, the federal government has been involved in setting market prices for U.S. crude oil for several decades. In 1935 the major producing states signed the Interstate Compact to Conserve Oil and Gas, an agreement that provided federal coordination for state imposed production controls. Though the regulators denied any intent to maintain crude oil prices, the actual effect of MER and market demand prorationing was to do exactly that. In the late 1950s, the imposition of voluntary and then mandatory controls on crude oil imports into the United States represented the second major component of an energy policy that kept domestic prices above international levels.[37] The result, according to one set of estimates, was a domestic price level that exceeded foreign price levels by 50 to 75% during the 1960s.[38] This high price policy was abruptly reversed in the early 1970s by a system of crude oil and refined product price controls, first instituted as part of an economy-wide price freeze, and then directed solely at the petroleum industry. Below, we will provide a brief history of federal price and allocation controls.

The U.S. experiment with crude oil and refined product controls began on August 15, 1971.[39] On that date President Nixon, acting under the powers granted in the Economic Stabilization Act of 1970, froze all prices and wages in the economy, except those involving first sale of agricultural or imported goods. This economy-wide freeze, later known as Phase I, lasted 90 days and was replaced, on November 14, 1971, by Phase II. The controls imposed under Phase II were more flexible and were intended to limit general price increases in the economy to no more than three percent per year. The ceiling prices of Phase I became "base prices" in Phase II. Increases above these base prices were permitted if justified by cost increases.

The Phase II controls generally allowed multiproduct manufacturers some latitude in pricing individual items and in making temporary price adjustments. Prices of individual products could generally be varied so long as weighted average price increases were in accordance with the guidelines. Likewise, under "term limit pricing," temporary price increases were allowed if offset by subsequent price reductions. Significantly, however, refined product producers were largely excluded from these flexibilities. Crude oil was specifically excluded from such arrangements; exclusions also applied to gasoline, heating oil, and residual fuel oil, products that collectively accounted for almost three-fourths of domestic refinery yields. Phase II controls nominally allowed costs of crude oil and refined product imports to be charged directly to consumers. In practice, however, requirements that such imports be physically segregated from domestic products prevented such pass-throughs.

In the absence of controls, prices of gasoline and heating oil tend to exhibit seasonal swings. The expectation of gasoline price increases in summer and heating oil price increases in winter provides refiners with incentives to build inventories of these products as periods of peak demand draw nearer. At the same time, such price swings act to ration available supplies among consumers during times when supplies are naturally strained. With this natural allocation mechanism suppressed by price regulations, shortages of heating oil began to develop in the winter of 1972–73.

In January 1973 Phase II controls were replaced by Phase III, a system of voluntary guidelines that were broadly similar in form. The removal of strict controls in an environment of substantial unmet demand led to sharp increases in heating oil prices during early 1973. This prompted federal regulators to impose, on March 6, 1973, a special set of mandatory price controls on the petroleum industry. Known as Special Rule 1, this policy allowed only limited price increases, and only if justified by costs, for the largest 24 integrated oil companies.

The majors were faced with a demand for crude oil and products that substantially exceeded amounts that could be profitably supplied at controlled prices. As a consequence, they were forced to allocate crude oil and products between their own integrated downstream operations and potential independent buyers. Claims of hardship and discrimination by both independent refiners, who sought access to price controlled crude from the majors, and by independent refined product marketers and distributors, generated pressure for a direct government role in the allocation of crude oil and products.

In June 1973, the voluntary Phase III guidelines were replaced by an economy-wide price freeze that lasted until mid-August. The Phase IV regulations that followed provided for a phase-out of controls on most sectors of the economy. For petroleum and refined product markets, however, mandatory controls were continued.

As the defects of petroleum industry controls under Phase IV were being worked out, the Arab-Israeli war of October 1973 broke out. Arab members of OPEC (OAPEC) immediately announced an embargo of crude oil shipments to the United States and the Netherlands. The same producing nations announced production cuts which, although partially offset by production increases elsewhere, resulted in a seven percent decrease in world supply in late 1973. During the same period, Arab and Iranian governments announced a series of price increases that raised the price of their crude oil exports from about $3.00 per barrel to over $11.00 per barrel. The embargo and its accompanying price increases hastened adoption of a comprehensive federal price and allocation policy. In November 1973, the Emergency Petroleum Allocation Act (EPAA) was signed into law. Although it was modified several times, the price and allocation apparatus of EPAA formed the foundation for the federal controls that remained into the early 1980s.

The centerpiece of the EPAA controls was a two-tier system of ceiling prices for crude oil. ''Old oil,'' defined to equal output from producing properties in amounts up to 1972 production levels, was subject to a ceiling price of about $5.25 per barrel. Output in excess of 1972 production levels was termed ''new oil'' and, together with oil from stripper wells and imported oil, was allowed to sell at uncontrolled prices.[40] During the period in which EPAA was in effect, prices for uncontrolled ''new oil'' averaged $10.00 to $12.00 per barrel. As a result of the price increases of late 1973, and the decontrol of new and imported oil under EPAA, petroleum refiners faced huge cost increases. Provisions of EPAA permitted refiners to base refined product prices on each firm's weighted average cost for old and new oil refined. These cost increases were then prorated across individual refined products on the basis of volumes of refined product outputs.

With the adoption of EPAA and its two-tier system of crude oil pricing, almost 40% of U.S. crude oil supply was decontrolled. However, the regulations failed to specify a mechanism whereby rights to price controlled old oil was to be allocated. Initially, the allocation problem was ''solved'' by freezing supplier-purchaser relationships in the industry. Accordingly, the 1972 patterns of trade between crude oil producers, refiners, wholesalers, distributors and refined product retailers, was required to continue. However, this approach did not guarantee specific quantities of crude oil to refiners, nor any particular mix of controlled versus uncontrolled prices. The latter issue was eventually resolved in December 1974, with implementation of the Entitlements Program, a policy that is discussed at length hereafter.

The regulations in force under EPAA were scheduled to lapse in early 1975. However, a series of extensions kept them in effect until December 1975, at which time they were replaced by the Energy Policy and Conservation Act (EPCA). This act further complicated the crude oil pricing program by defining three categories of crude oil, ''lower tier,'' ''upper tier,'' and ''highest tier.'' A target average price level for the three categories was prescribed, and the Federal Energy Administration was directed to establish prices for individual tiers that would achieve this average. The definition of lower tier oil was roughly equivalent to old oil as designated by EPAA, and upper tier oil included most of the balance of domestic production, except for the output of stripper wells. Highest tier oil was uncontrolled and consisted of imported oil and stripper oil (after September 1976). In effect, EPCA placed controls on much of the new oil that had been uncontrolled under EPAA. At the same time, however, EPCA called for the gradual decontrol of all domestic oil, to be accomplished by September 30, 1981.

Under EPCA controls, domestic crude oil prices during 1977–78 ranged from $5.00 to $5.50 per barrel for lower tier oil, to $11.00 to $12.00 per barrel for upper tier oil, and $12.00 to $13.00 for highest tier. After the Iranian revolution, and the wave of crude oil price increases that spread worldwide in 1979, price disparities became much more severe. In 1980, lower tier oil was priced at $6.00 to $7.00 per barrel, and upper tier oil sold for $13.00 to $14.00 per barrel; highest tier oil, however, traded for $35.00 to $36.00 per barrel. During the period that EPCA remained in effect, the proliferation of pricing categories continued; this took place as the process of gradual decontrol exempted additional portions of domestic crude oil from ceiling prices. Just prior to the expiration of controls in early 1981, the regulations distinguished 11 different categories of crude oil. (Distinctions were based on location, ownership, and method of production.)

EPCA continued the old EPAA refined product price controls with relatively minor modifications. Between 1976 and 1979, most refined products were exempted from controls. Motor gasoline, however, was not decontrolled until President Reagan eliminated all petroleum industry price controls in January 1981.

With the elimination of controls in 1981, it may have appeared that the deep federal involvement that had characterized the 1970s had finally ended. This was not, however, the case. Much of the differential treatment of crude oil prices that had been introduced under EPAA and EPCA was incorporated in the Crude Oil Windfall Profit Tax enacted in 1980. This tax was structured to capture, for the government, a portion of the price increases that domestic producers enjoyed when decontrol occurred. The economic effects of this tax are discussed elsewhere in this chapter.

PRICE CONTROLS AND PETROLEUM SUPPLY: ALLOCATIVE EFFECTS

The spectacular price increases that followed the October War of 1973 and the second wave of price jumps that accompanied the revolution and production cuts in Iran, generated strong political pressure in the United States to protect consumers from the massive wealth transfer that would otherwise have taken place. At the same time, the Arab embargo and threats that foreign oil supplies might be used to advance political aims made energy independence a popular cause in the United States. Energy independence, in turn, would be promoted by high domestic energy prices. Both Congress and the energy regulators apparently recognized the fundamental conflict in these two goals. Viewed from this perspective, the complex muddle of price and allocation controls that dominated U.S. energy policy in the 1970s is seen to be an attempt to reconcile or minimize this conflict.

Price controls affect resource allocation in two ways. First, they discourage current production and new investment in future productive capacity. In the case of petroleum, future capacity is secured by exploring and developing new reserves. If controls are expected to persist, the potential profitability of such activity is reduced.[41] Price controls can also alter the time profile of production from existing reserves. The second way price ceilings affect markets is by encouraging consumption in excess of levels that would prevail without controls. In combination the two effects tend to produce shortages, i.e. differences between demand and supply at the controlled price, and an allocation problem for the regulators. The net impact on consumers, the intended beneficiaries of the policy, is not entirely clear. Those consumers who manage to obtain supplies gain from the ability to buy at a lower price; however, some consumers' demands go unsatisfied. Depending on the allocation method used, access to supplies may be rendered uncertain, and such non-monetary costs as waiting lines at filling stations may arise.

The multi-tiered price system applied to domestic crude oil was an attempt to avoid the adverse supply effects that normally accompany price controls. Evidently, the regulators reasoned, "old oil" was profitably produced during 1972 at price levels equal to or below the ceiling prices imposed on such production under

EPAA. Hence, it seemed superficially plausible that price ceilings on such resources would not adversely affect supplies. Initially, at least, newly discovered oil and incremental production from existing reserves were not controlled. This feature appeared to preserve intact the industry's incentive to explore and increase production.

Despite these intentions, however, the crude oil controls did not simply replicate the supply incentives of an uncontrolled market setting.[42] As Kalt (1981) points out, the decontrol of stripper wells presented an incentive to reduce production of old oil to qualify a property for stripper status. Using 1975 price levels, an old oil output of 23 barrels would have yielded about the same revenue as 10 barrels of stripper production. Other analysts have pointed out that the "released oil" provisions of EPAA, whereby an increase in the production of new oil "released" a commensurate amount of old oil from controls, may have actually caused outputs from some properties to exceed levels that would have prevailed without controls. In general, then, the effect of controls on production from existing wells is difficult to determine.

Of primary importance, however, is the effect of controls on expectations of future profits and the return to the exploration and development of new reserves. By exempting the production of newly discovered oil, EPAA attempted to preserve this crucial set of incentives. However, the simple fact remained that the price of about 60% of the oil produced in the United States was held to a level below one-half of the world market price, and oil that was presently decontrolled might well be controlled by some future item of legislation. This thought could hardly have escaped those responsible for making investment decisions in the industry. Indeed, this fear has been borne out in every new set of regulations adopted since EPAA. With the passage of EPCA in 1975, ceiling prices were imposed on the new oil that had been uncontrolled under EPAA. EPCA also promised a phased decontrol of all oil prices, to be accomplished by late 1981. In 1980, however, that promise was effectively withdrawn with enactment of the Crude Oil Windfall Profit Tax, a levy that applies to newly discovered petroleum, as well as all other private supplies. The legislation that imposed the Windfall Profit Tax called for the tax to expire in 1991 or, if earlier, when a specific revenue target was attained. Given the recent history of petroleum price regulation, however, it seems unlikely that the firms now engaged in exploration and development place much confidence in the scheduled expiration of the tax.

According to estimates prepared by Kalt (1981), price controls reduced crude oil production in the United States by approximately 315 million barrels per year (about 900,000 barrels per day) during 1975–1979. This production impact resulted in a commensurate increase in imports from abroad. Moreover, because these foregone domestic supplies could have been made available at a lower cost than the foreign imports that replaced them, there was a "deadweight loss" to the U.S. economy. Kalt (1981) places this loss at about $1.0 billion per year during 1975–1978. When world prices rose in 1979 these deadweight losses rose accordingly, to an estimated $1.8 billion per year in 1979 and $4.6 billion per year in 1980.

Mandatory Allocations and the Entitlements Program

Regulations that are effective in reducing prices below equilibrium levels create shortages and necessitate the imposition of some "non-price" allocation mechanism. With adoption of EPAA, the primary problem was one of assigning rights to price controlled old oil among refiners. Initially old oil allocations were based on historic purchase patterns. However, this allocation method resulted in wide cost disparities among refiners, and eventually gave way to the "entitlements program." The entitlements program was designed to approximately equalize the effective cost of crude oil to all refiners at a level equal to the national weighted average of controlled and uncontrolled prices. In effect, the program distributed rights, or "entitlements," to old oil in proportion to individual refiners' use of uncontrolled oil. Thus, a refiner who paid the uncontrolled world price for a barrel of imported crude oil was granted an entitlement for a fraction of a barrel of low priced old oil. When EPCA was enacted, the entitlements program was amended to incorporate two categories of price controlled petroleum.

The entitlements program subsidized the use of crude oil imports. During 1974–1978, the price differential between old or lower tier oil, and imported oil ranged from $5.00 to $8.00 per barrel; during 1979 and 1980, this difference shot up to $17.00 to $30.00 per barrel. The fractional entitlement offered for each barrel of imports also varied over this period. As a result, the implicit subsidy for refining imported oil varied widely. During 1974–1978, the subsidy ranged from $1.60 to $2.85 per barrel; in 1979–1980, it grew to $3.08 to $5.16 per barrel.[43]

By subsidizing imports, the entitlements program worked in direct opposition to the goal of limiting the nation's reliance on foreign imports, and thus provided an ironic counterpoint to the rhetoric of "project independence." According to one set of estimates, entitlements resulted in additional crude oil imports of .305 to .491 million barrels per year during 1974–1980.[44] Also, because the entitlements subsidy made crude oil costs appear artificially low, the true cost of crude oil to the U.S. economy actually exceeded the value of some of the uses to which it was allocated. These deadweight losses on the consumption side have been estimated at about $750 million per year during 1975–1980.[45]

Aside from effects on resource allocation, the price regulations and entitlements program brought about a massive redistribution of wealth among various market participants.[46] Crude oil producers faced a loss in income that varied with the gap between controlled and uncontrolled crude oil prices. During the 1975–1980 period, these losses averaged an estimated $26.3 billion per year. The immediate beneficiaries were refiners, who faced lower crude oil input prices. However, a complete determination of the pattern of benefits is complicated by the fact that competition among refiners resulted in lower refined product prices for ultimate consumers. Also, some of the entitlements allocated for old oil were offered through a variety of special policies that offered preferential treatment to individuals and groups. Taking these factors into account, Kalt (1981) has estimated that the benefits captured by refiners averaged about $16.3 billion per year. Gains to consumers, as a result of lower product prices, are placed at $7.6 billion per year. Note that the sum of the gains to refiners and consumers falls short of the loss to producers. The difference, 2.4 billion per year, is an estimate of the major component of deadweight losses that these programs imposed on the economy.

Within the entitlements program there existed a variety of exemptions and special considerations. The most important of these was the "small refiners bias," a program that granted extra old oil entitlements, and hence a larger crude oil subsidy, to refiners with refining capacity less than 175,000 barrels per day. The degree of bias and the size of the subsidy were most pronounced for firms that had capacity below 10,000 barrels per day. At its peak in 1976, the program cut the crude oil acquisition cost for such firms by $3.00 per barrel below the industry average.

Quite naturally, the small refiner bias made the construction and operation of small refining plants artificially attractive. According to estimates presented by Boyce (1983), the small refiner subsidy was responsible for virtually all of the net entry of new firms and plants into refining during the existence of the policy. Minimum efficient scale in refining is generally believed to be about 200,000 barrels per day. In the 1975–1980 period, 54 firms (net) in the 0 to 30,000 barrel per day capacity range entered the industry. Within two years of the date the program terminated, an equal number of firms in that size range had exited the industry.[47]

Administration and Compliance Costs

The review of allocative effects and deadweight losses resulting from price and allocation controls would be incomplete without a discussion of the costs of administering and complying with the regulations. The enforcement of federal price and allocation policies was a major reason for the formation of the Federal Energy Office, and its successors, the Federal Energy Administration and the Department of Energy. Thus, a substantial portion of the budgets of these agencies must be attributed to the controls. In the private sector, the costs of interpreting and complying with the regulations were also substantial, though much more difficult to measure.

In one major attempt to estimate the public sector costs of administering crude oil price and allocation regulations, MacAvoy (1977) estimated costs in 1975 at over $60 million. In separate studies (U.S., President, various years), annual federal costs have been found to range from $63 million to $198 million, with an average level of $135 million per year during 1974–1980.

Private sector costs are much more difficult to estimate. As Kalt (1981, p. 208 ff.) points out, industry compliance involves three types of costs: the sheer burden of paperwork, the delays and disruption occasioned by the need to obtain government approval for various actions, and the rise of political acumen as a determinant of competitive viability. Only the first of these three has been documented to date. MacAvoy

(1977) surveyed the federal compliance program in 1975 and found that over 300,000 firms were affected, and that these firms were required to file over one million forms per year. To the cost of simply filling out such forms, one must add the costs of keeping records, obtaining legal advice, and seeking appeals and exceptions. Overall, MacAvoy (1977) estimated these paperwork costs at over $500 million in 1975. Given the nature of the problem, however, this estimate should clearly be viewed with caution.

The system of price and allocation controls that dominated U.S. energy policy during the 1970s represents a complex and costly chapter in the history of petroleum industry regulation. The apparent interest of the regulators was to redistribute the rents that accompanied crude oil price increases toward consumers of refined products, in the form of lower prices. To accomplish this, the regulators sought to limit crude oil price increases and to base refined product prices on crude oil costs. Although this may appear simple on the surface, it gave rise to a multitude of practical complications and inefficiencies. Among the end results were an increased dependence on foreign supplies at a time when they were exceedingly uncertain, and the rapid growth of a federal bureaucracy to manage the allocation problems the regulations created. At best, this episode demonstrated the inefficiency of price control as a tool for redistributing income.

REGULATORY ISSUES FOR THE FUTURE

The future, by definition, is uncertain. Energy problems will arise in the future which cannot be foreseen today, just as the approximately tenfold increase in the price of oil occurring in the 1970s was forecast by no one. Given perceived energy problems, governments may be compelled to legislate and regulate as a result of political pressures being brought to bear on it. The purpose of this section is to identify some future regulatory issues that can be foreseen at this time.

(A) What areas of the country should be reserved with oil exploration and production prohibited? Currently, environmental interests on one hand and oil industry and federal government interests on the other hand are debating which areas offshore from the Pacific Coast and California in particular should be leased for oil exploration and development. In the future, this debate will be extended to the Arctic National Wildlife Refuge located on Alaska's North Slope. This area is due east of Prudhoe Bay, bordering the Beaufort Sea to the north and Canada's Northwest Territory on the east. The U.S. Geological Survey and the oil industry believe that this national preserve is the most promising unexplored petroleum province remaining in the continental United States. As a matter of national energy policy, the nation must determine to what extent this and other reserved areas will be available for exploration and production where hydrocarbons are found.

(B) Currently, the U.S. government prohibits the export of oil from the prolific Prudhoe Bay oil field, also on Alaska's North Slope. This prohibition was legislated in the Export Administration Act partly in response to political pressure from environmental groups who opposed construction of the transAlaska oil pipeline from Prudhoe Bay to Valdez on the southern coast of Alaska. These environmental groups were joined by maritime unions wishing to avoid shipment of the oil to Japan on foreign-manned tankers. With the ban on exports, the Jones Act passed in 1926 requires that all marine shipments between any two U.S. ports be on ships (1) built in the United States, (2) owned by American firms, and most importantly, (3) manned by American crews.

The effect of the ban is, first, to require shipment to the U.S. West Coast with any excess over West Coast needs being transshipped to the Gulf and Atlantic coasts. The first problem that arises out of the export ban is that unnecessary transportation costs to the Gulf Coast amounting to approximately four dollars per barrel are incurred relative to export of some of this oil to Japan and substitute imports of oil into the Gulf and Atlantic areas of oil from the Middle East, Africa, and other areas. This is a social cost imposed on the nation. It unnecessarily uses scarce resources and inevitably brings about a lower standard of living for this nation than would otherwise be the case. Second, since the value of oil at wellhead is a residual after production and transportation costs are subtracted from market values, wellhead prices are reduced by approximately four dollars per barrel. The state of Alaska, through royalties and severance taxes, loses approximately 25% of the revenue which it would collect in the absence of the export ban. Similarly, the U.S. government loses revenues which would otherwise accrue in the form of windfall profit taxes and corporate income taxes on oil companies.

Finally, the export ban, by forcing all North Slope oil production into the West Coast, has brought about what is commonly referred to as the West Coast crude oil glut. The social costs of the export ban amounts to about $1,460 million per year. This is the result of about one million barrels per day of North Slope oil being shipped to the Gulf Coast at a social cost of approximately four dollars per barrel. Unless the people of the United States are willing to continue to bear this high cost, Congress must resolve this problem.

The West Coast crude oil glut resulting from the Alaskan North Slope oil export ban is exacerbated by production of crude oil from the Elk Hills California field. As indicated earlier, this field was formerly known as Naval Petroleum Reserve No. 1. The West coast glut is now being extended due to large discoveries and production from the prolific Santa Maria basin, offshore from Santa Barbara County, California. Production from this area is expected to reach 500 million barrels per day by 1990. Table 4.12.8 shows that currently approximately one million barrels per day of excess supply from the Pacific Coast are shipped via Panama to the Gulf Coast, the Atlantic Coast, or Caribbean regions. If the export ban is not removed, then lower cost methods of shipment to U.S. markets east of the Rockies must be considered. These methods might include extension of one existing pipeline, or construction of one or more totally new pipelines.

Table 4.12.8—West Coast Oil Supply-Demand Balance, April 1983 (Thousand bbls. per day)

Domestic Crude Oil Production	Bbls.
North Slope Alaska	1,599
Cook Inlet Alaska	66
California, onshore and offshore	1,187
Total West Coast	2,852
Imports	140
Total supply to West Coast	2,992
Refinery requirements on the West Coast	1,901
Excess shipped via Panama Canal, Panama pipeline to the Gulf Coast or Atlantic Coast or Caribbean	1,091

Source: Arlon Tussing, *ARTA Energy Insights*, No. 1, Summer 1983, p. 10.

(C) The giant Prudhoe Bay oilfield also contains an estimated 26 trillion cubic feet of natural gas. This is approximately 13% of the total U.S. natural gas reserves. The problem of a transportation system must be resolved. The estimated capital costs for the currently approved Alaska Natural Gas Transportation System (ANGTS) is estimated to be in excess of $35 billion. At current natural gas delivered prices and gas pipeline estimated operating costs, construction of the pipeline appears to be uneconomic. (Mead, 1978) The government must decide whether or not new and alternate transportation system proposals are to be invited, and whether or not an externality-justified subsidy to pipeline construction is merited.

(D) The Strategic Petroleum Reserve is now being filled to be drawn down in the event of a national emergency. However, regulations have not been adopted which prescribe procedures to be followed. Based upon past experience one can forecast that in the event of a national emergency and drawdown of these reserves, special interests will immediately assert their demand for special treatment. Small refineries will likely request a "small refiner bias" in the form of either prices below market or set-aside supplies, insulated from larger firm competition. Consumer groups will undoubtedly demand that oil be sold at below market prices in order to moderate petroleum product increases. Oil may be sold from the reserves at legislated or administered prices below competitive levels. Alternatively, oil may be sold impersonally to the highest bidder at competitive auctions. Congress should determine procedures to be followed in advance of a national emergency.

(E) Over the past couple decades, the "environmental revolution" has produced a multitude of new regulations governing energy exploration, development, production, transportation and use. Too often, the public has perceived regulations imposed on oil production, as protecting the environment, while "big oil companies" pay the costs. New regulations have been imposed, often with no consideration for their relative costs and benefits. From an economic perspective, the money costs may be borne by oil companies, if and only if markets do not allow higher costs to be passed on to consumers, but the resource costs are always borne by the nation at large, independent of who pays the money costs. Environmental regulations, like any other expenditure of a nation's scarce resources, should be justified by a showing that the benefits to society exceed the costs borne by society. Where the opposite relationship holds, such regulations are counterproductive of a conservation goal and lower the nation's standard of living. Congress should reexamine past environmental regulations in the energy industry, and should carefully examine future regulations before they are imposed to establish that the social benefits exceed the social costs.

SUMMARY AND CONCLUSIONS

As the analysis in this chapter demonstrates, the influence of government policy and regulation in the U.S. energy industry is pervasive. All of the policies discussed, whether intended to affect the distribution of wealth, to protect the environment, to promote the goals of national security or conservation, or to simply raise revenue for the public treasury, change the allocation of energy resources. They do so by altering the system of incentives and constraints that face the producers and consumers of these resources. As we stated in our opening section, there are sound economic reasons for government intervention in the marketplace. We also noted, however, that the costs of government involvement are often high. The most important of these costs do not show up in government budgets. Rather, they are borne by consumers and resource owners in the form of higher prices and costs. In recognition of these effects we offered the prescription that policy-makers ask whether the benefits of intervention exceed these costs before proceeding to implement policy. At present, this question is not being answered in any systematic way. Indeed, it is apparently seldom asked.

The principal rationales for government intervention include the presence of externalities, such as environmental pollution, an absence of competition, or an imperfection in the distribution of income. In reality, however, the pattern of actual public policy is only loosely related to these principles. Too often, policies promoted to correct for externalities serve, primarily, to enhance the wealth of a specific interest group. The market demand prorationing system, offered in the name of conservation, and the mandatory oil import quota program, ostensibly imposed to enhance national security, are important examples of this phenomenon. Similarly, the most significant effect of many income distribution measures is to alter the allocation of resources. The most important examples of this are in the systems of price controls for natural gas, crude oil, and refined products, and in the provisions of these and other policies that grant favorable tax and regulatory treatment to small producers. Finally, in those cases where the primary impact of policy coincides with an important public interest issue, as is the case with environmental protection, the form of government intervention is seldom efficiently structured.

In the coming decades the United States will face a variety of new energy policy and regulatory questions. Several of these items were raised in the preceding section, e.g., the debate over exploration and development on reserved public lands, policy toward exports of Alaskan crude oil and transportation of Alaskan natural gas, and management of the Strategic Petroleum Reserve. To this list one might add the issue of natural gas price decontrol, the question of whether the Windfall Profit Tax will be phased out as planned, and the controversy over the rights of states versus the federal government in the development of energy resources. If the past is any guide, the focus of public debate over these policies will often stray from, or overlook the principal economic issues involved. Yet, the costs of permitting history to be repeated are growing. As resources become increasingly scarce, the rewards for better management and more well-informed decision-making grow larger. By any sensible measure, America's energy resources are becoming increasingly scarce. From this perspective, then, the need for more objective and carefully considered energy policy can only become more urgent.

FOOTNOTES

* The authors would like to thank Margret Walls and Grant Gustafson for assistance in preparing this chapter.

1. The economic analysis of such issues is the subject matter of welfare economics. For an intermediate treatment of this subject, see Browning and Browning (1982), Chapter 17.
2. For an elegant statement of this proposition and a discussion of other results from the theory of general equilibrium and welfare economics, see Arrow (1977).

3. For a discussion of alternative policy approaches for dealing with environmental externalities, see Freeman, Haveman, and Kneese (1973), Chapter 5.
4. See Freeman, Haveman, and Kneese (1973), Chapters 6 and 7.
5. In the early 1970s the Nixon administration proposed the use of taxes to control lead additives in gasoline and emissions of sulfur oxides. These attempts were eventually abandoned, however, in favor of a traditional standards-regulation policy. For a description, see U.S. Council on Environmental Quality (1971), Chapter 1.
6. For an analysis, see Elliott (1973).
7. For the majority view see Federal Trade Commission (1974), Duchesneau (1975), Mancke (1976), Markham *et al.* (1977), Teece (1976), and Mitchell (1976). For a contrary view see Blair (1976), Allvine and Patterson (1972), and Measday (1977).
8. Adelman (1980) has argued in his paper "The Clumsy Cartel" that "Saudi Arabia, their neighbors, and others, are fine-tuning a cartel with coarse instruments" with the result that the OPEC price is highly unstable but generally short of unknown cartel maximizing price.
9. For alternative views on OPEC as a cartel with varying degrees of effectiveness, see Adelman (1978, 1980, and 1982), Teece (1982), Lichtblau (1981 and 1982), and Fesharaki, Feriedun, and Isaak (1982).
10. While the rate of return data referenced here are the best available, it should be noted that accounting profits do not correspond perfectly with economic concepts of profit. Fisher and McGowan (1983) observed that "Accounting rates of return, even if properly and consistently measured, provide almost no information about economic rates of return." Some of the measurement problems are cancelled out by the fact that both rates compared are accounting measures.
11. See, for example, the set of cases summarized in Harrington and Krupnick (1981), p. 542 ff. and in Council on Environmental Quality (1982), p. 60 ff.
12. See Council on Environmental Quality (1982) for a discussion of the extent of use of these approaches and for examples of cost savings attained.
13. See Congressional Research Service (1979), p. 105 ff.
14. For a more detailed survey of recent oil spills, see U.S. Council on Environmental Quality (1979), Chapters 2, 4, 5 and U.S. Council on Environmental Quality (1979), pp. 624 ff.
15. See U.S. Council on Environmental Quality (1977) for additional detail. This measure of emissions excludes natural oil seeps and pollution that settles from the atmosphere. Emissions from these sources are placed at 1.2 million tons per year.
16. *Ibid.*, p. 6 ff.
17. Most of the discussion of the Santa Barbara spill is drawn from Mead and Sorensen, 1970.
18. *Ibid.*
19. Descriptions presented in this section are drawn primarily from U.S. National Oceanic and Atmospheric Administration, 1983.
20. *Ibid.*, p. 31.
21. American Petroleum Institute, 1982.
22. *Ibid.* See, also, National Academy of Sciences (1979).
23. U.S. Internal Revenue Service, 1982, p. 40.
24. *Ibid.*
25. The net income limitation could, in principle, mitigate the tendency for production taxes to induce premature abandonment. The extent to which it does in practice would depend on how costs are reckoned, and on whether a marginal cost criterion is used in determining the economic limit. Regardless of these considerations, however, the WPT retains the disincentive to invest in exploration and development, an attribute that characterizes simpler production taxes.
26. Commerce Clearing House, Inc. (1982).
27. Much of the analysis presented in this section is discussed in greater detail in National Academy of Sciences (1980), Chapters 3, 4 and 6.
28. The result depends in part on the degree to which the exemption is applied to "marginal" as opposed to "average" net income. If, for example, the limit were applied uniformly to all producing wells in a field, then the tax would continue to adversely affect production from economically marginal wells in that field.
29. See Lehman (1981) for further analysis of this point. Sweeney (1977) presents a general framework in which one can analyze the effects of various market distortions, such as taxes and externalities, on the time pattern of production.
30. Several of the points developed in this discussion are adopted from National Academy of Sciences (1980), p. 51 ff.
31. See McDonald (1976) for further discussion of the corporate income tax. It should be pointed out that the effects of the corporate income tax apply to all corporate investments in the economy. Hence, the discussion in the text should not be interpreted to mean that the tax necessarily discriminates against investment in petroleum corporations relative to other corporations. The differential effects of the tax in individual sectors depends in part on the degree of risk and capital intensity in various sectors, but a thorough treatment of these issues is beyond the scope of the present chapter.
32. Several of the conclusions presented regarding the comparative effects of income and production taxes on resource allocations are discussed in more detail in National Academy of Sciences (1980), p. 56 ff.
33. *Ibid.*
34. Certain additional allowances are made from oil produced by qualified secondary or tertiary recovery methods.
35. See Schmalensee (1980).
36. For a subsequent economic evaluation of these measures, see McDonald (1979).
37. Both the prorationing policies and the mandatory oil import program are discussed at length elsewhere in this chapter.
38. See U.S. Cabinet Task Force on Oil Import Control, 1970.
39. This historical discussion is drawn primarily from Deacon, Mead and Agarwal (1980) Chapter 1, Kalt (1981) Chapter 1, and Lane (1981).
40. The regulations also permitted production at levels in excess of 1972 amounts to "release" a corresponding amount of old oil from controls. For an economic analysis of this "released oil" provision, see Kalt (1981), Chapter 3.
41. If price ceilings are expected to be temporary, or if a phased decontrol is in progress, short run supplies might fall dramatically, as producers curtail output in anticipation of future price increases. On the other hand,

announcement of a permanent ceiling could lead to increased short run production if the present value of future prices, and hence the profit from future production, is expected to fall. For a theoretical analysis, see Lee (1978) and Kalt (1981), Chapter 3.

42. See Kalt (1981), Chapter 3, for an extensive examination of the effects of crude oil controls on the decisions of producers.
43. The source of these estimates is Kalt (1981), p. 58 ff.
44. *Ibid.*
45. *Ibid.*
46. The source of estimates regarding these redistributions is Kalt (1981), p. 216 ff.
47. See Boyce (1981) Chapter 5.

References

1. Adelman, M.A., 1972, *The World Petroleum Market*, Johns Hopkins University Press, Baltimore, p. 8.
2. Adelman, M.A.,.1973, testimony, U.S. Senate, Committee on Interior and Insular Affairs, Subcommittee on Integrated Oil Operations, Hearings, December 20, 93rd Congress, 2nd Session, Part 4, p. 1338.
3. Adelman, M.A., 1978, "Constraints on the World Oil Monopoly Price," *Resources and Energy*, Sept., p. 4.
4. Adelman, M.A., 1980, "The Clumsy Cartel," *The Energy Journal*, Vol. 1, No. 1, January, 43–53.
5. Adelman, M.A., 1982, "OPEC as a Cartel," in Griffin, James M., and David J. Teece, *OPEC Behavior and World Oil Prices*, George Allen & Unwin, London, 37–63.
6. Allvine, F.C., and J.M. Patterson, 1972, *Competition Ltd.*, Indiana University Press, Bloomington, Ind.
7. American Petroleum Institute, 1982, "State and Local Oil and Gas Severance and Production Taxes," (mimeograph), Washington, D.C.
8. Blair, John M., 1976, *Control of Oil*, Pantheon Books, New York.
9. Bohi, Douglas R., and Milton Russell, 1978, *Limiting Oil Imports*, The Johns Hopkins University Press, Baltimore.
10. Boyce, Paul G., 1983, "The Small Refiner Bias: A Case Study in U.S. Federal Energy Regulation," unpublished dissertation, University of California at Santa Barbara, June, p. 71.
11. Brannon, Gerard M., 1974, *Energy Taxes and Subsidies*, Ballinger Publishing Co., Cambridge, Mass.
12. Browning, Edgar K., and Jacqueline M. Browning, 1983, *Public Finance and the Price System*, Macmillan, New York.
13. Deacon, Robert T., Walter J. Mead, and Vinod B. Agarwal, 1980, *Price Controls and International Petroleum Product Prices*, U.S. Department of Energy, Washington, D.C.: U.S. Government Printing Office.
14. DeChazeau, M.G., and A.E. Kahn, 1959, *Integration and Competition in the Petroleum Industry*, Yale University Press, New Haven, CT.
15. Duchesneau, Thomas D., 1975, *Competition in the U.S. Energy Industry*, Ballinger, Cambridge, Mass.
16. Elliott, R.D., 1973, "Economic Study of the Effect of Municipal Sewer Charges on Industrial Wastes and Water Use," *Water Resources Research*, October.
17. Erickson, Edward W., and R.M. Spann, 1971, "Supply Response in a Regulated Industry: The Case of Natural Gas," *Bell Journal of Economics and Management Science*, Volume 2, No. 1, pp. 94–121.
18. Fesharaki, Fereidun, and David T. Isaak, 1983, *OPEC, the Gulf, and the World Petroleum Market*, Westview Press, Boulder, Colo.
19. Fisher, Franklin M., and John J. McGowan, 1983, *American Economic Review*, Vol. 73, No. 1, March, 82–97.
20. Freeman, A. Myrick, Robert H. Haveman, and Allen V. Kneese, 1973, *The Economics of Environmental Policy*, John Wiley and Sons, New York.
21. Harrington, Winston, and Alan J. Krupnick, 1981, "Stationary Source Pollution Policy and Choices for Reform," *Natural Resources Journal*, Vol. 21, No. 3, July, pp. 539–564.
22. Hogan, William W., 1983, "Oil Stockpiling: Help Thy Neighbor," *The Energy Journal*, Vol. 4, No. 3, July, pp. 49–71.
23. Johany, A.D., 1978, "OPEC is Not a Cartel: A Property Rights Explanation of the Rise in Crude Oil Prices," Ph.D. dissertation, University of California, Santa Barbara.
24. Johany, A.D., 1980, *The Myth of the OPEC Cartel*, John Wiley & Sons, New York.
25. Joskow, Paul L., and Robert S. Pindyck, 1979, "Should the Government Subsidize Non-Conventional Energy Supplies?", Mimeographed Working Paper, M.I.T. Energy Laboratory, MIT-EL 79-003WP, January.
26. Kahn, A.E., 1974, testimony, Hearings before the Committee on Interior and Insular Affairs, U.S. Senate, 94th Congress, 1st session, part 1, pp. 317–323.
27. Kalt, Joseph P., 1981, *The Economics and Politics of Oil Price Regulations*, M.I.T. Press, Cambridge, Mass.
28. Kalter, Robert J. and William A. Vogely (eds.), 1976, *Energy Supply and Government Policy*, Cornell University Press, Ithaca, N.Y.
29. Lane, William C., 1981, *The Mandatory Petroleum Price and Allocation Regulations, A History and Analysis*, American Petroleum Institute, Washington, D.C.
30. Lehman, Dale E., 1981, "A Reexamination of the Crude Oil Windfall Profits Tax," Vol. 21, No. 4, October, pp. 583–589.
31. Lichtblau, J.H., 1981, "Factors Affecting World Oil Prices in the 1980's," *Oil and Gas Journal*, Vol. 79, November 9, p. xx.
32. Lichtblau, J.H., 1982, "The Limitation to OPEC's Pricing Policy," in Griffin, James M., and Teece, David J., *OPEC Behavior and World Oil Prices*, George Allen & Unwin, London, 131–174.
33. Liebeler, Wesley J., 1976, "Integration and Competition," in E.J. Mitchell (ed.)., *Vertical Integration in the Oil Industry*, American Enterprise Institute, Washington, D.C. pp. 5–34.
34. Lovejoy, Wallace F., and Paul T. Homan, 1967, *Economic Aspects of Oil Conservation Regulation*, Johns Hopkins University Press, Baltimore.
35. McDonald, Stephen L., 1971, *Petroleum Conservation in the United States: An Economic Analysis*, Johns Hopkins University Press, Baltimore.
36. McDonald, Stephen L., 1976, "Taxation System and Market Distortion," in Kalter, Robert J. and William A. Vogely (eds.), *Energy Supply and Government Policy*, Cornell University Press, Ithaca, N.Y. pp. 26–50.
37. McDonald, Stephen L., 1979, "The Energy Tax Act

of 1978," *The Energy Journal*, October, pp. 859–869.

38. McDonald, Stephen L., 1981, "The Incidence and Effects of the Crude Oil Windfall Profits Tax," *Natural Resources Journal*, Vol. 21, No. 2, April, pp. 331–340.
39. Mancke, Richard B., 1970, "The Long Run Supply Curve of Crude Oil Produced in the United States," *Antitrust Bulletin*, Vol. 15, pp. 727–756.
40. Mancke, Richard B., 1976, "Competition in the Oil Industry," in E.J. Mitchell (ed.), *Vertical Integration in the Oil Industry*, American Enterprise Institute, Washington, D.C., pp. 35–72.
41. Markham, J.W., A.P. Hourihan, and F.L. Sterling, 1977, *Horizontal Divestiture and the Petroleum Industry*, Lippincott, Cambridge, Mass.
42. Mead, Walter J., 1970, "The Economic Cost of the Santa Barbara Oil Spill," *Proceedings of the Santa Barbara Oil Spill Symposium*, December 16–18, 1970, University of California, Santa Barbara, pp. 183–226.
43. Mead, Walter J., 1971, "A National Defense Petroleum Reserve Alternative to Oil Import Quotas," *Land Economics*, Vol. XLVII, No. 3, August, pp. 211–224.
44. Mead, Walter J., 1978, "An Economic Appraisal of the Northwest Alcan Project," *An Overview of the Alaska Highway Gas Pipeline: The World's Largest Project*, American Society of Civil Engineers, pp. 23–40.
45. Mead, Walter J., 1979, "An Economic Analysis of Crude Oil Price Behavior in the 1970's," *The Journal of Energy and Development*, Vol. 9, No. 2, Spring, pp. 212–228.
46. Mead, Walter J., 1979, "The Natural Gas Policy Act of 1978: An Economic Evaluation," *Contemporary Economic Problems 1979*, William Fellner (editor), American Enterprise Institute, Washington, D.C., pp. 325–355.
47. Mead, Walter J., A. Moseidjord, and P.E. Sorensen, 1983, "The Rate of Return Earned by Lessees under Cash Bonus Bidding for OCS Oil and Gas Leases," *The Energy Journal*, Oct., Vol. 4, No. 4, pp. 37–52.
48. Measday, Walter S., 1977, "The Petroleum Industry," in Walter Adams (ed.), *The Structure of American Industry*, fifth edition, Macmillan, New York, 130–164.
49. Mitchell, Edward J., 1976, "Capital Cost Savings of Vertical Integration," in E.J. Mitchell (ed.), *Vertical Integration in the Oil Industry*, American Enterprise Institute, Washington, D.C. pp. 73–104.
50. Moran, T.H., 1978, *Oil Prices and the Future of OPEC*, Johns Hopkins University Press, Baltimore.
51. National Academy of Sciences, 1979, *A Taxonomy of Energy Taxes*, U.S. Department of Energy, DOE/EIA 10496, Washington, D.C.
52. National Academy of Sciences, 1980, *Energy Taxation: An Analysis of Selected Issues*, U.S. Department of Energy, DOE/EIA 0201/14, Washington, D.C.
53. Office of Defense Mobilization, 1956, *Hearings in the Matter of Petroleum Import Restrictions*, October 24, Washington, D.C.
54. Pindyck, R.S., 1978, "OPEC's Threat to the West," *Foreign Policy*, No. 10, Spring, pp. 36–52.
55. Salant, S.W., 1979, "Staving off the Backstop: Dynamic Limit-Pricing with a Kinked Demand Curve," *Advances in the Economics of Energy and Resources*, R.S. Pindyck, editor, Vol. 2, JAI Press, Greenwich, Conn, p. 187.
56. Schmalensee, Richard, 1980, "Appropriate Government Policy Toward Commercialization of New Energy Supply Technologies," *The Energy Journal*, Vol. 1, No. 2, April, 1–40.
57. Sweeney, James L., 1977, "The Economics of Depletable Resources: Market Forces and Intertemporal Bias," *Review of Economic Studies*, Vol. 44 (1), pp. 125–142.
58. Teece, David J., 1976, "Vertical Integration in the U.S. Oil Industry," in E.J. Mitchell (ed.), *Vertical Integration in the Oil Industry*, American Enterprise Institute, Washington, D.C. pp. 105–190.
59. Teece, David J., 1976, *Vertical Integration and Vertical Divestiture in the U.S. Oil Industry*, Institute for Energy Studies, Stanford, Calif.
60. U.S. Cabinet Task Force on Oil Import Control, 1970, *The Oil Import Question*, Washington, D.C. USGPO, February.
61. U.S. Cabinet Task Force on Oil Import Control, 1970, *The Oil Import Question: A Report on the Relationship of Oil Imports to National Security*, U.S. Government Printing Office, Washington, D.C.
62. U.S. Congressional Research Service, 1979, *Energy Development Project Delays: Six Case Studies*, in U.S. Senate Committee on Environment and Public Works, 96th Congress, 1st Session, U.S. Government Printing Office, Washington, D.C.
63. U.S. Council on Environmental Quality, 1971, *Environmental Quality, Second Annual Report*, U.S. Government Printing Office, Washington, D.C.
64. U.S. Council on Environmental Quality, 1977, *Oil and Gas in Coastal Lands and Waters*, U.S. Government Printing Office, Washington, D.C.
65. U.S. Energy Information Administration, 1978, *Oil and Gas Supply Curves for the Administrator's Annual Report*, U.S. Department of Energy, Washington, D.C.
66. U.S. Department of Energy, 1983, *Monthly Energy Review*, December.
67. U.S. Federal Trade Commission, 1952, *The International Petroleum Cartel*, Staff Report, Washington, D.C. pp. 22–23.
68. U.S. Federal Trade Commission, 1974, *Economic Report, Concentration Levels and Trends in the Energy Sector of the U.S. Economy, Staff Report*, Washington, D.C. March, p. 147.
69. U.S. Geologic Survey, 1980, *Future Supply of Oil and Gas from the Permian Basin of West Texas and Southeastern New Mexico*, U.S. Department of the Interior, Washington, D.C.
70. U.S. House of Representatives, 1974, Committee on Banking and Currency, Ad Hoc Committee on the Domestic and International Monetary Effect of Energy and Other Natural Resource Pricing, Report, 93rd Congress, 2nd Session, September, p. 48.
71. U.S. National Oceanic and Atmospheric Administration, 1983, *Assessing the Social Costs of Oil Spills: The Amoco Cadiz Case Study*, U.S. Department of Commerce, U.S. Government Printing Office, Washington, D.C.
72. Verleger, Phillip K., 1982, "The Determinants of Official OPEC Crude Prices," *The Review of Economics and Statistics*, Vol. 64, No. 2, May, pp. 177–183.
73. Wilcox, Susan M., 1975, "Joint Venture Bidding and Entry in the Market for Offshore Petroleum Leases," unpublished dissertation, University of California at Santa Barbara, March, p. 123.

Part 5

PUBLIC POLICY AND THE MINERAL INDUSTRIES

5.13

Energy Policy Issues

Richard L. Gordon

ENERGY POLICY—PRINCIPLES AND THEIR PRACTICE

Persistently, at least since the end of World War II, governments have intervened broadly, deeply, and incoherently in major energy markets. If we take a broad definition of energy that encompasses gas distribution and electric power, we can observe regulations of even older vintage.

This trend, of course, is not unique to energy. Expansion of government controls has occurred in many areas. The relevant basic question about energy regulation, therefore, is not why is energy special, but whether any general principles exist to explain the wisdom of existing regulations. This review takes place at a time, the middle 1980s, in which a massive onslaught is being undertaken against government regulation. However, the present chapter is influenced primarily by my observations of the failures of energy regulations and not by any inspiration from those ideologically opposed to government intervention.

Several subquestions arise in policy appraisal. A primary need is to establish clear objectives and determine the consistency between these goals and the policies imposed. Thus, the first step is to set objectives. The next is to outline ways to attain these goals.

However, it should be recognized that many apparently distinct options are economically equivalent. When the goal is to charge producers or consumers for something, it is possible to attain the same result through either government ownership or taxing private transactions. The tax can raise the price to the level the government would have charged if it were the owner. Creation of a Department of Energy is not necessarily any different from having separate energy agencies well coordinated by some small supervisory body.

Thus, a distinction should be made between substance and form in policymaking. The critical substance concerns such matters as whether the government imposes financial penalties, incentives, or rules to attain a goal. Then, there can be many equivalent ways to implement the basic policy form.

The second prior illustration suggests that one favorite question is how to attain proper integration of policy. At various times, voices on either (or even both) sides of the Atlantic have called for a coordinated energy policy. The U.S. Paley Commission made this a primary goal for future policy. The western European governments that were the founding members of the Coal and Steel, Economic, and Atomic Energy Communities spent much time in the middle fifties and much of the sixties trying futilely to agree on an energy policy. (See Gordon, 1970 for an annotated review). The energy price rises of the 1970s rekindled United States interest in coordination.

However, astute observers of the problem such as O. C. Herfindahl of the United States and Maurice Allais (1962) in France warned the discussions confused form with substance. The crucial issues were what were the goals and what were the best types of policies to attain such goals. Policy suffered, not from poor organization, but from discord about the goals and how to attain them.

All this suggests that we need principles to guide policy appraisal. The branch of economic theory known as welfare economics concentrates on providing guidance about the problems of policymaking. The theory examines both what can and what cannot readily be said about the wisdom of different policies. As such, the theory is often more a warning about the pitfalls of policy design than a definite guide. Nevertheless, the lessons are too important to ignore. The discussion here begins with an overview of welfare economics.

This overview proceeds in three steps. First, a warning is provided that neither welfare economics nor any other purely analytic tool can resolve questions of how the benefits of economic activity should be shared. Then, the demonstration of how competitive market economies promote better use of resources is presented and discussed. Finally, the problem of "market failure"—the inability of real markets to produce all the desirable changes in resource use—is reviewed.

With this background, the discussion turns to developing extensions relevant to a wide range of mineral policy issues. Two basic concepts are explained, interrelated, and used to develop models applicable to many questions. The first concept is that of economic surplus—various measures of the difference between the total benefits and the total costs of participating in the market economy. The other concept is scale economies—measures of whether larger scale raises or lowers marginal costs.

The key issue with surpluses is who should keep them. Generally, it is presumed by policymakers that households should keep their surpluses but firms should not. The key problem with scale is the effect on profitability. As discussed in Chapter 2.4, when costs fall with scale, a price equal to marginal costs of production produces losses. When costs rise with scale, a price equal to marginal cost causes economic surpluses.

The economies of scale-profitability relationship is the source of the principal exception to the rule that consumers should keep their surpluses. It has been suggested by public-utility economists that these surpluses can be tapped to promote output in decreasing cost industries. Conversely, the surplus arising from diseconomies of scale is widely viewed as unearned and, therefore, appropriately transferred elsewhere.

These points lead to consideration of various systems of charges, subsidies, price controls, and regulations to effect transfer of resources. In short, disposal of economic surplus proves the concept on which a surprisingly large portion of energy policy rests.

An additional area of concern is the proper way in which instability of oil and other energy supplies should be considered in policy appraisal. In particular, questions should be raised about what policies, if any, should be imposed to respond to such instability.

EQUITY—THE SHARING OF THE PIE

Perhaps the thorniest problem in the theory and practice of public policy is the determination of how the fruits of economic activity should be shared among members of society. The standard argument in textbooks is that the appraisal is a subjective one about which economic analysis can provide no guidance. Unfortunately, no one else can. A Harvard philosopher—John Rawls— tried to refute the contention that no obvious principles exist for determining uncontroversially the principles for sharing resources. He proposed a set of ethics that he contended would prove widely acceptable. The highly skeptical response to his efforts (see e.g., Nozick) confirmed the view that consensus was elusive.

A general accord may exist that whatever aid that should be provided should go from the richer to the poorer. However, this is insufficient to delineate fully the preferable policy options. Further questions remain about the amount of aid and the best way to provide it. The issues on the proper degree of aid have precluded accord on either intranational or international policies on income distribution.

(U.S. mineral leasing policy debate, for example, involves concern over the proper share of producing states in the rents but tend to favor returning the money to the state in which production occurs. Nominally, a 50/50 state-federal split prevails. As Commission member Donald Alexander pointed out in 1983 hearings of the Commission on Fair Market Value Policy for Federal Coal Leasing, the deductability of royalties and bonuses from taxable income and a 46% federal income tax means the 50% federal share is almost totally offset by the drop in income tax yield.) Controversies also abound about the best form in which to provide aid.

However, economic analysis can give guidance about the effects on income distribution of different policies. The theory can suggest the presumptions about the impacts. Observation can provide additional evidence.

Work on predicting and examining the distributional impacts of different policies has produced strong convictions about the proper way to proceed. In particular, a vigorous preference prevails in the economics literature for adopting the most direct possible means to achieve any policy goal.

Thus, monetary grants directly to the needy are the preferred way to provide assistance. The first problem with alternative measures is that they are likely to be costly. Monetary aid is preferred over grants of specific services because it allows the poor to decide how to use the benefits. It is argued that the poor are no worse able than anyone else to determine how best to spend their money.

Government-imposed choices of spending through providing aid in specific services is at best useless. So long as the government provides less of a service than would have been bought anyway, the effects of tied aid on consumption are nil. Otherwise, people get more than they want of some good and less of a more desired good. In addition, scattered programs of providing specific services can produce uneven and often perverse distribution effects.

Even if each program actually is limited to the truly needy, the total benefits to each family will vary with their tastes for the different forms of subsidized services. The U.S. legal aid program favors those with legal problems over those who do not. It is questionable that such difference in taste should affect the level of aid. (Critics of specific programs go on to suggest that they are designed more to aid those employed in program implementation than the poor).

Another problem is that each program may have different standards of eligibility. Programs with looser standards may divert money to more affluent groups—in some cases to groups that might be considered too affluent to merit aid. Food stamp allocations are widely criticized for this, and one complaint about legal aid is that it is used for social crusading of little or no benefit to the poor.

Economic analysis and practical experience suggest that an even worse alternative is to attempt to redistribute income by regulating individual markets. The experience with interference in individual markets designed to assist various participants suggests that practice can be highly unsatisfactory. Apologies for intervention of this type contend that tradeoffs are made to promote fairness at the cost of interfering with the smooth working of the markets. Critics counter that actually such intervention often proves unfair. The recipients of benefits are, not the needy, but the politically potent.

The most notorious examples of this sort of perverse distributional impacts of intervention are the policies to protect firms in an industry. The term aid to small business has incessantly been used to justify a wide variety of government controls. These have included farm subsidies in many countries, protection for small retailers, general aid to businesses, and numerous energy programs that protect smaller firms in the industry.

The basic complaint against such policies is that the owners of businesses tend to be more affluent than the rest of society. Government aid provides precisely the sorts of protection from competition that are illegal in the United States when attempted by private action. The sound reasons why such antitrust policies exist inspire skepticism over the government intervening to protect businessmen.

Questions even exist about whether the policies assist the least affluent in the business community. The size of a *business* is not necessarily a good measure of the wealth of its owners. Large publicly held companies with their diverse direct and indirect stockholders may have owners with lower average incomes than the owners of closely held enterprises. This likelihood is increased because aid to small business is rarely directed at the truly struggling. Small oil producers and refiners generally are rich men, but they are more typical of recipients of government aid than a poor family trying to establish a retail business.

Directing aid to consumers of a product may lessen, but certainly will not eliminate, the equity problems. Consumers of a given product are not necessarily a particularly worthy group. The problem is aggravated by a confusion about the nature of firms. Habitually, households are considered favored users and singled out for preference over firms. However, firms are merely organizations for transforming inputs into outputs (see Chapter 2.4). Only in those unusual cases in which aid to households does not affect the marginal cost to disfavored firms can the assistance not raise the cost of production and lead to higher prices of the goods whose producers were forced to transfer resources to households.

Such rises in prices, of course, are paid by the consumers of the products with increased prices. Thus, this group of consumers loses from favoring households over producers. For this loss to be offset, we must be confident that the households that benefited from market regulation are considerably more worthy than the consumers of the products whose prices are raised by the intervention. Were the groups equally meritorious (as in the case in which everyone consumes both goods in roughly similar amounts), the intervention is clearly harmful. We get no equity gains, complicate the world, and (as proven below) reduce the efficiency of resource use.

All of this suggests that intervention in individual markets in the interest of equity is undesirable. Better means exist to aid the deserving; at best, it is unclear whom intervention aids; experience suggests that often the wrong people get most of the benefits. The inappropriateness, moreover, is often so blatent that it can be readily inferred from the few widely accepted views of fairness we possess.

THE CONCEPT OF ECONOMIC EFFICIENCY

The economic theory of welfare stresses the question about which a definite answer can be given—are there any unambiguously better ways to use available resources. The theory of advantageous choices outlined in Chapter 2.4 provides the basis for defining a clearly preferable use of resources. Specifically, all arrangements that are profitable to at least one of the participants and produce gains, or at least no losses to others, should be undertaken. The seizing of all such opportunities is deemed economic efficiency or (after one of the pioneering writers on the subject) Pareto optimality.

Optimality proves remarkably similar to competitive market behavior. In fact, so long as all means to secure satisfaction can be attained in a marketplace, a competitive equilibrium is the same thing as an efficient allocation of resources. (As shown below, severe concerns exist about whether real markets are competitive or that all values can be reflected in the market).

A competitive equilibrium involves the maximization of satisfaction by consumers, the maximization of wealth by producers, and the establishment of prices for all goods such that every one can secure or be able to dispose of precisely the amount of good desired. The last condition, market clearing, means that when we sum up how much those who want more of any particular good desire at the prevailing prices for all goods and how much those who want less of that good are willing to surrender at those prices, the totals will be equal. The quantity supplied will equal the quantity demanded. (Recall Chapter 2.4).

Both profit maximization and satisfaction maximization are equivalent to each consumer and producer separately effecting every available deal that is profitable to them. In Chapter 2.4, I show that producers equate the marginal benefits of every possible action to the marginal costs. A comparable demonstration is available about consumer behavior. We simply develop the optimality rules by seeing the greatest amount of satisfaction possible for a given household, given its assets and prevailing market prices.

This is handled analogously to producer wealth maximization. Consumers maximize satisfaction of their needs (known in the jargon as utility) subject to the side condition that the lifetime value of consumption cannot exceed the present value of the resources each consumer owns. The sticking point of starting the analysis is resolving the question of a measure of satisfaction.

Chapter 2.4 pointed out that the concept of an observable measure of satisfaction has been abandoned as unnecessary to adequate analysis. (Similarly, the term utility often is misunderstood. In the theory, it is simply any kind of satisfaction be it from tangible goods or spiritual well-being. Utility, in the popular mind, has the narrower meaning of tangible direct benefit. This confusion of meaning leads to incorrect charges that economic analysis ignores the value of intangible benefits). However, it remains legitimate to talk about satisfaction maximization.

In the jargon of the theory, this is possible because only the ordinal or comparative properties of a satisfaction index matter. The critical consideration is a basic one about quantification. In many instances, the measuring rods are arbitrary, and any will do if it is employed consistently. This is well-illustrated by the many different systems of weights and measures used in the world. Each has its own, usually unknown, basis. So long as we stick to one system, we can get the critical results. Conversion tables allow us to find the equivalent relationships in another measurement system.

CONSUMER THEORY

This principle carries over into evaluating consumer behavior. It turns out that all we care about is that the consumer has some method for telling when she is better off. Since we are only interested in the consequences of this appraisal for choices, we end up cancelling out the index of satisfaction and dealing with the marginal rates of substitution implied. (This is analogous to the simplifications occurring in the derivations in Chapter 2.4).

In mathematical form, we have a utility function $U = U(x_i)$ where x_i are the goods consumed and a side condition

$$\Sigma\, p_i\, x_i \leqq \Sigma\, p_i\, \bar{x}_i$$

where the p_i's are prices and the $\bar{x}_i$'s are the consumer's initial endowment. The Lagrangean function for a consumer is

$$L = U + \lambda\, [\Sigma p_i\, x_i - \Sigma p_i\, \bar{x}_i]$$

The critical Kuhn-Tucker conditions for any good are

$$\frac{\partial U}{\partial x_i} + \lambda\, p_i \leqq 0$$

$$x_i\, [\frac{\partial U}{\partial x_i} + \lambda\, p_i] = 0 \text{ or } \frac{\partial U}{\partial x_i / p_i} \leqq -\lambda$$

For two goods i and j such that a strict equality prevails, we have

$$\frac{\partial U}{\partial x_i}/p_i = \frac{\partial U}{\partial x_j}/p_j \text{ or } \frac{\partial U}{\partial x_i}/\frac{\partial U}{\partial x_j} = p_i/p_j$$

By the implicit function rule, the left-hand term is the marginal rate of substitution

$$\frac{-\partial x_j}{\partial x_i} \quad \text{so} \quad \frac{-\partial x_j}{\partial x_i} = \frac{p_i}{p_j}$$

This says that consumption of any good is increased so that at the margin, what the consumer finds the worth of that good in terms of another good equals the market price of the first good in terms of the second. This is the consumer side analogy to the marginal benefit marginal cost rules discussed in Chapter 2.4.

GLOBAL OPTIMIZATION OUTLINED

The formal social optimization problem involves maximizing successively the utility of each individual in the economy and maximizing the net output of every good by every firm while (1) keeping all other utilities constant, (2) keeping all other outputs constant, (3) satisfying the limits on production imposed by production functions, and (4) the limits imposed by the available stock of existing goods and services. The condition (2) for firms considers whether it is possible to raise some outputs or lower inputs without changing other output or input levels.

Slight differences exist in the way that the problem is set up. One possibility is to define different Lagrangeans for each person and firm in the economy and separately optimize each. Another approach is to define a single equation that encompasses all the utility functions, all the availability constraints, and all the production functions.

The second approach is neater but creates problems in notation. The utility and production functions have a dual role. Most of the time they serve as constraints. However, when the time comes to consider variation in some person's utility function, it is allowed to vary, by separately varying the consumption of each good, restricted only by all the other expressions. Similarly, each production function is usually a constraint, but also varies when the variation of an output or input is considered. No simple way exists to write down this summation of terms, each of which has one set of turns at being the main function but otherwise remains a side condition on the main function.

If we consider each individual and each producer separately, we must add an optimality rule for production—an improvement is when someone can increase some output without decreasing the output of anything else. (Since use of inputs is treated as negative production, this rule suffices to include increased input use and implicitly says that the gain in output comes with neither decrease of other outputs nor increase in input use.)

The efficiency problem is formally described as determining the situation that insures that no one's utility can be increased without lowering someone else's utility.

This leads to the demonstration that an efficiency economy is equivalent to a competitive equilibrium. The prevalence of a single competitive market price then insures that the separate optimization by each actor in an ideal competitive market leads to efficiency.

The intuitive basis of this, as noted, was presented in Chapter 2.4. In the next sections, the formal optimization problem is presented and solved.

PARETO OPTIMALITY-EFFICIENCY IMPLIES COMPETITION

The satisfaction of a market clearing requirement is insured by a further property of the optimization process. One also differentiates with respect to each of the Lagrangeans and secures the original side conditions as optimizing conditions. The availability constraints are the formal expression of the market clearing conditions. This is precisely what we wanted to occur.

In sum, efficiency involves maximizing individual benefits and taking account of the impact on others. Competitive behavior involves each person maximizing his benefits without concern for the impacts on others. Specifically, the primary rules turn out to involve universal equality of marginal rates of substitution for a pair of goods. The marginal rate of substitution should be equal among all consumers *and* producers. Each consumer equates his marginal rate of substitution to that of every other consumer; each producer equates its rate to that of every other producer; producer rates are equated to the consumer rates.

The logic for such equalities producing efficiency is quite similar to that leading to their optimality in individual choice. In fact, my discussion in Chapter 2.4 of the advantage of trade is the basic explanation of why the universal attainment of marginal equalities is efficient. Thus, no further discussion of the common sense is needed here.

Chapter 2.4 also showed that, in a competitive equilibrium, there will be an equality of marginal rates of substitution among producers that will be equal to price ratios. The discussion above in this chapter shows that in the competitive markets marginal rates of substitution in consumption for goods are equated to price ratios by each consumer. This equality to the single price ratio prevailing in competitive equilibrium for each pair of goods leads to insurance of the basic efficiency condition of universal equality of marginal rates of substitution.

The consumer marginal rates of substitution being equal to the same price ratios are equal to each other. All the producer marginal rates of substitution, as Chapter 2.4 showed, are similarly equal to the price ratios and thus to each other and consumer rates. Thus, the requirement that the marginal rates of substitution be equal to attain efficiency implies that we need a competitive equilibrium to insure that these equalities are attained.

The formal proof is due to Malinvaud. In his most general model, he maximizes for any arbitrary consumer the utility function $U_1(x_1)$ where x_1 is a vector of goods consumed by consumer 1. (Note since the numbering is arbitrary, the analysis applies to any and thus all consumers). Similarly the profits for any arbitrary producer are maximized. Three sets of side conditions apply:

1. No other utilities should change.

$U_i(x_i) \geqq U_i(\bar{x}_i)$ for consumers 2 to m. U and x are defined analogously to the function for consumer 1. $\bar{x}_i$ refers to the fixed level of goods and thus satisfaction these others must enjoy. The index of consumers is i.

2. Production function should be satisfied as efficiently as possible.

$f_j(y_j) = 0$ for the n firms. Here $f_j(y_j)$ is a production function involving a vector y_j of net outputs (i.e., inputs are treated as negative outputs). J is the index of firms.

3. Satisfaction of availability constraints.

$$\sum_{i=1}^{m} x_{ih} = \sum_{j=i}^{n} y_{jh} + W_h$$

for the 1 goods indexed by the h where x is consumption, y is net output, and W is the initial stock.

It can be noted immediately that the critical characteristic of Lagrangean analysis that produces the constraints as part of the optimizing conditions leads to the most obvious way that a Pareto optimum has the characteristics of competitive equilibrium. First, the satisfaction of the last set of constraints is equivalent to market clearing as defined in Chapter 2.4. Second, the production function constraints express the limits on individual firms in competitive equilibrium.

Malinvaud adopts the symmetric form so that all m utilities are summed and the constants for each of the consumer's other than the first are omitted as are the constant initial stock terms.

Thus, his Lagrangean is

$$L = \sum_{i=1}^{m} \lambda_i U_i (x_i - \sum_{j=1}^{n} \mu_j f_j (y_j) - \sum_{h=1}^{e} \sigma_h [\sum_{i=1}^{m} x_{ih} - \sum_{j=1}^{n} y_{ih}]$$

For all h, i, and j, we have the basic conditions

$$\frac{\partial L}{\partial x_{ih}} = \lambda_i \frac{\partial U_i}{\partial x_{ih}} - \sigma_h = 0$$

$$\frac{\partial L}{\partial y_{jh}} = -\mu_j \frac{\partial f}{\partial y_{jh}} + \sigma_h = 0$$

Thus,

$$\lambda_i \frac{\partial \mu_i}{\partial x_{ih}} = \sigma_h \qquad \mu_j \frac{\partial f}{\partial y_{jh}} = \sigma_h$$

$$p\, x^{i1} \geqq p\, x^{i0}$$

Manipulations similar to those made in Chapter 2.4 lead to the postulated efficiency rules. Thus, for any two goods h and k actually consumed by any consumers, we get

$$\lambda_i \frac{\partial U_i}{\partial x_{ih}} = \sigma_h \qquad \lambda_i \frac{\partial U_i}{\partial x_{ik}} = \sigma_k$$

Dividing the two equations yields

$$\lambda_i \frac{\partial U_i}{\partial x_{ih}} \Big/ \lambda_i \frac{\partial U_i}{\partial x_{ik}} = \frac{\sigma_h}{\sigma_k}$$

The λ_i's cancel and the implicit function rule changes the ratio of marginal utilities to a marginal rate of substitution

$$-\frac{\partial x_{ik}}{\partial x_{ih}} = \frac{\sigma_h}{\sigma_k}$$

Since this requirement applies to every pair of goods for every consumer and the Lagrangean multipliers σ_h apply throughout the economy, this leads to the requirement of equal marginal rates of substitution among all actual consumers of goods h and k.

Malinvaud's net output concept collapses the three types of production efficiency criteria of Chapter 2.4 into one. For two net outputs j_h and j_k,

$$\frac{\partial f}{\partial y_{jh}} = \frac{\sigma_h}{\mu_j}; \frac{\partial f}{\partial y_{jk}} = \frac{\sigma_k}{\mu_j}$$

$$\frac{-\partial y_{jk}}{\partial y_{jh}} = \frac{\dfrac{\partial f}{\partial y_{jh}}}{\dfrac{\partial f}{\partial y_{jk}}} = \frac{\dfrac{\sigma_h}{\mu_j}}{\dfrac{\sigma_k}{\mu_j}} = \frac{\sigma_h}{\sigma_k}$$

The three interpretations—input-output, output-output, and input-input choices— are developed by considering how the equation behaves when one y is positive and the other negative (an input-output substitution), both positive (output-output), or both negative (input-input).

Thus, we get that Pareto optimality requires that marginal rates of substitution between all pairs of goods be equal to the ratio for all producers and consumers actually trading the goods. This is assured in competitive equilibrium because all these rates of substitution are equal to the prevailing competitive price ratios and thus to each other. (The rules to cover goods unconsumed by some households and not part of the production function of some firms follow the pattern of introducing Kuhn-Tucker conditions shown in Chapter 2.4 and need not be developed here. Again, a good is excluded because its cost in terms of all goods used exceeds the benefit).

PARETO OPTIMALITY—COMPETITION IMPLIES EFFICIENCY

A trickier proposition is to show that, not only does every efficient allocation satisfy the rules of competitive equilibrium, but every competitive equilibrium is efficient. The usual proof (also due to Malinvaud) is by contradiction. Suppose that there were some competitive equilibrium that was inefficient. This means that there is some consumer that could be made better off without harming another. However, it can be shown that this cannot occur because competitive equilibrium has, in fact, assigned all resources in a fashion that precludes aiding anyone without harming another.

The critical starting point is showing that any budget that could reallocate to benefit at least one person without harming anyone else must be more expensive than the allocations chosen in competitive equilibrium.

Formally, an improvement is possible, in state 1 alternative to state 0 which is the competitive equilibrium, if for all i

$$U^i (x^{i1}) \geqq U^i (x^{i0})$$

and for some i

$$U^i(x^{i1}) > U^i(x^{i0})$$

The assumptions about consumer choice imply that all assets are utilized. Having more of one good if no other good is lost is considered preferable. The consumer takes the best budget he can afford and more is better than less. Thus, the money value of post trade assets will equal the value of pretrade assets at the prevailing equilibrium market prices.

The assumptions combine to imply the alternative budget x^i for the consumer made better off in state 1 must be more expensive than the one chosen. The chosen budget maximized utility given trading all available assets. To get more satisfaction, more assets must be available. Thus, all budgets better than the one chosen in competitive equilibrium must exceed the last in cost. Different budgets producing equal satisfaction to the one chosen, however, can equal as well as exceed the chosen one in cost.

To see this, consider any budget that absorbs all a consumer's assets. Consider a second budget that is cheaper than the first and produces equal utility. Then neither would be chosen. A better third budget could be constructed by adding something to the cheaper budget. More is always better. Thus, any budget less valuable than available assets will be rejected for a better one. All chosen budgets will be at least as expensive as equally good unchosen ones.

Since firms maximize profits given prices, then profits will at best remain constant if we alter allocations. Using p to represent a vector of prices, the assumptions combine to imply that for some consumer

$$p\, x^{i1} > p\, x^{i0}$$

and for all others

$$p\, x^{i1} \geqq p\, x^{i0}$$

Then summing

$$p\, \Sigma\, x^{i1} > p\, \Sigma\, x^{i0}$$

(one inequality suffices to make the sum an inequality)

or

$$p\, [\, \Sigma x^{i1} - \Sigma x^{i0}\,] > 0$$

Now if y^{j1} is net production by firm j,

$$p\, \Sigma y^{j1} \leqslant p\, \Sigma y^{j0}$$

or

$$p\, [\, \Sigma y^{j1} - \Sigma y^{j0}\,] \leqslant 0$$

or

$$-p\, [\, \Sigma y^{j0} - \Sigma y^{j1}\,] \geqslant 0$$

Summing further

$$p\, [\, \Sigma x^{i1} - \Sigma x^{i0} - \Sigma y^{j1} + \Sigma y^{j0}\,] > 0$$

or

$$p\, [(\Sigma x^{i1} - \Sigma y^{j1}) - (\Sigma x^{i0} - \Sigma y^{j0})] > 0$$

or

$$\Sigma x^{i1} - \Sigma y^{j1} > \Sigma x^{i0} - \Sigma y^{j0}$$

However, the last is impossible because of availability limits competitive equilibrium implies

$$\Sigma x^{i0} - \Sigma y^{j0} = W$$

where W is previously existing supplies. This says you cannot consume more than existing supplies plus net production. What we have shown is that any allocation that would make at least one consumer better off than under competitive equilibrium without harming anyone else would require more resources than are available.

MARKET FAILURE PROBLEMS

This proof relies on several critical assumptions whose violation can upset the conclusion. The critical concerns are that there are markets for all goods, that each such market is competitive, that each person's utility depends only on his consumption, and that taxes are nondistortionary. This critical last requirement means that the only effect of a tax is reducing real income left for other things. Consumer and producer decisions are then supposed to be the same as they would have been had the lower income been produced by nontax impacts.

The principal market failures recognized in the literature are the impossibility of competition, the failure of competition to exist even if it were possible, the existence of publicness (the problem of *simultaneously* meeting the needs of many people), and the tendency of most feasible tax systems to cause violation of the conditions for attaining Pareto optimality. (If anything, the tradition has been for excess willingness to postulate market failures. This is a very dangerous approach because anything can be justified if we assume enough market failures. Thus, it is critical to be skeptical about both assertions of market failure and the desirability of intervention to correct them).

The problem of natural monopoly or oligopoly arises when it is impossible to operate an industry efficiently with a large enough number of firms to prevent any from attaining control over price. Considerable doubts can be raised about whether this problem is sufficiently prevalent that it justifies concern.

However, one rationalization of a widespread tendency to regulate or nationalize electric power companies, gas and water distribution companies, and firms in telephone and other forms of communication is that natural monopoly prevails. (Alternative theories are that regulation is created to create monopoly or to reallocate rents). The issues of how to deal with this problem is discussed below in the treatment of using levys on consumer surpluses to promote efficiency in natural monopolies and oligopolies.

CREATED MONOPOLY AND ITS CONTROL

Another possibility is that while an industry has room for a large enough number of efficiently-sized firms to be purely competitive, it has been organized in a fashion such that actually far fewer firms exist and avoidable oligopoly has developed. It was concern over this situation that was the core of the prohibition of monopolization in Section 2 of the U.S. Sherman Antitrust Act of 1890. (The literature on this is enormous. Scherer is probably the best review by an economist. Lawyers such as Bork, Posner, and Areeda with good background in economics have made some of the most interesting contributions).

The meager accomplishments of enforcers of the Act in actually producing such restructuring can be variously interpreted. The more optimistic view is that actually socially desirable breakups of industries are simply rarely worth doing and the enforcers have been wise to limit their efforts. An alternative view is that the process has been too timid. Enforcers have not sought aggressively to identify opportunities.

In fact, the literature on antitrust suggests the enforcers zealously seek to find and eliminate undesirable oligopoly. The courts have insisted that mere acquisition of monopoly power does not suffice to justify restructuring. So, a form of positive action to achieve market dominance must have been taken.

The importance of this last concern is, however, highly questionable. First, the courts have over the years modified its interpretation of what constitutes positive action to an extent which greatly limits the ability to be exonerated by claiming that the monopoly was inadvertently acquired. Moreover, Congress has the power to amend the Sherman Act to tighten the rules. Proposals have periodically been made to legislate rules to ease restructuring. However, they produce about as much political excitement as a Harold Stassen or George McGovern presidential campaign. Thus, I infer that the better presumption is that the opportunities for restructuring are limited.

PUBLIC GOODS

Considerably more interesting questions are associated with publicness and distortionary taxes. The essence of publicness is that certain activities unavoidably simultaneously benefit or harm large groups of individuals. For example, the provision of military or general police protection is available to everyone in the country. Each, in fact, is enjoying the same amount of total services of this sort. Conversely, everyone in an area suffers from air or water pollution or an eyesore.

Formidable problems exist in efficiently financing the provision of public benefits or in reducing harms to the socially optimal level. Basically, the problem is to find a mechanism to learn what the public collectively would pay to secure the collective good or what costs are being imposed in toto by the collective harm. The central problem is that any mechanism of learning the valuations would be expensive to operate. It is hard to imagine any cheap way to bring together all those involved and secure their views.

Writers on the subject, notably Samuelson, have added that a further problem is that powerful incentives exist to misstate the true preferences. Each person's contribution or gain is a small part of the total and each may feel that a slight misstatement of his true feelings would not materially affect the decision. A considerable literature exists on how to overcome this problem.

This may be a misplacement of emphasis. No matter how well, in principle, we can design a system for discerning true preferences, we cannot readily implement the system except at prohibitive cost. The crucial barrier seems to be the massive costs of communicating with all those involved. The state can intervene to reduce transaction costs and more economically provide goods. However, state action at best reduces administrative costs but never eliminates them.

A much neglected point in policy appraisal is that the reduction in administrative costs may still leave the costs of action greater than the benefits. Others would assert the government often administers so ineptly that it fails significantly to lower administrative costs. In any case, intervention can only be justified if government regulation lowers transactions costs sufficiently that the total costs of action are reduced below the benefits.

PUBLIC GOODS AND EXTERNALITIES

For many years, economists believed, on the basis of an analysis by Pigou, which currently is much discussed and little read, that a separate problem existed with the creation of inadvertent beneficial or detrimental effects (known technically as externalities or neighborhood effects). In a remarkable 1960 article, Ronald Coase revolutionized thinking on externalities. Among his most important and most fully accepted conclusion was that it was, not the side effect per se, but the fact that many involved publicness that was the policy problem.

Coase argued that where the number of affected parties was small, private negotiation could resolve the problem. From his long study of litigation over economic issues, he was able to reproduce many case-law examples to demonstrate that disputes involving a small number of parties could be resolved privately.

He added the further proposition that two means of reconciliation existed and that the choice between them was largely a matter of equity. Some of his critics (e.g., Rothbard 1982) seem to think Coase believes the equities do not matter. I read him as reiterating the point made above that equity is hard to determine.

The critical marginal condition for efficiency, here as elsewhere, is that the social cost equal the social benefit. Where, in addition to the direct cost to consumers, there is an inadvertent additional cost to bystanders, it is the sum of the direct and indirect costs that should be equated to the benefit. Coase noted that either charges for damage production or subsidies for abatement could cause the required adjustment in marginal costs.

Under the charge system, private marginal costs are increased by the amount of charge incurred as pollution is increased. Under the assistance system, marginal costs rise to the sum of direct marginal costs and the reduction of subsidies because increased output is reduced abatement.

Consider the situation in which a given marginal change in the absence of control produces a marginal benefit of \$10, a marginal cost to the recipient of the benefit of \$6, but a cost to others of \$6. In the absence of controls, the beneficiary gains \$10 − \$6 = \$4 and takes the step. With a tax of \$6, the cost rises to \$12 and the step is not taken. Similarly, under a scheme in which subsidy *falls* as harmful action is taken, making this step would also involve a \$12 cost—the \$6 direct cost and a \$6 drop in subsidy. Again the step would not be taken. Thus, suffering for external damages can be induced by raising a tax or lowering a subsidy as damage rises.

This conclusion so shocked economists that enormous effort was devoted to refute Coase. Only one major problem was unearthed. Where large numbers of firms are involved, the subsidy approach is administratively much more complicated to operate than the charge system. The primary concern is that it is much easier to identify who should be taxed than to determine who should be subsidized. Every actual creator of externalities should be taxed. Subsidies should go to anyone who might otherwise engage in the polluting activity. Such people could be difficult to identify and, in fact, one can imagine the creation of companies whose sole objective was to earn subsidies by threatening to engage in a polluting industry. (See Baumol and Oates for the best statement of this view).

A further, more technical administrative problem arises from the need also to consider social breakeven conditions. A full application of the optimality principle requires that we insure that every firm that operates produces social benefits at least equal to the social costs of its operation. This requirement is particularly problematic to implement in a subsidy system. The right subsidy system should insure that firms that are socially unprofitable should earn more money from taking their subsidies and not operating than they earn from operating. This may be difficult to insure.

Comparable problems arise under a charge system. Here too socially unprofitable companies should bear a total tax burden high enough to encourage shutdown and socially profitable companies should not be taxed so much that they go out of business. Baumol and Oates argue convincingly that a subsidy system that allows socially unprofitable firms to operate is a greater danger than a tax system that puts too many firms out of business (but tend to neglect the problem of optimum total taxes).

(However, Baumol and Oates' third objection to Coase's reasoning — an allegation that Coase seems to advocate taxes on the victims of pollution to force them to undertake abatement—seems less convincing. The actual passage by Coase seems actually to be making a much weaker point that certain government procedures for controlling environmental pollution might discourage socially efficient actions by the victims to adopt self-protective measures cheaper than the abatement measures imposed on the polluters by the regulators. A slightly different quite valid point developed by Buchanan and Stubblebine and summarized by Turvey is that when each party affects the other, each should be subject to some incentives to take corrective action).

An important aspect of Coase's analysis that has been neglected is the stress on the limitations on government intervention it implied. Coase himself concentrated on the problem that even in the cases in which high transactions costs precluded private negotiations, the alternative of government action might be unattractive for the reasons I discussed in reviewing the general argument about public goods.

He emphasized that government might not necessarily be able to effect the controls at an administrative cost lower than the net benefits of the policy. Society would lose because the direct payoff from the policy would be exceeded by administrative costs. He was further concerned about the additional costs of control imposed because the absence of direct financial benefits to governmental regulators lessened the pressures to adopt the most efficient controls. Thus, gains from intervention could be further attenuated by adoption of policies that produced less than maximizing benefit.

Clearly, this anticipated the nearly universal criticism economists specializing in environmental issues (e.g., Kneese, Mills, Lave, and Noll) about the specific policies chosen. Others stress that, in fact, the perverse incentive of seeking to maximize their authority further aggravates politicians' decision making.

Others have stressed that the political process of building coalitions by overweighting the strong preferences of various interest groups produces socially excessive intervention. Kalt (1981) characterized the situation as "*The political voting mechanism, then provides a means by which individuals secure private benefits at the expense of others.*" (Italics in the original).

Finally, others, notably Cheung, have pointed out that the social cost argument, in practice, has been pushed much too far. Many problems that could be solved in the marketplace are tackled by government. (To make his point, he conducted a test of the assertion J. E. Meade made in several widely cited works on social costs that apple growers benefiting from pollination by bees could not transact with beekeepers for the service. Cheung found that real apple growers do contract with beekeepers. His views appear in a pamphlet on social costs in which he lucidly lays out the issues and another author contributes a useful set of supplemental comments).

THE PRACTICE OF POLLUTION CONTROL

All this is in many senses moot since regulators, particularly in the United States, prefer to impose direct controls on the amount of pollution allowed (and in many cases also on the method of pollution control adopted). This preference apparently is based on a combination of considerations including reluctance to complicate an already complex tax system and belief that the goals are more surely attained by the regulations.

While the first argument has considerable merit, the second does not stand up under close scrutiny. The usual primary attack on taxes is that it is not clear that they will be high enough to produce the desired pollution reduction. However, a tax will impose immediate pressures to effect some pollution reduction while litigation can and often does allow substantial delays in the effecting of actual controls under a regulatory system. For that matter, it is not clear that in the longer run, the pressures to reduce pollution are any more attenuated by inadequately low taxes than by acquiescence in industry pressures to loosen regulation.

A blatant fallacy in the attack on taxes is that they may lead "only" to higher prices on a commodity instead of reducing the pollution per unit of production. However, the higher price reduces production and thus pollution and may turn out to produce more abatement than controlling emissions from a given output. Conversely, the imposition control devices also raise costs and prices so a price effect results from either approach. The virtue of the tax system is that it encourages selecting an optimum mix of reducing output and lowering pollution per unit of output. Regulation emphasizes only reducing per-unit-of-output pollution (which by raising costs indirectly leads, as noted, to lower output but not necessarily to the optimum degree).

Finally, it is often alleged that regulation is cheaper to monitor than taxation. This argument also is invalid. Either to tax or to be sure the regulations are met, it is necessary to have accurate estimates of emissions. The easier monitoring argument seems to mean that if you impose regulation, particularly ones where you specify how the pollution control is to be effected, you become willing to tolerate use of simpler monitoring measures. The existence of a pollution control device will be considered sufficient evidence of compliance.

Actually, considerable danger exists that you are monitoring inadequately. The mere installation of controls may not mean that the regulations are met. In fact, a chronic criticism of pollution control devices is that they do not live up to their design specifications. Conversely and more critically, the same measurement strategy could be adopted under a tax system. In principle, one could again accept the installation and use of a given pollution control device as a measure of the amount of pollution to tax.

IMPERFECT CAPITAL MARKETS

Inordinate attention is given in the economic literature to the alleged inability of capital markets to insure the fullest possible risk pooling. The point is that no one individual or even a small group dares absorb the losses from a significant failure but when the loss is shared widely, it is less burdensome to any one person and there is greater inclination to take risks.

The traditional view was simply that private institutions did not spread the risk out over enough people. Since the 1950s, the terminology has shifted to concern over the absence of complete futures markets. This concept recognizes that a well-functioning futures market pools effective risks for a commodity. If futures existed for all goods, all risks would be pooled.

This argument suffers from its failure to recognize three facts familiar to any observers of private and governmental behavior in the capital markets. First, private firms can be extremely innovative in developing devices (e.g., pension funds, mutual funds, joint ventures, and diversified corporations) for risk pooling. Second, largely because such new institutions upset politically influential existing organizations, governments often act to prevent the exercise of private initiative. A long struggle occurred in the United States to convince established banks to seek liberation from restraints instead of trying to prevent the creation of rival institutions.

It should be noted parenthetically that this provides an important lesson for those who consider it naive for economists relentlessly to advocate more efficient policies. If politicians cannot be reached directly, they can be convinced by the actions of astute businessmen who demonstrate the efficacy of more vigorous competition. Such businessmen may not need economic theory to see the benefits of innovative practices. However, the support of economic analysis can be helpful.

In any case, the third problem is the dismal record of government investment policy. A gigantic literature exists on the failures in this area. It is widely recognized that politicians believe public works attract votes and authorize many socially unprofitable ventures. This applies, for example, to much of the waterway improvements by the Army Corps of Engineers and irrigation projects. Government aid to industry is more likely to go into failing firms. The Chrysler success story is an exception to such experiences as the massive drains of farm subsidies around the world and the expensive European aid programs for coal and steel.

Thus, it is hard to accept the principle that increased government investment would lead to significantly more efficient investment decisions. All the defects of government policy could be reformed, but this would still not justify government aid to risky industries. Unfettered private institutions are likely to do the job more efficiently. The antigovernment argument that election pressures lead to shortsightedness appears the more valid appraisal. This is often epitomized by the argument that politicians have very short horizons, the time between elections, and seek investments with quick (i.e., high-interest-yielding) economic or political payoff.

However, there is a valid problem about capital markets—their behavior (and that of all parts of the economy) can be distorted by the tax laws. This is discussed below.

DISTORTIONARY TAXES

In any case, most of the market failures so widely stressed in the literature on welfare economics are trivial in impact compared to the problems resulting from imposing distortionary taxes. For reasons discussed below, the only known nondistortionary taxes that are feasible in practice are taxes on economic rents. Theorists have suggested another possibility of so-called lump sum taxes which would be levied upon individuals in some present amount that would not change with any economic decision. Therefore, no incentive would exist to alter any economic decision to alter the tax burden. (However, the reduction in private income would force a general reduction in personal expenditures).

The tax would be nondistortionary in the sense that at the margin no changes in decisions would be worth making to reduce the tax burden. The principal problem is that true lump sum taxes that would be nondistortionary are unlikely to be able to raise the substantial sums of money required by modern government.

Effective lump sum taxes do exist. Pennsylvania residents pay a number of municipal occupational privilege taxes that vary with the earning power of our occupation. (College presidents pay more than deans, who pay more than professors, who pay more than instructors). These appear to be largely nondistortionary taxes but only because their magnitude is so small that the cost in lower income of changing to a less heavily taxed profession is far greater than the tax saving.

The state and local governments must supplement these taxes with the usual array of income, sales, and property taxes that do distort decisions. The classic demonstration of a distortionary tax is that of a sales tax. Two well-established *alternative* presentations exist. In both cases, the principle is to show that the tax leads to the violation of the basic Pareto optimality conditions in exchange. One approach is to add the tax to the supply, the other to deduct the tax from the demand price.

In the first case, the distortion is expressed as the consumer paying a marginal cost in excess of the social marginal cost and equating marginal benefit to this tax inflated marginal cost and thus buying less than optimal. In the second method, it is shown that the producer receives a marginal receipt less than the true marginal benefit and thus ends up selling less than optimal.

Figure 5.13.1 presents both diagrams. The upper panel treats a tax as reduction in the demand price seen by the industry. The lower panel treats the tax as an increase in the supply price seen by consumers. In both cases, the basic demand curve D_c and S_c reflect respectively the true valuation and true resource cost of the good. This is simply another way of stating the earlier proposition that a competitive equilibrium with optimum provision of public goods is efficient. In the absence of a sales tax, the quantity demanded is equated to the quantity supplied at an output of Q_c selling for P_c (see Chapter 2.4).

In the demand-price reduction model, a tax reduces the value producers receive for a given quantity by the amount of the tax. Thus, we have an after-tax demand curve D_t which lies below D_c at each quantity by the amount of the tax.

Suppliers expand output to the point at which after-tax receipts equal marginal before-tax costs. This is determined by the intersection of D_t with S_c at the output Q_t, and the before-tax price P_p. Prices paid by consumers P_t are higher by the tax and can be found viewing the price on D_c associated with Q_t. Note Q_t is less than Q_c unless demand is totally price insensitive. P_p is lower than P_c unless supply price is independent of output. P_t must rise.

The same results arise if instead we generate an after-tax supply curve S_t that lies above the S_c at each output by the amount of the tax. Then the intersection of D_c and S_t determines Q_t and P_t and moving back to S_c gives us the associated P_p.

What is at work is a difference, because of the tax, between consumer payments and producer receipts. It is purely a matter of convenience whether we treat this difference as a reduction in receipts to producer or a rise in cost to consumers. All that is required is that some expression of the divergence is provided.

The various other major taxes can be treated

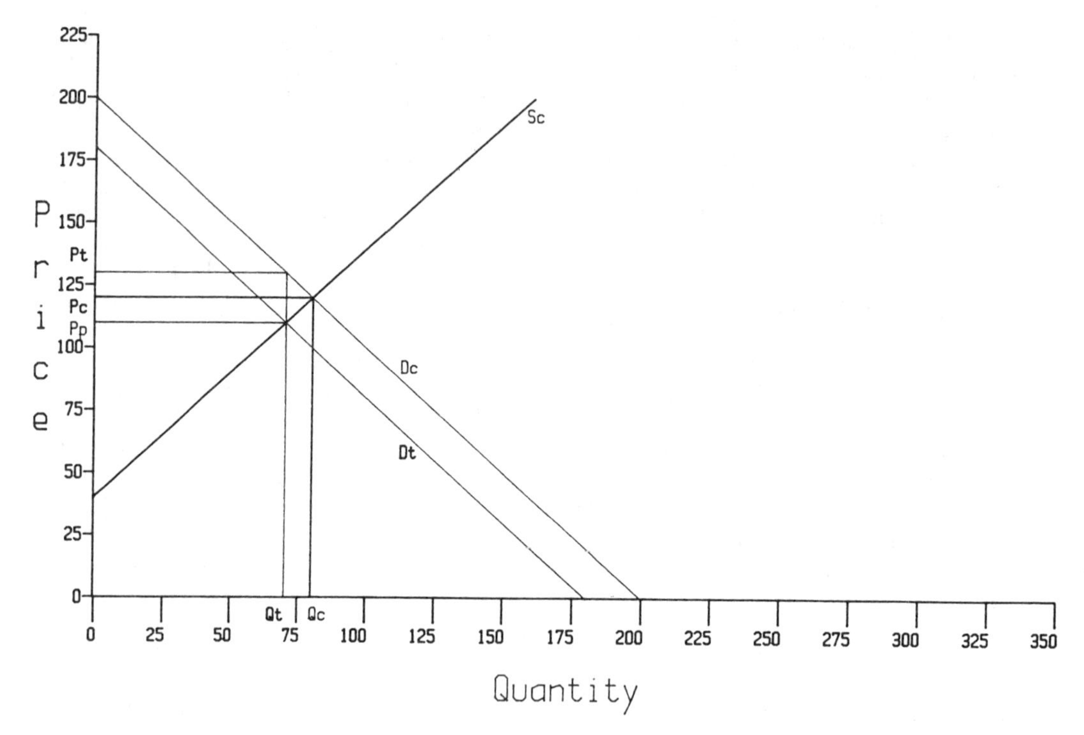

Fig.5.13.1a—Effects of a sales tax.

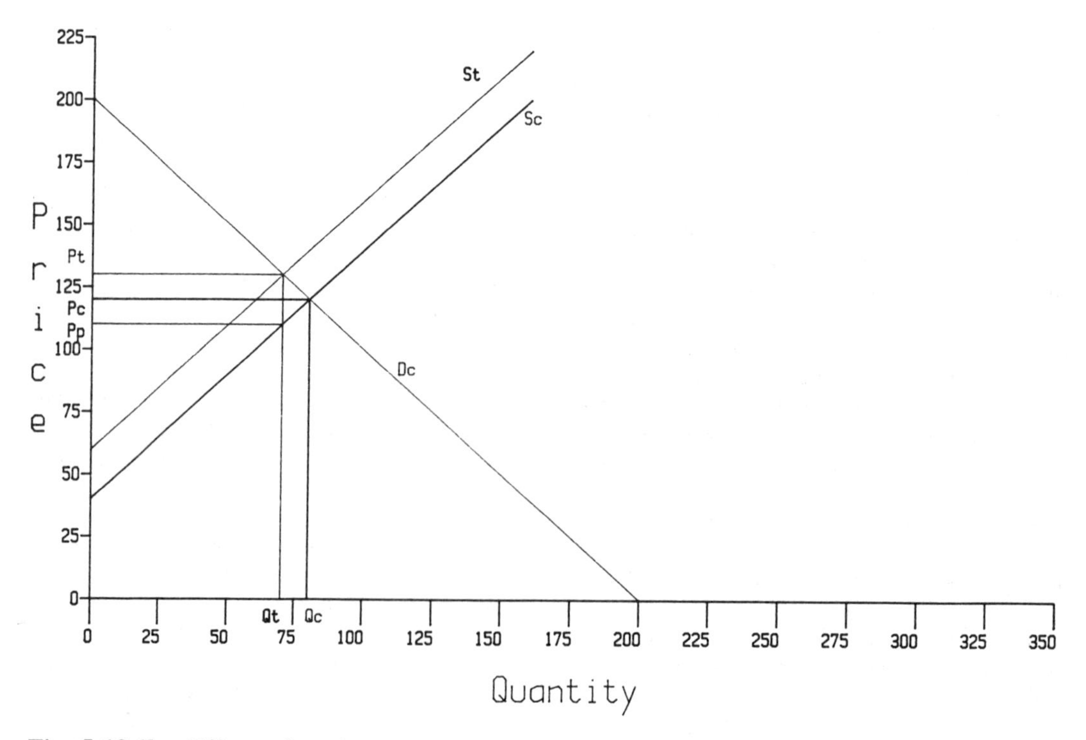

Fig. 5.13.1b—Effects of a sales tax.

equivalently. A general income tax is a tax on the sale of inputs and can be analyzed by exactly the same methods used for a commodity sales tax. We simply interpreted the demand curve as the input users' demand for input and the supply curve as the input owners' supplies of output. Taxes on special forms of income can be similarly analyzed by simply limiting the analysis to the market for the taxed inputs.

In every case, the result is that an inefficiently small amount of the good is sold. (The traditional analysis leaves out that the consumer recognizes that government services are received for the taxes. The implications of this are unclear, but certainly experience makes evident that the link to benefits is sufficiently weak that tax avoidance is widely practiced. Examples include the popularity of numerous schemes to encourage investment in ventures subject to lower taxes, tendencies to shop where sales taxes are lower, engaging in barter deals or cash transactions, and, where taxes are particularly onerous, dealing in smuggling).

A particularly disturbing form of distortionary tax is the heavy reliance on taxes on corporate income in addition to retaxing the income as it is distributed to stockholders. Such taxes greatly discourage the distribution of dividends to stockholders since then the second tax must be paid.

A critical and difficult-to-resolve issue associated with the existence of distortionary taxes is what interest rate to apply to governmental decisions. The problem lies in the taxation of returns to private investment. Therefore, a substantial difference exists between the before-tax yield and the after-tax yield. The literature suggests the range of possible appropriate interest rates for public investments is broader than the range defined by the before and the after-tax yield on private investments.

The appropriate rate, moreover, depends on a host of considerations. Some would challenge any use of market rates of interest as guides and rely on a concept of political preferences and considerations of intergenerational equity. Given the difficulties in implementing such an approach, most writers prefer to base determination of the appropriate interest rate on public investments on analysis of the market impacts of such investments.

In a valuable overview chapter in a volume of papers on the issue, Lind argues for a discount rate for public projects that is a weighted average of the before and after-tax rate. His conclusion arises from examination of the forces that must be considered in setting the appropriate rate of discount for a government investment. He builds on a framework developed by Bradford. The analysis centers on the disposition of its payoff on private savings and investment.

Bradford shows that, at one extreme, if public investment only displaces consumption and its payoff is all reinvested, the appropriate rate of interest is less than the after tax private return. Since only consumption is displaced, the initial cost is the lost interest that could have been received by consumers—namely the interest at the after-tax rate. A lower rate than this is needed to credit the government investment for creating private investment that otherwise would not have occurred.

For symmetric reasons, if government investment entirely displaces private investment and its yields create consumption that otherwise would not exist, the appropriate discount rate should exceed the before-tax rate. A weighted average should be chosen if financing lowers both consumption and investment in the private section and the repayment creates new consumption and investment. Lind argues for this approach even after developing a strong argument for using a rate at least as high as the before tax rate.

He suggests, implicitly following Martin Feldstein's controversial views on social security, that consumers may well consider public investment as obviating saving and that the investment displacement model is more realistic. It should be noted further that in one critical case for mineral economics—federal landholding, the displacement interpretation seems particularly valid.

Federal landowning is directly at the expense of private investment in land. It could be argued that this investment, in turn, displaces private investment elsewhere. However, the vigorous defense, by Hanke, of Reagan administration proposals for greater privatization of public lands suggests that the problem with public ownership is that it discourages investment by people more concerned than government officials in the profitable use of land. (See also Vernon Smith's even more extreme proposal for equitable disposal of all public land to individuals).

In any case, whatever the choice, formidable measurement problems exist. Great difficulties exist in measuring before and after tax yields. Fisher, McGowan, and Greenwood have effectively reminded us that accounting rates of return are quite different from economic rates of return. In addition, tax laws are so complex that direct application of tax rates to yields cannot accurately measure the divergence between before and after tax yields. Some claim tax laws have reached the point at which corporate income taxes have been virtually eliminated. If we accept the concept of weighting, determining the weights would be extremely difficult.

THE ECONOMICS OF SECOND BEST—CAN WE DECIDE ANYTHING

Because of all these departures from the conditions necessary to assure that a competitive equilibrium is equivalent to a Pareto optimum, a substantial economic literature has developed on the perils of imposing corrective policies. The usual terminology adopted is that of the theory of the second best. The term was adopted by J. R. Meade in his book, *Trade and Welfare*. Lipsey and Lancaster then prepared a journal article further developing the concept, and then numerous other contributions emerged. The Lipsey-Lancaster article presented the exceedingly pessimistic view that often one could not be sure what to do. The standard rules of welfare economics indicate what to do if the rules for a Pareto optimum are observed in all but one case. Where only one marginal benefit differs from marginal cost, that disparity should be eliminated. However, where many gaps between marginal cost and marginal benefit prevail, moving one benefit closer to cost may not necessarily be an improvement.

Various responses were developed. Some concluded that no policy advice could be given. Others, notably Graaf, reached the conclusion that the government should intervene in markets to insure that outputs were more fairly distributed. Just how the fairness of distribution could be any better determined than the efficiency was never made clear.

Others, such as Farrell, have attempted to determine what class of situations are such that a definite conclusion can be made. A critical thrust is that a wide variety of market failures lead to the underproduction of goods including public goods and underuse of inputs. Thus, policies that encourage greater input use and greater output of underproduced goods are desirable. More specifically, governments should avoid policies that produce output levels with marginal costs in excess of benefits. The general tendency for the value of output elsewhere to exceed its cost means that any output with marginal cost above price is even worse in an imperfectly competitive world than in a purely competitive world.

With pure competition and marginal costs everywhere else equal to price, a single output with a marginal cost in excess of price is causing an output loss equal exactly to the excess of marginal cost over price. In imperfect competition with prices of many goods in excess of marginal cost, the loss is larger.

Consider, for example, a situation in which it costs $15 to produce an increment of output of an item that sells for $10. With pure competition, an industry would use the $15 elsewhere to produce a product worth $15. Thus, in competition, the marginal loss is $5. With imperfect competition, the second industry might be securing $20 for the output with a marginal cost of $15. Thus, transfer of resources to the industry with a $10 price now produces a loss in goods to consumer of $10 in imperfect competition rather than the $5 cost that occurs in the pure competition. Overstimulation of industries thus is the one case in which the existence of market imperfections clearly reenforces the principle that prices below marginal costs are socially undesirable.

Of course, there are still questions here of exactly how much output reduction should occur in the industry producing at a marginal cost in excess of price. More generally, some cases may be far less clear-cut. Nevertheless, the theory of second best is more a warning that glib simple policy advice is inappropriate than proof that we should abandon marginal principles. The theory should be taken as a caution sign rather than a call to nihilism.

It should be noted parenthetically that a quite different use of second best has arisen in the literature—particularly that of international trade. This literature accepts the desirability of moving prices and marginal costs closer to equality. What is stressed is that there are better or worse ways of doing so. (See, e.g., Johnson, Bhagwati, or Corden).

In this approach, first best refers to a policy that removes the inefficiency in the least costly manner. A second-best policy is a higher-cost cure and in some situations, there may be third or even higher ranking (by cost) alternatives.

Thus, in the pollution control area, the first best policy would be to tax pollution (or subsidize abatement). This allows the firm to select the most efficient mix of responses to reduce pollution. A second-best method would be to tax the use of a particular fuel. This would discourage the use of that fuel and to the extent that the fuel use produced pollution, pollution would be lowered. No incentive would be provided to adopt other alternatives such as using abatement technologies that lead to less pollution per unit of fuel use even when these alternatives were cheaper than reducing fuel use. Thus, the policy produces abatement involving too much fuel use reduction and too little use of cleanup.

A third-best method would be to tax the output of the polluting industry. This reduces output and thus pollution. However, no incentive exists to alter the proportion of polluting inputs used or to engage in abating. Thus, here there is too much output reduction relative to changing input use or using cleanup techniques.

As is only implicitly noted, this theory also warns that certain popular policy tools cannot be used at all to accomplish a goal. Tariffs are a fourth-best way to discourage pollution arising in consumption. They discourage consumption but inefficiently stimulate domestic output and fail to stimulate efforts to consume using less polluting methods. However, where the problem is pollution in domestic industry, a tariff has the harmful effect of stimulating output and pollution. Import subsidies would be the foreign trade policy that would serve as the (fourth best) corrective.

ECONOMIC RENTS, THEIR NATURE, USE, AND ABUSE

Under various guises, the concepts of rent, surplus, economic profits, and the gains from trade have been important concepts in economic analysis. The discussion of the gains from exchange in Chapter 2.4 provides the basic explanation of where such gains come from. In this section, an effort is made to explain further the various concepts, their relationship, the implications of the concepts, and the basic policy considerations implied. Later sections deal with alternative approaches to the reallocation of economic rents.

A critical point to note is that the views of rent differ radically depending upon context of the discussion. Rents are simultaneously viewed as the critical fruits of engaging in a market economy and misgotten gains to be seized. The difference of view arises from deeply held opinions about who and for what reason some are worthy of earning rents. These opinions lead to alternative views about the desirability of attaining surplus income depending upon the identity of the recipient and the source of rent.

Whatever the terminology, the concept is a measure of the gain participants receive from engaging in transactions in a market economy. Gains arise in two basic forms. First, the market prices of goods are adjusted in a fashion that raises satisfaction. The implied adjustment process was outlined above in the discussion of optimization of economic efficiency. That process involves making the existing resources of the world more productive by using them to produce other goods and placing these goods with those who desire them most. On balance, goods become cheaper with the tendency of extant developed resources, principally labor, becoming more valuable and of the increased supply of produced goods lowering their scarcity. Since we have no way of knowing what prices might have been in the absence of specialization and trade, discussions of these benefits are confined to demonstrations, such as sketched above, that improvements result.

Given a set of prices, gains arise due to differences between the market prices and participants' valuation of commodities. Such surpluses, in fact, are the inevitable result of the two critical elements of moving to an economic optimum. First, as discussed in Chapter 2.4, we have the principle of diminishing marginal benefit—the more we transact, the less we gain on each additional unit traded. Second, if we trade at all, we ultimately extend trading to the point at which the marginal benefit equals the cost. Thus, when acquiring a good, we start out with the first unit we buy being worth more to us than its market price, the next also being worth more than market price but less than the first unit added. This process continues until the last unit bought has a value equal to the market price.

Then there is a surplus, namely, the difference between the market price and our valuation on the inframarginal purchases (i.e., those purchased prior to the last unit secured). A similar situation occurs with sales. The first item surrendered is worth less than the market price. Several more units of the same good are surrendered, each up to the last will be worth less to us than the market price. If several more units of the same good are surrendered, each up to the last will be worth less to us than the market price but the worth will rise closer to the market price. Again the gains are the difference on inframarginal deals between the market price and our value of the units.

The economics of the surplus of benefit over cost on inframarginal transactions differs radically depending upon whether households or firms are involved. The basic influence is the intrinsic difference in the economic concepts of firms and households.

Consumers are considered the center of the economic universe in that the key concern is that consumers benefit from economic activities. Thus, consumer profits or surplus are usually considered a good thing to be stimulated. (See below for an exception).

A firm is considered simply an inanimate organization that hires resources, uses them to produce goods, and then sells the goods. The firm as such has no tastes and is merely a conduit for resources. This concept is convenient to maintain whether we are concerned with something so intimately connected with its owner as a one-man consulting firm or huge corporation whose actual owners may be separated by several layers of intermediaries.

Much stock is nominally owned by firms which act as custodians for the actual legal owners. They might in turn be mutual or pension funds holding stocks for many individuals. Some of the individuals might be parents holding their children's money.

This separation is convenient because it handles all possible relationships between firms and their owners. Whatever these owners supply the firm is treated as a transaction between the owner as a person and the firm as an organization. The advantage of this approach is that it highlights that owners render services similar to others who deal with the firm and the analysis should recognize that the owner requires compensation as much as do other suppliers. A further advantage is that all the problems of dealing with feelings about the value of the outcome can be dealt with in analysis of consumer behavior. The firm is considered to be concerned only to make money; its owners then worry about the value of the money.

In any case, the excess of inframarginal benefits over costs gained by *firms* is equivalently called pure profit, excess profit, economic rent, or producers surplus. Its chief characteristics in economic analysis are (1) it is always appropriately measurable in money terms, (2) it is considered a bad thing to occur, and (3) vigorous competition is likely to eliminate the gain.

The first point follows from the concept of the firm as an insensate entity. It then becomes a device for making money. The more it has available (in terms of wealth, see Chapter 2.4) the better off are its owners. (This is a further illustration of why owners' services to the firm are best treated separately. The efforts owners make to produce income are best analyzed as a process of supplying managerial services). Greater wealth is simply unambiguously greater purchasing power in excess of the value of the inputs the owners provided.

In any case, most of the terms used are designed to denote that the net income is an excess over the market prices of *all* the inputs into production. The critical point to note is that economic costs usually exceed accounting costs. Accountants refuse to estimate costs that are not recorded in the marketplace. Stockholders expect some return on investment on their holdings, but no fully satisfactory means exists to determine the expectations (since the price of a stock reflects unknowable investor expectations about future income). Similarly, the value of owner-manager services is difficult to compute.

Accounting profit is usually defined as the difference between receipts and readily measured costs. However, a significant portion of these accounting profits consists of those costs that accountants refuse to estimate. Economic profit consists only of that portion which exceeds these omitted costs.

(A further complication to note is that accountants often use arbitrary rules to allocate even the measurable portions of cost. An enormous literature exists, for example, warning that accounting rules for depreciation—the allocating of the initial cost of investment among years—often are economically unsound. See for example Fisher, McGowan, and Greenwood). This difference between accounting and economic profits and more critically the practical problems of reconciling the two approaches is a critical one in policy analysis.

Considerable efforts are made to tax away economic surpluses. Controversy over what portion of accounting profits are truly economic profits is inevitable. Predictably, governments seeking to tax surpluses try to employ the most restrictive possible measures of the economic cost component of accounting profits. Conversely, companies seek to adopt the highest possible estimate of economic costs.

THE DISAPPEARANCE OF PRODUCER PROFITS

Given that firms ultimately return their profits to their owners, the argument that profits be taxed is, as the argument that consumer surpluses be retained, based on equities. It is presumed that it is generally a bad thing that stockholders keep gains from production but good that households retain most of their gains from trade. The presumption seems to be that on average consumers as a whole are more deserving than investors. This conclusion is based on the observation that obviously everyone is a consumer but few are substantial investors. Of course, the argument can be overdone. There may be specific cases such as the sale of luxury goods (say Tiffany jewelry, when it was a subsidiary of Avon Cosmetics) by a widely held company involves consumers less deserving than their suppliers.

A more fundamental aspect of economic surplus for firms is its tendency to disappear. Competition squeezes out rents. On the selling side, the existence of profits tends to attract new firms who augment supply and depress prices. The simplest case is that of perfect competition. This case normally is defined to mean that an unlimited number of firms can operate with identical costs. In that case, rents disappear entirely through lower prices. Specifically, entry occurs to the point at which prices are lowered to the point at which each firm and thus all firms exactly cover economic costs. The usual assumptions are that the cost function of the firm is characterized by an initial stage of economies of scale and then by a stage of diseconomies. Then each firm operated at exactly the point at which a transition from economies to diseconomies is made. Expansion of the industry is produced by increasing the number of firms. Prices in the long-run are at the point of minimum average cost for each of the identical operating firms.

As Samuelson has suggested, the assumption of variation of costs within but not among firms is made to simplify the analysis. If there were also no variation of costs within a firm with output, the firm could produce any output it desired and the division of output among firms would be undetermined. The basis for belief that costs rise as a firm gets too large is that firms do limit their size.

PROFIT ELIMINATION WHEN COSTS DIFFER— THE CREATION OF ECONOMIC RENTS

As Chapter 2.4 showed, when costs differ among firms, the process of output expansion with the entry of new firms involves increasing the price of the specialized resources used in the industry. This price for specialized resources is called its economic rent.

To reiterate, entry into competitive industry will occur when the *entrant* can operate profitably. That requires that prices be high enough that the entering firm can cover *its* average costs at the point of minimum average costs (or more generally can break even when operating at outputs at which economies of scale have been exhausted).

The new firm is higher cost than the old firms or else it would have entered sooner. Therefore, a price high enough to make the higher cost firm profitable makes the lower cost firms even more profitable. However, these rents generally are attributable to access to some superior resources. For example, a major source of cost differences in minerals production is variation in the economic attractiveness of deposits. Attractiveness, in turn, depends upon ease of mining, quality of the minerals, proximity to market, ease of transportation, and availability of infrastructure.

Another consequence of vigorous competition is that the cost saving due to access to such superior resources is captured, at least initially, by whoever possesses the property rights in these minerals, or other specialized resources. Thus, normally firms in highly competitive industries will not retain rents even if they are employing superior resources. Competition for the right to exploit these resources will encourage bidding up the payment to the resource owner to the point at which it will capture all the economic rent.

Interestingly, a strong tradition exists in public policy and the economics literature that it is always proper to insure that these rents are transferred to the government to finance public activities. The political and economic arguments are slightly different. Politicians simply assume that rent earners are probably undeserving people who should be heavily taxed. Economics adds the proposition that true rent taxes are the only easily-administered nondistortionary taxes available to governments. Rents, by definition, are incomes in excess of those needed to induce dedication of resources to an activity. Thus, their taxation still leaves enough income so that the resource owners will be willing to continue selling the same amount of resources for the same purposes.

It should be noted that the nondistortionary aspect of the tax is the more clearly defensible (however, recall my prior caveat that governments tend to overestimate rents, overtax, and create distortions). The view that rents from ownership of specialized resources deserves to be taxed while consumer surplus rarely should be taxed probably needs more scrutiny than it usually receives. The implicit assumption seems to be that most owners of specialized resources are likely to be sufficiently affluent that they can afford the loss of rents. This often is the case, but care must be taken to recognize that exceptions may exist.

Fundamentally, it would appear that this tradition is, nevertheless, well founded. Given the difficulties of measuring consumer surplus and in complicating the taxation of economic rents by trying to incorporate consideration of the neediness of the rent earners, it is administratively far more feasible to tax all rents than to try to tailor the tax to the neediness of the rent earners. Whatever effort we might make to transfer incomes from the rich to the poor are, as argued above, probably best done by special programs based on examination of the total incomes of each individual.

Measurement of consumer surplus is a particularly tricky problem because first, it requires unknowable information about the willingness to pay for inframarginal purchases and second, because the value of money to households varies with their income and no good way exists to adjust fully for this. This has led some economists—notably Paul Samuelson—to argue that economic analysis would be better off without use of the consumer surplus concept. Others, notably J. R. Hicks, have argued that the concept is too useful to abandon. Willig has developed a widely adopted technique for getting close approximations of the true consumer surplus, and this work has revived use of the concept.

In any case, the theme of rent collection and its utilization can be used as the basis for studying a broad range of mineral policy issues and much of this chapter is devoted to such a consideration.

The most obvious is to deal directly with the rent taxation issues and that is done first. An overview on this subject is followed by applications to U.S. coal leasing that I developed in the process of my work on a federal commission appraising aspects of coal leasing policy.

Then I turn to two other aspects of the issue. First I treat the reverse case of transferring consumer surpluses to producers to stimulate output in ''natural monopolies''—industries where economies of scale at the firm level make it efficient to concentrate output in a single firm. Then I develop the argument that most price control schemes in energy are devices to use rents to subsidize additional output. Here I develop analysis of why this is so and what results are produced by various approaches to using rents to subsidize output. In particular, I deal with the use of multitiered price controls such as have been used in U.S. oil and gas.

PRINCIPLES OF RENT TAXATION

Two issues of substance arise in designing government policies to transfer rents to the public treasury. First, given the many levels of government that prevail, questions can arise about the allocation of incomes among them. Second, many different bases for payment can be employed, and choices must be made among them. However, a considerable number of alternative institutional arrangements can be adopted to attain a given substantive goal.

Thus, at least three systems are used in the United States to transfer rents. First, much land has always been under private ownership under laws that include ownership of the mineral rights as part of the overall property rights. The federal government has several different approaches to the transfer of fuel mineral rights on land it owns. Basically, minerals are divided into locatable—principally metals including uranium—and the leasible—principally the fossil fuels. The system for locatable minerals transfers the nominal property rights to the finders.

They must put up signs marking the region in which they claim to have found valuable mineral deposits, provide adequate proof that such deposits really exist where the claim was staked, and undertake modest continued activity to maintain their rights.

With leasible minerals, the practice is to award a right to extract the minerals. For offshore oil and gas rights and for at least large coal fields, sealed competitive bids are taken and the award goes to the highest bidder. Usually, the bids involve offering a bonus on top of whatever royalties are required. However, for small onshore oil and gas leases, the leases are awarded by a lottery.

Several complications prevail. First, when federal land was sold for nonmineral uses such as farming and ranching, the federal government retained ownership of the minerals. (So did many private landsellers elsewhere). Problems have arisen about reconciling these retained public rights with the interests of the surface land owners.

Second, a complicated hodgepodge of ownership patterns developed in the western United States. In addition to fully federally owned lands and lands to which the federal government retains mineral rights, the states, Indian tribes, railroads, and probably many private parties have control of both the surface and the minerals beneath them. Thus, the states were given grants of land as part of their graduation from territorial to statehood status. The treaties with the Indians included land grants, and in recent years, some tribes have succeeded in securing court decisions that the prior grants had been inadequate and that the tribes were entitled to more land.

As part of the effort to encourage railroad development, the railroads were given land grants under a process usually referred to as checkerboarding. The land in the West was divided into square parcels and the railroads got alternating parcels around their right of ways. Checkerboarding is a metaphor for the analogy of this process of assigning all the red (or black) squares to one party and the other color to another party (in fact, actually keeping federal control).

As W. S. Gilbert pointed out, this is a case in which "Things are seldom what they seem; skim milk oft masquerades as cream." The choice not to impose any fees for access to locatable minerals can be and usually is offset by imposition of taxes on mineral production. Moreover, whether or not the federal government is the initial recipient of fees or taxes does not necessarily determine who ultimately receives the income. It is common for a substantial portion of the federal income to be transferred to the states in which the mining occurred. The states, in turn, could choose to contribute a substantial share of their receipts to the communities in which mining occurs.

What is probably a considerably more interesting question is how the claims of the different landowners get sorted out. The critical distinction is between situations in which a given landowner has undivided rights to the minerals and the surface of an area large enough to support an efficiently sized mine and those in which land must be purchased from several landowners.

Where a large number of landowners, each of whom can singlehandedly supply the resource needs of an entire mine, exists, the outcome is clear-cut. Under the circumstances, landowners can receive no more than the economic rents available from exploiting that property. Competition among potential mine operators will insure that at least that amount will be secured. Competition among landowners will insure that no more need be paid since it is possible to go to another landowner.

Another situation where the rent division is clear-cut is where the extant legal institutions clearly specify what must be paid. For example, in coal mining in the western United States, part of the rent collection system since 1976 includes a mandatory 12.5% federal royalty on surface mined coal and a tax at various rates by the states. (The 1976 law imposing the royalties on the coal added numerous additional requirements including limits on the amounts of land a company could lease on any one tract, in any one state, and nationally. Diligence requirements were imposed requiring that mining must start within ten years of lease grant but once production had been established advance payments of royalties could gain another ten years of delay). In addition, the 1977 act that established federal controls on the reclamation of land disturbed by coal mining included a provision that the owners of surface rights to land to which the federal right to minerals had been retained had veto power over exploitation of the land for surface mining.

The basic effect of all this could have been that all the rents left after the payment of federal royalties and state taxes would go to the surface landowners. The royalties and taxes were unavoidable and thus anyone dealing with surface owners had to reduce offers to insure that these royalty and tax obligations were met. However, there was no one else who had to be paid for access, the landowners legally were the first group with whom potential operations had to deal, and thus, it should be expected that these landowners could extract all the remaining rents. (A vague provision in the law allows the Interior Department to limit the payments to surface owners, but concern exists over the efficacy of the provision).

More complicated situations arise when it is necessary to deal with several landowners, none of whom have any special legal rights to secure enough land to support an efficiently sized mine. Complex bilateral bargaining situations can emerge in which because there is no universally accepted rule for rent sharing it may become difficult or impossible to effect satisfactory deals. Fortunately, this problem is most likely to lead only to an occasional anomaly in mine layout (akin perhaps to the thin building on New York's Herald Square where some landowner could not make a deal with Macy's whose huge store fills the rest of the block). The high cost to landowners of not being able to make deals can be a powerful force towards developing institutions for pooling interests and effecting deals. For example, consortiums of landowners may be formed. At worst, the states may choose to intervene, as many U.S. oil producing states have done, and pass laws (called unitization laws) requiring recalcitrant landowners to cooperate once the majority of their neighbors have agreed to cooperate in selling rights to extract oil and gas.

Another possible problem is that the owner of another critically scarce resource will attempt to extract part of the rents. It has been suggested that railroads with significant monopoly power might exercise that power by extracting a portion of the rent.

Thus, there are many claimants' possible problems of reconciling the claims and most critically a danger that the greed of these claimants may kill off too many ventures by attempting to charge too much. This danger presumably is greater when the claimants are diverse and uncoordinated so that none is fully aware of what the other is doing and thus cannot recognize the excess taxation.

Another fundamental problem is the application of the concept of distortionary taxes to rent collection.

AN OVERVIEW OF COAL-LEASING PROBLEMS

While coal-leasing policy and its goals have been widely discussed, an integrated, comprehensible review seems lacking. At least five requirements of leasing policy have been defined in the hearings and other material accumulated by the Commission on Fair Market Value Policy for Federal Coal Leasing.

First, the federal leases should produce a proper payment for its coal. (As noted above, the usual terminology that federal receipts should be optimal obscures that the operative law transfers the vast majority of the income to the mining states). Second, adequate supplies of coal should be made available. Third, undesirable environmental impacts of coal mining should be prevented. Fourth, the program must be administered within the budget available to the Department of the Interior. Fifth, the program must employ high standards of integrity.

For various reasons, the first two of these issues receives predominant emphasis here. Just as the Commission decided to limit its inquiry to that feasible given the limits of time and its expertise, this chapter defers to discussions elsewhere in this book. In this spirit, this discussion largely assumes that proper environmental controls are in place and provides limited comments about how defects in such controls might be remedied. (The review, however, draws heavily on the extensive literature on mineral leasing, much of which is cited in the bibliography to this chapter).

Administrative costs are best dealt with as part of the appraisal of the system. The integrity of the system obviously can be treated without economic analysis.

INFLUENCES ON LEASING

Governments can produce the effects of monopoly both by output restrictions and by sales taxes. Governmental output controls need no special treatment because they work exactly like private ones. As shown above, a sales tax or its equivalent—a royalty on production from government owned resources—has the effect of driving a wedge between buyer and seller. The payment by the buyer exceeds the receipt by the seller by the amount of the royalty of the tax. Instead of pushing trade to the point at which the value of additional purchases by the buyers equals the value of sales to buyer, the value to buyers exceeds that to sellers by the amount of the sale. Opportunities for additional mutually beneficial trade are lost in the same way as they are lost through monopolistic output restriction.

Two further points should be recalled. First is the difficulty in determining how much monopoly power exists. No satisfactory method exists to measure unambiguously whether a firm perceptibly affects price. Efforts to develop indirect measures based on shares of the market have proven no more satisfactory.

One problem is that the extent of the market is not clear-cut. Firms face varying degrees of competition from others. A coal producer from the Powder River basin clearly is closely rivaled by the coal producers in the region. Since some market potential exists as far east as Ohio and as far south as Texas, coal producers in other regions also have an impact. To some unclear extent, oil and natural gas producers also may be an influence. No methods exist to weight these influences to produce an acceptable measure of a firm's importance in the marketplace. To make matters worse, no accord has been reached by those who advocate use of market shares as indicators of monopoly power of what constitutes a share high enough to be dangerous.

Conversely, the admiration for competition expressed in economic theory is often criticized. Much of the criticism proves to be the self-serving rationalization for those who want to exercise monopoly power. Others fault the argument for not answering the question of whether the outcome is fair. Neither economists or anyone else can resolve this issue to everyone's satisfaction. All that can be said is that the system has the advantage of rewarding people for using their abilities to the fullest possible extent but cannot compensate for personal disadvantages that may be beyond the individual's control.

The concern most critical is fear that competition tends to be ruinous in the sense that the recognition of profits in an industry produces so much output that losses result. The ability of firms to survive for many decades in vigorously competitive industries suggests the concern for excess competition is misplaced. The survival implies firms can be profitable over extended periods.

However, the strength of competition is the critical question in setting leasing policy. Oil-and-gas leasing experience suggests that by organizing continuous large-scale leasing of tracts, each of interest to several bidders, the desired conditions can be created. Studies of competition in the coal industry suggest that a large number of well-financed participants operate. Given that coal resources are better delineated than those for oil and gas, it should be possible to make good valuation estimates when a large number of tracts are involved.

In practice, the number of leases offered is far smaller in coal than in oil and gas, and each attracts fewer bidders. Considerable concern, therefore, exists on whether competition *for leases* is adequate. A particular problem is the high proportion of offers that are maintenance or bypass tracts—properties near existing mines—that are more valuable to the operator of that mine than to anyone else. The operator often is the sole bidder, and this is widely interpreted as implying lack of competition. Similarly, a tract with no extant mine nearby may still be most attractive to a firm controlling adjacent nonfederal land and surface owner consent on the federal land.

This is not the only possible view. It could be that the fear of bidding from speculators forces the mine operator to bid as if it actually will encounter competition at the auction and that speculators have not entered the bidding because they can observe that the mine operators have acted to outdo potential rivals. An alternative view is that diligence requirements enable the mine operator to outwait speculators and discourage competition. (DOI civil servants feel actual bids reflect a tendency to underpay but it is possible these civil servants use appraisal methods that overstate the value of the coal).

The concept of intertract bidding has been proposed to stimulate competition. Several tracts of similar characteristics would be offered simultaneously. However, it would be announced that only a preset proportion of the leases would be granted. The separate bids would be compared, and awards would be made to those bidding highest. Thus, if six out of ten tracts were offered, the six highest bidders would get leases. The principal problem is finding comparable leases to offer together. The industry complaint that valid bids would be refused could be solved by varying the number of leases awarded depending upon the absolute level of the bids. If all exceeded some threshold, all might be granted.

Another important issue is that discussed above about whether or not the government should use a lower interest rate than the private sector. The critical implication of the interest-rate question is that leasing should be more limited if the public rate of interest is less than the private. The public sector then discounts the future less—i.e., gives future income a higher value—than does the private sector. Thus, private offers would be less than the value of the coal. Since these differences compound over time, the disparity is greatest for leases issued for coal used in the most distant future and to the extent possible these should be avoided. (Of course, if the public sector overdiscounts, rapid sale to the private sector is the optimal policy).

A final general point to discuss is market strength and how to recognize it. First, it is critical to recognize the difference between the current market for coal and the market conditions at the period starting several years after the grant of the lease during which the mine operates. Ideally, a temporary decline in the demand for coal should have no effect on the demand for leases. Foresighted investors will recognize the decline is temporary and ignore it in their bids. However, with poor foresight, investors may incorrectly interpret temporary drops as permanent ones and unwisely lower their bids.

The argument can go in the other direction. It is equally possible that bidders can overreact to temporary stimuli to demand and overpay for leases. A critical question about the drop in the demand for leases from 1980 to 1982 is its cause. Critics of federal leasing policy suggest that the problem is overreaction to the prevailing excess of capacity over sales. Defenders say that what had occurred is that investors have learned that the forecasts for coal use of even two years ago were far too high and adjusted their bids accordingly. (Examination of forecasts from three private services and the U.S. Energy Information Administration clearly shows such a decline).

Even those who claim the decline is temporary often accompany their statements with recognition of such things as slowdown in the growth of electric power, the unlikelihood that a substantial coal based synthetic fuel industry will arise, and diminished prospects that manufacturing firms will convert to coal. Such statements are inconsistent with the allegation that the decline in the market for leases is temporary.

A further problem is caused by the long lead times in leasing and the difficulties in evaluating the causes of changes in the leasing market and when, if ever, they will change again. Under these conditions, it may not be feasible to vary leasing with allegedly temporary shifts in the market. The requisite evaluation skills may be impossible for anyone to possess and the system may, in any case, be too rigid to allow such variations in lease levels. At most, leasing should vary with longer term trends.

THE EFFECT OF PUBLIC POLICY ON COAL SUPPLY AND FAIR MARKET VALUE

The prior discussion provides a framework for analysis of the impact of public policy on the markets for coal and the market for coal leases. In particular, it enables us to discuss what relationship, if any, exists between the market for produced coal and the market for coal leases. Further points can be made about the difference between the direct impact on the markets of leasing a given quantity of coal and the impact of particular lease conditions, other taxes, and legal claims by third parties such as surface-land owners.

The basic argument is that most problems stem from either the existence of monopolistic behavior by governments or lease buyers or from imperfections of foresight. Additional difficulties can arise from possible differences between the cost of capital to the public and private sector.

This can best be seen by combining the critical elements of the prior discussion. We begin by seeing what outcome would occur if we could be sure of the absence of monopoly, good foresight, and equality of public and private interest rates. Then we can examine the impacts of different forms of monopolistic behavior, poor foresight, and a public interest rate below the private.

The basic point to note about the value of a lease is that it is determined by the present value of the net income generated by the property. The property comes into operation when it becomes the lowest cost available. The expected timing of operation is determined by investor appraisal of how the property compares in economic attractiveness. Basically, varying the timing of the lease should be expected only to vary the present value to reflect the difference in the lag time between leasing and production. An earlier lease is worth less because of the longer wait for income but the loss in value should just equal the extra interest cost of waiting. The government can offset the loss by itself investing the funds.

If all the critical assumptions are met, it can be shown that increased offers of leases are always desirable as are any reductions in prices for leases or coal that result. The critical point is that in a competitive market increased leasing can only have a desirable effect on the coal market.

There are only two possible statuses for any new lease—it can be cheaper to exploit than some existing lease or it can be more expensive to exploit than all existing leases.

If the lease is cheaper to exploit than various existing leases, it will come in operation sooner. This would be an example of the mutually beneficial trades discussed above. Here we have a cheaper deposit displacing a dearer one.

The impact on lease prices and coal prices depends on the timing of the lease and the expectations of coal sellers. If the federal government makes all leases available well before the optimum time for it to start operations and all operators are confident that timely lease availability will prevail, the lease will have no effect on either lease values or market prices. The possibility that better tracts would be leased would have been considered both in bidding on past leases and in making commitments to develop them. Room would have been left for mines on leases yet to be issued.

The situation differs somewhat if, as must have been the case during the leasing moratorium, the prospects for timely leasing seem remote. Producers will develop an inferior property and pay more for it and other properties because they do not expect the superior one to become available.

The impact of belated availability depends on both the timing and the extent to which the delay has produced commitments to develop the inferior available properties. If large amounts of money have been invested in preparing for production at these inferior sites, it may be preferable to keep the mine operating and recover part of the investment than to shut down and recover nothing. This premature development of an inferior deposit will depress coal prices and make leases less valuable.

Two points should be noted. First, this is purely a transitional effect as one moves from a policy of dilatory leasing to one of timely leasing. Second, the net impacts on governments revenue are indeterminant, given the same leasing system over time. The government was paid too much for the earlier leases and too little for the later ones. Which error was greater is not clear.

Moreover, it is true here that it is better late than never. The mine will be developed and put downward pressure on prices only if it is profitable to do so; profitability is possible only if the development is socially beneficial.

Offering to lease properties that are inferior to existing leases is innocuous to both coal prices and prices on other leases. Inferiority means that this lease will await its turn until all the other leases are developed. Thus, the pattern of development of other leases will not be altered. Thus, supply, prices, and the willingness to pay for these other leases will not be affected. (An exception would arise if the level of leasing was so tight and uncertain that producers had made commitments to extend the life of existing deposits rather than develop the worse one in a timely fashion.)

The critical elements in this argument are that vigorous competition among producers in the coal market insures, as it should, that all the leased coal beneficial to mine is mined, that vigorous competition for leases insures that the government receives a payment equal on average to the present value of the expected rents, and foresight is good enough to insure that the expectations are reasonably correct.

Thus, satisfaction of the critical assumptions implies that increased offers of leases can only have desirable effects on the price of coal. Increased offers have an effect on the price of leases only when unexpected. However, the effect involves both overpayments before the leases and underpayment afterwards.

The implications of imperfect foresight and government interest rate under the private ones are simpler to treat than the many forms of monopoly that may exist. Thus, the other issues are disposed of first. Payments to the government will be higher or lower than the actual income stream depending on whether firms are overly or underly optimistic. As already noted, over optimistic firms do not survive so that a critical problem is more likely to be under-optimistic ones. It was already suggested that the best way to correct for pessimistic expectations is to make the payments somehow contingent on performance. Royalty or profit-sharing schemes may be developed.

When the public interest rate is lower than the private one, again more of the payments should be made at the time of production and less through lease bonuses than if the interest rates were the same. The federal government pays less of a cost for waiting and should therefore wait for its income. If federal interest costs are higher, lease bonuses become more attractive.

The first set of monopolistic practices to examine are those on the government side. As noted, interference with mutually beneficial trade can come from either restricting physical availability or by imposing sales taxes or royalties. A monopolistic restriction of leasing is one that systematically keeps off the market leases whose exploitation would be socially beneficial. This has major and clearly undesirable net social effects identical to those produced by private monopoly behavior.

Output is too low and prices too high. The Treasury gets more income, but like the monopolist's gain, this comes at a loss to coal consumers greater than the Treasury gain. Increasing leasing to prevent this monopolization would be desirable even if Treasury income fell. This constitutes an extension of the earlier conclusion that greater leasing can only have good effects on coal output. It adds a set of cases in which a decline in net Treasury revenue is certain but desirable.

We have already seen that a royalty or tax produces output reductions similar to those produced by monopoly. What remains to be added is discussion of the impact on bonus bids. To avoid combining effects, it is preferable to deal with the case in which the government tries to offer all the low cost leases. Here, all that is at work is the depressing effect the taxes and royalties have on output and the prices received by producers. This produces less after-tax rent and thus less is left for the lease bonus.

In short, the government can raise the *total* present value if its income by imposing a tax or royalty. However, the gain comes through the income for the tax or royalty. The amount left to pay a lease bonus is lower. Where we add limits on leasing, these can further raise prices and lease bonuses, possibly to above the level prevailing with unlimited leasing and no royalties. However, this is the effect of the leasing limits.

Most of the argument just stated applies even if a state tax is involved. There will be an output restriction which is larger, the larger is the tax. Governments collectively gain through the higher taxes and royalties offsetting the lower lease bonus. Direct state taxes give the state a bigger share of the revenues than if the states are content simply to share in federal royalties.

The impact on federal revenues is more complex. It would clearly fall if the tax were on before tax revenues of the firm. These fall more with two charges than with one. However, the Department of the Interior has chosen to tax revenues including taxes which effectively raises the tax rate as applied to before tax revenues. This could lead in some cases to higher federal revenues than if there were no state tax.

A final effect to note is that of the surface-owner protection provision of the Surface Mining Control and Reclamation Act. The provision requires written permission of the surface landowner before the underlying coal can be leased. It is accompanied by a provision that the requirement should not alter the value of private or federal property. Presumably, this means that compensation to the surface owner for permission should be limited to the compensation for the costs of having the land disturbed.

The difference between this cost and the present value of the coal lease should be kept by the federal government. The Surface Mining Act did not specify a mechanism for insuring this outcome, Interior appears not to have one, and, not surprisingly, testimony before the Commission indicates that surface owners actually succeed in extracting as much as possible of the rent remaining after taxes and royalties are paid.

With all this in mind, we can deal with the issue of "speculation" on coal that is frequently raised in discussions of the leasing program. Economists and financial specialists found the attacks confusing because the usual definition of speculation involves undertaking socially beneficial actions—namely smoothing out markets. Land speculators do the useful work of negotiating with numerous parties owning pieces of land so they can be aggregated into a package that is more attractive than the separate parts. For example, they can assemble all the rights needed to permit an efficiently-sized mine.

With a competitive market for speculative

services, the only excess profits any speculator can earn are those from being cleverer than the speculators. Close examination of the testimony before the Commission makes clear that this sort of speculation was not being criticized.

The value of activities to facilitate mining was recognized. The concern was over leaseholding not clearly intended to produce production. Such concern is essentially criticism of the absence of competitive bidding under pre-1976 coal policy. That policy made the holding of coal very cheap and encouraged the piling up of reserves without clear plans for their use. An effective bidding process imposes such penalties for lease holding and by itself insures that no one will bid unless a prospect for profitable use exists.

Thus, in a competitive market for leases, diligence requirements are at best redundant and, in many ways, undesirable. The competition for leasing suffices to limit holdings and provide adequate compensation to the federal government. (Of course, when thc government chooses to rely on some form of deferred compensation scheme such as royalties, it lessens the cost of holding unless some penalty is imposed. This could be done either by shifting towards greater reliance on bonus payments or by extending the provisions already in the act allowing progress payments in lieu of actual development). Diligence requirements have three drawbacks. First, they can inspire too early a start of production, too low prices, and too low lease payments. All this occurs when it is more profitable to start a mine before it is optimal to do so rather than lose the lease.

By definition, this produces more total output and less profits than would have prevailed without the diligence requirements. From this follows the indicated effects on lease values.

Another problem is the lessened incentive to bid. The time pressure increases the risk of not making a timely sale and makes the lease less attractive. Finally, the provision prevents third parties who believe other uses are more valuable than mining from buying the rights to prevent their use. A surface owner, an environmental group, local residents, or a coalition of such interests could outbid mining companies if they believed nonuse was worth more than mining. The diligence requirements preclude such bids because these third parties are not allowed to keep the rights if they do not mine.

The introduction of monopoly power by producers changes things greatly. Two types of monopoly power can be distinguished—that in selling coal and that in leasing. The analysis differs greatly depending upon whether or not government also is undertaking monopolistic practices. As would be expected, the more monopolistic forces that are present, the more complex is the situation.

Again, let us start with the simplest case—no public monopolistic practices, vigorous competition in coal sales, and weak competition in leasing. Here what all we get is that the government receives too little on leasing. The vigorous leasing and the willingness of firms to take the leases and vigorously compete in their use prevents output restriction.

The primary question about this case is how it can occur. The testimony before the Commission suggested that the problems of divided surface and coal ownership, the division of coal rights among federal, state, and private ownership, and preexisting leases, make it difficult to assemble lease tracts on which vigorous bidding is likely. It is further suggested that some of this could be cured in principle, by arranging for agreement among coal and surface owners to pool their interest and jointly offer tracts, or by swaps of lands so that each landowner exchanged a tract too small to mine by itself for one that when made added enough adjacent to another tract to constitute an attractive combined package.

It is also suggested that the combination of surface-owner right protection and land-use review provisions and the lack of resources to identify opportunities for such actions discourages the process.

However, it should be noted that the argument may be overstated. The absence of multiple bids does not always mean the absence of competition. The threat of competition may suffice to encourage adequate bids. Coal leasing is most likely what Baumol, Panzar and Willig call a contestable market in which the threat of competition suffices to insure a competitive outcome. Awareness of the tendency of the most eligible buyer always to offer at least as much as needed to win the bid would discourage rival bids. If low bids regularly occurred, speculators would begin to outbid the firm best able to use the property.

As with other problems that produce inadequate bids, one cure is reliance on royalties or profit sharing to get more income later. This produces two problems. First, these methods can introduce distortions or administrative problems and effectively the delay of payment can discourage development. Had the firm made all its payments in a lease bonus, there are greater benefits from operating as soon as possible than if the government must share in the operating profits. To offset the disincentive to operation caused by the actual reliance on royalties, the additional requirement of diligence (mainly involving loss of leases not developed within 10 years) has been introduced. This, as noted, makes leasing more risky and leads to depress bonus bids.

It should be further noted that this problem is independent of lease levels. The underpayment occurs separately on each lease. The only sense in which increased leasing is less remunerative is when the increase consists entirely of adding leases on which underpayment is likely to ones that can be sold at their full present value. (The effort rapidly to *increase* lease levels might cause lessened efforts to adding tracts that would attract competition).

The introduction of public monopolistic practices necessarily complicates things. The simpler case is where leasing levels are inadequate. The resulting output restriction problems remain and the key difference is that inadequate bidding again reduces government receipts. Again, the cure is approaches that increase competition or allow the government to share in the profits.

By turning to the case in which taxes and royalties are part of the output restricting activities of government, we can directly confront the previously undiscussed problems with the prior suggestions that royalties and profit sharing are ways to offset inadequate bidding. What was left unsaid is recalling that royalties produce higher revenues at the cost of retarding output. Thus, any reliance on royalties creates problems. The principle that higher taxes produce higher distortions means that addition of royalties to offset underpayment on leases to those imposed to raise revenues increases the undesirable effects on output. Thus, to the extent it is practical, it would be preferable to resort to profit sharing if competition for leases cannot be made more vigorous. (See below for further discussion).

The next case to consider is inadequate competition in both buying leases and selling coal. (It is unnecessary to consider the case when there is vigorous competition for leases but not for coal sales since it is unlikely to occur.) Here firms produce less as well as paying less. Varying leasing policy cannot help much. The cure would have to lie in increasing competition. (An alternative, discussed below, is so-called public utility regulation that forces the monopolist to produce more, but it is a difficult to administer policy whose practical utility is unclear.)

The combination of public and private monopolistic practices has the same implications as adding state taxes to federal—more retardation of output. Again, the only practical cure is increased competition. Fortunately, numerous studies by such diverse organizations as the Federal Trade Commission, the General Accounting Office, the Department of Justice, and the Department of Energy have concluded that competition in coal selling is vigorous. Western coal mining may be limited to large firms but many large firms operate in the West. This suffices to insure competition.

Thus, monopoly in coal selling is both something that would be difficult to cure through altered leasing policy and probably a nonproblem.

The upshot of this long discussion is that two critical deficiencies can exist in leasing—output restricting policies by government and inadequate payment by private firms. Care in appraisal must be taken to determine which problem or combination of problems is at work. It was further shown that increasing public leasing can only improve competition. The inadequacy of competition for leases is less specific and cannot be cured by not leasing. The only solution is to devise a method of increasing the yields on each lease.

The only sound reason for reducing leasing is when the government has lower interest costs than the private sector.

Thus, the critics of leasing policy have correctly identified the problems but have been incorrect about their effects and the cures.

METHODS FOR SECURING FAIR-MARKET VALUES

The method by which those who mine coal and other minerals should be charged for these rights has long been debated by both policy makers and economists. Three basic approaches to charges have been defined—fixed-sum bonus payments, royalties (or taxes) as a fixed amount per unit of output or percent of sales, and profit sharing. Each type has numerous variants, and actual charge systems may employ at least two of the basic systems. For example, U.S. fossil-fuel leasing policy involves requiring both bonus payments and royalties.

Given the many possible circumstances surrounding a given lease and the several, conflicting goals of setting a policy for charges, consensus on choice is difficult to attain. As discussed further below, the attractiveness of different approaches depends on how vigorous is the competition for leases and how well informed the potential leasers are on the economics of the lease.

BONUS BIDDING—THE ECONOMISTS' FIRST CHOICE

With vigorous competition, good foresight, access of a large number of bidders to the capital needed to finance bonus payments, and close correspondence between public and private interest rates, a system of leasing based entirely upon a bonus payment will extract, on average, the total present value of the economic rents from coal production. Therefore, the preferred method is to rely entirely on bonus payments. These could be a single payment at the time of leasing or a series of payments over some specified number of years.

The importance of foresight is that it critically affects the outcome of all investments including payments for coal leases. In the uncertain world in which we operate, it is impossible for anyone to be right about everything all the time. The best we can hope for is that in making a large number of choices, the dominant decision makers are correct on average. Their underestimates in some areas are offset by overestimates elsewhere. That many uninformed people operate in the market does not matter so long as there also exists a substantial group of well-informed participants. The informed will put the uninformed out of business. Competition among these knowledgeable participants will insure good performance.

Similarly, the definition of competition implies that firms are forced to pay the full expected value or lose out to rivals. A bid less than the full worth of the property would be topped by some rival. Observers of the problem suggest that such competition is possible only if many firms are both aware of the value of the resource and have access to the funds needed to finance the purchase.

The bonus-bid approach has the additional advantage of not influencing any of the post-lease decisions of the firm. The bonus, by definition, cannot be altered by any action of the firm, and thus no decisions can be affected by having paid a bonus. Of course, the continued willingness to participate in lease bidding depends upon the ability to insure that, on average, the firm does not overpay for leases and earns enough on its aggregate of leases to repay the bonus.

This requires both adequate foresight by the firm and the absence of rules such as a statutory minimum-required payment on all leases. Such a requirement would necessitate incurring losses on properties generating less value than the required payment.

The advantages of a bonus system persist only if the bonus payments are spread out over a predetermined fixed period with no possibility of default. Under these assumptions, the payments are still unaffected by any decisions of the firm. However, when the duration of payments is tied to the actual life of the mine, an unfavorable effect arises. As profits fall below the required annual payment, the ability to terminate such payments by closing the mine will be utilized.

To avoid this problem associated with bonuses spread over time, it is not sufficient merely to preset the number of years. Loopholes can arise particularly if leasers began to adopt the practice, common in Appalachia, of creating a separate corporation for each mine (or, in some cases, subsections of the mine). Then the firm could escape its obligation by declaring bankruptcy. This difficulty could be avoided if the term of payment was kept significantly shorter than the economic life of the mine or if parent corporations could be made liable for their subsidiaries' debts.

The financing problem is unlikely to be a major issue in federal coal leasing. However, spreading the payment schedule out over several years can ease any financing burden that exists.

The direct administrative costs of bonus bidding are the lowest of any system.

However, the perils of estimating values and providing assurance that competition is sufficiently vigorous make it difficult to secure acceptance of reliance solely on lease bonuses. Substantial additional administrative costs are incurred in attempting to provide sufficiently-convincing evidence that adequate compensation is secured. These problems of verification also arise with any system in which all the variation in federal receipts is produced by the lease bonus.

Failure of these assumptions to hold can make bonus payments diverge markedly from the present value to the government. Specifically, when competition is weak, future incomes underestimated and private interest rates above public, payments will be below the present value to the government.

It is fears of such underpayment on leases that cause governments to insist part of the payment be an annual one tied to income from operations. Because of its alleged administrative simplicity, a royalty is usually employed despite the inefficiency it causes. (In practice, monitoring production to determine the quantities on which royalties should be paid is difficult and complaints are rampant about fradulent reporting.) This need not be the case. Numerous alternatives have been proposed such as a tax on the excess profits. These would provide the desired additional payments without distorting output. However, the formidable accounting problems involved make governments reluctant to use profit sharing (see below).

ROYALTY SYSTEMS FURTHER CONSIDERED

The distinction between a royalty and a sales tax is purely legal (as is the practice of calling a tax on mineral sales a severance tax). Economically, the name chosen affects nothing. However, the method for laying the charges matters greatly. The traditional approach is to apply a single formula that applies to every unit of output of every producer. Thus, the royalty on surface mined federal coal is 12.5 % of sales revenue; Colorado has a tax of .708 cents per ton.

The advantage of taxes and royalties is that they adjust to two key uncertainties of the forecasting process—future prices and outputs. Government shares the risk of price fluctuation and of misestimates of the amount of economically recoverable coal in the leasehold. With limited exceptions, the charges are easily levied since the price paid can readily be determined and generally arises from transactions between independent entities.

Even when the coal is produced by subsidiaries of the consumer, valuation may not be difficult. The subsidiary may also be selling to independent buyers. Also, there are usually sales being made by nearby independent sellers of similar coals. Comparison to these selling prices is a well-established method to establish the market value of the coal. Moreover, where the coal producer is a regulated public utility, the desire to set prices higher to secure rate relief may predominate over the desire to set them lower to reduce taxes.

The classic objection to fixed royalty or tax charges or rates is that they fail to account for differences in the profitability of mines. The ability to pay the tax depends upon these profitability differences. Given the 12.5% federal royalty, mines making gains equal to 30% of sales fail to have all their excess profits taxed away, but the tax puts out of business any mine unable to generate profits equal to at least 12.5% of revenues.

Additionally on the disadvantage side, within any mine, some portions of the resource are more expensive to mine than others. The mines that average profits equal to 30% of sales can have areas where the costs are large enough to lower profits below 12.5% on mining this tonnage. A uniform tax prevents the production from these high cost sections of the mine. The importance of this effect depends on mining conditions. The problem is least for a mine with a single seam with highly uniform ratios between the amount of cover removed and the tons of coal produced per acre. Variations in the ratio and mining in multiple seams tend to create significant intramine cost differences.

These problems have inspired suggestions for more flexible royalty or tax systems. Such systems attempt to adjust for intermine, intramine cost differences, or both, Thus, a system gaged to intermine profitability would have a higher royalty rate for more profitable mines. Alternatively, formulas could be set to vary the tax burden for each mine in a fashion that lessened the burden placed on the more expensive-to-exploit portions of the deposit. Delicate balancing and extreme complexity would be required in a system to insure simultaneously higher charges on the low-cost portions of the output and lower charges on high-cost portions.

In principle then, properly-designed variable-royalty systems could be used to preserve some of the advantages and reduce the disadvantages of a preset uniform royalty. However, administration would be much more difficult.

The preferable alternatives for any variable royalty scheme appear to require setting the variations in advance. This is particularly true of setting the differences in the base royalties applied to each mine. An advanced commitment to a royalty rate that cannot be altered after the fact preserves incentives to lower costs, but varying the royalty rate once operating begins could make firms inadequately attentive to costs. Royalty reductions could remove the penalties of inadequate cost control. Royalty increases could remove the incentive to reduce costs. A preset royalty rate avoids this.

If we turn to the special problems of either intermine or intramine adjustments, even more difficulties arise.

A tricky problem with any system that imposes higher royalties on more profitable mines is how to insure that the system does not eliminate the competitive advantage of lower-cost mines. If the royalty system totally equalized costs, all mines would be on an equal footing. Several cures could be applied. The royalties could be set so that they left a cost advantage for the lower cost mines. This would preclude capture of the total excess profits. Addition of a bonus bid requirement to the system could capture the remaining profits.

Such a system would be hard to devise. We would want a system of bonuses that produce lower after tax costs for lower-cost firms. This is an advantage of the present system of a uniform tax and variable bonus. A fixed bonus (per ton) would not fully solve the problem because it would lower by the same amount what everyone was willing to pay as a royalty. Creation of a fixed bonus does discourage leasing and keeps out high cost mines that cannot cover the bonus. However, this leads to erring in the direction of allowing too little leasing by excluding release of land that could be profitably mined if the bonus were lower.

Thus, some complex set of rules that imposed higher bonuses as well as higher royalties on low-cost mines would be required. The higher bonus would insure that while the lower-cost mine paid higher royalties than a higher-cost one, the difference in royalties would be less than the cost difference. For example, one firm earning $20 profit and another earning $10 before royalties, might pay royalties of $18 and $10, respectively. This would leave a $2 cost advantage to the more profitable mine.

Another alternative would be to attempt to estimate future demands and the costs of exploiting specific tracts with sufficient accuracy that only the lowest cost ones needed would be leased—an estimation task far beyond our powers.

Still another problem is that a system of imposing higher royalties on a more profitable property means that more of the output cannot be mined than would be true with lower royalties. With a $10 price, coal costing $8 to mine can be recovered with a $2 royalty but not with a $3 levy.

Various other methods to impose higher charges on more profitable mines have been proposed but seem essentially variants on the variable royalty approach. It has been suggested that as output surged or prices began to exceed some threshold, the royalty rate be increased. The former approach assumes that output rises are indicators of rising profitability. Variation of royalty rates with output has also been proposed as a method of adjusting for intramine cost differences. This interpretation assumes that output declines are an indicator of rising cost.

A further variation is to secure the increase in royalties by receiving a share in the output in excess of some predetermined level. This is merely a roundabout way of imposing a higher tax (equal to the entire market value of the government's share of the additional output). It seems to have no advantage over directly imposing a high tax on additional production.

The approach, moreover, is less appropriate for coal than for many other minerals, particularly oil and gas. Coal is a particularly heterogeneous material, and transactions are made to match the coal characteristics needs of the buyers. Large western mines do this by signing long-term contracts with purchasers. The federal government would probably prefer to have the lease operator sell the federal coal share along with the other coal supplied by contract. Securing the revenue by a tax of equivalent amount would accomplish the same results as sharing in output at high levels.

The key question then is whether confiscating the total revenues on a portion of output is a good way to tap excess profit. Considerable concern exists about the severe administrative difficulties associated with such policies.

Turning to intramine cost variation problems, the literature suggests that some method be devised to reduce royalties on expensive-to-produce coal, but no satisfactory implementation procedures have been proposed. The rules would have to deal with the quite different problems of horizontal and vertical extensions of mining. One type of high-cost mining is that of going after a lower seam that is expensive to reach even when the higher seams and their cover are removed. This mining needs a concession when the upper seams are being stripped. Another sort of high-cost mining is when you reach a section that is generally high cost. When the tax break is needed depends upon whether it is more efficient to capture the coal when you reach it moving horizontally through the tract or to wait until later.

Again a system that lowered royalties as costs rose could discourage cost control. Administration, in any case, would require expensive audits of company costs. Alternatives would be to have either a universal rule for reducing the royalties as the mining neared some point in its operation at which costs rose or building mine-specific rules into the lease.

When output declines are an index of cost rises, royalties that rise as output is higher and drop if output declines would be such an approach. None would be easy to devise or administer. In fact, perhaps the only approach that would work would be to key the royalty to actual costs. This would produce a system virtually indistinguishable from profit sharing techniques discussed below. In short, variable royalties appear too complex to be practical policy tools.

PROFIT SHARING

Still a third approach is to levy charges on the profits of the firm. Profits can be defined in several ways. First, the tax-accounting concept could be used. However, this profit concept subjects to taxation part of the return on investment needed to attract stockholder funds. An alternative proposed approach is to add to accounting costs an estimate of this required return on investment, define profits as the difference between revenues and these adjusted costs, and tax profits defined in this fashion.

Advocates of profit-sharing approaches stress the potential of limiting the charges to exactly the amount of the excess profits. The only impact on the firm would be in possibly encouraging excessive investment because the government would be paying for these outlays by having agreed to reduce its receipts by an amount sufficient to recover all investments.

Incentives to reduce costs could be produced by allowing firms to share in the excess profits. However, this defeats the U.S. government goal of fully extracting the excess profits. A lease bonus could be employed to deal with that difficulty. The future gains effectively would be sold away in advance, but the firm would still have an incentive to cut costs and raise subsequent profits.

The system involves severe problems in administration and political acceptability. Given the chronic problems of seeking accord on the proper cost of capital to use in any form of financial analysis, it would be difficult to agree upon the appropriate formula for a scheme that included in costs capital recovery with an adequate return on investment.

A shift of the basis to a less controversial formula would involve the distortions of failing to recognize capital costs. Effectively coal mining would bear a higher corporate tax rate than other industries and tend to suffer greater distortion of its investment decisions.

However, an even greater problem would be the costs of administration. It is unlikely that, given the widespread concerns about insuring compliance, it would suffice to allow the firm to use the accounting system allowed by the Internal Revenue Service. More stringent rules and tighter audits are likely.

Moreover, the problems of cost attribution are more difficult at the mine level than at the firm level. These problems are quite vexsome even in the aggregate. Disputes arise over efforts to allocate profits so that as much as possible is earned in areas subject to lower tax rates. For example, vertically integrated mining companies try to get IRS to accept the highest possible valuation of their minerals to insure a higher depletion allowance.

However, greater problems arise when attempting to determine the costs of individual mines. Allocation of overheads are a perennial source of conflict.

Thus, none of the systems is ideal. Royalties seem particularly problematic. Variable royalties approaches seem to have problems of optimal design and administration that make them less desirable than a fixed-rate approach. Profit sharing with bonus bids avoids most of the problems with royalties and may prove no more administratively burdensome or intrusive to the firm than a royalty scheme that provides relief to high-cost production. As far as securing optimum payments in a less burdensome fashion, stimulating vigorous competition for leases, and being able to rely on timely leasing, bonus bidding is preferable. Whether it is feasible is less clear. Profit sharing seems second best; a shift to a lower royalty-higher bonus, third best. (DOI considers profit sharing administratively infeasible and would thus rank it lower, in practice, than I do).

SURPLUS TAXATION IN DECREASING COST INDUSTRIES

The case of industries with economies of scale is much simpler than that when diseconomies of scale prevail. In the former case, Euler's theorem indicates that losses arise if the efficiency rule requiring the equality of marginal cost to price is followed. One way to avoid this problem is by some method by which some consumer surplus is transferred to the firm to cover the losses.

Two basic methods are possible. Both involve attempting to charge each customer a price on marginal consumption equal to the socially efficient price for the commodity, i.e., prices such that all customers pay a price equal to the marginal cost of service. One way to cover the losses would be to impose some form of tax, service fee, or other charge on consumers that did not vary with consumption (known technically as a lump-sum payment). The total revenues from such lump sum payments must at least equal the firms' losses from prices equal to marginal costs.

An alternative, known in the jargon as declining block rates or Ramsey prices (after the economist who proposed them), is for prices to decline with consumption. Marginal consumption would be priced at marginal cost but lower units would be priced higher. This is illustrated by typical electric power rates which may charge 10 cents per kilowatt hour for the first 100 kw-hr per month used, 8 cents for the next hundred, six cents for the next hundred, and 4 cents for each kilowatt-hour after the first three hundred.

The ability either to raise lump-sum taxes or impose Ramsey prices rests on the existence of consumer surplus. If such surplus exists in sufficient amount, it provides a source of revenue to cover the losses. Thus, either method is potentially a method for attaining efficiency in industries with economies of scale without losses incurring.

The system is clearly feasible since public utilities have practiced it for years. However, several critical points ought to be made about the subject. First, obviously the workability (*and desirability*) of the approach depends upon the existence of enough consumer surplus to finance the losses. Only under extraordinary circumstances would that surplus exactly equal the loss.

However, if the surplus is not sufficient, this implies the venture is socially unprofitable and should not be opened. Otherwise, a residual surplus (i.e., an excess of gross consumer surplus over the losses that must be covered) remains. The sharing of this surplus is another of those equity questions that I noted early in this chapter on which we possess no agreed-upon rules. Thus, problems will arise about how to divide up the payments among various consumers.

A comparable, possibly worse problem arises in industries with diseconomies of scale and economic rents. One could transfer these rents in lump-sum payments to customers. Alternatively, what have been called lifeline or inverted block rates and are termed here reverse Ramsey prices could be (and have been) used to transfer rents. A low price is paid on the first few 100 kw-hr (e.g., three cents) and the charge rises with each increment with the last amounts costing the full marginal costs. The main problem with this approach compared to charges on consumer surpluses in industries in which economies of scale and losses occur is that the rationales are not reciprocal.

Consumers of the products of decreasing cost industries cannot get an efficient level of supply unless they aid the industry. The aid system is a way for the taxed to insure that *their* needs will be better satisfied. With diseconomies of scale, needs are efficiently satisfied by prices equal to marginal cost. Consumers need to do nothing to insure efficient output. In the industry with economies of scale, it was clear the beneficiaries were the most appropriate people to finance the extra output. The issue was exactly how to divide the burden among a basically well-defined group (but see below on some problem areas).

It is far less likely that the customers of an industry with diseconomies of scale and rents are most clearly those who should benefit from these rents. The problem then is not just one of allocating rents among customers (in itself a difficult process). A more basic question is whether customers should be favored at all or whether rents should be shared more broadly among various groups in society. (The next section deals more fully with models of rent sharing among customers including reverse Ramsey pricing and the pitfalls involved).

Finally, two types of errors can arise in Ramsey pricing. The first also can occur in financing by lump-sum payments. The socially efficient prices may be misestimated so that more or less than is proper is charged at the margin (and thus more or less than is socially efficient may be charged). With Ramsey pricing, a further problem is that the blocks may be of improper lengths. A given consumer may have *his* block rates drop too slowly so he consumes less than is efficient or too rapidly so that too much is consumed.

Thus, in Fig. 5.13.2a, we have the demand curve and marginal cost curve for our decreasing cost industry. (For simplicity, I assume a single product firm and no differences in the cost of serving customers as actually occurs in electric power). Then we have a socially efficient price of P_c and an efficient output of Q_c.

Even if each consumer was scheduled to pay P_o on what was presumed to be the block in which marginal consumption occurred, the blocks can be misdesigned. Thus, in Fig. 5.13.3, the demand curve is D_1 and an optimal price is P_o. However, the Ramsey price schedule is set so that for the optimal quantity Q_o this particular consumer demands, a price of $P_1 > P_o$ actually is charged. This consumer, as do all consumers facing a Ramsey price schedule, consumes to the point at which his quantity demanded at a given price equals what is supplied him at the (marginal Ramsey) price charged. In this case, the curve is drawn so that the consumer ends up paying P_1 but lowers consumption to Q_1 below Q_o. In Fig. 5.13.4, a worse situation is shown in which the charge for the optimum level of consumption is still Q_o but the consumer demand curve is such that the Q_2 actually chosen costs $P_2 > P_1 > P_o$.

As Fig. 5.13.5 and 5.13.6 show, the converse could occur. The optimum consumption for a particular consumer could be underpriced at say P_3. In Fig. 5.13.5, the consumer overconsumes to Q_3 but pays P_3. In Fig. 5.13.6, the overly generous schedule leads to paying the even lower marginal priced P_4 and leads to consumption of Q_4.

In short, the need to finance losses of industries enjoying economies of scale is the most clear-cut theoretic argument for using economic

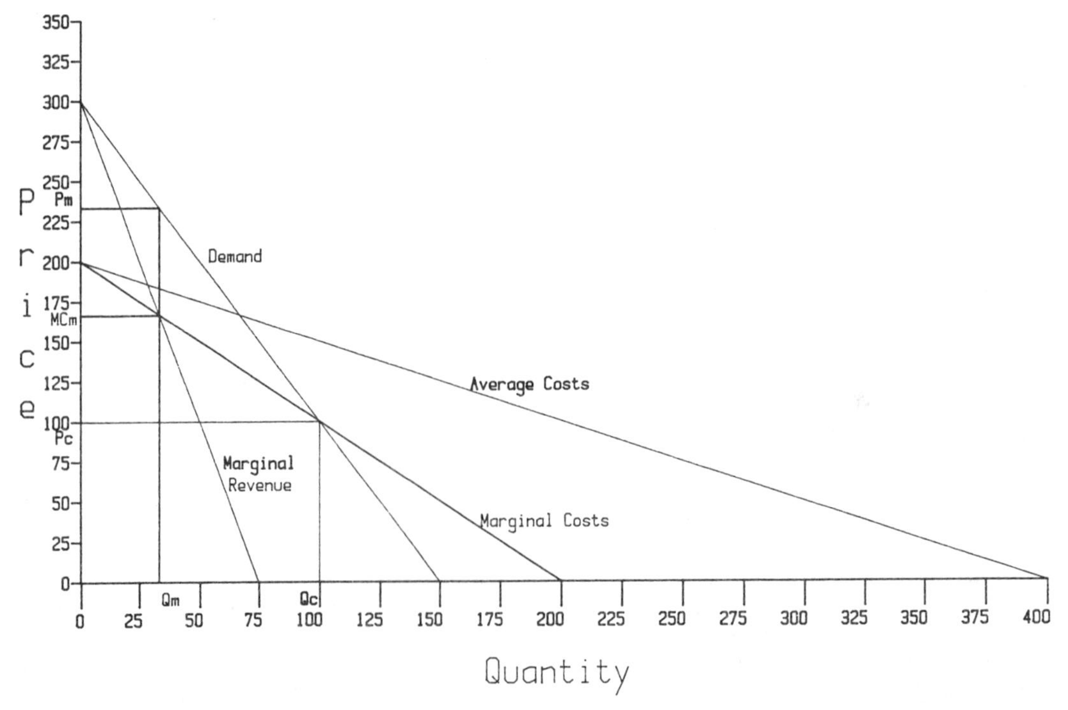

Fig. 5.13.2a—Decreasing cost efficiency.

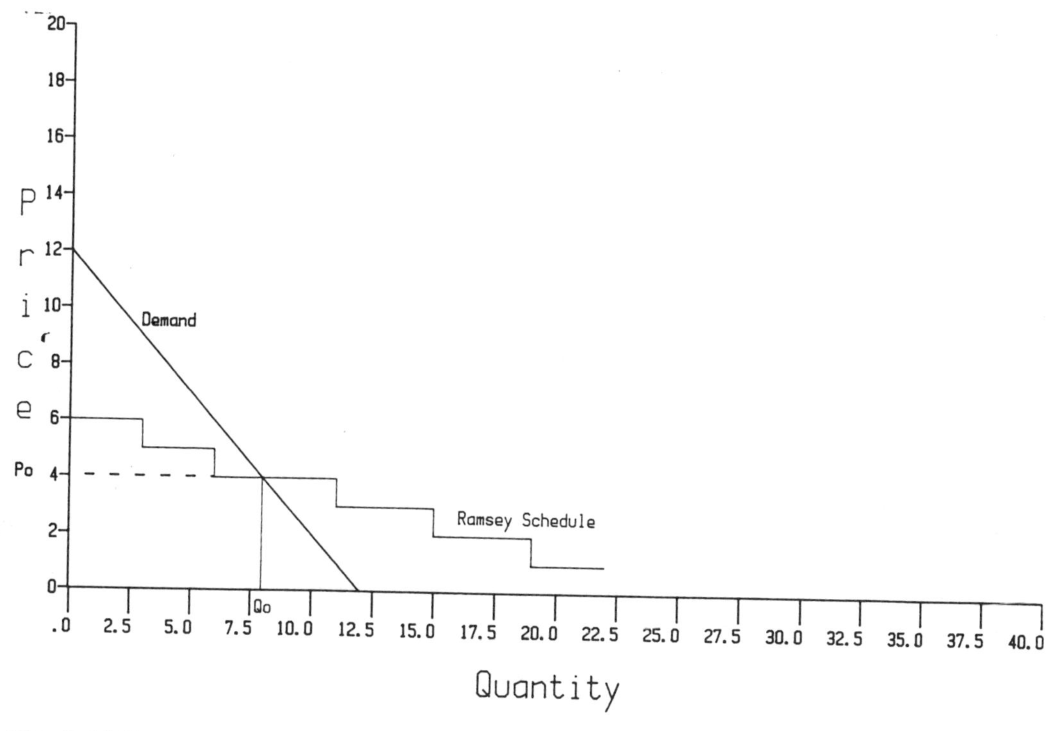

Fig. 5.13.2b—Optimum Ramsey schedule.

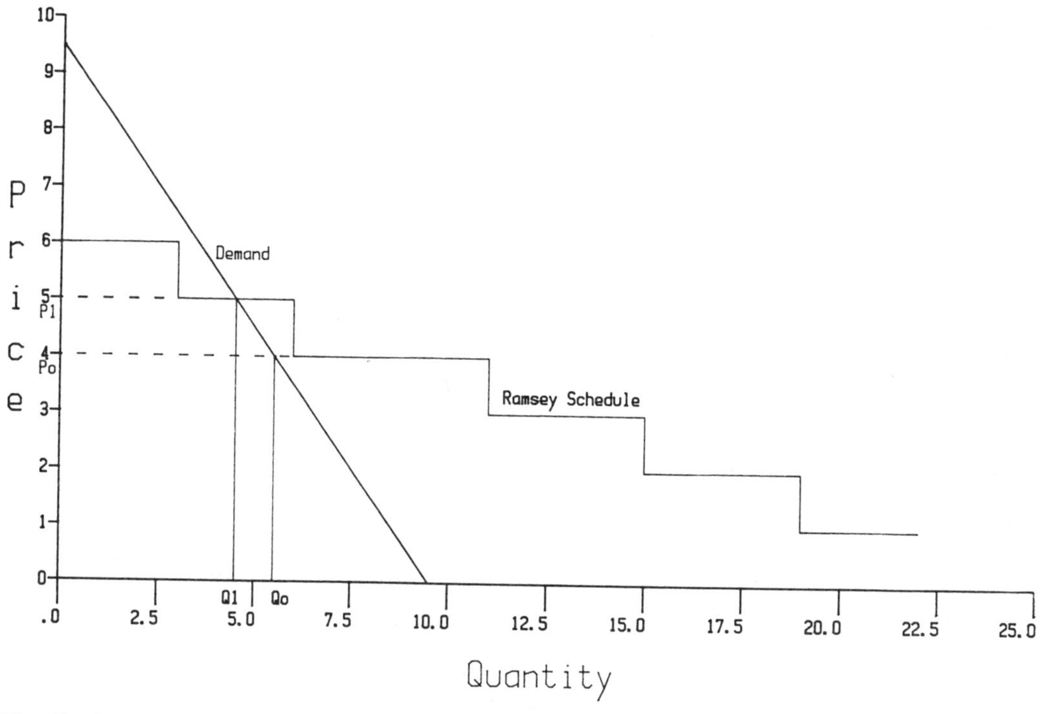

Fig. 5.13.3—Too high Ramsey schedule.

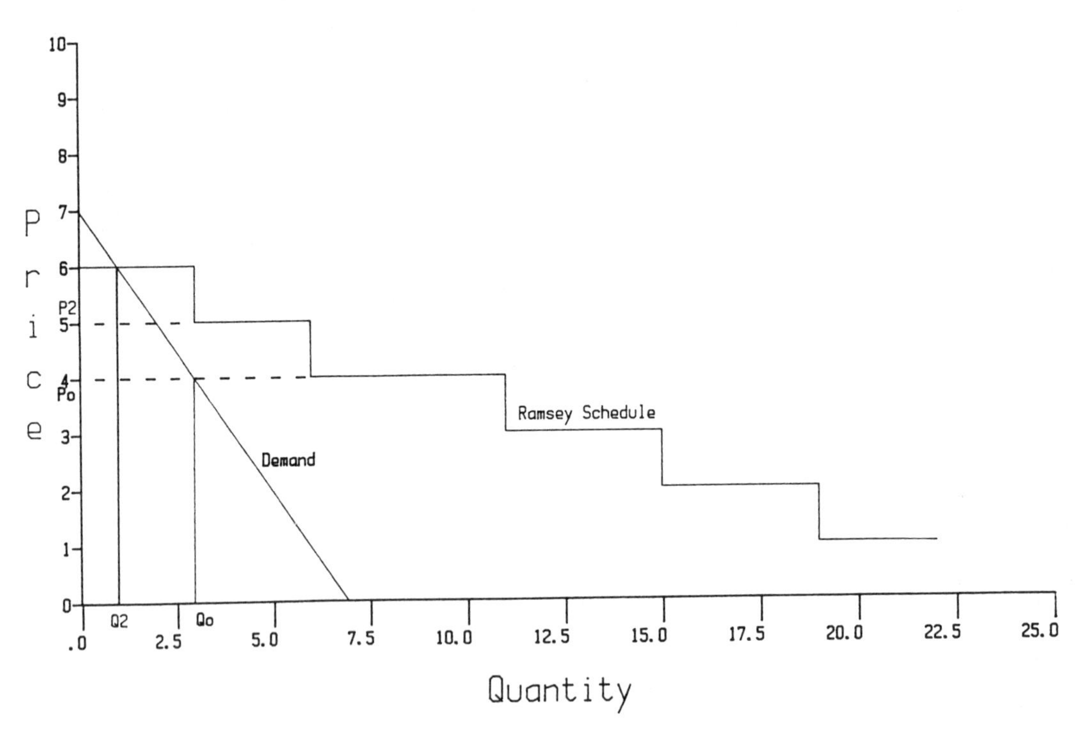

Fig. 5.13.4—Very high Ramsey schedule.

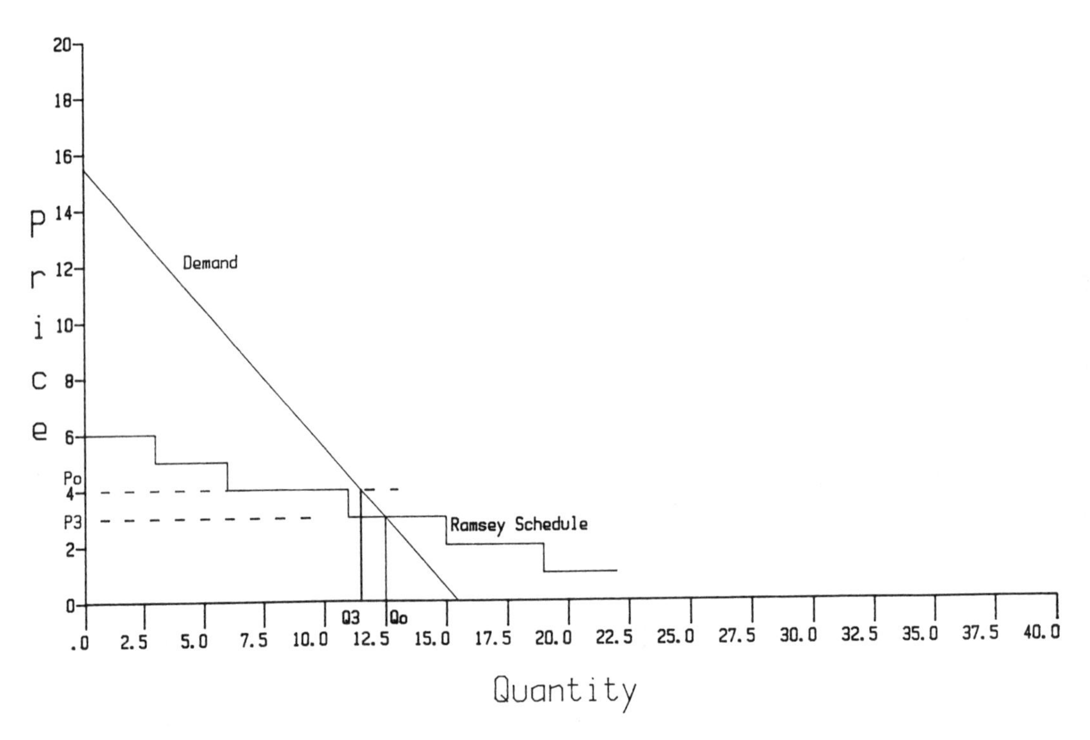

Fig. 5.13.5—Too low Ramsey schedule.

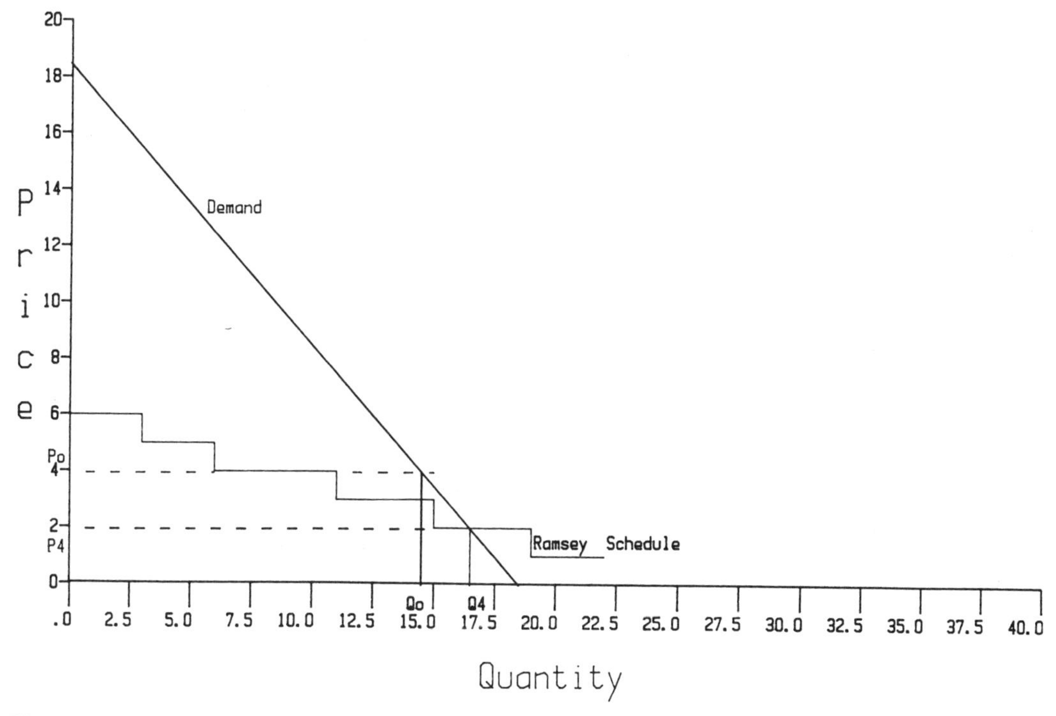

Fig. 5.13.6—Very low Ramsey schedule.

rents directly to affect the allocation of resources within an industry. However, problems arise. First, concern exists about how to allocate among consumers the cost of financing the subsidies. Second, problems arise in determining what the optimal price and output should be.

The prior discussion deliberately understated the difficulties. It was presumed that the same marginal cost applied to all consumers. This is patently untrue in practice. Different consumers costs quite different amounts to serve. For example, a large industry plant that is next door to a large electricity generating station and is capable of using the power without transformers to the lower voltage is cheaper to serve than a small household in New York City where lower voltage power must be delivered by underground wires. Determining the difference in the marginal costs among consumers is a formidable task that cannot easily be solved.

In regulated industries, the regulators often are accused of grossly underestimating the cost reduction involved with serving large customers and underestimating the cost of serving smaller ones. Thus, generally the prices charged to large customers tend to be "too high" under all pricing systems. This is most clearly explained in dealing with the third major problem of this form of rent use—the design of appropriate Ramsey price schedules for different users. The Ramsey schedule can be too high or too low for a given customer.

Industry complaints of mistreatment then can be interpreted as a tendency for imposing too high a Ramsey price schedule on industry and too low a one on households. Similarly, with the single price supplemented with lump sum payments, industry could complain of too high a price and argue that it was charged for too much of the lump sum payment. Where break-even is assured by a price greater than marginal cost, the industry complaint is harder to restate precisely. One possible interpretation is that the proportional loss of consumption is higher. Another is that on equity grounds industry paid too much.

In fact, a stronger conclusion is possible. It is difficult efficiently to apply Ramsey pricing to nonhousehold users. Ramsey prices applied to use of output of a decreasing cost industry by another industry would alter output decisions in the consuming industry. Consider first the case of a constant-cost consuming industry. There are no surpluses in the long run to transfer to the decreasing-cost industry. The surpluses can be created or maintained only by restricting output and prices in a fashion that generates the required payment. The introduction of increasing costs does not help matters much. Part of the cost might be financed by reductions of economic rents but there is no guarantee that this will happen and if it does, that it will provide all the extra income needed to pay for Ramsey prices.

The critical point is that competitive firms do not keep economic rents and thus do not possess a surplus to finance the Ramsey pricing scheme. The benefits have been transferred to the household consumers of the goods. An efficient method to subsidize the output of decreasing cost industries is best financed from levies on the ultimate beneficiaries.

It would be extremely difficult to devise any scheme such as a Ramsey-tax-rate schedule on the purchases of goods produced using the outputs of the decreasing cost industry. Thus, a more direct measure such as a lump-sum tax on consumers would be preferable to Ramsey pricing. This, of course, is a conclusion quite different from the frequent efforts to make nonresidential customers make as much a contribution as possible to the payment of subsidies to increased output.

All the theoretic problems are trivial compared to the practical ones. The first and by far the most important is whether industries with economies of scale are as widely extant as the public policies and theoretic discussions about them imply. Well-authenticated cases are rare. For example, a broad consensus has arisen that the generation and possibly the transmission of electricity are increasing cost industries. Scale economies are exhausted at low output levels. Many plants must serve the industry and increasing costs prevail among plants. Similar problems arise elsewhere.

Second, extremely severe problems are associated with administering optimal pricing. Such efforts would be difficult for any organization. Unfortunately, the United States assigns its regulations mainly to public agencies that have developed, among specialists on the subject, a reputation for massive ineptitude. At no political level is membership on such commissions considered highly attractive. This perception, moreover, is highly rational. The role of such commissions is to plow through masses of tedious accounting data on purported costs to determine the legally allowable rates.

The dullness of the task, of course, is the least concern. The decision will harm someone and that victim will react angrily. This could be consumers collectively or in some subclass (e.g., industry or households) who felt overcharged or utilities that considered themselves deprived of adequate return on investment. Actually, through the middle 1960s, falling real costs made ratemaking the more pleasant task of approving rate cuts. If any complaints were heard, they were from a few uninfluential specialists who feared rates did not fall fast enough. Companies allegedly reaped some of the gains of cost reduction through technical progress by taking advantage of "regulatory lag"—the ability to delay price cuts for some time after the cost decline.

However, subsequent cost escalations have led to the types of conflicts outlined above. It is thus little surprise that with rare exceptions, public utilities commissions are populated largely by either promising but still inexperienced young politicians or those of the party faithful whose reward is considered more important than seeking expertise on the commission. The situation is aggravated because state agencies tend to be small and the pay scale low compared to the federal civil service and even lower compared to private industry.

THE TRANSFER OF RENTS TO CONSUMERS—THE THEORY IN OUTLINE

As already noted, an alternative to rent taxation is transfer of the rent to consumers. A wide variety of transfer policies exist with various effects. Most schemes tend to produce market distortions but this is not necessarily the case. The concept of "lifeline" pricing of electricity noted above can be shown to be a system for nondistortionary transfer of rents from producers to consumers.

Lifeline rates invert the rate structure from the declining block approach discussed above. Rates are lowest on the first block of purchases (generally defined as covering the basic needs of households for electricity) and rise with various increments of purchases. This clearly turns Ramsey pricing on its head and might thus be termed reverse Ramsey pricing. In theory, the system could make transfers without distorting efficiency. In practice, counterparts of all the errors possible in ordinary Ramsey pricing can occur with reverse Ramsey pricing.

Most other systems for transfer of rents to consumers inevitably have distortionary effects on output. The distortions invariably raise the total social cost of energy. Some systems produce distortions by encouraging too much production by the industry from which rents are transferred. Others lower output of the regulated industry. Briefly, a simple price-control system applied to a competitive industry, will hold down rents at the cost of inefficiently reducing that industry's output. However, myriad means exist for using the rents to increase the output of the controlled industry or someone else. The simplest case is forcing an industry to use all its rents to finance expanded output. A more complex system involves imposing a multitier ceiling system and setting consumer prices at the weighted average of the ceilings. This last describes both natural gas and oil price controls in the United States.

Such systems can produce either more or less output than would occur in the absence of regulation. This can be shown to depend upon both the levels of ceilings allowed and how effectively the ceilings are granted to producers. Anticipating the subsequent discussion, a necessary but not sufficient condition for an output increase is ceilings on some portions of supply that exceed the competitive market clearing price.

The adequacy-of-grants problem is that the stimulus of above-market-clearing-price ceilings to some producers can be offset by failures to provide some producers with a ceiling high enough to produce. Some output that might have been produced in a free market might be lost because ceilings applied to it are below the marginal cost of production. Additionally, the stimulus of above-market-clearing-price ceilings could be attenuated because of failure to grant these ceilings to all who could take advantage of them. Ordinary ceilings, of course, produce inefficiently low outputs.

Many other uses of rents are possible. Historically, small oil refineries have been subsidized out of various rents produced by oil policies. Under the oil-import quota program, the rents available from access to imported oil were used to subsidize small refiners and some others. The overall oil price control system of the 1970s similarly was structured to subsidize small refiners. The system also involved a third type of impact—additional stimulus to the use of the closest substitute for the controlled fuel. As shown below, reducing output of the controlled fuel always stimulates use of the substitute fuels. The special aspect of oil price controls was that an additional stimulus was provided by effectively using some of the rents to subsidize imports.

REVERSE RAMSEY PRICING

The simplest case to treat is the concept of rising block rates. Overall, the analysis reverses the prior analysis of declining block rates. If any difference exists, they relate to possibly greater controversy about the transfers under reverse Ramsey pricing than under conventional Ramsey pricing. It may be easier to agree that all customers share in financing an efficient output level than to agree about who should benefit from rents. This is probably only a difference of degree. Many different patterns can be devised to divide up both the financing of extra output and the rents from an increasing cost industry.

As before the critical efficiency problem of designing an increasing block rate is to insure that each consumer faces a rate schedule that induces the same level of consumption as would prevail if an efficient price were charged (either because the industry is competitive or because regulators set a rate schedule such that the price charged equals the marginal cost of producing the output demanded at the regulated price).

Consumer equilibrium will come when the marginal value of consumption equals the marginal price. It is quite possible that consumers will face rate schedules producing either more or less consumption than is optimal. This is seen in Fig. 5.13.7. In each panel, we have a different consumer demand curve. In the first panel, the response to an efficient reverse Ramsey price schedule is shown. Under it or a policy of a flat rate schedule with rates at the efficient level, the customer pays a price, P and the consumption level occurs at the point, Q_o at which the demand price of the consumer is P_o. For reasons already discussed, this is the efficient consumption level.

In the second panel, I show a rising block rate schedule that increases so slowly that the consumer buys too much. In the third panel, I add a rising block schedule that rises too fast and the consumer buys too little. In both cases, the point is that with increasing rates, equilibrium comes with an intersection of demand with the rate schedule curve. A too low schedule

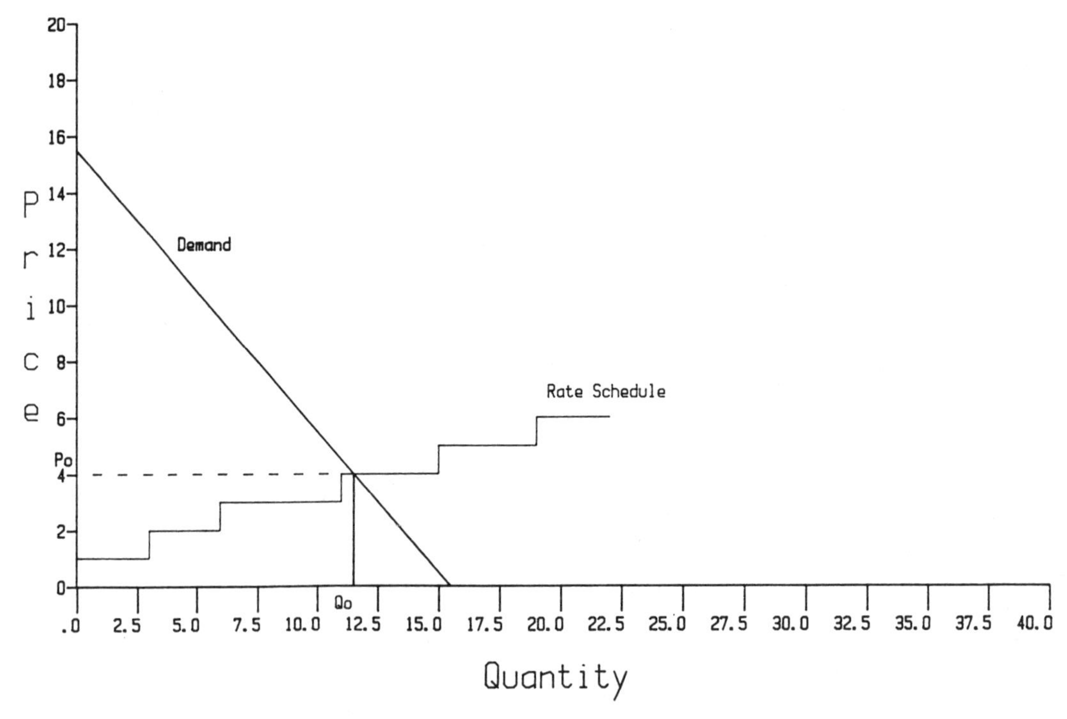

Fig. 5.13.7a—Optimum reverse Ramsey rates.

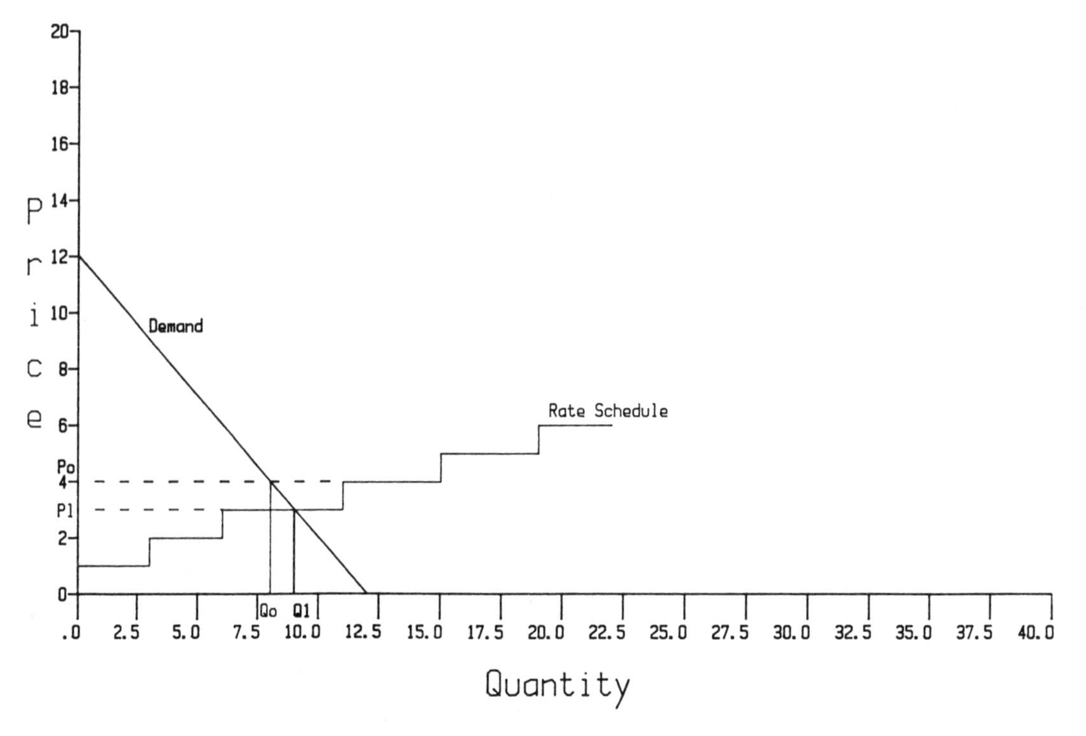

Fig.5.13.7b—Low reverse Ramsey rates.

causes an intersection at the undesirably high consumption Q_1 at a price P_1. A too high schedule produces an intersection with a too low quantity Q_h and a price P_h. In at least one special case, overconsumption is the more likely result. (Analogously to the prior argument, I could go on to show even more inefficiency could occur if the marginal price schedule produces a greater disparity between P_o and the marginal price paid).

There is one case in which the inefficiency problem could be avoided. Imagine a situation in which at some point the marginal cost function of the industry ceases to increase and the efficient output is in this range of constant marginal cost, P_o. (This case might apply to the electric utility industry under 1983 air pollution policy. The policy applies increasingly severe restrictions to newer coalfired plants. Three basic vintages exist—plants planned before 1971 governed by state rates, plants planned between 1971 and 1977 governed by 1971 U.S. government rules, and later plants have to comply with more expensive to meet standards imposed by the regulations implementing the 1977 Clear Air Act Amendments). In this case, the older plants are lower cost than the newer ones and the latest ones could be considered equal in cost. A reverse Ramsey price schedule that insured everyone paid P_E on marginal consumption would insure efficiency.

Actually, the shortage of accessible sites may cause increasing costs as it becomes necessary to move to less attractive sites. Proposals to reduce acid-rain could lead to expensive addition of pollution-control equipment to old plants and make them more expensive than new ones.

While Ramsey pricing raises profits and thus output, reverse Ramsey pricing lowers profits but such output effects as occur are produced by the inevitable errors in setting rate schedules. Thus, for reasons made clear in the debates (in the literature on economic theory) on the critical difference between increasing and decreasing cost industries, quite different efficiency concerns are involved between the two optimal demand prices.

RENT REALLOCATION AND OUTPUT—THE MAXIMUM STIMULUS CASE

As noted, many models can be formulated in which government policy involves removing the rents by market regulation. These included sys-

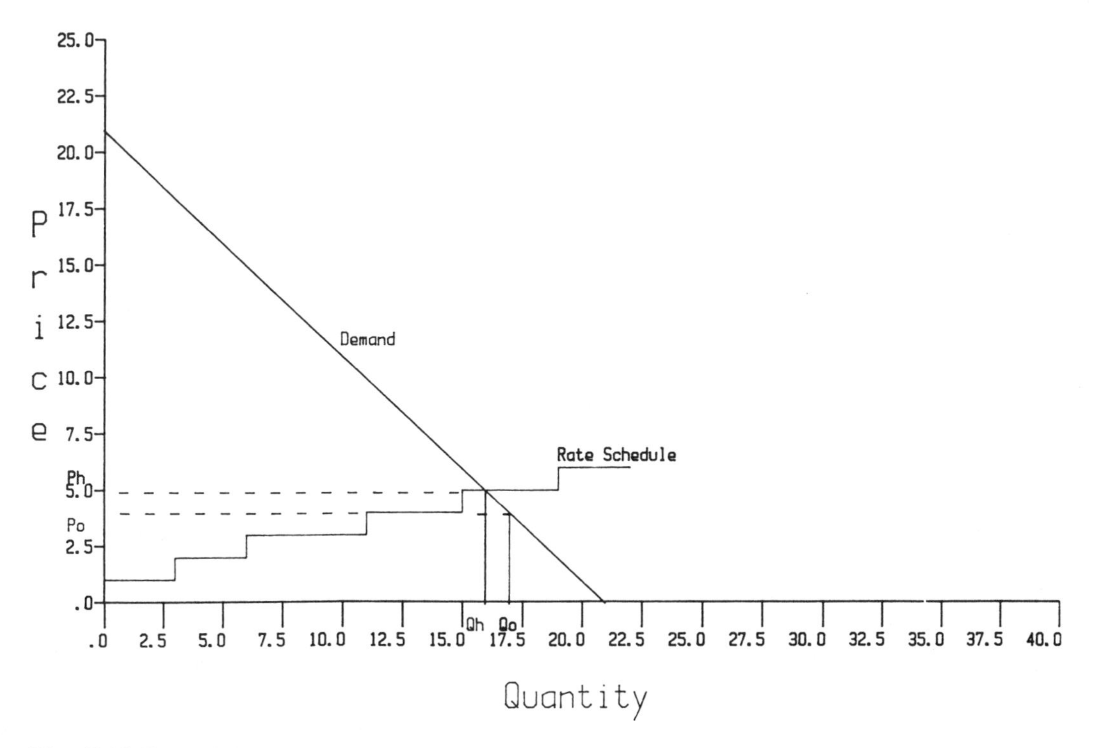

Fig. 5.13.7c—High reverse Ramsey rates.

tems that stress using the rents to stimulate output of the rent producing industry or its close competitors, policies that subsidize special interests, and policies that reduce output to hold down rents. In this section, I deal with systems that make the maximum possible use of rents to stimulate output of the rent producing industry.

The simplest way to visualize the process is to imagine first existence of some scheme for transferring all rents to the government. Government land ownership, combined with competitive bidding, reliance entirely on lease bonuses, and satisfaction of the critical assumptions noted above could produce this result. Then imagine establishment of an output subsidy fund.

The situation is diagrammed in Fig. 5.13.8. It is designed to treat either the domestic production of a commodity available in unlimited quantities at a fixed world price or a commodity facing a downward sloping demand curve that intersects the supply curve at output Q_c and price P_c which is also, by assumption, the price at which unlimited amounts can be sold at a prevailing world price. Rents of *ABC* are generated and also, by assumption, are taxed away.

In this case, the proceeds are used to subsidize domestic output. Given a world price, the need is to cover the difference between marginal costs and the world price. This subsidy starts at zero and rises steadily. The total funds are exhausted at an output of Q_w with an *MC* of MC_w. A subsidy *CEF* equal to *ABC* has been granted. Obviously the limit to the subsidy is set by the rent. Since the rent finances the subsidy, they are equal. The rent of *ABC* finances a subsidy of equal amount. Thus, the net effect of the scheme is to reduce rents to zero—i.e., have total revenues equal total costs or price equal average costs.

Therefore, one could label this as average-cost supply (see Gordon, 1970).The decision rule of equating average cost to price, if correctly implemented, produces the effect of the tax subsidy scheme outlined. No profit-maximizing firm would follow this rule but it could be, and in the case of nationalized European coal industries was, imposed on nationalized organizations. Similarly, public-utility regulation can be and probably is designed to force use of rents to subsidize further output.

A critical point to note is that the germane average costs are those of the industry. Marginal-cost rules must be used to minimize costs.

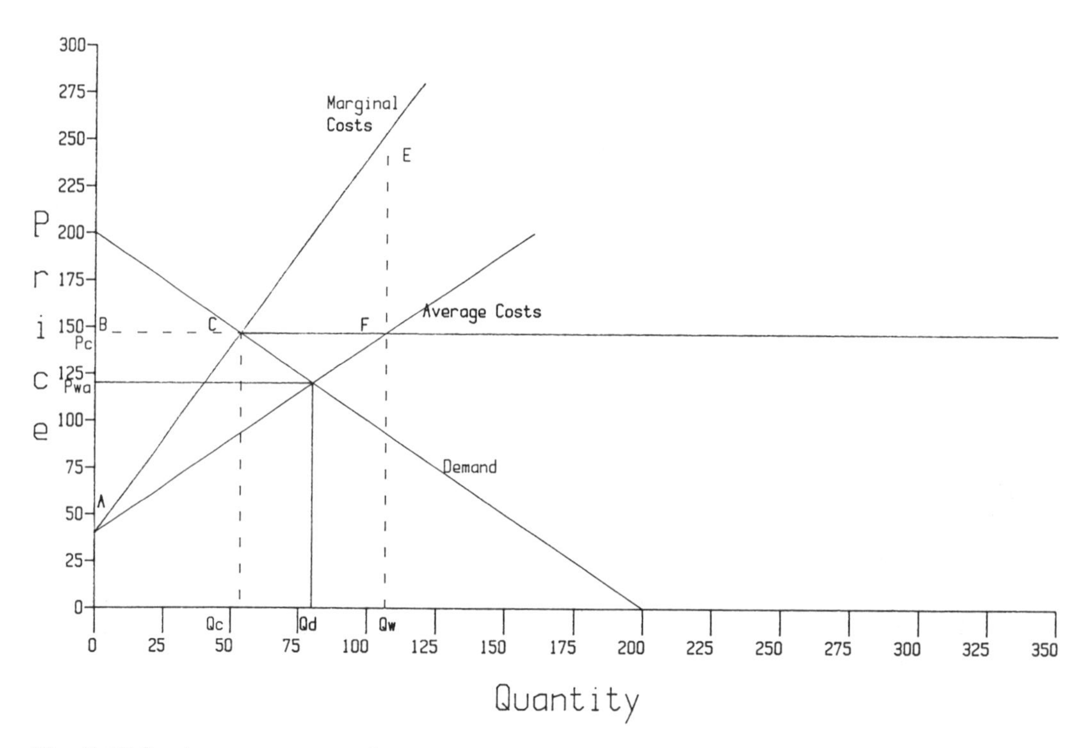

Fig. 5.13.8—Average cost supply.

The maximum amount of subsidy comes from first equating marginal costs among operating units to minimize costs and then, equating to price the average costs resulting from this cost minimization.

Since the European system was more a work relief than an output-stimulating system, behavior not surprisingly deviated radically from that modeled here. Any mine with average costs below allowable levels was kept operating. This could be true even if the marginal costs of that mine were higher than those at some other operation. The prevailing politics precluded transfer of resources to the mine with lower marginal costs. Workers were to be given protection of jobs in the mine in which they presently worked. Allais (1951) first documented this use of rents to subsidize preservation of high-cost French output. I (1970) extended his work to cover more years in France and applied it to Britain and Belgium where similar results were obtained.

Moreover, by the late 1950s, rising costs and falling oil prices eliminated the rents.

Therefore, the Europeans adopted other measures to provide further aid. Largely by use of administrative restrictions, coal imports were limited. A bewildering array of direct subsidies was granted. In addition to frankly labeled subsidies, the states lent money on favorable terms and often eventually wrote off the defaults. The industry's habit of paying pensions out of current receipts generated far less income than needed given falling employment and profits. The government, therefore, assumed the pension cost burden. Coal consumption was subsidized in various areas notably coking coal and by a complex German scheme in which electricity consumers were taxed to permit greater use of German coal. Electric utilities were generally pressured into using more domestic coal than was economic. The result was a very expensive effort that merely kept output and employment from shrinking as fast as they should have.

More generally, the output-stimulus effect obviously are attenuated if some of the rents are put to other uses such as subsidizing high-cost processors of the product (a special case of the earlier proposition that any cost increasing aspect of the policy will lessen output stimulus potential).

Dropping the assumption of a fixed world price leaves the basic argument unchanged. The main difference is that instead of being able to sell unlimited amounts at Q_c, it becomes necessary to cut prices to sell more. Thus, the subsidy now must offset both cost rises and price falls. Domestic prices fall to P_{wa} leading to output of only Q_d. Thus, subsidy costs rise more rapidly as output falls. This need to offset falling prices reduces the amount of output stimulus possible in the falling demand case compared to the case in which prices are set in the world market. In this case, rents are exhausted at Q_d at a price $P_{wa} = AC_{wa} < P_c$, the world price in the first case.

CLASSICAL PRICE CONTROL SCHEMES—NONMARKET CLEARING

The other methods used to affect the allocation of resources by altering the disposition of economic rents are associated with price control schemes. To understand such schemes, it is essential first to deal with the more traditional analyses of price controls. The rent-use models are extensions of the usual price-control models. Analysis begins with a system employing one ceiling for all producers in a competitive market, turns to discussions of how the excess demand spills over into other markets, goes on to the impacts of multiple ceilings, and then covers the case of price controls on monopoly.

Fig. 5.13.9 traces the effect of price controls in a competitive market. Without price controls equilibrium occurs at price P_o with Q_o produced. A ceiling P_c raises the quantity demanded to Q_w but reduces the quantity supplied to Q_c. Thus we have an excess of demand over supply at the ceiling price, what is popularly called a shortage.

The size of the shortage depends on the flexibility of supply and demand. The desire to impose price controls reflects a feeling that both supply and demand are highly inflexible (the supply and demand curves are more nearly vertical than is shown in Fig. 5.13.9. However, inflexibility alone does not insure low shortages. Inflexible supply and demand curves produce small shortages only if they are close together. When the demand curve is inflexible but well in excess of supply at the ceiling price, the shortage could be as large as in the Fig. 5.13.9 case (see Gordon, 1981, for a diagram). Greater inflexibility invariably implies that considerably greater price increases are needed to restore market equilibrium.

The reduction of price below the equilibrium level involves a reduction of economic rents. In particular, output Q_c would have been profitably produced whether prices were P_o or P_c. The ceiling lowers the income on this output to P_c and thus reduces revenues and rents on this output by $(P_o - P_c)Q_c$. (Further rent reductions consisting of the difference between marginal costs and P_o for the extra output produced without the ceilings occur; the triangle *DGE* in Fig. 5.13.9 measures this additional loss).

The last issue about the impacts of price control is what to do about the shortage. Many approaches and outcomes are possible. Any given demander may receive more than he demands at the ceiling price, but by the nature of the ceilings less will be available than is demanded at the ceiling. Thus, on average, consumers will have unsatisfied demands. A key problem with ceilings is how to allocate the dissatisfaction. This problem involves questions of equity about which no definite conclusions can be made. The policymakers must develop rules of thumb for sharing the shortages. This may involve, as has been true for natural gas, trying to rank end uses in order of desirability and favor the most desirable users. Alternatively, as in the oil products case, stress has been on allocations based on historical patterns of use. One also could favor the poor in energy allocations.

Deprivation, by definition, means that consumers have money that they wanted to spend on energy that could not be used as desired and so is available for other purposes. The most natural such use would be for sources of energy not subject to price controls. For some consumers, this might mean coal or nuclear power, but for many the most attractive alternative is imported oil. Thus, at least some of the excess demand may be expected to stimulate imports.

Given the excess demand problem in the competitive industry case, we may concentrate on the way that price controls of a domestically produced good will increase imports of that product. A ceiling creates excess demand and reduces domestic output. A step-by-step examination of the ways that the demand shortfall can be allocated shows why an increase in imports is normally presumed. I continue to assume a constant world oil price although the argument is independent of the assumption about world prices.

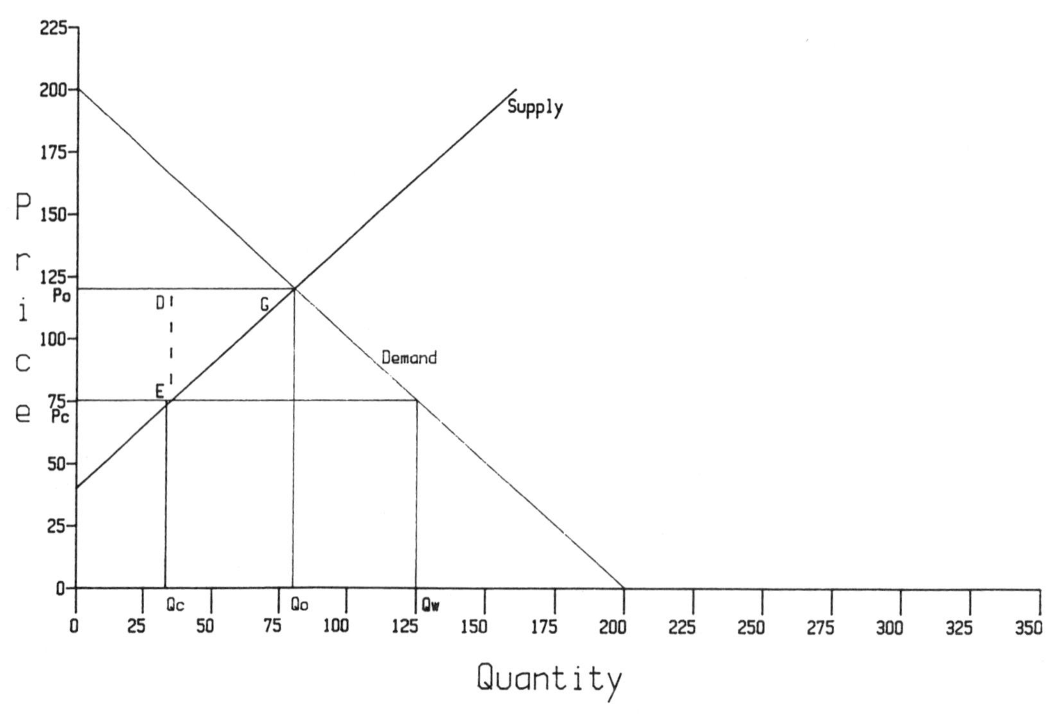

Fig. 5.13.9—Conventional price controls.

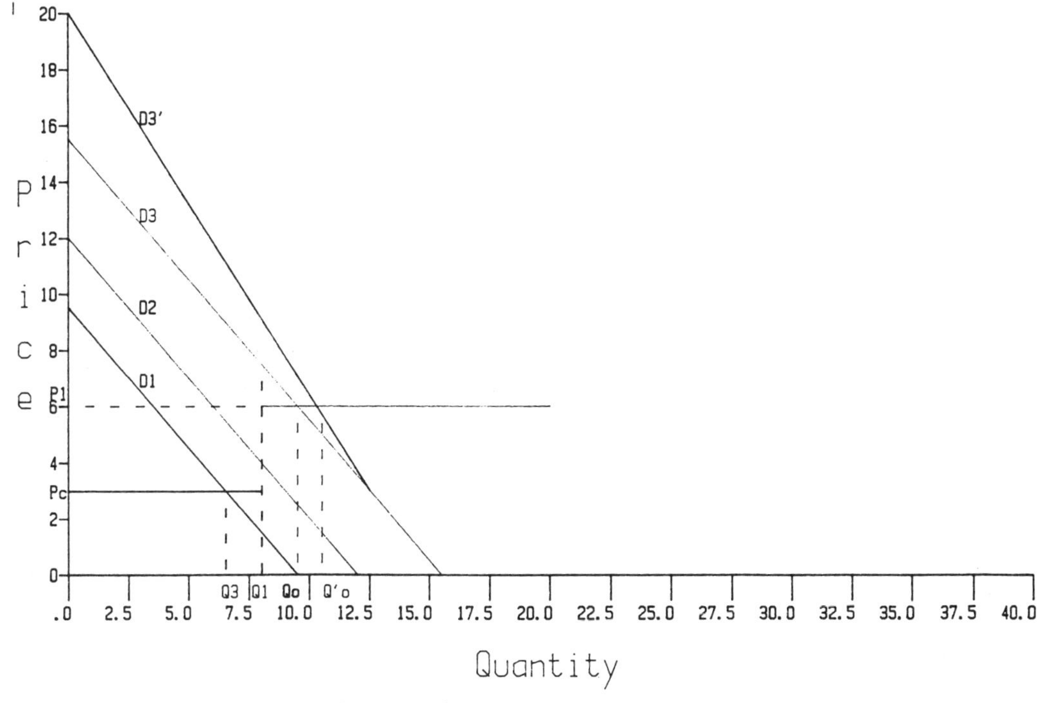

Fig. 5.13.10—Demand with price controls.

Consumers of the product all face a supply curve consisting of two segments. First the amount Q_1 is available at the domestic ceiling P_c; then all additional supplies must be purchased at the world price P_1. Any one consumer can be in one of three positions, represented in Fig. 5.13.10 by D_1, D_2, and D_3, respectively. In the first case (D_1) the allocation of domestic product exceeds demand at the domestic price, and the consumer does not use its full allocation but only consumes Q_3. In the D_2 case the quantity allocated is less than that demanded at the domestic price but more than demanded at the import price. All the available domestic output is worth more than it costs and is purchased, but imports cost more than the price that the D_2 consumer is willing to pay for additional consumption. Therefore, consumption is Q_1, the allocation of domestic fuel. Finally, consumers may have a greater demand at the import price than the allocation of domestic product that they receive and will meet the shortfall by imports consuming Q_o. This is the D_3 case. Further, the windfall profits of access to price-controlled domestic output raise real income, and if the product is a "normal good" (one for which demand increases with increases in real income), demand rises to the position represented by D_3 and consumption rises to Q_o.

The baseline case to which all other outcomes may be compared is that (1) every consumer is in the D_3 position, (2) no income effects occur, and (3) import prices are unaffected. Here imports clearly rise by an amount exactly equal to the fall in domestic output produced by a uniform ceiling. When every consumer has a D_3 type demand curve, total demand in the relevant range is simply the ordinary demand curve in the absence of a ceiling since everyone demands at P_1 exactly what is demanded under a free market. The demand for imports is, as usual, total demand at a price less domestic output. At prices above the ceiling domestic output is lower and thus import demand is higher under ceilings than under a free market.

Now if import prices are independent of demand, the entire demand increase can be satisfied. If import prices rise with imports, domestic consumption will be reduced below the level prevailing in the constant-price case. However, since demand for imports is increased, imports are above the free market level. Positive income effects raise the increase in import demands. Thus, positive income effects combined with constant import prices imply that imports rise more than domestic output falls. With a variable import price the positive income effect conflicts with the effect of higher import prices. Imports will be higher than in the absence of ceilings, but the change may be greater or less than the decline in domestic output. (The case in which the good is inferior—rising income reduces demand—is even more complex. We cannot be sure whether import demand rises or falls since the income effect now offsets the domestic output reduction effect. This case does not seem relevant for energy commodities).

Similarly, consumers who are in the D_1 or the D_2 position buy more than they would have bought at P_1. Their increased consumption must be met by increased imports, so the import demand curve is further raised. Thus imports are even higher at the uncontrolled equilibrium price. With infinitely elastic supply, imports increase by the amount of the total demand increase. Lesser supply elasticity, however, implies that rising prices lessen but do not eliminate the import rise.

The final complication of competitive cases is the impact of multiple ceilings. The cases in which multiple ceilings eliminate excess demand have been discussed. Therefore, only the cases in which excess demands persist are relevant.

The discussion of multiple ceiling suggests that several possibilities exist:

(1) Ideally the multiple ceiling could produce the domestic output that would have occurred under competition. This lessens the stimulus to imports. The only import demand stimulus effects would be produced by income effects from the grant of a supply of domestic energy equal to demand at the domestic weighted average price.

(2) Multiple ceilings could raise domestic output above their competitive levels and attenuate and possibly even reverse the import stimulating impacts of price controls.

(3) Multiple ceilings could reduce domestic output and aggravate the import stimulus problem.

THE CASE QUALIFIED: PRICE CEILINGS AND MONOPOLY

A basic element of monopoly is at least partially eliminated by price controls. Monopolists recognize the effect of their output changes on price and take advantage of this effect by restricting output to more profitable levels. Ceilings offset this by preventing monopolists from raising prices above the ceiling.

In particular, as Fig. 5.13.11 shows, the ceiling-affected marginal revenue curve is quite different from the marginal revenue curve for the unregulated monopolist. At a ceiling P_o the effective marginal revenue is, by the nature of a ceiling, P_o for outputs up to Q_o—the amount demanded at P_o. The ceiling is a rule that says prices cannot be raised above P_o but may be lower. Thus, the firm can sell its output at the lower of the ceiling or the price associated with the output on the demand curve. The ceiling is then the permissible price for outputs less than Q_o; the price indicated by the demand curve is the applicable one for outputs greater than Q_o. Thus, as output rises from zero to Q_o, marginal revenue remains constant at P_o; at Q_o further output increases can be made only if prices are cut and the conventional MR curve applies for higher outputs.

The effect of a ceiling depends on the relationship of the ceiling-shifted marginal revenue curve. There are four possibilities:

(1) No effect. Price controls have no effect if the ceiling is above the price that the monopolist would choose in the absence of control. Here the ceiling is so high that it pays to operate on the downward-sloping portion of the marginal revenue curve (see P_4 in Fig. 5.13.11).

(2) An increase in output to Q_o. This occurs as long as the ceiling is below the preferred price of an unregulated monopolist and no lower than the price P at which the demand curve intersects the marginal revenue curve. In this case outputs above Q_o, as usual, have a marginal revenue of P_o, and in every case this is by assumption greater than or equal to the marginal cost. By the assumption that Q_o is greater than the monopoly output, the marginal revenue for outputs above Q_o is less than the marginal cost. So Q_o is the preferred output. (See MC_1 in Fig. 5.13.11).

(3) An increase of output to levels above the monopoly level but with an excess demand. This occurs when the ceiling P_2 is below P_o. In this case the marginal cost curve intersects the ceiling-affected portion of the marginal revenue curve, and output is limited to that at which the marginal cost equals the ceiling. The quantity demanded exceeds the quantity supplied.

(4) Excess demand with output below the monopoly level. This result occurs when the

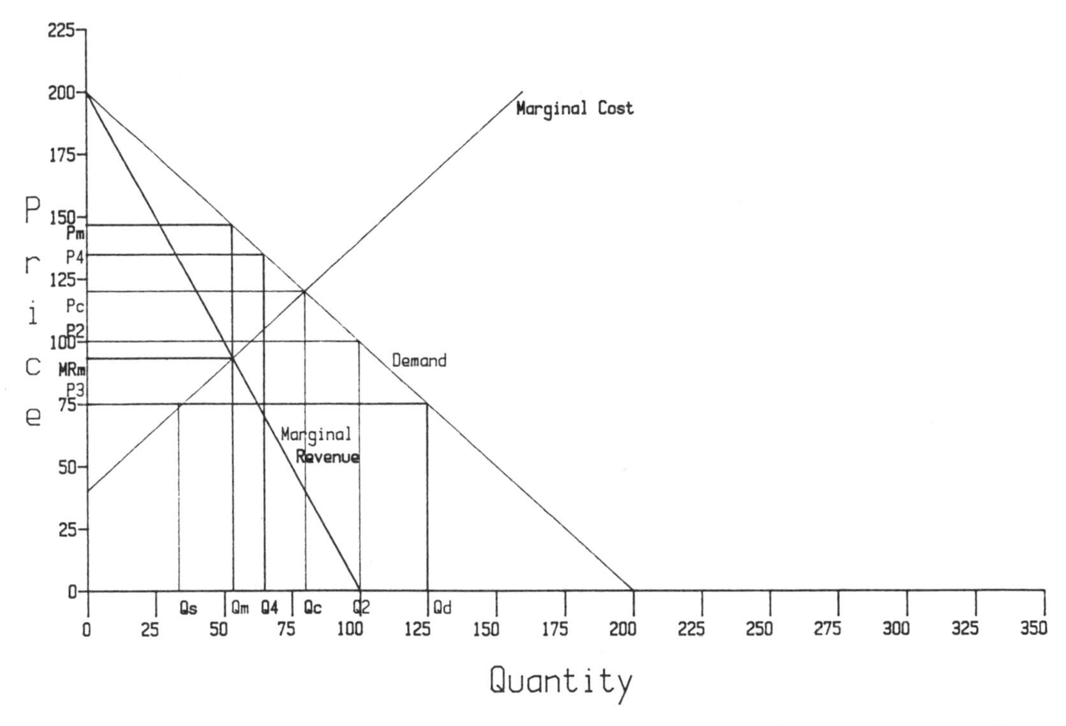

Fig. 5.13.11—Monopoly and price controls.

ceiling P_3 is so low that the output at which marginal cost equals the ceiling is below the monopoly level.

MULTIPLE CEILINGS AS AN ALTERNATIVE FORM OF OUTPUT STIMULUS

The United States price controls on oil and natural gas involved establishing a bewildering variety of vintages of oil or gas, each subject to its own ceiling. By itself, this is simply a form of reverse Ramsey pricing and has the same effects on allocative efficiency as any set of reverse Ramsey pricing. The primary interesting special questions relate to failures of the system to grant high enough ceilings to all those who would be competitive in a free market. Such failures reduce output and produce additional inefficiencies.

However, the actual U.S. schemes had additional features that produce output effects absent from reverse Ramsey prices. In particular, because of various regulatory schemes, the oil and gas was sold for the weighted average of the prices received by producers. In addition, in oil, domestic and import prices also were averaged.

The latter procedure can be shown to produce higher consumption, higher imports, and lower domestic prices than the absence of controls. The effects of averaging domestic prices is more complex. Its effects depend upon the basic system adopted and the skill with which it is administered. A critical distinction arises between systems that perpetuate the inherent problem of excess demand produced by imposing price controls on a competitive industry and those that succeed in eliminating the excess demand.

The critical step in ensuring elimination of excess demand is producing an output in excess of competitive equilibrium levels. This requires at a minimum that some producers receive ceilings above the competitive equilibrium price. To see this and other problems with multiple tiered ceilings, the possible effects of different ceilings should be reviewed.

As a baseline, it is desirable to assume initially that regulators have perfect knowledge of marginal costs and allocate ceilings accordingly. Thus, if we had two ceilings, P_1 and P_2, perfect knowledge would imply producers with marginal costs less than or equal to P_1 would get the ceiling, P_1; those with costs above P_1 but less than or equal to P_2 would rate a ceiling P_2.

Such a multiple ceiling is a variant of average-cost supply in which less rent is taxed and thus the potential for output stimulus is less. Consider the case in which P_1 is some price below the competitive equilibrium but $P_2 = P_c$, the competitive equilibrium price.

Compared to a rent tax, this system raises less "revenue." Producers of outputs subject to the ceiling P_1 lose, not their entire rents (*AFD* + *CEFD* in Fig. 5.13.12) but only *CEFD*—the difference between their revenues under competition and their revenues under the ceilings. They keep any difference between the ceiling and their marginal costs of producing (i.e., *AFD*). In this case, the other producers lose no rent at all since they get the competitive market price. Output is at the competitive level but is offered at a weighted average than is necessarily lower than P_c.

The last can be seen by formally developing the economics of weighted average prices. The supply price P_w for outputs up to Q_1 is P_1. For various levels of P_c the price is

$$\frac{P_1Q_1 + P_cQ_c}{Q_1 + Q_c} = \frac{(P_c - P_c + P_1)\,Q_1 + P_cQ_c}{Q_1 + Q_c}$$

$$= \frac{P_c(Q_1 + Q_c) - (P_c - P_1)\,Q_1}{Q_1 + Q_c} =$$

$$P_c - \frac{(P_c - P_1)}{Q_1 + Q_c} < P_c$$

$$\text{since } P_c - P_1 > O, \frac{Q_1}{Q_1 + Q_c} > O \text{ so}$$

$$\frac{(P_c - P_1)\,Q_1}{Q_1 + Q_c} > O, \; -\frac{(P_c - P_1)\,Q_1}{Q_1 + Q_c} < 0$$

The weighted average price P_w as expected lies below P_c. However, as the weight of Q_c in the total rises, the weighted average price P_w will rise closer to P_c.

Given that the same quantity that would be offered at the market clearing price P_c is now offered at a lower price at which the quantity demanded is higher and the quantities supplied

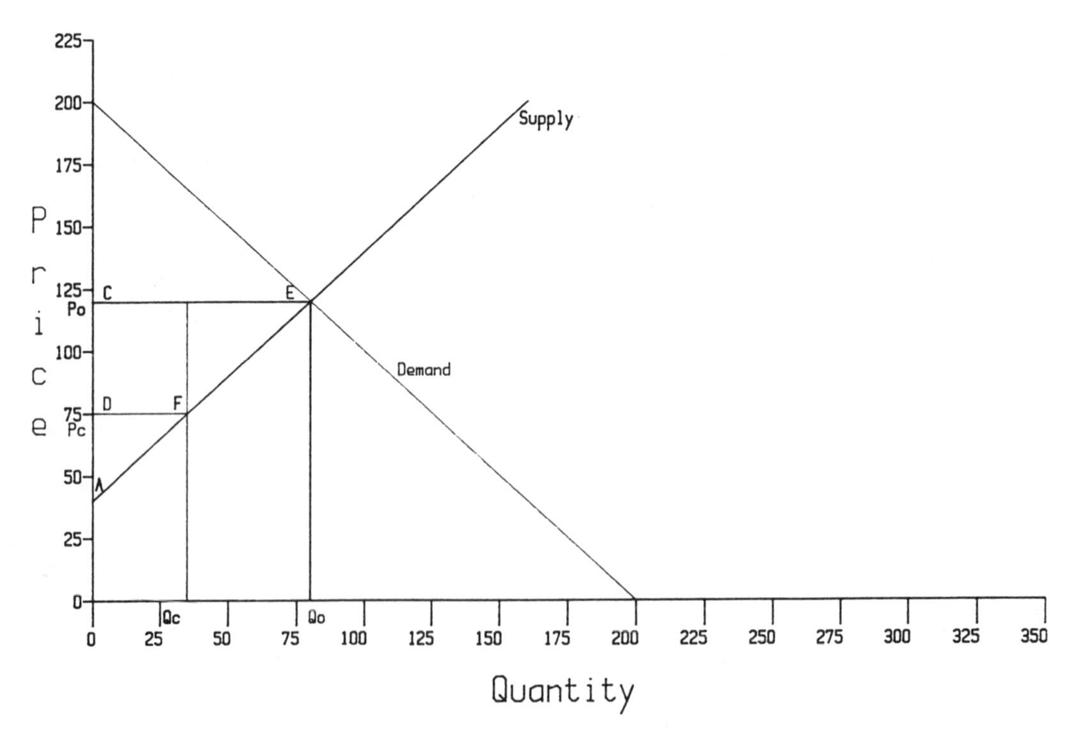

Fig. 5.13.12—Rent lost to price control.

of close substitutes are, if anything, lower, means excess demand prevails. A second tier ceiling below P_c would produce lower output and prices and thus an even higher excess demand. Thus, it can be seen that with perfect administration, excess demand cannot be eliminated if the maximum ceiling is P_c. However, if we created a higher ceiling that raised output above Q_c, we could expand output to its market clearing level.

In fact, a specific ceiling can be defined, given perfect administration and the ceilings on lower output levels, that will clear the market. Another two-tier system is the simplest one to treat the analysis. The first ceiling is one set below the competitive market level. The second is a price P_2 above P_c that leads to a weighted average price that can be calculated from a formula similar to that presented above.

Fig. 5.13.13 plots the process. Added to the conventional supply curve is a curve of the weighted average prices associated with every possible P_2. For any total quantity, the supply curve shows the ceiling required to produce it and the weighted-average price curve shows the cost to consumers of the regulated good. Then the intersection of the weighted average price curve with the demand curve defines the regulated market clearing price. Since the weighted average curve necessarily lies below the supply curve, the intersection with demand curve would occur with a higher output and a lower price than under competition.

If we shift to the case in which there is a fixed world price, the only difference in the analysis is that consumption is unchanged, the equilibrium weighted average is the world price and the required ceiling is that which raises the weighted average to the world price P_c.

Again by construction less than all the rent has been taken away so less output stimulus has been produced than if all the rents had been used to hold down prices. In particular, introducing a higher upper ceiling raise rents received by all producers subject to that ceiling. Turning to a three tier system where the second tier is P_c and the third P_2 is a market clearing and this leaves those getting P_2 and P_c in an unchanged position despite the rise in total output but giving a higher ceiling to all with marginal costs between P_c and P_2 has given them newly created rents.

The analysis can be extended in several di-

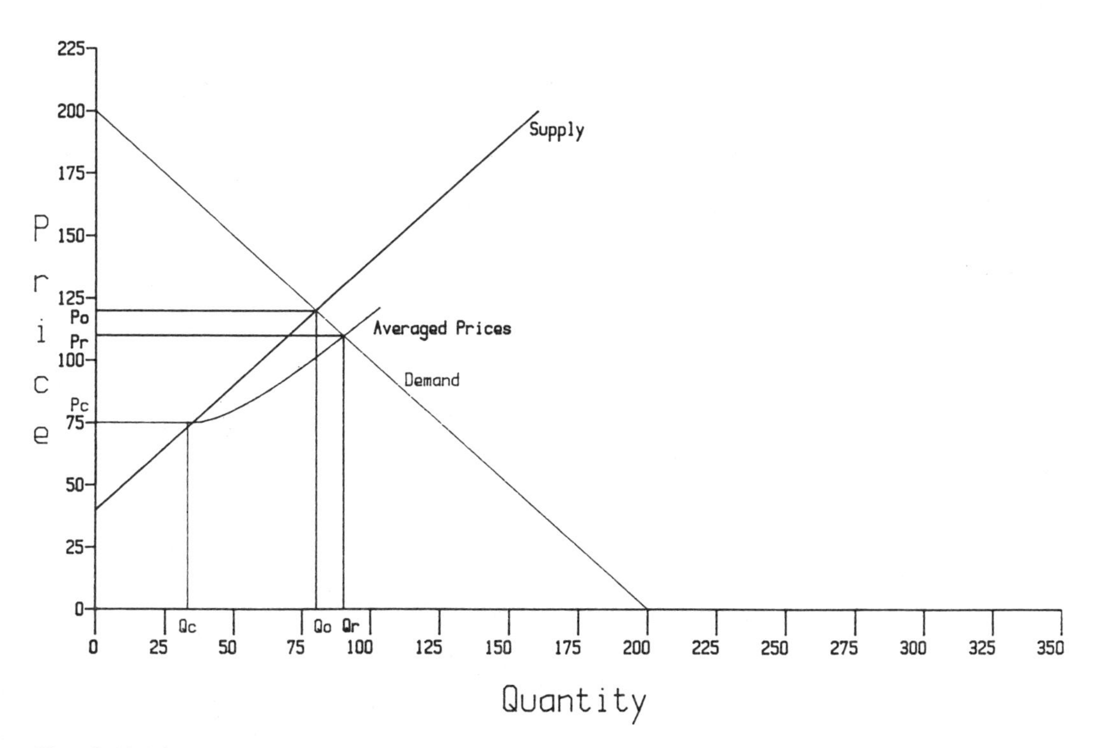

Fig. 5.13.13—Average regulated prices.

rections. Clearly, the tradeoff between administrative simplicity and maximum output subsidy leads to use of a multitiered system. The multitiers lessen rent collection compared to the full taxation system. However, the output impact increases as the number of tiers rises (assuming all output is assigned to the right tier). Setting say two ceilings P_0 and P_1 for the lower cost producers could lower rents on the first Q_0 of output. In fact, increasing the number of tiers leads inexorably to capturing more rent. Total rent taxation can be considered the ultimate in tier creation—one for each unit of output.

Another series of problems are associated with misassignment of ceilings. A given output can be classified into a higher or lower tier than necessary to insure output. Assignment to a higher tier involves raising the weighted average price by giving more rent than necessary and thus attenuates output stimulus. No low cost output is lost, but the average price rises.

Not giving a high enough ceiling necessarily reduces output. In the two tier case, this inevitably leads to lower weighted average prices for a given upper ceiling. By assumption a minimum ceiling of P_1 is available to all. No output costing less than P_1 can be lost. The only losses occur with outputs with *MC* between P_1 and P_c that get a ceiling of P_1.

If we add more tiers, this is not necessarily the case. Consider a four tier system involving P_0, P_1, P_c, and P_2. Now it is possible for some supplies in the P_0–P_1 marginal-cost range to be lost because their ceiling is only P_0. In this case, the effect on weighted average prices depends on whether more P_1 supplies are lost than supplies in the P_1–P_c range. With lower outputs at each weighted average price than if ceilings were assigned to all who could benefit from them, the result is less output stimulus than if all who could benefit actually receive ceilings. Combining cases, the errors reenforce each other and raise the curve further.

Various comparisons come to mind, the most critical of which is the relationship of equilibrium under multiple ceilings to competitive equilibrium. Clearly, if the only error is giving too high a ceiling to *some* producers, we still get a weighted average price below the competitive equilibrium, higher domestic output and, if a downward sloping demand curve prevails, a lower domestic price.

When the ceilings fail to be granted to all who need them, the outcome is indeterminant. We end up with a lower output and a weighted average price that may be higher or lower than if all who could use the higher ceilings actually received them. The critical question is whether the shifts in the weighted average price curve cause it to lie above or below the competitive supply curve at its point of intersection with the demand curve. Clearly, sufficiently modest output losses would leave output stimulus while large losses could lead to lower domestic output than under the free market.

Finally, the analysis can be used to deal with the partial deregulation prevailing under the U.S. Natural Gas Act. Assuming at first that deregulation applies only to outputs costing more than P_c, deregulation could be considered the simplest way to get prices up to P_2. This form of deregulation would have all the implications of a ceiling of P_2.

As gas costing less than Q_c also was deregulated, the weighted average price curve would swing upward producing a lower equilibrium level of all domestic output and, in the downward sloping demand case, raising prices above the levels prevailing if only high cost gas were deregulated. The main problem area would continue to be output losses because some producers continued to receive ceilings too low to cover their costs.

Obviously, total deregulation leads to the competitive solution. The effects of maintaining partial control will differ with circumstances. If the prior full control system increased output above competitive levels, the partial removal of controls will attenuate but not necessarily eliminate the excess. If full control lowered output, the effect of a shift to partial decontrol depends upon its nature. If decontrol only raises prices for gas that was produced under full control, then the rise in weighted average prices due to partial decontrol lowers supply further. If some lost supplies are recovered, this works in the direction of stimulating supply and could reverse the adverse effects of price controls. All this suggests that the logical next step after implementing the 1978 Natural Gas Act is to move to total decontrol.

MARKET CLEARING BY SUBSIDIZING CLOSE SUBSTITUTES

The U.S. oil entitlements program effectively represented a program in which rents were used to subsidize oil imports. Here I deal with an idealized system in which the process is direct and only devoted to import subsidy. This leaves out both an important matter of substance and the actual details. The process involved refiners viewing the relationship between their weighted average costs of crude oil acquisition and the national weighted average. Those who had lower weighted averages had to make payments for the entitlement to access to low cost crude. Those with higher costs received payments for their extra expense.

In practice, the payments did not cancel out. High cost refiners received greater compensation than lower cost refiners. This is another example of how rents can be expended in several ways and needs little further comment. The process reduces the rents available for other purposes.

The pure averaging system would cause the weighted average price of crude oil to be equal among refiners. As with the system discussed before, the supply curve seen by consumers would be the amount available at the weighted average price and the intersection of the weighted average price line and the demand curve determines domestic consumption.

Again, the only cases needing attention are those in which domestic price controls place weighted average prices below market clearing levels. If domestic prices are market clearing, no surpluses are available to subsidize imports and the analysis of the prior section applies. Given a U.S. supply Q_r with a weighted average price P_r, and an infinitely elastic, at price P_c, world oil supply, the weighted average is

$$\frac{P_c Q_I + P_r Q_r}{Q_r + Q_I}$$

where Q_I is the level of imports. As Fig. 5.13.14 shows, this curve lies below the competitive supply curve but approaches it as imports and the weighted average rises. Nevertheless, consumption rises because the intersection of the weighted average price curve S_1 with the demand curve occurs at a level Q_2 greater than Q_c because supply seen by consumers is greater.

As before, the impact on imports depends upon the ceiling price system adopted. A ceiling system that increases domestic output enough could lead to lower imports; otherwise imports are higher.

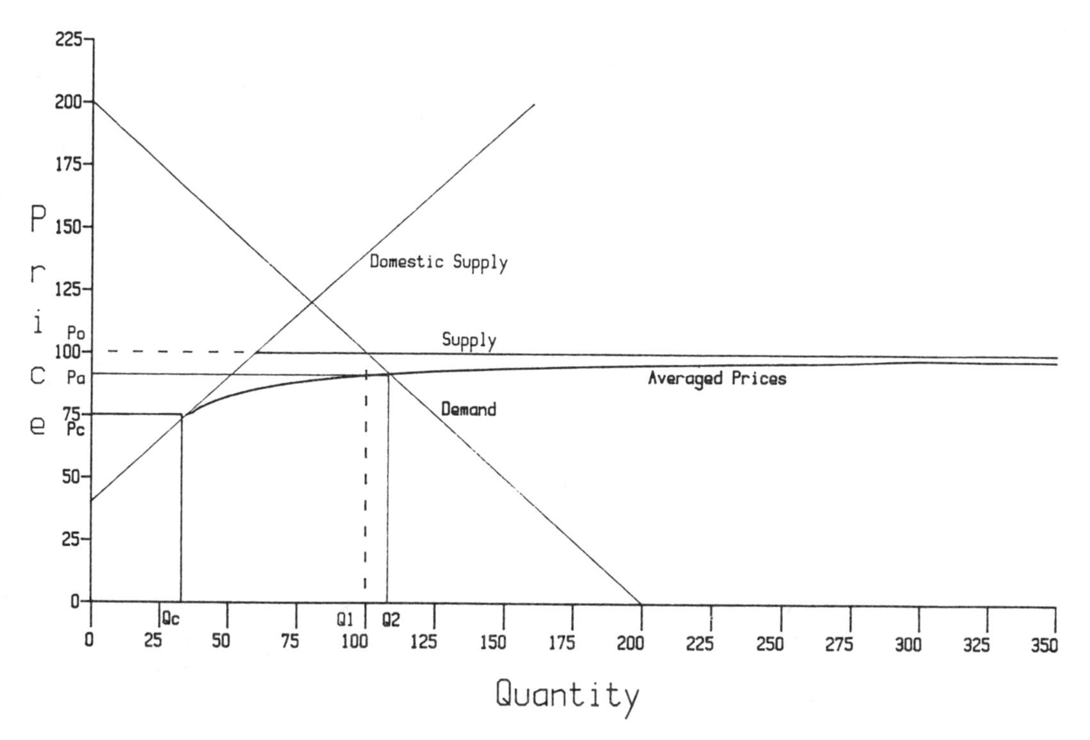

Fig. 5.13.14—U.S. oil price controls.

DEPENDENCE ON IMPORTED ENERGY

To deal with this issue, some further basic points must be made. First, the argument made above that monopoly is good for the monopolist can be elaborated further. Then, we can look at the effects on "macroeconomic" issues such as inflation and unemployment and of behavior in individual markets.

The most popular element of the economic analysis of monopoly is the suggestion that on an overall basis, monopoly is inefficient because it reduces the availability of goods. (The argument shows that whatever mix of output arises under monopoly, an unambiguously better one, namely one in which with every other output constant and the inputs to production unchanged, the output of some good can be higher if the monopoly is eliminated).

A less familiar but still well-developed element of the theory involves viewing the situation from the monopolist's point of view. The inefficiency arises because the monopolist has enriched itself. We have one person's gain at the cost of losses to other that exceed the gain. An extensive literature exists on the tension between the monopolist's gains and other peoples' losses and the circumstances under which it is desirable to favor a monopolist.

As is often the case in economic analysis, the most interesting work has been done in applications to international trade. It is noted that while one individual in any country has negligible impact on international trade, the country as a whole may have substantial influence. Given that the gains are internal and the losses foreign, a nationalistic government may decide to exploit national market power by restricting trade. This can be done limiting imports or exports.

The limitation process involves selecting the national-gain-maximizing trade limits. The easiest way to force up what is paid for exports or force down what is paid foreigners for imports is to tax them. The tax that produces the highest gain is called the optimal tariff. Associated with the optimal tariff will be an optimum (nonzero) trade level. As usual, balance must be attained. The price reduction gains from reduced imports and the price rise gains from reduced imports are at least partially offset by the impacts of reduced volumes. Trade restriction should stop at the point at which the additional price change benefit equals the additional (marginal) volume reduction loss.

The principal practical barrier to imposition of optimal tariffs is that in practice several nations possess some monopoly power. Its exercise by one nation may unleash a flood of efforts by others, and everyone may be worse off. These arguments can be and frequently, although usually implicitly, have been applied to intragroup policies within a country. The government creates a monopoly for one group after another at the expense of national economic efficiency. It was this problem that Thurow treated in his misnamed *Zero Sum Society*. As his actual text makes clear, the effects are not offsetting; the gainers get less than others lose.

The basic moral is that initially monopolization can do some people some good, but it is doubtful that in the long run, anyone can benefit from a monopoly-ridden world. In addition, it should be recognized that clever lobbyists can find many ways to disguise and thus rationalize the grant of monopoly power. The cries of job protection used to justify restrictions on steel and automobile imports conceal the reality that we are limiting competition in these industries by these trade restrictions.

The central issues of macroeconomics are the causes and cures of inflation and unemployment. Considerable accord exists that ultimately inflation results from expanding money supply faster than the expansion of productive capacity. However, the obvious solution of limiting monetary expansion is less universally accepted. Considerable fear exists that prices in the economy are so rigid that tight money will produce severe unemployment. It is presumed that prices are raised in anticipation of an inflationary expansion of the money supply, fail to decline if money is not expanded, and thus imply lower employment with the actual money supply. Although commentators (e.g., Patinkin) on John Maynard Keynes disagree about the importance of price rigidity to his analysis, the advocates of the excessive price rigidity argument, at least in the United States, are usually labeled Keynesians. This arises from the tendency of such people to express considerable enthusiasm for Keynes' analysis, their use of analytic tools suggested by Keynes, and to a lesser extent to their claims that their views were shared by Keynes. In any case, others would argue that price rigidity is a less serious problem, and inflation can be broken without *undue* cost by monetary restraint.

Examination of the arguments suggests that

a combination of ideological and factual disputes arise. Neither side claims that prices are either instantaneously or never responsive to monetary restraint. It is, moreover, agreed that in the long-run price stability can be maintained if the monetary authorities demonstrate in their actions a vigorous commitment to price stability.

The debate centers around different appraisals of the nature and consequences of both tight money and alternative policies. Opponents of tight money tend to expect more unemployment to result than do the advocates; in addition, there are different appraisals of the economic consequences of unemployment. One side stresses the immediate human misery and loss of output; the other, the economic distortions produced by inflation and the possible long-run efficiency gains from recession pressures on industries. They suggest that the diversion of resources into hedges against inflation is extremely costly to economic progress, as such devices spread they neutralized the employment effects of less stringent monetary policy and thus the last is totally ineffective in the long-run.

The advocates of fighting inflation by monetary restraints also tend to believe that recessions are usually the result of erratic government policies. They claim that governments tend to overreact. Expansions are overstimulated and then an overly abrupt shift to restriction is made, inducing a recession. Others suggest that the problem is private overreaction to stimuli. Nevertheless, the proponents of active government efforts to reduce macroeconomic instability have been forced to recognize that their critics were correct in noting that bad government policy has been responsible for much of the instability.

This is particularly true of inflation. Price rises in anticipation of inflation occur only after industry has experienced several years of inflationary government policies. What is rejected is the view that a more passive policy is preferable to improving the implementation of the prevailing commitment to active intervention. (In fact, some have gone so far as to suggest that careful examination of unregulated banking systems such as prevailed in the United States during the first half of the nineteenth century suggests that competition of private firms to provide credible monetary assets produced more stability than managed money and the supply of money ought to be turned over to the private sector).

The extensive literature on oil import policy has slowly moved towards relating the issues to the arguments presented above. (See Plummer, 1982, and Bohi and Montgomery, 1982). In particular, the badly neglected arguments of Newlon and Breckner that the case for intervention to offset the direct effects of supply crises is defective has finally become widely accepted. By now, it is agreed that the prointervention case is another dubious application of the imperfect capital market concept.

Private investors are capable of anticipating crises and stockpiling if they can be assured that they can profit from the higher prices during crises. Similarly, acceptance has also prevailed of Newlon and Breckner's further point that government through policies to prevent realization of profits on private stockpiles, is the principal barrier to efficient private stockpiles. Writers, notably Plummer, have become more willing to propose reform of intervention instead of relying on government stockpiles to offset the disincentives to private stockpiles. Clearly, the chronic failure of the U.S. government actually to stockpile has shaken faith that government stockpiles are politically more feasible than incentives to private stockpiles.

With the old pet argument tattered, advocates of intervention undauntedly have sought new ones. These turn out, as noted, to be applications of the optimal tariff and rigid price problem argument to oil. For reasons that are not totally clear, few writers explicitly used the optimum tariff concept as part of the debate until the early 1980s. The discussions still inadequately consider the full implications of the international trade literature. The original concept was simply that the buying power by the United States could force down world oil prices. The other concern is that oil supply disruptions have impacts outside the oil market that are not reflected in oil prices. The impacts consist of the simultaneous rise of inflation and unemployment due to the shock to the economy from supply disruption.

Despite the growing consensus on these views, they may be unsound. The prior arguments suggested why neither the optimum tariff nor the economic disruption argument may have relevance anywhere. Another consideration is that even if some application exists, it has not been proved that oil, minerals, or anything else is the best area in which to operate. It has already been suggested that a long-run movement to optimal tariffs can prove self-defeating.

As it becomes more evident that OPEC is an (imperfect) effort to monopolize world oil prices, particular attention must be given to the difficulties of applying an optimum tariff as a response to monopolization (which itself is OPEC's effort to charge an optimal tariff). The standard optimum tariff argument assumes imposition of tariffs on suppliers or customers without monopoly power. Such victims will clearly respond by lowering prices charged or accepting a higher price paid. A rival monopolist might act differently.

The earlier writings on the advantages of import limitations (e.g., Stobaugh and Yergin) demonstrated another defect of neglecting optimum tariff arguments. Such writers stressed controls on import quantities. The optimal tariff literature (see, e.g., Meade) has long warned that such controls on quantities are less satisfactory than tariffs because a distinct possibility exists that under quantitative controls monopoly suppliers may raise their prices and be the gainers. In short, it is unclear that there is any right area—oil, metals, or otherwise—in which it is wise to try to impose optimal tariffs. More conventional measures such as removing barriers to competition in energy—particularly the numerous government restrictions—may be more effective. Conversely, if the policy can succeed with one commodity, why can it not be applied elsewhere?

The crux of the concern over policies to offset supply disruptions because of their impacts on aggregative economic activity is whether the price-rigidity model is correct. Much of the work on the subject such as the analysis by Mork and Hall assumes the validity of the price rigidity argument. Given that there were an abundance of forces destabilizing the U.S. economy in the 1970s, it is hardly surprising that the evidence on the macroeconomic impact of OPEC is unclear (see, e.g., Darby). Further questions arise about the source of whatever impact arises. The contributors to the Plummer volume and Bohi and Montgomery concentrate upon the impacts of supply disruption and policies to offset it. Their advice that anticrisis stockpiles are socially efficient is correct if their premises, that it is supply disruption that causes the shocks and that the economy is not flexible enough to absorb them, are valid. However, it may be that it is the permanent price rise rather than the temporary supply disruption that is the problem. This would lead to quite different policies. In fact, there may be nothing worth doing aside from vigorous control of the money supply.

An optimum tariff would probably aggravate the macroeconomic situation whatever its effect on the cost to the economy of oil. While before-tax prices might fall, after-tax prices certainly would rise. The crux of the price-rigidity argument is that the economy is unable to effect the price adjustments elsewhere that reflect the real rise in purchasing power produced by the optimum tariff and the impact is inflationary pressures.

Whether inventory holding or anything else will prevent further price rises is unclear. The strongest case is that OPEC is so weak an organization that the only way it can be persuaded to cut output to sustain permanently higher prices is to view the benefits of crisis induced curtailments. Prices rise to the crisis-induced level and are kept there by conscious continuation of supply stringency. If inventories prevent crisis price rises from being so severe, they might limit or at least delay permanent changes. A major flaw in the argument is that OPEC may have enough cohesion to supplement the crisis caused supply restriction with conscious measures. In fact, Adelman has suggested that Saudi Arabia took precisely such actions during the 1979-1980 Iranian crisis.

In any case, severe limits exist in the extent to which OPEC can profitably raise prices. The essence of the optimal tariff and similar analyses of monopoly is that a specific optimal price exists at any given time. Once that optimum is reached, further changes are desirable only to the extent that underlying market conditions change markedly. The travails of OPEC in 1983 may indicate that the optimum for 1980 condition was actually reached and that subsequent conditions have lowered the optimum as investments to conserve on fuel and develop alternative energy supplies were completed. In any case, this speculation is less critical than the basic point that it is not clear how exercise of whatever permanent price raising power remains to OPEC can be prevented short of conceiving a strategy for forever breaking OPEC—a questionable prospect now that its members have seen the benefits of a decade of success.

Again a concern exists about whether oil, other minerals, or anything else is the right place to take action even if there is a severe price rigidity problem. Here too, it is possible that if action is taken, it should apply to many rather than few industries.

Finally, the imperfect-capital-market argument does serve the useful function of reminding us of the advantage of diversification. The argument generalizes to suggest that in addition to the familiar immediate gains from specialization and extensive trade, there are gains of diversification. We can ail, but others will be healthy. Considerations of supply instability thus suggest that greater diversification is desirable. More precisely, while we should optimally curtail dependence on unreliable suppliers, we should also seek a larger number of suppliers. (Recall, however, that there may not be a beneficial way to reduce dependence).

This argument is and, in fact, long has been established in energy debates. However, naive protectionist models still arise in some energy discussions, most metals policy debates, and, most blatently, in the soft energy path argument. The last two discussions incorrectly assume that greater self-sufficiency is the best way to avoid risks.

A NOTE OF THE ECONOMICS OF POLITICS

Economists dissatisfied with the explanations of why their advice is so often rejected have used economic analysis to devise an appraisal. The theory was first sketched in Schumpeter's appraisal of the future organization of the world economy and later extended by Downs. They argued that while firms sought profits, politicians sought voters. Votes are more easily won by granting the protectionist demands of many groups than by standing as an advocate for the general interest. The interest group feels much more strongly about its gains than others feel about their losses. Thus, politicians tend to stress the concerns of potent interest groups.

This argument has been criticized by some because certain political actions are taken because of their consistency with the ideology supported rather than visible voter concerns. Mitchell has suggested that many proponents of energy price controls did so because they were liberals and despite the probability that the policy harmed their constituents. Kalt has undertaken studies of stands on several energy issues and concluded that ideology and constituent concerns both are relevant. This suggests that much of the problem is the inability to tame support of interest groups but that another difficulty is that good economic performance is inadequately appreciated by politicians.

Still a further argument is that bureaucrats are paid in power, not profits. Therefore, the bureaucrats push power increasing policies—i.e., regulations—over ones that rely more on impersonal market mechanisms.

References

1. Adelman, M.A., 1976, "The World Oil Cartel: Scarcity, Economics, and Politics," *Quarterly Review of Economics and Business* 16:2, pp. 7–18.
2. Adelman, M.A., 1978, "International Oil," *Natural Resources Journal* 18:4 (October), pp. 725–730.
3. Adelman, M.A., 1979, Untitled transcript in *Seminar on Energy Policy: The Carter Proposals*, edited by Edward J. Mitchell, Washington, D.C.: American Enterprise Institute for Public Policy Research.
4. Adelman, M.A., 1980, "The Clumsy Cartel," *The Energy Journal* 1:43–53.
5. Allais, Maurice, 1951, *"La Gestion des Houilleres Nationalises et La Theorie Economique,"* Paris: Imprimerie Nationale.
6. Allais, Maurice, 1962, "Les Aspects Essential de la Politique de l'Energie—Synthese et Conclusions in M. Allais, ed., *La Politique de l'Energie*, Paris: Imprimerie Nationale (reprinted from *Annales des Mines*).
7. Areeda, Phillip, 1981, *Antitrust Analysis, Problems, Texts, Cases*, Boston: Little, Brown and Company.
8. Baumol, William J., and Wallace E. Oates, 1975, *The Theory of Environmental Policy: Externalities, Public Outlays, and the Quality of Life*, Englewood Cliffs, N.J.: Prentice-Hall Inc.
9. Baumol, William J., John C. Panzar, and Robert D. Willig, 1982, *Contestable Markets and the Theory of Industry Structure*, New York: Harcourt Brace and Jovanovich.
10. Bhagwati, Jagdish N., 1971, "The Generalized Theory of Distortions and Welfare," in Jagdish N. Bhagwati, Ronald W. Jones, Robert A. Mundell and Jaroslav Vanek, eds., *Trade, Balance of Payments and Growth*, papers in *International Economics in honor of Charles P. Kindleberger*, Amsterdam: North-Holland Publishing Co.
11. Bieniewicz, Donald J., 1980, *Fair Market Value and Unbiased Bid Rejection Procedures*, Washington, D.C.: U.S. Department of the Interior.
12. Bieniewicz, Donald J., 1982, *Options for Assuring the Receipt of Fair Market Value for Federal Coal Leases*, Washington, D.C.: U.S. Department of the Interior.
13. Bieniewicz, Donald J., 1983, *Options Analysis: Federal Coal Lease Sale/Bid Acceptance Procedures*, Washington, D.C.: U.S. Department of the Interior.
14. Bieniewicz, Donald J., and Robert H. Nelson, 1983, "Planning a Market for Federal Coal Leasing," *Natural Resources Journal*, 23 (July), pp. 593–604.
15. Bohi, Douglas R., and Milton Russell, 1978, *Limiting Oil Imports: An Economic History and Analysis*, Baltimore, Md.: The Johns Hopkins University Press for Resources for the Future.
16. Bohi, Douglas R., and W. David Montgomery, 1982,

Oil Prices, Energy Security, and Import Policy, Washington, D.C.: Resources for the Future, Inc.
17. Bork, Robert H., 1978, *The Antitrust Paradox, a Policy at War with Itself*, New York: Basic Books.
18. Bradford, David F., 1975, "Constraints on Government Investment Opportunities and the Choice of Discount Rate, *American Economic Review*, 65:5 (December), pp. 887–899.
19. Breyer, Stephen G., 1982, *Regulation and Its Reform*, Cambridge, Mass.: Harvard University Press.
20. Buchanan, James M., and William Craig Stubblebine, 1962, "Externality," *Economica*, NS 29, pp. 371–384. Reprinted in Kenneth J. Arrow and Tibor Scitovsky, eds., 1969, *Readings in Welfare Economics*, Homewood, Ill.: Richard D. Irwin.
21. Cass, Glen R., Robert W. Hahn, and Roger G. Noll, 1982, *Implementing Tradable Permits for Sulfur Oxide Emissions, A Case Study in South Coast Air*, Pasadena: Environmental Quality Laboratory, California Institute of Technology.
22. Cheung, Steven N., 1978, *The Myth of Social Cost*, London: The Institute of Economic Affairs. Reprinted with additional foreword, San Francisco: Cato Institute, 1980.
23. Clark, Richard A., Robert C. Lind, and Robert Smiley, 1976, *Enhancing Competition for Federal Coal Leases*, McLean, Va.: Science Application Inc.
24. Coase, Ronald H., 1960, "The Problem of Social Cost," *Journal of Law and Economics*, 3 (October), p. 1044.
25. Corden, W.M., 1974, *Trade Policy and Economic Welfare*, Oxford: Oxford University Press.
26. Darby, Michael R., 1982, "The Price of Oil and World Inflation and Recession," *American Economic Review*, 72:4 (September), pp. 738–751.
27. Deese, David A., and Joseph S. Nye, eds., 1981, *Energy and Security*, Cambridge, Mass.: Ballinger Publishing Company.
28. Downs, Anthony, 1957, *An Economic Theory of Democracy*, New York: Harper and Row.
29. Farrell, M.J., 1968, "In Defense of Public Utility Price Theory," in Ralph Turvey, ed., *Public Enterprise, Selected Readings*, Hammondsworth, England: Penguin Books. (Revised version of 1955 article in Oxford Economic Papers, NS 10, pp. 109–123).
30. Fisher, Franklin M., John J. McGowan, and Joen E. Greenwood, 1983, *Folded, Spindled, and Mutilated Economic Analysis and U.S. v. IBM*, Cambridge, Mass.: The MIT Press.
31. Gordon, Richard L., 1970, *The Evolution of Energy Policy in Western Europe: The Reluctant Retreat from Coal*, New York: Praeger Publishers.
32. Gordon, Richard L., 1974, "The Optimization of Input Supply Patterns in the Case of Fuels for Electric Power Generation," *Journal of Industrial Economics* 22:1 (September), pp. 19–37.
33. Gordon, Richard L., 1975, *U.S. Coal and the Electric Power Industry*, Baltimore, Md.: Johns Hopkins University Press for Resources for the Future.
34. Gordon, Richard L., 1978a, *Coal in the U.S. Energy Market: History and Prospects*, Lexington, Mass.: Lexington Books, D.C. Heath.
35. Gordon, Richard L., 1978b, "The Hobbling of Coal: Policy and Regulatory Uncertainties," *Science* 200 (14 April), pp. 153–158.
36. Gordon, Richard L., 1978c, "Hobbling Coal—Or How to Serve Two Masters Poorly," *Regulation* 2:4 (July-August), pp. 36–45.
37. Gordon, Richard L., 1981, *An Economic Analysis of World Energy Problems*, Cambridge, Mass.: The MIT Press.
38. Graaf, J. deV., 1967, *Theoretical Welfare Economics*, Cambridge: Cambridge University Press.
39. Hanke, Steve H., 1982, "The Privatization Debate: An Insider's View," *The Cato Journal*, 2:3 (Winter), pp. 652–662.
40. Heintz, H. Theodore, Jr., 1983, *OSC Oil and Gas Leasing: How the Country Benefits and How Interior Collects Revenues*, Washington, D.C.: U.S. Department of the Interior.
41. Hicks, John W., 1981, *Wealth and Welfare*, collected Essays on Economic Theory, Vol. 1, Cambridge, Mass.: Harvard University Press.
42. Herfindahl, Orris, 1974, *Resource Economics: Selected Works*, David B. Brooks, ed., Baltimore, Md.: Johns Hopkins University Press for Resources for the Future.
43. Johnson, Harry, 1965, "Optimal Trade Intervention in the Presence of Domestic Distortions," in Robert G. Baldwin (and 15 others) *Trade, Growth, and the Balance of Payments Essays in Honor of Gottfried Haberler*, Chicago: Rand McNally pp. 3–34.
44. Kalt, Joseph P., 1981, *The Economics and Politics of Oil Price Regulation: Federal Policy in the Post-Embargo Era*, Cambridge, Mass.: The MIT Press.
45. Kalt, Joseph P., 1981, "Public Goods and the Theory of Government," *The Cato Journal* 1:2 (Fall) pp. 565–584.
46. Keynes, John Maynard, 1936, *The General Theory of Employment, Interest, and Money*, London: MacMillan and Company.
47. Kneese, Allen V., and Charles L. Schultze, 1975, *Pollution, Prices and Public Policy*, Washington, D.C.: The Brookings Institution.
48. Lave, Lester B., and Gilbert S. Omenn, 1981, *Clearing the Air: Reforming the Clean Air Act*, Washington, D.C.: The Brookings Institution.
49. Leland, Hayne E., and Richard B. Norgaard, with Scott R. Pearson, 1974, *An Economic Analysis of Alternative Outer Continental Shelf Petroleum Leasing Policies*, Washington, D.C.: The National Science Foundation.
50. Lipsey, R.G., and K. Lancaster, "The General Theory of Second Best," *Review of Economic Studies*, 24:(195-7), pp. 11–32. Reprinted in M.J. Farrell, ed., 1973, *Readings in Welfare Economics*, London: MacMillan and Company.
51. Little, I.D.M., 1957, *A Critique of Welfare Economics*, second edition, Oxford: Oxford University Press.
52. Lund, Robert C., (with Kenneth J. Arrow, Gordon R. Corey, Partha S. Dasgupta, Amanta K. Sen, Thomas Stauffer, Joseph E. Stiglitz, J.A. Stockfisch, and Robert Wilson), 1982, *Discounting for Time and Risk in Energy Policy*, Washington, D.C.: Resources for the Future.
53. McDonald, Stephen L., 1979, *The Leasing of Federal Lands for Fossil Fuels Production*, Baltimore, Md.: Johns Hopkins University Press for Resources for the Future.
54. Malinvaud, E., 1972, *Lectures on Microeconomic Theory*, Amsterdam: North Holland Publishing Company.
55. Meade, J.E., 1955, *Trade and Welfare, The Theory of International Economic Policy*, vol. 2, London: Oxford University Press.

56. Mills, Edwin S., 1978, *The Economics of Environmental Quality*, New York: W.W. Norton & Company.
57. Mishan, E.J., 1976, *Cost-Benefit Analysis*, new and expanded edition, New York: Praeger Publishers.
58. Mitchell, Edward J., 1978, "Oil, Film, and Folklore," *Regulation*, July/August 1978, pp. 17–20.
59. Mork, Knut Anton, and Robert E. Hall, 1980a, "Energy Prices and the U.S. Economy in 1979–1981," *The Energy Journal*, 1:2 (April), pp. 41–53.
60. Mork, Knut Anton, 1980b, "Energy Prices, Inflation and the Recession in 1974–1975," *The Energy Journal*, 1:3 (July), pp. 31–63.
61. Nelson, Robert H., 1983, *The Making of Federal Coal Policy*, Durham, N.C.: Duke University Press.
62. Newlon, Daniel H., and Norman V. Breckner, 1975, *The Oil Security System: An Import Strategy for Achieving Oil Security and Reducing Oil Prices*, Lexington, Mass.: Lexington Books, D.C. Heath.
63. Nozick, Robert, 1974, *Anarchy, State, and Utopia*, New York: Basic Books.
64. Okun, Arthur M., 1975, *Equality and Efficiency, The Big Tradeoff*, Washington, D.C.: The Brookings Institution.
65. Patinkin, Don, 1965, *Money, Interest and Prices*, second edition, New York: Harper and Row.
66. Plummer, James L., ed., 1982, *Energy Vulnerability*, Cambridge: Ballinger Publishing Company.
67. Posner, Richard A., and Frank H. Easterbrook, 1981, *Antitrust: Cases, Economic Notes, and Other Materials*, St. Paul: West Publishing Company.
68. Rawls, John, 1971, *A Theory of Justice*, Cambridge, Mass.: Harvard University Press.
69. Rothband, Murry, 1982, "Law, Property Rights and Air Pollution," *The Cato Journal* 2 (Spring), pp. 55–99.
70. Samuelson, Paul A., 1966, *The Collected Scientific Papers of Paul A. Samuelson*, edited by Joseph E. Stiglitz, 2 vols., Cambridge, Mass.: The MIT Press. Contains earliest papers on public goods.
71. Samuelson, Paul A., 1969, "Pure Theory of Public Expenditures and Taxation," in J. Margolis, and H. Guitton, eds., *Public Economics*, New York: St. Martins Press, pp. 98–123. Reprinted in *Collected Scientific Papers of Paul A. Samuelson*, edited by Robert C. Merton, vol. 3, pp. 492–517, Cambridge, Mass.: The MIT Press, 1972.
72. Samuelson, Paul A., 1983, *Foundations of Economic Analysis*, expanded edition, Cambridge, Mass.: Harvard University Press.
73. Scherer, F.M., 1980, *Industrial Market Structure and Economic Performance*, second edition, Chicago, Ill.: Rand McNally College Publishing Company.
74. Schumpeter, Joseph A., 1950, *Capitalism, Socialism, and Democracy*, 3d ed., New York: Harper & Brothers.
75. Shone, R., 1976, *Microeconomics: A Modern Treatment*, New York: Academic Press.
76. Siebert, Horst, 1981, *Economics of the Environment*, Lexington, Mass.: Lexington Books, D.C. Heath.
77. Smith, Vernon L., 1982 "On Divestiture and the Creation of Property Rights in Public Lands," *The Cato Journal*, 2:3 (Winter), pp. 663–685.
78. Stobaugh, Robert, and Daniel Yergin, eds., 1979, *Energy Future, Report of the Energy Project at the Harvard Business School*, New York: Random House.
79. Thompson, Duane A., and Dennis Zimmerman, 1983, *An Economic Analysis of Federal Coal Leasing Policies*, Report 83-169ENR, Washington, D.C., The Library of Congress.
80. Thurow, Lester B., 1980, *The Zero Sum Society*, New York: Basic Books, Inc.
81. Turvey, Ralph, 1963, "On Divergencies Between Social Cost and Private Cost," *Economica*, 30:119 (August), pp. 309–313.
82. Tyner, Wallace E., and Robert J. Kalter, 1978, *Western Coal: Problem or Promise?*, Lexington, Mass.: Lexington Books, D.C. Heath.
83. U.S. Department of Energy, 1981, *Coal Competition Prospects for the 1980s*, Springfield, Va.: National Technical Information Service.
84. U.S. Department of Justice, annual since 1978, *Competition in the Coal Industry*, Washington, D.C.: U.S. Government Printing Office.
85. U.S. Federal Trade Commission, Bureau of Competition, Bureau of Economics, 1975, *Staff Report to the Federal Trade Commission on Federal Energy Land Policy*, Washington, D.C.
86. U.S. Federal Trade Commission, 1978, *Report to the Federal Trade Commission on the Structure of the Nation's Coal Industry 1964–1974*. Washington, D.C.: U.S. Government Printing Office. (Draft version available in FTC library).
87. U.S. General Accounting Office, 1977a, *The State of Competition in the Coal Industry*, Washington, D.C.
88. U.S. General Accounting Office, 1977b, *U.S. Coal Development: Promises, Uncertainties*, Washington, D.C.
89. U.S. President's Materials Policy Commission (The Paley Commission), 1952, *Resources for Freedom*, 5V, Washington, D.C.: U.S. Government Printing Office.
90. Varian, Hal R., 1978, *Microeconomic Analysis*, New York: W.W. Norton and Company.
91. Willig, R., 1976, "Consumer's Surplus Without Apology," *American Economic Review*, 66:589–597.

5.14

Non-Fuel Policy Issues

Walter R. Hibbard, Jr.

INTRODUCTION

Volatile markets and pricing, increasing imports, constraining but effective laws related to the environment, health and safety and price controls lead to concerns which generated pious laws stating non-fuel policy and authorizing studies but resulting in ineffective implementation. Several commissions, reports by the Secretary of Interior and even the President have led to many recommendations but little or no action. No single comprehensive policy could solve problems which are diverse, commodity specific and not categorizable (Landsberg 1982, 1983).

Thirteen laws administered by 27 agencies regulate or affect minerals and materials. Political instability created by communist regimes in Southern and Central Africa jeopardized sources of critical imported materials. This situation led to a Resource War mentality. It was feared that the U.S.S.R. was seeking to control this area and exclude the United States from resources (Vogely, 1980). The success of the Organization of Petroleum Exporting Countries with a petroleum cartel led to fears of an Organization of Minerals Exporting Countries which might create more cartels and/or make the existing ones more effective. Economic needs of developing countries required export to maintain employment and balance of payments have led to world surpluses, weak markets, low prices, and large loans from the World Bank which have undermined U.S. industry (Landsberg, 1982a). Uncooperative environmental laws, obsolete facilities and non-competitive costs have led to shipping U.S. ores to other countries for processing into ingot and bars which are imported back to the United States. Yet massive internal turmoil has interrupted supplies, and led to uncertain availability and higher prices. This situation could happen again. Included in this category of critical materials with uncertain sources are metals required in the manufacture of military aircraft, which has led to concern regarding the defense industry base (Morgan 1983).

Remedies have been proposed which include: updated stock piles, material substitution, international cooperation and coordination, foreign policy aimed at resolving racial and national stresses in South and Central Africa, relaxing U.S. environmental and public land regulations and action under the Defense Production Act (Goth 1982).

The President's report (1983) to Congress has stated a national materials policy which recognizes: that materials play a critical role in our economy, defense, and standard of living; that public lands contain a vast untapped mineral wealth; that the role of government is to identify mineral issues and act on them; and that there is a need for long term high potential pay off research (Bureau of Mines 1983).

There is a serious question as to whether or not a comprehensive national mining, minerals, and materials policy is feasible let alone necessary. One solution is to abandon the attempt at comprehensiveness and to target specific issues with realistic policies (Comptroller General 1979). In the following sections, the major studies of the 1970s will be reviewed, major issues yet to be faced will be identified, the principles involved in these resolutions will be reviewed, and each of several major issues will be discussed.

MAJOR STUDIES

The President's Commission on Materials Policy (1952) reviewed and forecast the availability of materials, projected problems with some, and came to the following conclusions and recommendations:

(1) No nation can be completely self-sufficient in raw materials.

(2) Long range demand may overrun domestic supply for some material.

(3) National Materials Policy should "insure an adequate and dependable flow of materials at lowest cost consistent with national security and the welfare of friendly nations."

(4) A National Security Resources Board should be established to report annually on the long term materials outlook with emphasis on emerging problems, changes in outlook and modifications of policy and program.

(5) The government's capacity to gather facts, analyze and evaluate problems should be strengthened.

The situation eased by the time the report was released and little government action resulted. However, the most important result was the establishment of Resources for the Future, Inc. by the Ford Foundation to study minerals and materials issues, which they have done, and are doing, effectively.

The Secretary of the Interior (1972–79) published eight annual reports to Congress in response to the Mining and Mineral Policy Act of 1970 (P.L. 91-631). These reports reviewed the state of the domestic mining, minerals, and minerals reclamations industries including; extensive statistics on the utilization and depletion of these resources, the viability of the industries and recommendations for legislative action. Each annual report, describing a deteriorating state of the industries, defined problems and recommended legislative solutions. Little, if any, action resulted.

The National Commission on Materials Policy (1973) was established pursuant to Section 201, Title II of PL 91-512, originally was concerned primarily with solid waste disposal and the recycling of materials but was extended to include all aspects of materials and the environment. An exhaustive multiyear study covered materials supply, consumption, recycling, management, land use, environmental effects and policy. The report entitled "Materials Needs and the Environment Today and Tomorrow" contained 198 recommendations in 12 principal categories which may be summarized as follows:

(1) National policy should contain the following elements: (a) provide adequate energy and materials supplies to meet the needs of a dynamic economy without waste; (b) rely on market forces to determine the mix of domestic and imported supply except where dangerous or costly; (c) conserve, protect, or enhance the environment while conserving resources; (d) recognize the interrelationships between materials, energy, and the environment in resource policy and planning.

(2) Modify the materials system so that all resources, including environmental, are paid by the user.

(3) Accelerate waste recycling and greater efficiency-of-use of materials.

(4) Manage materials policy so that laws, executive order and administration practices reinforce such policy and do not counter-act it.

Most of the recommendations were not implemented.

The Committee on the Survey of Materials Science and Engineering (1974) was established by the National Academy of Science and focused primarily on the state of science and engineering. Its extensive report entitled "Materials and Man's Needs" (five volumes) contributed the following concepts regarding policy: (a) materials are in a central position with respect to nature's processes, national economies, and in individual's daily life in the form of a global materials cycle; (b) materials, energy, and the environment are parts of the same vast system: policies and programs that deal with one part must respect the other two within the total materials system. The study produced 24 recommendations mostly related to research and education and mostly not implemented.

The National Commission on Supplies and Shortages (1976) was established in response to PL 93-426 to investigate shortages and historic high prices of many materials which occurred in 1973–74. The study focused on eight factors: (1) world resources, (2) market conditions, (3) 1973–74 shortages, (4) data analysis,

(5) government policy, (6) stockpiles, (7) recycle, (8) research and development.

The study covered four concerns: (1) resource exhaustion, (2) dependency on imports, (3) adequacy of government institutions for dealing with materials problems, (4) market ability to deal with shortages.

The findings were as follows: (1) resource exhaustion is not in the foreseeable future, (2) imports have increased only modestly, but threatened supply disruption must be counteracted, (3) improvements in government institutions are warranted and feasible, (4) the 1973–74 shortages were caused by a worldwide boom in 1972, inadequate production capacity, and a "shortage mentality" on behalf of purchasers. The market response took two years to solve the problem.

Twenty recommendations were made, including an economic stockpile, but none were implemented.

Non-Fuel Minerals Study (1977) was assigned to a cabinet level committee which studied nine areas of policy as follows: (1) supply, (2) imports, (3) relationship of environmental quality, health and safety standards to price and availability, (4) resource potential on public lands, (5) financial, capital formation and tax policies, (6) competitiveness of U.S. industry, (7) conservation, substitution and recycle, (8) adequacy of research and development, (9) adequacy of existing government capabilities to support policy making. A draft report, which was released in August 1979, identified the following items to be addressed: (1) slack markets, costs and lead times, industry instability and developing countries; (2) stockpiling; (3) mining industry's financial condition, low profits, increasing debt; (4) public lands with vast potential but little resource information; (5) environmental, safety, and health regulation impacts; (6) underinvestment in research and development; (7) the *ad hoc* and unstructured character of federal non-fuel minerals policy making; (8) the unresolved conflict between protecting and developing natural resources. In August 1980, the administration stated that the nature of the domestic policy review system does not require a final public report. No report was released. Apparently, the difference between the cabinet level participants could not be, or were not, resolved.

The President (1982) sent a report to Congress pursuant to the National Materials, and Minerals Policy, Research and Development Act of 1980, PL 96-479, which was entitled "National Materials and Minerals Program Plan" and stated the following elements of policy:

(1) to decrease minerals vulnerability
(2) to inventory minerals on public lands
(3) to prepare strategies and critical materials impact analysis on proposed land withdrawals
(4) to collect better data
(5) to concentrate government research on long term, high risk, high potential projects
(6) to update stockpile
(7) to coordinate natural materials policy through a cabinet council on National Resources and the Environment
(8) to identify potential critical materials
(9) to encourage private research and development
(10) to strengthen the data system
(11) to restore the current stockpile inventory
(12) to establish an amicable plan for stockpile acquisition or disposal
(13) consult and coordinate with industrial nations which are consumers of key materials.

The Bureau of Mines (1983) examined the domestic supply of 15 minerals which were considered critical in the following categories: (1) influence on the economy, (2) strategic to defense, (3) special properties not available elsewhere, (4) insufficient quantities produced in the United States, (5) important alloying elements. The report concluded that: (a) an adequate supply of minerals is essential to our economic well being and national defense, (b) demand will continue to grow especially for high technology minerals, (c) the world supply of critical minerals is adequate; however, the U.S. supply of some critical materials is inadequate, (d) some critical minerals (chromium, cobalt and platinum) are imported from insecure sources, (e) the domestic mineral processing industry is increasingly at a disadvantage.

Issues

From the numerous studies, the following major issues emerge from the rhetoric:

(1) the need for a supply of minerals adequate to support the economy and provide for national security;
(2) concern that imports required to meet this need may be unavailable from certain sources;

(3) concern for the quality of the defense stockpile and the deterioration of the U.S. minerals industry;

(4) inability of Congress to agree on the comprehensive materials policy action and the need to target specific issues with realistic policies and actions.

Specific Issues To Be Faced: include: (a) public land use, (b) the viability of the national industrial base, (c) sea bed mining, (d) international mineral trade, (e) extending U.S. supply through conservation, substitution, and recycling, (f) adequacy of government institutional capacity to execute minerals policies and plans, (g) the need for a statement of national non-fuel policy(s) to guide action in unrelated areas which have an impact on minerals—or should there be a requirement for a minerals impact statement with each proposed action. Each of these specific issues will be discussed in the following sections.

Public Land Use: The Public Land Law Review Commission report to the President in 1970 noted that economic concentrations of minerals must be mined where they are, and that mineral exploration and development should have preference over some other uses on much of our public lands. The Congress did not agree. Out of 741 million acres of public lands in 1967, approximately 120 million acres were withdrawn from such use. By 1974, the amount withdrawn increased to approximately 480 million acres (Goth 1982). A 1975 study indicated that approximately two thirds of the public lands were withdrawn from exploration and development. A 1977 study by the Department of Interior's Task Force on the Availability of Federal Owned Mineral Land reiterated these facts, but did nothing. A 1982 study noted that only 46 out of 244 known major mineral occurrences in Alaska are in areas open to mineral activity (Lee and Bennethum 1982).

The concept of wise and effective multiple public land use proposed by the 1970 commission has been replaced by a concept of preservation of public lands through an ecologically oriented public policy. Today about three million out of 2.3 billion acres of the United States are used for mineral activity, mostly for coal, sand, gravel and stone. Most of the land is used for agriculture and forestry. For society as a whole, non-mineral resources such as plants and species, scenic land forms, historical and archeological sites, wilderness and wild rivers are also only where you find them and alternative sites are not available at any cost. No generally accepted formula exists to identify mineral or social uniqueness. The latter is winning legislative favor (Anon. 1982).

National Industrial Base: During the last 20 years, the mining, materials, and many heavy industries have deteriorated because government regulations have added expenses not shared by foreign competitors and diverted research and development and investment efforts toward pollution, health and safety issues, have reduced profitability, discouraged re-investment in the U.S. facilities and restrained the availability of capital. Certain laws permit individual interveners to postpone, prolong, and even cancel industrial ventures aimed at improved competitiveness. As a result many materials and products are imported from foreign countries and with this trend, manufacturing capability and jobs are exported. As a result U.S. steel, aluminum, copper, automobile, applicances and communications industries have moved overseas and those remaining are being restructured to be more efficient, smaller, specialized, competitive and less labor intensive. Public policy often appears as an adversary of industry and, either intentionally or through unperceived side effects, has caused this change.

General Alvin Slay of the Air Force noted that the national industrial base had deteriorated to the point where it would not be possible to produce, on an emergency short term basis, 100 military aircraft. He presented the following information to Congress: (1) 90% of six metals required for jet engine manufacture are imported from Central or South Africa, (2) internal strife in producing countries has caused shortages and large price increases in five of these metals, (3) metal processing capacity has deteriorated so that only three companies can produce large landing gear forgings with lead times up to 116 weeks. He recommended actions most of which have not been implemented.

Sea-Bed Mining: The richest known sea-bed mineral resources are manganese nodules which lie in profusion on the ocean floor in the deep central Pacific Ocean and on the Blake Plateau in the Atlantic Ocean. The former are more attractive and contain rich amounts of copper, cobalt, nickel and manganese. Various consortia of U.S., Japanese, West German, and U.K. companies have developed technology to retrieve the nodules by several systems, but no commercial-sized equipment has been built because these consortia are awaiting clarification of legal access.

The Law of the Sea Treaty, approved by the United Nations, contains provisions which are unacceptable to the United States and other developed nations, which refused to sign the treaty. These provisions include ownership beyond national boundaries by the U.N. and administration and collections of royalties by that body. The United States, the United Kingdom, Japan, West Germany, and others are preparing mini-treaties, which permit them to lay claims to such deposits by discovery and to administer such claims on a national basis (Ratiner 1983).

Consortia which wished to mine under the Law of the Sea Treaty would have to obtain permits through countries which are signers of the treaty. Under the mini-treaties, consortia would have to be assured of their right and protection of the country involved.

No such venture has been launched by any consortia. Thus, supplies of minerals considered critical to the United States await resolution via international law and/or national policy and action.

Mineral Interdependence and International Trade: The economic rationale for unrestricted mineral trade is based on the efficiency of mineral activity occurring where costs are lowest. Cost savings generated by trade can be captured in such a way that most countries are not worse off and some are better off than if trade were restricted.

Trade restrictions do exist in other countries. In recent years, the U.S. liberal trade policy has been increasingly blamed for unemployment, imbalance of payments, and the demise of various mining, mineral, and materials industries. Large steel companies indicate that they can import ingots less expensively than they can make them. Groups who believe that they will be hurt by such imports are trying to block this trade. The arguments based on costs or risks beyond benefits include the following (Tilden 1980):

(1) Protection of special interests—trade tends to slightly improve the welfare of many people, but greatly harm a few people. The latter turn to the political system for relief.

(2) Security of supply—if the value of security of supply to the economy exceeds the value of security of supply to those firms purchasing materials, government intervention may be justified.

(3) Monopoly power of foreign producers—in general, cartels may not succeed, but concern for high prices, supply interruption, long lead times, and high capital expenditures to establish U.S. sources may justify government intervention.

(4) Balance of payments and the exchange value of the dollar—when imports exceed exports in an economically free system, the dollar is over valued and should adjust until imports and exports balance. Government intervention prevents this balance from being achieved.

(5) Dumping by foreign producers, i.e. selling below cost or domestic price—lower prices provide U.S. products with a cost advantage, but hurt domestic producers. Minimum imports price regulations have not helped.

(6) Protection of jobs and the tax rate—protection has an impact on other jobs, artificially props up the dollar, and has a negative impact on balance of payments. Sixty-five percent of the taxes depend on individual tax payers and, therefore, jobs.

(7) Environmental controls and other regulations—sudden changes in policy add unexpected costs, dilute earning power, and often are uneffective or unnecessary. Orderly regulations with known results are desired.

(8) National security and foreign policy—these questions can be treated by methods other than trade restrictions. Foreign sources do result in greater economic risk and political uncertainty than domestic sources. Other than trade restrictions, ameliorating policy can be: (a) subsidize domestic sources, (b) trade agreements, (c) stock piles, (d) substitution, (e) conservation, (f) recycling, (g) new technology (Berman 1980) (Bement 1983).

Materials Availability: World resources are more than adequate for the foreseeable future. In fact, many reserves have increased as a result of geological and technological discoveries. The principal questions are short time disruptions and higher prices. These factors are related to intensity of use, domestic capacity, pollution, energy and land use, and limits to growth. Such items are questions of values, preference, and ethics. Scarcity leads to higher prices, lower demand—conservation, substitution, and recycling until only high-value uses consume high-cost materials and a balance is achieved between supply, demand, and price. This imbalance requires time and is further distorted by regulations and shortage mentalities (Berman 1980).

Substitution: Demand for materials derives from demand for goods and services which are provided by products which function through design and materials selection. Selection is based on cost, availability, and performance which can be provided by several materials. Competition may take the form of the substitution of a new material in the existing design, a new material in a new design, a new or revised function with a different design, or a new material and elimination of the function. The driver for substitution is not simply cost and availability but all the factors comprising the total cost of the product and its relation to price and demand.

Conservation: Lowering demand is the form of conservation which involves using less material in providing the same function, such as eliminating waste, less scrap, zero defects, less material per unit product, longer-life products requiring fewer replacements, and less "disposable" products. For scarce or critical materials, conservation can take the form of recycling existing materials from discarded products and substituting available materials for scarce ones.

Recycling: Since materials are not created or destroyed (except by nuclear reactions), most of the materials used in the United States are somewhere in dumps, junk yards, and otherwise squirreled away. The economic problem is to find, collect, sort, transport, and reprocess these materials. If these costs plus the avoidance costs of disposal are less than the cost of similar materials produced from ore, recycling has occurred.

Iron and steel are recycled because they are readily identified, separated and recycled as the principal ingredient in junked cars.

Aluminum is recycled because it is readily identified in the form of beverage cans and packaging resulting in large energy savings.

Lead and antimony are recycled because auto storage battery dealers provide a trade-in allowance.

Municipal waste is recycled because of limitations on the availability of sanitary land fills (required by law) and technology approaching breakeven when avoidance costs and government subsidies are considered. Collection is already a service provided by municipal taxes.

In recent years, the technology for recycling critical materials such as chromium, cobalt, tantalum, and platinum (in automobile catalytic converters) has advanced to stimulate increased recycle of these materials (Hibbard 1982).

National Mining, Minerals and Materials Policy: Since 1970, several Congressional acts have stated the national mining, minerals, and materials policy in similar terms, which have been reconfirmed recently by the President. However, in the trade-offs with energy, environmental, labor, welfare, and defense policy, such mining, minerals, and materials policy statements have been ineffective and unable to generate supportive action in the Congress or the Administration. Efforts to achieve compromise have failed and the industries involved have deteriorated to an obsolete, over-regulated, noncompetitive state. These conditions must change if the industries are to survive viably in the United States. The most hopeful approach seems to abandon the concept of a broad comprehensive policy which is ineffective, and to target specific problems and seek realistic policies in those issues (Landsberg 1982a). Such an approach is also subject to attack from vocal single-issue groups who think only of their issue in their vested self-interest and threaten the political welfare of their political representatives. Excessive environmental regulations, public land use, coal slurry pipelines and other issues previously cited have escaped solution. The issue of critical materials and their relationship to economic and national security looms large in the picture, but no solution which satisfies all of the lobbies and preserves political longevity has emerged.

Government Institutions: Thirteen laws affecting the mining, minerals, and material industries are administered by 27 agencies but only one, or possibly two, agencies are charged with the welfare of these industries. Efforts to organize cabinet level councils, committees, and studies have failed to achieve the necessary balance to preserve these industries, and yet achieve the goals of the other 25 or so agencies which have prevailed. Presidents have chosen not to intervene. It is clear from the various studies that if an emergency related to critical materials arose, the response would be a conflicting maze of conflicting orders and regulations unless the President acted promptly and was blessed with wise advisors. In view of the fact that both the President and the Congress have not acted effectively on the recommendations of numerous studies, it is unlikely that any institutional remedies will occur in the government unless crises require precipitous reaction. The current government materials information system is probably inadequate to supply effective early warning of such crises.

Principles Underlying the Resolutions of Issues: Societal pressures and single-issue forces in a participatory democracy have expanded government intervention in the day-to-day life of our society. Such lobbies think only of themselves and indulge in selfish desires without reference to their impact on the economy or society. They expect little from themselves but a great deal from the government. Many government institutions which were originally established as scorekeepers, now act as intermediaries, policemen and tinkerers of the rules by which economic activity is conducted. The form in which we have addressed the betterment of economic disadvantaged people has created a seemingly unacceptable future for the nation. Our public education system has been ravaged. Our value judgments and priorities have swung far to one side.

If the principle of resolution by government intervention pervades, the issues will continue to be battered by single-issue lobbies and inadequate information (Charpie 1983).

An alternative principle for resolving these issues involves the economic free market approach. This approach requires the withdrawal of the government as an adversary of industry, a policeman and an intermediary of vocal single-issue groups. The success of cooperation between government and industry in Japan and West Germany, while retaining individual rights and welfare, is overwhelming.

In the long run, under these principles, it is the best product (mineral or material) produced at the lowest cost and sold at a competitive price which wins the race in the international market place. No government intervention can bias the competition long enough to create a meaningful competitive advantage unless the best product at the best price is about to be delivered.

No additional cost burdens, no matter how they are justified, can be tolerated in this competition. Burdens which are responsive to all the single-issue lobbies of the last 10 years are in place, including sharing benefits with disadvantaged, preserving the environment, and controlling prices. No additional costs of this type can be added to a product cost which is already too high competitively—and still permit the industry to prevail in the United States. Resolution of the issue on this basis must return to the basics of the free economy upon which our country was founded. New technology is based on scientific information which doubles every seven years. We must identify the next round of new technology which will make our mining, mineral, and materials industries competitive and pursue a strategy aggressively with heavy investment in technology and in managers willing to take risks to put that technology to work. We must avoid a legislated solution to our non-competitive problems. Only those issues which are frozen by legislative action can be resolved by legislative action.

It is probable that the actual resolutions of these issues, if they are resolved, will occur using hybrid principles from both approaches. The pendulum will be responsive to public outcry. Hopefully information, rather than emotion, will prevail. The media will play an important role, and not all voters can have it their way. By default, these industries may be abandoned to other countries along with their jobs, technology, and economic inputs.

References

1. Anon., 1981, "A Start Toward Welding a Minerals Policy," *Business Week*, April 6, pp. 54–55.
2. Anon., 1982, "Should Minerals Availability Really be the Top Priority in Federal Land-Use Policy?," *P.E.*, Sept., pp. 19–31.
3. Bement, A.L., Jr., 1983, "Utilization of Science and Technology to Reduce Materials Vulnerability," *Materials and Society*, Vol. 7, No. 1, pp. 87–92.
4. Berman, E.B., 1980, "The Comparative Risk of Domestic and Foreign Sources of Supply for Materials," AIME Annual Meeting, Las Vegas, Nev., AIME, NYC.
5. Brooks, D.B. and P.W. Andrews, 1974, "Minerals Resources, Economic Growth and World Population," *Science*, Vol. 185, No. 4145, July 5, pp. 13–20.
6. Bureau of Mines, 1983, "The Domestic Supply of Critical Materials," U.S. Government Printing Office, Washington, D.C.
7. Charpie, R.A. 1983, "Living in Interesting Times," *ASM News*, Nov., p. 6.
8. Committee on the Survey of Materials Science and Engineering (COSMAT), 1974, "Materials and Man's Needs," National Academy of Sciences, Washington, D.C.
9. Comptroller General, 1979, "Learning to Look Ahead: The Need for a National Materials Policy and Planning Process," G.A.O. Report EMD 79-30, April 19.
10. Dwyer, C.S., 1980, "The Politics of Import Dependence," AIME Annual Meeting, Las Vegas, Nev., AIME, NYC.
11. Goth, J.W., 1982, "Public Lands Must Be Opened for Minerals Exploration," *P.E.*, Sept., pp. 26–28.

12. Harwood, J.J. and R.W. Layman, 1983, "Some Industrial Viewpoints on National Materials Policy," *Materials and Society*, Vol. 7, No. 1, pp. 93–100.
13. Hibbard, W.R., Jr., 1982, "The Extractive Metallurgy of Old Scrap Recycle," *J. Metals*, July, pp. 50–53.
14. Kinner, W.K., 1979, "Our Developing National Materials Policy," *M.E.*, April, pp. 29–32.
15. Landsberg, H.H., 1982, "Is a U.S. Minerals Policy Really the Answer?," *P.E.*, Sept., pp. 17–21.
16. Landsberg, H.H., 1982a, "What Next for U.S. Minerals Policy," *Resources*, Oct., pp. 9–10.
17. Landsberg, H.H., 1983, "Key Elements Common to Critical Issues on Engineering Materials and Minerals," *Materials and Society*, Vol. 7, No. 1, pp. 101–114.
18. Lee, L.C. and G. Bennethum, 1982, "Is our Account Overdrawn? A Reassessment," *Materials and Society*, Vol. 6, No. 1, pp. 15–22.
19. Materials Policy Commission, 1952, "Resources for Freedom, A Report to the President." W.S. Paley, Chr., June, U.S. Government Printing Office, Washington, D.C.
20. Morgan, J.D., 1983, "Strategic Materials," AIME Annual Meeting, Atlanta, Ga., AIME NYC.
21. National Commission on Materials Policy, 1973, "Material Needs and the Environment Today and Tomorrow." A Report to the President and Congress pursuant to Section 201, Title II, PL 91-512. U.S. Government Printing Office, Washington, D.C.
22. National Commission on Supplies and Shortages, 1976, "Government and the Nation's Resources," A Report to the President and Congress in response to PL 93-426, U.S. Government Printing Office, Washington, D.C.
23. President of the United States, 1982, "National Materials and Minerals Program Plan," Report to Congress pursuant to the National Materials and Minerals, Policy Research and Development Act of 1980 (PL 96-479), U.S. Government Printing Office, Washington, D.C.
24. Ratiner, L., 1983, "The Future of Sea Bed Mining," *Materials and Society*, Vol. 7, No. 1, pp. 45–48.
25. Schwartz, M.A., 1983, "The National Materials Policy and Its Implementation," *Ceramic Bulletin*, Vol. 62, No. 7, pp. 765–6.
26. Secretary of the Interior, 1972, 1973, 1974, 1975, 1976, 1977, 1978, 1979. "Mining and Mineral Policy." Reports to the Congress under the Mining and Mineral Policy Act of 1970 (P.L. 91-631), U.S. Government Printing Office, Washington, D.C.
27. Strauss, S.D., 1980, "The Future of U.S. Mining," *Materials and Society*, Vol. 4, No. 1, pp. 5–10.
28. Tilton, J.E., 1980, "Issues in the Growing Controversy Over U.S. Mineral Trade," AIME Annual Meeting, Las Vegas, Nev.
29. Vogley, W.A., 1981, "Resource War?," *Materials and Society*, Vol. 6, No. 1, pp. 1–4.

5.15

Environmental Regulation and the Minerals Industry

David A. Gulley*

INTRODUCTION

The Scope of the Chapter

This chapter discusses environmental regulation and closely-related public policies associated with mineral development in the United States. This chapter outlines the nature, extent, and amelioration of secondary effects of mineral development. It sketches the policy foundations of regulation; discusses the implementation of environmental law; describes land use conflicts and land management regimes; and surveys the costs and benefits of environmental regulation. Since regulations are subject to frequent change, the chapter emphasizes basic factual and conceptual material, rather than specific regulations. The goal is to familiarize the reader with the general nature of environmental impacts and mitigation which arise in the minerals industry, and with the issues surrounding the regulation debate.

The chapter is organized as a series of subchapters. The first of these is a discussion of the conceptual foundations of regulatory policy. The second section summarizes the principal federal environmental statutes, and the implementation of such regulation. Following that, the third section summarizes the current state of knowledge regarding the types of environmental impacts associated with mineral development. This dicusssion, organized in a format similar to that used in environmental impact statements, can be thought of as a guide to reading and understanding such statements. Next, the fourth section surveys studies of the costs and benefits of environmental regulation. These empirical studies are evaluated in terms of the framework developed in the first section of the chapter. In the fifth section, the discussion turns to federal, state, and local approaches to land use control and siting, and the economic analysis of such conflicts. Included are discussions of the two principal forms of public lands mineral entry, the 1872 location patent system and the 1920 leasing approach.

CONCEPTUAL FOUNDATIONS OF REGULATORY POLICY

David A. Gulley

The Role of Regulation in a Market Economy

What is Regulation?: Regulation is the intervention of government in the activities of the private market. In a market society, the private sector is presumed to be the most appropriate economic agent. Given this presumption, regulation generally begins with a demonstration of persistent problems under laissez-faire in a given context. Environmental and related regulation occurs because the public believes the free market does not properly value environmental quality. In this section, concepts are advanced to aid the reader in understanding the government's role in this respect.

*Henry Krumb School of Mines, Columbia University, New York

The proper role of government in a market economy is the subject of ongoing discussion. According to a standard economics text (Musgrave and Musgrave, 1973), the public sector is required for several broad reasons. First, government responsibilities reflect political and social ideologies which depart from the premises of market economics. For instance, not everyone agrees with such economic tenets as consumer sovereignty and decentralized choice. Moreover, the market mechanism alone cannot perform all economic activities. For example, the marketplace needs a legal system, maintained by government, to enforce the property rights and contracts of the private sector. Also, the distribution of wealth and the level of economic activity are widely believed to be fit subjects for government involvement. Finally, economic society does not meet the idealized conditions of perfect competition. Thus, many people believe government can play a role in perfecting the marketplace. In this regard, market failure may exist with respect to certain goods, due to problems of externalities or collective consumption. This latter case is the justification of environmental regulation.

Government activities which address these problems can be organized into three broad groupings: allocation, distribution, and stabilization. The first of these refers to the provision of goods by the government and the apportionment of resources among the public and private sectors. The second refers to policies toward the sectors' distribution of wealth and income, and the third to macroeconomic conditions. This chapter is primarily concerned with the first function, allocation. However, most government programs involve all three missions. For example, environmental regulations redistribute the gains and losses associated with production, and they also can affect macroeconomic conditions such as inflation, international trade, and employment. However, most intervention by the government in resource markets, whether of a regulatory nature or as a landlord, primarily involves programs for correcting market failure.

In addition to the regulation of pollution, government intervention in the market for natural resources includes public lands management. The government is not only a policeman, but also a landlord. Depending upon one's interpretation, public land ownership can be viewed as either an anomaly of private enterprise (Nelson, 1982), or as another instance of a response to market failure (Krutilla, et. al., 1983). Although somewhat different in nature, public lands management and environmental issues are closely associated in the public mind, due to the attention focused on land use conflicts, such as mine siting versus wilderness.

What Conditions Justify Regulation?: Two broad categories of regulation exist, economic and social. For example, economic regulation is undertaken to ensure the smooth functioning of the marketplace, whereas social regulation involves broader social agendas. Economic regulation promotes the marketplace's efficiency, whereas social regulation addresses issues not usually thought of as economic, such as personal freedom. Not surprisingly, regulatory approaches encountered in the two cases can be quite different, a subject to be discussed in the next section.

Eleven different justifications have been given at one time or another for federal regulation (Committee on Governmental Affairs, 1978). However, the validity of these justifications is an open question. Most economists believe the first four items listed below have the greatest legitimacy. Taken as a whole, the 11 justifications embrace many circumstances in addition to environmental impacts. By mentioning them at this time, environmental regulation can be viewed in a larger policy context.

(1) A natural monopoly may need regulation to protect the public interest. Natural monopolies arise when scale economies make it inefficient for more than one firm to operate. Over the course of time, some industries may be natural monopolies for one period but not for others, as technology or economics change, or as substitute goods become available. Thus, during the recent past several ''natural monopolies'' in the area of transportation and communication have been deregulated.

(2) Regulations have been recommended to encourage the proper use of common property. For such resources the market may fail to work efficiently, since private property rights do not exist, leading to overexploitation. Petroleum reservoirs, radio frequency bands, and fisheries have been used as examples of common property resources. A variation on this argument is that the marketplace does not properly anticipate exhaustion or scarcity. From this perspective, economic resources are owned in common by the present and future generations. This question has been dealt with at some length elsewhere in

this volume.

(3) When the production and consumption of a good is governed by a price which does not reflect all associated costs, an external cost exists. A third justification for regulation is to require producers and consumers to take external costs into account, either by actually internalizing the cost by means of a governmental levy, or by directing the level of production and consumption to a preferred level. Later in this section, this concept will be related in detail to the environmental costs of mineral development.

(4) Government acts sometimes to correct information inadequacies. As society has become more complex, the old legal adage "let the buyer beware" has come to be seen as inadequate. It is one thing for a customer to be able to judge the health of a farm animal, and another thing for the customer to fully evaluate the safety of medicine, or a computer's conformance to the manufacturer's claims.

(5) Preventing destructive competition has been cited as a legitimate use of government powers. Regulation of transportation industries has often been justified with this doctrine.

(6) Price controls have been imposed to shield consumers from rapid price increases. For example, in the early 1970s wage and price controls were imposed to reduce inflation. Later, energy price regulations were used in hopes of containing price effects and preventing energy producers from profiting from a national crisis.

(7) Predatory and discriminatory pricing practices have been frequent targets of federal and state regulation. Unfair pricing practices may result from a firm's monopoly power; when monopoly is the cause, this type of regulation is subsumed by the first justification mentioned above. In such cases, regulation may be desirable. Where such regulation is used to restrict competition, however, it is harder to justify.

(8) Regulations are sometimes used to protect or promote a key industry. This may be done under various guises, such as unfair international competition (i.e., other countries are subsidizing their industry), strategic or other nonmarket considerations (in which case, this item may be analogous to the external cost justification), or to nurture a new industry. A problem commonly cited in regard to such policies is that industries which prosper under such a shield could continue to require the shield indefinitely.

(9) Guaranteeing services to certain groups has motivated regulation. An example of such regulations are those for transportation and communication industries which stipulate that services must be rendered to small communities at prices below marginal costs. In such cases, the guarantee of service is often better interpreted as something different, i.e. the subsidy of price.

(10) Cross-industry regulation has been imposed on some industries which compete with regulated industries. Cable television and trucking regulations were first created because in both cases a regulated industry was viewed as otherwise threatened. The argument was that it would be unfair to expect a constrained industry to compete with unregulated businesses. This same reason, however, may also be given to deregulate the already-regulated industry. One reason that some critics are leery of regulation is that intervention in one part of the economy can indeed create new distortions that require further intervention. This case illustrates such a regulatory spillover.

(11) Preservation of established privilege has been used to justify the continuation of regulatory and other policies. As a result of regulations, some firms or individuals may receive a windfall. The privilege to operate under such conditions may command a market price, and thus a type of property right is created. Grazing rights for domesticated animals on the public lands provide an example. Such rights are assigned to the party which has traditionally held the permit, and the cost of the permit is less than its true economic value. The permit thus becomes a productive asset. Such assets regularly command a market price. Culhane (1981), for example, has reported that grazing permits are used as loan collateral. Taxicab medallions, broadcast licenses, and stock exchange seats are other examples.

How Does One Regulate?: Having examined the range of American regulatory expression, the reader can appreciate that the subject of this chapter, regulation of environmental and related secondary effects, is but a subset of the general issue. Health, safety, and environmental issues have both economic and broader social ramifications. The degree to which such regulation is primarily social or economic is a matter of interpretation, depending on a person's orientation and on the specific case in question. This affects the form of regulation deemed to be suitable. Fines, licenses, quotas, and deregulation are all much discussed in economic regulation. But regulations for social issues often are intended not only to change behavior, but also to register moral or ethical standards of conduct and to reward or punish behavior accordingly.

For example, the primary purpose of air quality regulations may be to improve air quality, but one set of regulations might punish polluters while another set creates financial rewards to the same parties. The desirability of the controls would depend on not only the efficiency of meeting primary goals, but also the social statement each set is seen as making. The weight given to the social statement varies depending on whether the regulation is seen as economic or social. This in turn affects the nature of the regulatory instrument.

Thus economists may determine that a type of behavior, e.g. pollution abatement, can best be modified by the use of fines, subsidies, and other fiscal devices; but if this allows people to buy their way out of compliance or enables them to reap financial gain out of undesirable behavior, the abatement policy may be viewed as contrary to the social goals involved. In many cases it is clear whether the regulation is social or economic, but in the case of environmental residuals the situation is often more ambiguous and it is correspondingly more intricate to orchestrate the economically efficient and socially just aspects of regulation. (See Schultze, 1977, for an interesting discussion).

Several types of regulations exist. The most fundamental distinction is between command-control and decentralized forms of intervention. In the first case the government directly dictates the desired behavior, whereas in the latter case the government changes conditions within the marketplace but relies upon individualized decision making to achieve overall results. The two commonest forms of command-control are public enterprise and standards enforcement.

Public enterprise involves the direct participation of government in the production activity, either supplanting or acting in concert with private enterprise. For example, the federal government has retained ownership and control over some of the U.S. land area, and is actively involved in essentially entrepreneurial decisions about the desired mix of outputs from these lands. The standards-enforcement approach, probably the most commonly used form of intervention, can be further subdivided into two main forms. Performance standards may be used to require that a certain goal is achieved, such as an emission or ambient air quality level. A related approach uses specification standards, which require a particular means of achievement. An example is specifying a control technology, such as stack gas scrubbers or catalytic converters.

The decentralized approach is an alternative to command-control. It involves changes in the information or incentives given to private market participants. Information disclosure is a common example. Manufacturers or distributors can be required to provide consumers with information on energy efficiency, health effects, or other aspects of product use. Economic incentives and disincentives, such as direct subsidies, fees and penalties, quotas, and tax concessions, are an alternative to traditional regulation. Such approaches may channel compliance into least-cost avenues, but are also indirect and less certain of result. Of the various approaches the economic incentives approach is the least often employed. However, all types of regulatory methods are in current use. This subject is taken up in more detail later in the chapter.

Analyzing the Effects of Regulation

Economic science can make several contributions toward the intelligent consideration of proposed environmental stipulations: (1) by providing a framework for thinking about costs and benefits, (2) by empirically measuring costs and benefits, and (3) by assisting in the evaluation of alternate regulatory forms for attaining environmental quality. These contributions will be summarized below, but the reader should be aware that this is only an introduction to the substantial literature on this subject.

An Overview of the System: Regulation involves a complex system that embraces the interconnectedness of society and nature, the behavior of a given firm, industry, and market, and the relationship of one or more markets to the macroeconomy of that nation and the world. The dimensions involved are accordingly very broad and yet quite particular. If the interrelationships were well understood, the empirical measurements of effects and their costs and benefits would still represent a major challenge, and yet the interrelationships are anything but well understood.

A useful articulation of the relationship is :

(1) *Intervention:* regulation causes changes in the behavior of persons, and with it changes in the use of economic and environmental inputs.

(2) *Impacts:* intervention (1) leads to changes in the causal event, e.g. rate, time, and place of discharge of residuals, or changes in the use of environmental inputs.

(3) *Receptors:* Changes in impacts (2) lead to changes in ambient environmental quality.

(4) *Interface:* Changes affecting receptors (3) lead to changes in the flow of environmental services to people, also to direct effects on people as receptors, and perhaps in the way individuals use the environment.

(5) *Consumption:* Changes in (4) modify patterns of consumption of environmental amenities and goods and services, leading to changes in economic welfare or benefits.

Several of these stages involve economically quantifiable costs and benefits, but this framework also involves other disciplines. Freeman (1979) suggests that the phenomona included above as items (3) and (4) are the most poorly understood, and yet must be an essential prerequisite to benefit valuation. These, of course, rely on expertise other than economics.

It is worth noting that this framework is concerned with the measurement of costs and benefits accruing to people. Impacts on environmental systems are of concern, in this framework, only insofar as these impacts affect the well-being of people. Even more restrictively, in some instances the changes in well-being are inferred from the reaction of consumers, who are not necessarily wise judges of environmental priorities. The use of so anthropomorphic a value system is consonant with economic theory and methodology, as it must be for economics to cast light on the subject. However, this value system is disputed by some environmental scholars. Such a dispute can not be resolved here, but an active appreciation of it is essential for the intelligent consideration of results reported below and later in the chapter.

In employing the framework above, it is useful to systematize the routes taken by environmental changes in affecting people. (The exact nature of environmental impacts associated with mining and ancillary activities are discussed in detail later in the chapter.) The impacts can reach people through living systems (biological mechanisms) and through nonliving systems. The latter includes: economic costs due to materials damages, higher production costs, and the like; and changes in climate, aesthetics, and other environmental conditions. The former includes: human health effects; economic productivity of ecological systems (agriculture, forestry, fisheries); and other ecosystem impacts, such as recreation and research uses of the system, and more general effects associated with ecological diversity and stability. Economic estimates are easier to make for some categories than for others, and so it is not surprising that the frequency of treatment in the environmental economics literature gives an implicit weight to these effects, which is recognized as not necessarily coincident with the true importance of the various categories.

Evaluating Benefits: Any intervention in the market carries with it certain costs and benefits. In this section we consider how costs and benefits might be evaluated in principle, beginning with the latter. Such an evaluation must start with an understanding not only of the intervention, but of the intervention's actual affect upon behavior. Quite often intended effect differs from the actual effect, and an important task lies in the determination of what the regulation, statute, or other policy instrument actually accomplishes. This can be anything but easy, and if this alone can be determined (whether or not benefits and costs are numerically measured), it can constitute a major advance in knowledge. It is only after the effects of intervention on behavior are determined, that it is possible to perform a cost/benefit analysis.

Benefit-cost analysis of government programs has a long pedigree. For example, analysis of potential water resource projects was conducted with some regularity in the 1950s. The technique has been applied to regulations for nearly as long, and with regularity since the administration of President Ford. Moreover, similar techniques have been used for many years in the adjudication of torts, e.g. economic evaluation of such abstruse items as the value of human health and longevity. Nevertheless, the technique involves methodological difficulties, not only the obvious uncertainties of estimating specific numbers, but also conceptual controversies such as the choice of measures of consumer well-being. (For surveys, see Just, Hueth, and Schmitz, 1982; Freeman, 1979; and Sinden and Worrell, 1979). Nor is economic science the only source of uncertainty in these analyses.

Even at its best, such estimation is likely to involve controversies in the natural and biological sciences, too. If neither the numbers nor the concepts are easy to work with, why adopt the approach? For all the difficulties, economics offers more clarity about the tradeoffs involved. Other approaches may speak eloquently to particular values or to the urgency of a particular course of action, but comprehensive and balanced consideration is aided by the type of analysis discussed here.

The most widely applied test of economic welfare, although not the only such test, is the compensation principle, developed by Kaldor (1939) and Hicks (1939). This principle states that one alternative, A, is preferable to another, B, if all individuals could be made better off (or at least would be no worse off) in state A than in B. Stated differently, A is preferable to B if in moving from state A to state B the gainers can compensate the losers fully, with some extra benefit left over. This criterion does not require that the compensation actually be made. When actual compensation is required, the criterion is considerably more difficult to apply, and it goes by a special name, the Pareto criterion. Of course, society is interrelated and winners and losers are often hard to identify. Therefore, while society is not insensitive to the distribution of gains and losses, virtually all social change occurs without full compensation. The Kaldor-Hicks criterion formalizes the requirement that if change is for the better, then gains outweigh losses.

Having stated this principle, can we suggest how it might be applied? A lively controversy has revolved around this question, and some economists feel the verdict is still out. However, most economists believe the current state of the art must and does suffice for the time being, particularly given the rudimentary nature of the other scientific and empirical dimensions involved, and the considerably less rigorous approaches which are the alternative. Ordinarily benefits are measured by the changes in producer and consumer surplus. The former term can be interpreted as profits or quasi-rent, or alternatively as the "consumer's" surplus the producer enjoys as a purchaser of inputs. Consumer surplus also applies to the ultimate consumer, of course. Consumer surplus is not as airtight a measure of the consumer's welfare as was once thought, but is still commonly used in empirical work.

Two alternatives to consumer's surplus are the compensating and equivalent variations (Hicks, 1943), which are measures of the consumer's willingness to pay:

> "The compensating variation (C) is the amount of income which must be taken away from a consumer (possibly negative) after a price and/or income change to restore the consumer's original welfare level. Similarly, equivalent variation (E) is the amount of income that must be given to a consumer (again possibly negative) in lieu of price and income changes to leave the consumer as well off as with the change. (Just, Hueth, and Schmitz, 1982, p.85, emphasis deleted).

These concepts are also useful in examining welfare changes to resource owners, a particularly useful extension in the analysis of mineral policy. The compensating and equivalent variations can be theoretically developed using the indifference curves of consumer preference theory. However practicioners try to measure these concepts using market supply and demand relationships. Measurement techniques are discussed next.

Several classes of benefits might flow from environmental quality improvements. Health and longevity, recreation, and productivity benefits have been explored in the economic literature. Two general approaches to measuring the benefits of environmental improvement are the market and survey approaches. The market approach infers environmental benefits from whatever related markets exist. To use the market approach to valuing such benefits, a closely related market must be found, and a technique must be used to make inferences about environmental quality from the market. Environmental improvements may affect the production of a market good. For example, decreases in acid rain may increase crop harvests. In such cases, then, environmental quality is essentially a factor of production.

In other cases, environmental quality is an economic amenity, a good which produces utility for the individual. In such circumstances, the individual's demand for environmental quality may be estimated from the demand for closely related market items. We would expect that the recreational value of a fishing trip would at least equal the cost of the trip. Thus, direct costs including travel time provide a measure of benefits. The housing market is another possibility. The hedonic price technique, based on multiple regression of a cross-section of property values, can link the market with environmental quality. If two homes are identical except for their view, the market value of the view can be inferred. Of course consumers can not be expected to be knowledgeable about some aspects of environmental quality; scientists, after all, are uncertain about many potential environmentally-occuring health hazards. Therefore market-based valuation techniques can not be expected to provide an adequate measure of all possible benefits. However, for certain classes of benefits the technique has much to recommend it.

The other approach to benefit evaluation is the survey approach. This approach embraces various techniques, such as opinion sampling, questionnaires, bidding games, and voting. One method used in the survey approach is to ask people their willingness to pay for a particular environmental improvement. Another approach is to ask people how much of an environmental good or service they would demand at a given price. For example, how often would someone visit the Grand Canyon if there were to be a fifty dollar entrance fee? A third approach allows people to vote, either actually or hypothetically, for rival programs which differ in the environmental quality provided. The survey approach suffers from the problem of strategic behavior, i.e. people may believe that a less-than-candid response will further their goals better than a truthful response. It is not surprising that researchers using the survey approach often find inconsistent results. On the other hand, opinion research specialists have been laboring for years to perfect their techniques, and of course in the private marketplace, opinion-based market research is widely employed. However, most environmental economists prefer the market-based approach.

Evaluating Costs: In evaluating benefits and costs of compliance, care must be taken with terminology. For example, there are costs of surface restoration, and there are benefits of surface restoration, and alternatively there are the costs of unrestored surface damage. To avoid confusion, benefits and damages are ordinarily measured relative to a specified reference state, perhaps an existing, polluted state, or perhaps a hypothetical pristine state. Benefits are the favorable results of reduced pollution, damages the result of pollution, and benefits are a reduction in damages. While all this seems obvious, a moment's reflection should allow the reader to construct examples where these terms could have perverse or vague meanings. The term cost should be restricted to the sacrifices associated with compliance, and not include the environmental damages foregone as a result of compliance or damages remaining after compliance.

Typically, compliance costs fall into the following categories. Direct costs include expenditures by the regulatory agency, and expenditures by the private sector associated with interactions with the agency and the courts. Direct costs also include the capital and operating costs associated with compliance, such as the installation and operation of stack gas scrubbers. Indirect or unintended costs include: the costs of excess capacity, possibly due to idling older facilities rather than bringing them into compliance; present and future productivity losses, including regulatory impediments to technical change (which some economists regard as crucial, see Schultze, 1977); and effects on competition. Additional macroeconomic costs may be involved, such as the price level and employment. The costs of regulation therefore have an element of public and of private expenditures, and direct and indirect costs. Sometimes compliance costs are relatively easy to identify, since they involve so-called "bolt on" compliance technologies, such as scrubbers or end-of-the-pipe water treatment. At other times compliance costs are harder to estimate, such as when entire production processes have been changed for reasons of both greater productivity and greater compliance (e.g., a move to fluidized bed combustion of coal, or a change in chemical or plastics usage to reduce emissions as well as to improve the product).

In evaluating costs and benefits, some further points might be mentioned. In general, the determination of which costs are relevant to the analysis can be made on the basis of the usual rules of engineering economics (Barish and Kaplan, 1982; Riggs, 1982). Ordinarily, only relevant costs should be considered. These are costs which vary among the alternatives. They include future increases in capital and operating costs associated with compliance. Sunk costs associated with existing assets should be included when, and only when, compliance requires these assets to be diverted from other productive use (e.g., earth-moving equipment used in decommissioning or reclaiming a site, which would otherwise have been transferred to other uses or sold for salvage value).

Since a benefit-cost analysis of regulation obviously addresses nonmarket considerations, some costs not usually included in engineering economics might be considered as candidates for inclusion in the analysis. Care must be taken to avoid double-counting. As was already mentioned, mitigations to technological externalities, such as reduced emissions or other impacts, are ordinarily included in the damages or benefits-from-reduction category. To include the remaining level of impacts as a cost will distort the analysis of alternatives.

Another class of external costs exists, pecuniary externalities. These are secondary effects expressed in the price system, e.g. a change in copper prices due to smelter emissions control. Similar externalities might include macroeconomic effects, such as an increase in copper miners' unemployment rates, income transfers from mining regions to other regions, and so on. The Hicks-Kaldor criterion is, as mentioned earlier, insensitive to the distribution of gains and losses. Thus, these macro costs are legitimately included when and only when the displaced factors (underutilized capital or unemployed workers) represent a definite loss, and not a shift of productive inputs from one sector to another. For example, unemployment problems resulting from air pollution control will be offset to some degree by additional employment in the production of the air pollution control devices. If used correctly, the external cost would be computed from the net results of macroeconomic losses and gains.

In computing the losses, care must be taken in identifying the actual costs being borne. Ordinarily, the cost would be the lesser of avoiding, mitigating, or compensating the damage. For example, it could be the costs of reeducation or relocation, instead of the costs of unemployment. And to be consistent, the analysis should then also include the benefits of any economic stimulus the compliance might create. To continue the example, smelter emissions control might be bad for the copper industry, the smelter and mine workers, and the regional economy. However, it might be equally good for the manufacturers of the control technology, and their workers and their community. In a complex economy, tracing such multiplier effects involves considerable analytical effort. Ideally, the analysis would specify all such effects, but practically this may not be possible. As an alternative, one could ignore macro effects unless a clear reason exists to believe they are asymmetric. In a later subchapter, existing benefit and cost studies are surveyed.

It is clear from the foregoing that the economics perspective minimizes the equity and distributional dimensions, which are very real human and political dilemmas. While economists would argue against adding these to the cost category directly, as an alternative such considerations might be estimated and provided as a separate input for decision making. Justification for this may be provided by the immobility of factors and the lack of effectiveness of programs designed to aid hard-pressed populations. Separate consideration of such issues clarifies analysis.

The Economic Approach to Environmental Regulation

One way of viewing environmental problems is that they result from the failure of the marketplace to assign private property rights to environmental quality. Coase (1960) has shown that under ideal conditions a system of private property rights would lead to a court-enforced level of optimum abatement. That is, other than providing what it traditionally provides (a legal system), no further government action would be required once the rights are assigned. Moreover, environmental damages would ultimately reach the same equilibrium level, independent of whether the rights were initially assigned to the parties creating the damages or the parties suffering the damages. If polluters have the rights, sufferers would bribe them to clean up to a level where marginal abatement costs equal the reduction in marginal damages. Alternatively, if victims have the rights to a pristine environment, polluters would pay them for the rights to pollute, up to the level where it is cheaper to clean up emissions rather than pay for the (increasing) marginal cost of further damages. However, in both cases the level of emissions is the same level. Moreover, this level is the ideal, efficient level. The assignment of rights has an equity dimension, of course, since it determines who pays whom. However, both systems have the same result, an optimum or allocatively efficient level of damages, in this idealized case. The requirements of the idealized case are perfect information as to costs and benefits plus low transaction costs, i.e. the cost of identifying all parties and coming to agreement.

Since these conditions do not hold in real life, the principal value of the Coasian analysis is that it highlights the role of an economic system in directing behavior toward optimum discharge levels. Perhaps the most valuable insight to be gained from environmental economics is that the optimum level is that which minimizes all costs, and this is usually not the level of zero discharge. Given the fact that the law of torts, as outlined above, could not realistically be expected to function as sketched, economists have proposed a variety of other policy expressions which seek to capture as many of the advantages as possible. These regulatory instruments fall into the category of decentralized approaches, discussed earlier.

The simplest practicable approach is to add a surcharge of tax onto the price of a product, reflecting the marginal cost of damages associated with a unit of production. The higher price would be passed forward in whole or in part, so that the consumer would recognize the actual total cost of the item. Higher prices would reduce consumption, reducing the level of damages. Higher prices would also generate public revenues which could be used to reduce or compensate for the damages remaining. Finally, the higher prices would fine-tune economic performance, enhancing overall economic efficiency. Severance taxes, such as the Federal Black Lung Tax on coal production, are sometimes viewed as a means of internalizing such external costs. Page (1977) and Gulley (1982) have taken somewhat different positions on this issue. See Fig. 5.15.1

The product tax approach outlined above is somewhat crude. Since the product rather than the effluent is taxed, the charge does not directly respond to variations in damage costs, which might arise due to different locales, production technologies, or other features. The main virtue of the approach is its simplicity. An alternative is to meter and then tax the emission directly. Such an approach has often been advocated for the sulfur content of fuels, and in the early 1970s got as far, politically, as to receive the endorsement of the administration. However, not a single member of Congress would sponsor such a bill (Gulley, 1982). More recently, though, a fixed tax on some effluents was promulgated under the Comprehensive Environmental Response, Compensation, and Liability Act of 1980 (the "Superfund"), but this tax is set to expire in 1985 (Farrow, 1984).

The economics of this approach is similar to that of the Coasian case (both are summarized

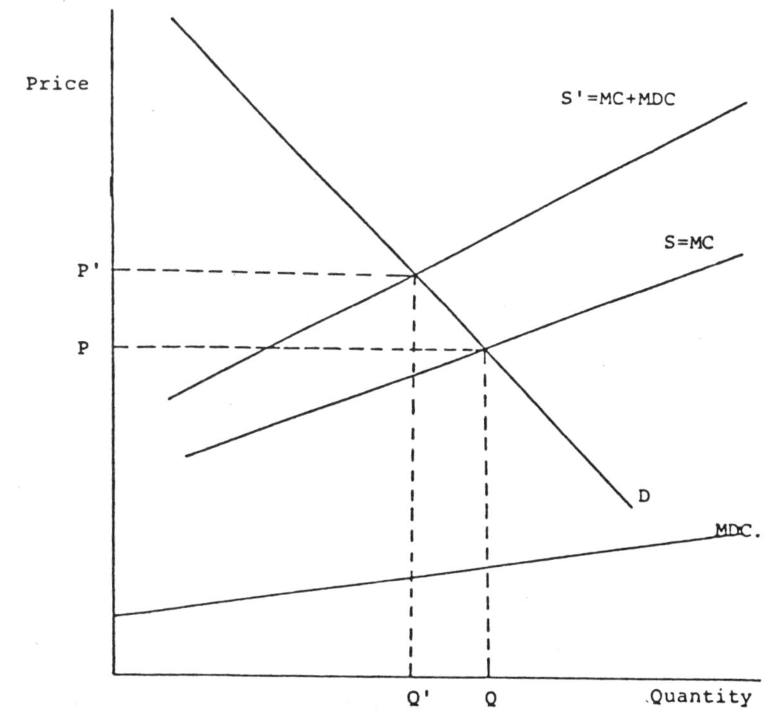

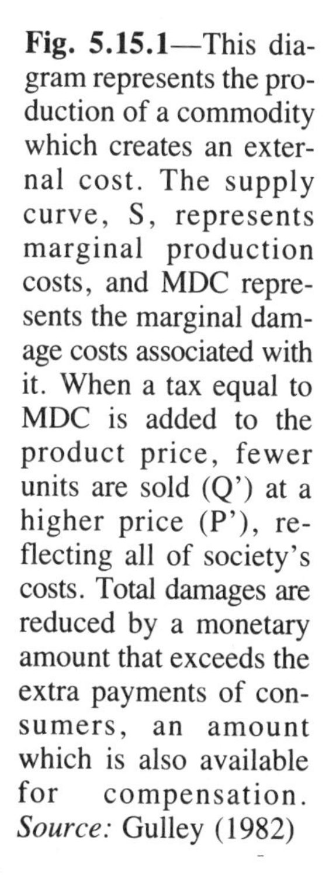

Fig. 5.15.1—This diagram represents the production of a commodity which creates an external cost. The supply curve, S, represents marginal production costs, and MDC represents the marginal damage costs associated with it. When a tax equal to MDC is added to the product price, fewer units are sold (Q') at a higher price (P'), reflecting all of society's costs. Total damages are reduced by a monetary amount that exceeds the extra payments of consumers, an amount which is also available for compensation. *Source:* Gulley (1982)

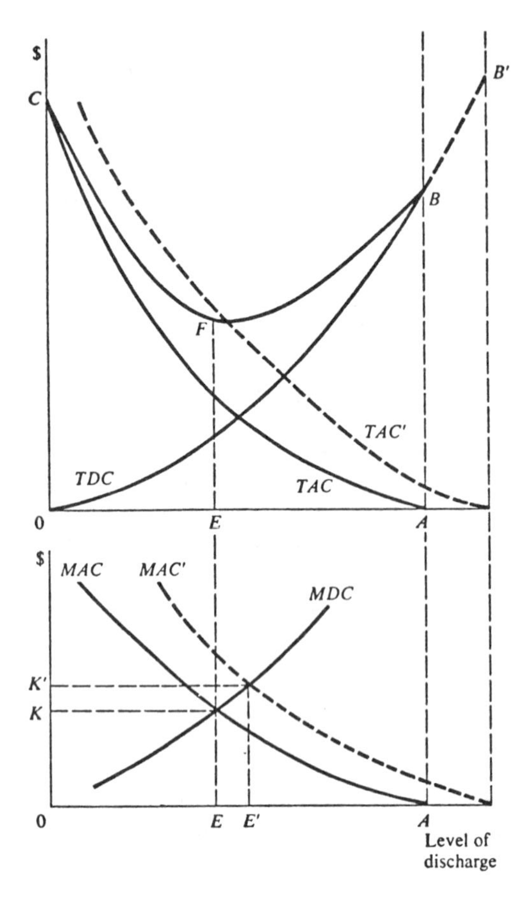

Pollution Control with Abatement Incentive

Fig. 5.15.2—The top diagram shows total damage (TDC) and abatement (TAC) costs. The lower diagram shows marginal costs. Discharges associated with two different production levels are indicated (TAC-MAC and TAC'-MAC'). A charge of K and K', respectively, could be levied as a means of bringing about an optimum discharge reduction, or alternatively, continuously varying charges equal to MDC could be levied. This diagram corresponds also to the Coasian analysis. Beginning at either O or A, depending on the initial assignment of rights, bargaining between agents could theoretically result in the optimum level. Where MDC intersects MAC or MAC', total costs are minimized. *Source:* Musgrave and Musgrave (1973).

in Fig. 5.15.2). The tax is set equal to the marginal damage cost, or if this is too difficult, the tax is set by trial and error to approximate the marginal damages at the likely level of emissions. Firms balance the tax against abatement costs, reducing emissions over that range where damage costs exceed abatement costs, and paying for damages over that range where abatement would not be worth the financial effort. This approach would lead to optimum abatement levels while also generating money to reduce or compensate for remaining damages. The approach has the added merit of creating incentives for further technological innovation designed to reduce pollution.

Both of these tax approaches suffer the drawback that estimates are needed of the costs of damages. As will be seen in a later subchapter, environmental benefit valuation is in a fairly primitive state. Thus economists have sought out strategies which minimize the government's need for information. One such strategy is the pollution quota system. In this system government determines an overall level of ambient environmental quality or a level of emissions, which it finds acceptable. Precisely how this might be done need not be specified. The system is an alternative to present command-control systems, which also must contend with the problem of setting permissable levels. Once the level is set, however, the quota system departs from command-control, in that the market is allowed to seek a solution which meets the goal. The preferred system presumably will be the least costly. Several different mechanisms have been considered. One approach is simply to auction permits for the tolerated level of emissions. Companies would pay up to the cost of abatement, so that those with the lowest abatement costs would reduce their emissions by the greatest extent. This is exactly what is needed to reach a given level of environmental quality at the lowest cost. The permits could be bought and sold, increasing flexibility.

Such a system has not been tried, but a somewhat related system, the "bubble" or "offset" system is under experimentation by the Environmental Protection Agency. Under this system, companies locating new facilities in a given location have the option of eliminating emissions from the new facility or of reducing emissions from older units by a corresponding amount. The company will choose the least-cost approach. Depending upon the specifics of the program and situation, the company may have the option of reducing emissions from another facility or another part of the same facility, or possibly of paying another company to reduce its emissions. The latter possibility raises antitrust issues of some complexity. This technique minimizes abatement costs but does not raise public revenues. Of course, all of these techniques do require a system of government monitoring and enforcement.

Economists have developed graphical and mathematical analyses of these systems, more systematic and detailed than the treatment provided here, The interested reader is directed to Musgrave and Musgrave (1973), Kneese and Schultze (1975), and Dorfman and Dorfman (1977). Fuller discussions of the pros and cons of the economic approach to environmental regulation, in theory and in practice, are found in Gulley (1982), Breyer (1979), and Russel (1979). An interesting paper by Blockman and Baumol (1980) suggests that fiscal incentives are not as underutilized as economists commonly suppose. For example, fines may be levied or performance bonds forfeited when firms are not in compliance with regulations. These and other provisions are real world approximations of the fiscal instruments conceived by economists.

Understanding the Regulatory Agency

The Federal System: Government in the United States is organized at several levels, in a federal system: the national or Federal government, 50 state governments, and many thousands of units of local government, the latter including governments of general purpose (townships, counties, cities) and focused purpose (special districts organized for tax collection or the provision of services, e.g. school or water districts). Roughly half of government expenditures occur at the Federal level, and local government spends roughly two-thirds of the remainder. Most government programs involve two or more levels of government. For example, Federal lands management must by law be coordinated with state and local land use planning, and surface mine control involves both Federal and state agencies.

Two distinct theories dominate discussions of public administration and decision making. The rationalist school of thought suggests that public agencies, like any organization, exist to fulfill certain purposes, and in seeking to do so they follow certain goals rationally. A central authority allocates resources and manages functions in pursuit of those goals. Individual workers' personal interests are subordinated to or harmonized with the organizational goals, The open systems approach suggests that administrative behavior is the result of various forces within and without the organization. In this view, an equilibrium state emerges from the tug and pull of individual interests and perceptions within the organization and from external parties at interest. A popular variant on this theory is clientelism, which suggests that just as private firms need customers, the public agency must identify, cultivate, and service one or more constituencies, who in turn will aid the agency in its dealing with executive and legislative authorities. Most observers favor one or another variation of open systems theory, although most agree that the rationalist model does express the outer form and many of the actual needs and conditions of some agencies. For more on this topic, see Simon (1965), Culhane (1981), and Wamsley and Zald (1973).

Organizations can be thought of as a vector of: task, structure, technology, and people. A simple but useful way of viewing a public agency is offered by simple elaborations on this abstraction. At it simplest and most general, the variables of task and structure can be thought of as coordinate axes, with the variables themselves suggesting a continuum of: agency interior vs. exterior (structure), and policy making vs. operations (task). This simple framework (see Fig. 5.15.3) suggests that in order to understand agency behavior, one should look at the following dimensions:

(1) Policy making tasks and the external environment—executive bodies (budget, personnel, other offices); legislative bodies (committees, auditors); independent review bodies (judiciary, courts); competitor agencies (rivals for jurisdiction and resources); interest groups and political parties; media; and citizenry.

(2) Policy making tasks and the internal environment—agency distribution of power; charter or constitution; dominant and rival coalitions or factions; systems for succession, recruitment, advancement; institutional norms, beliefs, values; patterns of communication.

(3) Operations and the external environment—demand characteristics; input characteristics; industry structure; and macroeconomic interactions.

(4) Operations and the internal environment—allocation rules; information systems; incentive system; authority structure.

For an elaboration, see Wamsley and Zald

	Environment structure and process	Internal structure and process
POLITICAL	*Superordinate and authoritative executive bodies and offices (and organized extensions—budget, personnel offices) *Superordinate and authoritative legislative bodies and committees (and organized extensions—ombudsman, inspectorates) *Independent review bodies—courts, judiciary *Competitors for jurisdiction and functions *Interest groups and political parties *Media-communications enterpreneurs *Interested and potentially interested citizenry	*Institutionalized distribution of authority and power Dominant coalition or faction Opposition factions, etc. *Succession system for executive personnel *Recruitment and socialization system for executive cadre *Constitution Ethos, myths, norms, and values reflecting institutional purpose *Patterns for aggregation and pressing demands for change by lower personnel
ECONOMIC	*Input characteristics: labor, material, technology, facilities, supply and cost factors *Output characteristics: demand characteristics and channels for registering demand *Industry structure (in and out of government) *Macro-economic effects on supply-demand characteristics	*Allocation rules Accounting and information systems *Task and technology related unit differentiation *Incentive system Pay, promotion, tenure, and fringes *Authority structure for task accomplishment *Buffering technological or task core

Fig. 5.15.3—Major Components of Political Economy in Public Organizations. *Source:* Wamsley and Zald (1973)

(1973). This framework can function as a checklist; to understand administrative behavior, one must understand dominant elements within these four quadrants, which will normally include some or all of the features indicated.

A related but less abstract model was developed by Wallace Sayre (Held, 1979). This model, displayed in Fig. 5.15.4, is intended to describe Federal administrative agencies, particularly bureau chief decisionmaking for domestic problems. It has been termed a model of the power structure at the national level. Sayre believed that the majority of domestic decisions are made at the bureau level, meaning at the level just below that of political appointments. This should be recognized as a generalization; in actual practice either political or career professional appointments can be make at several levels, and the degree of authority resting with the bureau chief will vary with agency and the personalities involved. The bureau is a major unit with all the powers necessary to further its programs, such as rule-making, budgeting, personnel management, and defining legislative requirements. The bureau chief is believed to have a central administrative role, dealing as a general manager to subordinates and as an expert to higher elected or appointed officials, who are generalists. Of course, a bureau may be named something else, a department, agency, service, or whatever; Sayre uses the term generically.

Nine sets of actors interact in the Sayre model, the bureau chief and eight other sets. The interplay of these defines the power structure. These are:

(1) The presidency—including the executive office, White House staff, Departmental secretaries, deputy, under, and assistant secretaries and administrators.

(2) Other bureaus—which compete with and/or cooperate with the bureau.

(3) The courts system—including departmental legal staffs, and the offices of legal counsel and solicitor general within the justice department.

(4) The media—both general and specialized.

(5) The career staff—particularly mid-level professionals whose experience and continuity allow the agency to function during changes in policy and appointments.

(6) Interest groups—both friendly and hos-

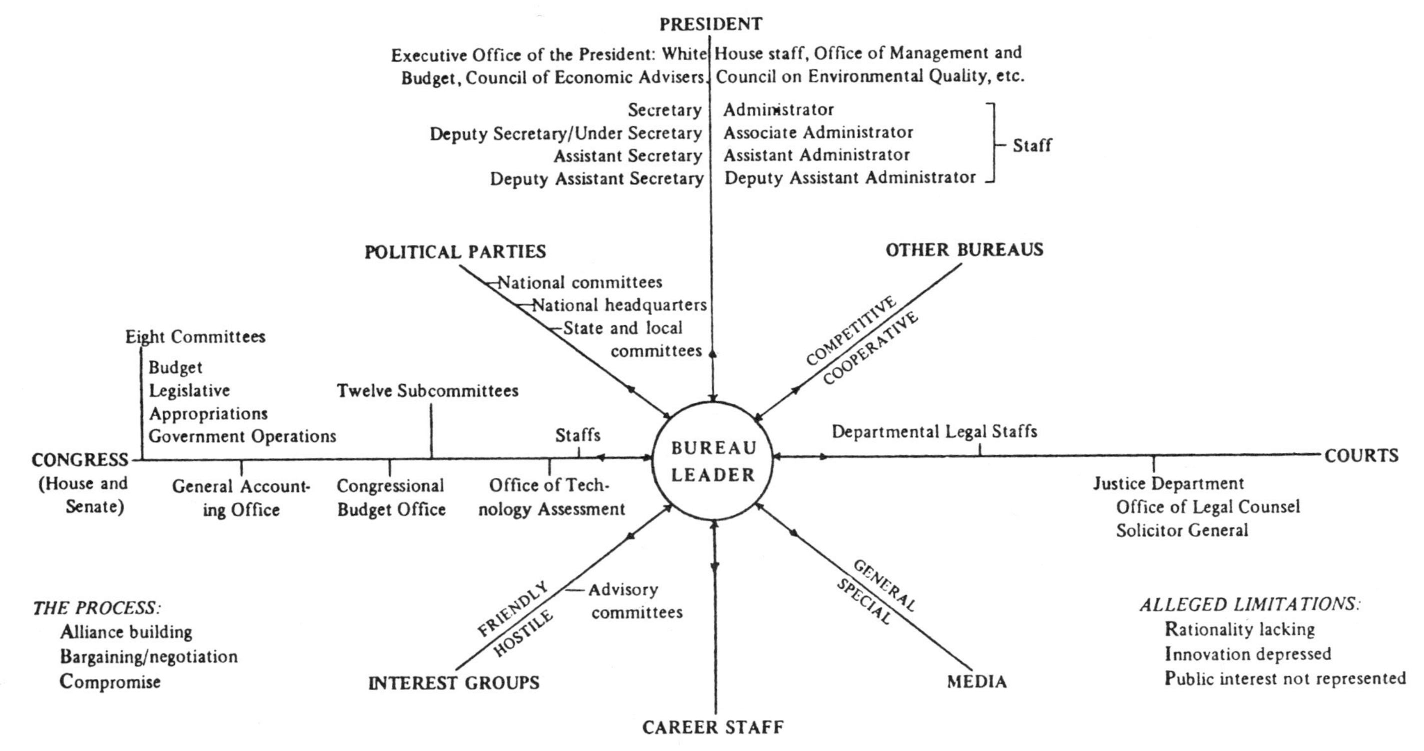

a. The presidential and congressional lines of influence have been modified to reflect organizational changes since the Sayre model was developed.

Fig. 5.15.4—*Source:* Held (1979)

tile, sometimes acting informally, sometimes organized into advisory committees or other groups.

(7) The Congress—including committees and subcommittees and congressional staffs and support offices.

(8) The political parties—at national, state, and local levels of organization.

The Congressional line of influence is particularly worthy of note. Generally speaking, eight committees are involved with any agency, four in each house of Congress. The four are: budget, appropriations, oversight, and legislative. Each committee appoints one or more subcommittees which are more narrowly focused on subjects involving the bureau. Although the subcommittee is composed of some committee members, it has its own personality, interests, and agendas.

References

1. Barish, N.N. and Seymour Kaplan, 1982, *Economic Analysis for Engineering and Managerial Decision Making*, 2d ed., McGraw-Hill, New York.
2. Blackman, S.A.B. and William J. Baumol, 1980 "Modified Fiscal Incentives in Environmental Policy," *Land Economics* v.56, pp. 417-431.
3. Coase, Ronald H., 1960, "The Problem of Social Cost," *Journal of Law and Economics*, v.3, p.1.
4. Committee on Government Regulation, U.S. Congress, 1978, *A Framework for Regulation*, U.S. Government Printing Office, Washington.
5. Culhane, Paul J., 1981, *Public Lands Politics: Interest Group Influence on the Forest Service and the Bureau of Land Management*, The John Hopkins Press for Resources for the Future, Baltimore.
6. Dorfman, Robert, and Nancy S. Dorfman, eds., 1977, *Economics of the Environment: Selected Readings, 2d. ed.*, W.W. Norton, New York.
7. Farrow, Scott, 1984, "An Empirical Method and Case Study to Test the Economic Efficiency of Extraction from a Stock Resource," paper presented at Eastern Economic Association meeting, April.
8. Freeman, A. Myrick III, 1979, *The Benefits of Environmental Improvement: Theory and Practice*, The John Hopkins Press for Resources for the Future, Baltimore.
9. Gulley, David A., 1982, "Severance Taxes and Market Failure," *Natural Resources Journal*, v.22, pp. 597-617.
10. Held, Walter G., 1979, *Decisionmaking in the Federal Government: The Walter S. Sayre Model*, The Brookings Institution, Washington.
11. Hicks, J. R., 1939, "The Foundations of Welfare Economics," *Economic Journal*, V.49, n.196, December, pp. 696-712.
12. Hicks, J. R., 1943, "The Four Consumers' Surpluses," *Review of Economic Studies*, v.6, n.1, pp. 31-41.
13. Hufschmidt, Maynard M., et. al., 1983, *Environment, Natural Systems, and Development: An Economic Valuation Guide*, The John Hopkins University Press, Baltimore.
14. Just, R. E., D.L. Hueth, and Andrew Schmitz, 1982, *Applied Welfare Economics*, Prentice Hall, Englewood Cliffs, NJ.
15. Kaldor, Nicholas, 1939, "Welfare Propositions of Economics and Interpersonal Comparisons of Utility," *The Economic Journal*, v.49, n.195, pp. 549-52.
16. Kneese, Allen V. and Charles L. Schultze, 1975, *Pollution, Prices, and Public Policy*, The John Hopkins Press for Resources for the Future and the Brookings Institution, Baltimore.
17. Krutilla, John V., et al., 1983, "Public versus Private Ownership: The Federal Lands Case," *Journal of Policy Analysis and Management*, v.2, n.4, pp. 548-558.
18. Musgrave, Richard A. and Peggy B. Musgrave, 1973, *Public Finance in Theory and Practice*, McGraw-Hill, New York.
19. Nelson, Robert C., 1982, "The Public Lands," *in* Paul R. Portney, ed., *Current Issues in Natural Resource Policy*, The John Hopkins Press for Resources for the Future, Baltimore.
20. Page, Talbot, 1977, *Conservation and Economic Efficiency: An Approach to Materials Policy*, The John Hopkins Press for Resources for the Future, Baltimore.
21. Riggs, J. L., 1982, *Engineering Economics, 2d. ed.*, McGraw-Hill, New York.
22. Schultze, Charles L., 1977, *The Public Use of Private Interest*, The Brookings Institution, Washington.
23. Simon, Herbert A., 1965, *Administrative Behavior*, The Free Press, New York.
24. Sinden, J. A. and A. C. Worrell, 1979, *Unpriced Values: Decisions without Market Prices*, John Wiley and Sons, New York.
25. Wamsley, Gary L., and Mayer N. Zald, 1973, *The Political Economy of Public Organizations: A Critique and Approach to the Study of Public Administration*, The University of Indiana Press, Bloomington, IN.

5.15A

The Implementation of Environmental Law

Lawrence J. MacDonnell*

This section begins with an introduction to several key environmental laws. This is followed by a discussion of some of the issues raised by such environmental regulation. Finally the subject of regulatory reform is addressed.

Overview of Selected Major Environmental Laws

The number, type and complexity of environmental laws and regulations affecting mineral resources development have grown at an astonishing rate since the 1960s. The most significant laws and regulations are those addressing the use of the public lands; those concerned with air quality, water quality, and surface reclamation; and those defining government responsibility for its actions having important environmental consequences. The laws and regulations governing activities on the public lands will be taken up in the section 5.15C, "Minerals and Land Management." In this section we will concentrate on five primary laws: the Clean Air Act, the Clean Water Act, the Surface Mining Control and Reclamation Act, the Endangered Species Act, and the National Environmental Policy Act.

The Clean Air Act: The legislative and regulatory framework for controlling air pollution has developed in a series of steps. Most important for present purposes are the Clean Air Act of 1970 (P.L. 91-604) and the 1977 Amendments (P.L. 95-95). The rather complex air quality control scheme includes four key elements—national ambient air quality standards, state implementation plans, the prevention of significant deterioration program, and the nonattainment program.

National Air Quality Standards. The 1970 Clean Air Act is a landmark law because it contains the first explicit recognition that pollution is a nationwide problem requiring a national control program. The federal government through the Environmental Protection Agency (EPA) was to establish national air quality standards for at least the six pollutants named in the Act (sulfur dioxide, carbon monoxide, total suspended particulates, photochemical oxidants, hydrocarbons, and nitrogen dioxide). These standards were to be established so as to protect public health, "allowing an adequate margin of safety." The cost of achieving these standards is not to be considered.

The 1970 Act unrealistically required that these standards be achieved in all areas of the country by 1975. As the program has developed, the air quality control regions are now to be designated as "attainment" or "nonattainment" with respect to each of the national standards.

State Implementation Plans. Under the Act the states are given the primary role in achieving the ambient standards. States are to develop and implement control plans governing emissions from all existing and new sources. These plans can include emission limitations for existing stationary sources, transportation control plans to reduce emissions from mobile sources, and preconstruction review of new stationary sources to assure compliance with new source performance standards.

The EPA must approve all state implementation plans. Sanctions are given to the EPA to be applied if states fail to submit acceptable

*Director, Natural Resources Law Center, University of Colorado School of Law

plans. In spite of these sanctions, development of acceptable state plans has been slow.

Most of the air pollution attributable to stationary sources (as opposed to mobile sources such as automobiles) comes from sources which have been in existence since 1970 or before. For such existing sources the 1977 Amendments require that state plans must incorporate all "reasonably available control measures as expeditiously as possible." Although existing stationary sources are a major cause of air pollution, the costs of clean-up for these sources are substantial and tend to impose a major economic burden. Generally the states have followed EPA control technique guideline documents as constituting acceptable levels of control. Enforcement has been uneven with most attention focused on the major sources.

State implementation plans also are required to include a permit program for major new or modified stationary sources. For facilities being constructed in areas not meeting one or more of the national standards, the requirements of the nonattainment program are in effect. This program is discussed next.

The Nonattainment Program. To allow economic development to occur in areas where the national air quality standards are not being met, the nonattainment program utilizes two key elements—the "offset policy" and the "lowest achievable emission rate" performance standard. The offset policy makes growth possible by allowing new sources of emissions in such areas but requiring that the quantity of new emissions introduced by a new or modified source be offset by a greater reduction of the same type of emissions from existing sources. This offset means that growth can actually be the impetus to improve air quality by eliminating emissions from older sources.

To assure that the technical performance from the new source in a nonattainment area will be as good as possible, the specified standard is the lowest achievable emission rate. This is defined as either the most stringent emission limitation included in any state implementation plan for the relevant source, or as the most stringent emission limitation achieved in practice.

The Prevention of Significant Deterioration Program. The 1977 Amendments formally adopted a program intended to maintain air quality in areas currently meeting or exceeding national standards. This program requires EPA to specifically designate all such areas. Three classes of such areas are recognized: Class I areas such as national parks where virtually no degradation is allowed, and Class II and III areas in which larger increments of new emissions are allowed.

Preconstruction review of major new and modified stationary sources is required. This review first determines if the source is subject to the PSD requirements. If it is so subject, then the review turns to determining the best available control technology (BACT) requirements to be applied. Best available control technology is defined as an emissions limitation based on the maximum degree of reduction which the permitting authority, on a case-by-case basis, taking into account energy, environmental and other costs, determines to be achievable. Out of this analysis come the expected emission levels which are included in the air quality analysis. This air quality analysis must demonstrate that the source can operate within the allowable PSD increment and that national air quality standards will not be violated.

One significant innovation within the PSD program has been the development of the "bubble" policy under which major modifications in an industrial facility may avoid PSD review. The plant is treated as if covered with a single bubble and only the net change in emissions is considered in determining whether the modification is subject to review. If emissions can be sufficiently reduced at other points in the facility, PSD review may be avoided.

Another significant provision added in the 1977 Amendments relates to protection of visibility in Class I areas. Since these are the areas where scenic views are especially valuable, the Clean Air Act requires the identification of the sources of visibility impairment in these areas and the implementation of measures to remedy such impairment and to prevent future impairment. Existing sources contributing to visibility impairment in designated Class I areas are required to incorporate the best available technologies.

The Clean Water Act: Like the Clean Air Act, the Clean Water Act has developed in a series of steps. Prior to passage of the Federal Water Pollution Control Act Amendments of 1972 (P.L. 92-500), water pollution control was largely dependent on state and local interest. General water quality standards were applied to surface waters according to their desired use (drinking water, body-contact recreation, fishing, agriculture, etc.). Control of individual sources was minimal.

The current scheme of water pollution control was established by the 1972 Amendments and the Clean Water Act of 1977 (P.L. 95-217). The 1972 legislation established a national goal of eliminating all pollutant discharges by 1985. The previous system of water quality standards was retained in a strengthened form. The most important change was to prohibit the discharge of pollutants into any public waters except as authorized by a permit issued under the National Pollutant Discharge Elimination System (NPDES). NPDES permits are issued either by the EPA or by a state with an approved NPDES program.

Effluent discharges from both new and existing sources are subject to control under the Clean Water Act. Two different standards of control may apply to existing sources—one for "conventional" pollutants (biochemical oxygen demand, suspended solids, fecal coliform, and pH) and one for toxic and nonconventional pollutants. Sources discharging conventional pollutants are required to use the "best conventional technology" by July 1, 1984. Sources discharging the other category of pollutants must use the "best available technology economically achievable" by July 1, 1984. New sources are subject to a stricter performance standard requiring the greatest degree of effluent reduction available through the use of the best available demonstrated control technology. The NPDES permit translates these performance standards into specific discharge limitations for each individual source. In addition, all sources are required to use "best management practices" to prevent pollution from nonpoint sources.

A separate provision in the Clean Water Act requires that a permit be obtained from the Army Corps of Engineers (or the state if its program has been approved) before any dredged or fill material may be discharged into any natural water. Under the Rivers and Harbors Act of 1899, a permit from the Army Corps of Engineers is required if a dam or dike is to be constructed that will span a navigable waterway of if a structure is to be built which will obstruct a navigable waterway. Finally, under the Safe Drinking Water Act (P.L. 93-523), a permit is required for the underground injection of fluid such as waste materials. Monitoring and mitigation measures are required to prevent the endangerment of the groundwater system.

The Surface Mining Control and Reclamation Act: The Surface Mining Control and Reclamation Act of 1977 (SMCRA) (P.L. 95-87) establishes uniform minimum federal standards for regulating surface coal mining and reclamation activities throughout the country on both public and private lands. No person may conduct a surface coal mining operation without first obtaining a permit from either the Federal Office of Surface Mining in the Department of the Interior or the state operating an approved program.

The permit applicant must be able to demonstrate ability to comply with a number of standards including:

—maximum utilization and conservation of the coal resource being recovered;
—restoration of disturbed land to support the same or better conditions;
—restoration of the approximate original surface contours;
—stabilization and protection of all surface areas;
—protection of prime farmland through specific reclamation techniques;
—minimization of disturbances to the existing hydrologic balance; and
—limitations on mining of steep slopes.

A performance bond sufficient to cover the full cost of reclamation is required.

Provision also is made for designating certain lands as unsuitable for surface mining. The Secretary of the Interior is given this responsibility with respect to the federal lands while states are responsible for designating nonfederal lands. Criteria for determining unsuitability are listed in the Act and include; incompatibility with state or local land use plans; the potential for significant damage to important historic, cultural, scientific, esthetics, or natural systems; the potential for significant loss of water supply or agricultural productivity from the land; and the potential to substantially endanger life and property due to natural hazards of the land, such as frequent flooding or geologic instability.

The Endangered Species Act: The Endangered Species Act of 1973 (P.L. 93-205) requires that federal agencies not undertake actions which may jeopardize the continued existence of an endangered or threatened species or harm the critical habitat of such species. Virtually every form of animal and plant life is potentially protected under this statute. The Fish and Wildlife Service in the Interior Department maintains a complete list of all designated endangered and threatened species. Amendments to the Act in 1978 created a procedure whereby an exemption from the requirements of the Act may be granted under specified circumstances.

The National Environmental Policy Act: The National Environmental Policy Act of 1969 (P.L. 91-190) is perhaps the best known of the many environmental laws passed in recent years. It is a deceptively short and simple piece of legislation. On its face it appears to be little more than a recognition, as a matter of policy, of the importance of the natural environment and the need for federal actions to take account of environmental consequences.

The provision which has made NEPA a major substantive environmental law contains the requirement that all federal agencies prepare an environmental impact statement (EIS) for any ". . . major Federal action significantly affecting the quality of the human environment." The intention of this provision is to establish a procedural requirement (a written statement) demonstrating that the federal agency has included environmental considerations in its decision-making process.

Legislation often includes such procedural requirements. Rarely do these requirements have the effect that the EIS requirements in NEPA have had. In this case the significant effect of NEPA on agency decision making resulted from a combination of the widespread environmental concern in the period following the passage of NEPA and the expansive interpretation given to this provision by the courts. The EIS requirement has been determined to apply to virtually all federal actions. The courts have actively reviewed EISs and have not hesitated to judge them as insufficient. At this point the NEPA requirements are well understood. New regulations for implementing NEPA, issued in 1979, have streamlined and simplified the EIS process. As a result, the NEPA is now well institutionalized.

Some Issues in Environmental Regulation

As the preceding survey has indicated, government intervention into private activities to achieve environmental quality objectives has largely taken the form of direct regulation in which specific, mandated requirements are established. Under this approach, the burden generally rests with the government to define these requirements, to cause their implementation, and then to assure compliance. There are a number of important issues raised under the regulatory approach. Here we discuss three areas —formulation of standards, enforcement, and economic effects.

Standards: Regulatory control of private activities has been established primarily through the use of standards. Standards have been utilized to specify the desired level of environmental quality. They have also been utilized to control the amounts of additional pollution allowed into the physical environment as well as to establish specific practices which must be employed to control environmental impacts. Without question, standards are the central feature of environmental regulation.

It is the responsibility of Congress first, and then the implementing department or agency, to establish standards in a manner which achieves environmental policy objectives. However, many environmental problems are scientifically complex and not well understood. We wish to control air pollution in order to protect public health. Thus, the Clean Air Act requires that ambient air quality standards be established at a level that protects public health "allowing an adequate margin of safety." Although such standards have been established by the EPA for seven specific pollutants, there is considerable disagreement regarding how meaningful these standards are (see, for example, the discussion in Landsberg, et al., *Energy: The Next Twenty Years* [1979], pp. 343 et seq.).

In addition to these uncertainties in establishing quality standards, there are a number of significant issues involved in developing a control strategy to achieve these quality standards. Kneese and Schultze (1975) have pointed out that the design of effective standards must account for three basic economic effects.

First, pollution reduction encounters the situation of increasing marginal costs. That is , as the performance standard moves closer to requiring total elimination of environmental damage, the costs of achieving that performance increase at a disproportionately rapid rate. Second, the costs of pollution reduction vary considerably from industry to industry and from firm to firm. To be economically efficient, standards should recognize these differences and seek to achieve the largest reductions of environmental damage from the lowest cost sources.

Third, optimal environmental controls may vary considerably from situation to situation. Allowances for these variations will permit more effective and economically efficient solutions.

The issue of increasing marginal costs of pollution control bears most directly on the legislative formulation for the general standards. The degree to which private economic costs are considered initially is a basic policy decision. In the Clean Air Act, for example, the primary ambient air quality standards are required to be determined strictly with reference to human health. The economic costs indicated by these standards are not to be considered. However, compliance costs are made an explicit part of the determination of performance standards for new stationary sources. Thus, economic considerations are to be included in developing specific control strategies.

Regulatory standard setting encounters considerable difficulty in accounting for variations in efficient approaches to pollution control. Appropriately, new sources are treated differently from existing sources. Performances standards for new operations generally are developed on the basis of a uniform, nationwide standard for distinguishable classes of operations. Such uniform standards obviously are easier to administer than determining standards on a case-by-case basis. There is an apparent equity in such uniform treatment, and it prevents competion for business among the states on the basis of environmental requirements.

However, uniform standards neglect entirely the fact that pollution problems often are site specific and that economically efficient solutions tend to vary considerably. Effluent going into the ocean is likely to be less environmentally degrading than effluent going into a river near large population centers. Moreover, the costs of pollution control vary considerably. Uniform requirements fail to seek out environmental improvements from the most cost-effective sources.

Control standards for new sources generally are based on the performance of specific technological approaches. Companies seeking to comply with these standards are likely to use this same approach rather then experiment with other approaches which may be more effective but are unproved. Moreover, once these standards are established it is likely that there will be considerable resistance to changing them.

Enforcement: Regulatory standards generally are implemented through a permit system. Individuals seeking to begin a regulated operation must demonstrate to the permitting agency that their operation will be able to comply with the established standards. This mechanism provides the permitting agency with a substantial degree of control over new sources. However, the permitting process often is time-consuming and costly. In some cases, it may be difficult to demonstrate compliance (see, for example, the discussion regarding the PSD permitting process by the National Commission on Air Quality [1981], pp. 3-5-34—3-5-39).

A regulatory system provides little incentive for voluntary compliance. The requirements often are detailed and complex so that individual firms may choose to await specific direction from the regulatory agency rather than pursue compliance on their own. Although environmental legislation normally provides for significant sanctions which may be applied against those operating in violation of the law, these sanctions may be less influential than they appear. According to Breyer (1982, p. 268):

> . . . the firm is likely to argue (1) that it was in compliance, (2) that the individual permit requirements were unreasonable, (3) that the general regulations and standards are unreasonable, and (4) that it needs more time. Given the informational difficulties facing the agency, the firm may well win. Even if it loses, the court will wish to work out a reasonable compliance schedule. Only repeat violators are likely to be penalized severely. A large violation may well receive no greater penalty than a small one, and firms will have little incentive to comply with particularized standards. They lose little —and may even gain —by waiting for court imposed compliance schedules and by using this threat as a weapon to obtain more relaxed requirements from the agency.

Moreover, effective enforcement requires periodic on-site inspection. The sheer number and variety of sources which must be inspected to insure compliance is a major problem. As a practical matter, such inspections tend to be concentrated on the major sources of pollution. When the complexities of the technologies involved and the technical requirements to determine compliance are considered, it is apparent that enforcement can be expensive and difficult.

Still another issue in enforcement concerns

the complexity and conflicting requirements in some regulatory schemes. For example, under the Clean Air Act, federal, state and local agencies are all involved in enforcement. The National Commission on Air Quality (1981, p. 3.8-2) noted that

> Stationary source requirements under the Act may be imposed under one or more of the following authorities: State implementation plan regulations, which generally apply to classes of sources rather than to individual sources; permits issued by a state or local control agency; federal regulations or permits; and consent orders or decrees between federal, state, local, or private parties and a source.
>
> Some sources have found that inconsistent or conflicting requirements imposed under the various authorities has led to confusion over what constitutes compliance.

Effective enforcement is made more difficult by these institutional and technical complexities.

Economic Effects: The consequences of a regulatory approach in terms of economic efficiency have already been mentioned. The inherent inflexibility in a legalistic regulatory scheme reduces opportunities for seeking least-cost solutions to environmental control. The costs of complying with environmental regulations are large (see section 5.15.D) and they are increasing. The macroeconomic consequences of these costs in terms of effects on capital investment decisions and inflation are by no means inconsequential. It is not a matter, however, of whether the nation is willing to pay for improving the quality of the physical environment. That fact is well established. At issue here is whether the current regulatory approach is achieving the desired environmental improvements in an economically efficient manner.

Another economic issue alluded to before is whether by regulating specific performance standards the effect is to freeze control technologies at their current levels. As Noll (1974) has pointed out, formalistic regulatory procedures following the rules of administrative law do not encourage flexibility and change. Some heavily regulated industries such as railroads have not produced the level of technological and managerial innovation apparent in other non-regulated industries. Yet it is clear that the achievement of environmental objectives will demand significant innovative developments.

On a more microeconomic level, the case has been made that uniform regulations tend to affect smaller businesses disproportionately (see for example Clarkson, Kadlec and Laffer [1979]). Simple compliance with paperwork may be more of a burden to smaller firms with fewer people to handle such tasks. Investments in pollution control equipment tend to be large and not necessarily in proportion to output. Thus, per unit costs of environmental compliance for a large firm may be substantially less than for a small firm in the same industry. The competitive effects are obvious. Some also have argued that regulatory requirements and related costs may create barriers to entry, thus removing potential competition in some industries.

Regulatory Reform and Alternatives to Regulation

The deluge of new regulations during the 1970s intended to achieve a broad spectrum of social objectives occurred with little regard for other implications —notably the costs of such regulations. Inevitably, this activity produced a strong reaction. The regulatory reform movement has produced considerable discussion, numerous proposals, and even a few changes. Not surprisingly, while the discussion and the proposed alternatives have covered a wide range of possibilities, actual changes have tended to be modest variations on the established theme. To this point at least, command-and-control regulation appears to be well established.

Directions for Reform: Proposals for regulatory reform can be grouped into three categories —those concerned with the regulatory agencies themselves and the way they currently operate, those concerned with improvements with the regulations themselves, and those proposing alternatives to traditional regulation.

Agency Reform. Some who express dissatisfaction with the present regulatory scheme believe that the problem resides with the agencies themselves. One concern commonly expressed is the quality and experience of regulatory agency personnel. The complexity of both devising and implementing regulations has been discussed previously, Sophisticated understanding of many matters including business and industrial activities is required. Many have questioned whether there are sufficient incentives available to attract and keep the kind of highly qualified individuals required to effectively administer existing programs. Moreover, the tendency for members of a regulatory agency to become identified with or "captured" by the industries which they regulate has been a subject of concern.

A more general concern is with the need to reduce and make more accountable the power of the regulatory agencies. Congress has found it necessary to delegate broad grants of authority to the implementing agencies in order to achieve the ambitious objectives of much of its legislation. Those unhappy with the way this expansive authority has been exercised often suggest either that Congress should be far more specific in the way it delegates authority or that it should more actively review and supervise the activities of the regulatory agencies. An alternative suggestion is to provide the executive branch with greater authority to review the regulatory activities of the various agencies.

Given the broad objectives contained in most environmental legislation, it is difficult to see how Congress could effectively specify the regulatory program to be used. If this task is difficult for an agency created for this very purpose, it is impossible for Congress which must and should be concerned with a far broader range of issues. Nor does a legislative veto process for individual new regulations seem to be a very effective approach. Active Congressional oversight, on the other hand, appears essential. Required periodic review of major authorizing legislation also seems appropriate as an opportunity to reconsider the regulatory program as it has developed as well as the continued validity of the policy objectives.

Beginning with the Ford administration, the executive agencies have been subject to a series of requirements to consider the impacts of new regulations. President Ford required the preparation of an "inflationary impact statement." President Carter replaced this requirement with his own requirement for the preparation of a "regulatory impact statement" which, in addition to considering the economic consequences, also required a full consideration of alternatives. President Reagan instituted his own program requiring the preparation of "regulatory impact analyses" which must describe the potential costs and benefits of proposed major regulations. New regulations may be issued only where potential social benefits outweigh potential social costs.

Such requirements, while potentially useful, appear not to have influenced significantly agency decision-making. Unlike the effect of environmental impact statements under NEPA, there is no strong popular sentiment for regulatory reform backed up by active judicial review. Moreover, review is required only for individual, newly proposed regulations. There is no mechanism for reviewing existing regulations nor for considering general regulatory approaches.

More comprehensive reform proposals seek to involve both the legislative and executive branches in a review process that would allow a cost and benefit evaluation of alternative programs. For example, the concept of a regulatory budget has received considerable attention in recent years. According to Liton and Nordhaus (1983, pp. 86-87), the need for a regulatory budget is indicated for at least four reasons:

> First, neither the Executive nor the Congress systematically determines the overall level of regulatory activity in a given period. Second, no office in the executive branch or committee in Congress is responsible for systematically establishing regulatory priorities across the government. Third, the Executive has not instituted any systematic process of submitting regulatory proposals of efforts to the Congress. Finally, there is no office to audit regulatory programs.

A regulatory budget procedure would impose a dollar limit on the private expenditures which could be mandated through new regulatory requirements. As in the federal budget process, the executive branch would submit a list of regulatory proposals including budgeted private expenditures which would be required. Congress would then review these proposals and authorize them with any modification it desires. Subsequent audit by an independent office would check to assure that the budget is being maintained.

Improved Regulations. Many observers of the regulatory process have concluded that major change of the regulatory schemes currently in place is unlikely in the forseeable future. This reality suggests to them that the only viable option for regulatory reform is to seek ways to make regulation better, to search for "smarter" regulations. Considerable progress was made during the Carter administration in this area—particularly in relation to environmental regulation. The major examples have developed in relation to the Clean Air Act.

The bubble policy is an especially good example of seeking regulatory solutions aimed at achieving the broader policy objectives rather than focusing narrowly on specific requirements. Originally the EPA had regulated emissions from each point source within a given facility. Under the bubble policy, new sources locating in a region where national ambient air quality standards are being met (or existing sources seeking to make a major modification) may treat the entire facility as a single point of emissions and develop the most cost-effective control strategy for all emission sources within the complex. The objective of maintaining air quality is met since the emissions are limited to the allowed increment. At the same time the individual firm is allowed to seek the lowest cost approach to achieve the objective.

A similar search for flexibility produced the offset policy under which new sources of regulated emissions may be introduced into nonattainment areas if existing emissions from other sources are reduced by more than the new source will add. In this case two often competing objectives are allowed to co-exist —growth and improved air quality. Indeed, growth becomes the cause of cleaner air by replacing existing dirty sources with new, cleaner sources.

As a general matter, many believe that the existing regulatory system would be improved by emphasizing more a result-oriented performance standard rather than the highly detailed standards generally in effect which operate more like design standards. By regulating on the basis of what is expected, rather than what is required to be done, there would be more incentive for innovation by individual firms as they seek their own best control strategy.

Still another approach which has been receiving attention recently seeks to include the parties-at-interest in a negotiated approach to developing new regulations. The intention is to clarify the concerns of these parties by engaging them in a constructive dialogue prior to the development of the regulation. This process can assist the agency by producing information needed to develop effective regulations. It may also promote more efficient implementation and reduce subsequent litigation.

Alternatives to Traditional Regulation. Many critics of regulation argue that the fundamental problem is with regulation itself. The only appropriate solution is to move to alternate approaches, primarily those which are incentive-based. Such alternatives tend to be favored by economists who stress the need to find market-oriented solutions rather than impose legalistic requirements. Two specific alternatives which have received the most attention in the environmental area are marketable permits and emission charges.

A system of marketable permits provides a means to control and limit an activity, but without the need for government to dictate the manner of performance. For example, in the case of pollution control, permits could be issued which, collectively, would limit the emissions of pollutants to the desired level. These permits would be saleable, thus allowing the exact pattern of pollution reduction to be determined by open-market trading. Sources able to inexpensively reduce emission would sell their extra permit rights to other sources whose clean-up would be more costly than the permits. Enforcement would still be a government responsibility and the problem of knowing how much pollution reduction is appropriate would still remain. But the burdensome chore for regulatory agencies of devising control requirements would be eliminated. Instead these decisions would be returned to individual firms with the necessary incentive to seek the least-cost solution.

Taxes applied directly to discharges of pollutants have long been advocated as an efficient approach to environmental control. In this most direct method of internalizing the external costs of environmental pollution, the price of the product being produced would then reflect the full cost of its production. If the environmental costs are substantial, the higher market price should tend to reduce demand —thus reducing the pollution output as well. A favorite of economists because of the use of the price mechanism to induce the desired change, pollution taxes have not found much political support in the United States.

Evidence from the use of market-based approaches to pollution control in several European countries and the theoretical support contained in a sizeable body of literature (see section 5.15) suggest that these are attractive alternatives to the present regulatory approach. Unfortunately, once systems are in place they are extremely difficult to change. At the national level, the pattern of seeking solutions to difficult problems through regulation became well established during the 1930s. With almost no serious consideration of the alternatives, Congress followed this pattern in its approach to environmental control.

The regulatory reform effort has succeeded in improving general awareness of the costs and benefits of regulation. Congressional zeal for creating complex new schemes of regulation has waned in recent years. Moreover, innovative approaches (such as the bubble policy) for achieving regulatory objectives suggest that improvements can be found within the existing structure. Although at this point the more ambitious proposals for reform appear unlikely to succeed, there is much that can be done to make the regulatory system more effective.

References

1. Breyer, Stephen G., 1982 *Regulation and Its Reform*, Harvard University Press, Cambridge, Mass.
2. Clarkson, Kenneth W., Charles W. Kadlec, and Arthur B. Laffer, 1979, "Regulating Chrysler Out of Business?" *Regulation*, September/October.
3. Kneese, Allen V. and Charles L. Schultze, 1975, *Pollution, Prices, and Public Policy*, The Brookings Institution, Washington, D.C.
4. Landsberg, Hans H., et al., 1979, *Energy: The Next Twenty Years*, Ballinger, Cambridge, Mass.
5. Litan, Robert E. and William D. Nordhaus, 1983, *Reforming Federal Regulation*, Yale University Press, New Haven, Conn.
6. National Commission on Air Quality, 1981, *To Breathe Clean Air*, Washington, D.C.
7. Noll, Roger G., et al., 1974, *Government Policies and Technological Innovation*, California Institute of Technology, Pasadena, Calif.

5.15B

The Secondary Effects of Mineral Development

Silver Miller* and J. C. Emerick†

After location and verification of exploitable mineral resources, mineral development requires capital goods, labor, and energy to extract the deposit. In addition, a growing social awareness of the need for a healthy environment requires inclusion of environmental burdens to the total costs incurred. This is because mineral development is detrimental to the environment in a number of ways. All mining operations involve the separation of valuable minerals from the surrounding waste rock, which in turn requires permanent disposal of unwanted or unused materials. Mine overburden, wastes, and tailings are often transferred to peripheral areas where they remain as unsightly features in the natural landscape. Some mine spoils contain harmful or contaminating substances as heavy metal salts or acid forming materials which subsequently are released to the environment.

Many types of mining operations handle significant quantities of water during extraction or beneficiation processes. When discharged, the quality of such water often is unsuitable for residential, agricultural, or recreational use. Since mining activity disturbs the natural vegetation in the project area, water may be allowed to transport soil materials into streams, causing erosion and siltation. In particular, surface mining operations disturb the top layers of soils, resulting in losses of productivity for long periods of time. Excavation, transportation, and further processing of large volumes of both ores and waste materials generate substantial quantities of particulate and gaseous emissions, some of which contain metal dusts or compounds of sulfur. Such fugitive and point source emissions can be transported beyond the confines of mineral operations to cause annoyance and damage at more distant locations.

Environmental degredation or pollution can be viewed as the alteration of the environment by man through the introduction of materials which represent potential or real hazards to human health, disruption to living resources and ecological systems, impairment to structures or amenity, or interference with socially desired uses of the environment.

The significance of environmental degradation is reflected in its effects on a range of receptors, including humans as well as the resources and ecological systems upon which we depend. Pollution also is judged in the social context of damage to environmental structures and amenities. The demand for restoration and maintenence of environmental quality presents the mineral industry with a challenge to successfully manage discharges and returns to the environment in order to avoid undesirable and harmful consequences.

AIR QUALITY

Impacts of Mining and Beneficiation Operations

Air quality generally is degraded by mining operations through the addition of dust or total suspended particulates (TSP), carbon monoxide (CO), sulfur dioxide (SO_2), nitrogen oxides (NO_x), and hydrocarbons (HC) to the atmos-

*Department of Mineral Economics, Colorado School of Mines, Golden
† Department of Environmental Sciences and Ecological Engineering, Colorado School of Mines, Golden

phere. The primary air emission in the mining industry is particulate matter, released in the form of fugitive dust.

Emission sources are categorized as fugitive or point sources and are present in all phases of the mining and on-site beneficiation process. Fugitive emission sources commonly include drilling activities, blasting , loading and hauling operations, stock and waste piles, mine roads, land reclamation, and wind erosion of unprotected surfaces. Diesel engines for equipment used in mining operations also are sources of fugitive air pollutants, most importantly particulates, and NO_x. Process point sources typically are crushing and grinding operations, screening installations, conveying systems, and drying units.

Ambient concentrations of various air pollutants which result from mining and beneficiation depend upon a variety of factors, the most significant of which are: moisture content, type, and amount of materials handled; type of equipment and operating practices employed; and, local geography and meteorology. If moisture content is reduced by evaporation, the release of fine particles increases appreciably. The handling and processing of softer materials with their lower resistance to fracture and crumbling creates a greater potential for emissions than operations involving harder materials. The type of material also determines the presence of toxic or hazardous constituents in particulate emissions.

Emissions from mining and processing equipment, such as those powered by diesel engines (crushers, screens, and conveyors) are generally a function of the total amount of materials handled, the amount of mechanically induced velocity applied to materials, and their frequency of handling. The crushing mechanism utilized (i.e. compression or impact) also affects emission. Due to the progressive movement of surface mining operations, the location of some pollutant sources shifts while other sources, such as roads or stationary installations emit continuously from well defined locations. Since fugitive dust emissions associated with mining occur primarily near the source, ambient TSP concentrations can decrease rapidly with distance.

Climate and seasonal changes affect emissions in several ways. The wind velocity, wind direction, amount and intensity of precipitation, and relative humidity can affect emissions significantly, particularly fugitive dust emissions. Topography and the type of vegetation surrounding the operations influence these meteorological parameters and thus indirectly affect ambient pollutant concentrations.

Despite the fact that they are intermittent operations of relatively short duration, drilling and blasting create large amounts of fugitive dust. Particulate emissions associated with drilling activities are generated by forced air flushing of cuttings and dust out of the bottom of the drill hole. These are transported up through the annular space between the hole wall and the drill rod by compressed air. Fugitive dust is an intrinsic characteristic of blasting operations. Blasting techniques, rock characteristics, and the size of the shot determine the degree and size of dust particles generated. The discharge of explosives also releases a number of gaseous chemical compounds into the atmosphere.

Loading and hauling operations represent a major source of fugitive dust emissions. Air motion across materials during hauling or conveying and vehicular transport over paved roads during mining activities are important sources of emissions from haulage operations. The type of road surface, the wetness of the surface, and the speed and volume of haulage vehicles and conveyors have been identified as major factors influencing emissions. Truck dumping also generates significant quantities of fugitive dust as the material cascades from the truck bed, striking the ground, the sides of receiving hoppers, or the edge of embankment slopes.

Specific sources of fugitive dust in descending order of importance include equipment and vehicle movement and operation in these areas, wind erosion of exposed embankments, loadout from the storage or spoil piles, and dumping onto piles. Tailings piles are subject to wind erosion and can emit mineral particles, as do waste or storage piles. However, due to the higher amount of fines, there is greater long-distance transport and deposition from tailings.

Both reclamation and clearing activities can produce significant fugitive dust emissions. All unvegetated surfaces that are exposed to wind erosion are typical fugitive dust sources. These emissions are influenced by wind speed, surface texture, and character of vegetation cover, where present.

Beneficiation processes, in contrast to most unit mining operations, can be characterized as point emission sources. Although significant, emissions from these sources can be more easily controlled as the processes are stationary and the emissions are generated from well defined points.

Emissions from crushing and grinding operations are principally affected by the moisture content of the material, the type of material processed and the type of crusher employed. The most important factor affecting the physical characteristics of emissions is whether compression or impact reduction is practiced. This has a substantial effect on the particle size distribution of the emissions, and the amount of mechanically induced energy imparted to the particles.

Dust is emitted from screening operations by the agitation of dry rock particles. The level of uncontrolled emissions is inversely related to the particle size of the material and directly related to the transmissions of mechanically induced energy to the material. As a general rule, the screening of finer particle sizes generates significantly higher levels of emissions than the screening of the coarser particle sizes. Correspondingly, screens agitated at high frequencies and at large amplitudes generate higher levels of dust emissions than those utilizing lower frequencies and lesser amplitudes.

All material hauling, conveyance, and transfer operations produce particulate emissions. The level of dust released depends on the size of the material and the degree of agitation to which it is subjected. The primary source is found at conveyor belt transfer points, both where materials leave the conveyor at the head pulley and where materials are received at the tail pulley. Factors influencing the amount of emissions produced are the conveyor belt speed and the free-fall difference in height between conveyor belts.

Dust generated by drying operations is of the same composition as the input concentrates. Small quantities of organic material in the concentrate can be decomposed or volatized in the drying operation. However, emission of metallic fumes or oxides of sulfur from concentrates are an unlikely occurence.

Impact Mitigation in Mining Operations

Dust suppression techniques and collection systems are used to reduce particulate emissions. Suppression techniques are applied to hinder the generation of fugitive or process dust. Collection systems are installed to capture or contain particulate matter after it is airborne. Preventing particulate matter from becoming airborne, whether through engineering or through the application of suppression agents, should be a primary goal. Once airborne, the collection of dust is generally more difficult and more costly than its initial suppression.

As a rule, the suppression of fugitive dust in mining operations is achieved by the application of water or chemical stabilizers to the surface of the dust source, and by the reduction of surface wind speed across exposed surfaces. While watering is the least expensive, it is also the least permanent in maintaining dust control. A thin coating of water creates an immediate cohesive force that binds surface particles together. This also results in a more compact and mechanically stable surface crust which after drying is less likely to produce emissions. However, repeated watering is required to maintain a film of moisture as the dried surface crust with its associated suppressive effects is easely ruined by vehicle disturbance or abrasion by loose particles blown across the surface.

Various dust suppression chemicals are available. Some are manufactured for use alone, while others are applied as additives to water. While many agents can re-encrust the surface after disturbance, the effect of natural weathering on the life of the treated surface can vary substantially. Although the primary use of chemical stabilizers in the minerals industry is confined to haul roads, tailings ponds, and land reclamation after mining operations, chemicals may also be applied successfully to piles of overburden and waste.

Control of particulate emissions from drilling operations is achieved through water injection and dry dust collection. Water injection is a wet drilling method in which water is introduced into the compressed air supplied to the drill bit for the flushing of cuttings from the drill hole. Wetting agents or surfactants can be added to increase efficiency by reducing the surface tension of water. A fine spray is produced by the injection of water into the stream of compressed air. The cuttings are thus agglomerated, dropping at the hole collar as pellets as they are blown out of the drill hole.

Dry collection systems utilize a shroud or hood to enclose the drill rod at the hole collar. Particulate matter is captured under vacuum and vented to a collection device. The most commonly used devices are cyclones or fabric filters preceded by a settling chamber. While cyclones are more suitable for medium-sized and coarse fractions, fabric filters are most efficient for fine particulates.

Techniques have not yet been developed to effectively suppress particulate emission from blasting. Still, multi-delay detonators that stage explosive charges in millisecond time intervals may reduce fugitive dust generation. Blasting operations also can be scheduled to take advantage of favorable meteorological conditions such as lower wind speed and low inversion potential to substantially reduce the impacts of emissions.

Fugitive dust suppression from loading operations has been limited to the wetting of materials prior to or during loading. This is generally done by water trucks or movable watering systems. Dumping operations pose similar difficulties. Wetting systems reduce particulates emitted during discharging operations. Otherwise, currently there are no additional methods available for controlling particulate emissions.

Methods of controlling particulate emissions resulting from hauling operations concentrate on road surfaces and traffic patterns. The movement of large, rubber-tired vehicles over dirt haulage roads is a major source of fugitive dust. The level of emissions is directly related to the condition of the road surface as well as to the frequency and speed of vehicles. Accordingly, suppression measures focus on maintaining desired road surfaces and regulating traffic. Although the diesel engines for mobile equipment are significant sources of particulates and NO_x, manufacturers' designs, including the appropriate pollution control devices of newer equipment, minimize the impact of exhausts.

Successful control of fugitive emissions from haulage roads includes the treatment of road surfaces with water, oil, or soil stabilizers. While the use of water is the most common, the beneficial effects are quickly destroyed by vehicle disturbance. Road dust can also be controlled by oiling, or oiling supplemented by watering. A potential adverse environmental effect of this treatment is the migration of oil into streams or aquifers. Soil stabilizers are agents which usually consist of water-based emulsions of either synthetic or petroleum resins which act as an adhesive or binder. Hygroscopic chemicals such as organic sulfonates or calcium chloride are also utilized. These dissolve in the moisture they absorb, forming an adhesive liquid which is resistant to evaporation.

Since the level of fugitive dust emissions is directly related to the speed of tire rotation and the total number of tires, reduced vehicle speed results in fewer emissions. Replacing smaller haulage units with equipment of larger capacity reduces the number of vehicle tires and haulage cycles and thus the total emissions per ton of material hauled.

In conveyor stacking of unconsolidated materials control of fugitive dust emissions can be achieved by the use of store ladders, telescoping chutes, and hinged-boom stacker conveyors. These are designed to minimize free-fall disturbances, thus reducing the impact of falling material onto the pile and protecting the stream of material from wind. An alternative is the installation of water sprays at conveyor discharge pulleys. A further control measure is the use of covered conveyors to provide protection from wind and to prevent particles from becoming airborne.

However, water sprays are not always practical, as added moisture may inhibit screening or grinding where agglomeration is undesired. This limitation also applies to post-drying operations. To compensate for limitations on moisture content, wetting agents or surfactants are mixed with water. These agents reduce the surface tension of water, minimizing the moisture necessary to suppress dust particles by improving wetting efficiency.

Spray headers equipped with pressure spray nozzles are used to distribute the dust suppressant liquid at desired treatment points. To prevent excessive wetting and to adjust to intermittent materials flow, spray rates and configuration must be appropriately regulated.

Dry dust collection systems consist of an exhaust system with hoods and enclosures to confine and capture emissions, and ducting and fans to convey contained emissions to a collection unit for particulate removal before the airstream is vented to the ambient air. Depending on the design of beneficiation facilities, point emission sources can be ventilated to one centrally located collection unit or to a number of strategically located units.

Effective control of fugitive dust includes the installation of hoods at conveyor transfer points, crushers, grinders, screens, and dryers. Ducts used to transport particulates must be designed so that adequate transport velocities are maintained to prevent dust particles from falling and settling en route to final collection. Fabric filters or baghouses currently are regarded as the most effective collection and particle removal devices in the minerals industry. These can be mechanical shaker type collectors or jet-pulse units. Other collection devices include cyclones and low-energy scrubbers. Although the latter collectors demonstrate high efficiency for coarse particles, their efficiency is lower for medium and fine particles. High-energy scrubbers and electrostatic precipitators approach the effectiveness of fabric filters, but tend to be more costly.

Dust from drying operations frequently consist of the fine particles present in the ore, their collection being complicated by the condensation of moisture in the warm emission stream. For dryers installed in beneficiation plants, dust control is best performed by collection and then removal by means of wet scrubbing, returning collected matter to the beneficiation process.

Impacts of Primary Metal Production

The primary metals industries discussed in this section include the nonferrous operations of primary aluminum production, copper smelters, lead smelters, and zinc smelters. Also included is primary iron production. These industries are characterized by large quantities of emitted sulfur oxides and particulates, some of which are considered toxic metal compounds.

In recent years regulatory and enforcement activities for the primary metals industry have been directed towards the abatement of sulfur dioxide and total particulate emissions. Besides these major pollutants, this industry has the potential to emit large quantities of airborne particulates composed of various metals. Although control systems have been developed by the industry to recover metal values, these have not been effective yet in capturing the volatile forms of these metals. In the case of both process and fugitive emissions the presence of significant levels of SO_2 requires greater margins of abatement success.

The major portion of aluminum produced in the United States is processed in prebaked reduction cells. The second commonly used pots are the horizontal-stud or vertical-stud Soderberg reduction cells. Emissions from aluminum reduction processes consist primarily of : gaseous hydrogen fluoride and particulate fluorides; alumina, hydrocarbons; and sulfur dioxide from the reduction cells and the anode baking furnaces. Large amounts of particulates are also generated during the calcining of aluminum hydroxide. However, only small quantities are emitted to the ambient air as the economic value of this dust encourages a high level of recovery.

The source of fluoride emissions from reduction cells is the fluoride electrolyte. Particulate emissions from the reduction cells consist of alumina and carbon from anode dusting, cryolite, aluminum fluoride, calcium chloride, chiolite, and ferric oxide. Moderate amounts of hydrocarbons derived from anode paste are emitted from horizontal-stud and vertical-stud Soderberg reduction cells.

Copper is produced primarily from lower-grade sulfide ores, which are concentrated by gravity and flotation methods. Copper sulfides normally are roasted in either multiple-hearth or fluidized-bed roasters to remove the sulfur and then are calcined in preparation for smelting in a reverberatory furnace. The high temperatures attained in roasting, smelting, and converting cause volatilization of a number of the trace elements present in copper ores and concentrates. The most significant among these are Hg, Zn, Cd, Pb, and As. The raw waste from these processes contains not only these pollutants but also dust and sulfur oxide. The volatile metals can appear in these dusts as sulfates, sulfides, oxides, chlorides, or fluorides. However, the value of volatilized metals dictates efficient collection of fumes and dusts, reducing pollutive emissions. Still, the copper smelting industry is a primary source of arsenic emissions and a major source of sulfur dioxide.

Lead is produced primarily from sulfide ores containing small quantities of copper, iron, zinc, and other trace elements. The normal practice of primary lead metal production from concentrate involves sintering. In this process, lead and sulfur are oxidized to produce lead oxide and sulfur dioxide. Following this, lead oxide is reduced and subsequently refined. Nearly 85% of the sulfur present in concentrates is eliminated in the sintering process. The primary lead metal industry emits fine particulates and sulfur dioxide to the atmosphere. The particulate emissions contain metals such as lead, cadmium, and arsenic. These originate from sinter machines, blast furnaces, slag fuming drossing, cadmium recovery, and antimony recovery processes. Control of sulfur dioxide emissions is a significant problem for this industry as only the gas stream from the sintering machine is amenable to sulfuric acid production, discussed in the following section.

Zinc is produced primarily from zinc and lead ores. The major products of the primary industry include zinc, zinc oxide, sulfuric acid, cadmium, and zinc sulfate. Cadmium generally is recovered at smelters from collected dusts and slags with sufficient metal content. Direct zinc oxide production uses the same ore concentrate as metallic zinc production. For efficient recovery of zinc, sulfur must be removed from concentrates to a level below two percent. This is accomplished by fluidized bed roasting, suspension roasting, and multiple-hearth roasting. In a zinc smelter the roasting process is responsible for more than 90% of potential sulfur dioxide emissions. Particulates emissions consist of fumes and dust composed of zinc concentrate elements. These include Zn, Pb, Cu, Mn, Hg. Distilled metal fumes constitute an appreciable amount of waste gas particulate carryover. Off gases also contain oxygen, carbon dioxide, and water-vapor.

Pig iron production requires the smelting of iron ore, iron-bearing materials, and fluxes with coke. These are charged into a blast furnace and allowed to react with large amounts of hot air to produce molten iron. Blast furnace by-products consist of slag, flue dust, and gas. Blast furnace combustion gas and gases that escape from bleeder openings constitute the major sources of particulate emissions. The dust in this gas contains iron, carbon, silicon dioxide, as well as small quantities of aluminum oxide, manganese oxide, and calcium oxide. All of the carbon monoxide generated in the gas is used for fuel, requiring the removal of particulates. The removal of useful flue dusts from blast furnace gas allows their recycling as blast furnace charge, eliminating most particulate emissions from the production process.

Impact Mitigation in Primary Metal Production

General selection methods for the application of particulate emissions control systems do not guarantee environmentally and economically successful installations. Proper selection of systems requires consideration of performance limits, emissions characteristics, and costs. Control devices should have the capacity to maintain continuous compliance regardless of short-term fluctuations in the composition of emissions, flow rates, and particle size-distribution. Particulate control systems which currently are available and employed in the primary metals industry include mechanical collectors, electrostatic precipitators, fabric filters, and wet scrubbers.

Mechanical collectors comprise a broad class of devices that utilize gravity settling, inertial, and dry impaction mechanisms. Settling chambers, elutriators, momentum separators, mechanically aided collectors, and inertial centrifugal collectors constitute the major classes of commercially available mechanical collectors. These are efficient for large particulate removal, but they cannot be expected to adequately remove fine particulates. In addition, they lack the ability to adjust efficiently to extremely variable flow rates. Mechanical collectors are tolerant of high dust loadings and operate efficiently at high gas flow rates.

Electrostatic precipitators are high efficiency particulate collection devices applicable to a wide variety of source categories and gas conditions. Particle collection is achieved through application of electrical energy for particle charging and gathering. Efficiencies of up to 99.9% are possible, depending on application, apparatus design, and gas and particulate characteristics. Electrostatic precipitators operate both dry and wet, the major difference being the manner by which particulates are removed from the collection electrodes. Dry systems may be installed for control of emissions from cement kilns and metallurgical furnaces. Wet precipitators are used primarily in the metallurgical industry at operating temperatures below 75°C. These are employed successfully for blast furnaces, acid mist, and sources with sticky and corrossive emissions.

Fabric filters are generally classified according to the method of fabric cleaning. The three major categories of cleaning methods are mechanical shaking, reverse air cleaning, and pulse jet cleaning. The particle collection mechanisms employed by fabric filters include inertial compaction, Brownian diffusion, interception, gravitational settling, and electrostatic attraction. Emission particles are physically collected on a dust cake supported on a fabric or on the fabric itself. At regular intervals a portion of this dust cake must be removed to enhance gas flow through the filter. A complete characterization of the effluent gas stream is important in the design of a fabric filter system. These would include: flow rate, temperature, acid dew point, moisture content, presence of large or sticky particulates, and total particulates. Properly sized filters operating under dry collections and within temperature limitations of the fabric can achieve extremely efficient collection over a wide range of particle sizes.

Wet scrubber technology is based primarily on the particle collection mechanisms on inertial impaction and Brownian diffusion. Scrubber liquids are used for particle collection in three ways. The most common method is to separate droplets, which are then mixed with the gas stream. Layers or sheets of water enveloping packing material are also used to collect particles by directing the gas stream around the individual packing elements. The least common method is to pass high-velocity gas through a vapor to create "jets" of liquids to collect particles.

The major types of wet scrubbers include spray scrubbers, packed-bed scrubbers, tray-type scrubbers, mechanically aided scrubbers, as well as venturi and orifice scrubbers. The performance of these scrubber systems generally exhibits strong particle size dependencies, with substantial differences of efficiency in the particle size range of 0.1 to 2μmA. Most wet scrubbers can efficiently collect large particulates (2μmA), while a few are also efficient for very small particulate collection. Thus many wet scrubbers exhibit a medium particle size window where collection is less efficient. Wet scrubbing of particulate matter is a two-step process, separation of the scrubbing liquid droplets from the gas stream being necessary after particulate collection. This step is important in the cumulative collection of particulates as poor liquid separation allows reentrainment of particulates.

Water-usage and waste disposal may become a major decisional factor in wet scrubber selection. Because scrubbing slurries are often corrosive and abrasive, liquid handling facilities can be extensive. Recirculation of scrubber liquors to prevent contamination of surface waters with the collected particulates tends to concentrate dissolved scrubbing liquor contaminants. With a view to minimizing contamination of the environment, slurries or liquors from wet scrubbing pollution control systems are treated by filteration or sedimentation to remove water and concentrate particulate matter. This permits the ultimate disposal of particulates as a solid waste, rather than a sludge or a slurry.

WATER RESOURCES

Impacts of Mining and Beneficiation Operations

The impact of mining operations on water quality is well documented. However, the emphasis has been on the impact of abandoned mines on water quality. While the major problems associated with water quality and surface mining are acid mine drainage and sedimentation, impacts on groundwater include disruption of groundwater movement, and changes in groundwater availability and quality.

Disturbance of groundwater resources can result from removal of portions of aquifers, dewatering on the perimeter of mining areas, and from backfill operations. Groundwater movement is modified by underground workings, and potential subsidence and fracturing of overlying rocks in underground mining.

Impacts due to aquifer disruption or mine dewatering relate in part to horizontal hydraulic conductivities of the strata in question. For very low conductivities, a locally limited drawdown can be expected. Aquifers may be affected in the vicinity of the mine site, but the net effects would not extend away from the mine since water contained in storage is unable to move quickly through the aquifer to seepage faces. Disruptions to the interaction between surface and groundwater occur where surface stream flows recharge aquifers and where aquifers discharge. Water recharge to both stream flows and aquifers can be altered and water quality affected.

Construction of subsurface drainage systems at the base of backfill materials in open pit mines, relocated overburden, or submarginal spoil materials are designed to collect inflows and transmit groundwater to surface diversions. Over very long periods of time, limited recharge may occur through the backfilled or relocated materials by infiltration and percolation of snowmelt, precipitation, or surface runoff. Such recharges may pick up additional ions in solution through geochemical interaction with spoil materials, resulting in an alteration of their loadings. Depending on the amount of recharge migrating through spoil or overburden materials, some affected water may reach the underdrain and eventually be discharged along with the original transmissions of groundwater flow. Potential groundwater impacts may result from interaction with both backfill and spoil materials as well as from the underdrain materials used.

A serious threat of aquifer depletion or contamination is posed by disturbance to, or removal of, impervious strata separating and protecting aquifers. The risk of inflows affecting underlying aquifers is significantly increased when interconnecting fractures become paths of hydraulic connection. Surface subsidence and related fractures resulting from underground mining provide such interconnections.

Effluent wastes from mining operations and beneficiation processes are comprised mostly of the following pollutants: total suspended solids (TTS), alkalinity or acidity (pH), settleable solids, iron in ferrous metal mining and dissolved metals in nonferrous mining. Suspended solids consist of small particles of solid pollutants that resist separation by conventional means.

Solids in suspension that will settle in one hour under quiescent conditions are settleable solids. Acid conditions prevalent in the minerals industry usually result from the oxidation of sulfides in mine waters or discharge from acid leach beneficiation processes. Alkaline leach beneficiation processes also contribute to effluent waste loading. Iron is often present in natural waters and is derived from common iron minerals in the substrata increasing iron levels present in process water and mine drainage. A number of dissolved metals are considered toxic pollutants. The major metal pollutants present in ore mining and beneficiation waste-waters are: arsenic, asbestos, cadmium, copper, lead, mercury, nickel, and zinc.

Surface water pollutants primarily originate from three major sources: mine dewatering, process waters, and precipitation runoff. For many mines mine dewatering is the only significant source of waste-water. It is usually low in suspended solids, but may contain dissolved minerals or metals. Process waters are those used in transportation, classification, washing, concentrating, and separation of ores. As an effluent it usually contains heavy loadings of suspended solids, and in nonferrous metals mining, dissolved metals. It also can contain process reagents, most notably cyanide. Since mining operations occupy large surface areas, precipitation runoff may contribute a major volume of waste-water and pollutant loading. This water typically contains suspended solids such as minerals, silt, clay, sand, and possibly dissolved toxic metals.

Other sources of possible water pollution include acid mine drainage and tailings pond leakage. Acid drainage can be produced by the leaching of water through any mine waste or structure containing sufficient pyrites or other sulfides. Tailings ponds are used for both the disposal of solid wastes and the treatment of waste-water streams. The supernatant decanted from these ponds contains suspended solids and at times process reagents introduced to the water during ore beneficiation. Percolation of waste-water from impoundments may occur if tailings ponds are not properly designed.

Water which accumulates in surface mine pits or underground mine workings is likely to be heavily loaded with sediment or dissolved salts, or both. As a result, mine dewatering can easily result in heavy pollution loads in the receiving waters. Pumping to maintain dry conditions may also lower the water table, reducing the yield of nearby wells, springs, and seeps. Mine waters originate from groundwater flows, as well as from rainfall and runoff. Thus, mine dewatering can significantly affect receiving waters. Water should be diverted from mine pits or workings when possible, as water pumped out of the mine will most likely require treatment before it can be discharged from the mine area.

The mineralogy of the ore and the beneficiation process determine the characteristics of the effluent stream. Beneficiation processes include gravity separation, froth flotation utilizing reagents, chemical extraction, and hydrometallurgy. Physical processes using water, such as gravity separation, discharge suspended solids. Froth flotation methods entail pH adjustments to increase flotation efficiency, which together with the fineness of ore grind substantially increase the solubility of mineral components. Ore leaching operations entail the use of large quantities of reagents such as strong acids and bases to achieve solubilization of ore components. This in turn results in a high percent of soluble metals content and hence availability of potential contaminants. The presence of reagents used in the flotation process can affect the quality of process waste-water. Surface mining can generate large volumes of sediment through precipitation runoff. If sediments cannot be contained but are allowed to reach adjacent waterways they become pollutants. Potential disturbed areas which can serve as sediment sources include active mine sites, future working sites, ore storage piles, overburden and spoils piles, and disturbed land adjacent to the mine. As sedimentation is directly related to erosion, the problem of sedimentation can be viewed as similar to that of erosion. Sources and mitigation measures for sedimentation are discussed in the section on Soil and Land Resources.

In areas where potential evaporation does not exceed the total of direct precipitation, collected runoff and tailings effluents, tailings ponds may inadvertently serve as infiltration pits. Although earth liners are often considered impermeable, leakage of leachates remains possible. Seepage rates and soil uptake of effluents depend on the chemical and physical properties of the soil, the design and construction of ponds, and prevalent geologic conditions. Soils and tailings deposits exhibiting high permeabilities may permit high seepage rates, especially if sandy soils underlie the pond.

Impact Mitigation in Mining and Beneficiation Operations

The selection of optimal control and treatment technology for mining and beneficiation operations is influenced by a number of factors. In the case of minewater, the operator has little control over the volume of water to be treated except for the diversion or runoff of surface mine areas. Differences in wastewater composition and treatability are caused by ore mineralogy, beneficiation techniques, and reagents used in the concentrating process. Geographic location, topography, and climatic conditions also influence treatment and control strategies.

Water control and treatment technologies fall into two general categories: in-process control, and end-of-pipe control. In-process control focuses on process changes available to existing operations to improve the quality or reduce the quantity of wastewater discharged. In-process recycling of waste streams to reduce wastewater volume can improve the performance of existing treatment systems. Recycling of concentrate thickener overflow and recycling of filtrate produced by concentrate filtering has been practiced more in recent years. Thickeners to reclaim water from tailings prior to their final discharge to ponds are also used. Water reclaimed in these ways can be used as beneficiation makeup water. In-process recycling of waste streams produced by concentrate dewatering conserves water and recovers metals that would normally be lost through discharge, improving the quality of treated water.

Many operations use mine drainage as beneficiation makeup water, and in some instances the entire water requirement is satisfied. This practice either eliminates the necessity of a mine water treatment system or greatly reduces the volume of wastewater discharged to a single system. In situations where mine water contains high concentrations of soluble metals, its use in beneficiation processes provides a more effective removal of these metals then generally could be expected by treatment of mine water alone. Thus the use of mine water as beneficiation makeup water can be considered a control practice which improves the quality of treated wastewater upon eventual discharge.

Pollutant discharges from mining and beneficiation operations may be reduced by limiting the total volume of discharge as well as by reducing pollutant concentrations in the waste stream. Techniques for reducing discharges of beneficiation wastewater include limiting water use, excluding incidental water from the waste stream, recycling of process water, and impoundment to evaporation. Apart from use as an input to the beneficiation process, mine waters generally are not amenable to control. The high cost of pumping the large volumes of mine water together with the cost of treatment facilities act as incentives to reduce process water demands.

Process water currently is recycled where it is necessary due to water shortage. By reducing the volume of discharge, this may not only reduce the gross pollutant load, but also allow treatment practices which would be uneconomical for the full waste stream.

By allowing concentrations to increase in some instances, the opportunity to recover desired waste components is improved. Recycling possibilities generally are limited to physical processing operations rather than to flotation operations as the latter can suffer reagent crossover and degradation. The development of resulting reagent mixes and contaminants can interfere with the extractive process. In general, flotation operations for sulfide ores can achieve a high degree of process water recycling with minimal difficulty or process modification. The flotation of non-sulfide ores and various oxide ores, however, has been found to be quite sensitive to input water quality.

Impoundment and evaporation are techniques practiced at many mining and beneficiation operations in arid regions to reduce the volume of discharges. Successful employment of this method requires a climate having low annual precipitation and high potential evaporation, as well as the availability of land adequate for the volumes of process waste waters. Where impoundment is not practical for the full process stream, impoundment and treatment of smaller, highly contaminated volumes of water from specific sources may be desirable.

End-of-pipe treatment techniques for wastewaters are utilized to improve quality characteristics that are not prevented through in-process control. A variety of treatment methods are available, including sedimentation, coagulation and flocculation, and pH neutralization. In conjunction with these methods, filtration, absorption, and iron oxidation also are practiced.

Sedimentation depends on the specific gravity of suspended solids to achieve their separation from wastewaters. The effectiveness of sedimentation systems is a function of settling velocities of the material in suspension. A sedimentation system must be designed to provide sufficient residence time to allow a desired amount of solids to settle. Residence time is controlled by the amount of flow and capacity of the impoundment. In theory, several ponds in a series will not remove more solids than one large pond of equal capacity, because the theoretical detention times are the same. However, short circuiting or too much depth in the primary pond, shock precipitation runoff, and an improper discharge structure in the primary pond are situations where secondary settling ponds can remove significant additional quantities of solids.

Coagulation and flocculation are terms often used interchangeably to describe the physiochemical process of suspended particle aggregation due to chemical additions to wastewater. Technically, coagulation involves the reduction of electrostatic surface charges and the formation of complex hydrous oxides. Flocculation is the physical process of solids aggregating into particles large enough to be separated from wastewaters by sedimentation, flotation, or filtration. Coagulants are added upstream of sedimentation ponds, clarifiers, or filter units to increase the efficiency of solids separation. This practice also has been shown to improve dissolved metals removal due to the formation of denser, rapidly settling flocs, which appear to be more effective in absorbing fine metal hydroxide precipitates. The major disadvantage of coagulant addition to wastewater streams is the generation of large quantities of sludge, which must remain in permanent storage within impoundments. A typical coagulation and flocculation system may consist of a mixing basin, then a clarifier or settling pond, and possibly a filtration unit.

Filtration is accomplished by the passage of wastewater through a physically restrictive medium with the resulting separation of suspended particulate matter. Filtration processes are categorized as surface filtration devices, including microscreens and diatomaceous-earth filters; and granular media filtration, or in-depth filtration devices such as rapid sand filters, and slow sand filters. Typical filtration applications include polishing units and pretreatment of input streams for reverse osmosis or ion exchange units.

Filtration is a versatile method in that it can be used to remove a wide range of suspended particle sizes. Next to gravity sedimentation, granular media filtration is the most widely used process for the separation of solids from wastewater. Most filter designs utilize a static bed with vertical flow, using gravity or pressure as the driving downward or upward force.

Absorption on solids, particularly activated carbon, has become a widely used operation for the purification of water and wastewater. Absorption involves the accumulation or concentration of substances at a surface or intersurface. The best known and most widely employed absorbent is activated carbon. This is because activated carbon has an extremely large surface area per unit of weight, which makes it a very efficient absorptive material. Compounds which are readily removed by activated carbon include aromatics, phenotics, and surfactants. The literature indicates that significant quantities of certain metals such as Cu, Cd, and Zn can be removed from wastewater by activated absorption.

Acid wastewaters are treated by neutralization. The most common neutralizing techniques include treatment with crushed limestone rock or with lime. Acid neutralization with limestone has been accomplished by means ranging from simple basins to complex pumping schemes which oscillate the waters through crushed limestone beds. Limestone treatment produces a dense, rapidly settling sludge which contains considerable amounts of unused limestone. Vigorous aeration is required to drive off the carbon dioxide formed in the neutralization process and to oxidize the ferrous iron. In the lime neutralization process, there are four basic steps: mixing of acidic water with lime slurry, aeration, diversion into clarifiers or settling basins for removal of solids, and finally the removal of sludge residues. In both treatment methods, concentrations of heavy metals in aqueous solution usually can be reduced by precipitation as insoluble hydroxides.

In a mine wastewater treatment system, such as any of the chemical neutralization methods, it is advantageous to oxidize any ferrous oxide present to the ferric form to be able to remove this as an insoluble hydroxide at near-neutral pH levels. Ferrous iron in mine drainage can be oxidized to the ferric form in the presence of oxygen. This oxidation is dependent on the pH level, proceeding rapidly at pH levels above eight. Above this point, the oxidation process becomes dependent on the availability of oxygen. As oxygen has a low solubility in water, vigorous aeration is necessary to achieve rapid intermixing.

SOIL AND LAND RESOURCES

Impacts of Mining and Beneficiation Operations

The quality of the covering over the land as well as the suitability of land for a particular variety of natural and human uses affects the value of soil and land resources. Depending on factors such as degree of slope, soil fertility, rockiness, erosiveness, vegetation, and climatic conditions the potential capabilities of land as a resource include cultivation, forestry, watershed, wildlife, and recreation. Whereas land as such is non-renewable, management of its physical configuration, soil characteristics, and vegetation determines its quality as a resource. In this context, the exclusion of land as a resource through disturbance or abandonment are major impacts of mineral development.

Erosion is the wearing away of the land surface by wind or water, with topsoil the first layer to be removed. Besides giving rise to sedimentation problems, erosion represents the irretrievable loss of native topsoils necessary for land rehabilitation. Although erosion occurs naturally from weather or runoff, it can be intensified dramatically by mining activities. The most important types of erosion associated with mining operations are those caused by storm water runoff. Three basic types are of major significance: sheet erosion, rill and gully erosion, and stream channel erosion. Sheet erosion results from rain striking a bare soil surface and displacing soil particles. Rill and gully erosion is caused by the removal of soil by water through small, well-defined channels. Stream channel erosion is a scouring of the stream channel by sediment reaching the stream from land surfaces.

The erosion potential of any area is determined by climate (precipitation, temperature, and wind); vegetative cover; soil characteristics (structure, texture, organic matter, moisture, and permeability); and topography. While these factors may be considered separately, it is their combined effects which determine total erosion potential. Sediment transportation and deposition are influenced by the flow patterns of the water and the characteristics of the particles being transported. As velocity and turbulence increase, more sediment is transported. Small, light particles such as fine sand, silt, and clay are easily transported, while coarser and heavier particles are more readily deposited.

The major sources of soil erosion in mining operations are areas disturbed by clearing, grubbing, or removal of stumps and roots, and scalping or removal of vegetation cover; roadways; spoil piles; and active mining areas. Clearing, grubbing, and scalping operations can cause soil losses through exposure of soil surfaces. In surface mining, clearing and grubbing too far above the unexcavated face of overburden or too far ahead of the open cut also facilitates erosion. Surface compaction and configurations created by operating equipment can significantly impede water infiltration or concentrate surface runoff. Improper placement or protection of salvaged or stockpiled overburdens can result in loss of topsoils or other useful materials.

Roadways are often a major source of soil erosion and may act as channels or conduits for sediment washing down from higher areas into natural drainage systems. Longer access roads can adversely disrupt natural drainage systems by intercepting, concentrating, and directing surface runoff. The result can be severe soil losses from roadway surfaces, ditches, cut slopes, and safety berms. This includes roadways within the mining area as well as access roads outside the actual mine site. Accelerated onsite and offsite erosion can occur after cessation of mining operations if measures are not taken to permanently stabilize exposed surfaces with vegetation and to otherwise alleviate disruption of normal drainage patterns.

Surface mine pit slopes and waste embankments represent large exposed surface areas with a great potential for erosion damage. Pit slope runoff can also carry erosive runoff from undisturbed areas above it. In contrast to embankment runoff which presents a problem for natural drainage systems, soil lost from mine pit slopes accumulates in the pit bottom where it can be effectively recovered and removed.

Surface mine pit areas, waste and stockpile dumps, tailings ponds, or logistical installations may render the land unfit for its original use, i.e. agriculture, rangeland, watershed, or forestry. Cumulative impacts to local topography can result from mass transfers of materials, construction of roads, excavation slopes and pits, dams, and reservoirs, resulting in the alteration of land use patterns or limited land accessibility.

The removal of overburden necessary in surface mining represents an important opportunity to conserve topsoils. The value of topsoils for revegetation cannot be overemphasized. Compared to subsoils, they are higher in organic matter and plant nutrients, and have better physical characteristics as a plant growth medium. If reused quickly, rather than being stockpiled for a period of time, seeds, rootstocks, and desirable soil organisms will remain alive and enhance the revegetation effort. The quality of topsoil cannot be duplicated by the use of mulches or soil amendments with spoil or subsoil, and topsoil reuse decreases the need and the expense for these additives. If care is not taken to identify and separately handle valuable topsoils, they often become lost through intermixing with waste materials or by being placed underneath spoil piles. Often times topsoils are inadvertently brought into contact with acidic spoils or leachate, causing serious contamination and impairing their utility for further use.

Another important environmental aspect of mining operations is the danger to human safety posed by unstable land surfaces, highwalls, exposed or abandoned adits, and property damage caused by subsidence. These are also important features toward which reclamation efforts are diverted, and may account for a substantial proportion of the reclamation cost, especially regarding abandoned mines.

Impact Mitigation in Mining and Beneficiation Operations

The choice of erosion and sediment control methods should be made during mine preplanning. Consideration should be given to local conditions pertaining to the erosiveness of disturbed materials, topography, rainfall, surface hydrology, and settling properties of sediments. Erosion and sediment control should constitute an integrated part of mining and reclamation planning, to be employed in conjunction with other abatement methods such as regrading, controlled mining, water infiltration control, and wastewater control. The most common methods of erosion control are runoff diversion, runoff control, channel protection, and sediment settling.

Runoff diversion isolates erosive material from moving water through channelization of these. Ditches are placed along the high end of a mine or elsewhere to intercept any significant amounts of water draining into the mine. Thus, water is collected before it can reach disturbed areas and is routed around or through the area to a receiving waterway. Cross sections and gradients of ditches are sized to accommodate estimated maximum storm runoff events. Dikes can be used in the same manner as ditches to deflect runoff waters. Complementary downslope dikes are often formed from ditch excavation materials. Diversion techniques can also be employed within surface mines to collect and convey incoming groundwater around erodible materials.

In contrast to runoff diversion, which is directed at preventing water from entering mining areas, runoff control is concerned with the management of erosion caused by water flowing through mine areas. Establishment of suitable vegetation is considered to be highly effective, economical, and the most universally applicable runoff control method. It is discussed later in this section.

Mulches can be successfully employed as a temporary measure to reduce erosion in areas where more permanent erosion controls are to follow. The function of a mulch is to provide surface protection from the impact of rain drops and reduce the velocity of subsequent water flow. Terracing of embankments to reduce water velocities on steep slopes is effective, as is reduction of steep slope areas. These are particularly applicable to tailings or spoil piles where steep and unstable slopes occur. A series of parallel diversion ditches placed in a configuration parallel to surface contours is an adaption of this technique. Surface modification through a series of closely spaced ridges parallel to surface contour lines reduces water velocity and allows sediment settling in troughs. As this method is short-lived due to erosion, it should be used in conjunction with revegetation. Finally, runoff control can be achieved by the use of surface stabilizers and compaction, as this reduces erosiveness of the surface. These techniques are temporary, since they are subject to weathering and physical damage.

Various methods are employed to control erosion in channels. As a rule, these involve the placement of a protective liner such as riprap, concrete, jute matting, flumes, or dumped rock in the channel. The effect is to reduce water velocities and to protect the underlying and sidewall materials. Protective cross channel dikes or energy dissipators also are used to reduce water velocities. Channel protection generally is used for diversion ditches through mine areas or for road drainage ditches. Sediments transported in runoff waters can be trapped through settling. Through internalization of drainage, sediments can be retained within the disturbed area. Water from collection and conveyance systems can be directed to settling ponds where the decrease in velocity lowers the competency of water to carry suspended matter. Settling can also be achieved without impoundments by distributing water discharge over a large area, reducing water velocity and depth, and allowing suspended material to settle. Area requirements for this would include a gentle slope and a rough, textured surface.

Significant contributions to erosion control can be achieved by minimizing disturbed areas and properly locating roads and waste piles. Removal of vegetation cover or topsoils should be limited as much as possible at any one time. Backfilling and grading should be carried out without delay, with reclamation instituted as quickly as possible. Roads can be located to minimize erosion and flooding. On road cuts, temporary erosion-control measures may be provided along with adequate drainage. Waste or storage piles can be placed to avoid interference with natural drainage. Selective placement of topsoil stockpiles or spoil piles on moderately sloping and naturally stable areas improves slope stability and lowers the potential for slope sliding and washout.

Reclamation measures maintain the quality and availability of land resources affected by mineral development. Reclamation can be characterized as the process of restoring disturbed land to levels of productivity equal to or greater than original levels. In its broadest sense reclamation returns land to prior or future uses in an ecologically stable state. Furthermore, this state should not contribute substantially to environmental degradation and should be compatible with surrounding aesthetic values.

The initiation of reclamation measures (which include topsoil preservation, soil erosion control, topographic reconstruction, and revegetation) should follow basic preliminary considerations: a predisturbance inventory of soils and vegetation, a review of mine plans to effectively and economically incorporate reclamation, and a decision concerning post-mining land use. Mining and reclamation practices required to attain post-mining land use goals need to be determined on a site-by-site basis. Climatic influences on the interaction of biological, physical, and chemical processes of the local environment must be understood. Reclamation practices selected for implementation should take into consideration both micro- and macroclimates at the project site. After the preparation of reclamation plans consistent with inventories and goals, rehabilitation of soils and the landscape are completed accordingly. This is followed by management of reclaimed areas and monitoring to determine the overall degree of success.

Topsoil preservation is generally practiced by stockpiling desired soils as they are removed during mining operations. However, this tends to disrupt both favorable physical soil properties and desirable biological components (including live seeds, root structures, microrhize, and soil invertebrates). Saturation with water or construction of storage piles deep enough to limit diffusion of oxygen can be employed to inhibit decomposition of organic matter and sustain water-holding properties. Still, this does not prevent soil compaction or maintain important living components. It is more advantageous if the topsoil can be reapplied immediately in areas undergoing reclamation.

Topographic reconstruction of disturbed land refers to regrading and reshaping earth to achieve a more desirable land configuration. An accompanying effect of grading in a reduction of infiltration capacity caused by compaction and breakdown of the soil structure, resulting in increased runoff, greater soil erosion, and reduced water availability in the soil profile. Poor infiltration is offset somewhat by reducing the slope of the reclaimed surface. Reestablishing landscapes after mining such that infiltration and runoff relationships for reclaimed soils are comparable to those for pre-mined soils remains a major problem.

Reclamation costs and revegetation difficulties are greater on steeper slopes. Revegetation of slopes greater than 2:1 often requires intensive manual labor and expensive mulches, such as jute or excensior matting. Revegetation success is highest on slopes of 4:1 or less, and most conventional form implements used in reclamation can operate on slopes of 5:1 or less.

Large amounts of waste materials are generated at both surface and underground mining and beneficiation operations. To avoid unsightly and unstable spoil piles or tailings, heavy erosion, possible landsliding, and water pollution, some degree of regrading and reshaping must be performed after mining. Regrading can be considered the technique most essential to the reclamation of disturbed surfaces. Implemented in conjunction with other reclamation techniques, it contributes significantly to the aesthetic improvement of land surfaces, the control or erosion and water pollution, and the provision of a suitable base for revegetation.

Revegetation techniques are used to encourage the establishment of a vegetative cover on disturbed lands. They are also some of the most effective pollution control methods for surface mined lands, waste piles, and tailings. Properly established vegetation provides efficient erosion control and contributes significantly to chemical contamination control. Revegetation results in aesthetic improvement and often returns land to agricultural, silvicultural, or recreational usefulness.

The various measures incorporated in revegetation encompass topsoil replacement, surface preparation, soil amelioration, plant species selection, and planting techniques. The replacement of topsoil improves the survival and growth of selected vegetation and its effectiveness in controlling erosion. The difficulties encountered in establishing plant cover on mine wastes or tailings can be reduced by the application of topsoil which provides an improved rooting medium and improved availability of moisture and nutrients, leading to more rapid and vigorous plant establishment.

Surface preparation and soil amelioration are generally done concurrently. Compacted surfaces not conducive to plant growth are scarified by discing or plowing, with the removal of as much rock as possible to decrease the average grain size of the remaining material. Soil supplements are often needed to rectify deficient nutrient levels or to adjust the pH to a tolerance range for the species to be planted. By increasing the organic content of surface soils with organic mulch, both water retention and cation exchange capacities can be improved.

Plant species selection and planting techniques are complementary measures designed to achieve an optimal level of vegetation success for the given conditions. Species should be selected according to a land use plan and the site environment. A dense ground cover of grasses and legumes provides good erosion control and soil development. Where needed, trees and shrubs provide for control of poor slope stability and landsliding. Seeds or transplants are the two methods available to initiate a vegetative cover. Depending on the species and prevailing weather conditions seeding can be accomplished by the use of a seed drill, by broadcast methods, or by hydroseeding. Transplanting insures rapid vegetation, but requires significant amounts of manual labor. However, the successful establishment of certain kinds of trees or shrubs may require transplantation. Federal, state, or local reclamation requirements imposed on mining operations may dictate the planting of specific plant species for wildlife habitat or other purposes.

References

1. Brookman, G.T. and Middlesworth, B.C., 1979, Assessessment of Surface Runoff from Iron and Steel Mills, USEPA-600/2-79-046, Research Triangle Park.
2. Cheremisinoff, Paul N. and Young, Richard A. (eds.), 1975, *Pollution Engineering Practice Handbook*, Ann Arbor Science, Ann Arbor.
3. Danielson, John A. (ed.), 1973, *Air Pollution Engineering Manual*, 2nd Ed.; USEPA-AP-40, Research Triangle Park.
4. Greber, J.S., et al., 1979, *Assessment of Environmental Impact of the Mineral Mining Industry*, USEPA-600/2-79-107, Cincinnati.
5. Jutze, George and Axetell, Kenneth, 1974, *Investigation of Fugitive Dust Volume I—Sources, Emissions, and Control*, USEPA-450/3-74-036-a, Research Triangle Park.
6. Katari, V., Isaacs, G. and Devitt, T.W., 1974, *Trace Pollutant Emissions from the Processing of Metallic Ores*, USEPA-650/2-74-115, Research Triangle Park.
7. Katari, V., Isaacs, G. and Devitt, T.W., 1974, *Trace Pollutant Emissions from the Processing of Non-Metallic Ores*, USEPA-650/2-74-122, Research Triangle Park.

8. Marple, Virginia, Rubow, Kenneth, and Lantto, Orville, 1980 *Fugitive Dust Study of an Open Pit Coal Mine*, U.S. Dept. of Interior, BOM J0295071, Washington.
9. Meek, Richard L. (ed.), 1979, *Control of Particulate Emissions in the Primary Nonferrous Metals Industries—Symposium Proceedings*, USEPA-600/2-79-211, Cincinnati.
10. Midwest Research Institute, 1984, *A Study of Waste Generation, Treatment and Disposal in the Metals Mining Industry*, USEPA-68-01-2665, Washington.
11. Nerkervis, R.J. and Hallowell, J.B., 1976, *Metals Mining and Milling Process Profiles with Environmental Aspects*, USEPA-600/2-76-167, Research Triangle Park.
12. PEDCo. Environmental, Inc., 1980, *Industrial Process Profiles for Environmental Use: Chapter 27—Primary Lead Industry*, USEPA-600/2/80/168, Cincinnati.
13. PEDCo. Environmental, Inc., 1980, *Industrial Process Profiles for Environmental Use: Chapter 28—Primary Zinc Industry*, USEPA-600/2-80-169, Cincinnati.
14. PEDCo. Environmental, Inc., 1980, *Industrial Process Profiles for Environmental Use: Chapter 29—Primary Copper Industry*, USEPA-600/2/80/170, Cincinnati.
15. PEDCo. Environmental, Inc., 1978, *Survey of Fugitive Dust from Coal Mines*, USEPA-600/1-78-003, Denver.
16. PEDCo. Environmental, Inc., 1977, *Technical Guidance for Control of Industrial Process Fugitive Particulate Emissions*, USEPA-450/3-77-010, Research Triangle Park.
17. Tourbler, J. Toby, and Westmacott, Richard, 1980, *Small Surface Coal Mine Operator's Handbook—Water Resources Protection Techniques*, U.S. Dept. of Interior: Office of Surface Mining, Grant No. 13-34-001-8900, University of Delaware.
18. TRW, 1979, *Evaluation of Best Management Practices for Mining Solid Waste Storage Disposal and Treatment*. USEPA-RFP-CI/79-0357, Cincinnati.
19. U.S. Environmental Protection Agency, 1977, *Compilation of Air Pollutant Emission Factors, 3rd Edition*, Publication No. AP-42, Parts A and B, Research Triangle Park.
20. U.S. Environmental Protection Agency, 1982, *Development Document for Final Effluent Limitations Guidelines and Standards for the Ore Mining and Dressing Point Source Category*, USEPA-440/1-82/061, Washington.
21. U.S. Environmental Protection Agency, 1976, *Erosion and Sediment Control*, USEPA-625/3-76-006, Washington.
22. U.S. Environmental Protection Agency, 1973, *Processes, Procedures, and Methods to Control Pollution from Mining Activities*, USEPA-430/9-73-011, Washington.
23. Williams, Roy E., 1975, *Waste Production and Disposal in Mining, Milling, and Metallurgical Industries*, Miller Freeman, San Francisco.
24. World Bank: Office of Environmental Affairs, 1981, *World Bank Environmentl Guidelines*, Washington.
25. World Bank: Office of Environmental Affairs, 1978, *Environmental Considerations for the Industrial Development Sector*, Washington.
26. World Bank: Office of Environmental Affairs, 1974, *Pollution Control Technology*, Research and Education Association, New York.

5.15C

Minerals and Land Management

L. J. MacDonnell* and D. A. Gulley**

INTRODUCTION

Environmental impacts are most commonly associated with air and water pollution. Notwithstanding this fact, to many people the primary disruption caused by mineral extraction is the change in the character of the locale. That is, the environmental impact of mining includes, but is not limited to, the generation of residuals. The dramatic transformation of wilderness or rural land often creates strong reactions to mineral development. Thus, the basic issues of land use management are a useful addition to environmental discussion. In this section, we take up a discussion of federal, state, and local land use management regimes, along with a brief introduction to the potential role of economic analysis. Elsewhere in this volume, land management issues are discussed from other orientations, and the interested reader may wish to look there as well.

Locatable Minerals on Federal Lands

Legislation and Administration: Access to hard-rock minerals on the public lands is governed by the General Mining Law of 1872. This law declares that "valuable mineral deposits in land belonging to the United States . . . are hereby declared to be free and open to exploration and purchase . . . ." In order to initiate rights, prospectors are required to physically stake claims of specified sizes and shapes following basic procedures set by federal law which may be supplemented by the individual states. Rights to these claims are protected at the exploration stage so long as actual possession is maintained and diligent exploration activity is underway. If a valuable mineral deposit is discovered on the claim, the claimholder earns the exclusive right of possession as against private parties so long as the claim is properly maintained. To maintain this exclusive right the claimholder must perform $100 worth of labor or improvements, known as assessment work, on each claim every year and make related annual filings with the U.S. Department of the Interior. If the existence of a valuable mineral deposit can be demonstrated, and other legal requirements are met, the claimholder may receive a patent giving full title to the claimed land, including the surface in most cases. The purchase price is either $2.50 or $5.00 per acre.

The Mining Law has remained largely unchanged by legislative amendment during its now more than 100-year existence. However, while the law itself has not changed, its interpretation and administration have changed considerably. Written at a time of rapid western expansion of the U.S. following the Civil War, the intent of the law was to promote mineral development by inviting unrestricted prospecting and by selling the lands containing mineral deposits at a minimal charge. As federal land policy shifted from one of rapid disposal into private ownership to one of retention and management for a number of uses, mineral patenting requirements have been administered more strictly. The most critical issue concerns what must be demonstrated to establish the existence of a valuable mineral deposit. The so-called "prudent man" standard was established in the case of *Castle v. Womble* (19 L.D. 455 (1894)):

*Natural Resources Law Center, University of Colorado School of Law
**Henry Krumb School of Mines, Columbia University.

> A mineral discovery, sufficient to warrant the location of a mining claim may be regarded as proven, where mineral is found and the evidence shows that a person of ordinary prudence would be justified in the further expenditure of his labor and means, with a reasonable prospect of success in developing a valuable mine.

More recently, it has been held (*United States v. Coleman*, 390 U.S. 599 (1968)) that a mineral deposit must be currently mineable at a profit for it to be considered "valuable." The increased stringency has resulted in far fewer patents granted than in earlier times.

Significant legislative and regulatory actions have changed the land use position of the mineral explorer on the public lands (see, generally, Marsh and Sherwood, 1980), Under the 1872 Mining Law the prospector is given an open invitation to explore the public lands. No limits are placed on these activities by the Mining Law. However, the Surface Resources Act of 1955 established the "multiple use" management philosophy for the surface resources of the public lands. Specifically, it subjected the surface use of unpatented mining claims to continued federal management and use.

In 1974 the Forest Service issued surface management regulations governing mining operations on national forest lands. With certain rather narrow exceptions, before activities potentially causing damage to surface resources can be conducted, a notice of intent must be filed with the district ranger. If the proposed activities are considered likely to cause "significant disturbance of surface resources," an acceptable plan of operations must be submitted. All activities must be conducted so as to minimize environmental damage. Surface reclamation of mined areas is required.

The Bureau of Land Management (BLM) also has issued regulations governing mining operations as they affect the public lands under its management. With the passage of the Federal Land Policy and Management Act (FLPMA) in 1976, the BLM was given a mandate for multiple-use management and required to undertake necessary actions to prevent "unnecessary or undue degradation of the lands." Under the BLM regulations, a notice of intent to operate is necessary before commencing activities involving more than "casual use" of land surface. If planned surface disturbance will exceed five acres on an annual basis, an approved operating plan is required. The operating plan must include appropriate environmental protection and reclamation procedures.

FLPMA also required the owners of existing unpatented mining claims to file a record of these claims with the appropriate BLM state office within three years. Notice of new claim locations must be given to the BLM within 90 days of location. Annual reporting of assessment work on claims also is required. Failure to make these filings is deemed to conclusively constitute an abandonment of a claim.

Approval of operating plans for mining by either the Forest Service or the BLM may be considered to be a "major federal action significantly affecting the human environment," thus requiring the preparation of an environmental impact statement (EIS) under NEPA. Other permit requirements associated with mining such as rights-of-way also may prompt the preparation of an EIS.

Mineral operations in designated wilderness areas or in areas under consideration for wilderness designation are restricted. The Wilderness Act of 1964 specifically allowed mineral exploration in wilderness areas until December 31, 1983. Following that time, holders of valid existing rights (mining claims or mineral leases) will be permitted to continue operations. However, no new claims or leases are now permitted in these areas. Surface management requirements designed to preserve the wilderness characteristics of lands in designated and prospective wilderness areas are, as might be expected, stringent.

Current Issues: Prior to the development of surface management regulations by the Forest Service and the BLM and passage of FLPMA by Congress, mining activities under the 1872 Mining Law were subject to virtually no direct federal control, although state reclamation requirements and the general federal environmental statutes were applicable. Indeed, mining and mineral exploration had been treated as a dominant land use based on the directive in the 1872 law that the public lands containing valuable mineral deposits be free and open to exploration and purchase. Now, however, it appears to be substantially clear that prospecting and mining activities may be regulated to allow protection for the surface values of the public lands. To this point, there has been relatively little conflict under the surface management regulations.

However, the potential for such conflict exists. A carefully balanced accommodation of interests is necessary if public land management is to function effectively while maintaining the needed incentive for mineral development.

The recordation requirements under FLPMA have enabled federal land management agencies to learn the geographic location of active claims and have cleared away the uncertainty associated with old, inactive claims. Accurate knowledge of existing mineral exploration activity should be of considerable value to land management agencies in their planning and management responsibilities. However, there has been some confusion in the administration of these recordation requirements and the mining industry generally has been unhappy with the conclusive presumption of abandonment upon failure to comply with the annual filing requirements (see, e.g. King, 1983).

The closing of wilderness areas to new claim location beginning in 1984 is likely to cause some conflicts. As discussed above, mining activities on valid existing claims may go forward subject to surface use regulations. However, claim locations not maintained in accordance with federal recordation requirements may no longer be restaked. Moreover, claim locations only become vested property rights with the claimholder when there has been a discovery of valuable minerals. As mentioned, current mineability at a profit must be shown to establish such a discovery. Because of the inherent conflict between wilderness and mining, the validity of mining claims in wilderness areas with respect to the existence of a valuable mineral deposit is likely to be actively challenged.

Before new wilderness areas are designated they must be examined by the Bureau of Mines and the Geological Survey to determine the extent of known mineralization. These surveys are based largely on available public information. Some field checking is carried out but there is limited geophysical and geochemical examination or drilling. With these areas now closed to new exploration, knowledge of their mineral potential is limited.

In addition to wilderness areas, other public lands withdrawn from mineral entry remain a source of controversy. FLPMA effected major changes regarding land withdrawals. Explicit statutory guidelines governing new withdrawals were established. Moreover, FLPMA requires a review of all existing withdrawals ''having a specific period'' to determine if the withdrawal is still required. Many withdrawals were made because of the absence of procedures for managing conflicting uses. With better management tools available, some of these areas should be made available for mineral entry once again. However, progress in reviewing withdrawn areas has been slow. Moreover, the withdrawal process continues to be a battleground for conflicting congressional and administration objectives. Public land withdrawals appear likely to remain a source of conflict for some time to come.

Finally, legislation passed by Congress in the 1970s requires the Bureau of Land Management and the Forest Service to develop comprehensive land management plans for virtually all public lands. The extent to which mineral development may be indirectly controlled or regulated through land management plans, especially with respect to restrictions on access to and across public lands, is developing as a major issue. A related area of recent conflict concerns the degree to which state and local laws may control mining activities on the public lands.

State and Local Government Controls

Traditionally, the ownership of land carries with it the right to the full use and enjoyment of that land. However, it has long been recognized that government, in its exercise of the ''police power,'' may interfere with this unrestricted use if necessary to protect public safety, health and welfare or for other valid public purposes. Thus, uses of private land which create public nuisances or hazards to public safety have been subjected to government regulation. Under the U.S. system, this police power exists with state government as a part of the inherent authority of the sovereign. In turn, state governments have made delegations of this authority to local government. Judicial review of the exercise of this authority by state or local government generally focuses on the ''reasonableness'' of the regulation.

Mineral development is a unique land use because it may occur only in the fortuitous circumstance where an economically valuable mineral deposit exists. As Justice Holmes noted in the case of *Pennsylvania Coal Co. v. Mahon* (260 U.S. 393, 414 [1922]): ''For practical purposes, the right to coal consists in the right to mine it.'' The mining of a given deposit is not a transportable activity. Regulation which prohibits or seriously impairs mining activity may amount to a taking of private land by government without just compensation in violation of the U.S. Constitution.

At the same time it is apparent that mineral development may create serious conflicts with adjacent land uses. Society's interest in obtaining the minerals is tempered by its interest in enjoying other values of the land. In general, the development of a valuable mineral deposit will be the highest use of the land containing that deposit. However, exceptions are recognized. For example, if that land is located in an already developed area such as a residential neighborhood or in an area identified as having unique ecological qualities, the disruption caused by mining may be considered undesirable or in need of strict control.

The major method of local government control of the use of land is through zoning. Zoning first was applied in urban areas to separate incompatible activities. In many places zoning has now been extended into non-urban areas as well. Local governments have used their zoning authority to prohibit or restrict mining activities. The decisions in legal cases involving a challenge to such zoning ordinances evidence an attempt to balance the competing interests in each specific situation (see, for example, the cases presented in 82 Am. Jur. 2d § 137-139). After reviewing the cases in this area, one author (Keppler, 1979, p. 814) has concluded:

> [P]erhaps all that can be said is that the probability that zoning which prohibits mining on a particular tract will be upheld increases the closer such tract is located to a municipality or concentrated residential development, the more disruptive the mining will be with respect to adjoining land uses, and the greater the alternative uses to which the land can be put. Conversely, the restriction is more likely to be struck down where the loss to the landowner is great, where society will noticeably suffer from not having the mineral extracted, where the mineral is extremely rare or the deposit unusually valuable, and where the mining will occur sufficiently distant from more fragile land uses so as to cause minimal disruption or disturbance.

Beginning in the 1960s several states began to take a more active role in land use decisions. (see Anderson, 1976) For example, Hawaii created a state agency with authority to zone the entire state using four classifications — rural, commercial, industrial, and recreational. Vermont established a system of state control of major development and made that development contingent upon obtaining a permit. Numerous states established explicit controls over development in specific areas such as coastal zones or wetlands.

This wave of activity at the state level grew out of a concern that local controls were too narrow and piecemeal and that rational land-use planning required at least a regional focus. Growing awareness of the environmental impacts of major development became an even more significant motivation for the states to get involved. An additional concern, especially among some of the sparsely populated western states, was with the so-called "socioeconomic" impacts associated with rapid development in rural areas — the effect on quality of life caused by inadequacies in housing and municipal services. As a result, some state and local governments now make the use of land for mineral development conditional upon implementing measures to mitigate environmental damages and, in some cases, socioeconomic impacts (see Barnhill, 1983).

Mineral development on federal land appears to be less subject to state and local control, especially where that control is deemed to significantly interfere with explicit federal objectives. For example, Ventura County, Calif. required the holder of a federal oil and gas lease on lands located within the county to obtain a permit for exploration and drilling. The Federal Circuit Court held that the county had no authority to interfere with the federally authorized use of these lands. (Ventura County v. Gulf Oil Corp., 601 F.2d 1080 [9th Cir. 1979], *aff'd without opinion*, 445 U.S. 947 [1980]).

Leasable Minerals on Federal Lands

Distinctive Features: Minerals not covered by the location-patent system described in the previous section fall into two general categories: those minerals not covered by the term "hard rock," and minerals on the so-called "acquired lands," which are not part of the original public domain. Public lands minerals in the first category are most notably the energy minerals coal, oil, and natural gas. An example of the second category is the lead deposits of Missouri. We will restrict our coverage to the energy minerals of the first category. These minerals are governed by the 1920 Mineral Leasing Act and ancillary legislation.

The leasing system embodies significant changes in public lands policy, reflecting changes in philosophy from 1872 to 1920, and also reflecting the unique circumstances of those minerals. The impetus for these changes came from a desire to retain ownership of public lands. It also reflects the desire to accrue public revenues from the development of federal lands, and to eliminate the acquisition of mineral lands indirectly, such as by homesteading. The change in philosophy corresponds to the ascendency of the progressive conservation movement (Hays, 1959).

In contrast to the location-patent system developed in the 1872 mining laws, leasable minerals are characterized by: higher payments to the federal treasury; retention of the surface estate by the government; and (in the case of coal) incorporation of mineral extraction into an overall multiple use framework. Payments can take several forms including a front-end payment (a bonus bid), production royalties, and annual acreage rentals. Sizeable sums of money are received by the government via energy mineral leases. In 1980, for example, total royalties under the Mineral Leasing Act were: petroleum, $2,961 million; natural gas, $1,653 million; and coal, $40 million (a moratorium on new leasing was in effect at the time) (U.S. Department of the Interior, 1982).

Unlike the patent-location system, in the leasing system the government retains ownership and some use privileges of the surface estate. However, in some instances the government does not own the surface, in which case resource development occurs only with surface owner consent. Since mineral leases are not intended to be permanent, private control is temporary, and in the case of undeveloped leases, public use of the surface may be ongoing. In contrast to the 1872 Mining Act, the 1920 leasing system does not elevate mineral extraction to the status of a dominant or preemptive land using activity. Instead, professional land managers are expected to balance the value of mineral production against the land's alternate uses. As mentioned previously, use privileges of the surface are governed by the Surface Resource Act of 1955, FLPMA, and NFMA. Each of the distinctive features of the leasing system has generated controversies, to be discussed shortly. Fuller discussions of the leasing system are provided by Gordon, elsewhere in this volume, and in McDonald (1979), Nelson (1983), and Gordon (1981).

The Leasing Process: The principal agency administering the leasing program at the present time is the Bureau of Land Management. However many other federal and state agencies are involved. In addition to the leasing acts, related legislation shapes the agency mission, the leasing procedures, and the terms of the lease. For example, FLPMA requires the BLM planning processes to be compatible with state and local plans. SMCRA establishes unsuitability criteria which constrain areas of leasing. NEPA has led to greater public involvement in the leasing process. Culhane (1981) provides an introduction to the milieu of the land agency.

Legislation and judicial review have created an overall leasing structure, although the details of the process often change. These details vary with the mineral, so that coal, on-shore oil and gas, and offshore oil and gas are handled somewhat differently. We restrict ourselves here to the general format used in coal leasing (and to a lesser extent, competitive oil and gas leasing). Four general components are involved: land use planning, activity planning, sales procedures and lease administration. In the land use planning stages, government officials assemble and analyze information on public lands parcels. It is at this stage that competing land uses are first evaluated. For example, coal-bearing lands will be identified, and information on the parcel's grazing, wildlife habitat, archaeological, and other forms of significance will be compiled.

In the second stage, activity planning, individual tracts are delineated and choices made as to which tracts will be offered for lease. Environmental statements are prepared, and public participation encouraged through series of hearings. In the third stage, sales are announced, bids evaluated, the bidder's qualifications reviewed, and lease documents are prepared. Mineral extraction must comply with various stipulations. Following the lease award comes the continuing monitoring of the lease activity referred to as lease administration. This phase also encompasses various other activities, such as exchange and exploration licensing. Readers will want to supplement this brief treatment with more detailed descriptions and with the specific regulations.

The most extensive and elaborate modifications to mineral leasing are those associated with federal coal. The principal legislation supplementing the Mineral Leasing Act is the Federal Coal Leasing Amendments act of 1976. This law together with several major judicial decisions led to a federal coal leasing program, initiated in 1981. For a description, see Nelson (1983), Gulley and Mei (1984) and U.S. Department of the Interior (1979). The coal program is noteworthy in several respects: the incorporation of state government and the public in leasing decisions; the length and complexity of the process; and a widespread belief that the coal format might serve as a prototype for the other minerals (Gulley and Mei, 1984).

Leasing and the Environment: Federal legislation and adjudication have led to the incorporation of environmental issues at several points in the leasing program. For example, in coal leasing the decision from Sierra Club v. Kleppe (427 U.S. 390, 1976) extensively dealt with NEPA environmental impact statement requirements and NRDC v. Huges (437 F. Supp. 981, D.D.C. 1977) extended the EIS requirements on a program-wide environmental statement. (That is, the types of federal actions requiring an EIS were expanded from site-specific to general federal programs). The Federal Coal Leasing Amendments Act (FCLAA) of 1976 addressed environmental and socioeconomic impacts not adequately dealt with in the 1920 act. Yet another law, FLPMA, further strengthened environmental orientation by endorsing multiple use, giving priority to the protection of areas of critical environmental concern, by requiring coordination with other federal and state and local plans. Sections 515 and 522 of SMCRA set environmental performance standards and require designations of areas unsuitable for surface coal mining. SMCRA also provides authority for the exchange of federal and private lands, where the exchange furthers the protection of alluvial valley floors, and requires surface owner consent in cases where federal coal is overlain by private-owned land. Nelson (1983) maintains that the surface owner consent provision transferred very lucrative property rights, i.e. permission to develop, from the public to private parties.

Within the coal leasing program, environmental impact mitigation is incorporated in a variety of ways. The key environmental feature is the use of unsuitability criteria as a screen to identify environmentally sensitive areas and prevent damaging development, either by prohibiting energy leasing on sensitive tracts or by requiring that energy developers obey restrictive conditions. Other mitigation measures include: (1) the extensive use of public participation to generate information about land use tradeoffs and environmental values and conditions; (2) the preparation of environmental impact statements and mitigation plans; (3) permit requirements and performance standards applicable to energy leases; and (4) lease tract ranking procedures that allow environmental and socioeconomic impacts for various potential tracts to be compared and used in deciding which tracts are to be offered for lease. See Department of Interior (1979) for more details.

The Preservation-Development Tradeoff

A Framework: Proper land management requires the determination of the highest and best use of the land. For some time now, it has been recognized that land has non-pecuniary value as well as commercial value. A need therefore exists to compare commercial and noncommercial values to determine highest and best use. Throughout this chapter, a variety of development constraints have been discussed which are intended to incorporate a variety of considerations in the land use decision. For example, zoning addresses the off-site impacts (external costs) of mining. In leasing, unsuitability criteria screen out otherwise leasable lands where the environmental costs are believed to be too high. Even in the location-patent system, where mining is presumed to be the highest and best use, the land's non-market values have a role in surface management requirements. For all this attention, however, public policy has seldom attempted to compare analytically a land parcel's amenity values (wilderness and recreation) with the parcel's commodity (mineral development) values. One reason is that such comparisons represent formidable analytical problems.

Recently, a number of economists have begun developing economic frameworks in which a site's commodity and amenity values could be compared directly. While it is safe to say the work is still in a preliminary state, the concepts (if not any specific numerical applications) provide a useful way of thinking about the general issue. The framework has been most fully articulated by Krutilla and Fisher (1975). Essentially, the approach is a modification of benefit-cost analysis (see Section 5.15). These authors believe the distribution of these benefits and costs is relevant. That is, the Kaldor-Hicks criterion is useful but cannot be the sole guide. (Krutilla and Fisher suggest that earlier writing, such as that of the Public Land Law Review Commission had implied that net benefits were an appropriate basis for decision, a much stronger statement.) The extensions of cost-benefit analysis lie in: first, subtracting environmental opportunity costs from the present value of development (a departure from the discussion in Section 5.15), and second, considering a time path of multiple (and irreversible) investments, instead of a single investment. Irreversibility is a central feature of the analysis. The development of mineral reserves embraces two aspects of irreversibility: a nonrenewable resource is depleted, so that the resource's future production value is lost (its "shadow price" or "user cost"); and, a loss of value in perpetuity arises, associated with the disruption of the wilderness amenities. An additional finding of Krutilla and Fisher is that under certain circumstances, the analysis need only consider two alternatives: immediate resource development or permanent preservation. The advantage of this finding is that it eliminates the need to consider the infinite intermediate cases of postponed development. The necessary condition, however, is restrictive — namely, the requirement that the returns to development relative to preservation are continuously decreasing or constant over time. Whether this is the case is of course a matter of judgment based on considerations of the availability of substitutes and so on. Krutilla and Fisher and also others have argued that given the nature of wilderness, this condition is likely to hold in a great many cases.

These conditions lead to the formulation of a model which makes the statement that one should maximize the discounted sum of development-related and production-related benefits, less investment costs. Environmental costs are incorporated implicitly as the foregone benefits from preservation. This need not always be correct, although it may be in areas qualifying as true wilderness. In areas of prior human use, new mine reclamation can make a net environmental improvement by cleaning up older problems. This has occurred in areas such as Appalachia, and also in the Rockies and Sierra Nevada where for example old tailings may create acid runoff problems, *prior* to development. However the analysis of Fisher and Krutilla is correct under the conditions they visualize. The formal mathematical model, stated as a simple problem in optimal control theory, is given in Krutilla and Fisher (1975, pp. 50-57). The benefits of preservation and development could be measured in terms of consumer's and producer's surplus, or the other concepts mentioned in section 5.15.

A well-known problem in benefit-cost analysis is the selection of a social discount rate. This issue has prompted lengthy discussions, and there is little need to air the issue in its entirety here. The approach taken by Krutilla and Fisher is to use the market opportunity cost of capital as a starting point, accompanied by sensitivity analysis. The authors recognize that virtually any benefit-cost approach reflects primarily the needs and interests of the present generation. The approach can not, therefore, be considered as the sole source of wilderness-development policy making.

A Case Study: Krutilla and Fisher (1975) applied their framework to a proposed open pit molybdenum mine-mill complex in the White Cloud Peaks area of Idaho. In brief, the options for this region are continued recreation and grazing, on the one hand, and on the other hand eventual development of the ore body. Due to the market conditions facing molybdenum producers, it appeared very unlikely that near term development was imminent. Interestingly, the analysis does not devolve to a mere comparison of the two alternatives' net present value, since in the case of deferred development the option also should include the near-term benefits from pre-mining land uses such as recreation and grazing.

Krutilla and Fisher were able to generate benefit values for wilderness recreation and as range resources. The analyses were complicated by the fact that further investment in the area's recreation and range potential would affect the level of benefits. Placing a dollar value on recreational benefits is obviously problematical. The authors' estimates of the present value ranged from about $1 million to $5 million. Range benefits were much more modest: about $100,000 to $200,000 in present value. By contrast, the mineral reserve's value might seem straightforward, but Krutilla and Fisher did not have the data to perform an adequate mine valuation study. They conjectured that the present value would be modest, perhaps in the area of $2 million to $3 million.

What these results seem to point to is that benefit-cost analysis might be a reliable guide only where one alternative is overwhelmingly attractive (or unattractive). Even in this case, which involves a small low-grade deposit of an alloying element for which present and future excess industrial capacity exists, no clear conclusions were possible. The drawback would seem to be that quantitative analysis is least necessary when the tradeoffs clearly favor one or the other. In such cases the best choice might be intuitively obvious; and in cases where a gray area exists, the uncertainty in the numbers will not clear up the matter. The benefit-cost approach is conceptually interesting, but not yet a practical guide to site-specific policy.

References

1. Anderson, Robert M., 1976, American Law of Zoning, 2nd ed., The Lawyers Co-operative Publishing Co., Rochester, NY.
2. Barnhill, Jr., Kenneth E. and Dianne Sawaya-Barnes, 1983, "The Role of Local Government in Mineral Development," *Proceedings of the Twenty-Eighth Annual Rocky Mountain Mineral Law Institute*, Mathew Bender & Company, New York.
3. Culhane, Paul J., 1981,*Public Lands Politics: Interest Group Influence on the Forest Service and the Bureau of Land Management*, The Johns Hopkins Press for Resources for the Future, Baltimore.
4. Gordon, R. L., 1981, *Federal Coal Leasing Policy Competition in the Energy Industries*, American Enterprise Institute for Public Policy Research, Washington, D.C.
5. Gulley, D. A. and D. M. Mei, 1984, "An Appraisal of Models as a Support of Decision Making in the Federal Coal Management Program," U.S. Bureau of Mines, Washington, Mining Research Contract Report J0133908.
6. Hays, S. P., 1959, *Conservation and the Gospel of Efficiency*, Harvard University Press, 297 pp. Cambridge, Mass.
7. Keppler, Peter, 1979, "County Regulation of Mineral Development," *Western Land Use Regulation and Mined Land Regulation*, Rocky Mountain Mineral Law Foundation.
8. King, James M., 1983, "The Validity and Interpretation of Section of 314 of FLPMA", *Proceedings of the Twenty-Eighth Annual Rocky Mountain Mineral Law Institute*, Mathew Bender & Company, New York.
9. Marsh, William R. and Don H. Sherwood, 1980, "Metamorphosis in Mining Law: Federal Legislative and Regulatory Amendment and Supplementation of the General Mining Law Since 1955," *Proceedings of the Twenty-Sixth Annual Rocky Mountain Mineral Law Institute*, Mathew Bender & Company, New York.
10. McDonald, S. L., 1979, *The Leasing of Federal Lands for Fossil Fuels Production*, The Johns Hopkins Press for Resources for the Future, Baltimore.
11. Nelson, Robert C., 1982, "The Public Lands," *in* Paul R. Portney, ed., *Current Issues in Natural Resource Policy*, The Johns Hopkins Press for Resources for the Future, Baltimore.
12. U.S. Department of the Interior, 1979, *Final Environmental Statement, Federal Coal Management Program*, U.S. Government Printing Office, Washington.
13. U.S. Department of the Interior, 1982, *Royalties: A Report on Federal and Indian Mineral Reserves for 1981*, U.S. Government Printing Office, 72 pp., Washington.

5.15D

Benefits and Costs of Environmental Compliance: A Survey

David A. Gulley* and Bonn J. Macy*

INTRODUCTION

Some of the most famous and graphic environmental incidents have arisen from mineral extraction, including oil spills, the Reserve Mining asbestos tailings case, Kaiparowitts surface mining and coal combustion, and other environmental causes celebre. The remote location of much of the mining industry to some extent reduces the human health hazards from pollution, but often times the pristine locale in which mining is conducted provides a sharp contrast to the impacting industry. This prompts questions as to what the research literature says about the true costs and benefits of environmental compliance. We survey the state of this knowledge. Where possible, we discuss studies focused on mining and related industries, such as primary metals production. In many cases, however, the studies pertain to the economy as a whole.

For present purposes, the mining industry is defined as companies engaged in the production and refinement of minerals, including metals, fuels, and nonmetallics (e.g., phosphate, sand and gravel). There are several stages in mining, with environmental effects occuring at every stage. The exploration stage comprises the period from initial prospecting to the point where the deposit has been located and delineated in sufficient detail to permit development decisions to be made. Geophysical and geochemical surveys, diamond drilling programs, and exploratory stripping and excavation are activities that may be employed during this stage. Some forms of prospecting are relatively benign, although even airborne surveys and motor vehicles generate noise pollution. Roads and the passage of vehicles and equipment affect land surfaces and vegetation, and attract heavier backcountry use. Clearings, stream crossings, and borrow pits can lead to increased erosion and sedimentation of streams. Geological survey grids disrupt vegetation and appearance. Moreover, prospecting can involve the excavation and stripping of overburden, resulting in soil and vegetation disruption as well as increased sedimentation of streams and lakes. The effects of exploration are often temporary, and frequently less severe than for other activities. Nevertheless, exploration-related impacts vary with the intensity and type of exploration effort, and can be significant.

During the development stage of mining, shafts are sunk, pits excavated, buildings erected, and transportation facilities established. Dust, noise, and other socioeconomic and environmental impacts are generated by the relatively large and sudden influx of people, equipment, and activity. Surface disturbance and solid waste are generated by development, and liquid wastes form when mines are dewatered.

The extraction phase is associated with a variety of environmental impacts. These impacts may differ for surface and underground mines. In surface mining, surface disturbance is obviously a major consequence. Piles of overburden and waste accumulation mar land surfaces. Wind and water erosion tend to spread the dis-

*Henry Krumb School of Mines, Columbia University

turbance to the surrounding areas, increasing the amount of area affected. Large quantities are involved in the aggregate: in 1975 open pit copper mining generated 686 million tons of mine waste. The mining and processing of iron and steel produced another 500 million tons of waste. Underground mines usually create little surface disturbance during extraction, although subsidence is an occasional problem. Both forms of mining generate hydrospheric residuals. Mine water drainage and the leaching of surface wastes are sources of acidic or alkaline effects, as well as sources of particulates and heavy metals. Mine drainage is particularly troublesome in Eastern and Midwestern coal mining. In other areas lead and zinc mining create serious acid mine drainage, and in still other areas silt and noxious materials from mining enter rivers and lakes.

The beneficiation and refining stages involve the separation of desired substances from the remainder of the raw ore or mineral. Crushing and grinding operations associated with this stage are a source of airborne particulates and noise. Major waste disposal problems are associated with the residue from the ore, called tailings. As with mine wastes, wind and water erosion tend to distribute these wastes over wider areas, and oxidation and leaching can produce various types of water pollution, including acid or alkaline effects, heavy metals, and other contaminants. Atmospheric pollution is also associated with this stage, such as the release of sulphur dioxide associated with metallic mineral smelting.

The nature and extent of environmental impacts vary with location, mineralization, activity, and other variables. Location affects both the nature and magnitude of the environmental impact, and also the effect of the impact upon human populations. Among the types of environmental impacts associated with the mining industry, mine drainage problems are particularly important. Mine drainage can threaten aquatic life and leisure-related human values; it can also create hazards to human health and cause property damage. Surface disturbances and air emissions likewise affect aesthetic and productive values, and can create hazards to health and safety.

The Benefits of Environmental Compliance

Morbidity and Mortality Valuation: Section 5.15 described benefit measurement concepts. An important benefit category is health. Improvements in health, due to decreased morbidity and mortality, are (as might be expected) difficult to quantify. Taken as a whole (for all industries and all environmental regulations), estimates of these benefits range from $3 billion to $43 billion annually (Crocker, 1979; Freeman, 1971a).

The two most important methodologies for health benefit estimation are the opportunity cost — productivity of human capital approach, and the willingness to pay method. The two methods differ considerably. Most morbidity benefit studies apply the productivity of human capital method (Ridker, 1967; Lave and Seskin 1970, 1977; Waddel, 1974; Sagan, 1974; and Liu and Yu, 1976). In this approach, health benefits are imputed by means of the loss of output measured by forgone earnings, plus medical costs, increased insurance costs, and other considerations. The human capital method assumes that earnings reflect the individual's marginal productivity or the individual's contribution to total economic output.

There are, however, a number of difficulties with this method. It fails to account for individual differences in attitudes toward risk. It oversimplifies the relationship between actual benefits and the probability of illness. It assumes both that a person is worth what the person earns, and that what the person earns is related to the person's output. The technique cannot be used for persons not in the workforce, such as volunteer workers, housewives, disabled or retired persons, or members of the structurally unemployed. Results are highly sensitive to the age, sex, and race of the individuals involved, since these factors correlate with earnings.

Because of the above problems, the willingness to pay approach is conceptually attractive. However, the willingness to pay approach is also much harder to put into practice. The method is based on the notion that health can be evaluated like any market commodity, and that the amount a person is willing to pay for better health is a measure of the total benefit he or she derives from it. Unlike the other method, in which benefits can be inferred from generally available data, the willingness to pay method requires that additional data be collected through surveys. Questions may be either direct, such as explicitly questioning someone on what they would be willing to pay, or indirect, where willingness to pay is inferred from related questions.

Willingness to pay can also be established through related market data, such as the market price for safety or health related products. Finally, willingness to pay is more consistent with the general principles of benefit-cost analysis than is the other procedure. However, the technique is less often used in practice because of data collection and processing difficulties.

Needless to say, there are further problems with any morbidity-mortality benefit valuation, not yet mentioned. These problems include the primitive state of knowledge regarding the relationships between pollution and health, and problems in controlling for other important variables affecting pollution-related illness (e.g., smoking, age, diet). Freeman (1979a), in a study for the Council on Environmental Quality, sought to reconcile the differences between various studies, and made adjustments in the results to bring them to a common set of assumptions.

Air and Water Quality Benefits: Freeman estimated the monetary value of health benefits accruing to the population in 1978 as a result of improved environmental quality from 1970 to 1978. During this period, Freeman estimated that environmental quality improved 20 percent. We consider first his analysis of air pollution. His findings, which contain the methodological difficulties discussed above, suggest a range of benefits of from $0.3 to $11.5 billion for morbidity reduction, with $3 billion for the most likely figure. Additionally, mortality reduction benefits were estimated as $3 billion to $28 billion, with a $14 billion most likely figure. These figures were based on the willingness to pay approach; the individual's apparent willingness to pay for the reductions in the risk of mortality was estimated as approximately $1 million per statistical life (Crocker, 1979).

Additional benefits accrue from the reduction of damages to materials, such as metals, fabrics, rubber, and building materials. Freeman estimated that reductions in the level of the three main pollutants, sulfur dioxide, ozone, and particulates, accounted for 97% of the cost savings in this category, which were estimated in the range of $0.5 to $1.4 billion. (See Waddell, 1974).

Additional benefits from decreased soiling and cleaning costs of households, associated with a reduction in suspended particulates in the atmosphere, ranged from $0.5 to $5 billion. (See also Watson and Jakesh, 1978; and Liu and Yu, 1976). Improved visibility and more pleasing surroundings led to estimates of $0.5 billion nationally (SRI, 1981). A study of the Four Corners region put the regional figure at $15.5 to $27.6 million (Randall, Ives, and Eastman, 1974). The aesthetic value of property together with other damage categories accounted for benefits to residential property of $1.1 to $8.9 billion.

Benefits from the control of water pollution consist mainly of recreational benefits, such as fishing, swimming, boating, and water fowl hunting, plus aesthetics. Using willingness to pay approaches, Freeman (1979a) estimated water pollution control benefits associated with full compliance in 1985 with Clean Water Act requirements (a hypothetical situation). Recreational benefits were estimated at $4.1 to $14.1 billion, and other benefits were estimated at $1 to $5 billion. See Davidson, Adams, and Seneca (1966); Heintz, Hershaft and Horak (1976); National Planning Association (1975); Bell and Canterbery (1975); and Unger (1975). Freeman estimated health improvement values at $3 million to $2 billion. Reductions in industrial and municipal water quality costs range from $1.5 billion to $3.7 billion. This category includes: municipal treatment costs, industrial treatment costs, effects on commercial fishing, and various effects on households.

The above discussion covers regulations affecting many industries. As a heavily polluting industry, mining figures fairly prominently in the above discussion, but it is not possible to separate figures for a given industry. However, certain regulations focus specifically on the mineral industries, and some benefit studies of these regulations have been undertaken. These are discussed in a later section.

Studies of the Costs of Environmental Compliance

Both direct and indirect costs are created by environmental regulation. Direct costs are the actual expenditures on pollution control equipment, both operating and capital costs; and administrative compliance costs (record-keeping, documentation, interactions between public agencies and industry). Indirect costs include: loss of income that is diverted from production and spent on abatement and control; productivity and efficiency losses introduced by changes in procedures or lags in development and construction; and losses due to decreased industry competitiveness and market share. Some costs affecting firms may not affect society generally, e.g. loss of mining jobs compensated by increased employment in the pollution control technology industry.

Direct Costs of Compliance: Aggregate cost data are typically large numbers, and interpreting such numbers requires care. Since the mining industry is a small proportion of GNP, numbers that are large relative to industry value added can be small relative to the economy as a whole. For example, total nonfarm business new plant and equipment expenditures amounted to $316 billion in 1982. In contrast, the primary metals industry spent $0.76 billion on pollution abatement, a large number but dwarfed by the total plant and equipment figure. The latter figure amounted to over 10% of that industry's total expenditures for plant and equipment (Russo and Rutledge, 1983). Such costs, no less than benefits, are distributed unevenly across regions and population cohorts. Since most economic analyses of this topic have used the Kaldor criterion of welfare, distributional aspects have been frequently subordinated to the major problem of estimating the magnitudes of the costs.

Cost impact studies fall into several categories: (1) surveys of actual expenditures; (2) engineering process compliance cost estimates; (3) evaluation of productivity effects or regulation; and (4) projections of industry and macroeconomic impacts due to cost, process, and related changes. Some studies deal with all sectors of the economy. Others focus on a particular set of regulations or legislation, or a specific industry. Finally, some studies are repeated at regular intervals, while others are one-time efforts.

Three annual surveys estimate the direct cost or actual expenditures on pollution abatement by industry. The surveys vary considerably, in part due to variations in methodology, in part to inherent uncertainties and data problems. The Bureau of the Census survey is undertaken at the level of the individual establishment or plant. The survey samples 20,000 plants, and makes inferences about the overall economy. The Bureau of Economic Analysis survey 15,000 units; the units are firms rather than plants. Finally, the McGraw-Hill survey surveys 350 firms.

The difference between survey units introduces an important source of discrepancy. A plant usually engages in only one type of activity, while firms often engage in many. Surveys of firms attribute all pollution control expenditures to the firm's primary product. Thus, pollution control expenditures for U.S. Steel's paintmaking operation are recorded under the category iron and steel (Portney, 1981). The differences between survey methodologies complicate comparisons among the studies, but at least offer several alternate approaches to the estimation of the same overall topic.

Turning to the survey findings for the minerals industry, the Bureau of Economic Analysis (BEA) survey provides the best data. The McGraw-Hill survey technique is less reliable, and the Census Bureau does not report results for mining specifically. In 1981, the BEA estimates total pollution abatement expenditures at $8.93 billion. Of this total, mining expenditures were $0.46 billion; primary metals expenditures were $0.78 billion; petroleum expenditures were $1.76 billion; and electric utilities spent $2.7 billion.

These data are subject to the usual problems associated with self-administered mail questionnaires. Additional difficulties are due to allocating expenditures to pollution control when the capital investment also improves productivity, and possibly problems of strategic responses, i.e., exaggerations for public relations purposes. See Portney (1981). Considerable variation in the above figures can occur from year to year.

The Census and BEA surveys estimate direct costs besides capital expenditures. The Census Bureau provides information on operating costs, and research and development expenditures. In 1981, abatement operating expenses for all industries was estimated by the Census Bureau at $9.1 billion. Primary metals accounted for $1.9 billion, nearly 20% of the total. Research and development expenditures in 1980 totaled $1.18 billion. Even more comprehensive data are generated by the BEA survey, which divides total pollution control compliance costs into categories for: current account spending, capital consumption allowance, net imputed return for pollution abatement capital, and research and development spending. The BEA data indicates that total industrial pollution control costs increased during the 1972-1981 decade at an average annual rate of 17.5% in current dollars (uncorrected for inflation). Total industrial costs in 1981 were estimated at $53.88 billion; total costs including government expenditures were $60.33 billion in the same year.

Another annual survey of pollution abatement costs was conducted by the Council on Environmental Quality (CEQ). The survey, no longer conducted, was based on the engineering process cost approach. Despite the differences in methodology, the CEQ and BEA studies arrived at similar cost figures. An interesting feature of the CEQ study was its recognition that pollution control expenditures include those made in response to regulation, and also expenditures which would have been made in the absence of federal regulation. Total incremental costs (those incurred as a result of regulation) were estimated at $36.9 billion in 1979, the last year the study was undertaken. That study forecast a growth in incremental expenditures to $69 billion in 1988, a growth rate roughly double that forecast for the economy as a whole.

Indirect Cost and Other Studies: Most studies of the indirect and macroeconomic costs of pollution control use large scale models of the macroeconomy. Typically, the model generates two scenarios, one which includes incremental abatement costs, and one without such costs. This leads to differences in various economic indices, such as GNP, and these differences (projected rather than actual) are taken as the aggregate cost of compliance.

A recent study of this type was conducted by Data Resources, Inc. (DRI, 1981). The study was based on the historical period 1970-1981, and projected results for 1981-1987. Four major effects of incremental abatement costs were documented. The first of these is the effect on the general price level. DRI found that the consumer price index was in its entirety 0.4% percentage points higher over the 1970-1987 period when these costs were added. Additionally, the producer's price index and GNP deflator were 0.5 and 0.4 percentage points higher, respectively. The second effect is upon GNP. This effect varied over time. Environmental compliance increased GNP in 1977 by $9.2 billion (0.5%), due to the added purchases of abatement equipment. However, by 1979 GNP had decreased, and was projected to be $21.8 billion (0.7%) lower by 1987. The third effect is upon employment; compliance stimulates employment throughout the period, creating 524,000 additional jobs by 1987 and reducing the unemployment rate by 0.3 percent. Finally, real business fixed investment is stimulated by mandated investment, with 1973 investment 2.6% higher, a figure that drops to 0.3% in 1987, due to higher prices, increased interest rates, and reduced aggregate demand.

Several other studies are worth mentioning at this time. Denison (1979) approaches the productivity growth question by means of a "growth accounting" framework, which uses a combination of ad hoc analysis and professional judgment (see also Haveman and Christainsen, 1981). Denison's results were similar to the DRI results, however. He estimated that environmental regulations reduced productivity 0.1% annually from 1969 to 1973, 0.22% annually during 1973-1975, and 0.08% per annum from 1975 to 1978. Crandale (1980) examined productivity growth in various industries, and suggested that "pollution control impacted industries" had experienced significantly larger declines in productivity during the 1970s than had other industries. However, all industries had experienced productivity declines.

Another study, by Ridker and Watson (1981), projected a variety of costs given differing air and water standards. Taking their estimates for relaxed and strict standards as a range (and recognizing the numbers are projections), the costs (in billions of 1971 dollars) include: for electric utilities, $1.2 to $1.4 (1975) and $1.7 to $4.0 (1985); for acid mine sediment control, under $0.7 in all cases; and for land reclamation, $0.4 (1975) and $0.3 (1985). Total estimated direct abatement costs are: $12.3 to $14.5 (1975) and $30.2 to $44.0 (1985). They also estimated abatement investment as a fraction of total investments for selected industries. These figures include the following, where the range includes both strict and relaxed standards, and projections for 1976 to 1990: for electric utilities, 8 to 16%; for petroleum refining, 8 to 28; for steel, 6 to 17; for aluminum, 4 to 11; and for cement, concrete, and gypsum, 4 to 20. The average for the economy was projected to be 3 to 5 percent.

In reviewing these figures, the reader should bear in mind that the accuracy of the model results is an open question. Sources of error may include basic assumptions about policy (e.g., some studies assumed an expansionary and accommodating monetary policy, which however did not materialize during part of the forecast period); additionally, there are always problems with baseline data, and with model forecast accuracy. Moreover, large scale macroeconomic models are useful in determining the effect of certain policy issues. Yet macroeconomic models are not amenable to the analysis of several classes of effects. For example, regulation not only increases costs, but can also discourage new activities from being undertaken. Project risks increase with regulatory uncertainty, biasing the selection of projects. Regulation also introduces distributional problems. For example, the 1977 Clean Air Act amendments prevent significant deterioration of air quality in relatively pristine locales, which many observors believe is slowing those regions' economic growth, and may also lead to a long term shift in the relative size of different components of GNP.

Benefits and Costs of Specific Mineral Operations

In previous sections, industry-wide benefit-cost analyses were summarized. The purpose of this section is to survey the literature pertaining to specific mineral technologies and operations. Cost and (to a lesser degree) benefit studies have been undertaken for a variety of processes and a variety of specific regulations, and the results of these studies are surveyed here.

While this information is provided in the belief it is a necessary point of departure, the reader is cautioned that the utility of the estimates is an open question. First, a variety of methods can be used to clear up emissions; since costs vary with the technology and with the circumstances in which the technology is used, it is very difficult to make generalizations. For example, a figure of X dollars per ton of ore is often almost meaningless. The reason for this is that many control costs vary not so much with tonnage extracted as with other variables (e.g., acreage disturbed, water discharged). Even where a correlation exists among such variables and production, wide variance among operating sites can be expected. Second, many companies are reluctant to reveal with much precision their actual costs. Furthermore, cost estimates are often produced for political purposes. Such statistics, generated by all sides of a debate, are sometimes used the way a drunk uses a lamppost—for support rather than illumination. Finally, even the best of cost estimates are based on the particular processes and regulations in effect at the time of the study, and quickly become obsolete. For all that, the interested reader may well find these sources to be useful.

Surface Mine Reclamation—Costs: In addition to being subject to the environmental regulations which face all U.S. industry, the mining and mineral industries are uniquely subject to another set of regulations. The most notable of these is the Surface Mine Control and Reclamation Act of 1977 (SMCRA), described in the section by MacDonnell. This controversial act mandates the restoration of land to its approximate original contour and/or the attainment of certain performance standards necessary for further use of the land, such as agricultural or recreational use. Moreover, the law requires uniform reclamation standards in all geographical regions, despite the fact that benefits and costs display considerable regional variations. The activities mandated under SMCRA require considerable resources, and the economic effects of this law have been frequently though only tentatively analyzed. Analyses have been undertaken by Kalt (1983), ICF (1977), Energy and Environmental Analysis (1977), and Misioleck and Noser (1982). Enough analysis has been undertaken to prompt the summary which follows.

Costs under SMCRA depend of course upon the act's implementing regulations, which vary from one time to another. For simplicity, we make no attempt to distinguish among alternate regulatory regimes. Moreover, it is doubtful if many of the cost estimates are precise enough to make such distinctions meaningful. Nevertheless, it is worth bearing in mind that several factors lead to significant cost variations; besides historical period, these include variations in mining and reclamation plan, and variations in geologic, topographic, and climatologic conditions.

Costs (and benefits) of surface reclamation are most directly related to the acreage involved, but are ultimately expressed in terms of coal tonnage. Some variation in costs and benefits per acre may occur in different parts of the country, but the major source of regional benefit/cost variation (expressed per ton of coal) arises from the large differences in per-acre reserve tonnage. For example, since western coal seams are much thicker than seams in the East and Midwest, compliance costs (and benefits) per ton are considerably lower in the West. For a large scale mine, the differences in costs between East and West can easily reach tens of millions of dollars. If the ratio of benefits to costs were invariant across regions, regional cost variations would be largely irrelevant; however, the benefit-to-cost ratio also varies regionally.

Per acre costs were found by ICF, Inc. (1977) to be fairly uniform in regions other than Appalachia, where the costs are higher. ICF based their estimates on mine simulation, using capital and operating costs for various operations associated with reclamation. The study concluded that the cost of coal mining is more greatly affected by SMCRA in the East than in the West. Apart from the higher costs, it might be impossible to reclaim land in parts of Appalachia, given the steep contours (Energy and Environmental Analysis, Inc. 1977). These considerations have led to projections that SMCRA results in a shift of production to the West, with the Appalachian area suffering the greatest production loss.

Estimates of the shift vary in magnitude. ICF estimated a 22 million tons per year reduction from Appalachia, while the EEA estimate was 13 million tons annual reduction. Schlottman (1977) suggested that $4000 per acre reclamation costs would lower Appalachian production by 10 percent. Actual surface mine production has dropped in the East and Midwest and increased in the West. Between 1977 and 1979, for example, production dropped 13.6% in Kentucky and 11.6% in Ohio; in the West, however, production increased 56.7% in Wyoming, 37.3% in New Mexico, and 59.1% in Colorado. These production changes reflect a variety of forces, including but not limited to SMCRA effects.

Other studies have yielded dissimilar results. Some of the discrepancy among cost studies are due to differences in methodology, and some are due to inherent data uncertainties. Many of the studies were prepared for ongoing policymaking, and the results are sometimes congenial to the viewpoint of the sponsoring agency. Kalt (1983) points out that federal environmental agencies sponsored a study concluding the average cost increase would be $1.10 per ton, whereas two mining industry committees argued the costs would more nearly be in the range of $10 to $22 per ton. Kalt synthesized various results, suggesting that the 1977 per ton costs of SMCRA (expressed in 1981 dollars) were: $5.44 in Appalachia, $2.56 in the Midwest, and $0.41 in the West. ICF (1977) estimated that SMCRA requirements would increase selling prices of coal by $0.27 to $4.80 per ton, depending on state, topography, and mine configuration. All of the above figures should be treated with caution, of course. A detailed mine-site study of the costs and benefits of reclaiming the Big Sky Mine in southeastern Montana was conducted by Julian (1979). At this site, the recreational, agricultural, hydrologic, aesthetic, atmospheric, and psychological benefits that would arise from reclamation had a present value (at 5%) of $1,288,610. However, the costs of filling, grading, replacing topsoil, and revegetation required by SMCRA amounted to $17,858,874. The resulting benefit/cost ratio equaled 0.07. Quite clearly, at this particular site in the West, reclamation is far from an efficient use of resources. On the other hand Randall, et al., 1978, determined that the benefits unambiguously exceed the costs of reclamation in the watershed area of the north fork of the Kentucky River in Central Appalachia. Howard (1971) in a study of another area in Kentucky determined that the benefits to be obtained under the much less stringent 1966 regulations were less than the reclamation cost.

Other regulatory programs create surface mine compliance costs, which we note here but do not quantify. (Kalt, 1983 provides estimates of some of these costs, compatible with the above figures for SMCRA, but for consistency these are omitted in this discussion, since we summarize studies of SMCRA.) SMCRA itself involves costs which strictly speaking are not reclamation costs. These include the need for performance bonding, which insures that reclamation can be completed in cases of forfeiture. Bonding costs have been reported to run as high as $6,000 per acre, depending upon the circumstances. Bonding costs can also be much lower. Also, SMCRA created a production tax whose revenues are earmarked for reclaiming abandoned land; this tax varies from $0.10 to $0.35 per ton. Non-SMCRA programs, such as state permitting processes, likewise create related costs not summarized here.

Increased coal costs can be expected to decrease consumption. In addition to the reclamation costs and regional production shifts characterized above, SMCRA-related costs will also shift production to underground mines and have employment effects. ICF, Inc. (1977) projected that SMCRA would reduce Appalachian jobs by 1400 in 1979. (The analysis assumed that the additional costs could not be passed forward; these figures are based on mine closings due to higher operating costs, not consumption shifts due to higher prices.) The shift to underground production has been estimated to be as much as 100 million tons (EEA, 1977). Schlottman (1977, 1979) estimated that underground production would rise roughly 14% in

Appalachia and 21% in the Midwest. The shift from surface to underground mining imposes additional and/or different external costs, such as subsidence and health and safety problems. These effects could be considered one category of cost associated with surface mine control. Under SMCRA, certain coal bearing lands are unmineable. These lands include: steeply-contoured and heavily-overburdened areas, primarily in the East, and alluvial valley floors. ICF, Inc. estimated that this effect would reduce total strippable reserves by 6 to 18 percent.

The effects of SMCRA vary not only among regions, but also with the age of the mine. Both old and new mines are subject to the same requirements. New mines, however, can amortize the expenses over a longer production horizon, lowering costs per ton.

Surface Mine Reclamation—Benefits: surface mining creates five major types of damages: acid mine drainage; silting damage; landslides; adjacent land costs; and surface and aesthetic damage. To date, several attempts have been made to measure benefits from a reduction in these types of damages. It is fair to say that more work is needed on the valuation of these benefits before they could be a reliable input to policy making. A study by Energy and Environmental Analysis, Inc. (1977) restricted the estimation to three classes of impacts, the value of restored land, reductions in acid drainage, and reductions in sediment yield. However, of these three, values were not attached to the latter two categories. For this reason, benefit/cost ratios for various study regions never exceeded 13 cents per dollar of reclamation cost. On the other hand, the Appalachian Resources Project in 1974 indicated that direct external damages for surface-mined Appalachian coal were $0.80 to $1.00 per ton or $1600 to $2000 per acre. Gordon (1973) estimated surface mine reclamation benefits at $50 to $150 per acre. An early estimate of strip mine damages to recreational usage was $35 million annually (Hechler, 1971). Central Appalachian coal strip mining is a particularly damaging activity. These mines account for 60% of all acid mine drainage effects, as well as 67% of total sedimentation and 52% of land disruption caused by coal mining (Energy and Environmental Associates, 1977). Original-contour reclamation, a controversial subject, has been estimated as having the potential to reduce total acid mine drainage by 96% and to reduce related sediment loading of streams and waterways by 80% nationally. Monetary benefits have not been estimated, however.

While approximate original contour reclamation regrading generally provides a high level of environmental protection, it may not achieve the best level of protection. Some studies have shown that terracing is often desirable in reducing runoff and conserving moisture. Original contours may be vulnerable to erosion and rain water runoff during the revegetation process. In many areas the haulage roads and mining benches, when properly reclaimed, have created new land-use opportunities (see ICF 1977).

Benefits are the reduction in external costs, particularly water quality damages, land damages, flooding, reduced recreation, and reduced aesthetics to locals and to non-users (i.e., existence value). As difficult as it is to measure compliance costs, valuation of benefits is even more tentative. In his survey of other research, Kalt (1983) found benefits from SMCRA to be roughly equivalent to costs, amounting to $4.49 per ton in Appalachia, $3.34 in the Midwest, and $1.03 in the West. In comparing these numbers to those reported in the cost section, the reader will note that benefits exceed costs in the Midwest and West, but not in Appalachia.

Interestingly, by far the greatest benefit category for the first two locales is that of non-user aesthetics (existence value). This category contributes $2.00 and $0.93 per ton for Midwest and West regional benefits. As Kalt acknowledges, these data were generated by a single study, and the authors of the study had cautioned against generalizing across sites. Existence value is also a major source of benefits in Appalachia, with land productivity and local aesthetics roughly equal in magnitude. Interestingly, damages to water quality and flooding were quite minor sources of benefits in all regions.

In addition to the basic questions of benefits and costs, there are the further questions of welfare gains and losses across income groups. Kalt (1983) has also estimated these effects. Welfare losses of roughly $400 million were estimated to accrue to local coal users. Appalachian coal producers experienced the greatest losses equaling $773 million while Midwestern and Western producers would experience losses of $142.4 and $122 million respectively. These figures refer to the year of implementation, 1977, and are expressed in 1981 dollars. Offsetting the losses to domestic coal users and surface coal producers are gains to environmental users and underground producers. Environmental users in Appalachia gain $720.7 million, in the Midwest $274.9 million, and in the West $140.8 million. The net overall effect is a loss to the nation of approximately $122 million. Kalt further suggests that since environmentalism and outdoor environmental amenities are positively related to income, SMCRA produces a regressive transfer of resources across income groups. This issue would, in all likelihood, persist regardless of an improvement in the efficiency of the regulation.

Bosselman (1967) argued that the control of unwanted land uses such as surface mining tend to be governed more by emotion than by sober comparisons of costs and benefits; if Kalt is correct, society has made some forward progress in this regard, as evidenced by the fact that surface mine control is not mismatched in a major way relative to benefits. Generalizability of these results to other forms of mineral environmental regulation is of course unwarranted.

Water Pollution Control Costs: Williams (1975) provides a detailed discussion of the Federal Water Pollution Control Act Amendments of 1972 (PL 92-500) and its effects on mining, milling, and metallurgy. Perhaps the major form of minerals-related impact addressed by that act was acid mine drainage, a widely-occurring problem especially associated with the Appalachian coal fields. The Appalachian Regional Commission once estimated that the elimination of acid mine drainage in that region alone would create annual benefits of over $4 million (1967 dollars), excluding benefits to recreation and fisheries. Most of the industrial savings were expected to occur in three minerals-based industries, primary metals, electric power generation, and chemical products. Practical control methods were: neutralization, drainage control, mine sealing, reverse osmosis/deep well injection, and ground-water flow system dewatering. Before listing the relevant costs (1967 dollars), it is important to note that the figures are based on many assumptions. Only the figures for conventional lime neutralization are given here, and only those based on actual mine operations. In the small sample reported, capital costs ranged from $120,000 (discharge of 2,725 cubic meters per day) to $1,094,000 (21,802 cubic meter discharge). Capital costs on a unit basis ranged from $43 to $382 per cubic meter. Annual operating costs ranged from $30,000 to $478,000; operating costs per thousand cubic meters ranged from $0.032 to $0.121.

Williams also reports cost data for: bauxite refining and aluminum smelting; phosphate fertilizer processes; the iron and steel industry and coke manufacturing; the ferroalloy industry; copper, lead, and zinc recovery (including smelting and refining); mill wastes; tailings and other solid waste disposal; and mined land reclamation. Most of the information pertains to water pollution control. Given the wide range of technologies and situations, it is exceedingly difficult to give a useful summary of this material. Much of the cost data is based on analyses which were conducted to help determine regulatory requirements, and do not therefore reflect actual operating experience with a given regulatory standard.

For illustrative purposes only, we mention a few of the numbers Williams reports. For bauxite refining: capital costs ranging from $0.7 to $9.5 million (1971 dollars) and annual costs of $76,000 to $571,000. For primary zinc refining plants: capital costs of $3.43 per annual ton and operating costs of $0.89 per ton. For primary lead: current plant treatment capital costs of $136,000 to $595,000 ($1.93 to $4.92 per annual metric ton) and operating costs of $56,000 to $212,000 ($0.41 to $1.64 per metric ton) (1971 dollars).

Coal Cleaning: Liu (1982) describes various techniques for cleaning coal, including a chapter on economic assessment made by S.P.N. Singh. Based on hypothetical analysis (rather than actual operating experience), Singh estimated coal beneficiation costs in the range of $7 to $28 per ton of coal processed (annualized costs in 1978 dollars). Sensitivity analyses showed up to 28% changes for individual adjustments in three variables: a 20% decrease in capital investment; 24 hour operations; 20% operating cost increases; and a 20% decrease in working capital requirements.

The Clean Air Act and Nonferrous Smelting: A recent study commissioned by a consortium of minerals companies (Everest and CRU, 1982) examined the effects of the clean air act on the competitive position of U.S. nonferrous smelting. The cost data was derived from actual operating experience as reported in a survey. The median value of recent (dates ranged from 1976 to 1980) expenditures reported for a small sample of smelters was about $1,500 per annual ton of capacity. The study mentions that the lowest costs in their survey works out to 9.6 cents per pound of copper, before adding the operating costs of compliance, as well as all other charges. The competitive world toll-smelting and refining charges are stated to be about 20 to 21 cents per pound. The study cites a report by the U.S. Department of Commerce, written in 1979, which estimated that combined compliance costs for the U.S. copper industry for environmental, health, and safety regulations for the period of 1974-1987 would be $4.5 billion. The study suggests that total smelter revenues in 1979 were roughly $615 million, and air pollution control costs are roughly $207 million annually, based on the government study.

Another study cited by Everest and CRU estimates the relationship between copper smelting costs and sulfur removal. The toll cost per pound of copper in 1977 dollars, based on a sample of eight smelters was: 13.5 cents for no removal, and 19.5 cents for 56% (the standard); if sulfur removal were to be raised to 90% or 98%, the costs were projected to rise to 28.4 and 36.5 cents, respectively. Similar data were not available for lead or zinc.

An early study, although no longer relevant, provides an interesting contrast. In 1973, the Environmental Protection Agency estimated annual costs in 1978 (1970 dollars) for implementing the Clean Air Act of 1970 in the various primary metals industries. They determined that the costs to the iron and steel industry would be $179 million, while the grey iron, aluminum, and copper industries would spend $168 million, $209 million, and $178 million, respectively. The estimated costs included interest on pollution abatement capital equipment, depreciation of equipment, and the costs of operation and maintenance. These figures were derived from an engineering cost approach based on composite plant data in existing (1973) situations. The variation in plants due to differences in location, technology, size, and age preclude the adaption of these results to the individual plants that comprise these aggregate figures.

Summary: The objective measurement of benefits and costs of environmental regulation is in its infancy. Actual capital expenditures for pollution control equipment can be surveyed easily enough, but even such objective numbers as these are subject to error and to misinterpretation. Indirect and macroeconomic costs depend upon projections from economic models; these projections are difficult to validate. Some of the costs are hard to measure because they are opportunity costs. For example, if environmental regulations prevent a plant from being built, no monetary expenditure can be recorded, although a sacrifice has occurred. Delays caused by siting requirements and other environmental regulations will increase costs (due to inflation and escalation) and decrease a project's present worth (due to the time value of money). Interestingly, even though these delays are widely cited as a cost of environmental regulation, we found no empirical estimation of their magnitude. It would be a very difficult estimate to make.

Benefits are perhaps even more problematical, although the difficulties vary with the type of benefit being measured. It may be, for example, that existence value (seemingly, the most intangible of benefits) is relatively easy to measure, in some circumstances. As with costs, it may be that benefit valuation excludes some elements. In the view of some environmentalists, the consumer sovereignty approach implicit in the economist's valuation of wilderness is simply an unrealistic guide to a pristine nature's true value. Indeed, true wisdom is in scarce supply, and economists would be content with reasonably accurate estimates of property damages and health problems avoided, and land and water recreational values to someone.

Environmental compliance costs are higher for the mineral industries than for the typical business enterprise. Nevertheless, minerals-related compliance costs do not have a major direct impact on the U.S. economy, simply because the minerals sector is a small part of GNP. However, there is some evidence that these costs sometimes have significant effects on the economics of particular firms and the economies of some regions.

References

1. Bell, Frederick W., and Canterbury, E. Ray, "An Assessment of the Economic Benefits Which will Accrue to Commercial and Recreational Fisheries from Incremental Improvements in the Quality of Coastal Waters," Florida State University, Tallahassee, 1975.
2. Bishoff, S.M., "Costs and Cost Effectiveness of Coal Mining Reclamation" *Mining Engineering*, Feb. 1984.
3. Bohm, Robert, "Benefits and Costs of Surface Mining in Appalachia," Appalachia Resources Project, The University of Tennessee, Knoxville 1974.
4. Brock, Samuel M., "Preservation of the Environment," in *Economics of the Mineral Industries*, Vogely,

W. ed., 3rd ed., AIME, Littleton, Colo. 1976.

5. Brookshire, David S., Ives, Berry C. and Schulze, William D., "The Valuation of Aesthetic Preferences," *Journal of Environmental Economics and Management*, Vol. 3 No. 4, Dec. 1976.
6. Bureau of the Census, "Pollution Abatement Costs and Expenditures, 1981," *Current Industrial Reports Series MA-200* (81)-1, Washington, D.C.
7. Cooper, Charles, *Economic Evaluation and the Environment*, Hodder and Stoughton, London 1981.
8. Council on Environmental Quality, *Annual Report 1979, 1980, 1981*; Washington, D.C.
9. Crandall, R., "Pollution Controls and Productivity Growth in Basic Industries," in *Productivity Measurements in Regulated Industries*; Cowing, T. and Stevenson, R. eds. 1980.
10. Crocker, Thomas D., et al., "Methods Development for Assessing Air Pollution Control Benefits," in *Experiments in the Economics of Epidemiology, Vol. 1*, U.S. E.P.A., Washington, D.C. 1979.
11. Data Resources Incorporated, *The Macroeconomic Impact of Federal Pollution Control Programs—1981 Assessment*, New York 1981.
12. Davidson, Paul F., Adams, F. Gerard and Seneca, Joseph, "The Social Value of Water Recreation Facilities Resulting From An Improvement in Water Quality: The Delaware Estuary" in Knesse, Allen V. and Smith, Steven eds. *Water Research*, Johns Hopkins University Press, Baltimore 1966.
13. Denison, Edward F., "Pollution Abatement Programs: Estimates of Their Effect Upon Output Per Unit of Input 1975–1978," *Survey of Current Business*, August 1979.
14. Department of Energy, Energy Information Administration, Washington, D.C. 1981.
15. Energy and Environmental Analysis Inc., "Benefit/Cost Analyses of Laws and Regulations Affecting Coal: Case Studies on Reclamation, Air Pollution, and Health and Safety Laws and Regulations; Report to the Office of Mineral Policy and Research Analysis," U.S. Dept. of Interior, Washington, D.C. 1977.
16. Environmental Protection Agency, *The Cost of Clean Air*, Annual Report of the Administrator of the EPA to the Congress of the United States, October 1973.
17. Evans, Robert J. and Bitler, John R., *Coal Surface Mining Reclamation Costs: Appalachian and Midwestern Coal Supply Districts*, U.S. Bureau of Mines Information Circular 8695, Washington, D.C., 1975.
18. Everest Consulting Associates and CRU Consultants, "The International Competitiveness of the U.S. Nonferrous Smelting Industry and the Clean Air Act, unpublished report, 1982.
19. Finklea, John F., et al., "The Role of Environmental Health Assessments in the Control of Air Pollution," in *Advances in Environmental Science and Technology*, Pitts, J.N., Metcalf, R.L. and Lloyd A.C. eds., Wiley, New York 1977.
20. Fisher, Anthony C. and Smith, V. Kerry, "Economic Evaluation of Energy's Environmental Costs with Special Reference to Air Pollution," *Annual Review of Energy, 1982*
21. *Fisher, Anthony C., Resource and Environmental Economics*, Cambridge University Press, New York, 1981.
22. Freeman, A. Myrick III, "The Benefits of Air and Water Pollution Control: A Review and Synthesis of Recent Estimates (Final Report)," Brunswick, Maine 1979.
23. Freeman, A. Myrick III, *The Benefits of Environmental Improvement, Theory and Practice*, Resources for the Future: Johns Hopkins University Press, Baltimore 1979.
24. Gordon, Richard, "Environmental Impacts of Energy Production and Use," Energy Supply Project, (unpublished manuscript on file, Resources for the Future, Washington, D.C. 1973).
25. Haveman, R.H. and Christainsen, G.B., "Environmental Regulations and Productivity Growth," in *Environmental Regulation and the U.S. Economy*, Peskin, H.M., Portney, P.R. and Knesse, A.V. eds., Resources for the Future: Johns Hopkins University Press, Baltimore 1981.
26. Hechler, Kenneth, Statement before the Subcommittee on Mines and Mining, *Congressional Record*, September 20, 1971.
27. Heintz, H.T., Jr., Hershaft, A. and Horak, G.C., "National Damages of Air and Water Pollution," a report to the U.S. Environmental Protection Agency, Enviro Control, Inc., Rockville, Md. 1976.
28. Howard, H. A., "A Measurement of the External Diseconomies Associated with Bituminous Coal Surface Mining, Eastern Kentucky 1962–1967," *Natural Resources Journal*, Jan. 1971.
29. Hueth, D., Just, R., and Schmitz, A., *Applied Welfare Economics*, Prentice-Hall, Englewood Cliffs, N.J. 1982.
30. ICF, Inc., "Energy and Economic Impacts of H.R. 13950 (Surface Mining Control and Reclamation Act)," Washington, D.C. 1977.
31. Julian, Edward L., *Big Sky Coal Mine—A Mine Site Study of Benefits and Costs of Reclaiming Surface-Mined Land in the West*, Pennsylvania State University, 1979.
32. Kaldor, Nicholas, "Welfare Propositions of Economics and Interpersonal Comparisons of Utility," *Economic Journal* Vol. 49, 1939.
33. Kalt, Joseph P., "The Costs and Benefits of Federal Regulation of Coal Strip Mining," *Natural Resources Journal*, Oct. 1983.
34. Lave, Lester B., and Seskin, Eugene P., "Air Pollution and Human Health," *Science*, Vol. 169, Aug. 21, 1970.
35. Lave, Lester B., and Seskin, Eugene P., *Air Pollution and Human Health*, Resources for the Future: Johns Hopkins University Press, Baltimore 1977.
36. Liu, Ben-chieh and Yu, Eden Siu-hung, "Physical and Economic Damage Functions for Air Pollution by Receptor," E.P.A., Corvallis, 1976.
37. Liu, Y.A., ed., *Physical Cleaning of Coal: Present and Developing Methods*, Marcel Dekker Inc., New York, 1982.
38. Mann, Charles E. and Heller, James N., "Coal and Profitability: An Investor's Guide," *Coal Week*, Hightstown, N.J. 1979.
39. Misioleck, W.S. and Noser, T.C., "Coal Surface Mine Land Reclamation Cost," *Land Economics* Vol. 58 No. 1, Feb. 1982.
40. National Commission on Air Quality, *To Breathe Clean Air*, Washington, D.C. 1981.
41. National Planning Association, "Water Related Recreation Benefits Resulting From P.L. 92-500," Washington, D.C. 1975.
42. O.E.C.D., "Costs of Coal Pollution Abatement: Results of An International Symposium (August 1983)," Paris 1983.
43. O.E.C.D., "Coal and Environmental Protection: Costs and Costing Methods," Paris 1983.

44. PEDCo Environmental, "Study of the Adverse Effects of Solid Wastes from Mining Activities and the Environment," March 1979.
45. Persse, F.H., Lockhard, D. W., and Lindquist, A.E., Coal Surface Reclamation Costs in the Western United States, U.S. Bureau of Mines Information Circular 8737, Washington, D.C. 1977.
46. Peskin, H.M., Portney, P.R., and Knesse, A.V., *Environmental Regulation in the U.S. Economy*, Resources for the Future: Johns Hopkins University Press, Baltimore, 1981.
47. Portney, Paul R., "The Macroeconomic Impacts of Federal Environmental Regulations," in *Environmental Regulation and the U.S. Economy*, Peskin, H.M., Portney, P.R. and Knesse, A.V. eds., Resources for the Future: Johns Hopkins University Press, Baltimore, 1981.
48. Randall, Alan, et al., "Reclaiming Coal Surface Mines in Central Appalachia: A Case Study of the Benefits and Costs," *Land Economics*, Vol. 54 No. 4, Nov. 1978.
49. Ridker, Ronald G., *Economic Costs of Air Pollution: Studies in Measurement*, Praeger, New York 1967.
50. Ridker, Ronald G., and Watson, W.D., "Long Run Effect of Environmental Regulation," *Natural Resources Journal*, Vol. 21, No. 3, pp. 565–587, 1981.
51. Ripley, Earle A., Redmann, Robert E., with Maxwell, James, *Environmental Impact of Mining in Canada*, Centre for Resource Studies, Queens University, Kingston, Ontario, 1978.
52. Russo, William J., Jr., and Rutledge, Gary L., "Plant and Equipment Expenditures by Business for Pollution Abatement, 1982 and Planned 1983," *Survey of Current Business*, June, 1983.
53. Rutledge, Gary L., and Lease-Trevathan, Susan, "Pollution Abatement and Control Expenditures, 1972–1981," *Survey of Current Business*, Feb. 1983.
54. Sagan, Leonard A., "Health Costs Associated with Mining, Transport and Combustion in the Steam-Electric Industry," *Nature*, Vol. 250, July 12, 1974.
55. Schlottman, Alan M., *Environmental Regulation and the Allocation of Coal: A Regional Analysis*, Praeger, New York 1977.
56. Schlottman, Alan M., "Economic Impacts of Surface Mining Reclamation," in *Coal Surface Mining: Impacts of Reclamation*, Rowe, James E., ed., Westview Press Inc., Boulder, Colo. 1979.
57. Schmidt-Bleak, F.K. and Moore, J.R., "Benefit–Cost Evaluation of Strip Mining in Appalachia," Appalachian Resources Project, University of Tennessee, Knoxville 1973.
58. Sinden, John A., and Worrell, Albert C., *Unpriced Values: Decisions Without Market Prices*, Wiley, New York 1979.
59. Sterling Hobe Corporation, "Survey of Methods Measuring the Economic Cost of Morbidity Associated with Air Pollution," Report for the National Commission on Air Quality, Washington, D.C., December 1980.
60. Unger, Samuel G., "National Benefits of Achieving the 1977, 1983, and 1985 Water Quality Goals," a report submitted to the U.S. Environmental Protection Agency, 1975.
61. Waddel, Thomas E., "The Economic Damages of Air Pollution," Office of Research and Development, U.S. E.P.A., May 1974.
62. Watson, William and Jakesch, John, "Household Cleaning Costs and Air Pollution," 1979.
63. Whitney, J.W. and Ramsey, J.B., "Effects of the Federal Regulatory Framework on Mineral Exploration and Mine Development in the U.S.," in *Economics of Exploration for Energy Resources*, Ramsey, J.B. ed., Jai Press, Greenwich, Conn. 1981.
64. Williams, Roy E., *Waste Production and Disposal in Mining, Milling, and Metallurgical Industries*, Miller Freeman Publications, 1975.